Holt Algebra with Trigonometry

Examination Guide

W9-CXY-472

A complete course for students of all abilities provides everything you need—

■ **The Step-by-Step problem-solving process** helps you teach algebraic concepts and methods.

> Chapters 1–4 include lessons on linear equations and linear inequalities (pp. 17, 42), and solving systems of two and three linear equations (pp. 119, 142). Chapters 5–8 and explore polynomial equations and their applications (pp. 170, 204, 258). Chapter 9 introduces students to complex numbers (p. 303). Chapters 11–16 focus on coordinate geometry and quadratic functions (p. 391), conic sections (p. 440), exponential and logarithmic functions (p. 486), probability (p. 517), and matrices (p. 607). Chapters 17–19 explore trigonometric functions and identities (pp. 619, 711).

■ **Exercises** progress in complexity to give you a choice of assignments.

> Exercises are plentiful with Classroom Exercises for oral discussion (p. 411) and Written Exercises that progress from easy to more challenging (pp. 411–412). Examples correspond to each type of written exercise for levels A and B.

■ **Reviews, quizzes, and tests** make it easy to integrate assessment with instruction.

> Each lesson includes a Prerequisite Quiz (p. 409) and a post-lesson Checkpoint (p. 411) in the side column of the *Annotated Teacher's Edition* and as blackline masters in the Quick Quizzes section of the testing package. Assessments in the textbook include Mixed Reviews (p. 412) and a Midchapter Review (p. 429) as well as Chapter Reviews, Tests, and College Prep Tests at the end of each chapter, and Cumulative Reviews in alternate chapters.

■ **Extensive features and worksheets** help you meet the NCTM *Standards* for the 90s!

> * **Technology**—Calculators and computers (pp. 7, 65, 378, 756)
> *Investigating Algebra with the Computer* Software–*Teacher's ResourceBank* ™
>
> * **Math connections**—In Teacher's side column for each lesson.
> Applications feature (pp. 12, 26, 57)
> Application Worksheets–*Teacher's ResourceBank* ™
>
> * **Communication about math**—
> Focus on Reading feature (pp. 43, 80)
> Written Exercises (pp. 37 #17, 371 #23)
>
> * **Cooperative learning**—Group Investigations begin each chapter
> Project Worksheets—*Teacher's ResourceBank* ™
>
> * **Problem solving**—Problem Solving Strategies feature (pp. 58, 112)
> Problem Solving Worksheets—*Teacher's ResourceBank* ™
>
> * **Other valuable timesavers for you**
> Reteaching and Practice Worksheets and Tests, Form A and Form B—*Teacher's ResourceBank* ™
> *Instructional Transparencies* and *Teacher's Notes*
> *Solution Key*

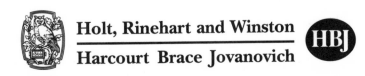

Holt, Rinehart and Winston
Harcourt Brace Jovanovich

HBJ

HOLT ALGEBRA

WITH TRIGONOMETRY

Annotated Teacher's Edition

Eugene D. Nichols

Mervine L. Edwards

E. Henry Garland

Sylvia A. Hoffman

Albert Mamary

William F. Palmer

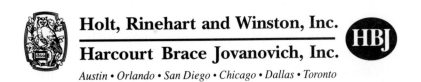

Holt, Rinehart and Winston, Inc.

Harcourt Brace Jovanovich, Inc.

Austin • Orlando • San Diego • Chicago • Dallas • Toronto

About the Authors

Eugene D. Nichols
Distinguished Professor of Mathematics Education
Florida State University
Tallahassee, Florida

Mervine L. Edwards
Chairman, Department of Mathematics
Shore Regional High School
West Long Branch, New Jersey

E. Henry Garland
Head of Mathematics Department
Developmental Research School
DRS Professor
Florida State University
Tallahassee, Florida

Sylvia A. Hoffman
Resource Consultant in Mathematics
Illinois State Board of Education
State of Illinois

Albert Mamary
Superintendent of Schools for Instruction
Johnson City Central School District
Johnson City, New York

William F. Palmer
Professor of Education and Director
Center for Mathematics and Science Education
Catawba College
Salisbury, North Carolina

Printed in the United States of America

123456   036   98765432

ISBN 0-03-005434-6

Acknowledgments

Reviewers

Jeanette Gann
Mathematics Supervisor
High Point Schools
High Point, North Carolina

Patrice Gossard, Ph.D.
Mathematics Teacher
Cobb County School District
Marietta, Georgia

Linda Harvey
Mathematics Teacher
Reagan High School
Austin, Texas

Gerald Lee
Chairman, Mathematics Department
McArthur High School
Lawton, Oklahoma

Janet Page
Mathematics Teacher
Hoffman Estates High School
Hoffman Estates, Illinois

For permission to reprint copyrighted material, grateful
acknowledgment is made to the following sources:

National Council of Teachers of Mathematics: Excerpts from
Curriculum and Evaluation Standards for School Mathematics,
prepared by the Working Groups of the Commission on Standards
for School Mathematics of the National Council of Teachers of
Mathematics. Copyright © 1989 by the National Council of Teachers
of Mathematics, Inc.

Photo Credits

Illustration

Computer Graphics

Chapter Contents

1 Linear Equations   xvi

Holt Algebra
with Trigonometry

MEETING THE STANDARDS OF TOMORROW— STARTING TODAY

The role that today's teachers play in the education of our students will affect our world for generations to come. At Holt, Rinehart and Winston, we recognize the importance of education and are dedicated to providing today's teachers and students with the finest textbooks available.

Holt Algebra with Trigonometry helps you move toward the NCTM *Standards,* while also providing you with a practical text that meets your standards. A variety of supplementary materials has been developed that incorporates the newest technology and current curriculum standards in mathematics education into a traditional mathematics program.

> "TODAY, WHAT HAPPENS in America's classrooms is being given lots of attention and scrutiny. The reactions and responses to the recent reports on education offer the mathematics education community a rare opportunity to shape school mathematics during the next decade. Public interest and concern, when combined with changing technology and a growing body of research-based knowledge, are the ingredients necessary for real reform. The NCTM *Standards* is a vehicle that can serve as a basis for improving the teaching and learning of mathematics in America's schools." (NCTM *Standards,* p. 254)

A COMPLETE SELECTION OF ANCILLARY MATERIALS ENRICHES THE PROGRAM.

The materials described below provide a wealth of activities that incorporate important topics in mathematics education. Computer activities help students feel comfortable using today's technology to perform real-life tasks.

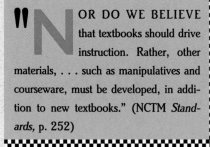

"NOR DO WE BELIEVE that textbooks should drive instruction. Rather, other materials, . . . such as manipulatives and courseware, must be developed, in addition to new textbooks." (NCTM *Standards*, p. 252)

Teacher's ResourceBank™

The *Teacher's ResourceBank™* contains a wealth of blackline master booklets, organized by chapter, that provide materials for practice, reteaching, review, enrichment, and an extensive testing program.

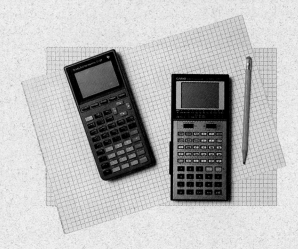

Chapter Worksheets and Tests

- ▲ Reteaching Worksheets (one per lesson)
- ▲ Practice Worksheets (one per lesson)
- ▲ Application Worksheets (one per chapter)
- ▲ Graphing Calculator Worksheets (Available for both Casio and Texas Instruments graphing calculators.)
- ▲ Problem Solving Worksheets (one per chapter)

- ▲ Project Worksheets (one per chapter)
- ▲ Quick Quizzes (one per lesson)
- ▲ Chapter Tests, Forms A and B (one test per chapter)
- ▲ Cumulative Tests, Forms A and B (one test for every four chapters)
- ▲ Semester Tests, Forms A and B (two tests for the school year)

Also Included in the *Teacher's ResourceBank*™
SAT®/ACT Practice Tests*
Investigating Algebra with the Computer software (Apple® and IBM®)*

* Also available for sale separately.

Additional Ancillary Materials

Instructional Transparencies and Teacher's Notes

The *Instructional Transparencies* provide over forty colorful transparencies (at least one transparency per chapter) for teachers who use an overhead projector in their classroom.

SAT®/ACT Practice Tests

The *SAT®/ACT Practice Tests* contains 4 sample SAT tests and 4 sample ACT tests which prepare students for the mathematics sections of these college entrance tests.

Investigating Algebra with the Computer software

The *Investigating Algebra with the Computer software* is designed to be used with the lessons that are found at the back of the *Pupil's Edition*. This software allows students to explore algebraic concepts visually.

Test Generator

The *Test Generator* uses a random number generator with a bank of algorithms to allow the teacher to create a virtually unlimited number of unique tests.

Solution Key

The *Solution Key* contains worked-out answers for all exercises and end-of-chapter activities in the *Pupil's Edition*.

FLEXIBLE TEACHING STRATEGIES ALLOW YOU TO CUSTOMIZE INSTRUCTION.

The teaching strategies, which are interleaved in the *Annotated Teacher's Edition* before each chapter, help you focus lessons on problem solving strategies, computer and calculator technology, and special features that enrich and extend students' learning.

In both the **Overview** and the **Objectives,** learning goals are defined in terms of what students will do—emphasizing the value of active, not passive, learning.

12 CONIC SECTIONS

OVERVIEW

In this chapter, equations of the circle, ellipse, parabola, and hyperbola are written and graphed. Vertices and foci of ellipses and hyperbolas are found, as well as equations of the asymptotes of the latter. The focus and directrix of a parabola are found. Linear-quadratic and quadratic-quadratic systems are solved and quadratic inequalities are graphed.

OBJECTIVES

- To write equations of conic sections
- To graph conic sections given their equations
- To find missing values in direct, inverse, joint, and combined variations
- To solve quadratic-quadratic systems of equations

PROBLEM SOLVING

In Lessons 12.5 and 12.6, several instances are provided that illustrate how inverse, joint, and combined variation are reflected in the physical world. In Lesson 12.8, the strategy *Using a Diagram* is employed to illustrate the conic sections introduced in this chapter. The problem solving strategies of *Using a Formula* and *Making a Graph* are used to solve problems in many of the lessons of this chapter.

READING AND WRITING MATH

A *Focus on Reading* appearing on page 438 asks students to fill in missing words to make a sentence true. Exercise 21 on page 442 asks students to explain a mathematical concept in their own words.

TECHNOLOGY

Calculator: Graphing calculators are now available that can accurately and rapidly display every graph shown in this chapter, including systems of quadratic equalities and quadratic inequalities.

SPECIAL FEATURES

Mixed Review pp. 412, 417, 422, 425, 434, 439, 442, 446, 451
Midchapter Review p. 429
Application: Eccentricity of an Ellipse p. 430
Focus on Reading p. 438
Application: LORAN p. 442
Brainteaser p. 446
Key Terms p. 452
Key Ideas and Review Exercises pp. 452–453
Chapter 12 Test p. 454
College Prep Test p. 455
Cumulative Review (Chapters 1–12) pp. 456–457

Reading and Writing Math identifies places in the chapter where students are asked to explain mathematical concepts in their own words.

Lessons in which the calculator can be used to explore mathematical concepts and facilitate routine computations are included in **Technology.**

A **Special Features** list allows you to decide in advance which applications and additional activities to incorporate into lessons.

> "THESE ALTERNATIVE methods of instruction will require the teacher's role to shift from dispensing information to facilitating learning, from that of director to that of catalyst and coach." (NCTM *Standards,* p. 128)

407A

Problem Solving identifies features and exercises in the chapter that help students develop strategies for solving problems.

TIME-SAVING PLANNING GUIDES HELP YOU ORGANIZE INSTRUCTION.

A quality curriculum requires versatile and complete lesson plans and the lesson plans in *Holt Algebra with Trigonometry* are organized in a format that is both functional and convenient.

The **Planning Guide** recommends appropriate classroom and written exercises for students of all ability levels, suggests when to incorporate special features into the lessons, and reminds you of the availability of other components.

> "A VARIETY OF INSTRUCtional methods should be used in classrooms in order to cultivate students' abilities to investigate, to make sense of, and to construct meanings from new situations; to make and provide arguments for conjectures; and to use a flexible set of strategies to solve problems . . ." (NCTM *Standards*, p. 125)

PLANNING GUIDE

Lesson	Basic	Average	Above Average	Resources
12.1 pp. 411–412	CE all WE 1–25 odd	CE all WE 7–31 odd	CE all WE 13–37 odd	Reteaching 82 Practice 82
12.2 pp. 416–417	CE all WE 1–19 odd	CE all WE 5–23 odd	CE all WE 9–29 odd	Reteaching 83 Practice 83
12.3 pp. 421–422	CE all WE 1–17 odd	CE all WE 5–21 odd	CE all WE 9–23 odd	Reteaching 84 Practice 84
12.4 p. 425	CE all WE 1–8	CE all WE 5–13	CE all WE 8–16	Reteaching 85 Practice 85
12.5 pp. 428–430	CE all WE 1–19 odd Midchapter Review Application	CE all WE 3–21 odd. Midchapter Review Application	CE all WE 5–25 odd Midchapter Review Application	Reteaching 86 Practice 86
12.6 pp. 433–434	CE all WE 1–8	CE all WE 3–10	CE all WE 6–16	Reteaching 87 Practice 87
12.7 pp. 438–439	FR all, CE all WE 1–17 odd	FR all, CE all WE 5–21 odd	FR all, CE all WE 7–23 odd	Reteaching 88 Practice 88
12.8 pp. 441–442	CE all WE 1–14 Application	CE all WE 6–19 Application	CE all WE 8–24 Application	Reteaching 89 Practice 89
12.9 pp. 445–446	CE all WE 1–13 odd Brainteaser	CE all WE 7–19 odd Brainteaser	CE all WE 11–23 odd Brainteaser	Reteaching 90 Practice 90
12.10 pp. 450–451	CE all WE 1–15 odd	CE all WE 7–21 odd	CE all WE 11–27 odd	Reteaching 91 Practice 91
Chapter 12 Review pp. 452–453	all	all	all	
Chapter 12 Test p. 454	all	all	all	
College Prep Test p. 455	all	all	all	
Cumulative Review pp. 456–457	1–47 odd	9–49 odd	15–51 odd	

CE = Classroom Exercises   WE = Written Exercises   FR = Focus on Reading
NOTE: For each level, all students should be assigned all Mixed Review exercises.

ATTRACTIVE CHAPTER OPENERS CONNECT MATH TO THE WORLD OUTSIDE THE CLASSROOM.

The **Chapter Opener** contains a colorful image that has been created using the computer. Each image illustrates how computer technology can be applied to mathematics, science, and the arts.

The **Investigation** is an activity that allows students to explore abstract mathematical concepts with the calculator. Students work cooperatively, in small groups, acquiring interactive problem-solving skills they will use for a lifetime. Follow-up questions help students form conjectures and draw conclusions about the activity.

◤◤◤ INVESTIGATION

The students are invited to explore the elliptical orbits of major and minor planets.

A planet orbiting about the Sun follows an elliptical path. The Sun will be at one of the points F_1 and F_2 in the figure.

The *eccentricity* e of an orbit is defined as $e = \frac{c}{a}$, which is a ratio that varies from 0 to 1. Given the eccentricity e and the semi-major axis a, it is easy to calculate c.

The values of a and e for the Earth and the asteroid Apollo are shown below. The value of the Earth's semi-major axis is *defined* as 1, which is called 1 astronomical unit (abbreviated 1 a.u.). Have the students calculate the value of c for the Earth and Apollo.

	Earth	Apollo
a	1.000	1.471
e	0.017	0.560

Earth: $c = 0.017$ Apollo: $c = 0.824$

The value of b, the semi-minor axis, can be calculated using the Pythagorean relation. In the above figure,

$$b^2 = a^2 - c^2.$$

Have the students calculate the value of b for the Earth and Apollo.
Earth: $b = 0.99986$, Apollo: $b = 1.218$

Have the students sketch the orbits of the Earth and the asteroid Apollo using the values for a and b. They should show the Sun at one focus of each ellipse.

Apollo is known as an *Earth-crossing asteroid*, because its orbit crosses inside the Earth's orbit.

408

12 CONIC SECTIONS

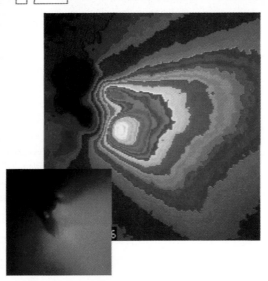

Comets that reappear on a regular basis, like Halley's Comet, have elliptical orbits. These images of Halley's Comet were produced from data gathered from the European Giotto space mission. Giotto is the name of an Italian pre-Renaissance artist who represented the Christmas star as a comet.

More About the Image

In 1986, a flotilla consisting of five separate spacecraft launched earlier by the USSR, Japan, and the European Space Agency (ESA) sent information back to the Earth about Halley's comet. ESA's satellite, called Giotto, flew closer to the comet than any of the other four spacecraft. It sent back valuable information, including the photo and the computer-enhanced image above. The data from the mission revealed that the nucleus of the comet is a velvety-black elongated object shaped like a peanut or a potato about the size of Manhattan Island.

More About the Image provides in-depth background information about the illustration featured in the **Chapter Opener.** This information can be used to start a class discussion about the value and usefulness of mathematics in our everyday lives.

EASY-TO-FOLLOW LESSONS COMMUNICATE
LEARNING GOALS CLEARLY AND INTELLIGIBLY.

Easy-to-follow lessons—featuring succinct objectives, brief intro-
ductions, and clear-cut examples—make mathematics accessible
to your students.

Teaching Resources provides a list of all supplementary
components available for use with the lesson. These
materials contain practice, review, reteaching and
enrichment activities.

12.1 The Circle

Objectives
To write equations of circles given certain conditions
To graph circles given their equations

People have been fascinated by circles
for thousands of years. Around 2000
B.C., an ancient people erected huge
standing stones in a large circular
ditch at Stonehenge in southern England. Most archaeologists believe
that Stonehenge served a religious
function, but some scientists suspect
that it also served as an astronomical
instrument for tracing the movements
of the sun and the moon and for
observing eclipses.

Definitions

A **circle** is the set of all points in a plane that are the same distance
from a given point, called the **center**. This distance is equal to the
length of any segment, called a **radius**, which joins the center to a
point on the circle.

The word *radius* can also denote the *length* of the radius. Thus,
radius 5 means "radius of length 5." In a coordinate plane, the
distance formula, $d = \sqrt{(x_2 - x_1)^2 + (y_2 - y_1)^2}$, can be used
to write an equation of the circle of radius 5 shown below, which has
its center at the origin. Let $P(x,y)$ be any point on the circle. Then use
the distance formula to express the distance between $C(0,0)$ and $P(x,y)$,
which is already known to be 5.

$$\sqrt{(x - 0)^2 + (y - 0)^2} = 5$$
$$\sqrt{x^2 + y^2} = 5$$
$$(\sqrt{x^2 + y^2})^2 = 5^2$$
$$x^2 + y^2 = 25$$

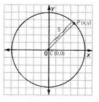

Thus, an equation of the circle
of radius 5 having its center at the
origin is $x^2 + y^2 = 5^2$, or
$x^2 + y^2 = 25$.

12.1 The Circle **409**

Teaching Resources

Problem Solving Worksheet 12
Quick Quizzes 82
**Reteaching and Practice
  Worksheets** 82

GETTING STARTED

Prerequisite Quiz

**Use the distance formula to find the
distance between the two given points.**

1. $P(0,0)$, $Q(3,4)$ 5
2. $P(0,0)$, $Q(5,-12)$ 13
3. $P(1,1)$, $Q(5,5)$ $4\sqrt{2}$

Simplify.

4. $(\sqrt{5})^2$ 5
5. $\sqrt{(x^2 + y^2)^2}$ $x^2 + y^2$

Motivator

Ask students why the equation $x^2 + y^2 = 25$
yields *two* values of y for $x = 3$.   Positive
and negative values will solve the equation
when squared.   What are they?   4, −4

Have students make a table of x- and
y-values for the equation above using the
following values of x: 5, 4, 3, 0, −3, −4,
−5. Plot the points corresponding to these
ordered pairs. Draw a smooth curve through
them. What does the shape appear to be?
A circle.

The **Prerequisite Quiz** helps
you assess what students
have learned in previous lessons and determine the overall readiness of the class for
the new lesson.

Through a series of questions,
the **Motivator** helps students
form a hypothesis. As students
complete the lesson, they will
discover if their conjectures
were correct and will be able
to draw mathematical conclusions about the concepts presented in the lesson.

Highlighting the Standards

Standard 8a: The circle has been studied
from various perspectives through different
levels of mathematics. This study can help
students grow in mathematical
understanding.

409

Highlighting the Standards
identifies how each lesson
addresses the NCTM *Standards*.

T7

TEACHING SUGGESTIONS INTEGRATE A VARIETY OF TOPICS INTO THE LESSONS.

The **Teaching Suggestions** relate each lesson to current issues in mathematics education. These features, located in the side-column of the text, provide a convenient way to enrich the lessons and make them meaningful to students.

Examples lead students through the problem-solving process by showing them how to formulate a **Plan** and then move step-by-step toward a solution.

The **Lesson Note** helps prepare students for the exercises by identifying concepts and skills from earlier lessons that may need to be reviewed.

Math Connections relates algebra to previously taught material, other areas of mathematics, real-life situations, and the history of mathematics.

Critical Thinking Questions develop higher-order thinking skills such as analysis, synthesis, and evaluation.

▰▰ TEACHING SUGGESTIONS

Lesson Note

For students who have already studied geometry, this lesson will be a good review of the properties of circles. Correlate the idea of locus, studied in geometry, with the set of points, $P(x,y)$, that satisfy an equation.

You can have students verify that the perpendicular bisector of a chord of a circle passes through the center of the circle. The students can find the equation of the perpendicular bisector of chord $\overline{AB}$, and then show that (0,0) satisfies this equation.

Math Connections

Circular Motion: Objects in circular motion often produce strange effects. In an experiment, a man stood at the center of a turntable holding his arms extended straight out to his side. When he folded his arms close to his body, his rotational speed more than doubled. The phenomenon illustrates the physical principle called the *conservation of angular momentum*.

Critical Thinking Questions

Analysis: Ask students to describe the set of points described by the equation of Theorem 12.2 for the case $r = 0$. Since the radius of a circle cannot be zero, some students may think that the set is empty. However, they should understand that the set of points in question consists of the single point (h,k). Thus, a point may be thought of as a "degenerate" circle, a circle with a radius of zero.

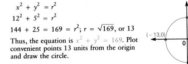

| **Theorem 12.1** | The standard form of the equation of a circle of radius r with its center at the origin is $x^2 + y^2 = r^2$. |

EXAMPLE 1 Write the equation of the circle centered at the origin and passing through the point $A(12,5)$. Graph the circle.

Plan Substitute 12 for x and 5 for y in the equation above to find r^2.

Solution
$$x^2 + y^2 = r^2$$
$$12^2 + 5^2 = r^2$$
$$144 + 25 = 169 = r^2; r = \sqrt{169}, \text{ or } 13$$
Thus, the equation is $x^2 + y^2 = 169$. Plot convenient points 13 units from the origin and draw the circle.

Circles can be *translated* (moved without rotation) so that their centers are not at the origin. In the figure at the right, circle S of radius 3 is centered at the origin. Circle T is a translation of circle S to $T(4,-2)$. The length of the radius, however, remains the same.

An equation of circle T can be found using the distance formula. Let $B(x,y)$ be any point on circle T. The distance between $T(4,-2)$ and $B(x,y)$ is 3. Therefore,
$$\sqrt{(x-4)^2 + [y - (-2)]^2} = 3$$
$$\sqrt{(x-4)^2 + (y+2)^2} = 3$$
$$(x-4)^2 + (y+2)^2 = 9 \quad \text{Square each side.}$$

Notice the following relationship: $\quad (x-4)^2 + (y+2)^2 = 9$

center: $(4,-2)$ radius: $\sqrt{9} = 3$

| **Theorem 12.2** | The standard form of the equation of a circle of radius r with its center at $C(h,k)$ is $(x-h)^2 + (y-k)^2 = r^2$. |

410   Chapter 12 Conic Sections

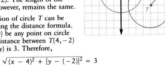

Additional Example 1

Write an equation of a circle in standard form with its center at the origin and passing through the point $A(-6,-8)$. $x^2 + y^2 = 100$

CLASSROOM AND WRITTEN EXERCISES CLARIFY AND REINFORCE NEWLY-LEARNED CONCEPTS AND SKILLS.

An extensive variety of exercises covers all topics from the lesson and challenges students of all ability levels.

Common Error Analysis diagnoses potential problem areas and prescribes corrective strategies that focus students' attention on correct procedures.

EXAMPLE 2   Write, in standard form, the equation of the circle centered at $C(-4,5)$ and passing through the point $P(2,10)$. Graph the circle.

Solution   Substitute 2 for x, 10 for y, -4 for h, and 5 for k in the standard form.

$$(x - h)^2 + (y - k)^2 = r^2$$
$$[2 - (-4)]^2 + (10 - 5)^2 = r^2$$
$$6^2 + 5^2 = r^2$$
$$61 = r^2$$
$$r = \sqrt{61} \approx 7.8$$

Thus, the standard form of the equation of this circle is $(x + 4)^2 + (y - 5)^2 = 61$

Plot several points 7.8 units from $C(-4,5)$. Draw the graph as shown.

EXAMPLE 3   Rewrite $x^2 - 8x + y^2 + 4y + 16 = 0$ in standard form. Identify the circle's center and radius.

Plan   Isolate the x- and y-terms. Then complete the squares in both x and y.

Solution
$$x^2 - 8x + \underline{\quad} + y^2 + 4y + \underline{\quad} = -16$$
$$x^2 - 8x + \left(\tfrac{1}{2}\cdot 8\right)^2 + y^2 + 4y + \left(\tfrac{1}{2}\cdot 4\right)^2 = -16 + 16 + 4$$
$$(x - 4)^2 + (y + 2)^2 = 4 \leftarrow r^2 = 4; r = 2$$

Standard form: $(x - 4)^2 + (y + 2)^2 = 4$; center: $(4, -2)$; radius: 2

Classroom Exercises

Give the radius and the coordinates of the center of each circle. Graph each circle.

1. $x^2 + y^2 = 49$ 2. $x^2 + y^2 = 19$
3. $(x - 4)^2 + (y - 1)^2 = 100$ 4. $(x - 1)^2 + y^2 = 144$

Written Exercises

Write, in standard form, the equation of the circle with center C and radius r given, or with center C and a point P on the circle given.

1. $C(0,0)$, $r = 13$ 2. $C(0,0)$, $r = 3.1$ 3. $C(2,3)$, $P(8,0)$ 4. $C(4,2)$, $P(0,8)$
5. $C(-2,8)$, $r = 9$ 6. $C(1,1)$, $r = \sqrt{3}$ 7. $C(1,-1)$, $P\left(\tfrac{1}{2},1\right)$ 8. $C\left(\tfrac{1}{4},\tfrac{1}{4}\right)$, $P(1,-1)$
9–16. Graph each circle in Written Exercises 1–8.

12.1 The Circle   **411**

Common Error Analysis

Error: When writing the equation of a circle in standard form, students tend to forget to square the radius. Thus, for the circle with center $(4,-2)$ and radius 3, they may incorrectly write the equation as $(x - 4)^2 + (y + 2)^2 = 3$.

Another typical error is to write the square of $\sqrt{5}$ as 25 rather than as 5.

Emphasize that $(\sqrt{a})^2$ is a, not a^2.

Checkpoint

Write an equation of each circle for the given conditions.

1. center: $(0,0)$, radius 12 $x^2 + y^2 = 144$
2. center: $(0,0)$; radius $\sqrt{7}$ $x^2 + y^2 = 7$
3. center: $(0,0)$; passing through $P(5,-2)$ $x^2 + y^2 = 29$
4. center: $(-2,-4)$; radius 5 $(x + 2)^2 + (y + 4)^2 = 25$

Closure

Ask students to give the standard form of the equation of a circle, the coordinates of the center, and the radius. $(x - h)^2 + (y - k)^2 = r^2$; (h,k); r Have students give the steps in writing an equation like $x^2 - 6x + y^2 - 14y - 42 = 0$ so that the circle can be graphed.   See Example 3.

Checkpoint provides a quick way to determine what students have learned from the lesson and what concepts and skills need to be reinforced.

Closure questions help students summarize, communicate, and form conclusions about the major concepts of the lesson.

Additional Example 2

Write an equation of the circle in standard form with center $C(2,-5)$ and passing through the point $P(7,7)$. Graph the circle.
$(x - 2)^2 + (y + 5)^2 = 169$

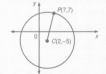

Additional Example 3

Rewrite $x^2 - 4x + y^2 + 8y - 16 = 0$ in standard form. Identify the center and radius.
$(x - 2)^2 + (y + 4)^2 = 36$; $c(2,-4)$; $r = 6$

411

Additional Examples may be used for reteaching or as a quiz to evaluate students' understanding of the lesson.

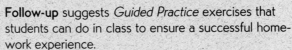

FREQUENT REVIEW OPPORTUNITIES HELP YOU EVALUATE STUDENTS' UNDERSTANDING.

The **Mixed Review**, which occurs at the end of most lessons, reinforces newly learned concepts and maintains skills from previous lessons. For the convenience of the student, exercises are referenced to the chapter and lesson where the related skill or concept was taught.

Follow-up suggests *Guided Practice* exercises that students can do in class to ensure a successful homework experience.

Independent Practice groups exercises according to A, B, and C difficulty levels for basic, average, and above-average students.

C-level exercises, numbered in red, require students to use critical-thinking skills to solve more challenging problems.

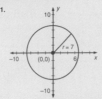

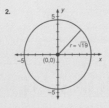

▬▬FOLLOW UP

Guided Practice
Classroom Exercises 1–4

Independent Practice
🅰 Ex. 1–24, 🅱 Ex. 25–31, 🅲 Ex. 32–38

Basic: WE 1–25 odd
Average: WE 7–31 odd
Above Average: WE 13–37 odd

Additional Answers
Classroom Exercises

1.

2.

See page 455 for the answers to Classroom Ex. 3 and 4 and pages 817–819 for Written Ex. 1–28, 30, 31, 33, 36, 38, and Mixed Rev. Ex. 1 and 2.

Graph, rewriting each equation in standard form if necessary. Identify the circle's center and radius.

17. $x^2 + y^2 = 4$ **18.** $x^2 + y^2 = 25$ **19.** $x^2 + y^2 = 64$ **20.** $x^2 + y^2 = 81$
21. $x^2 + y^2 = 8$ **22.** $x^2 + y^2 = 12$ **23.** $x^2 + y^2 = 21$ **24.** $x^2 + y^2 = 15$
25. $x^2 - 6x + y^2 - 10y - 2 = 0$ **26.** $x^2 + 8x + y^2 - 4y - 5 = 0$
27. $x^2 + 6x + y^2 - 14y - 42 = 0$ **28.** $x^2 - 12x + y^2 - 2y - 8 = 0$
29. A point $P(x,y)$ moves in a coordinate plane about a fixed point $C(-1,2)$ so that its distance from C is always $2\sqrt{3}$ units. Use the distance formula, $d = \sqrt{(x_2 - x_1)^2 + (y_2 - y_1)^2}$, to write an equation describing the path of point P. $(x + 1)^2 + (y - 2)^2 = 12$

Use the distance formula for Exercises 30 and 31.
30. Prove Theorem 12.1.   **31.** Prove Theorem 12.2.
32. The circle with equation $(x - 1)^2 + (y + 1)^2 = 9$ is translated 4 units to the right and 2 units downward. Find the equation of the resulting circle. $(x - 5)^2 + (y + 3)^2 = 9$
33. The circle with equation $x^2 - 8x + y^2 + 4y + 9 = 0$ is translated 2 units to the left and 5 units upward. Find the equation of the resulting circle. $(x - 2)^2 + (y - 3)^2 = 11$

Write, in standard form, the equation of the circle having the given properties.
34. center on the line $x = 5$, tangent to the y-axis at $P(0,9)$ $(x - 5)^2 + (y - 9)^2 = 25$
35. tangent to both rays of the graph of $y = |x|$, radius 4
(HINT: The slope of $y = x$ is one, so the triangle formed by the origin, the circle's center, and the point of tangency is isosceles.) $x^2 + (y - 4\sqrt{2})^2 = 16$

Use the circle $x^2 + y^2 = 100$ and triangle *ABC* for Exercises 36–38.
36. Show that the vertices $A(-6,8)$, $B(0,10)$, and $C(6,-8)$ of triangle *ABC* are points on the circle, and that the triangle is a right triangle.
37. Write an equation of the line tangent to the circle at $A(-6,8)$. $3x - 4y = -50$
38. Show that the perpendicular bisector of $\overline{AB}$ passes through the center of the circle.

Mixed Review

1. Graph $y = |x - 4| + 5.$ *11.3*   **2.** Graph $y = x^2 - 8x + 2.$ *11.5*

Simplify. *5.2, 8.2, 8.6, 9.2*
3. $(2x - 4)^0, x \neq 2$ 1 **4.** $\sqrt{32x^3}$ $4x^2\sqrt{2x}$ **5.** $27^{-\frac{2}{3}}$ $\frac{1}{9}$ **6.** $(3 + 4i)^2$ $-7 + 24i$

Enrichment

Challenge students to find the coordinates of the center of the circle that contains the points (1,2), (7,8), and (13,0).

You may wish to draw the figure in the center column on the chalkboard and give the hint that the perpendicular bisector of every chord of a circle passes through the center of the circle.

When the answer has been found, have the students check it by showing that the three points are equidistant from the center.

The center is $\left(\frac{50}{7}, \frac{13}{7}\right)$. The distance of each point from the center is $\frac{\sqrt{1850}}{7}$, or $\frac{5\sqrt{74}}{7}$ units.

The **Enrichment** problem is a challenging exercise for students who have mastered the lesson content.

IMPLEMENTING THE NCTM *STANDARDS* WITH HOLT ALGEBRA WITH TRIGONOMETRY

EVERYONE WHO TEACHES mathematics in the 1990s is concerned about how to implement the *Curriculum and Evaluation Standards for School Mathematics*, published by the National Council of Teachers of Mathematics. The professional articles in this section are designed to help you use *Holt Algebra with Trigonometry* to implement the NCTM *Standards* in your classroom.

NCTM's vision of the future "sees students studying much of the same mathematics currently taught but with quite a different emphasis" and builds on the premise that "*what* a student learns depends to a great degree on *how* he or she has learned it." (p. 5, NCTM *Standards*)

How will your students learn algebra? The six articles that follow are designed to help you apply these six new approaches in your classroom.

Applications and Connections
Cooperative Learning Groups
Problem-Solving Strategies
Critical Thinking Questions
Reading, Writing, and Discussing Math
Technology

Each of these articles can help you to achieve these five general goals for all students as listed in the NCTM *Standards* (pp. 5–6, NCTM *Standards*).

"
1. *Learning to value mathematics*
2. *Becoming confident in one's own ability*
3. *Becoming a mathematical problem solver*
4. *Learning to communicate mathematically*
5. *Learning to reason mathematically*"

These goals suggest new approaches to much of the traditional content of an algebra course. **Standard 5: Algebra** (p. 150, NCTM *Standards*) states (in part) that the curriculum should include topics such that the students can

"
• *represent situations that involve variable quantities with expressions, equations, inequalities, and matrices;*
• *use tables and graphs as tools to interpret expressions, equations, and inequalities;*
• *operate on expressions and matrices, and solve equations and inequalities;*
• *appreciate the power of mathematical abstraction and symbolism;*
• *use matrices to solve linear systems.*"

You will find ample material that relates directly to these topics as you survey *Holt Algebra with Trigonometry*, as well as content that relates to the other Standards. Pages 830 through 834 of the Annotated Teacher's Edition repeat the text of the applicable Standards and correlate each lesson in *Holt Algebra with Trigonometry* to the individual Standards. ■

APPLICATIONS AND CONNECTIONS

What are applications and connections?

A PERSON GATHERS, discovers, or creates knowledge in the course of some activity having a purpose. . . . Instructions should persistently emphasize 'doing' rather than 'knowing that.'" (p. 7, NCTM *Standards*) Applications and connections are the ways in which something is done with mathematics. Applications are the purposeful activities that students do in order to apply, or make use of, the algebra they have learned. Connections are the activities that relate the algebra learned today to that which was previously learned, or to other branches of mathematics, or to other disciplines such as chemistry, physics, or biology.

It is true that a person can keep a checkbook and fill out a tax form without using algebra. Students sometimes have difficulty seeing the relevance of what they are learning— "When am I ever going to use this?" They need to see how algebra and the skills they learn in class relate to getting a job, making decisions, and solving problems in the world outside the classroom.

Students in the United States drop out of mathematics classes at an alarming rate, and minorities drop out at a disproportionately greater rate. "Because mathematics is a key to leadership in our technological society, uneven preparation in mathematics contributes to unequal opportunity for economic power." (p. 18, *Educational Leadership*, Sept. 1989). Applications and connections give students a reason to continue their study of mathematics as they recognize its utility and importance.

How do I use applications and connections with my students?

Take every opportunity to relate mathematics and algebra to what is happening in the world today. Use newspapers, television programs, and magazines to relate mathematics to what is happening outside the classroom. Encourage students to bring in materials that show the use of mathematics in science and other disciplines. Invite them to share with the class the math they may be doing in some other course in school. Assign an interview so that students ask their parents and other adults what preparation in mathematics they had for their careers, and what mathematics they now wish they had taken in school. Use guest speakers to bring the applications of mathematics into the classroom.

For more formal experiences, use the worksheets in the *Teacher's ResourceBank*™. The *Application Worksheets* and the *Project Worksheets* guide the students in ways to use the skills they have learned. These help to convince the students that algebra is both real and useful.

Take a few minutes to discuss the images that open each chapter. Share with the students the additional information that is in the *Annotated Teacher's Edition* about how computer graphics are used in the world outside the classroom. The *Annotated Teacher's Edition* also gives Connections that you can share with the class. Use the Applications sections that occur within the chapters to stimulate discussion of how math relates to other topics.

How will applications and connections help me implement the *Standards*?

The *Standards* (p. 125, NCTM *Standards*)

" *call for a shift in emphasis from a curriculum dominated by memorization of isolated facts and procedures and by proficiency with paper-and-pencil skills to one that emphasizes conceptual understandings, multiple representations and connections, mathematical modeling, and mathematical problem solving.* **"**

Applications and connections help you and your students take the step from solving equations with paper and pencil to applying algebra and other mathematics to the world outside the classroom. Applications can help stimulate that all-important student comment, so rewarding to the teacher, "Oh, now I see!" ■

COOPERATIVE LEARNING GROUPS

What are cooperative learning groups?

THE *STANDARDS* POINT out that "instructional settings that encourage investigation, cooperation, and communication foster problem posing as well as problem solving." (p. 138, NCTM *Standards*) Cooperative learning groups are such an instructional setting.

Cooperative learning means more than simple working in teams or groups. Some of the distinguishing characteristics of true cooperative learning are these.

- Groups are heterogeneous to increase tolerance and understanding.
- Students are responsible for their own participation and for making sure that others participate.
- Help is sought from within the group and the teacher is asked for help only when the whole group agrees to ask a question.
- Consensus on the answer is required from the whole group.
- The group evaluates their own strategies and ideas rather than relying on the teacher for this evaluation.
- The teacher provides help by means of questions rather than giving hints.
- The students who learn at a faster pace do NOT do the task alone and then help other students.
- One student does NOT do all the work and then the others sign off on it.
- Participants encourage each other to explain answers and how they arrived at them.
- Ideas, not people, are criticized.
- The logic of an idea, not peer pressure or majority rule, determines its value.

Rather than working in competition with other students, or as an individual working for a personal best, students in cooperative learning groups are working together towards shared goals; the success of each student in the group depends on the success of the group as a whole, which in turn depends on the success of each member. Each individual is accountable for mastering the material, and also for helping everyone in the group to master the material. Thus cooperative learning groups foster interdependence.

The benefits of this learning structure incorporate those of peer tutoring (in which the tutor also learns by helping), increased understanding of diverse viewpoints, and simulation of the way work is done on the job.

How do I use cooperative learning groups with my students?

Students may not be familiar with interdependent groups in a classroom setting, though they will usually have experienced the team approach in games. The first time you organize cooperative learning groups, discuss a set of procedural rules for the groups based on the characteristics in the foregoing list. The first time, let students form their own groups of four or five. Later, after they have begun to learn how to work together, you can use random methods to form groups that are heterogeneous.

For the first try, assign a familiar task such as producing study notes for a test. Tell the students that the goal of the activity includes learning the process (cooperation) as well as producing the product (the notes). Encourage each student to take responsibility for

a) participating and contributing,
b) ensuring that all other members of the group participate,
c) staying on task, and
d) producing a high quality final product.

Once the procedures are understood (though the groups will probably need much practice before they can adhere effectively to them), use the appropriate activity worksheets (Projects, Applications, Graphing

Calculator, and Problem Solving), and the Investigations that begin each chapter, to provide algebra-related tasks for the groups.

You can also use cooperative learning groups with the Prerequisite Quizzes before the lesson to verify needed skills (Prerequisite Quiz, see also in the *Annotated Teacher's Edition*) and after the lesson to verify understanding (Checkpoint Quizzes, see also in the *Annotated Teacher's Edition*). The Classroom Exercises that follow each lesson can also be done in cooperative groups to prepare students for the Written Exercises.

Conclude each session by a discussion about how the procedure worked, and how they can improve the process.

How will groups help me implement the *Standards*?

Research indicates that "cooperative learning experiences tend to promote higher achievement than do competitive and individualistic learning experiences." (p. 15, *Circles of Learning: Cooperation in the Classroom,* by Johnson, Johnson, Holubec, and Roy.) This results in part because the discussion process promotes the development of critical thinking skills and the group setting increases student motivation. The peer support and the oral repetition of information that occurs in the group also contribute to efficient learning.

The use of cooperative learning groups clearly belongs in an effective variety of instructional methods as called for by the *Standards* (p. 125, 128, NCTM *Standards*).

" *A variety of instructional methods should be used in classrooms in order to cultivate students' abilities to investigate, to make sense of, and to construct meanings from new situations; to make and provide arguments for conjectures; and to use a flexible set of strategies to solve problems from both within and outside mathematics.* " ■

PROBLEM-SOLVING STRATEGIES

What are problem-solving strategies?

THE NCTM STANDARDS recommend increased attention to "problem solving as a means as well as a goal of instruction." (p. 129 of the *Standards*) As a means of instruction, problem-solving strategies are taught overtly so that students can choose from many possible ways to begin solving a given problem.

Some students, especially those who have been unsuccessful in mathematics, have developed, on their own, only one problem-solving strategy—guess and give up. Teaching problem-solving strategies overtly shows students that there are many procedures used by successful problem solvers to solve a problem. When you use problem solving as a means of instruction, students begin to realize that getting the right answer is not merely a matter of luck, innate talent, or magic. They begin to see that they too can experience success at mathematics.

In order to create an atmosphere of successful learning, change the emphasis in your classroom from correct answers to effective process and procedures. Pay attention to the reasons and steps that go into solving a problem. Give credit for the process. Ask for, and value, the explanations students give to tell how they reached a solution. Emphasize that there may be more than one way to arrive at a correct answer. Encourage discussions about the advantages (such as efficiency and generalizability) of various methods. For some students, it may be appropriate to point out that the easiest and most efficient way for a human to solve a problem may not be the most efficient way for a computer program to solve that same problem.

How do I use these strategies with my students?

To use problem-solving strategies with your students, follow these guidelines:

1. Teach problem-solving strategies directly by using the *Problem Solving Worksheets* and the Problem-Solving Strategies features in the text.
2. When you discuss the Examples in the Lesson, ask students how they plan to solve the problem.
3. Ask students to tell the class how they arrived at an answer.
4. Discuss with the class whether another strategy would have worked; compare different strategies that students may have used for a given problem.
5. Give recognition and credit for correct procedures as well as for correct answers; teach students that there may be only one correct answer but several correct ways to obtain that answer.
6. In reteaching students, ask them to explain their thoughts as they work the problem. This gives you a chance to help students to think logically and to suggest more effective problem-solving strategies.

How will problem solving help me implement the *Standards*?
(p. 139, NCTM *Standards*)

" *The importance of problem solving to all education cannot be overestimated. To serve this goal effectively, the mathematics curriculum must provide many opportunities for all students to meet problems that interest and challenge them and that, with appropriate effort, they can solve.* " ■

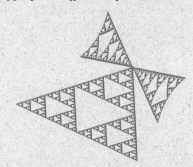

CRITICAL THINKING QUESTIONS

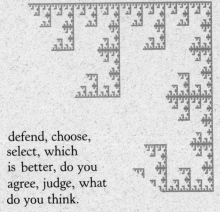

What are critical thinking questions?

CRITICAL THINKING questions ask the students to engage in mathematical thinking and to construct, symbolize, apply, and generalize mathematical ideas. This thinking can include investigations, constructing meanings from new situations, making and providing arguments for conjectures, working cooperatively, and creative and self-directed learning.

"Critical thinking" refers to types of thinking that are of a higher order in Bloom's taxonomy; these include application, analysis, synthesis, and evaluation.

Application involves applying concepts and ideas to new situations and is often signaled by these verbs: apply, build, choose, solve, plan, develop, construct, demonstrate, show.

Analysis means finding the underlying structure and breaking it down into stages or processes. This frequently involves looking for patterns and classifying examples and is often signaled by these verbs: relate, classify, compare, contrast, diagram, analyze, recognize.

Synthesis means bringing together data from various sources to come up with a new conclusion and is often signaled by these verbs and phrases: design, create, develop, make up, what happens if, invent, write a formula for.

Evaluation means judging the quality or worth of something, for example, which method works, or is more efficient. It is often signaled by these verbs and phrases: prove or disprove, evaluate, conclude,

defend, choose, select, which is better, do you agree, judge, what do you think.

For example, in your algebra class, you might ask the students to show that one expression is equivalent to another, and then discuss alternative approaches. You could ask questions such as these: "If substitution is used, how might you use another approach, such as simplifying? Can you think of several different ways to solve this problem? Which way is best, and why?" Critical thinking also includes investigating and exploring to derive properties, generalizations, and procedures.

How do I use these questions with my students?

Critical thinking questions appear throughout this *Holt Algebra with Trigonometry* text. The Investigation that introduces each chapter and the Motivator that introduces each lesson (in the *Annotated Teacher's Edition*) often include such questions.

The Teaching Suggestions for each lesson include Critical Thinking Questions pertinent to that lesson. Many of the worksheets include critical thinking questions.

How will critical thinking help me implement the *Standards*?

The *Standards* call for "an environment that encourages students to explore, formulate and test conjectures, prove generalizations, and discuss and apply the results of their investigations." (p. 128, NCTM *Standards*)

Critical thinking questions are a vital part of such an environment. ■

READING, WRITING, AND DISCUSSING MATH

What is the role of reading, writing, and discussing math?

THE FOURTH GENERAL goal set forth by the *Standards* is "Learning to communicate mathematically." In the past, we have sometimes allowed students to spell mathematical terms incorrectly, or to read aloud "x two" instead of "x-squared" or "the second power of x" because, as students hastened to explain, "This isn't English class." We now recognize that it is just as important that correct English be spoken in math class as it is that correct math be used in science class. These disciplines are interrelated and teach skills that will be integrated on the job. When we require students to use their newly-acquired skills only in a particular class, we participate in an artificial separation of knowledge into school courses.

We also now realize that precise and accurate communication about mathematics is closely connected to doing precise and accurate mathematics. For the classroom teacher, communication about mathematics brings an extra benefit. When a student writes or speaks about a problem, the teacher can often identify a mistaken or incomplete understanding that leads to an error in problem solving.

How do I encourage my students to communicate mathematically?

Ask a student to read an exercise (from the Classroom Exercises, for example) aloud and then answer the question. This helps you assess whether the student knows the vocabulary and the meaning of the various symbols.

Ask a student to put an exercise from the previous night's homework (in the Written Exercises) on the chalkboard and then explain the steps and process to the class. You could then ask if anyone else followed a different procedure to reach the same result. After the second student has explained the alternate method, ask the class to discuss the differences and the advantages and drawbacks of each method. This works particularly well with word problems and solving equations, as there are often two equally valid ways to reach the same conclusion.

Ask students frequently to justify their conclusions, *before* you say whether or not the answer is correct. This encourages them to look beyond a correct answer to think about and defend their procedures and reasoning.

Encourage students to verbalize conjectures *before* they read in the text about certain facts and relationships. Discuss the Closure questions that are given in *Holt Algebra with Trigonometry* at the end of each lesson in the *Annotated Teacher's Edition*. Ask students to keep a math journal in which they record their questions, ideas, and reactions to the activities and discussions in class and a summary of what they read in the text.

Students learn to value those activities that they see you value. If you take time to listen to them as they learn to communicate mathematically, and if you allow their reading, writing, and discussing mathematics to contribute to your on-going evaluation of their achievements, they will know that these things really do count.

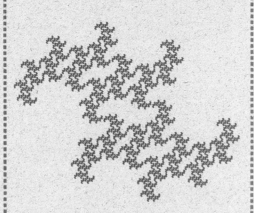

How will communicating mathematically help me meet the *Standards*?

The *Standards* (p. 6) put it this way:

“ *The development of a student's power to use mathematics involves learning the signs, symbols, and terms of mathematics. This is best accomplished in problem situations in which students have an opportunity to read, write, and discuss ideas in which the use of the language of mathematics becomes natural. As students communicate their ideas, they learn to clarify, refine, and consolidate their thinking.* ” ■

TECHNOLOGY

What is the role of technology?

THE NCTM STANDARDS speak of “removing the ‘computational gate’ to the study of high school mathematics” (p. 130, NCTM *Standards*):

“ *By assigning computational algorithms to calculator or computer processing, this curriculum seeks not only to move students forward but to capture their interest.* ”

It is particularly important to note that the non-college-intending student must have better preparation for the jobs of tomorrow (p. 130, NCTM *Standards*).

“ *The ever-increasing role of technology in our society further argues for a curriculum that moves all students beyond computation.* ”

Technology—the use of the calculator, computer, and graphing calculator—enables students to “study mathematics that is more interesting and useful and not characterized as remedial” and this in turn “will enhance students' self-concepts as well as their attitudes toward, and interest in, mathematics.” (p. 131, NCTM *Standards*)

How do I use technology with my students?

Almost everyone today uses a computer at work. The use of computers on the job and at home continues to grow. Make sure that *all* your students have access to whatever technology is available. The emphasis in today's curriculum is not on writing programs, but on investigation and foreshadowing of mathematical ideas and applications by means of the computer.

The Computer Investigation pages in *Holt Algebra with Trigonometry* pose problems that can be solved by using the BASIC programs on the disk called *Investigating Algebra with the Computer*. The *Graphing Calculator Worksheets* in the *Teacher's ResourceBank*™ provide students with problems that use either the Casio or Texas Instruments graphing calculator to explore various topics in algebra and related mathematics. Complete instructions help students to begin using this new technology.

In the *Holt Algebra with Trigonometry*, when exercises are especially suited to the calculator, the text shows what calculator keys to press. The Technology paragraph in the *Annotated Teacher's Edition* also points out appropriate places in the chapter for technology. However, you may want to encourage students to use hand-held calculators whenever they wish. This simulates mathematics as it is really used, both on the job and in scientific applications. Frequent use of calculators also helps students learn the importance of *estimating* before they calculate, and *checking* afterwards to be sure that answers are reasonable. "Appropriate use of calculators enhances children's understanding and mastery of arithmetic," according to *Everybody Counts, A Report to the Nation on the Future of Mathematics Education,* published by the National Academy of Sciences. The *Standards* (p. 8, NCTM *Standards*) also make this point:

> *Contrary to the fears of many, the availability of calculators and computers has expanded students' capability of performing calculations. There is no evidence to suggest that the availability of calculators makes students dependent on them for simple calculations.*

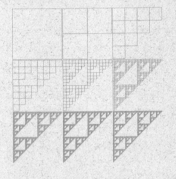

The Computer Investigations pages help students to graph equations and systems easily with the help of technology. With these programs, they can make predictions and draw inferences about patterns.

How will technology help me implement the *Standards*?

The *Standards* (page 129, NCTM *Standards*) propose increased attention to—

> *The use of calculators and computers as tools for learning and doing mathematics.*

The technology resources in *Holt Algebra with Trigonometry* and *Activity Worksheets* help you use the calculator and computer in this way. ∎

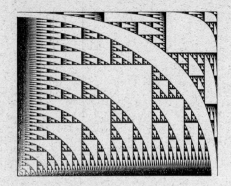

SUMMARY

AT THE BEGINNING of this section, we listed the five general goals for all students from the NCTM *Standards*.

1. Learning to value mathematics
2. Becoming confident in one's own ability
3. Becoming a mathematical problem solver
4. Learning to communicate mathematically
5. Learning to reason mathematically

These articles have explained how various instructional techniques used with the textbook, special features, and worksheets from *Holt Algebra with Trigonometry* can help you achieve those goals.

1. Cooperative learning groups help students value mathematics, as do the Applications, worksheets, and features in *Holt Algebra with Trigonometry*, and the Math Connections discussed in the *Annotated Teacher's Edition*.
2. Cooperative learning groups, problem-solving strategies, critical thinking questions, and technology used with *Holt Algebra with Trigonometry* and the accompanying worksheets all help students to become confident in their own abilities.
3. Problem-solving strategies, and the opportunities to practice and increase them provided by the Problem Solving Strategies features in the textbooks, and by the *Problem Solving Worksheets,* help students become mathematical problem solvers.
4. Cooperative learning groups, closure questions for each lesson in the *Annotated Teacher's Edition,* the questions on the worksheets, the Focus on Reading features, and the paragraphs in the chapter interleaf of the *Annotated Teacher's Edition* called Reading and Writing Math all help students learn how to communicate mathematically.
5. The entire *Holt Algebra with Trigonometry,* with its accompanying materials, helps the students learn to reason mathematically. The extensive practice with problem-solving strategies strengthens their reasoning abilities and their skills in communicating that reasoning to others. The chapter on geometry (Chapter 11: *Coordinate Geometry and Quadratic Functions*) and the lessons on graphing throughout the text help students learn to reason from a coordinate and an analytic perspective.

The *Standards* (p. 243, NCTM *Standards*) make this comment in the context of program evaluation:

> *The mathematics classroom envisioned in the Standards is one in which calculators, computers, courseware, and manipulative materials are readily available and regularly used in instruction.*

Holt Algebra with Trigonometry provides you with materials, in the textbook, in the *Annotated Teacher's Edition,* and in the worksheets, to help you meet these goals and standards as you teach your students. ∎

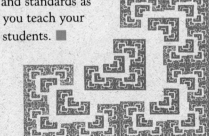

Symbol List

Symbol	Meaning	Symbol	Meaning
$\lvert x \rvert$	the absolute value of x	$\overrightarrow{AB}$	ray AB
$\overset{?}{=}$	is possibly equal to	$\vec{V}$	vector V
$\approx$	is approximately equal to	$\lVert \vec{v} \rVert$	the norm of the vector $\vec{v}$
$\neq$	is not equal to	$\angle A$	angle A
$\{\ \}$	set enclosures	$\mathbf{m}\angle A$	the measure of $\angle A$
$\mid$	such that	$^{\circ}$	degrees
$\varnothing$	empty set	$'$	minutes of arc
$<$	is less than	$\triangle ABC$	triangle ABC
$\leq$	is less than or equal to	i	the imaginary number $i = \sqrt{-1}$
$>$	is greater than		
$\geq$	is greater than or equal to	$\overline{a + bi}$	the conjugate of the complex number $a + bi$
$-x$	the additive inverse of x		
$\dfrac{1}{x}$	the multiplicative inverse of x	Σ	summation symbol
		$!$	factorial
$\pm$	plus-or-minus symbol		the number of combinations of n things taken r at a time;
(x,y)	ordered pair	$\dbinom{n}{r}$	
(x,y,z)	ordered triple		$\dbinom{n}{r} = \dfrac{n!}{r!(n-r)!}$
$f(x)$	f of x; the value of the function f at x	$P(E)$	the probability of event E
$g(f(x))$	composition; g of f of x	σ	standard deviation
f^{-1}	the inverse of the function f	$\log_b x$	the base-b logarithm of x
$\sqrt{x}$	the square root of x	A^{-1}	the inverse of matrix A
$\sqrt[n]{x}$	the nth root of x	$\det$	determinant
x^n	the nth power of x	$\mathrm{Sin}^{-1} x$	the principal values of the inverse of $\sin x$
$x^{\frac{p}{q}}$	the pth power of the qth root of x	π	the Greek letter pi
$\overline{AB}$	segment AB	θ	the Greek letter theta
$\overleftrightarrow{AB}$	line AB	φ	the Greek letter phi

1 LINEAR EQUATIONS

OVERVIEW

In this opening chapter, attention is given to several major areas of algebra such as order of operations with real numbers, linear equations, and problem solving. The first three lessons on fundamental real-number operations and linear equations serve as a background both for later chapters in general and for the three problem-solving lessons of this chapter in particular. This reflects one of the authors' main purposes, that of reaching problem-solving lessons very early in the course.

OBJECTIVES

- To identify properties of operations with real numbers
- To simplify and evaluate algebraic expressions
- To solve a linear equation in one variable
- To prove two algebraic expressions equivalent
- To solve word problems

PROBLEM SOLVING

Students should apply their algebraic skills to the solution of problems as soon as they have developed or reviewed those skills. For this reason, almost half of the lessons of the first chapter are devoted to problem solving. In each of the problem-solving lessons, students are led to understand some of the steps involved in the algebraic solution of a problem. This is done by suggesting, in the solutions to the Examples, some of the questions that students need to ask themselves as they solve a problem. For example, "What are you to find?" and "What is given?". Students are also shown specific strategies that they may apply to solve a problem, such as "Find an equation," or "Write the formula." The importance of checking the answer *in the original problem* is also emphasized.

READING AND WRITING MATH

In this chapter, as throughout the text, special attention is paid to helping students read mathematical symbols with understanding. This is seen particularly in the problem solving lessons. For example, in Lesson 1.6, English phrases are transformed into algebraic expressions and, in the Check of Example 1, the answers obtained are compared directly to the English wording of the problem. Throughout the text, similar careful attention is given to reading mathematical expressions with understanding.

TECHNOLOGY

Calculator: You may want to encourage students to use scientific calculators as an aid in computation for some of the exercises. These might include Written Exercises 28–32 on page 4, Exercises 16–17 on page 7 (using memory or the special keys), many of Exercises 1–25 on pages 20–21, Exercises 4–20 on pages 29–30, and the Application Exercises on page 12.

SPECIAL FEATURES

Mixed Review pp. 7, 12, 16, 25, 30, 35
Using the Calculator p. 7
Application: Insect Songs p. 12
Midchapter Review p. 21
Brainteaser p. 25
Application: Equilibrium p. 26
Key Terms p. 36
Key Ideas and Review Exercises
  pp. 36–37
Chapter 1 Test p. 38
College Prep Test p. 39

PLANNING GUIDE

Lesson	Basic	Average	Above Average	Resources
1.1 pp. 4–5	CE all WE 1–15, 16–33 odd	CE all WE 1–15 odd, 16–37 odd	CE all WE 1–15 odd, 16–39 odd	Reteaching 1 Practice 1
1.2 pp. 6–7	CE all WE 1–15, 18, 19, UC all	CE all WE 1–19 odd, 21, 22, UC all	CE all WE 1–21 odd UC all	Reteaching 2 Practice 2
1.3 pp. 11–12	CE all WE 1–14, 15–25 odd Application	CE all WE 1–25 odd, 26, 27 Application	CE all WE 1–29 odd Application	Reteaching 3 Practice 3
1.4 pp. 15–16	CE all WE 1–16, 17–27 odd	CE all WE 1–22 odd, 23–38	CE all WE 1–36 odd, 37, 39, 41	Reteaching 4 Practice 4
1.5 pp. 19–21	CE all WE 1–11, 13–21 odd Midchapter Review	CE all WE 1–21 odd, 23, 24 Midchapter Review	CE all WE 1–21 odd, 23, 25 Midchapter Review	Reteaching 5 Practice 5
1.6 pp. 23–26	CE all WE 1–8, 9–15 odd Brainteaser Application	CE all WE 1–16 odd, 18 Brainteaser Application	CE all WE 1–19 odd Brainteaser Application	Reteaching 6 Practice 6
1.7 pp. 29–30	CE all WE 1–9, 11–15 odd	CE all WE 1–15 odd, 18	CE all 1–19 odd	Reteaching 7 Practice 7
1.8 pp. 34–35	CE all WE 1–8, 9–15 odd	CE all WE 1–11 odd, 12, 14, 17, 20	CE all WE 1–15 odd 16, 19, 20	Reteaching 8 Practice 8
Chapter 1 Review pp. 36–37	all	1, 3, 5, 8, 11, 12, 15, 16,	1–20 odd	
Chapter 1 Test	1–13 odd	1–15 odd	1–17 odd	
College Prep Test p. 39	all	all	all	

CE = Classroom Exercises   WE = Written Exercises   UC = Using the Calculator
NOTE: For each level, all students should be assigned all Mixed Review exercises.

▮▮INVESTIGATION

Project: Working individually or in groups, the students use calculators to conduct an open-ended investigation of repeating decimal patterns within rational numbers.

Have the students divide integers on their calculators and observe the patterns of decimals that repeat. For instance, $1 \div 44 = .022727273$, where the pattern that repeats is 27. In this case, the last digit was rounded up. (Some calculators do not round the last digit.)

For some rational numbers, the pattern that repeats will be too long to show on the display. For example, the decimal expansion of $1 \div 17$ has a pattern 16 digits long.

The pattern begins to repeat here. ──────┐
                                       ↓

```
     .05882352941176470588235 29...
17)1.0000000000000000000000000...
   .99994
        600000
        599998
             200000
             199988
                 1200000
                 1199996
                      400000
                      399993
                           7
```

Challenge the students to find a calculator-assisted method for doing long division so that they can carry out the decimal expansion of $1 \div 17$.

For the first step, divide $1 \div 17 = .058823$. Discard the last digit and multiply $.05882 \times 17 = .99994$. Subtract as shown and bring down 5 zeros. Follow the pattern for as long as desired.

The number of digits that can be carried at each step depends on the calculator that is used and the number of digits in the divisor. The last digit on the display must always be discarded.

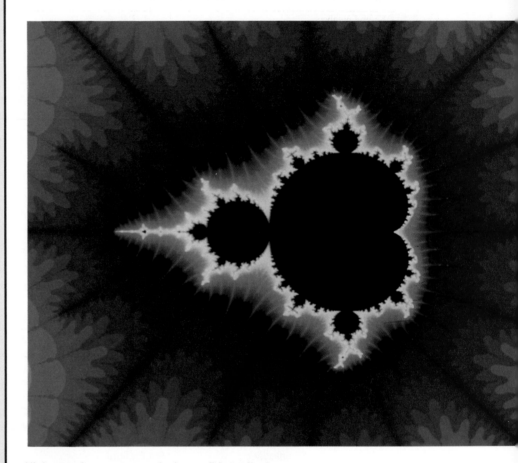

High-speed computers make it possible to display information graphically in mathematics, science, art, and design. The image above is an example of a purely mathematical object called a *fractal*. This particular fractal is known as the Mandelbrot set.

More About the Image

A **fractal** is a self-similar structure. The pattern of the whole repeats endlessly in the details of the parts. The overall pattern of the dark region of the Mandelbrot set occurs infinitely many times in the colored boundary regions. A computer is particularly useful in studying fractals. It can repeat a simple instruction over and over without further input from the user. Such a process is known as **iteration.**

1.1 Operations with Real Numbers

Teaching Resources

Problem Solving 1
Quick Quizzes 1
Reteaching and Practice
  Worksheets 1
Transparency 1

Objective To perform basic operations with pairs of real numbers

Listed below are examples of several kinds of real numbers.

Natural numbers	1, 2, 3, 4, . . .
Whole numbers: natural numbers and zero	0, 1, 2, 3, . . .
Integers: whole numbers and their opposites	. . . , -2, -1, 0, 1, 2, . . .
Rational numbers: quotients of integers	$\frac{3}{4}$, or 0.75; $\frac{-8}{1}$, or -8
(terminating decimals)	$\frac{-17}{5}$, or $-3\frac{2}{5}$, or -3.4
(repeating decimals)	$\frac{1}{3}$, or 0.33 . . . , or $0.\overline{3}$
Irrational numbers: nonterminating, nonrepeating decimals	$\sqrt{2}$, or 1.4142135 . . . 5.1511511151 . . . ; π

A **rational number** is a number that can be written as the quotient $\frac{a}{b}$ of two integers, a and b, where $b \neq 0$. The decimal form of a rational number either terminates (such as 0.75) or repeats (such as 0.33 . . .).

An **irrational number** is a nonterminating, nonrepeating decimal.

The set of **real numbers** is the set of all rational and irrational numbers. The sign rules for adding real numbers can be stated in terms of the *absolute values* of the numbers. The **absolute value** of a *positive* real number or zero is the number itself. The absolute value of a *negative* real number is its opposite. The absolute value of any real number x, symbolized by $|x|$, is never negative, as shown by the following examples.

$$|9| = 9 \quad |-3| = 3 \quad |0| = 0$$

Sign Rules for Addition
To add two nonzero real numbers with *like signs*, add their absolute values, and give the sum the sign of the original numbers.

To add two nonzero real numbers with *unlike signs*, subtract the lesser absolute value from the greater and give the difference the sign of the original number with the greater absolute value.

EXAMPLE 1 Simplify. **a.** $-4 + (-2.6)$ **b.** $2.4 + (-7.5)$

Solutions **a.** $-4 + (-2.6) = -(|-4| + |-2.6|) = -(4 + 2.6) = -6.6$
b. $2.4 + (-7.5) = -(|-7.5| - |2.4|) = -(7.5 - 2.4) = -5.1$

GETTING STARTED

Prerequisite Quiz

Find the value.

1. $3.82 + 7.4$ 11.22
2. $3.5(2.4)$ 8.4
3. $14.6 - 7.92$ 6.68
4. $18.72 \div 3.6$ 5.2
5. $\frac{1}{2} + \frac{2}{3}$ $\frac{7}{6}$
6. $\frac{2}{5} \cdot \frac{15}{8} \cdot \frac{1}{3}$ $\frac{1}{4}$

Motivator

Of the numbers your students will study in this lesson, the positive rational numbers are used in everyday situations. Irrational numbers are seldom used for everyday purposes. Have students identify some situations, and give a numerical example of each instance, where negative numbers can be useful.Two sample answers: debt, $-\$7.50$; degrees below zero, $-8°F$

Additional Example 1

Simplify.

a. $-4.7 + (-3.5)$ -8.2
b. $-9.6 + 7.1$ -2.5

Highlighting the Standards

Standard 5d: In Example 3, students review the connections between algebraic and arithmetic symbols. The exercises demonstrate the power of operational symbols.

1

Lesson Note

Students can think of the real numbers as all the numbers they have studied and used up to this time. The phrase "real number(s)" will be used frequently until complex numbers are introduced in Chapter 9.

Absolute value is introduced in order to express sign rules for addition and will be studied in depth in Chapter 2.

Math Connections

Measurement: Students should understand that real numbers are the basis of measurement for three fundamental objects of scientific interest: *displacement* (distance), *mass*, and *time*. The everyday use of real numbers as the basis of measurement began centuries ago, long before they were placed on a logically secure footing in the nineteenth century.

The operation of subtraction is defined as follows.

Definition

$a - b = c$ means that $c + b = a$.

For example, $3 - 7 = -4$ because $-4 + 7 = 3$.

Finding $a - b$ can be done by finding the opposite of b, that is $-b$, and then adding the real numbers a and $-b$.

Rule of Subtraction
$a - b = a + (-b)$ for all real numbers a and b.

EXAMPLE 2 Simplify.   **a.** $7.2 - 9.6$ **b.** $-12 - 3.5$

Solutions **a.** $7.2 - 9.6 = 7.2 + (-9.6)$ **b.** $-12 - 3.5 = -12 + (-3.5)$
$\qquad\qquad\qquad = -2.4$ $\qquad\qquad\qquad\qquad\qquad = -15.5$

The sign rules for multiplying or dividing a pair of nonzero real numbers fall into two cases, like signs and unlike signs, as illustrated below.

like $\begin{cases} 10(2.5) = 25 \\ -10 \div (-2.5) = 4 \end{cases}$ unlike $\begin{cases} 6.8(-2) = -13.6 \\ -6.8 \div 2 = -3.4 \end{cases}$
signs

Recall that *division by zero is undefined*. This can be explained by the definition of division that follows.

Definition

$a \div b = c$ means that $c \cdot b = a$.

For example, $15 \div 3 = 5$ because $5 \cdot 3 = 15$. However, $8 \div 0 = x$ has no solution because $x \cdot 0 = 8$ has no solution.

Sign Rules for Multiplication and Division
For all nonzero real numbers a and b with like signs, the product $a \cdot b$ and the quotient $a \div b$ are positive numbers.

For all nonzero real numbers a and b with unlike signs, the product $a \cdot b$ and the quotient $a \div b$ are negative numbers.

$a \cdot 0 = 0$ and $0 \cdot a = 0$ for each real number a.
$\frac{0}{a} = 0 \; (a \neq 0)$ and $\frac{a}{0}$ is undefined for each real number a.

2 Chapter 1 Linear Equations

Additional Example 2

Simplify.

a. $5.1 - 14.8$ -9.7
b. $-11 - 20.4$ -31.4

A **variable** is a symbol, usually a letter, that represents a number. An expression such as $c - d$ can be evaluated for given values of the variables c and d, such as 15 for c and -8 for d. To do this, substitute the values for the variables, and simplify the result as shown.

$$c - d = 15 - (-8) = 15 + 8 = 23$$

EXAMPLE 3 Evaluate each expression for $a = -16$, $b = 8$, $c = -4$, and $d = 0$, if possible.

Solutions

Expression	Result	Answer
$a + c$	$-16 + (-4)$	-20
$b - c$	$8 - (-4)$	12
$c - b$	$-4 - 8$	-12
$d - b$	$0 - 8$	-8
bc, or $b \cdot c$	$8(-4)$	-32
$-5c$, or $-5 \cdot c$	$-5(-4)$	20
$\frac{a}{b}$	$\frac{-16}{8}$	-2
$\frac{a}{c}$	$\frac{-16}{-4}$	4
$\frac{a}{d}$	$\frac{-16}{0}$	undefined
$\frac{d}{b}$	$\frac{0}{8}$	0
ad	$-16 \cdot 0$	0

Operations occur more than once in expressions such as $23 - 52 - (-14) - 7$ and $-5 \cdot 4(-10)(-3)$. Such expressions can be simplified by combining pairs of numbers in several steps.

EXAMPLE 4 Simplify. **a.** $23 - 52 - (-14) - 7$ **b.** $-5 \cdot 4(-10)(-3)$

Solutions

a. Change subtracting to *adding the opposite*. Group addends with the same sign and add.

$23 - 52 - (-14) - 7$
$23 + (-52) + 14 + (-7)$
$(23 + 14) + [-52 + (-7)]$
$37 + (-59)$
-22

b. Multiplication is the only operation. Work with pairs of factors.

$-5 \cdot 4(-10)(-3)$
$-20 \cdot 30$
-600

Critical Thinking Questions

Analysis: Ask students which would be easier, to prove that $\frac{1}{23}$ is a repeating decimal or that $\sqrt{2}$ is not a repeating decimal. They should see that, although it takes a few minutes of calculation to show that $\frac{1}{23}$ repeats (the period is 22 digits long), no amount of time will suffice to show that the digits of $\sqrt{2}$ do not repeat.

Common Error Analysis

Error: Students confuse operations like $\frac{0}{3}$ and $\frac{3}{0}$ even after studying the explanation of page 2. They know that one of the two is equal to 0 and the other is undefined, but cannot recall which is which.

A calculator can help, where $0 \div 3$ is 0 and $3 \div 0$ displays E, or Error.

Checkpoint

Simplify, if possible. If not possible, write *undefined*.

1. $(-22) + 15$ -7
2. $-30 - (-9)$ -21
3. $8(-12)$ -96
4. $9 \div 0$ Undef
5. $-28 \div (-7)$ 4
6. $-2 - 7 - 7$ -16

Evaluate each expression for $a = 30$, $b = 0$, $c = -6$, and $d = -3$, if possible.

7. $\frac{a}{d}$ -10
8. $c - b - a$ -36

Additional Example 3

Evaluate each expression for $a = 24$, $b = -8$, $c = 0$, and $d = 2$, if possible.

1. $a + b$ 16
2. $b - d$ -10
3. $d - b$ 10
4. $c - a$ -24
5. bd, or $b \cdot d$ -16
6. $-7b$, or $-7 \cdot b$ 56
7. $\frac{a}{b}$ -3
8. $\frac{a}{c}$ Undef
9. $\frac{a}{d}$ 12
10. $\frac{c}{d}$ 0
11. ca 0

Additional Example 4

Simplify.

a. $14 - (-6) - 18 - 46$ -44
b. $-9 \cdot 6(-8)(-5)$ $-2,160$

3

Closure

Have students recall the four basic operations with rational numbers. addition, subtraction, multiplication, and division Have them identify operations, different from the four basic operations, and give an example of each.

Three sample answers:
square root: $\sqrt{36} = 6$
raising to a power: $5^3 = 125$
averaging: The average of 6 and 9 is 7.5.

◤◤◤FOLLOW UP

Guided Practice

Classroom Exercises 1–15

Independent Practice

Ⓐ Ex. 1–27, Ⓑ Ex. 28–36, Ⓒ Ex. 37–39

Basic: WE 1–15, 16–33 odd
Average: WE 1–15 odd, 16–37 odd
Above Average: WE 1–15 odd, 16–39 odd

Additional Answers

Written Exercises

37. Because if $0 \div 0 = c$, then $c \cdot 0 = 0$, for which every number is a solution.

Classroom Exercises

Simplify, if possible. If not possible, write *undefined*.

1. $-15 + 45$ 30
2. $-17 + (-11)$ −28
3. $-18 + 18$ 0
4. $-11 - 30$ −41
5. $23 - (-5)$ 28
6. $-35 - (-15)$ −20
7. $-5(-7)$ 35
8. $-15 \div 3$ −5
9. $-24 \div (-6)$ 4
10. $0 \div \pi$ 0
11. $6 \div 0$ Undef
12. $-\sqrt{7} + 0$ $-\sqrt{7}$
13. $-5 - 3 - 3$ −11
14. $-7(-2)(-10)$ −140
15. $-5 \cdot 3(-4)$ 60

Written Exercises

Simplify, if possible. If not possible, write *undefined*.

1. $-45 + 175$ 130
2. $-73 + (-52)$ −125
3. $-62 - 32$ −94
4. $-18 - (-12)$ −6
5. $-8(-11)$ 88
6. $-5 \cdot 12$ −60
7. $-6 \cdot 0$ 0
8. $\frac{-84}{-21}$ 4
9. $\frac{0}{9}$ 0
10. $\frac{9}{0}$ Undef
11. $-9 + 9$ 0
12. $18 - 38 - 12$ −32
13. $-6 - (-31) - 15$ 10
14. $-9.3 + 5.7$ −3.6
15. $-18.2 - (-15.9)$ −2.3

Evaluate for $a = -24$, $b = 6$, $c = -3$, and $d = 0$, if possible.

16. $a + c$ −27
17. $b - c$ 9
18. bc −18
19. $-4c$ 12
20. ad 0
21. $\frac{a}{b}$ −4
22. $\frac{a}{c}$ 8
23. $\frac{b}{d}$ Undef
24. $a - b - c$ −27
25. $b - c - a$ 33
26. $c - a - b$ 15
27. $c - b - 9$ −18

Simplify, if possible.

28. $-8(-15)(-2)$ −240
29. $-6 \cdot 7 \cdot 5$ −210
30. $15(-6.5)$ −97.5
31. $-8.25(-400)$ 3,300
32. $-27 \div (-3.6)$ 7.5
33. $-5(-5)(-5)(-5)$ 625
34. $\frac{1}{2} - \frac{5}{6} - \frac{3}{4}$ $-\frac{13}{12}$
35. $-3 - \frac{2}{5} - \left(-\frac{3}{10}\right)$ $-\frac{31}{10}$
36. $-\frac{3}{2}\left(-\frac{8}{5}\right)\left(\frac{10}{9}\right)$ $\frac{8}{3}$

37. $a \div b = c$ if and only if $c \cdot b = a$. Thus, $3 \div 0 = c$ has no solution, and $\frac{3}{0}$ is undefined. Explain why $\frac{0}{0}$ must also be undefined.

In Exercises 38 and 39, x and y are any two *positive* real numbers with x less than y.

38. Is $(x - y)^{12}$ positive or negative?
Positive

39. If n is a positive integer, is $(x - y)^{2n}$ positive or negative? Positive

Enrichment

Have students experiment or use reasoning to find all the values of x that make each statement true.

1. The absolute value of x is 4. 4 or −4
2. The sum of x and 8 is −1. −9
3. The product of x and 4 is −20. 5
4. If 15 is divided by x, the expression is undefined. 0
5. If −15 is subtracted from x, the difference is −15. −30
6. The absolute value of x has a different sign than x has. All negative numbers

Have each student make up six sentences similar to these. Then pairs of students can exchange papers and solve the problems written by each other.

1.2 Order of Operations

Objective

To apply the rules for the order of operations

More than one operation may appear in a numerical expression, as in $18 - 12 \div 6 \cdot 2 + 4$. The value of such an expression will depend on the order in which the operations are performed. Thus, *only one* of the following can be a correct interpretation of $18 - 12 \div 6 \cdot 2 + 4$.

A. $18 - 12 \div 6 \cdot 2 + 4$
$ 18 - 2 \cdot 2 + 4$
$ 18 - 4 + 4$
$ 14 + 4$
$ 18 \text{ (correct)}$

B. $18 - 12 \div 6 \cdot 2 + 4$
$ 6 \div 6 \cdot 2 + 4$
$ 1 \cdot 2 + 4$
$ 2 + 4$
$ 6 \text{ (incorrect)}$

In the second interpretation (B), all operations are performed from left to right, but this is incorrect. The first interpretation (A) is the one mathematicians accept as correct. The rule is as follows.

> Multiply and divide in order from left to right. Then add and subtract, also in order from left to right.

On a calculator, enter $18 \boxminus 12 \boxdiv 6 \boxtimes 2 \boxplus 4 \boxequal$ and read the display. A *scientific* calculator will show 18, which is correct. Such a calculator automatically obeys the rule for the order for operations when there are no grouping symbols. If the calculator displays 6, which is incorrect, you will need to enter the data according to the rule for the order of operations:

$$\boxminus 12 \boxdiv 6 \boxtimes 2 \boxplus 18 \boxplus 4 \boxequal$$

EXAMPLE 1 Simplify: $-8 \div 4 - (-3)16 \div 4 + 5(-8)$

Solution Multiply and divide, left to right.
$-8 \div 4 - (-3)16 \div 4 + 5(-8)$
$-2 - (-48) \div 4 + (-40)$
$-2 - (-12) + (-40)$
Use $a - b = a + (-b)$. $-2 + 12 + (-40)$, or -30
The answer is -30.

The order-of-operations rule above does not include the operation of raising a number to a power. *Raising a number to a power* takes priority over all other operations, as illustrated below.

$$3 + 4^2 = 3 + (4^2) \qquad\qquad -5 \cdot 2^3 = -5(2^3)$$
$$ = 3 + 16, \text{ or } 19 \qquad\qquad = -5 \cdot 8, \text{ or } -40$$

Teaching Resources

Quick Quizzes 2
**Reteaching and Practice
Worksheets** 2

▰▰▰ GETTING STARTED

Prerequisite Quiz

Simplify.

1. 6^2 36
2. $(-5)^3$ −125
3. $1 - 6 + 2$ −3
4. $1 - (6 + 2)$ −7
5. $1 - (6 - 2)$ −3
6. $-1(5 - 7)$ 2

Motivator

Have your students compute $24 - 6 \cdot 3 + 2$ to obtain the answers −6, 4, 8, 56, and 90 using different orders of operations. Then show them that applying the standard order of operations will yield the answer 8.

▰▰▰ TEACHING SUGGESTIONS

Lesson Notes

Show students that the value of $18 - 12 \div 6 \cdot 2 + 4$ depends on the order in which operations are performed. State the rules of order of operations and emphasize that there is only one *correct* interpretation of the given numerical expression.

Additional Example 1

Simplify.

$50 \div (-10) - (-4) + 3 \cdot 4(-2) - (-5)4 - 5$ −10

Highlighting the Standards

Standard 14b: The order of operations shows students that initial decisions on procedures will have far reaching consequences.

5

Math Connections

Life Skills: Not all calculators automatically follow the rules for the order of operations. Students need to know if their calculator follows these rules and, if it does not, they will need to enter arithmetic expressions in a sequence that the calculator will process without error.

Critical Thinking Questions

Analysis: To be sure that students fully grasp the implications of the convention that, in the absence of parentheses, one should perform multiplication and division from left to right, ask whether $198 \times \frac{1000}{60} \times 17$ would be represented without the fraction bar as $198 \times 1{,}000 \div 60 \times 17$, as $(198 \times 1{,}000) \div (60 \times 17)$; as $198 \times [1{,}000 \div (60 \times 17)]$, or whether all of these are equivalent. The student will discover that only the first expression, $198 \times (1{,}000 \div 60) \times 17$ is equivalent to the original expression.

Common Error Analysis

Error: Emphasize the distinction between -3^2 and $(-3)^2$. Students often make errors in evaluating -3^2; they interpret the base as -3 and obtain 9 as the answer.

Point out that -3^2 indicates the negative of 3^2.

Parentheses and brackets are *grouping symbols*. They are often used to show an intended order of operations when the usual rules are not to be observed. Thus,

$$2 \cdot 5 - 7 = 10 - 7 = 3, \text{ but } 2(5 - 7) = 2(-2) = -4$$

When one expression is enclosed within another, simplify the innermost expression first. Thus,

$$5[7 - 3 - (2 + 6)] = 5[7 - 3 - 8] = 5[-4] = -20$$

EXAMPLE 2 Simplify each expression.

Solutions

a. $(3 + 4)^2$
7^2
49

b. $-5 + 3(2 - 7)(4 - 6)^3$
$-5 + 3(-5)(-2)^3$
$-5 + (-15)(-8)$
$-5 + 120$, or 115

A horizontal fraction bar serves as a grouping symbol. The intended order of operations is (1) simplify the numerator, (2) simplify the denominator, and (3) simplify the quotient, as shown below.

$$\frac{-18 - 3(-2)}{4 \cdot 6 - (-8)} = \frac{-18 + 6}{24 + 8} = \frac{-12}{32} = \frac{-3}{8} = -\frac{3}{8}$$

Rules for Order of Operations
1. First, simplify expressions within pairs of grouping symbols, beginning with the innermost pair.
2. Next, simplify all powers.
3. Then, multiply and divide in order from left to right.
4. Finally, add and subtract in order from left to right.

Classroom Exercises

Simplify, if possible. If not possible, write *undefined*.

1. $3 + 4 \cdot 2$ 11
2. $(3 + 4)2$ 14
3. $4 \cdot 2 - 3 \cdot 5$ −7
4. $4(2 - 3) \cdot 5$ −20
5. -5^2 −25
6. $(-5)^2$ 25
7. $6 \cdot 8 \div 4 \cdot 2$ 24
8. $6[8 \div (4 \cdot 2)]$ 6
9. $7 - 3^2$ −2
10. $(7 - 3)^2$ 16
11. $-5 \cdot 3^2$ −45
12. $(-5 \cdot 3)^2$ 225
13. $\frac{4 - 6 \cdot 2}{-2 \cdot 5 - 6}$ $\frac{1}{2}$
14. $\frac{-2 \cdot 3}{14 - 7 \cdot 2}$ Undefined

Additional Example 2

Simplify each expression.

a. $(5 + 1)^2$ 36
b. -7^2 −49
c. $(-7)^2$ 49
d. $4 - 4(3 - 8)(5 - 7)^3$ −156

Written Exercises

Simplify, if possible. If not possible, write *undefined*.

1. $-7 + 5 \cdot 3$ 8

2. $6(-8) - 10$ -58

3. $10(-4) - 5(-8)$ 0

4. $-12(-6) + 8(-4)$ 40

5. $\dfrac{30 - (-6)}{-5 - (-14)}$ 4

6. $\dfrac{8 + 2 \cdot 5}{3 \cdot 4 - 6 \cdot 2}$ Undef

7. $4^2 + 16 \cdot 9 - 2^4$ 144

8. $5(-3 - 4) + 35$ 0

9. $40 - 10^2$ -60

10. -10^2 -100

11. $-6 \cdot 2^3$ -48

12. $(-3 \cdot 2)^3$ -216

13. $7(-6 - 5^2)$ -217

14. $3[-18 - (6 - 14)]$ -30

15. $40 - 36 \div 18 \cdot 2$ 36

16. $\dfrac{-2 - 2[12 - 2(2.6 - 4.1)]}{(2.7 - 3.2)(4.4 - 8.4)^2}$ 4

17. $\dfrac{-7.16 - 5.42 \cdot 3}{3^2 - 6(1.5)}$ Undef

18. $(12 - 3^2)(12 - 3)^2$ 243

19. $-15 - 5[12 - 3(8.4 - 12.4)^2]$ 165

For each statement, determine whether it is true or false.

20. $(2^4)^2 = (2^2)^4$ True

21. $2^{(4^2)} = (2^4)^2$ False

22. $(5^2)^3 = 5^{(2^3)}$ False

Mixed Review

Evaluate each expression for $a = 6$, $b = -2$, and $c = -3$.   *1.1*

1. bc 6

2. $a - b$ 8

3. $5ac$ -90

4. $b - c - a$ -5

Using the Calculator

The example below shows how to use a *scientific* calculator with a minimum number of keystrokes.

$\dfrac{85}{3.9 - 5.6}$: $85 \boxed{\div} \boxed{(} 3.9 \boxed{-} 5.6 \boxed{)} \boxed{=} -50$

Second method: $3.9 \boxed{-} 5.6 \boxed{=} \boxed{1/x} \boxed{\times} 85 \boxed{=} -50$

Simplify using a calculator and a minimum number of keystrokes.

1. $-23 + 54 \cdot 83 - 31 \cdot 47$ 3,002

2. $75 - 15 \cdot 12 \div 30 + 36$ 105

3. $\dfrac{-75 - 5(65)}{50(3.2)}$ -2.5

4. $\dfrac{28.8 - 43.2}{-5.1 + 3.9}$ 12

Simplify, if possible. If not possible, write *undefined*.

1. $-5 - 3 \cdot 4 - 2(-7)$ -3

2. $\dfrac{18 - 2 \cdot 5}{15 + 3(-3)}$ $\dfrac{4}{3}$

3. $18 - 2 \cdot 5 + (6 - 8)^3$ 0

4. $\dfrac{-10 - 5 \cdot 2}{9 \cdot 4 - 6^2}$ Undef

Closure

Have students summarize the rules for order of operations.   See page 6.
Have them insert one pair of parentheses ( ) and one pair of brackets [ ] in $70 - 3 - 8 \cdot 4$ so that the value will be 90.
$70 - [(3 - 8) \cdot 4]$

◤◤◤◤ FOLLOW UP

Guided Practice

Classroom Exercises 1–14

Independent Practice

A Ex. 1–15, **B** Ex. 16–19, **C** Ex. 20–22

Basic: WE 1–15, 18, 19, UC
Average: WE 1–19 odd, 21,22, UC
Above Average: WE 1–21 odd, UC

Enrichment

Have the students determine whether each of the following statements is true or false. If a statement is false, have the students make it true by making no changes other than inserting or removing grouping symbols. Solutions may vary for statements that are false.

1. $5 + 3 \cdot 4 - 6 = 26$
F: $(5 + 3) \cdot 4 - 6 = 26$

2. $2 + 3 - 1 - 2 - 4 - 3 = 1$
F: $2 + 3 - 1 - 2 - (4 - 3) = 1$

3. $3 - (5 - 2) \cdot 6 + 5(6 - 4) = -5$ T

4. $2 \cdot (5 - 1) + 3 + 2 \cdot (4 - 1) = 19$
F: $2 \cdot 5 - 1 + 3 + 2 \cdot 4 - 1 = 19$

5. $(6 - 2^2) \cdot 5 + 3 - 2 \cdot (4 + 1) = -33$
F: $6 - 2^2 \cdot (5 + 3) - 2 \cdot 4 + 1 = -33$

▰▰GETTING STARTED

Prerequisite Quiz

True or false?

1. $3 + (-8)$ is an integer. T
2. $72 \cdot (9 \cdot 5) = (72 \cdot 9) \cdot 5$ T
3. $8(5 \cdot 4) = (8 \cdot 5) \cdot (8 \cdot 4)$ F
4. $-12 \cdot 0 = -12$ F
5. $-15 + 15 = 0$ T
6. $-7.63 + 2.8 = 2.8 + (-7.63)$ T

Motivator

Have students fill the blanks below to make true statements without making computations.

1. $7.62 + 4.89 = 4.89 + \underline{\ ?\ }$ 7.62
2. $32 \cdot 768 = \underline{\ ?\ } \cdot 768$ 32
3. $5(3 + 4) = \underline{\ ?\ } \cdot 3 + \underline{\ ?\ } \cdot 4$ 5, 5
4. $(512 \cdot 67) \cdot \underline{\ ?\ } = 512 \cdot (67 \cdot 9)$ 9

1.3 Algebraic Expressions

Objectives

To identify properties of operations with real numbers
To simplify algebraic expressions
To evaluate algebraic expressions

In Lessons 1.1 and 1.2, you reviewed the four major operations with real numbers. Real numbers are of great practical importance to scientists, engineers, and others who make and interpret measurements.

The set of real numbers is also of great interest to professional mathematicians. A main reason for this interest is that real numbers (together with the operations of addition and multiplication) provide one of the best examples of the mathematical structure known as a *field*.

Any set F is a **field** if the eleven properties listed below are true for the set. Unless otherwise stated, each property applies to all numbers a, b, and c in F.

Name of Property	Statement of Property
Closure for Addition	$a + b$ is in F.
Closure for Multiplication	$a \cdot b$ is in F.
Commutative of Addition	$a + b = b + a$
Commutative of Multiplication	$a \cdot b = b \cdot a$
Associative of Addition	$(a + b) + c = a + (b + c)$
Associative of Multiplication	$(a \cdot b) \cdot c = a \cdot (b \cdot c)$
Distributive of Multiplication over Addition	$a(b + c) = a \cdot b + a \cdot c$ and $(b + c)a = b \cdot a + c \cdot a$
Identity for Addition	There is a unique number 0 in F such that for all a in F, $a + 0 = a$.
Identity for Multiplication	There is a unique number 1 in F such that for all a in F, $a \cdot 1 = a$.
Additive Inverse	For each a in F, there is a unique number $-a$ in F such that $a + (-a) = 0$. a and $-a$ are **additive inverses**, or **opposites**.
Multiplicative Inverse	For each nonzero a in F, there is a unique number $\frac{1}{a}$ in F such that $a \cdot \frac{1}{a} = 1$. a and $\frac{1}{a}$ are **multiplicative inverses**, or **reciprocals**.

8 Chapter 1 Linear Equations

There are fields other than the set of real numbers. For example, the set of rational numbers (see Lesson 1.1) is a field. Also, the set of complex numbers (see Chapter 9) is a field.

EXAMPLE 1   Identify the property of operations with real numbers illustrated by the sentence.

Solutions

Sentence	Property
a. $v(4t) = (4t)v$	**a.** Commutative of Mult
b. $-a^2b + a^2b = 0$	**b.** Add Inverse
c. $7 \cdot \pi$ is a real number.	**c.** Closure for Mult
d. $-6r + 0 = -6r$	**d.** Identity for Add
e. $(8 + n) + (-n) = 8 + [n + (-n)]$	**e.** Associative of Add
f. $m - n = 1(m - n)$	**f.** Identity for Mult
g. $m(n^2 + n) = m \cdot n^2 + m \cdot n$	**g.** Distr of Mult over Add
h. $\frac{w}{4} \cdot \frac{4}{w} = 1$	**h.** Mult Inverse
i. $2c + 3c^2 = 3c^2 + 2c$	**i.** Commutative of Add
j. $-5(3t) = (-5 \cdot 3)t$	**j.** Associative of Mult
k. $8 + \sqrt{2}$ is a real number.	**k.** Closure for Add

Three examples of *algebraic expressions* are given below.

$$-4(5x) \qquad 7a^2 + 5a^2 \qquad -n + (3m + n)$$

An **algebraic expression** contains one or more variables (see Lesson 1.1) and one or more operations. The expression $-4(5x)$ contains three **factors**, -4, 5, and x. Factors are multiplied. The expression $7a^2 + 5a^2$ has two **terms**, $7a^2$ and $5a^2$. Terms are added. For the term $7a^2$, 7 is called the numerical **coefficient** of a^2.

All of the expressions above can be simplified. Steps in their simplification are shown in the table below.

Steps	Property or reason
$-4(5x) = (-4 \cdot 5)x$ $= -20x$	Associative Property of Multiplication (The factors -4, 5, and x were regrouped.)
$7a^2 + 5a^2 = (7 + 5)a^2$ $= 12a^2$	Distributive Property (The like terms, $7a^2$ and $5a^2$, were combined by adding their coefficients.)
$-n + (3m + n)$ $= (3m + n) + (-n)$ $= 3m + [n + (-n)]$ $= 3m + 0$ $= 3m$	Commutative Property of Addition Associative Property of Addition Additive Inverse Additive Identity (The three terms were reordered and regrouped.)

1.3 Algebraic Expressions

Lesson Note

The eleven properties of a field, listed on page 8, will be used throughout the book to (1) simplify algebraic expressions, (2) prove theorems, and (3) study fields other than the set of real numbers.

When asking students to evaluate algebraic expressions, encourage them to simplify the expression, if possible, before substituting for the variables.

Math Connections

Construction: The distinction between expressions that are *factors* and those that are *terms* can be illustrated by a contractor who is determining the perimeter of a rectangular foundation of a building he is constructing. If the dimensions of the building are m and n, then he can think of the perimeter as either the sum of two *unlike terms*, $2m$ and $2n$, or as the product of two *factors* 2, and $(m + n)$.

Additional Example 1

Identify the property of operations with real numbers illustrated by each sentence.

a. $2 + mn = mn + 2$ Comm Prop Add

b. $m + (n + 2) = (m + n) + 2$ Assoc Prop Add

c. $5 + \sqrt{3}$ is a real number.   Clos for Add

d. $3t + 0 = 3t$ Add Ident

e. $5(7x + 6) = 5(7x) + 5 \cdot 6$ Distr Prop

f. $-5(4y) = (-5 \cdot 4)y$ Assoc Prop Mult

g. $7t + t = 7t + 1t$ Mult Ident

h. $(\frac{3}{5} \cdot \frac{5}{3})c = 1c$ Mult Inv

Analysis: Ask students why there are no properties of subtraction and division in the Field Properties listed on page 8 of the Student text. Properties of subtraction and division are not necessary since subtraction is defined by the Additive Inverse and division is defined by the Multiplicative Inverse.

Checkpoint

Simplify. Evaluate each expression for $x = 2$ and $y = -3$, if possible.

1. $3x + 7 - y - 10 + x + 5y$
   $4x + 4y - 3$; -7

2. $6(2x - 3y) - (8x - 12y + 4) + 10$
   $4x - 6y + 6$; 32

3. $5x^2 - 2y - x^2 + xy - 9y$
   $4x^2 - 11y + xy$; 43

4. $\dfrac{x - (3y - 2x)}{3x + 2y}$ $\dfrac{3x - 3y}{3x + 2y}$; Undef

Closure

Have students produce an *equation* in one variable for which it would be appropriate to simplify one or both sides as a first step in solving the equation.   Answers will vary.

The expression $5x^2y + 3ab^2 + 2x^2y + 4a^2b$ contains four terms. $5x^2y$ and $2x^2y$ are called **like terms** because they differ only in their coefficients. Like terms can be combined using the Distributive Property.

$$5 \cdot x^2y + 2 \cdot x^2y = (5 + 2)x^2y = 7x^2y$$

The terms $3ab^2$ and $4a^2b$ are **unlike terms**. They differ in their variable factors, ab^2 and a^2b. Unlike terms cannot be combined.

The Distributive Property of Multiplication over Addition can be extended to subtraction.

For all real numbers a, b, and c,
$a(b - c) = a \cdot b - a \cdot c$ and $(b - c)a = b \cdot a - c \cdot a$

This statement is proved as a theorem in Lesson 1.8.

EXAMPLE 2   Simplify each expression.

Solutions   a. $-8x^2 + 5(4x^2 - 6)$
   $= -8x^2 + 5 \cdot 4x^2 - 5 \cdot 6$ Distribute 5 to $4x^2$ and 6.
   $= -8x^2 + 20x^2 - 30$
   $= 12x^2 - 30$ Combine like terms.

   b. $5n - (3n + 6)$ Use the multiplicative identity:
   $= 5n - 1(3n + 6)$ $(3n + 6) = 1(3n + 6)$.
   $= 5n + (-1 \cdot 3n) + (-1 \cdot 6)$ Distribute -1 to $3n$ and 6.
   $= 2n - 6$ Combine like terms.

Because in Example 2a, $-8x^2 + 5(4x^2 - 6)$ is simplified to $12x^2 - 30$, the two expressions are said to be *equivalent*. Similarly, in Example 2b, $5n - (3n + 6)$ and $2n - 6$ are shown to be equivalent expressions. Two algebraic expressions are **equivalent expressions** if they are equal for *all* values of the variables for which the expressions have meaning.

EXAMPLE 3   Evaluate $2x^2 - 6xy + 4x^2 - xy$ for $x = -3$ and $y = 2$.

Plan   Find a simpler equivalent expression for $2x^2 - 6xy + 4x^2 - xy$. Then evaluate the simpler expression.

Solution $2x^2 - 6xy + 4x^2 - xy = 2x^2 + 4x^2 - 6xy - xy$
   $= 6x^2 - 7xy$
   $= 6(-3)^2 - 7(-3)2$
   $= 6 \cdot 9 - (-42) = 96$

The value is 96.

Additional Example 2

Simplify each expression.

a. $-10x^3 + 4(2x^3 + 3)$ $-2x^3 + 12$
b. $16a - 5(2a + 3b - 4)$ $6a - 15b + 20$

Additional Example 3

Evaluate $4x^2 + 5xy - 3x^2 - xy$ for $x = 2$ and $y = 4$. $x^2 - 4xy$; -28

Summary (Simplifying Expressions)	Example
Factors are *multiplied* to give a product. Factors can be rearranged to simplify a product.	four factors $(3 \cdot x)(5 \cdot y) = (3 \cdot 5)(x \cdot y)$ $\qquad\qquad = 15xy$
Terms are *added* to give a sum. Terms can be rearranged, and like terms combined, to simplify a sum.	four terms $7a^2 + 5b^2 + 2a^2 + 8b$ $= (7a^2 + 2a^2) + (5b^2 + 8b)$ $= 9a^2 + 5b^2 + 8b$

Classroom Exercises

Identify the property of operations that is illustrated. All variables represent real numbers.

1. $-6.3 + 6.3 = 0$ Add Inv Prop
2. $6c + (5 - 2c) = (5 - 2c) + 6c$
3. $8.2 + \sqrt{10}$ is a real number.
4. $3c + c = 3c + 1c$ Identity Prop for Mult
5. $(7a)r = 7(ar)$ Associative Prop of Mult
6. $(5x + y)z = 5x \cdot z + y \cdot z$ Distr Prop

Simplify each expression. Then evaluate for $x = 3$ and $y = 4$.

7. $4x^2 - y + x^2$ $5x^2 - y$; 41 **8.** $2y - 3(y + 1)$ $-y - 3$; -7 **9.** $4 - 2(x - 1)$ $6 - 2x$; 0

Written Exercises

Simplify.

1. $8x + 1 - 2x$ $6x + 1$
2. $xy^2 + x^2y - xy^2$ x^2y
3. $3m - n + 15m - 14n + m - 12$
4. $16r - 11 - 18t - 15r + 17t + 10$
5. $9x^2 - 2y^2 + xy + 3x^2 + 2xy - y^2$
6. $4x^3 + 5x^2 - 6x^3 + 7x^2$ $-2x^3 + 12x^2$
7. $14 - 4(5x + 3) + 10x$ $-10x + 2$
8. $23 - (14 - 2n) - 12n$ $-10n + 9$

Evaluate each expression for $x = -2$ and $y = 5$, if possible.

9. $x - y + 2$ -5
10. $3(x + 2) + y$ 5
11. $8x + 3y + 4x - 5y - 6$ -40
12. $5(4x - y) - (15x + 2y) + 4$ -41
13. $x^2 - y^2 + xy$ -31
14. $3x^2y - 3xy - 3xy^2$ 240
15. $8.4x - 7.5y + 0.8 - 3.2x + 4.3y$
16. $5.7xy - 3.2x + 2.4y - 3.5xy + x - y$
17. $y(-5x) - 2(7 - 3x) - (4x + y)$ 27
18. $3.4x + 4.5(2x - y) - (10x - 3.3y)$ -10.8

FOLLOW UP

Guided Practice

Classroom Exercises 1–9

Independent Practice

A Ex. 1–14, **B** Ex. 15–26, **C** Ex. 27–30

Basic: WE 1–14, 15–25 odd, Application
Average: WE 1–25 odd, 26, 27, Application
Above Average: WE 1–29 odd, Application

Additional Answers

Classroom Exercises

2. Comm Prop of Add
3. Clos for Add

Written Exercises

3. $19m - 15n - 12$
4. $r - t - 1$
5. $12x^2 - 3y^2 + 3xy$
15. -25.6
16. -10.6
19. Rule of Subt
20. Mult Ident Prop
21. Distr Prop
22. Comm Prop Mult
23. Assoc Prop Mult
24. Add Inv Prop
25. Clos for Add
26. Add Ident Prop
30. $(a + (-b)) + b = a + ((-b) + b) =$ $a + 0 = a$. So, by definition, $a - b =$ $a + (-b)$.

Enrichment

Explain that the Distributive Property for Multiplication holds true if the parentheses contain more than two terms. Challenge the students to prove that
$r(x + y + z) = rx + ry + rz$.

Let $x + y = m$. Thus, $r(m + z) = rm + rz$. Now, substitute $x + y$ for m, giving $r(x + y) + rz$. Since $r(x + y) = rx + ry$, it follows that $r(x + y + z) = rx + ry + rz$.

Now have the students use a similar method to prove that $r(w + x + y + z) = rw + rx + ry + rz$.

Let $m = w + x$ and $n = y + z$. Thus, $r(w + x + y + z) = r(m + n) = rm + rn = r(w + x) + r(y + z) = rw + rx + ry + rz$.

11

Identify the rule or property of operations with real numbers that is illustrated. All variables represent real numbers.

19. $8(x - y) = 8[x + (-y)]$

20. $12t - (3t + 4) = 12t - 1(3t + 4)$

21. $6 + 3(c + d) = 6 + (3c + 3d)$

22. $d^2(-7c) = (-7c)d^2$

23. $(-7c)d^2 = -7(cd^2)$

24. $5m + [3n + (-3n)] = 5m + 0$

25. $15 + \sqrt{2}$ is a real number.

26. $(7a + 0) - 5b = 7a - 5b$

In Exercises 27–29, show that the given generalization is *false*. (HINT: A generalization is false if it is possible to find one *counterexample*— that is, one example for which the generalization is not true.) Answers may vary. Possible answers are given.

27. Subtraction is commutative ($a - b = b - a$) for all real numbers a and b. $9 - 7 \neq 7 - 9$

28. Subtraction is associative $[(a - b) - c = a - (b - c)]$ for all real numbers a, b, and c. $(10 - 8) - 1 \neq 10 - (8 - 1)$

29. Division is commutative for all real numbers. $8 \div 2 \neq 2 \div 8$

30. Use the definition of subtraction in Lesson 1.1 to prove the Rule of Subtraction, $a - b = a + (-b)$.

Mixed Review

Simplify, if possible. *1.2*

1. $-7(-13) - 3(-4)$ 103 **2.** $8 + (-2)(-3)(-4)$ −16 **3.** $36 - (-3)^2$ 27

Application: Insect Songs

Scientists have noticed a relationship between the temperature and the number of times a cricket chirps in a minute.

For one kind of cricket, a formula for the temperature T in degrees Fahrenheit in terms of the number of chirps C that a cricket makes in a minute is

$$T = 50 + \frac{C - 92}{4.7}.$$

1. What is the temperature if a cricket chirps 186 times per minute?
70°

2. What is the value of the fractional term in the formula for $C = 92$? What temperature would this be?
0; 50°

1.4 Linear Equations

Objective

To solve linear equations in one variable

The equation $6x - 15 - 3x + 5 = 9x + 6 - 4x$ is a **first-degree** or **linear equation** in one variable. (*Second-degree equations,* such as $6x^2 - 15 = 9x$, are discussed in Chapters 6 and 9.) Any value of x that makes the equation true is called a **solution** or **root** of the equation.

You can solve this equation by:
simplifying each side,
subtracting $5x$ from each side,
adding 10 to each side, and then
dividing each side by -2, the
coefficient of x. The root is -8.

$$6x - 15 - 3x + 5 = 9x + 6 - 4x$$
$$3x - 10 = 5x + 6$$
$$-2x - 10 = 6$$
$$-2x = 16$$
$$\frac{-2x}{-2} = \frac{16}{-2}$$
$$x = -8$$

In the above example, each of the equations ($6x - 15 - 3x + 5 = 9x + 6 - 4x$, $3x - 10 = 5x + 6$, and so on) has -8 as its only solution. Equations that have the same solution (or solutions) are called **equivalent equations**. A linear equation in one variable is solved by writing a sequence of equivalent equations, each one simpler than the one preceding it. To do this, one or more of the following properties are used.

Properties of Equality
For all real numbers a, b, and c,
if $a = b$, then $a + c = b + c$
     and $a - c = b - c$;   ← (Addition/Subtraction Properties)
if $a = b$, then $a \cdot c = b \cdot c$
     and $\frac{a}{c} = \frac{b}{c}$ ($c \neq 0$).   ← (Multiplication/Division Properties)

EXAMPLE 1   Solve $2(3c + 19) - 9c = 7 - 3(11 - 3c)$.

Solution
$$6c + 38 - 9c = 7 - 33 + 9c \qquad \leftarrow \text{First, the Distributive}$$
$$-3c + 38 = -26 + 9c \qquad\qquad \text{Property is used.}$$
$$-12c = -64$$
$$c = \frac{-64}{-12} = \frac{16}{3}$$
The solution is $\frac{16}{3}$, or $5\frac{1}{3}$.

Additional Example 1

Solve.

$3(12 - 2n) + 8n = 6 - (4n + 8)$
$-\frac{19}{3}$, or $-6\frac{1}{3}$

▬▬▬GETTING STARTED

Prerequisite Quiz

Simplify.

1. $5 + 4(9c - 3) - 20c$ $16c - 7$
2. $\frac{8.5}{3.4}$ 2.5
3. $\frac{5}{7} \cdot \frac{7}{5}x$ x
4. $12(\frac{3}{4}a + \frac{2}{3}b + 5)$ $9a + 8b + 60$
5. $\frac{2}{3}(3m + 15n - 6)$ $2m + 10n - 4$

Motivator

Ask students if -2 is a solution to the following equations.

1. $3x = 3x$ yes
2. $3x - 3 = 2x - 5$ yes
3. $\frac{x}{8} = \frac{5}{20}$ no

Highlighting the Standards

Standard 2a: Examples 3 and 4 show the kinds of equations that will recur in both theoretical and applied mathematics. In Exercises 23–32, students will clarify their thinking about fractional solutions.

Lesson Note

The students' first objective in solving an equation with decimals as coefficients, such as $2.1x + 3 = 51.3 - 1.35x$, should be to eliminate the decimals by multiplying each side by an appropriate power of 10 (100 in the case of this equation.)

Example 5 introduces literal equations. Exercises 21, 22, and 35–40 are typical equations that occur in solving word problems in the future lessons.

Math Connections

Life Skills: One of the most useful equation forms in everyday life is the *proportion*. Using a proportion, a cook can adapt a recipe to make a larger or smalller amount than the amount for which the recipe was written. When shopping, a consumer can verify whether a "large economy size" actually saves any money. The proportion is also used by scientists in their daily work.

EXAMPLE 2 Solve $2.1x + 3 = 51.3 - 1.35x$.

Plan To eliminate the decimals, multiply each side by an appropriate power of 10. Since the greatest number of decimal places is 2 (in 1.35), multiply each side by 10^2, or 100.

Solution
$$100(2.1x + 3) = 100(51.3 - 1.35x)$$
$$210x + 300 = 5{,}130 - 135x$$
$$345x = 4{,}830$$
$$x = 14$$

Check
$$2.1x + 3 = 51.3 - 1.35x$$
$$2.1(14) + 3 \overset{?}{=} 51.3 - 1.35(14)$$
$$29.4 + 3 \overset{?}{=} 51.3 - 18.9$$
$$32.4 = 32.4 \text{ True}$$

The solution is 14.

The technique demonstrated in Example 2 can also be used to solve an equation such as $\frac{3}{4}x + \frac{1}{2} = \frac{7}{6}x - 3$.

Use the Multiplication Property of Equality to eliminate the fractions. Recall that the *least common denominator* (LCD) of a set of fractions is the *least common multiple* (LCM) of the denominators. The least common denominator of $\frac{3}{4}$, $\frac{1}{2}$, and $\frac{7}{6}$ is 12. So, we multiply each side of the equation by 12.

EXAMPLE 3 Solve $\frac{3}{4}x + \frac{1}{2} = \frac{7}{6}x - 3$.

Solution Multiply each side by 12, the LCD of the fraction.
$$12\left(\frac{3}{4}x + \frac{1}{2}\right) = 12\left(\frac{7}{6}x - 3\right)$$
$$9x + 6 = 14x - 36$$
$$-5x = -42$$
$$x = \frac{42}{5}, \text{ or } 8\frac{2}{5}$$

The root is $8\frac{2}{5}$.

A **proportion** is an equation of the form $\frac{a}{b} = \frac{c}{d}$. Both $\frac{2}{3} = \frac{10}{15}$ and $2 \cdot 15 = 3 \cdot 10$ are true equations. This suggests a property of proportions that can be used to solve an equation such as $\frac{2x + 8}{3x - 2} = \frac{5}{4}$.

Proportion Property

If $\frac{a}{b} = \frac{c}{d}$, then $a \cdot d = b \cdot c$.

---

14 Chapter 1 Linear Equations

Additional Example 2

Solve.

$4.6x + 5 = 56.68 + 1.37x$. 16

Additional Example 3

Solve.

$\frac{2}{5}x - \frac{3}{4} = \frac{1}{2}x + 2$ $-27\frac{1}{2}$

EXAMPLE 4 Solve $\dfrac{2x + 8}{3x - 2} = \dfrac{5}{4}$.

Solution Use the Proportion Property.

$$\frac{2x + 8}{3x - 2} = \frac{5}{4}$$

$$(2x + 8)4 = (3x - 2)5$$
$$8x + 32 = 15x - 10$$
$$-7x = -42$$
$$x = 6$$

The solution is 6.

An equation with more than one variable, such as $px - 5 = n$, is called a **literal equation**. It is "solved for x" when x is alone on one side.

EXAMPLE 5 Solve each equation for x.

Solutions **a.** $px - 5 = n$ **b.** $y = \dfrac{7}{3}x + z$ **c.** $a(x + b) = d$

$$px = n + 5 \qquad y - z = \frac{7}{3}x \qquad\qquad x + b = \frac{d}{a}$$
$$x = \frac{n + 5}{p} \qquad \frac{3}{7}(y - z) = \frac{3}{7} \cdot \frac{7}{3}x \qquad x = \frac{d}{a} - b$$
$$\frac{3}{7}(y - z) = x$$

Classroom Exercises

Solve each equation for x.

1. $\dfrac{1}{3}x = 6$ 18
2. $7 - x = 9$ -2
3. $2 = 5 + x$ -3
4. $5 - 10x = 0$ $\dfrac{1}{2}$
5. $\dfrac{1}{5} = \dfrac{2}{x}$ 10
6. $\dfrac{x}{4} = \dfrac{x}{6}$ 0
7. $ax = b$ $\dfrac{b}{a}$
8. $x + n = m$ $m - n$
9. $5x + 14 = 6x$ 14
10. $x - c - d = 0$ $c + d$
11. $\dfrac{x}{a} = b$ ab
12. $5 - x = -2$ 7

Written Exercises

Solve each equation. Check.

1. $4a - 2 = 10$ 3
2. $10x = 2x - 32$ -4
3. $7y + 11 = 8 - 5y$ $-\dfrac{1}{4}$
4. $-43 = 7 - 8t$ $6\dfrac{1}{4}$
5. $2(x - 4) = -5$ $1\dfrac{1}{2}$
6. $15 = 4(6 - 3c)$ $\dfrac{3}{4}$

Checkpoint

Solve each equation.

1. $8c - (5 - c) = 6 + 2(6c - 4)$ -1
2. $\dfrac{3}{7}x = 6 - \dfrac{2}{7}x$ $8\dfrac{2}{5}$
3. $2.4x - 3.7 = 8.6x + 21.1$ -4
4. $\dfrac{2}{3} = \dfrac{2x - 5}{x + 6}$ $6\dfrac{3}{4}$
5. Solve for x. $t = \dfrac{5}{4}x - v$ $x = \dfrac{4}{5}(t + v)$

Additional Example 4

Solve.

$\dfrac{8x - 3}{3} = \dfrac{6x + 5}{2}$. $-10\dfrac{1}{2}$

Additional Example 5

Solve each equation for x.

a. $kx + t = w$ $x = \dfrac{w - t}{k}$
b. $y = \dfrac{2}{5}x + r$ $x = \dfrac{5}{2}(y - r)$
c. $5c(x - d) = 3cx + 4$ $x = \dfrac{4 + 5cd}{c}$

Closure

Have students define *linear equation*. An equation in one variable Have students give an example of equivalent equations. Answers will vary. Have them define *literal equation*. An equation with more than one variable. Have them select reasonable values for p and l in the perimeter formula $p = 2(l + w)$ and solve for w. Answers will vary.

◤▬▬FOLLOW UP

Guided Practice

Classroom Exercises 1–12

Independent Practice

A Ex. 1–19, B Ex. 20–36, C Ex. 37–42

Basic: WE 1–16, 17–27 odd

Average: WE 1–22 odd, 23–38

Above Average: WE 1–36 odd, 37, 39, 41

Additional Answers

Written Exercises

20. $\dfrac{7 + ab}{2a}$

21. $\dfrac{n - 4rt}{r}$

22. $2b + 3a$

39. No roots

40. Identity

41. No roots

42. Identity

Solve each equation.

7. $18x - 14 - 21x = 17 - x + 7$ -19 8. $14 + 5(6n - 2) = 25n - 16$ -4

9. $8y - 5.4 = 3y - 1.2$ 0.84

10. $2.9n - 5 = 3 - 0.3n$ 2.5

11. $\frac{3}{5}x + 11 = 31 - \frac{1}{5}x$ 25

12. $\frac{2}{7}y - 15 = \frac{6}{7}y - 9$ $-10\frac{1}{2}$

13. $12 - (7c - 11) = 3(5 - 3c)$ -4

14. $2.4(20 - x) + 1.8x = 43.8$ 7

15. $\dfrac{x - 3}{2} = \dfrac{x}{5}$ 5

16. $\dfrac{x + 2}{x - 4} = \dfrac{7}{4}$ 12

Solve each equation for x.

17. $ax - b = c$ $\frac{c + b}{a}$ 18. $5x - y = 3x + 3y$ $2y$ 19. $2tx + 7 = 8 - 3tx$ $\frac{1}{5t}$

20. $a(2x - b) = 7$ 21. $4r(x + t) = 3rx + n$ 22. $\dfrac{x - a}{2} = \dfrac{x + b}{3}$

Solve each equation.

23. $p + p(0.08)3 = 3{,}100$ $2{,}500$ 24. $p(0.09)4 + p(0.11)4 = 1{,}200$ $1{,}500$

25. $3(5.4 + r) = 5.4r$ 6.75 26. $4.2(r + 8.7) = 8.7r$ 8.12

27. $540 = v \cdot 6 - 5 \cdot 6^2$ 120 28. $242 = v \cdot 4 - 5 \cdot 4^2$ 80.5

29. $\frac{3}{4}n - \frac{7}{8} = \frac{3}{2}n - \frac{3}{4}$ $-\frac{1}{6}$ 30. $\frac{2}{3}y + \frac{5}{2} = \frac{4}{5}y + \frac{7}{6}$ 10

31. $\dfrac{8x - 3}{6x + 9} = \dfrac{-2}{3}$ $-\frac{1}{4}$ 32. $\dfrac{5x + 6}{3} = \dfrac{3x - 14}{5}$ $-4\frac{1}{2}$

33. $3(2y - 5) - (7y - 6) = -10$ 1 34. $-6(3 - 3x) = -9(2x - 30)$ 8

35. $\frac{3}{5}(10x - 15) + \frac{5}{4}(16x + 24) = 86$ $2\frac{1}{2}$ 36. $\frac{5}{6}(18 - 12t) - \frac{2}{3}(12t + 15) = 14$ $-\frac{1}{2}$

Equations such as $x + 3 = 3 + x$ or $2(y - 5) = y - 10 + y$ that have all real numbers as solutions are called *identities*. The equation $2x = 2x + 6$, however, has no root because $2x$ is less than $2x + 6$ for each real number x. Determine whether each equation is an identity or has no roots.

37. $5x + 2 = 2 + 5x$ Identity   38. $3a + 1 = 3a$ No roots

39. $5(4n + 6) - 10 = 2(15 + 10n)$ 40. $\frac{1}{3}y + \frac{1}{4} = \frac{1}{2}y + 0.25 - \frac{1}{6}y$

41. $2[8 - (15 - 2z)] = 4z - 15$ 42. $5(4y - 2) - 7 = 15y - 17 + 5y$

Mixed Review

Identify the property of operations with real numbers that is illustrated. *1.3*

1. $(m - n)3 = 3(m - n)$ Comm Prop of Mult   2. $-2(3x) = (-2 \cdot 3)x$ Assoc Prop of Mult

3. $-12c + 12c = 0$ Add Inverse Prop   4. $\frac{2}{7} \cdot \frac{7}{2}t = 1t$ Mult Inverse Prop

16   Chapter 1 Linear Equations

Enrichment

In each of the following equations, have the students determine the ratio $\frac{x}{y}$ in terms of the other variables in the equation. Give the hint that they must rewrite the equations as proportions in which the left side is $\frac{x}{y}$.

- $\dfrac{5x}{ay} = \dfrac{3c}{2d}$ $\dfrac{x}{y} = \dfrac{3ac}{10d}$

- $mnx - pqy = 0$ $\dfrac{x}{y} = \dfrac{pq}{mn}$

- $3ax = \dfrac{y}{b}$ $\dfrac{x}{y} = \dfrac{1}{3ab}$

- $\dfrac{ax}{2} + \dfrac{by}{3} = 0$ $\dfrac{x}{y} = -\dfrac{2b}{3a}$

- $\dfrac{abx}{cd} - \dfrac{mpy}{st} = 0$ $\dfrac{x}{y} = \dfrac{cdmp}{abst}$

1.5 Problem Solving: Using Formulas

Teaching Resources

Quick Quizzes 5
Reteaching and Practice
   Worksheets 5

Objective   To solve word problems using given formulas

Sometimes an equation is found to be useful in solving problems from business, industry, and everyday life. In such a case, the equation is often called a **formula**. Shown below is one formula from banking.

If you borrow $800 (principal) at 9% (rate) for 3 years (time), then at the end of three years you will owe money for the simple interest, which will be included in the total amount due (principal plus interest).

> You will owe $800(0.09)3, or $216 in simple interest, and
> $800 + $216, or $1,016 as the total amount.

In general, if you borrow or loan p dollars at an annual simple interest rate of r (expressed as a decimal) for t years, then

$$i = prt \qquad \text{and} \qquad A = p + prt$$

where p is the principal, r is the rate, t is the time, i is the amount of interest, and A is the total amount due.

The general rule, or *formula*, $A = p + prt$ contains four variables. The value of one of the variables can be found when values for the other variables are given, as shown in the example below.

EXAMPLE 1   At the end of 2 years 6 months, Sylvia owed a total of $3,637.50 on her junior-college loan. If she was being charged a simple interest rate of 8.5% per year, how much money did she originally borrow? Use the formula $A = p + prt$. (THINK: Did she borrow more or less than $3,637.50?)

Solution

What are you to find?	The principal, p
What is given?	$t = 2$ yr 6 mo, or 2.5 yr
	$A = \$3,637.50$
	$r = 8.5\%$, or 0.085
Write the formula.	$A = p + prt$
Substitute in the	$3{,}637.5 = p + p(0.085)2.5$
formula for A, r, and t.	$= 1p + 0.2125p$
	$= 1.2125p$
Solve the equation.	$\dfrac{3{,}637.5}{1.2125} = p$
	$3{,}000 = p$
State your answer.	Sylvia borrowed $3,000.

GETTING STARTED

Prerequisite Quiz

Use the formula $i = prt$ to solve for the indicated variable.

1. i; if $p = \$7,000$, $r = 8\%$, and $t = 5$ years
   $2,800
2. p; if $i = \$460$, $r = 11\frac{1}{2}\%$, and
   $t = 4$ years $1,000
3. r; if $i = \$4,320$, $p = \$8,000$, and
   $t = 6$ years   9%
4. Change $7\frac{1}{4}\%$ to a decimal.   0.0725
5. Use a decimal to express 5 years 9 months in years only.   5.75 yr

Motivator

Have students sketch a square. Ask them to represent the length of each side with the variable "s." Ask them to derive formulas for the perimeter and area of the square and solve for each if $s = 4$.   Perimeter = $4s$ = $4(4)$ = 16; Area = s^2 = 4^2 = 16.

Additional Example 1

The Warrens were charged 9% per year on a loan they took out 4 years 6 months ago. They now owe $2,810. How much money did they borrow?
Use the formula $A = p + prt$. $2,000

Highlighting the Standards

Standard 1d: Examples 1, 2, and 3 illustrate that similar kinds of equations are used in the mathematics of finance, in physics, and in computations based on geometric shapes.

Lesson Note

One objective of Example 3 is to solve the geometric formula $A = \frac{1}{2}h(b_1 + b_2)$ for b_1. This continues the skill of solving a literal equation for one of its variables. The formulas needed for Exercises 12–15 are given in the *Formulas from Geometry* section.

Students can use a graphing calculator to verify their answers to Exercises 16 and 17. Let $h = y$ and $t = x$ in the formula $h = vt - 5t^2$. Thus, $y = vx - 5x^2$ for a given value of v. Set the range to 0, 60, 10, 0, 5,000, 500.

16. GRAPH $300x - 5x^2$ EXE
        GRAPH 4500   EXE
17. GRAPH $240x - 5x^2$ EXE
        GRAPH 2900   EXE

Suppose an object is projected upward from the earth with an initial vertical velocity of 40 meters per second (40 m/s). While aloft, the object's height h in meters at the end of t seconds is given approximately by the formula $h = 40t - 5t^2$.

You can use this formula to verify the height for each time shown in the figure. For example, for $t = 3$ s,
$h = 40(3) - 5(3)^2 = 75$ m,
and for $t = 5$ s,
$h = 40(5) - 5(5)^2 = 75$ m.

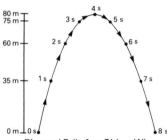

In general, if an object is shot upward with an initial vertical velocity of v meters per second, then its height h in meters at the end of t seconds will be given by the formula $h = vt - 5t^2$.

Rise and Fall of an Object When Initial Vertical Velocity is 40 m/s

EXAMPLE 2   Find the initial vertical velocity v (in meters per second) necessary for an object launched upward to reach a height of 800 m at the end of 4 s. Use the formula $h = vt - 5t^2$.

Solution

What are you to find?	The initial vertical velocity, v
What is given?	$h = 800, t = 4$
Write the formula.	$h = vt - 5t^2$
Substitute in the formula for h and t.	$800 = v \cdot 4 - 5 \cdot 4^2$
Solve the equation.	$880 = 4v$
State your answer.	$220 = v$

The necessary velocity is 220 m/s.

A formula can be solved for any of its variables in terms of the other variables. (Review Example 5 in Lesson 1.4.)

EXAMPLE 3   A trapezoid has an area of 144 square feet, or 144 ft^2. Find the length of one base given that the length of the other base is 7 ft and the height is 16 ft.

7 ft

$A = 144 \text{ ft}^2$ $h = 16$ ft

$b_1 = ?$

Plan   From the *Formulas from Geometry* at the back of the book, find the formula for the area of a trapezoid. Solve the formula for b_1 (read: "*b-sub-one*"). Then in the new formula, substitute 144 for A, 7 for b_2, and 16 for h. Compute the value of b_1, and label with the correct unit.

Additional Example 2

An object was shot upward to a height of 1,620 m in 6 sec. What was the initial velocity v? Use the formula $h = vt - 5t^2$.
300 m/s

Additional Example 3

A trapezoid has an area of 42 ft^2. Find the height of the trapezoid if the lengths of the bases are 6 ft and 8 ft.   6 ft

Solution   a.
$$A = \tfrac{1}{2}h(b_1 + b_2)$$
$$2 \cdot A = 2 \cdot \tfrac{1}{2} \cdot h(b_1 + b_2)$$
$$2A = h(b_1 + b_2)$$
$$2A = hb_1 + hb_2$$
$$2A - hb_2 = hb_1$$
$$\frac{2A - hb_2}{h} = b_1$$

b.
$$b_1 = \frac{2A - hb_2}{h}$$
$$= \frac{2 \cdot 144 - 16 \cdot 7}{16}$$
$$= \frac{288 - 112}{16}$$
$$= \frac{176}{16}, \text{ or } 11$$

The first base is 11 ft long.

Classroom Exercises

Change each percent to a decimal.

1. 7%   0.07   **2.** 12%   0.12   **3.** $9\tfrac{1}{2}$%   0.095   **4.** $11\tfrac{3}{4}$%   0.1175

Use a decimal to express each time period in years only.

5. 5 years 3 months
5.25 yr

6. 4 years 9 months
4.75 yr

7. May 15, 1987 to
Nov 15, 1993   6.5 yr

Find the simple interest earned in one year for the given principal (p)
and annual interest rate (r).

8. $p = \$500, r = 8\%$ $40   **9.** $p = \$900, r = 10\%$ $90   **10.** $p = \$1{,}000, r = 12\%$ $120

11. Find p in dollars given that $A = \$7{,}625$, $r = 10\tfrac{1}{2}\%$, and
$t = 5$ yrs. Use $A = p + prt$. $5,000

12. Find v in meters per second given that $h = 2{,}080$ m and $t = 8$ s.
Use $h = vt - 5t^2$.   300 m/s

Written Exercises

Use the formula $i = prt$ to find each of the following.

1. i given that $p = \$6{,}000$, $r = 7\%$,
and $t = 4$ years $1,680

2. p given that $i = \$500$, $r = 12\tfrac{1}{2}\%$,
and $t = 5$ years $800

3. r given that $i = \$990$, $p = \$1{,}200$, and $t = 10$ years $8\tfrac{1}{4}\%$

Solve each problem. Use the formula $i = prt$.

4. Edgar borrowed $3,500 for 4 years 9 months, and was charged
$1,330 in simple interest. What was the annual interest rate?   8%

5. Angela owes $1,806 in simple interest on a loan of $4,300 at 12%
per year. Find the time for which Angela has held the loan in years
and months.   3 yr 6 mo

Checkpoint

Solve each equation for r.

1. $4,560 = 3,000 + 3,000(r)(6.5)$ 0.08
2. $40(60 + r) = 60r$ 120
3. $175 = r \cdot 7 - 5 \cdot 7^2$ 60
4. $A = \frac{1}{3}rt - 4$ $r = \frac{3A + 12}{t}$
5. Find r_1 in ohms if $R = 12$ ohms and $r_2 = 18$ ohms. Use the formula $R(r_1 + r_2) = r_1r_2$.   36 ohms

Closure

Ask students to solve for t in the equation $i = prt$. $\frac{i}{pr} = t$ Have students use the formula $A = \sqrt{(s - a)(s - b)(s - c)}$ to find the area A of a triangle with sides of length $a = 6$, $b = 10$, $c = 12$, where s is one-half the perimeter.   Solution:

$s = \frac{1}{2}(6 + 10 + 12) = 14$ and
$A = \sqrt{(14 - 6)(14 - 10)(14 - 12)} = 8$

Solve each problem. Use the formula $A = p + prt$.

6. Find p in dollars given that $A = \$9,990$, $r = 9\frac{1}{2}\%$, and $t = 7$ years. $6,000

7. Mr. Worthy took out a loan on June 1, 1986, at $7\frac{1}{2}\%$ per year. How much did he borrow if he owed a total of $3,030 on December 1, 1989? $2,400

Solve each problem. Use the formula $h = vt - 5t^2$.

8. Find v in meters per second given that $h = 300$ m and $t = 6$ s. 80 m/s
9. Given that $h = 85$ m and $t = 5$ s, find v in meters per second. 42 m/s
10. At the end of 4 seconds, what will be the height of an object shot upward with an initial vertical velocity of 90 meters per second? 280 m
11. What initial vertical velocity is required to boost an object to a height of 168 m at the end of 4 s? 62 m/s

For each exercise, select the appropriate formula from the *Formulas from Geometry* on page 788. Then solve the formula for the appropriate variable. Use the new formula to solve the problem.

12. Find the length of a rectangle that is 6.5 cm wide and has a perimeter of 32 cm. $l = \frac{P - 2w}{2}$; 9.5 cm
13. The area of a triangle is 34.5 ft². Find the length of its base given that its height is 6 ft. $b = \frac{2A}{h}$; 11.5 ft
14. A trapezoid has an area of 111 cm². Find its height given that its bases are 17 cm and 20 cm in length. $h = \frac{2A}{b_1 + b_2}$; 6 cm
15. A cube has a volume of 50.653 m³. Find the length of an edge given that the area of a face is 13.69 m². $s = \sqrt{A}$; 3.7 m

Solve each problem. Use an appropriate formula.

16. If $h = vt - 5t^2$, then the maximum value of h is $\frac{v^2}{20}$. What initial vertical velocity is needed to launch an object to a maximum height of 4,500 m? 300 m/s
17. Find, to the nearest hundred meters, the maximum height reached by a projectile launched upward with an initial vertical velocity of 240 m/s (see Exercise 16). 2,900 m
18. Find t in years and months given that $A = \$11,400$, $p = \$7,500$, and $r = 8\%$ (simple interest). 6 yr 6 mo
19. Find the annual simple interest rate r given that $A = \$1,044$, $p = \$600$, and $t = 8$ years. 9.25%
20. At the end of 4 years 3 months, Mr. Benjamin owed a total of $4,594 on a loan made at $10\frac{1}{4}\%$ (simple interest). What was the amount of the loan? $3,200

Determine the time needed for each investment to be doubled in total value (investment plus simple interest earned).

21. \$3,000 at 8% per year $12\frac{1}{2}$ yr

22. *p* dollars at 10% per year 10 yr

If an object is sent upward from *d* meters above the surface of the earth with an initial vertical velocity of *v* meters per second, then its height *h* in meters at the end of *t* seconds is determined by the formula

$$h = d + vt - 5t^2.$$

Use the formula $h = d + vt - 5t^2$ for Exercises 23–25.

23. Find *h* for $d = 120$ m, $v = 80$ m/s, and $t = 6$ s. 420 m

24. Find *d* for $h = 275$ m, $v = 70$ m/s, and $t = 5$ s. 50 m

25. At the end of 4 s, what is the height above the earth of an object shot upward with an initial vertical velocity of 85 m/s from the bottom of a shaft that extends 40 m below the earth's surface? 220 m

Midchapter Review

Simplify, if possible. *1.1, 1.2*

1. $-7 - 3 \cdot 2(-5) + 6 - 21 \div 3$ 22

2. $-5 \cdot 2^3$ −40

3. $(-5 \cdot 2)^3$ −1,000

4. $\dfrac{6 - 9 \cdot 2}{-24 + 8 \cdot 3}$ Undef

Identify the property of operations with real numbers that is illustrated. *1.3*

5. $9 + \sqrt{2}$ is a real number.

6. $(3m + 2)n = 3m \cdot n + 2 \cdot n$

7. $(3m + 2)n = n(3m + 2)$

8. $-4c + 4c = 0$

9. Simplify $3(6x - 2y) - (8x - y) + xy$, and evaluate for $x = -5$ and $y = -2$. 10x − 5y + xy; −30

Solve. *1.4*

10. $3(2x + 11) - x = 7 - 2(12 - 5x)$ 10

11. $3.5x + 5 = 32.5 - 1.5x$ 5.5

12. Solve $t(3x - 4) = v$ for *x*. $\dfrac{v + 4t}{3t}$

13. Use the formula $A = p + prt$ to find *p* in dollars given that $A = \$1,120$, $r = 8\%$, and $t = 5$ years. *1.5* \$800

▰▰▰FOLLOW UP

Guided Practice
Classroom Exercises 1–12

Independent Practice
A Ex. 1–11, **B** Ex. 12–22, **C** Ex. 23–25

Basic: WE 1–11, 13–21 odd, Midchapter Review

Average: WE 1–21 odd, 23, 24, Midchapter Review

Above Average: WE 1–21 odd, 23 25, Midchapter Review

Additional Answers
Midchapter Review

5. Clos for Add
6. Distr Prop
7. Comm Prop Mult
8. Add Inv Prop

Enrichment

Have the students prove that if $a(b + c) = bc$, then $\dfrac{1}{a} = \dfrac{1}{b} + \dfrac{1}{c}$. Then have them check that each formula gives the same value for *a* when $b = 12$ and $c = 8$.

$\dfrac{1}{a} = \dfrac{1}{12} + \dfrac{1}{8} = \dfrac{5}{24}$, $a = \dfrac{24}{5} = 4.8$ ✔

$b + c = \dfrac{bc}{a}$; $\dfrac{b + c}{bc} = \dfrac{1}{a}$; $\dfrac{b}{bc} + \dfrac{c}{bc} = \dfrac{1}{a}$;

$\dfrac{1}{c} + \dfrac{1}{b} = \dfrac{1}{a}$, or $\dfrac{1}{a} = \dfrac{1}{b} + \dfrac{1}{c}$;

$a(12 + 8) = (12)(8)$, $a = \dfrac{96}{20} = 4.8$;

GETTING STARTED

Prerequisite Quiz

Write the three expressions in each list if $x = -4$.

1. $3x, 7 - 3x, 2(7 - 3x) + 5$ $-12, 19, 43$
2. $x, x + 6, x + 12$ $-4, 2, 8$
3. $2x, 2x - 8, 5(x - 4) + 32$ $-8, -16, -8$
4. $x - 12, 12 - x, 18 - (x - 12)$
   $-16, 16, 34$

Motivator

Have students create a word problem that would lead to writing and solving the equation $x + (x + 1) + (x + 2) = 63$.
Sample answers:
1. The sum of a number, one more than the number, and two more than the number is 63. Find the number.
2. The sum of three consecutive integers is 63. Find the three integers.

1.6 Problem Solving: One or More Numbers

Objective   To solve word problems about one or more numbers

When solving word problems, it is often necessary to translate word phrases and sentences into algebraic expressions or equations.

Word phrase or sentence	Algebraic expression or equation
Twelve decreased by a number	$12 - n$
Twelve less than a number	$n - 12$
Twice a number, increased by 12	$2a + 12$
Five decreased by the sum of a number and 12	$5 - (y + 12)$
Nine less than a number is 8 more than twice the number.	$n - 9 = 2n + 8$
Three divided by a number is the same as 6 divided by 2 more than the number.	$\dfrac{3}{y} = \dfrac{6}{y + 2}$

In some word problems, you are asked to find more than one number. To do this, begin by representing each number in terms of the same variable.

EXAMPLE 1   The second of three numbers is 4 times the first number. The third number is 5 less than the second number. If twice the first number is decreased by the third number, the result is the same as 23 more than the second number. What are the three numbers?

Solution

What are you to find?	Three numbers
Choose a variable.	Let f = the first number.
What does it represent?	Then $4f$ = the second number, and $4f - 5$ = the third number.

What is given?   Twice the 1st, minus the 3rd, equals the 2nd plus 23.

Write an equation. $2f - (4f - 5) = 4f + 23$

Solve the equation. $2f - 4f + 5 = 4f + 23$
$$-6f = 18$$
$$f = -3 \rightarrow \text{the first number} = -3$$

$f = -3$ $4f = -12$ $4f - 5 = -17$

Check in the original problem.   "Twice the 1st, minus the 3rd" is $2(-3) - (-17)$, or 11.
"23 more than the 2nd" is $-12 + 23$, or 11.

State the answer.   The three numbers are $-3, -12,$ and -17.

Highlighting the Standards

Standard 3c: Understanding that problem solving can be approached in a logical, sequential manner will help reduce students' anxiety.

Additional Example 1

The greatest of three numbers is 12 more than the smallest. The remaining number is twice the smallest number. The greatest number is 6 more than the sum of the other two numbers. Find the numbers.   3, 6, 15

Many problems require information that is not given directly in the problem. For example, a problem may involve consecutive integers or consecutive multiples of an integer. Four such word phrases, along with numerical examples and algebraic expressions, are given in the table below.

Word Phrase	Example	Algebraic Expression
Consecutive integers	$-2, -1, 0, 1$	$n, n + 1, n + 2, \ldots$ (n is an integer.)
Consecutive odd integers	$7, 9, 11, 13, 15$	$x, x + 2, x + 4, \ldots$ (x is odd.)
Consecutive even integers	$-4, -2, 0, 2$	$y, y + 2, y + 4, \ldots$ (y is even.)
Consecutive multiples of 3	$12, 15, 18, 21$	$t, t + 3, t + 6, \ldots$ (t is a multiple of 3.)

EXAMPLE 2 Find three consecutive odd integers such that twice the sum of the first two integers, decreased by the third integer, is the same as 20 less than the second integer.

Solution Let x, $x + 2$, and $x + 4$ represent the three odd integers.
$$2[x + (x + 2)] - (x + 4) = x + 2 - 20$$
$$4x + 4 - x - 4 = x - 18$$
$$2x = -18$$
$$x = -9 \leftarrow \text{the first odd integer} = -9$$

$$x = -9 \qquad x + 2 = -7 \qquad x + 4 = -5$$

Check Twice (the 1st plus the 2nd) minus the 3rd is $2[-9 + (-7)] - (-5)$, or -27.

20 less than the 2nd is $-7 - 20$, or -27.

The three consecutive odd integers are -9, -7, and -5.

Classroom Exercises

Represent each number in terms of one variable.

1. The second of three numbers is 5 times the first number. The third number is 4 less than the second number. $x, 5x, 5x - 4$

2. The greatest of three numbers is 40 more than the smallest. The remaining (middle) number is 5 less than 6 times the smallest number. $x, 6x - 5, x + 40$

Lesson Note

Note that the general sequence of the Polya problem-solving steps is shown in Example 1.

1. **Understand the problem** Read carefully. Identify what you are asked to find and the information needed to solve the problem. Choose a variable and represent the data.
2. **Develop a plan** Choose a strategy such as writing an equation, using guess and check, drawing a diagram, and so on.
3. **Carry out the plan** Apply the strategy by writing and solving an equation, and so on.
4. **Look back** Check the solution in the original problem. Consider whether the answer is reasonable. State the answer in a sentence.

The check is included since the solution of the equation may not be the answer to the problem.

Math Connections

Translation: This lesson connects English language sentences to mathematical sentences so that data given in paragraph form can be represented algebraically.

Additional Example 2

Find three consecutive even integers such that 5 times the sum of the first two integers, increased by 60, is the same as 9 times the third number. $-34, -32, -30$

Critical Thinking Questions

Analysis: Ask students whether two consecutive integers could both be prime numbers. If so, are there a limited number of such prime number pairs? The only such pair is 2, 3. If n and $n + 1$ were another such pair then either n or $n + 1$ would be even but since 2 is the only even prime number, such a second pair of consecutive prime numbers cannot exist.

Common Error Analysis

Error: To represent "$5n$ decreased by $n + 3$," students will often write $5n - n + 3$, rather than $5n - (n + 3)$.

Emphasize the need for parentheses when subtracting an expression that has more than one term.

Checkpoint

1. Find three consecutive multiples of 4 such that twice the sum of the last two numbers, decreased by the second number, is 4 less than 5 times the first number.   12, 16, 20
2. A Brand-A bottle holds a dozen more pills than a Brand-B bottle. A Brand-C bottle holds twice as many pills as a Brand-A bottle. Three Brand-A, 4 Brand-B, and 2 Brand-C bottles hold 150 pills. How many pills are in a Brand-C bottle?   36, or 3 doz

3. Write an equation for the three numbers of Classroom Exercise 1, given that their sum is 18. $x + 5x + (5x - 4) = 18$
4. Write an equation for the three numbers of Classroom Exercise 2, given that their sum is 3 times the middle number. $x + (6x - 5) + (x + 40) = 3(6x - 5)$
5–6. Solve the equations of Classroom Exercises 3 and 4. 2, 5
7. Eight less than 4 times a number is the same as 8 times the sum of the number and 2. What is the number? -6
8. Fifteen more than 7 times a number is twice the sum of the number and -6. What is the number? $-5\frac{2}{5}$

Written Exercises

1. If 18 is decreased by the sum of a number and 6, the result is 4 less than 3 times the number. Find the number. 4

2. If 3.5 times a number is increased by 2, the result is the same as 7.5 times 4 less than the number. Find the number. 8

3. The greater of two numbers is 10 less than 3 times the smaller. If the greater is increased by twice the smaller, the result is 8 less than 3 times the greater. What are the two numbers? 7, 11

4. A Brand X bottle contains one dozen fewer capsules than a Brand Y bottle. Five Brand X and 8 Brand Y bottles contain 486 capsules. How many capsules are in a Brand X bottle? 30

5. The second of three numbers is 4 times the first number. The third number is 12 less than the first. Find the three numbers given that their sum is 42. 9, 36, -3

6. Find three consecutive integers whose sum is 37 less than 5 times the third integer. 15, 16, 17

7. There are three consecutive even integers such that the sum of the second and third integers is 30 more than the first integer. Find these three even integers. 24, 26, 28

8. There are three consecutive odd integers such that twice their sum is 30 less than 8 times the third integer. Find these odd integers. 5, 7, 9

9. The greatest of three numbers is 7 times the smallest, and the middle number is 8 more than the smallest. Six more than the greatest number, decreased by 3 times the middle number, is 9 less than the smallest number. Find the three numbers. 3, 11, 21

10. The second of three numbers is twice the first, and the third is 5 less than the second. If 12 less than the first is decreased by the third, the result is -15. What are the three numbers? 8, 16, 11

11. Find three consecutive integers such that 6 times the first, decreased by the third, is 398. 80, 81, 82

12. Find three consecutive odd integers such that 5 times the second, decreased by twice the third, is 155. 51, 53, 55

13. Find three consecutive multiples of 15 such that twice their sum is 360. 45, 60, 75

14. One half of the sum of three consecutive multiples of 10 is 90. What are these multiples of 10? 50, 60, 70

15. A company's sales for June increased $600 over its May sales figure. July sales were twice June sales, and the sales for August doubled the figure for July. The sales total was $20,200 for the four months. Find the sales figure for each month.

16. A size *A* box holds twice as many oranges as a size *C* box, and a size *B* box holds one dozen fewer oranges than a size *A* box. Six size *A* boxes, 5 size *B* boxes, and 8 size *C* boxes will hold a total of 45 dozen oranges. Find the number of oranges that a size *B* box will hold. 28

17. Find all sets of three consecutive even integers such that 6 less than twice the third integer is the sum of the first and second integers.

18. A wire 50 in. long is bent to form a rectangle. Represent the rectangle's length given that $x - 3$ inches represents the width. $28 - x$ in

19. A first angle of measure *x* and a second angle of measure *y* are *complementary*—that is, the sum of their measures is 90. The second angle measures 4 times the measure of the first angle less 5*a*. Two angles are *supplementary* if the sum of their measures is 180. Represent the measure of the supplement of the first angle in terms of *a*. $162 - a$

Mixed Review

Solve each equation for *x*. *1.4*

1. $\dfrac{3x - 6}{x + 5} = \dfrac{3}{8}$ 3

2. $\dfrac{2}{3}x + \dfrac{1}{2} = \dfrac{5}{6}x$ 3

3. $a(5x + b) = 7$ $\dfrac{7 - ab}{5a}$

/ **Brainteaser**

The sum of two numbers is 1. Which is larger: the sum of the larger number and the square of the smaller number, or the sum of the smaller number and the square of the larger number? Both sums are the same.

Have students give an English phrase for each of the following algebraic expressions. $n - 3$? three less than a number $n, n + 1, n + 2,\dots$? Consecutive integers
Ask students to group the steps involved in the solution to Example 1 into four basic steps. See Lesson Note.

◤◤◤ FOLLOW UP

Guided Practice

Classroom Exercises 1–8

Independent Practice

Ⓐ Ex. 1–12, Ⓑ Ex. 13–16, Ⓒ Ex. 17–19

Basic: WE 1–8, 9–15 odd, Brainteaser, Application

Average: WE 1–16 odd, 18, Brainteaser, Application

Above Average: WE 1–19 odd, Brainteaser, Application

Additional Answers

Written Exercises

15. May: $2,000; June: $2,600; July: $5,200; August: $10,400

17. Any 3 consecutive even integers, {2*n*, 2*n* + 2, 2*n* + 4}, where *n* is an integer.

Enrichment

Ask the students to find this sum.
1 + 2 + 3 + 4 + · · · + 99 + 100
Challenge them to find a shortcut method.
(HINT: Start with 1 + 100 = 101.)

$$
\left.\begin{array}{r}
1 + 100 = 101 \\
2 + \ 99 = 101 \\
\cdot \qquad \cdot \qquad \cdot \\
49 + \ 52 = 101 \\
50 + \ 51 = 101
\end{array}\right\} 50 \cdot 101 = 5050
$$

Now have the students use a similar shortcut to find the sum of: (1) the first thousand positive integers, and (2) the multiples of 5, from 5 through 100. (1) 500,500 (2) 1,050

2. $2\frac{1}{4}$ ft toward the 260-lb shape

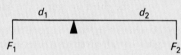

Application: *Equilibrium*

A sculptor is commissioned to create a large mobile for a public hall. For the mobile (shown below) to be balanced, the support cables must be attached at the precise points of equilibrium (*A* and *B*).

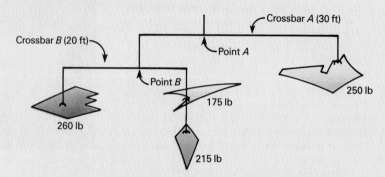

Crossbar *A* (30 ft)

Crossbar *B* (20 ft)

Point *A*

Point *B*

175 lb

250 lb

260 lb

215 lb

To calculate its point of equilibrium, think of each crossbar as a simple lever. A lever system is balanced when $F_1 \times d_1 = F_2 \times d_2$, where F_1 and F_2 are the forces (weights) and d_1 and d_2 are the distances.

d_1 d_2

F_1 F_2

EXAMPLE   Find the point of equilibrium for Crossbar *B* (point *B*). Assume that the weight of the crossbars and the cables can be disregarded.
Let x = distance from the 260-lb shape to the fulcrum (d_1).
Therefore, $d_2 = 20 - x$.

Solution
$$F_1 \times d_1 = F_2 \times d_2$$
$$260 \cdot x = (175 + 215)(20 - x)$$
$$x = 12 \leftarrow \text{ distance from 260-lb shape to fulcrum} = 12 \text{ ft}$$

The point of equilibrium for Crossbar *B* is 12 ft from the 260-lb shape.

1. Find the point of equilibrium for Crossbar *A*. $8\frac{1}{3}$ ft from Crossbar *B* cable, or $21\frac{2}{3}$ ft from the 250-lb shape

2. If a 150-lb shape were added to the cable holding the 260-lb shape, how far would point *B* move? Which way?

3. The construction team mistakenly attaches the main support cable to a point on Crossbar *A* 12 ft from the 250-lb shape. For the mobile to be balanced, how much weight must be added? To which end? 725 lb to the end with the 250-lb shape

4. The sculptor resolves the imbalance in Exercise 3 by adding another shape to Crossbar *A* at a point 4 ft from the cable holding the 250-lb shape. How much must this extra shape weigh in order to restore equilibrium? $1{,}087\frac{1}{2}$ lb

1.7 Problem Solving: Perimeter and Area

Teaching Resources

Quick Quizzes 7
Reteaching and Practice
  Worksheets 7

Objective   To solve word problems involving perimeter and area

If a word problem involves a geometric figure, you can represent the data by drawing and labeling the figure. Area and perimeter formulas can be found in the *Formulas from Geometry* at the back of the book.

EXAMPLE 1   Side *a* of a triangle is 9 cm longer than side *b*. Side *c* is 1.5 times as long as side *a*. Find the length of each side given that the perimeter of the triangle is 75 cm.

What are you to find?   The length of each side

What is given?
Represent the data as
  shown at the right.

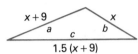

The **perimeter** is the sum of the lengths of the sides.

Write an equation. $(x + 9) + x + 1.5(x + 9) = 75$
Solve the equation. $x + 9 + x + 1.5x + 13.5 = 75$
$$3.5x = 52.5$$
$$x = 15$$

The three lengths are: $x = 15$
$x + 9 = 24$ $1.5(x + 9) = 36$

Check in the   The perimeter is $15 + 24 + 36 = 75$.
  original problem.
State the answer.   The lengths are 15 cm, 24 cm, and 36 cm.

The lengths of the sides of one geometric figure may be related to the lengths of the sides of another. In such cases, draw and label both figures to represent the data as shown below.

EXAMPLE 2   Rectangle I is 4 m longer than it is wide. Rectangle II is 3 m wider than, and twice as long as rectangle I. The difference between their perimeters is 24 m.

    a. Find the length of rectangle II.   **b.** Find the area of rectangle I.

Lesson Note

Students will generally find that a drawing helps clarify the data and relationships involved in a geometric word problem.

Note that the final step in solving each of these problems requires that the appropriate unit be included in the answer. Review the following abbreviations: ft, ft², cm, cm², in., in².

Math Connections

Architecture: When designing a building, an architect will be concerned with both its relative proportions and its absolute dimensions. Have students solve this problem, which is similar to the type of problem an architect must solve.

If the ratio of two sides of a building is designed to be 9:5 and the perimeter is to be 300 ft, what would be the length and width of the building? $l \approx 96.4$ ft; $w \approx 53.6$ ft

Plan   Draw and label the two rectangles. Use w to represent the width of rectangle I.

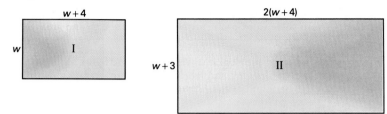

Solutions

a. Perimeter I $= 2w + 2(w + 4)$, or $4w + 8$
   Perimeter II $= 2(w + 3) + 2[2(w + 4)]$, or $6w + 22$
   Perimeter II $-$ Perimeter I $= 24$
   $$6w + 22 - (4w + 8) = 24$$
   $$6w + 22 - 4w - 8 = 24$$
   $$2w = 10$$
   $$w = 5 \leftarrow \text{width of rectangle I} = 5 \text{ m}$$
   The length of rectangle II, $2(w + 4)$, is 18 m.

b. The area of rectangle I, $w(w + 4)$, is $5 \cdot 9$, or 45 m².

The formulas for the circumference and area of a circle include the irrational number π. An approximate value of π is 3.14. This can be written as $\pi \approx 3.14$, where the symbol $\approx$ means "is approximately equal to."

Some scientific calculators have a key for π. Use such a calculator to find π to six decimal places.

$C = 2\pi r$
$A = \pi r^2$

EXAMPLE 3   The circumference of a circle is 47.1 in. Find the area of the circle to the nearest in².

Solution   Let $r =$ the length of the radius.
Use $C = 2\pi r$ and $\pi \approx 3.14$ to find r.
$$C = 2\pi r$$
$$47.1 \approx 2(3.14)r, \text{ or } 6.28r$$
$$\frac{47.1}{6.28} \approx r \quad \text{Thus, } 7.50 \approx r, \text{ or } r \approx 7.50$$
Use $A = \pi r^2$ to find the area with $r \approx 7.5$ and $\pi \approx 3.14$.
$$A \approx 3.14(7.5^2) = 3.14(56.25) = 176.625$$

The area is 177 in², to the nearest in².

Additional Example 3

The circumference of a circle is 94.2 cm. Find the area of the circle to the nearest cm². Use 3.14 for π.   707 cm²

Classroom Exercises

Find the perimeter or circumference of each figure. Include the appropriate unit of measure. Use $\pi \approx 3.14$.

1.

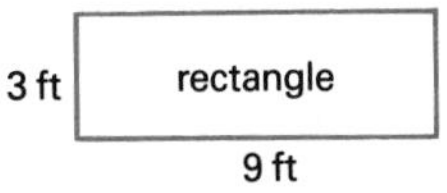

3 ft | rectangle
9 ft

24 ft

2.

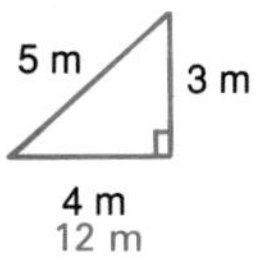

5 m / 3 m
4 m
12 m

3.

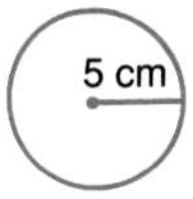

5 cm

31.4 cm

4–6. Find the area of each figure of Exercises 1–3. Include the appropriate unit of measure. Use $\pi \approx 3.14$. **4.** 27 ft^2 **5.** 6 m^2 **6.** 78.5 cm^2

7. Side a of a triangle is 5 ft longer than side b. Side c is twice as long as side b. Find the length of each side given that the perimeter of the triangle is 45 ft. $a = 15$ ft, $b = 10$ ft, $c = 20$ ft

Written Exercises

Write an expression for the perimeter of the figure shown below.

1.

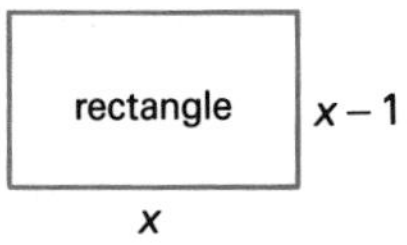

rectangle | $x - 1$
x

$4x - 2$

2.

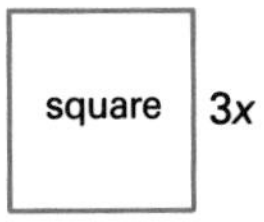

square | $3x$

$12x$

3.

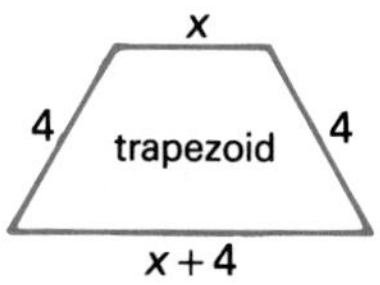

x
4 / trapezoid \ 4
$x + 4$

$2x + 12$

4. Side c of a triangle is 4 m longer than side a, and side b is 5 m shorter than side a. The perimeter is 59 m. What are the lengths of the sides?

5. The length of a rectangle is 6 yd more than 3 times its width. Find the length and the area of the rectangle given that its perimeter is 44 yd. $l = 18$ yd; $A = 72$ yd^2

6. The hypotenuse of a right triangle is 15 in. long and one leg is 3 in. longer than the other leg. Given that the perimeter is 36 in., find the area of the right triangle in square inches. 54 in^2

7. A square and a rectangle have the same width. The rectangle is 8 m longer than the square, and one perimeter is twice the other perimeter. Find the dimensions of the rectangle.

8. One side of an equilateral triangle is 6 ft shorter than one side of the square. The difference between the perimeters of the two figures is 28 ft. Find the perimeter of each figure.

9. The circumference of a circle is 31.4 cm. Find its area to the nearest 0.1 cm^2. Use $\pi \approx 3.14$. 78.5 cm^2

Critical Thinking Questions

Analysis: Ask students to solve this problem. The length of the longest side of a triangle is 3 more than twice the length of the shortest side. The remaining side is 2 more than the length of the shortest side. If the perimeter is 29, find the three lengths. After they find the three lengths, (6, 8, and 15) ask them to sketch the triangle. They will see that such a triangle cannot exist since $6 + 8 < 15$.

Checkpoint

1. The perimeter of a right triangle is 24 cm. One leg is 2 cm longer than the other leg and 2 cm shorter than the hypotenuse. Find the area of the right triangle. 24 cm^2
2. Rectangle I is 6 ft longer than it is wide. Rectangle II is 3 times as wide and twice as long as Rectangle I. The difference between the two perimeters is 42 ft. Find the area of Rectangle I and the length of Rectangle II. 55 ft^2; 22 ft

Closure

Have students tell how to represent data in solving perimeter and area problems.   **With a drawing**   Have students give the area formula for a rectangle, a square, a triangle, a circle, and a trapezoid. $A = bh$; $A = s^2$; $A = \frac{1}{2}bh$; $A = \pi r^2$; $A = \frac{1}{2}h(b_1 + b_2)$

▰▰▰FOLLOW UP

Guided Practice

Classroom Exercises 1–7

Independent Practice

A Ex. 1–9, **B** Ex. 10–16, **C** Ex. 17–20

Basic: WE 1–9, 11–15 odd
Average: WE 1–15 odd, 18
Above Average: WE 1–19 odd

Additional Answers

Written Exercises

4. 20 m, 15 m, 24 m
7. 4 m by 12 m
8. Triangle: 12 ft; square: 40 ft
16. Answers may vary. One possibility: Let x be the width. Adding twice the width to twice the length will give the perimeter, which is also given explicitly. Now solve for x, and then for the length. Multiplying these two will give the area in units squared.

Mixed Review

1. Clos for Add
2. Assoc Prop Add
3. Add Inv Prop
4. Mult Ident Prop

---

10. One base of a trapezoid is 1 in. shorter than the height, and the other base is 1 in. longer than twice the height. Find the area given that each leg is 5 in. long and the perimeter is 22 in.   24 in^2

11. The two longest sides of a pentagon (5-sided figure) are each 3.5 times as long as the shortest side. The remaining sides are each 9 m longer than the shortest side. Find the length of each side given that the perimeter is 78 m.   6 m, 15 m, 15 m, 21 m, 21 m

12. Each of the three longer sides of a hexagon (6-sided figure) is 2 in. less than 4 times the length of each of the three shorter sides. Find the length of a longer side given that the perimeter is 114 in.   30 in

13. One leg of a right triangle is 4 cm longer than twice the length of the other leg. The hypotenuse is 2 cm longer than the longer leg. Find the area given that the perimeter is 60 cm.   120 cm^2

14. A longer side of a parallelogram is 3 ft shorter than 4 times the length of a shorter side. One side of a rhombus (equilateral quadrilateral) and a longer side of the parallelogram are equal in length. The difference between the two perimeters is 24 ft. What is the perimeter of the rhombus?   68 ft

15. Find, to the nearest square meter, the area of a circular roof whose circumference is 78.5 m. Use $\pi \approx 3.14$.   491 m^2

16. Describe in your own words how to find the area of a rectangle, if its length is described in terms of its width, and its perimeter is given in feet.

17. The lengths of the sides of a quadrilateral have the ratio of 2 to 4 to 5 to 7. What is the length of each side if the perimeter is 117 in.?   13 in, 26 in, 32.5 in, 45.5 in

18. One side of an equilateral triangle is 4 cm longer than one side of a regular hexagon. The difference between their perimeters is 10.5 cm. Find the perimeter of the triangle.   34.5 cm

19. A right triangle has a perimeter of 30 ft, and one leg is 5 ft long. Find the area of the triangle. (HINT: $(m - n)^2 = m^2 - 2mn + n^2$)   30 ft^2

20. The perimeters x, y, and z of three figures are integers with the ratios $x{:}y = 2{:}3$ and $y{:}z = 5{:}4$. Find the three perimeters given that their sum is 74 ft. $x = 20$ ft; $y = 30$ ft; $z = 24$ ft

Mixed Review

Name the property of operations with real numbers that is illustrated.   *1.3*

1. $-4.2 + \sqrt{6}$ is a real number.
2. $(a + 3) + 5 = a + (3 + 5)$
3. $6t + [-(6t)] = 0$
4. $8x + x = 8x + 1x$

Enrichment

Challenge the students with this problem.

The number of linear units in the perimeter of a rectangle is the same as the number of square units in its area. Express the width w of the rectangle in terms of its length.

$w = \dfrac{2l}{l-2}$

Now have the students find the specific dimensions of one or two rectangles that meet the conditions of the problem.
Answers will vary. One answer is $l = 10$, $w = 2.5$.

1.8 Proof in Algebra

Objectives   To prove two algebraic expressions equivalent
To determine whether relations are equivalence relations

Statements that are assumed to be true are called **axioms** or **postulates**. The eleven properties of a field listed in Lesson 1.3 are postulates because they are accepted as true without proof. A **theorem** is a statement that has been logically deduced from a set of axioms or postulates.

The following additional postulates are needed to prove some theorems for all real numbers a, b, and c.

Zero Property for Multiplication
$a \cdot 0 = 0$

Properties of Equality

If $a = b$, then $a + c = b + c$.	(Add Prop of Eq)
If $a + c = b + c$, then $a = b$.	(Subt Prop of Eq)
If $a = b$, then $a \cdot c = b \cdot c$.	(Mult Prop of Eq)
If $a \cdot c = b \cdot c$ and $c \neq 0$, then $a = b$.	(Div Prop of Eq)
$a = a$	(Reflexive Prop of Eq)
If $a = b$, then $b = a$.	(Symmetric Prop of Eq)
If $a = b$ and $b = c$, then $a = c$.	(Transitive Prop of Eq)

Substitution

If $a = b$, then a can be replaced by b, and vice-versa, in any equation without changing the *truth-value* (truth or falsity) of the equation.

It was stated in Lesson 1.3 that the terms of an expression can be reordered and regrouped. One such rearrangement is stated in the following theorem. Notice that each statement in the proof of the theorem is justified by a postulate for operations with real numbers.

Theorem 1.1 $(a + b) + (c + d) = (a + c) + (b + d)$ for all real numbers a, b, c, and d.

Proof

Statement	Reason
1. $(a + b) + (c + d)$	
$= [(a + b) + c] + d$	Associative Prop of Add
2. $= [a + (b + c)] + d$	Associative Prop of Add
3. $= [a + (c + b)] + d$	Commutative Prop of Add
4. $= [(a + c) + b] + d$	Associative Prop of Add
5. $= (a + c) + (b + d)$	Associative Prop of Add
6. Thus, $(a + b) + (c + d) =$ $(a + c) + (b + d)$	

Teaching Resources

Quick Quizzes 8
Reteaching and Practice
  Worksheets 8

GETTING STARTED

Prerequisite Quiz

Name the property for real numbers or the rule of operations that is illustrated by each sentence, where x, y, and z are real numbers.

1. $x \cdot y$ is a real number.
Clos for Mult

2. $x \cdot y = y \cdot x$
Comm Prop Mult

3. $(x + y) + z = x + (y + z)$
Assoc Prop Add

4. $x(y + z) = xy + xz$
Distr Prop

5. $x = 1 \cdot x$
Mult Ident

6. $x + 0 = x$
Add Ident

7. $x + (-x) = 0$
Add Inv

8. $x - y = x + (-y)$
Rule of Subt

Highlighting the Standards

Standard 14b: Throughout the exercises, students learn that algebra, like geometry, is built on clearly defined, working postulates.

Motivator

Ask students if there are any statements in mathematics that are assumed to be true. Yes, axioms and postulates. Ask them if axioms and postulates can be used as reasons in a proof. Yes

TEACHING SUGGESTIONS

Lesson Note

To facilitate supplying reasons in a proof, students should keep a list of the Field Properties, the Properties of Equality, and each theorem as it is proved in this lesson.

Math Connections

Geometry: Equivalence relations can be found in plane geometry. Two examples are *congruence* and *similarity* when applied to the set of all triangles.

Statement 6 follows from the Transitive Property of Equality used several times in succession. (If $x = y$ and $y = z$, then $x = z$.)

Theorem 1.2 $-x = -1x$ for each real number x.

Proof

Statement	Reason
1. x is a real number.	Given
2. $-1x$ is a real number.	Closure Prop for Mult
3. $-1x + x$ is a real number.	Closure Prop for Add
4. $-1x + x = -1x + 1x$	Identity for Mult
5. $\quad\quad = (-1 + 1)x$	Distr Prop
6. $\quad\quad = 0 \cdot x$	Add Inverse Prop
7. $\quad\quad = 0$	Zero Prop for Mult
8. $-1x + x = 0$	Trans Prop of Eq (Steps 4–7)
9. $-x + x = 0$	Add Inverse Prop
10. $-1x + x = -x + x$	Sub (Steps 8–9)
11. $\quad -1x = -x$	Subt Prop of Eq
12. $\quad -x = -1x$	Sym Prop of Eq

It was stated in Lesson 1.3 that multiplication is distributive over subtraction: $(a - b)c = ac - bc$. To prove this, it is helpful to prove first that $(-a)b = -(ab)$.

Theorem 1.3 $(-a)b = -(ab)$ for all real numbers a and b.

Proof

Statement	Reason
1. $(-a)b = (-1a)b$	Theorem 1.2 $[-a = -1a]$
2. $\quad\quad = -1(ab)$	Associative Prop of Mult
3. $\quad\quad = -(ab)$	Theorem 1.2 $[-(ab) = -1(ab)]$
4. $(-a)b = -(ab)$	Trans Prop of Eq

Theorem 1.4 $(a - b)c = ac - bc$ for all real numbers a, b, and c.

Proof

Statement	Reason
1. $(a - b)c = [a + (-b)] c$	Rule of Subt
2. $\quad\quad = ac + (-b)c$	Distr Prop
3. $\quad\quad = ac + [-(bc)]$	Theorem 1.3 $[(-b)c = -(bc)]$
4. $\quad\quad = ac - bc$	Rule of Subt
5. $(a - b)c = ac - bc$	Trans Prop of Eq

A real-number fact, such as $-1 - 2 = -3$, can be used to replace one numeral with an equivalent numeral in a proof, as shown below.

Theorem 1.5

$-z - 2z = -3z$ for each real number z.

Proof

Statement	Reason
1. $-z - 2z = -1z - 2z$	Theorem 1.2 $[-z = -1z]$
2. $\quad\quad = (-1 - 2)z$	Theorem 1.4 $[(-1 - 2)z = -1z - 2z]$
3. $\quad\quad = -3z$	Number fact $[-1 - 2 = -3]$
4. $-z - 2z = -3z$	Trans Prop of Eq

You are asked to prove Theorems 1.6 and 1.7 below in Exercises 7 and 10.

Theorem 1.6

$-(-a) = a$ for each real number a.

Theorem 1.7

$(ab)(cd) = (ac)(bd)$ for all real numbers a, b, c, and d.

There are other relations besides $=$ that may be reflexive, symmetric, or transitive. In general, a relation $\Re$ is
(1) reflexive when: $a \, \Re \, a$ is true for all a.
(2) symmetric when: If $a \, \Re \, b$, then $b \, \Re \, a$ is true for all a and b.
(3) transitive when: If $a \, \Re \, b$ and $b \, \Re \, c$, then $a \, \Re \, c$ is true for all a, b, and c.

Definition

Any relation that is reflexive, symmetric, and transitive is called an **equivalence relation**.

EXAMPLE

Is the relation $<$ *(is less than)* for real numbers (1) reflexive, (2) symmetric, (3) transitive, or (4) an equivalence relation?

Solution

(1) $<$ is *not reflexive*: "$8 < 8$" is a false statement.
(2) $<$ is *not symmetric*: "If $5 < 6$, then $6 < 5$" is not true.
(3) $<$ *is transitive*: "If $7 < 9$ and $9 < 11$, then $7 < 11$" is true and "If $a < b$ and $b < c$, then $a < c$" is true for all real numbers a, b, and c.
(4) $<$ is *not an equivalence relation* because it does not satisfy all three conditions.

Critical Thinking Questions

Analysis: Ask students whether the theorem $-(a + b) = -a - b$ (Classroom Exercise 1) can be proved without using either the Identity Property for Multiplication or the Distributive Property. It is surprising that this can actually be done, which indicates that the relationship is "deeper" than it first appears. If a hint is needed, show the first two steps, as follows: $(a + b) + [-(a + b)] = 0$ (Def of Add Inv); $-a + (a + b) + [-(a + b)] = -a + 0$ (Add Prop Eq)

Checkpoint

Supply a reason for each step in the following proof.

Prove: $(a - b)c + bc = ca$
Proof:
1. $(a - b)c + bc = (ac - bc) + bc$
2. $\quad\quad = [ac + (-(bc))] + bc$
3. $\quad\quad = ac + [-(bc) + bc]$
4. $\quad\quad = ac + 0$
5. $\quad\quad = ac$
6. $\quad\quad = ca$
7. $(a - b)c + bc = ca$
   1. Thm 1.4; 2. Rule of Subt; 3. Assoc Prop Add; 4. Add Inv; 5. Add Ident; 6. Comm Prop Mult; 7. Trans Prop Eq

Closure

Have students explain why the relation *is a factor of*, for all positive integers, (1) is reflexive, (2) is *not* symmetric, (3) is transitive, and (4) is *not* an equivalence relation. Consider numerical examples and counterexamples as satisfactory explanations.   Answers will vary.

Additional Example

Is the relation "is the reciprocal of,"
(a) reflexive, **(b)** symmetric, **(c)** transitive, **(d)** an equivalence relation?

(a) It is not reflexive: 5 is not the reciprocal of 5. **(b)** It is symmetric: 5 is the reciprocal of $\frac{1}{5}$, and $\frac{1}{5}$ is the reciprocal of 5. **(c)** It is not transitive: 5 is the reciprocal of $\frac{1}{5}$ and $\frac{1}{5}$ is the reciprocal of 5, but 5 is not the reciprocal of 5. **(d)** It is not an equivalence relation: neither reflexive nor transitive.

Guided Practice

Classroom Exercises 1–4

Independent Practice

A Ex. 1–10, **B** Ex. 11–19, **C** Ex. 20

Basic: WE 1–8, 9–15 odd
Average: WE 1–11 odd, 12, 14, 17, 20
Above Average: WE 1–15 odd, 16, 19, 20

Additional Answers

Written Exercises

1. (1) yes; (2) no: $8 \geq 6$, but $6 \ngeq 8$; (3) yes; (4) no, not symmetric
2. (1) no: $3 \neq 3$ is false; (2) yes; (3) no: $4 \neq 3$ and $3 \neq 4$, but $4 \neq 4$ is false; (4) no, not reflexive or transitive
3. (1) yes; (2) no: $8 \not< 6$, but $6 < 8$; (3) yes; (4) no, not symmetric
4. (1) no: $5 \ngeq 5$ is false; (2) no: $6 \ngeq 8$ is true, but $8 \ngeq 6$ is false; (3) yes; (4) no, not reflexive or symmetric
5. (1) yes; (2) no: 3 is a divisor of 6, but 6 is not a divisor of 3; (3) yes; (4) no, not symmetric
6. (1) yes; (2) no: 4 is a multiple of 2, but 2 is not a multiple of 4; (3) yes; (4) no, not symmetric

Classroom Exercises

Justify each step in the proof.

1. $-(a + b) = -a - b$
   1. $-(a + b) = -1(a + b)$ Thm 1.2
   2. $= -1a + (-1)b$ Distr Prop
   3. $= -a + (-b)$ Thm 1.2
   4. $= -a - b$ Rule of Subt
   5. $-(a + b) = -a - b$ Trans Prop Eq

2. $(x + y) - y = x$
   1. $(x + y) - y = (x + y) + (-y)$ Rule of Subt
   2. $= x + [y + (-y)]$ Associative Prop Add
   3. $= x + 0$ Add Inverse Prop
   4. $= x$ Identity Prop Add
   5. $(x + y) - y = x$ Trans Prop Eq

Determine whether the relation is (1) reflexive, (2) symmetric, (3) transitive, or (4) an equivalence relation.

3. *is a multiple of,* for all numbers in $\{1, 2, 3, 4, 5, 6, 7, 8, 9, 10\}$

3. (1) yes, (2) no, (3) yes, (4) no

4. *is the same age as,* for all people   (1) yes, (2) yes, (3) yes, (4) yes

Written Exercises

Determine whether each relation is (1) reflexive, (2) symmetric, (3) transitive, or (4) an equivalence relation. If not (1), (2), or (3), supply a counterexample.

1. $\geq$, for real numbers
2. $\neq$, for real numbers
3. $\not<$, for real numbers
4. $\ngeq$, for real numbers
5. *is a divisor of,* for positive integers
6. *is a multiple of,* for positive integers

Supply a reason to justify each step in the following proofs. (Exercises 7–10)

7. $-(-a) = a$ (Theorem 1.6)
   1. $-(-a) = -1(-1 \cdot a)$ Thm 1.2
   2. $= [-1(-1)]a$ Associative Prop Mult
   3. $= 1a$ Number fact: $-1(-1) = 1$
   4. $= a$ Identity Prop Mult
   5. $-(-a) = a$ Trans Prop Eq

8. $(a - b) + b = a$
   1. $(a - b) + b = [a + (-b)] + b$ Rule of Subt
   2. $= a + [-b + b]$ Associative Prop Add
   3. $= a + 0$ Add Inverse Prop
   4. $= a$ Identity Prop Add
   5. $(a - b) + b = a$ Trans Prop Eq

9. $-(a - b) = b - a$
1. $-(a - b) = -1(a - b)$ Thm 1.2
2. $\quad = -1a - (-1)b$ Thm 1.4
3. $\quad = -a - (-b)$ Thm 1.2
4. $\quad = -a + [-(-b)]$ Rule of Subt
5. $\quad = -a + b$ Thm 1.6
6. $\quad = b + (-a)$ Commutative Prop Add
7. $\quad = b - a$ Rule of Subt
8. $-(a - b) = b - a$ Trans Prop Eq

10. $(ab)(cd) = (ac)(bd)$ (Theorem 1.7)
1. $(ab)(cd) = [(ab)c]d$ Associative Prop Mult
2. $\quad = [a(bc)]d$ Associative Prop Mult
3. $\quad = [a(cd)]d$ Commutative Prop Mult
4. $\quad = [(ac)b]d$ Associative Prop Mult
5. $\quad = (ac)(bd)$ Associative Prop Mult
6. $(ab)(cd) = (ac)(bd)$ Trans Prop Eq

11. Use Theorems 1.2 and 1.7, a real-number fact, the Identity Property for Multiplication, and the Transitive Property of Equality to prove that:
$(-x)(-y) = xy$, for all real numbers x and y.

Prove that each statement is true for all real numbers a, b, c, and d. Supply a reason for each step.

12. $(a + b) + c = (c + b) + a$ **13.** $a(-b) = -(ab)$
14. $a(b + c) + d = (d + ac) + ab$ **15.** $(ab + c) + ad = a(b + d) + c$

Determine whether each relation is (1) reflexive, (2) symmetric, (3) transitive, or (4) an equivalence relation.

16. *is due west of*, for all places in the Northern Hemisphere (1), (2), (3), (4)

17. *is a bisector of*, for all segments in a plane

18. *is the brother of*, for all people (3)

19. *is perpendicular to*, for all the lines in a plane (2)

20. Is the set of *irrational numbers* a field? Prove your answer. No.
Lacks closure: $\sqrt{3} + (-\sqrt{3}) = 0$. 0 is rational.

Mixed Review

Evaluate each expression for $x = -3$ and $y = 5$, if possible. If not possible, write *undefined*. *1.3*

1. $4(5x - 3y + 2) - 5(3x + 2y) + 4$ -113 **2.** $2x^3 - y^2 + xy - x$ -91

3. $\dfrac{y^2 - x^2}{(y - x)^2}$ $\dfrac{1}{4}$ **4.** $\dfrac{8}{x + y - 2}$ Undef

1.8 Proof in Algebra **35**

11. 1. $(-x)(-y) = (-1x)(-1y)$ (Thm 1.2)
2. $\quad = [-1(-1)](xy)$ (Thm 1.7, Ex. 10)
3. $\quad = 1(xy)$ (Number Fact: $-1(-1) = 1$)
4. $\quad = xy$ (Mult Ident Prop)
5. $(-x)(-y) = xy$ (Trans Prop Eq)

12. 1. $(a + b) + c = a + (b + c)$ (Assoc Prop Add)
2. $\quad = (b + c) + a$ (Comm Prop Add)
3. $\quad = (c + b) + a$ (Comm Prop Add)
4. $(a + b) + c = (c + b) + a$ (Trans Prop Eq)

13. 1. $a(-b) = (-b)a$ (Comm Prop Mult)
2. $-(ba)$ (Thm 1.3)
3. $-(ab)$ (Comm Prop Mult)
4. $a(-b) = -(ab)$ (Trans Prop Eq)

14. 1. $a(b + c) + d = d + a(b + c)$ (Comm Prop Add)
2. $\quad = d + (ab + ac)$ (Distr Prop)
3. $\quad = d + (ac + ab)$ (Comm Prop Add)
4. $\quad = (d + ac) + ab$ (Assoc Prop Add)
5. $a(b + c) + d = (d + ac) + ab$ (Trans Prop Eq)

15. 1. $(ab + c) + ad = ab + (c + ad)$ (Assoc Prop Add)
2. $\quad = ab + (ad + c)$ (Comm Prop Add)
3. $\quad = (ab + ad) + c$ (Assoc Prop Add)
4. $\quad = a(b + d) + c$ (Distr Prop)
5. $(ab + c) + ad = a(b + d) + c$ (Trans Prop Eq)

17. none

Enrichment

Challenge the students to prove that if $a = b$, then $a^2 = b^2$.

1. $a = b$ (Given)
2. $a \cdot a = b \cdot a$ (Mult Prop Eq)
3. $a^2 = b \cdot a$ (Def of exp)
4. $a^2 = b \cdot b$ (Subst Step 1)
5. $a^2 = b^2$ (Def of exp)

Next challenge the students to find the error in the following "proof" that $1 = -1$. (NOTE: Subt Prop of Eq and Div Prop of Eq, taught in Lesson 1.4, are used here in steps 3 and 5.)

1. $x = 1$ (Given)
2. $x^2 = 1$ (If $a = b$, $a^2 = b^2$.)
3. $x^2 - x = 1 - x$ (Subt Prop of Eq)
4. $x(x-1) = 1 - x$ (Distr. Prop)

5. $x = \dfrac{1 - x}{x - 1}$ (Div Prop of Eq)
6. $x = \dfrac{-(x - 1)}{x - 1}$ [See Exercise 9 in the Student text: $-(a - b) = b - a$]
7. $x = -1$ (Quotient Rule)
8. $1 = -1$ (Subst, using steps 1 and 8.)
In step 5, both sides of the equation are divided by 0 which is not possible. The Div Prop of Eq refers to nonzero numbers.

7. Add Ident Prop

9. $-2xy + 10x^2y$, 252

17. Answers may vary. One possibility is to use x, $x + 1$, and $x + 2$.

Chapter 1 Review

Key Terms

absolute value (p. 1)
Additive Inverse (p. 8)
algebraic expression (p. 9)
Associative Property (p. 8)
axiom (p. 31)
Closure Property (p. 8)
coefficient (p. 9)
Commutative Property (p. 8)
Distributive Property (p. 8)
equivalence relation (p. 33)
equivalent equations (p. 13)
equivalent expressions (p. 10)
factor (p. 9)
field (p. 8)
first-degree equation (p. 13)
formula (p. 17)
Identity for Addition (p. 8)
Identity for Multiplication (p. 8)
irrational number (p. 1)
like terms (p. 10)

linear equation (p. 13)
literal equation (p. 15)
Multiplicative Inverse (p. 8)
postulate (p. 31)
Properties of Equality (pp. 13, 31)
proportion (p. 14)
Proportion Property (p. 14)
rational number (p. 1)
real number (p. 1)
reciprocal (p. 8)
Reflexive Property (p. 31)
root (p. 13)
Substitution (p. 31)
Symmetric Property (p. 31)
term (p. 9)
theorem (p. 31)
Transitive Property (p. 31)
unlike terms (p. 10)
variable (p. 3)
Zero Property (p. 31)

Key Ideas and Review Exercises

1.1, 1.2 To simplify a numerical expression, follow the order of operations (summarized in Lesson 1.2), and recall that division by 0 is undefined.

Simplify, if possible. If not possible, write *undefined*.

1. $8 - 12 \cdot 3 + 4(-7) - (2 - 12)$ -46 **2.** $\dfrac{18 - 5(-4)}{2 \cdot 6 - 4 \cdot 3}$ Undef

1.3 The eleven properties of operations with real numbers are listed in Lesson 1.3. The rule of subtraction states that $a - b = a + (-b)$.

Identify the rule or property of operations with real numbers that is illustrated.

3. $4y - 7y = 4y + (-7y)$ Rule of Subt   **4.** $3(7t) = (3 \cdot 7)t$ Associative Prop Mult

5. $5(4c + 3) = 5(4c) + 5 \cdot 3$ Distr Prop   **6.** $3(7t) = (7t)3$ Commutative Prop Mult

7. $(6a - 2b) + 0 = 6a - 2b$ **8.** $4 \cdot \frac{1}{4}n = 1n$ Multiplicative Inverse

9. Simplify $xy + x^2y - 3xy + 9x^2y$, and evaluate for $x = -4$ and $y = 1.5$.

1.4 To solve a linear equation in one variable, first simplify all algebraic expressions. Then use the Properties of Equality listed in Lesson 1.4.

Solve each equation.

10. $3(4x - 6) - (2x + 7) =$
$10 - 5(4 - 3x)$ -3

11. $2.6y - 3.4 = 3.6 - 4.4y$ 1

12. $\frac{3}{4}w - \frac{1}{2} = \frac{5}{6}w + 2$ -30

13. $\dfrac{9x - 5}{6} = \dfrac{7x - 4}{5}$ $\frac{1}{3}$

14. Solve $8a(x - 2) = 5ax + 3$ for x. $\dfrac{16a + 3}{3a}$

1.5   Substitution in a given formula can be used to solve some problems.

15. Bernie borrowed some money at $8\frac{1}{2}\%$ (simple interest) per year. At the end of 4 years 3 months, he owed a total of $7,623 (in principal and interest). How much money did Bernie borrow? Use the formula $A = p + prt$. $5,600

1.6   To solve a word problem about more than one number, begin by representing the numbers in terms of one variable.

16. The greatest of three numbers is -8 times the smallest, and the remaining number is 10 more than the smallest. Find the three numbers, given that the greatest is 16 times the sum of the other two numbers. $-4, 6, 32$

17. Write in your own words how you would represent the numbers for a word problem in which you are to find three consecutive integers.

1.7   To solve a word problem involving geometric figures, begin by sketching and labeling the figures.

18. A rectangle is 4 m wider than a square and 3 times as long. The difference between the two perimeters is 36 m. Find the area of the square and the length of the rectangle. 49 m^2, 21 m

1.8   To prove algebraic theorems, use the field postulates in Lesson 1.3 and the postulates and theorems in Lesson 1.8.

19. Justify each step in the proof.

      Theorem: $(ab)c + d = d + b(ca)$

Proof:
1. $(ab)c + d = d + (ab)c$ Commutative Prop Add
2. $= d + c(ab)$ Commutative Prop Mult
3. $= d + (ca)b$ Associative Prop Mult
4. $= d + b(ca)$ Commutative Prop Mult
5. $(ab)c + d = d + b(ca)$ Trans Prop Eq

20. The properties of an equivalence relation are listed in Lesson 1.8. Determine whether, for all triangles, the relation $\cong$ (*has the same size and shape as*) is **a.** reflexive, **b.** symmetric, **c.** transitive, or **d.** an equivalence relation.   **a.** yes **b.** yes **c.** yes **d.** yes

Chapter Test

3. Comm Prop Add
4. Assoc Prop Mult
5. Add Inv Prop
15. (Sample answer) $3(5 \cdot 2) \neq$
$(3 \cdot 5) \cdot (3 \cdot 2)$, since $3(10) \neq 15 \cdot 6$,
since $30 \neq 90$
17. 1. $(x - y) - x = [x + (-y)] + (-x)$
(Rule of Subtr);
2. $= -x + [x + (-y)]$
(Comm Prop Add);
3. $= (-x + x) + (-y)$
(Assoc Prop Add);
4. $= 0 + (-y)$ (Add Inv Prop);
5. $= -y$ (Add Ident Prop);
6. $(x - y) - x = -y$ (Trans Prop Eq)

Chapter 1 Test

A Exercises: 1–9,11,14   B Exercises: 10,12,13
C Exercises: 15–17

Simplify, if possible. If not possible write *undefined*.

1. $12 - 6 \cdot 3 + 5(-2) - (-4)$ -12

2. $\dfrac{5 \cdot 8 - 4(-3)}{-4 \cdot 6 - (-24)}$ Undef

Identify the property of operations with real numbers that is illustrated.

3. $7 + 5x = 5x + 7$ **4.** $5(8y) = (5 \cdot 8)y$ **5.** $-7t + 7t = 0$

6. Simplify $2xy + 3x^2y - xy$, and evaluate for $x = 5$ and $y = -4$. $xy + 3x^2y, -320$

Solve each equation.

7. $3(2.4x - 1.2) - (5x + 4) =$ **8.** $\frac{2}{3}y - \frac{5}{4} = \frac{1}{2}y + 4$ $31\frac{1}{2}$
$7.2 - 2(x - 1)$ 4

9. Solve $5(mx + 3) = 2mx + c$ for x. $\frac{c - 15}{3m}$

10. Andrew borrowed some money at $9\frac{1}{2}\%$ (simple interest) per year.
At the end of 4 years 6 months, he owed a total of $11,420. How
much money did Andrew borrow? Use the formula $A = p + prt$. $8,000

11. The greatest of three numbers is -4 times the smallest number, and
the remaining number is 14 more than the smallest number. Find
the three numbers given that the greatest is 5 times the sum of the
other two numbers. $-5, 9, 20$

12. A rectangle is 5 ft wider than a square. The rectangle is twice as
long as it is wide. Find the area of the square given that the difference
between the two perimeters is 70 ft. 400 ft^2

13. Justify each step in the proof.

Theorem: $d + c(ba) = a(bc) + d$

Proof:
1. $d + c(ba) = c(ba) + d$ Commutative Prop Add
2. $= (ba)c + d$ Commutative Prop Mult
3. $= (ab)c + d$ Commutative Prop Mult
4. $= a(bc) + d$ Associative Prop Mult
5. $d + c(ba) = a(bc) + d$ Trans Prop Eq

14. Determine whether the relation $>$ is
   a. reflexive
   b. symmetric
   c. transitive
   d. an equivalence relation.   a. no b. no c. yes d. no

15. Use a numerical counterexample to prove that multiplication is *not*
distributive over multiplication.

16. Solve $3 - 2|4x - (3 - x)| = x - 2 - 11(x - 1)$.   All real numbers

17. Prove $(x - y) - x = -y$, for all real numbers x and y.

38   Chapter 1 Test

Strategy for Achievement in Testing

Try to obtain the following information before taking any important test.

(1) Is the test of the type that is intentionally made longer than most people can complete in the time provided? If so, knowing that fact may remove some of the tension and help your concentration.

(2) Is there a greater penalty for giving a wrong answer than for leaving a question unanswered? If not, it may pay to guess.

Directions: Choose the *one* best answer to each question or problem.

1. If $3a + 6b = 90$, then $2a + 4b = \underline{?}$.
(A) 120   (B) 75
(C) 60   (D) 45
(E) None of these
C

2. If x^* is defined to be $x + 2$, find the value of $(3^* + 5^*)^*$.   D
(A) 8   (B) 10
(C) 12   (D) 14
(E) None of these

3. A rectangle measures 9 in. by 12 in. Find the length of a diagonal.   B
(A) 21 in.   (B) 15 in.
(C) $10\frac{1}{2}$ in.   (D) 6 in.
(E) None of these

4. Each angle in the figure below is a right angle. Find the perimeter of the figure.   B
(A) 77 m (B) 36 m (C) 29 m (D) 18 m
(E) Not enough data to answer

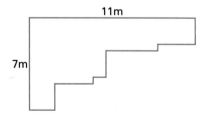

11m

7m

5. Which of the following rectangles has the greatest perimeter?   E
(A) 2 yd by 1 yd   (B) 7 ft by 2 ft
(C) 50 in. by 58 in.   (D) 5 ft 6 in. by 3 ft 6 in.   (E) They all have the same perimeter.

6. If $\dfrac{(x - 3)(y + 3)z}{3w} = 60$, which of the numbers x, y, z, and w *cannot* be 3?   A
(A) x (B) y (C) z
(D) w (E) All of these

7. If $\dfrac{15k}{3kx + 36} = 1$ and $x = 4$, what is the value of k?   E
(A) 2   (B) 3   (C) 4
(D) 8   (E) 12

8. Which number is a factor (divisor) of the product $15 \times 26 \times 77$?   D
(A) 4   (B) 9   (C) 36
(D) 55   (E) None of these

9. Which number is the sum of three consecutive integers?   B
(A) 158   (B) 258   (C) 358
(D) 458   (E) None of these

10. Find the number of fractions with two-digit numerators and two-digit denominators that are equivalent to $\frac{2}{9}$.   D
(A) 10   (B) 9   (C) 8
(D) 7   (E) 6

LINEAR INEQUALITIES

OVERVIEW

In this chapter, students find and graph the solution sets of simple linear inequalities. Compound inequalities are then introduced, followed by a lesson on sentences (equations as well as inequalities) that involve absolute value. The three problem-solving lessons continue the practice established in Chapter 1 of emphasizing problem solving early in the book. Algebraic principles are applied from both Chapters 1 and 2 in problem solving Lessons 2.4–2.6.

OBJECTIVES

- To find and graph the solution sets of simple linear inequalities
- To find and graph the solution sets of compound linear inequalities
- To solve equations involving absolute value
- To find and graph the solution sets of inequalities involving absolute value
- To solve word problems involving compound sentences

PROBLEM SOLVING

Throughout the text, problem solving lessons are organized in one of two ways. The first way is by *mathematical concept*, for example, inequalities, as in Lesson 2.4. The second way is by *problem type*, for example, the age problems of Lesson 2.5. In addition to problem solving lessons, attention is given throughout the book on the use of *strategies* to solve problems. This is done in the form of the full page feature, *Problem Solving Strategies*. The first of these strategies, on constructing a table to represent data, appears on page 58.

READING AND WRITING MATH

In this chapter, the importance of reading is evident throughout, perhaps most noticeably in Lesson 2.2, in which compound sentences are introduced.

The tables for conjunction and disjunction illustrate how the meanings that mathematicians and logicians assign to *and* and *or* compare to the everyday meanings of these words. Exercises that require students to write explanations of concepts and procedures occur once or twice in each chapter. In this chapter, such questions can be found on both page 50 and page 66.

TECHNOLOGY

Calculator: Some exercises in Lesson 2.3 on *absolute value* may be solved using a calculator, for example, Exercises 16–18 on page 53. See *Using the Calculator* on page 65.

SPECIAL FEATURES

Mixed Review pp. 44, 50, 53, 61, 65
Focus on Reading p. 43
Statistics: Stem-and-Leaf Plots p. 45
Midchapter Review p. 57
Application: Circumference of the Earth p. 57
Problem Solving Strategies: Making a Table p. 58
Application: Etching p. 61
Using the Calculator p. 65
Key Terms p. 66
Key Ideas and Review Exercises pp. 66–67
Chapter 2 Test p. 68
College Prep Test p. 69
Cumulative Review (Chapters 1–2) pp. 70, 71

PLANNING GUIDE

Lesson	Basic	Average	Above Average	Resources
2.1 pp. 43–45	FR CE all WE 1–13 odd Statistics	FR CE all WE 9–23 odd Statistics	FR CE all WE 13–29 odd Statistics	Reteaching 9 Practice 9
2.2 pp. 49–50	CE all WE 1–15 odd	CE all WE 5–19 odd	CE all WE 9–25 odd	Reteaching 10 Practice 10
2.3 p. 53	CE all WE 1–17odd	CE all WE 5–23 odd	CE all WE 9–27 odd	Reteaching 11 Practice 11
2.4 pp. 55–58	CE all WE 1–11 odd Midchapter Review Application Problem Solving	CE all WE 3–13 odd Midchapter Review Application Problem Solving	CE all WE 5–17 odd Midchapter Review Application Problem Solving	Reteaching 12 Practice 12
2.5 pp. 60–61	CE all WE 1–7 odd Application	CE all WE 3–9 odd Application	CE all WE 5–13 odd Application	Reteaching 13 Practice 13
2.6 pp. 63–65	CE all WE 1–9 odd UC	CE all WE 3–11 odd UC	CE all WE 7–15 odd UC	Reteaching 14 Practice 14
Chapter 2 Review pp. 66–67	all	all	all	
Chapter 2 Test p. 68	all	all	all	
College Prep Test p. 69	all	all	all	
Cumulative Review pp. 70–71	all	1–37 odd	1–17 odd	

*CE = Classroom Exercises   WE = Written Exercises   FR = Focus on Reading   UC = Using the Calculator
NOTE: For each level, all students should be assigned all Mixed Review exercises.

▬ INVESTIGATION

Project: Students are asked to consider properties of sets defined by strict inequality statements.

Ask the students to consider the difference between the following questions:

1. If you could spend any amount less than $50.00, what is the greatest amount you could spend?
2. What is the greatest number less than 50?

The first question has a definite answer in terms of today's US currency: $49.99.

The second question has no definite answer. For example, if we were to say that the greatest number less than 50 was 49.99, someone might reply that 49.999 is a larger number less than 50.

In mathematics we make the assumption that there is no smallest number. One consequence of this assumption is that between any two numbers there is always another number. For instance, given 49.999 and 50, 49.9995 is a number that is between them.

Have students consider the set defined as the set of all numbers greater than 0 and less than 50. In set notation this is $\{x \mid 0 < x < 50\}$. On the number line, this is represented as the set of numbers between the points 0 and 50.

Ask the students whether they think the portion of the line between the two points has endpoints. The answer is no, for the reasons given above. For a point to be an endpoint, it would have to be the greatest number less than 50 or the smallest number greater than zero. No number could qualify.

Sets such as the above are called **open sets.** They have many interesting mathematical properties. Have students draw other open sets and describe them in set notation.

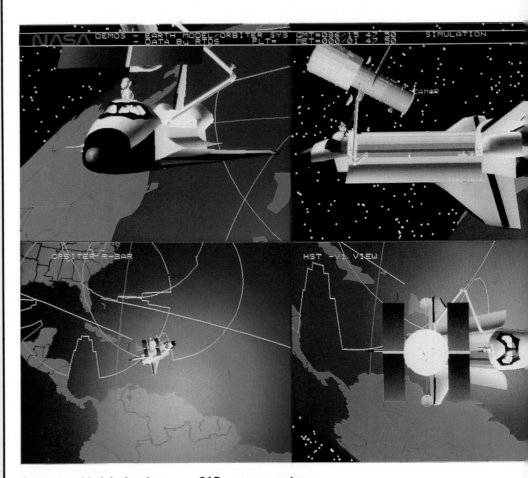

Computer-aided design, known as CAD, uses computer graphics to create a three-dimensional image of an object. CAD enables NASA engineers to gather information about the space shuttle to predict how it will perform under a given set of conditions.

More About the Image

Computer graphics are generated by programs that control the pixels of a computer display screen. A CAD program may be used to create a three-dimensional image of an object on the computer screen, and then show it in its full range of configurations—independent of limitations normally imposed by gravity.

2.1 Linear Inequalities

Objective

To find and graph the solution sets of linear inequalities

In Chapter 1, you learned how to solve problems with exactly one answer. You did this by writing a linear equation and finding its only solution, or root. Sometimes problems have a *range* of answers. To solve these problems, the relations < *(is less than)* and > *(is greater than)* are often used. These relations are called **order relations.**

```
      A     B                    C          D
   ◄──●─────●─────────────────────●──────────●──────►
     -5  -4  -3  -2  -1   0   1   2   3   4   5   6
```

Use the number line above to verify the statements below.

$-3 < 2$ because B is to the *left* of C
$5 > -5$ because D is to the *right* of A

Several properties are developed below using numerical examples.

1. $-9 < 6$
    $-9 + 4 < 6 + 4$
    $-5 < 10$

2. $-9 < 6$
    $-9 - 2 < 6 - 2$
    $-11 < 4$

3. $-9 < 6$
    $\frac{-9}{3} < \frac{6}{3}$
    $-3 < 2$

4. $-9 < 6$
    $-9 \cdot 2 < 6 \cdot 2$
    $-18 < 12$

In the two cases below, the order of the relation changes.

5. $-9 < 6$
    $-9(-2) > 6(-2)$
    $18 > -12$

6. $-9 < 6$
    $\frac{-9}{-3} > \frac{6}{-3}$
    $3 > -2$

Multiplying or dividing each side of an inequality by the same *negative* number *reverses* the order of the inequality.

Properties of Inequality
For all real numbers a, b, and c,

if $a < b$, then $a + c < b + c$ and $a - c < b - c$
(Addition/Subtraction)

if $c > 0$ and $a < b$, then $a \cdot c < b \cdot c$ and $\frac{a}{c} < \frac{b}{c}$

if $c < 0$ and $a < b$, then $a \cdot c > b \cdot c$ and $\frac{a}{c} > \frac{b}{c}$
(Multiplication/Division)

2.1 Linear Inequalities   **41**

Teaching Resources

Ouick Quizzes 9
Reteaching and Practice
   Worksheets 9
Transparency 3

▬▬ GETTING STARTED

Prerequisite Quiz

Solve.

1. $3(14 - 4x) = 8 - 7x$ 6.8
2. $10x - (12x + 6) = x$ -2
3. $\frac{7x - 2}{4} = \frac{5x + 1}{3}$ 10
4. $\frac{5}{7}x = 35$ 49

Motivator

Ask students what is meant by the phrase x is less than 5.   The solution set includes all values of x less than 5.   Ask students to compare the solution(s) of the expressions $x = 3$ and $x > 3$.   The equality has exactly one solution; the inequality has more than one solution.

▬▬ TEACHING SUGGESTIONS

Lesson Note

Students can recall the inequality properties by their relation to the equation properties as follows.

The inequality properties are similar to the equation properties with one exception: *multiplying* (or *dividing*) each side of an inequality by a *negative* number *reverses* the order symbol < (or >) to > (or <).
In answering Classroom Exercises 5–12, students should state the solution sets using set-builder notation.

Highlighting the Standards

Standard 3b, 10a: In Written Exercises 24–29, students examine the truth of mathematical statements and provide counterexamples. In the Statistics section, they construct and draw inferences from a stem-and-leaf plot of data on the Pollution Index.

Math Connections

Laboratory Research: Laboratory experiments often are better expressed as inequalities than as equalities. An example is an experiment designed to determine the temperature and pressure conditions that ensure that a compound will be in a completely gaseous state. For water, these conditions are $t > 374$ and $p < 218$, where t is the Celsius temperature and p is the pressure in atmospheres.

Critical Thinking Question

Analysis: Ask students whether the sentence "$x \neq 3.5$" can be expressed using other symbols that they know. The answer anticipates somewhat the content of Lesson 2.2, since the answer is clearly "$x > 3.5$ or $x = 3.5$."

Common Error Analysis

Error: Some students will forget to reverse the symbol $<$ after dividing both sides of an inequality such as $-2x < 10$ by -2. Ask them to recall the operations that reverse the order symbol for an inequality.

Students may draw the graph of $4 < x$ as a ray that extends to the left because $<$ points to the left. Have the students rewrite $4 < x$, with x on the left, as $x > 4$.

Checkpoint

Find the solution set. Graph it.

1. $8 - 3x + 9 > 5x - 3 + 2x$ $\{x \mid x < 2\}$

2. $\frac{3}{4}(12y - 8) \geq \frac{2}{3}(9y - 18)$ $\{y \mid y \geq -2\}$

The properties stated for $a < b$ can easily be restated for $a > b$. For example, if $c < 0$ and $a > b$, then $a \cdot c < b \cdot c$.

You can use inequality properties to find the set of all solutions, or the **solution set,** of a *linear inequality*, as shown below.

EXAMPLE 1   Find the solution set of $5x - 4 < 8x + 2$.

Solution

$$5x - 4 < 8x + 2$$

Add 4 to each side. $5x < 8x + 6$
Subtract $8x$ from each side. $-3x < 6$
Divide each side by -3 $x > -2$
and reverse the order.

The solution set is the set of all real numbers greater than -2.

As Example 1 illustrates, a solution set may have an infinite number of solutions. For this reason, such sets are called **infinite sets**. The solution set of Example 1 is an infinite set written as follows.

$$\{x \mid x > -2\} \leftarrow \text{Read: "The set of all } x \text{ such that } x > -2.\text{"}$$

The number-line graph of this solution set is shown below. The open dot indicates that -2 is not a solution.

$$\overset{\longleftrightarrow}{\underset{-6 \quad -4 \quad -2 \quad 0 \quad 2 \quad 4}{}}$$

The order relations can be combined with the *equals* relation ($=$) to form the relations $\leq$ *(is less than or equal to)* and $\geq$ *(is greater than or equal to)*. Notice that the following inequalities are both true.

$$2 \leq 5 \text{ (because } 2 < 5) \qquad -3 \geq -3 \text{ (because } -3 = -3)$$

EXAMPLE 2   Find the solution set of $2(15 - 3x) \geq 4x - 5$. Graph it on a number line.

Solution

$$2(15 - 3x) \geq 4x - 5$$

Use the Distributive Property. $30 - 6x \geq 4x - 5$
Subtract 30 from each side. $-6x \geq 4x - 35$
Subtract $4x$ from each side. $-10x \geq -35$
Divide each side by -10. $x \leq 3.5 \leftarrow$ NOTE: "$\geq$" changes to "$\leq$."

The solution set is $\{x \mid x \leq 3.5\}$. The graph is the ray to the left of 3.5, including its endpoint. The endpoint is represented by a closed dot.

$$\overset{\longleftrightarrow}{\underset{-2 \quad 0 \quad 2 \quad 3.5 \; 4}{}}$$

Additional Example 1

Find the solution set of
$3x - 12 > 7x + 8$.
$\{x \mid x < -5\}$

Additional Example 2

Find the solution set of
$3(12 - 4x) \leq 6x + 9$.
Graph the number line.
$\{x \mid x \geq 1.5\}$

$$\overset{\longleftrightarrow}{\underset{1.5}{}}$$

If $5 > 3$, then $3 < 5$, and in general, if $a > b$, then $b < a$. This property allows an inequality such as $4 > n$, with the variable on the right, to be written as $n < 4$, with the variable on the left. When the variable is written on the left, the symbols $<$ and $>$ will "match" the arrows on the graphs of the relations as shown below.

| $n<4$ | $<$ and both point to the left. | $n>6$ | $>$ and both point to the right. |

EXAMPLE 3 Find and graph the solution set of $\frac{1}{6}x - \frac{3}{2} < \frac{2}{3}x$.

Solution Multiply each side of the inequality by 6, the LCD.

$$6\left(\frac{1}{6}x - \frac{3}{2}\right) < 6 \cdot \frac{2}{3}x$$
$$x - 9 < 4x$$

Subtract x from each side. $\quad -9 < 3x$

Divide each side by 3, $\quad -3 < x$

and write the variable on the left side. $\quad x > -3$

The solution set is $\{x \mid x > -3\}$. The graph is the ray (excluding its endpoint) that extends to the right of -3.

Focus on Reading

Match each sentence with exactly one graph.

1. $6 < x$ c
2. $6 \leq x$ a
3. $x < 6$ d
4. $x \leq 6$ b

a.
b.
c.
d.

Classroom Exercises

For each statement, determine whether it is true or false.

1. $-18 < -10$ T   **2.** $-12 > 10$ F   **3.** $5 \leq 7$ T   **4.** $8 \geq 8$ T

State the solution set of each inequality. Graph it on a number line.

5. $-4y \leq 12$ **6.** $x + 7 > 5$ **7.** $28 > 7n$ **8.** $-t < -3$
9. $\frac{1}{2}c < 6$ **10.** $-6a \geq -24 + 2a$ **11.** $x - 4 < 21$ **12.** $2x + 5 \geq 1$

Ask students to name four relations studied in this lesson. $<, >, \leq, \geq$ Ask students how they would draw a picture of the solution set of an inequality. Graph the solutions on a number line. Have students state the properties of inequality for which the order relation must be reversed. Properties of multiplication and division.

FOLLOW UP

Guided Practice

Classroom Exercises 1–12

Independent Practice

A Ex. 1–14, **B** Ex. 15–23, **C** Ex. 24–29

Basic: FR, WE 1–13 odd, Statistics
Average: FR, WE 9–23 odd, Statistics
Above Average: FR, WE 13–29 odd, Statistics

Additional Answers

Classroom Exercises

5. $\{y \mid y \geq -3\}$

6. $\{x \mid x > -2\}$

7. $\{n \mid n < 4\}$

8. $\{t \mid t > 3\}$

Additional Example 3

Find the solution set of $\frac{4}{15}x + \frac{1}{2} > \frac{3}{5}x$.
Graph it. $\{x \mid x < 1.5\}$

43

9. $\{c \mid c < 12\}$

10. $\{a \mid a \le 3\}$

11. $\{x \mid x < 25\}$

12. $\{x \mid x \ge -2\}$

Written Exercises

1. $\{x \mid x \le 5\}$

2. $\{y \mid y \ge -3\}$

3. $\{n \mid n > 2\}$

4. $\{c \mid c > -3\frac{1}{2}\}$

5. $\{y \mid y > -4\}$

6. $\{x \mid x \le 5\}$

See page 45 for the answers to Ex. 7–16 and page 57 for the answers to Ex. 17–23, 27.

Written Exercises

Find the solution set. Graph it on a number line.

1. $2 \ge x - 3$ {x | x ≤ 5}

2. $4y - 12 \le 8y$ {y | y ≥ −3}

3. $7 - 2n < 3n - 3$ {n | n > 2}

4. $11c + 4 > 9c - 3$ {c | c > −3½}

5. $18y + 36 > 8y - 4$ {y | y > −4}

6. $9x + 10 - x \le 2x + 40$ {x | x ≤ 5}

7. $3(4x - 5) \le 8x + 3$ {x | x ≤ 4.5}

8. $8 - 2(3x + 7) \ge 12$ {x | x ≤ −3}

9. $5x > 12x - (x - 24)$ {x | x < −4}

10. $30 - (12 - x) < 7x$ {x | x > 3}

11. $7x - 15 - 3x < 14 - 6x + 11$

12. $5(2 - 3x) \ge 4 - 3(4x + 7)$

13. $5p - (7p + 2) < 29 + 3(2p - 5)$

14. $-x > 0$ {x | x < 0}

15. $-\frac{2}{3}n > 6$ {n | n < −9}

16. $\frac{5}{3} < \frac{1}{6}y$ {y | y > 10}

17. $\frac{3}{4}a - \frac{1}{2} > a + \frac{2}{3}$

18. $\frac{3x - 8}{5} \le \frac{x + 4}{3}$

19. $\frac{2x + 7}{3} \le \frac{x + 6}{2}$

20. $\frac{5}{6}a - \frac{3}{8}a \ge \frac{1}{2}a - 2$ {a | a ≤ 48}

21. $\frac{3x + 1}{-4} < 5$ {x | x > −7}

22. $\frac{2y + 3}{-5} > 3 - y$ {y | y > 6}

23. $-3(a + 5) > \frac{a - 5}{-2}$ {a | a < −7}

Find a numerical counterexample that proves each statement false.

24. If $a < b$ and $b \ne 0$, then $a^2 < b^2$ for *all* real numbers a and b. −5 < −3, but 25 ≮ 9

25. If $a < b$, then $\frac{1}{a} > \frac{1}{b}$ for *all* real numbers $a \ne 0$ and $b \ne 0$. −2 < 2, but $-\frac{1}{2} \not> \frac{1}{2}$

For each statement, determine whether it is true or false. If it is false, provide a numerical counterexample to show it is false.

26. If $a < b$ and $b < c$, then $a < c$ for all real numbers a, b, and c. T

27. If $a < b$ and $b < c$, then $a + b < c$ for all real numbers a, b, and c.

28. If $\frac{a}{b} < \frac{c}{d}$, then $ad < bc$ for all positive real numbers a, b, c, and d. T

29. If $\frac{a}{b} < \frac{c}{d}$, then $\frac{a}{b} < \frac{a+c}{b+d}$, and $\frac{a+c}{b+d} < \frac{c}{d}$ for all positive real numbers a, b, c, and d. T

Mixed Review

Evaluate each expression for $x = -4$ and $y = 3$. *1.3*

1. $8(3x - 4y) - (20x - 30y)$ −22

2. $5xy - x^2y + 2xy^2$ −180

Solve each equation for x. *1.4*

3. $\frac{3x - 2}{4} = \frac{2x + 5}{3}$ 26

4. $2a(x - m) = am - c$ $\frac{3am - c}{2a}$

Enrichment

Challenge students to decode the following message. Each letter could be thought of as an integer on the number line (A = 1, B = 2, etc.). If you arrange the letters so that all the inequalities are true, the message will be clear.

$S < R$	$F < O$	$R < L$
$T > S$	$N > U$	$C < H$
$L < U$	$T < A$	$N < C$
$H > L$	$O < H$	$U > A$
$Y < F$	$Y > A$	$R > O$

Stay for lunch

Statistics: *Stem- and-Leaf Plots*

The data below are values of the Pollution Index for Austin, Texas for 23 weekdays of May, 1990. Each value is the highest average hourly concentration of the air pollutant ozone in parts per million (ppm) for that day. A score of 0–50 is considered good, 50–100 moderate, and 100–199 unhealthful. Readings approaching or exceeding 200 warrant advisories or warnings.

37 27 36 38 63 47 45 39 46 33 30 38
54 24 74 49 62 36 33 29 47 37 33

In their present form, the numbers appear to be haphazard. To organize them, construct a **stem-and-leaf plot** as shown below.

1. Divide each number into two parts: a stem and a leaf. For the given data, it is convenient to take the tens digit as the stem and the units digit as the leaf.

2. Write the stem of the smallest number. Then in a column below it, list all of the possible stems up to the stem of the largest number, as shown at the right above. Draw a vertical line to separate the stems from the leaves.

2	4, 7, 9
3	0, 3, 3, 3, 6, 6, 7, 7, 8, 8, 9
4	5, 6, 7, 7, 9
5	4
6	2, 3
7	4

3. The first ozone reading, 37, has a leaf of 7. Write this leaf opposite its corresponding stem, 3. Do this for every reading in the table, listing each leaf in the row opposite its stem.

It is now easy to make observations about the data. For example, the plot shows at a glance that the greatest number of readings lies in the thirties.

Exercises

For Exercises 1–3, refer to the stem-and-leaf plot above.

1. Give the three readings represented by the first row of the plot. 27, 24, 29
2. For how many days was the air quality moderate? unhealthful? 4; 0
3. What was the most common quality assessment of the Pollution Index? Good
4. Construct a stem-and-leaf plot of the following test scores.

   78 55 94 89   92 85 69 98 86 89
   86 67 74 46 100 48 71 79 67 87

   Which of the stems contain most of the test scores? 6–9

7. {x | x ≤ 4.5}

4.5

8. {x | x ≤ −3}

−3

9. {x | x < −4}

−4

10. {x | x > 3}

3

11. {x | x < 4}

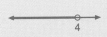
4

12. {x | x ≤ 9}

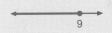
9

13. {p | p > −2}

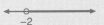
−2

14. {x | x < 0}

0

15. {n | n < −9}

−9

16. {y | y > 10}

10

▰▰▰GETTING STARTED

Prerequisite Quiz

Write the solution set.

1. $6x - 12 < 2x$ $\{x \mid x < 3\}$
2. $8 < 5y + 23$ $\{y \mid y > -3\}$
3. $9 - n > 4$ $\{n \mid n < 5\}$
4. $5 \geq 3c + 17$ $\{c \mid c \leq -4\}$

Motivator

Ask students what it means to say "We will go to the library *and* to class." They will go to both places. What does it mean to say, "We will go to the library *or* to the park." They might go both places, the library and not the park, or the park and not the library.

▰▰▰TEACHING SUGGESTIONS

Lesson Note

One procedure covers all cases for graphing the conjunction or disjunction of two linear inequalities. Graph the two inequalities on parallel number lines. For their disjunction (*or* sentence), place *both* graphs on a third number line. For their conjunction (*and* sentence), graph only their *overlap* on a third number line.

Highlighting the Standards

Standard 3f: In Exercise 2 of the Mixed Review, students supply the reasons for an algebraic proof.

2.2 Compound Sentences

Objective   To find and graph the solution sets of compound sentences

Two **simple sentences** such as "John is shopping" and "John is at the supermarket" can be combined into a **compound sentence** such as "John is shopping *and* John is at the supermarket."

For a mathematical example of a compound sentence, consider first the simple sentences "$4 + 1 = 5$" and "$6 < 3$." These can be connected by *and* or by *or* to form the two compound sentences below.

$$4 + 1 = 5 \text{ } and \text{ } 6 < 3 \qquad 4 + 1 = 5 \text{ } or \text{ } 6 < 3$$

The first compound sentence, which contains *and*, is called a **conjunction** of two simple sentences. The second sentence, which contains *or*, is called a **disjunction** of the simple sentences.

The conjunction (*and* sentence) "$4 + 1 = 5$ and $6 < 3$" is *false* because at least one of the simple sentences ($6 < 3$) is false.

The disjunction (*or* sentence) "$4 + 1 = 5$ or $6 < 3$" is *true* because at least one of the simple sentences ($4 + 1 = 5$) is true.

> **Truth-Value of a Compound Sentence**
> If p and q are a pair of statements, then
> 1. the conjunction p *and* q is true if and only if both p is true and q is true.
> 2. the disjunction p *or* q is false if and only if both p is false and q is false.

Next, consider the conjunction $x > 3$ *and* $x < 6$. Note that $x > 3$ *and* $x < 6$ is true only for numbers between 3 and 6, as shown by the table below.

x	$x > 3$ *and* $x < 6$	
1	$1 > 3$ *and* $1 < 6$	False
3	$3 > 3$ *and* $3 < 6$	False
4	$4 > 3$ *and* $4 < 6$	True (4 is between 3 and 6.)
8	$8 > 3$ *and* $8 < 6$	False

46   Chapter 2 Linear Inequalities

The graph of the solution set of a conjunction is the intersection (set of all points in common), or *overlap* of the graphs of its parts.

Graph of $x > 3$:

Graph of $x < 6$:

Graph of $x > 3$ *and* $x < 6$:

The solution set is $\{x \mid 3 < x < 6\}$, where $3 < x < 6$ means that $3 < x$ *and* $x < 6$. Read $3 < x < 6$ as: "x is between 3 and 6."

Finally, consider the disjunction $x > 3$ *or* $x < 6$. This compound sentence is true for *all* real numbers, as suggested by the table.

x	$x > 3$ *or* $x < 6$	
1	$1 > 3$ *or* $1 < 6$	True
3	$3 > 3$ *or* $3 < 6$	True
4	$4 > 3$ *or* $4 < 6$	True
8	$8 > 3$ *or* $8 < 6$	True

The graph of the solution set of a disjunction is the *union* of the graphs of its parts.

Graph of $x > 3$:

Graph of $x < 6$:

Graph of $x > 3$ *or* $x < 6$:

The solution set is the set of all real numbers.

EXAMPLE 1   Graph the solution set of $5x - 10 \le 3x$ *and* $7 < 4x + 15$ on a number line. State the solution set.

Plan   Simplify and graph each part. Then graph their intersection.

Solution

$$
\begin{array}{lll}
5x - 10 \le 3x & and & 7 < 4x + 15 \\
2x \le 10 & and & -8 < 4x \\
x \le 5 & and & -2 < x
\end{array}
$$

Graph $x \le 5$.

Graph $x > -2$.

Graph the intersection.

Graph of the solution set.

The solution set is $\{x \mid -2 < x \le 5\}$.

Life Skills: Students should appreciate the fact that the word "or" may have two slightly different meanings, both in mathematical work and in everyday life. The *exclusive* "or" which means *one or the other, but not both*, and the *nonexclusive* "or" which means *one or the other or both*. It is the nonexclusive "or" that is used in the lesson and throughout the text.

Critical Thinking Questions

Analysis: Ask students to prepare a table like the one on page 47 for the sentence "$x > 3$ or $x < 6$" but with "or" used in the exclusive sense rather than the nonexclusive sense. Then have them graph the sentence. The result will be oppositely directed arrows that start from closed dots at 3 and 6 with an empty region between 3 and 6.

Common Error Analysis

Error: Students forget that in graphing a conjunction, they should graph only the intersection (or overlay) of the two graphs.

Review the logical meaning of the word *and*. Then show how the graph of a conjunction includes only the set of points in common.

Additional Example 1

Graph the solution set of $6x - 12 \le 2x$ *and* $8 < 5x + 28$ on a number line. Find the solution set.

$\{x \mid -4 < x \le 3\}$

47

Checkpoint

Find the solution set.
Graph the solution set of each compound sentence on a number line.

1. $x < 7$ or $x < 4$ $\{x \mid x < 7\}$

$$\longleftarrow\!\!\!\!\underset{7}{\circ}\!\!\longrightarrow$$

2. $x > 3$ and $x > -2$ $\{x \mid x > 3\}$

$$\longleftarrow\!\!\!\!\underset{3}{\circ}\!\!\longrightarrow$$

3. $5y - 4 < 3y$ and $4 - y < 7$
   $\{y \mid -3 < y < 2\}$

$$\longleftarrow\!\!\!\underset{-3}{\circ}\!\!-\!\!\underset{2}{\circ}\!\!\longrightarrow$$

4. $3n \geq n + 6$ or $0 \geq n + 4$
   $\{n \mid n \leq -4 \text{ or } n \geq 3\}$

$$\longleftarrow\!\!\!\underset{-4}{\bullet}\!\!-\!\!\underset{3}{\bullet}\!\!\longrightarrow$$

Closure

Have students explain, in general terms, how they would graph a disjunction and a conjunction of two linear inequalities. Have students give an example of a compound sentence that has no solution.

▰▰▰FOLLOW UP

Guided Practice

Classroom Exercises 1–14

Independent Practice

A Ex. 1–10, **B** Ex. 11–19, **C** Ex. 20–27

Basic: WE 1–15 odd
Average: WE 5–19 odd
Above Average: WE 9–25 odd

EXAMPLE 2   Graph the solution set of $3x + 11 > 2$ *or* $8 - x > 4$ on a number line. State the solution set.

Plan   Simplify and graph each part. Then graph their union.

Solution

$$
\begin{array}{ccc}
3x + 11 > 2 & or & 8 - x > 4 \\
3x > -9 & or & -x > -4 \\
x > -3 & or & x < 4
\end{array}
$$

Graph $x > -3$.

Graph $x < 4$.

Graph the union.

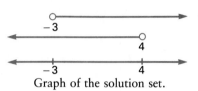

Graph of the solution set.

The solution set is the set of all real numbers.

Sometimes the graphs of the parts of a compound sentence have no points in common, or no overlap. For example, consider the disjunction $4x < 12$ *or* $2x > 12$ and its solution set.

Simplify each part. $x < 3$ *or* $x > 6$

Graph $x < 3$.

Graph $x > 6$.

Graph their union.

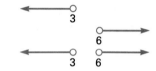

The solution set is $\{x \mid x < 3 \text{ or } x > 6\}$.

The conjunction $x < 3$ *and* $x > 6$, however, has *no* solutions. Because there is no overlap, the graph of the conjunction has no points, and the solution set has no numbers. This solution set is called the **empty set**. It is represented by the symbol $\varnothing$.

EXAMPLE 3   Graph the solution set of $5x + 21 \leq 6$ *or* $17 - 3x < 5$ on a number line. State the solution set.

Solution

$$
\begin{array}{ccc}
5x + 21 \leq 6 & or & 17 - 3x < 5 \\
5x \leq -15 & or & -3x < -12 \\
x \leq -3 & or & x > 4
\end{array}
$$

Graph the union because the disjunction is used.

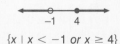

The solution set is $\{x \mid x \leq -3 \text{ or } x > 4\}$.

Additional Example 2

Graph the solution set of $6x - 12 \leq 2x$ *or* $8 < 5x + 28$ on a number line. Find the solution set.

Set of all real numbers

Additional Example 3

Graph the solution set of $3x + 8 < 5$ *or* $12 - x \leq 8$ on a number line. Find the solution set.

$$\longleftarrow\!\!\!\underset{-1}{\circ}\!\!-\!\!\underset{4}{\bullet}\!\!\longrightarrow$$

$\{x \mid x < -1 \text{ or } x \geq 4\}$

A compound inequality such as $-5 < 2y - 9 \leq 5$ can be solved using the Properties of Inequality on each of the three parts: -5, $2y - 9$, and 5.

EXAMPLE 4 Find the solution set of $-5 < 2y - 9 \leq 5$. Graph it on a number line.

Solution
$$-5 < 2y - 9 \leq 5$$
$$4 < 2y \leq 14$$
$$2 < y \leq 7$$

The solution set is $\{y \mid 2 < y \leq 7\}$. The graph is shown below.

Classroom Exercises

Match each compound sentence with the graph of its solution set.

1. $x < -4$ or $x < 2$ b
2. $x < -4$ and $x < 2$ a
3. $x < -4$ or $x > 2$ g
4. $x < -4$ and $x > 2$ h
5. $x > -4$ or $x > 2$ c
6. $x > -4$ and $x > 2$ d
7. $x > -4$ or $x < 2$ f
8. $x > -4$ and $x < 2$ e

a. -4

b. ⟵————○ 2

c. ○——————→ -4

d. ○——————→ 2

e. ○————○ -4 2

f. ⟵——————→

g. ⟵—○ ○—→ -4 2

h. ∅

For each compound sentence, determine whether it is true or false. Justify your answer.

9. $-5 < 3$ and $(-5)^2 < 3^2$

10. $-2 \cdot 6 > 2 \cdot 5$ or $-2(-1) > 0$

Graph the solution set of each compound sentence on a number line. State the solution set.

11. $x < 1$ or $x > 2$

12. $x > -1$ and $x < 2$

13. $x < 2$ or $x < 5$

14. $x < 2$ and $x < 5$

Additional Answers

Classroom Exercises

9. False; T *and* F = F
10. True; F *or* T = T
11. $\{x \mid x < 1 \text{ or } x > 2\}$

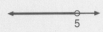
   1 2

12. $\{x \mid -1 < x < 2\}$

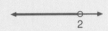
   -1 2

13. $\{x \mid x < 5\}$

⟵————○——→
   5

14. $\{x \mid x < 2\}$

⟵————○——→
   2

Written Exercises

1. $\{x \mid x < -4 \text{ or } x > 3\}$

⟵—○ ○—→
   -4 3

2. $\{x \mid x \leq 5\}$

⟵————●——→
   5

3. $\{y \mid y \geq 3\}$

⟵——●————→
   3

4. $\{y \mid 3 < y < 6\}$

⟵—○——○—→
   3 6

5. {Real numbers}

⟵————+————→
   0

Additional Example 4

Find the solution set of $-2 \leq 3y + 4 < 16$.
Graph it on a number line.
$\{y \mid -2 \leq y < 4\}$

   -2 4

6. ∅

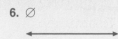

7. {a | −4 ≤ a < 2}

    ●———○
   −4   2

8. {t | −3 < t ≤ 5}

    ○———●
   −3   5

9. {x | −3 < x < 12}

    ○———○
   −3   12

10. {x | x ≤ 3 or x ≥ 5}

    ●   ●
    3   5

11. {x | x < 3}

         ○
         3

12. {x | x < −3 or x > 4}

    ○———○
   −3   4

13. {x | x ≥ 2}

    ●
    2

14. {x | −4 ≤ x < 4}

    ●———○
   −4   4

See page 61 for the answers to Written Exercises 15–19.

Written Exercises

Graph the solution set of each compound sentence on a number line. State the solution set.

1. $x + 9 < 5 \text{ or } 4x > 12$

2. $x - 3 \le 2 \text{ or } 5x \le 10$

3. $5y \ge 15 \text{ and } y + 8 > 2$

4. $2y > 6 \text{ and } y - 2 < 4$

5. $c - 8 < 2 \text{ or } 6c > -18$

6. $7x < -14 \text{ and } x + 5 > 8$

7. $3 \le a + 7 < 9$

8. $-12 < 4t \le 20$

9. $-15 < 2x - 9 < 15$

10. $3x + 11 \le 20 \text{ or } 4x \ge 20$

11. $5x + 12 < 2 \text{ or } 5x - 12 < 3$

12. $14 - 3x < 2 \text{ or } 5 - 4x > 17$

13. $7 - 6x \le 19 \text{ and } 14 - 5x \le 4$

14. $-2 \le 3x + 10 \text{ and } 5 > 2x - 3$

15. $2x < 7x - 10 \text{ or } 8x \le 3x - 15$

16. $2x - 7 < 5x + 8 \text{ or } 8 - 2x > 0$

17. $-7.5 < 4x + 2.5 < 11.3$

18. $-0.7 \le 3x - 4.3 \le 8$

19. Write in your own words how to draw the graph of the solution set of the disjunction (*or* sentence) of two inequalities.

Find the solution set.

20. $8 < 4x - 7 < 22 \text{ and } x \text{ is an odd integer.}$ {5, 7}

21. $18 < 5y + 6 < 50 \text{ and } y \text{ is an even integer.}$ {4, 6, 8}

22. $2w + 2(3w - 6) \le 40 \text{ and } w > 0 \text{ and } 3w - 6 > 0$ {w | 2 < w ≤ 6.5}

23. $4a + 2(6a - 8) < 100 \text{ and } a > 0 \text{ and } 4a - 8 > 0$ {a | 2 < a < 7.25}

24. $-8 < -4x < 12$ {x | −3 < x < 2}   **25.** $1 \le 6 - 5x \le 16$ {x | −2 ≤ x ≤ 1}

26. $3x - 10 < 5x + 2 < 3x + 4$ **27.** $-5 < 3x - 2 < 4 \text{ or } 3x - 2 > 10$

  {x | −6 < x < 1}   {x | −1 < x < 2 or x > 4}

Mixed Review

1. A rectangle is 6 ft longer than it is wide. A second rectangle is twice as wide and 3 times as long as the first rectangle. The sum of the two perimeters is 104 ft. Find the area of the first rectangle and the length of the second rectangle. *1.7* 40 ft²; 30 ft

2. Supply a reason for each step in the proof shown below. *1.3, 1.8*
1. $a(x + b) + c = c + a(x + b)$ Comm Prop Add
2. $\qquad\qquad = c + (ax + ab)$ Distr Prop
3. $\qquad\qquad = c + (ax + ba)$ Comm Prop Mult
4. $\qquad\qquad = (c + ax) + ba$ Assoc Prop Add
5. $a(x + b) + c = (c + ax) + ba$ Trans Prop Eq

50   Chapter 2 Linear Inequalities

Enrichment

Introduce the concept of the *negation* of a statement. Two statements are negations of each other if one must be true when the other is false. A common mistake is to think that $x > 5$ is the negation of $x < 5$. These statements are not negations of each other, since they are both false when $x = 5$.

Challenge the students to write the negations of these statements.

- $x < 2$ $x \ge 2$
- $x > 3 \text{ or } x < -3$ $-3 \le x \le 3$
- $0 < x \le 1$ $x \le 0 \text{ or } x > 1$
- $|x| \le 3$ $x < -3 \text{ or } x > 3$
- $|x - 2| > 2$ $0 \le x \le 4$

2.3 Sentences with Absolute Value

Objectives
To solve equations involving absolute value
To find and graph the solution sets of inequalities involving absolute value

Definition

The **absolute value** of any real number x, written as $|x|$, is the distance on a number line between its location and the origin. Thus, the absolute value of 4 is 4 ($|4| = 4$), and the absolute value of -4 is also 4 ($|-4| = 4$), as shown in the figure below.

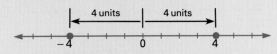

$|x| = -x$, if and only if x is a negative number ($x < 0$)

$|x| = x$, if and only if x is a nonnegative number ($x \geq 0$)

It follows from this definition that if $|y| = 10$,
then $y = -10$ or $y = 10$.

Equation Property for Absolute Value
If $|x| = k$ and $k \geq 0$, then $x = -k$ or $x = k$.

If $|x| = k$ and $k < 0$, then there is no solution.

EXAMPLE 1   Solve the equation $|3n + 4| = 19$.

Plan   Use the Equation Property above. Replace x with $3n + 4$ and k with 19.

Solution

$$|3n + 4| = 19$$
$$3n + 4 = -19 \text{ or } 3n + 4 = 19$$
$$3n = -23 \text{ or } \quad 3n = 15$$
$$n = \frac{-23}{3} \text{ or } \quad n = 5$$

Checks: $|3 \cdot \frac{-23}{3} + 4| \stackrel{?}{=} 19$
$$19 = 19 \text{ True}$$
$$|3 \cdot 5 + 4| \stackrel{?}{=} 19$$
$$19 = 19 \text{ True}$$

The roots are $-7\frac{2}{3}$ and 5.

Additional Example 1

Solve the equation.

$|5x - 3| = 22$ $-3\frac{4}{5}$ or 5

Teaching Resources

Project Worksheet 2
Quick Quizzes 11
**Reteaching and Practice
   Worksheets** 11
Transparency 5

GETTING STARTED

Prerequisite Quiz

Write the solution set.

1. $3n - 6 = -9$ or $3n - 6 = 9$ $\{-1, 5\}$
2. $4a + 2 < -10$ or $4a + 2 > 10$
   $\{a \mid a < -3 \text{ or } a > 2\}$
3. $-7 < 2y - 5 < 7$ $\{y \mid -1 < y < 6\}$

Motivator

The solutions of absolute value equations and inequalities locate the points at a given *distance* from a given point on a number line. Have students label the point at 0 on a number line and the two points that are 5 units from that point. Ask them to write an equation that represents the graph. $|x| = 5$ Then have them graph all points between 5 and -5. Ask them to write an inequality that represents the graph. $|x| < 5$

TEACHING SUGGESTIONS

Lesson Note

Students should examine an absolute value sentence first to see whether the relation is $=$, $<$, or $>$ with respect to k where $k > 0$. Then the appropriate transformation property should be selected and applied.

Highlighting the Standards

Standard 2f: In this lesson, students expand their ideas about absolute value notation and its alternative form, a compound equation.

Math Connections

Geometry: Point out to students that the graph of the solution set of an equation or inequality involving absolute value can form a geometric figure. Have them give sentences where the graph of the solution set will form a point, line segment, ray, or line.

Critical Thinking Questions

Analysis: Ask students whether the first half of the definition of the *absolute value* of x, "| x | = −x, if and only if x is a negative number" could be "rephrased" as "| −x | = x, if and only if x is a negative number." Students should see that if x is negative, it cannot equal the absolute value of a number. So the answer is "No." Ask whether the rephrasing would be correct if the word "negative" were replaced by "positive", or by "nonnegative". no; yes

Checkpoint

1. Solve the equation |x + 5| = 17.
−22 or 12

Find the solution set of each inequality.

2. |2y − 3| ≤ 7   {y | −2 ≤ y ≤ 5}
3. |n + 5| > 9
   {n | n < −14 or n > 4}

Closure

Ask students what is meant by the phrase, "*absolute value* of any real number x". Distance on number line between its location and the origin. Have students use the equation property and the inequality properties to rewrite the sentences below as compound sentences without absolute value symbols.

1. |x − 3| = 5   x = −2 or x = 8
2. |x − 3| < 5   x < 8 and x > −2
3. |x − 3| > 5   x > 8 or x < −2

The solutions of $|x| < 4$ are the numbers whose graphs are *less than* 4 units from the origin.

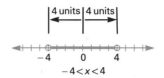
$-4 < x < 4$

The solutions of $|x| \geq 4$ are those numbers whose graphs are *at least* 4 units from the origin.

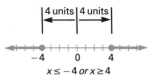
$x \leq -4 \, or \, x \geq 4$

Inequality Properties for Absolute Value
If $|x| \leq k$ and $k > 0$, then $-k \leq x \leq k$.
If $|x| \geq k$ and $k \geq 0$, then $x \leq -k \, or \, x \geq k$.

EXAMPLE 2   Find the solution set of $|4y - 2| \leq 10$. Graph it.

Plan   Use the property: If $|x| \leq k$, then $-k \leq x \leq k$.

Solution
$$|4y - 2| \leq 10$$
$$-10 \leq 4y - 2 \leq 10$$
$$-8 \leq 4y \leq 12$$
$$-2 \leq y \leq 3$$

The solution set is $\{y \mid -2 \leq y \leq 3\}$. The graph is shown below.

-2 3

EXAMPLE 3   Find the solution set of $|2n + 1| > 7$. Graph it.

Solution   Restate the property above as: If $|x| > k$, then $x < -k \, or \, x > k$.

$$|2n + 1| > 7$$
$$2n + 1 < -7 \, or \, 2n + 1 > 7$$
$$2n < -8 \, or \qquad 2n > 6$$
$$n < -4 \, or \qquad n > 3$$

The solution set is $\{n \mid n < -4 \, or \, n > 3\}$.

-4 3

Additional Example 2
Find the solution set of
$|2y + 7| < 15$. Graph it.
$\{y \mid -11 < y < 4\}$

−11   4

Additional Example 3
Find the solution set of
$|4n - 6| \geq 10$. Graph it.
$\{n \mid n \leq -1 \, or \, n \geq 4\}$

−1   4

Classroom Exercises

Match each sentence on the left with a sentence at the right.

1. $|x + 5| = 3$ b
2. $|x + 5| < 3$ d
3. $|x + 5| < -3$ e
4. $|x + 5| > 3$ c

a. $x + 5 = -3$ and $x + 5 = 3$
b. $x + 5 = 3$ or $x + 5 = -3$
c. $x + 5 < -3$ or $x + 5 > 3$
d. $x + 5 > -3$ and $x + 5 < 3$
e. There is no solution.

Solve each equation.

5. $|n + 4| = 7$ −11, 3
6. $|y - 3| = 4$ −1, 7
7. $|12 - x| = 12$ 0, 24
8. $|4 + c| = 22$ −26, 18
9. $|11 - r| = 3$ 8, 14
10. $|s - 6| = 17$ −11, 23

Written Exercises

Solve each equation.

1. $|x - 4| = 9$ −5, 13
2. $|a + 5| = 3$ −8, −2
3. $|6 - y| = 8$ −2, 14
4. $|2t - 17| = 11$ 3, 14
5. $|3n + 12| = 18$ −10, 2
6. $|10 - 4a| = 32$ −5.5, 10.5

Find the solution set of each inequality. Graph it.

7. $|x - 4| > 1$
8. $|2c| > 6$
9. $|y + 3| \leq 7$
10. $|n - 2| \leq 5$
11. $|3c| \leq 12$
12. $|t + 8| > 4$

Solve each equation.

13. $|4y + 5| = 17$ −5.5, 3
14. $|5x - 7| = 3$ 0.8, 2
15. $|10 - 3n| - 5 = 0$ $1\frac{2}{3}$, 5
16. $|2x + 7.5| = 19.5$ −13.5, 6
17. $|2.5x - 8| = 10$ −0.8, 7.2
18. $|7.5 - 5n| = 2.5$ 2, 1

Find the solution set of each inequality. Graph it.

19. $|2y - 3| < 11$
20. $|4t + 6| \leq 14$
21. $|3y + 6| > 15$
22. $|4n - 5| \geq 15$
23. $|2x + 3.5| > 11.5$
24. $14 > |4z - 2|$

Solve each equation. (HINT: $|a - b| = |-a + b| = |b - a|$)

25. $|n - 3| + |3 - n| = 14$ −4, 10
26. $|2y + 7| + |3 - 2y| = 20$ −6, 4
27. $|x + 5| + |-x - 9| = 30$ −22, 8
28. $|3 - 2c| + |-c - 9| = 27$ −11, 7

Mixed Review

Solve for x. 1.4

1. $3x - 6 = 0$ 2
2. $400(0.09)x = 180$ 5
3. $2[x + (x + 2)] = 23$ 4.75

▰▰▰FOLLOW UP

Guided Practice

Classroom Exercises 1–10

Independent Practice

🅰 Ex. 1–12, 🅱 Ex. 13–24, 🅲 Ex. 25–28

Basic: WE 1–17 odd
Average: WE 5–23 odd
Above Average: WE 9–27 odd

Additional Answers

Written Exercises

7. $\{x \mid x < 3 \text{ or } x > 5\}$

3   5

8. $\{c \mid c < -3 \text{ or } c > 3\}$

−3   3

9. $\{y \mid -10 \leq y \leq 4\}$

−10   4

10. $\{n \mid -3 \leq n \leq 7\}$

−3   7

11. $\{c \mid -4 \leq c \leq 4\}$

−4   4

12. $\{t \mid t < -12 \text{ or } t > -4\}$

−12   −4

See page 58 for the answers to Written Exercises 19–24.

Enrichment

Introduce the concept of an absolute value function. These functions have V-shaped angles for their graphs. For example, the graph of the function $y = |x|$ is an angle with its vertex at (0, 0), having (−1,1) on one side, and (1,1) on the other.

After analyzing a few variations of the graphs of absolute value functions, challenge the students to find the function whose graph is a "V" having its vertex at (1,−2), one side containing (0,0), and the other side containing (2,0).

$y = 2|x - 1| - 2$

▬▬GETTING STARTED

Prerequisite Quiz

Write the solution set.

1. $x - 2 < 6$ *and* $x - 3 > 0$
   $\{x \mid 3 < x < 8\}$
2. $50 \le 20t \le 160$
   $\{t \mid 2.5 \le t \le 8\}$
3. $3.5 < n < 9.5$ *and* n is an even integer.
   $\{4, 6, 8\}$
4. $0 < 7(x - 3) < 14$ $\{x \mid 3 < x < 5\}$

Motivator

Have students express these phrases algebraically:

1. "Any time between 30 seconds and 90 seconds." $30 < t < 90$
2. "The sum of three consecutive integers is greater than 40." $n + (n + 1) + (n + 2) > 40$

Ask students how many answers are possible for Question 2.   An infinite number

▬▬TEACHING SUGGESTIONS

Lesson Note

A compound sentence must be written for each example and exercise in this lesson since a *second condition* is implied in each situation. For example, if $4x - 8$ represents the *length* of a rectangle, or if it represents a positive real number, then $4x - 8 > 0$, $4x > 8$, and $x > 2$.

2.4 Problem Solving: Using Inequalities

Objective   To solve word problems involving a compound sentence

In Chapter 1 you used algebra to solve word problems with exactly one answer. As you have learned since, however, some problems have several answers, and others none. The problem below, for example, has many answers, yet algebra can still be used to solve it.

Find the time period during which \$400 invested at 9% per year will earn from \$135 to \$180, inclusive, in simple interest.

$i = prt \leftarrow$ formula for simple interest
$135 \le prt \le 180$
$135 \le 400(0.09)t \le 180$
$135 \le 36t \le 180$
$3.75 \le t \le 5$

Thus, for any time between 3 years 9 months and 5 years, inclusive, \$400 invested at 9% per year will earn between \$135 and \$180, inclusive, in simple interest.

Sometimes, the realistic scope, or *domain*, of a variable must also be considered. In the problem below, the length $(3x - 6)$ and width $(2x)$ of the rectangle must always be positive.

EXAMPLE 1   A rectangle is twice as wide as a square. The rectangle's length is 6 ft less than 3 times the square's length. The sum of the two perimeters is less than 86 ft. Find the set of all possible lengths for the square.

Solution

Perimeter of square: $4x$
Perimeter of rectangle: $10x - 12$
Sum of perimeters: $4x + (10x - 12)$

The length of each side must be positive.

$4x + 10x - 12 < 86$ *and* $3x - 6 > 0$ *and* $2x > 0$ *and* $x > 0$
$\qquad\quad 14x < 98$ *and* $\qquad 3x > 6$ *and* $x > 0$
$\qquad\qquad\quad x < 7$ *and* $\qquad\quad x > 2$
So, $2 < x < 7$.

Check

Check a number between 2 and 7:

If $x = 5$, then $4x = 20$, $10x - 12 = 38$, $20 + 38 = 58$, and $58 < 86$. The length of the square is between 2 ft and 7 ft.

In checking problems, such as Example 1, it is also a good idea to check numbers that are not between 2 and 7 ($x \le 2$ *or* $x \ge 7$):
If $x = 8$, then $4x = 32$, $10x - 12 = 68$, $32 + 68 = 100$, and $100 \not< 86$ (100 is not less than 86). If $x = 1$, then the length of the rectangle $(3x - 6)$ would be -3 ft, which is not possible.

54   Chapter 2 Linear Inequalities

Highlighting the Standards

Standard 4d, 6b: In the Application, students connect geometry to early measurements of the earth. In the Problem Solving Strategies, they represent data in a table.

Additional Example 1

A rectangle is 2 m wider than a side of a square. The rectangle's length is 12 m less than 4 times the square's length. The sum of the two perimeters is less than 50 m. Find the set of possible widths for a side of the square.   The width of the square is between 3 m and 5 m.

EXAMPLE 2 Find three consecutive odd integers such that twice the sum of the first two is greater than 23, and 3 times the sum of the last two is less than 61.

Solution Let $x, x + 2, x + 4$ be the three consecutive odd integers.

x is odd *and* $2[x + (x + 2)] > 23$ *and* $3[(x + 2) + (x + 4)] < 61$

$4x + 4 > 23$ *and* $\qquad\qquad 6x + 18 < 61$

$x > 4\frac{3}{4}$ *and* $\qquad\qquad\qquad x < 7\frac{1}{6}$

Thus, x is odd *and* $x > 4\frac{3}{4}$ *and* $x < 7\frac{1}{6}$. In other words, x is an odd integer between $4\frac{3}{4}$ and $7\frac{1}{6}$. So, x is either 5 or 7.

There are two answers: 5, 7, 9 and 7, 9, 11.

Recall (pg. 18) that $vt - 5t^2$ is the height h an object will reach at the end of t seconds if it is launched upward with an initial vertical velocity of v meters per second.

EXAMPLE 3 What is the range of the initial vertical velocity that will be needed to launch an object to a height of 810 m $\pm$ 30 m at the end of 6 s? Use the formula $h = vt - 5t^2$.

Solution
$$h = vt - 5t^2$$
$$810 - 30 \le vt - 5t^2 \le 810 + 30 \leftarrow 810 \pm 30 \text{ refers to}$$
$$780 \le 6v - 5 \cdot 6^2 \le 840 \qquad \begin{array}{l}\text{numbers } between \text{ 810} - \text{30}\\ \text{and 810} + \text{30, inclusive.}\end{array}$$
$$960 \le 6v \le 1{,}020$$
$$160 \le v \le 170$$

An initial vertical velocity between 160 m/s and 170 m/s, inclusive, will be needed.

Classroom Exercises

1. Find three consecutive even integers given that x is the first even integer and $6\frac{2}{3} < x < 9\frac{1}{2}$. 8, 10, 12

2. Find three consecutive multiples of 5 given that y is the first multiple and $12 < y < 24$. 15, 20, 25 and 20, 25, 30

In Exercises 3 and 4, the interval for n is $6 < n < 10$.

3. What is the interval for $n - 4$?
$2 < n - 4 < 6$

4. What is the interval for $5(n - 4)$?
$10 < 5(n - 4) < 30$

In Exercises 5–7, a certain rectangle has a width x and a length $4(x - 5)$. Can x be:

5. equal to -2? No

6. equal to 2? No

7. equal to 5? No

Math Connections

Mechanics: Algebraic expressions are often used to describe mechanical forces upon an object. In Example 3, the coefficient 5 appears in the formula $h = vt - 5t^2$ for the height an object attains in a trajectory. This value is used to represent the acceleration experienced by a freely falling body under the influence of gravity.

Critical Thinking Questions

Analysis: Ask students how the solution and the answer to Example 2 would be affected if three consecutive even integers were considered rather than three consecutive odd integers. The algebraic steps of the solution would be the same, but there would be only one answer: 6, 8, 10.

Checkpoint

1. Find all the time periods in which $800 invested at 7% simple interest per year will earn between $112 and $140, inclusive ($i = prt$). All time periods from 2 yr through 2.5 yr

2. If 8 less than a positive number is multiplied by 3, the result is less than 32. Find the set of all such positive numbers. All positive numbers less than $18\frac{2}{3}$

Closure

Ask students to tell what relations they would use to represent problems with a range of answers. $<, >, \le, \ge$ Ask students if $x < 4$ is a realistic scope of x-values for a rectangle $x + 2$ by $2x + 6$, with a perimeter less than 40. No, length, width, and perimeter must be positive.

Additional Example 2

Find three consecutive even integers such that 3 times the sum of the first two is greater than 50, and 4 times the sum of the last two is less than 110.
There are two possible answers:
8, 10, 12 and 10, 12, 14.

Additional Example 3

What set of initial velocities is possible if an object is to reach a height of 260 m $\pm$ 20 m at the end of 4 s when it is launched vertically? ($vt - 5t^2$ represents the height.)
$\{v \mid 80 \text{ m/s} \le v \le 90 \text{ m/s}\}$

FOLLOW UP

Guided Practice

Classroom Exercises 1–7

Independent Practice

A Ex. 1–9, **B** Ex. 10–13, **C** Ex. 14–17

Basic: WE 1–11 odd, Midchapter Review, Application, Problem Solving

Average: WE 3–13 odd, Midchapter Review, Application, Problem Solving

Above Average: WE 5–17 odd, Midchapter Review, Application, Problem Solving

Additional Answers

Written Exercises

1. Between 2 yr 6 mo and 6 yr 3 mo, inclusive
2. Between $1,500 and $2,000, inclusive
6. $\{x \mid 2.5 \text{ cm} < x < 10 \text{ cm}\}$
11. Between $18,000 and $24,000, inclusive

Midchapter Review

6. $\{x \mid -2 < x < 5\}$

7. $\{x \mid x > -2\}$

Written Exercises

1. Find the time period during which $600 invested at 8% per year will earn between $120 and $300, inclusive, in simple interest.

2. How much principal must be invested at 10% per year for 3 years 6 months to earn between $525 and $700, inclusive, in simple interest?

3. Find the set of real numbers for which 18 more than 5 times the sum of a number and 4 is greater than 98. $\{x \mid x > 12\}$

4. If 6 less than a positive number is multiplied by 2, the result is less than 17. Find the set of all such positive numbers. $\{x \mid 0 < x < 14\frac{1}{2}\}$

5. A rectangle's length is 12 in. less than 3 times its width. Find the set of all the possible widths given that the perimeter must be less than 48 in. $\{w \mid 4 \text{ in.} < w < 9 \text{ in.}\}$

6. Each leg of an isosceles triangle is 5 cm shorter than twice the length of the base. Find the set of possible lengths of a leg given that the perimeter is less than 40 cm.

7. Find three consecutive integers such that the sum of the first two is greater than 12, and the sum of the last two is less than 18. 6, 7, 8 and 7, 8, 9

8. Find three consecutive even integers for which the sum of the first two integers is greater than 9 more than the third integer, and the sum of the first and third integers is less than 17 more than the second integer. 12, 14, 16 and 14, 16, 18

9. What is the range of the initial vertical velocity needed for an object to reach a height of 760 m ± 120 m at the end of 4 s? (Use the formula: $h = vt - 5t^2$.) 180 m/s ≤ v ≤ 240 m/s

10. A rectangle is 3 times as wide as a square. The rectangle's length is 15 ft shorter than 5 times the square's length. Given that the sum of the two perimeters must be less than 150 ft, find the set of all possible lengths for the rectangle. $\{l \mid 0 \text{ ft} < l < 30 \text{ ft}\}$

11. Mrs. McTier wants to invest some money at 11% per year in order to have a total amount (investment plus interest) between $27,900 and $37,200, inclusive, at the end of 5 years. Find the amount of money she must invest. (HINT: Let $p + prt$ represent the total amount.)

12. Mr. McTier invests $600 at 9% per year, and $800 at 12% per year. During what time period will the $1,400 earn between $525 and $975, inclusive, in simple interest? 3.5 yr ≤ t ≤ 6.5 yr

13. Find three consecutive multiples of 5 such that 3 times the sum of the first two multiples is less than 200, and twice the sum of the last two is greater than 125. 25, 30, 35, and 30, 35, 40

14. The second of three negative numbers is 6 less than the first negative number. The third negative number is 3 times the second number. Find the set of all possible third numbers given that twice the sum of the first two numbers is greater than the third number. $\{x \mid -36 < x < -18\}$

15. Given that the base of a triangle is 5 ft, find all the whole-number values for its height that will result in an area between 8 ft² and 18 ft². 4 ft, 5 ft, 6 ft, 7 ft

56 Chapter 2 Linear Inequalities

A rectangular solid (box) has a height h of 4 in. and a width w that is 8 in. less than the length l. Use this information for Exercises 16 and 17.

16. Find all the combinations of whole-number values of l, w, and h for which the resulting volume V will be less than 140 in.^{3}. ($V = lwh$) (l,w,h): (9,1,4),(10,2,4),(11,3,4)

17. Find all the combinations of whole-number values of l, w, and h for which the resulting surface area T will be between 636 in.^{2} and 936 in.^{2}. ($T = 2lw + 2lh + 2wh$) (l,w,h): (19,11,4),(20,12,4), (21,13,4),(22,14,4)

Midchapter Review

Find the solution set. *2.1, 2.3*

1. $3(4 - 2x) < 18 - 3x$ $\{x \mid x > -2\}$
2. $\frac{3}{4}n - \frac{1}{6}n \geq \frac{1}{3}n + 3$ $\{n \mid n \geq 12\}$
3. $|4n - 2| = 14$ $\{-3, 4\}$
4. $|4n - 2| < 14$ $\{n \mid -3 < n < 4\}$
5. $|4n - 2| > 14$ $\{n \mid n < -3 \text{ or } n > 4\}$

Graph the solutions. State the solution set. *2.2*

6. $x + 5 > 3x - 5$ and $4x + 4 > x - 2$ $\{x \mid -2 < x < 5\}$
7. $3x - 4 > x + 6$ or $4x > x - 6$ $\{x \mid x > -2\}$

Application: *Circumference of the Earth*

About 240 B.C., Eratosthenes, then the head of the Library at Alexandria, made the first scientific measurement of the earth's circumference. He knew that in the city of Syene at noon on Midsummer's Day, the sun shone directly into a well, while in Alexandria, a camel's ride of 5,000 stadia (1 stadium = 606 ft) to the north, the sun cast a shadow of 7.2 degrees from the vertical.

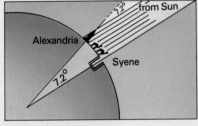

Using alternate interior angles to show that 7.2 was also the central angle between Alexandria and Syene, Eratosthenes then concluded that the measure 7.2 was to a circle as 5,000 stadia is to the earth's circumference.

1. Use this data to find the earth's circumference to the nearest mile. (Use 1 stadium = 606 ft.) 28,693 mi

2. Explain why Eratosthenes had to assume that all the sun's rays are parallel to one another.

2.4 Problem Solving: Using Inequalities **57**

Application

2. To use alternate interior angles, one must first have parallel lines.

Additional Answers, page 44

17. $\{a \mid a < -4\frac{2}{3}\}$

$-4\frac{2}{3}$

18. $\{x \mid x \leq 11\}$

11

19. $\{x \mid x \leq 4\}$

4

20. $\{a \mid a \leq 48\}$

48

21. $\{x \mid x > -7\}$

-7

22. $\{y \mid y > 6\}$

6

23. $\{a \mid a < -7\}$

-7

27. False; Example: $2 < 5$ and $5 < 6$, but $2 + 5 \not< 6$.

Enrichment

Students sometimes tend to ignore or overlook the importance of inequalities. Have the students use reference sources to list practical cases in which inequalities are essential. For example, in a machine shop, *tolerance* is the allowable range of variation from a standard. In a cooking recipe, the amount of an ingredient to be used is sometimes placed within an acceptable range. Also, nutritionists speak of ranges of values for daily requirements of vitamins and minerals.

Each student should be encouraged to provide several similar examples.

3.

Ages	Amos	Beth	Carlo
Now	$3x$	x	$x + 4$
5 yrs ago	$3x - 5$	$x - 5$	$x - 1$

Additional Answers, page 53

19. $\{y \mid -4 < y < 7\}$

-4   7

20. $\{t \mid -5 \le t \le 2\}$

-5   2

21. $\{y \mid y < -7 \text{ or } y > 3\}$

-7   3

22. $\{n \mid n \le -2.5 \text{ or } n \ge 5\}$

-2.5   5

23. $\{x \mid x < -7.5 \text{ or } x > 4\}$

-7.5   4

24. $\{z \mid -3 < z < 4\}$

-3   4

 Problem Solving Strategies

Making a Table

Objective   To construct tables to represent the data in word problems

EXAMPLE   José is 9 years younger than Maria. In 4 years, she will be twice as old as he will be. Construct a table that represents their ages, both now and in 4 years, in terms of one variable.

Plan   The table must distinguish between:
(1) José and Maria, and
(2) ages now and ages in 4 years.

Table: Let m = Maria's age now.

Ages	José	Maria
Now	$m - 9$	m
In 4 yr	$m - 9 + 4$	$m + 4$

Notice that the information "she will be twice as old as he" has not yet been used in representing the data. This fact would be used in the next problem-solving step, that is, writing an equation.

Exercises

1. Bus A left a terminal and averaged 35 mi/h. Two hours later, Bus B left the terminal and traveled the same route at 45 mi/h. Construct a table that represents the rate, time, and distance for each bus. Let t = the time in hours for Bus A. (Use the formula: $r \times t = d$.)

	rate	time	distance
Bus A	35	t	$35t$
Bus B	45	$t - 2$	$45(t - 2)$

2. Some corn worth 75¢/lb is mixed with oats worth $1.10/lb to make 25 lb of an animal feed that is worth 96¢/lb. Construct a table that represents the weight, value per pound, and total value for the corn, oats, and mixture. Let x = number of pounds of corn.

	No. of lb	¢/lb	Value in ¢
corn	x	75	$75x$
oats	$25 - x$	110	$110(25 - x)$
mixture	25	96	$96(25)$

3. Amos is 3 times as old as Beth and Carlo is 4 years older than Beth. Five years ago, the sum of their ages was 29 years. Construct a table that represents their ages now and 5 years ago.

2.5 Problem Solving: Ages

Objective

To solve word problems involving ages

You will often find that a word problem becomes easier to solve once its data has been represented in a table. Note that in the example below, each entry in the bottom row of the table is clearly 6 greater than the corresponding entry in the row above it.

EXAMPLE

Charlene is 19 times as old as Enrique. Sonya is 30 years younger than Charlene. In 6 years, Charlene's age will be twice the sum of Enrique's and Sonya's ages then. How old is each person now?

Solution

Let E = Enrique's age now. Then, Charlene's age now is represented by $19E$ and Sonya's age by $19E - 30$.

Ages	Enrique	Charlene	Sonya
Now	E	$19E$	$19E - 30$
In 6 years	$E + 6$	$19E + 6$	$(19E - 30) + 6$, or $19E - 24$

In 6 years, Charlene's age	will be	twice the sum of the others' ages (in 6 years)
$19E + 6$	$=$	$2[(E + 6) + (19E - 24)]$

$$19E + 6 = 40E - 36$$
$$42 = 21E$$
$$E = 2 \qquad 19E = 38 \qquad 19E - 30 = 8$$

Check

So, Enrique is 2 years old now, Charlene is 38 years old now, and Sonya is 8 years old now. In 6 years, Charlene will be $38 + 6$, or 44, and twice the sum of the others' ages then will be $2[(2 + 6) + (8 + 6)]$, or 44.

An age problem can also involve an inequality and have many answers in its solution set. In such a problem, be sure to consider the realistic domain of the variable. For example, if you are x years old and 6 years ago you were $x - 6$ years old, then $x - 6 > 0$ and $x > 6$. So, you must now be more than 6 years old.

2.5 Problem Solving: Ages **59**

Teaching Resources

Quick Quizzes 13
Teaching and Practice Worksheets 13

▰ GETTING STARTED

Prerequisite Quiz

Solve each equation.

1. $15m + 6 = 3[(m + 6) + (15m - 26)]$
2

2. $3x - 10 + 4x - 10 + x + 6 - 10 = 88$
14

Motivator

Ask students where they would find information represented in a table. Newspapers, encyclopedias, etc. Ask them why they might use a table to organize information. Organizing data in a table often reveals patterns that help solve problems.

▰ TEACHING SUGGESTIONS

Lesson Note

The A-level exercises provide age problems involving equations. Age problems involving inequalities are included in B-level exercises.

Additional Example

Carol is 4 times as old as Andrew, and Byron is 12 years younger than Carol. Four years ago, Carol's age was twice the sum of Andrew's and Byron's ages. How old is Byron now? 12 yr

Highlighting the Standards

Standard 6b, 4d: In the Example, students continue to explore the representation of a relationship by means of a table which helps them solve a problem. They practice doing this in the Exercises. In the Application, they connect mathematics to the etching process.

59

Math Connections

Life Skills: Students use tables in their daily lives; for example, to find out the amount of postage for a letter or perhaps, as a part-time clerk, to calculate the sales tax on an item. Tables such as these are useful in organizing information.

Critical Thinking Questions

Analysis: In the Example on page 59, ask students what changes would occur in the table if E were to represent Enrique's age six years from now rather than his age now. The top row would begin with $E - 6$ and other positions would change accordingly.

Common Error Analysis

Error: In writing an equation for "she was 3 times as old as he," students may write 3(her age) = his age."

Such students can be shown the following:

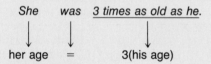

Checkpoint

1. Ben is 5 years older than Tammy. In 7 years, the sum of their ages will be 31. Find their ages now.
Tammy: 6; Ben: 11

2. House A is 4 times as old as House B and House C is 12 years newer than House A. Two years ago, the age of House A was twice the sum of the ages of Houses B and C. Find the present age of each house. A:20; B:5; C:8

Classroom Exercises

Represent each age now in terms of one variable.

1. A horse is 15 years older than twice the age of a colt. $c, 2c + 15$

2. A redwood tree is 100 years older than 3 times the age of a pine tree. $p, 3p + 100$

3. A second city was chartered 15 years before a first city. A third city is 3 times as old as the first city. $f, f + 15, 3f$

4. A bald eagle is twice as old as a falcon. A sparrow was hatched 10 years after the eagle was hatched. $f, 2f, 2f - 10$

In the table shown, find the entry for:

5. Mary's age 10 yr ago. x

6. Sue's age now. $4x + 10$

Ages	Mary	Sue
now		
10 years ago	x	$4x$

7. If Sue is now 2 times as old as Mary, how old is Sue? 30

8. In Classroom Exercise 2, suppose that in 115 years the redwood tree will be twice as old as the pine tree. Find the age of each now. Redwood: 145 Pine: 15

9. Mrs. Adams is 26 years older than her son. Ten years ago, she was 3 times as old as he was. How old is each now? Mother: 49, Son: 23

Written Exercises

1. Today, Selma is 5 times as old as Frank, but in 10 years she will be only 3 times as old. How old are they now?
Frank: 10; Selma: 50

2. Andre is 4 years younger than Charlene. Six years ago, the sum of their ages was 16. Find their ages now.
Charlene: 16; Andre: 12

3. Barbara is 6 years older than Matt, and Christi is twice as old as Barbara. Three years ago, the sum of all their ages was 41. Find their present ages.
Matt: 8; Barb: 14; Christi: 28

4. An oak tree is 30 years older than a pine tree. In 8 years, the oak will be 3.5 times as old as the pine. Find their present ages. Pine: 4; Oak: 34

5. Truck 1 is 3 times as old as Truck 2, and Truck 3 is 16 years younger than Truck 1. One year ago, the age of Truck 1 was twice the sum of the ages of Trucks 2 and 3. Find their present ages. Truck 1: 21; Truck 2: 7; Truck 3: 5

6. Building A was built 30 years before Building B, and 20 years after Building C. In 10 years, Building C will be 40 years younger than the combined ages of Buildings A and B. What is the present age of each building?
A: 80; B: 50; C: 100

7. City B is 3 times as old as City A. Six years ago, the sum of their ages was less than 32 years. Find the possibilities for the present age of City B. Between 18 yr and 33 yr

8. Mary is 10 years younger than her brother. In 3 years, the sum of their ages will be less than 24. Find their possible ages now.
Brother: between 10 and 14; Mary: less than 4

9. The ratio of the present ages of three paintings is 1 to 3 to 4. In three years, the sum of their ages will be less than 81 years. Find the possible present ages of each painting. (HINT: The ratio 1:3:4 can be written more generally as $1x:3x:4x$.) P1 < 9 yr, P2 < 27 yr, P3 < 36 yr

10. The ratio of the ages of three companies is 2 to 4 to 5. Twenty years ago, the age of the oldest company was greater than the combined ages of the other two companies. Find the possibilities for the present age of the oldest company. Between 50 and 100 yr old

11. A person who is n years old now was p years old 8 years ago. Represent the person's age f years from now in terms of p and f. $p + 8 + f$

12. In x years, a painting will be y years old. Represent the painting's age z years ago in terms of x, y and z.
$y - x - z$

13. Company A was formed n years after Company B was started, and p years before Company C was organized. Two years ago, the combined age of the companies was 50 years. Express the present age of Company B in terms of n and p. $\dfrac{56 + 2n + p}{3}$

14. City A is half as old as City B, and City B is one-third as old as City C. City B is now b years old. In 6 years, the combined age of Cities A and B will be $3t$ years less than the age of City C. Express the present age of City C in terms of t. $6t + 12$

Mixed Review

Solve each sentence for x.

1. $5.3x - 1.2(5x - 6) = 2.3x - 1.8$ **1.4**
   1.4 **3**

2. $\frac{2}{3}x - \frac{3}{4} = \frac{5}{8}x - \frac{7}{4}$ **1.4** **−24**

3. $-5 < 3x + 1 < 16$ **2.2**
   $\{x \mid -2 < x < 5\}$

4. $-6x + 4 > -8$ or $2 < x - 3$ **2.2**
   $\{x \mid x < 2 \text{ or } x > 5\}$

Application: *Etching*

In one method of etching, artists use acid baths to cut images into metal plates. Once etched, the plates are inked and used to print the artwork. Control of the bath's acid content is crucial. For example, a printer has 64 fluid ounces (fl oz) of 10% nitric acid solution. To cover some zinc plates, she adds 32 fl oz of water. The calculations below show, to the nearest tenth of an ounce, the amount of acid (x) she now has to add to bring the acid content up to 8%.

$$10\% \text{ of } 64 \text{ fl oz} + x \text{ fl oz} = 8\% \text{ of } (64 + 32 + x) \text{ fl oz}$$
$$0.1(64) + x = 0.08(64 + 32 + x)$$
$$x = 1.4 \text{ fl oz of acid}$$

1. To the nearest tenth, how much acid must be added to 128 fl oz of a 12% solution to make it a 15% solution?
   4.5 fl oz of acid

2. A printer mixed a 40% acid solution with water to make 64 fl oz of an 8% solution. How much of each did he use?
   12.8 fl oz of solution; 51.2 fl oz of water

2.5 Problem Solving: Ages **61**

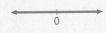

▬▬GETTING STARTED

Prerequisite Quiz

Solve each equation.

1. $36t + 40(t - 2.5) = 356$ 6
2. $20t = 52(t - 4)$ 6.5
3. $20t = 52(t - 4) + 32$ 5.5

Motivator

The formula to find the distance traveled by an object in motion is $d = rt$. Ask students how they would solve the equation for r.
$r = \frac{d}{t}$ Ask students how they would represent the rate of speed of two objects if one is twice as fast as the other.
r = slower object; $2r$ = faster object

2.6 Problem Solving: Two Objects in Motion

Objectives   To solve word problems involving two objects moving in opposite directions

To solve word problems involving two objects moving in the same direction

A bus driven at an average rate of 45 mi/h for 4 h travels 45 · 4, or 180 mi. A sprinter who runs at an average rate of 11 m/s for 5 s travels 11 · 5, or 55 m. The general relationship among distance, rate, and time for an object in motion is stated below.

The distance d traveled by a moving object is equal to its average rate of speed r multiplied by its time t in motion: $d = rt$.

Sometimes two objects travel in *opposite* directions, as shown in Example 1 below.

EXAMPLE 1   Two trains are moving toward each other on parallel tracks from stations 332 mi apart. One train is averaging 48 mi/h and the other 56 mi/h. If the faster train leaves 1 h 30 min after the slower train, how long will it take the faster train to meet the slower one?

Solution   Change 1 h 30 min to $1\frac{1}{2}$ h, or 1.5 h.

Let t = time for the faster train.
Then $t + 1.5$ = time for the slower train.

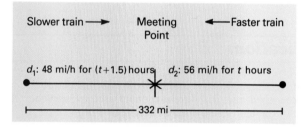

Use the formula $d = rt$.

Substitute for r_1, t_1, r_2, and t_2.

$$d_1 + d_2 = 332$$
$$r_1t_1 + r_2t_2 = 332$$
$$48(t + 1.5) + 56t = 332$$
$$48t + 72 + 56t = 332$$
$$104t = 260$$
$$t = 2.5$$

It will take the faster train $2\frac{1}{2}$ h (2 h 30 min) to meet the slower one.

Highlighting the Standards

Standard 4a, 1b: In the Classroom Exercises, students recognize equivalent representations of the same concept and in the Written Exercises they apply this to various problems about transportation.

Additional Example 1

Bus A and Bus B traveled toward each other from terminals 362 mi apart. Bus A averaged 52 mi/h and Bus B averaged 56 mi/h. If Bus B left its terminal 1 h 15 min after Bus A left, how long did it take Bus B to meet Bus A? $2\frac{3}{4}$ h, or 2 h 45 min

In the next example, the two objects start from the same point and move in the *same* direction.

EXAMPLE 2 An ocean liner, traveling at 30 km/h, leaves port at 7:00 A.M. At 10:00 A.M., a helicopter leaves the same port and travels the same route at 66 km/h. At what time will the helicopter overtake the liner? (THINK: Do the liner and the helicopter travel at the same speed? for the same time? for the same distance?)

Solution Let t = time for the helicopter. Then $t + 3$ = time for the liner.

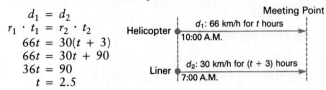

$$\begin{aligned} d_1 &= d_2 \\ r_1 \cdot t_1 &= r_2 \cdot t_2 \\ 66t &= 30(t + 3) \\ 66t &= 30t + 90 \\ 36t &= 90 \\ t &= 2.5 \end{aligned}$$

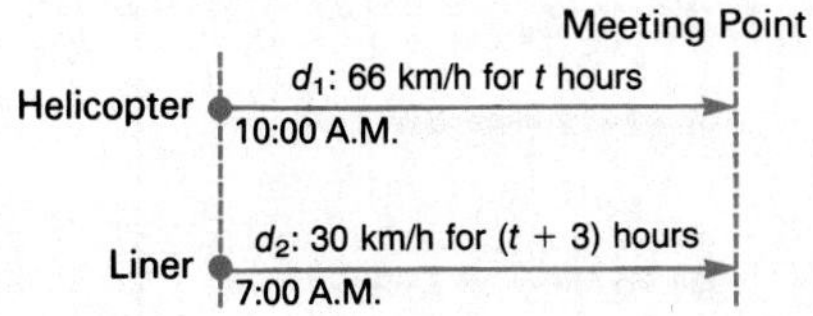

The helicopter will fly for $2\frac{1}{2}$ h, or 2 h 30 min, and overtake the liner at 10:00 A.M. plus 2 h 30 min, or 12:30 P.M.

EXAMPLE 3 Marcia averaged 8 mi/h and André averaged 10 mi/h during a long-distance race. If they began the race at 7:30 A.M., at what time were they 2.5 mi apart?

Solution Let t = both Marcia's time and André's time in hours after 7:30 A.M.

$$\begin{aligned} d_1 + 2.5 &= d_2 \\ r_1 \cdot t_1 + 2.5 &= r_2 \cdot t_2 \\ 8t + 2.5 &= 10t \\ 2.5 &= 2t \\ 1.25 &= t \end{aligned}$$

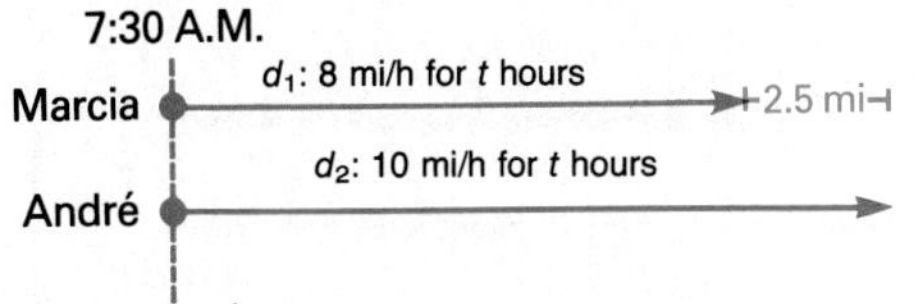

Marcia was 2.5 mi behind André at $1\frac{1}{4}$ h, or 1 h 15 min, after 7:30 A.M. So it was at 8:45 A.M. that they were 2.5 mi apart.

Classroom Exercises

Match each sentence on the left with a diagram on the right.

1. $d_1 = d_2$ b
2. $d_1 + d_2 = 100$ a
3. $d_1 + 100 = d_2$ c
4. $r_1t_1 + 100 = r_2t_2$ c
5. $r_1t_1 + r_2t_2 = 100$ a
6. $r_1t_1 = r_2t_2$ b

a.

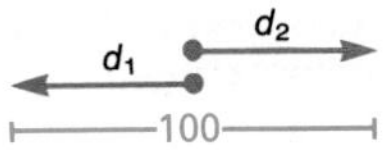

b.

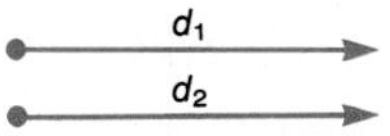

c.

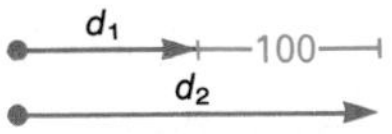

TEACHING SUGGESTIONS

Lesson Note

The drawings in Example 1–3 are used to help students write the appropriate equation, $d_1 + d_2 = k$, $d_1 = d_2$, or $d_1 + k = d_2$. Then the distance formula, $d = rt$, is used to rewrite the equation in the form $r_1t_1 + r_2t_2 = k$, $r_1t_1 = r_2t_2$, or $r_1t_1 + k = r_2t_2$.

Math Connections

Orbiting Spacecraft: The examples of this lesson can be regarded as simplified versions of the problem of launching a spacecraft under conditions that will allow it to meet a satellite already in orbit. In both situations, it is important to have information about the location and velocity of both moving bodies at several specified moments of their flight paths.

Additional Example 2

A truck traveling at 35 mi/h left a warehouse at 6:00 A.M. At 8:00 A.M., a motorcyclist left the warehouse and traveled the same route at 55 mi/h. At what time did the cyclist overtake the truck? 11:30 A.M.

Additional Example 3

Samuel averaged 30 mi/h and Teresa averaged 26 mi/h in a long-distance bicycle race. If they began the race at 11:30 A.M., at what time were they 6 mi apart? 1:00 P.M.

Analysis: Ask students how they would organize the examples of the lesson into a tabular format, such as the one used in Lesson 2.5. As a hint, have them refer to the *Problem Solving Strategy* on page 58.

Checkpoint

1. A bus left a terminal at 7:30 A.M. and averaged 75 km/h. Another bus, traveling in the opposite direction, left the terminal at 9:30 A.M. and averaged 60 km/h. At what time were the buses 420 km apart? 11:30 A.M.

2. Two trains left a terminal at the same time and in the same direction on parallel tracks. One train averaged 48 mi/h and the other averaged 32 mi/h. In how many hours were they 20 mi apart? $1\frac{1}{4}$ h, or 1 h 15 min

Closure

Have students (1) create a word problem with two objects moving from the same point in the same direction with different starting times and (2) sketch and label a diagram to represent the data.
Answers will vary.

Mr. Adams left his house at 7:00 A.M., and traveled for t hours. Mrs. Adams left the house at 9:00 A.M., and traveled for $t - 2$ hours along the same road until she overtook Mr. Adams. (Exercises 7–10)

7. If $t = 5$ h, for how many hours did Mr. Adams travel? 5 h

8. If $t = 5$ h, for how many hours did Mrs. Adams travel? 3 h

9. If $t = 8$ h, for how many hours did Mrs. Adams travel? 6 h

10. If $t = 6$ h, at what time did Mrs. Adams overtake Mr. Adams? 1:00 P.M.

Written Exercises

1. Two buses start toward each other at the same time from towns 190 mi apart. If one bus travels at 40 mi/h and the other travels at 55 mi/h, in how many hours will they meet? 2 h

2. From a buoy in a lake, Sharon swims toward the near shore at 4.2 m/s, while George swims in the opposite direction at 2.8 m/s. If they begin at the same time, in how many seconds will they be 105 m apart? 15 s

3. A ship leaves port traveling at 15 km/h. Two hours later, a motor boat leaves the same port traveling at 75 km/h in the same direction as the ship. How long will it take the motor boat to overtake the ship? 0.5 h

4. Two trains leave a terminal at the same time traveling in the same direction on parallel tracks. One train averages 57 mi/h, and the other averages 42 mi/h. In how many hours will they be 30 mi apart? 2 h

5. A bus left the city at 8:00 A.M., traveling at 70 km/h. Another bus, traveling in the opposite direction at 55 km/h, left at 11:00 A.M. At what time were the buses 460 km apart? 1:00 P.M.

6. A van left City A at 7:30 A.M., and traveled toward City B at 65 km/h. At 11:30 A.M., a bus left City B and headed toward the van at 75 km/h. At what time did the bus pass the van given that the cities are 470 km apart? 1:00 P.M.

7. An aircraft carrier and a destroyer leave the same port at 6:00 A.M., sailing in the same direction. The carrier averages 16 mi/h and the destroyer averages 40 mi/h. At what time will they be 54 mi apart? 8:15 A.M.

8. Amy drove her car the same distance in 6 h that Juan drove his car in $5\frac{1}{2}$ h. Find the speed (rate) of Amy's car, given that she drove 5 mi/h slower than Juan. 55 mi/h

9. It took a truck 5 h 30 min to cover the same distance that a car covered in 4 h 15 min. Find the speed of the car, given that the car was driven 10 mi/h faster than the truck. 44 mi/h

10. Cities A and B are 360 mi apart on an east-west highway. A bus left City A for City B at 10:00 A.M.. Two hours later, a car left City B for City A. The bus's speed was 8 mi/h slower than the car's and the car passed the bus at 3:00 P.M. Find the speed of the bus. 42 mi/h

11. A plane left an airfield at 9:00 P.M. cruising at 240 mi/h. At 11:30 P.M., a pursuit plane left the same airfield flying at 440 mi/h in the same direction. At what time did the pursuit plane overtake the slower plane? 2:30 A.M.

12. A cruiser and a submarine leave the same port at the same time and travel eastward. The cruiser averages 18 mi/h, and the submarine 42 mi/h. For what period of time will the cruiser be 12 mi or less behind the submarine? For the first 30 min

13. A plane left an airfield and flew eastward at 250 mi/h. Two hours later, a second plane left the same airfield and flew 450 mi/h in pursuit of the first plane. For which period after take-off was the second plane 100 mi or less behind the first plane? Between 2 and $2\frac{1}{2}$ h

14. An ocean freighter left port at 8:00 A.M. traveling 20 mi/h. At 10:00 A.M. a speedboat left the same port traveling the same route at 70 mi/h. As the speedboat caught and then passed the freighter, how long were they 5 miles or less from each other? 12 min

15. An automobile traveled x miles in a hours, then y miles in b hours, and finally z miles in c hours. Represent the average speed for the total trip in terms of the variables.
$$\frac{x + y + z}{a + b + c}$$

Mixed Review

Find the solution set of each sentence. *2.2, 2.3*

1. $3x - 2 < 16$ *or* $11 > 21 - x$

2. $5x - 4 < 7x$ *and* $6 - 2x > -4$

3. $|2x - 3| = 9$ $\{-3, 6\}$

4. $|2x - 3| > 9$ $\{x \mid x < -3 \text{ or } x > 6\}$

▰▰▰/ *Using the Calculator*

Use a calculator to solve each equation or inequality. Round each answer to the nearest hundredth.

Example   Solve. $|150x + 1.1| = 5.6$

$$150x + 1.1 = -5.6 \quad or \quad 150x + 1.1 = 5.6$$

$$x = \frac{-5.6 - 1.1}{150} \quad or \quad x = \frac{5.6 - 1.1}{150}$$

The roots are -0.04 and 0.03.

Exercises

1. $5.43x - 72.8 = 6.92$ 14.68

2. $\frac{2}{3.75}y + 4.03 < 10.54$ $\{y \mid y < 12.21\}$

3. $\frac{3.6}{4.5}t = \frac{48}{7.5}$ 8

4. $\left|\frac{2.4}{3.2}n - \frac{6.2}{4.5}\right| = \frac{58}{4.5}$ $-15.35, 19.02$

5. $\frac{x}{350} = \frac{9}{140}$ 22.5

6. $\frac{5.4}{1.5} = \frac{63}{y}$ 17.5

2.6 Problem Solving: Two Objects in Motion **65**

Guided Practice

Classroom Exercises 1–10

Independent Practice

A Ex. 1–7, **B** Ex. 8–11, **C** Ex. 12–15

Basic: WE 1–9 odd, UC

Average: WE 3–11 odd, UC

Above Average: WE 7–15 odd, UC

Additional Answers

Mixed Review

1. $\{x \mid x < 6 \text{ or } x > 10\}$

2. $\{x \mid -2 < x < 5\}$

Enrichment

Ask the students to solve this problem. Mr. Mherabi drove 100 mi to the city at an average rate of 60 mi/h. Because of heavy traffic, he made the 100-mi return trip at an average rate of 40 mi/h. What was his average rate for the entire round trip?

48 mi/h

1. $\{x \mid x > 2\}$

2

2. $\{a \mid a \le -10\}$

−10

3. $\{x \mid x < -2\}$

−2

4. $\{y \mid y \ge 6\frac{1}{2}\}$

$6\frac{1}{2}$ 7

5. $\{n \mid n \le 2\}$

2

6. $\{a \mid a < 32\}$

32

7. $\{x \mid x < 4\}$

4

8. $\{x \mid x \ge -2\}$

−2

9. Answers will vary.

Chapter 2 Review

Key Terms

absolute value (p. 51)
compound sentence (p. 46)
conjunction (p. 46)
disjunction (p. 46)
empty set (p. 48)
infinite set (p. 42)

linear inequality (p. 42)
order relations (p. 41)
Properties of Inequality (p. 41)
simple sentence (p. 46)
solution set (p. 42)

Key Ideas and Review Exercises

2.1 When solving a linear inequality, recall that multiplying or dividing each side by a *negative* number reverses the order.

Find the solution set. Graph it on a number line.

1. $3x - 2(4x - 7) < 8 - 2x$ $\{x \mid x > 2\}$ **2.** $\frac{5}{6}a - \frac{2}{3} \le \frac{1}{2}a - 4$ $\{a \mid a \le -10\}$

3. $-3x > 6$ $\{x \mid x < -2\}$ **4.** $21 - 6y \le 8 - 4y$ $\{y \mid y \ge 6\frac{1}{2}\}$

5. $6(2n - 4) \ge 5(3n - 4) - 10$ **6.** $\frac{3}{4}a < 24$ $\{a \mid a < 32\}$
$\qquad \{n \mid n \le 2\}$

2.2 To graph the solution set of the disjunction (*or* sentence) or conjunction (*and* sentence) of two simple sentences, first graph the two simple sentences on separate number lines. Then for the disjunction, graph the *union* of the parts. For the conjunction, graph the *intersection* of the parts.

Graph the solution set on a number line.

7. $6x + 3 < 15 \text{ or } 2x < 12 - x$ **8.** $3 - 4x \le 11 \text{ and } 19 \ge 7 - 2x$

9. Write in your own words how to draw the graph of the solution set of the conjunction of two linear inequalities.

2.3 To solve a sentence involving absolute value, use one of the following properties.

$\qquad$ If $|x| = k$ and $k \ge 0$, then $x = -k \text{ or } x = k$.

$\qquad$ If $|x| \ge k$ and $k \ge 0$, then $x \le -k \text{ or } x \ge k$.

$\qquad$ If $|x| \le k$ and $k > 0$, then $-k \le x \le k$.

10. Solve $|3y - 5| = 22$. $-5\frac{2}{3}, 9$

11. Solve $|x - 8| = 12$. $-4, 20$

Find the solution set. Graph it on a number line.

12. $|2y + 7| < 15$ {$y \mid -11 < y < 4$}   **13.** $|4t - 6| \geq 10$ {$t \mid t \leq -1 \text{ or } t \geq 4$}

2.4   Solving a word problem can sometimes involve writing and solving a compound sentence.

14. Find three consecutive odd integers such that the sum of the first two is greater than 28, and the sum of the last two is less than 42. 15, 17, 19 or 17, 19, 21

2.5   To solve a word problem involving the ages of several objects, represent the ages now and at another time using one variable.

15. An oak tree is 5 times as old as a pine tree, and an elm tree is 6 years younger than the oak. In 3 years, the age of the elm will be twice the age of the pine. How old is the oak tree now? 15 yr old

16. The ratio of the present ages of two houses is 2 to 5. Ten years ago, the sum of their ages was less than 43 years. Find the possibilities for the present age of the younger house. Between 10 yr old and 18 yr old

17. Building B was built 10 years before Building A and 20 years after Building C. Twenty years ago, the age of Building C was the same as the combined ages of Buildings A and B. What is the present age of each building? A: 40, B: 50, C: 70

2.6   To solve a problem involving objects A and B traveling distances d_1 and d_2 at rates r_1 and r_2 for times t_1 and t_2, respectively:

   (1) If $d_1 + d_2 = k$, use $r_1t_1 + r_2t_2 = k$.
   (2) If $d_1 = d_2$, use $r_1t_1 = r_2t_2$.
   (3) If $d_1 = d_2 + k$, use $r_1t_1 = r_2t_2 + k$.

18. A car left City A at 9:00 A.M. driving 52 mi/h toward City B. A truck left City B at 11:00 A.M. traveling 43 mi/h toward City A. At what time did the car and truck meet given that Cities A and B are 294 mi apart? 1:00 P.M.

19. A cargo plane left an airfield at 8:30 A.M. flying due west at 150 mi/h. At 11:30 A.M. a pursuit plane left the same field flying 350 mi/h to overtake the cargo plane. At what time did the pursuit plane overtake the cargo plane? 1:45 P.M.

20. In 2 h, Alice drove her car the same distance that Gene drove his car in 2 h 30 min. Find the speed of Gene's car given that he drove 10 mi/h slower than Alice drove. 40 mi/h

21. A bus left City A at 9:00 A.M. heading toward City B at 50 mi/h. Then, a car left City B at 10:00 A.M. traveling 40 mi/h toward City A. At what time did they meet if the cities are 230 miles apart? 12:00 noon

22. Sam began a marathon run at 6:30 A.M. and averaged 12 km/h, while Pam began the run at 7:00 A.M. and averaged 15 km/h. At what time did Pam pass Sam? 9:00 A.M.

12. {$y \mid -11 < y < 4$}

13. {$t \mid t \leq -1 \text{ or } t \geq 4$}

Chapter Test

1. $\{x \mid x \le 2\}$

2

2. $\{a \mid a > 6\}$

6

3. $\{y \mid y < -4 \text{ or } y > 1\}$

−4 1

4. $\{x \mid x < 5\}$

5

5. $\{x \mid 1 \le x \le 5\}$

1   5

6. $\{x \mid -2 < x < 3\}$

−2   3

7. $\{a \mid 2 < a < 8\}$

2   8

16. $\{a \mid -3 < a < 11 \text{ or } a < -6 \text{ or } a > 14\}$

−6 −3 11 14

For Exercises 1–7, find the solution set and graph it on a number line.

1. $5x - 3(5x - 8) \ge 16 - 6x$ {x | x ≤ 2}

2. $3a > 18 \text{ and } a - 2 > 1$ {a | a > 6}

3. $4y + 6 < -10 \text{ or } 2y + 5 > 9 - 2y$ {y | y < −4 or y > 1}

4. $5x - 3 < 7 \text{ or } 8 - x > 3$ {x | x < 5}

5. $-4 \le 3x - 7 \le 8$ {x | 1 ≤ x ≤ 5}

6. $6x - 1 < 17 \text{ and } 2 - x < 4x + 12$ {x | −2 < x < 3}

7. $|a - 5| < 3$ {a | 2 < a < 8}

8. Solve $|3n - 15| = 6$. 3, 7

9. Find three consecutive even integers such that the sum of the first two is greater than 27 and the sum of the last two is less than 39. 14, 16, 18, and 16, 18, 20

10. An oak tree is 6 times as old as a pine tree and an elm tree is 6 years younger than the oak tree. In 3 years, the age of the elm will be 3 times the age of the pine. How old is the oak tree now? 24 yr old

11. The ratio of the present ages of two museums is 3 to 5. Twelve years ago, the sum of their ages was less than 56 years. Find the possibilities for the present age of the younger museum. Between 12 yr and 30 yr

12. Cities A and B are located on an east-west highway at a distance of 244 mi from each other. A truck left City A at 8:00 A.M. driving 47 mi/h toward City B. A car left City B at 10:00 A.M. traveling 53 mi/h toward City A. At what time did the car pass the truck? 11:30 A.M.

13. What is the range of the initial vertical velocity that will be needed to launch an object to a height of 800 m ± 40 m at the end of 8 s? Use the formula $h = vt - 5t^2$. Between 135 m/s and 145 m/s, inclusive

14. A ship leaves port at 9:30 A.M. sailing 50 km/h due east. At 11:00 A.M., a helicopter leaves the same port flying 80 km/h to overtake the ship. At what time does the helicopter reach the ship? 1:30 P.M.

15. In 3 h, Andrea drove her car the same distance that Covey drove his car in 3 h 30 min. Find the speed of Covey's car given that he drove 8 mi/h slower than Andrea drove. 48 mi/h

16. Find the solution set of $|a - 4| < 7 \text{ or } |a - 4| > 10$. Graph it on a number line. {a | a < −6 or −3 < a < 11 or a > 14}

17. Solve $|3y + 2| + |8 - 3y| = 30$. −4, 6

18. An antique automobile is now n years old. In 5 years, it will be f years old. Represent its age p years ago in terms of f and p. $f - 5 - p$

In each item, you are to compare a quantity in Column 1 with a quantity in Column 2. Write the letter of the correct answer from these choices.

A—The quantity in Column 1 is greater than the quantity in Column 2.
B—The quantity in Column 2 is greater than the quantity in Column 1.
C—The quantity in Column 1 is equal to the quantity in Column 2.
D—The relationship cannot be determined from the given information.

NOTE: Information centered over both columns refers to one or both of the quantities to be compared.

Column 1	Column 2	
Sample Question 1		**Answer**
$x = 3$ *and* $y = 4$		The answer is B, since $3 \cdot 4 > 3 + 4$.
$x + y$	xy	
Sample Question 2		**Answer**
$5a$	$7a$	The answer is D, since $5a > 7a$ if $a = -1$, $7a > 5a$ if $a = 1$, and $5a = 7a$ if $a = 0$.

Column 1	Column 2		Column 1	Column 2	
1. $5 \cdot 629 \cdot 6$	$10 \cdot 629 \cdot 3$	C	**6.** $\|x\| = 3$ *and* $\|y\| = 4$		D
			$4x$	$3y$	
2. $a \cdot a = 16$ *and* $b \cdot b = 25$		D	**7.** $p < q$ *and* $q < r$		B
a	b		$p - r$	2	

3.

Items 8–9 use the following definition:

For all positive integers x,
$\text{(x)} = x$, if x is even, and
$\text{(x)} = x + 1$, if x is odd.

Column 1	Column 2	
4. $58 \cdot 2^3 \cdot 47$	$47 \cdot 3^2 \cdot 58$	B
5. $a < c$ *and* $b < c$		B
$a + b$	$2c$	
8. $\text{(61)} + \text{(65)}$	$\text{(62)} + \text{(64)}$	A
9. $\text{(y)} + \text{(y)} + \text{(1)}$	$\text{(2y)} + \text{(1)}$	D

Simplify, if possible. Choose the correct response among the choices A, B, C, D, and E.

1. $-5 - (-4) + (-8)$ D   *1.1*
   (A) -17 (B) -7 (C) 7
   (D) -9 (E) 9

2. $\dfrac{5 - 4 \cdot 3 + 9}{-9 \cdot 4 + (-6)^2}$ E   *1.2*
   (A) $-\frac{1}{36}$ (B) $-\frac{1}{6}$ (C) $\frac{2}{9}$
   (D) 0 (E) Undefined

3. $8 - 3(4x - 6) - (x - 2) + 5$ D   *1.3*
   (A) $19x - 23$ (B) $19x - 3$
   (C) $-13x + 5$
   (D) $-13x + 33$
   (E) None of these

Solve the equation. Choose the correct solution(s) among the choices A, B, C, and D.

4. $\frac{2}{3}x - \frac{5}{2} = \frac{3}{4}x - 4$ D   *1.4*
   (A) $\frac{48}{31}$ (B) $9\frac{1}{2}$
   (C) 12 (D) 18

5. $3(5x - 4) - 4(4x + 3) =$ *1.4*
   $8 - (5 - 2x)$ C
   (A) -1 (B) -4
   (C) -9 (D) 3

6. $\dfrac{6}{3x - 2} = \dfrac{4}{4x + 3}$ A   *1.4*
   (A) $-2\frac{1}{6}$ (B) $-\frac{5}{12}$
   (C) $-\frac{5}{6}$ (D) 15

7. $|2x + 5| = 11$ D   *2.3*
   (A) $3, -3$ (B) $-8, 8$
   (C) $8, -3$ (D) $3, -8$

For Exercises 8–14, match the sentence with a property or rule (A–G) listed below the sentences. All the variables represent real numbers.   *1.3*

8. $3t + (4t + 6) = (3t + 4t) + 6$ C

9. $-5 - 8 = -5 + (-8)$ E

10. $-5.4 + \sqrt{3}$ is a real number. D

11. $2.4(5n + 15) =$
    $2.4(5n) + 2.4(15)$ F

12. $(x + 3)4 = 4(x + 3)$ G

13. $-5y + 5y = 0$ B

14. $8c + c = 8c + 1c$ A
    (A) Multiplicative Identity
    (B) Additive Inverses
    (C) Associative of Addition
    (D) Closure for Addition
    (E) Rule of Subtraction
    (F) Distributive
    (G) Commutative of Multiplication

Match the sentence with its solution set. Choose the correct response among the choices A–D.

15. $4x - 3(2x - 6) < 24$ B   *2.1*
    (A) $\{x \mid x < -3\}$
    (B) $\{x \mid x > -3\}$
    (C) $\{x \mid x < -21\}$
    (D) $\{x \mid x > -21\}$

16. $3x < 18$ or $5 > x - 4$ B   *2.2*
    (A) $\{x \mid x < 6\}$
    (B) $\{x \mid x < 9\}$
    (C) $\{x \mid 6 < x < 9\}$
    (D) $\{x \mid x < 6 \text{ or } x > 9\}$

17. $6(x - 4) < x + 6$ *and*   *2.2*
$5 - 3x < x - 7$ C

(A) $\{x \mid x < 3\}$
(B) $\{x \mid x < 6\}$
(C) $\{x \mid 3 < x < 6\}$
(D) $\{x \mid 6 < x < 3\}$

18. $|x - 5| < 6$ C   *2.3*

(A) $\{x \mid x > 11\}$
(B) $\{x \mid x > -1\}$
(C) $\{x \mid -1 < x < 11\}$
(D) $\{x \mid x < -1 \text{ or } x > 11\}$

Simplify, if possible.   *1.2*

19. $\dfrac{17 + 3 \cdot 5}{6 + 2 \cdot 4 + 2}$ 2

20. $\dfrac{3 \cdot 8 - 7 \cdot 8}{(-6)^2 + 3(-12)}$ Undef

21. $-8 - 7(-2)$ 6

22. $6(9 - 11)^3$ -48

23. $-8 + 6 \div 2 - 4(-3) - (-20)$ 27

24. $9 - 2[3(6 - 10) + 8]$ 17

Simplify.   *1.3*

25. $8c - 4 - d + 12 + 9d - c$
$7c + 8d + 8$

26. $12x - 3(8x - 5) + 6$ $-12x + 21$

27. $17 - (6 - 11n) - 8n$
$11 + 3n$

Solve. (Exercises 28–31)   *1.4*

28. $5(3x + 4) - (x + 12) =$
$15 - 3(8 - 4x)$ -8.5

29. $\dfrac{5}{6}m - \dfrac{7}{2} = \dfrac{5}{4}m + 4$ -18

30. $3.2y - 5.6 = 7.9 - 5.8y$ 1.5

31. $\dfrac{8x - 4}{5} = \dfrac{7x - 3}{4}$ $-\dfrac{1}{3}$

32. Solve $6ax + 3c = 5c - 3ax$ *1.4*
for x. $\dfrac{2c}{9a}$

33. Justify each step in the proof.   *1.8*

Theorem: $a(bc + d) =$
$\qquad\qquad ad + b(ca)$
Proof:

1. $a(bc + d) = a(bc) + ad$ Distr Prop
2. $\qquad\quad = (bc)a + ad$ Comm Prop Mult
3. $\qquad\quad = b(ca) + ad$ Assoc Prop Mult
4. $\qquad\quad = ad + b(ca)$ Comm Prop Add
5. $a(bc + d) = ad + b(ca)$ Trans Prop Eq

34. Ann borrowed some money   *1.5*
at $9\frac{1}{2}\%$ per year in simple
interest. At the end of 5
years 6 months, she owed a
total amount of $6,090,
including principal and
interest. Use the formula
$A = p + prt$ to find how
much money Ann borrowed. $4,000

35. The greatest of three num-   *1.6*
bers is -3 times the smallest
number, and the remaining
(middle) number is 15 more
than the smallest. Find the
three numbers given that the
greatest is 3 times the sum of
the other two numbers. -5, 10, 15

36. A rectangle is 5 in. wider   *1.7*
than a square and 4 times as
long. The difference between
the two perimeters is 52 in.
Find the area of the square. 49 in^2

37. The age of an antique auto-   *2.5*
mobile is 10 years less than
the age of an antique table.
In 15 years the table's age
will be 45 years less than
twice the auto's age. How
old is the auto now? 40 yr old

38. Mario drove his racing car   *2.6*
the same distance in 6 h that
Bobby drove his car in 5 h 45
min. Find the speed at which
Mario drove, given that he
drove 10 mi/h slower than
Bobby. 230 mi/h

3 RELATIONS, FUNCTIONS: GRAPHING LINEAR FUNCTIONS

OVERVIEW

In this chapter, students will concentrate on linear functions, with emphasis on such topics as the equation of a straight line, slope of a line, and parallel and perpendicular lines. They will examine *Direct Variation* as a special case of a linear function, and graph linear inequalities.

OBJECTIVES

- To determine the domain and range of relations and functions, whether a relation is a function, and $f(x)$ given x
- To find the slope and slant of a line
- To write an equation of a line
- To graph a linear function or linear inequality in two variables
- To solve a problem involving direct variation

PROBLEM SOLVING

In this chapter, attention to problem solving continues in the regular lessons. It also appears in many of the features, for example in the application on the Doppler Effect and in a Problem Solving Strategy, *Insufficient Data and Contradictions*. The latter deals with the matter of having inadequate or conflicting data in a problem situation.

The *Mixed Problem Solving* feature that appears on page 113, and also in the remaining odd-numbered chapters throughout the rest of the book, gives students the opportunity to practice problem-solving skills.

READING AND WRITING MATH

The *Problem Solving Strategy* on page 112 can be used to stress the importance of reading a problem not only to understand its meaning, but also to determine whether there is enough information given to solve the problem. Students should also be encouraged to reflect on the solution to a problem and ask themselves whether it makes sense in the given situation.

The Focus on Reading on page 80 asks students to translate English phrases into algebraic notation or to identify definitions and algebraic equations.

TECHNOLOGY

Calculator: A graphing calculator can be used by students to verify the accuracy of their graphs.

SPECIAL FEATURES

Mixed Reviews pp. 77, 81, 87, 90, 99, 102, 109, 111
Brainteaser pp. 77, 102
Focus on Reading p. 80
Application: The Doppler Effect p. 82
Statistics: Box-and-Whisker Plots p. 83
Midchapter Review p. 94
Application: Familiar Linear Relationships p. 95
Application: Möbius Strip p. 99
Extension: Logical Reasoning pp. 103–104
Problem Solving Strategies: Insufficient Data and Contradictions p. 112
Mixed Problem Solving p. 113
Key Terms p. 114
Key Ideas and Review Exercises pp. 114–115
Chapter 3 Test p. 116
College Prep Test p. 117

PLANNING GUIDE

Lesson	Basic	Average	Above Average	Resources
3.1 pp. 76–77	CE all WE 1–15 Brainteaser	CE all WE 3–17 Brainteaser	CE all WE 7–22 Brainteaser	Reteaching 15 Practice 15
3.2 pp. 80–83	FR all, CE all WE 1–13 odd Application Statistics	FR all, CE all WE 3–15 odd Application Statistics	FR all, CE all WE 3–19 odd Application Statistics	Reteaching 16 Practice 16
3.3 p. 87	CE all WE 1–15 odd	CE all WE 3–19 odd	CE all WE 5–21 odd	Reteaching 17 Practice 17
3.4 p. 90	CE all WE 1–14	CE all WE 5–18	CE all WE 8–21	Reteaching 18 Practice 18
3.5 pp. 93–95	CE all WE 1–27 odd Midchapter Review Application	CE all WE 3–29 odd Midchapter Review Application	CE all WE 5–23 odd Midchapter Review Application	Reteaching 19 Practice 19
3.6 pp. 98–99	CE all, WE 1–9 Application	CE all, WE 3–11 Application	CE all, WE 3–13 Application	Reteaching 20 Practice 20
3.7 pp. 101–104	CE all WE 1–19 odd Extension	CE all WE 3–21 odd Extension	CE all WE 5–23 odd Brainteaser, Extension	Reteaching 21 Practice 21
3.8 pp. 107–109	CE all WE 1–31 odd	CE all WE 7–37 odd	CE all WE 9–41 odd	Reteaching 22 Practice 22
3.9 pp. 111–113	CE all WE 1–13 odd Problem Solving Mixed Problem Solving	CE all WE 3–15 odd Problem Solving Mixed Problem Solving	CE all WE 5–17 odd Problem Solving Mixed Problem Solving	Reteaching 23 Practice 23
Chapter 3 Review pp. 114–115	all	all	all	
Chapter 3 Test p. 116	all	all	all	
College Prep Test p. 117	all	all	all	

CE = Classroom Exercises   WE = Written Exercises   FR = Focus on Reading
NOTE: For each level, all students should be assigned all Mixed Review exercises.

■■■ INVESTIGATION

Project: In this activity, the students relate the form of a linear equation to an equation which should already be familiar to them.

The formula for conversion from Fahrenheit to Celsius measure is

$$F = \frac{9}{5}C + 32.$$

The students should recognize that this is the slope-intercept form of a linear equation, where the variables x and y have been replaced by C and F.

To graph this equation, the students should draw a set of axes and label them as C and F. The vertical F-axis should extend to at least 250 degrees in the positive direction. Locate the F-intercept and draw a straight line with a slope of $\frac{9}{5}$ through the intercept.

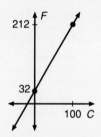

At $C = 0$, $F = 32$ (the F-intercept), which agrees with the fact that 0° C corresponds to 32° F, the freezing point of water. Also, 100° C corresponds to 212° F on the graph, the boiling point of water on the two scales.

Point out to the students that these known values of the freezing and boiling points of water can be used to plot two points on the graph. The straight line through the two points is the complete graph.

Using their graphs, the students should be able to give approximate answers to the following questions. If it is 22° C outside, what is the temperature in Fahrenheit degrees? 38° C?   72° F, 100°F

By rewriting the equation as $\frac{9}{5}C - F = -32$, the students should recognize the standard form of a linear equation, $ax + by = c$. Ask the students to explain the relation of the conversion formula to the standard form.

x and y are replaced by F and C, and $a = \frac{9}{5}$, $b = -1$, and $c = 32$.

3 RELATIONS, FUNCTIONS: GRAPHING LINEAR FUNCTIONS

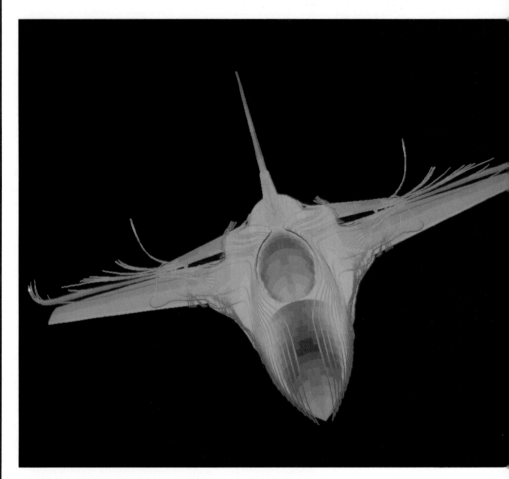

Research and development engineers use computer graphics to create new designs without building expensive models. The image above shows an aircraft being tested for wind resistance. The red lines simulate the movement of air around the hull and wings.

More About the Image

Computers are used by aviation engineers to simulate conditions that affect the design of their product. Using certain physical data of an airplane, such as the shape and weight of different components, a computer creates a three-dimensional model that tests how well the craft will fly. Mathematically speaking, the model consists of a large, complex set of equations that express relationships between various parts of the airplane. Engineers can test alternative designs for the plane by changing the values for certain components and observing the way the changes affect the plane's flight.

3.1 Relations and Functions

Objectives

To graph relations and determine their domains and ranges
To determine relations from their graphs
To determine whether a relation is a function

Sentences with one variable, such as $3x - 1 < x$, are graphed on a number line. Sentences with two variables, such as $x + y = 1$, are graphed on a *coordinate plane*.

The figure at the right shows a vertical number line (**y-axis**) perpendicular to a horizontal number line (**x-axis**). The two axes intersect in a point called the **origin**, where they form a **coordinate plane** separated into four **quadrants**. The quadrants are numbered counterclockwise beginning with Quadrant I on the upper right.

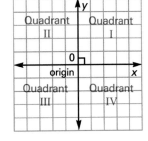

Each point in a coordinate plane corresponds to exactly one **ordered pair** of real numbers, (x, y). The point is the graph of the ordered pair. The x-coordinate is called the **abscissa**. The y-coordinate is called the **ordinate**.

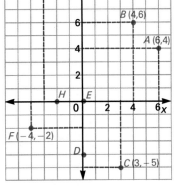

The coordinates of point A in the graph at the right are written as the ordered pair $(6, 4)$. Order is important because $A(6, 4)$ and $B(4, 6)$ are not the same point.

To graph, or *plot*, the point $C(3, -5)$, proceed as follows: Start at the origin. Move 3 units to the *right* and 5 units *down*. Place a dot in the coordinate plane. Label the point. C is the graph of the ordered pair $(3, -5)$.

The sign of the coordinate indicates the direction in which to move. The absolute value of the coordinate indicates the distance. The x-coordinate (abscissa) indicates the direction and distance from the origin along the x-axis. The y-coordinate (ordinate) indicates the direction and distance from the origin along the y-axis.

Teaching Resources

Quick Quizzes 15
Reteaching and Practice Worksheets 15

▰▰▰ GETTING STARTED

Prerequisite Quiz

Use set builder notation to describe each set of points graphed on the number line.

1.
   -2 0 3
   $\{x \mid -2 < x \le 3\}$

2.
   4 8
   $\{x \mid 4 \le x < 8\}$

3.
   -3 -1
   $\{x \mid -3 < x < -1\}$

4.
   1 5
   $\{x \mid 1 \le x \le 5\}$

Motivator

Ask students to tell how many coordinates are needed to locate a point on a number line. one Then ask them how many coordinates are needed to locate points on a plane. two Ask them how they would describe where a student is seated in the classroom. row number and seat number

Highlighting the Standards

Standard 5b: The coordinate system, as described in the opening paragraphs, is a powerful concept. Students will constantly return to graphical representations of algebraic ideas.

■ TEACHING SUGGESTIONS

Lesson Note

Have students describe how to locate the following points on the graph on page 73.

- A (6,4) right 6, up 4
- B (4,6) right 4, up 6
- F (−4,−2) left 4, down 2
- G (−3,8) left 3, up 8

Point out that points H, D, and E are part of one or both axes and, therefore, not in any quadrant.

Drawings such as the following are useful in helping students to determine the domain and the range of a relation from a graph.

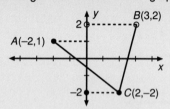

For the range, draw dashed lines from A, B, and C perpendicular to the y-axis. Shade the distance between the *extremes* on the y-axis from −2 to 2. The range is $\{y \mid -2 \le y < 2\}$.

For the domain, draw dashed lines from A, B, and C perpendicular to the x-axis.

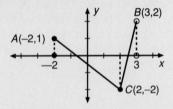

Shade the distance between the *extremes* on the x-axis from −2 to 3. The domain is $\{x \mid -2 \le x < 3\}$.

The graph at the right consists of four points. It can be described by a set of ordered pairs called a *relation*:

$$R = \{(-3, 1), (-3, 5), (0, -5), (4, -2)\}$$

This relation can also be shown by a **mapping**.

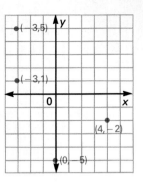

The set of x-values is called the *domain* of the relation.
Domain: $\{-3, 0, 4\}$

The set of y-values is called the *range* of the relation.
Range: $\{-5, -2, 1, 5\}$

Notice that −3 is listed only once in the domain.

Definitions

A **relation** is a set of ordered pairs.

The **domain** of a relation is the set of all first coordinates of the ordered pairs.

The **range** of a relation is the set of all second coordinates of the ordered pairs.

EXAMPLE 1 Write the relation P whose graph is shown at the right. List its domain and range.

Solution

$$P = \{(-2, 1), (-1, -2), (1, 1), (2, 4), (3, 2)\}$$

Domain of P: $\{-2, -1, 1, 2, 3\}$

Range of P: $\{-2, 1, 2, 4\}$

(In the range, list the 1 only once.)

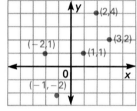

In Example 1, the domain and the range are finite sets because the relation consists of only a finite number of ordered pairs. For many relations, however, this is not the case. As illustrated in Example 2, some relations consist of an infinite number of ordered pairs.

74 Chapter 3 Relations, Functions: Graphing Linear Functions

Additional Example 1

Write the relation P whose graph is shown. List the domain and the range.

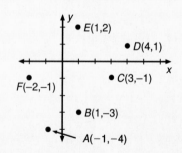

Relation $P = \{(-1,-4), (1,-3), (3,-1), (4,1), (1,2), (-2,-1)\}$

Domain of P: $\{-2, -1, 1, 3, 4\}$

Range of P: $\{-4, -3, -1, 1, 2\}$

EXAMPLE 2 Determine the domain and range of the relation Q, whose graph is shown at the right.

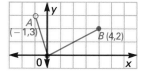

Solution The open circle at $A(-1, 3)$ indicates that this point is not a point of the graph. Therefore, -1 is not included in the domain of the relation, and 3 is not in the range.

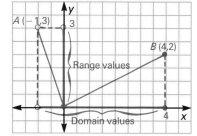

Domain: $\{x \mid -1 < x \le 4\}$

Range: $\{y \mid 0 \le y < 3\}$

Look back at the relation P in Example 1. Notice that no two first coordinates are the same. This is not true for all relations, but when it is true, the relation is called a *function*.

Definition

A **function** is a relation such that no two ordered pairs have the same first coordinate.

EXAMPLE 3 Determine whether each of the following relations is a function.

a. $S = \{(-5, 3), (4, 3), (11, -1)\}$ b.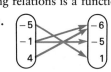

Solutions a. No two ordered pairs have the same first coordinate. The relation is a function.

b. The ordered pairs $(-1, -6)$ and $(-1, -5)$ have the same first coordinate. The relation is not a function.

You can also determine whether a relation is a function by applying the following **vertical-line test** to its graph: *If a vertical line can be drawn that crosses the graph more than once, then the relation is not a function. Otherwise, the relation is a function.*

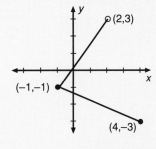
75

Checkpoint

Graph each relation. Determine its domain and its range. Is it a function?

1. {(2,2), (1,0), (0,1), (−1,1)}
D: {−1, 0, 1, 2}
R: {0, 1, 2}; function

2. {(1,2), (0,2), (1,0), (−1,0)}
D: {−1, 0, 1}
R: {0, 2}; not a function

Closure

Ask students how the signs of an ordered pair of numbers indicate the quadrant in which the corresponding point is located. Gives direction   What is a relation?   See page 74.   What is the domain and the range of a relation?   See page 74.   How do you use a graph to determine if a relation is a function?   Use the vertical-line test.

Classroom Exercises

State the domain and range of each relation.

1. {(2, −1), (4, 2)}   **2.** {(−3, −3), (8, −6), (1, −6)}   **3.** {(1, 1), (−1, 3), (1, −3)}

4–6. Graph each relation in Exercises 1–3.

7–9. Tell which relations in Exercises 1–3 are functions.   **7.** Yes, **8.** Yes, **9.** No

Written Exercises

Graph each relation and determine its domain and range. Is it a function?

1. {(2, 1), (1, 2)}   **2.** {(0, 0), (−1, 4)}
3. {(−1, 6), (−1, 2)}   **4.** {(3, 1), (−1, 1)}
5. {(4, 2), (3, −2), (−2, 4), (−3, −3), (0, 3)}
6. {(9, 8), (−6, 9), (−9, 8), (6, −1)}
7. {(−1, −1), (7, −1), (8, −1), (−5, −1)}
8. {(6, 2), (6, −1), (6, −5), (6, 0)}

Write the relations whose graphs are given. List the domain and range of each. Determine whether each relation is a function.

9. 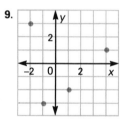   **10.** 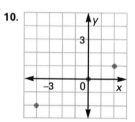   **11.**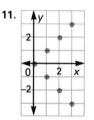

Determine whether each mapping represents a function.

12. 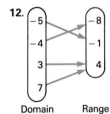   **13.** 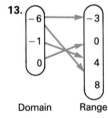   **14.**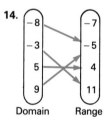

Domain   Range   Domain   Range   Domain   Range

Determine the domain and range of each relation. Use the vertical-line test to determine whether the relation is a function.

15.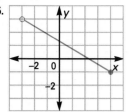

16.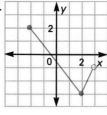

17.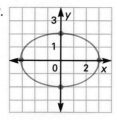

18.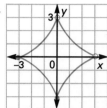

19.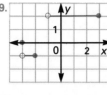

20.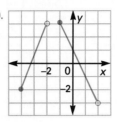

Give the domain and range of each relation. Is the relation a function?

21. $\{\ldots(-7,-5), (-4,-3), (-1,-1), (2, 1), (5, 3), \ldots\}$

22. $\{(1, 5), (4, 10), (9, 17), (16, 26), (25, 37), \ldots\}$

Mixed Review

Find the solution set. Graph it on a number line. *2.1, 2.3*

1. $2x - 15 < 8x + 3$ **2.** $\dfrac{3x - 8}{5} < \dfrac{x + 4}{3}$ **3.** $|4x - 2| = 8$ **4.** $|2x - 3| < 11$

Brainteaser

Three partners in an appliance store appear before a bankruptcy judge. Each partner makes four statements. Says Mr. Houlihan: "Washington owes me $5,000. Steinberg owes me $10,000. Everything Steinberg says is true. Nothing Washington says is true." Mr. Washington replies: "Everything Houlihan says is wrong. I don't owe him anything. Steinberg owes me $5,000. I should know because I was the bookkeeper." Finally, Mr. Steinberg says: "Washington was not the bookkeeper. And only two of his statements are true. Furthermore, I don't owe anyone anything. And I always tell the truth." Only one partner is telling the complete truth. Which one is it? Washington

FOLLOW UP

Guided Practice

Classroom Exercises 1–9

Independent Practice

A Ex. 1–14, **B** Ex. 15–17, **C** Ex. 18–22

Basic: WE 1–15, Brainteaser
Average: WE 3–17, Brainteaser
Above Average: WE 7–22, Brainteaser

Additional Answers

Classroom Exercises

1. $D = \{2, 4\}$, $R = \{-1, 2\}$
2. $D = \{-3, 1, 8\}$, $R = \{-6, -3\}$
3. $D = \{-1, 1\}$, $R = \{-3, 1, 3\}$

4.

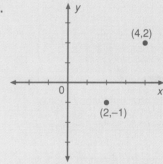

See pages 803–804 for the answers to Classroom Ex. 5, 6, Written Ex. 1–22, and Mixed Review Ex. 1–4.

Enrichment

Write this function on the chalkboard.

$\{(0,-4), (1,-2), (5,6), (6,10), (8,12), (10,16)\}$

Tell the students that five of the ordered pairs fit a pattern, but that one of the ordered pairs does not fit the pattern. Challenge the students to describe the pattern and to identify the ordered pair that does not fit the pattern.

The pattern is that the y-coordinate is 4 less than twice the x-coordinate. The ordered pair (6,10) does not fit the pattern.

■■■■GETTING STARTED

Prerequisite Quiz

Solve each equation for y.

1. $3x - 4y = 8$
   $y = \frac{3}{4}x - 2$
2. $x^2 - y = 7$
   $y = x^2 - 7$
3. $2x - y = 4$
   $y = 2x - 4$

Find the value of y for the given value of x.

4. $y = 3x + 5; x = -4$ -7
5. $y = -x^2 - 2x + 3; x = -2$ 3

Motivator

Ask students how they would solve $4x - 2y = 8$ for y. First subtract $4x$ from each side; then divide each side by -2. Have them solve for y. $y = 2x - 4$ Have them use the result to find y for $x = 3$. Plot the resulting ordered pair (x,y). Now have them find y for each of the following values of x: 4, -2, 0, 5. Plot the resulting ordered pairs. What appears to be true about the graph?   The graph is a line.

Objectives

To find functions described by equations
To find values of functions
To find the ranges of functions for given domains

A function composed of an infinite number of points cannot be described by listing all the ordered pairs corresponding to those points. This type of function is usually described by writing an *equation in two variables*.

A function that has a line as its graph is called a **linear function**. Its equation is called a *linear equation in two variables*. The graph at the right describes a linear function. Its equation, $2x - y = 2$, is a linear equation in two variables. Each solution is an ordered pair of real numbers.

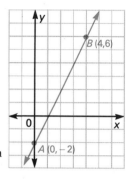

Definition

The *standard form* of a **linear equation in two variables** is $ax + by = c$, where a and b are *not both* equal to 0.

EXAMPLE 1   Graph $x - 2y = -8$.

Solution

1. Solve for y in terms of x:
   $y = \frac{1}{2}x + 4$.
2. Find two ordered pairs by choosing two values for x; find the corresponding values for y.
3. Plot the points; draw the graph.

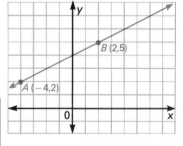

x	$\frac{1}{2}x + 4$	y	ordered pair
2	$\frac{1}{2} \cdot 2 + 4$	5	$(2, 5)$
-4	$\frac{1}{2} \cdot (-4) + 4$	2	$(-4, 2)$

78   Chapter 3 Relations, Functions: Graphing Linear Functions

Highlighting the Standards

Standard 6c: Example 2 and the material that follows show that a function is best understood when seen as a mapping, a set of ordered pairs, and a set of corresponding points.

Additional Example 1

Graph $3x - 2y = 6$.

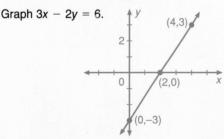

The y-values in Example 1 can also be found using a scientific calculator. For example, if $x = -4$, the steps are as shown below.

$$0.5 \; \boxed{\times} \; 4 \; \boxed{+/-} \; \boxed{+} \; 4 \; \boxed{=} \; 2$$

In Example 2, the equation that is graphed is *not* linear.

EXAMPLE 2 Graph the equation $x^2 - y = 4$. Use the following values for x to make a table of ordered pairs: $\{-3, -2, -1, 0, 1, 2, 3\}$.

Solution Since the equation is not linear, more than 2 points need to be plotted.

  1. Solve the equation for y: $y = x^2 - 4$.

  2. Make a table of ordered pairs using the given values for x.

  3. Plot the points and draw a smooth curve through them.

x	$x^2 - 4$	y	ordered pair
-3	$(-3)^2 - 4$	5	$(-3, 5)$
-2	$(-2)^2 - 4$	0	$(-2, 0)$
-1	$(-1)^2 - 4$	-3	$(-1, -3)$
0	$(0)^2 - 4$	-4	$(0, -4)$
1	$(1)^2 - 4$	-3	$(1, -3)$
2	$(2)^2 - 4$	0	$(2, 0)$
3	$(3)^2 - 4$	5	$(3, 5)$

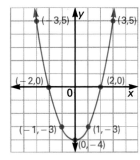

Recall that a function may be described by a mapping, as shown at the right.

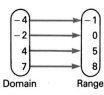

Domain   Range

Each member of the range is a *value* of the function. The statement, "The value of f at -4 is -1," can be written as $f: -4 \to -1$ or, more commonly, as $f(-4) = -1$. Similarly, $f(-2) = 0$ is read as "The value of f at -2 is 0."

In general, the notation $f(x)$ means "the value of f at x," and it is read either as "f of x" or as "f at x." Thus, $f(-2)$ can be read either as "f at -2," or as "f of -2."

Definition For any ordered pair (x, y) of a function f, the **value of the function** at x is equal to y—that is, $f(x) = y$.

■■■ **TEACHING SUGGESTIONS**

Lesson Note

In graphing an equation such as the one shown in Example 1, you may wish to have students graph a third point as a check that it is on the same line as the other two. Following Example 2, the text introduces $f(x)$ notation for functions.

Math Connections

Language: The word *function* is used in everyday speech in a sense that often suggests its mathematical meaning. For example, compare the mathematical statement "y is a function of x" to a nonmathematical statement such as "Success is a function of hard work." In each case, you are saying that something *is dependent upon* something else.

Critical Thinking Questions

Analysis: Ask students to extend Example 4 to the evaluation of the fraction

$$\frac{f(a + h) - f(a)}{h}.$$

The value will be 6. Then ask whether the fraction would have the value 6 if $h = 0$. No, since the fraction would be undefined. Have them consider a case in which h were very close to 0, but not equal to 0. yes

Common Error Analysis

Error: When evaluating $f(x)$ in equations like $f(x) = -x^2 + 2$, students may apply the negative of x^2 to x. For example, at $x = -3$, they get 9 instead of -9 for $-x^2$.

If necessary, have the students rewrite $-x^2$ as $-1(x^2) = -1(-3)^2 = -1(9) = -9$.

Additional Example 2

Graph the equation $x^2 + y = 2$. Use the following values for x: $\{-3, -2, -1, 0, 1, 2, 3\}$

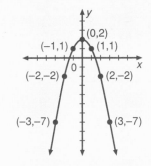

Checkpoint

Let $f(x) = -x^2 - 2x + 5$ and $g(x) = -4x - 7$. Find each of the following.

1. $f(4)$ -19
2. $g(-2)$ 1
3. $f(-3) - g(-3)$ -3
4. $g(a + h)$ $-4a - 4h - 7$

Closure

Ask students what is meant by $f(x)$?
Value of f at x.   Ask them how to find $f(2)$
for $f(x) = x^2 - 4x + 2$?   Substitute 2 for x
and evaluate.   Have them give the
standard form of a linear equation in two
variables?
$ax + by = c$

▬▬▬FOLLOW UP

Guided Practice

Classroom Exercises 1–8

Independent Practice

A Ex. 1–11, **B** Ex. 12–17, **C** Ex. 18, 19

Basic: FR, WE 1–13 odd, Application,
Statistics

Average: FR, WE 3–15 odd, Application,
Statistics

Above Average: FR, WE 3–19 odd,
Application, Statistics

Additional Answers

Classroom Exercises

1. $\{-11,-8,-5,-2,1,4,7\}$
2. $\{4,3,2,1,0,-1,-2\}$
3. $\{10,8,6,4,2,0,-2\}$
4. $\{-12,-7,-2,3,8,13,18\}$

EXAMPLE 3   Given that $f(x) = -4x + 8$, find each of the following values.

  a. $f(-2)$ b. $f\left(\frac{3}{8}\right)$ c. $f(3a + 1)$

Solutions

a. $f(x) = -4x + 8$
   $f(-2) = -4(-2) + 8$
        $= 16$

b. $f(x) = -4x + 8$
   $f\left(\frac{3}{8}\right) = -4\left(\frac{3}{8}\right) + 8$
        $= -1\frac{1}{2} + 8$
        $= 6\frac{1}{2}$

c. $f(x) = -4x + 8$
   $f(3a + 1) = -4(3a + 1) + 8$
        $= -12a - 4 + 8$
        $= -12a + 4$

EXAMPLE 4   Given that $f(x) = 6x - 2$, find $f(a + h) - f(a)$.

Plan   Substitute $a + h$ for x in $f(x)$. Then substitute a for x in $f(x)$. Subtract.

Solution

$f(x) = 6x - 2$ $f(x) = 6x - 2$
$f(a + h) = 6(a + h) - 2$ $f(a) = 6a - 2$

$f(a + h) - f(a) = [6(a + h) - 2] - [6a - 2]$
$= 6a + 6h - 2 - 6a + 2 = 6h$

EXAMPLE 5   A function g is defined for all numbers in its domain by $g(x) = -x^2 - 4$. The domain of g is $\{-3, 0, 1.2\}$. Find the range of g.

Solution

$g(x) = -x^2 - 4$ $g(x) = -x^2 - 4$ $g(x) = -x^2 - 4$
$g(-3) = -(-3)^2 - 4$ $g(0) = -(0)^2 - 4$ $g(1.2) = -(1.2)^2 - 4$
$= -9 - 4$ $= 0 - 4$ $= -1.44 - 4$
$= -13$ $= -4$ $= -5.44$

The range is $\{-13, -5.44, -4\}$.

▬▬ *Focus on Reading*

Match each item at the left with its corresponding item at the right.

1. set of first coordinates of a relation e   a. $f(x)$
2. the value of f at x a   b. range
3. set of second coordinates of a relation b   c. $3x - 2y = 8$
4. set of ordered pairs of real numbers d   d. relation
5. linear equation c   e. domain

Additional Example 3

If $f(x) = -5x + 1$, find each of the following.

a. $f(-3)$ 16

b. $f\left(-\frac{10}{3}\right)$ $17\frac{2}{3}$

c. $f(3a - 1)$ $-15a + 6$

Additional Example 4

If $f(x) = -4x - 5$, find $f(a + 3) - f(a)$.
-12

Additional Example 5

A function is defined for all numbers in its domain by $h(x) = -x^2 - 5x + 4$. The domain of h is $\{-4, -3, -1, 0, 4\}$. Find the range of h.
$R: \{-32, 4, 8, 10\}$

Classroom Exercises

Find the values of y for $x = -3, -2, -1, 0, 1, 2, 3$.

1. $y = 3x - 2$ **2.** $y = -x + 1$ **3.** $y = -2x + 4$ **4.** $y = 5x + 3$

5–8. Graph each equation in Exercises 1–4.

Written Exercises

Graph each equation.

1. $3x - y = 6$ **2.** $6x - y = 3$ **3.** $f(x) = -x - 1$

Make a table of ordered pairs using the following values
for x: $-3, -2, -1, 0, 1, 2, 3$. Graph each equation.

4. $x^2 - y = 8$ **5.** $x^2 - y = 3$ **6.** $f(x) = 3x^2 - 10$

Let $f(x) = -6x + 8$ and $g(x) = -x^2 + 6$. Find each of the following values.

7. $f(-4)$ 32   **8.** $f(0.2)$ 6.8   **9.** $g(2a)$ $-4a^2 + 6$

Find the range of each function for the given domain D.

10. $f(x) = -3x + 5; D = \{-4, 6, 8\}$ **11.** $g(x) = \frac{2}{3}x - 4; D = \{-6, 2, 3\}$

Let $f(x) = 4x - 3$ and $g(x) = x^2 - 2$. Find each of the following values.

12. $f(4) - f(8)$ -16 **13.** $g(2) - g(-3)$ -5 **14.** $f(4) - g(5)$ -10

15. $f(3a - 4)$ $12a - 19$ **16.** $g(a) - g(3)$ $a^2 - 9$ **17.** $f(3a) - g(2a)$

18. If $h(a + b) = h(a) + h(b)$ for all real numbers a and b, prove that
$h(2a) = 2h(a)$. $h(2a) = h(a + a) = h(a) + h(a) = 2h(a)$

19. Let W be a function such that for all positive real numbers x, y,
and z, $W(xyz) = W(x) + W(y) + W(z)$. Prove that $W(t^3) = 3W(t)$
for all positive real numbers t.

Mixed Review

Identify the property of operations with real numbers.   *1.3*

1. $(x - 7)4 = 4(x - 7)$ **2.** $-8y + 8y = 0$ **3.** $a(b + c) = ab + ac$

4. Simplify $xy^2 + xy - 7xy^2 + 3xy$. *1.3*   **5.** Solve $\frac{5x - 2}{3} = \frac{3x - 5}{2}$.   *1.4*

5.

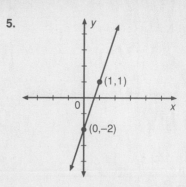

6.

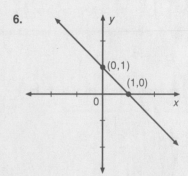

7.

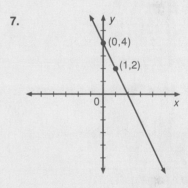

See pages 82–83 for the answers to
Classroom Ex. 8 and Written Ex. 1–5, and
page 804 for Written Ex. 6, 10, 11, 17, 19,
and Mixed Review Ex. 1–5.

Enrichment

Write the following equations for functions
on the chalkboard.

$y = \frac{1}{x}$ $y = \frac{1}{x + 1}$ $y = \frac{1}{x - 1}$

$y = x^2$ $y = \sqrt{x}$ $y = \frac{x}{x - 1}$

$y = \sqrt{x + 10}$ $y = \sqrt{10 - x}$

Challenge the students to match each of the
following restrictions on domain and range
with one of the equations.

1. The domain includes all real numbers,
but the range excludes negative
numbers. $y = x^2$

2. Neither the domain nor the range
includes negative numbers. $y = \sqrt{x}$

3. Both the domain and range include all
real numbers except 0. $y = \frac{1}{x}$

4. Both the domain and range include all
real numbers except 1. $y = \frac{x}{x - 1}$

5. The domain excludes all real numbers
greater than 10, and the range excludes
all negative numbers. $y = \sqrt{10 - x}$

8.

Written Exercises

1.

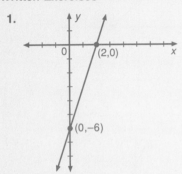

2.

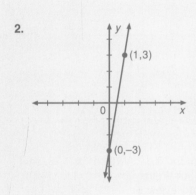

Application: *The Doppler Effect*

The next time a train passes by, stop for a moment and listen. Doesn't the pitch of its whistle seem higher as the train moves toward you and lower as it moves away? The frequency of the whistle never changes, of course, but the pitch of the sound reaching you does. This change in observed pitch, caused by the motion of the source, is known as the Doppler Effect.

The Doppler Effect for a moving source and a stationary observer can be described using the formula shown below, where f' is the frequency (in vibrations per second) of the sound heard, f is the frequency of the sound produced, and v is the velocity of the source of sound (in mi/h). Use positive values for v when the source is moving toward you, and negative values when it is moving away from you.

$$f' = \frac{f}{1 - \dfrac{v}{760}}$$

EXAMPLE If a train were moving away from you at 19 mi/h, and the frequency of its whistle was 820 vib/s, what frequency would you hear?

Use -19 for v because the train is moving away.

$$f' = \frac{f}{1 - \dfrac{v}{760}} = \frac{820}{1 - \dfrac{(-19)}{760}} = \frac{820}{1.025} = 800$$

You would hear a frequency of 800 vib/s.

1. A train is moving toward you at 87.4 mi/h. The frequency of its whistle is 3,540 vib/s. What frequency do you hear? 4,000 vib/s

2. A race car is leaving the pit at 171 mi/h. To the pit crew, the pitch of its engine seems to be 440 vib/s. What is the engine's actual pitch? 539 vib/s

3. The frequency produced by a siren is 1,540 vib/s. You hear a frequency of 1,400 vib/s. What is the speed of the source? Is it coming or going? 76 mi/h; going

4. In the formula above, what does the 760 represent? (HINT: In any formula, all the measures must cancel out. So, 760 must be a measure of miles per hour.) Speed of sound

Statistics: *Box-and-Whisker Plots*

The table at the right lists the 1989 batting averages for each of the New York Mets with at least 30 "at bats". A **box-and-whisker** plot gives a quick graphical display of the data.

Batting Averages

.183	.228	.233	.270	.291
.183	.229	.247	.272	.308
.205	.231	.256	.286	
.225	.231	.258	.287	

A. List the values in order, as in the table above. Take the first and last numbers. These 2 numbers determine the **range** of the batting averages.

B. Find the **median** of the values. For an *odd* number of values, the median is the middle value. For an *even* number of values, as in the table above, the median is the average of the 2 middle values. The median here is the average of the 9th and 10th values: (.233 + .247) ÷ 2 = .240.

C. Find the first and third *quartiles*. The **first quartile** is the median of the lower half of the values, and the **third quartile** is the median of the upper half. (For an odd number of values, the middle value is counted in both halves.) Taking the medians of the upper and lower 9 values gives a first quartile of .228 and a third quartile of .272.

You are now ready to draw a box-and-whisker plot for the data.

D. Select a scale appropriate for the range of the values, and draw a box from the first quartile to the third quartile. Draw a bar at the median.

E. Add the "whiskers" by drawing segments to the boxes from the highest and lowest values as shown.

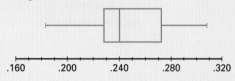

Exercises   1. Oakland   2. Atlanta: ≈ .161, ≈ .315; Oakland: ≈ .193, ≈ .336

1. Referring to the plots on the right, which team had the higher median?

2. Estimate each of the team's best and worst batting averages.

3. What was the range of the middle half of the averages for each team?

4. In 1989, the Oakland Athletics won 61% of their games. The Atlanta Braves won only 39% of theirs. Do the plots show greater batting strength for the Athletics? Discuss.

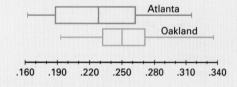

3. Atlanta: ≈ .074; Oakland: ≈ .039   4. Yes; All comparable values are higher for Oakland.

Statistics **83**

3.

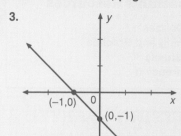

4. (−3,1), (−2,−4), (−1,−7), (0,−8), (1,−7), (2,−4), (3,1)

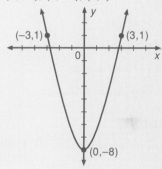

5. (−3,6), (−2,1), (−1,−2), (0,−3), (1,−2), (2,1), (3,6)

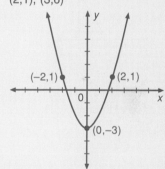

▰▰▰GETTING STARTED

Prerequisite Quiz

Given the coordinates of two points *P* and *Q* on a number line, find the distance *PQ*.

1. $P: -5; Q: 7$ 12
2. $P: 4; Q: 6$ 2
3. $P: -3\frac{1}{2}; Q: 4\frac{1}{2}$ 8
4. Evaluate $\frac{5}{x-4}$ for $x = 4$.   Undef
5. Evaluate $\frac{x-5}{x-2}$ for $x = 5$.   0
6. Evaluate $\frac{x-6}{x-8}$ for $x = 10$.   2

Motivator

Draw the following figure on the chalkboard.

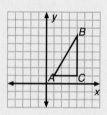

What are the lengths *AC* and *CB*?   3, 5
What is the value of the ratio $\frac{CB}{AC}$? $\frac{5}{3}$

3.3 Slope

Objectives   To find the slope of a line given two of its points
To determine whether three given points lie on the same line

A mountain road that rises rapidly is said to have a steep *slope*. Linear functions can also be described using this same concept of slope.

The *slope*, or degree of steepness, of a line is described by the ratio slope $= \frac{\text{rise}}{\text{run}}$, as illustrated in the figures below.

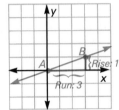   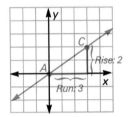   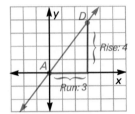

slope of $\overleftrightarrow{AB} = \frac{1}{3}$ slope of $\overleftrightarrow{AC} = \frac{2}{3}$ slope of $\overleftrightarrow{AD} = \frac{4}{3}$

The steeper the line, the greater is the absolute value of $\frac{\text{rise}}{\text{run}}$.

In the figure at the right, a right triangle is drawn to illustrate the slope of $\overleftrightarrow{PQ}$.

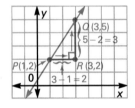

$\overline{RQ}$: rise $= 5 - 2 = 3$ (change in y)

$\overline{PR}$: run $= 3 - 1 = 2$ (change in x)

So, the slope $= \frac{\text{rise}}{\text{run}} = \frac{5-2}{3-1} = \frac{3}{2}$.

Notice that the slope of $\overleftrightarrow{PQ}$ is

$$\frac{\text{change in } y}{\text{change in } x} = \frac{5-2}{3-1} = \frac{3}{2}, \text{ or } \frac{2-5}{1-3} = \frac{-3}{-2} = \frac{3}{2}.$$

Definition

The **slope**, *m*, of a nonvertical line containing the points $A(x_1, y_1)$ and $B(x_2, y_2)$ is given by the following formula.

$$\text{slope of } \overleftrightarrow{AB} = m(\overleftrightarrow{AB}) = \frac{\text{change in } y}{\text{change in } x} = \frac{y_2 - y_1}{x_2 - x_1}, \text{ or } \frac{y_1 - y_2}{x_1 - x_2}$$

84   Chapter 3 Relations, Functions: Graphing Linear Functions

Highlighting the Standards

Standard 2a: Slope is an algebraic topic in which concept and skill are closely related. Students gain a fresh understanding of ratio and immediately put it to use.

EXAMPLE 1   Find the slope of $\overleftrightarrow{AB}$.

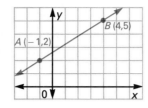

Solution
$$m(\overleftrightarrow{AB}) = \frac{y_2 - y_1}{x_2 - x_1} = \frac{5 - 2}{4 - (-1)} = \frac{3}{5}$$

The slope is $\frac{3}{5}$.

In Example 1, the slope of $\overleftrightarrow{AB}$ is $\frac{3}{5}$, a positive number, and the line slants up to the right.

The sign of the slope of a line tells whether the line slants up to the right or down to the right. In the figure at the right, the slope of $\overleftrightarrow{AB}$ is

$$\frac{2 - 5}{3 - (-1)} = \frac{-3}{4}, \text{ or } -\frac{3}{4}.$$

Notice that the slope of the line is negative and that the line slants down to the right.

In general, a line with a *positive* slope slants *up* to the right. A line with a *negative* slope slants *down* to the right.

When different pairs of points on a line are used to compute the slope of the line, the slope is the same. This suggests the following theorem which can be proved using the properties of similar triangles from geometry.

Theorem 3.1   The slope of a nonvertical line is the same for any two points on the line.

Slope can also be used to determine whether three points are *collinear*. Three points are **collinear** if they all lie on the same line.

3.3 Slope   **85**

TEACHING SUGGESTIONS

Lesson Note

Emphasize that the slope of $\overleftrightarrow{PQ}$ for $P(x_1, y_1)$ and $Q(x_2, y_2)$ can be expressed *two* ways,

$$\frac{y_2 - y_1}{x_2 - x_1} \text{ or } \frac{y_1 - y_2}{x_1 - x_2},$$

but not as $\frac{y_2 - y_1}{x_1 - x_2}$.

Subscripts must be similarly placed in the numerator and the denominator.

Signs of slopes can sometimes be visualized in terms of walking up or down a staircase, from left to right.

positive slope   *negative* slope

Math Connections

Rate of Speed: Slope may be interpreted physically as the *speed* of a moving object if the horizontal and vertical axes are respectively labeled as t (for time) and d (for distance). Then the three slopes near the top of page 84 illustrate the distance formula $d = rt$ (Lesson 2.6) for the three cases $r = \frac{1}{3}, \frac{2}{3}$, and $\frac{4}{3}$. If the rate is measured in mi/min, then a slope of $\frac{1}{3}$ will be equivalent to 20 mi/h, a slope of $\frac{2}{3} = $ 40 mi/h, and a slope of $\frac{4}{3} = $ 80 mi/h.

Additional Example 1

Find the slope of $\overleftrightarrow{AB}$ for the given coordinates. Tell whether the line slants up or down to the right.
$m(\overleftrightarrow{AB}) = -\frac{6}{1} = -6$; down to the right.

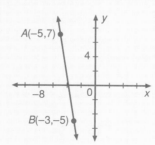

85

Application: Ask students to give interpretations of *slope*. Any constant rate of change, especially one measured with respect to time, can serve as an answer. One example is a simple annual interest rate of 6% on one dollar. The equation is $i = 0.06t$ (see Lesson 1.5).

Checkpoint

Find the slope of $\overleftrightarrow{PQ}$ for the given coordinates. Determine whether the line slants up to the right, down to the right, is horizontal, or is vertical.

1. $P(-5,1)$, $Q(4,4)$ $\frac{1}{3}$; up
2. $P(3,-3)$, $Q(5,-6)$ $-\frac{3}{2}$; dn
3. $P(-5,-1)$, $Q(-5,6)$ Undef; vert
4. $P(-5,-1)$, $Q(8,-1)$ 0; horiz

Closure

Ask students how they would find the slope of a line given the coordinates of two points on a line? Find the value of the difference of *y*-coordinates divided by the difference of *x*-coordinates.

What is the slope of a horizontal line?   0
Ask them to give the slope of a vertical line.
Undef   Ask them how they would determine whether three points lie on the same line.
If any two pair of the three points have equal slopes, then the three points are collinear.

EXAMPLE 2   Determine, without graphing, whether the following three points lie on the same line: $A(0, 1)$, $B(-4, 4)$, and $C(8, 3)$.

Plan   If the three points lie on the same line, then the slopes for any two of them must be the same. Determine whether $m(\overleftrightarrow{AB}) = m(\overleftrightarrow{BC})$.

Solution   For $A(0, 1)$, $B(-4, 4)$, $m(\overleftrightarrow{AB}) = \dfrac{4 - 1}{-4 - 0} = -\dfrac{3}{4}$ ← different slopes

For $B(-4, 4)$, $C(8, 3)$, $m(\overleftrightarrow{BC}) = \dfrac{3 - 4}{8 - (-4)} = -\dfrac{1}{12}$

Because the slopes are different, the three points do not lie on the same line. They are not collinear.

In Example 3, the slopes of horizontal or vertical lines are considered.

EXAMPLE 3   Find the slope of each line.

a. 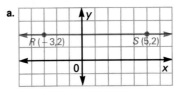   b.

Solutions   a. $m(\overleftrightarrow{RS}) = \dfrac{2 - 2}{5 - (-3)} = \dfrac{0}{8} = 0.$ b. $m(\overleftrightarrow{AB}) = \dfrac{3 - (-1)}{-1 - (-1)} = \dfrac{4}{0}$

The slope is 0.   Since division by zero is undefined, the slope is undefined.

Theorem 3.2   The slope of any horizontal line is 0.

Theorem 3.3   The slope of any vertical line is undefined.

Some important ideas about slope are summarized below.
(1) Lines with positive slopes slant up to the right.
(2) Lines with negative slopes slant down to the right.
(3) The steeper the line, the greater is the absolute value of its slope.

Additional Example 2

Determine, without graphing, whether the three following points are on the same line: $A(0,-3)$, $B(4,-5)$, and $C(6,-8)$.

$m(\overleftrightarrow{AB}) = -\frac{1}{2}$; $m(\overleftrightarrow{BC}) = -\frac{3}{2}$
No; slopes are not equal.

Additional Example 3

Find the slope of each line.
a. horizontal line

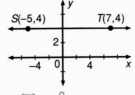

$m(\overleftrightarrow{ST}) = \frac{0}{12} = 0$ Slope is 0.

b. vertical line

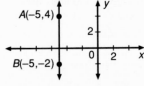

$m(\overleftrightarrow{AB}) = \frac{-6}{0}$, undef
Slope is undefined.

Classroom Exercises

Find the slope of each line whose graph is shown.

1. 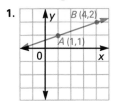 $\frac{1}{3}$

2.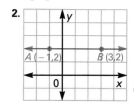

3. 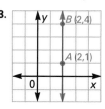 Undefined

Describe the slant of the line with the given slope.

4. $\frac{2}{3}$ Up to right   **5.** -4 Down to right   **6.** $\frac{0}{4}$ Horiz   **7.** Undefined Vert

Written Exercises

Find the slope of $\overleftrightarrow{PQ}$. Describe the slant of the line.

1. $P(0, 0)$, $Q(3, 6)$ **2.** $P(0, 0)$, $Q(-2, 11)$ **3.** $P(4, 2)$, $Q(6, 8)$

4. $P(7, -5)$, $Q(-4, -5)$ **5.** $P(-3, 6)$, $Q(-3, 10)$ **6.** $P(7, 5)$, $Q(-5, 14)$

7. $P(9, 2)$, $Q(5, 8)$ **8.** $P(-8, -2)$, $Q(-4, 8)$ **9.** $P(-3.5, -7)$,
$-\frac{3}{2}$; down to right $\frac{5}{2}$; up to right $Q(-1.5, -7)$ 0; horizontal

10. $P(-6, -2)$, $Q(-6, -8)$ **11.** $P(-7, -5)$, $Q(-3, -6)$ **12.** $P(31, 1)$, $Q(-2, -8)$

Determine, without graphing, whether points A, B, and C are collinear.

13. $A(0, -7)$, $B(2, -3)$, $C(5, 3)$ Yes   **14.** $A(-4, 4)$, $B(8, -5)$, $C(0, -4)$ No

Find the slope of $\overleftrightarrow{PQ}$ for the given coordinates.

15. $P(-2, -3)$, $Q\left(-\frac{1}{2}, -\frac{1}{3}\right)$ **16.** $P\left(-\frac{2}{3}, \frac{1}{2}\right)$, $Q\left(\frac{5}{6}, 5\right)$ 3 **17.** $P\left(\frac{1}{7}, \frac{1}{4}\right)$, $Q\left(\frac{5}{7}, \frac{3}{4}\right)$ $\frac{7}{8}$

18. $P(3a, -b)$, $Q(-a, 2b)$ **19.** $P(7a, -3b)$, $Q(-a, -b)$ **20.** $P(-10a, 12b)$,
$-\frac{3b}{4a}$ $-\frac{b}{4a}$ $Q(-2a, 7b)$ $-\frac{5b}{8a}$

21. A line contains $A(6, y)$ and $B(9, -1)$. Find y given that $m(\overleftrightarrow{AB}) = \frac{2}{3}$. -3

22. The points $P(-2, 4)$ and $Q(0, 2)$ determine a line. Suppose that the point $R(3, a - 1)$ is a point on the same line. Find the value of a. 0

Mixed Review

Solve each equation for x.   *1.4*

1. $6x - 1 = 9 - 2x$ $\frac{5}{4}$ **2.** $\frac{1}{2}x + \frac{1}{3} = x$ $\frac{2}{3}$ **3.** $ax - 2 = c$ $\frac{c + 2}{a}$

◼◼FOLLOW UP

Guided Practice

Classroom Exercises 1–7

Independent Practice

🅐 Ex. 1–14, 🅑 Ex. 15–20, 🅒 Ex. 21, 22

Basic: WE 1–15 odd
Average: WE 3–19 odd
Above Average: WE 5–21 odd

Additional Answers

Written Exercises

1. 2; up to right
2. $-\frac{11}{2}$; down to the right
3. 3; up to right
4. 0; horizontal
5. Undef; vertical
6. $-\frac{3}{4}$; down to the right
10. Undef; vertical
11. $-\frac{1}{4}$; down to the right
12. $\frac{3}{11}$; up to the right
15. $\frac{16}{9}$

Enrichment

Have the students solve this problem.
On a camping trip, Barbara and Wali leave their camper and walk 50 meters up a section of a hill with a slope of $\frac{3}{4}$. The slope of the hill then changes to $\frac{5}{12}$, and they walk another 130 meters. How many meters higher than the level of the camper is their present position? 80 m

◼◼◼◼GETTING STARTED

Prerequisite Quiz

Find the slope of $\overleftrightarrow{AB}$ for the given coordinates of A and B.

1. $A(-4,0)$, $B(-1,2)$ $\frac{2}{3}$
2. $A(1,-5)$, $B(-2,-1)$ $-\frac{4}{3}$
3. $A(2,1)$, $B(2,-3)$ Undef
4. $A(-8,-4.5)$, $B(6,-4.5)$ 0

Motivator

Ask students to plot the points $A(3,7)$ and $B(6,9)$. Draw a line through the two points and extend it in both directions. Label a general point $G(x,y)$ somewhere on the line so that B is between A and G.

Have students answer these questions:

What is the slope of $\overline{AB}$? $\frac{2}{3}$

What is the slope of $\overline{BG}$? $\frac{2}{3}$

Why are these two slopes equal?
The three points are collinear.

Write an equation setting these two slopes equal. $\frac{2}{3} = \frac{y-9}{x-6}$

3.4 Equation of a Line

Objective

To write an equation of a line, given its slope and one of its points, or given two of its points

Recall that the *standard form* of a linear equation in two variables is $ax + by = c$. Also, recall that the slope of a nonvertical line is the same for any two points on that line. (Theorem 3.1)

The line in the figure at the right contains the point $A(x_1, y_1)$. Let $G(x, y)$ represent any other point on the line. Let the slope of the line be m. Then, the slope of $\overleftrightarrow{AG} = m$.

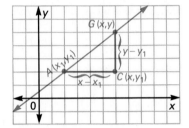

$$\frac{y - y_1}{x - x_1} = m$$

$$y - y_1 = m(x - x_1)$$

Equation of a Line (*Point-Slope Form*)
If a nonvertical line has slope m and contains the point $A(x_1, y_1)$, then the **point-slope form** of the equation of that line is given by:
$$y - y_1 = m(x - x_1).$$

EXAMPLE 1

Write an equation, in standard form, of the line containing the point $P(2, -3)$ and having the slope $-\frac{4}{3}$.

Solution

Write the point-slope form.

$$y - y_1 = m(x - x_1)$$
$$y - (-3) = -\frac{4}{3}(x - 2)$$
$$y + 3 = -\frac{4}{3}(x - 2)$$

Multiply each side by 3.

$$3y + 9 = -4(x - 2)$$
$$3y + 9 = -4x + 8$$

Write in standard form. $4x + 3y = -1$

Highlighting the Standards

Standard 5b: Equations are always easier to understand when accompanied by graphs. The basics contained in this lesson form an important prerequisite to the use of computer graphing.

Additional Example 1

Write an equation, in standard form, of the line containing the point $A(3,1)$ and having the slope $\frac{3}{8}$. $3x - 8y = 9$

EXAMPLE 2   Write an equation, in standard form, of the line containing the points $P(3, -1)$ and $Q(-6, -13)$.

Plan   First find the slope of the line $\overleftrightarrow{PQ}$.

Solution
$$m = \frac{-13 - (-1)}{-6 - 3} = \frac{-12}{-9} = \frac{4}{3}$$

Then write the point-slope form of the equation. Use either point $P(3, -1)$ or $Q(-6, -13)$.

$$
\begin{array}{ll}
y - y_1 = m(x - x_1) & \text{or} \qquad y - y_1 = m(x - x_1) \\
y - (-1) = \frac{4}{3}(x - 3) & \qquad y - (-13) = \frac{4}{3}[x - (-6)] \\
y + 1 = \frac{4}{3}(x - 3) & \qquad y + 13 = \frac{4}{3}(x + 6) \\
3y + 3 = 4x - 12 & \qquad 3y + 39 = 4x + 24 \\
-4x + 3y = -15 & \qquad -4x + 3y = -15
\end{array}
$$

An equation in standard form is $-4x + 3y = -15$, or $4x - 3y = 15$.

Recall that the slope of a vertical line is undefined (Theorem 3.3). Therefore, the point-slope form cannot be used to write an equation of a vertical line, for example, one containing the points $(3, 3)$ and $(3, 1)$. The value of the x-coordinate of every point on this line remains *constant* (always 3). Thus, an equation of the line is $x = 3$.

Equation of a Vertical Line
The standard form of the equation of a vertical line is $x = c$, where c is a constant.

Recall that the slope of any horizontal line is 0 (Theorem 3.2).

Although the point-slope form could be used to write an equation of the line containing the points $(-1, 3)$ and $(5, 3)$, it is not needed because the y-coordinate of every point on the line remains *constant* (always 3). So, an equation of the line is $y = 3$.

Equation of a Horizontal Line
The standard form of the equation of a horizontal line is $y = c$, where c is a constant.

A function whose graph is a horizontal line is called a *constant linear function*, or a **constant function**.

Lesson Note

The point-slope form is used to write the equation of a line when the coordinates of a point on the line and its slope are known. The equation can then be put in standard form, $ax + by = c$. Emphasize that the development of the point-slope form hinges on Theorem 3.1 of Lesson 3.3 that the slope of a nonvertical line is the same for any two points of the line.

Math Connections

Construction: The graph on page 88 suggests the slant of a roof of a house, which might lead one to conclude that architects use the slope of the roof to indicate its degree of steepness. However, architects prefer another measure, the *pitch* of a roof, which is defined as follows.

$$\text{pitch} = \frac{\text{rise}}{\text{span}},$$

where the span is *twice* the run. Thus, a roof with a pitch of $\frac{1}{3}$ would have a slope $\frac{2}{3}$.

Critical Thinking Questions

Analysis: The text indicates that $y = c$ is an equation of a horizontal line because the y-coordinate of every point on the line remains constant. Ask students to show how they could prove this directly using the point-slope form of a line and the fact that the slope of a horizontal line is 0.
$y - y_1 = 0(x - x_1) = 0$
Thus, $y = y_1 = c$.

Additional Example 2

Write an equation, in standard form, of the line containing the points $A(3,1)$ and $B(0,3)$.
$2x + 3y = 9$

Common Error Analysis

Error: Students often make errors with signs, particularly in cases that involve a double negative. For example, if a line contains points $A(-4,-6)$ and $B(-2,-1)$, the slope is $\frac{-1-(-6)}{-2-(-4)}$, or $\frac{5}{2}$, not $\frac{-1-6}{-2-4}$.

Have students use parentheses to clarify cases involving a double negative.

Checkpoint

Write an equation of the line, in standard form, that contains the given points.

1. $A(-3,6)$, $B(6,-6)$ $4x + 3y = 6$
2. $A(-4,5)$, $B(-4,-19.2)$ $x = -4$
3. $A(2,-6)$, $B(-3,-11)$ $x - y = 8$
4. $A(4,-3)$, $B(-6,-3)$ $y = -3$

Closure

Have students write an equation of a nonvertical line containing two given points (x_1, y_1) and (x_2, y_2) in point-slope form.

Find slope $\frac{y_2 - y_1}{x_2 - x_1} = m$. Use either point in the equation: $y - y_1 = m(x - x_1)$.

▰▰▰FOLLOW UP

Guided Practice

Classroom Exercises 1–6

Independent Practice

Ⓐ Ex. 1–12, Ⓑ Ex. 13–18, Ⓒ Ex. 19–21

Basic: WE 1–14
Average: WE 5–18
Above Average: WE 8–21

Additional Answers

See page 99 for the answers to Written Ex. 1–12, 15, 16, 19.

Classroom Exercises

Tell whether the line containing the points A and B is vertical or horizontal. Also, give an equation of the line.

1. $A(-3, 6)$, $B(-3, 8)$
   Vert; $x = -3$
2. $A(-4, -6)$, $B(11, -6)$
   Horiz; $y = -6$
3. $A(3, 7)$, $B(3, 0)$
   Vert; $x = 3$

Write an equation, in standard form, of the line containing the given point and having the given slope.

4. $P(4, 1)$, $m = \frac{2}{3}$
   $2x - 3y = 5$
5. $P(1, 3)$, $m = 3$
   $3x - y = 0$
6. $P(-1, -3)$, $m = -\frac{2}{3}$
   $2x + 3y = -11$

Written Exercises

In Exercises 1–18, you are given either the coordinates of two points on a line or the slope of a line and one point on it. Write an equation for the line in standard form.

1. $R(1, 3)$, $m = \frac{2}{3}$
2. $A(4, -3)$, $m = 4$
3. $L(2, 7)$, $m = -\frac{3}{4}$
4. $P(6, 0)$, $Q(-3, -6)$
5. $S(8, -3)$, $T(-4, 3)$
6. $G(-4, 1)$, $H(8, 8)$
7. $A(-4, 7)$, $B(-4, -1)$
8. $M(9, 3\frac{1}{4})$, $N(-5, 3\frac{1}{4})$
9. $W(2.3, -5)$, $U(2.3, -4)$
10. $T(-1, -6)$, $U(6, -8)$
11. $R(\frac{1}{6}, -\frac{1}{2})$, $C(0, 1)$
12. $Y(2, -\frac{1}{3})$, $Z(6, 1)$
13. $G(0.2, -0.4)$, $H(1.2, 2.6)$ $3x - y = 1$
14. $S(0.1, -0.19)$, $T(0, -0.4)$ $2.1x - y = 0.4$
15. $R(2.1, 14.42)$, $S(-1.2, 4.52)$
16. $J(1.03, 6.3)$, $K(-0.07, 1.9)$
17. $I(5, -4)$, undefined slope $x = 5$
18. $A(-4, -3)$, horizontal line $y = -3$

19. Write how to find an equation of a line containing any pair of given points. Discuss these cases: (1) lines that are neither vertical nor horizontal, (2) vertical lines, (3) horizontal lines.

20. Write an equation of the line containing the points $A(4, 5)$ and $B(12, -1)$. Then use the equation to find y when $x = 2$. $3x + 4y = 32, 6\frac{1}{2}$

21. Write an equation of the line containing the points $A(4, 5)$ and $B(0, 8)$. The point $(t, -7)$ is a point on the line. Find t. $3x + 4y = 32; 20$

Mixed Review

Find the solution set.   *2.1, 2.2, 2.3*

1. $4x - 3(2x - 6) < 24$ $\{x \mid x > -3\}$
2. $3x - 2 > 7$ *and* $-x - 2 > -8$ $\{x \mid 3 < x < 6\}$
3. $|2x - 6| = 8$ $\{-1, 7\}$
4. $|x + 6| \leq 4$ $\{x \mid -10 \leq x \leq -2\}$

Enrichment

Have the students solve this problem.
A professor gives an experimental test, and the scores range from 40 to 80. The professor decides to "scale" the test to make the scores range from 60 to 90. Let x represent an original score and y, the converted scale score. Find the equation the professor must use to convert the scores. (HINT: Use three ordered pairs, (x,y), $(40,60)$, and $(80,90)$.)

$$m = \frac{90 - 60}{80 - 40} = \frac{3}{4}$$
$$y - 60 = \frac{3}{4}(x - 40)$$
$$y = \frac{3}{4}x + 30$$

Now have the students convert the following original scores to scaled scores.

45	64	70	83
60	75	75	86

3.5 Graphing Linear Relations

Teaching Resources

Problem Solving Worksheet 3
Quick Quizzes 19
Reteaching and Practice
  Worksheets 19
Transparency 7

Objectives

To write equations of lines given their slopes and y-intercepts
To find the slopes and y-intercepts of lines given their equations
To graph lines using the slope-intercept method
To determine whether given points lie on given lines

The line $\overleftrightarrow{PQ}$ at the right crosses, or *inter-cepts*, the y-axis at $P(0,4)$. The *y-intercept* is 4. The slope of $\overleftrightarrow{PQ}$ is

$$m = \frac{8 - 4}{6 - 0} = \frac{4}{6} = \frac{2}{3}.$$

An equation of $\overleftrightarrow{PQ}$ can be written using the point-slope method.

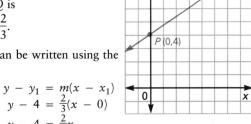

$$y - y_1 = m(x - x_1)$$
$$y - 4 = \frac{2}{3}(x - 0)$$
$$y - 4 = \frac{2}{3}x$$

Now solve for y. $\qquad y = \frac{2}{3}x + 4$

Notice the pattern. $\qquad$ slope $\qquad$ y-intercept

When you solve an equation of a nonvertical line for y, the result is the *slope-intercept form* of the equation. The *y-intercept* is the y-coordinate of the point at which the line intersects the y-axis.

In general, if a curve crosses the y-axis at the point $(0, b)$, then b is called the **y-intercept** of the curve.

Equation of a Line *(Slope-Intercept Form)*
If a nonvertical line has a slope m and y-intercept b, then the **slope-intercept form** of the equation of that line is given by $y = mx + b$.

EXAMPLE 1 Write an equation, in slope-intercept form, of the line whose slope is $-\frac{4}{5}$ and whose y-intercept is 2.

Plan Use the slope-intercept form.

Solution $\qquad y = mx + b \qquad y = -\frac{4}{5}x + 2$

TEACHING SUGGESTIONS

Lesson Note

Students should see that if a line is drawn through two points P and Q, shown on page 91, the coordinates of P and Q will satisfy the equation of the line, $y = \frac{2}{3}x + 4$. For $Q(6,8)$, it is true that $8 = \frac{2}{3}(6) + 4$. For $P(0,4)$, it is true that $4 = \frac{2}{3}(0) + 4$. Conversely, any point (x,y) with coordinates that satisfy the equation $y = \frac{2}{3}x + 4$ must lie on the line. If $x = 9$, for example, then $y = \frac{2}{3}(9) + 4 = 10$.

Math Connections

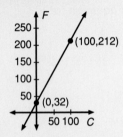

Meteorology: The formula that is used to convert temperatures in Celsius degrees to temperatures in Fahrenheit degrees is $F = \frac{9}{5}C + 32$. From the graph of this equation, you can see that this is a linear relationship expressed in slope-intercept form; that is, the y-intercept is 32 and the slope is $\frac{9}{5}$.

Critical Thinking Questions

Analysis: Have students compare the standard form of a linear equation, $ax + by = c$ (Lesson 3.2), with the slope-intercept form, $y = mx + b$. Then ask them whether the coefficent b has the same meaning in the two forms.   It does not.

EXAMPLE 2 Find the slope and y-intercept of the line whose equation is given.
  a. $5x + 3y = 4$ **b.** $-x - y = 8$

Plan Solve each equation for y to obtain the form $y = mx + b$.

Solutions
  a. $5x + 3y = 4$
  $3y = -5x + 4$
  $y = -\frac{5}{3}x + \frac{4}{3}$

  slope $= -\frac{5}{3}$ y-intercept $= \frac{4}{3}$

  b. $-x - y = 8$
  $-1y = 1x + 8$
  $y = -1x - 8$

  slope $= -1$ y-intercept $= -8$

The slope-intercept form of an equation of a line can also be used to graph a line such as $2x - 5y = 20$.

Slope-Intercept Method of Graphing a Linear Equation

1. Solve the equation for y. Identify the slope and y-intercept.

   $2x - 5y = 20$
   $-5y = -2x + 20$
   $y = \frac{2}{5}x - 4$

   slope $= \frac{2}{5}$ y-intercept $= -4$

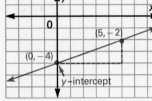

2. Place a dot at -4 on the y-axis [the point $(0,-4)$]. Use the slope, $\frac{2}{5}$, to find a second point as follows: From the point $(0,-4)$, move 5 units to the right and 2 units up, to the point $(5,-2)$.
3. Place a dot at the point $(5,-2)$. Draw a line through the two points.

EXAMPLE 3 Graph the function $f(x) = -\frac{3}{4}x + 6$.

Solution
  1. Find the slope and y-intercept.
     $$f(x) = -\frac{3}{4}x + 6$$
     $$m = -\frac{3}{4} \qquad b = 6$$
  2. Plot the point $(0, 6)$. From this point, move 4 units to the right and 3 units down ($m = \frac{-3}{4}$), or 4 units left and 3 units up ($m = \frac{3}{-4}$).
  3. Plot the second point. Draw a line through the two points. The graph is shown above.

Additional Example 2

Find the slope and y-intercept of the line whose equation is given.

a. $4x - 3y = -6$ $\frac{4}{3}; 2$
b. $3x + 7y = 12$ $-\frac{3}{7}; \frac{12}{7}$

Additional Example 3

Graph the function $f(x) = \frac{2}{3}x - 5$.

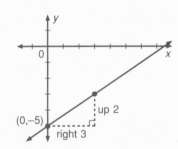

Recall that $y = c$ is the standard form of an equation of a horizontal line, and that $x = c$ is the standard form of an equation of a vertical line. Vertical and horizontal lines can be graphed without making tables or using special methods.

EXAMPLE 4   Graph $-7x - 14 = 0$.

Plan   Solve for x. Compare to the standard form $x = c$.

Solution
$$-7x - 14 = 0$$
$$-7x = 14$$
$$x = -2$$
The graph is shown at the right.

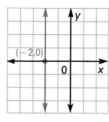

You can determine whether a point such as $P(15,2)$ lies on the line $2x - 5y = 20$ by substituting the coordinates of the point into the equation and determining whether the result is a true equation. The procedure is illustrated in Example 5.

EXAMPLE 5   Determine whether the given point lies on the line described by the given equation.
  a. $P(15, 2)$, $2x - 5y = 20$ **b.** $R(7, -3)$, $5y - x - 12 = 0$

Solutions   **a.** Substitute 15 for x and 2 for y.

$$2 \cdot 15 - 5 \cdot 2 \stackrel{?}{=} 20$$
$$30 - 10 \stackrel{?}{=} 20$$
$$20 = 20 \text{ (True)}$$

The pair $(15, 2)$ *satisfies* the equation. So, P is a point on the line.

b. Substitute 7 for x and -3 for y.

$$5(-3) - 17 - 12 \stackrel{?}{=} 0$$
$$-15 - 7 - 12 \stackrel{?}{=} 0$$
$$-34 = 0 \text{ (False)}$$

The pair $(7, -3)$ does *not satisfy* the equation. So, R is not a point on the line.

Classroom Exercises

Find the slope and y-intercept, if any, of the line whose equation is given.

1. $y = \frac{3}{4}x - 7$ $\frac{3}{4}, -7$
2. $y = -\frac{4}{7}x + 9$ $-\frac{4}{7}, 9$
3. $y = 3x + 7$ 3, 7
4. $x = 8$ Undef, none
5. $y = -1$ 0, -1
6. $x = 4$ Undef, none
7. $y = \frac{3}{2}x - 1$ $\frac{3}{2}, -1$
8. $y = -5$ 0, -5
9. $-3x = 12$ Undef, none
10. $y = -x - 7$ $-1, -7$
11. $y = 4 + x$ 1, 4
12. $3x = -15$ Undef, none

13–24. Graph each equation in Classroom Exercises 1–12.

Common Error Analysis

Error: When graphing lines by the slope-intercept method, students may become confused by equations such as $y = 3x + 4$. They are not sure how to treat 3 as the value of the slope.

Have students rewrite the slope as a fraction, $\frac{3}{1}$. Also, emphasize the correct order for using the slope to plot a second point after plotting the y-intercept. For instance, a slope of $\frac{2}{3}$ means 3 right, 2 up.

Also show that a slope of $-\frac{3}{4}$ can be interpreted two ways:
$\frac{-3}{4}$ means 4 right, 3 down
$\frac{3}{-4}$ means 4 left, 3 up.
Either will give another point on the line.

Checkpoint

Graph each equation.

1. $3x - 2y = 2$
2. $x + y = 3$

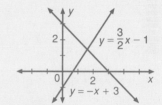

Closure

Ask students how they would write an equation of a line given the slope and the y-intercept? $y = mx + b$ For the equation $y = mx + b$, what are the roles of m and b in graphing the line?   Slope and y-intercept

Additional Example 4

Graph $3y + 9 = 0$.

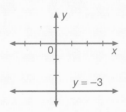

Additional Example 5

Determine whether the point having the given coordinates is on the line described by the given equation.

a. $P(5,3)$; $3x - 2y = 11$ $3(5) - 2(3) = 11$ (False); P is not a point on the line.
b. $R(-4,-1)$; $-4x - 3y = 19$ $-4(-4) - 3(-1) = 19$ (True); R is a point on the line.

◼️ FOLLOW UP

Guided Practice

Classroom Exercises 1–24

Independent Practice

🅰️ Ex. 1–26, 🅱️ 27–30, 🅲️ Ex. 31–34

Basic: WE 1–27 odd, Midchapter Review, Application

Average: WE 3–29 odd, Midchapter Review, Application

Above Average: WE 5–23 odd, Midchapter Review, Application

Additional Answers

Classroom Exercises

13.

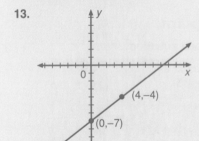

(4,−4)
(0,−7)

14.

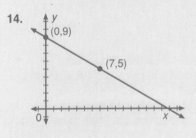

(0,9)
(7,5)

See page 104 for Classroom Ex. 15–17, and pages 804–807 for Classroom Ex. 18–24, Written Ex. 9–26, 31, 32, and Midchapter Review Ex. 1–6.

Written Exercises

Write an equation of the line with the given slope and y-intercept.

1. slope: $-\frac{2}{3}$, y-intercept: 5 $y = -\frac{2}{3}x + 5$
2. slope: -2, y-intercept: -5 $y = -2x - 5$

Find the slope and y-intercept of the line whose equation is given.

3. $y = -\frac{4}{9}x + 1$ $-\frac{4}{9}, 1$
4. $y = \frac{5}{4}x - 9$ $\frac{5}{4}, -9$
5. $y = -\frac{2}{3}x - 1$ $-\frac{2}{3}, -1$
6. $y = -x$ $-1, 0$
7. $4x + 5y = -3$ $-\frac{4}{5}, -\frac{3}{5}$
8. $y = 4$ $0, 4$

Graph each equation.

9. $y = \frac{2}{3}x - 1$
10. $y = -\frac{4}{5}x + 1$
11. $y = -x + 5$
12. $y = x - 6$
13. $2x - 5y = 15$
14. $4x + 3y = 18$
15. $2x - 7y = 21$
16. $-x - y = 6$
17. $f(x) = 2x - 6$
18. $f(x) = -x - 3$
19. $f(x) = \frac{3}{4}x - 3$
20. $f(x) = -\frac{5}{4}x + 8$
21. $-5y = -20$
22. $8 = -4y + 20$
23. $-14 = -7x$
24. $12 = 9 + 3x$
25. $-6x - 12y = 36$
26. $f(x) = -2x - 1$

Determine whether the point lies on the line described by the equation.

27. $P(1, 4)$; $3x + 4y = 19$ Yes
28. $P(3, 4)$; $x - 4y = -13$ Yes
29. $P(6, 8)$; $\frac{2}{3}x - \frac{1}{4}y = 2$ Yes
30. $P(10, -6)$; $\frac{2}{5}x + \frac{5}{3}y = 6$ No

Graph the line described by each equation.

31. $3(x - 4) - (4 - 4y) = 8$
32. $-(x - 5y) - 2(3x + y) = 12$
33. For what value of a will the graph of $3ax - 4y = 8$ have the same slope as the line whose equation is $3x - 2y = 14$?   2
34. Point P is a point on the line whose equation is $4x - 3y = 6$. The abscissa of P is 1 less than twice its ordinate. Find the coordinates of P.   (3, 2)

Midchapter Review

Graph each relation or equation.   *3.1, 3.2, 3.5*

1. $\{(-2, 0), (-1, 2), (0, 1), (1, 0)\}$
2. $\{(2, 3), (-1, 4), (0, 4), (2, 4)\}$
3. $x + y = 2$
4. $f(x) = -x + 4$
5. $y = -\frac{1}{4}x + 2$
6. $2x - y = 3$

Find the slope of $\overleftrightarrow{PQ}$. Describe the slant of the line.   *3.3*

7. $P(5, 1)$, $Q(0, 2)$ $-\frac{1}{5}$, down to the right
8. $P(-3, 4)$, $Q(-2, 8)$ 4, up to the right
9. Write an equation of the line through $P(4, -1)$ that has a slope of $\frac{3}{5}$.   *3.4* $3x - 5y = 17$

Enrichment

Have the students write the standard form of the linear equation, $ax + by = c$, then solve for y. $y = -\frac{a}{b}x + \frac{c}{b}$
Point out that the slope is now represented as $-\frac{a}{b}$, and the y-intercept is $\frac{c}{b}$. Thus, the slope and the y-intercept of an equation can be seen easily in an equation in standard form.

Have the students find the slope and y-intercept of each of the following equations without first changing the equation to slope-intercept form.

1. $3x - 2y = -1$ $\frac{3}{2}; \frac{1}{2}$
2. $-x + y = -5$ $1; -5$
3. $5x - 10y = 20$ $\frac{1}{2}; -2$
4. $\frac{1}{2}x + 2y = \frac{1}{4}$ $-\frac{1}{4}; \frac{1}{8}$
5. $0.1x - 0.5y = -1.5$ $\frac{1}{5}; 3$

Application: *Familiar Linear Relationships*

The linear relations you have been studying are readily observed in many familiar situations in your own life. The relation between degrees Fahrenheit and degrees Celsius, for example, is linear. Constant rates, such as constant distance per time (speed), are also linear relations.

Use the following exercises to explore linear relationships.

1. Use the slope-intercept form $y = mx + b$ and the two points given in the figure at the right to show that $C = \frac{5}{9}(F - 32)$, the relationship between Fahrenheit and Celsius, is truly a linear relation. Let $y = C$, $m = \frac{5}{9}$, $x = F$, $b = -\frac{160}{9}$.

2. Using the points in the second figure, show that $F = \frac{9}{5}C + 32$ is a linear relationship. Let $y = F$, $m = \frac{9}{5}$, $x = C$, $b = 32$.

3. At what temperature (point) do the Celsius and Fahrenheit scales register the same (intersect)? Show your derivation. -40

4. Thermometers sell for \$3.50 each. Let n represent the number of thermometers and C the cost of n thermometers. Write an equation for representing C in terms of n. Draw the graph showing this relationship. $C = 3.50n$

5. A car traveled at an average speed of 70 km/h. How far did it travel in 3 h? in 5 h? Draw the graph showing this relation. 210 km, 350 km

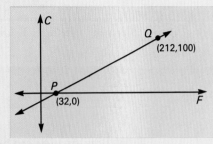

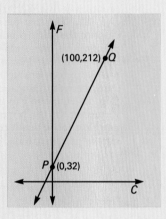

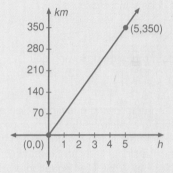

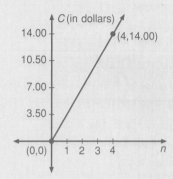

Application

3. Since you want C° = F°, we can substitute C for F (or vice versa) in one of the equations above. So we have C = $\frac{9}{5}$C + 32, or $\frac{4}{5}$C = -32, or C = -40.

4. C = 3.50n;

5. 210 km, 350 km;

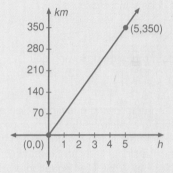

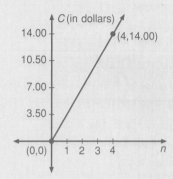

Prerequisite Quiz

1. Find the slope of a line that contains the points (3,420) and (7,900). $\frac{120}{1}$, or 120

Use the equation $y = 14x + 2$ for Exercises 2–4.

2. Find y when $x = 20$.   282
3. Find x when $y = 100$.   7
4. When x increases from 4 to 5, by what amount does y increase?   14

Motivator

Ask the class to plot the following points: $A(2,4)$, $B(3,7)$, $C(4,11)$, $D(5,13)$, $E(6,15)$, $F(7,19)$.

Ask students the following questions: Can a line be drawn to contain all the points?   No
What is the greatest number of these points through which a single line can be drawn?
4   Draw this line. What is an equation of this line? $y = 3x - 2$

3.6 Linear Models

Objectives
To construct linear models for sets of data
To use the equations of linear models to predict new data values

Scientists, economists, sociologists, and business managers all record data in order to find patterns by which future results may be predicted. For example, a naturalist might record the number of ants in an anthill at the end of a number of months in order to predict insect population changes over time. Below is a sample of such data and its graph.

Months: x	2	3	4	5	6	7
Ant population: y	8,000	9,500	10,000	11,000	13,000	13,500

Notice that a line can be drawn through two of the given points, say $A(2, 8,000)$ and $D(5, 11,000)$, that passes very close to the other plotted points. This line is called a **linear model**, and it approximates the relationship in the study between time and population. Its equation can be used to predict the ant population at any time for which recorded data is not available.

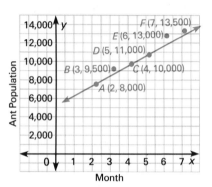

EXAMPLE 1   Use points $A(2, 8,000)$ and $D(5, 11,000)$ to write the equation of the linear model above.

Plan   Find the slope. Use the point-slope form to write the equation.

$$m = \frac{11,000 - 8,000}{5 - 2} = \frac{3,000}{3} = 1,000$$

Solution
$$y - y_1 = m(x - x_1)$$
$$y - 8,000 = 1,000(x - 2)$$
$$y - 8,000 = 1,000x - 2,000$$
$$y = 1,000x + 6,000$$

96   Chapter 3 Relations, Functions: Graphing Linear Functions

Highlighting the Standards

Standard 6a: The opening example models a real-world situation with a function. Slopes and graphs that use large numbers provide a new context.

Additional Example 1

A manufacturer constructed a linear model for the cost y (in dollars) of producing x cameras. It cost about $12,500 to produce 300 cameras, and $30,000 to produce 800 cameras. Use this information to write an equation in slope-intercept form of the linear model. $y = 35x + 2,000$

The equation of the linear model $y = 1,000x + 6,000$ describes a linear function where y, the number of ants, is a function of x, the number of months. $y = 1,000x + 6,000$ can therefore be written as

$$f(x) = 1,000x + 6,000.$$

This function can be used to predict the approximate number of ants at any time, as shown in Example 2.

EXAMPLE 2   Use the formula above to predict the number of ants after:

a. 4 months.   **b.** 8 months.

Solutions   **a.** 4 months
Find $f(4)$.
$f(x) = 1,000x + 6,000$
$f(4) = 1,000 \cdot 4 + 6,000$
$f(4) = 10,000$ (This confirms the value in the table.)

b. 8 months
Find $f(8)$.
$f(x) = 1,000x + 6,000$
$f(8) = 1,000 \cdot 8 + 6,000$
$f(8) = 14,000$

So, there would be 10,000 ants after 4 months and 14,000 ants after 8 months.

The linear model used in Examples 1 and 2 also illustrates the following physical interpretation of slope.

$A(2, 8,000)$ corresponds to 8,000 ants after 2 months. $D(5, 11,000)$ corresponds to 11,000 ants after 5 months.

change of 3,000
in ant population
↓
$$\text{Slope of } \overleftrightarrow{AD} = \frac{11,000 - 8,000}{5 - 2} = \frac{3,000}{3} = \frac{1,000}{1}$$
↑
change of 3 months

Lesson Note

In discussing the data involving ants, emphasize that the line is an *approximation* that represents the set of all data for the population study. Have the students look in newspapers for graphs that approximate straight lines. Encourage them to write equations for the graphs; then discuss the meaning of slope and possible limitations for the domain and range.

Math Connections

Research: The examples of the lesson illustrate how a mathematical model may be devised to fit recorded data. Students should understand that a straight line is often an approximation.

Critical Thinking Questions

Analysis: The linear model for Example 1 on page 96 shows that the ant population increases proportionately as time passes. The slope of the line is positive. Ask students to describe the slope of a linear model for which an increase in one variable is associated with a proportionate decrease in the slope of a second variable.   The slope will be negative.   Suppose that the slope of a linear model showing the relationship between business aptitude scores and music aptitude scores is negative. Ask students to interpret this relationship.   Persons with higher music aptitude scores will have lower business aptitude scores, and vice versa.

Additional Example 2

Use the formula for Additional Example 1 to find the cost of producing each number of cameras.

a. 500 $19,500
b. 1,200 $44,000

Checkpoint

The value of a certain painting appears to increase each year according to the equation $y = 75x + 1,500$, where y is the dollar value and x is the number of years after it was first sold. Find the value of the painting at the times given.

1. after 5 years $1,875
2. after 10 years $2,250
3. at the time it was sold $1,500
4. How many years after the painting was first sold will its value have doubled?
   20
5. How much does the value of the painting increase each year? $75

Closure

Ask students how they would construct a linear model for a given set of data?   Plot points and draw a line through two of the given points that passes close to the other points.   Will all points of the model satisfy the equation?   No   How can you predict more members of the linear model?   Find rate of change; use as slope. Use known data and slope to write an equation. Use the equation to estimate more values.

■■■FOLLOW UP

Guided Practice

Classroom Exercises 1–8

Independent Practice

A Ex. 1–6, **B** Ex. 7–11, **C** Ex. 12–13

Basic: WE 1–9, Application
Average: WE 3–11, Application
Above Average: WE 3–13, Application

EXAMPLE 3   A researcher constructs a linear model for a set of data that shows how dosages of x grams of an appetite stimulant caused rats to gain y grams of weight. Over a restricted dosage range, the equation of the linear model is found to be $y = 2x + 50$, or $f(x) = 2x + 50$.

Predict each of the following:
a. the weight gain for 30 g of stimulant,
b. the dosage necessary for a weight gain of 120 g, and
c. the average gain in weight with respect to the dosage of stimulant.

Solutions

a. $y = 2x + 50$
   $y = 2 \cdot 30 + 50$
   $y = 60 + 50$, or 110

   The weight gain is 110 g.

b. $y = 2x + 50$
   $120 = 2x + 50$
   $70 = 2x$
   $35 = x$

   The necessary dosage is 35 g.

c. $y = 2x + 50$, so the slope is 2.

   The average gain in weight is 2 g for every 1 g of stimulant.

Classroom Exercises

For Exercises 1–8, suppose that the equation $y = -5x + 100$ represents the number of bacteria y (in thousands) left x hours after an antibacterial spray is initiated. Predict the number of bacteria left after the given number of hours.

1. 2   90,000
2. 5   75,000
3. 10   50,000
4. 7   65,000
5. 8   60,000
6. 12   40,000

7. If the linear model is correct, in how many hours will there be no bacteria left?   20 h

8. Give the average rate of change in the number of bacteria with respect to time (in hours).   −5,000 bacteria per hour

Written Exercises

A linear model for the cost y (in dollars) of producing x radios contains the points $A(5, 160)$ and $B(10, 220)$.

1. Write an equation for the linear model. $y = 12x + 100$

2. Find the cost of producing 15 radios. $280

A linear model for the total number y of a particular brand of television set sold over x years contains the points $A(2, 2,600)$, $B(5, 3,500)$, and $C(10, 5,000)$.

3. How many television sets were sold after 2 years?   2,600 sets

---

Additional Example 3

A leak was discovered in a water storage tank. A linear model was made to describe the number of gallons y in the tank x days after the leak was discovered. Over a restricted time span, the equation of the linear model was found to be
$y = 5,400 - 80x$.

Find each of the following.

a. the number of gallons left after 5 days
   5,000
b. the number of days for the water to decrease to 3,000 gallons   30
c. the average loss of water per day
   80 gal

4. How many years did it take to sell 5,000 sets? 10 yr

5. Write the equation of the linear model. $y = 300x + 2,000$

6. Find the average rate of change in the total number of sets sold with respect to time (in years). 300 sets per year

The table at the right shows the number of chirps per minute that a cricket makes at various temperatures. Use the table for Ex. 7–13.

Temperature (°C): x	6	8	10	15	20
Number of chirps per minute: y	12	28	40	74	110

7. Graph a linear model using the data for 10°C and 20°C.

8. Write the equation of the linear model. $y = 7x - 30$

9. Use the equation to predict the number of chirps at 40°C. 250 chirps

10. Write the average rate of change in the number of chirps with respect to the temperature. 7 chirps per °C rise

11. Predict at what temperature the number of chirps will be 180. 30°C

12. Predict the number of chirps at a temperature of 4°C. Explain the significance of your answer. 0 chirps, negative chirps not possible

13. Predict the number of chirps at temperature a°C. $7a - 30$ chirps

Mixed Review

1. Simplify $5xy - 7x - 6xy - 4x$. *1.3* $-xy - 11x$

2. Solve $0.06x = 1.02$. *1.4* 17

3. Solve $|x - 2| = 5$. *2.3* $-3, 7$

4. Given that $f(x) = 4 - x - x^2$, find $f(-3)$. *3.2* -2

Application: *Möbius Strip*

Look at the bands shown below. The band on the left is the familiar ring-shaped sort with two sides and two edges. The band on the right, however, has a twist. This type of band is called a **Möbius strip**.

1. How many sides does a Möbius strip have? How many edges? 1 side, 1 edge

2. If a "Möbius twist" were added to an endless answering-machine cassette, what would happen to the time available for a message? It would double.

Written Exercises

7.

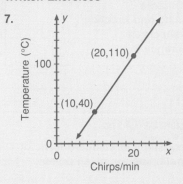

Additional Answers, page 90

1. $-2x + 3y = 7$
2. $4x - y = 19$
3. $3x + 4y = 34$
4. $2x - 3y = 12$
5. $x + 2y = 2$
6. $-7x + 12y = 40$
7. $x = -4$
8. $y = 3\frac{1}{4}$
9. $x = 2.3$
10. $2x + 7y = -44$
11. $9x + y = 1$
12. $x - 3y = 3$
15. $-3x + y = 8.12$
16. $-4x + y = 2.18$
19. Answers may vary. One possible answer: For nonvertical, nonhorizontal lines, use the two points to find the slope, then use the point-slope form and one of the points to find the equation. For vertical lines, use the form $x = k$, where k is the x-coordinate of either point. For horizontal lines, use the form $y = k$, where k is the y-coordinate of either point.

Enrichment

Have the students use dictionaries to look up the meanings of *interpolate* and *extrapolate* and have them write the definitions that seem most applicable to the mathematics of this lesson. (A brief way to recall the distinction between the terms is to think: *inter*polate means place *between* and *extra*polate means place *outside*.) Discuss with the students these two forms of estimation.

Refer to Example 2. Finding the number of ants after 8 months is an example of extrapolation, since 8 months is outside the range of observed data (2 months through 7 months). On the other hand, estimating the number of ants at $4\frac{1}{2}$ months is an example of interpolation, since $4\frac{1}{2}$ lies between elements of the observed data.

Have each student create a graph similar to those in this lesson and give examples of interpolation and extrapolation.

3.7 Parallel and Perpendicular Lines

Objectives   To determine whether two given lines are perpendicular, parallel, or neither

To write an equation of a line that contains a given point and that is parallel (or perpendicular) to a given line

■■■■GETTING STARTED

Prerequisite Quiz

Write an equation, in standard form, of the line containing the given points.

1. $P(0,-4)$, $Q(2,2)$ $3x - y = 4$
2. $P(7,1)$, $Q(-3,-9)$ $x - y = 6$
3. $P(-5,-1)$, $Q(-5,4)$ $x = -5$
4. $P(-4,-3)$, $Q(0,-3)$ $y = -3$

Motivator

Ask the class what is meant by parallel lines. See page 100. Ask the class to graph each of the following on the same set of axes.

$$3x - 2y = 6$$
$$5x - 2y = 10$$
$$6x - 4y = 2$$

What special geometric relationship exists for two of these lines? parallel What is the slope of each line? $\frac{3}{2}, \frac{5}{2}, \frac{3}{2}$, respectively
What generalization seems possible for parallel lines and their equations?
Slopes are equal.

In the figure at the right, $\overleftrightarrow{PQ}$ is *parallel* to $\overleftrightarrow{RS}$. (Parallel lines are lines in the same plane that never meet.) Notice that their slopes are the same.

$$m(\overleftrightarrow{PQ}) = \frac{5 - 3}{3 - 0} = \frac{2}{3}$$
$$m(\overleftrightarrow{RS}) = \frac{1 - (-1)}{5 - 2} = \frac{2}{3}$$

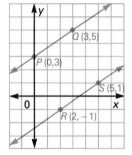

Theorem 3.4   **Slopes of Parallel Lines:** In a plane, if two nonvertical lines are **parallel,** then they have the same slope. Conversely, two lines in a plane that have the same slope are parallel.

EXAMPLE 1   Write an equation, in standard form, of the line containing the point $P(8, 3)$ that is parallel to the line whose equation is $4y - 3x = 8$.

Plan   Because the lines are parallel, their slopes are the same.

Solution   (1) Solve $4y - 3x = 8$ for y. (2) Write an equation of the parallel line.
$$4y = 3x + 8 \qquad\qquad y - y_1 = m(x - x_1)$$
$$y = \frac{3}{4}x + 2 \qquad\qquad y - 3 = \frac{3}{4}(x - 8)$$
The slope of the given line is $\frac{3}{4}$. $\qquad 12 = 3x - 4y$

An equation of the line in standard form is $3x - 4y = 12$.

In the figure, $\overleftrightarrow{PQ}$ is *perpendicular* to $\overleftrightarrow{RS}$. (Perpendicular lines meet at right angles.)
$$m(\overleftrightarrow{PQ}) = \frac{-1 - 3}{4 - (-2)} = \frac{-4}{6}, \text{ or } -\frac{2}{3}$$
$$m(\overleftrightarrow{RS}) = \frac{4 - 1}{3 - 1} = \frac{3}{2}$$

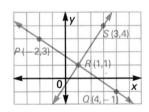

Highlighting the Standards

Standard 14b: The logic of algebraic procedures is just as compelling as the logic in geometry. The theorems of parallel and perpendicular lines illustrate this logic.

Additional Example 1

Write an equation, in standard form, of the line that contains the point $P(3,-1)$ and that is parallel to the line with equation $2x + 3y = 7$. $2x + 3y = 3$

Notice that the product of the two slopes is $-\frac{2}{3} \cdot \frac{3}{2}$, or -1.

Therefore, the slope of one line is the opposite of the reciprocal of the slope of the other. This suggests the following theorem.

Theorem 3.5

Slopes of Perpendicular Lines
Given two lines that are neither vertical nor horizontal, if the two lines are **perpendicular**, then the product of their slopes is -1.

Conversely, if the product of the slopes of two lines is -1, then the lines are perpendicular. Therefore, Theorem 3.5 can be rewritten as follows: Two nonvertical, nonhorizontal lines are perpendicular *if and only if* the product of their slopes is -1.

EXAMPLE 2   Write an equation, in standard form, of the line containing the point $(-5, 4)$ that is perpendicular to the line with equation $3x - 2y = 8$.

Solution

1. Find the slope of the given line.

Solve for y. $3x - 2y = 8$
$$-2y = -3x + 8$$
$$y = \frac{3}{2}x - 4$$

The slope is $\frac{3}{2}$. So, the slope of a perpendicular line is $-\frac{2}{3}$.

2. Use the point-slope form to write an equation.
$$y - y_1 = m(x - x_1)$$
$$y - 4 = -\frac{2}{3}(x + 5)$$
$$3y - 12 = -2x - 10$$
$$2x + 3y = 2$$

An equation in standard form is $2x + 3y = 2$.

You can tell whether two lines are parallel or perpendicular by comparing their slopes. For example, consider the slopes of $y = \frac{5}{2}x + 2$ and $y = -\frac{2}{5}x + \frac{7}{10}$. The lines are perpendicular because the product of their slopes, $\frac{5}{2}$ and $-\frac{2}{5}$, is -1.

Classroom Exercises

Tell whether the given lines are parallel, perpendicular, or neither.

1. $y = \frac{2}{3}x - 1$; $y = -\frac{3}{2}x + 5$ **2.** $y = 3x - 5$; $y = 3x + 4$ **3.** $x = 6$; $y = 8$
    perpendicular   parallel   perpendicular

Give the slope of a line parallel to the given line.

4. $y = 3x - 5$ 3   **5.** $y = -3x$ -3 **6.** $y = 4$ 0

Lesson Note

If students have studied geometry, use similar triangles to prove that parallel lines have the same slope.

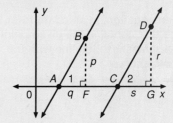

$\overleftrightarrow{AB} \parallel \overleftrightarrow{CD}$, so $\angle 1 \cong \angle 2$. (Corr $\angle$s of $\parallel$ lines are $\cong$.) Also, $\angle F$ and $\angle G$ are $\cong$ rt $\angle$s. So $\triangle ABF \sim \triangle CDG$ by AA. Then $\frac{p}{r} = \frac{q}{s}$ (Def $\sim \triangle$s.) Now rewrite the proportion as $\frac{p}{q} = \frac{r}{s}$. Therefore, the slopes are equal.

Math Connections

Logic: Have students review the logical structure of a theorem such as Theorem 3.4 in which a statement and its converse are both stated as being true. The biconditional form *p if* and *only if q* combines the two forms *p if q*, which is equivalent to *if q then p*, and *p only if q* (if *p* then *q*).

Critical Thinking Questions

Analysis: Ask students to reword Theorem 3.4 to bring out its logical structure as indicated in *Math Connections* above. The result should be along the following lines: *If two nonvertical lines in the same plane have the same slope, then they are parallel*, and *if two nonvertical lines in a plane are parallel, then they have the same slope.* The combined form then becomes *two nonvertical lines in the same plane have the same slope if and only if they are parallel.*

Additional Example 2

Write an equation, in standard form, of the line that contains the point $(-4, 2)$ and is perpendicular to the line with equation
$5x - 2y = 10$.
$2x + 5y = 2$

Checkpoint

Write an equation, in standard form, of the line that contains the given point and is parallel to the line with the given equation.

1. $P(1, -3)$; $2x - y = -3$ $2x - y = 5$
2. $P(-4, -1)$; $3y + 4 = 13$ $y = -1$

Write an equation, in standard form, of the line that contains the given point and is perpendicular to the line with the given equation.

3. $P(1, -3)$; $2x - y = -3$ $x + 2y = -5$
4. $P(-4, -1)$; $3y + 4 = 13$ $x = -4$

Closure

Given the equation of two lines, ask students how they determine if the lines are:
(1) parallel   Slopes are =.
(2) perpendicular
Slopes are opp reciprocals.

▬▬▬FOLLOW UP

Guided Practice

Classroom Exercises 1–6

Independent Practice

A Ex. 1–18, **B** Ex. 19–21, **C** Ex. 22,23

Basic: WE 1–19 odd, Extension
Average: WE 3–21 odd, Extension
Above Average: WE 5–23 odd, Brainteaser, Extension

Additional Answers

Written Exercises

1. $3x - y = 2$

Written Exercises

Write an equation, in standard form, of the line that contains the given point and is parallel to the given line.

1. $P(1, 1)$; $y = 3x + 2$ **2.** $Q(0, -2)$; $y = -\frac{1}{2}x - 1$ **3.** $R(-1, 2)$; $y = -x$
                          $x + 2y = -4$ $x + y = 1$
4. $P(4, 3)$; $x + 2y = 7$ **5.** $A(5, 1)$; $4x + 3y = 15$ **6.** $C(-2, -3)$; $3x - 5y = 9$
   $x + 2y = 10$ $4x + 3y = 23$ $3x - 5y = 10$
7. $D(3, -1)$; $x = 4$ **8.** $F(-3, -6)$; $4y - 6 = 2$ **9.** $P(-5, 3)$; $3x - 7 = 14$
   $x = 3$ $y = -6$ $x = -5$

Write an equation, in standard form, of the line that contains the given point and is perpendicular to the given line.

10. $P(4, 2)$; $3x + 2y = 16$ **11.** $A(-5, -1)$; $4x - 3y = -17$ **12.** $J(3, 5)$; $x + y = 8$
   $y = \frac{2}{3}x - 4$ $4y + 3x = 8$ $y - x = 11$
13. $R(4, -3)$; $x = 4$ **14.** $A(-5, -4)$; $y = -4$ **15.** $T(3, 2)$; $3x - y = 7$
   $8 = -4y - 4$ $3x - 4 = 5$ $-x - 3y = 15$

Determine whether the given lines are parallel, perpendicular, or neither.

16. $y = 2x$;   **17.** $y = 3x + 4$;   **18.** $4x - 5y = 10$;
   $y = -2x$ Neither $y = 3x - 5$ parallel $15x + 12y = 36$
                                               perpendicular

Write an equation of the line satisfying the given conditions.

19. perpendicular to the line $-x - y = 9$ at $P(-6, -3)$ $y = x + 3$
20. containing $T(3, 0)$ and parallel to the line through $R(5, 4)$ and $S(3, 1)$. $3x - 2y = 9$
21. perpendicular to the line containing $A(2, -1)$ and $B(3, 1)$ at B $x + 2y = 5$
22. For what value of k will the line $3x + ky = 8$ be perpendicular to the line $4x - 3y = 6$?   4
23. For what value of k will the line $2x - ky = 4$ be parallel to the line $x - y = -1$?   2

Mixed Review

Let $f(x) = 3x - 4$ and $g(x) = x^2 - 2$. Find each of the following.   *3.2*

1. $f(-4)$ -16 **2.** $g(-3)$ 7   **3.** $f(-5)$ -19 **4.** $f(2) - g(-3)$ -5

▬▬▬/ **Brainteaser**

Find a natural number n such that $n(n - 1)(n - 2)(n - 3)(n - 4) = 95,040$.   12

Enrichment

Draw this figure on the chalkboard and explain to the students that *ABCD* represents any square.

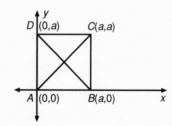

Challenge the students to use information learned in this lesson to prove that the diagonals of a square are perpendicular.

Slope of $\overleftrightarrow{AC} = \dfrac{a - 0}{a - 0} = \dfrac{a}{a} = 1$

Slope of $\overleftrightarrow{DB} = \dfrac{a - 0}{0 - a} = \dfrac{a}{-a} = -1$

Since $(1)(-1) = -1$, $\overleftrightarrow{AC}$ and $\overleftrightarrow{DB}$ are $\perp$ by Thm 3.5.

Extension

Logical Reasoning

When the equation $2x + 6 = 14$ is solved, the result is $x = 4$. Recall from Chapter 1 that if you give a reason for each step and write them in statement-reason format, you are presenting the proof of your conclusion. Thus, if you are given $2x + 6 = 14$, you can *conclude* $x = 4$ from the following proof. It is *assumed* that the given is true.

Statement	Reason
1. $2x + 6 = 14$	1. Given
2. $2x = 8$	2. Subt Prop of Eq
3. $\therefore\ x = 4$	3. Div Prop of Eq

A statement written in the following form is a **conditional** statement.

"If p, then q."

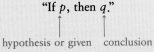

hypothesis or given   conclusion

Theorems can also be stated in the form of a conditional. For example, Theorem 3.2 of this chapter states that the slope of a horizontal line is 0.

If *a line is horizontal*, then *the slope of the line is 0*.

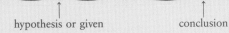

hypothesis or given   conclusion

EXAMPLE 1 Write as a conditional: The slope of a vertical line is undefined.

Solution Hypothesis: a line is vertical. Conclusion: slope is undefined.

Conditional: If *a line is vertical*, then *its slope is undefined*.

Interchanging the hypothesis and the conclusion of a conditional produces another conditional, called the *converse*. The converse of a true conditional may or may not be true.

EXAMPLE 2 Conditional: If you live in Ohio, then you live in the United States. (True)

Converse: If you live in the United States, then you live in Ohio. (False)

Definition

The **converse of a conditional** is the statement formed by interchanging the hypothesis and conclusion.

Extension

1. If a line is horizontal, then its slope is 0.
2. If lines are parallel, then their slopes are the same.
3. If angles are supplementary, then the sum of their measures is 180.
4. Hypothesis: $3x - 7 = 8$
   Conclusion: $x = 5$
5. Hypothesis: $a > b$ and $b > c$
   Conclusion: $a > c$
6. Hypothesis: $x = 2$; conclusion: $x^2 = 4$
7. True; converse: If $x = 10$, then $6x - 6 = 4x + 14$. True.
   Biconditional: $6x - 6 = 4x + 14$ if and only if $x = 10$.
8. True; converse: If $|x| = 4$, then $x = -4$. False
   Biconditional $x = -4$ if and only if $|x| = 4$.
9. Conditional: If the square of a number is 9, then the number is 3.
   Converse: If a number is 3, then its square is 9.
   The biconditional is false since the conditional is false.

15.

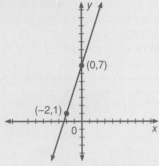

16.

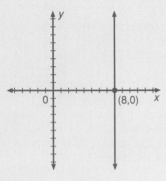

17.

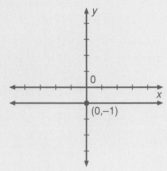

Recall that when $2x + 6 = 14$ was solved, the result was $x = 4$. However, you should not conclude that 4 is the solution of the equation without checking.

Substitute 4 for x.

$$2x + 6 = 14$$
$$2 \cdot 4 + 6 \stackrel{?}{=} 14$$
$$14 = 14 \text{ True}$$

Thus, 4 is a solution of $2x + 6 = 14$, since the conditional and its converse are *both* true.

In this case the conditional and its converse can be combined into a single statement called a *biconditional*.

Conditional:	If $2x + 6 = 14$, then x = 4.	True
Converse:	If $x = 4$, then $2x + 6 = 14$.	True
Biconditional:	$2x + 6 = 14$ *if and only if* $x = 4$.	

Definition

When a conditional statement and its converse are combined by "if and only if," the resulting statement is called a **biconditional**. The biconditional, "*p if and only if q*," is true only when the conditional, "if p, then q," and its converse, "if q, then p," are *both* true.

A conditional or its converse can be shown false by finding one example where it is false. Such an example is called a **counterexample**.

Exercises

Write the following statements as conditionals.

1. The slope of a horizontal line is 0.
2. The slopes of parallel lines are the same.
3. Supplementary angles are angles the sum of whose measures is 180.

Identify the hypothesis and conclusion of each conditional.

4. If $3x - 7 = 8$, then $x = 5$.
5. If $a > b$ and $b > c$, then $a > c$.
6. If $x = 2$, then $x^2 = 4$.

Determine whether each conditional is true. Write the converse of the conditional. Is it true? Write a biconditional for the conditional and its converse.

7. If $6x - 6 = 4x + 14$, then $x = 10$.
8. If $x = -4$, then $|x| = 4$.
9. Write the conditional and converse that produces the following biconditional: The square of a number is 9 if and only if the number is 3. Is the biconditional true? Why?

104 Extension

3.8 Direct Variation

Objectives

To determine whether relations are direct variations
To find missing values in direct variations
To solve word problems involving direct variations

Recall that the function $f(x) = mx + b$ (or $y = mx + b$) is a linear function. If the y-intercept of the graph of a linear function (b) is 0, then the function is a special type of function called a *direct variation*.

The figure at the right is the graph of the direct variation $f(x) = \frac{2}{3}x$, or $y = \frac{2}{3}x$. Notice that for any pair (x, y) other than $(0, 0)$, the ratio $\frac{y}{x}$ is *constant*.

The table below illustrates this.

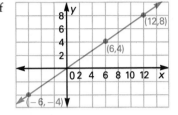

x	y
−6	−4
6	4
12	8

$$\frac{y}{x} = \frac{-4}{-6} = \frac{4}{6} = \frac{8}{12} = \frac{2}{3}$$

The ratio of y to x is always the same, or constant: $\frac{2}{3}$.

Definition

A linear function defined by an equation of the form $y = kx$ ($k \neq 0$) is called a **direct variation**. The constant k is called the **constant of variation**.

The equations $\frac{y}{x} = k$ and $y = kx$ can be read as "y varies directly as x."

EXAMPLE 1 Determine whether the table below expresses a direct variation. If so, find the constant of variation and an equation of the function.

Solution

x	y
2	−8
5	−20
6	−24

$$\frac{y}{x} = \frac{-8}{2} = \frac{-20}{5} = \frac{-24}{6} = -4$$

$$\frac{y}{x} = -4, \text{ or } y = -4x$$

So, y varies directly as x. The constant of variation is -4, and an equation of the function is $y = -4x$.

Additional Example 1

Determine whether the table below expresses a direct variation. If so, find the constant of variation and an equation of the function.

x	y
3	9
−2	−6
5	15

Yes; 3 is the constant of variation. $\frac{y}{x} = 3$, or $y = 3x$

Lesson Note

Emphasize the equation form $\frac{y}{x} = k$ as an interpretation of "y varies directly as x." This will help in Chapter 12 when students study inverse variation and direct and inverse variation combined. For example, $yx = k$ is read "y varies inversely as x," and $\frac{yz}{x} = k$ is read "y varies directly as x and inversely as z."

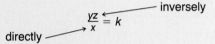

$$\underset{\text{directly}}{\longrightarrow} \frac{yz}{x} = k \underset{\text{inversely}}{\longleftarrow}$$

Math Connections

Electrical Engineering: There are many examples of direct variation in science and engineering. One of the most familiar is *Ohm's Law*, which expresses the relationship between a variable direct current and variable voltage in an electric circuit that has a constant resistance. The formula is $E = IR$, where E is the voltage in volts, I is the current in amperes, and R is the constant resistance in ohms.

EXAMPLE 2 Determine whether the equation represents a direct variation. If so, find the constant of variation.

  a. $-9y = 27x$ **b.** $y = 7x - 4$

Solutions

  a. Solve for y. $y = -3x$
  The equation is of the form $y = kx$, where $k = -3$. So, $y = -3x$ represents a direct variation, and -3 is the constant of variation.

  b. The equation is not of the form $y = kx$.
  So, $y = 7x - 4$ does not represent a direct variation.

EXAMPLE 3 The bending of a beam varies directly as the mass of the load it supports. A beam is bent 20 mm by a mass of 40 kg. How much will the beam bend when supporting a mass of 100 kg?

Solution

1. Write a formula to find k.
   $\underbrace{\text{bending}}\ \underbrace{\text{varies directly}}\ \text{as}\ \underbrace{\text{mass}}$

   $b = km$
   $20 = k \cdot 40$
   $\frac{1}{2} = k$

2. Substitute $\frac{1}{2}$ for k in the formula.
   $b = km$ becomes $b = \frac{1}{2}m$.

3. Find b for $m = 100$.
   $b = \frac{1}{2}m = \frac{1}{2} \cdot 100 = 50$

So, the beam will bend 50 mm when supporting a mass of 100 kg.

A direct variation problem can also be solved using a proportion (see Lesson 1.4). If (x_1, y_1) and (x_2, y_2) are any two ordered pairs of a direct variation, and neither is $(0, 0)$, then $\frac{y_1}{x_1} = k$, $\frac{y_2}{x_2} = k$, and by the Transitive Property of Equality, $\frac{y_1}{x_1} = \frac{y_2}{x_2}$.

In Example 3, the bending of the beam *is directly proportional* to the mass of the load. The solution below shows how a proportion can be used to solve the problem.

The beam is bent 20 mm by a mass of 40 kg.

(b_1, m_1)
$(20, 40)$

Find the bending for a mass of 100 kg.

(b_2, m_2)
$(b_2, 100)$

$$\frac{b_1}{m_1} = \frac{b_2}{m_2}$$
$$\frac{20}{40} = \frac{b_2}{100}$$

Solve the proportion for b_2. $40b_2 = 2{,}000$
$$b_2 = 50 \text{ mm}$$

Additional Example 2

Determine whether the equation represents a direct variation. If so, find the constant of variation.

a. $3x - y = 9$ No
b. $-4y = 16x$ Yes; -4 is the constant of variation.

Additional Example 3

A stone-and-gravel center charges $24 for enough white rocks to cover a square 4 ft on a side. At this rate, how much would it cost to cover a square 25 ft on a side?

$\frac{24}{4^2} = \frac{x}{25^2}$; $x =$ $937.50

Furthermore, direct variation need not be restricted to linear functions, as shown in the next example.

EXAMPLE 4 A nursery charges $48 for enough plants to cover a square patch 8 ft on a side. At that rate, how much would it cost to cover a square patch 6 ft on a side?

Solution The cost of the ground cover (C) is directly proportional to the area of the patch, which is the square of one side (s).

$$\frac{C_1}{(s_1)^2} = \frac{C_2}{(s_2)^2}$$

$$\frac{48}{8^2} = \frac{C_2}{6^2}$$

$$64C_2 = 1{,}728$$

$$C_2 = 27$$

Calculator Solution

Solve the proportion for C_2.

$$C_2 = \frac{(48)(6^2)}{8^2}$$

Then use a calculator as follows.

48 ⊠ 6 ⌈xʸ⌉ 2 ÷ 8 ⌈xʸ⌉ 2 ⌈=⌉ 27

So, it would cost $27 to cover a square patch 6 ft on a side.

Classroom Exercises

Give an equation that describes each direct variation. Use k for the constant of variation.

1. y varies directly as z. $\frac{y}{z} = k$, or $y = kz$
2. m is directly proportional to t. $\frac{m}{t} = k$, or $m = kt$
3. p varies directly as w. $\frac{p}{w} = k$, or $p = kw$
4. a varies directly as the square of g.
5. e is directly proportional to f. $\frac{e}{f} = k$, or $e = kf$
6. g varies directly as h. $\frac{g}{h} = k$, or $g = kh$

Find the constant of variation for each direct variation.

7. $\frac{y}{x} = \frac{4}{7}$ $\frac{4}{7}$
8. $y = \frac{2}{3}x$ $\frac{2}{3}$
9. $-4y = -20x$ 5
10. $-5y = -15x$ 3
11. $\frac{y}{2x} = 7$ 14
12. $\frac{3y}{x} = -6$ -2

3.8 Direct Variation **107**

Critical Thinking Questions

Analysis: Ask students to consider a formula that they have used earlier in the course and to discuss the conditions under which it might represent direct variation. For example, the distance formula $d = rt$ would represent direct variation if $r = 40$ mi/h (a constant value) and also if the time t were 2 h or some other constant value.

Checkpoint

y varies directly as x. Find the value as indicated. (Exercises 1–2)

1. $y = 18$ when $x = 6$. Find y when $x = 7$. 21
2. $y = 3$ when $x = 10$. Find x when $y = \frac{1}{2}$. $1\frac{2}{3}$
3. y is directly proportional to the square of x. $y = 8$ when $x = 3$. Find y when $x = 4$. $\frac{128}{9}$, or $14\frac{2}{9}$

Closure

Ask students how they can determine if a table of x and y values expresses a direct variation. Find k, the constant of variation. Then ask them to find the value of y if (14, 3.5) and (8.4, y) are two ordered pairs of a direct variation. 2.1

Additional Example 4

A skydiver jumps from a plane and falls 88.2 meters in 3 seconds. The distance the skydiver falls varies as the square of the number of seconds falling. How far will the sky diver fall in 8 seconds? 627.2 meters

FOLLOW UP

Guided Practice

Classroom Exercises 1–12

Independent Practice

A Ex. 1–29, **B** Ex. 30–38, **C** Ex. 39–42

Basic: WE 1–31 odd

Average: WE 7–37 odd

Above Average: WE 9–41 odd

Additional Answers

Classroom Exercises

4. $\frac{a}{g^2} = k$, or $a = kg^2$

Written Exercises

42. $f(x) = kx$

$$f(a) + f(b) = ka + kb$$
$$= k(a + b)$$
$$= f(a + b)$$

Written Exercises

Determine whether the table or equation expresses a direct variation. If so, find the constant of variation, and for the tables, write an equation of the function.

1.

x	y
5	1
15	3
25	5
45	9

$\frac{1}{5}$, $y = \frac{1}{5}x$

2.

x	y
4	7
12	21
−8	−14
−16	−28

$\frac{7}{4}$, $y = \frac{7}{4}x$

3.

x	y
−18	6
12	−4
−45	15
54	−18

$-\frac{1}{3}$, $y = -\frac{1}{3}x$

4.

x	y
5	4
9	8
−3	−4
−9	−10

No

5.

x	y
7	7
−8	−8
13	13
−3	−3

1, $y = x$

6.

x	y
4	5
6	7
−3	−2
0	1

No

7.

x	y
2	3
8	12
14	21
−4	−6

$\frac{3}{2}$, $y = \frac{3}{2}x$

8.

x	y
2	−1
4	1
9	6
11	8

No

9.

x	y
−4	5
8	−10
24	−30
−12	15

$-\frac{5}{4}$, $y = -\frac{5}{4}x$

10. $y = -4x$ -4

11. $y = -x$ -1

12. $y = 3x - 5$ No

13. $\frac{y}{x} - 4 = 0$ 4

14. $y = -3x$ -3

15. $y = 2x$ 2

16. $7y = 4x$ $\frac{4}{7}$

17. $-4y = -8x$ 2

18. $y = 4 - 3x$ No

19. $y = \frac{3}{x}$ No

20. $y = x + 3$ No

21. $xy = 4$ No

In Exercises 22–25, t varies directly as u. Find the value as indicated.

22. Given that $t = 21$ when $u = 7$, find t when $u = 4$.   12

23. Given that $t = -27$ when $u = 3$, find t when $u = 18$.   −162

24. Given that $t = 16$ when $u = -8$, find t when $u = 4$.   −8

25. Given that $t = 2.6$ when $u = 20.8$, find u when $t = 3.9$.   31.2

26. Given that y varies directly as x, and $y = 0.18$ when $x = -6$, find y when $x = -9$.   0.27

27. Given that v varies directly as w, and $v = 24$ when $w = -6$, find v when $w = 9$.   −36

28. Given that p is directly proportional to q, and $p = -81$ when $q = 9$, find p when $q = -3$.   27

29. Given that y is directly proportional to x, and $y = -6$ when $x = -30$, find x when $y = 22.2$.   111

108   Chapter 3 Relations, Functions: Graphing Linear Functions

30. A number y varies directly as the square of the number x. If $y = 32$ when $x = 4$, find y when $x = 6$. 72

31. A number y varies directly as the cube of a second number x. If $y = 54$ when $x = 3$, find y when $x = 5$. 250

32. The current I in amperes in an electric circuit varies directly as the voltage V. When 24 volts are applied, the current is 8 amperes. Find the current when 36 volts are applied. 12 amperes

33. The cost of gold varies directly as its mass. Given that 6 g of gold cost $102, find the cost of 14 g of gold. $238

34. Gas consumption of a car is directly proportional to the distance traveled. A car uses 20 gallons of gas to travel 300 mi. How much gas will the car consume on a trip of 650 mi? $43\frac{1}{3}$ gal

35. The distance required to stop a car varies directly as the square of its speed. It requires 144 ft to stop a car at 60 mi/h. What distance is required to stop a car at 45 mi/h? 81 ft

36. The height reached by a ball thrown vertically upward is directly proportional to the square of its initial velocity. If a ball reaches a height of 46 m when it is thrown upward with an initial velocity of 30 m/s, what height will the ball reach if the initial vertical velocity is 40 m/s? $81\frac{7}{9}$ m

37. Hooke's Law says that the distance a spring is stretched by a hanging object is directly proportional to the weight w of the object. If the distance is 80 cm when the weight is 6 kg, what will be the distance when the weight is 10 kg? $133\frac{1}{3}$ cm

38. Given that A is directly proportional to B, and $(A_1, B_1) = (36, 5)$, find B_2 when $A_2 = -9$. $-\frac{5}{4}$

39. If y varies directly as x, and the value of y is positive and held constant, what happens to the value of k as x increases? Decreases

40. If y varies directly as x, what effect will tripling x have on y? Triples y

41. If y varies directly as the square of x, what effect will doubling x have on y? Quadruples y

42. Write an equation of a function f given that f is a direct variation whose constant of variation is k. Then show that $f(a) + f(b) = f(a + b)$ for all real numbers a and b.

Mixed Review

Find the solution set. *2.1, 2.2*

1. $3x - 2 < 5x + 6$ $\{x \mid x > -4\}$
2. $3x - 8 > x + 6$ *and* $2x - 5 < 23$ $\{x \mid 7 < x < 14\}$
3. $x + 5 \leq 8$ *or* $2x - 3 > 1$ {Real numbers}
4. $-5 < 2x - 9 \leq 5$ $\{x \mid 2 < x \leq 7\}$

For each compound sentence, determine whether it is true or false. *2.2*

5. $2^4 = 4^2$ *or* $4 - 7 = -(7 - 4)$ T
6. $8 > 4 \cdot 2$ *and* $-6 \leq -3 \cdot 2$ F

Enrichment

Discuss the fact that direct variation has countless applications in everyday living. To take one example, the hanging spring scale found in grocery stores is a practical application of Hooke's law, mentioned in Exercise 37.

Have the students use library resources to discover similar practical applications of direct variation, and have them write reports on their findings.

As an extension of this project, have the students organize a bulletin-board display on the applications of direct variation.

▰▰GETTING STARTED

Prerequisite Quiz

Solve for x.

1. $6 - 2x < 8$ $x > -1$
2. $5x - 4 \geq 7x - 8$ $x \leq 2$

Graph on a number line.

3. $x < -4$ ⟵———○———⟶
                    -4

Motivator

Ask the class to graph $x > 4$ on a number line.   Shade all points to the right of $x = 4$. Ask the class to graph $y > 2x$ on a coordinate plane.   First graph $y = 2x$ as a dotted line and then shade to the left and above the dotted line.

▰▰TEACHING SUGGESTIONS

Lesson Note

Compare the graphing of inequalities on a line with graphing inequalities in a plane. Instead of shading a line to the left or right of a point, the student will shade a half-plane above, below, to the left, or to the right of a line.

Math Connections

Models: The half-planes of this lesson can be compared to the one-dimensional inequalities of Chapter 2. The two-dimensional *plane* of this lesson is the counterpart of the one-dimensional *line* of Lesson 2.1.

3.9 Graphing Inequalities

Objective   To graph linear inequalities in two variables

In Chapter 2, you graphed linear inequalities in *one* variable as *rays*.

The graphs of linear inequalities in two variables are *half-planes*.

The graph of $y = \frac{1}{2}x - 2$ separates the coordinate plane into two regions called **half-planes**. One half-plane is above the line, and the other is below the line. The line is called the **boundary** of the two half-planes. One half-plane is described by $y > \frac{1}{2}x - 2$ and the other by $y < \frac{1}{2}x - 2$.

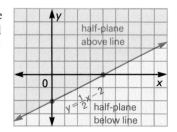

EXAMPLE 1   Graph the inequality $-x - 5y \geq 15$. Check.

Solution   1. Solve the inequality for y.
$$-x - 5y \geq 15$$
$$-5y \geq x + 15$$
$$y \leq -\frac{1}{5}x - 3$$

2. Graph the equation $y = -\frac{1}{5}x - 3$. Because the inequality sign is $\leq$, the points on the line are part of the graph. Use a *solid* line to show this.

3. Shade the region *below* the line.

Check   Pick one point, such as $Q(2, -5)$, in the region below the line, and another point, such as $P(3, 1)$, above the line. Substitute.

$$-x - 5y \geq 15 \qquad\qquad -x - 5y \geq 15$$
$$-2 - 5(-5) \overset{?}{\geq} 15 \qquad -3 - 5 \cdot 1 \overset{?}{\geq} 15$$
$$23 \geq 15 \text{ True} \qquad\qquad -8 \geq 15 \text{ False}$$

Because $23 \geq 15$ is true, the point $Q(2, -5)$ is a part of the graph. Thus, the graph of $-x - 5y \geq 15$ is the line and the shaded region below the line.

Highlighting the Standards

Standard 5b: The equation, table, and graph offer three different perspectives for viewing an inequality. Together, they lead to fuller understanding than any can provide alone.

Additional Example 1

Graph the inequality $x - y > -4$. Check. $y < x + 4$
Check (0,0) in the shaded region.
$x - y > -4;\ 0 - 0 > -4$ (True)
Check (0,5) above the line.
$x - y > -4;\ 0 - 5 > -4$ (False)

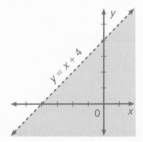

EXAMPLE 2   Graph the inequality $3x - 4 < 11$.

Solution   Solve for x.
$$3x - 4 < 11$$
$$3x < 15$$
$$x < 5$$

Graph the equation $x = 5$. Use a *dashed* line because the inequality symbol is $<$, *not* $\leq$. The points on the line are not part of the graph. Shade the region to the left of the line because the x-coordinate of every point to the left of the line is less than 5. The graph is the shaded region, not including the vertical dashed line.

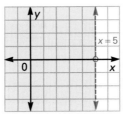

Classroom Exercises

Solve each inequality for *y*. Then, graph each inequality.

1. $2y > 4x - 6$ $y > 2x - 3$
2. $-3y < 6x - 3$ $y > -2x + 1$
3. $-y > x - 2$ $y < -x + 2$
4. $-5y \leq -10x + 20$ $y \geq 2x - 4$

Written Exercises

Graph each inequality. Check.

1. $y < 3x - 4$
2. $x > 4$
3. $y > -4x - 8$
4. $y < 5$
5. $-4x \leq 8$
6. $y \leq -4x - 3$
7. $3y > 18$
8. $y > -x + 6$
9. $y < 2x - 5$
10. $y > \frac{2}{3}x - 1$
11. $x \leq -4$
12. $y < -4$
13. $3x - 2y \leq 8$
14. $5x + 2y < 10$
15. $3x - 7y \geq 21$
16. $4 < y < 6$
17. $8 > 2x - 6 > 4$
18. $4 < -2x - 8 \leq 10$

Mixed Review

Write an equation, in standard form, of the line satisfying the given conditions.   *3.4, 3.7*

1. containing the points $A(0, -5)$ and $B(9, 1)$ $2x - 3y = 15$
2. containing the point $T(-6, 8)$ and having slope $-\frac{2}{3}$ $2x + 3y = 12$
3. containing $P(1, 2)$ and parallel to the line $3x - 2y = 6$ $3x - 2y = -1$

2. No solution; $36\frac{1}{3}$, $37\frac{1}{3}$, and $38\frac{1}{3}$ are not integers.

3. Insufficient data; need to know width, perimeter, or area

4. No solution; no triangle has sides 7 in., 8 in., and 15 in. since 7 in. + 8 in. = 15 in.

5. Insufficient data; need to know the width or area of the floor

Additional Answers, page 111

Classroom Exercises

1.

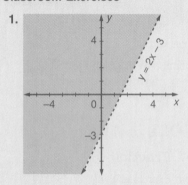

2.

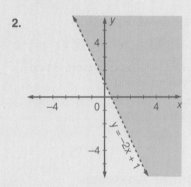

Problem Solving Strategies

Insufficient Data and Contradictions

Sometimes, a problem may not provide enough data to obtain an answer.

Example 1

How much change would you receive from a 10-dollar bill if you purchased 6 quarts of motor oil?

Example 1, as stated, cannot be solved. There is *insufficient data*. That is, to solve this problem, you would need to know the price of 1 quart and the amount of sales tax, if any.

On the other hand, there can be enough data to solve a problem but the solution leads to a contradiction. This is illustrated in Example 2.

Example 2

A rectangle and a square have the same width and the rectangle's length is 5 ft more than twice its width. Find the perimeter of the rectangle if the difference between the two perimeters is 8 ft.

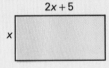

$$2x + 2(2x + 5) - 4x = 8$$
$$2x + 10 = 8$$
$$2x = -2$$
$$x = -1$$

The width x of the square cannot be -1 ft because the width of a square must be a positive number. Thus, there is *no solution*.

Exercises

Solve each problem if possible. If there is insufficient data, state this and tell what additional data is needed. If the solution leads to a contradiction, state "no solution" and describe the contradiction.

1. How much change would you receive from $20 if you purchased 3 neckties at $2.20 each, 2 belts at $5.80 each, and paid a 5% sales tax. 89¢

2. The sum of three consecutive integers is 112. Find the three integers.

3. A rectangle is twice as long as it is wide. Find the length of the rectangle.

4. Side a of a triangle is 1 in. shorter than side b, and side c is 7 in. longer than side b. Find the length of each side if the triangle's perimeter is 30 in.

5. A rectangular floor, 18 ft long, is to be covered by a carpet that costs $45 per sq yd, tax included. Find the cost of the carpet.

Mixed Problem Solving

Solve each problem.

1. Find the two numbers whose sum is 85, given that the greater number is 16 times the smaller number. 5, 80

2. Karen is 7 years older than Bob. Three years ago, Karen's age was twice Bob's age then. How old is Karen now? 17 yr old

3. Rita is 24 years older than Manuel, and 2 years from now she will be 5 times as old as he will be. How old is Rita now? 28 yr old

4. Find three consecutive multiples of 5 such that the sum of the two smaller numbers is 50 more than the greatest number. 55, 60, 65

5. Find three consecutive odd integers such that twice the sum of the first two integers is 73 more than 3 times the third integer. 81, 83, 85

6. What initial vertical velocity is required to boost an object to a height of 375 m at the end of 3 s? Use the formula $h = vt - 5t^2$. 140 m/s

7. Find three consecutive multiples of 9 such that 5 times the sum of the first two multiples is greater than 600, and 4 times the sum of the last two multiples is less than 660. 63, 72, 81

8. The perimeter of a triangle is 71 ft. The length of the first side is 11 ft less than twice the length of the second side. The third side is 14 ft longer than the second side. Find the length of the third side. 31 ft

9. An aircraft carrier traveling 25 km/h left a port at 6:30 A.M. At 10:00 A.M., a helicopter left the same port, traveling the same route at 60 km/h. At what time did the helicopter overtake the carrier? 12:30 P.M.

10. The stock of women's shoes in a shoe store is 40 pairs more than twice the stock of men's shoes. The stock of children's shoes is 80 pairs less than twice the stock of women's shoes. Given that the total stock is 600 pairs of shoes, find the number of pairs of children's shoes. 320 pairs

11. Machine A is 3 times as old as Machine B. Machine C was constructed 20 years after Machine A was built. In 12 years, the age of the oldest machine will be the same as the combined ages of the other two machines now. How old is each machine now?
A: 96 yr old
B: 32 yr old
C: 76 yr old

12. The length of a rectangle is 5 cm less than 3 times its width, and one side of an equilateral triangle is as long as the length of the rectangle. Given that the sum of the two perimeters is 60 cm, find the length of the rectangle. 10 cm

13. At 8:00 A.M., a van left City A and traveled 60 km/h toward City B. At 10:00 A.M., an automobile left City B and traveled at 80 km/h toward City A. At what time did the automobile and van meet, given that the cities are 330 km apart? 11:30 A.M.

Additional Answers, page 111

3.

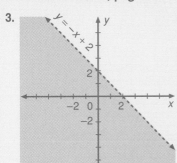

4.

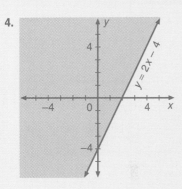

Written Exercises

1.

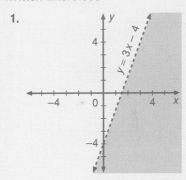

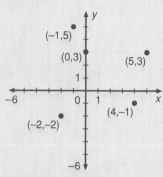
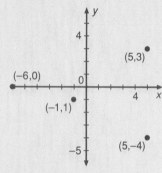
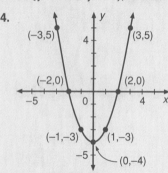
Chapter 3 Review

Key Terms

abscissa (p. 73)	linear function (p. 78)	relation (p. 74)
collinear (p. 85)	linear model (p. 96)	slope (p. 84)
constant function (p. 89)	mapping (p. 74)	slope-intercept form (p. 91)
constant of variation (p. 105)	ordered pair (p. 73)	value of a function (p. 79)
coordinate plane (p. 73)	ordinate (p. 73)	vertical-line test (p. 75)
direct variation (p. 105)	origin (p. 73)	x-axis (p. 73)
domain (p. 74)	point-slope form (p. 88)	y-axis (p. 73)
function (p. 75)	quadrant (p. 73)	y-intercept (p. 91)
half-plane (p. 110)	range (p. 74)	

Key Ideas and Review Exercises

3.1 To verify that a relation is a function, show either that (1) no two ordered pairs have the same first coordinate, or that (2) no vertical line crosses the graph of the relation more than once.

Graph each relation and determine its domain and range. Is it a function?

1. {(5,3), (4,−1), (−1,5), (−2,−2), (0,3)}   **2.** {(5,3), (−6,0), (−1,−1), (5,−4)}

3. Determine the domain and range of the relation whose graph is given at the right. Use the vertical-line test to determine whether the relation is a function.

3.2 To graph a function given its equation and domain, solve for y, and construct a table of values for x and y. Plot the points; draw the graph.

4. Graph $x^2 - y = 4$ for the following values of x: −3, −2, −1, 0, 1, 2, 3.

Let $f(x) = x^2 - 4x - 2$ and $g(x) = -5x + 2$. Find each of the following.

5. $f(-4)$ 30   **6.** $g(-3)$ 17

7. The range of f for the domain {−6, −2, 5}.   **8.** $g(3 + h)$ $-5h - 13$
{3, 10, 58}

3.3 To determine the slant of a line, find the slope. If the slope > 0, then the line slants up to the right. If the slope < 0, then the line slants down to the right. If the slope = 0, then the line is horizontal. If the slope is undefined, then the line is vertical. (See p. 84 for the slope formula.)
To verify that three points are collinear, show that the slope is the same for any two pairs of the points.

Find the slope of $\overleftrightarrow{PQ}$. Describe the slant of the line.

9. $P(-4, 2)$, $Q(-3, -2)$ **10.** $P(-6, -5)$, $Q(-6, -1)$ **11.** $P(-3, 4)$, $Q(7, 4)$ 0, horiz

12. Determine whether the points $A(2, 5)$, $B(0, -1)$, and $C(1, 2)$ are collinear. Yes

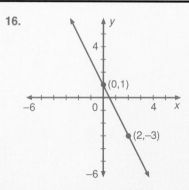

3.4 To find an equation of a line,

1. Use $y - y_1 = m(x - x_1)$ if the line is nonvertical and m and (x_1, y_1) are given. If two points are given, first use the slope formula to find m.
2. Use $y = c$ if the line is horizontal and contains $P(x, c)$.
3. Use $x = c$ if the line is vertical and contains $P(c, y)$.

Find an equation, in standard form, of the line containing the given points.

13. $P(0, -1)$, $Q(-3, 1)$ **14.** $P(4, 5)$, $Q(4, -2)$ $x = 4$ **15.** $P(7, 2)$, $Q(1, -7)$
    $2x + 3y = -3$ $3x - 2y = 17$

3.5 To graph a line using the slope-intercept method, write an equation in $y = mx + b$ form. Plot the point $(0, b)$, where b is the y-intercept. Use this point and the slope, m, to plot a second point. Draw the graph.

Graph each equation.

16. $f(x) = -2x + 1$ **17.** $3x - 5y = 15$ **18.** $3y - 4 = 11$

3.6 To write an equation of a linear model for a given set of data, first plot the points corresponding to the data. Then draw the straight line that best fits the points. Finally, use two points on the line to write the equation.

19. Use the table to construct a linear model. Write an equation of the model.

Number of months: x	2	3	4	5	6
Ants (thousands): y	9	10	13	18.5	19

20. Describe in your own words the procedure for writing an equation of a linear model corresponding to a given set of data.

3.7 To find an equation of a nonvertical line parallel to a given line, use the fact that parallel lines have the same slope.

To find an equation of a nonvertical, nonhorizontal line perpendicular to a given line, use the fact that the product of the slopes of two perpendicular lines is -1.

Find an equation of the line satisfying the given conditions.
    $4x - y = -21$
21. containing $P(-5, 1)$ and parallel to the line whose equation is $4x - y = 8$

22. containing $Q(-2, 5)$ and perpendicular to the line whose equation is $-3x - 5y = 20$ $5x - 3y = -25$

3.8 To verify that a function is a direct variation, show that for all $x \neq 0$ and for all y, $\frac{y}{x} = k$ or $y = kx$, where k is a nonzero constant.

Determine whether the equation expresses a direction variation.

23. $y = 4x + 2$ No   **24.** $c = 5s$ Yes   **25.** $6y = 11x$ Yes

17.

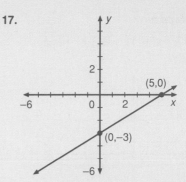

18.

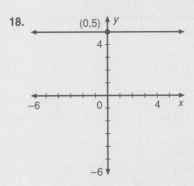

19. Answers may vary. A good fit is the line $y = \frac{5}{2}x + 4$.

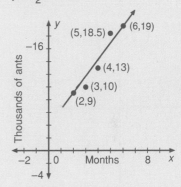

20. Choose two points on the linear model. Calculate the change in y over the change in x to find the slope. Substitute the slope and one of the points into the point-slope form of the equation of a line to find the equation of the linear model.

1. D: $\{-4, -1, 3\}$; R: $\{-2, 1, 6\}$; not a function

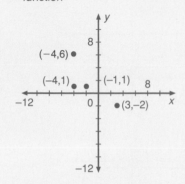

2. D: $\{-5, 2, 6, 8\}$; R: $\{-4, -3, 0\}$; function

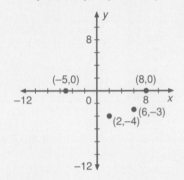

14.

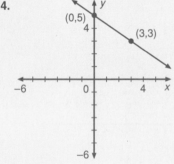

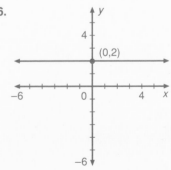

Chapter 3 Test   A Exercises: 1–5, 7–23   B Exercises: 6, 24–26
C Exercise: 27

Graph each relation and determine its domain and range. Is each a function?

1. $\{(-4, 1), (3, -2), (-4, 6), (-1, 1)\}$ **2.** $\{(-5, 0), (2, -4), (6, -3), (8, 0)\}$

Let $f(x) = -x^2 - x + 4$ and $g(x) = -4x + 5$. **Find each of the following.**

3. $g(-3)$ 17   **4.** $f(4)$ -16

5. the range of f for the domain $\{-2, 0, 5\}$ **6.** $g(2 + h)$ $-4h - 3$

$\{-26, 2, 4\}$

Find the slope of $\overleftrightarrow{AB}$. Describe the slant of the line.

Down to right   Undef; vert   0, horiz

7. $A(-2, 5), B(3, -1)$ $-\frac{6}{5}$;   **8.** $A(-4, -1), B(-4, 6)$ **9.** $A(-6, -3), B(2, -3)$

10. Determine whether the points $A(0, 1), B(2, -3), C(4, 5)$ are points on the same line. No

Find an equation, in standard form, of the line containing the given points.

$3x + 4y = 28$ $y = -1$ $4x - y = 23$

11. $P(8, 1), Q(-4, 10)$ **12.** $A(-3, -1), B(4, -1)$ **13.** $A(5, -3), B(3, -11)$

Graph each equation.

14. $f(x) = -\frac{2}{3}x + 5$ **15.** $-4x - 3y + 12 = 0$ **16.** $-4 = 6 - 5y$

17. Use the table to construct a linear model. Write an equation of the model.

Number of hours: x	4	6	8	10	12	14
Thousands of bacteria: y	80	70	58	52	36	33

18. Predict the number of bacteria after 18 h.

Write an equation, in standard form, of the line satisfying the given conditions.

19. containing $P(-1, 2)$ and parallel to the line whose equation is $-2x - 4y = 8$ $x + 2y = 3$

20. containing $T(3, -1)$ and perpendicular to the line whose equation is $x - y = 5$ $x + y = 2$

Determine whether each equation represents a direct variation. If so, find the constant of variation.

21. $-6y = 12$ No   **22.** $x - y = 4$ No   **23.** $\frac{x}{y} = -4$ $-\frac{1}{4}$

Graph each inequality.

24. $y < -3x + 1$ **25.** $4x + 2y \le 8$ **26.** $3x - 1 > 5$

27. Graph $-4 < 2x - 8 \le 6$ in a coordinate plane.

15.

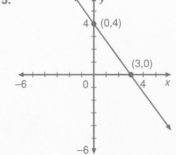

16.

College Prep Test

Choose the *one* best answer to each question or problem.

1. Which ordered pair (x,y) is not a solution of the equation $-13 + 28 = xy - 13$? C

(A) $(4,7)$ (B) $(14,2)$ (C) $(14,14)$
(D) $\left(\frac{1}{2},56\right)$ (E) $(-7,-4)$

2. If y varies directly as the square of x, what will be the effect on y of doubling x? C

(A) y will double.
(B) y will be half as large.
(C) y will be four times as large.
(D) y will decrease in size.
(E) None of these.

3. If $5x + 4y - xy + 8 = 0$ and $x + 3 = 9$, then $3 - y =$ __?__ B
(A) -19 (B) -16 (C) 8
(D) 19 (E) 22

4. Segments of the lines $x = 4$, $x = 9$, $y = -5$, and $y = 4$ form a rectangle. What is the area of this rectangle? E
(A) 6 (B) 10 (C) 20
(D) 18 (E) 45

5. In the figure below, what is the area of the shaded region of the rectangle? E

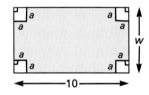

(A) $10w$ (B) $4a^2$ (C) $10w - 4a$
(D) $10w - a^2$ (E) $10w - 4a^2$

6. For any integer k, which of the following represents three consecutive even integers? E
(A) $2k, 4k, 6k$
(B) $k, k + 1, k + 2$
(C) $k, k + 2, k + 4$
(D) $4k, 4k + 1, 4k + 2$
(E) $2k, 2k + 2, 2k + 4$

7. Six oranges cost as much as 3 pears. If 2 oranges cost 50¢, what will 2 pears cost? A
(A) $\$1.00$ (B) 50¢   (C) $\$3.00$
(D) 75¢   (E) $\$1.50$

8. Dividing a number by $\frac{1}{5}$ gives the same result as multiplying by which one of the following? E
(A) $\frac{1}{5}$ (B) 0.5 (C) 20%
(D) 3 (E) 5

9. What is the arithmetic mean (average) of $\frac{1}{3}$ and $\frac{1}{4}$? E
(A) $\frac{1}{6}$ (B) $\frac{1}{7}$ (C) $\frac{1}{12}$
(D) $\frac{7}{12}$ (E) $\frac{7}{24}$

10. Given that $x = 5a$ and $y = \dfrac{1}{15a + 2}$, what is y in terms of x? A
(A) $\dfrac{1}{3x + 2}$ (B) $\dfrac{1}{3x}$

(C) $\dfrac{1}{x + 2}$ (D) $\dfrac{5}{x + 2}$

(E) None of these

College Prep Test   **117**

Additional Answers, page 116

17. Answers may vary. One possible answer; $y = -5x + 100$

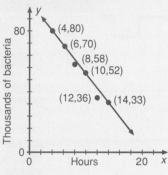

18. Answers may vary. One possible answer: 16,000 bacteria

24.

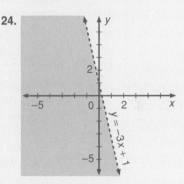

25.

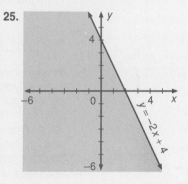

26.

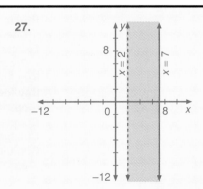

27.

4 LINEAR SYSTEMS IN TWO VARIABLES

OVERVIEW

In this chapter, students solve *linear systems*. Systems of two equations in two variables and of three equations in three variables are explored. The methods of solution of the systems include graphing, substitution, and the linear combination method. The chapter closes with a lesson on systems of linear inequalities in two variables followed by an *Extension* on linear programming.

OBJECTIVES

- To solve systems of two linear equations in two variables.
- To solve systems of three linear equations in three variables
- To solve word problems using a system of two linear equations in two variables
- To solve systems of two linear inequalities by graphing

PROBLEM SOLVING

Special emphasis to problem solving is given in Lessons 4.4 and 4.5 where students are asked to solve money/mixture problems and wind/current/angle problems, respectively. In addition to the regular lessons, there are two *Applications*, a *Statistics* lesson, and a *Problem Solving Strategies* lesson on the role of guessing and checking in the solution of problems.

READING AND WRITING MATH

This chapter continues the practice established earlier of developing problem-solving skills through questions that follow the Polya model for problem-solving. These questions guide students through a word problem in a manner that encourages them to read the problem carefully. The first question is usually, "What are you to find?" and is intended to focus a student's attention on the words of the problem that answer that question. Two exercises are included in this chapter in which students are asked to explain, in their own words, how to apply mathematical concepts. The first is in Lesson 4.1 (Exercise 14), and the second is in the Chapter Review (Exercise 19).

TECHNOLOGY

Calculator: In a few instances, problem checks are done using a calculator. The calculators employed should follow the arithmetic order of operations discussed on page 5 of Lesson 1.2.

SPECIAL FEATURES

PLANNING GUIDE

Lesson	Basic	Average	Above Average	Resources
4.1 pp. 121–123	CE all WE 1–11 Application Statistics	CE all WE 3–13 Application Statistics	CE all WE 5–15 Application Statistics	Reteaching 24 Practice 24
4.2 pp. 126–127	CE all WE 1–12	CE all WE 3–14	CE all WE 5–16	Reteaching 25 Practice 25
4.3 pp. 130–132	CE all WE 1–19 odd Midchapter Review Application	CE all WE 7–25 odd Midchapter Review Application	CE all WE 9–29 odd Midchapter Review Application	Reteaching 26 Practice 26
4.4 pp. 136–137	CE all WE 1–8	CE all WE 3–10	CE all WE 5–12	Reteaching 27 Practice 27
4.5 pp. 140–141	CE all WE 1–8 Brainteaser	CE all WE 3–10 Brainteaser	CE all WE 5–12 Brainteaser	Reteaching 28 Practice 28
4.6 pp. 144–146	CE all WE 1–11 odd Problem Solving	CE all WE 5–15 odd Problem Solving	CE all WE 7–17 odd Problem Solving	Reteaching 29 Practice 29
4.7 pp. 150–153	CE all WE 1–10 Extension	CE all WE 3–12 Extension	CE all WE 5–15 Extension	Reteaching 30 Practice 30
Chapter 4 Review pp. 154–155	all	all	all	
Chapter 4 Test p. 156	all	all	all	
College Prep Test p. 157	all	all	all	
Cumulative Review pp. 158–159	1–45 odd	5–41 odd	11–43 odd	

CE = Classroom Exercises   WE = Written Exercises

NOTE: For each level, all students should be assigned all Mixed Review exercises.

▬▬INVESTIGATION

Project: This investigation invites the students to explore the idea that a graph is a solution set of an equation.

To reinforce the idea that every point on the graph of an equation is a solution to the equation, ask the students to determine several points of a graph of a linear function. Start with the equation $x + y = 0$. Have the students make a table of values that satisfy the equation. Then have the students plot all the points on a coordinate plane before connecting them with a straight line.

x	−2	−1	0	1	2
y	2	1	0	−1	−2

Lead the students to the realization that any point they could put in the table would be on the straight line graph.

Conversely, the student should realize that any point on the graph could be a value in the table. Thus, the complete graph would be a complete solution set, while the points in the table are a partial solution set.

Have the students work in groups. Have some of the students in the group write linear equations, while the others graph them. Then, on the same coordinate plane, have the students graph the equation $x + y = 0$, and ask them to describe the points of intersection with the other graphs.

Discuss the fact that a point of intersection of two graphs lies on *both* of the graphs. Thus, it is in the solution set of both equations. Some pairs of linear equations will have one solution in common, others will have none, and others will have infinitely many.

A three-dimensional model of Hurricane Gilbert is created from two sets of data transmitted via satellite. The computer uses the data to calculate the height of cloud tops and to track the movement of the storm. Images can be generated to view the hurricane from any height or angle.

More About the Image

Computer graphics enables meteorologists to plot information about storm systems in three-dimensional coordinate space. As new data is transmitted from satellites, the heights of the cloud tops are recalculated and an updated image is plotted. Over time, image after image can be created. When the images are viewed one after the other, the storm is set in motion—giving meteorologists an accurate model of its size as well as the speed and direction it is moving.

4.1 Graphing Linear Systems

Objectives

To solve systems of two linear equations in two variables by graphing

To determine whether a system of two linear equations is consistent, inconsistent, dependent, or independent

Teaching Resources

Problem Solving Worksheet 4
Project Worksheet 4
Quick Quizzes 24
Reteaching and Practice
  Worksheets 24
Transparencies 9A, 9B

Recall that equations such as $2x + y = 3$ and $2y - x + 4 = 0$ describe *linear* functions. The graphs of the two functions are shown below. Notice that their *point of intersection*, $P(2, -1)$, is the only point on both lines. Therefore, $(2, -1)$ is the only ordered pair (x, y) that satisfies *both* equations.

Recall (Lesson 2.2) that the conjunction p *and* q is true if and only if both p is true and q is true. Therefore, the conjunction

$$2x + y = 3 \quad and \quad 2y - x + 4 = 0$$

is true for $(2, -1)$ because $(2, -1)$ is a solution of both equations.

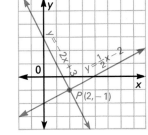

$$
\begin{array}{ll}
2x + y = 3 & 2y - x + 4 = 0 \\
2 \cdot 2 - 1 \overset{?}{=} 3 & 2(-1) - 2 + 4 \overset{?}{=} 0 \\
4 - 1 \overset{?}{=} 3 & -2 - 2 + 4 \overset{?}{=} 0 \\
3 = 3 & -4 + 4 \overset{?}{=} 0 \\
\text{True} & 0 = 0 \\
& \text{True}
\end{array}
$$

So, the ordered pair $(2, -1)$ is the only solution of both parts of the conjunction

$$2x + y = 3 \quad and \quad 2y - x + 4 = 0,$$

and of the following **system of equations:** $\begin{array}{l} 2x + y = 3 \\ 2y - x + 4 = 0 \end{array}$

To solve a system of two equations in two variables by graphing:
1. Graph both equations on the same coordinate plane.
2. Find the coordinates of the point of intersection of the two lines.

Definitions

> If a system of linear equations has *at least* one solution, the system is called a **consistent system.**
>
> If a system of linear equations has *at most* one solution, the equations are called **independent**, and the system is called an **independent system.**

Prerequisite Quiz

Find the slope and y-intercept of the line described by the given equation.

1. $y = \frac{2}{3}x - 4$ $\frac{2}{3}; -4$
2. $3x + 4y = 8$ $-\frac{3}{4}; 2$
3. $2x - 5y = 10$ $\frac{2}{5}; -2$
4. $4x - y = 1$ $4; -1$

Motivator

Ask students how the slopes of two parallel lines are related. They are the same. Ask students if two lines can have more than one point in common. Yes If so, ask them how many points the two lines have in common. An infinite number

■■■ **TEACHING SUGGESTIONS**

Lesson Note

Emphasize that solving systems by graphing is not very accurate when the solutions are not integers. This provides the motivation for teaching the algebraic solutions in the next two lessons.

Highlighting the Standards

Standard 4a: The intersection of two lines and the common ordered pair are two ways of looking at the solution to a system of linear equations.

Nutrition: Nutritionists and animal scientists can use systems of equations to ensure proper diets. For example, a horse breeder might need to ensure that a horse get a given amount of protein and a given number of calories from a mix of two types of feed. She can then write two linear equations, one for the total amount of protein in a mix of the two feeds and one for the total calories in the mix. The intersection of the two equations gives the proper combination.

Critical Thinking Questions

Analysis: Ask students these questions. If the slopes of the lines in a system of linear equations are 0 and undefined, respectively, is the system consistent?   Yes Is the system dependent?   No. It has exactly one solution.   If both slopes are undefined, is the system dependent? Possibly. If the solution set is infinite, then the system is dependent.

Common Error Analysis

Error: Students tend to think that an infinite number of solutions for a system means that any ordered pair (x,y) is a solution. Show that this is not true by using a counterexample. For example, for $2x + 3y = 6$ and $-4x - 6y = -12$, $(2,5)$ is not a solution.

Emphasize that each equation has an infinite number of solutions since its graph is a line and a line contains an infinite number of points. Then, the coordinates of any of this infinite number of points will be a solution of the other equation.

For example, the system $\begin{array}{l} 2x + y = 3 \\ 2y - x + 4 = 0 \end{array}$ is both *consistent* and *independent* because it has exactly one solution, $(2, -1)$.

EXAMPLE 1 Solve the system $\begin{array}{l} 5y - 2(x + y) = 15 \\ 2x - 3y = 12 \end{array}$ by graphing.

Solution First, write each equation in slope-intercept form, then graph it.

$$5y - 2(x + y) = 15$$
$$5y - 2x - 2y = 15$$
$$3y = 2x + 15$$
$$y = \tfrac{2}{3}x + 5$$
$$2x - 3y = 12$$
$$-3y = -2x + 12$$
$$y = \tfrac{2}{3}x - 4$$

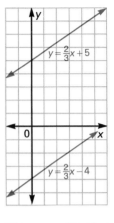

Notice that the slopes of the two lines are the same, but their y-intercepts are different. Therefore, the lines are parallel, and because parallel lines never meet, there is no point of intersection. So, the system has no solution.

Definition A system of equations that has *no* solution is an **inconsistent system**.

EXAMPLE 2 Solve by graphing. $0.2x - 0.3y = 0.9$ *and* $\dfrac{y}{2} = \dfrac{x}{3} - \dfrac{3}{2}$

Solution First, write each equation in slope-intercept form and graph it.

$$0.2x - 0.3y = 0.9$$
$$10(0.2x - 0.3y) = 10(0.9)$$
$$2x - 3y = 9$$
$$-3y = -2x + 9$$
$$y = \tfrac{2}{3}x - 3$$

$$\frac{y}{2} = \frac{x}{3} - \frac{3}{2}$$
$$2 \cdot \frac{y}{2} = 2\left(\frac{x}{3} - \frac{3}{2}\right)$$
$$y = \tfrac{2}{3}x - 3$$

The slopes and y-intercepts of the two equations are the same. Therefore, the graphs coincide, and *all* ordered pairs (x,y) such that $y = \tfrac{2}{3}x - 3$ are solutions of the system. The solution set, which is infinite, can be written as $\{(x,y) \mid y = \tfrac{2}{3}x - 3\}$.

Definition If a system of equations has an infinite number of solutions, the equations are **dependent**, and the system is a **dependent system**.

Additional Example 1

Solve this system by graphing:

$$4y - (3x - 2y) = 12$$
$$-2y + x = 8$$

The lines are parallel. There is no solution.

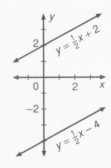

As the examples of this lesson have shown, a system of linear equations may have one solution, no solution, or infinitely many solutions.

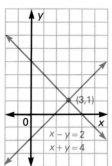 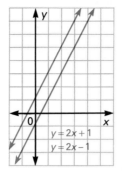 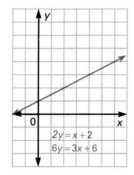

One solution
Consistent
Independent

No solution
Inconsistent
Independent

Infinitely many solutions
Consistent
Dependent

Classroom Exercises

Determine whether the given ordered pair is a solution of the given system of equations.

1. $(3, -1)$;
$2x + y = 5$
$x - y = 1$ No

2. $(3,4)$;
$x + y = 7$
$2x - y = 1$
No

3. $(5,1)$;
$x + 2y = 7$
$2x - y = 9$
Yes

4. $(-3,1)$;
$2x + y = -5$
$y = -\frac{1}{3}x$ Yes

Solve each equation for *y*.

5. $2y = 4x - 8$ $y = 2x - 4$

6. $-3y = 6x - 12$ $y = -2x + 4$

7. Solve Exercise 1 by graphing. $(2,1)$

Written Exercises

Solve each system by graphing. Indicate whether the system is *consistent* or *inconsistent* and *dependent* or *independent*.

1. $y = x + 4$
$y = -x + 2$

2. $y = 2x - 4$
$y = 2x + 5$

3. $y = 3x - 2$
$2y = 6x - 4$

4. $y = 2x$
$y = -2x + 4$

5. $-x + y = 6$
$-2x + 2y = 10$

6. $x + y = 6$
$-3x - 3y = -18$

4.1 Graphing Linear Systems   **121**

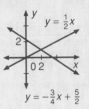

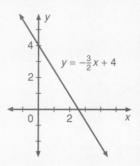

Independent Practice

A Ex. 1–9, **B** Ex. 10–14, **C** Ex. 15,16

Basic: WE 1–11, Application, Statistics
Average: WE 3–13, Application, Statistics
Above Average: WE 5–15, Application, Statistics

Additional Answers

Classroom Exercises

7.

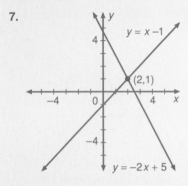

See page 123 for the answers to Written Ex. 1–3, and pages 808–809 for Written Ex. 4–14, Mixed Review Ex. 4, Application Ex. 2, and Statistics Ex. 4.

Solve each system by graphing.

7. $y = 2 - 3x$
$\frac{x}{2} = \frac{1}{3} - \frac{y}{6}$

8. $y = 8 - x$
$\frac{y}{2} = x + \frac{5}{2}$

9. $2x = 4 - y$
$-\frac{3}{2}x + \frac{1}{4}y = -1$

10. $5x - 3(y + x) = -6$
$2x - 3y = 3$

11. $6x - 2(5y - 2x) + 12y = 10$
$-5x - (y - 8x) + 2y - 3 = 0$

12. $0.4x - 0.2y = 0.6$
$0.6x + 0.3y = -0.9$

13. $0.2x + 0.2y = 0.4$
$0.4x - 0.4y = 0.8$

14. Write, in your own words, how to solve a system of linear equations by graphing.

Find the value(s) of k that satisfy the given condition for each system.

15. $kx - 5y = 8$
$7x + 5y = 10$ is consistent. $k \neq -7$

16. $5x - 2y = 7$
$-20x + ky = -28$ has infinitely many solutions. $k = 8$

Mixed Review

Find the solution set. *2.1, 2.3, 3.5*

1. $3x - 2(4x + 1) \geq x - 8$ $\{x \mid x \leq 1\}$

2. $|2x - 4| = 10$ $\{-3,7\}$

3. $|x - 6| < 1$ $\{x \mid 5 < x < 7\}$

4. Graph $3x - 5y = 15$.

Application: *Supply and Demand*

Supply and demand equations are used by economists to describe, for a particular time, the relationship between price y and quantity x. Supply equations have positive slopes because, as price increases, so does production. Demand equations, however, have negative slopes because, as price increases, demand decreases. The solution of a system of supply and demand equations is the equilibrium price for that item.

1. Supply of and demand for rag rugs can be described by $2y - x = 18$ and $2y + 2x = 36$, where $x =$ thousands of rugs and $y =$ price in dollars. Find the equilibrium price.
$12.00

2. What are some of the factors that could cause supply to fall below the equilibrium point? What are some of the factors that could cause the equilibrium point to shift?

Enrichment

Discuss this situation. A department store had an 8-day sales promotion. The graph shows income and expenses during this period. The break-even point occurs when income equals expenses. Note that although lines are drawn through each set of points, it is the points that give the daily totals.

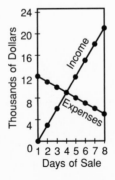

Days of Sale

Ask the student these questions.

1. Which day represents the break-even point? *fourth day*

2. Estimate the net profit (+) or loss (−) for each day of the sale.

Day 1: −$12,000   Day 5: +$4,000
Day 2: −$8,000   Day 6: +$8,000
Day 3: −$4,000   Day 7: +$12,000
Day 4: break-even day   Day 8: +$16,000

Statistics: *Median Line to Fit*

The table below shows gas mileage ratings for different cars reported in a recent year. The engine size x is in liters of displacement. The mileage y is in highway miles per gallon.

x	1.0	1.3	1.5	1.6	1.8	1.9	2.0	2.2	2.3	2.5
y	58	47	49	38	41	32	40	32	33	31

Statisticians sometimes use a method called the **median line of fit** to determine a line through a set of data points like the one above.

A. Graph the points on a coordinate plane. This is called a **scatter diagram.**

B. Partition the data into three nearly equal sets such that there is an odd number of points in the first and third sets. In this example, the obvious choice is sets of 3, 4, and 3 points.

C. Locate the *median x* and *y* values in the first and third sets. *The median is the value with as many values above it as below it.*

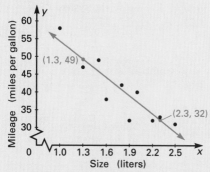

The median of the x values in the first group of 3 points is 1.3. The median of the y values is 49. Combine these 2 values as the point (1.3, 49) and add it to the diagram. Do the same with the median x and y values from the 3rd group to locate a second point, (2.3, 32). Connect the 2 points with a straight line.

Exercises

1. Find the slope of the line using the points (1.3, 49) and (2.3, 32). -17

2. Using the slope from Exercise 1 and one of the two points, determine the equation of the line. $y = -17x + 71.1$

3. Predict the mileage for an engine with a 1.7 l engine. 42.2 mpg

4. Use the data in the table below to construct a scatter diagram, using "pages" for the x-variable. Draw the median line of fit and determine its equation. Predict the number of hours to type 70 pages.

Pages	5	8	14	20	23	29	36	40	44	49
Hours to Type	.5	1.1	1.4	2.8	1.8	3.8	4.7	3.9	5.0	4.8

Additional Answers, page 122

Written Exercises

1. Consistent, independent; $(-1,3)$

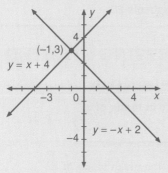

2. Inconsistent, independent; No solution

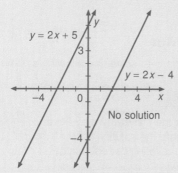

3. Consistent, dependent; $\{(x,y) \mid y = 3x - 2\}$

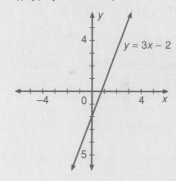

4.2 The Substitution Method

GETTING STARTED

Prerequisite Quiz

Solve for _y_ in terms of _x_.

1. $4x - y = 6$ $y = 4x - 6$
2. $3x = 5 + y$ $y = 3x - 5$

Find _y_ for the given value of _x_.

3. $y = 3x - 2; x = -4$ -14
4. $y = 8 - x; x = -3$ 11

Motivator

Ask the class to solve the following system by graphing.
$$6x - y = 1$$
$$2x + y = 1$$
Then ask the class why they have to approximate the solution. The solution does not have integer coordinates.

TEACHING SUGGESTIONS

Lesson Note

Point out that substitution is not the best method for solving systems of equations unless a variable has a coefficient of 1 or −1. Indicate that another, more general method will be taught in the next lesson.

Objectives To solve systems of two linear equations using the substitution method
To solve word problems using systems of equations

Because solving a system of equations by graphing involves estimating the coordinates of the point of intersection, the graphing method works poorly when the coordinates are not integers. There are, however, algebraic methods for solving systems of equations that produce exact results. One of these is the *substitution method*.

EXAMPLE 1 Solve the system $\begin{array}{l} 2x - y = 8 \\ -4x = -23 - 3y \end{array}$ by substitution.

Plan First, solve one of the equations for one of its variables. Notice that it is easier to solve for a variable whose coefficient is already 1 or −1, such as _y_ in $2x - y = 8$.

Solution Solve $2x - y = 8$ for _y_.
$$2x - y = 8$$
$$-y = 8 - 2x$$
$$y = -8 + 2x$$

Substitute $-8 + 2x$ for _y_ in the other equation.
$$-4x = -23 - 3y$$
$$-4x = -23 - 3(-8 + 2x)$$
$$-4x = -23 + 24 - 6x$$
$$2x = 1$$
$$x = \frac{1}{2}$$

To find the value of _y_, substitute $\frac{1}{2}$ for _x_ in $y = -8 + 2x$. (This equation is used because it is already solved for _y_.)
$$y = -8 + 2x$$
$$= -8 + 2 \cdot \frac{1}{2}$$
$$= -8 + 1$$
$$= -7$$

Check
$$2x - y = 8$$
$$2(\tfrac{1}{2}) - (-7) \stackrel{?}{=} 8$$
$$8 = 8 \quad \text{True}$$

$$-4x = -23 - 3y$$
$$-4(\tfrac{1}{2}) \stackrel{?}{=} -23 - 3(-7)$$
$$-2 = -2 \quad \text{True}$$

The solution is $\left(\frac{1}{2}, -7\right)$.

In Example 1, the system $\begin{array}{l} 2x - y = 8 \\ -4x = -23 - 3y \end{array}$ was simplified to the system: $\begin{array}{l} x = \frac{1}{2} \\ y = -7 \end{array}$

Because these two systems have the same solution set, $\{(x,y) \mid x = \frac{1}{2}$ *and* $y = -7\}$, the two systems are referred to as **equivalent systems.**

Highlighting the Standards

Standard 5c: Any algebra course must continually return to the solution of equations. This chapter is entirely devoted to the use, and solution, of linear equations.

Additional Example 1

Solve this system by substitution.

$$-2y = 5 - 3x$$
$$3y - x = 3$$

Check the solution.   (3, 2)

Sometimes the equations of a linear system contain fractions or decimals. When this occurs, it helps to eliminate the fractions or decimals first.

To accomplish this, multiply each side of the equation by the LCM of the denominators or, for decimals, the appropriate power of 10, as illustrated in Example 2 below.

EXAMPLE 2 Solve the conjunction $\frac{x}{3} - \frac{y}{6} = \frac{1}{2}$ and $0.3x + 0.7y = 4.7$ using the substitution method.

Solution

First, eliminate the fractions. Multiply each side of the first equation by the LCD, 6.

$$6\left(\frac{x}{3} - \frac{y}{6}\right) = 6 \cdot \frac{1}{2}$$
$$6 \cdot \frac{x}{3} - 6 \cdot \frac{y}{6} = 6 \cdot \frac{1}{2}$$
$$2x - y = 3$$

Next, eliminate the decimals. Multiply each side of the second equation by 10.

$$10(0.3x + 0.7y) = 10(4.7)$$
$$10(0.3x) + 10(0.7y) = 10(4.7)$$
$$3x + 7y = 47$$

Now solve the equivalent system.

(1) $2x - y = 3$
(2) $3x + 7y = 47$

Solve Equation (1) for y, because it has the convenient coefficient -1.

(1) $2x - y = 3$
$-y = 3 - 2x$
$y = 2x - 3$

Substitute $2x - 3$ for y in Equation (2).

(2) $3x + 7y = 47$
$3x + 7(2x - 3) = 47$
$3x + 14x - 21 = 47$
$17x = 68$
$x = 4$

To find the value of y, substitute 4 for x in $y = 2x - 3$, because this equation is already solved for y.

$y = 2x - 3$
$= 2 \cdot 4 - 3$
$= 8 - 3$, or 5

Check

$$\frac{x}{3} - \frac{y}{6} = \frac{1}{2}$$
$$\frac{4}{3} - \frac{5}{6} \overset{?}{=} \frac{1}{2}$$
$$\frac{1}{2} = \frac{1}{2} \text{ True}$$

$$0.3x + 0.7y = 4.7$$
$$0.3(4) + 0.7(5) \overset{?}{=} 4.7$$
$$1.2 + 3.5 \overset{?}{=} 4.7$$
$$4.7 = 4.7 \text{ True}$$

The solution is $(4,5)$.

Sometimes a word problem can be solved using a system of equations.

Math Connections

Ancient Greece: Archimedes (287–212 B.C.) is said to have solved a system of linear equations to find out how much gold and how much silver was in a crown made of a mixture of the two. Knowing the weight of the crown and the amounts of water displaced by a given weight of gold and of silver, respectively, led to a system equivalent to the following, where g is the amount of gold in the crown and s is the amount of silver.

$$3g + \frac{3}{2}s = 20$$
$$g + s = 10$$

Then he solved the system algebraically.

Critical Thinking Question

Analysis: Have students answer these questions.

1. Are the equations $2x + 7 = 8$ and $-4x = -23 + 21$ equivalent? Explain.
   Yes, because they have the same solution, $x = \frac{1}{2}$.

2. Are these systems equivalent? Explain.
   $2x - y = 8$ and $4x = -19 - 3y$
   $x = \frac{1}{2}$ $y = -7$
   Yes, because they have the same solution set, $\{(\frac{1}{2}, -7)\}$.

Additional Example 2

Solve this conjunction.

$\frac{2}{5}x - \frac{1}{10}y = 9$ and $0.5x + 0.4y = 6$
$(20, -10)$

Common Error Analysis

Error: After using substitution to solve for the first variable, students do not find the value of the second variable.

Emphasize the need to find the value of the other variable also.

Checkpoint

Solve by the substitution method.

1. $2x - y = 5$
   $-x + y = -3$ $(2, -1)$

2. $2x + y = 6$
   $5x = -y$ $(-2, 10)$

3. $0.2x + 0.1y = 0.7$
   $\frac{x}{2} - \frac{y}{4} = \frac{1}{4}$ $(2, 3)$

Closure

Ask students to give the steps in solving a system of linear equations by the substitution method. See Example 1. Ask students what they should do if the system contains an equation involving fractions or decimals. Eliminate the fractions or decimals by multiplying each side of the equation by the LCM of the denominators or by the appropriate power of 10, respectively.

EXAMPLE 3   At a senior-citizens center, 35 tables were set up for games of checkers (2 players per table) and bridge (4 players per table). There were 92 players occupying 35 tables. Find the number of checker players and the number of bridge players.

Solution

What are you to find?	the number of each type of player
Choose two variables. What do they represent?	Let c = the number of checker tables. Let b = the number of bridge tables. Then $2c$ = the number of checker players, and $4b$ = the number of bridge players.

What is given?   The number of tables is 35.   The number of players is 92.

Write a system.   (1) $c + b = 35$ (2) $2c + 4b = 92$

Solve the system.   Solve (1) for one variable. $c = 35 - b$

Substitute $35 - b$ for c in (2).
   (2) $2(35 - b) + 4b = 92$

Solve for b. $70 - 2b + 4b = 92$
                     $b = 11 \leftarrow$ number of bridge tables

Substitute 11 for b in (1).
   (1) $c + 11 = 35$
Solve for c. $c = 24 \leftarrow$ number of checker tables

Number of checker players = $2c = 2 \cdot 24 = 48$

Number of bridge players = $4b = 4 \cdot 11 = 44$

Check in the original problem.   35 tables were set up: $11 + 24 = 35$

There were 92 players: $48 + 44 = 92$

State the answer.   There were 48 checker players and 44 bridge players.

Classroom Exercises

Tell which equation is easier to solve. Which variable would you solve for?

1. $2x + 3y = 7$
   $3x - y = 5$
   Second, y

2. $2y = 6 - x$
   $4x + 5y = 18$
   First, x

3. $7x = 6 - 2y$
   $4y = 9 - x$
   Second, x

4. $3x - y = 9$
   $5x + 2y = 4$
   First, y

Solve using the substitution method.

5. $y = 3x$ $(2, 6)$
   $4x + 3y = 26$

6. $x = 4y$ $(8, 2)$
   $2x + 4y = 24$

7. $y = x - 4$ $(5, 1)$
   $y + x = 6$

8. $x = y + 1$ $(3, 2)$
   $3x + y = 11$

Additional Example 3

The sum of two numbers is 7. Twice the second number is 6 less than 3 times the first number. Find the two numbers.   4, 3

Written Exercises

Solve using the substitution method.

1. $y = 2x$
$x + y = 9$ (3,6)

2. $x = y + 1$
$x + 4y = 11$ (3,2)

3. $y = 3x - 1$
$2x + y = 14$ (3,8)

4. $x = 4y$
$2x + 3y = 22$ (8,2)

5. $2x - y = 5$
$-x + y = -3$ (2,-1)

6. $15x + 4y = 23$
$10x - y = -3$ $\left(\frac{1}{5}, 5\right)$

7. $3x - y + 2 = 0$
$5x - 3y + 4 = 0$ $\left(-\frac{1}{2}, \frac{1}{2}\right)$

8. $9m + 8n = 21$
$2m = 7 - n$ (5,-3)

9. $5a + 4b = 7$
$2a - b = 8$ (3,-2)

Solve each conjunction.

10. $\frac{1}{2}x + \frac{1}{4}y = \frac{3}{4}$ and $0.5x - 0.2y = 0$ $\left(\frac{2}{3}, \frac{5}{3}\right)$

11. $0.12x + 0.15y = 132$ and $x = 1{,}000 - y$ (600,400)

Solve each problem by using a system of two linear equations and the substitution method.

12. The sum of two numbers is 35. Twice the first number is equal to 5 times the second number. Find the two numbers. 25, 10

13. Martha's age is twice that of her brother Ned. The sum of their ages is 24. Find the age of each. Martha: 16 yr old, Ned: 8 yr old

14. An apartment building contains 200 units. Some of these are one-bedroom units that rent for $435 each month. The rest are two-bedroom units that rent for $575 per month. When all the units are rented, the total monthly income is $97,500. How many apartments are there of each type? 1-br: 125, 2-br: 75

15. The units digit of a two-digit number is $\frac{1}{2}$ the tens digit. If the digits are reversed, the sum of the new number and the original number is 132. Find the original number. (HINT: Write the number as $10t + u$, where t and u are the tens and units digits, respectively.) 84

16. The graph of $y = ax^2 + bx$ contains the points $P(2,5)$ and $Q(1,-1)$. Find a and b. $\frac{7}{2}, -\frac{9}{2}$

17. Find the coordinates of the vertices of the triangle formed by the graphs of the lines $2x - y = -1$, $x - y = -1$, and $2x - 5y = -29$. (0,1), (3,7), (8,9)

Mixed Review

1. Find $f(-3)$ for $f(x) = -x^3$. **3.2** 27

2. Graph $f(x) = 2x - 4$. **3.5**

3. Find the slope of $\overline{AB}$ for $A(5,-1)$ and $B(-2,3)$. **3.3** $-\frac{4}{7}$

4. If y varies directly as x, and if $y = 12$ when $x = 3$, what is y when $x = 15$? **3.8** 60

FOLLOW UP

Guided Practice

Classroom Exercises 1–8

Independent Practice

A Ex. 1–9, B Ex. 10–15, C Ex. 16,17

Basic: WE 1–12
Average: WE 3–14
Above Average: WE 5–16

Additional Answers

Mixed Review

2.

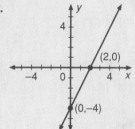

Enrichment

Have the students solve this system for x in terms of a, b, c, d, and e.

$ax + by = c$
$y = dx + e$

$x = \frac{c - be}{a + bd}$

Now have the students verify this formula for x by applying it to find the value of x in the system below. Then have them find y by using $y = dx + e$.

$2x + 3y = 21$
$y = x + 2$

$x = 3, y = 5$

Prerequisite Quiz

1. Find y if $x = 2$ in $3x + 2y = 14$. 4
2. Write the resulting equation when each side of $3x - 2y = 5$ is multiplied by -2.
   $-6x + 4y = -10$

Find the additive inverse.

3. -6 6
4. $7x$ $-7x$

Motivator

Ask students why they would find the substitution method of the last lesson difficult to apply to the system below.

$$8x + 5y = 22$$
$$6x + 2y = 13$$

When solving for x or y, fractional coefficients occur. Ask students if each of the following equations is true.

$$3 + 4 = 7$$
$$7 - 2 = 5 \quad \text{Yes}$$

Ask them what equation will result from adding the two equations above.
$10 + 2 = 12$ Is the third equation true?
Yes

4.3 The Linear Combination Method

Objective   To solve systems of two linear equations using the linear combination method

Another algebraic method for solving a system of linear equations, the *linear combination method*, applies the following theorem.

Theorem 4.1   If $a = b$ and $c = d$, then $a + c = b + d$.

Proof

Statement	Reason
1. $a = b, c = d$	1. Given
2. $a + c = b + c$	2. Add Prop of Eq (Lesson 1.4)
3. $a + c = b + d$	3. Subst (Lesson 1.8)

In the system at the right, the y-terms, $2y$ and $-2y$, are *additive inverses*. So, when the equations are added, the resulting equation has only one variable term, $7x$. Next, solving for x, and then solving for y by substituting for x in either equation, gives the solution $(2, 1)$.

$$
\begin{array}{ll}
(1) & 3x + 2y = 8 \\
(2) & 4x - 2y = 6 \\
\hline
 & 7x \quad\;\; = 14 \\
 & x = 2
\end{array}
$$

$$
\begin{array}{ll}
(1) & 3x + 2y = 8 \\
 & 3 \cdot 2 + 2y = 8 \\
 & 6 + 2y = 8 \\
 & 2y = 2 \\
 & y = 1
\end{array}
$$

Verify that the ordered pair $(2, 1)$ is a solution of the system: $\begin{array}{l} 3x + 2y = 8 \\ 4x - 2y = 6 \end{array}$

Also, $(2, 1)$ is a solution of the equation formed from the sum of *any* two constant multiples of these equations in the manner illustrated below (see Mult Prop of Eq in Lesson 1.4).

Use 5 as a multiplier for (1).   Use 3 as a multiplier for (2).

$$
\begin{array}{ll}
(1) & 3x + 2y = 8 \\
 & 5(3x + 2y) = 5 \cdot 8 \\
(3) & 15x + 10y = 40
\end{array}
\qquad
\begin{array}{ll}
(2) & 4x - 2y = 6 \\
 & 3(4x - 2y) = 3 \cdot 6 \\
(4) & 12x - 6y = 18
\end{array}
$$

Add the corresponding sides of the resulting equations.

$$
\begin{array}{ll}
(3) & 15x + 10y = 40 \\
(4) & 12x - \;\; 6y = 18 \\
\hline
(5) & 27x + \;\; 4y = 58
\end{array}
$$

Verify that $(2, 1)$ is a solution of Equation (5).

$$27 \cdot 2 + 4 \cdot 1 \overset{?}{=} 58$$
$$54 + 4 = 58 \quad \text{True}$$

Thus, if $(2, 1)$ is a solution of (1) $3x + 2y = 8$ and (2) $4x - 2y = 6$, then $(2, 1)$ is also a solution of (3) and (4). The resulting equation,

Highlighting the Standards

Standard 14b: The method of solving a system of equations through linear combination provides practice in precise thinking and careful manipulation.

(5) $27x + 4y = 58$, is called a *linear combination* of Equations (1) and (2).

A **linear combination** of two equations may be obtained in the following manner.
1. Multiply the first equation by a nonzero constant.
2. Multiply the second equation by a nonzero constant.
3. Add the two resulting equations to obtain a new equation.

If the constants in Steps (1) and (2) above are chosen properly, the new equation obtained in Step (3) can be solved for one of the variables. The other variable will have been eliminated in the addition step. This is illustrated in Example 1.

EXAMPLE 1 Solve the system $\begin{aligned} 8x + 5y &= 22 \\ 6x + 2y &= 13 \end{aligned}$ using the linear combination method.

Plan Solve the system by eliminating either the *x*- or *y*-terms. For example, to make the *y*-terms additive inverses of each other, multiply each side of the first equation by 2, and each side of the second equation by -5.
The result is the equivalent system: $\begin{aligned} 16x + 10y &= 44 \\ -30x - 10y &= -65 \end{aligned}$

Solution $\begin{aligned} 8x + 5y &= 22 \\ 6x + 2y &= 13 \end{aligned} \rightarrow \begin{aligned} 2(8x + 5y) &= 2(22) \\ -5(6x + 2y) &= -5(13) \end{aligned} \rightarrow \begin{aligned} 16x + 10y &= 44 \\ -30x - 10y &= -65 \end{aligned}$

Add the resulting equations. $-14x = -21$

Solve for *x*. $x = \dfrac{3}{2}$

To find the value of *y*, substitute $\frac{3}{2}$ for *x* in either equation. Equation (1) is used here.

$\begin{aligned} (1) \quad 8x + 5y &= 22 \\ 8\left(\tfrac{3}{2}\right) + 5y &= 22 \\ 12 + 5y &= 22 \\ 5y &= 10 \\ y &= 2 \end{aligned}$

So, the solution is $\left(\frac{3}{2}, 2\right)$. The check is left for you.

In Example 1, the system $\begin{aligned} 8x + 5y &= 22 \\ 6x + 2y &= 13 \end{aligned}$ was reduced to the equivalent system: $\begin{aligned} x &= \tfrac{3}{2} \\ y &= 2 \end{aligned}$

At the right, the two systems are represented together graphically, as is their common solution $\left(\frac{3}{2}, 2\right)$.

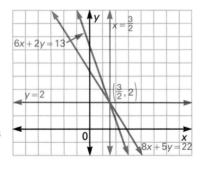

Lesson Note

Students have used the LCM of two numbers to simplify an equation with fractional coefficients. Now they will use the LCM to find appropriate multipliers when solving a system of equations by the linear combination method. Point out that it may not be necessary to multiply both equations by a constant in order to obtain additive inverses for a pair of terms having the same variable.

Math Connections

Business: The linear combination method may be applied to business problems to find the break-even point from a given set of cost conditions. For example, in the *Application: Profit and Loss* (page 132), the two equations $y = 30,000 + 4x$ and $y = 7x$ may be added using the linear combination method.

Critical Thinking Questions

Analysis: Ask students to provide examples of equations in one variable that are counterparts of inconsistent systems in two variables. They might come up with examples of equations with no roots at all, such as $x = x + 1$, $|x| = -3$, and $0 \cdot x = 2$. These equations are equivalent, since they have the same solution set, $\varnothing$.

Additional Example 1

Solve this system by the linear combination method.

$5x + 2y = 19$
$x - 3y = -3$

$(3, 2)$

Common Error Analysis

Error: When multiplying an equation by the necessary multiplier, some students tend to forget to multiply the right side of the equation also by that multiplier.

Emphasize that *both* sides of the equation must be multiplied by the same number.

Checkpoint

Solve by the linear combination method. Is the system inconsistent or dependent?

1. $4x + 2y = 2$
   $10x + 5y = 5$
   Dependent

2. $3x - 6y = 12$
   $-2x + 4y = 1$
   Inconsistent

3. $\dfrac{y}{2} = \dfrac{13 - 3x}{4}$
   $4x + 5y = 29$ $(1,5)$

Closure

Ask students if it is always necessary to use a multiplier for each equation in solving a system by the linear combination method. No   What are the steps in solving a system of two equations by the linear combination method if neither the *x* nor the *y* coefficients are additive inverses of each other?   See Example 1.

EXAMPLE 2   Solve each system using the linear combination method.

 a. $3x - 6y = 7$ b. $10x - 6y = 4$
    $-x + 2y = 6$ $-5x + 3y = -2$

Solutions

 a. (1) $3x - 6y = 7$ b. (1) $10x - 6y = 4$
    (2) $-x + 2y = 6$ (2) $-5x + 3y = -2$
         $3(-x + 2y) = 3 \cdot 6$ $2(-5x + 3y) = 2(-2)$
    (3) $-3x + 6y = 18$ (3) $-10x + 6y = -4$

 Add Equations (1) and (3).   Add Equations (1) and (3).
 (1) $3x - 6y = 7$ (1) $10x - 6y = 4$
 (3) $\underline{-3x + 6y = 18}$ (3) $\underline{-10x + 6y = -4}$
        $0 + 0 = 25$ False $0 + 0 = 0$ True

 Thus, there is *no* solution.   Thus, there are an *infinite num-*
 The system is *inconsistent*.   *ber* of solutions. The system is
 The solution set is $\varnothing$.   *dependent*. The solution set is
                                  $\{(x,y) \mid -5x + 3y = -2\}$.

The graph of the system in Example 2a is a pair of parallel lines (see Example 1 of Lesson 4.1). The graph of the system in Example 2b is a pair of coincident lines (see Example 2 of Lesson 4.1).

Classroom Exercises

For each given system, select a multiplier for one of the equations so that the sum of the equations of the new system is an equation in one variable. Answers may vary. One possible answer is given.

1. $3x - 5y = 1$ 2. $4x - 5y = 13$ 3. $3x + 7y = 4$ 3   4. $3x + y = 4$
   $x + y = 7$ 5 $3x + y = 6$ 5 $-x + 5y = 2$ $x + 2y = 3$ -3

Solve using the linear combination method.

5. $x + y = 4$ (3,1)   6. $x + y = 2$ (5,-3)   7. $-x + y = 5$ 8. $3x - y = 8$ (2,-2)
   $x - y = 2$ $x - y = 8$ $x + y = 7$ (1,6) $2x + y = 2$

Indicate whether each system has exactly one solution, no solution, or infinitely many solutions.

9. $3x - 4y = -1$ 10. $3x - 5y = 9$ 11. $7x - 8y = 5$ 12. $3x = 2y + 3$
   $x + 4y = 5$ $-3x + 5y = -9$ $-7x + 8y = 5$ $6x + 4y = 3$
   Exactly one   Infinitely many   No solution   Exactly one

Solve using the linear combination method. If the system is inconsistent or dependent, indicate this.

13. $3x + 2y = 4$ 14. $5x - 3y = 8$ 15. $6x + y = 12$ (2,0) 16. $4x - 2y = 1$
    $5x - 2y = 28$ $10x - 6y = 18$ $x - 2y = 2$ $-8x + 4y = -2$
    $(4, -4)$ Inconsistent   Dependent

130   Chapter 4 Linear Systems in Two Variables

Additional Example 2

Solve each system by the linear combination method.

a. $5x - 2y = 4$
   $-10x + 4y = -8$
   Infinite number of solutions

b. $4x + 2y = 6$
   $6x + 3y = 5$
   No solution

Written Exercises

Solve using the linear combination method. If the system is inconsistent or dependent, indicate this.

1. $5x - 2y = 30$ (6,0)
   $x + 2y = 6$

2. $-7x + y = 10$ (−1,3)
   $7x + 2y = -1$

3. $-8x + 7y = 17$ (−3,−1)
   $8x + 11y = -35$

4. $5x + 3y = 4$ (2,−2)
   $4x - 3y = 14$

5. $6x - y = 5$ (1,1)
   $8x + 2y = 10$

6. $4x - 3y = 1$ (1,1)
   $2x + y = 3$

7. $7y - 2x = -4$ (9,2)
   $4y = x - 1$

8. $2x = 1 + 3y$ (2,1)
   $5x + 6y = 16$

9. $12x + 10y = 4$ (−3,4)
   $9x = 21 - 12y$

10. $3x + 2y = 5$ Dependent
    $-6x - 4y = -10$

11. $2x - 4y = 5$ Inconsistent
    $-x + 2y = 8$

12. $3x + 2y = 5$ (3,−2)
    $4x = 22 + 5y$

13. $5x - 2y = 3$ (1,1)
    $2x + 7y = 9$

14. $3x + 2y = 12$ (4,0)
    $2x + 5y = 8$

15. $3x = -2y + 13$ (7,−4)
    $2x + 3y = 2$

16. $\dfrac{x - 2y}{8} = \dfrac{1}{2}$ (2,−1)
    $3x + 2y = 4$

17. $\dfrac{6}{7}x - y = \dfrac{3}{7}$ $\left(\dfrac{1}{3}, -\dfrac{1}{7}\right)$
    $\dfrac{9}{7}x + 2y = \dfrac{1}{7}$

18. $\dfrac{x}{2} = \dfrac{y + 4}{3}$ (2,−1)
    $\dfrac{x - y}{6} = \dfrac{1}{2}$

19. $10t + u = 7(t + u) + 3$ and $3u = t + 2$ $(t,u) = (7,3)$

20. $2(b + c) = 16$ and $3(b - c) = 16$ $(b,c) = \left(\dfrac{20}{3}, \dfrac{4}{3}\right)$

21. $12x - 11y = 7$ and $13x - 10y = 21$ (7,7)

22. $27x - 165y = 105$ and $11x + 55y = 165$ (10,1)

Solve each problem using a system of linear equations and the linear combination method.

23. The larger of two numbers is 3 more than twice the smaller. If twice the larger is decreased by 5 times the smaller, the result is 0. Find the two numbers. 6, 15

24. Twice one number is 15 less than a second number. When 13 is added to the second number, the result is 7 less than 9 times the first number. Find the numbers. 5, 25

25. The perimeter of a rectangle is 22 m. The length of the rectangle is 1 m less than 3 times the width. Find the length and the width. $l = 8$ m, $w = 3$ m

26. Twice one number is 20 more than a second number. One-fourth of the first number is 2 less than $\dfrac{1}{4}$ of the second. Find each number. 28, 36

27. If a theater added 5 seats to each row, it would need 20 fewer rows to seat the same number of people. If each row had 3 fewer seats, it would take 20 more rows to seat the same number. How many people does the theater seat now? [HINT: $(a + b)(c + d) = ac + bc + ad + bd$] 1,200 people

FOLLOW UP

Guided Practice

Classroom Exercises 1–16

Independent Practice

A Ex. 1–15, **B** Ex. 16–26, **C** Ex. 27–30

Basic: WE 1–19 odd, Midchapter Review, Application

Average: WE 7–25 odd, Midchapter Review, Application

Above Average: WE 9–29 odd, Midchapter Review, Application

Additional Answers

Midchapter Review

1. $(4, -1)$; consistent, independent

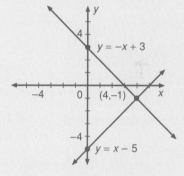

2. $(-1, 1)$; consistent, independent

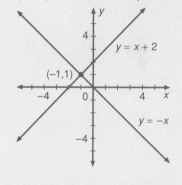

Enrichment

Have the students solve this problem: On a recent math test containing 25 problems, a certain number of points were given for each correct answer, and a certain number of points were deducted for each wrong answer. Paula answered 20 questions correctly and received a score of 50.

Werner answered 22 questions correctly and received a score of 60. What is the maximum score that could have been achieved on this test? 75

3. Inconsistent, independent

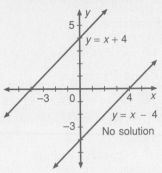

$y = x + 4$

$y = x - 4$

No solution

4. $\{(x,y) \mid y = 3x - 6\}$ consistent, dependent

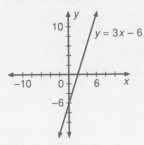

$y = 3x - 6$

Solve for x and y.

28. $ax + by = n$ $\left(\frac{2n}{a}, -\frac{n}{b}\right)$ **29.** $\frac{x}{y} = \frac{a}{b}$ $\left(1 - \frac{2a}{b}, \frac{b}{a} - 2\right)$ **30.** $\frac{8}{x} + \frac{15}{y} = 33$ $\left(-\frac{1}{9}, \frac{1}{7}\right)$

$2ax + 3by = n$ $\frac{x+1}{2a} = \frac{y+1}{b}$ $\frac{4}{x} + \frac{5}{y} = -1$

Midchapter Review

Solve each system by graphing. Indicate whether the system is consistent or inconsistent and dependent or independent. **4.1**

1. $y = x - 5$ **2.** $y = -x$ **3.** $x - y = 4$ **4.** $3x - y = 6$
$y = 3 - x$ $y = x + 2$ $y - x = 4$ $2y = 6x - 12$

Solve using either the substitution method or the linear combination method. **4.2, 4.3**

5. $y = -3x$ $(3, -9)$ **6.** $\frac{1}{3}t + \frac{1}{5}u = \frac{8}{15}$ **7.** $2x - y = -4$ **8.** $7x - 2y = 4$
$x - y = 12$ $3t = 4 - u$ $(1,1)$ $4x + 2y = 12$ $\left(\frac{1}{2}, 5\right)$ $3x - 3y = 21$
$(-2, -9)$

9. The total annual cost T of operating a new car is $f + cm$, where f is the fixed cost (such as depreciation and insurance), c is the operating cost per mile, and m is the number of miles driven. The total cost for 10,000 mi is $1,800, and the total cost for 15,000 mi is $2,300. Find the fixed cost and the operating cost per mile. **4.3** $800, 10¢ per mile

---

▨/ **Application:** *Profit and Loss*

For any business, profit (or loss) is simply the difference between costs and revenues. The equations below show costs and revenues for a blanket production run of x blankets. Fixed costs are $30,000; variable costs are $4 per blanket; revenues are $7 per blanket.

Costs = fixed + variable
$y = 30,000 + 4x$

Revenues = price × sales
$y = 7x$

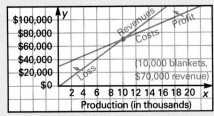

(10,000 blankets, $70,000 revenue)

Production (in thousands)

The solution to any cost-revenue system is the business's break-even point (where revenues = costs). In this example, the revenue from sales of 10,000 blankets equals the cost of their production.

Calculate the break-even point for The Bagel Bakery. Fixed costs are $24,000; variable costs are $1/dozen bagels; revenues are $5/dozen. 6,000 dozen bagels

4.4 Problem Solving: Money Problems/Mixture Problems

Objectives

To solve word problems involving coins
To solve word problems involving unit price
To solve word problems about investments
To solve word problems involving mixtures

A system of linear equations can be used to solve such problems as finding the numbers of coins having a given total value, finding the number of tickets sold at a given price, or finding the amount of money invested at a given rate of return.

These problems involve *numbers* of items and *monetary values* of those items. For example, in problems involving coins, it is often necessary to distinguish the *number* of coins from the *total value* of each in cents. Thus, the total *number* of coins in 3 dimes and 8 nickels is $3 + 8$, or 11, while the total *value* of these 11 coins is $3 \cdot 10 + 8 \cdot 5 = 30 + 40$, or 70 cents.

Similarly, the total value in cents of q quarters and d dimes is $q \cdot 25 + d \cdot 10$, or $25q + 10d$.

EXAMPLE 1

Renee has $3.75 in dimes and quarters. The number of quarters is 2 less than 3 times the number of dimes. Find the number of coins of each type.

What are you to find?	the number of dimes and quarters
Choose two variables.	Let d = the number of dimes.
What do they represent?	Let q = the number of quarters.
	Then $10d$ = the value of dimes in cents,
	and $25q$ = the value of quarters in cents.
What is given?	No. of quarters is 2 less than 3 times number of dimes.

$$q = 3d - 2$$

Value of quarters plus value of dimes is $3.75.

$$25q \quad + \quad 10d \quad = 375$$

Write a system.	(1) $q = 3d - 2$ (2) $25q + 10d = 375$
Solve the system.	Substitute $3d - 2$ for q in (2). Solve for d.

$$(2)\ 25(3d - 2) + 10d = 375$$
$$75d - 50 + 10d = 375$$
$$d = 5 \leftarrow \text{number of dimes}$$

Teaching Resources

Application 4
Quick Quizzes 27
Reteaching and Practice
   Worksheets 27

GETTING STARTED

Prerequisite Quiz

Solve.

1. $x + y = 30,000$
   $475x + 595y = 15,450,000$
   (20,000, 10,000)
2. $q = 3d - 2$
   $25q + 10d = 375$
   $q = 13, d = 5$

Write each percent as a decimal.

3. 59%   0.59
4. 5%   0.05
5. $6\frac{1}{2}$%   0.065
6. $7\frac{1}{4}$%   0.0725

Motivator

Ask the class why they need to distinguish between the total number of coins a person has and the value of the coins.   To find the number of coins that have a given total value.   Ask them the following question: If Joe has 5 quarters and 3 dimes,

1. How many coins does he have altogether?   8
2. What is their total value in cents? $1.55

Additional Example 1

Bill has $2.15 in quarters and nickels. The number of quarters is 1 less than 3 times the number of nickels. Find the number of coins of each type.   8 quarters, 3 nickels

Highlighting the Standards

Standard 5a: Familiarity with the process of building equations increases student confidence in both using symbols and solving equations.

Lesson Note

Point out that coin and interest problems are special cases of mixture problems. Coin and interest problems are similar in the two equations that are used to solve them: one involving amounts invested—or numbers of coins; the other involving interest on investments—or values of coins. It may be necessary to review the conversion of percents to decimals and the simple interest formula, $i = prt$.

Math Connections

Dimensional Analysis: Measures of the form $\frac{p}{q}$ (eg., $\frac{mi}{h}$, $\frac{dollars}{lb}$) can be manipulated as if they were fractions. This principle can be used in this lesson to ensure that the expressions in an equation make dimensional sense. Thus, in Example 1, students should understand that

$25 \frac{cents}{quarter} \cdot q \text{ quarters} = 25q \text{ cents}$,

$10 \frac{cents}{dime} \cdot d \text{ dimes} = 10d \text{ cents}$,

and

$100 \frac{cents}{dollars} \cdot 3.75 \text{ dollars} = 375 \text{ cents}$.

Substitute 5 for d in (1). Solve for q.
$$(1)\ q = 3 \cdot 5\ -2$$
$$= 13 \leftarrow \text{number of quarters}$$

Check in the original problem.

13 quarters is 2 less than 3 times 5 dimes.
$$13 = 3 \cdot 5 - 2$$
Renee has \$3.75: $10 \cdot\ \ 5 = 50$ cents in dimes
$$25 \cdot 13 = \underline{325 \text{ cents in quarters}}$$
$$375 \text{ cents, or } \$3.75$$

Calculator check using decimals:
0.1 ⊗ 5 ⊕ 0.25 ⊗ 13 ⊜ 3.75

State the answer. Renee has 5 dimes and 13 quarters.

EXAMPLE 2

The receipts from a ball game were \$154,500, paid by 30,000 spectators. Bleacher seats sold for \$4.75 each, and reserved seats sold for \$5.95 each. How many seats of each type were sold?

Solution

What are you to find? the number of bleacher seats and reserved seats sold

Choose two variables. What do they represent?

Let x = the number of bleacher tickets.
Let y = the number of reserved tickets.

Then $4.75x$ = the value of the bleacher tickets and $5.95y$ = the value of the reserved tickets.

What is given?

Value of bleacher tickets + value of reserved tickets = \$154,500

$$(1)\ 4.75x + 5.95y = 154,500$$

No. of bleacher tickets + no. of reserved tickets = 30,000

$$(2)\ x + y = 30,000$$

Write a system. (1) $4.75x + 5.95y = 154,500$ (2) $x + y = 30,000$

Solve the system. Multiply each side of (1) by 100 and (2) by -595.

$$(3)\quad 475x + 595y = 15,450,000 \leftarrow (1) \cdot 100$$
$$(4) -595x - 595y = -17,850,000 \leftarrow (2) \cdot (-595)$$

Add (3) and (4). $-120x = -2,400,000$
$$x = 20,000 \leftarrow \text{number of bleacher seats}$$

Substitute 20,000 for x in (2).
$$(2)\ 20,000 + y = 30,000$$
$$y = 10,000 \leftarrow \text{number of reserved seats}$$

Additional Example 2

Mr. Lead wants to mix pencils costing 25¢ each with pencils costing 20¢ each. The mixture will cost \$3.60. If the number of 20¢ pencils is 9 more than the number of 25¢ pencils, how many pencils of each type should be in the package? 4 at 25¢; 13 at 20¢

State the answer. 20,000 bleacher seats and 10,000 reserved seats were sold. The check is left for you.

EXAMPLE 3 Julio invested some money at $8\frac{1}{2}$% interest, and a second amount at $6\frac{3}{4}$%. The amount at $6\frac{3}{4}$% was $1,200 more than the amount at $8\frac{1}{2}$%. Find the amount invested at each rate, given that the total income after 2 years was $711. Use the simple-interest formula $i = prt$.

Solution Let x = the amount invested at $6\frac{3}{4}$%, or 6.75%.

Let y = the amount invested at $8\frac{1}{2}$%, or 8.5%.

Write an equation relating the amounts invested.

Amount at $6\frac{3}{4}$% was $1,200 more than $8\frac{1}{2}$% amount.

(1) $x = y + 1,200$

Equation for total income using $i = prt$:	income on x dollars at $6\frac{3}{4}$% for 2 years	+	income on y dollars at $8\frac{1}{2}$% for 2 years	=	total income

$$x \cdot 0.0675 \cdot 2 \quad + \quad y \cdot 0.085 \cdot 2 \quad = 711$$
$$0.135x \quad + \quad 0.17y \quad = 711$$

Multiply each side by 1,000. (2) $135x + 170y = 711,000$

Solve by substitution. $135(y + 1,200) + 170y = 711,000$
Substitute $y + 1,200$ for x $135y + 162,000 + 170y = 711,000$
in Equation (2).

$$305y = 549,000$$
$$y = 1,800$$
↑
amount at 8.5%

Substitute 1,800 for y $x = 1,800 + 1,200$
in Equation (1). $= 3,000$ ← amount at 6.75%

The check is left for you.

The amounts invested were $3,000 at $6\frac{3}{4}$% and $1,800 at $8\frac{1}{2}$%.

handwritten: 5,000 .09 45,0.00 4000 .08 320 450 770

EXAMPLE 4 A chemist mixes a solution containing 18% alcohol with a second solution containing 45% alcohol. The result is 12 oz of solution that is 36% alcohol. How many ounces of each solution are mixed? (THINK: Which is greater, the number of ounces of 18% solution or of 45% solution? How do you know?)

Critical Thinking Questions

Analysis: Ask students how many ounces of 20% rubbing alcohol should be added to 40% rubbing alcohol to obtain 50% rubbing alcohol. No amount of 20% rubbing alcohol can bring a solution of 40% rubbing alcohol to a strength higher than 40%. Such a mixture must be weaker than 40%. Ask students what the minimum strength of rubbing alcohol required to bring 40% rubbing alcohol up to 50% is. A solution stronger than 50% is required.

Common Error Analysis

Error: With coin problems, students tend to forget to convert the values of the coins to cents in the equation that represents the values of the coins.

Have the students find the total value in cents of the following:

1. 5 quarters and 3 dimes 155¢
2. 6 nickels and 7 quarters 205¢
3. 4 half-dollars and 8 dimes 280¢

Checkpoint

1. Mark has 6 more quarters than dimes. He has $3.95 in all. How many coins of each type does he have? 7 dimes, 13 quarters
2. The cost of an adult ticket to a play was $2.00. A student ticket costs $1.50. The number of student tickets sold was 50 more than twice the number of adult tickets sold. The total income from the tickets was $800. How many tickets of each type were sold? 340 student tickets, 145 adult tickets

Additional Example 3

Ted borrowed $9,000, part from a bank at $9\frac{3}{4}$% a year and the rest from his father at $8\frac{1}{2}$% a year. How much did he borrow from each if he owed $1,630 in interest at the end of two years? Assume that he had made no payments during this time.
$4,000 from the bank, $5,000 from his father

Additional Example 4

A solution that contains 25% acid is mixed with a second solution that contains 40% acid to obtain a 15-liter solution which is 30% acid. How many liters of each solution are used? 10 liters of the 25% solution, 5 liters of the 40% solution

handwritten:
$$x + y = 9000$$
$$.09x + .08y = 770$$
$$.09(9,000 - y) + .08y = 770$$
$$810 - .09y + .08y = 770$$
$$-.01y = -40$$

Closure

Joe has *n* nickels and *d* dimes. Have students write an equation representing a total collection of 12 coins. $n + d = 12$ Have them write an equation representing a total value of $0.85. $5n + 10d = 85$

FOLLOW UP

Guided Practice

Classroom Exercises 1–4

Independent Practice

A Ex. 1–6, B Ex. 7–10, C Ex. 11,12

Basic: WE 1–8
Average: WE 3–10
Above Average: WE 5–12

Additional Answers

Written Exercises

4. $8,000 at $8\frac{1}{2}$%, $6,000 at $10\frac{3}{4}$%

6. 45%: 8 kg; 70%: 12 kg

7. $3,600 at $15\frac{1}{2}$%, $2,400 at $14\frac{1}{4}$%

10. $4,000 each

Solution

Let x = the number of ounces containing 18% alcohol.

Let y = the number of ounces containing 45% alcohol.

	Number of ounces of solution	Number of ounces of alcohol
18% solution	x	$0.18x$
45% solution	y	$0.45y$
Total	12	$0.36(12) = 4.32$

Write two equations.

Total ounces of solution = 12   Total ounces of alcohol = 4.32
$$x + y = 12$$
$$0.18x + 0.45y = 4.32$$
$$18x + 45y = 432$$

Solve the system $\begin{array}{l}(1)\ x + y = 12 \\ (2)\ 18x + 45y = 432\end{array}$ using the substitution method.

The above system can be simplified to the equivalent system: $\begin{array}{l} x = 4 \\ y = 8 \end{array}$

Thus, there are 4 ounces of the 18% solution and 8 ounces of the 45% solution. The solution and check are left for you.

Classroom Exercises

Give an algebraic expression for the income on each investment. Use $i = prt$.

1. x dollars at 5% for 3 years 0.15x

2. y dollars at 4% for 5 years 0.2y

3. x dollars at $9\frac{1}{2}$% for 6 months 0.0475x

4. The Hayakawas invested $20,000, part at $11\frac{1}{4}$% and the rest at $12\frac{1}{2}$%. At the end of one year, the return was $2,350. How much money was invested at each rate? $12,000 at $11\frac{1}{4}$%, $8,000 at $12\frac{1}{2}$%

Written Exercises

Use a linear system of two equations to solve each problem.

1. Jon has $1.15 in dimes and nickels. The number of nickels is 5 less than twice the number of dimes. Find the number of each type of coin. 7 dimes, 9 nickels

2. Megan has twice as many quarters as half-dollars. The total value is $5.00. How many of each type of coin are there? 10 quarters, 5 half-dollars

Enrichment

Challenge the students to solve this puzzle. Mrs. Kelly had a full 36-oz pitcher of pure orange juice in the refrigerator. Patti came home from school first, poured 9 oz of juice for herself and then added water to make the pitcher full again. Shortly afterwards, Dennis came home, poured 9 oz of the mixture for himself and then added water to make the pitcher full again. Finally, Sheila came home, drank a 9-oz glass of the mixture, and went out to play. How many ounces of orange juice was in the 9-oz mixture that Sheila drank?

$\frac{81}{16}$, or 5.0625 oz

3. The cost of an adult ticket to a football game is $2.00, and a student ticket is $1.50. The total receipts from 300 tickets were $550. How many tickets of each kind were sold?
200 adult, 100 student

4. Hank invested one sum of money at 8.5%, and another at 10.75%. The amount invested at 8.5% was $4,000 less than twice the amount at 10.75%. If his total income from simple interest for one year was $1,325, how much did he invest at each rate?

5. A chemist has one solution containing 20% acid, and a second containing 30% acid. How many liters of each solution must be combined to obtain 80 liters of a mixture that is 28% acid? 16 l of 20%, 64 l of 30%

6. A scientist wants to combine two metal alloys into 20 kg of a third alloy which is 60% aluminum. He plans to use one alloy with 45% aluminum content, and a second alloy with 70% aluminum content. How many kilograms of each alloy must be combined?

7. Julio borrowed $6,000, part at $15\frac{1}{2}\%$ and the rest at $14\frac{1}{4}\%$. At the end of 3 years and 6 months, he had paid $3,150 in simple interest. How much did he borrow at each rate?

8. An $8.50 assortment of pads contains $0.65 pads and $0.50 pads. If the number of $0.50 pads is 1 less than $\frac{1}{2}$ the number of $0.65 pads, how many of each type of pad are in the assortment? 10 $0.65 pads, 4 $0.50 pads

9. Jake invested one-third of his money at $9\frac{3}{4}\%$, and the rest at $11\frac{1}{2}\%$. At the end of 18 months he had earned $4,912.50 in simple interest. Find the total amount of money invested. $30,000

10. Bill invested a total of $8,000, part at 10% and the remainder at 15%. The total simple interest income for one year is $12\frac{1}{2}\%$ of the total investment. Find the amount of each investment.

11. The Easy Kar Rental Agency charges a fixed amount for each day a car is rented and a fixed amount for each mile the car is driven. Mr. Wong spent $159 to rent a car for 5 days and 600 mi. Ms. Weintraub spent $130 to rent a car for 4 days and 500 mi. Find the daily rental charge and the charge per mile. $15 per day, 14¢ per mile

12. A metallurgist has 30 grams of a metal alloy that is 60% silver. How much pure silver must she add to the alloy to obtain a mixture that is 70% silver? 10 g

Mixed Review

1. Solve $\frac{4}{5}x - 2 = \frac{1}{3}x + 5$. *1.4* 15

2. Solve $\frac{3x - 2}{2x + 1} = \frac{2}{3}$. *1.4* $\frac{8}{5}$

3. Write an equation of the line containing the point $P(4,10)$ and parallel to the line described by $3x + 2y = 4$. *3.7* $3x + 2y = 32$

4. Suppose y varies directly as the square of x. If $y = 4$ when $x = 2$, what is y when $x = 5$? *3.8* 25

4.5 Problem Solving: Wind and Current Problems/Angle Problems

GETTING STARTED

Prerequisite Quiz

1. Write a formula for distance in terms of rate and time. $d = rt$
2. Solve. $3x + 3y = 900$
   $4\frac{1}{2}(x - y) = 900$ 250, 50

Motivator

Ask students to answer the following questions.

1. Suppose that you are in a motorboat that can go 10 mi/h in still water. If you are headed downstream, in the direction of the current, and the current is moving at a speed of 3 mi/h, how far will you actually travel in one hour?   13 miles
2. What then is your effective speed?
   13 mi/h
3. If m represents the speed of the boat in still water, c represents the speed of the boat due to the current, and the boat traveled 13 miles in an hour, write an equation to represent the effective speed. $m + c = 13$ mi/h

Objectives   To solve motion problems involving wind or current

To solve word problems involving complementary or supplementary angles

Recall (see Lesson 2.6) that the distance d traveled by an object is given by the formula $d = rt$, where r is the object's rate (speed) and t is the elapsed time. The rate of a plane in still air is called the *airspeed* of the plane. The rate of the plane with respect to the ground is its *ground speed*. Ground speed is affected by the wind as shown below.

Rate of plane *with* wind is airspeed + tail wind = ground speed.

Rate of plane *against* wind is airspeed − head wind = ground speed.

tail wind   head wind

EXAMPLE 1   A plane flies 900 mi in 3 h with a tail wind, and returns the same distance in $4\frac{1}{2}$ h against the wind. Find the wind speed. Assume a constant airspeed and a constant wind speed.

Solution   Let p = the airspeed of the plane.

Let w = the wind speed.

Then $p + w$ = the ground speed with the wind, and $p - w$ = the ground speed against the wind.

	rate (r) $\times$	time (t) $=$	distance (d)
	rate	time	distance
With wind	$p + w$	3	$3(p + w)$
Against wind	$p - w$	$4\frac{1}{2}$	$4\frac{1}{2}(p - w)$

Write two equations. Use the fact that both distances are 900 mi.

$3(p + w) = 900$ $\frac{9}{2}(p - w) = 900$

$\frac{1}{3} \cdot 3(p + w) = \frac{1}{3} \cdot 900$ $\frac{2}{9} \cdot \frac{9}{2}(p - w) = \frac{2}{9} \cdot \frac{900}{1}$

$p + w = 300$ $p - w = 200$

Now solve the system of equations.   (1) $p + w = 300$
                                           (2) $p - w = 200$

Add equations (1) and (2). $2p = 500$
                                $p = 250$ ← airspeed = 250 mi/h

Highlighting the Standards

Standard 1a: The problem solving in this lesson is a means to increase students' understanding of how linear equations work. Exercises like the ones here could be used in regular reviews.

Additional Example 1

Bill flew his ultralight plane to the next city against a headwind of 10 mi/h. The trip took $2\frac{1}{2}$ hours. The return trip with the help of a 10 mi/h–tailwind took $1\frac{1}{2}$ hours. Find the distance to the next city and the airspeed of the plane. (Assume a constant airspeed.)   75 mi; 40 mi/h

To find the value of w,   (1) $250 + w = 300$
substitute 250 for p in (1). $w = 50 \leftarrow$ wind speed $= 50$ mi/h

Check   Distance *with* the wind: Distance *against* the wind:
$3(250 + 50) = 3(300) = 900$ $4\frac{1}{2}(250 - 50) = 4\frac{1}{2}(200) = 900$

So, the wind speed is 50 mi/h.

Systems of equations can also be applied to problems involving relationships between angles. The diagrams below illustrate the properties of complementary and supplementary angles.

Two angles are *complementary* if the sum of their degree measures is 90.

Two angles are *supplementary* if the sum of their degree measures is 180.

  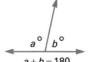 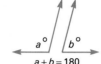

$x + y = 90$ $x + y = 90$ $a + b = 180$ $a + b = 180$

EXAMPLE 2   The degree measure of the larger of two complementary angles is 30 less than twice the measure of the smaller. Find the measure of each angle.

What are you to find?   The measures of two angles.

Choose two variables.   Let x = the measure of the smaller angle.
What do they represent?   Let y = the measure of the larger angle.

What is given?   Larger is 30 less than twice smaller.
Write two equations.   (1) y = 30 less than $2x$, or $2x - 30$

        Angles are complementary.
        (2) $x + y = 90$

Write a system.   (1) $y = 2x - 30$
               (2) $x + y = 90$

Solve the system.   Substitute $2x - 30$ for y in (2).

        (2) $x + (2x - 30) = 90$
              $3x - 30 = 90$
                  $3x = 120$
                    $x = 40 \leftarrow$ measure of
                            smaller angle

◾ TEACHING SUGGESTIONS

Lesson Note

Motion problems have already been studied in Chapter 2. Problems with winds or currents introduce only a slight variation. Therefore this lesson includes a second type of problem.

Review the definitions of complementary and supplementary angles. Use two board compasses to demonstrate that adjacent complementary angles form a right angle and adjacent supplementary angles form a straight angle.

Math Connections

Navigation: Problems that involve *airspeed*, *wind speed* and *ground speed* ordinarily need to be solved using trigonometric methods. Thus, Example 1 may be thought of as a simple instance of the interaction of two velocity vectors. Vectors are considered in greater detail in Chapter 9 and also in Chapter 19.

Critical Thinking Questions

Analysis: Ask students how they would set up the equations of a wind/current problem when the airspeed or wind speed changes in a known way. For example, in Example 1, how would the answer be affected if the wind speed on the return trip were half the wind speed on the first half of the trip? The system of equations would be changed to
$$3(p + w) = 900$$
$$\frac{9}{2}\left(p - \frac{1}{2}w\right) = 900$$
The new answers are: $233\frac{1}{3}$ mi/h for the airspeed, $66\frac{2}{3}$ mi/h for the initial wind speed.

Additional Example 2

The degree measure of the larger of two complementary angles is 15 less than twice the measure of the smaller. Find the measure of each angle. The degree measures are 55 and 35.

139

Checkpoint

1. The degree measure of the larger of two supplementary angles is 60 less than twice the measure of the smaller. Find the measure of each angle. 100, 80
2. A jet flew 2,100 km with the jet stream in $2\frac{1}{2}$ hours. The return flight against the jet stream took $3\frac{3}{4}$ hours. Find the speed of the jet stream and the airspeed of the plane. Jet stream: 140 km/h, airspeed: 700 km/h

Closure

Ask students the following questions on how to set up a table to solve problems involving wind currents.

1. What does each of the three columns represent? rate, time, and distance
2. What does each of the two rows represent? with the wind, against the wind

FOLLOW UP

Guided Practice

Classroom Exercises 1–7

Independent Practice

A Ex. 1–6, **B** Ex. 7–11, **C** Ex. 12

Basic: WE 1–8, Brainteaser
Average: WE 3–10, Brainteaser
Above Average: WE 5–12, Brainteaser

Substitute 40 for x in (1).

$$
\begin{aligned}
(1)\; y &= 2x - 30 \\
&= 2 \cdot 40 - 30 \\
&= 50 \leftarrow \text{measure of larger angle}
\end{aligned}
$$

Check in the original problem.

Larger is 30 less than twice smaller.
$50 = 2 \cdot 40 - 30$ True
Angles are complementary.
$50 + 40 = 90$ True

State the answer.

The measures are 40 and 50.

Classroom Exercises

Using x and y to represent the measures of the two angles, write a system of equations to represent each pair of sentences.

1. Two angles are complementary. The measure of one is twice the measure of the other. $x + y = 90, y = 2x$
2. Two angles have equal measure. The angles are supplementary. $x = y, x + y = 180$

Find the ground speed of each plane.

3. airspeed = 500 mi/h, head wind = 50 mi/h 450 mi/h
4. airspeed = 660 mi/h, tail wind = 40 mi/h 700 mi/h
5. airspeed = A mi/h, head wind = H mi/h $(A - H)$ mi/h
6. airspeed = A mi/h, tail wind = T mi/h $(A + T)$ mi/h
7. A plane flew 2,400 mi in 5 h with a tail wind. The return trip took 6 h against the wind. Find the wind speed. Assume a constant airspeed and a constant wind speed. 40 mi/h

Written Exercises

1. A plane flew 1,800 mi in 4 h with a tail wind, and flying at the same airspeed, returned the same distance in $4\frac{1}{2}$ h against the wind. Find the wind speed. 25 mi/h
2. A plane flew 720 mi in 3 h against the wind, and flying at the same airspeed, returned the same distance in 2 h with the wind. Find the airspeed of the plane. 300 mi/h
3. Flying with the wind, a plane can travel 1,500 mi in $2\frac{1}{2}$ h. Flying at the same airspeed, the return trip against the wind requires $\frac{1}{2}$ h more to make the same trip. Find the airspeed. 550 mi/h
4. The measure of the larger of two complementary angles is twice the measure of the smaller. Find the measure of each angle. 30, 60

Enrichment

Have the students solve this problem. During a recent school competition, 3 points were awarded to the winner of each event, and 2 points were awarded to the individual who placed second. At the end of the competition, Raj had accumulated 32 points. List all the possible combinations of first and second place awards that might have accounted for Raj's score. Use ordered pairs (a,b), in which a, the first number, is the number of first places, and b is the number of second places. (10,1), (8,4), (6,7), (4,10), (2,13), (0,16)

5. Two angles are supplementary. The measure of the larger is 60 less than twice the measure of the smaller. Find the measure of each angle. 80, 100

6. Two angles are supplementary. The measure of the larger is 60 more than 3 times the measure of the smaller. Find the measure of each angle. 30, 150

7. Two angles are complementary. The measure of one is $\frac{2}{3}$ the measure of the other. Find the measure of the supplement of the larger angle. 126

8. The measure of the supplement of angle A is 3 times that of its complement. Find the measure of angle A. 45

9. An Everglades air boat moves 78 mi downstream (with the current) in the same amount of time it requires to move upstream (against the current) a distance of 48 mi. The boat's engines drive it in still water at an average rate of 16 mi/h greater than the speed of the current. Find the speed of the current. (HINT: Current is to water as wind is to air.) 5 mi/h

10. Traveling downstream (with the current), a boat covers 54 mi in 3 hours. Traveling upstream (against the current), it takes the boat twice as long to cover $\frac{2}{3}$ of the distance. Find the rate of the boat in still water and the speed of the current. (See the HINT offered in Exercise 9). Boat: 12 mi/h, Current: 6 mi/h

11. Four times the measure of the complement of angle B is 12 more than twice the difference between the measures of its supplement and complement. Find the measure of angle B. 42

12. A woman can row 11 mi downstream in the same time it takes her to row 7 mi upstream. She rows downstream for 3 h, then turns and rows back for 4 h, but finds that she is still 5 mi from where she started the trip. How fast does the stream flow (see the HINT offered in Exercise 9)? 2 mi/h

Mixed Review

Identify the property of operations with real numbers. *1.3*
Assoc Prop Add   Comm Prop Mult

1. $3x + (5x + 2) = (3x + 5x) + 2$
2. $-7x(-4y) = -4y(-7x)$
3. Find $f(-2)$ for $f(x) = -x^5$. *3.2* 32
4. Write an equation of the line that contains $A(4,3)$ and $B(4,-2)$. *3.4* $x = 4$

▰▰▰ Brainteaser

A ticket to a chamber-music concert costs twice as much as a ticket to a band concert. Suppose that you have a choice of inviting 40 friends to one band concert or inviting two friends to 10 chamber music concerts. Which is the more expensive choice? They cost the same.

4.6 Systems of Three Linear Equations

Objectives   To solve systems of three linear equations in three variables
To solve word problems involving systems of three linear equations

Prerequisite Quiz

1. Solve by substitution.
   $y = 2x + 3$
   $4x + 3y = 59$ (5, 13)
2. Solve by the linear combination method.
   $3x + 5y = 11$
   $7x + 4y = 18$ (2, 1)
3. Given: $3x + 2y - 4z = -8$, $x = 2$,
   and $y = 3$. Find z.   5

Motivator

Consider the following system of equations.

1. $-x + 3y + z = -10$
2. $3x + 2y - 2z = 3$
3. $2x - y - 4z = -7$

Ask students how they can eliminate z from the last two equations.   Multiply equation (2) by −2 and add to equation (3).   Ask them how they can eliminate z from equations (1) and (2).   Multiply equation (1) by 2 and add.   These operations will result in two equations in how many variables? two

At the right, direct substitution shows that the given equation is true when $x = 2$, $y = 1$, and $z = 3$.

$$2x + 3y - z = 4$$
$$2 \cdot 2 + 3 \cdot 1 - 3 = 4$$
$$4 + 3 - 3 = 4 \text{ True}$$

The **ordered triple** (2, 1, 3) is a solution of the equation $2x + 3y - z = 4$. Any equation of the form $ax + by + cz = d$ (with a, b, and c not all zero) is a **first-degree equation in three variables**. Such equations are also called *linear* even though their graphs are not lines but *planes* in three-dimensional space.

As with systems of two linear equations in two variables, systems of three linear equations in three variables may have exactly one solution, no solution, or infinitely many solutions. You can also use the same methods you learned earlier to solve systems of three equations in three variables.

EXAMPLE 1   Solve for x, y, and z.   (1) $2x - 3y - z = 12$
                                        (2) $y + 3z = 10$
                                        (3) $z = 4$

Plan   Since Equation (3) is already solved for z, substitute 4 for z in Equation (2) to find y. Then the values of y and z can be substituted in Equation (1) to find x.

Solution   Substitute 4 for z in Equation (2).   (2) $y + 3z = 10$
                                        $y + 3 \cdot 4 = 10$
                                        $y = -2$

Substitute -2 for y and 4 for z in Equation (1).   (1) $2x - 3y - z = 10$
                                        $2x - 3(-2) - 4 = 12$
                                        $2x + 6 - 4 = 12$
                                        $2x = 10$
                                        $x = 5$

Check
$$2x - 3y - z = 12 \qquad y + 3z = 10 \qquad z = 4$$
$$2 \cdot 5 - 3(-2) - 4 \stackrel{?}{=} 12 \qquad -2 + 3 \cdot 4 \stackrel{?}{=} 10 \qquad 4 = 4 \text{ True}$$
$$10 + 6 - 4 \stackrel{?}{=} 12 \qquad -2 + 12 \stackrel{?}{=} 10$$
$$12 = 12 \text{ True} \qquad 10 = 10 \text{ True}$$

Thus, $(5, -2, 4)$ is the solution of the system.

Highlighting the Standards

Standard 3c: The method for solving three linear equations follows the same logic as the method for two. But students must thoroughly understand what they are doing to use this method.

Additional Example 1

Solve for x, y, and z.

$3x + 2y = 7$
$x = 1$
$2x - 3y + 4z = 8$ 1, 2, 3

EXAMPLE 2

Solve.
(1) $-x + 3y + z = -10$
(2) $3x + 2y - 2z = 3$
(3) $2x - y - 4z = -7$

Plan

Choose any two equations, say the first two, and eliminate one of the variables, say x, by linear combination. Then repeat the procedure, eliminating the same variable x from a different pair of equations. This leaves two equations in two variables.

Solution

Multiply (1) by 3 and add to (2) to eliminate x.

(1) $-x + 3y + z = -10$ → $\quad -3x + 9y + 3z = -30$
(2) $3x + 2y - 2z = 3$ → $\quad \underline{3x + 2y - 2z = 3}$
$$(4) \qquad\qquad 11y + z = -27$$

Multiply (1) by 2 and add to (3).

(1) $-x + 3y + z = -10$ → $\quad -2x + 6y + 2z = -20$
(3) $2x - y - 4z = -7$ → $\quad \underline{2x - 1y - 4z = -7}$
$$(5) \qquad\qquad 5y - 2z = -27$$

Solve the resulting system, (4) and (5). Multiply (4) by 2 and add.

(4) $11y + z = -27$ → $\quad 22y + 2z = -54$
(5) $5y - 2z = -27$ → $\quad \underline{5y - 2z = -27}$
$$27y = -81$$
$$y = -3$$

Substitute -3 for y in either (4) or (5) to find the value of z.

(4) $\quad 11y + z = -27$
$\quad 11(-3) + z = -27$
$\quad\quad\quad\quad z = 6$

Substitute 6 for z and -3 for y in one of the original equations to find the value of x.

(1) $\quad -x + 3y + z = -10$
$\quad -x + 3(-3) + 6 = -10$
$\quad -x - 9 + 6 = -10$
$\quad\quad\quad -x = -7$, or $x = 7$

Check

A check of Equation (1) is shown below using a scientific calculator. The rest of the check is left for you.

Calculator check of (1): 7 [+/−] [+] 3 [×] 3 [+/−] [+] 6 [=] -10

Thus, the solution of the system is $(7, -3, 6)$.

EXAMPLE 3

Joe has $4.70 in half-dollars, quarters, and dimes. The number of quarters is 10 less than twice the number of dimes. The number of half-dollars decreased by the number of quarters is 2. Find the number of coins of each type.

■■■ **TEACHING SUGGESTIONS**

Lesson Note

Use the Prerequisite Quiz to review both the substitution and the linear combination methods for solving systems of two equations. Solving systems with three variables frequently involves a combination of the two methods. As shown in the lesson development, numbering the equation is helpful for ease of reference in solving the system.

Math Connections

Geometry: The graph of a first-degree equation in three variables forms a plane. The graph of the solution set of a system of three linear equations in three variables can form a point, a line, a plane, or can have no solution.

Critical Thinking Questions

Analysis: In a plane, the equation $x = 4$ is the graph of a line parallel to the y-axis. Sketch the three axes of the three-dimensional coordinate system on the chalkboard (or project the system on a screen) and ask students to describe the graph of $z = 4$ in Example 1. *The graph is a plane parallel to the xy-plane.*

Common Error Analysis

Error: A typical error is failure to keep in mind which variable is being eliminated first. For example, if x is eliminated from one pair of equations, then x must also be eliminated from a second pair of equations.

Emphasize that the objective at this stage is to derive a pair of linear equations in two variables—the same two variables in each equation.

Additional Example 2

Solve.

$x + y + z = 1$
$x + 3y + 7z = 13$
$x + 2y + 3z = 4$ $\quad (1, -3, 3)$

Additional Example 3

Find the number of nickels, dimes, and quarters in a collection of 80 such coins if the nickels and quarters are worth $4.50 and the quarters and dimes are worth $5.50. *40 nickels, 30 dimes, 10 quarters*

Checkpoint

Solve each system of equations.

1. $z = 5$
$3x + 2z = 13$
$2x - y + z = 9$ $(1, -2, 5)$

2. $3x + y - z = -2$
$x + 2y + 4z = 9$
$2x - 3y + z = 11$ $(1, -2, 3)$

3. Bob has 12 coins in dimes, nickels, and quarters. The dimes and nickels are valued at $0.85. The dimes and quarters have a total value of $1.20. Find the number of coins of each type.   7 dimes, 3 nickels, 2 quarters

Closure

Ask students to give the steps for solving a system of equations such as the one below.

$x + 2y - z = 1$
$2x + y + 3z = 5$
$3x + y + 2z = 8$

See Example 2.

◤◤◤FOLLOW UP

Guided Practice

Classroom Exercises 1–9

Independent Practice

Ⓐ Ex. 1–9. Ⓑ Ex. 10–15, Ⓒ Ex. 16–18

Basic: WE 1–11 odd, Problem Solving

Average: WE 5–15 odd, Problem Solving

Above Average: WE 7–17 odd, Problem Solving

Solution   Let h = number of half-dollars, q = number of quarters and d = number of dimes. Total value is 470 cents: $50h + 25q + 10d = 470$.

The number of quarters is 10 less than twice the number of dimes: $q = 2d - 10$. The number of half-dollars decreased by the number of quarters is 2: $h - q = 2$.

So, the system is: (1) $50h + 25q + 10d = 470$
(2) $q = 2d - 10$
(3) $h - q = 2$

Equation (2) is already solved for q in terms of d. Substitute $2d - 10$ for q in each of the other two equations.

When we substitute $2d - 10$ for q in (1) and (3) we obtain:

(4) $50h + 60d = 720$ and (5) $h - 2d = -8$.

When we solve the system (4) and (5) by multiplying Equation (5) by 30 and adding, we obtain $h = 6$. Substituting 6 for h in Equation (5) we get $d = 7$. Substituting 7 for d in Equation (2), we get $q = 4$.

Thus, there are 6 half-dollars, 4 quarters, and 7 dimes

Classroom Exercises

State whether $(3, 4, -1)$ is a solution of the given equation.

1. $x + y + z = 6$ Yes   **2.** $2x + 3y - z = 13$ No   **3.** $x - y + 3z = 4$ No
4. $3x + y - 2z = 15$ Yes   **5.** $-x - y - 2z = -5$ Yes **6.** $3x = 2y - z$ Yes

Solve each system of equations for x, y, and z.

7. $x + 2y + z = 14$ (1,5,3)   **8.** $x = 3$ (3,4,-1)   **9.** $y = 6$ $\left(-2, 6, \frac{28}{3}\right)$
$y + z = 8$ $2x + y = 10$ $x - y + 3z = 20$
$z = 3$ $2x + y + z = 9$ $y - x = 8$

Written Exercises

Solve each system of equations.

1. $5x - 4y - z = 25$ **2.** $3x + 2y - z = 16$ **3.** $-3x - 2y = 14$
$2y + 7z = 17$ $x = 4$ $x - 2y + z = -1$
$z = 3$ (4,-2,3) $2x - y = 5$ (4,3,2) $y = -4$ (-2,-4,-7)

4. $x + 2y - z = 1$ **5.** $a + b + c = 6$ **6.** $2x + 3y - 2z = 4$
$2x + y + 3z = 5$ $3a - b + c = 8$ $3x - 2y + 2z = 16$
$3x + y + 2z = 8$ (3,-1,0) $2a + 3b - 2c = 10$ (3,2,1) $-x - 12y + 8z = 5$ $\left(3, 5, \frac{17}{2}\right)$

Enrichment

A system of two equations in three variables has no unique solution. However, solutions can be found if one of the variables is controlled. Consider the following system.

$2x + 2y - z = 4$ (1)
$y - z = 1$ (2)
Let (r, s, t) be a solution. Thus,
$2r + 2s - t = 4$ (3)
$s - t = 1$ (4)
From (4), $s = t + 1$;

from (4) and (3),
$2r + 2(t + 1) - t = 4$
$2r + 2t + 2 - t = 4$
$r = \dfrac{2 - t}{2}$
Thus, a solution of the system is
$\left(\dfrac{2 - t}{2}, t + 1, t\right)$.
For example, if $t = 2$, then a solution is $(0, 3, 2)$. If $t = -6$, a solution is $(4, -5, -6)$. Have the students find solutions (r, s, t) to

equations (1) and (2) when $t = 0$, $t = 4$, and $t = -1$. $(1, 1, 0)$; $(-1, 5, 4)$; $\left(\frac{3}{2}, 0, -1\right)$

You may wish to explain that a controlled variable, such as t in this illustration, is called a *parameter*.

7. $x + 2y = 2$
$2y - 3z = 4$
$2x + 3z = -1$ $\left(1, \frac{1}{2}, -1\right)$

8. $x + 3y - z = 8$
$2x - y + 2z = -9$
$2x + z = -5$ $(-2, 3, -1)$

9. $3x + 5y = -3$
$10y - 2z = 2$
$x + 4z = 6$ $\left(-2, \frac{3}{5}, 2\right)$

10. Moira has $2.95 in half-dollars, dimes, and nickels. The number of half-dollars is 1 less than twice the number of dimes. The sum of the number of half-dollars and the number of nickels is 8. Find the number of coins of each type. 5 half-dollars, 3 dimes, 3 nickels

11. Find three numbers in decreasing order such that their sum is 20, the difference of the first two numbers is 1, and the sum of the smallest number and the greatest number is 13. 8, 7, 5

12. A three-digit number is 198 less than the number when its digits are reversed. Twice the sum of the digits is 5 more than 7 times the tens digit. The tens digit is 5 less than twice the hundreds digit. Find the original number. (HINT: Write the number as $100h + 10t + u$, where h, t and u are the hundreds, tens, and units digits, respectively.) 436

13. The tens digit of a three-digit number is 3 less than 5 times the units digit. Three times the sum of the digits is 2 more than 4 times the hundreds digit. If the digits are reversed, the new number is 594 less than the original number. Find the original number (see the HINT offered in the previous exercise). 721

Solve each system.

14. $\dfrac{2x + 4}{3} - (y + 4) + \dfrac{z + 1}{3} = 0$

$\dfrac{x - 4}{6} + \dfrac{y + 1}{8} - \dfrac{z - 2}{4} = -\dfrac{1}{2}$

$\dfrac{x + y + 1}{2} = \dfrac{3}{4} - \dfrac{z - 1}{4}$ $(1, -1, 2)$

15. $0.2x - 0.3y + 0.1z = 0.3$
$0.5x - 0.5y + 0.1z = 0.8$
$0.6x - 0.7y + 0.2z = 0.9$ $(1, -1, -2)$

16. $\dfrac{1}{x} + \dfrac{2}{y} + \dfrac{2}{z} = 16$

$\dfrac{1}{y} + \dfrac{2}{z} + \dfrac{2}{x} = 15$

$\dfrac{1}{z} + \dfrac{2}{x} + \dfrac{2}{y} = 14$ $\left(\frac{1}{2}, \frac{1}{3}, \frac{1}{4}\right)$

17. $x + y + z + w = 10$
$2x - y + z - 3w = -9$
$3x + y - z - w = -2$
$2x - 3y + z - w = -5$
$(w, x, y, z) = (4, 1, 2, 3)$

18. Find a, b, and c such that the solution of the equation $ax + by + cz = 6$ contains the ordered triples $\left(\frac{1}{2}, \frac{3}{8}, \frac{3}{2}\right)$, $\left(\frac{2}{3}, \frac{1}{2}, 1\right)$, and $\left(1, \frac{1}{2}, \frac{1}{2}\right)$. $(3, 4, 2)$

Mixed Review

Solve. *1.4, 2.1, 2.3*

1. $5 - \frac{2}{3}(6 - 9x) = \frac{1}{2}x + 4$ $\frac{6}{11}$

2. $0.6t - 0.2 = 0.12t - 0.08$ 0.25

3. $4(y + 1) < 7y - 2(3 - y)$ $y > 2$

4. $8 - 2|2 - a| = -2$ $-3, 7$

Problem Solving Strategies

$2x(x + 4) + 16 = (x + 8)^2$;

$-4,0,4$; $12,16,20$

Problem Solving Strategies

Guess and Check

A problem can sometimes be solved by a guess-and-check method. This method can also be used to suggest an equation whose solution(s) will help in solving the problem. Both of these objectives are achieved in the example below and practiced in the exercise that follows.

Example The sum of the areas of a square and a rectangle is 240 ft^2. The rectangle is 3 ft wider than and twice as long as the square. Find the length of a side of the square.

Plan Choose (guess) a value of s for a side of the square. Check in the problem as shown below. Repeat the process until the sum of the areas is 240 ft^2.

Solution

Square		Rectangle			Sum of Areas	?	Sum of Areas
side	area	width	length	area			
s	s^2	w	l	lw	$s^2 + lw$		240
5	25	$5 + 3$	$2 \cdot 5$	80	$25 + 80 = 105$	<	240
10	100	$10 + 3$	$2 \cdot 10$	260	$100 + 260 = 360$	>	240
8	64	$8 + 3$	$2 \cdot 8$	176	$64 + 176 \rightarrow$	=	240
x	x^2	$x + 3$	$2x$	$2x(x + 3)$	$x^2 + 2x(x + 3)$	=	240

Notice above that 5 for s is too small, since $105 < 240$, and 10 is too large, since $360 > 240$. Try 7 for s; then try 9 for s. This process continues until the value of s narrows down to 8, the correct value of s. Also, the patterns developed in each column are generalized, using the variable x, to provide the equation $x^2 + 2x(x + 3) = 240$, whose positive root, 8, is the length of the side in feet. Now, try the following exercise in a similar way. Include the equation and its solution(s).

Find three consecutive multiples of 4 such that twice the product of the first two, increased by 16, is the square of the third.

1st	2nd	3rd	$2 \cdot$ 1st $\cdot$ 2nd $+ 16$	?	$(\text{3rd})^2$
4	8	12	$2 \cdot 4 \cdot 8 + 16 = 80$	<	$12^2 = 144$
16	20	24			
.					
.					
.					
x					

4.7 Solving Systems of Linear Inequalities

Objectives   To solve systems of two linear inequalities by graphing
To graph polygonal regions described by conjunctions of several linear inequalities

To solve a system such as $2x + 3y > 12$ *and* $-3y \geq 6 - 4x$, first graph both inequalities on the same coordinate plane. The solution will be the region where the two graphs overlap, as shown in Example 1.

EXAMPLE 1   Solve by graphing. $2x + 3y > 12$
$-3y \geq 6 - 4x$

Solution   First, solve each inequality for y.

$$2x + 3y > 12 \qquad\qquad -3y \geq 6 - 4x$$
$$3y > -2x + 12 \qquad\qquad -3y \geq -4x + 6$$
$$y > -\tfrac{2}{3}x + 4 \qquad\qquad y \leq \tfrac{4}{3}x - 2 \leftarrow \text{Change } \geq \text{ to } \leq.$$

Now, graph each inequality on the same coordinate plane.

First graph $y = -\tfrac{2}{3}x + 4$.
    ↗ ↑
   slope *y*-intercept

Graph $y = \tfrac{4}{3}x - 2$ on the same
    ↑ ↑ coordinate plane.
  slope *y*-intercept

Use a dotted line because $>$ means *is greater than*. Shade the region above the line.

Use a solid line because $\leq$ means *is less than or equal to*. Shade the region below the line.

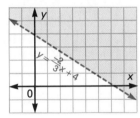   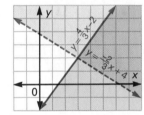

The solution is the set of points in the double-shaded region, including the upper boundary of that region.

Check   Pick a point in the double-shaded region, say $P(6, 2)$. Show that its coordinates satisfy both inequalities.

▬▬▬GETTING STARTED

Prerequisite Quiz

Graph the solution set of each inequality on a number line.

1. $2x - 4 \leq 8$

    ←————————
                 6

2. $5x + 2 > 12$

    ○————————→
    2

3. $-4 - 2x < 6$

    ○————————→
   -5

Motivator

Ask the class to recall how to solve the compound inequality $x \geq 5$ *and* $x \leq 10$ on a number line.   Graph both inequalities on the same number line. The solution will be the region where the two graphs overlap.

Additional Example 1

Solve by graphing.

$y + x \geq 1$
$-2y < 8 - 3x$

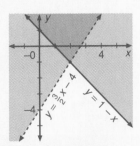

Highlighting the Standards

Standard 4c: Solving inequalities builds on the work done with equations. The extension takes students to new ground and a wider perspective of algebra.

Lesson Note

Point out the similarity between solving systems of inequalities and solving systems of equations by graphing. Once the lines are drawn—solid or dashed—simply decide whether to shade above or below the line (right or left if the line is vertical).

Example 2 paves the way for the Introduction to Linear Programming which follows this lesson. For this reason, it is important to find the coordinates of the vertices of the resulting polygon.

Math Connections

Logic: The sentences of this lesson can be used to review the language of symbolic logic. Inequalities such as $-3y \geq 6 - 4x$ in Example 1 can be written as the disjunction $-3y > 6 - 4x$ or $-3y = 6 - 4x$.

Critical Thinking Questions

Analysis: Ask students whether the shortcut mentioned at the bottom of page 149 will work under all circumstances.

They should see that although the short cut will always provide the vertices of a polygon, it is still necessary to check all of the inequalities to be sure that their graphs *all* intersect in the polygon's interior.

$$2x + 3y > 12 \qquad\qquad -3y \geq 6 - 4x$$
$$2 \cdot 6 + 3 \cdot 2 \overset{?}{>} 12 \qquad -3 \cdot 2 \overset{?}{\geq} 6 - 4 \cdot 6$$
$$12 + 6 \overset{?}{>} 12 \qquad\qquad -6 \overset{?}{\geq} 6 - 24$$
$$18 > 12 \text{ True} \qquad -6 \geq -18 \text{ True}$$

The pair $(6, 2)$ satisfies both inequalities.

Thus, the graph of the solution set is the double-shaded region *below* and *including* the graph of $-3y = 6 - 4x$ and *above* the graph of $2x + 3y = 12$.

Example 1 illustrates the *intersection* of the graphs of the two given inequalities. The **intersection** of two graphs is the set of all points in *both* the first graph *and* the second graph. This set of points is the graph of the solution set of Example 1.

The **union** of two graphs, such as those of the inequalities in Example 1, is the set of all points in either the first graph *or* the second graph, including all points in both graphs.

EXAMPLE 2 Graph the solution set of the following conjunction. Then find the coordinates of the vertices of the polygonal region that is formed.

$$3 \leq x \leq 6 \text{ and } y \geq 0 \text{ and } x + y \geq 5 \text{ and } 21 - 3y \geq 2x$$

Solution 1. Rewrite $3 \leq x \leq 6$ as the conjunction $x \geq 3$ *and* $x \leq 6$. Then graph both inequalities on the same coordinate plane.

Graph of $x \geq 3$:

Graph the vertical line $x = 3$. Use a solid line. Shade the half-plane to the right.

Graph of $x \leq 6$:

Graph the vertical line $x = 6$. Use a solid line. Shade the half-plane to the left.

Graph of $3 \leq x \leq 6$:

The graph is the *intersection* of the graphs of $x \geq 3$ and $x \leq 6$.

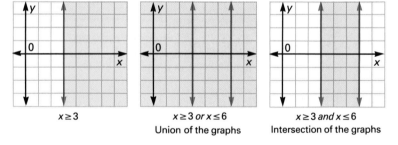

$x \geq 3$

$x \geq 3 \text{ or } x \leq 6$
Union of the graphs

$x \geq 3 \text{ and } x \leq 6$
Intersection of the graphs

Additional Example 2

Graph the solution set of the following conjunction of inequalities. Then find the coordinates of the vertices of the polygon formed.

$y \leq x + 4$ *and* $y \geq -x + 1$ *and* $0 \leq x \leq 4$ *and* $y \geq 0$

The solution set is the polygonal region with vertices (0,4), (0,1), (1,0), (4,0), and (4,8).

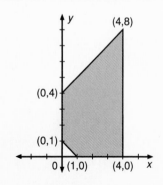

2. Graph $y \geq 0$ on the same coordinate plane as $3 \leq x \leq 6$. Graph the horizontal line $y = 0$. Use a solid line. Shade the half-plane above the line.

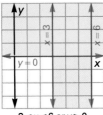 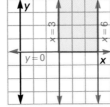

$3 \leq x \leq 6$ or $y \geq 0$ $3 \leq x \leq 6$ and $y \geq 0$

3. Continue in a similar fashion with the two remaining inequalities.

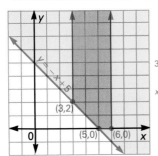

$3 \leq x \leq 6$
$y \geq 0$
$x + y \geq 5$

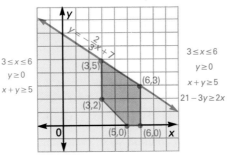

$3 \leq x \leq 6$
$y \geq 0$
$x + y \geq 5$
$21 - 3y \geq 2x$

Thus, the graph of the solution set of the conjunction is the polygonal region with vertices $A(3, 2)$, $B(5, 0)$, $C(6, 0)$, $D(6, 3)$, and $E(3, 5)$. The graph of the region is shown at the right.

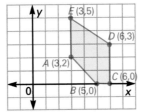

There is, however, a shorter and more efficient way to graph the system of linear inequalities in Example 3.

First graph the boundary lines defined by each inequality.

$$3 \leq x \leq 6 \rightarrow x = 3, x = 6$$
$$y \geq 0 \rightarrow \qquad y = 0$$
$$x + y \geq 5 \rightarrow \qquad y = -x + 5$$
$$21 - 3y \geq 2x \rightarrow \qquad y = -\tfrac{2}{3}x + 7$$

Then shade the interior of the polygonal region bounded by the lines.

Checkpoint

1. Graph the solution set.
   $3x - 2 \leq y < 3$

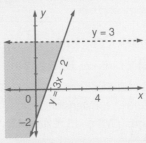

2. Graph the solution set of the conjunction of inequalities. Then find the coordinates of the vertices of the polygonal region formed.
   $x + 2y \leq 12$ and $2 \leq x \leq 6$ and $y \geq 2$

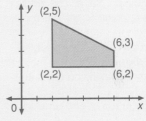

The solution set is the polygonal region with vertices (2,2), (6,2), (6,3), and (2,5).

Closure

Ask students to give the steps in solving a system of linear inequalities in two variables by graphing. See Example 1. Ask students what is meant by a conjunction of several linear inequalities. The intersection of all the graphs

Enrichment

This Enrichment activity provides an extension of the type of activity presented in the Enrichment for Lesson 4.1.

The shaded regions in the graph show *loss* when expense exceeds income, and *profit* when income exceeds expense.
Have the students determine the net profit or loss during the 6 months of this business operation. If you wish, you can give the hint

that the students should compare the areas of the two triangles.

$\tfrac{1}{2}(40)(4) - \tfrac{1}{2}(20)(2) = 60$

net profit = $60,000

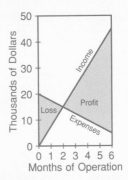

Guided Practice

Classroom Exercises 1–7

Independent Practice

A Ex. 1–9, **B** Ex. 10–12, **C** Ex. 13–15

Basic: WE 1–10, Extension
Average: WE 3–12, Extension
Above Average: WE 5–15, Extension

Additional Answers

Classroom Exercises

3. Answers may vary. One possible answer: Graph each inequality on the same coordinate plane. The solution is the set of all points in the doubly-shaded region.

4. Answers may vary. One possible answer: Solve $-x + 2y \geq 4$ for y, then graph each inequality on the same coordinate plane. The solution is the set of all points in the doubly-shaded region.

5. Answers may vary. One possible answer: Solve both inequalities for y, then graph each on the same coordinate plane. The solution is the set of all points in the doubly-shaded region.

6. Answers may vary. One possible answer: Rewrite $4 \leq x \leq 8$ as the conjunction $x \geq 4$ and $x \leq 8$, solve $y + 2x \geq 6$ for y, then graph each inequality on the same coordinate plane. The solution is the set of all points in the triple-shaded region.

See pages 151–153 for the answers to Classroom Ex. 7 and Written Ex. 1–9. See page 809 for the answers to Written Ex. 10–15.

Classroom Exercises

State each inequality as a conjunction.

1. $3 \leq x \leq 7$ $x \geq 3$ and $x \leq 7$

2. $3x - 2 \leq y \leq 3x + 8$
   $y \geq 3x - 2$ and $y \leq 3x + 8$

Describe each step in solving the following systems of linear inequalities.

3. $y \geq x$
   $y \leq -x + 4$

4. $-x + 2y \geq 4$
   $y \geq 3$

5. $x + y \leq 6$
   $x - y \geq 4$

6. $4 \leq x \leq 8$
   $y + 2x \geq 6$

7. Graph the solution set of Classroom Exercise 3.

Written Exercises

Graph the solution set.

1. $y > x + 5$
   $y < -x + 1$

2. $y \leq 3x - 4$
   $y \geq 6 - x$

3. $y < x + 4$
   $x \leq 4$

4. $y \geq \frac{1}{2}x - 1$
   $y < 2$

5. $x + y > -1$
   $3x - 2y > 4$

6. $4x - y \leq -3$
   $y \leq -1$

7. $2x - 1 \leq y < 5$

8. $4 \leq x \leq 8$

9. $6 \leq y < -2x + 1$

Graph the solution set of each conjunction of inequalities. Then find the coordinates of the vertices of the polygonal region formed.

10. $y \leq 2x - 4$ and $x \leq 6$ and $y \geq 2$ **11.** $2x + y \leq 5$ and $y \geq 1$ and $x \geq 1$

12. $x \geq 0$ and $y \geq 0$ and $x + 2y \leq 7$ and $2x + y \leq 8$

Write a system of inequalities describing each polygon and its interior.

13.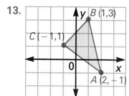

14.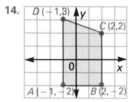

15.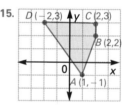

Mixed Review

1. Solve $\dfrac{2x - 4}{x - 3} = \dfrac{2}{3}$.   **1.4** $\frac{3}{2}$

2. Solve $|2x - 4| < 8$.   **2.3** $-2 < x < 6$

3. Given that $f(x) = -2x^3$, find $f(-4)$.   **3.2** 128

4. Solve $\begin{aligned}2x - y &= 4 \\ 3x + y &= 1\end{aligned}$.   **4.3** $(1, -2)$

Extension

Introduction to Linear Programming

A branch of mathematics called **linear programming** deals with the graphs of systems of linear inequalities.

Suppose that the sum S of two numbers, x and y, is to be maximized. Suppose, moreover, that x and y must also be the coordinates of a point in the convex polygonal region defined in Example 2 of Lesson 4.7.

Because the polygonal region contains an infinite number of points, it would be impossible to find the maximum value of S by checking all of these points. The following theorem, however, reduces to a manageable number the list of points that must be checked.

$$3 \leq x \leq 6$$
$$y \geq 0$$
$$x + y \geq 5$$
$$21 - 3y \geq 2x$$

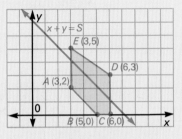

Theorem 4.2

Suppose that x and y are the coordinates of points in a convex polygonal region defined by a system of linear inequalities. Let P represent the linear expression $ax + by + c$ (with a and b not both zero). Then P attains its maximum value at one or more of the vertices of the region, and its minimum value at one or more of the remaining vertices.

The sum-of-two-numbers formula, $S = x + y$, is an example of an expression that can be maximized or minimized using Theorem 4.2. To find the maximum and minimum values of S, calculate the sum of the coordinates for each vertex of the polygonal region.

$$x + y = S$$

Test $(3, 2)$ $3 + 2 = 5$ ⎫ minimum of 5
Test $(5, 0)$ $5 + 0 = 5$ ⎭
Test $(6, 0)$ $6 + 0 = 6$
Test $(6, 3)$ $6 + 3 = 9$ ← maximum of 9
Test $(3, 5)$ $3 + 5 = 8$

So, S has a maximum value of 9 and a minimum value of 5 under the given constraints on x and y.

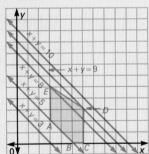

7.

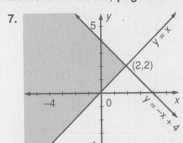

Written Exercises

1.

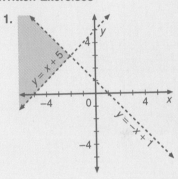

2.

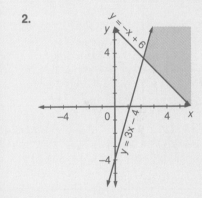

3.

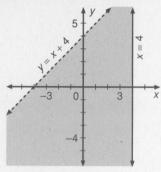

4.

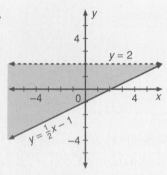

5.

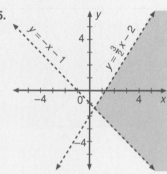

Example 1   Find the maximum and minimum values of *P* for $P = 2x + 7y$ given the following constraints: $x + y \leq 4$, $x - y \leq 2$, $x \geq 0$, and $y \geq 0$.

Solution
1. Graph the polygonal region and label the vertices.
2. Test the coordinates of each vertex in the formula $2x + 7y = P$.
   $(2, 0)$: $2 \cdot 2 + 7 \cdot 0 = 4 + 0 = 4$
   $(3, 1)$: $2 \cdot 3 + 7 \cdot 1 = 6 + 7 = 13$
   $(0, 4)$: $2 \cdot 0 + 7 \cdot 4 = 0 + 28 = 28$
   $(0, 0)$: $2 \cdot 0 + 7 \cdot 0 = 0 + 0 = 0$

So, the maximum value of *P* is 28, and the minimum value is 0.

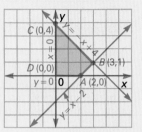

Example 2   A company manufactures television sets and video cassette recorders (VCRs). It must produce at least 20 TV sets per month, but cannot make more than 60 of them. The company also cannot produce more than 100 VCRs per month. Total production cannot exceed 140 TV sets and VCRs combined. The profit for a TV set is $45, and $175 for a VCR. Find the number of each that should be manufactured in order to maximize profit.

Solution   Let x = the number of TV sets produced each month.

Let y = the number of VCRs produced each month.

At least 20 TV sets means 20 or more TV sets. $x \geq 20$
No more than 60 TV sets means 60 or fewer TV sets. $x \leq 60$
No more than 100 VCRs means 100 or fewer VCRs. $y \leq 100$
The number of VCRs cannot be negative. $y \geq 0$
Total production *is less than or equal to* 140. $x + y \leq 140$

Total profit $P = 45x + 175y$.

Now find the maximum value of *P* under the above constraints.

1. Graph the polygonal region and label the vertices.
2. Test the coordinates of each vertex in $45x + 175y = P$.

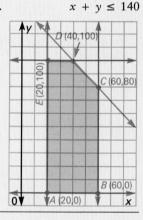

The maximum value occurs at $D(40,100)$: $45 \cdot 40 + 175 \cdot 100 = 19,300$.

So, the maximum profit is $19,300, which corresponds to $D(40,100)$, or 40 TV sets and 100 VCRs.

Exercises

The formula $P = 2x + 3y$ has constraints on x and y. The graph of these constraints is shown at the right. Find the value of P at the given vertex.

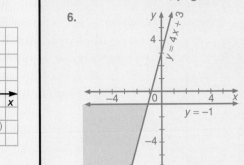

1. A -5
2. B 2
3. C 14
4. D 18
5. E 11
6. Give the maximum value of P. 18
7. Give the minimum value of P. -5
8. In the graph of the constraints, find a point (x, y) for which $P = 0$. Answers may vary. One possible answer: $(3, -2)$

Find the maximum and minimum values of P under the given constraints. Begin by graphing a polygonal region and determining the coordinates of the vertices of the polygon.

9. $P = 4x + 7y$
$x + y \leq 8$
$x - y \leq 2$
$x \geq 0$
$y \geq 0$ 56, 0

10. $P = 3x + 5y$
$x - 2y \geq 0$
$x + 2y \leq 8$
$1 \leq x \leq 6$
$y \geq 0$ 23, 3

11. $P = 2x + 7y$
$x \geq 1$
$0 \leq y \leq 4$
$4x - 2y \leq 8$ 36, 2

12. $P = 3x + 2y$
$x + y \leq 5$
$y - x \geq 5$
$4x + y \geq -10$ 10, -5

Solve each problem using linear programming.

13. A farmer uses two types of fertilizer on his bean field. Each 100-lb bag of 5-10-10 fertilizer ($10 per bag) contains 5 lb of nitrogen. Each 100-lb bag of 10-10-10 fertilizer ($12 per bag) contains 10 lb of nitrogen. The bean field needs at least 160 lb of nitrogen, and the farmer's fertilizer budget for the field is $240. If he already has 6 bags of 5-10-10 fertilizer on order, how many more bags of each type should he order to minimize the cost? 13 bags of 10-10-10 fertilizer

14. At radio station QXYZ, 6 min of each hour are devoted to news and the remainder to music, commercials, and other programming. Station policy requires at least 30 min of music per hour, as well as at least 3 min of music for each minute of commercial time. Find the maximum number of minutes of commercial time available each hour. $13\frac{1}{2}$ min

15. The manager of the Heinz bicycle factory, which has to make 50 frames per day to show a profit, has 240 man-hours available to him each day. It takes 2 h for one man, or 2 man-hours, to make a frame. It takes $\frac{1}{3}$ man-hour to make each wheel, and at least two wheels must be made for each frame. Find the maximum number of completed bicycles that the Heinz factory can make in one day. 90

Extension **153**

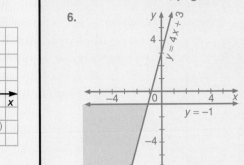

Additional Answers, page 150

6.
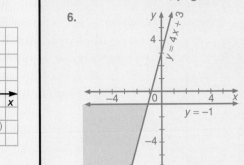

7.
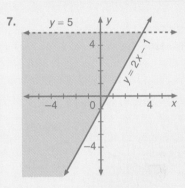

8.
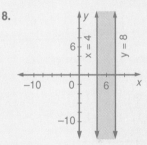

9.
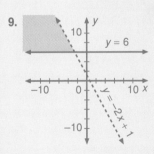

153

Chapter 4 Review

1. (5, −1); Consistent, independent

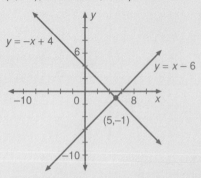

$y = -x + 4$
$y = x - 6$
(5, −1)

2. Inconsistent, independent

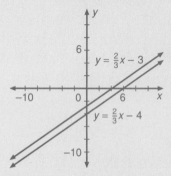

$y = \frac{2}{3}x - 3$
$y = \frac{2}{3}x - 4$

3. $\{y \mid y = -2x + 6\}$; Consistent, dependent

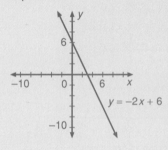

$y = -2x + 6$

Key Terms

consistent system (p. 119)
dependent system (p. 120)
equivalent systems (p. 124)
first-degree equation in three variables (p. 142)
inconsistent system (p. 120)
independent system (p. 119)

intersection (p. 148)
linear combination (p. 129)
linear programming (p. 151)
ordered triple (p. 142)
system of equations (p. 119)
union (p. 148)

Key Ideas and Review Exercises

4.1   To solve a system of two linear equations in two variables by graphing, graph both equations on the same coordinate plane.

Three situations can occur. (1) There is one point of intersection. The coordinates of the point are the solution. The system is *consistent* and *independent*. (2) The lines are parallel. There is no solution. The system is *independent* and *inconsistent*. (3) The lines coincide. There are infinitely many solutions. The system is *consistent* and *dependent*.

Solve each system by graphing. Indicate whether the system is *consistent* or *inconsistent* and *dependent* or *independent*.

1. $x + y = 4$
$x - y = 6$

2. $y = \frac{2}{3}x - 4$
$2x - 3y = 9$

3. $4x + 2y = 12$
$6x + 3y = 18$

4.2   To solve a system of two linear equations by substitution, look for an equation with a variable whose coefficient is 1 or −1. Solve the equation for that variable. Then substitute for that variable in the remaining equation.

Solve using the substitution method.

4. $y = 2x - 6$ (4,2)
$3x + 2y = 16$

5. $2x - y = 5$ (3,1)
$3x + 5y = 14$

6. $\dfrac{y}{2} + \dfrac{x - 3}{3} = 3$
$2x - y = 8$ (6,4)

4.3   To solve a system of two linear equations using the linear combination method, choose multipliers for one or both equations so that one pair of terms will be additive inverses of each other. Add both equations, and solve using substitution.

Solve using the linear combination method. If the system is inconsistent or dependent, indicate this.

7. $2x + 5y = 16$ (3,2)
$4x + 3y = 18$

8. $6x + 3y = 9$ Inconsistent
$4x + 2y = 5$

9. $3x - 2(y - 4) = 2$ (4,9)
$\dfrac{x}{2} = \dfrac{y - 1}{4}$

4.4, 4.5 To solve special types of word problems, use systems of equations as follows. *Coin, Unit price, Investment*: One equation relates *numbers* of coins (items) or *amounts* invested. The second relates cent or dollar *values*.

Mixture: One equation relates total amounts of a substance. The second relates amounts of a particular ingredient.

Wind: Equations have the form *rate* × *time* = *distance*, where
(1) ground speed = airspeed + tail wind, and
(2) ground speed = airspeed − head wind.

Complementary/Supplementary Angles: For one of the equations, use the sum of the degree measures. The sum of the measures of two complementary angles is 90. The sum of the measures of two supplementary angles is 180.

Use a system of equations to solve each problem.

10. Mr. Ruiz has $1.10 in dimes and nickels. The number of nickels is 2 more than $\frac{1}{2}$ the number of dimes. Find the number of coins of each type.   8 dimes, 6 nickels

11. Tina invested $4,000, part at 8% and part at 11%. The total amount of simple interest earned on her investments in two years was $790. How much did she invest at each rate? $1,500 at 8%, $2,500 at 11%

12. A small plane flew 800 mi in 5 h against the wind, and returned the same distance with the wind in 4 h. Find the plane's average airspeed.   180 mi/h

4.6   To solve a system of three linear equations in three variables, you can use the substitution or the linear combination method, or both methods together.

Solve each system of equations for x, y, and z. $(4, -3, -\frac{5}{2})$

13. $x = 4$ (4,−1,4)
$2x - 3y = 11$
$3x - 4y - 2z = 8$

14. $x + 2y = -6$ (4,−5,8)
$y + 2z = 11$
$2x + z = 16$

15. $2x + 3y - 2z = 4$
$3x - 3y + 2z = 16$
$6x - 2y + 8z = 10$

4.7   To solve a system of linear inequalities, graph both inequalities on the same coordinate plane and shade the region common to both.

Graph the solution set.

16. $y > x - 4$ *and* $y < -x + 2$

17. $y < x - 6$ *and* $x \geq -3$

18. Graph the solution set of $2x - y \leq 4$ *and* $y \leq 2$ *and* $x \geq 1$. Then find the coordinates of the vertices of the polygonal region formed.

19. Describe in a paragraph the difference in what is being represented by each of the two equations written to solve a problem dealing with coins. How is the strategy for solving coin problems similar to that for solving investment problems?

16.
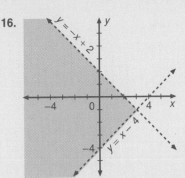

17.
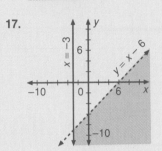

18. (1,2), (1,−2), (3,2)

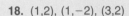
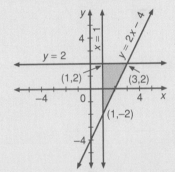

19. Answers may vary. One possible answer: In coin problems, one equation represents the number of coins, while the other represents their relative value. In investment problems, the number equation represents the number of dollars invested, while the value equation represents the value in interest income of each investment.

Chapter 4 Test

1. (4,2); Consistent, independent

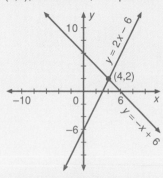

2. Inconsistent, independent

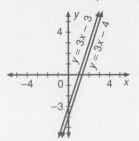

14.

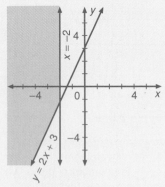

Solve each system by graphing. Indicate whether the system is consistent or inconsistent and dependent or independent.

1. $x + y = 6$
   $-2x + y = -6$

2. $6x - 2y = 6$
   $y = 3x - 4$

3. Solve by substitution.
   $x = 8 - y$ (2,6)
   $2x + 3y = 22$

4. Solve by linear combination.
   $4x + 2y = 10$
   $3x + 5y = 11$ (2,1)

Solve the conjunction by any method.

5. $3x + y = -5$ and $\frac{x}{3} + \frac{y}{2} = -\frac{1}{6}$ (−2,1)   **6.** $2x - (3 - 2y) = 5$ and $\frac{3x + 4y}{2} = 9$
(−2,6)

Use a system of equations to solve each problem.

7. The perimeter of a rectangle is 38 cm. The length is 9 cm less than 3 times the width. Find the length and the width. $l = 12$ cm, $w = 7$ cm

8. Bill has $3.00 in nickels and quarters. The number of nickels is 4 more than twice the number of quarters. Find the number of coins of each type. 8 quarters, 20 nickels

9. Alicia invested a total of $6,400, part at 9% per year and the rest at 8% per year. At the end of 2 years 6 months, the money had earned a total of $1,320 in simple interest. How much did she invest at each rate? $1,600 at 9%, $4,800 at 8%

10. The ratio of the measures of two complementary angles is 4:5. Find the measure of each angle. 40, 50

11. Flying with a tail wind, a plane can travel 3,600 mi in 6 h. Flying against the wind, the plane can return in 7.2 h. Find the average airspeed of the plane. 550 mi/h

Solve each system of equations for x, y, and z.

12. $2x + 3y + z = 11$ (1,2,3)
   $3x - y - 2z = -5$
   $4x + 3y + 3z = 19$

13. $3x + 4y = 3$ (5,−3,−6)
   $5y - 2z = -3$
   $x + 3z = -13$

14. Graph the solution set of $2x - y \le -3$ and $x \le -2$.

15. Graph the solution set of $x + 2y \le 8$ and $3x + 2y \le 12$ and $x \ge 0$ and $y \ge 0$. Then find the coordinates of the vertices of the polygonal region formed. (0,0), (0,4), (4,0), (2,3)

16. Find the value(s) of k for which the following conjunction has infinitely many solutions: $3x - 4y = 8$ and $-6(x - 3) + ky = -10$. None

15.

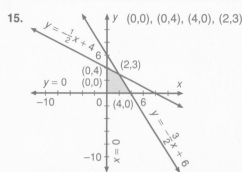

(0,0), (0,4), (4,0), (2,3)

In each item, you are to compare a quantity in Column 1 with a quantity in Column 2. Write the letter of the correct answer from the following choices.

A—The quantity in Column 1 is greater than the quantity in Column 2.
B—The quantity in Column 2 is greater than the quantity in Column 1.
C—The quantity in Column 1 is equal to the quantity in Column 2.
D—The relationship cannot be determined from the given information.

NOTE: Information centered over both columns refers to one or both of the quantities to be compared.

Sample Items	Answers
Column 1 **Column 2**	
S1. 4,570 rounded to 4,523 rounded to the nearest the nearest hundred hundred	The answer is A: $4{,}600 > 4{,}500$.
$x > -6$ *and* $y < 6$ S2. x y	The answer is D: For example, if $x = 7$ and $y = 5$, then $x > y$. If $x = 0$ and $y = 5$, then $x < y$.

Column 1	Column 2
1. B $\dfrac{8}{12}$	$\dfrac{23}{24}$
$-1 < y < 1$, and $y \neq 0$	
2. D $3y$	$\dfrac{3}{y}$
3. C $\frac{1}{3}(30 + 90)$	$\frac{1}{6}(60 + 180)$
C $6 \leq x \leq 12$ *and* $4 \leq y \leq 11$	
4. Maximum value of $y - x$ 5	
$x < y < 0$	
5. A $\dfrac{y}{x}$	x
k is a positive integer.	
6. A $(-1)^{2k}$	$(-1)^{2k+1}$

The figure below shows a circle of diameter x in a circle of radius x.

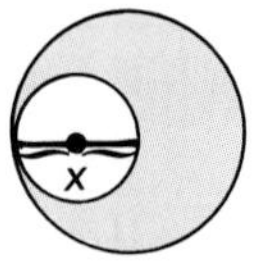

Column 1	Column 2
7. C Area of the shaded region	$\dfrac{3\pi x^2}{4}$
$x > 0$	
8. A x plus an increase of 75% of x	$0.75x$
$4(x + y) = 24$ *and* $3x - y = 6$	
9. C x	y

For additional standardized test practice, see the SAT/ACT Practice Test for Cumulative Test, Chapters 1–4.

18. $-\frac{1}{2}$, down to right
19. 0, horiz
20. Undef, vert

23.

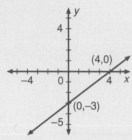

24.

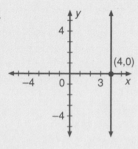

25.

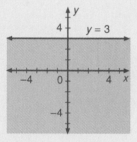

26.

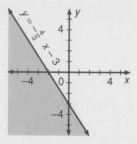

Cumulative Review (Chapters 1–4)

Simplify, if possible.   *1.2*

1. $7(6 - 5^2)$ –133

2. $\dfrac{-8.34 - 1.53 \cdot 2}{3 \cdot 0.27 - (0.9)^2}$ Undef

Identify the property of operations with real numbers. (Exercises 3–6)   *1.3*

3. $5a + (3a + 6) =$
$(5a + 3a) + 6$ Assoc Prop Add

4. $\frac{2}{3}(6x + 12) =$
$\frac{2}{3} \cdot 6x + \frac{2}{3} \cdot 12$ Distr Prop

5. $-4a + 4a = 0$ Add Inverse Prop

6. $9c + 5d = 5d + 9c$ Comm Prop Add

7. Simplify $5ab - a^2b - ab + 2a^2b$, and then evaluate for $a = 3$, $b = 2$. $4ab + a^2b$, 42

For Exercises 8–11, choose the root of the equation from among the five choices.   *1.4*

8. $2(3x + 8) - 3(x - 2) =$
$9x - (4x - 6)$ A
(A) 8   (B) –8   (C) 14
(D) –2   (E) 2

9. $x - 0.01 = 12.2 - 2.3x$ E
(A) 37   (B) 0.37   (C) 3,700
(D) 370   (E) 3.7

10. $\frac{3}{2}x - 1 = \frac{4}{5}x + 6$ B
(A) –10   (B) 10   (C) 1
(D) –1   (E) $-\frac{50}{7}$

11. $\dfrac{2x + 8}{3x - 2} = \dfrac{5}{4}$ C
(A) –6   (B) $\frac{10}{7}$ (C) 6
(D) $\frac{6}{7}$ (E) $3\frac{1}{7}$

Find the solution set. $\{x \mid x \le 6\}$

12. $5x - 2(3x - 1) \ge x - 10$ *2.1*

13. $3x - 2 > 7$ *and*   *2.2*
$2x - 4 \le 18$ $\{x \mid 3 < x \le 11\}$

14. $|2x - 8| = 12$ $\{-2, 10\}$ *2.3*

15. $|x - 7| < 4$ $\{x \mid 3 < x < 11\}$

Let $f(x) = -3x - 5$ and $g(x) = -2x^2 - x$. Find each value.   *3.2*

16. $f(-4)$ 7
17. $g(-4)$ –28

Find the slope of $\overleftrightarrow{PQ}$. Describe its slant as *up to the right, down to the right, horizontal,* or *vertical.*   *3.3*

18. $P(-4,5)$, $Q(-8,7)$
19. $P(3,-1)$, $Q(-5,-1)$
20. $P(-6,-7)$, $Q(-6,8)$

Write an equation of the line satisfying the given conditions. $y = \frac{2}{3}x + 1$

21. The line containing the   *3.4*
points $A(0,1)$ and $B(-6,-3)$.

22. The line containing $P(3,-4)$ *3.7*
and parallel to the line
$x - y = 4.$ $x - y = 7$

Graph each equation or inequality on a coordinate plane. (Exercises 23–26).

23. $3x - 4y = 12$ *3.5*
24. $4x - 3 = 13$
25. $6y - 4 \le 14$ *3.9*
26. $4x + 3y \le -9$
27. If y varies directly as the   *3.8*
square of x, and $y = 6$ when
$x = 4$, what is y when $x = 2$? $\frac{3}{2}$

Solve each system by graphing. Determine whether the system is *consistent* or *inconsistent* and *dependent* or *independent*. *4.1*

28. $3x - 2y = 4$
$x + y = 3$

29. $4x - 2y = 6$
$-6x + 3y = 9$

Solve algebraically.

30. $2x - y = 2$
$3x + 2y = 10$ (2,2) *4.2*

31. $4x + 6y = 6$
$6x + 9y = 5$ Inconsistent *4.3*

32. $3x - 2(y - 1) = 7$
$\dfrac{2x}{3} = \dfrac{y + 1}{2}$ (9,11)

33. $x = 4$
$2x + y = 11$
$3x - y + z = 10$ (4,3,1) *4.6*

34. $x + y + z = 2$
$2x + y + 2z = 3$
$3x - y + z = 4$ (2,1,-1)

35. Graph the solution set. *4.7*
$y > 2x - 4$
$y < -x + 2$

36. Jack borrowed $4,000, and paid $585 in simple interest after 1 year and 6 months. Find the annual rate of interest. Use the formula $i = prt$. $9\frac{3}{4}$% *1.5*

37. The sum of three consecutive odd integers is 69 more than twice the third odd integer. Find the three odd integers. 71, 73, 75 *1.6*

38. A rectangle and a square have the same width. The rectangle is 6 cm longer than a side of the square. One perimeter is twice the other. Find the dimensions of the rectangle. l = 9 cm, w = 3 cm *1.7*

39. Two trains started toward each other from stations 732 km apart. One train traveled at 96 km/h, and the other at 148 km/h. After how many hours did they meet? 3 h *2.6*

40. Al is 4 times as old as Bill. Ten years from now, he will be twice as old as Bill will be. Find the age of each now. *2.5*

41. The sum of two numbers is 14. Twice the larger number, increased by 3 times the smaller number, is 34. Find the two numbers. 6, 8 *4.2*

42. Mr. Brown invested a total of $6,400, part at 9% per year and the rest at 8% per year. At the end of 2 years 6 months, he had earned a total of $1,380 in simple interest. How much was invested at each rate? $4,000 at 9%, $2,400 at 8% *4.4*

43. Angelo has $1.90 in dimes and nickels. The number of nickels is 4 fewer than 5 times the number of dimes. How many of each type does he have? 6 dimes, 26 nickels

44. The cost of an adult ticket to a football game is $1.75. The cost of a student ticket is $1.25. Total receipts last week from ticket sales were $1,700. If the number of student tickets sold was twice the number of adult tickets, how many of each type were sold? 400 adult, 800 student

45. An airplane flies 840 mi in 3 h with a tail wind. The return trip at the same airspeed takes $3\frac{1}{2}$ h. Find the wind speed and the airspeed. *4.5*
Wind speed: 20 mi/h, airspeed: 260 mi/h

28. (2,1); Consistent, independent

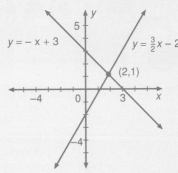

29. Inconsistent, independent

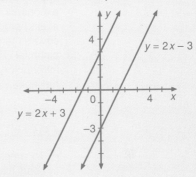

35.

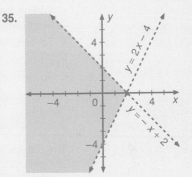

40. Al: 20 yr; Bill: 5 yr

5 POLYNOMIALS

OVERVIEW

This chapter introduces students to polynomials together with operations on polynomials. The chapter begins with two lessons on the rules of exponents, with special attention given to zero and negative exponents. The polynomial lessons focus on finding products and factors. Lesson 5.8 covers the division of a polynomial by a binomial. The chapter closes with an introductory lesson on composition of functions.

OBJECTIVES

- To simplify and evaluate expressions containing integral exponents
- To find the sum, difference, or product of two polynomials
- To factor polynomials
- To divide polynomials by binomials
- To find values of composite functions

PROBLEM SOLVING

The Extension on page 168 enables students to use the problem-solving strategy *Guess and Check* to estimate products and quotients. Guessing and checking is a useful strategy to use when factoring a trinomial into two binomials. Problem solving continues to be reinforced in the *Mixed Problem Solving* page that precedes the *Chapter Review*. In addition, there is an *Application* on page 192 that allows students to solve applied problems.

READING AND WRITING MATH

Exercise 39 on page 183 asks students to explain, in their own words, why the sum of two squares will always be greater than or equal to twice the product of the two numbers.

TECHNOLOGY

Calculator: A scientific calculator can be used to illustrate algebraic rules such as the rules for operating with exponents. This is shown in Lesson 5.2.

SPECIAL FEATURES

Mixed Review pp. 164, 167, 173, 177, 183, 186, 189, 192
Brainteaser pp. 164, 189
Extension: Estimating Products and Quotients p. 168
Midchapter Review p. 180
Application: Measures of Motion p. 192
Mixed Problem Solving p. 193
Key Terms p. 194
Key Ideas and Review Exercises pp. 194–195
Chapter 5 Test p. 196
College Prep Test p. 197

PLANNING GUIDE

Lesson	Basic	Average	Above Average	Resources
5.1 pp. 163–164	CE all WE 1–29 odd Brainteaser	CE all WE 5–33 odd Brainteaser	CE all WE 13–41 odd Brainteaser	Reteaching 31 Practice 31
5.2 pp. 166–169	CE all WE 1–35 odd Extension	CE all WE 7–41 odd Extension	CE all WE 9–43 odd Extension	Reteaching 32 Practice 32
5.3 pp. 172–173	CE all WE 1–31 odd	CE all WE 5–35 odd	CE all WE 9–39 odd	Reteaching 33 Practice 33
5.4 pp. 176–177	CE all WE 1–31 odd	CE all WE 5–35 odd	CE all WE 7–39 odd	Reteaching 34 Practice 34
5.5 pp. 179–180	CE all WE 1–17 odd Midchapter Review	CE all WE 1–27 odd Midchapter Review	CE all WE 3–29 odd Midchapter Review	Reteaching 35 Practice 35
5.6 pp. 182–183	CE all WE 1–23 odd	CE all WE 1–35 odd	CE all WE 5–39 odd	Reteaching 36 Practice 36
5.7 pp. 185–186	CE all WE 1–23 odd	CE all WE 1–39 odd	CE all WE 11–49 odd	Reteaching 37 Practice 37
5.8 pp. 188–189	CE all WE 1–12 Brainteaser	CE all WE 4–16 Brainteaser	CE all WE 1–23 Brainteaser	Reteaching 38 Practice 38
5.9 pp. 191–193	CE all WE 1–29 odd Application Mixed Problem Solving	CE all WE 9–37 odd Application Mixed Problem Solving	CE all WE 15–43 odd Application Mixed Problem Solving	Reteaching 39 Practice 39
Chapter 5 Review pp. 194–195	all	all	all	
Chapter 5 Test p. 196	all	all	all	
College Prep Test p. 197	all	all	all	

CE = Classroom Exercises   WE = Written Exercises

NOTE: For each level, all students should be assigned all Mixed Review exercises.

■■■INVESTIGATION

Project: The students explore the properties of exponents graphically and numerically.

Students should be familiar with graphs of functions of a variable with an exponent, such as $y = x^2$ or $y = x^3$. It should not be difficult for them to graph functions like $y = 2^x$ or $y = 3^x$, where a number has a *variable exponent*. Have the students make a table of values for $y = 2^x$.

x	−3	−2	−1	0	1	2	3
y	.125	.250	.500	1	2	4	8

Have the students graph the function. Then have them graph other similar functions on the same coordinate plane. Ask them to discuss their findings.

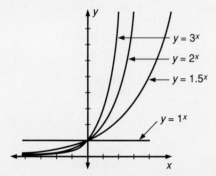

As the base numbers of the power get larger, the curves become more and more steep. Each curve passes through the point (0,1). Other answers are possible.

Since the points are connected by a continous curve, the graph represents exponents other than integers. For instance, $2^{1.5}$ is shown by the graph to be a number a little less than 3. The students can give meaning to this by noting that $2^{1.5} = 2^{\frac{3}{2}} = \sqrt{2^3} = 2.828427125$.

Students can use their calculators to verify the rules for multiplying numbers with exponents in the cases of non-integer values. For example,

$$2^{1.3} \times 2^{1.7} \overset{?}{=} 2^3$$
$$2.462288827 \times 3.249009585 \overset{?}{=} 8$$
$$8 \overset{\checkmark}{=} 8$$

This computer-generated drawing of a fern is an example of a fractal. Fractals are *self-similar structures,* containing patterns within patterns. Fractal-like structures are found in nature in clouds, mountain ranges, coastlines—and in the "fern" shown here.

More About the Image

Many natural growth processes often involve the repetition of a simple pattern over and over. The same patterns, with variation, occur many times at different levels in the three-dimensional **fractal models** of the fern shown above. Several levels of branching, from large to small, are visible. Notice that the curl of the tips of the branches is repeated for every branch at all visible levels of branching.

5.1 Positive Integral Exponents

Teaching Resources

Quick Quizzes 31
Reteaching and Practice
Worksheets 31

Objectives

To simplify expressions containing positive integral exponents
To solve exponential equations

Over three hundred years ago, the mathematician René Descartes (1596–1650) economized by writing $b \cdot b$ as b^2, $b \cdot b \cdot b$ as b^3, $b \cdot b \cdot b \cdot b$ as b^4, and so on. This notation is now universally used.

In the expression b^3, 3 is the **exponent**, and b is the **base**. Exponents that are positive integers signify the number of times the base is used as a factor. Thus, b^3 means $b \cdot b \cdot b$. Read b^3 as "the third power of b," or "b-cubed." When no exponent is written, the exponent is understood to be 1.

The meaning of the exponent suggests laws of exponents that can be used to simplify products and quotients of powers with the same base.

Product or Quotient	Expanded Form	Simplified Form	Suggested Relationship
$a^4 \cdot a \cdot a^2$	$aaaa \cdot a \cdot aa$	a^7	$a^4 \cdot a^1 \cdot a^2$ $= a^{4+1+2} = a^7$
$\dfrac{a^6}{a^2}$	$\dfrac{aa \cdot aaaa}{aa}$	a^4	$\dfrac{a^6}{a^2} = a^{6-2} = a^4$
$\dfrac{b^2}{b^5}$	$\dfrac{bb}{bb \cdot bbb}$	$\dfrac{1}{b^3}$	$\dfrac{b^2}{b^5} = \dfrac{1}{b^{5-2}} = \dfrac{1}{b^3}$
$\dfrac{c^3}{c^3}$	$\dfrac{c \cdot c \cdot c}{c \cdot c \cdot c}$	1	$\dfrac{c^3}{c^3} = 1$

Laws of Exponents
For all positive integers m and n and all real numbers b,

1. $b^m \cdot b^n = b^{m+n}$ (Product of powers)

2. $\dfrac{b^m}{b^n} = \begin{cases} b^{m-n}, & \text{where } m > n \text{ and } b \neq 0 \\ \dfrac{1}{b^{n-m}}, & \text{where } n > m \text{ and } b \neq 0 \\ 1, & \text{where } m = n \text{ and } b \neq 0 \end{cases}$ (Quotient of powers)

◾◾◾ **GETTING STARTED**

Prerequisite Quiz

Find the value of each expression in the list.

1. $\left(\dfrac{1}{2}\right)^2$, $\left(\dfrac{1}{2}\right)^3$, $\left(\dfrac{1}{2}\right)^4$ $\dfrac{1}{4}, \dfrac{1}{8}, \dfrac{1}{16}$

2. $\left(\dfrac{2}{3}\right)^2$, $\left(\dfrac{2}{3}\right)^3$, $\left(\dfrac{2}{3}\right)^4$ $\dfrac{4}{9}, \dfrac{8}{27}, \dfrac{16}{81}$

3. $\left(\dfrac{3}{4}\right)^2$, $\left(\dfrac{3}{4}\right)^3$ $\dfrac{9}{16}, \dfrac{27}{64}$

Solve each equation for x.

4. $2^x = 2^5$ 5
5. $3^{x-1} = 3^4$ 5
6. $5^{2x} = 5^6$ 3

Motivator

Have students suggest a simpler way to write $5 \cdot 5 \cdot 5$. 5^3 Have students use a scientific calculator to verify statements 1-3 below.

1. $3^4 \cdot 3^2 \cdot 3^5 = 3^{11}$

2. $\dfrac{2^{11}}{2^6} = 2^5$

3. $(3^2)^4 = 3^8$

Highlighting the Standards

Standard 5d: Students have worked with positive exponents since the middle grades. This return to familiar ground will reinforce the continuity of mathematical learning.

Lesson Note

In discussing the first two laws of exponents, emphasize that these laws apply only when the bases are the same number. Note that following Example 1, the text indicates that expressions such as $a^7 \cdot b^4$ and $\frac{a^3}{c^5}$ cannot be simplified.

Math Connections

Geometry: The area of a geometric figure is given in *square* units. Since the area of a *square* can be represented as $A = s^2$, a polygon with an area of 9 *square* centimeters is written as 9 cm^2.

Critical Thinking Questions

Analysis: Ask students what they can say about the exponential equation $5^{4x-2} = -25$ (obtained by modifying Example 3). They should quickly see that there is no real-number solution since the left side of the equation is the power of a positive number, which must be positive, while the right side of the equation is negative.

EXAMPLE 1   Simplify.

a. $-8a^5b \cdot 4a^2b^3$

b. $\dfrac{-12a^5b^4c}{18a^2b^4c^6}$

Solutions

a. $-8 \cdot a^5 \cdot b^1 \cdot 4 \cdot a^2 \cdot b^3$
$= -8 \cdot 4 \cdot a^5a^2 \cdot b^1b^3$
$= -32a^7b^4$

b. $\dfrac{-12}{18} \cdot \dfrac{a^5}{a^2} \cdot \dfrac{b^4}{b^4} \cdot \dfrac{c^1}{c^6}$

$= \dfrac{-2}{3} \cdot \dfrac{a^3}{1} \cdot \dfrac{1}{1} \cdot \dfrac{1}{c^5} = \dfrac{-2a^3}{3c^5}$

Notice in the answers for Example 1 that $a^7 \cdot b^4$ and $\dfrac{a^3}{c^5}$ cannot be simplified because the bases are different.

Exponents can also be used to show the power of a power, such as $(b^5)^3$, the power of a product, such as $(5b)^3$, and the power of a quotient, such as $\left(\dfrac{5}{b}\right)^3$.

$(b^5)^3 = b^5 \cdot b^5 \cdot b^5$
$= b^{15}$, or $b^{5 \cdot 3}$

$(5b)^3 = 5b \cdot 5b \cdot 5b$
$= 125b^3$, or $5^3 \cdot b^3$

$\left(\dfrac{5}{b}\right)^3 = \dfrac{5}{b} \cdot \dfrac{5}{b} \cdot \dfrac{5}{b}$

$= \dfrac{125}{b^3}$, or $\dfrac{5^3}{b^3}$

Laws of Exponents

For all positive integers m and n and all real numbers a and b,

(1) $(a^m)^n = a^{m \cdot n}$ (Power of a power)

(2) $(ab)^n = a^n \cdot b^n$ (Power of a product)

(3) $\left(\dfrac{a}{b}\right)^n = \dfrac{a^n}{b^n}$, where $b \neq 0$ (Power of a quotient)

EXAMPLE 2   Simplify.

a. $8(a^3)^5$
$= 8 \cdot a^{3 \cdot 5}$
$= 8a^{15}$

b. $2a^2(5a^3b)^4$
$= 2a^2 \cdot 5^4(a^3)^4b^4$
$= 2a^2 \cdot 625a^{12}b^4$
$= 1{,}250a^{14}b^4$

c. $\left(\dfrac{-2a^2}{5b^4c}\right)^3$

$= \dfrac{(-2a^2)^3}{(5b^4c)^3}$

$= \dfrac{(-2)^3(a^2)^3}{5^3(b^4)^3c^3}$

$= \dfrac{-8a^6}{125b^{12}c^3}$

Additional Example 1

Simplify.

a. $-5x^3y \cdot 2xy^4(-3x^2y^2)$ $30x^6y^7$

b. $\dfrac{-16m^2np^6}{12m^8np^3}$ $\dfrac{-4p^3}{3m^6}$

Additional Example 2

Simplify.

a. $5(x^4)^2$ $5x^8$

b. $3m(2mp^2)^3$ $24m^4p^6$

c. $\left(\dfrac{-3a^3}{4b^4}\right)^2$ $\dfrac{9a^6}{16b^8}$

Equations such as $3^x = 81$, $2^{3x} = 32$, and $5^{4x-2} = 125$ are called **exponential equations**. To solve these equations, you need to know the various powers of 3, 2, and 5.

EXAMPLE 3 Solve $5^{4x-2} = 125$.

Plan The left side, 5^{4x-2}, is in exponential form and the base is 5. So, express the right side, 125, as a power of 5.

Solution $5^{4x-2} = 5^3$

Because the bases are equal, the exponents must also be equal.

$$4x - 2 = 3$$
$$4x = 5$$
$$x = \frac{5}{4} \quad \text{Check: } 5^{4x-2} = 5^{4 \cdot \frac{5}{4} - 2} = 5^3 = 125$$

Thus, the solution is $\frac{5}{4}$.

Classroom Exercises

Simplify.

1. $a^3 \cdot a^3$ a^6

2. $\dfrac{b^{16}}{b^4}$ b^{12}

3. $(c^2)^3$ c^6

4. $(de)^3$ d^3e^3

5. $(f^2g)^4$ f^8g^4

6. $\dfrac{10x^3y^5}{-2xy^2}$ $-5x^2y^3$

7. Solve $2^x = 128$ for x. 7

Written Exercises

Simplify each expression.

1. $5x^2 \cdot x^4$ $5x^6$

2. $y(6y^3)$ $6y^4$

3. $2z(-6yz)$ $-12yz^2$

4. $7xy^3 \cdot 4x^2y^4$ $28x^3y^7$

5. $-11mn^6 \cdot 2m^3n$ $-22m^4n^7$

6. $-6c^2d \cdot 4c^4d^6$ $10c^3d^2$

7. $\dfrac{18a^3b^7}{-9a^9b^2}$ $-\dfrac{2b^5}{a^6}$

8. $\dfrac{8x^6y^4z^3}{6x^2y^4z^7}$ $\dfrac{4x^4}{3z^4}$

9. $\dfrac{-12a^5bc^2d^5}{-8ac^{10}d^5}$ $\dfrac{3a^4b}{2c^8}$

10. $n(n^4)^3$ n^{13}

11. $(2x)^4$ $16x^4$

12. $(-4c)^3$ $-64c^3$

13. $5(-3a)^2$ $45a^2$

14. $(5c^2)^3$ $125c^6$

15. $5x(-3x^3)^4$ $405x^{13}$

16. $\left(\dfrac{-2a}{5}\right)^4$ $\dfrac{16a^4}{625}$

17. $\left(\dfrac{-7c^3}{10d^5}\right)^2$ $\dfrac{49c^6}{100d^{10}}$

18. $\left(\dfrac{3b^3}{2c^2}\right)^3$ $\dfrac{27b^9}{8c^6}$

Common Error Analysis

Error: Students may confuse $(x^4)^3$ with $x^4 \cdot x^3$ and write x^7 for $(x^4)^3$.

Ask the students to find the base for the exponent 3 in $(x^4)^3$ and write the base three times as a factor, $x^4 \cdot x^4 \cdot x^4$. Students may also confuse $5x^3$ with $(5x)^3$. Again, ask the students to find the base for the exponent 3 and write it three times as a factor.

$$5x^3 \text{ means } 5 \cdot x \cdot x \cdot x$$

but

$$(5x)^3 \text{ means } 5x \cdot 5x \cdot 5x.$$

Checkpoint

Simplify.

1. $-4a^2b \cdot 5ab^3 \cdot 2a^3b$ $-40a^6b^5$

2. $(-2y^2)^3$ $-8y^6$

3. $10c(3c^3)^4$ $810c^{13}$

4. $\dfrac{-15x^6y^4z}{10x^2y^8z}$ $\dfrac{-3x^4}{2y^4}$

Solve for x.

5. $3^{x-1} = 27$ 4

6. $5^{2x+1} = 125$ 1

Additional Example 3

Solve $2^{3n+2} = 32$. 1

Closure

Have students simplify

$$\left(\frac{-3a^2a^4b^3}{6a^3b^6}\right)^3 \cdot -\frac{a^9}{8b^9}$$

Ask them how many of the Laws of Exponents they used to simplify the expression.   all five

◤◤◤FOLLOW UP

Guided Practice

Classroom Exercises 1–7

Independent Practice

A Ex. 1–27, **B** Ex. 28–33, **C** Ex. 34–41

Basic: WE 1–29 odd, Brainteaser

Average: WE 5–33 odd, Brainteaser

Above Average: WE 13–41 odd, Brainteaser

Additional Answers

Written Exercises

6. $-240c^9d^9$

Solve each equation.

19. $3^{x-1} = 81$ 5

20. $10^{n+2} = 1,000$ 1

21. $2^{3y} = 64$ 2

22. $5^{-x} = 25$ −2

23. $3^{2n+1} = 243$ 2

24. $5^{3x-1} = 625$ $\frac{5}{3}$

25. $4^{6-y} = 256$ 2

26. $2^{6x+6} = 1,024$ $\frac{2}{3}$

27. $4^{5-3x} = 64$ $\frac{2}{3}$

Simplify each expression.

28. $(-5a)^3b^7c \cdot (2a)^4bc^5$ $-2,000a^7b^8c^6$

29. $(10x)^2y^2 \cdot (5x)^2z^4(-4yz^2)$ $-10,000x^4y^3z^6$

30. $8m(-5m^3n^4)^3$ $-1,000m^{10}n^{12}$

31. $4c^2d(2a^3cd^4)^4$ $64a^{12}c^6d^{17}$

32. $\left|\dfrac{5a^2b^3}{-2c^4d^2}\right|^2$ $\dfrac{25a^4b^6}{4c^8d^4}$

33. $\left|\dfrac{-2mnp^3}{3x^3y^2}\right|^3$ $\dfrac{-8m^3n^3p^9}{27x^9y^6}$

34. $x^ay^{2b} \cdot x^{3c}y^{4d}$ $x^{a+3c}y^{2b+4d}$

35. $x^{3a+2}y^{3b} \cdot x^{4a}y^{2b-1}$ $x^{7a+2}y^{5b-1}$

36. $(x^ay^{b+2})^2$ $x^{2a}y^{2b+4}$

37. $(x^{2a}y^{b-1})^c$ $x^{2ac}y^{bc-c}$

38. $x^a(x^by^c)^d$ $x^{a+bd}y^{cd}$

39. $x^{2a}y^b(x^{a+3}y^b)^3$ $x^{5a+9}y^{4b}$

40. $\dfrac{x^{8a}y^{4b}}{x^{2a}y^b}$ $x^{6a}y^{3b}$

41. $\dfrac{x^{5a+2b}y^{3a-5b}}{x^{a+b}y^{2a-3b}}$ $x^{4a+b}y^{a-2b}$

Mixed Review

Find the solution set.   *2.1, 2.2, 2.3*   1. $\{x \mid x > 1\}$

1. $8 - 5(4x - 3) < 6x - (7 - 4x)$

2. $|2x - 7| < 11$ $\{x \mid -2 < x < 9\}$

3. $|4x + 6| > 14$ $\{x \mid x < -5 \text{ or } x > 2\}$

4. $3x - 2 < 10 \text{ or } 3 - 5x > 18$ $\{x \mid x < 4\}$

5. $3x + 6 > 2x - 4 \text{ and } -2x + 2 < 2x - 10$ $\{x \mid x > 3\}$

6. $x + 3 > 2x - 6 \text{ or } -x - 3 > 2x - 9$ $\{x \mid x < 9\}$

◤◤◤/Brainteaser

A group of blocks is stacked in a corner as shown at the right. There are no empty spaces behind any of the blocks.

a. How many blocks cannot be seen? 20

b. How many blocks will not be seen if another layer of blocks is added to the bottom of the stack? 35

Enrichment

Exponential equations may be of the form $2^{x+5} = 4^{3x}$. Explain to students that they cannot say $x + 5 = 3x$ because the bases are different. However, since $4 = 2^2$, the equation can be rewritten as $2^{x+5} = (2^2)^{3x}$, or $2^{x+5} = 2^{6x}$. Then $x + 5 = 6x$ and $x = 1$. Now challenge the students to solve the following equations.

1. $3^{x-1} = 27^x$ $x = -\frac{1}{2}$

2. $8^{x+2} = 4^x$ $x = -6$

3. $25^{2-x} = 125^x$ $x = \frac{4}{5}$

5.2 Zero and Negative Integral Exponents

Objectives To evaluate expressions containing negative integral exponents
To simplify expressions containing negative integral exponents
To use scientific notation

The domain of the exponent x in the function $y = 2^x$ can be expanded to include zero and negative integers. That is, expressions such as 2^0 and 2^{-3} can be defined so that the Laws of Exponents (see Lesson 5.1) cover all integral exponents—positive, negative, and zero.

First, consider 2^0.

$$\frac{2^5}{2^5} = \frac{32}{32} = 1 \text{ and } \frac{2^5}{2^5} = 2^{5-5} = 2^0.$$

Therefore, for the Laws of Exponents to hold for all integral exponents, 2^0 must equal 1.

This suggests the following definition of *zero exponent*.

Definition

> **Zero Exponent:** $b^0 = 1$, for each real number b where $b \neq 0$.

Similarly, 2^{-3} can be defined to mean $\frac{1}{2^3}$, and $\frac{1}{5^{-4}}$ defined to mean 5^4.
$\frac{2^0}{2^3} = 2^{0-3} = 2^{-3}$ and $\frac{2^0}{2^3} = \frac{1}{2^3}$. Thus, $2^{-3} = \frac{1}{2^3}$.

(By calculator, $2 \boxed{x^y} 3 \boxed{+/-} \boxed{=} 0.125$ and also $2 \boxed{x^y} 3 \boxed{=} \boxed{1/x}$ = 0.125.)

$\frac{1}{5^{-4}} = \frac{5^0}{5^{-4}} = 5^{0-(-4)} = 5^4$. Thus, $\frac{1}{5^{-4}} = 5^4$.

Definition

> **Negative Exponent:** For each real number b, $b \neq 0$, and each positive integer n, $b^{-n} = \frac{1}{b^n}$ and $\frac{1}{b^{-n}} = b^n$.

When simplifying expressions such as $5a^{-2}$ and $(5a)^{-2}$, note that the base in $5a^{-2}$ is a while the base in $(5a)^{-2}$ is $5a$.
Thus, $5a^{-2} = 5 \cdot a^{-2} = 5 \cdot \frac{1}{a^2} = \frac{5}{a^2}$ and $(5a)^{-2} = \frac{1}{(5a)^2} = \frac{1}{25a^2}$.

5.2 Zero and Negative Integral Exponents **165**

Teaching Resources

Application Worksheet 5
Project Worksheet 5
Quick Quizzes 32
**Reteaching and Practice
Worksheets** 32

GETTING STARTED

Prerequisite Quiz

Simplify.

1. $5 \cdot 2^3$ 40
2. $8.36 \times 10,000$ 83,600

Determine the base for the exponent.

3. $2b^3$ b 4. $7c(4c^2)^3$ $4c^2$

Simplify.

5. $4a^2 \cdot 6a^4 \cdot a$ $24a^7$ 6. $3n(5n^3)^2$ $75n^7$

Motivator

Have students simplify the following expression. $\frac{3^2}{3^2}$ 1 Ask them if $\frac{3^2}{3^2} = 3^{2-2}$.
Yes Then ask if $\frac{3^2}{3^2} = 3^0$. Yes

TEACHING SUGGESTIONS

Lesson Note

Encourage students to use the "easy" way to evaluate expressions as shown in Example 3, rather than take steps like the following.

a. $\frac{1.2 \times 10^{-2}}{4.8 \times 10^{-5}} = \frac{0.012}{0.000048} = 250$

In fact, it may be well to discourage the use of a calculator in this lesson.

Highlighting the Standards

Standard 3c: The definitions of zero and negative exponents are consistent with the definition of positive exponents. Students appreciate the logic involved in the algebra of exponents.

Math Connections

Astronomy: Scientific notation is used for work in some branches of science that routinely deal with enormous numbers. As an example, the number of stars in an average galaxy has been estimated to be about 1,000,000,000, or 10^9 stars. The number of galaxies in the universe has been estimated to be about the same number, that is, 10^9. If these numbers are taken to be correct, then the number of stars in the universe is easily obtained using scientific notation by writing $(1 \times 10^9)^2 = 1 \times 10^{18}$.

Critical Thinking Questions

Analysis: How would you obtain the product $(3.2 \times 10^3)(4.5 \times 10^5)$ on a calculator that does not follow the rules for order of operations? $3.2 \boxed{\times} 4.5 \boxed{\times} 10 \boxed{y^x}$ $8 \boxed{=}$ What is the product $a \times 10^b$ when $b = 0$? a

Checkpoint

Simplify.

1. $3a^{-5} \cdot a^2 \cdot 4a^0$ $\quad \dfrac{12}{a^3}$

2. $\dfrac{2c^{-3}}{6c^2}$ $\quad \dfrac{1}{3c^5}$

3. $\dfrac{8m^{-1}}{4m^7}$ $\quad \dfrac{2}{m^8}$

4. $5(2a^2)^{-3}$ $\quad \dfrac{5}{8a^6}$

5. $5(2a^{-2})^3$ $\quad \dfrac{40}{a^6}$

Closure

Ask students to identify the base in the expression $4x^{-4}$. x in $(4x)^{-4}$. $4x$ Ask students if the number $7.98 \times 10^{\frac{1}{2}}$ is written in scientific notation. No. The exponent is not an integer.

EXAMPLE 1 Find the value of each expression.

a. $4 \cdot 5^{-2} = \dfrac{4}{5^2} = \dfrac{4}{25}$

b. $\dfrac{3}{(4 \cdot 5)^{-2}} = 3(4 \cdot 5)^2$
$= 3 \cdot 20^2 = 1{,}200$

EXAMPLE 2 Simplify and write each expression using positive exponents.

a. $-6x^3 \cdot y^{-4} \cdot 5x^{-7} \cdot y^6$
$= -6 \cdot 5 \cdot x^3 x^{-7} \cdot y^{-4} y^6$
$= -30 \cdot x^{3+(-7)} \cdot y^{-4+6}$
$= -30x^{-4}y^2$
$= \dfrac{-30y^2}{x^4}$

b. $\dfrac{-10a^7b^{-2}c^{-3}}{15a^{-2}b^{-5}c^5}$
$= \dfrac{-10}{15} \cdot \dfrac{a^7}{a^{-2}} \cdot \dfrac{b^{-2}}{b^{-5}} \cdot \dfrac{c^{-3}}{c^5}$
$= \dfrac{-2}{3} \cdot \dfrac{a^7 a^2}{1} \cdot \dfrac{b^5}{b^2} \cdot \dfrac{1}{c^3 c^5}$
$= \dfrac{-2a^9b^3}{3c^8}$

The numbers 7.62×10^5 and 7.62×10^{-3} are written in *scientific notation*. They are rewritten in ordinary notation below.

$$7.62 \times 10^5 = 762{,}000.0 \qquad 7.62 \times 10^{-3} = 0.00762$$

Notice that the exponents 5 and -3 tell the number of places and the direction in which to move the decimal point in order to obtain ordinary notation. A number is written in **scientific notation** when it is in the form $a \times 10^c$, where $1 \le a < 10$ and c is an integer.

EXAMPLE 3 Find the value of the expression and state it in ordinary notation.
$(8.2 \times 10^4) \times (1.5 \times 10^{-7})$
$= (8.2 \times 1.5) \times (10^4 \times 10^{-7})$
$= 12.3 \times 10^{-3}$
$= 0.0123$

Classroom Exercises

Simplify each expression.

1. 3^{-2} $\quad \dfrac{1}{9}$

2. $5 \cdot 2^{-3}$ $\quad \dfrac{5}{8}$

3. $(5 \cdot 2)^{-3}$ $\quad \dfrac{1}{1{,}000}$

4. 2.34×10^2 $\quad 234$

5. $6c^{-3}$ $\quad \dfrac{6}{c^3}$

6. $(6c)^{-3}$ $\quad \dfrac{1}{216c^3}$

7. $(7 \cdot 4)^0$ $\quad 1$

8. $a^{-4} \cdot a^7$ $\quad a^3$

166 Chapter 5 Polynomials

Additional Example 1

Find the value of each expression.

a. $(-3)^{-2} \cdot 4$ $\quad \dfrac{4}{9}$

b. $\dfrac{5}{(-3 \cdot 4)^{-2}}$ $\quad 720$

Additional Example 2

Simplify and write each expression using positive exponents.

a. $-4x^2 \cdot 3x^0 \cdot 2x^{-5}$ $\quad \dfrac{-24}{x^3}$

b. $\dfrac{x^{-3}y^{-6}z^4}{x^4y^{-2}z^{-2}}$ $\quad \dfrac{z^6}{x^7y^4}$

c. $\left(\dfrac{2c^3}{3d^{-2}}\right)^3$ $\quad \dfrac{8c^9d^6}{27}$

Additional Example 3

Find the value of each expression and state it in ordinary notation.

a. $\dfrac{7.5 \times 10^{-11}}{2.5 \times 10^{-8}}$ $\quad 0.003$

b. $(1.2 \times 10^7) \times (5 \times 10^{-4})$ $\quad 6{,}000$

166

Written Exercises

Find the value of each expression.

1. 5^{-3} $\frac{1}{125}$ **2.** 4^{-3} $\frac{1}{64}$ **3.** $6 \cdot 3^{-2}$ $\frac{2}{3}$ **4.** $(4 \cdot 3)^{-2}$ $\frac{1}{144}$

5. $8^4 \cdot 8^{-6}$ $\frac{1}{64}$ **6.** $8^{-4} \cdot 8^6$ 64 **7.** 3.8×10^{-3} 0.0038 **8.** 842×10^{-2} 8.42

9. $\dfrac{4}{5^{-3}}$ 500 **10.** $\dfrac{2^{-3}}{3^{-2}}$ $\frac{9}{8}$ **11.** $\dfrac{4.6}{10^{-3}}$ $4{,}600$ **12.** $\dfrac{5.79}{10^2}$ 0.0579

Simplify each expression. Use positive exponents.

13. $8x^{-3}$ **14.** $(5y)^{-4}$ **15.** $-2a^{-3}$ **16.** $(-2a)^{-4}$

17. $x^{-3} \cdot x^5$ **18.** $5y^2 \cdot 2y^{-5}$ **19.** $-2n^{-3} \cdot 7n^{-5}$ **20.** $5c^{-4} \cdot 2c^0$

21. $(x^{-3})^4$ **22.** $(5a^3)^{-2}$ **23.** $(2c^{-4})^{-3}$ **24.** $2(-3x^{-2})^4$

25. $\dfrac{12c^6}{15c^{-2}}$ $\frac{4c^8}{5}$ **26.** $\dfrac{10n^{-6}}{15n^{-8}}$ $\frac{2n^2}{3}$ **27.** $\left(\dfrac{x^2}{y^{-3}}\right)^5$ $x^{10}y^{15}$ **28.** $\left(\dfrac{2a^{-3}}{d^2}\right)^4$ $\frac{16}{a^{12}d^8}$

Find the value of each expression and state it in ordinary notation.

29. $\dfrac{8.8 \times 10^2}{4.4 \times 10^{-2}}$ $20{,}000$ **30.** $\dfrac{3.3 \times 10^{-3}}{2.2 \times 10^{-5}}$ 150 **31.** $\dfrac{1.4 \times 10^{-1}}{2.8 \times 10^3}$ 0.00005 **32.** $\dfrac{2.1 \times 10^{-8}}{8.4 \times 10^{-6}}$ 0.0025

33. $(4.6 \times 10^{-2}) \times (1.5 \times 10^{-1})$ 0.0069 **34.** $(5.6 \times 10^{-3}) \times (3.8 \times 10^7)$ $212{,}800$

Simplify each expression. Use positive exponents.

35. $\dfrac{2a^6b^{-5}}{6c^{-3}d^4}$ $\frac{a^6c^3}{3b^5d^4}$ **36.** $\dfrac{15x^7y^{-12}}{10x^{-3}y^{10}}$ $\frac{3x^{10}}{2y^{22}}$ **37.** $\left(\dfrac{12x^{-4}}{y^3z^{-3}}\right)^2$ $\frac{144z^6}{x^8y^6}$ **38.** $\left(\dfrac{3a^{-2}b^3}{2c^{-1}}\right)^4$

39. $3x^{-8}y^2 \cdot 5x^3y^{-4}$ **40.** $(-3a^3d^{-2})^4$ $\frac{81a^{12}}{d^8}$ **41.** $\left(\dfrac{-4x^{-3}y}{3z^2w^{-4}}\right)^3$ **42.** $\left(\dfrac{2a^3}{3b^2}\right)^{-2}$ $\frac{9b^4}{4a^6}$

43. x^{m-n}, where $m < 0 < n$ $\frac{1}{x^{n-m}}$ **44.** y^{a-b}, where $0 < a < b$ $\frac{1}{y^{b-a}}$

Mixed Review

Find each value given that $f(x) = 6x - 2$. *3.2*

1. $f(7)$ 40 **2.** $f(-2c^2)$ $-12c^2 - 2$ **3.** $f(a) + f(5)$ $6a + 26$ **4.** $f(c + 3) - f(c)$ 18

5. Find the set of real numbers for which 3 less than 3 times the sum of a number and 6 is greater than 12. *2.4* $\{x \mid x > -1\}$

FOLLOW UP

Guided Practice

Classroom Exercises 1–8

Independent Practice

A Ex. 1–34, **B** Ex. 35–42, **C** Ex. 43, 44

Basic: WE 1–35 odd, Extension
Average: WE 7–41 odd, Extension
Above Average: WE 9–43 odd, Extension

Additional Answers

Written Exercises

13. $\frac{8}{x^3}$

14. $\frac{1}{625y^4}$

15. $\frac{-2}{a^3}$

16. $\frac{1}{16a^4}$

17. x^2

18. $\frac{10}{y^3}$

19. $\frac{-14}{n^8}$

20. $\frac{10}{c^4}$

21. $\frac{1}{x^{12}}$

22. $\frac{1}{25a^6}$

23. $\frac{c^{12}}{8}$

24. $\frac{162}{x^8}$

38. $\frac{81b^{12}c^4}{16a^8}$

39. $\frac{15}{x^5y^2}$

41. $\frac{-64y^3w^{12}}{27x^9z^6}$

Enrichment

Challenge the students to evaluate these expressions.

1. $\{[2(3^0 - 2)^{-2}]^0\}^{-1}$ 1

2. $\{[(3^{-1})^{-2}]^{-1}\}^{-2}$ 81

3. $\left\{2\left[2\left(\dfrac{1}{2^{-1}}\right)^{-2}\right]^{-1}\right\}^{-2}$ $\frac{1}{16}$

4. $\{[(10^{-2})(10)]^{-2}\}^2$ $10{,}000$

5. $\left\{\dfrac{2(2^{-2})^{-1}}{[(2^{-1})(2)^0]^{-2}}\right\}^{-1}$ $\frac{1}{2}$

6. $\left\{\dfrac{\left\{\left(\dfrac{5^{-1}}{5^{-2}}\right)^{-2}\right\}^{-1}}{\left\{\left(\dfrac{1}{5^{-3}}\right)^{-1}\right\}^{-2}}\right\}^{-1}$ 625

1. 4; 78,000; 7.8×10^4
2. 3; 460,000; 4.6×10^5
3. 3; 38,000,000; 3.8×10^7
4. 3; 0.0046; 4.6×10^{-3}
5. 4; 0.75; 7.5×10^{-1}
6. 4; 21×10^5; 2.1×10^6
7. 3; 32×10^{-3}; 3.2×10^{-2}
8. 3; 0.072×10^{-2};
   7.2×10^{-4}
9. 6; 30,000,000,000 cm/s;
   3.0×10^{10} cm/s

Extension

Estimating Products and Quotients

Scientific notation can be used to express very large or very small numbers in a compact form. For example, 982,000,000 ft/s, the speed of light, can be written as 9.82×10^8 ft/s. When a measurement is written in scientific notation, $a \times 10^c$, the *significant* digits are placed in the first factor, a, where $1 \le a < 10$.

Each of the nine nonzero digits, 1 through 9, in a measurement is a **significant digit**. Zero (0) is a significant digit when its *only* purpose is *not* to place the decimal point as shown below.

Number	Number of Significant Digits	
0.0507	3	0.0<u>507</u>
5.007	4	<u>5.007</u>
5.70	3	<u>5.70</u>
57,000	2	<u>57</u>,000
57,000 to the nearest hundred	3	<u>570</u>,00

Example 1 Round 0.0006080 to two significant digits, and write the result in scientific notation.

0.0006080 = 0.00061 to two significant digits, and
0.00061 = 6.1×10^{-4} in scientific notation.

One strategy to be used when solving problems is to determine if your answer is reasonable. Scientific notation can be used to *estimate* a quotient or product to be compared with a calculated answer.

Example 2 Estimate, to one significant digit, the time needed for light to travel 10 mi at a rate of 982,000,000 ft/s.

Plan Convert 10 mi to feet. Round the distance and rate to two significant digits, and write each in scientific notation.

Use $\dfrac{\text{distance}}{\text{rate}} = \text{time} \left(\dfrac{d}{r} = t \right)$ to find the time in seconds.

Solution $d = 10$ mi $= 10 \times 5{,}280$ ft $= 52{,}800$ ft $\approx 53{,}000$ ft $= 5.3 \times 10^4$ ft
$r = 982{,}000{,}000$ ft/s $\approx 980{,}000{,}000$ ft/s $= 9.8 \times 10^8$ ft/s

$$t = \frac{d}{r} \approx \frac{5.3 \times 10^4}{9.8 \times 10^8} = \frac{5.3}{9.8} \times 10^4 \times 10^{-8}$$
$$\approx 0.5 \times 10^{-4} = 5 \times 10^{-5}$$

Thus, it will take approximately 5×10^{-5}, or 0.00005 s. This compares favorably with the calculated answer, 0.00005376782 s.

Example 3   If $r = 0.0000718$ km/h and $t = 232$ h, estimate $r \cdot t$ to one significant digit.

$$(7.2 \times 10^{-5}) \times (2.3 \times 10^2) = (7.2 \times 2.3) \times (10^{-5} \times 10^2)$$
$$\approx 20 \times 10^{-3} = 2 \times 10^{-2}$$

Thus, $r \cdot t$ is approximately 2×10^{-2}, or 0.02 km.

This compares favorably with the calculated answer, 0.0166576 km.

Exercises

Find the number of significant digits. Then round the number to two significant digits, and write the result in scientific notation.

1. 78,350   **2.** 457,000   **3.** 37,800,000   **4.** 0.00462
5. 0.7506   **6.** 20.64×10^5 **7.** 32.4×10^{-3} **8.** 0.0718×10^{-2}
9. the velocity of light: 29,979,300,000 cm/s

Estimate the time t in seconds to one significant digit. Use $\frac{d}{r} = t$.

10. $d = 564,000,000$ mi, $r = 28,100$ mi/s   20,000 s
11. $d = 1,000$ mi, $r = 982,000,000$ ft/s   0.005 s
12. $d = 38,600,000$ km, $r = 640$ m/s 6×10^7 s

Estimate the distance d to one significant digit. Use $r \cdot t = d$.

13. $r = 0.00818$ mi/s, $t = 72,000$ s   600 mi
14. $r = 186,000$ mi/s, $t = 0.0023$ s   400 mi

Choose the correct estimate of the value for each expression to one significant digit. Do not use a calculator.

15. $382,124 \times 5,012$ c
   a. 20,000,000
   b. 200,000,000
   c. 2,000,000,000

16. $39.46 \div 81,460$ b
   a. 0.05
   b. 0.0005
   c. 0.00005

Prerequisite Quiz

Simplify.

1. $3x - 5y + 2z + (3z - 5x - 2y)$
   $-2x - 7y + 5z$
2. $4x - 6y + 3z - (4z - 6x - 4y)$
   $10x - 2y - z$
3. $5(3x^2 - 2x + 6)$ $15x^2 - 10x + 30$
4. $-3(-2a + 5b - c)$ $6a - 15b + 3c$

Motivator

Ask students if the following expressions can be simplified.

1. x^2 no
2. $x^2 - 5x^2$ yes
3. $ac + cd + ca - bd$ yes

5.3 Polynomials

Objectives   To classify polynomials by the number of terms and their degree
To simplify sums and differences of polynomials
To multiply polynomials

A **monomial** is a numeral, or a variable, or the product of a numeral and one or more variables. Five examples of monomials are shown below.

$$\frac{2}{3} \qquad z \qquad -4t \qquad x^3y^2 \qquad \frac{1}{2}c^2d$$

A **polynomial** is a monomial or the sum of two or more monomials. The monomials in a polynomial are called the **terms** of the polynomial. The expressions below are examples of polynomials.

$$-5c^4 \text{ (monomial)}$$
$$3x^2 + \frac{1}{2}x \text{ (binomial)}$$
$$2x^3y - 5xy - 3y^2 \text{ (trinomial)}$$

Polynomials can be classified by the number of terms they contain. A **binomial** has exactly two terms; a **trinomial** has exactly three terms. A polynomial such as $5x - 3y + 2z - 1.5$ is classified as a polynomial with four terms. Expressions such as $\frac{3y}{x}$ (division by a variable) and $\sqrt{x}$ (the square root of a variable) are not polynomials.

The **degree of a monomial** is the sum of the exponents of its variables. For example, x^2yz is of degree 4 because $x^2yz = x^2y^1z^1$ and $2 + 1 + 1 = 4$. The degree of a nonzero number such as 6 is 0, because $6 = 6 \cdot 1 = 6x^0$.

The **degree of a polynomial** with more than one term is the same as that of its term with the greatest degree. The polynomial $5xy + 6x^2yz^2 - 2x + 6$ is of degree 5. This can be determined as shown below.

$$\underset{\substack{\text{degree} \\ 2}}{5x^1y^1} \quad + \underset{\substack{\text{degree} \\ \circledS}}{6x^2y^1z^2} \quad - \underset{\substack{\text{degree} \\ 1}}{2x^1} \quad + \underset{\substack{\text{degree} \\ 0}}{6x^0}$$

First-degree polynomials such as $3x - 7$ and $4x + 5y - 6$ are also called **linear polynomials** (see Lesson 3.2). Second-degree polynomials such as $5y^2 - 3y + 2$ and $x^2 - 6xy + 9y^2$ are called **quadratic polynomials**.

EXAMPLE 1 Classify each polynomial by number of terms, degree, and whether the polynomial is linear, quadratic, or neither.

Polynomial	Number of terms	Degree	Linear/Quadratic
8	Monomial	0	Neither
$-3x^2$	Monomial	2	Quadratic
$5x + 7$	Binomial	1	Linear
$x^2 - 4xy + 4y^2$	Trinomial	2	Quadratic
$a^2 + 3ab^3 + a^3bc - a$	Four terms	5	Neither

Polynomials can be added, subtracted, multiplied, and divided. To simplify the sum of two or more polynomials, combine like terms.

EXAMPLE 2 Find the simplified form of the sum of the polynomials $5a^2b - 2ab + b^2$ and $7ab - 8a^2b + 2b^2 - 6$.

Plan Use the Rule of Subtraction, $x - y = x + (-y)$, to convert all subtractions to additions. Then use the Commutative and Associative Properties of Addition to regroup the terms.

Solution
$(5a^2b - 2ab + b^2) + (7ab - 8a^2b + 2b^2 - 6)$
$= 5a^2b + (-2ab) + b^2 + 7ab + (-8a^2b) + 2b^2 + (-6)$
$= 5a^2b - 8a^2b - 2ab + 7ab + b^2 + 2b^2 - 6$
$= -3a^2b + 5ab + 3b^2 - 6$

The product of a monomial and a polynomial can be simplified using the Distributive Property.

$-2x^2(3x - 4x^3 + 5) = -2x^2 \cdot 3x - 2x^2(-4x^3) - 2x^2 \cdot 5$
$= -6x^3 + 8x^5 - 10x^2$
$= 8x^5 - 6x^3 - 10x^2$

To simplify $(a + b)(c + d)$, the product of two binomials, you can use the Distributive Property three times as shown below.

$(a + b)(c + d) = a(c + d) + b(c + d) = ac + ad + bc + bd$

The product $ac + ad + bc + bd$ can be found using the "First-Outer-Inner-Last Terms" method (FOIL method).

5.3 Polynomials **171**

Lesson Note

Emphasize that the degree of a polynomial is found by referring to each term, determining its degree, and taking the greatest of these. The exponents on the variables in a polynomial must be nonnegative integers. Thus, $\frac{3y}{x}$ or $3yx^{-1}$, is not a polynomial.

Later, $\sqrt{x}$ will be written as $x^{\frac{1}{2}}$, and thus $\sqrt{x}$ is not a polynomial. However, 7 can be written as $7 \cdot 1$, or $7x^0$, and so 7 is a polynomial (monomial). Use the words *linear* and *quadratic* frequently in the classroom since these words occur often in this book. Be sure that students associate *linear* with degree 1 and *quadratic* with degree 2.

Math Connections

Geometry: A geometric figure such as a circle with its center at the origin can be described by the equation $x^2 + y^2 = r^2$, where the binomial $x^2 + y^2$ is equal to the monomial r^2.

Additional Example 1

Classify each polynomial according to its number of terms and its degree. Tell whether the polynomial is linear, quadratic, or neither of these. $x^2y + 2x^3y - 5x$
Trinomial, 4, neither $3ab - b^2$ Binomial, 2, quadratic -4 Monomial, 0, neither $2c - 3d + r + 9$ Polynomial with 4 terms, 1, linear

Additional Example 2

For the polynomials $5a - 6 + a^2$ and $8 - 3a^2 - 6a$, find the simplified form of their sum. $-2a^2 - a + 2$

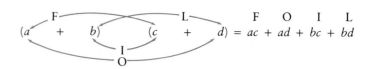

$$(a \quad + \quad b)(c \quad + \quad d) = ac + ad + bc + bd$$

EXAMPLE 3   Simplify $(5x + 4)(2x - 3)$. Use the FOIL method.

Solution

$$\begin{aligned}(5x + 4)(2x - 3) &= 5x \cdot 2x + 5x(-3) + 4 \cdot 2x + 4(-3) \\ &= 10x^2 - 15x + 8x - 12 \\ &= 10x^2 - 7x - 12\end{aligned}$$

To simplify the product of a monomial and two binomials, it is usually easier to begin with the binomials. For example,

$$\begin{aligned}7y(2y^2 - 5)(3y^2 - 2) &= 7y[2y^2 \cdot 3y^2 + 2y^2(-2) - 5 \cdot 3y^2 - 5(-2)] \\ &= 7y(6y^4 - 4y^2 - 15y^2 + 10) \\ &= 7y(6y^4 - 19y^2 + 10) \\ &= 42y^5 - 133y^3 + 70y\end{aligned}$$

The product $(3a - b)(2a + 4b - 5)$ can be simplified by distributing each term of the binomial to each term of the trinomial, as shown in the next example.

EXAMPLE 4   Simplify $(3a - b)(2a + 4b - 5)$.

Solution

$$\begin{aligned}(3a - b)(2a + 4b - 5) &= 3a(2a + 4b - 5) - b(2a + 4b - 5) \\ &= 6a^2 + 12ab - 15a - 2ab - 4b^2 + 5b \\ &= 6a^2 + 10ab - 15a - 4b^2 + 5b\end{aligned}$$

Classroom Exercises

Classify each polynomial by the number of its terms.

1. $x^2 - 4$ Binomial   **2.** $x^2 - 2xy + y^2$ **3.** $9a^2y^2$ Monomial   **4.** $x - y + z - w$
                              Trinomial   Polynomial, 4 terms

Simplify.

5. $(2x^2 + x - 5) - (x - y^2 + 2)$ **6.** $3ab(b^2 - ab + 2)$
    $2x^2 + y^2 - 7$ $3ab^3 - 3a^2b^2 + 6ab$

Written Exercises

Give the degree of each polynomial.

1. $x^2 + x - 6$ 2   **2.** $2x^3y^2$ 5   **3.** $-4abc$ 3   **4.** $a + b + c$ 1

5. 8 0   **6.** $7x$ 1   **7.** $x^3 - 2x$ 3   **8.** $x^2y^2 + xy^2$ 4

Simplify.

9. $(x^2 - 4x) + (2x^2 + 1)$ $3x^2 - 4x + 1$ **10.** $(3m^2 - 8) - (m^2 - 3)$ $2m^2 - 5$

11. $(6x^2 + 4x - 3) - (3x^2 - 3x - 4)$ **12.** $(3n - 2n^2 + 10) - (6 + 2n^2 - 7n)$

13. $-3x(7x + 2y - 5)$ $-21x^2 - 6xy + 15x$ **14.** $5x^2(3x^2 - 2x - 4)$ $15x^4 - 10x^3 - 20x^2$

15. $(x + 1)(x - 2)$ $x^2 - x - 2$ **16.** $(y - 3)(y - 4)$ $y^2 - 7y + 12$

17. $(3t + 4)(2t - 3)$ $6t^2 - t - 12$ **18.** $(5n - 3)(2n - 1)$ $10n^2 - 11n + 3$

19. $(5x - y)(2x + 3y)$ $10x^2 + 13xy - 3y^2$ **20.** $(3m + 4n)(5m + 2n)$ $15m^2 + 26mn + 8n^2$

21. $(x^2 + 8)(x^2 + 10)$ $x^4 + 18x^2 + 80$ **22.** $(n^3 - 5)(n^3 - 7)$ $n^6 - 12n^3 + 35$

23. $(2y^3 - 7)(5y^3 - 2)$ $10y^6 - 39y^3 + 14$ **24.** $(8x^2 + 3)(3x^2 + 5)$ $24x^4 + 49x^2 + 15$

25. $4(x + 3)(2x - 5)$ $8x^2 + 4x - 60$ **26.** $-4(3x + 10)(3x - 10)$ $-36x^2 + 400$

27. $(3x + 4)(x^2 - 3x + 2)$ **28.** $(4y - 3)(2y^2 + 5y - 1)$

29. $(3x^2y^2 - 2xy - xy^2) + (5xy - 3x^2y^2 - xy^2)$ $3xy - 2xy^2$

30. $(6x^3y - 8x^2y^2) - (4xy^3 - x^2y^2 + 6x^3y)$ $-7x^2y^2 - 4xy^3$

Simplify.

31. $(5x^2 + y^2)(3x^2 - 4y^2)$ **32.** $(c^3 - 2d^2)(2c^3 + 6d^2)$

33. $(2a - 3b)(4a - 5b + 3)$ **34.** $(5x + 2y)(x - y + 4)$

35. $-3y(4y - 3)(5y + 2)$ **36.** $(3.2x + 0.4)(0.7x + 0.3)$

37. $(x^n + 4)(x^n - 2)$ $x^{2n} + 2x^n - 8$ **38.** $(2x^{2a} + 3)(x^{2a} + 1)$ $2x^{4a} + 5x^{2a} + 3$

39. $(x^a + y^{2a})(x^a + 3y^{2a})$ **40.** $(5x^{2a} + 2y^b)(4x^{2a} - 3y^b)$

Mixed Review

Write an equation in standard form, $ax + by = c$, for the line that contains the given point(s) and/or satisfies other conditions that are stated. *3.4, 3.7*

1. $P(0,3)$; slope $= \frac{4}{5}$ $-4x + 5y = 15$

2. $Q(-2,4)$; slope $= -\frac{2}{3}$ $2x + 3y = 8$

3. $R(4,-2)$; $S(-2,6)$ $4x + 3y = 10$

4. $T(5,-4)$; parallel to the line described by $2x + 3y = 6$ $2x + 3y = -2$

5.3 Polynomials   **173**

Guided Practice

Classroom Exercises 1–6

Independent Practice

Ⓐ Ex. 1–26, Ⓑ Ex. 27–36, Ⓒ Ex. 37–40

Basic: WE 1–31 odd
Average: WE 5–35 odd
Above Average: WE 9–39 odd

Additional Answers

Written Exercises

11. $3x^2 + 7x + 1$

12. $-4n^2 + 10n + 4$

27. $3x^3 - 5x^2 - 6x + 8$

28. $8y^3 + 14y^2 - 19y + 3$

31. $15x^4 - 17x^2y^2 - 4y^4$

32. $2c^6 + 2c^3d^2 - 12d^4$

33. $8a^2 - 22ab + 6a + 15b^2 - 9b$

34. $5x^2 - 3xy + 20x - 2y^2 + 8y$

35. $-60y^3 + 21y^2 + 18y$

36. $2.24x^2 + 1.24x + 0.12$

39. $x^{2a} + 4x^ay^{2a} + 3y^{4a}$

40. $20x^{4a} - 7x^{2a}y^b - 6y^{2b}$

Enrichment

Discuss the prefixes of the terms *monomial*, *binomial*, and *trinomial*. *Mon* is from the Greek *monos*, meaning *alone*. *Bi* and *tri* are from the Latin, meaning *two* and *three*, respectively. Have the students write three lists of nonmathematical words that contain these prefixes. Each list should contain definitions of words. Suggest the use of a dictionary.

5.4 Factoring

Objectives

To factor a trinomial into two binomials
To factor a polynomial whose terms contain a common factor
To solve literal equations by factoring

▰▰▰GETTING STARTED

Prerequisite Quiz

1. Factor 180 into prime factors.
   $2 \cdot 2 \cdot 3 \cdot 3 \cdot 5$
2. Simplify $2x(3x + 2)(2x - 3)$.
   $12x^3 - 10x^2 - 12x$

Find the greatest common factor, other than 1, for each list of numbers.

3. 6, 12, 18, 10   2
4. 4, 4, 9, 16, 25   None

Solve each equation for x.

5. $ax - b = 5 - 3ax$ $x = \dfrac{b + 5}{4a}$
6. $4(rx + 2) = 2rx + 11$ $x = \dfrac{3}{2r}$

Motivator

Have students factor expressions 1–4 below as the indicated product of two positive integers.

1. 51   (17)(3)
2. $3 \cdot 23 + 2 \cdot 23$ (5)(23)
3. $11 \cdot 6 + 11 \cdot 3 - 11 \cdot 2$ (7)(11)
4. $6(5 + 2) - 4(5 + 2) + 3(5 + 2)$ (7)(5)

Sometimes you need to find the *factors* of a whole number in order to solve a problem in arithmetic. In algebra, too, it is often necessary to factor a polynomial in order to solve a problem.

You can reverse the FOIL method to help you factor $3x^2 + 14x - 5$ by trial-and-error. Try $3x$ and $1x$ for the first terms because $3x \cdot 1x$ is $3x^2$.

$$3x^2 + 14x - 5 = (\underline{\ \ ?\ \ })(\underline{\ \ ?\ \ })$$

$$3x^2 + 14x - 5 = (3x \underline{\ \ ?\ \ })(1x \underline{\ \ ?\ \ })$$

To find the last terms, notice first that because their product is negative, their signs must be opposite. So, the last terms are either $+5$ and -1, or -5 and $+1$. After trying several possible pairs of factors, you may find a pair that works. In this case, the pair $3x - 1$ and $x + 5$ give the correct middle term, $+14x$. So, $3x^2 + 14x - 5 = (3x - 1)(x + 5)$.

EXAMPLE 1   Factor $5n^2 - 42n + 16$ into two binomials, if possible.

Solution $5n^2 - 42n + 16 = (5n - \underline{\ \ })(n - \underline{\ \ })$ ← (THINK: Why must the missing terms be negative?)

Possible Factors	Middle Term
$(5n - 1)(n - 16)$	$-80n - 1n = -81n$
$(5n - 16)(n - 1)$	$-5n - 16n = -21n$
$(5n - 4)(n - 4)$	$-20n - 4n = -24n$
$(5n - 8)(n - 2)$	$-10n - 8n = -18n$
$(5n - 2)(n - 8)$	$-40n - 2n = -42n$ ← (Correct middle term)

So, $5n^2 - 42n + 16 = (5n - 2)(n - 8)$.

Highlighting the Standards

Standard 14c: Students are comfortable factoring whole numbers. They should see the factoring of polynomials as an extension of this process.

Additional Example 1

Factor $18a^2 - 5a - 2$ into two binomials, if possible. $(9a + 2)(2a - 1)$

Some polynomials cannot be factored when only integers are allowed as coefficients of the factors. In $12y^2 + 9y + 1$, no pair of factors gives the correct middle term.

Possible Factors	Middle Term
$(12y + 1)(y + 1)$	$13y$
$(6y + 1)(2y + 1)$	$8y$
$(4y + 1)(3y + 1)$	$7y$

Thus, $12y^2 + 9y + 1$ cannot be factored using integers.

The product of a monomial and a polynomial is a polynomial whose terms contain a *common monomial factor*.

$$4x^2(6x^2 - 11x - 10) = 24x^4 - 44x^3 - 40x^2$$

Each monomial term of $24x^4 - 44x^3 - 40x^2$ has $4x^2$ as a common factor. So, $24x^4 - 44x^3 - 40x^2$ can be factored into a monomial and a trinomial.

$$24x^4 - 44x^3 - 40x^2 = 4x^2(6x^2 - 11x - 10)$$

Because $4x^2$ is a common factor, 2, 4, x, x^2, $2x$, $4x$, and $2x^2$ must also be common factors. However, $4x^2$ is the *greatest common factor* (GCF). The **greatest common factor** of a polynomial is the common factor that has the greatest degree and the greatest constant factor.

The factoring of $24x^4 - 44x^3 - 40x^2$ into $4x^2(6x^2 - 11x - 10)$, however, is not "complete" because the factor $6x^2 - 11x - 10$ can be factored into two binomials. The *complete* factoring is shown below.

$$24x^4 - 44x^3 - 40x^2 = 4x^2(6x^2 - 11x - 10)$$
$$= 4x^2(3x + 2)(2x - 5)$$

To factor a polynomial completely: **1.** factor out the GCF of its terms, if any, and **2.** factor any resulting factor, if possible.

EXAMPLE 2   Factor $18a^3b^4 + 24a^2b^2 - 12a^2b$ completely.

Solution   Find the GCF of the terms.

The GCF of 18, 24, and 12 is 6.

The GCF of a^3, a^2, and a^2 is a^2.

The GCF of b^4, b^2, and b is b.

So, the GCF of the terms is $6a^2b$.

Thus, $18a^3b^4 + 24a^2b^2 - 12a^2b = 6a^2b(3ab^3 + 4b - 2)$.

■ TEACHING SUGGESTIONS

Lesson Note

Polynomials of degree 2, 3, or 4 are factored using integers as coefficients, if possible. Rational coefficients are used in the C-level exercises.

The use of a common monomial factor in solving a literal equation is shown in Example 4. This is an extension of Lesson 1.4, where literal equations were introduced.

Math Connections

Coordinate Geometry: Many demonstrations in coordinate geometry rely on the ability to factor polynomials. For example, to show that $x^2 - 8x + 16 + y^2 - 25 = 0$ is the equation of a circle, it is necessary to show that the given expression can be written in the form $(x - h)^2 + (y - k)^2 = r^2$. Since the given expression can be written in this form as $(x - 4)^2 + y^2 = 5^2$, it is the equation of a circle.

Critical Thinking Questions

Analysis: Ask students to devise trinomials that can be dismissed as nonfactorable on immediate inspection, that is, without any trial-and-error work at all. Examples are $x^2 + 50x + 1$ and $x^2 - 30x + 1$ with middle terms that are clearly too large to have emerged from the product of binomials such as $x + 1$ and $x - 1$.

Additional Example 2

Factor completely.

$12m^3n^2 - 8m^2n^2 + 4mn^3$

$4mn^2(3m^2 - 2m + n)$

Checkpoint

Factor completely, if possible.

1. $x^2 - 8x - 48$ $(x - 12)(x + 4)$
2. $4y^2 + 3y - 10$ $(4y - 5)(y + 2)$
3. $6n^3 - 27n^2 + 12n$ $3n(2n - 1)(n - 4)$
4. $12a^2 - 2a - 3$ Not possible
5. Solve $m(x - 7) = 5x + 2m$ for x.
   $x = \dfrac{9m}{m - 5}$

Closure

Ask students to define Greatest Common Factor. See page 175. Ask students if the expression $4x(12x^4 - 8x^3 - 24x^2 + 2)$ is factored completely. No.

█████FOLLOW UP

Guided Practice

Classroom Exercises 1–19

Independent Practice

A Ex. 1–27, **B** Ex. 28–36, **C** Ex. 37–39

Basic: WE 1–31 odd

Average: WE 5–35 odd

Above Average: WE 7–39 odd

Additional Answers

Classroom Exercises

1. $(x - 10)(x - 2)$
2. $(y + 3)(y - 8)$

EXAMPLE 3   Factor $45x^2y - 21xy^2 - 6y^3$ completely.

Solution   The GCF of the terms is $3y$. Factor out $3y$.
$$45x^2y - 21xy^2 - 6y^3$$
$$3y(15x^2 - 7xy - 2y^2)$$

Factor $15x^2 - 7xy - 2y^2$, if possible. $3y(5x + y)(3x - 2y)$

$5x + y$ and $3x - 2y$ cannot be factored using integers.

Thus, $45x^2y - 21xy^2 - 6y^3 = 3y(5x + y)(3x - 2y)$.

Recall that an equation such as $a(x - b) = 2 - cx$ is called a *literal equation* (see Lesson 1.4). Sometimes factoring can be used to solve such equations.

EXAMPLE 4   Solve $a(x - b) = 2 - cx$ for x.

Solution

	$a(x - b) = 2 - cx$
Use the Distributive Property.	$ax - ab = 2 - cx$
Rewrite the equation with the x-terms alone on one side.	$ax + cx = ab + 2$
Factor out the common factor.	$(a + c)x = ab + 2$
Solve for x.	$x = \dfrac{ab + 2}{a + c}$

Classroom Exercises

Factor into two binomials, if possible.

1. $x^2 - 12x + 20$ 2. $y^2 - 5y - 24$ 3. $45c^3 + 5c$
4. $x^2 + 2x - 3$ 5. $x^2 + 3x + 1$ 6. $x^2 - 5x + 4$
7. $4d^3 + 2d^2$ 8. $x^2 - x - 6$ 9. $3n^2 + 8n + 4$
10. $5t^2 - 17t + 6$ 11. $4y^2 - 5y - 6$ 12. $4y^2 + 2y - 12$
    $(5t - 2)(t - 3)$ $(4y + 3)(y - 2)$ $(2y - 3)(2y + 4)$

Find the GCF of each group of expressions.

13. $12x^2, 36x, 18$ 6   14. $15x^2, 20x^3$ $5x^2$
15. a^3b^2, a^2b^3, a^2b a^2b 16. $x(a - b), 2y(a - b)$ $a - b$

Factor completely.

17. $a^3 + 2a^2 - 8a$ 18. $m^3 - 4m^2 + 3m$ 19. $y^4 + 7y^3 + 10y^2$
    $a(a + 4)(a - 2)$ $m(m - 3)(m - 1)$ $y^2(y + 2)(y + 5)$

176   Chapter 5 Polynomials

Additional Example 3

Factor completely.

$16x^2y + 12xy - 18y$ $2y(4x - 3)(2x + 3)$

Additional Example 4

Solve for x.

$2m(3x + 1) = tx + 8$ $x = \dfrac{8 - 2m}{6m - t}$

Written Exercises

Factor into two binomials, if possible.

1. $x^2 + x - 20$ $(x + 5)(x - 4)$ **2.** $x^2 - 8x - 20$ $(x - 10)(x + 2)$

3. $y^2 - 5y + 6$ $(y - 3)(y - 2)$ **4.** $y^2 - 5y - 6$ $(y - 6)(y + 1)$

5. $a^2 + 10a + 36$ Not possible   **6.** $b^2 - 7b - 24$ Not possible

7. $3n^2 - 5n - 1$ Not possible   **8.** $2a^2 - 8a + 1$ Not possible

9. $7y^2 - 36y + 5$ $(7y - 1)(y - 5)$ **10.** $3c^2 + 14c - 5$ $(3c - 1)(c + 5)$

11. $10t^2 - 11t - 6$ $(5t + 2)(2t - 3)$ **12.** $4n^2 + 7n - 15$ $(4n - 5)(n + 3)$

Factor completely, if possible.

13. $9m^2 - 18m + 8$ **14.** $8c^2 - 34c + 21$ **15.** $3n^2 + 6n - 45$

16. $4x^2 - 28x + 48$ **17.** $6t^3 - 11t^2 - 10t$ **18.** $8a^3 - 18a^2 + 9a$

19. $6x^3 - 28x^2 - 10x$ **20.** $24y^3 - 44y^2 - 40y$ **21.** $12x^2 + 4xy - y^2$

     $2x(3x + 1)(x - 5)$ $4y(3y + 2)(2y - 5)$ $(2x + y)(6x - y)$

Solve each equation for x.

22. $tx + vx = 7$ **23.** $mx - 5x = f$ **24.** $rx - a = tx + 5$

25. $c - 4x = d - ax$ **26.** $c(x + 4) = 8x + 7c$ **27.** $a(6x + 5) = 2nx - 4a$

28. Find two integers for k such that $x^2 + kx + 3$ can be factored into two binomials. $-4, 4$

29. There are four integers for k such that $y^2 + ky - 10$ can be factored. Find these four integers. $-9, -3, 3, 9$

30. Find one positive integer and at least three negative integers for k such that $x^2 + 2x + k$ can be factored. $1, -3, -8, -15, \ldots$

Factor completely.

     $(6y + 5)(2y - 3)$ $(8x - 9)(3x + 4)$ $5xy(3x^2 + xy - 2y)$

31. $12y^2 - 8y - 15$ **32.** $24x^2 + 5x - 36$ **33.** $15x^3y + 5x^2y^2 - 10xy^2$

34. $15c^3 - 9c^2d + 3c^3d$ **35.** $36y^5 - 46y^3 - 12y$ **36.** $30n^5 + 93n^3 + 72n$

37. $x^{6m} - 8x^{3m} + 12$ **38.** $6y^{4m} + 19y^{2m} + 15$ **39.** $5x^{2a} - 23x^ay^b + 12y^{2b}$

     $(x^{3m} - 6)(x^{3m} - 2)$ $(3y^{2m} + 5)(2y^{2m} + 3)$ $(5x^a - 3y^b)(x^a - 4y^b)$

Mixed Review

Find the solution set. Graph it on a number line. (Exercises 1 and 2)
2.2, 2.3

1. $3x - 2 < 10$ or $2 - 3x > 14$ **2.** $|2x - 7| < 15$

3. Solve $\dfrac{6}{2x + 3} = \dfrac{10}{3x - 5}$.   *1.4* -30 **4.** Solve $m(tx - 6) = m + 5$ for x.   *1.4*

---

3. $5c(9c^2 + 1)$
4. $(x + 3)(x - 1)$
5. Not possible
6. $(x - 4)(x - 1)$
7. $2d^2(2d + 1)$
8. $(x - 3)(x + 2)$
9. $(3n + 2)(n + 2)$

Written Exercises

13. $(3m - 4)(3m - 2)$
14. $(4c - 3)(2c - 7)$
15. $3(n + 5)(n - 3)$
16. $4(x - 3)(x - 4)$
17. $t(3t + 2)(2t - 5)$
18. $a(4a - 3)(2a - 3)$
22. $\dfrac{7}{t + v}$
23. $\dfrac{f}{m - 5}$
24. $\dfrac{5 + a}{r - t}$
25. $\dfrac{d - c}{a - 4}$
26. $\dfrac{3c}{c - 8}$
27. $\dfrac{-9a}{6a - 2n}$

See page 196 for the answers to Written Ex. 34–36 and Mixed Review Ex. 1, 2, 4.

Enrichment

Students sometimes have difficulty factoring trinomials in which the lead coefficient is a number other than 1. Discuss the following strategy that can be used in factoring such a trinomial.

$9m^2 - 18m + 8$ (Ex. 13)

$\dfrac{1}{9} \cdot 9(9m^2 - 18m + 8)$

$\dfrac{1}{9} \cdot [(9m)^2 - 18(9m) + 72)]$

Let $a = 9m$.

$\dfrac{1}{9}(a^2 - 18a + 72)$

$\dfrac{1}{9}(a - 12)(a - 6)$

Replace a with $9m$.

$\dfrac{1}{9}(9m - 12)(9m - 6)$

$\dfrac{1}{3}(9m - 12) \cdot \dfrac{1}{3}(9m - 6)$

$(3m - 4)(3m - 2)$

Have the students try the strategy with several similar exercises.

▬GETTING STARTED

Prerequisite Quiz

Simplify.

1. $(5x + 2)(3x - 4)$ $15x^2 - 14x - 8$
2. $(4a - 3b)(4a + 3b)$ $16a^2 - 9b^2$
3. $(5a + 3)(5a + 3)$ $25a^2 + 30a + 9$

Let $f(x) = 3x + 5$. Find the following.

4. $f(-4)$ -7
5. $f(c + 2)$ $3c + 11$
6. $f(a + h) - f(a)$ $3h$

Motivator

Ask students to find the product of the following:

1. $(x + 4)(x - 4)$ $x^2 - 16$
2. $(y + 2)(y - 2)$ $y^2 - 4$
3. $(a + b)(a - b)$ $a^2 - b^2$

Have students devise a rule based on their observations.   See page 178.

5.5 Special Products

Objectives   To simplify products of the form $(x + y)(x - y), (x + y)^2$, and $(x - y)^2$
To evaluate quadratic functions

Products of the form $(a + b)(a - b)$ are readily simplified.

$$(a + b)(a - b) = a^2 - ab + ab - b^2, \text{ or } a^2 - b^2$$

Product of Sum and Difference of Two Terms
For all real numbers a and b, $(a + b)(a - b) = a^2 - b^2$. That is, the product of the sum and difference of two terms is the difference of the squares of the two terms.

EXAMPLE 1   Simplify.

Solutions
a. $(4x + 3y)(4x - 3y)$
$= (4x)^2 - (3y)^2$
$= 16x^2 - 9y^2$

b. $(5n^2 - 8)(5n^2 + 8)$
$= (5n^2)^2 - 8^2$
$= 25n^4 - 64$

The second special product, $(a + b)^2$, is the square of a binomial. Notice that $(a + b)^2 = (a + b)(a + b) = a^2 + ab + ab + b^2 = a^2 + 2ab + b^2$. So, the square of a binomial is the sum of (1) the square of the first term, (2) twice the product of the terms, and (3) the square of the last term. In a similar way, you can find that $(a - b)^2 = a^2 - 2ab + b^2$.

Square of a Binomial
For all real numbers a and b, $(a + b)^2 = a^2 + 2ab + b^2$
and $(a - b)^2 = a^2 - 2ab + b^2$.

The trinomials $a^2 + 2ab + b^2$ and $a^2 - 2ab + b^2$ are called **perfect-square trinomials.**

EXAMPLE 2   Simplify.

Solutions
a. $(3n + 5)^2$
$= (3n)^2 + 2 \cdot 3n \cdot 5 + 5^2$
$= 9n^2 + 30n + 25$

b. $(4x^2 - 5)^2$
$= (4x^2)^2 - 2 \cdot 4x^2 \cdot 5 + 5^2$
$= 16x^4 - 40x^2 + 25$

178   Chapter 5 Polynomials

Highlighting the Standards

Standard 2d: This would be a good lesson to have students study the presentation themselves. Most will be pleasantly surprised to find that they can read mathematics.

Additional Example 1

Simplify.

a. $(5x - 2y)(5x + 2y)$ $25x^2 - 4y^2$
b. $(3n^2 + 4)(3n^2 - 4)$ $9n^4 - 16$

Additional Example 2

Simplify.

a. $(4n - 3)^2$ $16n^2 - 24n + 9$
b. $(3x^2 + 7)^2$ $9x^4 + 42x^2 + 49$

Recall that if $g(x) = 4x - 7$, then $g(-2) = 4(-2) - 7 = -15$ and $g(a + h) = 4(a + h) - 7 = 4a + 4h - 7$. The function g is called a *linear function* because $4x - 7$ is a linear (first-degree) polynomial. Now, let $f(x) = 3x^2 - 2x + 5$. Then the function f is a *quadratic function* because $3x^2 - 2x + 5$ is a quadratic (second-degree) polynomial. To find $f(a - 4)$, substitute the binomial as shown below.

$$f(x) = 3x^2 - 2x + 5$$
$$f(a - 4) = 3(a - 4)^2 - 2(a - 4) + 5$$
$$= 3(a^2 - 8a + 16) - 2a + 8 + 5$$
$$= 3a^2 - 24a + 48 - 2a + 13$$
$$= 3a^2 - 26a + 61$$

EXAMPLE 3   Let $f(x) = 2x^2 - 5x + 12$. Find the following values.
   a. $f(-3)$ **b.** $f(4c)$ **c.** $f(x + h) - f(x)$

Solutions   Given that $f(x) = 2x^2 - 5x + 12$:
   a. $f(-3) = 2(-3)^2 - 5(-3) + 12 = 2 \cdot 9 + 15 + 12 = 45$
   b. $f(4c) = 2(4c)^2 - 5 \cdot 4c + 12 = 2 \cdot 16c^2 - 20c + 12$
       $= 32c^2 - 20c + 12$
   c. $f(x + h) - f(x)$
       $= [2(x + h)^2 - 5(x + h) + 12] - [2x^2 - 5x + 12]$
       $= 2(x^2 + 2hx + h^2) - 5x - 5h + 12 - 2x^2 + 5x - 12$
       $= 2x^2 + 4hx + 2h^2 - 5h - 2x^2$
       $= 4hx + 2h^2 - 5h$

Classroom Exercises

Choose one of the expressions a–d that is equivalent to the given expression.

1. $(3x - 5)(3x + 5)$ d
   a. $3x^2 - 10$
   b. $9x^2 - 30x + 25$
   c. $3x^2 - 25$
   d. $9x^2 - 25$

2. $(2y - 7)^2$ c
   a. $4y^2 - 49$
   b. $2y^2 - 14y - 49$
   c. $4y^2 - 28y + 49$
   d. $4y^2 - 14y + 49$

3. $(2c^2 + d)(2c^2 - d)$ d
   a. $4c^2 - d^2$
   b. $2c^4 - d^2$
   c. $4c^4 + 4c^2d + d^2$
   d. $4c^4 - d^2$

Simplify.

4. $(a + 2)(a - 2)$ $a^2 - 4$ **5.** $(x^2 - 1)(x^2 + 1)$ $x^4 - 1$ **6.** $(y + 2)^2$ $y^2 + 4y + 4$
7. $(b - 2)(b - 2)$ **8.** $(6x + 3)(6x - 3)$ **9.** $(4x - 3)^2$
   $b^2 - 4b + 4$ $36x^2 - 9$ $16x^2 - 24x + 9$

Lesson Note

The special products, $(a + b)(a - b) = a^2 - b^2$ and $(a + b)^2 = a^2 + 2ab + b^2$ will be reversed in the next lesson to give the special factors $a^2 - b^2 = (a + b)(a - b)$ and $a^2 + 2ab + b^2 = (a + b)^2$. Functional notation, introduced in Chapter 3, is extended to polynomials of degree 2. For example, if $f(x) = 4x^2 - 5$, then $f(a - 3) = 4(a - 3)^2 - 5$, where the students must be successful in squaring a binomial.

Math Connection

History: The ancient Babylonians had a well-developed algebra as early as 2000 B.C. It was closely related to practical measurements, which involved verbal rather than symbolic representations. There is evidence that the Babylonians solved quadratic equations by substituting into a general formula (pattern) and completing the square. Cubic equations of the form $x^3 + x^2 = b$ were solved by using an $n^3 + n^2$ table.

Critical Thinking Questions

Analysis: Ask students to square the binomial $a + b$ under the assumption that all of the field postulates (see Lesson 1.3) hold except for the Commutative Property of Multiplication. What result is obtained? Students should obtain $a^2 + ba + ab + b^2$.

Common Error Analysis

Error: Some students will simplify expressions of the form $(a + b)^2$ to $a^2 + b^2$.

Ask these students to find the base for the exponent 2, write the base twice as a factor and use the FOIL method to simplify.

Additional Example 3

Let $g(x) = 3x^2 + 4$. Find:

a. $g(7)$ 151
b. $g(-2c)$ $12c^2 + 4$
c. $g(x - 5)$ $3x^2 - 30x + 79$
d. $g(x + a) - g(x)$ $6ax + 3a^2$

179

Checkpoint

Simplify.

1. $(6x + 5a)(6x - 5a)$ $36x^2 - 25a^2$
2. $(3x - 2y)^2$ $9x^2 - 12xy + 4y^2$

Let $f(x) = 2x^2 + x$. Find the following.

3. $f\left(-\frac{1}{2}\right)$ 0
4. $f(4a)$ $32a^2 + 4a$
5. $f(2x + 3)$ $8x^2 + 26x + 21$

Closure

Ask students to give an example of a *perfect square trinomial.* Answers will vary. Ask students to factor the expression $4x^2 - 49$. $(2x + 7)(2x - 7)$

▰▰▰FOLLOW UP

Guided Practice

Classroom Exercises 1–9

Independent Practice

🅐 Ex. 1–16, 🅑 Ex. 17–28, 🅒 Ex. 29, 30

Basic: WE 1–17 odd, Midchapter Review

Average: WE 1–27 odd, Midchapter Review

Above Average: WE 13–29 odd, Midchapter Review

Additional Answers

Written Exercises

23. 57
24. $130c^2 + 8c - 30$
25. $-15c^2 + 62c + 168$
26. $10xh + 5h^2 - 2h$
27. $-10a^2 + 14a + 12$
28. $-4xh - 2h^2 + 6h$

Written Exercises

Simplify.

1. $(n + 11)(n - 11)$ $n^2 - 121$
2. $(7x - 6)(7x + 6)$ $49x^2 - 36$
3. $(2x + 5)^2$ $4x^2 + 20x + 25$
4. $(4y - 5)^2$ $16y^2 - 40y + 25$
5. $(3 - 7a)^2$ $49a^2 - 42a + 9$
6. $(6c + 2)^2$ $36c^2 + 24c + 4$
7. $(x + c)^2$ $x^2 + 2cx + c^2$
8. $(2a^2 - 6)(2a^2 + 6)$ $4a^4 - 36$
9. $(3x + 2a)(3x - 2a)$ $9x^2 - 4a^2$
10. $(x - 2y)^2$ $x^2 - 4xy + 4y^2$

Let $f(x) = 4x^2 - 10$ and $g(x) = 3x^2 - 4x$. Evaluate.

11. $f\left(-\frac{1}{2}\right)$ -9
12. $g\left(\frac{1}{3}\right)$ -1
13. $f(3c)$ $36c^2 - 10$
14. $g(-2a)$ $12a^2 + 8a$
15. $f(2x - 5)$ $16x^2 - 80x + 90$
16. $g(x + h) - g(x)$ $6xh - 4h + 3h^2$

Simplify.

17. $(3m^2 + 4n)(3m^2 - 4n)$ $9m^4 - 16n^2$
18. $(6c - 5d^2)(6c + 5d^2)$ $36c^2 - 25d^4$
19. $(x + 0.1)^2$ $x^2 + 0.2x + 0.01$
20. $(5x - 4y)^2$ $25x^2 - 40xy + 16y^2$
21. $(3x + 2h^2)^2$ $9x^2 + 12xh^2 + 4h^4$
22. $(4a^2 - k^2)^2$ $16a^4 - 8a^2k^2 + k^4$

Let $f(x) = 5x^2 - 2x - 15$ and $g(x) = -2x^2 + 6x + 10$. Evaluate.

23. $f(4)$
24. $f(-5c) + f(c)$
25. $f(c + 6) - f(2c)$
26. $f(x + h) - f(x)$
27. $g(2a - 1) + g(-a)$
28. $g(x + h) - g(x)$
29. $(x^m + y^n)(x^m - y^n)$ $x^{2m} - y^{2n}$
30. $(x^{3a} - y^{2b})(x^{2a} + y^b)$

$x^{5a} + x^{3a}y^b - x^{2a}y^{2b} - y^{3b}$

Midchapter Review

Simplify each expression. Use positive exponents. *5.1, 5.2, 5.3, 5.5*

1. $\dfrac{-10a^3b^2c^7}{15a^6b^2c^2}$ $\dfrac{-2c^5}{3a^3}$
2. $\left(\dfrac{-3a}{2c^2}\right)^4$ $\dfrac{81a^4}{16c^8}$
3. $\dfrac{10a^6c^{-3}d^{-1}}{5a^{-2}c^2d^{-3}}$ $\dfrac{2a^8d^2}{c^5}$
4. $(-2x^{-2})^3$ $\dfrac{-8}{x^6}$

5. $(8x^3 - 5x^2 + 4x) - (5x^3 - 2x^2 - 3x - 2)$ $3x^3 - 3x^2 + 7x + 2$
6. $3c(5c + 4)(2c - 3)$ $30c^3 - 21c^2 - 36c$
7. $(4x - 3y)(4x + 3y)$ $16x^2 - 9y^2$
8. Solve $3^{2x-1} = 81$ for x. *5.1* $\frac{5}{2}$

Factor completely. *5.4*

9. $6n^2 - 14n - 12$ $2(3n + 2)(n - 3)$
10. $60m^2n - 9mn^2 - 6n^3$

$3n(4m + n)(5m - 2n)$

Enrichment

There are five children in the Fernandez family, and the age of each is expressed as a natural number. If you choose any one or more of the children, the total of the ages selected will be different from any other possible selection of one or more. The sum of all five ages is the least possible total. What are the ages of the Fernandez children? 1, 2, 4, 8, 16

5.6 Special Factors

Objectives
To factor the difference of two squares
To factor perfect-square trinomials
To factor the sum or difference of two cubes

Teaching Resources

Quick Quizzes 36
Reteaching and Practice Worksheets 36

When the special products $(a + b)(a - b)$ and $(a + b)^2$ are simplified, the results are $a^2 - b^2$, a difference of two squares, and $a^2 + 2ab + b^2$, a perfect-square trinomial. These procedures can be reversed to find the special factors of $a^2 - b^2$ and $a^2 + 2ab + b^2$.

Special Factors
For all real numbers a and b,
$a^2 - b^2 = (a + b)(a - b)$ (Difference of two squares)
$\left.\begin{array}{l}a^2 + 2ab + b^2 = (a + b)^2 \\ a^2 - 2ab + b^2 = (a - b)^2\end{array}\right\}$ (Perfect-square trinomials)

The trinomial $16n^2 + 40n + 25$ is a perfect-square trinomial because it can be written as the square of a binomial.

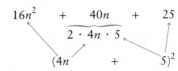

Check: The square of $4n$ is $16n^2$, the first term. The square of 5 is 25, the third term. Twice $4n \cdot 5$ is $40n$, the middle term.

EXAMPLE 1
a. Factor $16c^2 - 49$.
b. Factor $16c^4 - 8c^2d + d^2$.

Solutions
a. $16c^2 - 49 = (4c)^2 - 7^2$
$= (4c + 7)(4c - 7)$
b. $16c^4 - 8c^2d + d^2 = (4c^2)^2 - 2 \cdot 4c^2 \cdot d + d^2$
$= (4c^2 - d)^2$

The special products $a^2 - b^2$ and $(x + y)^2$ may sometimes occur in the same expression, as in $(c + d)^2 - 5^2$. Factor as follows.

Use $a^2 - b^2$ to factor $(c + d)^2 - 5^2$. $\leftarrow a^2 - b^2 = (a + b)(a - b)$
$(c + d)^2 - 5^2 = [(c + d) + 5][(c + d) - 5]$
$= (c + d + 5)(c + d - 5)$

5.6 Special Factors **181**

GETTING STARTED

Prerequisite Quiz

Simplify.

1. $(3x - 2)(9x^2 + 6x + 4)$ $27x^3 - 8$
2. Factor -1 out of $-4y^2 + 12y - 9$.
   $-1(4y^2 - 12y + 9)$
3. Write $25m^2 - 16n^2$ in the form $(a)^2 - (b)^2$. $(5m)^2 - (4n)^2$

Factor if possible.

4. $4x^2 - 12x + 9$ $(2x - 3)(2x - 3)$
5. $9y^2 + 4$ Not possible

Motivator

Ask students if $x^2 + 4x - 4$ is a perfect square trinomial. No. Ask students if they think $x^3 + y^3$ can be factored into the product of two binomial terms. No. Have students multiply $(x + 4)(x^2 - 4x + 16)$ $x^3 + 64$.

TEACHING SUGGESTIONS

Lesson Note

Polynomials of the form $a^2 - b^2$ or $a^2 + 2ab + b^2$ can be factored by the trial-and-error method rather than using the formulas. The factors of $a^3 - b^3$ and $a^3 + b^3$ should be memorized. Students should rewrite a sum or difference of two cubes such as $8x^3 + 125y^3$ in the form $(2x)^3 + (5y)^3$ before using the formulas for factoring $(a)^3 + (b)^3$.

Additional Example 1

Factor $36n^2 - 25$. $(6n + 5)(6n - 5)$

Highlighting the Standards

Standard 3d: This lesson is an opportunity for students to use their algebraic reasoning. The many forms of special factors lead to an understanding of the underlying principles.

181

Math Connections

Mental Arithmetic: Sometimes special factors can be used to perform arithmetic calculations mentally. However, to do this with confidence and a certain amount of regularity, students should know the perfect squares through 625 (25^2).

Critical Thinking Questions

Analysis: Ask students to consider the polynomials $a^4 - b^4$ and $a^6 - b^6$. Are these factorable? They should see that both polynomials are factorable. The first, $a^4 - b^4$, can be considered as a difference of two squares by writing it as $(a^2)^2 - (b^2)^2$. The second, $a^6 - b^6$, can be thought of as either a difference of two squares, $(a^3)^2 - (b^3)^2$, or as a difference of two cubes, $(a^2)^3 - (b^2)^3$.

Checkpoint

Factor if possible.

1. $x^4 - 36$ $(x^2 + 6)(x^2 - 6)$
2. $(m - n)^2 - 36$
   $(m - n + 6)(m - n - 6)$
3. $x^2 + 8x + 16 - y^2$
   $(x + 4 + y)(x + 4 - y)$
4. $27x^3 + 8$ $(3x + 2)(9x^2 - 6x + 4)$

Closure

Ask students to give a second factor to make a perfect square trinomial if the first factor is $(x - y)$. $x - y$ Ask students if the expression $(x + y)^2 - 4^2$ is, in fact, the difference of two squares.   yes

---

EXAMPLE 2   Factor.   **a.** $25x^2 - 40x + 16 - 9y^2$ **b.** $16x^2 - 4y^2 - 12y - 9$

Solutions   **a.** Use $25x^2 - 40x + 16 = (5x - 4)^2$.
$$(25x^2 - 40x + 16) - 9y^2 = (5x - 4)^2 - (3y)^2$$
$$= (5x - 4 + 3y)(5x - 4 - 3y)$$

b. Use $4y^2 + 12y + 9 = (2y + 3)^2$.
$$16x^2 - 4y^2 - 12y - 9 = 16x^2 - (4y^2 + 12y + 9)$$
$$= (4x)^2 - (2y + 3)^2$$
$$= [4x + (2y + 3)][4x - (2y + 3)]$$
$$= (4x + 2y + 3)(4x - 2y - 3)$$

The sum of two cubes, $a^3 + b^3$, and the difference of two cubes, $a^3 - b^3$, can be factored into the product of a binomial and a trinomial. To understand this factoring, start with the two factors.

$$(a + b)(a^2 - ab + b^2) = a^3 - a^2b + ab^2 + a^2b - ab^2 + b^3$$
$$= a^3 + b^3$$
$$(a - b)(a^2 + ab + b^2) = a^3 + a^2b + ab^2 - a^2b - ab^2 - b^3$$
$$= a^3 - b^3$$

> **Special Factors**
> For all real numbers a and b,
> $a^3 + b^3 = (a + b)(a^2 - ab + b^2)$ (Sum of two cubes)
> $a^3 - b^3 = (a - b)(a^2 + ab + b^2)$ (Difference of two cubes)

EXAMPLE 3   Factor $64x^3 - 27y^3$.

Solution $64x^3 - 27y^3$ can be written as $(4x)^3 - (3y)^3$, a difference of cubes.

$$a^3 - b^3 = (a - b)(a^2 + ab + b^2)$$
$$64x^3 - 27y^3 = (4x)^3 - (3y)^3 = (4x - 3y)(16x^2 + 12xy + 9y^2)$$

Classroom Exercises

Determine whether each polynomial is a *difference of two squares*, a *perfect-square trinomial*, a *sum of two cubes*, a *difference of two cubes*, or *none of these*.

1. $4c^2 - 9$ D.S.
2. $y^2 + 16$ None
3. $x^2 + 8x + 16$ P.S.
4. $x^2 - 12x - 36$ None
5. $8x^3 + 1$ S.C.
6. $a^2 + 2a - 1$ None

7–12. Factor each polynomial in Classroom Exercises 1–6, if possible.

Additional Example 2

a. Factor $9x^2 + 30x + 25 - 4t^2$.
   $(3x + 5 + 2t)(3x + 5 - 2t)$
b. Factor $25x^2 - 9t^2 + 24t - 16$.
   $(5x - 3t + 4)(5x + 3t - 4)$

Additional Example 3

Factor $m^3 + 125n^3$.
   $(m + 5n)(m^2 - 5mn + 25n^2)$

Written Exercises

Factor each polynomial, if possible. Write each perfect-square trinomial as the square of a binomial.

1. $4c^2 - 25$ $(2c + 5)(2c - 5)$ 2. $49 - 36d^2$ 3. $x^2 - 8x + 16$ $(x - 4)^2$
4. $100a^2 + 20a + 1$ 5. $c^2 + 1$ Not possible   6. $x^2 + 10x + 16$
7. $y^4 - 49$ $(y^2 + 7)(y^2 - 7)$ 8. $25 - d^4$ $(5 + d^2)(5 - d^2)$ 9. $16n^2 + 24n + 9$ $(4n + 3)^2$
10. $4a^2 - 20a + 25$ 11. $(x + y)^2 - 16$ 12. $(c - d)^2 - 25$
13. $x^2 - (y - 2)^2$ 14. $m^2 - (n + 3)^2$ 15. $x^2 + 6x + 9 - y^2$
16. $m^2 - 2mn + n^2 - 9$ 17. $a^4 - 16a^2 + 64$ 18. $y^3 - 8$
19. $x^3 + 27$ 20. $t^2 - v^2 + 8v - 16$ 21. $64d^3 + 1$
   $(x + 3)(x^2 - 3x + 9)$ $(t + v - 4)(t - v + 4)$ $(4d + 1)(16d^2 - 4d + 1)$

22. Find the two values of k for which $x^2 + kx + 16$ is a perfect-square trinomial. $-8, 8$

23. For what value of k is $y^2 - 12y + 9k$ a perfect-square trinomial? 4

Factor each polynomial, if possible.

24. $49a^2b^2 - 100$ 25. $36x^6 - y^4$ 26. $4c^2 - 20cd + 25d^2$
27. $25a^4 + 60a^2b^2 + 36b^4$ 28. $25c^2 + 9$ Not possible   29. $4x^2 - 10xy + 25y^2$
30. $9x^2 + 30xy + 16y^2$ 31. $m^3 - 64n^3$ 32. $125c^3 + d^3$
33. $8t^3 + 125v^3$ $(2t + 5v)(4t^2 - 10tv + 25v^2)$ 34. $27x^6 - 1{,}000y^3$
35. $4x^2 + 12xy + 9y^2 - 25$ 36. $9x^2 - 6xy + y^2 - 16$
37. $x^{2n} - y^{4n+6}$ $(x^n + y^{2n+3})(x^n - y^{2n+3})$ 38. $4x^{6m+4} + 12x^{3m+2}y^n + 9y^{2n}$
39. Is $x^2 + y^2 \geq 2xy$ true for all real numbers x and y? Explain.
40. Prove that $[(a + b) + c]^2 = a^2 + b^2 + c^2 + 2ab + 2ac + 2bc$ for all real numbers a, b, and c.

Mixed Review

1. Betty is 3 years older than Alex, and Carlo is twice as old as Betty. In 5 years, Carlo's age will be 7 times the difference in Betty's and Alex's ages. Find Carlo's present age. *2.5* 16 yr

2. Two twins leave their house at the same time, traveling in the same direction. One drives a moped at 30 mi/h and the other rides a bike at 20 mi/h. In how long will they be 5 mi apart? *2.6* 30 min

Find the slope and describe the slant of the line determined by the two points. *3.3*

3. $P(-3,5)$, $Q(2,-4)$ 4. $M(2,7)$, $N(-3,7)$ 5. $A(6,1)$, $B(6,-4)$
   $-\frac{9}{5}$; down to right   0; horizontal   Undef; vertical

▰▰FOLLOW UP

Guided Practice

Classroom Exercises 1–12

Independent Practice

A Ex. 1–21, **B** Ex. 22–36, **C** Ex. 37–40

Basic: WE 1–23 odd
Average: WE 1–35 odd
Above Average: WE 5–39 odd

Additional Answers

Classroom Exercises

7. $(2c + 3)(2c - 3)$
8. Not possible
9. $(x + 4)^2$
10. Not possible
11. $(2x + 1)(4x^2 - 2x + 1)$
12. Not possible

Written Exercises

2. $(7 + 6d)(7 - 6d)$
4. $(10a + 1)^2$
6. $(x + 8)(x + 2)$
10. $(2a - 5)^2$

See pages 809–810 for Written Ex. 11–18, 24–27, 29–32, 34–36, 38–40.

Enrichment

Explain that the factorization of polynomials in this lesson assumes that all numbers are integral. If the process were open to all real numbers, the factorizations below would be valid.

$x^2 - \frac{1}{4}x - \frac{1}{8} = (x - \frac{1}{2})(x + \frac{1}{4})$; $x^2 - 2 = (x + \sqrt{2})(x - \sqrt{2})$

Under this rule, challenge students to factor the following expressions.

1. $x^2 + \frac{1}{2}x - \frac{1}{2}$ $(x + 1)(x - \frac{1}{2})$
2. $x^2 - 10$ $(x + \sqrt{10})(x - \sqrt{10})$
3. $2x^2 - 9$ $(\sqrt{2}x + 3)(\sqrt{2}x - 3)$
4. $\sqrt{15}x^2 + \sqrt{10}x + \sqrt{6}x + 2$
   $(\sqrt{5}x + \sqrt{2})(\sqrt{3}x + \sqrt{2})$

▰▰GETTING STARTED

Prerequisite Quiz

Find the greatest common factor of the expressions in each list.

1. 16, 80, 32, 48   16
2. x^2y^2, x^3y, xy^2 xy
3. $15x^3$, $10x$, $25x^2$ $5x$

Factor completely.

4. $3xy - 12x^2 + 6xy^2$ $3x(y - 4x + 2y^2)$
5. $12x^2 + 14x - 6$ $2(3x - 1)(2x + 3)$
6. $8x^3 + 27$ $(2x + 3)(4x^2 - 6x + 9)$

Motivator

Have students factor 1,800 into prime numbers. $2 \cdot 2 \cdot 2 \cdot 3 \cdot 3 \cdot 5 \cdot 5$ Have students identify the common binomial factor and then factor $a(x - 3) + b(x - 3) - c(x - 3)$. $(a + b - c)(x - 3)$

▰▰TEACHING SUGGESTIONS

Lesson Note

Emphasize that the first step in factoring is to identify the GCF.

5.7 Combined Types of Factoring

Objectives   To factor polynomials completely
To factor polynomials by grouping pairs of terms

Some integers can be factored into the product of more than two integers. For example, $18 = 2 \cdot 9 = 2 \cdot 3 \cdot 3$. In a similar way, some polynomials can be factored into the product of more than two polynomials. This can be illustrated by factoring $75x^4 - 27x^2$ completely as shown below.

Recall that the first step in factoring a polynomial completely is to find the GCF of the terms of the polynomial. The GCF of the terms in $75x^4 - 27x^2$ is $3x^2$. So, $75x^4 - 27x^2 = 3x^2(25x^2 - 9)$.

Second, factor $25x^2 - 9$, if possible. Recall that $25x^2 - 9$ is a difference of squares. So, $3x^2(25x^2 - 9) = 3x^2(5x + 3)(5x - 3)$.

Some fourth-degree trinomials can be factored into two second-degree binomials. Then one or both of the binomials can be factored into the product of two first-degree binomials. This case is shown in Example 1.

EXAMPLE 1   Factor $4y^4 - 17y^2 + 4$ completely.

Plan   The terms of $4y^4 - 17y^2 + 4$ have no common factor, and $4y^4 - 17y^2 + 4$ is not a special product. Therefore, try to factor it into two binomials.

Solution
$$4y^4 - 17y^2 + 4 = (4y^2 - 1)(y^2 - 4)$$

Each factor, $4y^2 - 1$ and $y^2 - 4$, is a difference of squares.

$$(4y^2 - 1)(y^2 - 4) = (2y + 1)(2y - 1)(y + 2)(y - 2)$$

Thus, $4y^4 - 17y^2 + 4 = (2y + 1)(2y - 1)(y + 2)(y - 2)$.

Because the coefficient of the first term in $-18x^2 + 24x - 8$ is negative, it is helpful to factor out -2 rather than 2.

$$
\begin{aligned}
-18x^2 + 24x - 8 &= -2(9x^2 - 12x + 4) \\
&= -2(3x - 2)(3x - 2) \\
&= -2(3x - 2)^2
\end{aligned}
$$

Highlighting the Standards

Standard 14b: In proceeding from one type of factoring to another, students should be taught to rely on general methods rather than isolated mechanics.

Additional Example 1

Factor $9y^4 - 40y^2 + 16$ completely.
$(3y + 2)(3y - 2)(y + 2)(y - 2)$

EXAMPLE 2 Factor $-24y^3 + 81$ completely.

Plan Factor out the GCF, -3. Then factor the resulting polynomial.

Solution
$$-24y^3 + 81 = -3(8y^3 - 27)$$

Factor $8y^3 - 27$ as a difference of cubes: $(2y)^3 - 3^3$.

$$8y^3 - 27 = (2y)^3 - 3^3 = (2y - 3)(4y^2 + 6y + 9)$$

Thus, $-24y^3 + 81 = -3(2y - 3)(4y^2 + 6y + 9)$.

Some polynomials with four terms can be factored by grouping the terms into pairs of terms. The *common monomial factor z* can be factored out in $5xz - 3z$. In a similar way, the *common binomial factor $y - 2$* can be factored out in $5x(y - 2) - 3(y - 2)$.

This technique can be used to factor $5xy - 10x - 3y + 6$ by grouping pairs of terms as follows.

$$(5xy - 10x) + (-3y + 6) = 5x(y - 2) - 3(y - 2)$$
$$= (5x - 3)(y - 2)$$

EXAMPLE 3 Factor $6x^2 + 8x - 15xy - 20y$ by grouping pairs of terms.

Solution
$$6x^2 + 8x - 15xy - 20y = (6x^2 + 8x) + (-15xy - 20y)$$
$$= 2x(3x + 4) - 5y(3x + 4)$$
$$= (2x - 5y)(3x + 4)$$

Here are the steps for factoring a polynomial completely.
1. Factor out the greatest common factor (GCF).
2. Factor the resulting polynomial, if possible.
3. Factor each polynomial factor, if possible.

Classroom Exercises

Determine whether each expression is factored completely. If not, complete the factoring.

1. $11 \cdot 17 \cdot 21$
2. $19 \cdot 29 \cdot 37$ Yes
3. $7(3x - 6y)$ No; $21(x - 2y)$
4. $-7(a^3 + 2ab)$
   No; $-7a(a^2 + 2b)$
5. $4(x^2 - 2x + 3)$ Yes
6. $5c(9c^2 + 1)$ Yes

Factor out a negative integer; then factor completely.

7. $-12a + 10b - 6c$
   $-2(6a - 5b + 3c)$
8. $-6x^2 + 45xy - 21y^2$
   $-3(2x - y)(x - 7y)$
9. $-4x^2 + 9$
   $-1(2x + 3)(2x - 3)$

Math Connections

Geometric Models: Some factoring situations can be visualized geometrically. For example, $2a^2 - 8$ can be thought of as the volume of a rectangular solid of dimensions $2 \times a \times a$ from which a cube of dimensions $2 \times 2 \times 2$ has been removed. In its factored form, the expression is $2(a - 2)(a + 2)$, which can be represented by a rectangular solid of dimensions $2 \times (a - 2) \times (a + 2)$. Thus, the same volume is represented by two different solids that correspond to the two different algebraic expressions.

Critical Thinking Questions

Analysis: Ask students to make geometrical interpretations of selected algebraic expressions in their factored and unfactored forms. In addition to the example shown above in *Math Connections*, you can suggest examples based on factoring relationships such as $a^2 + 2ab + b^2 = (a + b)(a + b)$ and $a^3 - b^3 = (a - b)(a^2 + ab + b^2)$.

Checkpoint

Factor completely.

1. $-9n^2 + 30n - 25$ $-1(3n - 5)^2$
2. $10x^2 - 5x - 15$ $5(2x - 3)(x + 1)$
3. $27y^4 + 64y$ $y(3y + 4)(9y^2 - 12y + 16)$
4. $x^4 - 22x^2 - 75$ $(x - 5)(x + 5)(x^2 + 3)$
5. $15x^2 - 6xy - 20x + 8y$
   $(3x - 4)(5x - 2y)$

Additional Example 2

Factor $-4x^3 - 32$ completely.
$-4(x + 2)(x^2 - 2x + 4)$

Additional Example 3

Factor $6x^2 - 4xy - 15x + 10y$ by grouping pairs of terms. $(2x - 5)(3x - 2y)$

185

Closure

Ask students to give the steps for factoring a polynomial completely.   See page 185.

See page 185.

■■■FOLLOW UP

Guided Practice

Classroom Exercises 1–9

Independent Practice

A Ex. 1–22, **B** Ex. 23–40, **C** Ex. 41–50

Basic: WE 1–23 odd

Average: WE 1–39 odd

Above Average: WE 11–49 odd

Additional Answers

Classroom Exercises

1. No; $11 \cdot 17 \cdot 3 \cdot 7$

Written Exercises

1. $4(x - 4)(x - 3)$
2. $3(n + 5)(n - 3)$
3. $3(2y + 1)(2y - 1)$
4. $2(5c + 3)(5c - 3)$
5. $4(n - 5)^2$
6. $5(x + 4)^2$
7. $(x + 2)(x - 2)(x + 3)(x - 3)$

See pages 194–195 for answers to Written Ex. 8–15, 19–24, 31, 32, 41–44.

Written Exercises

Factor each polynomial completely.

1. $4x^2 - 28x + 48$
2. $3n^2 + 6n - 45$
3. $12y^2 - 3$
4. $50c^2 - 18$
5. $4n^2 - 40n + 100$
6. $5x^2 + 40x + 80$
7. $x^4 - 13x^2 + 36$
8. $y^4 - 5y^2 + 4$
9. $x^4 - 5x^2 - 36$
10. $y^4 - 3y^2 - 4$
11. $2y^3 + 54$
12. $3n^3 - 375$
13. $-a^2 + 10a - 25$
14. $-x^2 - 8x - 16$
15. $-4n^2 - 4n + 3$
16. $-6m^2 + 11m - 3$
    $-1(3m - 1)(2m - 3)$
17. $-y^3 + 1$
    $-1(y - 1)(y^2 + y + 1)$
18. $-x^3 - 64$
    $-1(x + 4)(x^2 - 4x + 16)$

Factor each polynomial by grouping pairs of terms.

19. $8xy - 20x + 6y - 15$
20. $6cd + 14c - 15d - 35$
21. $8x^2 - 2xy + 12x - 3y$
22. $20c^2 - 15c - 4cd + 3d$
23. $3x^3 - 4x^2 + 15x - 20$
24. $10y^3 - 25y^2 - 16y + 40$
25. $6c^3 - 4c^2 - 3cd + 2d$
    $(2c^2 - d)(3c - 2)$
26. $8a^4 + 28a^2 - 6a^2b - 21b$
    $(4a^2 - 3b)(2a^2 + 7)$

Factor each polynomial completely.

27. $10a^3 - 5a^2 - 50a$ $5a(2a - 5)(a + 2)$
28. $2a^3 - 50ab^2$ $2a(a + 5b)(a - 5b)$
29. $3x^3y - 6x^2y + 3xy$ $3xy(x - 1)^2$
30. $18a^3b + 60a^2b^2 + 50ab^3$ $2ab(3a + 5b)^2$
31. $9y^4 - 40y^2 + 16$
32. $25y^4 - 101y^2 + 4$
33. $9y^4 + 32y^2 - 16$ $(3y + 2)(3y - 2)(y^2 + 4)$
34. $25y^4 - 21y^2 - 4$ $(y + 1)(y - 1)(25y^2 + 4)$
35. $24x^3 + 375$ $3(2x + 5)(4x^2 - 10x + 25)$
36. $-54y^3 + 128$ $-2(3y - 4)(9y^2 + 12y + 16)$
37. $-y^4 + 64y$ $-y(y - 4)(y^2 + 4y + 16)$
38. $2x^4 + 2,000x$ $2x(x + 10)(x^2 - 10x + 100)$
39. $-8ac^2 - 8ac - 2a$ $-2a(2c + 1)^2$
40. $-3xy^2 + 27x$ $-3x(y + 3)(y - 3)$

Factor completely.

41. $a^3 - a^2b - a^2b^2 + ab^3$
42. $6c^3 + 12c^2d - 6cd^2 - 12d^3$
43. $4c^8 - 13c^4 + 9$
44. $36d^8 - 13d^4 + 1$
45. $x^{a+5} + x^3$ $x^3(x^{a+2} + 1)$
46. $x^{6c} - x^{5c}$ $x^{5c}(x^c - 1)$
47. $y^{n+6} + y^{n+4}$ $y^{n+4}(y^2 + 1)$
48. $x^{6c} - 9x^{4c}$ $x^{4c}(x^c + 3)(x^c - 3)$
49. $x^{4n+2} - 8x^{2n+2} + 16x^2$
    $x^2(x^n - 2)^2(x^n + 2)^2$
50. $y^{4n} - y^n$ $y^n(y^n - 1)(y^{2n} + y^n + 1)$

Mixed Review

Find the slope of the line described by each equation.   *3.5*

1. $3x - 5y = 4$ $\frac{3}{5}$
2. $8y - 16 = 0$ 0
3. $-3x = 6$
   Undefined
4. $\frac{1}{2}x + \frac{1}{3}y = 1$ $-\frac{3}{2}$

Enrichment

Explain this strategy for factoring trinomials in the form $ax^2 + bx + c$.

1. Find the product ac.
2. Factor ac into two numbers whose sum is b.
3. Using these two numbers as coefficients, rewrite the middle term as two terms.
4. Factor by grouping.
   For example, factor $8x^2 + 10x + 3$.

Think: $8 \cdot 3 = 24$
$6 \cdot 4 = 24$ and
$6 + 4 = 10$
Write: $8x^2 + 4x + 6x + 3$
$4x(2x + 1) + 3(2x + 1)$
$(2x + 1)(4x + 3)$

Now have the students use this strategy to factor the following polynomials.

1. $x^2 + 10x + 21$ $(x + 3)(x + 7)$
2. $6x^2 - 29x + 35$ $(3x - 7)(2x - 5)$
3. $12x^2 + 23x + 5$ $(4x + 1)(3x + 5)$

5.8 Dividing Polynomials

Objectives To divide polynomials in one variable by binomials
To determine whether a binomial is a factor of a polynomial

Teaching Resources

Quick Quizzes 38
**Reteaching and Practice
Worksheets** 38

A divide-multiply-subtract cycle can be used to divide 678 by 32.

Ordinary form	Divide-multiply-subtract (D-M-S) cycle		Expanded form
21	$600 \div 30 = 20$	(D)	$20 + 1$
$32\overline{)678}$	$20(30 + 2) = 600 + 40$	(M)	$30 + 2\overline{)600 + 70 + 8}$
$\underline{640}$	$(600 + 70) - (600 + 40) = 30$	(S)	$\underline{600 + 40}$
38	$30 \div 30 = 1$	(D)	$30 + 8$
$\underline{32}$	$1(30 + 2) = 30 + 2$	(M)	$\underline{30 + 2}$
6	$(30 + 8) - (30 + 2) = 6$	(S)	6

So, $678 \div 32 = 21$, remainder 6. Check: $32 \times 21 + 6 = 678$.

32 is not a factor of 678 because there is a nonzero remainder.

The divide-multiply-subtract cycle shown in expanded form above can be used to divide a polynomial in one variable by a binomial.

EXAMPLE 1 Divide $5x^2 - 11x - 21$ by $x - 3$ to find the polynomial quotient and remainder. Is $x - 3$ a factor of $5x^2 - 11x - 21$?

Solution

$$\begin{array}{r} 5x + 4 \\ x - 3\overline{)5x^2 - 11x - 21} \\ \underline{5x^2 - 15x} \\ 4x - 21 \\ \underline{4x - 12} \\ -9 \end{array}$$

$5x^2 \div x = 5x$	(D)
$5x(x - 3) = 5x^2 - 15x$	(M)
$(5x^2 - 11x) - (5x^2 - 15x) = 4x$	(S)
$4x \div x = 4$	(D)
$4(x - 3) = 4x - 12$	(M)
$(4x - 21) - (4x - 12) = -9$	(S)

The polynomial quotient is $5x + 4$, and the remainder is -9.

Check $(x - 3)(5x + 4) + (-9) = 5x^2 - 11x - 21$

So, $x - 3$ is not a factor of $5x^2 - 11x - 21$ because there is a nonzero remainder.

■■■ **GETTING STARTED**

Prerequisite Quiz

Simplify.

1. $10x^3 \div 5x$ $2x^2$
2. $12x^3 - 4x^2 - (12x^3 - 6x^2)$ $2x^2$
3. $14n^3 - 5n^2 - (14n^3 + 7n^2)$ $-12n^2$
4. $6y^3 + 0 - (6y^3 + 2y^2)$ $-2y^2$

Motivator

Have students explain how to verify that $657 \div 32$ is 20 with remainder 17 without dividing.   Show that $32 \cdot 20 + 17 = 657$. Ask students if 32 is a factor of 657.   No. Ask them why 32 is not a factor of 657. Because there is a nonzero remainder.

■■■ **TEACHING SUGGESTIONS**

Lesson Note

An understanding of the terms *divisor*, *dividend*, *quotient*, and *remainder* is important in this work.

Additional Example 1

Divide $3x^3 + 4x^2 + 3x + 14$ by $x + 2$ to find the polynomial quotient and remainder.
Is $x + 2$ a factor of $3x^3 + 4x^2 + 3x + 14$?
$3x^2 - 2x + 7$; 0; yes

Highlighting the Standards

Standard 4d: Once again, the connections between arithmetic and algebra are presented. Division of polynomials will lose much of its mystery once students see that the process is similar to division with whole numbers.

Math Connections

Algorithms: The divide-multiply-subtract cycle that is central to the process of dividing one polynomial by another is an example of an *algorithm*, a fixed set of steps that are used to solve an arithmetic or algebraic problem. Other examples of algorithms are the Euclidean Algorithm for finding the greatest common divisor of two numbers and the algorithm for obtaining the square root of a number to any desired number of decimal places.

Critical Thinking Questions

Analysis: Ask students how to write an algebraic expression for the answer to Example 1.   By thinking of similar division problems from arithmetic, such as $\frac{5}{3} = 1 + \frac{2}{3}$, or $1\frac{2}{3}$, they should be able to reason that the expression in question must be $5x + 4 + \frac{-9}{x - 3}$.

Checkpoint

Simplify.

1. $(35x^4 + 14x^2) \div (7x)$ $5x^3 + 2x$

2. $\dfrac{-12cd^3 + 15c^4d^2 - 21cd}{3cd}$
   $-4d^2 + 5c^3d - 7$

Divide to find the polynomial quotient and remainder. Is the divisor a factor of the dividend?

3. $(6y^2 - 8y - 24) \div (y - 3)$ $6y + 10$; 6; no

4. $(12a^3 - 14 - 35a - 13a^2) \div (3a + 2)$
   $4a^2 - 7a - 7$; 0; yes

EXAMPLE 2   Divide $6x^3 - 8 - 7x^2 - 26x$ by $3x + 4$. Is $3x + 4$ a factor of the dividend?

Plan   Arrange the terms of the dividend in descending order of the exponents.

Solution

$$
\begin{array}{r}
2x^2 - 5x - 2 \\
3x + 4\overline{)6x^3 - 7x^2 - 26x - 8} \\
\underline{6x^3 + 8x^2} \\
-15x^2 - 26x \\
\underline{-15x^2 - 20x} \\
-6x - 8 \\
\underline{-6x - 8} \\
0
\end{array}
$$

Check $(3x + 4)(2x^2 - 5x - 2) + 0 = 6x^3 - 8 - 7x^2 - 26x$

Thus, the polynomial quotient is $2x^2 - 5x - 2$, the remainder is 0, and $3x + 4$ is a factor of the dividend.

If there are "missing terms" in the dividend, use the coefficient 0 and replace the missing terms. To divide $4x^3 - 5x - 35$ by $2x - 5$, first replace the missing x^2-term with $0x^2$.

$$
\begin{array}{r}
2x^2 + 5x + 10 \\
2x - 5\overline{)4x^3 + 0x^2 - 5x - 35} \\
\underline{4x^3 - 10x^2} \\
10x^2 - 5x \\
\underline{10x^2 - 25x} \\
20x - 35 \\
\underline{20x - 50} \\
15
\end{array}
$$

Classroom Exercises

Simplify each quotient.

1. $(16x^5 - 24x^3) \div (8x)$ $2x^4 - 3x^2$

2. $(7cd^2 + 9c^2d) \div (cd)$ $7d + 9c$

3. $(15a^2b^2 - 12a^2b) \div (3a^2b)$ $5b - 4$

4. $\dfrac{25m^4 - 20m^3 - 15m^2}{5m^2}$ $5m^2 - 4m - 3$

5. $\dfrac{-8cd^4 + 10c^5d^2 - 14c^3d^3}{2cd}$ $-4d^3 + 5c^4d - 7c^2d^2$

6. $\dfrac{28x^3y^3 + 16x^2y^2 - 24xy^3}{4xy^2}$ $7x^2y + 4x - 6y$

7. Divide $(2c^3 - 3c^2 + 2c - 1)$ by $(c - 1)$. $2c^2 - c + 1$; R:0

8. Is the divisor in Classroom Exercise 7 a factor of the dividend?   Yes

---

Additional Example 2

Divide $8y^3 + 15 - 26y^2 + 5y$ by $2y - 5$ to find the polynomial quotient and remainder. Is $2y - 5$ a factor of the dividend?
$4y^2 - 3y - 5$; -10; no

Written Exercises

Divide to find the polynomial quotient and the remainder. Is the divisor a factor of the dividend?

1. $(3c^2 + c - 3) \div (c + 1)$ $\quad 3c - 2, -1;$ no
2. $(y^2 + 5y + 5) \div (y + 2)$ $\quad y + 3, -1;$ no
3. $(8x^2 - 6x - 20) \div (x - 2)$
4. $(2n^2 + 23n + 45) \div (2n + 5)$
5. $(4y^3 + 9y^2 + y + 36) \div (y + 3)$
6. $(9x - 17x^2 - 9 + 5x^3) \div (x - 3)$
7. $(16n^3 - 33n + 20) \div (4n - 5)$
8. $(9y^4 + 5y^2 - 12) \div (3y + 2)$
9. $(3a^4 + a + 20 - 49a^2) \div (a + 4)$
10. $(8y^4 - 22y^2 + 9) \div (2y - 3)$
11. $(y^3 - 8) \div (y - 2)$ $\quad y^2 + 2y + 4, 0;$ yes
12. $(125x^3 + 64) \div (5x + 4)$
13. $(3c^4 - 2c^3 + 8c - 48) \div (c^2 - 4)$
14. $(4x^4 + 6x^3 - 15x - 5) \div (2x^2 - 5)$
15. $(y^6 - 3y^4 + y^2 - 8) \div (y^4 + 1)$
16. $(9x^6 + 3x^5 + x^2 - 4) \div (3x^3 + 1)$
17. $\left(3x^3 + \frac{1}{2}x^2 - 5x + 2\right) \div \left(x - \frac{1}{2}\right)$ $\quad 3x^2 + 2x - 4, 0;$ yes
18. $\left(x^3 - \frac{7}{3}x^2 - x + \frac{1}{2}\right) \div \left(x + \frac{2}{3}\right)$ $\quad x^2 - 3x + 1, -\frac{1}{6};$ no
19. $(x^5 + y^5) \div (x + y)$ $\quad x^4 - x^3y + x^2y^2 - xy^3 + y^4, 0;$ yes
20. $(x^7 - y^7) \div (x - y)$ $\quad x^6 + x^5y + x^4y^2 + x^3y^3 + x^2y^4 + xy^5 + y^6, 0;$ yes
21. $(x^{3m} - 4x^{2m} + 3x^m - 12) \div (x^m - 4)$ $\quad x^{2m} + 3, 0;$ yes
22. $(24x^3y + 7x^2y^2 - 6x^2y - 6xy^3 - 4xy^2) \div (3x + 2y)$ $\quad 8x^2y - 3xy^2 - 2xy, 0;$ yes
23. Find the value of m that will make $x - 2$ a factor of
   $3x^4 + 10x^3 - 40x^2 + mx - 12.$ $\quad 22$

Mixed Review

1. A stack of 20 coins contains dimes and quarters. Given that the stack has a face value of \$2.75, find the number of coins of each type. $\quad$ *4.4* 5 quarters, 15 dimes
2. One of two supplementary angles measures 45 less than twice the measure of the other angle. Find the measure of each angle. $\quad$ *4.5* 75, 105
3. If $f(x) = 3x + 7$, what is $f\left(a + \frac{2}{3}\right)$? $\quad$ *3.2* $3a + 9$
4. If $g(x) = 2x^2 + 1$, what is $g(c - 3)$? $\quad$ *5.5* $2c^2 - 12c + 19$

◤ Brainteaser

The sum of the volumes of two cubes is equal to the sum of the lengths of all their edges. For each cube, the length of an edge is a whole number. Find the dimensions of each cube. $\quad$ 2 by 2 by 2, 4 by 4 by 4

Enrichment

Mrs. Vlahos left a bowl of grapes in the refrigerator for after-school snacks for her three children. Clara came home first and gave one grape to Spot, their dog. Then she took a third of the remaining grapes for herself. A few minutes later, Jorge came home, gave one grape to Spot, and took a third of the remaining grapes for himself. Finally, Stella came home, gave one grape to Spot, and took a third of the remaining grapes for herself. If the bowl originally held fewer than 100 grapes, what are the possible numbers of grapes that Mrs. Vlahos left for the children? $\quad$ 25, 52, 79

Teaching Resources

Problem Solving Worksheet 5
Quick Quizzes 39
Reteaching and Practice
   Worksheets 39
Transparency 13

▰▰▰ GETTING STARTED

Prerequisite Quiz

Let $f(x) = 2x + 5$ and $g(x) = x^2 - 3$.
Find each value.

1. $f\left(-\frac{1}{2}\right)$ 4

2. $g(-5)$ 22

3. $f(c + 6)$ $2c + 17$

4. $g(a - 5)$ $a^2 - 10a + 22$

Motivator

Refer to the Prerequisite Quiz above. Have
students determine the value of $g(f(x))$.

$g(f(x)) = g(2x + 5)$
$\quad\quad = (2x + 5)^2 - 3$
$\quad\quad = 4x^2 + 20x + 25 - 3$
$\quad\quad = 4x^2 + 20x + 22$
$\quad\quad = 2(2x^2 + 10x + 11)$

▰▰▰ TEACHING SUGGESTIONS

Lesson Note

Call to the students' attention that two
functions f and g can be described by
equations like $f(x) = x + 4$ and $g(x) =$
$x^2 - 3$, respectively.

However, the composition of f with g is a new
function, where $f(g(x)) = f(x^2 - 3) = x^2 -$
$3 + 4 = x^2 + 1$. Thus, the composition of f
with g is described by one equation, $f(g(x))$
$= x^2 + 1$, which can also be written $fg(x)$.

Highlighting the Standards

Standard 6b: The composition of functions
is a step toward a higher degree of
abstraction. It offers a taste of what lies
ahead in advanced mathematics.

5.9 Composition of Functions

Objectives   To find composite functions
To evaluate composite functions

Two functions can often be combined into one. The resulting function
is called a *composite function.*

Consider a function f: *adding* 6, followed by a second function g: *multiplying* by 3. Apply the function f to the set $\{1, 2, 3\}$ and then apply
the function g to the resulting set.

x	f: adding 6	$f(x)$	g: multiplying by 3	$g(f(x))$
1	$1 + 6 = 7$	7 $\longrightarrow$	$7 \times 3 = 21$	21
2	$2 + 6 = 8$	8 $\longrightarrow$	$8 \times 3 = 24$	24
3	$3 + 6 = 9$	9 $\longrightarrow$	$9 \times 3 = 27$	27

Notice that the result of applying g to $f(x)$ is written as $g(f(x))$. Another
way of writing the combined function is $(f \circ g)(x)$, or $f \circ g$.

Definition   If f and g are functions such that the range of g is in the domain of f,
then the function whose value at x is $g(f(x))$ is called the **composite** of
the functions f and g. The operation of combining the two is called
composition.

It is important that the results of applying the first function be values
that will work for the second function. For instance, if the second function is the square root function, then negative values will not work for
it. In other words, the *range* of the first function must be in the
domain of the second function.

EXAMPLE 1   Let $f(x) = x + 5$ and $g(x) = x^2 - 10$.
   a. Define the composite function $g(f(x))$ as an
   expression in terms of the variable x.
   b. Evaluate the expression for $x = 5$.

Solutions   **a.** $g(f(x)) = g(x + 5)$ $\longleftarrow$ Substitute $x + 5$ for $f(x)$.
   $g(f(x)) = (x + 5)^2 - 10$ $\longleftarrow$ THINK: The function g
   $\quad\quad = x^2 + 10x + 25 - 10$ squares a number and
   $\quad\quad = x^2 + 10x + 15$ subtracts 10.

Additional Example 1

Given functions f and g such that $f(x) =$
$x + 3$ and $g(x) = x^2 - 5$, determine:

a. $f(g(x))$ $f(g(x)) = x^2 - 2$
b. $g(f(x))$ $g(f(x)) = x^2 + 6x + 4$
c. $f(g(-4))$ 14
d. $g(f(-4))$ -4

b. $g(f(x)) = x^2 + 10x + 15$ ← Substitute $x = 5$.

$g(f(5)) = 5^2 + 10(5) + 15$

$= 25 + 50 + 15 = 90$

EXAMPLE 2 Let $f(x) = x - 10$ and $g(x) = 2x + 3$. Define the composite functions $g(f(x))$ and $f(g(x))$. Evaluate each for $x = 4$. Are the results the same?

Solutions

$g(f(x)) = g(x - 10)$
$= 2(x - 10) + 3$
$= 2x - 20 + 3$
$= 2x - 17$

$g(f(4)) = 2(4) - 17$
$= 8 - 17$
$= -9$

$f(g(x)) = f(2x + 3)$
$= (2x + 3) - 10$
$= 2x - 7$

$f(g(4)) = 2(4) - 7$
$= 8 - 7$
$= 1$

The results in this case are not the same.

Classroom Exercises

Given the functions $t = \{(7,5), (5,7), (4,-6), (-3,-2)\}$, $s = \{(5,-6), (4,-3), (-3,8), (-6,5)\}$, and $f(x) = 2x$, find each value if it exists.

1. $s(t(4))$ 5
2. $t(s(4))$ -2
3. $s(t(7))$ -6

4. $t(s(-6))$ 7
5.   None
6. $t(s(-3))$ None

7. $s(s(4))$ 8
8. $t(t(7))$ 7
9. $f(t(5))$ 14

Written Exercises

Given $f(x) = 3x - 5$ and $g(x) = x^2 + 2$, find each value.

1. $g(f(2))$ 3
2. $f(g(2))$ 13
3. $g(f(-3))$ 198
4. $f(g(-3))$ 28

5. $f(g(\frac{1}{3}))$ $\frac{4}{3}$
6. $f(f(-\frac{1}{3}))$ -23
7. $g(g(2))$ 38
8. $g(f(a))$

9. $f(g(a))$ $3a^2 + 1$
10. $f(f(a))$ $9a - 20$
11. $g(g(-a))$
12. $f(g(5c))$

Given $s(x) = x^2 + 4x$ and $t(x) = -2x + 3$, find each value.

13. $s(t(2))$ -3
14. $t(s(-5))$ -7
15. $t(s(2.5))$ -29.5
16. $s(t(4.5))$ 12

17. $t(s(a))$ $-2a^2 - 8a + 3$
18. $s(t(a))$ $4a^2 - 20a + 21$
19. $t(s(3c))$
20. $s(t(-5c))$

21. $t(s(c - 5))$ $-2c^2 + 12c - 7$
22. $s(s(-4))$ 0
23. $s(s(2))$ 192
24. $t(t(a + b))$ $4a + 4b - 3$

Math Connections

Application: In 1988, parts of the United States suffered a severe drought. The lack of rain was said to be a function of unusual upper air movement. The poor crops in Iowa in 1988 were said to be a function of the lack of rain. Thus, it can be concluded that the poor crops in Iowa were a function (result) of unusual upper air movement. This is an illustration of a composite function.

Critical Thinking Questions

Analysis: Ask students whether they can devise two functions f and g such that $f(g(x)) = g(f(x))$. One example is $f(x) = x + 4$, $g(x) = x - 4$. If f and g are inverses of one another, then $fg = gf$.

Checkpoint

Let $f(x) = 4x - 3$, $g(x) = x^2 + 4$, $h = \{(6,9), (2,-1), (9,6)\}$ and $j = \{(9,2), (-1,4), (6,2)\}$. Find each value.

1. $h(j(9))$ -1
2. $j(h(6))$ 2
3. $g(j(-1))$ 20
4. $f(g(3.5))$ 62

Closure

Have students describe how to find $f(g(x))$ where $f(x) = x^2 - 2x + 4$ and $g(x) = 3x - 1$.

1. Replace the variables in $f(x)$ by $3x - 1$.
$f(g(x)) = f(3x - 1)$
$= (3x - 1)^2 - 2(3x - 1) + 4$
2. Simplify.
$f(g(x)) = 9x^2 - 6x + 1 - 6x + 2 + 4$
$= 9x^2 - 12x + 7$

Additional Example 2

For $f(x) = \frac{1}{2}x + \frac{1}{2}$ and $g(x) = 2x - 1$, show that $f(g(x)) = g(f(x))$.
$f(g(x)) = x$ and $g(f(x)) = x$; therefore, $f(g(x)) = g(f(x))$.

Guided Practice

Classroom Exercises 1–9

Independent Practice

A Ex. 1–28, **B** Ex. 29–38, **C** Ex. 39–43

Basic: WE 1–29 odd, Application, Mixed Problem Solving

Average: WE 9–37 odd, Application, Mixed Problem Solving

Above Average: WE 15–43 odd, Application, Mixed Problem Solving

Additional Answers

Written Exercises

8. $9a^2 - 30a + 27$
11. $a^4 + 4a^2 + 6$
12. $75c^2 + 1$
19. $-18c^2 - 24c + 3$
20. $100c^2 + 100c + 21$
25. $g(f(x)) = 3x + 24$
    $f(g(x)) = 3x + 8$
26. $g(f(x)) = x + 3$
    $f(g(x)) = x + 3$
27. $g(f(x)) = 35x - 19$
    $f(g(x)) = 35x + 41$
28. $g(f(x)) = x - 24$
    $f(g(x)) = x + 24$

See page 193 for the answers to Written Ex. 29–40.

For each pair of functions, f and g, determine $g(f(x))$ and $f(g(x))$.

25. $f(x) = x + 8$, $g(x) = 3x$
26. $f(x) = x + 7$, $g(x) = x - 4$
27. $f(x) = 5x - 4$, $g(x) = 7x + 9$
28. $f(x) = 12 - x$, $g(x) = -x - 12$
29. $f(x) = x + 3$, $g(x) = x^2 - 1$
30. $f(x) = 2x - 1$, $g(x) = x^2 + x$
31. $f(x) = 2x^2 + 6$, $g(x) = 3x - 2$
32. $f(x) = 3x^2 - 4x$, $g(x) = 2x + 5$
33. $f(x) = 6x$, $g(x) = \frac{1}{3}x$
34. $f(x) = 4x + 12$, $g(x) = \frac{1}{4}x - 3$
35. $f(x) = \frac{1}{5}x - 7$, $g(x) = 5x + 35$
36. $f(x) = \frac{2}{3}x + 6$, $g(x) = \frac{3}{2}x - 9$

Determine $s(t(x))$ and $t(s(x))$. Then use $s(t(x))$ and $t(s(x))$ to find $s(t(10))$ and $t(s(10))$.

37. $s(x) = x^2 - x - 12$, $t(x) = x + 3$
38. $s(x) = x + 2$, $t(x) = (2x - 3)^2$

For Exercises 39–43, $f(x) = x^2 - 3$, $g(x) = 5x + 2$, and $h(x) = 4 - x$. Write an equation for each composition.

39. $f \circ (g \circ h)$ (HINT: Find $f(g(h(x)))$.)
40. $(f \circ g) \circ h$ (HINT: Find $f \circ g$ first.)
41. $f \circ (h \circ g)$ $25x^2 - 20x + 1$
42. $h \circ (g \circ f)$ $-5x^2 + 17$
43. $(h \circ g) \circ f$ $-5x^2 + 17$

Mixed Review

Find the solution set for each sentence. *2.2, 2.3*

1. $-7 \leq 5x - 27 < 18$ $\{x \mid 4 \leq x < 9\}$
2. $|x + 5| > 2$ $\{x \mid x < -7 \text{ or } x > -3\}$
3. $2x - 1 < 5$ or $1 - x < 5$
   {Real numbers}
4. $2x - 1 < 5$ and $1 - x < 5$
   $\{x \mid -4 < x < 3\}$

Application: *Measures of Motion*

Scientists often express measures of motion using negative exponents. For instance, velocity in units of meters per second (m/s) can be expressed as ms^{-1}, and acceleration can be written as units of ms^{-2}. When solving motion problems, exponential notation makes it easy to find the units of the answer. For example, if velocity is divided by acceleration, the result is time (in seconds).

$$\frac{m}{s} \div \frac{m}{s^2} = \frac{m}{s} \cdot \frac{s^2}{m} = ms^{-1} \cdot m^{-1}s^2 = s$$

Find the units of the answer. Use meters and seconds.

1. acceleration divided by velocity multiplied by the square of time Seconds
2. velocity squared multiplied by time divided by distance Meters per second

---

Enrichment

Write the functions $f(x) = x - 1$ and $g(x) = x + 1$ on the chalkboard and have the students find:

$f(g(3))$ 3
$f(g(5))$ 5
$f(g(x))$ x
$g(f(3))$ 3
$g(f(5))$ 5
$g(f(x))$ x

Stress that for any value of x, when these conditions are satisfied, $f(x)$ and $g(x)$ are called *inverse functions*. Not all functions have inverses, but the following functions do.

Challenge the students to find the inverse, $g(x)$, for each $f(x)$.

1. $f(x) = 10x$ $g(x) = \frac{x}{10}$
2. $f(x) = 5x - 2$ $g(x) = \frac{x + 2}{5}$
3. $f(x) = 2 - 3x$ $g(x) = \frac{2 - x}{3}$
4. $f(x) = x^3$ $g(x) = \sqrt[3]{x}$

Mixed Problem Solving

1. How many liters of water must be added to 30 liters of a 28% salt solution to dilute it to a 12% solution? 40 l

2. How much water must be evaporated from 15 liters of an 18% salt solution to obtain a 27% solution? 5 l

3. Find four consecutive even integers such that 8 times the smallest integer is the same as 5 times the largest integer. 10, 12, 14, 16

4. Timothy was 7 years old when Ada was born. Hal, who is 5 years older than Timothy, is now 3 times as old as Ada. How old is each person now?
   Ada: 6, Tim: 13, Hal: 18

5. How many milliliters of water must be evaporated from 200 ml of an 18% iodine solution to obtain a 30% solution? 80 ml

6. Skim milk contains no butterfat. How many liters of skim milk must be added to 800 liters of milk that is 3.2% butterfat to obtain milk that is 2.0% butterfat? 480 l

7. If y varies directly as the square of x, and $y = 12$ when $x = 0.6$, what is y when $x = 0.9$? 27

8. A triangle has a perimeter of 58 cm, and one side is 22 cm long. Find the lengths of the other two sides given that their ratio is 4:5. 16 cm, 20 cm

9. The measures of two supplementary angles are in the ratio of 3:5. Find the measure of the complement of the smaller angle. $22\frac{1}{2}$

10. If an object is sent upward, what initial vertical velocity is required to reach a height of 275 m at the end of 5 s? Use the formula $h = vt - 5t^2$.
    80 m/s

11. An airplane flew 480 mi in 2 h aided by a tailwind. It would have flown the same distance in $2\frac{1}{2}$ h against the same wind. Find the wind speed. 24 mi/h

12. A car left a rest area and drove south on a highway at 45 mi/h. Ten minutes later, a motorcycle left the same rest area and headed south at 60 mi/h. How long did it take the motorcycle to overtake the car? 30 min

13. A collection of 35 coins contains dimes and quarters. If there were 7 more dimes, the face value of the dimes would equal the face value of the quarters. How many quarters are in the collection? 12

14. A dealer planned to buy 30 radios of a popular model. When the price of each radio was reduced by $3, the dealer was able to buy 3 more radios for the same total cost. What was the total cost of the radios? $990

15. If $8,000 were invested, part at 9% and the remainder at 11%, the amount of simple interest earned at the end of 5 years would be $4,150. Find the amount invested at 9%. $2,500

16. Some almonds worth $5.40/kg are mixed with some peanuts worth $3.60/kg to make 24 kg of a mixture worth $4.80/kg. How many kilograms of peanuts are in the mixture? 8 kg

Chapter Review

17. $6xh + 3h^2 - 2h$

18. $(5xy + 6)(5xy - 6)$

19. $(4c - 5)^2$

21. $(x + 2)(x^2 - 2x + 4)$

22. $(5y - 3)(25y^2 + 15y + 9)$

23. $(x^2 + 3)(x^2 - 3)$

Additional Answers, page 186

8. $(y + 2)(y - 2)(y + 1)(y - 1)$

9. $(x + 3)(x - 3)(x^2 + 4)$

10. $(y + 2)(y - 2)(y^2 + 1)$

11. $2(y + 3)(y^2 - 3y + 9)$

12. $3(n - 5)(n^2 + 5n + 25)$

13. $-1(a - 5)^2$

14. $-1(x + 4)^2$

15. $-1(2n - 1)(2n + 3)$

19. $(4x + 3)(2y - 5)$

20. $(2c - 5)(3d + 7)$

21. $(4x - y)(2x + 3)$

Chapter 5 Review

Key Terms

base (p. 161)
binomial (p. 170)
composite function (p. 190)
degree of a monomial (p. 170)
degree of a polynomial (p. 170)
exponent (p. 161)
exponential equation (p. 163)
greatest common factor (p. 175)

linear polynomial (p. 170)
monomial (p. 170)
perfect-square trinomial (p. 178)
polynomial (p. 170)
quadratic polynomial (p. 170)
scientific notation (p. 166)
trinomial (p. 170)

Key Ideas and Review Exercises

5.1, *Laws of Exponents* (For restrictions, see Lessons 5.1 and 5.2.)
5.2

$$b^m \cdot b^n = b^{m+n} \qquad (a^m)^n = a^{m \cdot n} \qquad \left(\frac{a}{b}\right)^n = \frac{a^n}{b^n}$$

$$\frac{b^m}{b^n} = b^{m-n} \qquad (ab)^n = a^n \cdot b^n$$

For negative exponents and the zero exponent:

$$b^{-n} = \frac{1}{b^n} \qquad\qquad \frac{1}{b^{-n}} = b^n \qquad\qquad b^0 = 1$$

Simplify each expression. Use positive exponents.

1. $10c(2c^2)^3$ $80c^7$

2. $5x^{-5} \cdot 3x^3 \cdot 2x^0$ $\dfrac{30}{x^2}$

3. $\dfrac{18x^{-5}y^4}{12x^{-2}y^{-1}}$ $\dfrac{3y^5}{2x^3}$

Find the value of each expression and state it in ordinary notation.

4. $\dfrac{7.5 \times 10^2}{2.5 \times 10^{-3}}$ 300,000

5. $(8.4 \times 10^3) \times (2.5 \times 10^{-7})$ 0.0021

5.3   Polynomials can be added, subtracted, and multiplied.

Simplify. Then give the degree of the polynomial in the answer.

6. $(7xy - xy^2 - x^2y) - (2xy + 5xy^2 - 6x^2y)$ $5xy - 6xy^2 + 5x^2y$; degree: 3

7. $(7x^2 + 3y)(2x^2 - y)$ $14x^4 - x^2y - 3y^2$; degree: 4

5.4   Some trinomials can be factored into binomials by reversing the FOIL method for multiplying.

Factor each trinomial into two binomials, if possible.

8. $x^2 - 5xy - 36y^2$
$(x + 4y)(x - 9y)$

9. $12x^4 + x^2 - 20$
$(3x^2 + 4)(4x^2 - 5)$

10. $c^2 - 3c - 16$
Not possible

194   Chapter 5 Review

11. Solve $ax - t = 5x + 6$ for x. $\dfrac{t + 6}{a - 5}$

5.5,   *Special Products and Special Factors*
5.6,
5.7 $a^2 - b^2 = (a + b)(a - b)$ $a^3 + b^3 = (a + b)(a^2 - ab + b^2)$
     $a^2 + 2ab + b^2 = (a + b)^2$ $a^3 - b^3 = (a - b)(a^2 + ab + b^2)$

Common binomial factor: $a(x + y) + b(x + y) = (a + b)(x + y)$

Simplify.

12. $(5m + 3n)(5m - 3n)$ **13.** $(6c - 5d)^2$ **14.** $(a + 1)(a^2 - a + 1)$
     $25m^2 - 9n^2$ $36c^2 - 60cd + 25d^2$ $a^3 + 1$

Let $f(x) = 3x^2 - 2x$. Evaluate.

15. $f(-4)$ 56   **16.** $f(3c - 2)$ $27c^2 - 42c + 16$ **17.** $f(x + h) - f(x)$

Factor, if possible. Write each perfect-square trinomial as the square of a binomial.

18. $25x^2y^2 - 36$ **19.** $16c^2 - 40c + 25$ **20.** $9a^2 + 4$ Not possible
21. $x^3 + 8$ **22.** $125y^3 - 27$ **23.** $x^4 - 9$

The steps for *factoring completely* are listed on p. 185.

Factor each polynomial *completely*.

24. $27y^2 - 12$ **25.** $2n^3 - 24n^2 + 72n$ **26.** $x^4 - 26x^2 + 25$
     $3(3y + 2)(3y - 2)$ $2n(n - 6)^2$ $(x + 5)(x - 5)(x + 1)(x - 1)$

Factor each polynomial by grouping pairs of terms.

27. $4xy + 2x - 6y - 3$ $(2x - 3)(2y + 1)$ **28.** $6x^3 - 15x^2 - 4x + 10$ $(3x^2 - 2)(2x - 5)$

5.8   If $P(x) \div B(x) = Q(x) + R$ for polynomials $P(x)$, $B(x)$, $Q(x)$, and some
     number R, then $Q(x)$ is the *polynomial quotient* and R is the *remainder*.
     If $R = 0$, then $B(x)$, the *divisor*, is a factor of $P(x)$, the *dividend*.

Divide to find the polynomial quotient and the remainder. Is the divisor a factor of the dividend?

29. $(15c^3 - 28c^2 + 15c - 8) \div (3c - 2)$ $5c^2 - 6c + 1, -6$; no
30. $(2y^4 + 2y + 15 - 19y^2) \div (y + 3)$ $2y^3 - 6y^2 - y + 5, 0$; yes

5.9   To find $f(g(x))$ and $g(f(x))$, given equations for $f(x)$ and g(x): substitute
     g(x) for x in $f(x)$ and substitute $f(x)$ for x in g(x).

Let $f(x) = 2x + 1$ and $g(x) = x^2 - 4$.

31. Find $g(f(-3))$.   21   **32.** Find $f(g(3c))$. $18c^2 - 7$
33. Write an equation that determines the function $g(f(x))$, the composition of
     g with f. $g(f(x)) = 4x^2 + 4x - 3$

22. $(5c - d)(4c - 3)$
23. $(x^2 + 5)(3x - 4)$
24. $(5y^2 - 8)(2y - 5)$
31. $(3y + 2)(3y - 2)(y + 2)(y - 2)$
32. $(5y + 1)(5y - 1)(y + 2)(y - 2)$
41. $a(a - b^2)(a - b)$
42. $6(c + d)(c - d)(c + 2d)$
43. $(2c^2 + 3)(2c^2 - 3)$
     $(c^2 + 1)(c + 1)(c - 1)$
44. $(2d^2 + 1)(2d^2 - 1)(3d^2 + 1)(3d^2 - 1)$

Chapter Test

5. $15a^4 - a^2b - 2b^2$
6. $14t^3 - 27t^2 - 24t + 10$
25. $c^2 + 6c + 15$
26. $c^2 + 2cd$

Additional Answers, page 177

34. $3c^2(5c - 3d + cd)$
35. $2y(9y^2 + 2)(2y^2 - 3)$
36. $3n(5n^2 + 8)(2n^2 + 3)$

Mixed Review

1. $\{x|x < 4\}$

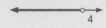

4

2. $\{x|-4 < x < 11\}$

-4 11

4. $x = \dfrac{7m + 5}{mt}$

Chapter 5 Test

A Exercises: 1–4, 10–16, 19–30   B Exercises: 5–19, 17, 18
C Exercises: 31, 32

Simplify each expression. Use positive exponents.

1. $-5ac \cdot 7a^3bc^2 \cdot 2b^2c^3$ $-70a^4b^3c^6$

2. $\dfrac{15x^8y^3z^5}{10x^4y^3z^7}$ $\dfrac{3x^4}{2z^2}$

3. $(6a^{-3})^2$ $\dfrac{36}{a^6}$

4. $(2b^0c^2)^{-3}$ $\dfrac{1}{8c^6}$

Simplify.

5. $(5a^2 - 2b)(3a^2 + b)$

6. $(2t - 5)(7t^2 + 4t - 2)$

7. $(7ab^2 - a^2b - 5ab) - (6a^2b - ab^2 - 2ab)$ $8ab^2 - 7a^2b - 3ab$

8. $3y(4y - 5)(4y + 5)$ $48y^3 - 75y$

9. $(8m + 3n)^2$ $64m^2 + 48mn + 9n^2$

10. $(x - 1)(x^2 + x + 1)$ $x^3 - 1$

Find the value of each expression and state it in ordinary notation.

11. $\dfrac{8.8 \times 10^2}{2.2 \times 10^{-3}}$ 400,000

12. $(4 \times 10^6) \times (2.1 \times 10^{-8})$ 0.084

Give the degree of each polynomial.

13. $-7x^3y$ 4

14. $5x^3 - 4x + 6$ 3

Factor completely, if possible.

15. $5x^2 + 10x - 40$ $5(x - 2)(x + 4)$

16. $9c^3 - 30c^2 + 25c$ $c(3c - 5)^2$

17. $25x^4 - y^2$ $(5x^2 + y)(5x^2 - y)$

18. $4x^4 + 1$ Not possible

19. $n^3 - 64$ $(n - 4)(n^2 + 4n + 16)$

20. $y^4 - 29y^2 + 100$
$(y + 5)(y - 5)(y + 2)(y - 2)$

21. Solve $nx + 4 = px + r$ for x. $\dfrac{r-4}{n-p}$

22. Factor $15xy + 6y - 20x - 8$ by grouping pairs of terms. $(3y - 4)(5x + 2)$

23. Solve $2^{4x-3} = 32$ for x.   2

Let $f(x) = x^2 + 6$ for Exercises 24–26.

24. Find $f(-4)$.   22

25. Find $f(c + 3)$.

26. Find $f(c + d) - f(d)$.

27. Divide $(2x^3 - 15x + 15)$ by $(x + 3)$ to find the polynomial quotient and the remainder. Is the divisor a factor of the dividend? $2x^2 - 6x + 3$, 6; no

For Exercises 28–30, let $f(x) = 3x - 2$ and $g(x) = x^2 + 4$.

28. Find $g(f(-1))$.   29

29. Find $f(g(-1))$.   13

30. Find $f(g(c))$. $3c^2 + 10$

31. Simplify $(3x^{2a} - y^b)(x^{2a} + y^b)$. $3x^{4a} + 2x^{2a}y^b - y^{2b}$

32. Factor $16x^{2n} + 24x^n + 9$. $(4x^n + 3)^2$

Strategy for Achievement in Testing

If guessing is not penalized on the test, you may be able to improve your score by eliminating the obviously wrong choices and guessing one of the remaining choices.

Sample: If $3^{x+1} = 27^{x-1}$, then $x =$ ___?___
(A) 0 (B) -2 (C) 1 (D) 3 (E) None of these

Even if you do not know how to solve the problem directly, you might still notice that choices (A) and (C) are incorrect. By eliminating these two choices, you have improved your chances of guessing the correct answer, (E).

Choose the *one* best answer to each question or problem.

1. $\overline{AB}$ is a line segment in the figure above. Find $\dfrac{x+y}{x-y}$. D

   (A) $\dfrac{5}{3}$ (B) $\dfrac{8}{5}$ (C) $\dfrac{8}{3}$
   (D) 4   (E) 24

2. What is the greatest monomial factor of $2(6x^2y - 9xy^3)(15a^3x + 10ay^2)$?
   (A) 2   (B) $6xy$ (C) $10a$
   (D) $30axy$ (E) None of these D

3. If $4^{x-1} = 8^{2x+2}$, then $x =$ ___?___
   (A) 2   (B) 1   (C) 0
   (D) -1 (E) -2 E

4. If $f(x) = (3x - 1)^2 - 5x + 2x^2$, then $f(0.1) =$ ___?___
   (A) 0.01   (B) 0.1   (C) 1
   (D) 1.21   (E) None of these A

5. $x\star$ means $4(x - 2)^2$. Find the value of $(3\star)\star$.
   (A) 8   (B) 12   (C) 16
   (D) 36   (E) None of these C

6. A function f is described by $f(x) = 3x - 6$ and a function g is described by $g(x) = 12 - 6x$. Which statement is true?
   (A) $g(f(x)) = f(g(x))$
   (B) $g(f(x)) = 2 \cdot f(g(x))$
   (C) $g(f(x)) = -2 \cdot f(g(x))$
   (D) $g(f(x)) = f(g(x)) + 18$
   (E) None of these is true. D

7. If $f(x) = 2x - 6$ and $g(f(x)) = x$, then ___?___
   (A) $g(x) = \dfrac{x+6}{2}$
   (B) $g(x) = \dfrac{1}{2}(x - 6)$
   (C) $g(x) = -2x + 6$
   (D) $g(x) = -2x - 6$
   (E) $g(x) = \dfrac{1}{2x - 6}$ A

8. If $f(x) = x^{-2}$, then $f(f(2)) =$ ___?___
   (A) $\dfrac{1}{16}$ (B) $\dfrac{1}{4}$ (C) 4
   (D) 16   (E) None of these D

9. $9[4^{-2}(-2)^4 - 3^{-2}]^{-1} =$ ___?___
   (A) 8   (B) $\dfrac{1}{8}$ (C) $-\dfrac{1}{8}$
   (D) -8 (E) None of these E

6 HIGHER-DEGREE EQUATIONS AND INEQUALITIES

OVERVIEW

In Chapter 6, students focus on solving equations of degree two and higher using the Zero-Product Property. The same property is used also in a later lesson to solve second- and third-degree inequalities. Students also explore techniques for solving higher-degree equations by applying the Integral Zero Theorem, which shows how to test potential integral zeros of a polynomial until a second-degree factor is found.

OBJECTIVES

- To solve equations using the Zero-Product Property
- To find and graph solution sets of second- or third-degree inequalities in one variable
- To divide a polynomial by a binomial using synthetic division
- To solve word problems involving quadratic equations

PROBLEM SOLVING

This chapter continues the strong emphasis on problem solving that was established at the beginning of the text. Two full lessons are devoted to the solving of word problems that can be modelled by second-degree equations. In addition, the chapter contains an *Application* feature, *Projectile Motion* on page 207, and a *Statistics* feature, *Population Sampling* on page 212, in which students solve real-world problems. A *Problem Solving Strategies* activity teaches the strategy *Solving a Simpler Problem*.

READING AND WRITING MATH

In this chapter there are two *Focus on Reading* features, on pages 201 and 220. There are two exercises (Exercise 33 on page 216 and Exercise 13 on page 227) in which students are asked to explain mathematical concepts in their own words.

TECHNOLOGY

Calculator: A calculator greatly cuts down on the time required to evaluate a polynomial for a given value of x using synthetic substitution. The details are found in the feature *Using the Calculator* on page 222.

SPECIAL FEATURES

PLANNING GUIDE

Lesson	Basic	Average	Above Average	Resources
6.1 pp. 201–203	FR all CE all WE 1–26 Problem Solving	FR all CE all WE 14–40 Problem Solving	FR all CE all WE 23–49 Problem Solving	Reteaching 40 Practice 40
6.2 pp. 205–207	CE all WE 1–14 Application	CE all WE 3–16 Application	CE all WE 5–18 Application	Reteaching 41 Practice 41
6.3 pp. 210–212	CE all WE 1–13 odd Midchapter Review Statistics	CE all WE 3–15 odd Midchapter Review Statistics	CE all WE 7–19 odd Midchapter Review Statistics	Reteaching 42 Practice 42
6.4 pp. 215–216	CE all WE 1–17 odd	CE all WE 9–25 odd	CE all WE 23–39 odd	Reteaching 43 Practice 43
6.5 pp. 220–222	FR all CE all WE 1–13 odd Using the Calculator	FR all CE all WE 3–15 odd Using the Calculator	FR all CE all WE 5–19 odd Using the Calculator	Reteaching 44 Practice 44
6.6 p. 225	CE all WE 1–9	CE all WE 3–11	CE all WE 7–15	Reteaching 45 Practice 45
Chapter 6 Review pp. 226–227	all	all	all	
Chapter 6 Test p. 228	all	all	all	
College Prep Test p. 229	all	all	all	
Cumulative Review pp. 230–231	1–49 odd	3–45 odd	11–45 odd	

CE = Classroom Exercises   WE = Written Exercises   FR = Focus on Reading
NOTE: For each level, all students should be assigned all Mixed Review exercises.

▬ INVESTIGATION

Project: The students investigate the relation of a quadratic equation to its factors graphically. The investigation may be extended to cubic or other higher degree equations.

Have the students factor and solve the quadratic equation $x^2 - 4 = 0$.

Factors: $x + 2$, $x - 2$
Solutions: $x = 2$, $x = -2$

Emphasize the idea that to solve the equation means to find values of x for which the expression on the left equals 0. This is sometimes called *finding the zeros* of the equation.

A graph of the equation displays the values of the expression for a range of values of x. The values are displayed in the y-direction. The choices of x that give a value of $y = 0$ are the solutions to the equation.

Have the students graph the quadratic equation above. Then have them graph the equations of the two factors, $y = x + 2$ and $y = x - 2$, on the same coordinate plane. Have them discuss their results.

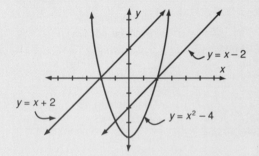

The graphs of the two factors are each linear equations with a single solution or zero. Students should notice that the zero of each linear equation corresponds to one of the zeros of the quadratic equations.

The students should be able to see that for a given x-value, the corresponding y-values of the linear equations multiplied together equal the y-value of the quadratic equation. For example, for $x = -2$ the y-values of the two linear equations are 0 and -4. The product of these y-values is $0 \cdot (-4) = 0$, which is the value of the quadratic equation for $x = -2$. Other values of interest are $x = 0$ and $x = 2$.

This is a magnification of a region of the Mandelbrot Set, the fractal that appears on page XVI. Notice that the shape of the dark region is very much like the overall shape of the set. The basic shape occurs infinitely many times in the set, an example of the self-similarity of fractals.

More About the Image

The Mandelbrot Set is a fractal. The actual Mandelbrot Set consists of the dark regions of the graph. It is made up of the points on the complex plane whose absolute value remains finite under repeated iterations of a single operation. The basic shape, the dark "snowman" or "bug" occurs infinitely many times in a structure connected by tiny dark threads. However, the color-coded areas at the edges of the set are very interesting both mathematically and artistically. Each color represents a different property for numbers close to, but not actually in, the Mandelbrot Set.

6.1 The Zero-Product Property

Teaching Resources

Problem Solving Worksheet 6
Quick Quizzes 40
Reteaching and Practice
  Worksheets 40

Objective   To solve equations of degree 2, 3, or 4 for their rational roots by factoring

You already know that the solution, or *root*, of a linear (first-degree) equation in one variable can be found by solving for that one variable (see Lesson 1.4). Thus, $2x - 1 = 0$ can be solved in two steps to give $x = \frac{1}{2}$. Equations of a higher degree, however, such as $x^2 + x = 0$ and $x^4 = 3x^3 - 8$, can rarely be solved in this same direct manner. In fact, higher-degree equations can often be solved only approximately. In this chapter, though, you will work with higher-degree equations that *can* be solved exactly. The simplest of these is the second-degree, or *quadratic* equation.

Equations such as $3x^2 - 10x - 8 = 0$, $0 = 9x^2 - 16$, and $5x^2 + 15x = 0$ are called **quadratic equations** in one variable because each equation contains a quadratic (second-degree) term.

> **Standard Form of a Quadratic Equation in One Variable**
> The standard form of a quadratic equation in one variable is either $ax^2 + bx + c = 0$ or $0 = ax^2 + bx + c$, where a, b, and c are real numbers and $a \neq 0$.

Some quadratic equations can be solved by writing the equation in standard form, factoring the polynomial, and then setting each factor equal to zero. This method is based on the *Zero-Product Property*.

> **Zero-Product Property**
> If $m \cdot n = 0$, then $m = 0$ or $n = 0$, for all real numbers m and n.

EXAMPLE 1   Solve $3x^2 = 6 - 7x$.

Plan   Write the equation in standard form and factor the polynomial. Then use the Zero-Product Property.

Solution
$$3x^2 = 6 - 7x$$
$$3x^2 + 7x - 6 = 0$$
$$(3x - 2)(x + 3) = 0$$
$$3x - 2 = 0 \quad \text{or} \quad x + 3 = 0$$
$$x = \tfrac{2}{3} \quad \text{or} \quad x = -3$$

GETTING STARTED

Prerequisite Quiz

Solve.
1. $0 = x + 7$ -7
2. $5n - 4 = 0$ $\frac{4}{5}$
3. $6x = 0$ 0

Factor completely.

4. $4y^2 - 4y - 15$ $(2y - 5)(2y + 3)$
5. $y^4 - 13y^2 + 36$
   $(y + 2)(y - 2)(y + 3)(y - 3)$
6. $x^3 - 36x$ $x(x - 6)(x + 6)$

Motivator

Have students find the solutions of $5(x + 3)(x - 4)(2x + 1) = 0$ by substitution from the replacement set $\{0, 3, -3, 4, -4, 5, \frac{1}{2}, -\frac{1}{2}\}$. -3; 4; $-\frac{1}{2}$

Ask students to describe a common feature when each of the solutions was substituted into the equation. One of the operations in parenthesis resulted in 0.

Additional Example 1

Solve $5x^2 = 6 + 7x$. $-\frac{3}{5}$, 2

Highlighting the Standards

Standard 14b: Without an understanding of the algebraic structure supporting the zero principle, students may try to use the procedure when an equation is equal to something other than zero.

Lesson Note

In solving quadratic equations by factoring, students can use either of the standard forms, $ax^2 + bx + c = 0$ or $0 = ax^2 + bx + c$. Usually finding the factors is easier when the polynomial is expressed with a as a positive number. The Zero Product Property is extended to four factors in Example 3.

Math Connections

Matrix Algebra: Students should realize that there are mathematical systems in which the Zero-Product Property does not hold, that is, it is *not* the case that if $m \cdot n = 0$ then either $m = 0$ or $n = 0$. For example, if m and n are *matrices* of numbers rather than numbers (see Chapter 16) and "$\cdot$" represents matrix multiplication, then it may happen that $m \cdot n = 0$ (the "zero matrix") but neither m nor n is 0.

Critical Thinking Questions

Analysis: Can the Zero-Product Property be used to solve the equation $x(x + y) + 3 = 0$. If so, how? If not, why? No. When the Distributive Property is used, the resulting equation, $x^2 + xy + 3$, does not contain a product of factors.

Checks

$$x = \frac{2}{3}: \qquad 3x^2 = 6 - 7x$$
$$3\left(\frac{2}{3}\right)^2 \stackrel{?}{=} 6 - 7 \cdot \frac{2}{3}$$
$$3 \cdot \frac{4}{9} \stackrel{?}{=} \frac{18}{3} - \frac{14}{3}$$
$$\frac{4}{3} = \frac{4}{3} \quad \text{True}$$

$$x = -3: \qquad 3x^2 = 6 - 7x$$
$$3(-3)^2 \stackrel{?}{=} 6 - 7(-3)$$
$$3 \cdot 9 \stackrel{?}{=} 6 + 21$$
$$27 = 27 \quad \text{True}$$

The solutions, or *roots*, are $\frac{2}{3}$ and -3.

Sometimes it is more convenient to use the form $0 = ax^2 + bx + c$.

EXAMPLE 2   **a.** Solve $-5x^2 = 20x$.   **b.** Solve $16 = 9n^2$.

Solutions

a. $-5x^2 = 20x$
$$0 = 5x^2 + 20x$$
$$0 = 5x(x + 4)$$
$$5x = 0 \quad or \quad x + 4 = 0$$
$$x = 0 \quad or \quad x = -4$$
The roots are 0 and -4.

b. $16 = 9n^2$
$$0 = 9n^2 - 16$$
$$0 = (3n + 4)(3n - 4)$$
$$3n + 4 = 0 \quad or \quad 3n - 4 = 0$$
$$n = -\frac{4}{3} \quad or \quad n = \frac{4}{3}$$
The roots are $-\frac{4}{3}$ and $\frac{4}{3}$.

A fourth-degree equation can be solved for its *rational roots*.

EXAMPLE 3   Find the rational roots of each equation.
   a. $x^4 - 13x^2 + 36 = 0$ **b.** $4x^4 + 35x^2 - 9 = 0$

Plan   Factor each trinomial completely.

Solutions   **a.**
$$x^4 - 13x^2 + 36 = 0$$
$$(x^2 - 4)(x^2 - 9) = 0$$
$$(x + 2)(x - 2)(x + 3)(x - 3) = 0$$

Use the Zero-Product Property extended to four factors.

$$x + 2 = 0 \quad or \quad x - 2 = 0 \quad or \quad x + 3 = 0 \quad or \quad x - 3 = 0$$
$$x = -2 \qquad\qquad x = 2 \qquad\qquad x = -3 \qquad\qquad x = 3$$

There are four rational roots: $-2, 2, -3$, and 3.

b.
$$4x^4 + 35x^2 - 9 = 0$$
$$(4x^2 - 1)(x^2 + 9) = 0$$
$$(2x + 1)(2x - 1)(x^2 + 9) = 0$$
$$2x + 1 = 0 \quad or \quad 2x - 1 = 0 \quad or \quad x^2 + 9 = 0$$
$$x = -\frac{1}{2} \qquad\qquad x = \frac{1}{2} \qquad\qquad x^2 + 9 \text{ cannot be factored.}$$
$$x^2 = -9 \text{ has no real roots.}$$

There are two rational roots: $-\frac{1}{2}$ and $\frac{1}{2}$.

Additional Example 2

a. Solve $-6x = 18x^2$. $0, -\frac{1}{3}$

b. Solve $25 = 4n^2$. $-\frac{5}{2}, \frac{5}{2}$

Additional Example 3

Find the rational roots of each equation.

a. $-x^4 = 49 - 50x^2$ $-7, 7, -1, 1$

b. $0 = x^4 - 48x^2 - 49$ $-7, 7$

Recall the Division Property of Equality (see Lesson 1.4).

$$\text{If } a = b, \text{ then } \frac{a}{c} = \frac{b}{c} \ (c \neq 0).$$

To solve $4x^2 - 36 = 0$, divide each side of the equation by 4, and factor. This is done at the right.

The roots are -3 and 3.

$$4x^2 - 36 = 0$$
$$x^2 - 9 = 0$$
$$(x + 3)(x - 3) = 0$$
$$x + 3 = 0 \text{ or } x - 3 = 0$$
$$x = -3 \qquad x = 3$$

However, to solve $x^3 - 9x = 0$, do *not* divide each side by x because x may equal 0.

Correct method

$$x^3 - 9x = 0$$
$$x(x^2 - 9) = 0$$
$$x(x + 3)(x - 3) = 0$$
$$x = 0 \text{ or } x = 3 \text{ or } x = -3$$

The roots are 0, -3, and 3.

Incorrect method

$$\frac{x^3}{x} - \frac{9x}{x} = \frac{0}{x}$$
$$x^2 - 9 = 0$$
$$(x + 3)(x - 3) = 0$$
$$x = -3 \text{ or } x = 3$$

The roots are -3 and 3.

Dividing each side by x caused the loss of the root 0.

Focus on Reading

For each statement, determine whether it is true or false.

1. $0 = 3x^2 + 6x$ is a quadratic equation. T

2. $4x - 8 = 0$ is a quadratic equation. F

3. A quadratic equation may have two rational roots. T

4. $x^2 - 6x + 9 = 0$ is a quadratic equation with exactly one root. T

5. A fourth-degree equation may have four rational roots. T

6. A fourth-degree equation may have exactly two rational roots. T

Classroom Exercises

For each statement, determine whether it is true or false.

1. If $x(x - 5) = 0$, then $x = 0$ or $x - 5 = 0$. T

2. If $(3n - 2)(n - 4) = 0$, then $n = 2$ or $n = 4$. F

3. If $(y - 6)(y - 6) = 0$, then $y = 6$. T

4. $x^2 + 4 = 0$ has no real-number roots. T

Find the rational roots of each equation.

5. $-3x = 0$ 0

6. $x^2 = 0$ 0

7. $(x - 1)x = 0$ 1, 0

8. $(x - 2)(x - 3) = 0$ 2, 3

9. $(5n - 3)(n^2 + 4) = 0$ $\frac{3}{5}$

10. $y^2 + 15y + 54 = 0$ $-9, -6$

Checkpoint

Find the rational roots, if any, of each equation.

1. $2x^2 = 5 - 9x$ $\frac{1}{2}, -5$

2. $-10a^2 = 4a$ $0, -\frac{2}{5}$

3. $x^2 + 49 = 0$ None

4. $10x^2 - 45x - 25 = 0$ $-\frac{1}{2}, 5$

5. $2y^3 - 32y = 0$ $0, -4, 4$

Closure

Ask students if the equation, $ax^2 + bx = c$, is in standard form.   No   Ask students if the Zero-Product Property can be used to find all the roots of equations of degree 2.   No.   Some equations of degree 2 may not have real roots.

FOLLOW UP

Guided Practice

Classroom Exercises 1–10

Independent Practice

A Ex. 1–24, **B** Ex. 25–42, **C** Ex. 43–49

Basic: FR all, WE 1–26, Problem Solving

Average: FR all, WE 14–40, Problem Solving

Above Average: FR all, WE 23–49, Problem Solving

Enrichment

Challenge the students to determine under what conditions the sum of two numbers will equal the product of the two numbers. That is, under what conditions is it true that $a + b = ab$?

$$a + b = ab$$
$$b - ab = -a$$
$$b(1 - a) = -a$$
$$b = \frac{-a}{1 - a}$$

$$b = \frac{a}{a - 1}, a \neq 1$$

Thus, if $b = \frac{a}{a - 1}$, then $a + b = ab$.

Example:
Choose $a = 7$. Then $b = \frac{7 - 1}{7} = \frac{7}{6}$.

Check: $7 + \frac{7}{6} \overset{?}{=} 7 \cdot \frac{7}{6}$
$$\frac{49}{6} = \frac{49}{6} \ ✔$$

1. $\frac{1}{3}$, -2

2. $-\frac{5}{2}$, 4

4. $\frac{2}{3}$, $-\frac{3}{2}$

6. -10, -4

17. ± 1, ± 5

22. -6, 2

23. -5

27. -5, 0, 2

35. $\pm\frac{1}{3}$, ± 2

36. $\pm\frac{3}{5}$, ± 1

38. $\pm\frac{3}{5}$

40. $\frac{4}{3}$, 4

41. -6, 0, 1

43. $3a$, $-2a$

44. $2a$, $4a$

45. $\pm a$, $\pm 3a$

49. Converse: For all real numbers m and n, if $m = 0$ or $n = 0$, then $m \cdot n = 0$.
True

Written Exercises

Find the rational roots, if any, of each equation.

1. $(3x - 1)(x + 2) = 0$ 2. $(2a + 5)(a - 4) = 0$ 3. $y(y + 5) = 0$ 0, -5

4. $(3c - 2)(2c + 3) = 0$ 5. $x^2 - 13x + 40 = 0$ 5, 8   6. $0 = x^2 + 14x + 40$

7. $y^2 = y + 12$ -3, 4   8. $6a = 16 - a^2$ -8, 2   9. $0 = 3c^2 - 15c$ 0, 5

10. $7x^2 = -28x$ -4, 0   11. $z^2 - 49 = 0$ ± 7 12. $16 = y^2$ ± 4

13. $x^2 + 25 = 0$ None   14. $0 = n^2 + 36$ None   15. $0 = a^2 - 4a + 4$ 2

16. $x^2 + 8x + 16 = 0$ -4 17. $y^4 - 26y^2 + 25 = 0$ 18. $a^4 = 29a^2 - 100$ ± 2, ± 5

19. $y^4 - 24y^2 - 25 = 0$ ± 5 20. $a^4 = 21a^2 + 100$ ± 5 21. $6x^2 - 42x + 60 = 0$ 2, 5

22. $0 = 5n^2 + 20n - 60$ 23. $0 = 4n^2 + 40n + 100$ 24. $60c - 180 = 5c^2$ 6

25. $a^3 - 9a = 0$ 0, ± 3 26. $0 = c^3 - 16c$ 0, ± 4 27. $x^3 + 3x^2 - 10x = 0$

28. $2x^2 + 8x = x^3$ -2, 0, 4   29. $2c^2 = 7c - 6$ $\frac{3}{2}$, 2   30. $15 - 4y = 3y^2$ -3, $\frac{5}{3}$

31. $16n^2 - 9n = 0$ 0, $\frac{9}{16}$ 32. $30c = 12c^2$ 0, $\frac{5}{2}$ 33. $30t^2 = 125t - 120$ $\frac{3}{2}$, $\frac{8}{3}$

34. $2m + 70 = 24m^2$ $-\frac{5}{3}$, $\frac{7}{4}$ 35. $9y^4 - 37y^2 + 4 = 0$ 36. $25b^4 - 34b^2 + 9 = 0$

37. $9y^4 + 35y^2 - 4 = 0$ $\pm\frac{1}{3}$ 38. $25b^4 + 16b^2 - 9 = 0$ 39. $(n - 7)^2 = 29 - n^2$ 2, 5

40. $2(24 - n^2) = (8 - n)^2$ 41. $5x^3 = 30x - 25x^2$ 42. $3y^3 + 36y^2 = 3y^4$ -3, 0, 4

Solve each equation for x in terms of a.

43. $x^2 - ax - 6a^2 = 0$ 44. $3x^2 + 24a^2 = 18ax$ 45. $x^4 - 10a^2x^2 + 9a^4 = 0$

46. Solve $|x^2 - 5| = 4$. (HINT: There are four roots.) ± 1, ± 3

47. Find the two roots of $|a^2 + 2a - 8| = |a^2 - 4|$. -3, 2

48. Find the three roots of $|2y^2 - 11y + 5| = |y^2 - 3y - 10|$. $-\frac{1}{3}$, 3, 5

49. The Zero-Product Property is stated in "If . . ., then . . ." form (see page 199). Write the converse of this property. Is the converse true? (See Extension, pages 103–104.)

Mixed Review

1. Find three consecutive even integers such that 2 more than 5 times the third integer is the same as 3 times the sum of the first two integers.   *1.6*   16, 18, 20

2. Supply a reason for each step in the proof that $a(b + c) + d$ is equivalent to $d + (ba + ca)$.   *1.8*

    1. $a(b + c) + d = d + a(b + c)$ Comm Prop Add
    2. $= d + (ab + ac)$ Distr Prop
    3. $= d + (ba + ca)$ Comm Prop Mult
    4. $a(b + c) + d = d + (ba + ca)$ Trans Prop Eq

Problem Solving Strategies

Solving a Simpler Problem

A service engineer is asked to install telephone lines connecting 10 buildings. Every pair of buildings is to have a separate line. How many lines are needed?

Use a decagon (10-sided polygon) to **model** the situation. Let each vertex represent a building. The problem is to find the total number of *sides and diagonals* of the figure.

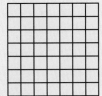

Consider a number of **simpler problems.** Draw several polygons and find the sum of the diagonals and sides for each.

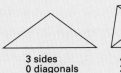

| 3 sides | 4 sides | 5 sides | 6 sides |
| 0 diagonals | 2 diagonals | 5 diagonals | 9 diagonals |

Make a **table** to organize the results. Look for a **pattern.** Use the pattern to extend the table.

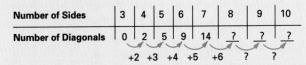

Number of Sides	3	4	5	6	7	8	9	10
Number of Diagonals	0	2	5	9	14	?	?	?

+2 +3 +4 +5 +6 ? ?

A decagon will have 35 diagonals and 10 sides. So, 45 lines will be needed to connect the 10 buildings.

Exercises

1. There are 7 students at a party. Each student shakes hands once with every other student in the room. How many handshakes are there? 21

2. How many different squares are there in the 7 by 7 grid at the right? 140

▰▰GETTING STARTED

Prerequisite Quiz

Write an algebraic expression for each word phrase.

1. three consecutive odd integers
   $x, x + 2, x + 4$
2. three consecutive multiples of 6
   $n, n + 6, n + 12$

Solve.

3. $(2f - 3)(f + 2) = 15$ $3, -\frac{7}{2}$
4. $2m(m + 4) + 16 = (m + 8)^2 - 4(m + 4)$ $8, -4$

Motivator

Have students name the dimensions of all rectangles whose interior can be covered by 12 one-foot square tiles.
$1 \times 12; 2 \times 6; 3 \times 4; 4 \times 3; 6 \times 2; 12 \times 1$
Is -3×-4 a reasonable dimension for the rectangle? Why or why not?   No; dimensions must be positive numbers.

6.2 Problem Solving: Using Quadratic Equations

Objective   To solve word problems using quadratic equations

In the following example, you see that a given word problem may have two sets of answers.

EXAMPLE 1   The product of two consecutive integers is 12. Find all such consecutive integers.

What are you to find?	Two consecutive integers
Choose a variable.	Let m = the first consecutive integer.
What does it represent?	Then $m + 1$ is the second consecutive integer.
What is given?	The product of m and $m + 1$ is 12.
Write an equation.	$m(m + 1) = 12$
Solve the equation.	$m^2 + m = 12$
	$m^2 + m - 12 = 0$
	$m + 4 = 0$ or $m - 3 = 0$
	$m = -4$ $m = 3$
	If $m = -4$, then $m + 1 = -3$.
	If $m = 3$, then $m + 1 = 4$.
Check in the original problem.	The check is left for you.
State the answer.	There are two pairs of consecutive integers, $-4, -3$ and $3, 4$.

When a quadratic equation is used to solve a problem, it is necessary to determine whether both solutions provide reasonable answers to the problem. For example, if x apples are placed in each of $x - 6$ boxes and there is a total of 40 apples, then $x(x - 6) = 40$ and the roots are 10 and -4.

If $x = 10$, then $x - 6 = 4$ and there are 10 apples in each of 4 boxes. If $x = -4$, then $x - 6 = -10$, and there are -4 apples in each of -10 boxes, which is not reasonable. The root -4 is therefore not a reasonable solution of the problem.

204   Chapter 6 Higher-Degree Equations and Inequalities

Highlighting the Standards

Standard 1a: The first example in this lesson takes students through the sequential thinking that should accompany the solving of a quadratic equation.

Additional Example 1

Find three consecutive odd integers such that twice the product of the first two is 11 less than the square of the third integer.

$[2x(x + 2) = (x + 4)^2 - 11]$
5, 7, 9, and $-1, 1, 3$

EXAMPLE 2   One hundred cubes are placed in rows so that the number of cubes in each row is 1 more than 6 times the number of rows. Find the number of cubes in each row.

Plan   Number × Number of cubes = Total number
of rows   in one row   of cubes

Let r = the number of rows. Then $6r + 1$ = the number of cubes per row.

Solution $r(6r + 1) = 100$ ← (THINK: The answer must be a factor of what number?)

$$6r^2 + r = 100$$
$$6r^2 + r - 100 = 0$$
$$(6r + 25)(r - 4) = 0$$

$6r + 25 = 0$ or $r - 4 = 0$
$\quad r = -\frac{25}{6}$ or $\quad r = 4$

The number of rows cannot be $-\frac{25}{6}$, so $r = 4$ and $6r + 1 = 6 \cdot 4 + 1$, or 25. Check: 4 rows × 25 cubes per row = 100 cubes. There are 25 cubes in each row.

Classroom Exercises

Tell which of the following values substituted for x in Exercises 1–8 give results that are *not* reasonable: $-3, 2\frac{1}{2}, 8$.

1. a first number x and a second number $4x - 2$ None
2. three consecutive integers beginning with x $2\frac{1}{2}$
3. four consecutive odd integers beginning with x $2\frac{1}{2}, 8$
4. x nickels, $3x$ quarters, $3x - 2$ dimes $-3, 2\frac{1}{2}$
5. x males and $x - 10$ females $-3, 2\frac{1}{2}, 8$
6. a rectangle, x units wide and $2x + 3$ units long -3
7. a triangle with a base x units long and a height of $2x - 5$ units $-3, 2\frac{1}{2}$
8. x pounds of rice in each of $4x$ bags -3
9. Find three consecutive multiples of 5 so that the square of the third number, decreased by 5 times the second number, is the same as 25 more than twice the product of the first two numbers. $-5, 0, 5,$ and $10, 15, 20$
10. Some light bulbs are placed in boxes, and the boxes are packed in cartons. The number of bulbs in each box is 4 less than the number of boxes in each carton. Find the number of bulbs in each box given that a full carton contains 60 light bulbs.   6 bulbs

TEACHING SUGGESTIONS

Lesson Note

Students must realize that a root of an equation may not give a reasonable answer to a word problem. This is demonstrated in Example 2 and Classroom Exercises 1–8. The Plan step provided in Example 2 is necessary in solving Written Exercises 3, 4, 9, 10, 15, and 16.

Math Connections

Nonlinear Problems: Students should now begin to see that some problems, including problems in science and engineering, cannot be solved by using linear equations. Fortunately, many such problems can be solved using quadratic equations, which often are relatively easy to solve using methods that have been demonstrated in Lesson 6.1 or by methods shown later in this textbook.

Critical Thinking Questions

Analysis: Ask students whether they can devise a problem that has two answers or sets of answers but leads to an equation that is not quadratic.   One example is the following: "The difference between 12 and a number is 7. Find the number." An equation for the problem is $|x - 12| = 7$, which leads to the two answers 5 and 19.

Additional Example 2

Sixty lemons are placed in bags so that the number of lemons in each bag is 2 more than twice the number of bags. Find the number of lemons in one bag.

$[b(2b + 2) = 60]$; 12 lemons

Checkpoint

Determine which of the numbers −5, 4½, and 6, are not reasonable values for x (Ex. 1–3).

1. $x + 9$ boys and $2x$ girls −5, 4½
2. two consecutive odd integers, beginning with x 4½, 6
3. three numbers: x, $2x$, $x − 10$ None
4. Twenty-four band members are arranged in rows so that the number of rows is 5 less than the number of members in each row. Find the number of rows.
   3 rows

Closure

Have students create a word problem that would lead to solving the equation $x(x + 4) = 6(x + 8)$. Sample answer: Find three consecutive multiples of 4 such that the product of the first two is 6 times the third.

FOLLOW UP

Guided Practice

Classroom Exercises 1–10

Independent Practice

A Ex. 1–12, **B** Ex. 13–16, **C** Ex. 17, 18

Basic: WE 1–14, Application
Average: WE 3–16, Application
Above Average: WE 5–18, Application

Written Exercises

1. Find three consecutive even integers such that the product of the first two integers is 48. −8, −6, −4; 6, 8, 10

2. Find three consecutive integers so that the product of the second and third integers is 56. −9, −8, −7; 6, 7, 8

3. Seventy chairs are placed in rows so that the number of chairs in each row is 3 less than the number of rows. What is the number of chairs in each row? 7 chairs

4. Thirty bulbs are placed in some boxes where the number of bulbs in each box is 7 more than the number of boxes. Find the number of bulbs in each box. 10 bulbs

5. One number is 5 less than another number, and their product is 24. Find all such pairs of numbers. −3, −8; 8, 3

6. The product of two integers is 44. One is 3 more than twice the other. Find both numbers. 4, 11

7. Find three consecutive multiples of 5 so that the product of the first two numbers is 150. −15, −10, −5; 10, 15, 20

8. Find three consecutive multiples of 10 so that the product of the first two numbers is 15 times the third number. 20, 30, 40

9. Eighty peaches were packed in boxes so that the number of boxes was 6 less than twice the number of peaches in each box. Find the number of boxes used. 10 boxes

10. A complete album holds 480 stamps. How many pages are in the album given that the number of pages is 8 more than 4 times the number of stamps on each page? 48 pages

11. Find three consecutive integers so that the square of the first, increased by the square of the third, is 74. −7, −6, −5; 5, 6, 7

12. Find three consecutive odd integers such that the sum of the squares of the second and third integers is 130. −11, −9, −7; and 5, 7, 9

13. The product of two numbers is 4, and one number is 9 times the other number. Find all such pairs of numbers. $-\frac{2}{3}, -6; \frac{2}{3}, 6$

14. Find two numbers whose product is −9 given that one number is −16 times the other number. $-\frac{3}{4}, 12; \frac{3}{4}, -12$

15. Some tiles are arranged in rows so that the number of tiles in each row is 8 more than the number of rows. The same number of tiles can be arranged in 3 more rows than in the first pattern with 16 tiles in each row. Find the total number of tiles. 240 tiles

16. Some chairs are placed in rows so that the number of rows is 2 less than the number of chairs in each row. The same number of chairs can be placed in 6 more rows than in the first case with 15 chairs per row. Find the total number of chairs. 360 chairs

17. Find four consecutive multiples of 0.5 for which the product of the first and third numbers is 0.25 less than 3 times the second number. −0.5, 0, 0.5, 1; 2.5, 3, 3.5, 4

18. Find four consecutive multiples of π in which the product of the second and third multiples is $2\pi^2$ more than 3π times the fourth multiple. $-3\pi, -2\pi, -\pi, 0; 3\pi, 4\pi, 5\pi, 6\pi$

Enrichment

Ask the students to find these quotients.

$\frac{x^3 - 1}{x - 1}$ $x^2 + x + 1$

$\frac{x^5 - 1}{x - 1}$ $x^4 + x^3 + x^2 + x + 1$

$\frac{x^6 - 1}{x - 1}$ $x^5 + x^4 + x^3 + x^2 + x + 1$

Then have them use the pattern to state the general quotient for $\frac{x^n - 1}{x - 1}$.

$x^{n-1} + x^{n-2} + x^{n-3} + \cdots + x^2 + x + 1$

Mixed Review

1. The length of a rectangle is 6 units more than its width. Represent the perimeter and the area of the rectangle in algebraic terms. *1.7* $4x + 12; x^2 + 6x$

2. The base of a triangle is 6 units shorter than 4 times its height. Represent the area of the triangle in algebraic terms. *1.7* $2x^2 - 3x$

3. What is the perimeter of the right triangle whose sides are $5x$, $12x$, and $13x$ and whose area is 30? *1.7* 30; 30

4. Find the area of the rectangle whose dimensions are $6a$ by $8a$, and whose perimeter is 70. *1.7* 300

Application: *Projectile Motion*

From observation, Aristotle (384-322 B.C.) and Galileo (1564-1642) knew that an object thrown horizontally off a cliff took a curved path to the ground. Galileo was able to show that the projectile traveled forward and downward simultaneously, and that the horizontal and vertical components of its motion were independent of each other.

The distance x (in feet) traveled horizontally can be expressed using $d = rt$. If the initial horizontal velocity is 8 ft/s, then $x = 8t$. The height y (in feet) can also be expressed in terms of time (in seconds): $y = h_o - 16t^2$, where h_o is the original height. If the cliff is 100 ft high, then $y = 100 - 16t^2$. At $t = 1$ s for example, $x = 8$, $y = 84$, and the object is at a point 8 ft away from the cliff and 84 ft high.

To find the spot where this projectile hits the ground $(y = 0)$, set the polynomial $100 - 16t^2$ equal to zero and solve for t by factoring. Then substitute this value for t into $x = 8t$ to find how far the projectile will travel horizontally in that time.

$$100 - 16t^2 = 0 \qquad\qquad x = 8t$$
$$(10 + 4t)(10 - 4t) = 0 \qquad = 8(\tfrac{5}{2})$$
$$t = \frac{10}{4} = \frac{5}{2}s \qquad\qquad = 20 \text{ ft}$$

1. A cannon is fired from a fort 144 ft high. The shell is shot horizontally with an initial velocity of 900 ft/s. How far away from the fort does the shell hit the ground? 2,700 ft

2. A motorcycle daredevil prepares to jump a canyon from the North Rim to the South Rim. The canyon is 352 ft wide, and the South Rim is 64 ft lower than the North Rim. What is the minimum speed (in ft/s) necessary, given a horizontal takeoff, for the jump to work?

176 ft/s

▰▰GETTING STARTED

Prerequisite Quiz

Find the area of each figure.

1. rectangle, 7 ft by 5 ft   35 ft²
2. triangle, base 9 cm and height 10 cm
   45 ft²

Solve each equation.

3. $x^2 + (x + 7)^2 = (x + 8)^2$ 5, −3
4. $w(2w + 5) = 33$ 3, $-\dfrac{11}{2}$
5. $\dfrac{1}{2}b(3b + 4) = 10$ 2, $-\dfrac{10}{3}$

Motivator

Have students find a^2, b^2, and c^2, for each set of data below. Then have them write an equation relating the three in each case.

1. $a = 12$, $b = 5$, $c = 13$ 144, 25, 169
2. $a = 6$, $b = 8$, $c = 10$ 36, 64, 100
3. $a = 15$, $b = 8$, $c = 17$ 225, 64, 289
   In each case, $a^2 + b^2 = c^2$.

6.3 Problem Solving: Area and Length

Objectives   To find the lengths of sides of right triangles using the Pythagorean relation
To solve area problems using quadratic equations

For each right triangle ABC with right angle C, the lengths of the two legs are a and b and the length of the hypotenuse is c. For any such right triangle, the Pythagorean relation states that
$$a^2 + b^2 = c^2.$$

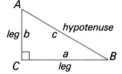

EXAMPLE 1   One leg of a right triangle is 14 m longer than the other leg and 2 m shorter than the hypotenuse. Find the length of each side of the triangle.

Solution   What are you to find?   the length of each side
What is given?

Represent the data as shown at the right.

Write an equation. $a^2 + b^2 = c^2$
Substitute in the formula for a, $x^2 + (x + 14)^2 = (x + 16)^2$
b, and c. $x^2 + x^2 + 28x + 196 = x^2 + 32x + 256$
Solve the equation. $x^2 - 4x - 60 = 0$
    $(x - 10)(x + 6) = 0$
    $x = 10$ or $x = -6$
The length of a side cannot be − 6 m.
If $x = 10$, then $x + 14 = 24$ and
$x + 16 = 26$.

Check in the original problem.   24 m is 14 m longer than 10 m.

24 m is 2 m shorter than 26 m.

$10^2 + 24^2 \stackrel{?}{=} 26^2$
    $676 = 676$ True

State the answer.   The legs measure 10 m and
24 m. The hypotenuse is 26 m long.

Recall that the area of a triangle is found by multiplying half the length of the base by the height of the triangle.

Highlighting the Standards

Standard 1b: Once again, the Polya model is used to lead students through the kind of thinking that mathematicians use as problem solvers. The method presented will at least offer all students a way of getting started.

Additional Example 1

The length of a rectangle is 1 ft longer than the width and 1 ft shorter than a diagonal. Find the length of the diagonal.

$[x^2 + (x + 1)^2 = (x + 2)^2]$; 5 ft

EXAMPLE 2   The height of a triangle is 9 cm less than 3 times the length of its base. Find the length of the base and the height of the triangle given that the area of the triangle is 60 cm^2.

Solution   Let b = the length of the base.
Then $3b - 9$ = the height.

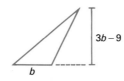

$$\tfrac{1}{2} \times \text{base} \times \text{height} = \text{area}$$
$$\tfrac{1}{2}b(3b - 9) = 60$$
$$b(3b - 9) = 120$$
$$3b^2 - 9b - 120 = 0$$
$$b^2 - 3b - 40 = 0$$
$$(b - 8)(b + 5) = 0$$
$$b = 8 \quad or \quad b = -5 \quad \text{(The base cannot be } -5 \text{ cm.)}$$

If $b = 8$, then $3b - 9 = 15$.

Check $\tfrac{1}{2} \times 8 \text{ cm} \times 15 \text{ cm} = 60 \text{ cm}^2$

The base is 8 cm long, and the height is 15 cm.

EXAMPLE 3   The length of a rectangular floor is 4 m shorter than 3 times its width. The width of a rectangular carpet on the floor is 2 m shorter than the floor's width, and the carpet's length is 2 m more than twice its own width. Find the area of the floor given that 31 m^2 of the floor is not covered by the carpet.

Solution   Draw and label both rectangles.

Let x = the width of the floor.

Then $3x - 4$ = the length of the floor; $x - 2$ = the width of the carpet; and $2(x - 2) + 2$, or $2x - 2$ = the length of the carpet.

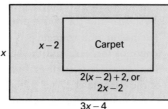

Use the formula for the area of a rectangle: $A = lw$.

$$(\text{Area of floor}) - (\text{Area of carpet}) = \text{Area not covered}$$
$$x(3x - 4) - (x - 2)(2x - 2) = 31$$
$$3x^2 - 4x - (2x^2 - 6x + 4) = 31$$
$$x^2 + 2x - 35 = 0$$
$$(x - 5)(x + 7) = 0$$
$$x = 5 \quad or \quad x = -7 \quad \text{(Width cannot be } -7 \text{ m.)}$$

If $x = 5$, then $3x - 4 = 11$, and $x(3x - 4) = 55$.

The area of the floor is 5 m $\times$ 11 m, or 55 m^2.

Lesson Note

Remind students that the following are important things to remember when solving geometric word problems.

1. Represent the data by drawing and labeling a figure.
2. Examine the roots of an equation to determine whether each one provides a reasonable answer.
3. Include the correct unit (linear, square, or cubic) in the answer.

Math Connections

History: The Pythagorean relation, one of the most familiar relationships in all of geometry, was used as early as 2000 B.C. by the ancient Egyptians. There have also been records of the relationship being used long ago by the Babylonians and the Chinese. About 500 B.C., the Greek mathematician Pythagoras became well known for proving the theorem and teaching many of its applications. Thereafter, it was named in honor of Pythagoras.

Additional Example 2

The base of a triangle is 7 ft more than twice the height. Find the length of the base if the area of the triangle is 11 ft^2.

$\left[\tfrac{1}{2}(2h + 7)h = 11\right]$; 11 ft

Additional Example 3

A square is twice as wide as a rectangle. The rectangle is 1 in. longer than it is wide. The difference in the two areas is 24 in^2. Find the area of the square.

$[(2w)^2 - w(w + 1) = 24]$; 36 in^2

Critical Thinking Questions

Analysis: In Example 3, ask students to suppose that it is not given that 31 m^2 of the floor is uncovered by carpet. Is it nevertheless true that the carpet's area must be less than that of the floor? The answer is "Yes." Since it is known that the carpet's width is 2 m less than the floor's width, it is sufficient to show only that the carpet's length is less than the floor's length: $2x - 2 < 3x - 4$ if and only if $-2 + 4 < 3x - 2x$; that is, if and only if $2 < x$, or $x > 2$. This last inequality is easily shown to be true. Since the carpet's width is given as $x - 2$, it follows that $x - 2 > 0$, or $x > 2$.

Common Error Analysis

Error: Students may think of the Pythagorean relation as $a^2 + b^2 = c^2$ rather than $(\text{leg}_1)^2 + (\text{leg}_2)^2 = (\text{hypotenuse})^2$.

This sometimes leads to the incorrect identification of a or b with the hypotenuse. Emphasize the verbal approach rather than the formula $a^2 + b^2 = c^2$.

Checkpoint

1. Find the length of each side of a right triangle if one leg is 3 m longer than the other leg and 3 m shorter than the hypotenuse. $[x^2 + (x + 3)^2 = (x + 6)^2]$; 9 m, 12 m, 15 m

2. The base of a triangle is 6 cm less than the height. Find the length of the base if the area of the triangle is 20 cm^2. $\left[\frac{1}{2}h(h - 6) = 20\right]$; 4 cm

Classroom Exercises

Which figure has the greater area?

1. a square 8 m wide or a rectangle 10 m by 6 m? square

2. a triangle with base 7 m and height 9 m or a rectangle 4 m by 8 m? rectangle

3. a rectangle x units by $4(x - 2)$ units or a rectangle $4x$ units by x units? rectangle $4x$ by x

4. a square $4x$ units wide or a rectangle $8x$ units by $2x - 2$ units? square

Write an expression for the area of each figure.

5. 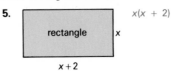 $x(x + 2)$

6. 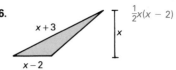 $\frac{1}{2}x(x - 2)$

7. In a right triangle, the hypotenuse is 8 in. longer than one leg and 4 in. longer than the other leg. Find the length of each side. 12 in, 16 in, 20 in

Written Exercises

1. One leg of a right triangle is 2 ft shorter than the other leg. The hypotenuse is 10 ft long. Find the length of each leg. 6 ft, 8 ft

2. The length of a rectangle is 4 cm less than twice its width. Find the length and the width given that the area is 70 cm^2. l = 10 cm, w = 7 cm

3. The length of a rectangle is 5 m more than 4 times its width. Find the length and the width given that the area is 51 cm^2. l = 17 m, w = 3 m

4. The height of a triangle is 8 times the length of its base. Find the length of the base and the height given that the area is 36 ft^2. b = 3 ft, h = 24 ft

5. The base of a triangle is 3 yd shorter than its height, and the area of the triangle is 27 yd^2. Find its height and the length of its base. h = 9 yd, b = 6 yd

6. One square is 3 times as wide as another square. Find the area of each square given that the difference between the two areas is 200 cm^2. 25 cm^2, 225 cm^2

7. The width of one square is twice the width of another square. The sum of their areas is 180 m^2. Find the area of the larger square. 144 m^2

8. The length of a rectangle is 3 cm more than 3 times the width and 1 cm less than the length of a diagonal. Find the length of the rectangle. 24 cm

9. A square and a rectangle have the same width. The rectangle's length is 4 times its width. Find the area of each figure given that the sum of the two areas is 80 ft^2. 16 ft^2, 64 ft^2

10. The area of a triangle is 57 cm². Find the height of the triangle given that the height is 5 cm less than 4 times the length of the base. 19 cm

11. A rectangle is 7 m longer than it is wide, and each diagonal is 8 m longer than the rectangle's width. What is the length of each diagonal? 13 m

12. The length of a rectangle is 5 in. more than twice its width. The width of a square is the same as the length of the rectangle. Find the area of each figure given that one area, decreased by the other area, is 88 in². 33 in², 121 in²

13. A rectangular wall is 3 m longer than twice its height. The width of a picture on the wall is 1 m less than the wall's height. The picture is 2 m longer than it is wide. Find the area of the wall given that 19 m² of the wall are not covered by the picture. 27 m²

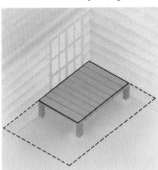

14. A rectangular patio floor has a length that is 1 m less than twice its width. The floor is extended by adding an additional 3 m to the original patio's length and an additional 2 m to its width. Find the area of the original floor given that the total area of the patio floor is 40 m² after the extension is built. 15 m²

15. A rectangle's length is 4 times its width. A triangle's height is 4 times the length of its base. The triangle's height is the same as the rectangle's length. Find the area of each figure given that the sum of the areas is 150 m². 100 m², 50 m²

16. A triangle's base is 2 ft shorter than a rectangle's width. The rectangle's length is 6 times its width, and the triangle's height is the same as the rectangle's length. Find the area of each figure given that the difference between the areas is 105 ft². 45 ft², 150 ft²

17. A rectangular lawn measures 80 ft by 104 ft and is surrounded by a uniform sidewalk. The outer edge of the sidewalk is a rectangle with an area of 10,260 ft². Find the width of the uniform sidewalk. 5 ft

18. A rectangular picture measures 30 cm by 52 cm. It is surrounded by a uniform frame whose outer edge is a rectangle with an area of 2,280 cm². Find the width of one side of the frame. 4 cm

19. A rectangular piece of cardboard is 4 in. longer than it is wide. A 3-inch square is cut from each corner, and the four flaps are turned up to form an open box with a volume of 351 in³. Find the length and the width of the original piece of cardboard. l = 19 in; w = 15 in

20. Find the value of x given that the *area* of the block-letter H at the right is 5,400 square units. 20

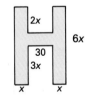

Closure

Have students explain the Pythagorean relation in terms of the legs and the hypotenuse. See *Common Error Analysis* above. Have students give the area formulas for a triangle and a rectangle. See page 209.

■■■ **FOLLOW UP**

Guided Practice

Classroom Exercises 1–7

Independent Practice

A Ex. 1–9, B Ex. 10–16, C Ex. 17–20

Basic: WE 1–13 odd, Midchapter Review, Statistics

Average: WE 3–15 odd, Midchapter Review, Statistics

Above Average: WE 7–19 odd, Midchapter Review, Statistics

Enrichment

Have students solve this puzzle. The solution involves the concept of *limit*, a topic they will encounter in Chapter 15.

A 10-cm square has a square inscribed in it, and a third square is inscribed in the second square, as shown. If this process could be continued forever, what would be the total area of all the squares? 200 cm²

Midchapter Review

Solve. *6.1*

1. $2x^2 = 5x + 12$ $-\frac{3}{2}, 4$ **2.** $12x = 4x^2$ 0, 3

3. $16 = 25x^2$ $\pm\frac{4}{5}$ **4.** $9y^4 - 40y^2 + 16 = 0$ $\pm\frac{2}{3}, \pm2$

5. $5n^2 + 40n + 80 = 0$ -4 **6.** $a^3 - 25a = 0$ 0, ±5

7. Find three consecutive even integers such that the square of the third integer decreased by 5 times the first integer is 7 times the third integer.   *6.2*   −2, 0, 2, and 6, 8, 10

8. A rectangle is 3 m longer than it is wide. The rectangle's length is 3 m less than a diagonal's length. Find the length of a diagonal.   *6.3*   15 m

9. The base of a triangle is 3 ft shorter than twice the triangle's height. Find the length of the base given that the area of the triangle is 27 ft².   *6.3*   9 ft

---

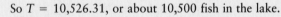

Statistics: *Population Sampling*

Biologists often use **tag sampling** to estimate the total population of fish in a lake. The fish from a *random sample* are counted, tagged, and released. At a later time, another random sample is taken and the following formula is used to predict the total fish population, T.

$$
\begin{array}{ccccc}
\text{Tagged} & \rightarrow & \dfrac{s_1}{S} = \dfrac{F}{T} & \leftarrow & \text{1st Sample} \\
\text{2nd sample} & \rightarrow & & \leftarrow & \text{Population}
\end{array}
$$

A random sample of 500 fish is taken from a lake, tagged, and released. Later, another random sample is taken. This sample contains 800 fish, 38 of which have tags. Substituting in the formula,

$$s_1 = 38, \; S = 800, \; \text{and} \; F = 500.$$

$$\frac{38}{800} = \frac{500}{T} \qquad 38T = 400{,}000$$

So $T = 10{,}526.31$, or about 10,500 fish in the lake.

Exercises

Compute an estimate for the total population T to the nearest hundred.

1. $F = 900, S = 400, s_1 = 73$ **2.** $F = 850, S = 250, s_1 = 25$

     4,900   8,500

6.4 Inequalities of Degree 2 and Degree 3

Objective

To find and graph the solution sets of second- or third-degree inequalities in one variable

In Chapter 2, you solved and graphed linear inequalities by using your knowledge of linear *equalities* as well as that of compound sentences involving *and* and *or*. Now you can solve *quadratic inequalities* in a similar fashion, using your knowledge of quadratic equalities.

An inequality such as $x^2 - 2x - 8 > 0$ is called a **quadratic inequality** in one variable. The standard form of a quadratic inequality is either

$$ax^2 + bx + c > 0 \quad \text{or} \quad ax^2 + bx + c < 0, \text{ where } a \neq 0.$$

When $ax^2 + bx + c$ can be factored as in $x^2 - 2x - 8 > 0$, or $(x + 2)(x - 4) > 0$, the following property involving the positive product of two factors can be used to solve the inequality.

Positive Product of Two Factors

If $m \cdot n > 0$, then [$m < 0$ *and* $n < 0$] *or* [$m > 0$ *and* $n > 0$]. (If the product of two numbers is positive, then the numbers must be either both negative or both positive.)

EXAMPLE 1

Find the solution set of $x^2 - 2x - 8 > 0$. Graph the solutions.

Plan

Factor the polynomial. Use the property of the Positive Product of Two Factors.

Solution

$$x^2 - 2x - 8 > 0$$
$$(x + 2)(x - 4) > 0$$

$(x + 2 < 0 \text{ and } x - 4 < 0)$	*or*	$(x + 2 > 0 \text{ and } x - 4 > 0)$
$(x < -2 \text{ and } x < 4)$	*or*	$(x > -2 \text{ and } x > 4)$
$x < -2$	*or*	$x > 4$

The solution set is $\{x \mid x < -2 \text{ or } x > 4\}$.

The quadratic inequality $x^2 - 9 < 0$, or $(x + 3)(x - 3) < 0$, can be solved using the property of the *Negative Product of Two Factors*.

Additional Example 1

Find the solution set of $x^2 - 3x - 18 > 0$. Graph the solutions.

$$\{x \mid x < -3 \text{ or } x > 6\}$$

Teaching Resources

Quick Quizzes 43
Reteaching and Practice Worksheets 43
Transparency 14

▮▮▮GETTING STARTED

Prerequisite Quiz

Graph the solution set.

1. $x < 2$ *or* $x > 5$

2. $x < 2$ *and* $x > 5$ No points, $\varnothing$
3. $x < 3$ *and* $x < 6$

4. $x > 4$ *and* $x < 7$

Motivator

Ask students the following questions.

1. Does the graph of $x < -2$ *and* $x < 4$ include all points less than 4?   No.
2. What points are included in the graph of Exercise 1?
   The solution set is $\{x \mid x < -2\}$.
3. How will the graph change if the "*and*" in Exercise 1 is changed to "*or*"?   The solution set will be $\{x \mid x < 4\}$.

Highlighting the Standards

Standard 3c: Solving inequalities is an exercise in logic rather than in the use of an algorithm.

Lesson Note

For inequalities of degree 2, the properties of the positive or negative product of two factors are used, as in Examples 1 and 2. For inequalities of degree greater than 2, the method of Example 3 should be used. Students may discover that the method of Example 3 can also be used with quadratic inequalities.

Math Connections

Linear Inequalities: The inequalities of this lesson have graphs that closely resemble the graphs of compound inequalities found in Lessons 2.2 and 2.3. This occurs because, although the technique for solving quadratic inequalities requires one initially to write a compound sentence of some complexity, it is possible to simplify such a sentence to a simple conjunction or disjunction of the kind found in Chapter 2.

Negative Product of Two Factors
If $m \cdot n < 0$, then $[m < 0 \ and \ n > 0]$ or $[m > 0 \ and \ n < 0]$.
(If the product of two numbers is negative, then one of the numbers must be positive and the other must be negative.)

If the coefficient of the square term in a quadratic inequality is negative, it may be helpful to divide each side by -1 as shown below.

EXAMPLE 2   Find the solution set of $-x^2 + 9 > 0$. Graph the solutions.
Divide each side by -1 and reverse the order of the inequality.

Solution
$$-x^2 + 9 > 0$$
$$x^2 - 9 < 0$$
$$(x + 3)(x - 3) < 0$$

Next, use the property of the Negative Product of Two Factors.

$(x + 3 < 0 \ and \ x - 3 > 0)$ or $(x + 3 > 0 \ and \ x - 3 < 0)$
$\underbrace{(x < -3 \ and \ x > 3)}_{\text{no solution}}$ or $\underbrace{(x > -3 \ and \ x < 3)}_{-3 < x < 3}$

The solution set is $\{x \mid -3 < x < 3\}$.

For a *polynomial inequality* of degree 3 or higher, it is usually easier to graph the solutions *first*, and then write the solution set. Begin by factoring the polynomial completely. Then solve the equation that corresponds to the inequality, as shown below.

EXAMPLE 3   Graph the solutions of $x^3 - x^2 - 20x > 0$. Then write the solution set.

Solution   Factor the polynomial completely.
$$x^3 - x^2 - 20x > 0$$
$$x(x^2 - x - 20) > 0$$
$$x(x + 4)(x - 5) > 0$$

Let $x(x + 4)(x - 5) = 0$. The roots of this *equation* are 0, -4, and 5. Plot these numbers with *open* dots on a number line. (The open dots indicate that these numbers are *not* solutions of the inequality $x(x + 4)(x - 5) > 0$.)

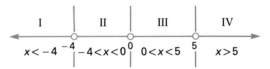

Additional Example 2

Find the solution of $-y^2 + 2y + 15 > 0$. Graph the solutions.

$$\{y \mid -3 < y < 5\}$$

○——○
−3   5

Additional Example 3

Find the solutions of $a^3 + a^2 - 12a < 0$. Then graph the solution set.

$$\{a \mid a < -4 \ or \ 0 < a < 3\}$$

◄—○ ○——○
−4   0   3

The points at -4, 0, and 5 separate the preceeding number line into four parts: I, II, III, and IV. Choose a number for x in each part and evaluate $x(x + 4)(x - 5)$ to find where its value is positive.

Part	x	$x(x + 4)(x - 5)$	Value	Positive(?)
$x < -4$	-5	$-5(-1)(-10)$	-50	No
$-4 < x < 0$	-2	$-2(+2)(-7)$	$+28$	Yes
$0 < x < 5$	$+1$	$+1(+5)(-4)$	-20	No
$x > 5$	$+6$	$+6(+10)(+1)$	$+60$	Yes

Notice in the above table that the value of the polynomial *changes sign* whenever one of the boundary points, -4, 0, or 5, is crossed.

The graph of the solutions is parts II and IV, a segment and a ray.

The solution set is $\{x \mid -4 < x < 0 \quad or \quad x > 5\}$.

If the inequality $x^3 - x^2 - 20x > 0$ were combined with its related equation to form the compound sentence $x^3 - x^2 - 20x \geq 0$, then the numbers 0, -4, and 5, would be solutions of the compound sentence.

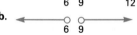

Classroom Exercises

Match each inequality to exactly one graph.

1. $6 < x < 9$ e
2. $x < 6$ or $x > 9$ b
3. $x < 6$ or $9 < x < 12$ a
4. $6 < x < 9$ or $x > 12$ d
5. $x^2 + 9 > 0$ f
6. $x^2 + 6 < 0$ c

a. ◄————○ ○————————○—→
         6 9   12

b. ◄————○ ○—————————→
         6 9

c. no points

d. ○—————————○ ○——→
     6   9 12

e. ○—————————○
    6   9

f. ◄——●———●———●——→
      6   9   12

Match each sentence at the left with one compound sentence at the right.

7. $(x + 3)(x - 5) < 0$ b

8. $(x + 3)(x - 5) > 0$ c

9. $(x - 1)(x - 4) \leq 0$ a

a. $[x - 1 \leq 0 \ and \ x - 4 \geq 0]$ or $[x - 1 \geq 0 \ and \ x - 4 \leq 0]$

b. $[x + 3 < 0 \ and \ x - 5 > 0]$ or $[x + 3 > 0 \ and \ x - 5 < 0]$

c. $[x + 3 < 0 \ and \ x - 5 < 0]$ or $[x + 3 > 0 \ and \ x - 5 > 0]$

Analysis: Ask students to show that in Example 3 (or in another problem of the same kind), whenever a boundary point is crossed, the value of the factored polynomial must change signs.   This occurs because an interval around a boundary point can always be found that is small enough so that the factor with that boundary point is negative on one side of the point and positive on the other, while all the other factors have the same sign throughout the interval. For example, for all values of x, when $-5 \leq x \leq -3$, both x and $x - 5$ are negative. However, $x + 4$ is negative when $-5 \leq x < -4$, and positive when $-4 < x \leq -3$. So, in any small interval around -4, $x + 4$ changes sign while the other factors remain negative.

Common Error Analysis

Error: Some students will incorrectly extend the Zero Property (if $mn = 0$, then $m = 0$ or $n = 0$) to situations involving inequalities. They may think mistakenly: If $mn > 0$, then $m > 0$ or $n > 0$.

Have the students write a true inequality like $-4(-3) > 0$ and discuss the signs of the factors.

Checkpoint

Find the solution set of each inequality. Graph the solutions.

1. $-x^2 - 2x + 24 \geq 0$ $\{x \mid -6 \leq x \leq 4\}$

●————————●
-6 4

2. $y^3 - 64y < 0$
   $\{y \mid y < -8 \ or \ 0 < y < 8\}$

Closure

Ask students how to graph the solution set of a quadratic inequality.   See Examples 1 and 2.   Ask students to describe the first step in graphing the solution set of a polynomial inequality of degree 3 or higher.
Factor completely

▰▰▰FOLLOW UP

Guided Practice

Classroom Exercises 1–9

Independent Practice

Ⓐ Ex. 1–12, Ⓑ Ex. 13–33, Ⓒ Ex. 34–42

Basic: WE 1–17 odd
Average: WE 9–25 odd
Above Average: WE 23–39 odd

Additional Answers

Written Exercises
 1. $\{n \mid n < 2 \text{ or } n > 6\}$

 2   6

 2. $\{x \mid -2 < x < 5\}$

 -2   5

 3. $\{c \mid -5 \le c \le -2\}$

 -5   -2

 4. $\{n \mid n \le 1 \text{ or } n \ge 6\}$

 1   6

See pages 220–222 for answers to Written Ex. 5–12, 15–42, and Mixed Review Ex. 1–4.

Written Exercises

Find the solution set of each inequality. Graph the solutions.

1. $n^2 - 8n + 12 > 0$ **2.** $x^2 - 3x - 10 < 0$ **3.** $c^2 + 7c + 10 \le 0$
4. $n^2 - 7n + 6 \ge 0$ **5.** $x^2 - 16 \ge 0$ **6.** $y^2 - 36 \le 0$
7. $-y^2 + 25 < 0$ **8.** $-a^2 + 4 > 0$ **9.** $n^2 - 4n > 0$
10. $c^2 + 6c < 0$ **11.** $-3c^2 - 12c \ge 0$ **12.** $-4x^2 + 12x \le 0$

13. Find the set of all numbers whose squares, decreased by 49, result in a negative number. $\{x \mid -7 < x < 7\}$

14. Find the set of all numbers such that 5 times the square of a number, decreased by 20 times the number, is a positive number. $\{n \mid n < 0 \text{ or } n > 4\}$

Graph the solutions of each inequality. Then write the solution set.

15. $y^3 - y^2 - 12y < 0$ **16.** $x^3 + 6x^2 - 7x > 0$ **17.** $x^3 - 16x \ge 0$
18. $y^3 - 25y \le 0$ **19.** $-y^3 - y^2 + 20y > 0$ **20.** $-x^3 + 49x < 0$

Find the solution set of each inequality. Graph the solutions.

21. $2x^2 + x < 15$ **22.** $4a^2 > 9a + 9$ **23.** $2y^2 \ge 9y$
24. $4c^2 \le 10c$ **25.** $-c^2 - c + 12 < 0$ **26.** $-x^2 + 6x \ge 8$
27. $0 < -a^2 + 4a - 3$ **28.** $0 \ge 35 - 2y - y^2$ **29.** $2x^3 - x^2 - 21x > 0$
30. $3a^3 + 5a^2 - 28a \le 0$ **31.** $a^2 < 4 \text{ or } a^2 > 36$ **32.** $c^2 \ge 4 \text{ and } c^2 \le 36$

33. Write in your own words the property of the Negative Product of Two Factors.

Graph the solutions and write the solution set.

34. $4x^2 + 1 > 3x^2$ **35.** $y^2 - 3 > 2y^2 + 3$ **36.** $|x^2 - 5| < 4$
37. $|y^2 - 10| > 6$ **38.** $a^4 - 10a^2 + 9 < 0$ **39.** $x^4 - 26x^2 + 25 > 0$
40. $a^4 - 16 \ge 0$ **41.** $n^4 - 81 \le 0$ **42.** $y^4 - 20y^2 + 64 > 0$

Mixed Review

Divide to find the polynomial quotient and the remainder. Is the divisor a factor of the dividend?   *5.8*

1. $(2x^3 - 4x - 10) \div (x - 2)$ **2.** $(y^3 + 4y^4 - y - 2 - 2y^2) \div (y + 1)$
3. $(3x^3 + 8x^2 - 4x - 3) \div (x + 3)$ **4.** $(y^3 - 10) \div (y - 2)$

5. An 18% alcohol solution is mixed with a 45% alcohol solution to produce 12 ounces of a 36% alcohol solution. How many ounces of each solution must be used? *4.4* 18%: 4 oz; 45%: 8 oz

Enrichment

Challenge the students to prove that the sum of any positive number and its reciprocal is greater than or equal to 2. Give the hint that they should use an indirect proof. That is, in this case, they should assume that $x + \frac{1}{x} < 2$, then show that this leads to an impossibility.

Assume $x + \frac{1}{x} < 2$ (1).

$\frac{x^2 + 1}{x} < 2$ (2)

$\frac{x^2 + 1}{x} - 2 < 0$ (3)

$\frac{x^2 - 2x + 1}{x} < 0$ (4)

$\frac{(x - 1)^2}{x} < 0$ (5)

Since $(x - 1)^2 \ge 0$ for any real number and x is a positive number, statement (5) is impossible. So, $x + \frac{1}{x} \ge 2$ for any positive number x.

6.5 Synthetic Substitution and the Remainder Theorem

Objectives

To divide a polynomial $P(x)$ by $x - a$ using synthetic division
To evaluate a polynomial $P(x)$ using synthetic substitution
To determine whether a binomial $x - a$ is a factor of a polynomial $P(x)$

For brevity, a polynomial such as $3x^3 - 10x^2 + 14x - 7$ is sometimes represented as $P(x)$ (read "P of x" or "P at x"). Similarly, a *polynomial function* can be represented as $y = P(x)$. Thus, the two polynomial functions $y = 3u^2 + 1$ and $y = 2w^5 + w^2 - 1$ might be represented as $y = P_1(u)$ and $y = P_2(w)$, respectively.

Recall from Lesson 5.8 the long-division technique for dividing a polynomial by any binomial. In this lesson, a polynomial $P(x)$ is divided by a binomial in the special form $x - a$. It is also determined whether the divisor $x - a$ is a factor of $P(x)$.

Here, the polynomial
$P(x) = 3x^3 - 10x^2 + 14x - 7$
is divided by $x - 2$ using
long division.

$$
\begin{array}{r}
3x^2 - 4x + 6 \\
x - 2 \overline{)\,3x^3 - 10x^2 + 14x - 7} \\
\underline{3x^3 - 6x^2} \\
-4x^2 + 14x \\
\underline{-4x^2 + 8x} \\
6x - 7 \\
\underline{6x - 12} \\
5
\end{array}
$$

Shown below is a more compact form of this division called *synthetic division*. Note that the divisor $x - a = x - 2$, and that $a = 2$. To begin, write 2 and the coefficients of $P(x)$.

Bring down the 3. Then, use the "multiply by 2 and add" cycle as shown.

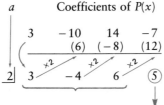

Quotient: $3x^2 - 4x + 6$ Remainder: 5

Notice that both methods yield the same results: (1) the polynomial quotient $Q(x)$ is $3x^2 - 4x + 6$, (2) the remainder R is 5, and (3) the divisor $x - 2$ is *not* a factor of the dividend $P(x)$ because the remainder is not 0.

▮▮▮ GETTING STARTED

Prerequisite Quiz

Divide to find the polynomial quotient and the remainder. Is the divisor a factor of the dividend?

1. $(4x^3 + 5x^2 - 8x - 4) \div (x + 2)$
   $4x^2 - 3x - 2$; 0; yes
2. $(2y^3 - 22y + 18) \div (y - 3)$
   $2y^2 + 6y - 4$; 6; no

Motivator

Have students use long division to divide $3x^3 + x^2 - 14x - 6$ by $x + 2$ to determine the polynomial quotient. $3x^2 - 5x - 4$ Is there a remainder?   Yes.   Ask students if $x + 2$ is a factor of $3x^3 + x^2 - 14x - 6$.   No.

▮▮▮ TEACHING SUGGESTIONS

Lesson Note

As mentioned earlier, this lesson is part of a strand that includes the Integral Zero Theorem and the Rational Zero Theorem and their use in solving equations and graphing polynomial functions of degree greater than 2. The calculator activities at the bottom of page 222 tell students how to use synthetic substitution and a calculator with the ⌐+/−⌐ function to evaluate $P(x)$ for $x > 0$ and for $x < 0$.

Highlighting the Standards

Standard 5d: The theorems in this lesson are landmarks in algebra. Students should work through the numerical introduction to appreciate the reasoning behind the theorems.

It can also be shown that the value of $P(x) = 3x^3 - 10x^2 + 14x - 7$ is 5 when $x = 2$. That is, $P(2) = 5$.

$$\text{Dividend} = \text{Divisor} \times \text{Quotient} + \text{Remainder}$$

$$P(x) = (x - 2)(3x^2 - 4x + 6) + 5$$
$$P(2) = (2 - 2)(3 \cdot 2^2 - 4 \cdot 2 + 6) + 5$$
$$= 0 + 5$$
$$= 5$$

Thus, the value of the polynomial at $x = 2$ is 5, which is the same as the remainder when the polynomial is divided by $x - 2$. This suggests the following theorem.

Theorem 6.1

> **Remainder Theorem:** When a polynomial in x is divided by $x - a$, the remainder equals the value of the polynomial when $x = a$.

One consequence of the Remainder Theorem is that $x - a$ and $Q(x)$ are factors of $P(x)$ when $R = 0$. This suggests the **Binomial Factor Theorem** (also referred to as the *Factor Theorem*).

Theorem 6.2

> **Binomial Factor Theorem:** The binomial $x - a$ is a factor of the polynomial $P(x)$ if and only if $P(a) = 0$.

If $x - a = x + 5$, then $x - a = x - (-5)$ and $a = -5$. Thus, if a polynomial $P(x)$ is divided by $x + 5$, the remainder will be the value of $P(-5)$.

EXAMPLE 1   Given $P(x) = 2x^4 + 11x^3 - 27x - 10$, use synthetic division to divide $P(x)$ by $x + 5$. (1) Find the polynomial quotient $Q(x)$ and the remainder R. (2) Find the value of $P(x)$ for $x = -5$. (3) Determine whether $x + 5$ is a factor of $P(x)$.

Plan   There is no x^2 term. Write 0 for the coefficient of the missing x^2 term.

Solution

$$x + 5 = x - (-5)$$

$$-5\big|\ \begin{array}{ccccc} 2 & 11 & 0 & -27 & -10 \\ & (-10) & (-5) & (25) & (10) \\ \hline 2 & 1 & -5 & -2 & 0 \end{array}$$

(1) $Q(x) = 2x^3 + x^2 - 5x - 2$ and $R = 0$
(2) $P(-5) = 0$, because $P(-5) = R$ and $R = 0$
(3) $x + 5$ is a factor of $P(x)$ because $R = 0$.

Additional Example 1

Given $P(x) = 3x^4 + 5x^3 - 4x^2 + 8$, use synthetic division to divide $P(x)$ by $x + 2$. (1) Find the polynomial quotient and remainder, (2) find the value of $P(x)$ if $x = -2$, and (3) determine whether $x + 2$ is a factor of $P(x)$.
$Q(x) = 3x^3 - x^2 - 2x + 4, R = 0$; $P(-2) = 0$; yes

Finding a value of a polynomial by synthetic division is called **synthetic substitution.** For example, $P(x) = 6x^3 + 7x^2 - 8x - 9$ is evaluated for $x = 2$ by synthetic substitution as follows.

$$
2 \left| \begin{array}{cccc}
6 & 7 & -8 & -9 \\
 & (12) & (38) & (60) \\
\hline
6 & 19 & 30 & \boxed{51}
\end{array} \right.
$$

The last number, 51, is the value of $P(x)$ when $x = 2$. $P(2) = 51$.

A more compact form of the synthetic substitution above is shown below. In this form, some of the operations are performed mentally. Thus, the multiples of 2 (12, 38, and 60) are not shown.

$$
2 \left| \begin{array}{cccc}
6 & 7 & -8 & -9 \\
\hline
6 & 19 & 30 & \boxed{51}
\end{array} \right.
$$

The "multiply by 2 and add" cycle above may be better understood by seeing $P(x)$ expressed in a different way. First, rewrite $P(x)$ as follows.

$$P(x) = 6x^3 + 7x^2 - 8x - 9 = (6x^2 + 7x - 8)x - 9$$
$$= [(6x + 7)x - 8]x - 9$$

Next, use direct substitution to see the "multiply by 2 and add" cycle.

$$P(2) = [(6 \cdot 2 + 7)2 - 8]2 - 9 = \boxed{51}$$

Multiply by 2.   Then add.

EXAMPLE 2   Evaluate $P(x) = 2x^4 - x^3 + 4x - 12$ for $x = -2$ and for $x = 3$.

Solution

$$
\begin{array}{c}
-2 \\
\\
3
\end{array} \left| \begin{array}{ccccc}
2 & -1 & 0 & 4 & -12 \\
 & (-4) & (10) & (-20) & (32) \\
\hline
2 & -5 & 10 & -16 & \boxed{20} \\
 & (6) & (15) & (45) & (147) \\
\hline
2 & 5 & 15 & 49 & \boxed{135}
\end{array} \right.
$$

Compact Form

$$
\begin{array}{c}
-2 \\
3
\end{array} \left| \begin{array}{ccccc}
2 & -1 & 0 & 4 & -12 \\
\hline
2 & -5 & 10 & -16 & \boxed{20} \\
\hline
2 & 5 & 15 & 49 & \boxed{135}
\end{array} \right.
$$

Thus, $P(-2) = 20$ and $P(3) = 135$.

Additional Example 2

Evaluate.

$P(x) = 2x^4 - 51x^2 + 3x + 10$ if $x = 5$ and again if $x = -1$.
$P(5) = 0$; $P(-1) = -42$

5. $\{x \mid x \le -4 \text{ or } x \ge 4\}$

$$\begin{array}{c} \bullet\!\!\!\!\!-\!\!\!\!\!\longleftrightarrow\!\!\!\!\!\longleftrightarrow\!\!\!\!\!-\!\!\!\!\!\bullet \\ -4 \qquad 4 \end{array}$$

6. $\{y \mid -6 \le y \le 6\}$

$$\begin{array}{c} \bullet\!\!\!\!\!\longleftrightarrow\!\!\!\!\!\bullet \\ -6 \qquad 6 \end{array}$$

7. $\{y \mid y < -5 \text{ or } y > 5\}$

$$\begin{array}{c} \circ\!\!\!\!\!\longleftrightarrow\!\!\!\!\!\circ \\ -5 \qquad 5 \end{array}$$

8. $\{a \mid -2 < a < 2\}$

$$\begin{array}{c} \circ\!\!\!\!\!\longleftrightarrow\!\!\!\!\!\circ \\ -2 \qquad 2 \end{array}$$

9. $\{n \mid n < 0 \text{ or } n > 4\}$

$$\begin{array}{c} \circ\!\!\!\!\!\longleftrightarrow\!\!\!\!\!\circ \\ 0 \qquad 4 \end{array}$$

10. $\{c \mid -6 < c < 0\}$

$$\begin{array}{c} \circ\!\!\!\!\!\longleftrightarrow\!\!\!\!\!\circ \\ -6 \qquad 0 \end{array}$$

11. $\{c \mid -4 \le c \le 0\}$

$$\begin{array}{c} \bullet\!\!\!\!\!\longleftrightarrow\!\!\!\!\!\bullet \\ -4 \qquad 0 \end{array}$$

12. $\{x \mid x \le 0 \text{ or } x \ge 3\}$

$$\begin{array}{c} \bullet\!\!\!\!\!\longleftrightarrow\!\!\!\!\!\bullet \\ 0 \qquad 3 \end{array}$$

15. $\{y \mid y < -3 \text{ or } 0 < y < 4\}$

$$\begin{array}{c} \circ\!\!\!\!\!\longleftrightarrow\!\!\!\!\!\circ\!\!\!\!\!-\!\!\!\!\!\circ \\ -3 \quad 0 \quad 4 \end{array}$$

16. $\{x \mid -7 < x < 0 \text{ or } x > 1\}$

$$\begin{array}{c} \circ\!\!\!\!\!\longleftrightarrow\!\!\!\!\!\circ\!\!\!\!\!-\!\!\!\!\!\circ \\ -7 \quad 0 \; 1 \end{array}$$

17. $\{x \mid -4 \le x \le 0 \text{ or } x \ge 4\}$

$$\begin{array}{c} \bullet\!\!\!\!\!\longleftrightarrow\!\!\!\!\!\bullet\!\!\!\!\!-\!\!\!\!\!\bullet \\ -4 \quad 0 \quad 4 \end{array}$$

18. $\{y \mid y \le -5 \text{ or } 0 \le y \le 5\}$

$$\begin{array}{c} \bullet\!\!\!\!\!\longleftrightarrow\!\!\!\!\!\bullet\!\!\!\!\!-\!\!\!\!\!\bullet \\ -5 \quad 0 \quad 5 \end{array}$$

EXAMPLE 3 Evaluate $P(x) = 3x^4 + 3x^3 - 20x^2 - 2x + 12$ for -3, -2, 2, and 3. Name the binomial factors found, if any.

Plan To evaluate $P(x)$ for a or $-a$, write $\underline{a}\rfloor$ or $\underline{-a}\rfloor$, respectively. If $P(a) = 0$, then $x - a$ is a factor of $P(x)$. If $P(-a) = 0$, then $x + a$ is a factor of $P(x)$.

Solution

$$
\begin{array}{r|rrrrr}
 & 3 & 3 & -20 & -2 & 12 \\
 &   & (-9) & (18) & (6) & (-12) \\
\hline
-3\rfloor & 3 & -6 & -2 & 4 & \boxed{0} \\
 &   & (-6) & (6) & (28) & (-52) \\
\hline
-2\rfloor & 3 & -3 & -14 & 26 & \boxed{-40} \\
 &   & (6) & (18) & (-4) & (-12) \\
\hline
2\rfloor & 3 & 9 & -2 & -6 & \boxed{0} \\
 &   & (9) & (36) & (48) & (138) \\
\hline
3\rfloor & 3 & 12 & 16 & 46 & \boxed{150} \\
\end{array}
$$

Compact form

$$
\begin{array}{r|rrrrr}
 & 3 & 3 & -20 & -2 & 12 \\
\hline
-3\rfloor & 3 & -6 & -2 & 4 & \boxed{0} \\
\hline
-2\rfloor & 3 & -3 & -14 & 26 & \boxed{-40} \\
\hline
2\rfloor & 3 & 9 & -2 & -6 & \boxed{0} \\
\hline
3\rfloor & 3 & 12 & 16 & 46 & \boxed{150} \\
\end{array}
$$

Thus, $P(-3) = 0$, $P(-2) = -40$, $P(2) = 0$, and $P(3) = 150$. Two factors of $P(x)$ are $x + 3$ and $x - 2$.

Focus on Reading

Use the example of synthetic substitution shown at the right and the list of expressions below to complete the paragraph.

$$
\begin{array}{r|rrrr}
 & 2 & 9 & 0 & -6 \\
-4\rfloor & 2 & 1 & -4 & \boxed{10} \\
\end{array}
$$

The dividend is __k__, the divisor is __d__, 10 is the __a__, and the polynomial quotient is __i__. The value of $P(x)$ is __c__ when the value of x is __h__. The binomial $x + 4$ is not a __g__ of $2x^3 + 9x^2 - 6$.

a. remainder
b. -10
c. 10
d. $x + 4$
e. $x - 4$
f. $2x^3 + x^2 - 4x + 10$
g. factor
h. -4
i. $2x^2 + x - 4$
j. $2x^2 + 9x - 6$
k. $2x^3 + 9x^2 - 6$

220 Chapter 6 Higher-Degree Equations and Inequalities

Additional Example 3

Evaluate $P(x) = 3x^4 - 6x^3 - 25x^2 + 2x + 8$ using each of -2, -1, 1, and 4. Name the binomial factors found, if any. $P(-2) = 0$, $P(-1) = -10$, $P(1) = -18$, $P(4) = 0$; $x + 2$, $x - 4$

Classroom Exercises

Determine whether each statement is true or false.

1. If $P(x) = (x + 3)(x^2 - 2x - 5) + 4$, then $x + 3$ is a factor of $P(x)$. F
2. If $P(x) = (x - 5)(2x^2 + 7)$, then $x - 5$ is a factor of $P(x)$. T
3. If $P(x) = (x + 4)(x^2 - 5x + 6)$, then $P(x) = 0$ when $x = -4$. T
4. If $P(x) = (x - 6)(x^2 + 4x + 4) + 5$, then $P(6) = 5$. T

Evaluate $P(x) = x^2 + x - 2$ for each value of x.

5. $x = 0$ -2   6. $x = -1$ -2   7. $x = 1$ 0   8. $x = -2$ 0   9. $x = 2$ 4
10. Use the results of Exercises 5–9 to name the binomial factors of $x^2 + x - 2$. $(x - 1)(x + 2)$

Written Exercises

Use synthetic division to find the polynomial quotient and the remainder. Is the divisor a factor of the dividend?

1. $(2x^3 - 11x^2 + 18x - 15) \div (x - 3)$ $2x^2 - 5x + 3, -6$; no
2. $(3x^3 + 17x^2 - 8x - 12) \div (x + 6)$ $3x^2 - x - 2, 0$; yes
3. $(y^4 + 2y^3 - 3y^2 + 2y + 16) \div (y + 2)$ $y^3 - 3y + 8, 0$; yes
4. $(3y^4 - 12y^3 - 20y^2 + 30y + 10) \div (y - 5)$ $3y^3 + 3y^2 - 5y + 5, 35$; no
5. $(4n^5 - 2n^3 + 6n^2 - 9n + 1) \div (n - 1)$ $4n^4 + 4n^3 + 2n^2 + 8n - 1, 0$; yes

Evaluate each polynomial for the given values using synthetic substitution. Name the binomial factors found, if any.

6. $x^3 + 2x^2 - 8x - 21$ for $x = 3$ and again for $x = -2$ $0, -5; x - 3$
7. $x^3 - 4x^2 - 4x - 5$ for $x = -2$ and again for $x = 5$ $-21, 0; x - 5$
8. $3y^4 + y^3 - 14y^2 + 10y - 60$ for $y = 2$ and again for $y = -3$ $-40, 0; y + 3$
9. $2x^5 - 6x^4 - 3x^2 + 44$ for $x = 3$ and again for $x = 2$ $17, 0; x - 2$

Evaluate each polynomial for $-3, -2, -1, 0, 1, 2,$ and 3. Name the binomial factors found, if any.

10. $2x^4 - 4x^3 - 5x^2 - 2x - 3$ 11. $x^4 + x^3 - 4x - 16$

Use synthetic division to find the polynomial quotient and the remainder. Is the divisor a factor of the dividend?

12. $(c^3 - 8,000) \div (c - 10)$ 13. $(c^3 + 8,000) \div (c + 20)$
14. $(x^6 + x^4 + x^2 - 84) \div (x + 2)$ 15. $(x^5 - x^3 - 210) \div (x - 3)$

6.5 Synthetic Substitution and the Remainder Theorem   **221**

Enrichment

Write this polynomial on the chalkboard.

$x^4 + ax^3 + 5x^2 + 5x - 6$

Ask the students to find the value of a such that $x - 2$ will be a factor of the polynomial. Tell them that they should use synthetic substitution and the Remainder Theorem to find the answer. $a = -5$

Now ask each student to write three similar problems of their own, and to challenge each other with them.   Problems and solutions will vary.

30. $\left\{a \mid a \le -4 \text{ or } 0 \le a \le 2\frac{1}{3}\right\}$

$-4 \quad 0 \quad \frac{7}{3}$

31. $\{a \mid -2 < a < 2 \text{ or } a < -6 \text{ or } a > 6\}$

$-6 \; -2 \quad 2 \quad 6$

32. $\{c \mid -6 \le c \le -2 \text{ or } 2 \le c \le 6\}$

$-6 \; -2 \quad 2 \quad 6$

33. Answers will vary. Possible answer:
The product of two factors is negative if
and only if the factors have opposite
signs. Therefore, one and only one of
the factors must be negative.

34. {Real Numbers}

35. $\varnothing$

36. $\{x \mid -3 < x < -1 \text{ or } 1 < x < 3\}$

$-3 \; -1 \quad 1 \quad 3$

37. $\{y \mid y < -4 \text{ or } -2 < y < 2 \text{ or } y > 4\}$

$-4 \; -2 \quad 2 \quad 4$

38. $\{a \mid -3 < a < -1 \text{ or } 1 < a < 3\}$

$-3 \; -1 \quad 1 \quad 3$

39. $\{x \mid x < -5 \text{ or } -1 < x < 1 \text{ or } x > 5\}$

$-5 \; -1 \quad 1 \quad 5$

40. $\{a \mid a \le -2 \text{ or } a \ge 2\}$

$-2 \qquad 2$

41. $\{n \mid -3 \le n \le 3\}$

$-3 \qquad 3$

42. $\{y \mid y < -4 \text{ or } -2 < y < 2 \text{ or } y > 4\}$

$-4 \; -2 \quad 2 \quad 4$

Use synthetic substitution to find the value of k in each polynomial $P(x)$, given a factor of $P(x)$.

16. $x - 2$ is a factor of $P(x) = x^3 + 4x^2 - 15x + k$. 6
17. $x + 3$ is a factor of $P(x) = 2x^4 + 5x^3 + kx^2 + 10x - 6$. 1

Use synthetic division to find the polynomial quotient and the remainder. Is the divisor a factor of the dividend?

18. $(2x^3 - 7x^2 + 7x - 2) \div (x - \frac{1}{2})$ **19.** $(2y^3 - \frac{5}{2}y^2 + \frac{15}{4}y - \frac{3}{2}) \div (y - \frac{1}{2})$

Mixed Review

Write an equation in standard form, $ax + by = c$, for each line described below.

1. crossing the y-axis at $P(0, -2)$ with a slope of $\frac{2}{3}$ *3.5* $2x - 3y = 6$
2. passing through $P(4, -2)$ and parallel to the line described by $2x - y = 5$ *3.7* $2x - y = 10$
3. passing through $P(4, -2)$ and perpendicular to the line described by $2x - y = 5$ *3.7* $x + 2y = 0$

▰ Using the Calculator

You can use synthetic substitution and a calculator with the $\boxed{+/-}$ function to evaluate any polynomial $P(x)$ for a given value of x. For example, to evaluate $P(x) = 5x^4 + 4x^3 - 6x^2 - 7x + 12$ for $x = 8$, a positive number, use the sequence of keystrokes below.

$5 \boxed{\times} 8 \boxed{+} 4 \boxed{=} \boxed{\times} 8 \boxed{-} 6 \boxed{=} \boxed{\times} 8 \boxed{-} 7 \boxed{=} \boxed{\times} 8 \boxed{+}$
$12 \boxed{=} 22,100$

Thus, $P(8) = 22,100$.

Notice that the keystrokes use the multiply-and-add cycle of synthetic substitution.

The keystrokes for $x = -8$, a negative number, are as follows.

$5 \boxed{\times} 8 \boxed{+/-} \boxed{+} 4 \boxed{=} \boxed{\times} 8 \boxed{+/-} \boxed{-} 6 \boxed{=} \boxed{\times} 8 \boxed{+/-} \boxed{-}$
$7 \boxed{=} \boxed{\times} 8 \boxed{+/-} \boxed{+} 12 \boxed{=} 18,116$

Thus, $P(-8) = 18,116$.

Evaluate $P(x) = 6x^4 - 5x^3 + 4x^2 - 7x - 9$ for the given value of x.

1. $x = 2$ 49 **2.** $x = 3$ 357 **3.** $x = -2$ 157 **4.** $x = -3$ 669

Mixed Review

1. $2x^2 + 4x + 4, -2$; no
2. $4y^3 - 3y^2 + y - 2, 0$; yes
3. $3x^2 - x - 1, 0$; yes
4. $y^2 + 2y + 4, -2$; no

6.6 The Integral Zero Theorem

Objectives

To find the rational zeros of integral polynomials

To factor integral polynomials into first-degree factors

To find the rational roots of integral polynomial equations of degree greater than 2

The polynomial $P(x) = 2x^4 - 7x^3 - 5x^2 + 28x - 12$ is an **integral polynomial** because all of its coefficients are integers. By synthetic substitution, or otherwise, you can show that $P(3) = 0$. The number 3 is called a *zero* of $P(x)$.

Definition

A number r is called a **zero of a polynomial function** $P(x)$ if and only if $P(r) = 0$.

A *zero of a polynomial function* may be referred to more briefly as a *zero of a polynomial*.

EXAMPLE 1 Find the four rational zeros of $P(x) = 2x^4 - 7x^3 - 5x^2 + 28x - 12$, given that $x - 3$ and $x + 2$ are factors.

Plan Use synthetic division and the Factor Theorem.

1. Divide by one of the factors, say $x - 3$, to obtain a quotient, $Q_1(x)$, with a zero remainder ($R_1 = 0$).

2. Divide $Q_1(x)$ by $x + 2$ to find $Q_2(x)$ with $R_2 = 0$. Because $x + 2$ is a factor of $Q_1(x)$, it must also be a factor of $P(x)$.

3. Factor $Q_2(x)$.

Solution

1.
$$3 \begin{array}{|ccccc} 2 & -7 & -5 & 28 & -12 \\ \hline 2 & -1 & -8 & 4 & \boxed{0} \end{array}$$

$Q_1(x) = 2x^3 - x^2 - 8x + 4$; $R_1 = 0$

2.
$$-2 \begin{array}{|cccc} 2 & -1 & -8 & 4 \\ \hline 2 & -5 & 2 & \boxed{0} \end{array}$$

$Q_2(x) = 2x^2 - 5x + 2$; $R_2 = 0$

The *second-degree* factor $2x^2 - 5x + 2$ can be factored into two first-degree factors: $2x^2 - 5x + 2 = (2x - 1)(x - 2)$.

Teaching Resources

Quick Quizzes 45
Reteaching and Practice Worksheets 45

GETTING STARTED

Prerequisite Quiz

Solve.

1. $(2x - 3)(4x + 1)(x + 4)(x - 5) = 0$
   $\frac{3}{2}, -\frac{1}{4}, -4, 5$

2. $(y - 4)(y - 4)(3y + 1)(3y + 1) = 0$
   $4, -\frac{1}{3}$

Use synthetic substitution to determine whether the binomial is a factor of the polynomial $P(x)$.

3. $x + 2$, $P(x) = 4x^3 + 5x^2 - 8x - 4$
   Yes, ($R = 0$)

Motivator

Have students factor 11,067 completely given that 17 and 31 are two of the factors. $17 \times 31 \times 7 \times 3$ Ask: If 11,067 is divided by any of these factors, will there be a remainder other than zero?   No.

TEACHING SUGGESTIONS

Lesson Note

Students should understand that the Integral Zero Theorem is used to factor an integral polynomial into first-degree factors so that the rational zeros of the polynomial can be found. To find the zeros of an integral polynomial, students should begin by listing potential integral zeros.

Additional Example 1

Given that $x + 2$ is a factor of $P(x) = 3x^3 - 2x^2 - 12x + 8$ and that $P(x)$ has three rational zeros, find the three zeros.
$-2, 2, \frac{2}{3}$

Highlighting the Standards

Standard 3a: The Factor Theorem is one that is quickly put to use. Students can test their abilities to identify zeros of an equation.

Advanced Algebra: When classified by the nature of its coefficients, an integral polynomial is one of the simplest types of polynomial. Even so, there is no simple and sure method, such as an algorithm, that can be used to find all the real zeros of an integral polynomial. However, there are techniques for limiting the work required. The *Integral Zero Theorem* shows one way to cut down on the effort. In Chapter 10, other techniques will be studied that can be used to find the zeros of a polynomial.

Critical Thinking Questions

Analysis: Ask students whether a polynomial must have *any* real zeros.   It is easy to construct a very simple polynomial, such as $x^2 + 1$, that clearly has none.

Common Error Analysis

Error: Some students will list $x - 3$ as a factor if the value of $P(x)$ is 0 when $x = -3$.

Emphasize that $x - 3$ is a factor when $P(3) = 0$ and $x + 3$ is a factor when $P(-3) = 0$.

Checkpoint

1. List the potential integral zeros of $2x^3 - x^2 - 40$. $\pm 1, \pm 2, \pm 4, \pm 5, \pm 8, \pm 10, \pm 20, \pm 40$
2. If 6 is a zero of a polynomial $P(x)$, what is one first-degree factor of $P(x)$? $x - 6$
3. If $P(x) = x(x - 7)(x + 2)(x - 7)(x - 7)$, find the multiplicity for the zero 7.   3

Thus, $P(x) = 2x^4 - 7x^3 - 5x^2 + 28x - 12$
$= (x - 3)(2x^3 - x^2 - 8x + 4)$
$= (x - 3)(x + 2)(2x^2 - 5x + 2)$
$= (x - 3)(x + 2)(2x - 1)(x - 2)$

Thus, the four rational zeros are $3, -2, \frac{1}{2},$ and 2 because $P(3) = 0$, $P(-2) = 0$, $P\left(\frac{1}{2}\right) = 0$, and $P(2) = 0$.

In Example 1, three of the four *rational zeros* $(3, -2, 2)$ are **integral zeros**. The four zeros are also the roots of the polynomial equation $2x^4 - 7x^3 - 5x^2 + 28x - 12 = 0$.

For any given integral polynomial, the number of *potential* integral zeros is finite. The positive integral zeros are integral factors of the constant term. From this, the **Integral Zero Theorem** follows:

Theorem 6.3

> **Integral Zero Theorem:** If an integer r is a zero of the integral polynomial $P(x)$, then r is a factor of the constant term in $P(x)$. That is, if r is an integral zero of
> $$a_0 x^n + a_1 x^{n-1} + a_2 x^{n-2} + \cdots + a_{n-1} x^1 + a_n,$$
> then r is a factor of a_n.

An integral polynomial may have *no* integral zeros. For example, the potential integral zeros of $x^3 - x + 1$ are 1 and -1, but neither of these numbers is a zero of $x^3 - x + 1$.

EXAMPLE 2   Solve the equation $2x^4 - 5x^3 - 12x^2 - x + 4 = 0$.

Plan   Let $P(x) = 2x^4 - 5x^3 - 12x^2 - x + 4$. The roots of the equation are the zeros of $P(x)$. The potential integral zeros are $\pm 1, \pm 2,$ and ± 4. Test each potential zero until a second-degree factor is found.

Solution

	2	−5	−12	−1	4
−1	2	−7	−5	4	⓪
−1	2	−9	4	⓪	

$2x^2 - 9x + 4 = (x - 4)(2x - 1)$

The zeros of $P(x)$ are $-1, 4,$ and $\frac{1}{2}$.

Successive Factorizations of $P(x)$

$(x + 1)(2x^3 - 7x^2 - 5x + 4)$
$(x + 1)(x + 1)(2x^2 - 9x + 4)$
$(x + 1)(x + 1)(x - 4)(2x - 1)$

In the example above, the number -1 is called a zero with a **multiplicity** of 2 because the factor $x + 1$ appears *twice* in the factorization of $P(x)$. Each of the remaining zeros, 4 and $\frac{1}{2}$, has a multiplicity of 1.

Additional Example 2

Solve the equation.

$3x^4 + 5x^3 - 22x^2 - 52x - 24 = 0$.

$-2, 3, -\frac{2}{3}$ (The number -2 is a zero with a multiplicity of 2.)

Classroom Exercises

State the potential integral zeros of each polynomial.

1. $2x^2 + 14x + 3$ $\pm 1, \pm 3$

2. $2x^5 + 9$ $\pm 1, \pm 3, \pm 9$

Given that -3 is a zero of the polynomial $P(x)$, find each of the following.

3. a first-degree factor of $P(x)$ $x + 3$

4. a root of $P(x) = 0$ -3

5. Find each root of $(x + 7)(x - 5)(3x - 1)(2x + 3) = 0$. $-7, 5, \frac{1}{3}, -\frac{3}{2}$

Find each zero and its multiplicity for the given polynomial $P(x)$.

6. $P(x) = (x + 3)(x - 5)(x + 3)$
-3(mult 2), 5(mult 1)

7. $P(x) = (x - 1)^3(x + 2)$
1(mult 3), -2(mult 1)

Written Exercises

Factor $P(x)$ into first-degree factors. Find the zeros of $P(x)$.

1. $P(x) = x^3 - 2x^2 - x + 2$

2. $P(x) = x^3 + 3x^2 - x - 3$

3. $P(x) = 2x^3 + 3x^2 - 17x + 12$

4. $P(x) = 6x^3 + 25x^2 + 21x - 10$

5. $P(x) = 10x^3 - 41x^2 + 2x + 8$

6. $P(x) = 10x^3 + 51x^2 + 3x - 10$

7. $P(x) = x^4 - 15x^2 + 10x + 24$

8. $P(x) = 6x^4 + 23x^3 - 36x^2 - 3x + 10$

Solve each equation. Find the multiplicity m for each root whose $m > 1$.

9. $x^3 - 6x^2 + 12x - 8 = 0$

10. $3x^3 + 29x^2 + 65x - 25 = 0$

11. $4a^4 - 4a^3 - 43a^2 - 26a + 24 = 0$

12. $5c^4 + 6c^3 - 79c^2 + 84c + 20 = 0$

13. Write a fourth-degree polynomial $P(x)$ whose zeros are $2, -2, -1$, and $\frac{1}{2}$. $2x^4 + x^3 - 9x^2 - 4x + 4$

14. Write a fourth-degree polynomial $P(x)$ whose zeros are 3 and $-\frac{1}{2}$, each with a multiplicity of 2. $4x^4 - 20x^3 + 13x^2 + 30x + 9$

15. The length of a rectangular solid (box) is 3 ft longer than twice its width. The height is 2 ft less than the width. Find the 3 dimensions given that the volume is 195 ft^3. Use $V = lwh$.   13 ft by 5 ft by 3 ft

16. Solve the equation $x^6 - 3x^4 + 3x^2 - 1 = 0$. -1(mult 3), 1(mult 3)

Mixed Review

Find the slope of $\overleftrightarrow{PQ}$ and describe its slant.   *3.3*

1. $P(-3, -4)$, $Q(1, 2)$
$\frac{3}{2}$, up to the right

2. $P(7, 3)$, $Q(7, -2)$
Undefined, vertical

3. $P(1, 5)$, $Q(-2, 5)$
0, horizontal

Closure

Ask students to describe under what conditions a number n is a zero of a polynomial. For polynomial $P(x)$, n is a zero if $P(n) = 0$. Have students explain in their own words the meaning of the Integral Zero Theorem. Answers will vary.

◤FOLLOW UP

Guided Practice

Classroom Exercises 1–7

Independent Practice

A Ex. 1–8, **B** Ex. 9–12, **C** Ex. 13–16

Basic: WE 1–9

Average: WE 3–11

Above Average: WE 7–15

Additional Answers

Written Exercises

1. $(x - 2)(x - 1)(x + 1)$; zeros: 2, 1, -1
2. $(x - 1)(x + 1)(x + 3)$; zeros: 1, -1, -3
3. $(x - 1)(x + 4)(2x - 3)$; zeros: 1, -4, $\frac{3}{2}$
4. $(x + 2)(3x - 1)(2x + 5)$; zeros: -2, $\frac{1}{3}$, $-\frac{5}{2}$
5. $(x - 4)(5x + 2)(2x - 1)$; zeros: 4, $-\frac{2}{5}$, $\frac{1}{2}$
6. $(x + 5)(5x - 2)(2x + 1)$; zeros: -5, $\frac{2}{5}$, $-\frac{1}{2}$
7. $(x - 3)(x - 2)(x + 1)(x + 4)$; zeros: 3, 2, -1, -4
8. $(x - 1)(x + 5)(3x - 2)(2x + 1)$; zeros: 1, -5, $\frac{2}{3}$, $-\frac{1}{2}$
9. 2 (mult of 3)
10. -5 (mult of 2), $\frac{1}{3}$
11. 4, -2, $-\frac{3}{2}$, $\frac{1}{2}$
12. 2 (mult of 2), -5, $-\frac{1}{5}$

Enrichment

Write this polynomial on the chalkboard.

$x^5 + 5x^4 + 10x^3 + 10x^2 + 5x + 1$

Ask students to find all the zeros of the polynomial, and the multiplicity of each. The only zero is -1, and it has a multiplicity of 5. Emphasize that the given polynomial is equivalent, therefore, to $(x + 1)^5$. Explain that the original polynomial is an example of a *binomial expansion*, a topic covered in Chapter 15.

Chapter Review

6. $\{x \mid x < -2 \text{ or } x > 8\}$

$-2\ \ 8$

7. $\{y \mid -2 \le y \le 2\}$

$-2 \qquad 2$

8. $\{a \mid -5 < a < 0 \text{ or } a > 5\}$

$-5\ \ 0\ \ 5$

13. Answers will vary. One possible answer: The zero of a polynomial is that value of the variable for which the polynomial equals zero.

Key Terms

Binomial Factor Theorem (p. 218)
integral polynomial (p. 223)
integral zero (p. 224)
Integral Zero Theorem (p. 224)
multiplicity (p. 224)
polynomial function (p. 217)
polynomial inequality (p. 214)

quadratic equation (p. 199)
quadratic inequality (p. 213)
Remainder Theorem (p. 218)
synthetic division (p. 217)
synthetic substitution (p. 219)
zero of a polynomial (p. 223)
Zero-Product Property (p. 199)

Key Ideas and Review Exercises

6.1 To find the rational roots of an equation such as $16x^4 = 17x^2 - 1$, write the equation in standard form. Then factor the polynomial completely, set each factor equal to 0, and solve.

Find the rational roots of each equation. $-3, 0, \frac{3}{5}$

1. $6x^2 = 5x + 4$ $-\frac{1}{2}, \frac{4}{3}$ **2.** $16x^4 = 17x^2 - 1$ $\pm\frac{1}{4}, \pm 1$ **3.** $20y^3 + 48y^2 = 36y$

6.2 To find a pair of numbers or consecutive multiples of a number, represent the numbers in terms of one variable and follow the six-step procedure for problem solving.

4. Find three consecutive multiples of 3 so that twice the square of the second number is 90 more than the product of the first and third numbers. $-12, -9, -6;\ 6, 9, 12$

6.3 To solve problems involving the areas of rectangles, triangles, and right triangles, first sketch and label each figure. Then use an appropriate formula: for a rectangle, $A = lw$; for a triangle, $A = \frac{1}{2}bh$; for a right triangle, $a^2 + b^2 = c^2$ and $A = \frac{1}{2}ab$.

5. The length of the base of a triangle is 10 cm less than 4 times its height. Find the length of the base and the height given that the area of the triangle is 25 cm^2. $b = 10$ cm, $h = 5$ cm

6.4 To solve a second-degree inequality such as $-y^2 + 4 \ge 0$, divide each side by -1 and reverse the order. Then use the appropriate property of the product of two factors:

If $m \cdot n > 0$, then $[m < 0 \text{ and } n < 0]$ or $[m > 0 \text{ and } n > 0]$.
If $m \cdot n < 0$, then $[m < 0 \text{ and } n > 0]$ or $[m > 0 \text{ and } n < 0]$.

Find the solution set of each inequality. Graph the solutions.

6. $x^2 - 6x - 16 > 0$ **7.** $-y^2 + 4 \ge 0$

6.4   To find the solution set of a third-degree inequality such as $a^3 - 25a > 0$:
1. factor the polynomial completely,
2. plot the roots of the related equation on a number line to separate the line into parts,
3. from each part, test a number in the factorization, and
4. graph the parts for which the test is true.

8. Graph the solutions of $a^3 - 25a > 0$. Then write the solution set.

6.5   To divide a given polynomial $P(x)$ by a binomial of the form $x - a$ using synthetic division:
1. Use synthetic substitution to evaluate $P(x)$ for $x = a$.
2. Use the resulting row of numbers to obtain the coefficients of the quotient and the remainder.

Use synthetic division to find the polynomial quotient and the remainder. Is the divisor a factor of the dividend?

9. $(x^6 + 2x^4 + x^2 - 900) \div (x + 3)$ $x^5 - 3x^4 + 11x^3 - 33x^2 + 100x - 300, 0;$ yes

6.5   To evaluate a given polynomial $P(x)$ for $x = -m$ and again for $x = n$, and to name any binomial factors found:
1. include 0 for the coefficient of any missing terms,
2. use synthetic substitution with $-m\rfloor$ and $n\rfloor$,
3. then use the Binomial Factor Theorem: if $P(-m) = 0$, then $x + m$ is a factor of $P(x)$; if $P(n) = 0$, then $x - n$ is a factor of $P(x)$.

10. Use synthetic substitution to find the value of $P(x) = 2x^4 - 17x^2 + x - 12$ for $x = -2$ and again for $x = 3$. Name the binomial factors found, if any. $-50, 0; x - 3$

6.6   To factor a given polynomial $P(x)$ into first-degree factors and find the zeros of $P(x)$:
1. list the integral factors of the constant term to find the potential integral zeros,
2. then use synthetic substitution to factor $P(x)$ completely.

11. Factor $P(x) = 3x^3 + 2x^2 - 37x + 12$ into first-degree factors. Find the zeros of $P(x)$. $(x - 3)(x + 4)(3x - 1);$ zeros: $3, -4, \frac{1}{3}$

12. Solve $4x^4 + 15x^3 + 9x^2 - 16x - 12 = 0$. Find the multiplicity m for each root whose $m > 1$. -2(multiplicity of 2), $1, -\frac{3}{4}$

13. Write in your own words what is meant by a *zero* of a polynomial.

7. $\{x \mid -4 < x < 3\}$

8. $\{y \mid y \le -5 \text{ or } y \ge 5\}$

9. $\{a \mid -6 < a < 0 \text{ or } a > 6\}$

16. $\{x \mid -5 \le x \le -2 \text{ or } 2 \le x \le 5\}$

Chapter 6 Test A Exercises: 1–5, 7, 9–11 B Exercises: 6, 8, 12–13
C Exercises: 14 – 16

Find the rational roots of each equation.

1. $5x^2 + 18x = 8$ $-4, \frac{2}{5}$

2. $4y^4 - 37y^2 + 9 = 0$ $\pm\frac{1}{2}, \pm 3$

3. $6a^3 - 33a^2 + 36a = 0$ $0, \frac{3}{2}, 4$

4. Find three consecutive multiples of 5 so that twice the square of the second number is 250 more than the product of the first and third numbers. $-20, -15, -10; 10, 15, 20$

5. The length of the base of a triangle is 2 m less than twice its height. Find the length of the base and the height given that the area of the triangle is 56 m^2. $b = 14$ m, $h = 8$ m

6. A rectangle is twice as wide as a square. The rectangle is 5 ft longer than it is wide. The sum of the areas of the two figures is 240 ft^2. What is the area of the square? 36 ft^2

Find and graph the solution set of each inequality.

7. $x^2 + x - 12 < 0$

8. $-y^2 + 25 \le 0$

9. Graph the solutions of $a^3 - 36a > 0$. Then write the solution set.

Use synthetic division to find the polynomial quotient and the remainder. Is the divisor a factor of the dividend? (Exercise 10)

10. $(3x^4 + 10x^3 + 2x^2 - 5x - 3) \div (x + 3)$ $3x^3 + x^2 - x - 2; 3;$ no

11. Use synthetic substitution to find the value of $P(x) = 2x^4 - 33x^2 + 3x + 4$ for $x = 4$ and again for $x = -1$. Name the binomial factors found, if any. $0; -30; x - 4$

12. Factor $P(x) = 2x^3 + 7x^2 + 2x - 3$ into first-degree factors. Find the zeros of $P(x)$. $(x + 1)(x + 3)(2x - 1); -1, -3, \frac{1}{2}$

13. Solve $2x^4 - 5x^3 - 21x^2 + 45x + 27 = 0$. Find the multiplicity m for each root whose $m > 1$. 3(mult of 2), $-3, -\frac{1}{2}$

14. Solve $|x^2 - 20| = 16$. $\pm 2, \pm 6$

15. Solve $x^4 - 10a^2x^2 + 9a^4 = 0$ for x in terms of a. $\pm a, \pm 3a$

16. Graph the solutions of $x^4 - 29x^2 + 100 \le 0$. Write the solution set.

In each item, you are to compare a quantity in Column 1 with a quantity in Column 2. Write the letter of the correct answer from these choices:

A—The quantity in Column 1 is greater than the quantity in Column 2.
B—The quantity in Column 2 is greater than the quantity in Column 1.
C—The quantity in Column 1 is equal to the quantity in Column 2.
D—The relationship cannot be determined from the given information.

NOTE: Information centered over both columns refers to one or both of the quantities to be compared.

Sample Question	Answer
$x \neq 2$ Column 1 Column 2 $\dfrac{x^2 - 4}{x - 2}$ x	$\dfrac{x^2 - 4}{x - 2} = \dfrac{(x - 2)(x+2)}{x - 2} = x + 2,$ if $x \neq 2$, $x + 2 > x$ for each value of x. The answer is A.

Column 1	Column 2
1.	$y > 0$
$\dfrac{y^2 - 25}{y + 5}$	$y - 1$ B
2. $P(x) = 18x^6 + 12x^4 - 8x^2$	
$P(7)$	$P(-7)$ C
3. $(x - 6)^2$	$(x + 6)^2$ D
4. $(x + 1)^2$	$x(x + 2)$ A
5. B	

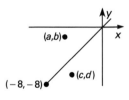

$a + d$	$b + c$

Column 1	Column 2
6. Number of minutes in two weeks	Number of seconds in 7 hours B
7.	x is an integer.
$\dfrac{x}{-2}$	$2x$ D

8.

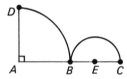

B is the midpoint of $\overline{AC}$.
$\overset{\frown}{BC}$ and $\overset{\frown}{BD}$ are arcs of circles with centers at E and A, respectively.

Length of $\overset{\frown}{BC}$	Length of $\overset{\frown}{BD}$ C
9.	x^* means x^{-2}.
$(3^*)^*$	3^* A

28.

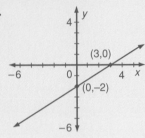

The line passes through (3,0) and (0,−2).

29.

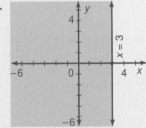

The line is horizontal through (0,3).

30.

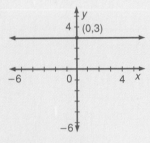

$x = 3$

31.

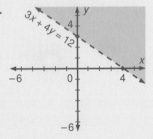

$3x + 4y = 12$

Identify the property of operations with real numbers. (Exercises 1–3)   *1.3*

1. $(a + 5)c = ac + 5c$ Distr Prop

2. $-7t + 7t = 0$ Add Inverse Prop

3. $n = 1n$ Mult Identity Prop

4. Simplify   *1.3*
$5(3x + 2) - (7x - 4y^3)$
and evaluate the result for
$x = 1.5$ and $y = -2$.
$8x + 4y^3 + 10, -10$

For Exercises 5–8, solve the   *1.4*
equation. Then, select the
root from the choices
A, B, C, D, and E.

5. $5x + 7 - 4(2 - x) =$
$7x - (2x + 9)$ B
   (A) $-\frac{5}{7}$ (B) -2 (C) 2
   (D) -6 (E) 6

6. $x - 0.04 = 8.6 - 2.2x$ B
   (A) 0.27 (B) 2.7 (C) 27
   (D) 270 (E) $2,700$

7. $\frac{3x + 8}{5x - 2} = \frac{2}{5}$ A
   (A) $-8\frac{4}{5}$ (B) $-\frac{5}{44}$ (C) $\frac{25}{36}$
   (D) $1\frac{11}{25}$ (E) $-1\frac{11}{25}$

8. $\frac{1}{2}x - 2 = \frac{3}{5}x + 4$ B
   (A) -6 (B) -60 (C) $-\frac{6}{11}$
   (D) 20 (E) -20

Solve each equation for x.
(Exercises 9–16)

9. $10x - (8 - 2x) =$ *1.4*
$12 + 4(x - 7)$ -1

10. $\frac{5x - 2}{5} = \frac{2x + 6}{3}$ $7\frac{1}{5}$

11. $\frac{2}{3}x - \frac{1}{2} = \frac{3}{4}x + 2$ -30

12. $a(2x - b) = 6ab - 5$ $\frac{7ab - 5}{2a}$

13. $|2x - 7| = 3$ 2,5   *2.3*

14. $5^{2x-1} = 125$ 2   *5.1*

15. $m(x - a) = 3x + 4am$ $\frac{5am}{m - 3}$ *5.4*

16. $2x^2 - 7x = 15$ $-\frac{3}{2}, 5$ *6.1*

17. Supply the reason for each   *1.8*
step in the proof of the theorem.

Theorem:
$(ab + c) + ad = a(b + d) + c$

Proof:
1. $(ab + c) + ad$
   $= ab + (c + ad)$ Assoc Prop Add
2. $= ab + (ad + c)$ Comm Prop Add
3. $= (ab + ad) + c$ Assoc Prop Add
4. $= a(b + d) + c$ Distr Prop
5. $(ab + c) + ad$
   $= a(b + d) + c$ Trans Prop Eq

Find the solution set.

18. $7 - 5x < 2(x - 7)$ $\{x \mid x > 3\}$ *2.1*

19. $-4 \le 3x - 7 \le 8$ *2.2*
   $\{x \mid 1 \le x \le 5\}$

20. $|2x + 5| < 11$ $\{x \mid -8 < x < 3\}$ *2.3*

21. $|x - 3| > 5$ $\{x \mid x < -2 \text{ or } x > 8\}$

22. $x^2 - 2x - 8 < 0$ *6.4*
   $\{x \mid -2 < x < 4\}$

For Exercises 23–25, $f(x) = 3x - 2$
and $g(x) = x^2 + 4$. Find the following.

23. $g(-4) - f(4)$ 10   *3.2*

24. $g(c + 5) - g(c)$ $10c + 25$ *5.5*

25. $f(g(2))$ 22   *5.9*

Find the slope and describe the   *3.3*
slant of the line determined by
the points P and Q.

26. $P(5,2)$, $Q(-3,4)$ $-\frac{1}{4}$, down to right

27. $P(-2,4)$, $Q(-5,4)$ 0, horizontal

Draw the graph on a coordinate plane. (Exercises 28–31)

28. $2x - 3y = 6$ *3.5*

29. $4y + 2 = 14$

30. $2x + 3 \leq 9$ *3.9*

31. $3x + 4y > 12$

32. If y varies directly as x, and   *3.8*
    $y = 84$ when $x = 3.5$, what
    is x when $y = 60$? 2.5

33. Given that y varies directly as
    the square of x and $y = 12$
    when $x = 3$, find y when
    $x = 6$. 48

Solve each system for (x, y). (Exercises 34–35)

34. $3x - 4y = 7$ (5,2)   *4.2*
    $y = 2x - 8$

35. $5x - 3y = 1$ (2,3)   *4.3*
    $3x + 4y = 18$

Simplify. Use positive exponents.

36. $(4n^{-2})^3$ $\dfrac{64}{n^6}$ *5.2*

37. $\dfrac{7.2 \times 10^{-6}}{3.6 \times 10^{-8}}$ 200

38. $5a(3a + 4)(a - 6)$ $15a^3 - 70a^2 - 120a$ *5.3*

39. $(5x - 3)^2$ $25x^2 - 30x + 9$ *5.5*

Factor completely.   *5.7*

40. $9x^4 - 37x^2 + 4$ $(x + 2)(x - 2)(3x + 1)(3x - 1)$

41. $2x^3 + 16$ $2(x + 2)(x^2 - 2x + 4)$

42. $16x^3 - 100x$ $4x(2x + 5)(2x - 5)$

For Exercises 43 and 44, $P(x) =$ *6.6*
$6x^4 - 11x^3 - 22x^2 + x + 6$.

43. Factor $P(x)$ into first-degree
    factors. $(x - 3)(x + 1)(2x - 1)(3x + 2)$

44. List the zeros of $P(x)$. $3, -1, \frac{1}{2}, -\frac{2}{3}$

45. The second of three numbers   *1.6*
    is 5 times the first number,
    and the third number is 4
    less than the second number.
    Twice the first number, de-
    creased by the third number,
    is the same as 20 less than
    the second number. Find the
    three numbers. 3, 15, 11

46. Maria is 5 years older than   *2.5*
    Andre. In 4 years, the sum of
    their ages will be less than
    45 years. Find the possible
    ages for Maria. Between 5 and 21 yr

47. A rectangle is 6 in. longer   *6.3*
    than twice its width. A sec-
    ond rectangle is 3 times as
    wide and one-half as long as
    the first rectangle. The sum
    of the areas of the two rec-
    tangles is 140 in^2. Find the
    length of the first rectangle. 14 in

**Write a system of two equations with
two variables to solve each problem.**

48. A collection of 35 coins con-   *4.4*
    tains nickels and dimes. Find
    the number of dimes given
    that the total face value of
    the coins is $2.55. 16 dimes

49. Teresa invested $600, part at   *4.4*
    7% and the rest at 9%. At
    the end of 4 years, the total
    simple interest earned was
    $184. How much money
    did she invest at 9%? $200

50. A plane flies 800 m in 4 h   *4.5*
    with a tail wind, and returns
    the same distance at the same
    airspeed in 5 h against the
    wind. Find the wind speed. 20 mi/h

51. The ratio of the measures of   *4.5*
    two supplementary angles is
    4:5. Find the measure of
    each angle. 80, 100

7 RATIONAL EXPRESSIONS

OVERVIEW

In this chapter, students focus on adding, subtracting, multiplying, dividing, and simplifying rational expressions. Equations that involve rational expressions are covered in Lesson 7.6. A lesson on dimensional analysis (Lesson 7.3) is provided in which units of measurement are manipulated as if they were rational expressions. Although the manipulations are not true mathematical operations, they serve a practical purpose by assuring that the results obtained using actual operations make dimensional sense.

OBJECTIVES

- To simplify and evaluate rational expressions
- To find products and quotients of rational expressions
- To find sums and differences of rational expressions
- To solve equations containing rational expressions
- To solve word problems that involve rational expressions

PROBLEM SOLVING

Although most of Chapter 7 is devoted to algebraic skills, much is done also in the field of problem solving. One lesson (Lesson 7.7) is devoted completely to problem solving and two *Application* features are included on photography and Archimedes' Principle, respectively. A *Problem Solving Strategies* lesson that teaches *Using a Table to Find a Pattern* appears on page 244 and the regular *Mixed Problem Solving* page that occurs in odd-numbered chapters is on page 261.

READING AND WRITING MATH

The *Focus on Reading* on page 254 enables students to use the vocabulary taught in the lesson in a grammatical context. Exercise 11 on page 263 asks students to explain, in their own words, how to add rational expressions with unlike denominators.

TECHNOLOGY

Calculator: Because of the emphasis on algebraic operations, arithmetic operations occur less frequently in Chapter 7 than in most other chapters, with a corresponding decrease in a need for calculators. Even here, however, calculators can be used to check the equations of Lesson 7.6 and in the evaluation of formulas such as those of Exercises 27–30 on page 256.

SPECIAL FEATURES

Mixed Review pp. 237, 241, 243, 251, 256, 260
Application: Photography p. 241
Problem Solving Strategies: Using a Table to Find a Pattern p. 244
Midchapter Review p. 248
Focus on Reading p. 254
Brainteaser p. 256
Application: Flotation p. 257
Mixed Problem Solving p. 261
Key Terms p. 262
Key Ideas and Review Exercises pp. 262–263
Chapter 7 Test p. 264
College Prep Test p. 265

PLANNING GUIDE

Lesson	Basic	Average	Above Average	Resources
7.1 pp. 236–237	CE all WE 1–27 odd	CE all WE 3–29 odd	CE all WE 7–33 odd	Reteaching 46 Practice 46
7.2 pp. 240–241	CE all WE 1–19 odd Application	CE all WE 5–23 odd Application	CE all WE 7–25 odd Application	Reteaching 47 Practice 47
7.3 pp. 243–244	CE all WE 1–8 Problem Solving	CE all WE 1–9 Problem Solving	CE all WE 3–11 Problem Solving	Reteaching 48 Practice 48
7.4 pp. 247–248	CE all WE 1–19 odd Midchapter Review	CE all WE 1–19 odd Midchapter Review	CE all WE 3–21 odd Midchapter Review	Reteaching 49 Practice 49
7.5 pp. 250–251	CE all WE 1–9 odd	CE all WE 3–11 odd	CE all WE 9–17 odd	Reteaching 50 Practice 50
7.6 pp. 254–257	FR all CE all WE 1–21 odd Brainteaser Application	FR all CE all WE 11–31 odd Brainteaser Application	FR all CE all WE 15–35 odd Brainteaser Application	Reteaching 51 Practice 51
7.7 pp. 259–261	CE all WE 1–7 Mixed Problem Solving	CE all WE 2–8 Mixed Problem Solving	CE all WE 3–9 Mixed Problem Solving	Reteaching 52 Practice 52
Chapter 7 Review pp. 262–263	all	all	all	
Chapter 7 Test p. 264	all	all	all	
College Prep Test p. 265	all	all	all	

CE = Classroom Exercises   WE = Written Exercises   FR = Focus on Reading

NOTE: For each level, all students should be assigned all Mixed Review exercises.

■■■ INVESTIGATION

Project: The students explore what happens at points where a rational expression is undefined. If the students do not have graphing calculators or a computer with graphing software, ordinary calculators can be used to locate y coordinates for given values of x.

Write the following rational expression on the chalkboard.

$$\frac{x + 1}{x - 2}$$

Ask students to give the value of the expression when $x = 2$.   Undefined.   Have the students graph the equation

$$y = \frac{x + 1}{x - 2}$$

and discuss what happens to the graph as x approaches 2.   Starting from the left, the expression approaches negative infinity as x increases toward 2. At $x = 2$ there is no value for the graph. Starting from the right and moving toward 2, the expression approaches positive infinity. Other answers are possible.   Have students graph the expression

$$\frac{10x - 20}{x^2 - 16}$$

and discuss their findings.   Students should notice the zero at $x = 2$, corresponding to $10x - 20 = 0$ for the numerator, and the undefined points at $x = -4$ and $x = 4$.

Architects use CAD technology to create three-dimensional drawings of a building project. The image above shows the central space of a building in exact detail. The computer enables architects to view the interior from any perspective or change the style of any design element.

More About the Image

Architects use computers in a number of different ways. As a design tool, the graphics capability of the computer produces drawings in three dimensions. Once the original design is created, the architect can rotate the image to view the structure from any perspective. Sophisticated CAD programs integrate the drafting function with formulas for engineering calculations. Thus, CAD technology can be used to test materials against external forces such as hurricanes or earthquakes.

7.1 Rational Expressions and Functions

Objectives

To find the domains of rational functions
To find the zeros of rational functions
To simplify rational expressions
To find the values of rational functions

Teaching Resources

Problem Solving Worksheet 7
Quick Quizzes 46
Reteaching and Practice
   Worksheets 46

Each expression below is a quotient of two polynomials.

$$\frac{3x + 2}{12x + 7} \qquad \frac{8}{x^2 - x - 6} \qquad \frac{x^2 - 2}{9}$$

Such expressions are called *rational expressions*. Thus, the expression $\frac{x^2 + 4x + 3}{\sqrt{3x - 4}}$ is not a rational expression because $\sqrt{3x - 4}$ is not a polynomial.

Definition

> A **rational expression** is a quotient of two polynomials.

Recall that the quotient $\frac{a}{b}$ of two numbers is undefined if the denominator b is 0. Similarly, a rational expression is undefined if its denominator is 0.

EXAMPLE 1 For what values of x is $\dfrac{2x - 3}{x^2 - 8x + 12}$ undefined?

Plan Find the values of x for which the denominator, $x^2 - 8x + 12$, is 0.

Solution Solve the quadratic equation $x^2 - 8x + 12 = 0$.
$$(x - 6)(x - 2) = 0$$
$$x - 6 = 0 \quad or \quad x - 2 = 0$$
$$x = 6 \qquad\qquad x = 2$$

Thus, $\dfrac{2x - 3}{x^2 - 8x + 12}$ is undefined for $x = 6$ and for $x = 2$.

A **rational function** is a function described by a rational expression. For example, the function described by the rule $f(x) = \dfrac{2x - 3}{x^2 - 8x + 12}$ is a rational function.

7.1 Rational Expressions and Functions   **233**

GETTING STARTED

Prerequisite Quiz

Find the zeros of each function.

1. $f(x) = x^2 - 7x + 10$ 2, 5
2. $f(x) = 3x - 12$ 4

Find each indicated function value.

3. $f(x) = x^2 - 4x + 3$
   Find $f(-2)$.   15
4. $f(x) = -x^3 - 5x - 7$
   Find $f(-1)$. -1

Simplify, if possible.

5. $\frac{0}{4}$ 0
6. $\frac{8}{0}$ Undef

Motivator

Ask students how $\frac{12}{4}$ can be written as a division problem.   Emphasize that $\frac{12}{4}$ means $12 \div 4$ or 3.   Ask students how they check a division problem.   Recall $12 \div 4 = 3$ only if $4 \times 3 = 12$.   Ask them if $\frac{4}{0}$ can be written as a division problem. No. There is no quotient which, when multiplied by zero, will equal 4.   Ask them what this indicates about any fraction with a denominator of zero.   It is undefined.

Additional Example 1

For what values of x is $\frac{3x - 4}{x^3 - 9x^2}$ undefined?
$x = 0, x = 9$

Highlighting the Standards

Standard 6b: Work with algebraic expressions remains a useful skill. But unifying themes such as functions should be woven into the treatment of manipulative skills as is done in this lesson.

Lesson Note

It may be helpful to review simplifying fractions such as $\frac{18}{30}$. Show the numerator and denominator factored into primes, as is done with algebraic expressions:

$$\frac{18}{30} = \frac{3 \cdot \cancel{3} \cdot \cancel{2}}{\cancel{3} \cdot \cancel{2} \cdot 5} = \frac{3}{5}$$

Math Connections

Calculus: The expression $f(x + h) - f(x)$ has appeared in two earlier lessons (Lesson 3.2, Example 4 and Lesson 5.5, Example 3). This is the numerator of one of the two expressions that, as mentioned at the top of page 236, are often encountered in calculus. Each expression represents the slope of a line that is tangent to the curve $y = f(x)$ at a point (x, y) of the curve.

Critical Thinking Questions

Analysis: Students already know that a ratio of two integers, such as $\frac{7}{9}$, is a rational number. Ask them whether it is also a rational *expression*.   It is, by the definition on page 233, since 7 and 9 can each be thought of as a polynomial of degree zero. That is, $\frac{7}{9}$ is the ratio of the *polynomials* 7 and 9 and thus is also a rational expression.

Recall that the domain of a function f is the set of all values of x for which the function f is defined. For example, the domain of f in Example 1, where $f(x) = \dfrac{2x - 3}{x^2 - 8x + 12}$, is the set of all real numbers except 2 and 6; that is, $\{x \mid x \neq 2 \; and \; x \neq 6\}$.

Unless otherwise stated, the domain of any rational function is assumed to be the set of all real numbers for which the value of any denominator is not zero.

Recall that $\frac{a}{b} = c$ means that $c \cdot b = a$ (see Lesson 1.1). This definition is the basis of the following theorem.

Theorem 7.1 | For all real numbers a and b, where $b \neq 0$, if $\frac{a}{b} = 0$, then $a = 0$.

The **zero of a function** f is a number x such that $f(x) = 0$ (see Lesson 6.6). In the next example, Theorem 7.1 is used to find the zeros of a rational function.

EXAMPLE 2 Find the zeros of the function $f(x) = \dfrac{2x - 3}{x^2 - 8x + 12}$.

Plan Find the values of x for which $2x - 3 = 0$.

Solution $2x - 3 = 0, \; 2x = 3, \; x = \frac{3}{2}$

Thus, $\frac{3}{2}$ is the only zero of f.

A useful property of fractions is the *Cancellation Property*.

Cancellation Property of Fractions
For all real numbers a, b, and c, where $b \neq 0$ and $c \neq 0$, $\dfrac{ac}{bc} = \dfrac{a}{b}$.

For example, $\dfrac{6}{10} = \dfrac{3 \cdot 2}{5 \cdot 2} = \dfrac{3}{5}$

This property can be used to simplify rational expressions. A rational expression is said to be in *simplest form* when it is written as a polynomial or the quotient of two polynomials whose greatest common factor is 1. A rational expression can be simplified by:
1. factoring the numerator and denominator completely, and
2. applying the Cancellation Property of Fractions.

234   Chapter 7 Rational Expressions

Additional Example 2

Find the zeros of the function

$$f(x) = \frac{x^2 - 11x + 18}{x^2 - 25}. \quad 9, 2$$

EXAMPLE 3   Simplify $\dfrac{2x}{6x^4 - 18x^3}$.

Solution   Factor the numerator and denominator completely.

$$\dfrac{2x}{6x^4 - 18x^3} = \dfrac{2x}{6x^3(x - 3)}$$

$$= \dfrac{2x \cdot 1}{2x \cdot 3x^2(x - 3)}$$

Use the Cancellation Property of Fractions.

$$= \dfrac{1}{3x^2(x - 3)} \leftarrow \dfrac{ac}{bc} = \dfrac{a}{b}$$

A shorter method of simplification, *dividing out common factors*, is based on the Cancellation Property of Fractions. It is illustrated in Example 4 below. In order to divide out *common* factors, it is sometimes useful to change the numerator or denominator to a more convenient form.

EXAMPLE 4   Simplify $\dfrac{y^2 - y - 42}{14 + 5y - y^2}$.

Plan   First rewrite $14 + 5y - y^2$ in a more convenient form as follows.
(1) Write in descending order of exponents. $-1y^2 + 5y + 14$
(2) Factor out -1. $-1(y^2 - 5y - 14)$

Solution
$$\dfrac{y^2 - y - 42}{14 + 5y - y^2} = \dfrac{y^2 - y - 42}{-1(y^2 - 5y - 14)}$$

$$= \dfrac{\overset{1}{\cancel{(y + 6)}}(y - 7)}{-1(y + 2)\cancel{(y - 7)}\underset{1}{}} = \dfrac{y + 6}{-1(y + 2)}, \text{ or } -\dfrac{y + 6}{y + 2}$$

EXAMPLE 5   Simplify $f(t) = \dfrac{t^3 - t^2 - 4t + 4}{t^2 + t - 2}$. Then find $f(9.3)$.

Solution   Factor the numerator by grouping pairs of terms.

$$\dfrac{t^2(t - 1) - 4(t - 1)}{t^2 + t - 2} = \dfrac{(t - 1)(t^2 - 4)}{t^2 + t - 2} = \dfrac{\overset{1}{\cancel{(t - 1)}}\overset{1}{(t + 2)}(t - 2)}{\underset{1}{\cancel{(t - 1)}}\underset{1}{(t + 2)}} = t - 2$$

So, $f(t) = t - 2$ and $f(9.3) = 9.3 - 2$, or 7.3.

7.1 Rational Expressions and Functions   **235**

Common Error Analysis

Error: Many students will simplify an expression such as

$$\dfrac{3x(x - 7)}{6x(x - 7)(x + 2)}$$

as $2(x + 2)$. This occurs when they "cancel" mechanically.

Emphasize that they are dividing out common factors. Writing the *ones* is sometimes helpful, as shown in Examples 4 and 5.

Checkpoint

Find the domain of the function for the given rule.

1. $f(x) = \dfrac{3x - 18}{2x - 10}$ $\{x \mid x \neq 5\}$

2. $f(x) = \dfrac{x^2 - 7x}{x^2 - 25}$ $\{x \mid x \neq 5 \text{ and } x \neq -5\}$

3. Simplify. $\dfrac{b^2 + 2b - 24}{12 + b - b^2}$ $\dfrac{-b + 6}{b + 3}$

4. Simplify the rational expression in $f(t) = \dfrac{2t^2 + 3t - 2}{t^2 - 4}$. Then find $f(4)$.
$f(t) = \dfrac{2t - 1}{t - 2}$; $f(4) = \dfrac{7}{2}$

Closure

Ask students to give the two steps for simplifying a rational expression. See page 234. Ask students to tell when a rational expression is undefined. When the denominator is 0. Have students describe the domain of a rational function. The set of all real numbers for which the value of any denominator is not zero.

Additional Example 3

Simplify.

$\dfrac{3x^2 - 21x}{6x^3 - 30x^2 - 84x}$ $\dfrac{1}{2(x + 2)}$

Additional Example 4

Simplify.

$\dfrac{y^2 - 5y - 24}{16 + 6y - y^2} - \dfrac{y + 3}{y + 2}$
$\dfrac{-2(y + 3)}{y + 2}$

Additional Example 5

Simplify.

$f(t) = \dfrac{t^3 + 5t^2 - 4t - 20}{t^2 + 3t - 10}$
Then find $f(-1.3)$.
$f(t) = t + 2$; $f(-1.3) = 0.7$

235

Rational expressions of the form $\dfrac{f(a) - f(b)}{a - b}$ or $\dfrac{f(x + h) - f(x)}{h}$ are encountered often in the study of calculus. When the function f is a polynomial function, such rational expressions can be simplified.

EXAMPLE 6

a. Given that $f(x) = 5x + 4$, simplify $\dfrac{f(a) - f(7)}{a - 7}$.

b. Given that $g(x) = x^2 - 2$, simplify $\dfrac{g(x + 3) - g(x)}{3}$.

Solutions

a.
$$\dfrac{f(a) - f(7)}{a - 7}$$
$$= \dfrac{(5a + 4) - (5 \cdot 7 + 4)}{a - 7}$$
$$= \dfrac{5a - 35}{a - 7}$$
$$= \dfrac{5(a - 7)}{a - 7} = 5$$

b.
$$\dfrac{g(x + 3) - g(x)}{3}$$
$$= \dfrac{[(x + 3)^2 - 2] - (x^2 - 2)}{3}$$
$$= \dfrac{x^2 + 6x + 9 - 2 - x^2 + 2}{3}$$
$$= \dfrac{6x + 9}{3} = 2x + 3$$

Classroom Exercises

Determine whether each expression is a rational expression. If not, explain.

1. $\dfrac{4x^2 - 7x + 12}{x - 8}$ Yes

2. $\dfrac{0}{7x - 5}$ Yes

3. $\dfrac{\sqrt{3x + 5}}{2x - 9}$ No, $\sqrt{3x + 5}$ is not a polynomial.

For what value(s) of x is the rational expression undefined?

4. $\dfrac{5x^2 - 25x}{2x}$ 0

5. $\dfrac{3x - 15}{x^2 - 25}$ ± 5

6. $\dfrac{3x - 6}{10 - 5x}$ 2

7–9. Simplify each rational expression in Classroom Exercises 4–6.

7. $\dfrac{5(x - 5)}{2}$ 8. $\dfrac{3}{x + 5}$ 9. $-\dfrac{3}{5}$

Written Exercises

Find the domain of the function with the given rule.

1. $f(x) = \dfrac{3x - 15}{2x - 6}$ $\{x \mid x \neq 3\}$

2. $f(x) = \dfrac{x^2 - 9}{3x + 15}$ $\{x \mid x \neq -5\}$

3. $f(x) = \dfrac{x^2 - 9x}{x^2 - 7x}$ $\{x \mid x \neq 0 \text{ and } x \neq 7\}$

4. $f(x) = \dfrac{x^2 - 5x}{x^2 - 7x + 12}$ $\{x \mid x \neq 3 \text{ and } x \neq 4\}$

5–8. Find the zeros of each function in Written Exercises 1–4.   **5.** 5 **6.** ± 3 **7.** 0,9 **8.** 0,5

Additional Example 6

a. Simplify $\dfrac{f(b) - f(3)}{b - 3}$ if $f(b) = 3b - 2$.   3

b. Simplify $\dfrac{g(x + 4) - g(x)}{4}$ if $g(x) = x^2 - 5$.
$2x + 4$

Simplify each rational expression.

9. $\dfrac{x^5(x+4)}{x^7(x+4)}$ $\dfrac{1}{x^2}$

10. $\dfrac{(2x+5)(x-4)}{(x-4)(5+2x)}$ 1

11. $\dfrac{x}{x^2+5x}$ $\dfrac{1}{x+5}$

12. $\dfrac{3x-15}{6x-30}$ $\dfrac{1}{2}$

13. $\dfrac{x^2-x}{x^2-1}$ $\dfrac{x}{x+1}$

14. $\dfrac{4x}{8x^2-8x}$ $\dfrac{1}{2(x-1)}$

15. $\dfrac{5x}{10x^2-20x}$ $\dfrac{1}{2(x-2)}$

16. $\dfrac{2x^3-4x^2}{6x^3}$ $\dfrac{x-2}{3x}$

17. $\dfrac{6x^2-12x}{8x^3}$ $\dfrac{3(x-2)}{4x^2}$

18. $\dfrac{a^2+4a-12}{3a^2-12a+12}$ $\dfrac{a+6}{3(a-2)}$

19. $\dfrac{6m^2-4m}{9m^2-12m+4}$ $\dfrac{2m}{3m-2}$

20. $\dfrac{27a^4-75a^2}{27a^3-18a^2-45a}$ $\dfrac{a(3a+5)}{3(a+1)}$

21. $\dfrac{x^2-8x-20}{12x-x^2-20}$ $-\dfrac{x+2}{x-2}$

Simplify the expression that describes each function. Then find the indicated value of the function.

22. $f(x)=\dfrac{x^2-25}{2x^2-7x-15}$

Find $f(-4)$. $\dfrac{x+5}{2x+3}, -\dfrac{1}{5}$

23. $f(x)=\dfrac{-x^2+8x-12}{3x^2-2x-8}$

Find $f(-3)$. $\dfrac{-(x-6)}{3x+4}, -\dfrac{9}{5}$

In Exercises 24–26, $f(x)=3x-8$ and $g(x)=x^2+5$. Simplify the following.

24. $\dfrac{f(c)-f(10)}{c-10}$ 3

25. $\dfrac{g(2a)-g(3)}{2a-3}$ $2a+3$

26. $\dfrac{f(x+12)-f(x)}{12}$ 3

Simplify each rational expression.

27. $\dfrac{5a^4-a^4m}{a^4m^2-7a^4m+10a^4}$ $\dfrac{-1}{m-2}$

28. $\dfrac{x^4-10x^2+9}{27-3x^2}$ $\dfrac{(x-1)(x+1)}{-3}$

29. $\dfrac{x^2a^2-9a^2-4x^2+36}{xa-3a+2x-6}$ $(a-2)(x+3)$

30. $\dfrac{-3x-x^2+4}{x^3+4x^2-x-4}$ $\dfrac{-1}{x+1}$

31. $(3+2a-a^2)(a^4-10a^2+9)^{-1}$

32. $(ax+ay-xb-yb)(a^2-b^2)^{-1}$

33. Prove Theorem 7.1.

Mixed Review

Solve each equation or system of equations. *1.4, 2.3, 4.2*

1. $\dfrac{2x-3}{2}=\dfrac{4x+1}{3}$ $-\dfrac{11}{2}$

2. $|2x+4|=8$ $-6, 2$

3. $\begin{aligned} y&=2x-6 \\ 3x-2y&=2 \end{aligned}$ $(10,14)$

Enrichment

Challenge students to find what is wrong with the following rational expression and its restriction on the domain: $\dfrac{xa+xb}{x+xd}$, where $d \neq -1$. The denominator factors into $x(1+d)$, so the restriction on the domain should include $x \neq 0$.

GETTING STARTED

Prerequisite Quiz

Simplify.

1. $\dfrac{x^2 - 4}{5x - 10}$ $\dfrac{x + 2}{5}$

2. $\dfrac{a^2 - 5a + 6}{a^2 - 6a + 9}$ $\dfrac{a - 2}{a - 3}$

3. $\dfrac{6 + 7x - 3x^2}{3x - 9}$ $\dfrac{-(3x + 2)}{3}$

4. $\dfrac{x^2 - 2x - 35}{21 + 11x - 2x^2}$ $\dfrac{x + 5}{-(2x + 3)}$

Motivator

Have students give the prime factorization of the numerator and denominator of the expression $\dfrac{42}{105}$. $\dfrac{(7)(2)(3)}{(7)(3)(5)}$ Have students divide out the common factors and write the expression in its simplified form. $\dfrac{2}{5}$

TEACHING SUGGESTIONS

Lesson Note

Rules for multiplying and dividing fractions have been justified in earlier studies. However, it may be helpful to justify the rule for dividing fractions again.

Think: $a \div b$ means $\dfrac{a}{b}$. Therefore,

$\dfrac{x}{y} \div \dfrac{m}{n}$ means $\dfrac{\frac{x}{y}}{\frac{m}{n}} = \dfrac{\frac{x}{y} \cdot \frac{n}{m}}{\frac{m}{n} \cdot \frac{n}{m}} = \dfrac{\frac{x}{y} \cdot \frac{n}{m}}{1} = \dfrac{x}{y} \cdot \dfrac{n}{m}$.

Highlighting the Standards

Standard 5a: Virtually every area of science and social science now uses quantitative methods. These methods are often algebraic, using procedures such as the ones in this lesson.

7.2 Products and Quotients of Rational Expressions

Objective   To find products and quotients of rational expressions

The product of two rational expressions is a rational expression of the form $\dfrac{\text{product of the numerators}}{\text{product of the denominators}}$.

Rule for Multiplying Rational Expressions
If $\dfrac{a}{b}$ and $\dfrac{c}{d}$ are rational expressions, where $b \neq 0$ and $d \neq 0$,

then $\dfrac{a}{b} \cdot \dfrac{c}{d} = \dfrac{a \cdot c}{b \cdot d}$.

The above rule is illustrated below.

1. Apply the Rule for Multiplying Rational Expressions.

$$\dfrac{6}{5x + 20} \cdot \dfrac{x^2 + 4x}{9} = \dfrac{6(x^2 + 4x)}{(5x + 20) \cdot 9}$$

2. Factor the numerator and denominator.

$$= \dfrac{\overset{1}{\cancel{3}} \cdot 2 \cdot x(\cancel{x + 4})^{1}}{5(\cancel{x + 4})_{1} \cdot \cancel{3}_{1} \cdot 3}$$

3. Divide out common factors.

4. Simplify.

$$= \dfrac{2x}{15}$$

It is usually convenient to divide out common factors *before* using the Rule for Multiplying Rational Expressions. This is shown in Example 1.

EXAMPLE 1   Simplify $\dfrac{6a^3b}{15xy^4} \cdot \dfrac{10x^2y^3}{9ab^4}$.

Solution

$$\dfrac{6a^3b}{15xy^4} \cdot \dfrac{10x^2y^3}{9ab^4} = \dfrac{\overset{1}{\cancel{3}} \cdot 2 \cdot \overset{a^2}{\cancel{a^3}} \cdot \overset{1}{\cancel{b}}}{\cancel{5} \cdot 3 \cdot \cancel{x} \cdot \cancel{y^4}} \cdot \dfrac{\overset{1}{\cancel{5}} \cdot 2 \cdot \overset{x}{\cancel{x^2}} \cdot \overset{1}{\cancel{y^3}}}{3 \cdot 3 \cdot \cancel{a} \cdot \cancel{b^4}} = \dfrac{4a^2x}{9yb^3}$$

Additional Example 1

Simplify.

$\dfrac{4a^5b}{27m^8n^2} \cdot \dfrac{9m^7n^2}{8ab^2}$ $\dfrac{a^4}{6mb}$

EXAMPLE 2 Simplify $\dfrac{2x^2 - 12x - 14}{x^3 - 16x} \cdot \dfrac{-16 - 4x}{6x - 42}$.

Plan Write $-16 - 4x$ in a more convenient form: $-16 - 4x = -4x - 16 = -4(x + 4)$. Then factor the other numerators and denominators completely, divide out the common factors, and multiply.

Solution

$$\dfrac{2x^2 - 12x - 14}{x^3 - 16x} \cdot \dfrac{-16 - 4x}{6x - 42} = \dfrac{2(x^2 - 6x - 7)}{x(x^2 - 16)} \cdot \dfrac{-4(x + 4)}{6(x - 7)}$$

$$= \dfrac{2(x - 7)(x + 1)}{x(x + 4)(x - 4)} \cdot \dfrac{-2 \cdot 2(x + 4)}{2 \cdot 3(x - 7)}$$

$$= \dfrac{-4(x + 1)}{3x(x - 4)}$$

To divide a rational number by a nonzero rational number, multiply the first number by the *reciprocal*, or *multiplicative inverse*, of the second number. Rational expressions are divided in a similar manner.

Rule for Dividing Rational Expressions

If $\dfrac{a}{b}$ and $\dfrac{c}{d}$ are rational expressions where $b \neq 0$, $c \neq 0$, and $d \neq 0$, then $\dfrac{a}{b} \div \dfrac{c}{d} = \dfrac{a}{b} \cdot \dfrac{d}{c}$.

EXAMPLE 3 Simplify $\dfrac{3x^2 - 2xy}{8x^3y} \div \dfrac{9x^2 - 4y^2}{12x^2y^5}$.

Solution

$$\dfrac{3x^2 - 2xy}{8x^3y} \div \dfrac{9x^2 - 4y^2}{12x^2y^5} = \dfrac{3x^2 - 2xy}{8x^3y} \cdot \dfrac{12x^2y^5}{9x^2 - 4y^2}$$

$$= \dfrac{x(3x - 2y)}{8 \cdot x^3 \cdot y} \cdot \dfrac{12 \cdot x^2 \cdot y^5}{(3x + 2y)(3x - 2y)}$$

$$= \dfrac{3y^4}{2(3x + 2y)}$$

Cancellation Property: The Cancellation Property of Lesson 7.1 is a special case of the Rule for Multiplying Rational Expressions on page 238. To see that this is so, let $d = c$ and read the Rule from right to left. The result is $\dfrac{ac}{bc} = \dfrac{a}{b} \cdot \dfrac{c}{c} = \dfrac{a}{b} \cdot 1 = \dfrac{a}{b}$.

Critical Thinking Questions

Analysis: Ask students to use the Rule for Multiplying Rational Expressions on page 238 and the *definition of division* in Lesson 1.1 of the text ($x \div y = z$ means that $z \cdot y = x$) to prove the Rule for Dividing Rational Expressions on page 239. As a hint, have them first write the equation $\dfrac{a}{b} \div \dfrac{c}{d} = z$. Then use the *definition of division* to rewrite this equation in another form and solve it for z.

Common Error Analysis

Error: When products and quotients are mixed in the same set of exercises, some students will multiply even when division is indicated. This is particularly true in an exercise with three rational expressions, as in Example 4, where multiplication is indicated first, then division.

Emphasize that students should multiply by the reciprocal of the divisor as shown in the solution of Example 4.

Additional Example 2

Simplify.

$$\dfrac{4 - x}{3x^3 - 6x^2} \cdot \dfrac{6x - 12}{4x^2 - 20x + 16} \qquad \dfrac{-1}{2x^2(x - 1)}$$

Additional Example 3

Simplify.

$$\dfrac{4c + 6}{49d^3e^9} \div \dfrac{2c^2 + 5c + 3}{7d^3e^4} \qquad \dfrac{2}{7e^5(c + 1)}$$

Checkpoint

Simplify.

1. $\dfrac{y^5}{7} \cdot \dfrac{21}{y^4}$ $3y$

2. $\dfrac{25a^6}{16b^4} \cdot \dfrac{8b^5}{15a^7}$ $\dfrac{5b}{6a}$

3. $\dfrac{5-a}{a^2-a-20} \cdot \dfrac{a^2+8a+16}{a^2+7a+12}$ $-\dfrac{1}{a+3}$

4. $\dfrac{y^3+3y}{y^2-9} \div \dfrac{y^2+5y-14}{21-4y-y^2}$ $-\dfrac{y(y^2+3)}{(y+3)(y-2)}$

Closure

Ask students to name the four steps in simplifying the product of rational expressions. See page 238. Ask them to give the reciprocal of a non-zero number. $\dfrac{1}{n}$ Have them give another name for reciprocal. Multiplicative inverse Ask students to give the rule for dividing rational expressions. See page 239.

◾◾◾FOLLOW UP

Guided Practice

Classroom Exercises 1–8

Independent Practice

A Ex. 1–18, **B** Ex. 19–23, **C** Ex. 24, 25

Basic: WE 1–19 odd, Application

Average: WE 5–23 odd, Application

Above Average: WE 7–25 odd, Application

Additional Answers

Classroom Exercises

1. $\dfrac{14}{2b} \cdot \dfrac{2y}{7x}$ 2. $\dfrac{5}{x-y} \cdot \dfrac{4}{x+y}$

3. $\dfrac{x-2}{x+3} \cdot \dfrac{x-3}{2x-4}$ 4. $\dfrac{5}{6x-3} \cdot \dfrac{1-2x}{15x}$

EXAMPLE 4   Simplify $\dfrac{-2x-5}{x^2-1} \cdot \dfrac{x^2-1}{2x^2-x-15} \div \dfrac{x+1}{x-3}$.

Solution

$$\dfrac{-2x-5}{x^2-1} \cdot \dfrac{x^2-1}{2x^2-x-15} \cdot \dfrac{x-3}{x+1}$$

$$= \dfrac{-1(2x+5)}{(x-1)(x+1)} \cdot \dfrac{(x-1)(x+1)}{(x-3)(2x+5)} \cdot \dfrac{(x-3)}{(x+1)} = \dfrac{-1}{x+1}, \text{ or } -\dfrac{1}{x+1}$$

Classroom Exercises

Express each quotient as the product of rational expressions.

1. $\dfrac{14}{2b} \div \dfrac{7x}{2y}$ 2. $\dfrac{5}{x-y} \div \dfrac{x+y}{4}$ 3. $\dfrac{x-2}{x+3} \div \dfrac{2x-4}{x-3}$ 4. $\dfrac{5}{6x-3} \div \dfrac{15x}{1-2x}$

5–8. Simplify the quotients in Classroom Exercises 1–4, if possible.

Written Exercises

Simplify.

1. $\dfrac{x^3}{5} \cdot \dfrac{10}{x^4}$ $\dfrac{2}{x}$ 2. $\dfrac{4}{x^5} \cdot \dfrac{x^2}{12}$ $\dfrac{1}{3x^3}$ 3. $\dfrac{x-3}{14x^2} \cdot \dfrac{7x^3}{x-3}$ $\dfrac{x}{2}$

4. $\dfrac{x^3}{x^5(2x+5)} \cdot (2x+5)$ 5. $\dfrac{2a}{3b} \cdot \dfrac{b^2}{4a^3}$ $\dfrac{b}{6a^2}$ 6. $\dfrac{10x^3}{9y^5} \cdot \dfrac{6y^4}{15x^4}$ $\dfrac{4}{9xy}$

7. $\dfrac{3}{4x+8} \cdot \dfrac{x^2+2x}{9}$ $\dfrac{x}{12}$ 8. $\dfrac{x^2+6x}{10} \cdot \dfrac{4}{x^2-36}$ 9. $\dfrac{7}{a^2-9a+20} \cdot \dfrac{a^2-4a}{14}$

10. $\dfrac{x^2y^4}{6a^2} \div \dfrac{x^4y}{10a^2}$ $\dfrac{5y^3}{3x^2}$ 11. $\dfrac{8x+40}{6x^3} \div \dfrac{4x+20}{8x^5}$ 12. $\dfrac{x^2-9}{6} \div \dfrac{3-x}{8}$

13. $\dfrac{x^2-y^2}{5x^3y^2} \cdot \dfrac{15x^2y^5}{4x+4y}$ $\dfrac{3y^3(x-y)}{4x}$ 14. $\dfrac{4a+8}{5a-20} \div \dfrac{10+3a-a^2}{a^2-4a}$ $\dfrac{-4a}{5(a-5)}$

15. $\dfrac{3x^2+10x-8}{3x^2-17x+10} \cdot \dfrac{5+9x-2x^2}{x^2+3x-4}$ 16. $\dfrac{2y^2-18}{24-6y} \div \dfrac{3y^2+24y+45}{2y^2-9y+4}$

17. $\dfrac{4ab^3}{3a^2-a-10} \div \dfrac{6a^5b^7}{3a^2+17a+20}$ 18. $\dfrac{3x^2+14x-5}{x^2+2x-15} \div \dfrac{3x^2-25x+8}{8+15x-2x^2}$

19. $\dfrac{6y^2-13y-5}{-3x-x^2} \cdot \dfrac{2xy+6y+5x+15}{4y^2-25}$ 20. $\dfrac{x^2+2x-3}{x^2-2x+1} \div \dfrac{x^2+5x+6}{2-x-x^2}$ -1

Additional Example 4

Simplify.

$$\dfrac{x}{x+4} \cdot \dfrac{x^2+6x+8}{x+10} \div \dfrac{x^2}{2x+20} \quad \dfrac{2(x+2)}{x}$$

$3y^2 + 15y - y - 5$
$3y(y+5) - 1(y+5)$
$15y - y$

21. $\dfrac{x^3 - 27}{8x^3} \cdot \dfrac{20x - 4x^2}{x^2 - 8x + 15}$

22. $\dfrac{a^2 + 3a - 18}{3 + 2a - a^2} \cdot \dfrac{a^3 + 1}{2a^2 + 7a - 30}$

23. $\dfrac{x^2 + 4x - 32}{x^2 - 12x + 35} \cdot \dfrac{x^2 - 4x - 21}{16x - 4x^2} \cdot \dfrac{x^2 - 10x}{x^2 + 11x + 24}$ $\dfrac{-(x-10)}{4(x-5)}$

24. $\dfrac{2x^2y - x^3}{a^3 - 27b^3} \cdot \dfrac{x^3 + 8y^3}{x^2 - 4y^2} \div \dfrac{x^5 - 2x^4y + 4x^3y^2}{ax + 6by - 2ay - 3bx}$ $\dfrac{2y - x}{x(a^2 + 3ab + 9b^2)}$

25. $\dfrac{x^3y^3 - 8y^3 - x^3 + 8}{x^2 - 10x + 25 - 49y^2} \cdot \left(\dfrac{xy + 2 - 2y - x}{x^2 - 5x - 7yx}\right)^{-1}$ $\dfrac{x(x^2 + 2x + 4)(y^2 + y + 1)}{x - 5 + 7y}$

Mixed Review

Find the solution set.   *2.1, 2.2, 2.3, 6.4*

1. $6 - 2x \geq 4(6 - x)$ $\{x \mid x \geq 9\}$

2. $-6 \leq 2x - 4 \leq 8$ $\{x \mid -1 \leq x \leq 6\}$

3. $|x - 5| > 4$ $\{x \mid x < 1 \text{ or } x > 9\}$

4. $x^2 + 4x - 12 < 0$ $\{x \mid -6 < x < 2\}$

Application: *Photography*

When light passes through a camera lens, the curvature of the lens causes the light rays to bend so that they all pass through a point F, called the **focal point** of the lens. The distance from the center of the lens to the vertical plane through F is f, the *focal length* of the lens.

A photographic image is in focus when

$$\frac{1}{p} + \frac{1}{q} = \frac{1}{f},$$

where p is the distance of the object from the lens, q is the image distance, and f is the focal length.

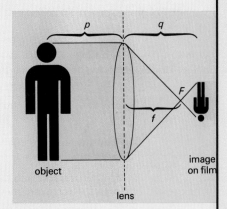

For a lens of focal length 50 mm, an image distance of 52 mm, and an image size of 12 mm, find the following.

1. the distance of the object from the lens in meters   1.3 m

2. the size of the object given that

$$\frac{\text{image size}}{\text{object size}} = \frac{\text{image distance}}{\text{object distance}}$$ 300 mm

---

5. $\dfrac{2y}{bx}$

6. $\dfrac{20}{(x-y)(x+y)}$

7. $\dfrac{x-3}{2(x+3)}$

8. $-\dfrac{1}{9x}$

Written Exercises

4. $\dfrac{1}{x^2}$

8. $\dfrac{2x}{5(x-6)}$

9. $\dfrac{a}{2(a-5)}$

11. $\dfrac{8x^2}{3}$

12. $\dfrac{-4(x+3)}{3}$

15. $\dfrac{-(2x+1)}{x-1}$

16. $\dfrac{-(y-3)(2y-1)}{9(y+5)}$

17. $\dfrac{2(a+4)}{3a^4b^4(a-2)}$

18. $\dfrac{-(2x+1)}{x-3}$

19. $\dfrac{-(3y+1)}{x}$

21. $\dfrac{-(x^2+3x+9)}{2x^2}$

22. $\dfrac{-(a^2-a+1)}{2a-5}$

Enrichment

Students may find it surprising that π can be approximated in the unusual way described here.

The mathematician John Wallis (1616–1703), proved that a product formed by pairs of factors

$$\frac{2n}{2n - 1} \cdot \frac{2n}{2n + 1}$$

for $n = 1, 2, 3, \cdots$ approaches closer and closer to $\dfrac{\pi}{2}$ as the number of factors increases.

Have the students verify that for $n = 1, 2, 3$, the product is:

$$\frac{\pi}{2} \approx \frac{2}{1} \cdot \frac{2}{3} \cdot \frac{4}{3} \cdot \frac{4}{5} \cdot \frac{6}{5} \cdot \frac{6}{7}$$

Next, have the students use a calculator to find approximations to $\dfrac{\pi}{2}$ using 8 and 9 factors ($n = 1, 2, 3, 4$). Then have them compare the sum of these approximations to a familiar value for π.

Using 8 factors: $\dfrac{\pi}{2} \approx 1.486$

Using 9 factors: $\dfrac{\pi}{2} \approx 1.651$

So, $\pi \approx 3.137 \approx 3.14$.

The students may notice that the approximation is high if the number of factors used is odd, and the approximation is low if the number of factors is even. By taking other approximations for larger successive values of n, the sum of the two approximations should be closer to π.

GETTING STARTED

Prerequisite Quiz

Complete.

1. 1 mi = _?_ ft   5,280
2. 1 m = _?_ cm   100
3. 1 h = _?_ s   3,600

Motivator

Recall that 12 in. = 1 ft. Since these are the same, $\frac{12 \text{ in.}}{1 \text{ ft}} = \frac{1 \text{ ft}}{12 \text{ in.}} = 1$.
Have students convert 15 inches to feet by multiplying 15 in. by *one* of the two fractions above. Ask students which choice will *eliminate* inches. $\frac{1 \text{ ft}}{12 \text{ in.}}$

TEACHING SUGGESTIONS

Math Connections

Electricity: In physics, the *joule* is a unit of energy, and the *coulomb* a unit of electric charge. The rate of flow of electric charge in a circuit is measured in $\frac{coulombs}{second}$, or *amperes*, while the potential difference in volts is the amount of energy per coulomb, or $\frac{joules}{coulomb}$. The power of an electric circuit is the product.
$$\frac{coulombs}{s} \times \frac{joules}{coulombs} = \frac{joules}{s} \text{ or } watts$$

7.3 Dimensional Analysis

Objective

To convert from one unit of measurement to another using dimensional analysis

The equation 12 in = 1 ft relates feet to inches. This relation can be written as a fraction that is equal to 1 in two ways.

Divide each side by 1 ft.   Divide each side by 12 in.
$$\frac{12 \text{ in}}{1 \text{ ft}} = 1 \qquad\qquad 1 = \frac{1 \text{ ft}}{12 \text{ in}}$$

To convert 18 in to feet, multiply 18 in by $\frac{1 \text{ ft}}{12 \text{ in}}$
so that the common factor *in.* will divide out and leave the new unit *ft*.

$$18 \text{ in} = \frac{18 \text{ in}}{1} \cdot \frac{1 \text{ ft}}{12 \text{ in}} = \frac{18}{12} \text{ ft} = 1\tfrac{1}{2} \text{ ft}$$

To convert 18 ft to inches, multiply 18 ft by $\frac{12 \text{ in}}{1 \text{ ft}}$.

This conversion procedure is called **dimensional analysis.**
Fractions such as $\frac{12 \text{ in}}{1 \text{ ft}}$ that are used to make the conversions are called *conversion factors* or **conversion fractions.**

EXAMPLE 1   Convert 60 miles per hour (mi/h) to feet per second (ft/s).

Plan   Write 60 mi/h as a fraction: $\frac{60 \text{ mi}}{1 \text{ h}}$. Then note that 5,280 ft = 1 mi, 60 min = 1 h, and 60 s = 1 min. Use these relationships to write conversion fractions.

Solution
$$\frac{60 \text{ mi}}{1 \text{ h}} \cdot \frac{5{,}280 \text{ ft}}{1 \text{ mi}} \cdot \frac{1 \text{ h}}{60 \text{ min}} \cdot \frac{1 \text{ min}}{60 \text{ s}} = \frac{60 \cdot 5{,}280 \text{ ft}}{60 \cdot 60 \text{ s}}$$
$$= \frac{88 \text{ ft}}{\text{s}}, \text{ or } 88 \text{ ft/s}$$

Thus, 60 mi/h is equivalent to 88 ft/s.

Conversion fractions can be used with area data to convert measurements with square units. Recall that 3 ft = 1 yd and 12 in = 1 ft.

Square each side. $9 \text{ ft}^2 = 1 \text{ yd}^2$ and $144 \text{ in}^2 = 1 \text{ ft}^2$

This leads to the following conversion fractions.

$$\frac{9 \text{ ft}^2}{1 \text{ yd}^2} = 1 \qquad \frac{1 \text{ yd}^2}{9 \text{ ft}^2} = 1 \qquad \frac{144 \text{ in}^2}{1 \text{ ft}^2} = 1 \qquad \frac{1 \text{ ft}^2}{144 \text{ in}^2} = 1$$

Additional Example 1

Convert 30 mi/h to yd/min.   880 yd/min

EXAMPLE 2   A rug measures 12 ft by 10 ft and costs $27.90 per square yard. Find the cost of the rug.

Solution   The area of the rug is 12 ft · 10 ft, or 120 ft^2. Use a conversion fraction.

$$\frac{120 \text{ ft}^2}{1} \cdot \frac{1 \text{ yd}^2}{9 \text{ ft}^2} \cdot \frac{\$27.90}{1 \text{ yd}^2} = \frac{120 \cdot \$27.90}{9} = \$372$$

Thus, the cost is $372.

Classroom Exercises

Determine the conversion fraction that should be used for each conversion indicated.

1. 18 yd to feet $\frac{3 \text{ ft}}{1 \text{ yd}}$ **2.** 26 pt to quarts $\frac{1 \text{ qt}}{2 \text{ pt}}$ **3.** 90 in. to yards $\frac{1 \text{ yd}}{36 \text{ in.}}$

4. 4.5 lb to ounces $\frac{16 \text{ oz}}{1 \text{ lb}}$ **5.** 300 g to kilograms $\frac{1 \text{ kg}}{1000 \text{ g}}$ **6.** 18 qt to gallons $\frac{1 \text{ gal}}{4 \text{ qt}}$

Written Exercises

1–6. Convert each measure in Classroom Exercises 1–6 as indicated.

Use dimensional analysis to convert each measure.

7. 12 oz/pt to pounds per gallon   6 lb/gal   **8.** 972 lb/ft^2 to ounces per square inch   108 oz/in.^{2}

9. A test car is driven for 5 h at an average speed of 40 mi/h. The car averages 24 mi/gallon. Find the cost of the gas used given that gas costs $1.35/gallon. $11.25

10. A construction crew uses 125 ft^3 of sand each work week. Given that sand weighs 1.215 tons per cubic yard (T/yd^3), find the number of pounds of sand the crew uses in 3 work weeks.   33,750 lb

11. A city has 320 youths involved in its softball program. Twenty players are assigned to each team. Each team is provided with 30 bats for the season. The cost is $42.60 per carton, with a dozen bats in each carton. Find the total cost of the bats given that the city receives a 15% discount. $1,448.40

Mixed Review

Find the slope and describe the slant of the line.   *3.5*

1. $3x - 2y = 10$ **2.** $5x - 10 = 0$ **3.** $2y - 5 = 3$ **4.** $y = 2 - x$

7.3 Dimensional Analysis   **243**

243

Problem Solving Strategies

Using a Table to Find a Pattern

An important problem solving strategy is to use a table to organize data and find a pattern. The data in the table may help you to discover a pattern so that you can solve the problem.

The Italian astronomer Galileo (1564–1642) discovered the relationship between the length of a pendulum and the time required for it to make a complete swing. Thus, given the swing time, he could predict the length of the pendulum. A student, trying an experiment similar to Galileo's, constructs the following table based on her results.

Time of Swing	Length of Pendulum
1 s	0.25 m
2 s	1 m
3 s	2.25 m
5 s	6.25 m

Using this data, the student discovers a pattern and then writes an equation relating the time of the swing to the length of the pendulum.

Exercises

1. By noting a pattern in the data, predict the length of a pendulum with a swing time of 4 seconds. (HINT: Express the length of the pendulum in fourths of a meter, as $\frac{1}{4}$, $\frac{4}{4}$, and so on.) 4 m

2. Write a formula for the length given the time for one swing. $l = \frac{1}{4}s^2$

Tables can also be used to indicate possible solutions to problems involving geometric patterns. For example, the three circles below show the number of regions formed by drawing all possible segments between a given number of points on a circle.

3. Use the pattern to guess the missing numbers in the table.

Number of points connected	2	3	4	5	6
Number of regions	2	4	8	16	32

4. Draw all possible segments between 5 points on a circle and then 6 points on a circle and count the number of regions formed. Are the predictions made from the table accurate in these cases? Only for 5 points

7.4 Sums and Differences of Rational Expressions

Objective To find sums and differences of rational expressions

$$x(y + z) = xy + xz \qquad x(y - z) = xy - xz$$

The Distributive Properties above can be used to simplify the sum $\frac{3}{7} + \frac{2}{7}$ and the difference $\frac{3}{7} - \frac{2}{7}$.

$$\frac{3}{7} + \frac{2}{7} = \frac{1}{7} \cdot 3 + \frac{1}{7} \cdot 2 \qquad\qquad \frac{3}{7} - \frac{2}{7} = \frac{1}{7} \cdot 3 - \frac{1}{7} \cdot 2$$
$$= \frac{1}{7}(3 + 2) = \frac{1}{7} \cdot 5, \text{ or } \frac{5}{7} \qquad = \frac{1}{7}(3 - 2) = \frac{1}{7} \cdot 1, \text{ or } \frac{1}{7}$$

The above results can be generalized into the following theorem.

Theorem 7.2

For all real numbers x, y, and z, where $z \neq 0$,
$$\frac{x}{z} + \frac{y}{z} = \frac{x + y}{z} \text{ and } \frac{x}{z} - \frac{y}{z} = \frac{x - y}{z}.$$

EXAMPLE 1 Simplify $\dfrac{3a}{a^2 - 16} - \dfrac{a + 8}{a^2 - 16}$.

Plan Use $\dfrac{x}{z} - \dfrac{y}{z} = \dfrac{x - y}{z}$ and simplify the resulting rational expression.

Solution
$$\frac{3a}{a^2 - 16} - \frac{a + 8}{a^2 - 16} = \frac{3a - (a + 8)}{a^2 - 16} \qquad \leftarrow \text{Be sure to include the parentheses in the numerator.}$$
$$= \frac{3a - a - 8}{a^2 - 16}$$
$$= \frac{2a - 8}{a^2 - 16} = \frac{2(a - 4)}{(a + 4)(a - 4)}, \text{ or } \frac{2}{a + 4}$$

In the sum $\dfrac{3y}{10} + \dfrac{7y - 2}{5}$, the denominators are not the same.
The denominators would be the same, however, if you multiplied the numerator and denominator of $\dfrac{7y - 2}{5}$ by 2.

$$\frac{3y}{5 \cdot 2} + \frac{(7y - 2)2}{5 \cdot 2} = \frac{3y}{10} + \frac{14y - 4}{10} = \frac{17y - 4}{10}$$

Teaching Resources

Quick Quizzes 49
Reteaching and Practice Worksheets 49

▰▰ GETTING STARTED

Prerequisite Quiz

Find the sum or difference.

1. $\frac{2}{3} + \frac{1}{2} + \frac{5}{6}$ 2
2. $7 - \frac{4}{5}$ $6\frac{1}{5}$

Factor.

3. $7x - 21$ $7(x - 3)$
4. $4x^2 - 25$ $(2x + 5)(2x - 5)$

Motivator

To add the fractions $\frac{3}{5} + \frac{7}{2(5)}$, the denominators must be the same.

Ask students to give the factor that is missing from the first denominator to make it the same as the second denominator.   2
Ask them what the numerator and denominator of the first fraction should be multiplied by to have the same denominator as the second fraction. $\frac{2}{2}$

Additional Example 1

Simplify.

$\dfrac{4a}{a^2 - 7a + 10} - \dfrac{2a + 10}{a^2 - 7a + 10}$ $\dfrac{2}{a - 2}$

Highlighting the Standards

Standard 3c: Computation with algebraic fractions follows the rules of arithmetic. Students will be less likely to get lost in the details if they understand the parallels.

Lesson Note

When finding sums and differences of rational expressions, students must first learn how to determine their LCD. The most efficient method is to factor each denominator into primes and then decide what is needed in each denominator so that all the denominators will contain the same factors. Point out that in this lesson, the slash marks used to show the common factors that have been divided out will be eliminated.

Math Connections

Arithmetic Fractions: Students should notice the similarity between the addition of rational expressions and the addition of fractions. In each case, to find the sum or difference, express each fraction in terms of the LCD. Then perform the operation.

Critical Thinking Questions

Analysis: Ask students why parentheses must be inserted in the first step of Example 1 (and also in other similar examples). The parentheses preserve the grouping formerly provided by the fraction bar. Specifically, "$a + 8$", which is grouped by the fraction bar, must have that grouping preserved in some way after the single numerator is formed.

For the sum $\dfrac{3y}{10} + \dfrac{7y - 2}{5}$, 10 is the *Least Common Denominator* (LCD) of the fractions, or the *Least Common Multiple* (LCM) of the denominators.

To find the LCM of several algebraic expressions such as

$$x^2 - 6x + 9, \; 2x^2 - 6x, \text{ and } 4x + 12:$$

1. factor the expressions completely, and

2. form the LCM by writing the product of the factors, with each factor counted only the greatest number of times it occurs in any one expression.

EXAMPLE 2   Find the LCM of $x^2 - 6x + 9$, $2x^2 - 6x$, and $4x + 12$.

Solution   Factor each expression completely.

$$\begin{array}{lll} x^2 - 6x + 9 & 2x^2 - 6x & 4x + 12 \\ (x - 3)(x - 3) & 2 \cdot x(x - 3) & 2 \cdot 2(x + 3) \end{array}$$

There are at most *two* $(x - 3)$s, *one* x, *two* 2s, and *one* $(x + 3)$ in any expression.

The LCM is $(x - 3)(x - 3) \cdot x \cdot 2 \cdot 2(x + 3)$,
or $4x(x - 3)^2(x + 3)$.

EXAMPLE 3   Simplify $\dfrac{5}{4x - 16} + \dfrac{7}{2x - 8} + \dfrac{1}{3x - 12}$.

Solution   Factor each denominator completely.

$$\frac{5}{2 \cdot 2(x - 4)} + \frac{7}{2(x - 4)} + \frac{1}{3(x - 4)}$$

Each denominator contains at most two 2s, one 3, and one $(x - 4)$.

$$\frac{5}{2 \cdot 2(x - 4)} + \frac{7}{2(x - 4)} + \frac{1}{3(x - 4)}$$
$$\quad\;\text{needs 3} \quad\;\; \text{needs 2 and 3} \quad \text{needs two 2s}$$

Multiply each numerator and denominator by the factors needed.

The LCD is $2 \cdot 2(x - 4) \cdot 3$.

$$\frac{5 \cdot 3}{2 \cdot 2(x - 4) \cdot 3} + \frac{7 \cdot 2 \cdot 3}{2(x - 4) \cdot 2 \cdot 3} + \frac{1 \cdot 2 \cdot 2}{3(x - 4) \cdot 2 \cdot 2} =$$

$$\frac{15}{12(x - 4)} + \frac{42}{12(x - 4)} + \frac{4}{12(x - 4)} = \frac{61}{12(x - 4)}$$

Additional Example 2

Find the LCM.

$x^2 - 10x + 25$, $4x + 20$, and $3x^2 - 15x$.
$12x(x - 5)^2(x + 5)$

Additional Example 3

Simplify.

$\dfrac{7}{9x - 18} + \dfrac{5}{3x - 6} + \dfrac{3}{5x - 10}$ $\dfrac{137}{45(x - 2)}$

EXAMPLE 4 Simplify $\dfrac{3y - 5}{2y - 6} - \dfrac{4y - 2}{5y - 15}$.

Solution

$$\dfrac{3y - 5}{2y - 6} - \dfrac{4y - 2}{5y - 15} = \dfrac{3y - 5}{2(y - 3)} - \dfrac{4y - 2}{5(y - 3)}$$

Express each fraction in terms of the LCD, $2 \cdot 5(y - 3)$.

$$= \dfrac{(3y - 5) \cdot 5}{2(y - 3) \cdot 5} - \dfrac{(4y - 2) \cdot 2}{5(y - 3) \cdot 2}$$

$$= \dfrac{15y - 25}{2 \cdot 5(y - 3)} - \dfrac{8y - 4}{2 \cdot 5(y - 3)}$$

Use $\dfrac{a}{b} - \dfrac{c}{b} = \dfrac{a - c}{b}$.

$$= \dfrac{15y - 25 - (8y - 4)}{2 \cdot 5(y - 3)}$$

$$= \dfrac{15y - 25 - 8y + 4}{2 \cdot 5(y - 3)}$$

Factor $7y - 21$. Divide out common factors.

$$= \dfrac{7y - 21}{2 \cdot 5(y - 3)} = \dfrac{7(y - 3)}{2 \cdot 5(y - 3)} = \dfrac{7}{10}$$

EXAMPLE 5 Simplify $\dfrac{3a - 39}{a^2 - 7a + 10} + \dfrac{3}{2 - a}$.

Solution

$$\dfrac{3a - 39}{a^2 - 7a + 10} + \dfrac{3}{2 - a} = \dfrac{3a - 39}{(a - 2)(a - 5)} + \dfrac{3}{2 - a}$$

$a - 2$ is the opposite of $2 - a$.

$$= \dfrac{3a - 39}{(a - 2)(a - 5)} + \dfrac{-3}{a - 2}$$

Multiply the numerator and denominator of the second fraction by -1.

$$= \dfrac{3a - 39}{(a - 2)(a - 5)} + \dfrac{-3(a - 5)}{(a - 2)(a - 5)}$$

$$= \dfrac{3a - 39 - 3a + 15}{(a - 2)(a - 5)}$$

$$= \dfrac{-24}{(a - 2)(a - 5)}$$

Classroom Exercises

Find the LCM of each pair of numbers or expressions.

1. 3, 13   39
2. 4, 12   12
3. $6x^2$, $x(x - 1)$ $6x^2(x - 1)$
4. $x^2 - 1$, $x + 1$ $(x + 1)(x - 1)$

Simplify.

5. $\dfrac{2b}{7} + \dfrac{3b}{14}$ $\dfrac{b}{2}$
6. $\dfrac{7x}{10} + \dfrac{3x}{2} - \dfrac{x}{5}$ $2x$
7. $\dfrac{7}{x - 3} + \dfrac{4}{(x - 3)(x + 1)}$ $\dfrac{7x + 11}{(x - 3)(x + 1)}$

Common Error Analysis

Error: Students tend to forget to use the Distributive Property in exercises such as Example 4. They do not multiply *both* terms of $(8y - 4)$ by -1. Therefore, they mistakenly write $-8y - 4$ instead of $-8y + 4$ when they combine numerators.

Emphasize that students should always perform the operation of multiplying each term by -1 and changing the sign accordingly.

Checkpoint

Simplify.

1. $\dfrac{7x}{24} + \dfrac{5x}{24}$ $\dfrac{x}{2}$

2. $\dfrac{7b - 3}{b^2 - 25} - \dfrac{5b + 7}{b^2 - 25}$ $\dfrac{2}{b + 5}$

3. $\dfrac{2x + 1}{7x + 21} + \dfrac{3x - 2}{5x + 15}$ $\dfrac{31x - 9}{35(x + 3)}$

4. $\dfrac{2}{t + 5} + \dfrac{20}{t^2 - 25}$ $\dfrac{2}{t - 5}$

Closure

Ask students to give the steps in adding rational expressions with unlike denominators.

1. Factor each denominator completely.
2. Multiply each numerator and denominator by the factors needed for the LCD.
3. Add the numerators.

Additional Example 4

Simplify.

$\dfrac{4y - 1}{3y - 12} - \dfrac{y + 6}{2y - 8}$ $\dfrac{5}{6}$

Additional Example 5

Simplify.

$\dfrac{a^2 - 3a}{a^2 + a - 12} + \dfrac{4}{a + 4} + \dfrac{5}{3 - a}$ $\dfrac{a - 8}{a - 3}$

Guided Practice

Classroom Exercises 1–7

Independent Practice

A Ex. 1–18, **B** Ex. 19, 20, **C** Ex. 21, 22

Basic: WE 1–19 odd, Midchapter Review

Average: WE 1–19 odd, Midchapter Review

Above Average: WE 3–21 odd, Midchapter Review

Additional Answers

Written Exercises

9. $\dfrac{3(5a + 6)}{10(a + 4)}$

21. 1. $\dfrac{x}{z} + \dfrac{y}{z} = x \cdot \dfrac{1}{z} + y \cdot \dfrac{1}{z}$
    (Mult Ident Prop)

  2. $= (x + y)\dfrac{1}{z}$ (Distr Prop)

  3. $= \dfrac{x+y}{z}$ (Ident Prop Mult)

  4. $\dfrac{x}{z} + \dfrac{y}{z} = \dfrac{x+y}{z}$ (Trans Prop Eq)

22. 1. $\dfrac{x}{z} - \dfrac{y}{z} = \dfrac{x}{z} + \dfrac{(-y)}{z}$ (Rule of Subtr)

  2. $= x \cdot \dfrac{1}{z} + (-y) \cdot \dfrac{1}{z}$
    (Mult Ident Prop)

  3. $= [x + (-y)]\dfrac{1}{z}$ (Distr Prop)

  4. $= (x - y)\dfrac{1}{z}$ (Rule of Subtr)

  5. $= \dfrac{x - y}{z}$ (Ident Prop Mult)

  6. $\dfrac{x}{z} - \dfrac{y}{z} = \dfrac{x - y}{z}$ (Trans Prop Eq)

Written Exercises

Simplify.

1. $\dfrac{3a}{25} + \dfrac{2a}{25}$ $\dfrac{a}{5}$

2. $\dfrac{7}{2a} + \dfrac{5}{2a}$ $\dfrac{6}{a}$

3. $\dfrac{11a}{15} - \dfrac{2a}{15}$ $\dfrac{3a}{5}$

4. $\dfrac{2x + 5}{4x + 8} + \dfrac{1 + x}{4x + 8}$ $\dfrac{3}{4}$

5. $\dfrac{4b - 1}{b^2 - 9} - \dfrac{14 - b}{b^2 - 9}$ $\dfrac{5}{b + 3}$

6. $\dfrac{7x + 15}{x^2 + 7x + 10} - \dfrac{2x + 5}{x^2 + 7x + 10}$ $\dfrac{5}{x + 5}$

7. $\dfrac{7x}{10} + \dfrac{3x}{20} - \dfrac{x}{5}$ $\dfrac{13x}{20}$

8. $\dfrac{2a}{7} + \dfrac{3a}{14} + \dfrac{7a}{2}$ $4a$

9. $\dfrac{2a}{3a + 12} + \dfrac{5a}{6a + 24} + \dfrac{9}{5a + 20}$

10. $\dfrac{7}{9a - 27} - \dfrac{4}{15a - 45}$ $\dfrac{23}{45(a - 3)}$

11. $\dfrac{2}{9x + 45} + \dfrac{7}{6x + 30}$ $\dfrac{25}{18(x + 5)}$

12. $\dfrac{5}{6a + 15} - \dfrac{5}{8a + 20}$ $\dfrac{5}{12(2a + 5)}$

13. $\dfrac{5}{m - 7} - \dfrac{55}{m^2 - 3m - 28}$ $\dfrac{5}{m + 4}$

14. $\dfrac{3x - 10}{x^2 - 8x + 12} - \dfrac{2}{x - 6}$ $\dfrac{1}{x - 2}$

15. $\dfrac{-2x - 3}{x^2 - 3x} + \dfrac{x}{x - 3}$ $\dfrac{x + 1}{x}$

16. $\dfrac{-9b - 27}{b^2 - 3b - 18} - \dfrac{b}{6 - b}$ $\dfrac{b - 9}{b - 6}$

17. $\dfrac{-9y - 3}{y^2 - 11y + 18} + \dfrac{y + 3}{y - 9}$ $\dfrac{y + 1}{y - 2}$

18. $\dfrac{-24}{a^2 - 7a + 10} + \dfrac{a + 3}{a - 5}$ $\dfrac{a + 6}{a - 2}$

19. $\dfrac{2a - 3}{3a^2 - 13a - 10} + \dfrac{2a + 1}{5 - a} + \dfrac{1}{3a + 2}$ $\dfrac{-2(3a^2 + 2a + 5)}{(3a + 2)(a - 5)}$

20. $\dfrac{5}{xy + 3y - 2x - 6} + \dfrac{4}{x + 3} - \dfrac{2}{2 - y}$ $\dfrac{3 + 4y + 2x}{(x + 3)(y - 2)}$

21. Prove the first part of Theorem 7.2.

22. Prove the second part of Theorem 7.2.

Midchapter Review

Simplify.   7.1, 7.2, 7.4

1. $\dfrac{4x + 2}{10x + 5}$ $\dfrac{2}{5}$

2. $\dfrac{a + 4}{9y^3} \cdot \dfrac{3y^4}{a + 4}$ $\dfrac{y}{3}$

3. $\dfrac{6x + 2}{x^2 - 2x - 8} + \dfrac{2}{8 - 2x}$ $\dfrac{5x}{(x + 2)(x - 4)}$

Convert each measure as indicated.   7.3

4. 60 mg to grams   0.06 g
5. 10 yd^2 to square ft   90 ft^2
6. 20 gal to quarts   80 qt

Enrichment

Explain that the statement below represents a common error.

$$\dfrac{a}{b + c} = \dfrac{a}{b} + \dfrac{a}{c}$$

Challenge the students to *prove* that the statement is false. Proof by counterexample: As one possibility, let $a = 30$, $b = 3$, and $c = 2$. This results in the statement $6 = 25$, which clearly disproves the given statement.

Now challenge the students to prove that *no* positive values of a, b, and c can make the statement true.

If $\dfrac{a}{b + c} = \dfrac{a}{b} + \dfrac{a}{c}$,

then $\dfrac{a}{b + c} = \dfrac{ab + ac}{bc}$

$\dfrac{a}{b + c} = \dfrac{a(b + c)}{bc}$

$$(b + c)^2 = bc$$
$$b^2 + 2bc + c^2 = bc$$
$$b^2 + c^2 = -bc$$

But this last equation is impossible since $-bc < 0$ and $b^2 + c^2 > 0$.

7.5 Complex Rational Expressions

Objective   To simplify complex rational expressions

When a quotient has fractions in the numerator or the denominator (see the fractions at the right), it is called a *complex rational expression*.

$$\frac{\frac{1}{10} + \frac{3x}{5}}{\frac{x}{2} - \frac{1}{5}} \qquad \frac{a^{-1} - b^{-1}}{a^{-2} - b^{-2}}$$

Definition   A **complex rational expression** is a rational expression in which the numerator, the denominator, or both contain at least one rational expression.

There are two methods for simplifying complex rational expressions.

Method 1: Express the complex rational expression as a quotient using the division symbol ($\div$). Then simplify the result.

Method 2: Use the property $\dfrac{a}{b} = \dfrac{a \cdot c}{b \cdot c}$, where c is the LCM of all the denominators of the individual expressions. Multiply the numerator and the denominator of the complex rational expression by the LCM.

EXAMPLE 1   Simplify $\dfrac{\frac{1}{10} + \frac{3x}{5}}{\frac{x}{2} - \frac{1}{5}}$.

Solution   The LCM of 10, 5, and 2 is 10, or $5 \cdot 2$.

Method 1

$$\left(\frac{1}{10} + \frac{3x}{5}\right) \div \left(\frac{x}{2} - \frac{1}{5}\right)$$

$$= \left(\frac{1}{10} + \frac{3x \cdot 2}{5 \cdot 2}\right) \div \left(\frac{x \cdot 5}{2 \cdot 5} - \frac{1 \cdot 2}{5 \cdot 2}\right)$$

$$= \frac{1 + 6x}{10} \div \frac{5x - 2}{10}$$

$$= \frac{1 + 6x}{10} \cdot \frac{10}{5x - 2}$$

$$= \frac{1 + 6x}{5x - 2}$$

Method 2

$$\frac{5 \cdot 2\left(\frac{1}{10} + \frac{3x}{5}\right)}{5 \cdot 2\left(\frac{x}{2} - \frac{1}{5}\right)}$$

$$= \frac{5 \cdot 2 \cdot \frac{1}{10} + 5 \cdot 2 \cdot \frac{3x}{5}}{5 \cdot 2 \cdot \frac{x}{2} - 5 \cdot 2 \cdot \frac{1}{5}}$$

$$= \frac{1 + 2 \cdot 3x}{5 \cdot x - 2 \cdot 1}, \text{ or } \frac{1 + 6x}{5x - 2}$$

Additional Example 1

Simplify.

$$\frac{\frac{2}{15} - \frac{7x}{3}}{\frac{x}{3} + \frac{4}{5}} \qquad \frac{2 - 35x}{5x + 12}$$

Prerequisite Quiz

Find the LCM of each set of numbers or algebraic expressions.

1. 9, 2, 3   18
2. a^2, a^3, a^4 a^4
3. $(x - 2), (x + 2), (x^2 - 4)$ $x^2 - 4$

Simplify.

4. $\dfrac{1}{10} + \dfrac{3x}{5}$ $\dfrac{6x + 1}{10}$
5. $\dfrac{x}{2} - \dfrac{1}{5}$ $\dfrac{5x - 2}{10}$

Motivator

Ask students to identify the numerator of the fraction below. $\dfrac{2}{3}$ Ask them to identify the denominator. $\dfrac{5}{6}$

Ask students to find the LCD for each of these two separate fractions.   6

$$\frac{\frac{2}{3}}{\frac{5}{6}}$$

Ask them to multiply the numerator and denominator of the **entire** fraction by the LCD of the two separate fractions and give the result. $\dfrac{4}{5}$

Highlighting the Standards

Standard 14b: At the high school level, a teacher cannot point to the structure of mathematics. Rather, students will gradually grasp underlying principles as they expand their skill base.

Lesson Note

Although this lesson presents two methods of simplifying complex fractions, some teachers may prefer to select only one for class illustration to avoid confusing some students. Then the students can have the option of using the other method if they want to learn it on their own.

Math Connections

Harmonic Mean of Two Numbers: The definition of the harmonic mean of two numbers is "the reciprocal of the arithmetic mean of the reciprocals of the two numbers." The practical effect of this definition is to obtain a complex rational expression. If a and b are the two numbers then the resulting expression is as follows.

$$\frac{2}{\frac{1}{a} + \frac{1}{b}}$$

Critical Thinking Questions

Average Speed: Have students find the average speed of an automobile that travels for 3 mi at 40 mi/h and immediately returns along the same route at 20 mi/h. Of course, the "obvious" answer, 30 mi/h, is not correct. Students should first use the distance formula $d = rt$ to solve for the time needed for each part of the trip and only then calculate the rate. After they have done so, they should notice that the answer is the harmonic mean of 20 and 40, or $26\frac{2}{3}$ mi/h.

EXAMPLE 2 Simplify $\dfrac{\dfrac{3x}{x^2 - 3x - 10} - \dfrac{2}{x + 2}}{\dfrac{4}{3x - 15} + \dfrac{2x}{x^2 - 3x - 10}}$.

Plan Factor each denominator. Then multiply the numerator and denominator of the complex expression by the LCM of all the denominators.

Solution

$$\dfrac{3(x - 5)(x + 2)\left[\dfrac{3x}{(x - 5)(x + 2)} - \dfrac{2}{x + 2}\right]}{3(x - 5)(x + 2)\left[\dfrac{4}{3(x - 5)} + \dfrac{2x}{(x - 5)(x + 2)}\right]}$$ ← The LCM is $3(x - 5)(x + 2)$.

$$= \dfrac{3(x - 5)(x + 2) \cdot \dfrac{3x}{(x - 5)(x + 2)} + 3(x - 5)(x + 2) \cdot \dfrac{-2}{x + 2}}{3(x - 5)(x + 2) \cdot \dfrac{4}{3(x - 5)} + 3(x - 5)(x + 2) \cdot \dfrac{2x}{(x - 5)(x + 2)}}$$

$$= \dfrac{3 \cdot 3x + 3(x - 5) \cdot (-2)}{(x + 2)4 + 3 \cdot 2x} = \dfrac{9x + (-6x + 30)}{4x + 8 + 6x}, \text{ or } \dfrac{3x + 30}{10x + 8}$$

EXAMPLE 3 Simplify $\dfrac{1 - 7x^{-1} - 18x^{-2}}{1 - 4x^{-2}}$.

Plan Rewrite the rational expression so that it contains only positive exponents.

Solution

$$\dfrac{1 - \dfrac{7}{x} - \dfrac{18}{x^2}}{1 - \dfrac{4}{x^2}} = \dfrac{x^2\left(1 - \dfrac{7}{x} - \dfrac{18}{x^2}\right)}{x^2\left(1 - \dfrac{4}{x^2}\right)} = \dfrac{x^2 \cdot 1 - x^2 \cdot \dfrac{7}{x} - x^2 \cdot \dfrac{18}{x^2}}{x^2 \cdot 1 - x^2 \cdot \dfrac{4}{x^2}}$$

$$= \dfrac{x^2 - 7x - 18}{x^2 - 4} = \dfrac{(x - 9)(x + 2)}{(x - 2)(x + 2)}, \text{ or } \dfrac{x - 9}{x - 2}$$

Classroom Exercises

Find the LCD for each complex rational expression.

1. $\dfrac{\dfrac{x}{3} + \dfrac{1}{2}}{\dfrac{5}{6} - \dfrac{x}{2}}$ 6

2. $\dfrac{\dfrac{1}{y} + \dfrac{1}{3}}{\dfrac{1}{y} - \dfrac{1}{3}}$ 3y

3. $\dfrac{\dfrac{5}{a} - 2}{\dfrac{3}{a} + 1}$ a

4. $\dfrac{2a + \dfrac{1}{2}}{\dfrac{a}{3} + \dfrac{2}{5}}$ 30

5–8. Simplify the rational expressions in Classroom Exercises 1–4 above.

Additional Example 2

Simplify.

$$\dfrac{\dfrac{2}{x} - \dfrac{10}{x^2 + 7x}}{\dfrac{5}{x + 7} + \dfrac{2}{x}}$$ $\dfrac{2}{7}$

Additional Example 3

Simplify.

$$\dfrac{1 + 12x^{-1} + 27x^{-2}}{x^{-1} + 9x^{-2}}$$ $x + 3$

Written Exercises

Simplify.

1. $\dfrac{\dfrac{7}{x^2-4}+\dfrac{2}{x-2}}{\dfrac{6}{x-2}+\dfrac{5}{x+2}}$ $\dfrac{2x+11}{11x+2}$

2. $\dfrac{\dfrac{3}{2a-10}+\dfrac{6}{a+2}}{\dfrac{9}{a^2-3a-10}+\dfrac{1}{a-5}}$ $\dfrac{3(5a-18)}{2(a+11)}$

3. $\dfrac{\dfrac{4}{x^2-6x-16}-\dfrac{3}{5x-40}}{\dfrac{5}{2x-16}+\dfrac{3}{x+2}}$ $\dfrac{-2(3x-14)}{5(11x-38)}$

4. $\dfrac{1-5m^{-1}+4m^{-2}}{1-16m^{-2}}$ $\dfrac{m-1}{m+4}$

5. $\dfrac{1-25a^{-2}}{1-3a^{-1}-10a^{-2}}$ $\dfrac{a+5}{a+2}$

6. $\dfrac{1-2x^{-1}+x^{-2}}{1+2x^{-1}-3x^{-2}}$ $\dfrac{x-1}{x+3}$

7. $\dfrac{\dfrac{2}{x}-\dfrac{10}{x^2+7x}}{\dfrac{5}{x+7}+\dfrac{2}{3x}}$ $\dfrac{6(x+2)}{17x+14}$

8. $\dfrac{\dfrac{2}{x-3}-\dfrac{1}{x-5}}{\dfrac{ax-7a-x+7}{x^2-8x+15}}$ $\dfrac{1}{a-1}$

9. $\dfrac{\dfrac{3}{a}-\dfrac{9}{a^2+3a}}{\dfrac{4}{a+3}-\dfrac{1}{a}}$ $\dfrac{a}{a-1}$

10. $\dfrac{\dfrac{10a}{a^2+6a+8}}{\dfrac{7}{a+4}+\dfrac{3}{a+2}}$ $\dfrac{5a}{5a+13}$

11. $\dfrac{\dfrac{1}{2}+\dfrac{1+\dfrac{1}{x}}{2}}{\dfrac{1}{x}+\dfrac{1}{2}}$ $\dfrac{2x+1}{x+2}$

12. $1+\dfrac{3+\dfrac{1}{x}}{1+\dfrac{1}{x}}$

Simplify $\dfrac{f(3+h)-f(3)}{h}$ for each function in Exercises 13–16.

13. $f(x)=\dfrac{1}{x}$ 14. $f(x)=\dfrac{1}{x-1}$ 15. $f(x)=\dfrac{1}{x^2}$ 16. $f(x)=\dfrac{1}{2x+1}$

Simplify.

17. $\dfrac{(x+y)y^{-1}-2x(x+y)^{-1}}{(x-y)y^{-1}+2x(x-y)^{-1}}$ $\dfrac{x-y}{x+y}$

Mixed Review

Graph each sentence. *3.5, 3.9*

1. $3x-2y=4$ 2. $4x-4=8$ 3. $2x-3y\ge 9$
4. Write an equation of the line containing the points $A(3,7)$ and $B(-2,-3)$. *3.4* $y=2x+1$

Enrichment

Write the complex expression shown below on the chalkboard. Challenge students to evaluate it. HINT: Have students begin by simplifying the expression from the bottom.

Have each student make up a similar complex expression and challenge each other with their evaluations.

$$\cfrac{1}{1+\cfrac{1}{1+\cfrac{1}{1+\cfrac{1}{1+1}}}}\quad \dfrac{5}{8}$$

▮▮▮ GETTING STARTED

Prerequisite Quiz

Solve.
1. $7x - 8 = 9x - 10$ 1
2. $1 - 4x = 9 + x$ $-\dfrac{8}{5}$
3. $3(2x - 4) = 5 - (6 - x)$ $\dfrac{11}{5}$
4. $x^2 - 8x + 12 = 0$ 2, 6
5. $4x^2 - 8x = 0$ 0, 2

Motivator

Suppose that your school has a population of 600 and 3 out of 5 students take Algebra. Have students use a **proportion** to find the number of students taking Algebra.
Let x = the number taking Algebra.
$\dfrac{3}{5} = \dfrac{x}{600}$; $3(600) = 5(x)$; $x = 360$
Ask students if they could solve the problem by multiplying each side of the equation by the LCD of all the fractions.   Yes.

▮▮▮ TEACHING SUGGESTIONS

Lesson Note

Students sometimes find it contradictory that, in adding rational expressions, the LCD is retained as the denominator. However, in solving equations with rational expressions, the denominators "vanish." Emphasize that an expression like $\dfrac{a}{b} + \dfrac{c}{d}$ is *not* an equation; the multiplication property of equations cannot be applied as it can in $\dfrac{a}{b} = \dfrac{c}{d}$.

7.6 Equations Containing Rational Expressions

Objective   To solve equations containing rational expressions

The easiest way to solve equations that contain rational expressions is to multiply each side of the equation by the LCD of all the fractions.

EXAMPLE 1   Solve $\dfrac{2x - 3}{4} + 2 = \dfrac{2x + 1}{3}$.

Plan   Factor each denominator to find the LCD.

Solution $\dfrac{2x - 3}{2 \cdot 2} + \dfrac{2}{1} = \dfrac{2x + 1}{3}$ ← The LCD is $2 \cdot 2 \cdot 3$.

Multiply each side by the LCD.

$$2 \cdot 2 \cdot 3 \cdot \left(\dfrac{2x - 3}{2 \cdot 2} + \dfrac{2}{1}\right) = 2 \cdot 2 \cdot 3 \cdot \left(\dfrac{2x + 1}{3}\right)$$

$$2 \cdot 2 \cdot 3 \cdot \left(\dfrac{2x - 3}{2 \cdot 2}\right) + 2 \cdot 2 \cdot 3 \cdot \dfrac{2}{1} = 2 \cdot 2 \cdot 3 \cdot \left(\dfrac{2x + 1}{3}\right)$$

$$3(2x - 3) + 2 \cdot 2 \cdot 3 \cdot 2 = 2 \cdot 2(2x + 1)$$
$$6x + 15 = 8x + 4$$
$$-2x = -11$$
$$x = 5\tfrac{1}{2}$$

Thus, the solution is $5\tfrac{1}{2}$. The check is left for you.

Possible solutions of a fractional equation may not check in the original equation. Therefore, the check is especially important.

EXAMPLE 2   Solve $\dfrac{x + 1}{x - 3} = \dfrac{3}{x} + \dfrac{12}{x^2 - 3x}$ and check.

Solution $\dfrac{x + 1}{x - 3} = \dfrac{3}{x} + \dfrac{12}{x(x - 3)}$

$$x(x - 3) \cdot \dfrac{x + 1}{x - 3} = x(x - 3) \cdot \dfrac{3}{x} + x(x - 3) \cdot \dfrac{12}{x(x - 3)}$$

$$x^2 + x = 3x - 9 + 12$$
$$x^2 - 2x - 3 = 0$$
$$(x - 3)(x + 1) = 0$$
$$x = 3 \quad or \quad x = -1$$

Highlighting the Standards

Standard 2a: Example 2 shows the possibility of introducing extraneous roots, a new experience for students. This teaches students that checking is an essential part of the algebraic process.

Additional Example 1

Solve.
$\dfrac{3a}{4} - \dfrac{2a - 1}{2} = \dfrac{a - 7}{6}$ 4

Additional Example 2

Solve and check.
$\dfrac{4}{y^2 - 8y + 12} = \dfrac{y}{y - 2} + \dfrac{1}{y - 6}$
-1; 6 is extraneous.

Checks

Substitute 3 for x.

$$\frac{x+1}{x-3} = \frac{3}{x} + \frac{12}{x^2 - 3x}$$

$$\frac{3+1}{3-3} \overset{?}{=} \frac{3}{3} + \frac{12}{3^2 - 3 \cdot 3}$$

$$\frac{4}{0} \overset{?}{=} 1 + \frac{12}{0}$$

Substitute -1 for x.

$$\frac{x+1}{x-3} = \frac{3}{x} + \frac{12}{x^2 - 3x}$$

$$\frac{-1+1}{-1-3} \overset{?}{=} \frac{3}{-1} + \frac{12}{(-1)^2 - 3(-1)}$$

$$\frac{0}{-4} \overset{?}{=} -3 + 3$$

$$0 = 0 \quad \text{True}$$

Because the symbols $\frac{4}{0}$ and $\frac{12}{0}$ are undefined, 3 is not a solution of the original equation.

Thus, -1 is the only solution of the original equation.

In Example 2 above, 3 is a solution of the *derived* equation, $x^2 - 2x - 3 = 0$. However, 3 is not a solution of the *original* equation. In this case, 3 is called an *extraneous solution*, or *extraneous root*.

Definition

An **extraneous root** is a solution of a derived equation that is not a solution of the original equation.

The formula $\frac{1}{R} = \frac{1}{x} + \frac{1}{y}$ is a literal equation that contains rational expressions. It can be simplified by multiplying each side by the LCD. Then it can be solved for any one of its variables.

EXAMPLE 3

Solve $\frac{1}{R} = \frac{1}{x} + \frac{1}{y}$ for x. Then find the value of x for $R = 4.5$ and $y = 6.0$.

Solution

$$\frac{1}{R} = \frac{1}{x} + \frac{1}{y}$$

$$Rxy \cdot \frac{1}{R} = Rxy \cdot \frac{1}{x} + Rxy \cdot \frac{1}{y}$$

$$xy = Ry + Rx$$

$$xy - Rx = Ry$$
$$x(y - R) = Ry$$

$$x = \frac{Ry}{y - R}$$

Next, substitute 4.5 for R and 6.0 for y.

$$x = \frac{Ry}{y - R}$$

$$= \frac{4.5(6.0)}{6.0 - 4.5}$$

$$= \frac{27.0}{1.5}$$

$$= 18$$

Thus, $x = \frac{Ry}{y - R}$, and $x = 18$ when $R = 4.5$ and $y = 6.0$.

Math Connections

Parallel Circuits: The equation of Example 3 can be interpreted physically as the relationship between the resistances of two pieces of electrical equipment that are connected in a parallel circuit. If x is the resistance of one piece and y the resistance of the other, then the total resistance R of the circuit is as given by the equation. In other words, the sum of the reciprocals of the two resistances in parallel equals the reciprocal of the total resistance of the circuit.

Critical Thinking Questions

Analysis: Each of the equations of the lesson has at least one root and perhaps an extraneous root or two as well. Ask students whether they think that an equation with rational expressions might have *only* extraneous roots, that is, no actual roots at all. Ask them to try to devise such an equation. Here is an example of such an equation.

$$\frac{1}{x-2} = \frac{3}{x} + \frac{2}{x^2 - 2x}$$

Checkpoint

Solve. Check for extraneous roots.

1. $\frac{2x-3}{5} + \frac{x-8}{2} = \frac{-1}{10}$ 5

2. $\frac{6}{a+2} + \frac{3}{a^2-4} = \frac{2a-7}{a-2}$ $-\frac{1}{2}$, 5

3. $\frac{2x-1}{x+4} - \frac{x}{x+3} = \frac{-1}{x^2+7x+12}$ -2, 1

4. $\frac{a}{a-4} + \frac{2}{a} = \frac{16}{a^2-4a}$

   -6; 4 is extraneous.

Additional Example 3

Solve $\frac{p - af}{q} = t$ for f. Then find the value of f if $p = 72.0$, $a = 24.0$, $q = 1.2$, and $t = 140$. $f = \frac{p - qt}{a}$; $f = -4$

253

Closure

Ask students to give the steps in solving an equation containing rational expressions.

1. Factor each denominator to find the LCD.
2. Multiply each side by the LCD.
3. Check possible solutions in the original equation.

Ask students to define an *extraneous* solution.   See page 253.

**FOLLOW UP**

Guided Practice

Classroom Exercises 1–12

Independent Practice

A Ex. 1–20, **B** Ex. 21–34, **C** Ex. 35, 36

Basic: FR, WE 1–21 odd, Brainteaser, Application

Average: FR, WE 11–31 odd, Brainteaser, Application

Above Average: FR, WE 15–35 odd, Brainteaser, Application

Additional Answers

Classroom Exercises

7. -2 8. 15

9. 3   10. $\frac{5}{2}$

11. $-\frac{9}{4}$ 12. 4

Written Exercises

21. No solution (5 is extraneous.)
22. No solution (1 is extraneous.)
28. $x_2 = \frac{Rx_1}{x_1 - R}$; 9.6
29. $b = \frac{a}{1 + cd}$; 0.5

EXAMPLE 4   The numerator of a fraction is 7 less than twice its denominator. If the numerator is doubled and the denominator is increased by 4, the resulting fraction is $\frac{2}{3}$. Find the original fraction.

What are you to find?	the original fraction
Choose two variables. What do they represent?	Let $n =$ the numerator. Let $d =$ the denominator. So, the original fraction is $\frac{n}{d}$.

What is given?   (1) The numerator is 7 less than twice the denominator.

(2) When the numerator is doubled and the denominator increased by 4, the result is $\frac{2}{3}$.

Write a system.   (1) $n = 2d - 7$ (2) $\dfrac{2n}{d + 4} = \dfrac{2}{3}$

Solve the system.   (2) $\dfrac{2(2d - 7)}{d + 4} = \dfrac{2}{3}$ ← The substitution method is used.

Use the Proportion Property: $2(2d - 7) \cdot 3 = (d + 4) \cdot 2$
If $\frac{a}{b} = \frac{c}{d}$, then $a \cdot d = b \cdot c$.

$$12d - 42 = 2d + 8$$
$$10d = 50$$
$$d = 5 \leftarrow \text{denominator}$$
$$(1)\ n = 2d - 7$$
$$= 2(5) - 7$$
$$= 3 \leftarrow \text{numerator}$$

Check in the original problem.   The numerator is 7 less than twice the denominator.
$$3 = 2 \cdot 5 - 7$$

If the numerator is doubled and the denominator is increased by 4, the resulting fraction is $\frac{2}{3}$.
$$\frac{3 \cdot 2}{5 + 4} = \frac{6}{9} = \frac{2}{3}$$

State the answer.   The original fraction $\frac{n}{d}$ is $\frac{3}{5}$.

Focus on Reading

1. A solution of a derived equation that is not a solution of the original equation is called _____.   extraneous
2. One method for solving a fractional equation is to multiply each side of the equation by the _____ of the _____.   LCD, fractions

---

Additional Example 4

The denominator of a fraction is 1 more than twice the numerator. If the numerator is doubled and the denominator is decreased by 2, the resulting fraction is $\frac{6}{5}$. Find the original fraction. $\frac{3}{7}$

Classroom Exercises

Find the least common denominator (LCD) for each equation.

1. $\dfrac{x}{2} + \dfrac{2}{3} = \dfrac{x}{6}$ 6

2. $\dfrac{3a}{5} + \dfrac{3}{2} = \dfrac{7a}{10}$ 10

3. $\dfrac{1}{m} + \dfrac{2}{3} = 1$ 3m

4. $\dfrac{4}{5} + \dfrac{3}{a} = 2$ 5a

5. $\dfrac{2a - 3}{7} - \dfrac{a}{2} = \dfrac{a + 3}{14}$ 14

6. $\dfrac{3x}{4} - \dfrac{2x - 1}{2} = \dfrac{x - 7}{6}$ 12

7–12. Solve the equations in Classroom Exercises 1–6 above. Check for extraneous roots.

Written Exercises

Solve each equation. Check for extraneous roots.

1. $\dfrac{2}{5} + \dfrac{2}{y} = 1$ $3\frac{1}{3}$

2. $\dfrac{6}{x} + \dfrac{9}{2x} = 3$ $3\frac{1}{2}$

3. $\dfrac{4a + 3}{3} = \dfrac{2a + 5}{4}$ $\dfrac{3}{10}$

4. $\dfrac{x}{10} + \dfrac{x}{6} + \dfrac{x}{15} = 1$ 3

5. $\dfrac{2}{3n^2} = \dfrac{1}{4n^2} + \dfrac{5}{6n}$ $\dfrac{1}{2}$

6. $\dfrac{2n - 3}{2} = \dfrac{3}{4} + \dfrac{n - 4}{8}$ 2

7. $\dfrac{3x - 2}{2} + \dfrac{x - 4}{3} = \dfrac{1}{4}$ $1\frac{9}{22}$

8. $\dfrac{2x - 3}{5} + 1 = \dfrac{x + 3}{3}$ 9

9. $\dfrac{3n - 7}{n - 5} + \dfrac{n}{2} = \dfrac{8}{n - 5}$ -6

10. $\dfrac{a - 4}{a + 3} = \dfrac{3a + 2}{a + 3} + \dfrac{a}{4}$ -8

11. $\dfrac{3}{x - 4} + \dfrac{2}{x + 4} = \dfrac{14}{x^2 - 16}$ 2

12. $\dfrac{-3a}{a^2 - 4a - 32} = \dfrac{2}{a - 8} + \dfrac{3}{a + 4}$ 2

13. $\dfrac{14}{x^2 - 3x} - \dfrac{8}{x} = \dfrac{-10}{x - 3}$ -19

14. $\dfrac{2m - 1}{m^2 - 9m + 20} = \dfrac{7}{m - 5} - \dfrac{4}{m - 4}$ 7

15. $\dfrac{x + 5}{x - 4} = \dfrac{3}{x} + \dfrac{36}{x^2 - 4x}$ -6

16. $\dfrac{2a - 9}{a - 7} + \dfrac{a}{2} = \dfrac{5}{a - 7}$ -4

17. $\dfrac{5}{2y + 6} - \dfrac{3}{y - 4} = \dfrac{2y - 4}{y^2 - y - 12}$ -6

18. $\dfrac{4}{y^2 - 8y + 12} = \dfrac{y}{y - 2} + \dfrac{1}{y - 6}$ -1

19. $\dfrac{6}{x + 2} + \dfrac{3}{x^2 - 4} = \dfrac{2x - 7}{x - 2}$ $-\dfrac{1}{2}$, 5

20. $\dfrac{2}{a + 2} + \dfrac{a}{2 - a} = \dfrac{13}{4 - a^2}$ -3, 3

21. $\dfrac{2x - 3}{x - 5} = \dfrac{x}{x + 4} + \dfrac{20x - 37}{x^2 - x - 20}$

22. $\dfrac{2b + 3}{b - 1} - \dfrac{10}{b^2 - 1} = \dfrac{2b + 3}{b + 1}$

23. $\dfrac{x - 2}{x + 1} = \dfrac{x - 3}{x^2 - 5x - 6} - \dfrac{2x - 7}{x - 6}$ $\dfrac{2}{3}$, 4

24. $\dfrac{a - 3}{3a} = \dfrac{1}{3a^2 + 9a} + \dfrac{1}{a + 3}$ -2, 5

25. $\dfrac{7y - 20}{y^2 - 7y + 12} = \dfrac{y}{y - 3} - \dfrac{2}{4 - y}$ 2, 7

26. $\dfrac{8}{12 + 4x - x^2} + \dfrac{x + 1}{6 - x} = \dfrac{5}{x + 2}$ -10, 2

 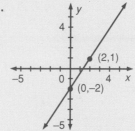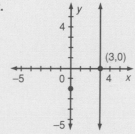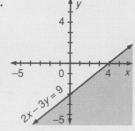
Enrichment

Challenge students with this puzzle.
In a certain math class, each student has a textbook, every 2 students share a book of tables, every 3 students share a problem book, and every 4 students share a mathematics dictionary. If the total number of books is 75, how many students are there in the class?

$$x + \dfrac{x}{2} + \dfrac{x}{3} + \dfrac{x}{4} = 75$$
$$12x + 6x + 4x + 3x = 75 \cdot 12$$
$$25x = 75 \cdot 12$$
$$x = 36$$

There are 36 students in the class.

Solve each formula for the specified variable. Then find the value of that variable for the given data.

27. Solve $\dfrac{a + b}{ab} = \dfrac{2}{x}$ for x.

$a = 8.12$; $b = 8.7$ $x = \frac{2ab}{a + b}$; 8.4

28. Solve $\dfrac{1}{R} = \dfrac{1}{x_1} + \dfrac{1}{x_2}$ for x_2.

$R = 3.2$; $x_1 = 4.8$

29. Solve $\dfrac{a - b}{b} = cd$ for b.

$a = 48.35$; $c = 5.8$; $d = 16.5$

30. Solve $\dfrac{t + mf}{m} = g$ for m. $m = \dfrac{t}{g - f}$; 2

$t = 65.8$; $g = 46.3$; $f = 13.4$

Solve each problem.

31. The denominator of a fraction is 1 more than twice its numerator. If the numerator is increased by 1 and the denominator is decreased by 1, the resulting fraction is $\frac{2}{3}$. Find the original fraction. $\frac{3}{7}$

32. The denominator of a fraction is 14 less than the square of its numerator. The fraction can be simplified to $\frac{1}{5}$. What could the original fraction be?
$\frac{7}{35}$, $\frac{-2}{-10}$

33. The numerator of a fraction is 6 less than twice its denominator. If the numerator is increased by 1 and the denominator is doubled, the resulting fraction is $\frac{1}{2}$. Find the value of the original fraction. $\frac{4}{5}$

34. The denominator of a fraction is the same as the numerator increased by its square. If the numerator and denominator are each increased by 2, the resulting fraction, simplified, is $\frac{1}{2}$. Find the original fraction. $\frac{2}{6}$

Solve each equation. Check for extraneous roots.

35. $\dfrac{3x^2}{x^3 - 8} = \dfrac{2x - 5}{x^2 + 2x + 4}$ $-10, 1$

36. $\dfrac{x^3 + x^2 - 7x - 1}{x - 1} = x^2 + x + 1$ $0, 7$

Mixed Review

Given the points $A(-2, -8)$, $B(6,8)$, and $C(6, -2)$, find the following slope or equation. *3.3, 3.4*

1. slope of $\overleftrightarrow{AB}$ 2

2. equation of $\overleftrightarrow{AB}$
$y = 2x - 4$

3. equation of $\overleftrightarrow{BC}$ $x = 6$

Brainteaser

Starting 3 mi away, a boy walks home at 3 mi/h. His dog, starting with him, runs home at 5 mi/h, then immediately turns around and runs back to the boy, then back home, back to the boy, and so on until the boy reaches home. How far does the dog run?   5 mi

Application: Flotation

Two thousand years ago, Archimedes discovered that objects placed in water are buoyed up by a force working against gravity that is equal to the weight of the water the object displaces. The key is the object's density, or mass per unit volume. For example, lead is more dense than wood because a block of lead weighs more than a block of wood of the same size. Objects denser than water sink, although more slowly than they would in air. Objects less dense than water float.

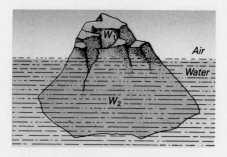

In algebraic terms, the weight w of an object *in water* is $w_1 - w_2$, where w_1 is the weight of the object and w_2 is the weight of the displaced water. For objects denser than water, $w_1 > w_2$ (the volume being equal), and w is positive. Therefore, the objects weigh more than an equal volume of water and sink. For objects less dense than water, however, w is negative, and the objects float.

To find what part of a floating object is submerged, first set $w = 0$. You can do this because floating is an equilibrium state, and therefore the pull of gravity downward is exactly offset by the buoyant force of the water acting upward. Next, divide the object into two parts, one part above the water line and the other below.

In the case of an iceberg, for example, let W_1 = the weight of the portion of the iceberg above water and W_2 = the weight of the portion submerged. The pull of gravity on the iceberg is then equal to the total weight of the iceberg ($W_1 + W_2$), and the equivalent buoyant force is the weight of the displaced water. The weight of the displaced water is 1.08 times the weight of the submerged part of the iceberg because water is 1.08 times as dense as ice. That is, a given volume of water weighs 1.08 times as much as an equal volume of ice.

To find the relationship between W_1 and W_2, use the equation below.

$$\text{pull of gravity} \rightarrow W_1 + W_2 = 1.08\,W_2 \leftarrow \text{buoyant force}$$

1. How many times larger is the submerged part of an iceberg than the part above the water line? 12.5

2. A large pine log floats in water. Given that water is 2.5 times more dense than pine, what part of the log is below the water? $\frac{2}{5}$

7.6 Equations Containing Rational Expressions **257**

GETTING STARTED

Prerequisite Quiz

Solve.

1. $\frac{x}{5} + \frac{x}{10} = 1$ $3\frac{1}{3}$

2. $\frac{4}{7} + \frac{10}{x} + \frac{20}{x} = 1$ 70

3. $\frac{1}{2} + \frac{4}{x} = 1$ 8

4. $\frac{x}{6} + \frac{x}{8} = 1$ $3\frac{3}{7}$

5. $\frac{16}{x} + \frac{2}{5} = 1$ $26\frac{2}{3}$

Motivator

Ask students the following questions. Suppose it takes you 4 hours to get a refreshment stand ready to open for a school football game if you do all the work by yourself. What part of the work or job will you complete at the end of 3 hours? $\frac{3}{4}$
What part of the work will remain to be done by someone else? $\frac{1}{4}$

TEACHING SUGGESTIONS

Lesson Note

Emphasize that the sum of the fractional parts of a job done by people working together must add up to 1.

7.7 Problem Solving: Rates of Work

Objective   To solve work problems

Wanda can mow a lawn in 5 h. Therefore, in 1 h she can mow $\frac{1}{5}$ of the lawn. Her work rate is $\frac{1}{5}$ of the job per hour. At this rate, in 4 h she should complete $4 \cdot \frac{1}{5}$, or $\frac{4}{5}$ of the job. This suggests the following relationship:

(number of hours worked) · (rate per hour) = fractional part of job completed

$$4 \cdot \frac{1}{5} = \frac{4}{5}$$

Suppose two painters work together to paint a house. One of them can do $\frac{3}{5}$ of the work while the other is doing $\frac{2}{5}$ of the work.
Working together, they complete the job, because $\frac{3}{5} + \frac{2}{5} = 1$. The sum of the fractional parts of the job done by each worker is 1.

EXAMPLE 1   It takes Fay 5 h to paint a room. Rob can do the job in 10 h. How long will it take them to do the job if they work together? (THINK: Will it take more or less time than 5 h?)

Solution   Let x = the number of hours Fay and Rob work together.

Use a table to represent the part done by each person.

	Part of job done in 1 h	Number of hours working together	Part of job completed
Fay	$\frac{1}{5}$	x	$x \cdot \frac{1}{5} = \frac{x}{5}$
Rob	$\frac{1}{10}$	x	$x \cdot \frac{1}{10} = \frac{x}{10}$

Fay's part + Rob's part = total job

$$\frac{x}{5} + \frac{x}{10} = 1$$
$$10 \cdot \frac{x}{5} + 10 \cdot \frac{x}{10} = 10 \cdot 1$$
$$2x + 1x = 10$$
$$3x = 10$$
$$x = \frac{10}{3}, \text{ or } 3\frac{1}{3} \text{ hours working together}$$

Highlighting the Standards

Standard 1a: As with many algebraic problems, the ones in this lesson are a means to reinforce mathematical concepts and skills. It is important to check answers.

Additional Example 1

Sam can rake the leaves in 4 hours. His sister Heather can do the same job in 6 hours. How long will it take to do the job if they work together? $2\frac{2}{5}$ hours

Check

To check, add the parts done in $3\frac{1}{3}$ h.

$$x \cdot \frac{1}{5} + x \cdot \frac{1}{10} = \frac{10}{3} \cdot \frac{1}{5} + \frac{10}{3} \cdot \frac{1}{10} = \frac{2}{3} + \frac{1}{3} = 1$$

Thus, the job will take $3\frac{1}{3}$ h if Fay and Rob work together.

EXAMPLE 2 Working together, Pat, Frank, and Pam can wallpaper an apartment in 20 h. Pam can do the job alone in 35 h, and Pat can do the job alone in twice the time it would take Frank working alone. Find the time it would take Frank to do the job alone.

Solution Let x = the number of hours for Frank to do the job alone.

Then $2x$ = the number of hours for Pat to do the job alone.

	Part of job done in 1 hr	Number of hours working together	Part of job completed
Pam	$\frac{1}{35}$	20	$20 \cdot \frac{1}{35} = \frac{20}{35} = \frac{4}{7}$
Pat	$\frac{1}{2x}$	20	$20 \cdot \frac{1}{2x} = \frac{20}{2x} = \frac{10}{x}$
Frank	$\frac{1}{x}$	20	$20 \cdot \frac{1}{x} = \frac{20}{x}$

Pam's part + Pat's part + Frank's part = total job

$$\frac{4}{7} + \frac{10}{x} + \frac{20}{x} = 1$$

$$7x \cdot \frac{4}{7} + 7x \cdot \frac{10}{x} + 7x \cdot \frac{20}{x} = 7x \cdot 1$$

$$4x + 70 + 140 = 7x$$

$$70 = x$$

Thus, it would take Frank 70 h working alone. The check is left for you.

Classroom Exercises

It takes Bill 18 h to build a cabinet. Represent the part built in the given period of time.

1. 5 h $\frac{5}{18}$ 2. 8 h $\frac{4}{9}$ 3. x h $\frac{x}{18}$ 4. $2x$ h $\frac{x}{9}$

5. 10 h $\frac{5}{9}$ 6. $3x$ h $\frac{x}{6}$ 7. $(2x + 1)$ h $\frac{2x+1}{18}$ 8. $\frac{x}{2}$ h $\frac{x}{36}$

9. If it takes Ed 12 h to build Bill's cabinet (see above), how long would it take Bill and Ed to build the cabinet working together? $7\frac{1}{5}$ h

Math Connections

Mathematical Models: The work problems of this lesson illustrate the advantages and disadvantages of a mathematical model. Partitioning the total work into manageable shared parts illustrates one way to think about such problems. However, a mathematical model will not reflect practical difficulties such as the extra time needed to coordinate the wallpapering efforts of three people in Example 2. Students may benefit from a class discussion of both the advantages and the limitations of mathematical models.

Critical Thinking Questions

Analysis: Ask students to think of examples where people are working together but the "work problem" model of the lesson does not apply. One case would be where the participants do not share all the required skills in equal measure.

Checkpoint

Solve each problem.

1. It takes Ronald 4 h to prepare a dinner. Monica can do it in 5 h. How long will it take them to do the job if they work together? $2\frac{2}{9}$ h

2. Working together, Mr. Ramirez and his daughter can paint a house in 4 weeks. It takes Mr. Ramirez twice as long as it takes his daughter to do the job alone. How long would it take each to do it alone? Mr. Ramirez: 12 weeks; his daughter: 6 weeks

Additional Example 2

Working together, Tina, Tony and Andrea can paint a house in 35 h. Andrea can do the job alone in 70 h. Tina can do the job alone in twice the time it would take Tony working alone. Find the time it would take Tony working alone. 105 h

In this lesson students learned to use a table to organize the data of a work problem. Ask them to name the labels for each of the three columns of such a table. See page 258 or page 259. Ask them how each row is labeled. By the names of the participants

◼◼◼FOLLOW UP

Guided Practice

Classroom Exercises 1–9

Independent Practice

A Ex. 1–6, **B** Ex. 7, 8, **C** Ex. 9–11

Basic: WE 1–7, Mixed Problem Solving
Average: WE 2–8, Mixed Problem Solving
Above Average: WE 3–9, Mixed Problem Solving

Additional Answers

Mixed Review

2. $6x^3 - 33x^2 - 18x$

6. $\dfrac{7 - 2x}{(x + 2)(x - 2)}$

Written Exercises

Solve each problem.

1. It takes Don 6 h and Joyce 8 h to paint a room alone. How long would it take them to paint the room if they worked together? $3\frac{3}{7}$ h

2. It takes Regina 3 h to prepare dinner. Her husband, Rich, can prepare the same dinner in 2 h. How long would it take them working together? $1\frac{1}{5}$ h

3. Working together, it takes Dick and Gail 16 h to tile a floor. It would take Gail 40 h to do it alone. How long would it take Dick working alone? $26\frac{2}{3}$ h

4. It takes Lois 3 times as long as Richie to mow a lawn. How long would it take each of them alone, if together they can do it in 5 h? Lois: 20 h; Richie: $6\frac{2}{3}$ h

5. Working together, Andy and Sal can build a fence in 7 h. Alone, it takes Andy twice the time it takes Sal. How long does it take each working alone? Andy: 21 h; Sal: $10\frac{1}{2}$ h

6. Mary can wallpaper a room in 6 h. It takes Jackie 2 h and Kevin 3 h to do the same job alone. How long would it take to do the job if they all worked together? 1 h

7. Dick, Florence, and Betty can paint a house in 12 h if they work together. If each worked alone, Dick would take 30 h, and Betty twice as long as Florence. How long would it take each girl working alone? Florence: 30 h; Betty: 60 h

8. A large pipe can fill a tank in 5 h, and a smaller one can fill the tank in 8 h. A drain pipe can empty the tank in 10 h. Find the total time needed to fill the tank when all three pipes are left open. $4\frac{4}{9}$ h

9. Holly can harvest a strawberry patch in 12 h, and Evelyn can do the job in 8 h. Given that Evelyn starts 2 h after Holly has begun working, find the total time needed to do the job. 6 h

10. It takes a man 6 days less than his son to build a garage. If they work together, they can do it in $\frac{1}{3}$ the time it takes the son. How long would it take each working alone? Man: 6 d; son: 12 d

11. If 5 men and 2 boys work together, a job can be completed in one day. If 3 men and 6 boys work together, the job can also be completed in one day. How long would it take 1 boy working alone to do the work? 12 d

Mixed Review

Simplify. Use positive exponents. *5.2, 5.3, 5.5, 7.1, 7.2, 7.4*

1. $(5n^2)^{-3}$ $\dfrac{1}{125n^6}$

2. $3x(2x + 1)(x - 6)$

3. $(8x - 3)^2$ $64x^2 - 48x + 9$

4. $\dfrac{x^3 - 8}{2 - x}$ $-(x^2 + 2x + 4)$

5. $\dfrac{5x^3}{2a^4} \cdot \dfrac{6a}{25x}$ $\dfrac{3x^2}{5a^3}$

6. $\dfrac{3}{x^2 - 4} - \dfrac{2}{x + 2}$

Enrichment

Have the students try this problem. Norman and Carl are stacking books in a library. The books have been stored in cartons, with each carton containing the same number of books. Norman can stack a carton of books in 25 min, while Carl takes 20 min to stack the same number of books. If they work together for $2\frac{1}{2}$ h, how many cartons of books can they stack? $13\frac{1}{2}$

1. How many milliliters of water must be added to 400 ml of a 77% salt solution to dilute it to a 55% solution?
   160 ml

2. How many liters of a 75% sulfuric acid solution must be added to 30 liters of a 20% sulfuric acid solution to obtain a 45% acid solution? 25 L

3. Given that y varies directly as x and $y = 9.6$ when $x = 4.5$, find y when $x = 15$. 32

4. Convert 4 ounces per square inch to pounds per square foot. 36 lb/ft^2

5. Mr. Brown can carpet a floor in 3 hours. If his apprentice helps him, the job can be done in 2 hours. How long would it take the apprentice to do the job working alone? 6 h

6. The hypotenuse of a right triangle is 1 cm longer than one leg, and 4 cm longer than 3 times the length of the other leg. Find the length of the hypotenuse. 25 cm

7. The height of a triangle is 7 ft less than 5 times the length of the base. Find the length of the base and the height of the triangle given that its area is 12 ft^2. $b = 3$ ft; $h = 8$ ft

8. The length of a rectangle is 7 ft more than 3 times its width. Find the set of all possible lengths given that the perimeter is less than 102 ft. $\{l \mid 7\text{ ft} < l < 40\text{ ft}\}$

9. How many milliliters of a 72% alcohol solution should be added to 300 ml of a 27% alcohol solution to make a 36% alcohol solution? 75 ml

10. The measures of two supplementary angles have the ratio 2:3. Find the measure of the complement of the smaller angle. 18

11. Find three consecutive multiples of 5 such that 7 times the second number is the same as 55 increased by 4 times the third number. 20, 25, 30

12. Find three consecutive multiples of 3 given that twice the square of the first number is 18 more than the product of the other two numbers. $-3, 0, 3$ or 12, 15, 18

13. A second number is twice a first number. The product of the first number and 2 more than the second number is 60. Find all such pairs of numbers.
    $-6, -12$ or 5, 10

14. Mrs. Montero invested $10,000, part at 9% and the remainder at 10% per year. If at the end of 5 years the two investments had earned $4,800 in simple interest, what amount was invested at 9%? $4,000

15. One bus left a terminal at 6:00 A.M. and headed east at 80 km/h. A second bus left the same terminal at 7:30 A.M. and traveled west at 90 km/h. At what time were the buses 460 km apart? 9:30 A.M.

16. If 90 ml of a 40% citrus solution and 20 ml of a 25% citrus solution were mixed with a 60% citrus solution to make a 50% solution, how many milliliters of the 60% solution were used?
    140 ml

1. Zeros: 1,8

   domain: $\left\{x \mid x \neq -\frac{2}{3}\right\}$

2. Zeros: 3,4

   domain: $\{x \mid x \neq -7 \text{ and } x \neq 7\}$

3. $\frac{-(x+4)}{3x^2(2x+5)}$; $f(-2) = -\frac{1}{6}$

4. $\frac{-2(2x-1)}{3x-1}$; $f(3) = -\frac{5}{4}$

11. Answers may vary. One possible answer: Find the LCD of the rational expressions. Then multiply the numerator and denominator of each by the factors needed to make that denominator equal the LCD. Then add the numerators, retaining the LCD for the denominator.

Chapter 7 Review

Key Terms

Cancellation Property (p.234)
complex rational expression (p.249)
conversion fraction (p.242)
dimensional analysis (p.242)

extraneous root (p.253)
rational expression (p.233)
rational function (p.233)
zero of a function (p.234)

Key Ideas and Review Exercises

7.1   For a function such as $f(x) = \dfrac{x^2 - 9x + 8}{3x + 2}$:

  1. To find the zeros, set the numerator equal to 0 and solve.
  2. To find the domain, find the values of x for which the denominator is *not* 0.

Find the zeros and the domain of each function.

1. $f(x) = \dfrac{x^2 - 9x + 8}{3x + 2}$

2. $f(x) = \dfrac{x^2 - 7x + 12}{x^2 - 49}$

To simplify a rational expression such as $\dfrac{2x^3 + 2x^2 - 24x}{6x^3(15 + x - 2x^2)}$:

  1. Put $15 + x - 2x^2$ into a more convenient form by factoring out -1.
  2. Factor the numerator and denominator completely.
  3. Divide out the common factors.

Simplify the expression that describes each function. Then find the indicated value of the function.

3. $f(x) = \dfrac{2x^3 + 2x^2 - 24x}{6x^3(15 + x - 2x^2)}$; $f(-2)$

4. $f(x) = \dfrac{-10 + 22x - 4x^2}{3x^2 - 16x + 5}$; $f(3)$

7.2   To simplify products or quotients of rational expressions, first apply $\dfrac{a}{b} \cdot \dfrac{c}{d} = \dfrac{ac}{bd}$ or $\dfrac{a}{b} \div \dfrac{c}{d} = \dfrac{a}{b} \cdot \dfrac{d}{c}$. Then simplify the result.

Simplify.

5. $\dfrac{n^2 + 3n - 18}{2n^2 - 5n - 3} \cdot \dfrac{1 - 4n^2}{n^2 - 36}$ $\dfrac{-(2n-1)}{n-6}$

6. $\dfrac{x}{x + 4} \div \dfrac{6x^2}{3x + 12}$ $\dfrac{1}{2x}$

7.3   To convert from one unit of measurement to another, use pairs of equivalent measures.

7. Convert 144 ft^2 to square yards.   16 yd^2 **8.** Convert 90 mi/h to feet per second.
   132 ft/s

7.4   To add or subtract rational expressions with unlike denominators, see Examples 3–5 in Lesson 7.4.

Simplify.

9. $\dfrac{6a - 23}{a^2 - 7a + 12} + \dfrac{a - 5}{a - 4}$ $\dfrac{a + 2}{a - 3}$ **10.** $\dfrac{5a}{a^2 - 9} - \dfrac{4}{a + 3} + \dfrac{2}{3 - a}$ $\dfrac{-a + 6}{(a + 3)(a - 3)}$

11. Write in your own words how to add several rational expressions with unlike denominators.

7.5   To simplify a complex rational expression, use either of the two methods shown in Lesson 7.5.

Simplify.

12. $\dfrac{1 - 6x^{-1} + 5x^{-2}}{1 - 3x^{-1} - 10x^{-2}}$ $\dfrac{x - 1}{x + 2}$ **13.** $\dfrac{\dfrac{1}{x + 6} + \dfrac{1}{x + 2}}{\dfrac{x^2 + 11x + 28}{x^2 + 8x + 12}}$ $\dfrac{2}{x + 7}$

7.6   To solve an equation containing rational expressions,
    1. factor each denominator if possible,
    2. multiply each side of the equation by the LCD of all the fractions,
    3. solve the resulting equation, and
    4. check for extraneous solutions. (No denominator can be 0.)

Solve each equation for x.

14. $\dfrac{2x + 3}{x - 1} - \dfrac{2x - 3}{x + 1} = \dfrac{10}{x^2 - 1}$ $\varnothing$ **15.** $\dfrac{5a}{x + b} = \dfrac{a}{x - b}$ $x = \dfrac{3b}{2}$

To solve word problems involving the numerator and denominator of a fraction, use two equations as shown in Example 4 of Lesson 7.6.

16. The numerator of a fraction is 5 less than the denominator. If the numerator were increased by 3 and the denominator increased by 4, the resulting fraction would be $\frac{1}{2}$. Find the original fraction. $\frac{3}{8}$

7.7   To solve work problems,
    (1) represent the fractional part of the work done by each worker, and
    (2) set the sum of the fractional parts done by each worker equal to 1 and solve the equation.

17. Bill can do a job in 15 h. Jane can do the same job in 10 h. If Jane were to join Bill 3 h after he had begun, what would be the total time needed to do the job? $7\frac{4}{5}$ h

1. Zero: $\frac{4}{3}$

   domain: $\{x \mid x \neq -5\}$
2. Zeros: 5, 6

   domain: $\{x \mid x \neq -2 \text{ and } x \neq 1\}$
3. $\frac{x-2}{x-1}$; $f(3) = \frac{1}{2}$
4. $\frac{2}{x+2}$: $f(-6) = -\frac{1}{2}$
6. $\frac{2(3c+1)}{3c}$
8. $\frac{x-3}{x}$

Chapter 7 Test

A Exercises: 1–14   B Exercises: 15, 17   C Exercises: 16, 18

Find the zeros and the domain of each function.

1. $f(x) = \dfrac{3x - 4}{2x + 10}$

2. $f(x) = \dfrac{x^2 - 11x + 30}{x^2 + x - 2}$

Simplify the expression that describes each function. Then find the indicated value of the function.

3. $f(x) = \dfrac{x^2 - 12x + 20}{x^2 - 11x + 10}$; $f(3)$

4. $f(x) = \dfrac{2x^2 - 4x}{x^3 - 4x}$; $f(-6)$

Simplify.

5. $\dfrac{3n^2 - 15n + 18}{45 - 5n^2}$ $\dfrac{-3(n-2)}{5(n+3)}$

6. $\dfrac{3c^2 - 5c - 2}{6c^2} \cdot \dfrac{4c^2 - 8c}{c^2 - 4c + 4}$

7. $\dfrac{x}{x+4} \div \dfrac{6x^2}{3x + 12}$ $\dfrac{1}{2x}$

8. $\dfrac{-2x^2 - 5x}{x^2 + 7x} + \dfrac{x - 2}{x + 7} + \dfrac{2x - 3}{x}$

9. $\dfrac{2x + 7}{x^2 - 5x + 4} - \dfrac{2x - 3}{x - 4}$ $\dfrac{-(2x+1)}{x-1}$

10. $\dfrac{1 - 4x^{-1} - 45x^{-2}}{2 + 7x^{-1} - 15x^{-2}}$ $\dfrac{x-9}{2x-3}$

11. Convert 288 in^2 to square feet. 2 ft^2

12. Convert 30 T/h to pounds per minute.
    1,000 lb/min

Solve each equation. Check for extraneous roots.

13. $\dfrac{5}{x - 4} - \dfrac{x}{2} = \dfrac{x + 1}{x - 4}$ -2

14. $\dfrac{n - 6}{n^2 - 2n - 8} + \dfrac{3}{n - 4} = \dfrac{2}{n + 2}$ -4

15. Solve $A = \dfrac{h}{2}(x + y)$ for x. Then find x when $h = 10$, $A = 70$, and $y = 8$. $x = \dfrac{2A}{h} - y$, or $x = \dfrac{2A - hy}{h}$; 6

16. Bill can do a job alone in 8 days. After he works alone for 2 days, Marie joins him. They finish the job together in 2 more days. How long would it take Marie to do the entire job alone? 4 d

17. The denominator of a fraction is 3 less than twice the numerator. If the numerator were doubled and the denominator increased by 8, the resulting fraction would be $\frac{2}{3}$. Find the original fraction. $\frac{5}{7}$

Simplify.

18. $\dfrac{ab + 5ac - 2bd - 10cd}{b^2 + 7bc + 10c^2} \cdot \left(\dfrac{a^2 + 2ad - 8d^2}{b^2 - 4c^2} \right)^{-1}$ $\dfrac{b - 2c}{a + 4d}$

Choose the *one* best answer to each question or problem.

1. Given that $y = \frac{1}{xz}$, what is the value of z when $x = \frac{1}{4}$ and $y = 2$? (A)
   (A) 2   (B) $\frac{1}{2}$ (C) 8   (D) 6
   (E) None of these

2. The figure below represents the scale drawing of a rectangular living room. The scale used is 3 cm = 1 m. Find the actual area of the room. (D)

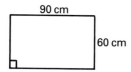

   (A) 50 m^2 (B) 100 m^2
   (C) 150 m^2 (D) 600 m^2
   (E) 5,400 m^2

3. *ABCD* is a square, and the shaded region is the interior of a circle. What is the ratio of the shaded area to the area of the square? (A)

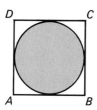

   (A) $\frac{\pi}{4}$ (B) $\frac{4}{\pi}$ (C) π
   (D) $\frac{1}{\pi}$ (E) $\frac{\pi}{8}$

4. If 5 more than x is 2 less than y, what is y in terms of x? (E)
   (A) $x + 3$ (B) $y - 7$ (C) $y - 3$
   (D) $x - 7$ (E) $x + 7$

5. $\frac{1}{10^{15}} - \frac{1}{10^{16}} = \underline{\ ?\ }$ (B)
   (A) $-\frac{9}{10^{16}}$ (B) $\frac{9}{10^{16}}$ (C) $\frac{1}{10}$
   (D) $-\frac{1}{10}$ (E) $\frac{1}{10^{16}}$

6. If $x + y = 4$ and $x + 2y = 5$, then $x - y = \underline{\ ?\ }$ (B)
   (A) 4   (B) 2   (C) -2
   (D) $3\frac{1}{3}$ (E) None of these

7. If $a \neq 0$, then $\frac{(-4a)^3}{-4a^3} = \underline{\ ?\ }$ (C)
   (A) -16 (B) 1   (C) 16
   (D) -1 (E) 3

8. If $\frac{1}{a} + \frac{1}{a} = 12$, then $a = \underline{\ ?\ }$ (B)
   (A) 6   (B) $\frac{1}{6}$ (C) $\frac{1}{12}$
   (D) 12   (E) $\frac{1}{24}$

9. Three lines intersect as shown. What is the sum of the degree measures of the marked angles? (C)

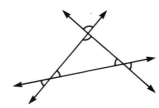

   (A) 90   (B) 180   (C) 360
   (D) 540   (E) It cannot be determined from the given information.

10. If $\frac{4}{a} = \frac{5}{b} + \frac{1}{c}$, then $ab - 4bc = \underline{\ ?\ }$ (A)
    (A) $-5ac$ (B) $-5bc$ (C) $5ac$
    (D) $5bc$ (E) $5abc$

8 RADICALS AND RATIONAL EXPONENTS

OVERVIEW

In this chapter, students focus on operations with radicals and radical expressions, especially square roots of numbers and of algebraic expressions. They also perform operations with radicals in which the index of the radical is greater than two. The last two lessons of the chapter are devoted to the study of rational exponents.

OBJECTIVES

- To add, subtract, multiply, divide, and simplify expressions containing square roots
- To solve equations of the form $x^2 = k$
- To simplify expressions containing indices greater than 2
- To simplify and evaluate expressions containing rational-number exponents

PROBLEM SOLVING

In this chapter, students should be reminded of the problem-solving strategy *Solving a Simpler Problem.* In many instances, to solve exponential equations, it is necessary to rewrite the equation so that all terms have the same positive base. Once this is done, the problem can be solved.

READING AND WRITING MATH

A *Focus on Reading* appears on page 274. There are also two exercises, Exercise 33 on page 274 and Exercise 9 on page 296, in which students are asked to explain or describe in their own words how to simplify expressions involving radicals.

TECHNOLOGY

Calculator: The feature *Using the Calculator* on page 291 shows several ways in which a scientific calculator can be used to evaluate radicals (or numbers that are equivalent to radicals but are expressed as powers using rational exponents). The calculator can also be used to check that the simplification of a radical has been done correctly, as illustrated in Example 1 on page 271.

SPECIAL FEATURES

Mixed Review pp. 270, 275, 278, 287, 291, 295
Focus on Reading p. 274
Application: Inventory Modeling p. 275
Brainteaser p. 278
Midchapter Review p. 283
Using the Calculator p. 291
Key Terms p. 296
Key Ideas and Review Exercises
   pp. 296–297
Chapter 8 Test p. 298
College Prep Test p. 299
Cumulative Review (Chapters 1-8)
   pp. 300–301

PLANNING GUIDE

Lesson	Basic	Average	Above Average	Resources
8.1 p. 270	CE all WE 1–16	CE all WE 3–18	CE all WE 5–20	Reteaching 53 Practice 53
8.2 pp. 274–275	FR all CE all WE 1–23 odd Application	FR all CE all WE 11–33 odd Application	FR all CE all WE 17–39 odd Application	Reteaching 54 Practice 54
8.3 p. 278	CE all WE 1–27 odd Brainteaser	CE all WE 3–29 odd Brainteaser	CE all WE 5–31 odd Brainteaser	Reteaching 55 Practice 55
8.4 pp. 282–283	CE all WE 1–27 odd Midchapter Review	CE all WE 7–33 odd Midchapter Review	CE all WE 13–39 odd Midchapter Review	Reteaching 56 Practice 56
8.5 pp. 287	CE all WE 1–33 odd	CE all WE 5–37 odd	CE all WE 7–39 odd	Reteaching 57 Practice 57
8.6 pp. 290–291	CE all WE 1–27 odd Using the Calculator	CE all WE 5–31 odd Using the Calculator	CE all WE 7–33 odd Using the Calculator	Reteaching 58 Practice 58
8.7 pp. 294–295	CE all WE 1–31 odd	CE all WE 7–37 odd	CE all WE 15–45 odd	Reteaching 59 Practice 59
Chapter 8 Review pp. 296–297	all	all	all	
Chapter 8 Test p. 298	all	all	all	
College Prep Test p. 299	all	all	all	
Cumulative Review pp. 300–301	1–39 odd	5–47 odd	7–51 odd	

CE = Classroom Exercises   WE = Written Exercises   FR = Focus on Reading

NOTE: For each level, all students should be assigned all Mixed Review exercises.

◼◼◼INVESTIGATION

In this investigation, the students are introduced to processes of *iteration*. Ask students to give the meaning of the word "iteration."   Repetition, repeating a process

Ask students to consider the following iteration process.

> Take a number z. Square it.
> Let the result be the new number.
> Then square the new number.

Have students use their calculators to follow the procedure above for $z = 41$. After six repetitions, ask them to describe what happens to the number.   The iteration process results in a number greater than the capacity of the calculator.

Then have students follow the procedure above for $z = 0.23$. After 8 repetitions, have the students describe what happens to the number. The iteration results in a number that approaches zero.

Ask students to summarize the results of the iteration process.   When $z > 1$, the repetition of the function results in numbers that get larger and larger. When $0 < z < 1$, the repetition of the function results in numbers that get closer and closer to zero.

Have students form groups and use their calculators to do an iteration in which they take the square root of z. After they have explored possibilities for $0 < z < 1$, and $z > 1$, have them summarize their results.   The students should discover that any number greater than zero goes to 1.

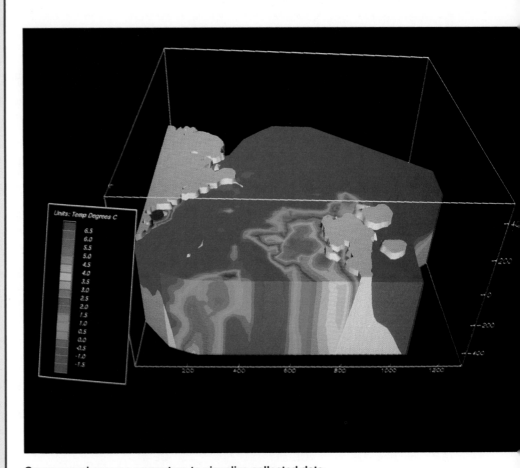

Oceanographers use computers to visualize collected data and gain a greater understanding of the physical environment they are studying. The images above show the 11-year mean temperature field in the upper 500 m of Fram Strait. The red and orange colors outline the warmer Atlantic waters.

More About the Image

Computer-generated models give scientists a way of representing and understanding collected data. The image above is an example of *volumetric modeling*. It provides a three-dimensional picture of an environment based on data collected over a specific period of time for a given set of coordinates in space. Models of this type often combine the visualization of three-dimensional space with quantitative analysis to calculate and represent volumes accurately.

8.1 Square Roots and Functions

Teaching Resources

Quick Quizzes 53
Reteaching and Practice
Worksheets 53

Objectives
To find square roots of positive numbers
To find domains of functions involving square roots
To solve simple quadratic equations

Every positive real number has two *square roots*, one positive and the other negative. To find the two square roots of 64, find the two numbers that have 64 as their square. That is, find the values of x such that $x^2 = 64$. Because $8 \cdot 8 = 64$ and $(-8) \cdot (-8) = 64$, the square roots of 64 are 8 and -8.

Definitions

> The **square roots** of any nonnegative number b are the solutions of the equation $x^2 = b$. The **principal square root** is the positive square root of that number, written as $\sqrt{b}$. The negative square root is written $-\sqrt{b}$.

From the above definitions, it follows that

$$\sqrt{b} \cdot \sqrt{b} = (\sqrt{b})^2 = b.$$

The symbol $\sqrt{b}$ is called a **radical**, and $\sqrt{}$ is the **radical sign**. The number b under the radical sign is called the **radicand**.

Notice that the above definitions do not include negative radicands because the square root of a negative number is not a real number.

For example, $\sqrt{-9}$ does not name a real number because there is no real number whose square is -9.

Because $\sqrt{}$ indicates the principal square root, which is positive, $\sqrt{a^2} = |a|$ for any real number a. That is, the principal square root of a^2 is the absolute value of a. (Notice that when $a \geq 0$, $\sqrt{a^2} = a$.) For example, $\sqrt{64} = \sqrt{8^2} = |8| = 8$, and $\sqrt{64} = \sqrt{(-8)^2} = |-8| = 8$. On a calculator, only the principal square root is displayed.

EXAMPLE 1 Simplify.

a. $\sqrt{0.25}$
b. $\sqrt{\dfrac{9}{49}}$

Solutions

a. $\sqrt{0.25} = 0.5$ because $(0.5)^2 = 0.25$

b. $\sqrt{\dfrac{9}{49}} = \dfrac{3}{7}$ because $\left(\dfrac{3}{7}\right)^2 = \dfrac{9}{49}$

Additional Example 1

Simplify.

a. $\sqrt{0.0081}$ 0.09

b. $\sqrt{\dfrac{64}{121}}$ $\dfrac{8}{11}$

GETTING STARTED

Prerequisite Quiz

Simplify.

1. $(-3)^2$ 9
2. $8 - 5^2$ -17
3. $\left(-\dfrac{2}{7}\right)^2$ $\dfrac{4}{49}$
4. $(0.2)^2$ 0.04

Evaluate for $x = -2$.

5. $(8 - x)^2$ 100
6. $-3(x^2)$ -12

Motivator

Ask students what positive number multiplied by itself equals 36.   6   Ask students to give the meaning of $\sqrt{\dfrac{4}{9}}$ in their own words. The square root of four-ninths. Ask them to simplify $\sqrt{\dfrac{4}{9}}$. $\dfrac{2}{3}$

Highlighting the Standards

Standard 5d: The radical sign and related definition of principal square root have an important effect in computing and in solving equations.

Lesson Note

Use specific examples to show that *finding a square root of a number* and *squaring a number* are inverse operations. Emphasize that $\sqrt{b} \cdot \sqrt{b} = (\sqrt{b})^2 = b$, ($b > 0$). This will be important in future lessons on simplifying radicals.

Notice that the function concept is continued in this lesson in the discussion of the domain of $f(x) = \sqrt{g(x)}$.

Math Connections

The Pendulum: Formulas that include radicals are often just another form of a formula that is basically quadratic. For example, the formula for the period T of a pendulum of length L could be written in quadratic form as follows.

$$T^2 = 4\pi^2 \frac{L}{g}$$

(The quantity g is the acceleration of an object under the influence of gravity.) However, the formula is usually written explicitly for T in the following form.

$$T = 2\pi\sqrt{\frac{L}{g}}$$

Critical Thinking Questions

Analysis: Ask students whether the fact that $\sqrt{-9}$ is not a real number necessarily implies that $\sqrt{-9}$ does not exist at all.
Such a number may conceivably exist as a number but not as a real number. The matter is discussed more fully in Lesson 9.1.

In Example 1, the radicands 0.25 and $\frac{9}{49}$ are *perfect squares* because their principal square roots, 0.5 and $\frac{3}{7}$, are rational numbers.

Definition

> A rational number is a **perfect square** if and only if its principal square root is also a rational number. That is, a is a perfect square if and only if $\sqrt{a}$ is a rational number.

Although a number such as $\sqrt{43}$ is not rational, there are some rational numbers that have squares that are very close to 43. For instance,

$$(6.557)^2 = 42.994249 \text{ and } (6.558)^2 = 43.007364.$$
$$\text{Notice that } 6.557 < \sqrt{43} < 6.558.$$

Calculators, however, cannot be used reliably to determine whether the square root of a number is or is not rational. For example, if the rational number 6.5574385 is squared on some scientific calculators, the answer will be displayed as exactly 43. If correct, this would imply that $\sqrt{43}$ is 6.5574385, which is rational. But, in fact, the displayed answer is rounded and thus is slightly too large.

EXAMPLE 2 Evaluate each of the following expressions for $x = 8$. Approximate irrational results to the nearest hundredth using a calculator or the Square Root Table on page 789.

a. $\sqrt{\dfrac{5x + 9}{16}}$
b. $\sqrt{4x + 3}$

Solutions
a. $\sqrt{\dfrac{5 \cdot 8 + 9}{16}} = \sqrt{\dfrac{49}{16}} = \dfrac{7}{4}$
b. $\sqrt{4 \cdot 8 + 3} = \sqrt{35} \approx 5.92$

Recall that the *domain* of a function f is the set of all values of x for which the function is defined. For a function involving square roots, such as

$$f(x) = \sqrt{3x - 15} \text{ or } f(x) = -\frac{x}{\sqrt{6 - x}},$$

$f(x)$ is not defined when a radicand is negative (or when a denominator is zero). Therefore, unless otherwise stated, the domain of a function that involves square roots is assumed to be the set of all real numbers for which the radicand is nonnegative, and for which the denominator is not zero.

268 Chapter 8 Radicals and Rational Exponents

Additional Example 2

Evaluate each of the following expressions for $x = 6$. Indicate whether the resulting square root is rational or irrational. Approximate irrational results to the nearest hundredth using a square-root table or calculator.

a. $\sqrt{\dfrac{7x - 6}{2x + 13}}$ $\dfrac{6}{5}$; rational

b. $\sqrt{5x - 8}$ $\sqrt{22}$ is irrational; $\sqrt{22} \approx 4.69$

EXAMPLE 3 Find the domain of each function.

  a. $f(x) = \sqrt{3x - 15}$ **b.** $f(x) = -\dfrac{x}{\sqrt{6 - x}}$

Solutions   **a.** Find the values of x **b.** Find the values of x for which
       for which $3x - 15$ $6 - x$ is nonnegative *and* for which
       is nonnegative. $\sqrt{6 - x} \neq 0$.

   Solve $3x - 15 \geq 0$.   Solve $6 - x \geq 0$ *and* $\sqrt{6 - x} \neq 0$.
              $3x \geq 15$ $6 \geq x$ *and* $6 - x \neq 0$
               $x \geq 5$ $x \leq 6$ *and* $x \neq 6$

   The domain is   The domain is
   $\{x \mid x \geq 5\}$. $\{x \mid x < 6\}$.

The definition of square root can also be applied to simple quadratic equations, as shown below.

$$x^2 = 25$$
$$x = \sqrt{25} \quad or \quad x = -\sqrt{25}$$
$$x = 5 \quad or \quad x = -5$$

In general, if $x^2 = b$, and $b \geq 0$, then $x = \sqrt{b}$ or $x = -\sqrt{b}$.

EXAMPLE 4 Solve each equation. Approximate irrational results to the nearest hundredth.

  a. $4x^2 + 9 = 45$ **b.** $3x^2 - 86 = x^2 + 86$

Solutions   **a.** $4x^2 + 9 = 45$ **b.** $3x^2 - 86 = x^2 + 86$
             $4x^2 = 36$ $2x^2 = 172$
              $x^2 = 9$ $x^2 = 86$

   $x = \sqrt{9}$ *or* $x = -\sqrt{9}$ $x = \sqrt{86}$ *or* $x = -\sqrt{86}$
   $x = 3$ *or* $x = -3$ $x \approx 9.27$ *or* $x \approx -9.27$

   The roots are 3 and -3,   The roots are 9.27 and -9.27,
   or ± 3.   or ± 9.27 to the nearest hundredth.

Sometimes, as in Example 4, a number and its opposite are represented by a single symbol. Thus, the symbol ± 3 represents the two numbers 3 and -3 and is read as "plus or minus 3."

Closure

Ask students what is meant by the principal square root of a number. The principal square root of a^2 is the absolute value of a. Ask students to describe the conditions in which a rational number is a perfect square. See definition, page 268. Ask students how they can find the domain of a function like $f(x) = \sqrt{2x - 8}$. The domain is the set of real numbers for which the radicand is nonnegative. Thus, the domain is $x \geq 4$.

◤◤◤◤FOLLOW UP

Guided Practice

Classroom Exercises 1–6

Independent Practice

Ⓐ Ex. 1–15, Ⓑ Ex. 16–18, Ⓒ Ex. 19–22

Basic: WE 1–16
Average: WE 3–18
Above Average: WE 5–20

Additional Answers

Written Exercises

19. $\{x \mid x \leq 3 \text{ or } x \geq 4\}$
20. $\{x \mid 1 \leq x \leq 4 \text{ or } x > 8\}$

Classroom Exercises

Indicate if each number is rational.

1. $\sqrt{25}$ Yes 2. $\sqrt{53}$ No 3. $-\sqrt{36}$ Yes 4. $\sqrt{\frac{4}{9}}$ Yes 5. $\sqrt{0.04}$ Yes 6. $\sqrt{0.9}$ No

Written Exercises

Simplify if the indicated square root is a real number. Otherwise, write *not real*.

1. $\sqrt{81}$ 9 2. $-\sqrt{49}$ −7 3. $\sqrt{1.44}$ 1.2 4. $\sqrt{\frac{25}{49}}$ $\frac{5}{7}$ 5. $\sqrt{-\frac{16}{81}}$ Not real 6. $\sqrt{0.0121}$ 0.11

Evaluate each of the following expressions for the given value of x. Approximate irrational results to the nearest hundredth using the Square Root Table on page 789 or a calculator.

7. $\sqrt{2x + 6}$; $x = 5$ 4
8. $\sqrt{\dfrac{7 - 5x}{9}}$; $x = -8$ 2.29
9. $\sqrt{\dfrac{3x + 1}{2x}}$; $x = 8$ $\frac{5}{4}$

Find the domain of each function.

10. $f(x) = \sqrt{2x - 6}$
$\{x \mid x \geq 3\}$
11. $f(x) = \sqrt{8 - 4x}$
$\{x \mid x \leq 2\}$
12. $f(x) = -\dfrac{1}{\sqrt{2x + 3}}$
$\{x \mid x > -\frac{3}{2}\}$

Solve each equation. Approximate irrational results to the nearest hundredth.

13. $x^2 = 144$ ± 12
14. $9a^2 = 2a^2 + 294$ ± 6.48
15. $3a^2 - 4 = a^2 + 68$ ± 6
16. $y^2 = 0.25$ ± 0.5
17. $x^2 - 0.0009 = 0.004$ ± 0.07
18. $5x^2 - 1 = 4x^2 + 0.21$ ± 1.1

Find the domain of each function.

19. $f(x) = \sqrt{x^2 - 7x + 12}$
20. $f(x) = \sqrt{\dfrac{x^2 - 5x + 4}{6x - 48}}$
21. $f(x) = \sqrt{\dfrac{x^4 - 16}{2x - 6}}$
$\{x \mid -2 \leq x \leq 2 \text{ or } x > 3\}$
22. Simplify $\sqrt{x^2 + 6x + 9}$. (HINT: $\sqrt{a^2} = |a|$ for each real number a.) $|x + 3|$

Mixed Review

Simplify. 5.1, 5.2, 7.1

1. $(3x^2y^3)^4$ $81x^8y^{12}$
2. $5a^0$ 5
3. $\dfrac{4x^{-3}y^{-5}}{16x^{-5}y^{-2}}$ $\dfrac{x^2}{4y^3}$
4. $\dfrac{x^2 - 7x + 12}{9 - x^2}$ $-\dfrac{x - 4}{x + 3}$

Enrichment

Tell the students that ancient mathematicians were not familiar with the $\sqrt{}$ symbol. They usually wrote the word "root." It evolved that this was abbreviated to the letter "r." The symbol $\sqrt{}$ came into use because it resembles the letter "r." As a research project, have the students find out who it was that first used the root symbol. The first use of the root symbol is attributed to a German named Christoff Rudolph in 1525. The symbol was not universally accepted until the 17th century.

8.2 Simplifying Square Roots

Teaching Resources

Project Worksheet 8
Quick Quizzes 54
Reteaching and Practice
  Worksheets 54
Transparency 18

Objectives   To simplify square roots of positive numbers
To simplify algebraic expressions containing square roots

The following illustration suggests a theorem for products of square roots.

$$\sqrt{16 \cdot 9} \overset{?}{=} \sqrt{16} \cdot \sqrt{9}$$
$$\sqrt{144} \overset{?}{=} 4 \cdot 3$$
$$12 = 12 \quad \text{True}$$

So, $\sqrt{16 \cdot 9} = \sqrt{16} \cdot \sqrt{9}$. In general, the square root of the product of two nonnegative real numbers is the product of the square roots of the numbers.

Theorem 8.1   For all real numbers a and b, where $a \geq 0$ and $b \geq 0$, $\sqrt{a} \cdot \sqrt{b} = \sqrt{ab}$.

Proof

Statement	Reason
1. $(\sqrt{a} \cdot \sqrt{b})^2 = (\sqrt{a})^2 (\sqrt{b})^2$	1. Power of a product: $(ab)^n = a^n b^n$
2. $\phantom{(\sqrt{a} \cdot \sqrt{b})^2} = ab$	2. $(\sqrt{a})^2 = \sqrt{a} \cdot \sqrt{a} = a$; $(\sqrt{b})^2 = \sqrt{b} \cdot \sqrt{b} = b$
3. $(\sqrt{ab})^2 = ab$	3. Def of principal square root
4. $(\sqrt{a} \cdot \sqrt{b})^2 = (\sqrt{ab})^2$	4. Trans and Sym Prop of Eq
5. $\sqrt{a} \cdot \sqrt{b} = \sqrt{ab}$	5. Take the square root of both sides.

A radical such as $\sqrt{108}$ is not in simplest form as long as its radicand contains a factor that is a perfect square. Example 1 below illustrates a method for simplifying square roots whose radicands contain perfect square factors.

EXAMPLE 1   Simplify $5\sqrt{108}$. Use a calculator to check the result.

Plan   Find the greatest perfect square factor of 108. Then use Theorem 8.1.

Solution

$$\begin{aligned} 5\sqrt{108} &= 5\sqrt{36 \cdot 3} \\ &= 5\sqrt{36} \cdot \sqrt{3} \\ &= 5 \cdot 6\sqrt{3} = 30\sqrt{3} \end{aligned}$$

Calculator check:

5 ✕ 108 √ = 51.961524
30 ✕ 3 √ = 51.961524

Additional Example 1

Simplify $6\sqrt{80}$. $24\sqrt{5}$

GETTING STARTED

Prerequisite Quiz

Simplify.

1. $\sqrt{49}$ 7
2. $\sqrt{36}$ 6
3. $\sqrt{64}$ 8
4. $\sqrt{25}$ 5
5. $\sqrt{1,600}$ 40
6. $\sqrt{1.44}$ 1.2

Motivator

Ask students to simplify the following.

1. $\sqrt{9} \cdot \sqrt{4}$ 6
2. $\sqrt{9 \cdot 4}$ 6
3. $\sqrt{16} \cdot \sqrt{25}$ 20
4. $\sqrt{16 \cdot 25}$ 20

TEACHING SUGGESTIONS

Lesson Note

Example 2 illustrates the process for determining whether an absolute value symbol is needed in simplifying a square root. This process is difficult for many students, especially if both odd and even exponents are involved. Some teachers choose to demonstrate the formal technique and then tell students to assume on class tests that all variables represent positive numbers.

Highlighting the Standards

Standard 3c: The rules for working with radicals are logical and consistent but students often make mistakes. The Standard referred to urges students to follow the rules of mathematical logic.

Math Connections

History: The word radical comes from the Latin word, radix, meaning root. The radical symbol $\sqrt{}$, comes from an old European lower case letter "r".

Critical Thinking Questions

Analysis: Ask students whether, in Example 1, they can use divisibility rules from arithmetic to determine that the perfect square, 36, is a factor of 108.   Yes; a number is divisible by 4 if its last two digits are divisible by 4 and a number is divisible by 9 if the sum of its digits is divisible by 9. Since 108 meets both of these tests, it follows that 4×9, or 36, is a factor of 108.

Common Error Analysis

Error: Students may not completely simplify a radical that has numerical coefficients. For example, students may write $\sqrt{72a^4} = 3a^2\sqrt{8}$, without noticing that 8 has a perfect square factor.

Emphasize that students should look further for perfect square factors even when they think that a radical is simplified. Thus,

$$3a^2\sqrt{8} = 3a^2 \cdot \sqrt{4} \cdot \sqrt{2}$$
$$= 3a^2 \cdot 2 \cdot \sqrt{2}$$
$$= 6a^2\sqrt{2}.$$

In Example 1, the greatest perfect square factor of 108 is 36. Smaller perfect square factors (4 and 9) may also be used, as shown below.

$$5\sqrt{108} = 5\sqrt{4 \cdot 9 \cdot 3} = 5\sqrt{4} \cdot \sqrt{9} \cdot \sqrt{3} = 5 \cdot 2 \cdot 3\sqrt{3} = 30\sqrt{3}$$

By definition, $\sqrt{b}$ is a solution of $x^2 = b$. Therefore, the positive number $\sqrt{3^{10}}$ is a solution of the equation $x^2 = 3^{10}$. But 3^5 is also the positive solution of this equation because $(3^5)^2 = 3^{10}$ is also true. It therefore follows that $\sqrt{3^{10}}$ and 3^5 are the same number. In general,

$$\sqrt{b^{2n}} = b^n, \text{ if } n \text{ is a natural number and } b \geq 0.$$

If n is an *even* number, the above formula is also true for $b < 0$, as illustrated below for $b = -2$ and $n = 2$.

$$\sqrt{(-2)^{2 \cdot 2}} = \sqrt{(-2)^4} = \sqrt{16} = 4 = (-2)^2$$

If, however, n is odd and $b < 0$, then the formula must be modified to $\sqrt{b^{2n}} = |b^n|$ to ensure that the right side of the equation is a positive number. For example, if $b = -2$ and $n = 3$, then

$$\sqrt{(-2)^{2 \cdot 3}} = \sqrt{(-2)^6} = \sqrt{64} = 8 = |(-2)^3|.$$

So, if b is any real number, then

$$\sqrt{b^{2n}} = b^n \text{ if } n \text{ is an } even \text{ natural number, and}$$
$$\sqrt{b^{2n}} = |b^n| \text{ if } n \text{ is an } odd \text{ natural number.}$$

EXAMPLE 2 Simplify each radical.
  a. $\sqrt{28a^{16}}$ **b.** $2\sqrt{45b^{10}}$

Solutions   **a.** $\dfrac{16}{2} = 8$, an even number.   **b.** $\dfrac{10}{2} = 5$, an odd number.

So, no absolute value symbol is needed.

So, the absolute value symbol is necessary.

$\sqrt{28a^{16}}$

$= \sqrt{4 \cdot 7 \cdot a^{16}}$

$= \sqrt{4} \cdot \sqrt{7} \cdot \sqrt{a^{16}}$

$= 2\sqrt{7} \cdot a^8$

$= 2a^8\sqrt{7}$

$2\sqrt{45b^{10}}$

$= 2\sqrt{9 \cdot 5 \cdot b^{10}}$

$= 2\sqrt{9} \cdot \sqrt{5} \cdot \sqrt{b^{10}}$

$= 2 \cdot 3\sqrt{5}|b^5|$

$= 6|b^5|\sqrt{5}$

If it can be assumed that $b \geq 0$, then absolute value symbols are not necessary when simplifying $\sqrt{b^{2n}}$, regardless of whether n is even or odd. Thus, $\sqrt{b^6} = b^3$, if $b \geq 0$.

Additional Example 2

Simplify each radical.

  a. $\sqrt{40a^{14}}$ $2|a^7|\sqrt{10}$
  b. $8\sqrt{24b^{12}}$ $16b^6\sqrt{6}$

Radicals such as $\sqrt{b^1}$ and $\sqrt{b^7}$, whose radicands have exponents that are odd numbers, are not real numbers if b is negative. For example, $\sqrt{(-4)^3} = \sqrt{-64}$, and there is no real number whose square is -64. However, if $b \geq 0$, then a radical such as $\sqrt{b^7}$ can be simplified by first rewriting the radicand as the product of two factors, one of which is b^1.

Thus, $\sqrt{b^7}$, where $b \geq 0$, can be rewritten as

$$\begin{aligned}\sqrt{b^7} &= \sqrt{b^6 \cdot b^1} \\ &= \sqrt{b^6} \cdot \sqrt{b^1} = b^3\sqrt{b}\end{aligned}$$

No absolute value symbol is needed because b is given as nonnegative. In general, if $b \geq 0$ and n is a natural number, then $2n + 1$ is odd and

$$\sqrt{b^{2n+1}} = \sqrt{b^{2n} \cdot b^1} = \sqrt{b^{2n}} \cdot \sqrt{b} = b^n\sqrt{b}$$

EXAMPLE 3 Simplify $\sqrt{27a^9}$, where $a \geq 0$.

Solution
$$\begin{aligned}\sqrt{27a^9} = \sqrt{9 \cdot 3 \cdot a^8 \cdot a^1} &= \sqrt{9 \cdot a^8 \cdot 3 \cdot a} \\ &= \sqrt{9} \cdot \sqrt{a^8} \cdot \sqrt{3a} = 3a^4\sqrt{3a}\end{aligned}$$

Sometimes simplifying a radical involves radicands with both odd and even exponents, as illustrated in Example 4.

EXAMPLE 4 Simplify $4a^3b\sqrt{72a^7b^{10}}$, where $a \geq 0$ and $b \geq 0$.

Solution There is no need for the absolute value symbol because $a \geq 0$ and $b \geq 0$.

$$\begin{aligned}4a^3b\sqrt{72a^7b^{10}} &= 4a^3b\sqrt{36 \cdot 2 \cdot a^6 \cdot a^1 \cdot b^{10}} \quad \leftarrow \text{Write } a^7 \text{ as } a^6 \cdot a^1 \\ &\qquad\qquad\qquad\qquad\qquad\qquad\qquad\qquad \text{because 7 is odd.} \\ &= 4a^3b\sqrt{36 \cdot a^6 \cdot b^{10} \cdot 2 \cdot a} \\ &= 4a^3b\sqrt{36} \cdot \sqrt{a^6} \cdot \sqrt{b^{10}} \cdot \sqrt{2a} \\ &= 4a^3b \cdot 6 \cdot a^3 \cdot b^5 \cdot \sqrt{2a} = 24a^6b^6\sqrt{2a}\end{aligned}$$

When simplifying a square root, keep the following points in mind.

(1) A square root is not in simplest form if its radicand contains a factor that is a perfect square.

(2) If the exponent of b is *even*, then:
$\sqrt{b^{2n}} = |b^n|$ if n is an *odd* natural number,
$\sqrt{b^{2n}} = b^n$ if n is an *even* natural number, and
$\sqrt{b^{2n}} = b^n$ if n is a natural number and $b \geq 0$.

(3) If the exponent of b is *odd*, then:
$\sqrt{b^{2n+1}} = b^n\sqrt{b}$ if n is a natural number and $b \geq 0$.

Checkpoint

Simplify. Assume that all variables are positive real numbers.

1. $\sqrt{36a^4}$ $6a^2$
2. $\sqrt{40m^5}$ $2m^2\sqrt{10m}$
3. $\sqrt{24x^6y^7}$ $2x^3y^3\sqrt{6y}$
4. $3xy^3\sqrt{28x^3y^4}$ $6x^2y^5\sqrt{7x}$

Closure

Ask students how they would simplify the square root of a non-prime positive integer. Find the greatest perfect square factor, then use Theorem 8.1. Ask students to simplify each of the following for $a > 0$.

$\sqrt{a^n}$ where n is even   Use the formula $\sqrt{b^{2n}} = b^n$ when n is even.

$\sqrt{a^n}$ where n is odd   Use the formula $\sqrt{b^{2n+1}} = b^n\sqrt{b}$.

Additional Example 3

Simplify $\sqrt{60t^{11}}$, $t > 0$. $2t^5\sqrt{15t}$

Additional Example 4

Simplify $7xy^3\sqrt{48x^{13}y^{14}}$, $x > 0$, $y > 0$.

$28x^7y^{10}\sqrt{3x}$

FOLLOW UP

Guided Practice

Classroom Exercises 1–10

Independent Practice

A Ex. 1–20, **B** Ex. 21–33, **C** Ex. 34–39

Basic: FR all, CE all, WE 1–23 odd, Application

Average: FR all, CE all, WE 11–33 odd, Application

Above Average: FR all, CE all, WE 17–39 odd, Application

Additional Answers

Written Exercises

17. $m^3\sqrt{m}$
18. $b^7\sqrt{b}$
19. $b^8\sqrt{b}$
20. $a^3\sqrt{a}$
29. $-10x^7y^5\sqrt{5y}$
30. $9x^3y^6\sqrt{2x}$
31. $12x^4y^4\sqrt{2y}$
32. $8y^9z^{12}\sqrt{11y}$
33. Answers may vary. One possible answer: Find the greatest perfect square factor of each factor in the radicand. Then use Theorem 8.1 to simplify.
39. Either $\sqrt{x^2+y^2} \leq x+y$ or $\sqrt{x^2+y^2} > x+y$. Assume $\sqrt{x^2+y^2} > x+y$. Then $x^2+y^2 > (x+y)^2$, $x^2+y^2 > x^2+2xy+y^2$, $0 > 2xy$. But this is a contradiction. Because x and y are nonnegative numbers, $2xy$ must also be nonnegative. Therefore, the assumption is false, and $\sqrt{x^2+y^2} \leq x+y$.

Focus on Reading

Indicate whether each of the following statements is always true, sometimes true, or never true.

1. $\sqrt{a^n}$ is defined for all real numbers a, if n is even. Always true
2. $\sqrt{a^3} = a\sqrt{a}$, if $a < 0$. Never true
3. $\sqrt{a^{2n}} = |a^n|$ Always true
4. $\sqrt{a} \cdot \sqrt{b} = \sqrt{ab}$ Sometimes true

Classroom Exercises

Find the greatest perfect square factor of each radicand.

1. $\sqrt{8}$ 4   2. $\sqrt{12}$ 4   3. $\sqrt{18}$ 9   4. $\sqrt{32}$ 16   5. $\sqrt{200}$ 100

Simplify each radical.

6. $\sqrt{t^{32}}$ t^{16} 7. $\sqrt{9a^2}$ $3|a|$ 8. $\sqrt{c^{19}}, c \geq 0$ $c^9\sqrt{c}$ 9. $\sqrt{25y^8}$ $5y^4$ 10. $\sqrt{t^{22}}, t \geq 0$ t^{11}

Written Exercises

Simplify.

1. $\sqrt{24}$ $2\sqrt{6}$ 2. $\sqrt{40}$ $2\sqrt{10}$ 3. $\sqrt{45}$ $3\sqrt{5}$ 4. $\sqrt{44}$ $2\sqrt{11}$
5. $\sqrt{50}$ $5\sqrt{2}$ 6. $\sqrt{48}$ $4\sqrt{3}$ 7. $\sqrt{72}$ $6\sqrt{2}$ 8. $\sqrt{98}$ $7\sqrt{2}$
9. $3\sqrt{32}$ $12\sqrt{2}$ 10. $-5\sqrt{20}$ $-10\sqrt{5}$ 11. $0.75\sqrt{80}$ $3\sqrt{5}$ 12. $-\frac{1}{14}\sqrt{490}$ $-\frac{1}{2}\sqrt{10}$
13. $\sqrt{16a^4}$ $4a^2$ 14. $\sqrt{49b^6}$ $7|b^3|$ 15. $\sqrt{36c^{10}}$ $6|c^5|$ 16. $\sqrt{144x^{16}}$ $12x^8$
17. $\sqrt{m^7}, m \geq 0$ 18. $\sqrt{b^{15}}, b \geq 0$ 19. $b^4\sqrt{b^9}, b \geq 0$ 20. $a\sqrt{a^5}, a \geq 0$
21. $4\sqrt{12a^{18}}$ $8|a^9|\sqrt{3}$ 22. $-3\sqrt{18m^{12}}$ $-9m^6\sqrt{2}$ 23. $\sqrt{32a^3}, a \geq 0$ $4a\sqrt{2a}$ 24. $\sqrt{45m^9}, m \geq 0$ $3m^4\sqrt{5m}$

Simplify. For Exercises 25–32 assume that x and y are positive real numbers.

25. $\sqrt{12x^4y^5}$ $2x^2y^2\sqrt{3y}$ 26. $\sqrt{24x^7y^{14}}$ $2x^3y^7\sqrt{6x}$ 27. $\sqrt{80x^{16}y^{11}}$ $4x^8y^5\sqrt{5y}$ 28. $2xy^2\sqrt{75x^7y^6}$ $10x^4y^5\sqrt{3x}$
29. $-5x^3y\sqrt{20x^8y^9}$ 30. $3xy\sqrt{18x^5y^{10}}$ 31. $3xy^3\sqrt{32x^6y^3}$ 32. $4y^3z^5\sqrt{44y^{13}z^{14}}$

33. Write how to simplify an expression such as $\sqrt{50a^7b^6}$, where $a \geq 0$ and $b \geq 0$.

Enrichment

Recall that a rational number is defined as one that can be expressed in the form $\frac{a}{b}$, where a and b are integers and $b \neq 0$. Then show the following proof that $\sqrt{2}$ is not a rational number. The proof is indirect. It starts with the assumption that $\sqrt{2}$ can be written as a fraction in reduced form. Assume that $\sqrt{2} = \frac{a}{b}$, and that a and b have no common factor ($GCF = 1$).

1. $2 = \frac{a^2}{b^2}$ 2. $2b^2 = a^2$
3. Since $2b^2$ is even, a^2 is even and a is even. Thus, a can be written as $2c$.
4. $2 = \frac{4c^2}{b^2}$ 5. $2b^2 = 4c^2$ 6. $b^2 = 2c^2$
7. Since $2c^2$ is even, then b^2 is even and b is even.
8. Thus, both a and b have 2 as a factor, but this contradicts the assumption.
9. Therefore, $\sqrt{2}$ is not a rational number.

Note: Some students may see that Steps 3 and 7 require another proof, that if the square of an integer is even, then the integer is even. See whether the students can prove this number property. Have the students write a proof to show that $\sqrt{3}$ is irrational. Ask: How will steps 3 and 7 have to be changed? If the square of an integer is a multiple of 3, then the integer is a multiple of 3. Thus, a can be written as $3c$.

Simplify, assuming that m and n are positive integers and that x and y are positive real numbers.

34. $\dfrac{\sqrt{x^{6m}}}{x^{2m}}$ x^m

35. $\dfrac{\sqrt{x^{4m-2}}}{x^{m-1}}$ x^m

36. $x^m\sqrt{x^{4m+7}}$ $x^{3m+3}\sqrt{x}$

37. $x^{m+3}y^{2n-5}\sqrt{x^{8m-2}y^{6n+5}}$ $x^{5m+2}y^{5n-3}\sqrt{y}$

38. Simplify $\sqrt{8x^2 - 16x + 8}$. $2|x - 1|\sqrt{2}$

39. Prove that for all nonnegative real numbers x and y, $\sqrt{x^2 + y^2} \le x + y$.
(HINT: Assume that $\sqrt{x^2 + y^2} > x + y$. Then show that this leads to a contradiction.)

Mixed Review

Simplify. *5.2, 5.5*

1. $(3x - 4)^0$ 1

2. $\left(\dfrac{2a^{-2}}{b^{-1}}\right)^3$ $\dfrac{8b^3}{a^6}$

3. $(4x - 3)(4x + 3)$ $16x^2 - 9$

Application: *Inventory Modeling*

To maximize sales, retailers must maintain adequate inventories of the items they sell. To maximize profit, however, they need to maintain those inventories at the lowest possible cost. Inventory modeling balances these objectives by considering the following variables:

k = set-up cost per order h = monthly holding cost per item
c = purchase cost per item s = monthly sales (30-day months)

If Q = the number of items ordered each cycle, then the total cost per order includes the set-up cost k, the purchasing cost $c \cdot Q$, and the holding cost. The average inventory during a cycle is $Q \div 2$ items. The length of a cycle is $Q \div s$ 30-day months. So, the holding cost per order is $hQ^2 \div 2s$.

Therefore, the total cost per month, T, for Q items ordered per cycle is

$$T(Q) = \left(k + cQ + \frac{hQ^2}{2s}\right) \div \frac{Q}{s} = \frac{sk}{Q} + sc + \frac{hQ}{2}.$$

The value of Q that minimizes T is $Q = \sqrt{(2sk) \div h}$.

Each month, a retailer sells 106 radios for which she pays $12 each. Set-up cost is $35 per order, and her holding cost is $1 per radio. Find the number of radios per cycle that minimizes her inventory cost, the minimum inventory cost, and the length of each cycle. 86 radios, $1,358.14, 24 days

▮▮▮GETTING STARTED

Prerequisite Quiz

Simplify.

1. $\sqrt{12}$ $2\sqrt{3}$
2. $\sqrt{18}$ $3\sqrt{2}$
3. $\sqrt{27}$ $3\sqrt{3}$
4. $\sqrt{50}$ $5\sqrt{2}$
5. $\sqrt{20x^3}$ $2|x|\sqrt{5x}$

Motivator

Ask students how they would simplify an
expression such as $3a + 5b + 7a$.
Combine like terms, $10a + 5b$. Ask them
to extend this idea to simplifying an
expression like $5\sqrt{12} + 2\sqrt{3} + \sqrt{12}$.
Adding like radicals resembles adding
expressions with like terms. The simplified
expression is $14\sqrt{3}$.

▮▮▮TEACHING
   SUGGESTIONS

Lesson Note

When multiplying radicals, students should
be aware of shortcuts. A radical such as
$\sqrt{36 \cdot 7}$ can be written directly as $6\sqrt{7}$,
without the intermediate step, $\sqrt{36} \cdot \sqrt{7}$.
When radicals are equal, students can use
$\sqrt{a} \cdot \sqrt{a} = a$. For example, $\sqrt{375} \cdot \sqrt{375}$
$= 375$. In a product such as $\sqrt{45} \cdot \sqrt{539}$,
students can simplify before multiplying:
$3\sqrt{5} \cdot 7\sqrt{11} = 21\sqrt{55}$.

8.3 Sums, Differences, and Products of Square Roots

Objectives

To simplify expressions containing sums or differences of square roots
To simplify expressions containing products of square roots

The radical expressions $5\sqrt{3}$ and $-7\sqrt{3}$ are called *like radicals* because
their radical factors are the same. Recall that the Distributive Property
allows like terms to be added.
So, for like radicals such as $a\sqrt{b}$ and $c\sqrt{b}$, $a\sqrt{b} + c\sqrt{b} = (a + c)\sqrt{b}$.
Thus, adding like radicals resembles adding like terms:

$$8\sqrt{3} + 5\sqrt{3} = 13\sqrt{3}$$

Sometimes radicals have to be simplified before like radicals can be
recognized and combined. This is illustrated below.

EXAMPLE 1   Simplify $4\sqrt{12} - 2\sqrt{18} - 2\sqrt{27} + \sqrt{50}$.

Solution
$$4\sqrt{12} - 2\sqrt{18} - 2\sqrt{27} + \sqrt{50}$$
$$= 4\sqrt{4 \cdot 3} - 2\sqrt{9 \cdot 2} - 2\sqrt{9 \cdot 3} + \sqrt{25 \cdot 2}$$
$$= 4\sqrt{4} \cdot \sqrt{3} - 2\sqrt{9} \cdot \sqrt{2} - 2\sqrt{9} \cdot \sqrt{3} + \sqrt{25} \cdot \sqrt{2}$$
$$= 4 \cdot 2\sqrt{3} - 2 \cdot 3\sqrt{2} - 2 \cdot 3\sqrt{3} + 5\sqrt{2}$$
$$= 8\sqrt{3} - 6\sqrt{3} - 6\sqrt{2} + 5\sqrt{2} \leftarrow \text{Group like radicals.}$$
$$= 2\sqrt{3} - 1\sqrt{2}, \text{ or } 2\sqrt{3} - \sqrt{2} \leftarrow \text{Combine like radicals.}$$

EXAMPLE 2   Simplify $3x^2\sqrt{20x^3y} - x\sqrt{45x^5y} - x^3\sqrt{5xy}$, where
$x \geq 0$ and $y \geq 0$.

Solution
$$3x^2\sqrt{4 \cdot 5 \cdot x^2 \cdot x \cdot y} - x\sqrt{9 \cdot 5 \cdot x^4 \cdot x \cdot y} - x^3\sqrt{5xy}$$
$$= 3x^2\sqrt{4 \cdot x^2 \cdot 5xy} - x\sqrt{9 \cdot x^4 \cdot 5xy} - x^3\sqrt{5xy}$$
$$= 3x^2 \cdot 2 \cdot x\sqrt{5xy} - x \cdot 3 \cdot x^2\sqrt{5xy} - x^3\sqrt{5xy}$$
$$= 6x^3\sqrt{5xy} - 3x^3\sqrt{5xy} - x^3\sqrt{5xy}$$
$$= 2x^3\sqrt{5xy}$$

When simplifying the product of two square roots, first determine
whether the radicands are the same. If they are, use $\sqrt{a} \cdot \sqrt{a} = a$.
Thus, $(5\sqrt{x})^2$ becomes $5\sqrt{x} \cdot 5\sqrt{x}$, or $25x$.

276   Chapter 8 Radicals and Rational Exponents

Highlighting the Standards

Standard 14b: This lesson, showing the
application of whole number, exponential,
and polynomial skills to radicals, continues
the work on the logical manipulation of
radicals.

Additional Example 1

Simplify.

$4\sqrt{45} - 7\sqrt{12} - 6\sqrt{20} - \sqrt{3}$. $-15\sqrt{3}$

Additional Example 2

Simplify.

$4x^2\sqrt{18x^3y^2} + 6x\sqrt{98x^5y^2} - 8x^3\sqrt{2xy^2}$,
$x > 0, y > 0$. $46x^3y\sqrt{2x}$

If the product of two square roots results in a large integral radicand, it is often easier to factor the radicands first before multiplying them. Both methods are shown in Example 3.

EXAMPLE 3 Simplify $\sqrt{15} \cdot \sqrt{10}$.

Solution

Method 1

$$\sqrt{15} \cdot \sqrt{10} = \sqrt{150}$$
$$= \sqrt{25 \cdot 6}$$
$$= 5\sqrt{6}$$

Method 2

$$\sqrt{15} \cdot \sqrt{10} = \sqrt{5} \cdot \sqrt{3} \cdot \sqrt{5} \cdot \sqrt{2}$$
$$= \sqrt{5} \cdot \sqrt{5} \cdot \sqrt{3} \cdot \sqrt{2}$$
$$= 5\sqrt{6}$$

EXAMPLE 4 Simplify $4\sqrt{3x^3} \cdot 5\sqrt{15x^8}$, where $x \geq 0$.

Solution

$$4\sqrt{3x^3} \cdot 5\sqrt{15x^8} = 20\sqrt{45x^{11}}$$
$$= 20\sqrt{9 \cdot 5 \cdot x^{10} \cdot x^1}$$
$$= 20 \cdot 3 \cdot x^5\sqrt{5 \cdot x} = 60x^5\sqrt{5x}$$

The following example utilizes two extensions of the Distributive Property: $a(b + c + d) = ab + ac + ad$ and $(a + b)(c + d) = ac + ad + bc + bd$. The second of these duplicates the FOIL method (see Lesson 5.3).

EXAMPLE 5 Simplify each expression.

    a. $2\sqrt{5}\,(6\sqrt{5} - \sqrt{10} + 4\sqrt{7})$ **b.** $(4\sqrt{3} + \sqrt{2})(2\sqrt{3} - 5\sqrt{2})$

Solutions

    a. Apply the Distributive Property to $2\sqrt{5}\,(6\sqrt{5} - \sqrt{10} + 4\sqrt{7})$.

$$2\sqrt{5} \cdot 6\sqrt{5} - 2\sqrt{5} \cdot \sqrt{10} + 2\sqrt{5} \cdot 4\sqrt{7}$$
$$= 12\sqrt{5} \cdot \sqrt{5} - 2\sqrt{50} + 8\sqrt{35}$$
$$= 12 \cdot 5 - 2\sqrt{25 \cdot 2} + 8\sqrt{35}$$
$$= 60 - 2 \cdot 5\sqrt{2} + 8\sqrt{35}$$
$$= 60 - 10\sqrt{2} + 8\sqrt{35}$$

    b. Apply the FOIL method to $(4\sqrt{3} + \sqrt{2})(2\sqrt{3} - 5\sqrt{2})$.

$$4\sqrt{3} \cdot 2\sqrt{3} - 4\sqrt{3} \cdot 5\sqrt{2} + \sqrt{2} \cdot 2\sqrt{3} - \sqrt{2} \cdot 5\sqrt{2}$$
$$= 8\sqrt{3} \cdot \sqrt{3} - 20\sqrt{6} + 2\sqrt{6} - 5\sqrt{2} \cdot \sqrt{2}$$
$$= 8 \cdot 3 - 20\sqrt{6} + 2\sqrt{6} - 5 \cdot 2$$
$$= 24 - 20\sqrt{6} + 2\sqrt{6} - 10 = 14 - 18\sqrt{6}$$

Math Connections

The Field Properties: In Lesson 1.3, it was pointed out that the set of real numbers together with the operations + and × constitute a *field*. It was also pointed out that the set of *rational* numbers is also a field. However, it is not the case that the set of *irrational* numbers (of which radicals make up a significant portion) is a field (See *Critical Thinking Questions*.).

Critical Thinking Questions

Analysis: Ask students to show that the set of irrational numbers is not a field (see *Math Connections* above). It suffices to show that any one of the eleven field properties fails for some suitably chosen subset of irrational numbers. For example, students could show that the Closure Property of Multiplication fails for $\sqrt{2}$ and $\sqrt{8}$ since $\sqrt{2} \times \sqrt{8}$ (= 4) is not irrational.

Common Error Analysis

Error: When simplifying products such as $(2\sqrt{7} + 3\sqrt{2})\,(3\sqrt{7} - \sqrt{2})$, students tend to forget the middle term.

It may help to review briefly products such as $(3x - 5)(2x + 1)$, without radicals, and then make the transition to binomials with radicals.

Additional Example 3

Simplify $\sqrt{12} \cdot \sqrt{150}$. $30\sqrt{2}$

Additional Example 4

Simplify $6\sqrt{8x^5} \cdot 7\sqrt{6x^{10}}$, $x > 0$.
$168x^7\sqrt{3x}$

Additional Example 5

Simplify each expression.

  a. $4\sqrt{3}(7\sqrt{3} - \sqrt{6} + 8\sqrt{5})$
    $84 - 12\sqrt{2} + 32\sqrt{15}$
  b. $(2\sqrt{7} + 3\sqrt{2})(3\sqrt{7} - \sqrt{2})$
    $36 + 7\sqrt{14}$

Checkpoint

Simplify. Assume that variables represent nonnegative real numbers.

1. $4\sqrt{2} + \sqrt{8}$ $6\sqrt{2}$
2. $4\sqrt{7} - 3\sqrt{63} - 2\sqrt{28}$ $-9\sqrt{7}$
3. $(\sqrt{5x})^2$ $5x$
4. $4\sqrt{6x^5} \cdot 2\sqrt{3x^3}$ $24x^4\sqrt{2}$
5. $(4\sqrt{3} - \sqrt{5})(2\sqrt{3} + 3\sqrt{5})$
   $9 + 10\sqrt{15}$

Closure

Ask students to give the steps in simplifying a sum or difference of two radicals.

Simplify to find like radicals, group like radicals, then combine like radicals. Ask students how they would simplify a product of two radicals that are the same. If the radicands are the same, then use the formula $\sqrt{a} \cdot \sqrt{a} = a$.

◼◼◼◼FOLLOW UP

Guided Practice

Classroom Exercises 1–8

Independent Practice

Ⓐ Ex. 1–26, Ⓑ Ex. 27–30, Ⓒ Ex. 31, 32

Basic: WE 1–27 odd, Brainteaser

Average: WE 3–29 odd, Brainteaser

Above Average: WE 5–31 odd, Brainteaser

Additional Answers

Classroom Exercises

3. $2\sqrt{5y}$

See page 299 for the answers to Written Ex. 7, 8, 11, 12, 16–21, 27, 30, and Brainteaser.

Classroom Exercises

Simplify. Assume that variables represent nonnegative real numbers.

1. $5\sqrt{3} - 7\sqrt{3}$ $-2\sqrt{3}$ 2. $7\sqrt{y} + \sqrt{y}$ $8\sqrt{y}$ 3. $9\sqrt{5y} - 7\sqrt{5y}$ 4. $9\sqrt{a} - 9\sqrt{a}$ 0
5. $\sqrt{5} \cdot \sqrt{11}$ $\sqrt{55}$ 6. $3\sqrt{2} \cdot 4\sqrt{3}$ $12\sqrt{6}$ 7. $\sqrt{7} \cdot \sqrt{7}$ 7 8. $3\sqrt{2} \cdot 3\sqrt{2}$ 18

Written Exercises

Simplify. Assume that variables represent nonnegative real numbers.

1. $2\sqrt{3} + \sqrt{27}$ $5\sqrt{3}$ 2. $\sqrt{20} - 4\sqrt{5}$ $-2\sqrt{5}$ 3. $\sqrt{2} - \sqrt{8}$ $-\sqrt{2}$ 4. $\sqrt{12} - 2\sqrt{3}$ 0
5. $\sqrt{98} + \sqrt{50}$ $12\sqrt{2}$ 6. $6\sqrt{18} + \sqrt{32}$ $22\sqrt{2}$ 7. $3\sqrt{99} - 5\sqrt{44}$ 8. $5\sqrt{45} - \sqrt{80}$
9. $7\sqrt{20} + 8\sqrt{5} - 2\sqrt{45}$ $16\sqrt{5}$ 10. $\sqrt{28} - 2\sqrt{7} - 4\sqrt{63}$ $-12\sqrt{7}$
11. $6\sqrt{8} - \sqrt{24} + 3\sqrt{72} - \sqrt{54}$ 12. $8\sqrt{5} - 2\sqrt{48} - \sqrt{45} + 4\sqrt{75}$
13. $(\sqrt{8x})^2$ $8x$ 14. $(9\sqrt{x})^2$ $81x$ 15. $(\sqrt{4-x})^2$ $4 - x$ 16. $(-2\sqrt{x-1})^2$
17. $\sqrt{14} \cdot \sqrt{32}$ 18. $-\sqrt{15} \cdot \sqrt{35}$ 19. $4\sqrt{5} \cdot 6\sqrt{10}$ 20. $3\sqrt{6} \cdot (-4\sqrt{8})$
21. $2\sqrt{3}(7\sqrt{3} - \sqrt{8} + 2\sqrt{5})$ 22. $4\sqrt{2}(\sqrt{12} - 3\sqrt{2} + 4\sqrt{8})$ $40 + 8\sqrt{6}$
23. $(4\sqrt{2} - 2\sqrt{3})(5\sqrt{2} - \sqrt{3})$ $46 - 14\sqrt{6}$ 24. $(4\sqrt{5} + \sqrt{6})(3\sqrt{5} - 2\sqrt{6})$ $48 - 5\sqrt{30}$
25. $(3\sqrt{2} - 4\sqrt{6})^2$ $114 - 48\sqrt{3}$ 26. $(7\sqrt{5} + 3\sqrt{11})^2$ $344 + 42\sqrt{55}$
27. $(\sqrt{3x} - \sqrt{5y})(2\sqrt{3x} + \sqrt{5y})$ 28. $(3\sqrt{x} - 4\sqrt{y})^2$ $9x - 24\sqrt{xy} + 16y$
29. $\sqrt{49ab^3} - \sqrt{ab^3} + 4b\sqrt{ab}$ $10b\sqrt{ab}$ 30. $y\sqrt{12x^5y} - x^2y\sqrt{3xy} + x^2\sqrt{27xy^3}$
31. $\sqrt{4y - 12} + \sqrt{9y - 27}$ $5\sqrt{y-3}$ 32. $\sqrt{4x - 4} - \sqrt{x^3 - x^2}$ $(2 - x)\sqrt{x-1}$

Mixed Review

Solve for x.   *1.4, 2.3, 5.4, 6.4*

1. $4x - (7 - x) = 12$ $3\frac{4}{5}$ 2. $|2x - 8| = 4$ $2, 6$ 3. $ax - b = cx + m$ $\dfrac{b + m}{a - c}$ 4. $x^2 - 4 < 0$ $-2 < x < 2$

◼◼◼◼ Brainteaser

Given identical cups of cocoa and milk, a teaspoon of milk is taken from the milk cup and stirred into the cocoa cup. Then a teaspoon is taken from the cocoa cup and returned to the milk cup. Is there more milk in the cocoa or more cocoa in the milk? Explain.

278   Chapter 8 Radicals and Rational Exponents

Enrichment

Explain how to find a square root such as $\sqrt{8 + 2\sqrt{15}}$.

Let $\sqrt{8 + 2\sqrt{15}} = \sqrt{a} + \sqrt{b}$. Square: $8 + 2\sqrt{15} = a + 2\sqrt{ab} + b$. Then $8 = a + b$ and $15 = ab$, so $15 = a(8 - a)$. Solutions for a and b are 3 and 5. Have the students check that $(\sqrt{3} + \sqrt{5})^2 = 8 + 2\sqrt{15}$.

Now have the students find these roots.

1. $\sqrt{7 + 2\sqrt{10}}$ $\sqrt{2} + \sqrt{5}$
2. $\sqrt{14 + 2\sqrt{33}}$ $\sqrt{3} + \sqrt{11}$

8.4 Quotients of Square Roots

Objective   To simplify expressions containing quotients of square roots

In the study of trigonometry, one occasionally encounters such equations as

$$\tan x = \frac{\dfrac{\sqrt{3}}{3} + 1}{1 - \dfrac{\sqrt{3}}{3}}.$$

To solve for x in these equations, it is necessary first to simplify the complex fractions. In this lesson, you will learn how to do this.

First notice that $\sqrt{\dfrac{9}{16}} = \dfrac{3}{4}$ and $\dfrac{\sqrt{9}}{\sqrt{16}} = \dfrac{3}{4}$.

Thus, $\sqrt{\dfrac{9}{16}} = \dfrac{\sqrt{9}}{\sqrt{16}}$.

This suggests the following theorem, the proof of which is similar to that of Theorem 8.1.

Theorem 8.2   For all real numbers a and b, where $a \geq 0$ and $b > 0$, $\dfrac{\sqrt{a}}{\sqrt{b}} = \sqrt{\dfrac{a}{b}}$.

Theorem 8.2 can be used to simplify such expressions as $\dfrac{\sqrt{48}}{\sqrt{6}}$, where the radicand of the numerator is exactly divisible by the radicand of the denominator. Example 1 illustrates such a simplification.

EXAMPLE 1   Simplify each expression.

a. $\dfrac{\sqrt{48}}{\sqrt{6}}$ b. $\dfrac{\sqrt{54x^7}}{\sqrt{2x^2}}$, $x > 0$

Solutions   a. $\dfrac{\sqrt{48}}{\sqrt{6}} = \sqrt{\dfrac{48}{6}}$ b. $\dfrac{\sqrt{54x^7}}{\sqrt{2x^2}} = \sqrt{\dfrac{54x^7}{2x^2}}$

$\qquad\qquad = \sqrt{8}$ $\qquad\qquad = \sqrt{27x^5}$

$\qquad\qquad = \sqrt{4 \cdot 2}$ $\qquad\qquad = \sqrt{9 \cdot 3 \cdot x^4 \cdot x}$

$\qquad\qquad = 2\sqrt{2}$ $\qquad\qquad = 3x^2\sqrt{3x}$

Teaching Resources

Quick Quizzes 56
Reteaching and Practice Worksheets 56

GETTING STARTED

Prerequisite Quiz

Simplify.

1. $\sqrt{27}$ $3\sqrt{3}$
2. $\sqrt{x^5}$ $x^2\sqrt{x}$
3. $\sqrt{12a^3}$ $2|a|\sqrt{3a}$
4. $\sqrt{5} \cdot \sqrt{5}$ 5
5. $(\sqrt{5} + \sqrt{3})(\sqrt{5} - \sqrt{3})$ 2

Motivator

Students have seen that $\sqrt{ab} = \sqrt{a} \cdot \sqrt{b}$. Ask them what similar property would hold for quotients. $\sqrt{\dfrac{a}{b}} = \dfrac{\sqrt{a}}{\sqrt{b}}$

TEACHING SUGGESTIONS

Lesson Note

Point out that Example 1 shows the easiest way to simplify a quotient of square roots if the radicand of the numerator is divisible by the radicand of the denominator. However, this is seldom the case. In general, a practical first step is to simplify the radical of the denominator.

Additional Example 1

Simplify each expression.

a. $\dfrac{\sqrt{60}}{\sqrt{5}}$ $2\sqrt{3}$

b. $\dfrac{\sqrt{80x^5}}{\sqrt{20x^2}}$, $x > 0$ $2x\sqrt{x}$

Highlighting the Standards

Standard 2a: Work with quotients of square roots is an extension of work with products. The more complex problems help students clarify their thinking.

Trigonometry: Students who have had some elementary trigonometry in their previous year will be able to understand the following simpler version of the opening example on page 279. If the lengths of the three sides of a right triangle are 1, 2, and $\sqrt{3}$, then the tangent of the smallest angle of the triangle (30°) is $\frac{1}{\sqrt{3}}$ or $\frac{\sqrt{3}}{3}$.

Critical Thinking Questions

Analysis: After discussing the criteria for determining whether a radical expression is in simplest form (see page 282), ask students whether *simplest form* has been as precisely spelled out for earlier topics of the book such as polynomials or rational expressions.   They will discover that it has not.   Ask them why they think this is so. One possible answer is that for the earlier topics, intuition and common sense are usually adequate guides for judging when an expression is in "simplified" form. In the case of radicals, common sense is not always the best guide. Thus, $\frac{1}{\sqrt{3}}$ and $\frac{\sqrt{3}}{3}$ may appear equally simple, but the latter is more useful for most purposes.

The denominator of $\frac{2}{3\sqrt{5}}$ contains a radical.

This expression can be rewritten so that the *denominator is rational* as follows:

$$\frac{2}{3\sqrt{5}} = \frac{2 \cdot \sqrt{5}}{3\sqrt{5} \cdot \sqrt{5}} = \frac{2\sqrt{5}}{3 \cdot 5} = \frac{2\sqrt{5}}{15}$$

The above procedure, known as *rationalizing the denominator*, makes use of the property $\frac{a}{b} = \frac{a \cdot c}{b \cdot c}$, where $b \neq 0$ and $c \neq 0$.

EXAMPLE 2   Simplify. Rationalize the denominator.

a. $\dfrac{15}{\sqrt{18}}$

b. $\dfrac{6x^2y^4}{\sqrt{24x^3y^{10}}}$, $x > 0$, $y > 0$

Plan   Simplify each radical. Then rationalize the denominator.

a.
$$\frac{15}{\sqrt{18}} = \frac{15}{\sqrt{9 \cdot 2}}$$
$$= \frac{15}{3\sqrt{2}}$$
$$= \frac{5}{\sqrt{2}}$$
$$= \frac{5 \cdot \sqrt{2}}{\sqrt{2} \cdot \sqrt{2}}$$
$$= \frac{5\sqrt{2}}{2}$$

b.
$$\frac{6x^2y^4}{\sqrt{24x^3y^{10}}} = \frac{6x^2y^4}{\sqrt{4 \cdot 6 \cdot x^2 \cdot x \cdot y^{10}}}$$
$$= \frac{6x^2y^4}{2xy^5\sqrt{6x}}$$
$$= \frac{3x}{y\sqrt{6x}}$$
$$= \frac{3x \cdot \sqrt{6x}}{y\sqrt{6x} \cdot \sqrt{6x}}$$
$$= \frac{3x\sqrt{6x}}{6xy} = \frac{\sqrt{6x}}{2y}$$

Example 2a can be checked using a calculator, as shown below.

15 ÷ 18 √ = 3.5355339
5 × 2 √ ÷ 2 = 3.5355339

EXAMPLE 3   Simplify $\sqrt{\dfrac{27a^3}{20m^5n^2}}$, $a \geq 0$, $m > 0$, and $n > 0$.

Solution
$$\sqrt{\frac{27a^3}{20m^5n^2}} = \frac{\sqrt{27a^3}}{\sqrt{20m^5n^2}} \leftarrow \text{Rewrite as the quotient of two radicals.}$$
$$= \frac{3a\sqrt{3a}}{2m^2n\sqrt{5m}} = \frac{3a\sqrt{3a} \cdot \sqrt{5m}}{2m^2n\sqrt{5m} \cdot \sqrt{5m}} = \frac{3a\sqrt{15am}}{2m^2n \cdot 5m} = \frac{3a\sqrt{15am}}{10m^3n}$$

Additional Example 2

Simplify. Rationalize the denominator.

a. $\dfrac{10}{\sqrt{20}}$ $\sqrt{5}$

b. $\dfrac{8x^2y^5}{\sqrt{40x^5y^{12}}}$, $x > 0$, $y > 0$ $\dfrac{2\sqrt{10x}}{5xy}$

Additional Example 3

Simplify.

$\dfrac{8m^5}{75b^3c^4}$, $m > 0$, $b > 0$, $c > 0$

$\dfrac{2m^2\sqrt{6mb}}{15b^2c^2}$

The denominator of $\dfrac{4}{\sqrt{5} - \sqrt{3}}$ contains a difference of radicals.

Multiplying both the numerator and denominator by $\sqrt{5}$ or $\sqrt{3}$ will *not* rationalize the denominator. Instead, to rationalize the denominator use the *conjugate* of $\sqrt{5} - \sqrt{3}$, which is $\sqrt{5} + \sqrt{3}$. Notice that the product of these two irrational numbers is a rational number.

$$(\sqrt{5} + \sqrt{3})(\sqrt{5} - \sqrt{3}) = (\sqrt{5})^2 - (\sqrt{3})^2 = 5 - 3, \text{ or } 2$$

The above procedure is based upon the special product $(a + b)(a - b) = a^2 - b^2$ (see Lesson 5.5). Here, the expressions $a + b$ and $a - b$ are sometimes called a pair of **conjugates**. The conjugate of $a + b$ is $a - b$, and the conjugate of $a - b$ is $a + b$.

EXAMPLE 4   Simplify the given expressions. Rationalize the denominators.

a. $\dfrac{14}{\sqrt{5} + \sqrt{3}}$

b. $\dfrac{\sqrt{x} + \sqrt{y}}{\sqrt{x} - \sqrt{y}}, \; x > 0, y > 0, x \neq y$

Solutions

a. $\dfrac{14}{\sqrt{5} + \sqrt{3}}$

$= \dfrac{14(\sqrt{5} - \sqrt{3})}{(\sqrt{5} + \sqrt{3})(\sqrt{5} - \sqrt{3})}$

$= \dfrac{14(\sqrt{5} - \sqrt{3})}{5 - 3}$

$= \dfrac{14(\sqrt{5} - \sqrt{3})}{2}$

$= 7(\sqrt{5} - \sqrt{3})$

b. $\dfrac{\sqrt{x} + \sqrt{y}}{\sqrt{x} - \sqrt{y}}$

$= \dfrac{(\sqrt{x} + \sqrt{y})(\sqrt{x} + \sqrt{y})}{(\sqrt{x} - \sqrt{y})(\sqrt{x} + \sqrt{y})}$

$= \dfrac{(\sqrt{x})^2 + 2 \cdot \sqrt{x} \cdot \sqrt{y} + (\sqrt{y})^2}{(\sqrt{x})^2 - (\sqrt{y})^2}$

$= \dfrac{x + 2\sqrt{xy} + y}{x - y}$

To simplify such complex expressions as

$$\dfrac{\dfrac{\sqrt{3}}{3} + 1}{1 - \dfrac{\sqrt{3}}{3}},$$

which contain one or more radicals, first multiply the numerator and denominator of the complex expression by their LCD. Then rationalize the denominator.

Checkpoint

Simplify. Assume that no denominator has a value of zero and that, for all values of the variables, the radicands are positive.

1. $\dfrac{\sqrt{28}}{\sqrt{7}}$ 2

2. $\dfrac{\sqrt{y^{13}}}{\sqrt{y^5}}$ y^4

3. $\dfrac{6}{\sqrt{3}}$ $2\sqrt{3}$

4. $\dfrac{6x}{\sqrt{12x^5}}$ $\dfrac{\sqrt{3x}}{x^2}$

5. $\dfrac{\sqrt{5} + \sqrt{7}}{\sqrt{5} - \sqrt{7}}$ $\dfrac{12 + 2\sqrt{35}}{-2} = -6 - \sqrt{35}$

Additional Example 4

Simplify the given expressions. Rationalize the denominators.

a. $\dfrac{20}{\sqrt{7} - \sqrt{2}}$ $4(\sqrt{7} + \sqrt{2})$

b. $\dfrac{2\sqrt{x} - 3\sqrt{y}}{2\sqrt{x} + 3\sqrt{y}}$ $\dfrac{4x - 12\sqrt{xy} + 9y}{4x - 9y}$

Closure

Ask students how they would *rationalize* the denominator of an expression with a radical in the denominator. If the denominator contains only one radical, multiply the numerator and denominator by that radical. If the denominator contains the sum or difference of radicals, multiply the numerator and denominator by the conjugate of the denominator.

◼◼◼FOLLOW UP

Guided Practice

Classroom Exercises 1–8

Independent Practice

A Ex. 1–24, **B** Ex. 25–36, **C** Ex. 37–40

Basic: WE 1–27 odd , Midchapter Review

Average: WE 7–33 odd , Midchapter Review

Above Average: WE 13–39 odd , Midchapter Review

Additional Answers

Written Exercises

21. $14(\sqrt{3} + \sqrt{2})$

22. $2(\sqrt{7} - \sqrt{2})$

23. $3(2 - \sqrt{3})$

24. $-2(\sqrt{5} + 3)$

29. $5 - 2\sqrt{6}$

30. $\dfrac{7 + \sqrt{15}}{2}$

31. $\dfrac{6x + 8\sqrt{xy}}{9x - 16y}$

32. $\dfrac{a - 2\sqrt{ab} + b}{a - b}$

EXAMPLE 5

Simplify $\dfrac{\frac{\sqrt{3}}{3} + 1}{1 - \frac{\sqrt{3}}{3}}$.

Plan   Multiply the numerator and denominator by 3.

Solution

$$\frac{3\left(\frac{\sqrt{3}}{3} + 1\right)}{3\left(1 - \frac{\sqrt{3}}{3}\right)} = \frac{\sqrt{3} + 3}{3 - \sqrt{3}}$$

$$= \frac{(3 + \sqrt{3})(3 + \sqrt{3})}{(3 - \sqrt{3})(3 + \sqrt{3})}$$

$$= \frac{9 + 2 \cdot 3\sqrt{3} + 3}{9 - 3} \quad \leftarrow \text{From the special product}$$
$$\hspace{5cm} (a + b)^2 = a^2 + 2ab + b^2$$

$$= \frac{12 + 6\sqrt{3}}{6} = \frac{6(2 + \sqrt{3})}{6} = 2 + \sqrt{3} \quad .$$

A radical expression is in simplest form if:
(1) no radicand contains a perfect-square factor other than 1,
(2) no denominator contains a radical,
(3) no radicand contains a fraction, and
(4) the expression is not a complex expression.

Classroom Exercises

Find the conjugate of each expression.

1. $4 + \sqrt{5}$ $4 - \sqrt{5}$ 2. $2\sqrt{3} - 6$ $2\sqrt{3} + 6$ 3. $4\sqrt{2} + 6\sqrt{7}$ $4\sqrt{2} - 6\sqrt{7}$ 4. $-\sqrt{11} - \sqrt{3}$ $-\sqrt{11} + \sqrt{3}$

Simplify. Assume that the variables represent positive real numbers.

5. $\dfrac{\sqrt{24}}{\sqrt{6}}$ 2 6. $\dfrac{\sqrt{a^7}}{\sqrt{a}}$ a^3 7. $\dfrac{6}{\sqrt{12}}$ $\sqrt{3}$ 8. $\dfrac{12}{\sqrt{5} - 1}$ $3(\sqrt{5} + 1)$

Written Exercises

Simplify. Assume that no denominator has a value of zero and that all variables represent positive real numbers.

1. $\dfrac{\sqrt{24}}{\sqrt{3}}$ $2\sqrt{2}$ 2. $\dfrac{\sqrt{140}}{\sqrt{7}}$ $2\sqrt{5}$ 3. $\dfrac{\sqrt{60}}{\sqrt{5}}$ $2\sqrt{3}$ 4. $\dfrac{\sqrt{96}}{\sqrt{2}}$ $4\sqrt{3}$

282   Chapter 8 Radicals and Rational Exponents

Additional Example 5

Simplify.

$\dfrac{1 + \frac{\sqrt{2}}{2}}{1 - \frac{\sqrt{2}}{2}}$ $3 + 2\sqrt{2}$

5. $\dfrac{\sqrt{x^{11}}}{\sqrt{x^3}}$ x^4 **6.** $\dfrac{\sqrt{b^9}}{\sqrt{b^3}}$ b^3 **7.** $\dfrac{\sqrt{56a^7}}{\sqrt{2a^3}}$ $2a^2\sqrt{7}$ **8.** $\dfrac{\sqrt{80x^3}}{\sqrt{2x}}$ $2x\sqrt{10}$

9. $\dfrac{4}{\sqrt{2}}$ $2\sqrt{2}$ **10.** $\dfrac{12}{\sqrt{6}}$ $2\sqrt{6}$ **11.** $\dfrac{4}{\sqrt{8}}$ $\sqrt{2}$ **12.** $\dfrac{8}{\sqrt{20}}$ $\dfrac{4\sqrt{5}}{5}$

13. $\dfrac{-20d}{\sqrt{28d^3}}$ $\dfrac{-10\sqrt{7d}}{7d}$ **14.** $\dfrac{15x^2}{\sqrt{27x^5}}$ $\dfrac{5\sqrt{3x}}{3x}$ **15.** $\dfrac{6a^2}{\sqrt{18a^8}}$ $\dfrac{\sqrt{2}}{a^2}$ **16.** $\dfrac{10y^4}{\sqrt{20y^7}}$ $\sqrt{5y}$

17. $\sqrt{\dfrac{4}{a^3}}$ $\dfrac{2\sqrt{a}}{a^2}$ **18.** $\sqrt{\dfrac{8}{3}}$ $\dfrac{2\sqrt{6}}{3}$ **19.** $\sqrt{\dfrac{x^5}{y^3}}$ $\dfrac{x^2\sqrt{xy}}{y^2}$ **20.** $\sqrt{\dfrac{4a^2}{125b^3}}$ $\dfrac{2a\sqrt{5b}}{25b^2}$

21. $\dfrac{14}{\sqrt{3}-\sqrt{2}}$ **22.** $\dfrac{10}{\sqrt{7}+\sqrt{2}}$ **23.** $\dfrac{3}{2+\sqrt{3}}$ **24.** $\dfrac{8}{\sqrt{5}-3}$

25. $\dfrac{4xy^4}{\sqrt{32x^5y^8}}$ $\dfrac{\sqrt{2x}}{2x^2}$ **26.** $\dfrac{2a^3b^4}{\sqrt{48a^6b^7}}$ $\dfrac{\sqrt{3b}}{6}$ **27.** $\dfrac{-18x^3y}{\sqrt{27x^7y^4}}$ $\dfrac{-2\sqrt{3x}}{xy}$ **28.** $\dfrac{8x^5y^4}{\sqrt{72x^2y^5}}$ $\dfrac{2x^4y\sqrt{2y}}{3}$

29. $\dfrac{\sqrt{3}-\sqrt{2}}{\sqrt{3}+\sqrt{2}}$ **30.** $\dfrac{2\sqrt{5}-\sqrt{3}}{\sqrt{5}-\sqrt{3}}$ **31.** $\dfrac{2\sqrt{x}}{3\sqrt{x}-4\sqrt{y}}$ **32.** $\dfrac{\sqrt{a}-\sqrt{b}}{\sqrt{a}+\sqrt{b}}$

33. $\dfrac{1+\dfrac{\sqrt{2}}{5}}{1-\dfrac{\sqrt{2}}{5}}$ **34.** $\dfrac{\dfrac{\sqrt{2}}{2}+4}{1-\dfrac{\sqrt{2}}{2}}$ **35.** $\dfrac{\dfrac{\sqrt{6}-\sqrt{2}}{2}}{\dfrac{\sqrt{6}}{2}+\sqrt{2}}$ **36.** $\dfrac{\dfrac{\sqrt{5}}{2}-\dfrac{\sqrt{3}}{3}}{\dfrac{\sqrt{5}}{2}+\dfrac{\sqrt{3}}{3}}$

37. $\dfrac{\sqrt{a+b}+\sqrt{a-b}}{\sqrt{a+b}-\sqrt{a-b}}$ **38.** $\dfrac{\sqrt{4x^2+1}+2x}{\sqrt{4x^2+1}-2x}$ **39.** $\sqrt{\dfrac{\sqrt{20}+\sqrt{12}}{\sqrt{5}+\sqrt{3}}}$

40. Prove Theorem 8.2.

Midchapter Review

Simplify. Assume that variables represent nonnegative real numbers.   *8.1, 8.2, 8.3, 8.4*

1. $-\sqrt{144}$ -12 **2.** $\sqrt{90x^6}$ $3x^3\sqrt{10}$ **3.** $-2\sqrt{45x^4}$ $-6x^2\sqrt{5}$ **4.** $\sqrt{44a^7b^8}$ $2a^3b^4\sqrt{11a}$

5. $\sqrt{35}\cdot\sqrt{21}$ $7\sqrt{15}$ **6.** $2\sqrt{27}+4\sqrt{75}$ $26\sqrt{3}$ **7.** $\dfrac{\sqrt{65}}{\sqrt{5}}$ $\sqrt{13}$ **8.** $\dfrac{-2}{\sqrt{5}-\sqrt{7}}$ $\sqrt{5}+\sqrt{7}$

Solve each equation. Approximate irrational results to the nearest hundredth.   *8.1*

9. $y^2=196$ ± 14 **10.** $x^2=7$ ± 2.65 **11.** $b^2-8=12$ ± 4.47 **12.** $a^2-0.36=0$ ± 0.6

8.4 Quotients of Square Roots **283**

---

33. $\dfrac{27+10\sqrt{2}}{23}$

34. $9+5\sqrt{2}$

35. $3\sqrt{3}-5$

36. $\dfrac{19-4\sqrt{15}}{11}$

37. $\dfrac{a+\sqrt{a^2-b^2}}{b}$

38. $8x^2+4x\sqrt{4x^2+1}+1$

39. $\sqrt{2}$

40. 1. $\left(\dfrac{\sqrt{a}}{\sqrt{b}}\right)^2=\dfrac{(\sqrt{a})^2}{(\sqrt{b})^2}$ (Power of a quotient)

2. $=\dfrac{a}{b}$ (Def of principal square root)

3. $\left(\sqrt{\dfrac{a}{b}}\right)^2=\dfrac{a}{b}$ (Def of principal square root)

4. $\left(\dfrac{\sqrt{a}}{\sqrt{b}}\right)^2=\left(\sqrt{\dfrac{a}{b}}\right)^2$ (Trans and Sym Prop Eq)

5. $\dfrac{\sqrt{a}}{\sqrt{b}}=\sqrt{\dfrac{a}{b}}$ (Each is the unique positive solution of $x^2=\dfrac{a}{b}$.)

Enrichment

Write this expression on the chalkboard.

$2\sqrt{6}+3\sqrt{15}$

Challenge the students to rewrite the expression as a fraction whose numerator is a binomial made up of two irrational numbers and whose denominator is a monomial irrational number.

$2\sqrt{6}+3\sqrt{15}=2\sqrt{3}\cdot\sqrt{2}+3\sqrt{3}\cdot\sqrt{5}=$

$\sqrt{3}(2\sqrt{2}+3\sqrt{5})=\dfrac{3(2\sqrt{2}+3\sqrt{5})}{\sqrt{3}}=$

$\dfrac{6\sqrt{2}+9\sqrt{5}}{\sqrt{3}}$ Have students challenge each other with similar exercises of their own creation.

▰▰▰ GETTING STARTED

Prerequisite Quiz

Simplify. Assume that all radicands are nonnegative.

1. $(\sqrt{3x - 4})^2$ $3x - 4$
2. $\sqrt{28a^3}$ $2a\sqrt{7a}$
3. $3\sqrt{2} + 5\sqrt{8} - \sqrt{50}$ $8\sqrt{2}$
4. $3\sqrt{6x^3} \cdot 5\sqrt{8x^7}$ $60x^5\sqrt{3}$

Motivator

Sometimes $\sqrt{25}$ is written as $\sqrt[2]{25}$. Ask students to discuss the significance of the 2. It gives the index of the root. Ask them to guess the meaning of an expression like $\sqrt[3]{27}$ or $\sqrt[4]{16}$. The cube root of 27, and the fourth root of 16.

Objective | To simplify expressions containing cube, fourth, or fifth roots

---

To solve the equation $x^2 = 36$, the definition of square root is used: $x = \sqrt{36} = 6$ or $x = -\sqrt{36} = -6$. The *principal square root* is 6.

Equations such as $x^4 = 81$ can be solved in a similar way. Because $3^4 = 81$ and $(-3)^4 = 81$, $x = \sqrt[4]{81} = 3$ or $x = -\sqrt[4]{81} = -3$. The numbers 3 and -3 are *fourth roots* of 81. Of the two real fourth roots of 81, the *principal fourth root* is defined as the positive root 3, symbolized as $\sqrt[4]{81}$.

Definition

> The **nth roots** of a real number b are the solutions of the equation $x^n = b$, where n is a positive integer.

In general, the *principal nth root* of a number b is the real-number solution of the equation $x^n = b$ and is written as $\sqrt[n]{b}$. As with square roots, the number b is the *radicand*, and the positive integer n, where $n > 1$, is the **index**. For square roots, the index 2 is understood.

The principal nth root of a real number b is *not* necessarily positive, and may not exist as a real number. Consider the following.

(1) n even, radicand positive: $\sqrt[4]{16} = \sqrt[4]{2^4} = 2$
$$\sqrt[4]{16} = \sqrt[4]{(-2)^4} = |-2| = 2$$

(2) n even, radicand negative: $\sqrt[4]{-16}$ is not a real number because there is no real number whose fourth power is negative.

(3) n odd, radicand positive: $\sqrt[3]{8} = \sqrt[3]{2^3} = 2$

(4) n odd, radicand negative: $\sqrt[3]{-8} = \sqrt[3]{(-2)^3} = -2$

Thus, if n is odd and the radicand is negative, the principal root is negative.

Definition

> The **principal nth root** of a real number b, symbolized by $\sqrt[n]{b}$, represents either
>
> (1) the nonnegative nth root of b if $b \geq 0$ and n is even, or
>
> (2) the one real nth root of b if n is odd.

284   Chapter 8 Radicals and Rational Exponents

Highlighting the Standards

Standard 3e: It is important to stress the reasonableness of procedures with radicals. Students should be able to give reasons for the steps they take in simplifying radicals.

Illustrations (1)–(4) and the definition just given also suggest the following generalizations:

$(\sqrt[n]{b})^n = b$, because $\sqrt[n]{b}$ is a solution of $x^n = b$.

$\sqrt[n]{b^n} = |b|$ if n is even.

$\sqrt[n]{b^n} = b$ if n is odd. (Do not use absolute value symbols if n is odd.)

EXAMPLE 1 Simplify each radical.

    a. $\sqrt[4]{(4a)^4}$ **b.** $(\sqrt[4]{x-1})^4$ **c.** $(2\sqrt[3]{2x-1})^3$

Solutions   **a.** $\sqrt[4]{(4a)^4} = |4a|$ **b.** $(\sqrt[4]{x-1})^4 = x - 1$
                    $= 4|a|$

    c. $(2\sqrt[3]{2x-1})^3 = 2^3(\sqrt[3]{2x-1})^3$
                        $= 8(2x - 1)$
                        $= 16x - 8$

Theorem 8.3

> For all odd integers $n > 1$ and all real numbers a and b, and for all even integers $n > 1$ and all real numbers $a \geq 0$ and $b \geq 0$,
> $\sqrt[n]{a \cdot b} = \sqrt[n]{a} \cdot \sqrt[n]{b}$.

To simplify radicals such as $\sqrt[5]{-128}$, first find the greatest fifth power that is a factor of -128.

EXAMPLE 2 Simplify.

    a. $\sqrt[5]{-128}$ **b.** $\sqrt[4]{160}$ **c.** $\sqrt[3]{27x^{12}}$

Solutions   **a.** $\sqrt[5]{-128}$ **b.** $\sqrt[4]{160}$ **c.** $\sqrt[3]{27x^{12}}$
       $= \sqrt[5]{(-2)^7}$ $= \sqrt[4]{2^4 \cdot 10}$ $= \sqrt[3]{27} \cdot \sqrt[3]{x^{12}}$
       $= \sqrt[5]{(-2)^5} \cdot \sqrt[5]{(-2)^2}$ $= \sqrt[4]{2^4} \cdot \sqrt[4]{10}$ $= \sqrt[3]{3^3} \cdot \sqrt[3]{(x^4)^3}$
       $= -2\sqrt[5]{4}$ $= 2\sqrt[4]{10}$ $= 3x^4$

EXAMPLE 3 Simplify $\sqrt[3]{250a^{11}}$.

Solution $\sqrt[3]{250a^{11}} = \sqrt[3]{125 \cdot 2 \cdot a^9 \cdot a^2}$ $\leftarrow$ Use $125 = 5^3$ and $(a^3)^3 = a^9$.
                   $= \sqrt[3]{5^3 \cdot a^9 \cdot 2a^2}$
                   $= \sqrt[3]{5^3} \cdot \sqrt[3]{(a^3)^3} \cdot \sqrt[3]{2a^2}$
                   $= 5a^3\sqrt[3]{2a^2}$

8.5 Simplifying Radicals With Indices Greater Than 2   **285**

Lesson Note

Students should memorize the third, fourth, and fifth powers of some bases. In this lesson, we again discuss the simplification of radicals with positive and negative radicands, and with odd and even indices. In checks, however, we shall indicate that radicands with variables are assumed to be nonnegative. Even in advanced mathematics courses, such as calculus, this assumption is implied in the exercises.

Math Connections

The Zero-Product Property: The nth root of a real number is defined as a solution of $x^n = b$. When this is written in the form $x^n - b = 0$, it suggests the Zero-Product Property of Lesson 6.1. It is of interest to compare the results obtained using the two different approaches. For example, $x^4 = 81$, which is solved using the definition of nth roots at the top of page 284, can also be solved by rewriting the equation as $x^4 - 81 = 0$, factoring the left side to obtain $(x^2 + 9)(x + 3)(x - 3) = 0$, and then using the Zero-Product Property to obtain the two real fourth roots of 81, 3 and -3.

Additional Example 1

Simplify each radical.

a. $\sqrt[6]{(2m)^6}$ $2|m|$

b. $(\sqrt[6]{x-8})^6$ $x - 8$

c. $(4\sqrt[3]{3x+5})^3$ $64(3x + 5)$ or
    $192x + 320$

Additional Example 2

Simplify.

a. $\sqrt[5]{-160}$ $-2\sqrt[5]{5}$

b. $\sqrt[4]{162}$ $3\sqrt[4]{2}$

c. $\sqrt[3]{216x^6}$ $6x^2$

Additional Example 3

Simplify $\sqrt[4]{80a^{13}}$. $2|a^3|\sqrt[4]{5a}$

Critical Thinking Questions

Analysis: Ask students whether they can think of a way to rationalize $\dfrac{1}{\sqrt[3]{5}-1}$ using an approach similar to that shown in Lesson 8.4 using conjugate pairs. As a hint, remind them that:

$a^3 - b^3 = (a - b)(a^2 + ab + b^2)$

(See also Exercise 44 of Lesson 8.7.)

Checkpoint

Simplify.

1. $(3\sqrt[3]{2x-3})^3$ $54x - 81$
2. $\sqrt[3]{x^{12}}$ x^4
3. $8\sqrt[4]{2} - 9\sqrt[4]{32}$ $-10\sqrt[4]{2}$
4. $\sqrt[3]{3b^5} \cdot \sqrt[3]{18b^2}$ $3b^2\sqrt[3]{2b}$

Closure

Ask students to give the steps in simplifying an expression like $\sqrt[3]{40x^9y^6}$. See Example 2. Ask them how they would simplify the following expression.

$$\frac{8x}{\sqrt[4]{2}}$$

Rationalize the denominator using the property $(\sqrt[n]{b})^n = b$. Multiply the numerator and denominator by $\sqrt[4]{8}$ to get $\sqrt[4]{16}$, or 2, as the denominator.

Like radicals can be added or subtracted just as square roots are. Similarly, products and quotients of nth roots can also be simplified in the same manner as square roots.

EXAMPLE 4   Simplify $5\sqrt[4]{2} + 6\sqrt[4]{32} - \sqrt[4]{162}$.

Solution

$$5\sqrt[4]{2} + 6\sqrt[4]{32} - \sqrt[4]{162}$$
$$= 5\sqrt[4]{2} + 6\sqrt[4]{2^4 \cdot 2} - \sqrt[4]{3^4 \cdot 2} \; \leftarrow 32 = 16 \cdot 2 = 2^4 \cdot 2 \text{ and}$$
$$= 5\sqrt[4]{2} + 6 \cdot 2\sqrt[4]{2} - 3\sqrt[4]{2} \qquad 162 = 81 \cdot 2 = 3^4 \cdot 2$$
$$= 5\sqrt[4]{2} + 12\sqrt[4]{2} - 3\sqrt[4]{2} = 14\sqrt[4]{2}$$

EXAMPLE 5   Simplify $\sqrt[4]{3y^3} \cdot \sqrt[4]{6y} \cdot \sqrt[4]{9y^2}$, where $y \geq 0$.

Solution

$$\sqrt[4]{3y^3} \cdot \sqrt[4]{6y} \cdot \sqrt[4]{9y^2} = \sqrt[4]{3y^3 \cdot 6y \cdot 9y^2}$$
$$= \sqrt[4]{162y^6}$$
$$= \sqrt[4]{81 \cdot 2 \cdot y^4 \cdot y^2}$$
$$= 3y\sqrt[4]{2y^2}$$

The property of $\left(\sqrt[n]{b}\right)^n = b$ may be used to rationalize a denominator.

$$\frac{10}{\sqrt[3]{25}} = \frac{10}{\sqrt[3]{5 \cdot 5}} = \frac{10}{\sqrt[3]{5} \cdot \sqrt[3]{5}} = \frac{10 \cdot \sqrt[3]{5}}{\sqrt[3]{5} \cdot \sqrt[3]{5} \cdot \sqrt[3]{5}} = \frac{10\sqrt[3]{5}}{5} = 2\sqrt[3]{5}$$

Theorem 8.4   For all odd integers $n > 1$ and all real numbers a and b, $b \neq 0$, and for all even integers $n > 1$ and all real numbers a and b, $a \geq 0$ and $b > 0$,

$$\sqrt[n]{\frac{a}{b}} = \frac{\sqrt[n]{a}}{\sqrt[n]{b}} \; .$$

EXAMPLE 6   Simplify $\sqrt[4]{\dfrac{3}{8a^2}}$, $a > 0$.

Solution

$$\sqrt[4]{\frac{3}{8a^2}} = \frac{\sqrt[4]{3}}{\sqrt[4]{2^3a^2}}$$
$$= \frac{\sqrt[4]{3}}{\sqrt[4]{2^3a^2}} \cdot \frac{\sqrt[4]{2a^2}}{\sqrt[4]{2a^2}} \; \leftarrow \text{Choose the radicand } 2a^2 \text{ to get the fourth root}$$
$$\text{of a fourth power: } 2^3a^2 \cdot 2a^2 = 2^4a^4 = (2a)^4.$$
$$= \frac{\sqrt[4]{3 \cdot 2a^2}}{\sqrt[4]{2^4a^4}} = \frac{\sqrt[4]{6a^2}}{2a}$$

Additional Example 4

Simplify $7\sqrt[4]{3} - 3\sqrt[4]{48} + 5\sqrt[4]{243}$.

$16\sqrt[4]{3}$

Additional Example 5

Simplify $\sqrt[5]{2m^3} \cdot \sqrt[5]{16m^2} \cdot \sqrt[5]{4m}$, $m > 0$.

$2m\sqrt[5]{4m}$

Additional Example 6

Simplify $\sqrt[3]{\dfrac{5}{4a^2}}$, $a > 0$. $\dfrac{\sqrt[3]{10a}}{2a}$

Classroom Exercises

Simplify.

1. $\sqrt[3]{8}$ 2   **2.** $\sqrt[4]{16}$ 2   **3.** $\left(\sqrt[4]{3}\right)^4$ 3   **4.** $\left(\sqrt[5]{a}\right)^5$ a **5.** $\sqrt[3]{5} \cdot \sqrt[3]{5} \cdot \sqrt[3]{5}$ 5

6. $\sqrt[3]{-27}$ -3 **7.** $\sqrt[3]{\dfrac{8}{27}}$ $\dfrac{2}{3}$ **8.** $\sqrt[6]{a^6}$ $|a|$ **9.** $\sqrt[5]{64}$ $2\sqrt[5]{2}$ **10.** $3\sqrt[4]{64} + \sqrt[4]{4}$ $7\sqrt[4]{4}$

Written Exercises

Simplify. Assume that no denominator is 0.

1. $\left(\sqrt[4]{7x}\right)^4$ $7x$ **2.** $\left(\sqrt[4]{x+2}\right)^4$ $x+2$ **3.** $\left(-2\sqrt[3]{3y}\right)^3$ $-24y$ **4.** $\left(4\sqrt[3]{2x-1}\right)^3$ $128x-64$

5. $\sqrt[4]{81}$ 3   **6.** $\sqrt[3]{125}$ 5   **7.** $\sqrt[5]{-32}$ -2 **8.** $\sqrt[5]{243}$ 3

9. $\sqrt[4]{64}$ $2\sqrt[4]{4}$ **10.** $\sqrt[3]{-54}$ $-3\sqrt[3]{2}$ **11.** $\sqrt[3]{-250}$ $-5\sqrt[3]{2}$ **12.** $\sqrt[5]{128}$ $2\sqrt[5]{4}$

13. $\sqrt[3]{x^{21}}$ x^7 **14.** $\sqrt[4]{(2x)^4}$ $2|x|$ **15.** $\sqrt[6]{a^{30}}$, $a \geq 0$ a^5 **16.** $\sqrt[4]{b^{16}}$ b^4

17. $\sqrt[4]{16a^{23}}$, $a \geq 0$ **18.** $\sqrt[3]{27x^9}$ $3x^3$ **19.** $\sqrt[5]{32b^{10}}$ $2b^2$ **20.** $\sqrt[4]{81x^{12}}$, $x \geq 0$ $3x^3$

21. $7\sqrt[4]{2} + 8\sqrt[4]{32}$ $23\sqrt[4]{2}$ **22.** $3\sqrt[3]{54} - 5\sqrt[3]{2}$ $4\sqrt[3]{2}$

23. $4\sqrt[4]{80} - 2\sqrt[4]{5}$ $6\sqrt[4]{5}$ **24.** $\sqrt[4]{8a^5} \cdot \sqrt[4]{4a^7}$, $a \geq 0$ $2a^3\sqrt[4]{2}$

25. $\sqrt[3]{9b^7} \cdot \sqrt[3]{12b^5}$ $3b^4\sqrt[3]{4}$ **26.** $\sqrt[4]{8x^5} \cdot \sqrt[4]{20x^2}$, $x \geq 0$ $2x\sqrt[4]{10x^3}$

27. $\dfrac{14}{\sqrt[3]{7}}$ $2\sqrt[3]{49}$ **28.** $\dfrac{6x}{\sqrt[4]{2}}$ $3x\sqrt[4]{8}$ **29.** $\sqrt[4]{\dfrac{7}{8}}$ $\dfrac{\sqrt[4]{14}}{2}$ **30.** $\sqrt[4]{\dfrac{5}{6}}$ $\dfrac{\sqrt[4]{1,080}}{6}$

31. $5\sqrt[3]{3} + \sqrt[3]{24} - 7\sqrt[3]{81}$ $-14\sqrt[3]{3}$ **32.** $7\sqrt[4]{3} + 8\sqrt[4]{243} - \sqrt[4]{48}$ $29\sqrt[4]{3}$

33. $\sqrt[4]{7d^5} \cdot \sqrt[4]{8d^3} \cdot \sqrt[4]{2d^3}$, $d \geq 0$ $2d^2\sqrt[4]{7d^3}$ **34.** $\sqrt[5]{10a^7} \cdot \sqrt[5]{2a} \cdot \sqrt[5]{16a^4}$ $2a^2\sqrt[5]{10a^2}$

35. $\sqrt[3]{\dfrac{7}{25y^2}}$ $\dfrac{\sqrt[3]{35y}}{5y}$ **36.** $\dfrac{4h^3}{\sqrt[5]{8h^7}}$ $2h\sqrt[5]{4h^3}$ **37.** $\sqrt[5]{\dfrac{3}{16a^4b^2}}$ $\dfrac{\sqrt[5]{6ab^3}}{2ab}$ **38.** $\dfrac{6a^2}{\sqrt[4]{8a^2}}$ $3|a|\sqrt[4]{2a^2}$

39. Prove Theorem 8.3.   **40.** Prove Theorem 8.4.

Mixed Review

Solve.   *5.1, 6.4, 6.6*

1. $2^{3n-1} = 32$ 2   **2.** $x^2 < 9$ $-3 < x < 3$ **3.** $2x^3 + 3x^2 - 8x + 3 = 0$ $-3, \frac{1}{2}, 1$

■ **FOLLOW UP**

Guided Practice

Classroom Exercises 1–10

Independent Practice

A Ex. 1–30, **B** Ex. 31–38, **C** Ex. 39, 40

Basic: WE 1–33 odd
Average: WE 5–37 odd
Above Average: WE 7–39 odd

Additional Answers

Written Exercises

17. $2a^5\sqrt[4]{a^3}$

39. 1. $\left(\sqrt[n]{a} \cdot \sqrt[n]{b}\right)^n = \left(\sqrt[n]{a}\right)^n\left(\sqrt[n]{b}\right)^n$
    (Power of a product)

   2. $= ab$ (Def of principal n^{th} root)

   3. $\left(\sqrt[n]{ab}\right)^n = ab$
    (Def of principal n^{th} root)

   4. $\left(\sqrt[n]{a} \cdot \sqrt[n]{b}\right)^n = \left(\sqrt[n]{ab}\right)^n$
    (Trans and Sym Prop Eq)

   5. $\sqrt[n]{a} \cdot \sqrt[n]{b} = \sqrt[n]{ab}$ (If n is negative, each is the unique solution of $x^n = ab$. If n is positive, each is the unique positive solution.)

40. 1. $\left(\dfrac{\sqrt[n]{a}}{\sqrt[n]{b}}\right)^n = \dfrac{\left(\sqrt[n]{a}\right)^n}{\left(\sqrt[n]{b}\right)^n}$ (Power of a quotient);

   2. $= \dfrac{a}{b}$ (Def of principal n^{th} root)

   3. $\left(\sqrt[n]{\dfrac{a}{b}}\right)^n = \dfrac{a}{b}$ (Def of principal n^{th} root)

   4. $\left(\dfrac{\sqrt[n]{a}}{\sqrt[n]{b}}\right)^n = \left(\sqrt[n]{\dfrac{a}{b}}\right)^n$
    (Trans and Sym Prop Eq)

   5. $\dfrac{\sqrt[n]{a}}{\sqrt[n]{b}} = \sqrt[n]{\dfrac{a}{b}}$ (If n is negative, each is the unique solution of $x^n = \dfrac{a}{b}$. If n is positive, each is the unique positive solution.

Enrichment

Ask the students to find a number that is both a perfect square and a perfect cube. If they have difficulty, point out that 64 is the square of 8 and the cube of 4.

Next, challenge them to find a formula which will enable them to find the entire family of such numbers.

Let $x =$ a positive number such that $\sqrt{x}$ and $\sqrt[3]{x}$ are rational numbers. Then $\dfrac{\sqrt{x}}{\sqrt[3]{x}} = a$,

where $a > 0$ and a is a rational number. Therefore, $\sqrt{x} = a\sqrt[3]{x}$

$$x^{\frac{1}{2}} = ax^{\frac{1}{3}}$$
$$\left(x^{\frac{1}{2}}\right)^6 = \left(ax^{\frac{1}{3}}\right)^6$$
$$x^3 = a^6x^2$$
$$x = a^6$$

Thus, for any positive rational number a, a^6 will be both a perfect square and a perfect cube. For example, if $a = 2$, then $a^6 = 64$.

Teaching Resources

Application Worksheet 8
Quick Quizzes 58
Reteaching and Practice
    Worksheets 58

▮ GETTING STARTED

Prerequisite Quiz

Simplify.

1. $(x^3)^2$ x^6
2. $(\sqrt{x})^2$ x
3. 3^{-2} $\frac{1}{9}$
4. $\sqrt{x^4}$ x^2
5. $(\sqrt{x})^4$ x^2
6. $(\sqrt[3]{8})^{-2}$ $\frac{1}{4}$

Motivator

Have the class discover the meaning of a fractional exponent with the aid of calculators. Ask students to find 6^2 with a calculator. 36 Then ask them to find $36^{\frac{1}{2}}$ using a calculator. 6 Ask them what this suggests about the meaning of an exponent like $\frac{1}{2}$. $b^{\frac{1}{2}}$ is the same as $\sqrt{b}$.

▮ TEACHING SUGGESTIONS

Lesson Note

Emphasize that while $a^{mn} = (a^m)^n = (a^n)^m$, the order of computations with rational-number exponents can make a practical difference. For example, it is much easier to evaluate $32^{\frac{3}{5}}$ as $(\sqrt[5]{32})^3$ than as $\sqrt[5]{32^3}$.

Objectives

To write expressions with rational-number exponents in radical form, and vice versa

To evaluate expressions containing rational-number exponents

Recall that b^0, where $b \neq 0$, is defined as 1, and b^{-n}, where $b \neq 0$ and n is a positive integer, is defined as $\frac{1}{b^n}$.

These two definitions allowed the Laws of Exponents to be extended to nonpositive integral exponents. In this lesson, a further extension will be made to cover exponents such as $\frac{1}{5}$ and $-\frac{2}{3}$ that are rational but not integral.

Recall the power-of-a-power law, $(b^m)^n = b^{mn}$, where m and n are integers and $b \neq 0$. If this law is to be extended to rational-number exponents, then the following statement must be true.

$$\left(4^{\frac{1}{5}}\right)^5 = 4^{\frac{1}{5} \cdot 5} = 4^1, \text{ or } 4.$$

Thus, the fifth power of $4^{\frac{1}{5}}$ must be 4, and the fifth power of $\sqrt[5]{4}$ is also 4, so $4^{\frac{1}{5}}$ can be *defined* as $\sqrt[5]{4}$.

This definition of the meaning of the exponent $\frac{1}{5}$ can be extended to the following general definition.

Definition

> For all integers $n > 1$ and all real numbers b,
> $b^{\frac{1}{n}} = \sqrt[n]{b}$, provided that $\sqrt[n]{b}$ is a real number.

The definition of negative rational exponents such as $-\frac{1}{2}$ and $-\frac{1}{4}$ is somewhat similar to the definition of negative integer exponents given in Chapter 5.

Definition

> For all integers $n > 0$ and all nonzero real numbers b,
> $b^{-\frac{1}{n}} = \dfrac{1}{b^{\frac{1}{n}}}$, provided that $b^{\frac{1}{n}}$ is a real number.

From this definition, $(-8)^{-\frac{1}{3}} = \dfrac{1}{(-8)^{\frac{1}{3}}} = \dfrac{1}{\sqrt[3]{-8}} = \dfrac{1}{-2}, \text{ or } -\dfrac{1}{2}$.

Highlighting the Standards

Standard 14a: At the school level, structure cannot be learned in itself; it can only be appreciated gradually through the accumulation of mathematical experiences.

EXAMPLE 1   Evaluate each of the following.

     a. $49^{\frac{1}{2}}$ **b.** $(-64)^{\frac{1}{3}}$ **c.** $81^{-0.25}$

Solutions   **a.** $49^{\frac{1}{2}} = \sqrt{49} = 7$

     b. $(-64)^{\frac{1}{3}} = \sqrt[3]{-64} = \sqrt[3]{(-4)^3} = -4$

     c. $81^{-0.25} = 81^{-\frac{1}{4}} = \dfrac{1}{81^{\frac{1}{4}}} = \dfrac{1}{\sqrt[4]{81}} = \dfrac{1}{3}$

The Laws of Exponents can also be extended to rational exponents of the form $\frac{m}{n}$. To see how this is done, assume that the power-of-a-power law holds in the two parallel examples below.

$$8^{\frac{2}{3}} = 8^{2 \cdot \frac{1}{3}} = (8^2)^{\frac{1}{3}} \qquad \text{or} \qquad 8^{\frac{2}{3}} = 8^{\frac{1}{3} \cdot 2} = (8^{\frac{1}{3}})^2$$
$$= \sqrt[3]{8^2} \qquad\qquad\qquad\qquad = (\sqrt[3]{8})^2$$
$$= \sqrt[3]{64} = 4 \qquad\qquad\qquad\quad = 2^2 = 4$$

Thus, $8^{\frac{2}{3}} = \sqrt[3]{8^2}$, or $(\sqrt[3]{8})^2$.

This development suggests the following definition.

Definition

> For all integers m and n, where $n > 1$, and all real numbers $b \neq 0$,
> $$b^{\frac{m}{n}} = (\sqrt[n]{b})^m = \sqrt[n]{b^m}$$ provided that $\sqrt[n]{b}$ is real.

With these definitions it can be shown that all the Laws of Exponents stated in Chapter 5 do hold true for rational exponents.

EXAMPLE 2   Write in exponential form.

     a. $\sqrt[7]{x^4}$ **b.** $(\sqrt[3]{13})^2$

Solutions   **a.** $\sqrt[7]{x^4} = x^{\frac{4}{7}}$ **b.** $(\sqrt[3]{13})^2 = 13^{\frac{2}{3}}$

EXAMPLE 3   Write in radical form and simplify.

     a. $2^{\frac{4}{5}}$ **b.** $(-8)^{\frac{2}{3}}$ **c.** $-8^{\frac{2}{3}}$

Solutions   **a.** $2^{\frac{4}{5}} = \sqrt[5]{2^4}$ **b.** $(-8)^{\frac{2}{3}} = \sqrt[3]{(-8)^2}$ **c.** $-8^{\frac{2}{3}} = -\sqrt[3]{8^2}$

            $= \sqrt[5]{16}$ $= \sqrt[3]{64} = 4$ $= -\sqrt[3]{64} = -4$

Math Connections

Group Theory: A set of numbers forms a group under multiplication if it satisfies these four properties: closure, associativity, existence of an identity element, and existence of a multiplicative inverse for each number. It can be shown that the set of nonzero numbers of the form $a + b\sqrt{2}$, where a and b are rational numbers, forms a group under multiplication.

Additional Example 1

Evaluate each of the following.

a. $\dfrac{1}{64^{-\frac{1}{2}}}$ $\frac{1}{8}$

b. $(-27)^{\frac{1}{3}}$ -3

c. $32^{-0.2}$ $\frac{1}{2}$

Additional Example 2

Write in exponential form.

a. $\sqrt[4]{x^3}$ $x^{\frac{3}{4}}$

b. $(\sqrt[5]{17^2})$ $17^{\frac{2}{5}}$

Additional Example 3

Write in radical form and simplify.

a. $3^{\frac{2}{5}}$ $\sqrt[5]{3^2} = \sqrt[5]{9}$

b. $(-32)^{\frac{4}{5}}$ $(\sqrt[5]{-32})^4 = (-2)^4 = 16$

c. $-32^{\frac{4}{5}}$ $-(\sqrt[5]{32})^4 = -2^4 = -16$

Critical Thinking Questions

Analysis: In the Examples of this lesson, rational-number exponents are limited to values that are less than 1. Ask students whether rational-number exponents must be restricted so that they do not have values greater than 1. They should see that there are no restrictions on m and n other than those given in the definition on page 289. See also Exercises 22, 26, 29, 33, and 34.

Common Error Analysis

Error: Students tend to misinterpret a negative rational exponent as meaning the negative of the power rather than as the reciprocal.

Review the meaning of negative integral exponents. Thus, 3^{-2} means $\frac{1}{3^2}$, and not $-(3^2)$.

Emphasize that in simplifying a power such as $8^{-\frac{2}{3}}$, it is useful to write $\frac{1}{8^{\frac{2}{3}}}$ as the first step.

Checkpoint

Evaluate.

1. $4^{\frac{1}{2}}$ 2

2. $32^{-\frac{2}{5}}$ $\frac{1}{4}$

3. $16^{0.75}$ 8

Evaluate for $x = 25$ and $y = -27$.

4. $x^{\frac{3}{2}}$ 125

5. $\dfrac{y^{-\frac{1}{3}}}{y^{\frac{1}{3}}}$ $\frac{1}{9}$

6. Write $\sqrt[7]{a^3}$ in exponential form. $a^{\frac{3}{7}}$

When finding the value of a power with a rational-number exponent, such as $b^{\frac{m}{n}}$, it is easier to find the nth root first and then to find the mth power of the result.

EXAMPLE 4   Find the value of each of the following expressions.

  a. $x^{-\frac{4}{5}}$, for $x = 32$ b. $5y^{\frac{3}{4}}$, for $y = 16$

Solutions

a. $x^{-\frac{4}{5}} = 32^{-\frac{4}{5}}$

$= \dfrac{1}{32^{\frac{4}{5}}}$

$= \dfrac{1}{(\sqrt[5]{32})^4} = \dfrac{1}{2^4}$, or $\dfrac{1}{16}$

b. $5y^{\frac{3}{4}} = 5 \cdot 16^{\frac{3}{4}}$

$= 5(\sqrt[4]{16})^3$

$= 5 \cdot 2^3$

$= 5 \cdot 8$, or 40

Classroom Exercises

Give each expression in exponential form.

1. $\sqrt[3]{43}$ $43^{\frac{1}{3}}$
2. $\sqrt{19}$ $19^{\frac{1}{2}}$
3. $\sqrt[7]{32}$ $32^{\frac{1}{7}}$
4. $\sqrt[5]{x}$ $x^{\frac{1}{5}}$
5. $\sqrt[n]{a}, a > 0$ $a^{\frac{1}{n}}$

Give each expression in radical form. Simplify, if possible.

6. $27^{\frac{1}{3}}$ $\sqrt[3]{27}; 3$
7. $5^{\frac{1}{4}}$ $\sqrt[4]{5}$
8. $28^{\frac{1}{2}}$ $\sqrt{28}; 2\sqrt{7}$
9. $4^{\frac{2}{3}}$ $\sqrt[3]{4^2}; 2\sqrt[3]{2}$
10. $5^{-\frac{1}{2}}$ $\frac{1}{\sqrt{5}}; \frac{\sqrt{5}}{5}$

Written Exercises

Evaluate.

1. $36^{\frac{1}{2}}$ 6
2. $8^{\frac{1}{3}}$ 2
3. $81^{-\frac{1}{4}}$ $\frac{1}{3}$
4. $(-27)^{-\frac{1}{3}}$ $-\frac{1}{3}$
5. $400^{-0.5}$ $\frac{1}{20}$
6. $4 \cdot 16^{\frac{1}{2}}$ 16
7. $5 \cdot 4^{-\frac{1}{2}}$ $2\frac{1}{2}$
8. $-7 \cdot 27^{\frac{1}{3}}$ -21
9. $-4 \cdot 81^{0.25}$ -12
10. $-3 \cdot 16^{0.25}$ -6

Write each expression in exponential form.

11. $\sqrt[5]{a^3}$ $a^{\frac{3}{5}}$
12. $\sqrt[4]{b^3}, b \geq 0$ $b^{\frac{3}{4}}$
13. $(\sqrt[3]{11})^4$ $11^{\frac{4}{3}}$
14. $(\sqrt{13})^5$ $13^{\frac{5}{2}}$
15. $\sqrt[5]{23^4}$ $23^{\frac{4}{5}}$

Write each expression in radical form and simplify.

16. $3^{\frac{2}{3}}$ $\sqrt[3]{3^2}, \sqrt[3]{9}$
17. $5^{\frac{3}{4}}$ $\sqrt[4]{5^3}, \sqrt[4]{125}$
18. $(-7)^{\frac{2}{3}}$ $\sqrt[3]{(-7)^2}; \sqrt[3]{49}$
19. $3^{-\frac{4}{5}}$
20. $a^{\frac{4}{5}}$ $\sqrt[5]{a^4}$

Evaluate.

21. $16^{\frac{3}{4}}$ 8
22. $25^{-\frac{5}{2}}$ $\frac{1}{3,125}$
23. $(-27)^{\frac{2}{3}}$ 9
24. $-27^{\frac{2}{3}}$ -9
25. $32^{\frac{2}{5}}$ 4
26. $6 \cdot 4^{-\frac{5}{2}}$ $\frac{3}{16}$
27. $-4 \cdot 343^{-\frac{2}{3}}$
28. $8 \cdot 16^{-\frac{3}{4}}$ 1
29. $6^0 \cdot 27^{-\frac{5}{3}}$
30. $8^{\frac{2}{3}} \cdot 2^{-1}$ 2

290   Chapter 8 Radicals and Rational Exponents

Additional Example 4

Find the value of each of the following.

a. $x^{-\frac{5}{3}}$, for $x = 27$ $\frac{1}{243}$

b. $3y^{\frac{5}{4}}$, for $y = 16$ 96

31. The price p, in dollars, of a certain type of solar heater is approximated by $p = 3x^{\frac{1}{2}} + 5x^{\frac{2}{3}}$, where x is the number of units. Find the price of 64 units. $104

Simplify.

32. $\sqrt[3]{x^3 - 6x^2 + 12x - 8}$ $x - 2$ **33.** $[1^m - (-27)^{\frac{1}{3}}]^{-\frac{3}{2}}$ $\frac{1}{8}$

34. Given that $f(x) = (x^3 + 31)^{-\frac{3}{2}}$, find $f(\sqrt[3]{5})$. $\frac{1}{216}$

Mixed Review

1. Graph $3x - 4y = 12$.   *3.5*

2. Simplify $(3x^{-2}y^5)^{-2}$.   *5.2* $\dfrac{x^4}{9y^{10}}$

3. Solve: $\begin{array}{l} 3x - 2y = 4 \\ 4x + 3y = 11 \end{array}$ *4.3* (2,1)

4. Write an equation of the line containing the point $(2, -1)$ and parallel to the line whose equation is $4x - y = 8$.   *3.7* $4x - y = 9$

/ Using the Calculator

Though most scientific calculators do not have special keys for roots with indices greater than 2, many do have an $\boxed{x^y}$ (or $\boxed{y^x}$) key and other special-function keys. To find $\sqrt[6]{64}$ using such a calculator, rewrite $\sqrt[6]{64}$ as $64^{\frac{1}{6}}$ because the exponent $\frac{1}{6}$ can be entered directly.

If you have the reciprocal key ($\boxed{1/x}$), enter: 64 $\boxed{x^y}$ 6 $\boxed{1/x}$ $\boxed{=}$ 2.

Without a reciprocal key, use the left and right parentheses keys to express the exponent as $(1 \div 6)$: 64 $\boxed{x^y}$ $\boxed{(}$ 1 $\boxed{\div}$ 6 $\boxed{)}$ $\boxed{=}$ 2.

If your calculator has an inverse key ($\boxed{INV}$ or $\boxed{2nd}$), $\sqrt[6]{64}$ can be found using the fact that finding the 6th root is the inverse (reverse) of finding the 6th power: 64 $\boxed{INV}$ $\boxed{x^y}$ 6 $\boxed{=}$ 2.

The change-of-sign key $\boxed{+/-}$ can be used to show that $243^{-\frac{1}{5}} = \frac{1}{3}$.

   Enter: 243 $\boxed{x^y}$ 5 $\boxed{1/x}$ $\boxed{+/-}$ $\boxed{=}$ 0.33333333

To find $64^{\frac{5}{6}}$, enter $\frac{5}{6}$ as $(5 \div 6)$ or find $64^{\frac{1}{6}}$ and raise the result to the 5th power, because $64^{\frac{5}{6}} = \left(64^{\frac{1}{6}}\right)^5$.

Use a scientific calculator to check your answers to Written Exercises 1–10 and 21–25. (Many calculators do not accept a negative number as a base. If yours does not, compute the result with a positive base, and then determine the sign of the result.)

Ask students to explain the significance of the 3 and the significance of the 2 in the expression $27^{\frac{2}{3}}$.   See the definition on page 289.   Ask students if it is easier to work with the numerator or the denominator first when simplifying.   The denominator.   Then ask them why.   The cube root of 27 is 3. Thus, by applying the denominator first, students get $(3)^2 = 9$.

FOLLOW UP

Guided Practice

Classroom Exercises 1–10

Independent Practice

A Ex. 1–25, **B** Ex. 26–31, **C** Ex. 32–34

Basic: WE 1–27 odd, Using the Calculator
Average: WE 5–31 odd, Using the Calculator
Above Average: WE 7–33 odd, Using the Calculator

Additional Answers

Written Exercises

12. $b^{\frac{3}{4}}$

19. $\dfrac{1}{\sqrt[5]{3^4}}$; $\dfrac{1}{\sqrt[5]{81}}$

27. $-\dfrac{4}{49}$

29. $\dfrac{1}{243}$

See page 295 for the answer to Mixed Review Ex. 1.

Enrichment

Write this expression on the chalkboard and challenge students to rewrite it using a single root symbol.

$$\sqrt[3]{\sqrt[4]{\sqrt[5]{\sqrt{a}}}} \quad \left(\left(\left((a)^{\frac{1}{2}}\right)^{\frac{1}{5}}\right)^{\frac{1}{4}}\right)^{\frac{1}{3}} = a^{\frac{1}{120}} = \sqrt[120]{a}$$

▬▬GETTING STARTED

Prerequisite Quiz

Simplify.

1. $4^{\frac{3}{2}}$ 8
2. $27^{-\frac{2}{3}}$ $\frac{1}{9}$
3. $a^2 \cdot a^{-4}$ $\frac{1}{a^2}$
4. $n^5 \cdot n^{-5}$ 1

Complete.

5. $\frac{3}{5} + \frac{?}{} = 1$ $\frac{2}{5}$
6. $\frac{7}{8} + \frac{?}{} = 1$ $\frac{1}{8}$

Motivator

Ask students what number must be added
to $\frac{3}{5}$ to equal 1. $\frac{2}{5}$
Write the following expression on the
chalkboard.

$$\frac{1}{7^{\frac{3}{5}}}$$

Ask students to find a value for x that will
make this equation true.

$\frac{1}{7^{\frac{3}{5}}} \cdot \frac{7^x}{7^x} = \frac{7^x}{7}$ Since $7^{\frac{3}{5}} \cdot 7^{\frac{2}{5}} = 7^{\frac{5}{5}}$, or 7,

x must equal $\frac{2}{5}$.

8.7 Applying Rational-Number Exponents

Objectives   To simplify expressions containing rational-number exponents
To solve exponential equations

It has now been demonstrated that the Laws of Exponents hold for all
rational-number exponents. Several of these laws are shown below.

$$b^r \cdot b^s = b^{r+s} \qquad \frac{b^r}{b^s} = b^{r-s} \qquad (b^r)^s = b^{rs}$$
$$(ab)^r = a^r b^r \qquad \left(\frac{a}{b}\right)^r = \frac{a^r}{b^r} \qquad b^{-r} = \frac{1}{b^r}$$
$$(\sqrt[n]{b})^n = b$$
$$b^{\frac{m}{n}} = (\sqrt[n]{b})^m = \sqrt[n]{b^m}$$

All of the above laws are true when a and b are positive real numbers, r
and s are rational numbers, m is an integer, and n is a natural number.
Under some conditions, the laws also hold for negative values of a and
b. Review Chapter 5 and earlier lessons of this chapter for the various
restrictions.

These laws can be used to simplify expressions that involve rational
exponents.

EXAMPLE 1   Simplify $6^{-\frac{3}{5}}$.

Plan   First rewrite $6^{-\frac{3}{5}}$ as $\frac{1}{6^{\frac{3}{5}}}$. Then rationalize the denominator by
obtaining a power of 6 that is a positive integer.

Solution   The least positive integer greater than $\frac{3}{5}$ is 1.

$\frac{3}{5} + \frac{2}{5} = \frac{5}{5} = 1$, so multiply numerator and denominator of
$\frac{1}{6^{\frac{3}{5}}}$ by $6^{\frac{2}{5}}$.

$$6^{-\frac{3}{5}} = \frac{1}{6^{\frac{3}{5}}} = \frac{1 \cdot 6^{\frac{2}{5}}}{6^{\frac{3}{5}} \cdot 6^{\frac{2}{5}}} = \frac{6^{\frac{2}{5}}}{6^{\frac{5}{5}}} = \frac{6^{\frac{2}{5}}}{6^1}, \text{ or } \frac{6^{\frac{2}{5}}}{6}$$

Note that an expression such as $6^{-\frac{3}{5}}$ is considered to be in the simplest
form when its exponent is positive and its denominator is rational,

as in $\frac{6^{\frac{2}{5}}}{6}$.

Highlighting the Standards

Standard 4c: The Laws of Exponents now
bring together a number of concepts that
have been studied separately. Students will
begin to see the unity linking various
computations.

Additional Example 1

Simplify.

$4^{-\frac{2}{3}}$. $\frac{4^{\frac{1}{3}}}{4}$

Additional Example 2

Simplify.

$3y^{-1} \cdot -4y^{\frac{1}{2}} \cdot 6y^{-\frac{1}{4}}, y > 0$ $\frac{-72y^{\frac{1}{4}}}{y}$

EXAMPLE 2 Simplify $4x^{-1} \cdot 7x^{\frac{1}{3}} \cdot 2x^{-\frac{2}{3}}$, where $x > 0$.

Solution
$$4x^{-1} \cdot 7x^{\frac{1}{3}} \cdot 2x^{-\frac{2}{3}} = 56x^{-1+\frac{1}{3}-\frac{2}{3}}$$
$$= 56x^{-\frac{4}{3}}$$
$$= \frac{56}{x^{\frac{4}{3}}} = \frac{56 \cdot x^{\frac{2}{3}}}{x^{\frac{4}{3}} \cdot x^{\frac{2}{3}}} = \frac{56x^{\frac{2}{3}}}{x^2}$$

EXAMPLE 3 Simplify $(27x^{12}y^{-6})^{-\frac{2}{3}}$, where $x > 0$ and $y > 0$.

Plan Use $(ab)^r = a^r b^r$ and $(b^r)^s = b^{rs}$.

Solution
$$(27x^{12}y^{-6})^{-\frac{2}{3}} = 27^{-\frac{2}{3}} \cdot x^{12\left(-\frac{2}{3}\right)} \cdot y^{-6\left(-\frac{2}{3}\right)}$$
$$= 27^{-\frac{2}{3}}x^{-8}y^4 = \frac{y^4}{27^{\frac{2}{3}}x^8} = \frac{y^4}{(\sqrt[3]{27})^2 x^8} = \frac{y^4}{3^2 x^8} = \frac{y^4}{9x^8}$$

The laws $\frac{b^r}{b^s} = b^{r-s}$ and $\left(\frac{a}{b}\right)^r = \frac{a^r}{b^r}$ are used in Examples 4 and 5.

EXAMPLE 4 Simplify $\dfrac{8^{\frac{3}{4}}x^{-\frac{1}{2}}y^{\frac{1}{4}}}{8^{\frac{1}{12}}x^{\frac{3}{4}}y^{-\frac{1}{6}}}$, where $x > 0$ and $y > 0$.

Plan Rewrite the exponents of like bases so that they have the same denominator.

Solution
$$\frac{8^{\frac{9}{12}}x^{-\frac{2}{4}}y^{\frac{3}{12}}}{8^{\frac{1}{12}}x^{\frac{3}{4}}y^{-\frac{2}{12}}} = 8^{\frac{8}{12}}x^{-\frac{5}{4}}y^{\frac{5}{12}} = \frac{8^{\frac{2}{3}}y^{\frac{5}{12}}}{x^{\frac{5}{4}}} = \frac{4y^{\frac{5}{12}} \cdot x^{\frac{3}{4}}}{x^{\frac{5}{4}} \cdot x^{\frac{3}{4}}} = \frac{4y^{\frac{5}{12}}x^{\frac{3}{4}}}{x^2}$$

EXAMPLE 5 Simplify $\left(\dfrac{4c^{-6}}{25d^{-4}}\right)^{\frac{3}{2}}$, where $c > 0$ and $d > 0$.

Solution
$$\left(\frac{4c^{-6}}{25d^{-4}}\right)^{\frac{3}{2}} = \frac{4^{\frac{3}{2}}c^{-6\left(\frac{3}{2}\right)}}{25^{\frac{3}{2}}d^{-4\left(\frac{3}{2}\right)}} = \frac{4^{\frac{3}{2}}c^{-9}}{25^{\frac{3}{2}}d^{-6}} = \frac{4^{\frac{3}{2}}d^6}{25^{\frac{3}{2}}c^9} = \frac{8d^6}{125c^9}$$

The Laws of Exponents can also be used to simplify the products and quotients of radicals with unequal indices.

Lesson Note

Point out that rational-number exponents are used to rationalize denominators. Compare the methods of rationalizing denominators using radicals and rational-number exponents. For example,

$$\frac{1}{\sqrt{2}} = \frac{1 \cdot \sqrt{2}}{\sqrt{2} \cdot \sqrt{2}} = \frac{\sqrt{2}}{2}$$

New method, with rational-number exponents:

$$\frac{1}{2^{\frac{1}{2}}} = \frac{1 \cdot 2^{\frac{1}{2}}}{2^{\frac{1}{2}} \cdot 2^{\frac{1}{2}}} = \frac{2^{\frac{1}{2}}}{2^1}, \text{ or } \frac{2^{\frac{1}{2}}}{2}$$

Math Connections

Logarithms: The Laws of Exponents summarized at the beginning of the lesson will reappear in connection with the introduction to logarithms in Chapter 13.

Critical Thinking Questions

Analysis: Ask students under what conditions exponential equations such as that of Example 6 on page 294 cannot be solved by using the methods studied so far. They will probably see that such equations cannot be solved if it turns out not to be possible to write a common base for the two sides of the equation. Students may be told that such equations can be solved using the logarithm techniques of Chapter 13.

Additional Example 3

Simplify.

$(-32x^{10}y^{-15})^{-\frac{3}{5}}$, $x > 0$, $y > 0$

$\dfrac{-y^9}{8x^6}$

Additional Example 4

Simplify.

$\dfrac{16^{\frac{5}{12}}a^{-\frac{1}{3}}b^{\frac{3}{5}}}{16^{\frac{1}{6}}a^{\frac{1}{2}}b^{\frac{1}{10}}}$, $a > 0$, $b > 0$

$\dfrac{2a^{\frac{1}{6}}b^{\frac{1}{2}}}{a}$

Additional Example 5

Simplify.

$\left(\dfrac{8x^{-3}}{27y^9}\right)^{-\frac{4}{3}}$ $\dfrac{81x^4y^{12}}{16}$

Checkpoint

Simplify each expression. Assume that values of the variables are positive.

1. $2^{-\frac{4}{5}}$ $\dfrac{2^{\frac{1}{5}}}{2}$

2. $(a^8b^{10})^{\frac{3}{2}}$ $a^{12}b^{15}$

3. $\dfrac{a}{a^{\frac{4}{5}}}$ $a^{\frac{1}{5}}$

4. Simplify $\sqrt[4]{x^3} \cdot \sqrt[3]{x^2}$ using the least number of radicals. $x\sqrt[12]{x^5}$

Closure

Ask students to describe the conditions in which an expression like a^p, where p is a negative fraction, is said to be in simplest form.   When its exponent is positive and its denominator is rational. Ask students to give examples of the laws of exponents that hold for rational numbers as well as for integers. See page 292.

▰▰▰FOLLOW UP

Guided Practice

Classroom Exercises 1–4

Independent Practice

Ⓐ Ex. 1–28, Ⓑ Ex. 29–40, Ⓒ Ex. 41–46

Basic: WE 1–31 odd

Average: WE 7–37 odd

Above Average: WE 15–45 odd

To simplify an expression such as $\sqrt[3]{x^2} \cdot \sqrt[4]{x^3}$, $x > 0$, first write the expression with rational-number exponents.

$$\sqrt[3]{x^2} \cdot \sqrt[4]{x^3} = x^{\frac{2}{3}} \cdot x^{\frac{3}{4}} = x^{\frac{17}{12}} = x^{1+\frac{5}{12}} = x \cdot x^{\frac{5}{12}} = x\sqrt[12]{x^5}.$$

To solve exponential equations involving rational-number exponents, set the exponents equal to each other and solve the resulting equation.

EXAMPLE 6   Solve $32^{3x-2} = \dfrac{1}{16}$ and check.

Plan   First, rewrite 32 and $\dfrac{1}{16}$ as powers of the same positive base.

Solution

$$32^{3x-2} = \frac{1}{16}$$
$$32^{3x-2} = 16^{-1}$$
$$(2^5)^{3x-2} = (2^4)^{-1}$$
$$2^{15x-10} = 2^{-4}$$
$$15x - 10 = -4$$
$$15x = 6$$
$$x = \frac{2}{5}$$

Check:
$$32^{3x-2} = \frac{1}{16}$$
$$32^{3 \cdot \frac{2}{5} - 2} \overset{?}{=} \frac{1}{16}$$
$$32^{-\frac{4}{5}} \overset{?}{=} \frac{1}{16}$$
$$\frac{1}{32^{\frac{4}{5}}} \overset{?}{=} \frac{1}{16}$$
$$\frac{1}{16} = \frac{1}{16} \quad \text{True}$$

Classroom Exercises

Simplify each expression. Assume that $x > 0$.

1. $7^{-\frac{1}{3}}$ $\dfrac{7^{\frac{2}{3}}}{7}$

2. $x^{-\frac{1}{2}}$ $\dfrac{x^{\frac{1}{2}}}{x}$

3. $x^{-\frac{2}{5}} x^{\frac{7}{5}}$ x

4. $\dfrac{x^{\frac{4}{7}}}{x^{\frac{2}{7}}}$ $x^{\frac{2}{7}}$

Written Exercises

Simplify each expression. Assume that the values of the variables are positive.

1. $5^{-\frac{3}{4}}$ $\dfrac{5^{\frac{1}{4}}}{5}$

2. $4^{-\frac{2}{3}}$ $\dfrac{4^{\frac{1}{3}}}{4}$

3. $30x^{-\frac{4}{3}}$ $\dfrac{30x^{\frac{2}{3}}}{x^2}$

4. $2a^{-\frac{7}{3}}$ $\dfrac{2a^{\frac{2}{3}}}{a^3}$

5. $4^{\frac{2}{9}} \cdot 4^{-\frac{1}{9}} \cdot 4^{\frac{4}{9}}$ $4^{\frac{5}{9}}$

6. $a^{\frac{5}{4}} \cdot a^{\frac{3}{4}} \cdot a^{-\frac{7}{4}}$ $a^{\frac{1}{4}}$

7. $t^4 \cdot t^{-\frac{1}{4}}$ $t^{\frac{15}{4}}$

8. $7a^3 \cdot 8a^{-\frac{2}{3}}$ $56a^{\frac{13}{5}}$

9. $(x^{\frac{2}{3}}y^{\frac{1}{4}})^{12}$ x^8y^3

10. $(x^{\frac{3}{4}} \cdot y^{\frac{1}{2}})^8$ x^6y^4

11. $(16x^4y^6)^{-\frac{1}{2}}$ $\dfrac{1}{4x^2y^3}$

12. $(16a^8b^4)^{\frac{5}{4}}$ $32a^{10}b^5$

Additional Example 6

Solve.

$4^{2x-3} = \dfrac{1}{32}$ $x = \dfrac{1}{4}$

Check: $4^{-\frac{5}{2}} = \dfrac{1}{4^{\frac{5}{2}}} = \dfrac{1}{2^5} = \dfrac{1}{32}$

13. $\dfrac{a^{\frac{4}{7}}}{a^{\frac{3}{7}}}$ $\;a^{\frac{1}{7}}$

14. $\dfrac{x^{\frac{2}{5}}}{x^{-\frac{1}{5}}}\,x^{\frac{3}{5}}$

15. $\dfrac{a^{-\frac{1}{4}}}{a^{-\frac{5}{4}}}\,a$

16. $\dfrac{b^{4}}{b^{\frac{2}{5}}}\,b^{\frac{18}{5}}$

17. $\left(\dfrac{x^{\frac{2}{5}}}{y^{\frac{1}{2}}}\right)^{10}\dfrac{x^{4}}{y^{5}}$

18. $\left(\dfrac{a^{\frac{2}{5}}}{b^{\frac{1}{3}}}\right)^{15}\dfrac{a^{6}}{b^{5}}$

19. $\left(\dfrac{49m^{8}}{16m^{4}}\right)^{\frac{1}{2}}\dfrac{7m^{2}}{4}$

20. $\left(\dfrac{125c^{9}}{27z^{15}}\right)^{-\frac{2}{3}}\dfrac{9z^{10}}{25c^{6}}$

Simplify. Write each expression using *one* radical. Assume that $x > 0$.

21. $\sqrt{x}\cdot\sqrt[3]{x}\;\sqrt[6]{x^{5}}$

22. $\sqrt[5]{x^{2}}\cdot\sqrt[3]{x}\;\sqrt[15]{x^{11}}$

23. $\dfrac{\sqrt[3]{x}}{\sqrt[4]{x}}\;\sqrt[12]{x}$

24. $\dfrac{\sqrt[4]{x^{3}}}{\sqrt[6]{x}}\;\sqrt[12]{x^{7}}$

Solve each equation. Check.

25. $8^{x}=16$ $\frac{4}{3}$

26. $3^{6x}=\dfrac{1}{27}$ $-\frac{1}{2}$

27. $3^{8x}=\dfrac{1}{81}$ $-\frac{1}{2}$

28. $4^{6x}=32$ $\frac{5}{12}$

29. $4^{2x-1}=8$ $\frac{5}{4}$

30. $2^{2x+4}=\dfrac{1}{16}$ -4

31. $10^{3x-2}=\dfrac{1}{1{,}000}$ $-\frac{1}{3}$

32. $3^{1-x}=9^{1-x}$ 1

Simplify each expression. Assume that the values of the variables are positive.

33. $x^{\frac{3}{7}}\cdot x^{\frac{2}{7}}\cdot x^{-\frac{6}{7}}$

34. $b^{-3}\cdot b^{\frac{3}{4}}$

35. $-5a^{-\frac{3}{4}}\cdot 6a^{-\frac{5}{8}}$

36. $3x^{-2}\cdot\left(-5x^{-\frac{1}{4}}\right)$

37. $(16x^{20}y^{-8})^{-\frac{3}{4}}$

38. $\dfrac{27^{\frac{7}{12}}x^{-\frac{1}{10}}y^{\frac{1}{2}}}{27^{\frac{1}{4}}x^{\frac{1}{5}}y^{-\frac{1}{3}}}$

39. $\dfrac{t^{\frac{1}{2}}v^{-\frac{1}{5}}\,v^{\frac{3}{5}}t^{\frac{5}{6}}}{t^{\frac{2}{3}}v^{-\frac{4}{5}}\,t}$

40. $\left(\dfrac{64m^{-3}}{27m^{-9}}\right)^{\frac{2}{3}}\dfrac{16m^{4}}{9}$

Simplify. Write each expression with the least number of radicals.
Assume that the values of the variables are positive.

41. $(\sqrt[4]{x^{3}}+\sqrt[6]{y})(\sqrt[4]{x^{3}}-\sqrt[6]{y})$

42. $(\sqrt[4]{a}-\sqrt[4]{b})(\sqrt[4]{a}+\sqrt[4]{b})$

43. $(2\sqrt{x}-3\sqrt[6]{y})^{2}$

44. Rationalize the denominator of $\dfrac{21}{5^{\frac{1}{3}}+2^{\frac{1}{3}}}$. (HINT: Use the factors of $a^{3}+b^{3}$.)

45. Solve $8x^{-\frac{2}{3}}-x^{\frac{1}{3}}=0$. 8

46. Solve $8(2x-6)^{-\frac{1}{2}}-(2x-6)^{\frac{1}{2}}=0$. 7

Mixed Review

Solve. *2.2, 6.1, 6.6, 7.6*

1. $-6<2x-4\le 8$ $-1<x\le 6$

2. $2x^{2}-15=-7x$ $-5,\frac{3}{2}$

3. $2x^{3}-13x^{2}+23x-12=0$ $1,\frac{3}{2},4$

4. $\dfrac{x}{5}=\dfrac{2}{x+3}$ $-5,2$

Written Exercises

33. $\dfrac{x^{\frac{6}{7}}}{x}$

34. $\dfrac{b^{\frac{3}{4}}}{b^{3}}$

35. $\dfrac{-30a^{\frac{5}{8}}}{a^{2}}$

36. $\dfrac{-15x^{\frac{3}{4}}}{x^{3}}$

37. $\dfrac{y^{6}}{8x^{15}}$

38. $\dfrac{3x^{\frac{7}{10}}y^{\frac{5}{6}}}{x}$

41. $x\sqrt{x}-\sqrt[3]{y}$

42. $\sqrt{a}-\sqrt{b}$

43. $4x-12\sqrt[6]{x^{3}y}+9\sqrt[3]{y}$

44. $3(5^{\frac{2}{3}}-10^{\frac{1}{3}}+2^{\frac{2}{3}})$

Additional Answer, page 291

Mixed Review

1.

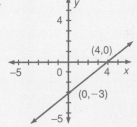

Enrichment

Discuss the fact that a scientific calculator has a key for finding square roots, but no keys for finding higher roots. To find higher roots, students can use the $\boxed{y^{x}}$ key and express exponents in decimal form. For example, $\sqrt[5]{x}=x^{\frac{1}{5}}=x^{0.2}$.

To find the fifth root of 32, press the following keys.

$32\;\boxed{y^{x}}\;0.2\;\boxed{=}$

Have the students use a scientific calculator to find these roots.

1. $\sqrt[5]{243}$ $243\;\boxed{y^{x}}\;0.2\;\boxed{=}$; 3

2. $\sqrt[4]{1{,}296}$ $1296\;\boxed{y^{x}}\;0.25\;\boxed{=}$; 6

3. $\sqrt[3]{729}$

HINT: Use memory keys. $1\;\boxed{\div}\;3\;\boxed{=}$

$\boxed{M+}\;729\;\boxed{y^{x}}\;\boxed{MR}\;\boxed{=}$; 9

295

9. Answers may vary. One possible answer: Use Theorem 8.1 to factor out the greatest even power of x. Then simplify.

10. $21\sqrt{3} - 29\sqrt{2}$

Chapter 8 Review

Key Terms

conjugate (p. 281)
index (p. 284)
nth root (p. 284)
perfect square (p. 268)
principal nth root (p. 284)

principal square root (p. 267)
radical (p. 267)
radical sign (p. 267)
radicand (p. 267)
square root (p. 267)

Key Ideas and Review Exercises

8.1   To identify the domain of a function such as $f(x) = \sqrt{4x - 12}$, find the values of x for which $4x - 12 \geq 0$.

To solve an equation such as $3x^2 - 2 = 25$, use the fact that if $x^2 = a$, then $x = \pm\sqrt{a}$.

Find the domain of each function.

1. $f(x) = \sqrt{4x - 32}$ $\{x \mid x \geq 8\}$
2. $f(x) = \sqrt{21 - 7x}$ $\{x \mid x \leq 3\}$

Solve each equation. Approximate irrational results to the nearest hundredth.

3. $3x^2 - 2 = 25$ ± 3
4. $c^2 + 6 = 19$ ± 3.61

8.2   To simplify the square root of a positive number, find a, the greatest perfect-square factor of the radicand, and use $\sqrt{ab} = \sqrt{a} \cdot \sqrt{b}$.

To simplify square roots when the sign of the variable is not indicated, use $\sqrt{x^{2n}} = x^n$ when n is even and $\sqrt{x^{2n}} = |x^n|$ when n is odd.

Simplify.

5. $\sqrt{50}$ $5\sqrt{2}$
6. $-2\sqrt{108}$ $-12\sqrt{3}$
7. $\sqrt{16b^4}$ $4b^2$
8. $\sqrt{25a^{10}}$ $5|a^5|$

9. Write a brief paragraph describing how to simplify $\sqrt{x^n}$ if x is positive and n is an odd integer.

8.3   To simplify sums and differences of square roots, simplify each radical and combine like radicals.

To simplify products of square roots, use $\sqrt{a} \cdot \sqrt{b} = \sqrt{ab}$, $a \geq 0$, $b \geq 0$.

Simplify. Assume that the values of the variables are positive.

10. $5\sqrt{27} - 6\sqrt{32} + 3\sqrt{12} - \sqrt{50}$
11. $7\sqrt{3} \cdot 8\sqrt{6}$ $168\sqrt{2}$
12. $4\sqrt{3}(7\sqrt{2} - \sqrt{6} + 2\sqrt{3})$
$28\sqrt{6} - 12\sqrt{2} + 24$
13. $6\sqrt{8y^5} \cdot 2\sqrt{3y^4}$ $24y^4\sqrt{6y}$

8.4   To simplify quotients of square roots, use $\dfrac{\sqrt{a}}{\sqrt{b}} = \sqrt{\dfrac{a}{b}}$, $a \geq 0$, $b > 0$.

If a denominator contains a radical, rationalize the denominator as shown in Lesson 8.4.

To simplify a complex fraction involving square roots, multiply the numerator and denominator by the LCD of the fractions in the expression and simplify.

Simplify. Assume that the values of the variables are positive.

14. $\dfrac{\sqrt{72}}{\sqrt{6}}$ $2\sqrt{3}$ **15.** $\dfrac{6}{\sqrt{8}}$ $\dfrac{3\sqrt{2}}{2}$ **16.** $\dfrac{4xy^4}{\sqrt{20x^5y^8}}$ $\dfrac{2\sqrt{5x}}{5x^2}$ **17.** $\dfrac{24z^5}{\sqrt{45y^4z^3}}$ $\dfrac{8z^3\sqrt{5z}}{5y^2}$

18. $\dfrac{10}{\sqrt{7} - \sqrt{2}}$ **19.** $\dfrac{5\sqrt{2}}{3 - \sqrt{2}}$ **20.** $\dfrac{1 + \dfrac{\sqrt{2}}{2}}{1 - \dfrac{\sqrt{2}}{2}}$ $3 + 2\sqrt{2}$ **21.** $\dfrac{1 - \dfrac{\sqrt{5}}{5}}{2 + \sqrt{5}}$

$2(\sqrt{7} + \sqrt{2})$ $\dfrac{15\sqrt{2} + 10}{7}$ $\dfrac{14\sqrt{5} - 30}{10}$

8.5   To simplify radicals with index n, for all integers $n > 1$, use Theorem 8.3. For sums, differences, products, and quotients of such radicals, see Examples 4, 5, and 6 in Lesson 8.5.

Simplify. Assume that the values of the variables are positive.

22. $\sqrt[3]{50y^5}$ $y\sqrt[3]{50y^2}$ **23.** $\sqrt[3]{56a^{11}}$ $2a^3\sqrt[3]{7a^2}$

24. $\dfrac{21}{\sqrt[3]{7}}$ $3\sqrt[3]{49}$ **25.** $\sqrt[4]{\dfrac{5}{8x}}$ $\dfrac{\sqrt[4]{10x^3}}{2x}$

26. $5\sqrt[3]{16} - 8\sqrt[3]{2}$ $2\sqrt[3]{2}$ **27.** $\sqrt[4]{8x^3} \cdot \sqrt[4]{4x^9}$ $2x^3\sqrt[4]{2}$

8.6, 8.7   To write expressions with rational exponents in radical form, or vice versa, use $b^{\frac{m}{n}} = (\sqrt[n]{b})^m = \sqrt[n]{b^m}$.

To simplify expressions containing rational exponents, apply the Laws of Exponents. An expression is in simplest form when all its exponents are positive and any denominators are rational.

28. Write $\sqrt[7]{x^2}$ in exponential form. $x^{\frac{2}{7}}$ **29.** Write $2^{\frac{4}{5}}$ in radical form. Simplify.

                                                      $\sqrt[5]{2^4};\ \sqrt[5]{16}$

Simplify. Assume that the values of the variables are positive.

30. $(-32)^{\frac{1}{5}}$ -2 **31.** $8^{-\frac{2}{3}}$ $\dfrac{1}{4}$ **32.** $16^{0.25}$ 2 **33.** $5^{\frac{3}{7}} \cdot 5^{-\frac{1}{7}} \cdot 5^{\frac{4}{7}}$ $5^{\frac{6}{7}}$

34. $4x^{-\frac{3}{5}}$ $\dfrac{4x^{\frac{2}{5}}}{x}$ **35.** $\dfrac{x}{x^{\frac{5}{8}}}$ $x^{\frac{3}{8}}$ **36.** $\left(\dfrac{8y^{-9}}{125x^{-6}}\right)^{\frac{2}{3}}$ $\dfrac{4x^4}{25y^6}$ **37.** $\dfrac{x^{\frac{1}{4}}y^{-\frac{2}{7}}}{x^{\frac{3}{2}}y^{-\frac{4}{7}}}$ $\dfrac{y^{\frac{2}{7}}x^{\frac{3}{4}}}{x^2}$

38. Solve and check $5^{2x-1} = \dfrac{1}{25}$. $-\dfrac{1}{2}$

1. Evaluate $\sqrt{\dfrac{5x + 11}{4}}$ for $x = 5$. 3

2. Give the domain of the function $f(x) = \sqrt{4x - 20}$. $\{x \mid x \geq 5\}$

Solve each equation. Approximate irrational results to the nearest hundredth.

3. $a^2 - 7 = 29$ ± 6 **4.** $5x^2 + 17 = 47$ ± 2.45

5. Simplify $4\sqrt{24x^6}$ using absolute value symbols, if necessary. $8|x^3|\sqrt{6}$

Simplify. Assume that the values of the variables are positive.

6. $\sqrt{81x^4}$ $9x^2$ **7.** $8\sqrt{44}$ $16\sqrt{11}$ **8.** $\sqrt[3]{-40b^{13}}$ $-2b^4\sqrt[3]{5b}$ **9.** $\sqrt[4]{16x^8}$ $2x^2$

10. $\dfrac{\sqrt{48}}{\sqrt{6}}$ $2\sqrt{2}$ **11.** $\dfrac{4}{\sqrt{12}}$ $\dfrac{2\sqrt{3}}{3}$ **12.** $\dfrac{20x^6}{\sqrt{8x^3}}$ $5x^4\sqrt{2x}$ **13.** $\sqrt[4]{\dfrac{3}{2x}}$ $\dfrac{\sqrt[4]{24x^3}}{2x}$

14. $25^{\frac{3}{2}}$ 125 **15.** $32^{\frac{4}{5}}$ 16 **16.** $(-27)^{-\frac{2}{3}}$ $\dfrac{1}{9}$ **17.** $3y^{-\frac{2}{5}}$ $\dfrac{3y^{\frac{3}{5}}}{y}$

18. $\dfrac{6}{\sqrt{5} - \sqrt{2}}$
$2(\sqrt{5} + \sqrt{2})$

19. $\dfrac{1 + \dfrac{\sqrt{3}}{3}}{1 - \dfrac{\sqrt{3}}{3}}$ $2 + \sqrt{3}$

20. $\dfrac{x}{x^{\frac{3}{4}}}$ $x^{\frac{1}{4}}$

21. $\dfrac{x^{\frac{1}{2}}y^{-\frac{2}{3}}}{x^{\frac{3}{4}}y^{-\frac{4}{3}}}$ $\dfrac{y^{\frac{2}{3}}x^{\frac{3}{4}}}{x}$

22. $7\sqrt{12} - 5\sqrt{48} - 2\sqrt{75}$ $-16\sqrt{3}$ **23.** $5\sqrt{2}(8\sqrt{2} - 3\sqrt{6} - \sqrt{8})$ $60 - 30\sqrt{3}$

24. $(3\sqrt{2} + \sqrt{5})(6\sqrt{2} - \sqrt{5})$ $31 + 3\sqrt{10}$ **25.** $6\sqrt[4]{32} - 8\sqrt[4]{2}$ $4\sqrt[4]{2}$

26. $\sqrt[3]{4x^{10}} \cdot \sqrt[3]{40x^2}$ $2x^4\sqrt[3]{20}$ **27.** $8\sqrt{5a^3b^2} - 4ab\sqrt{20a}$ 0

28. Write $\sqrt[5]{x^3}$ in exponential form. $x^{\frac{3}{5}}$ **29.** Write $3^{\frac{3}{4}}$ in radical form.
Simplify. $\sqrt[4]{3^3}; \sqrt[4]{27}$

30. Solve and check $4^{3x-2} = \dfrac{1}{64}$. $-\dfrac{1}{3}$

31. Four less than the square of a number is 8. Find the number to the nearest hundredth. ± 3.46

Simplify.

32. $\sqrt{\dfrac{\sqrt[10]{32} \cdot \sqrt[4]{4}}{2^{-3} \cdot 3^{-2}}}$ 12

33. $\dfrac{a^2 + b^2}{a^{\frac{2}{3}} + b^{\frac{2}{3}}}, a \neq 0, b \neq 0$ $a^{\frac{4}{3}} - a^{\frac{2}{3}}b^{\frac{2}{3}} + b^{\frac{4}{3}}$

34. Solve $-3y^{-\frac{1}{3}} + 7y^{\frac{2}{3}} = 0, y > 0$. $\dfrac{3}{7}$

College Prep Test

Choose the *one* best answer to each question or problem.

1. Which of the following products has the greatest value less than 100? E
   (A) $2 \times 4 \times 6$ (B) $2 \times 4 \times 9$
   (C) $4 \times 4 \times 9$ (D) $3 \times 3 \times 9$
   (E) $4 \times 4 \times 6$

2.

   The figure above represents a drawer with compartments in a chest of identical drawers. If each compartment in each drawer contains 30 screws, and if the chest contains 9,000 screws, how many drawers are in the chest? C
   (A) 12   (B) 300   (C) 15
   (D) 60   (E) None of these

3. On a map, 1 in. represents 2 mi. A circle on the map has a circumference of 5π in. This represents a circular region with what area? B
   (A) 10π mi^2 (B) 25π mi^2
   (C) 5π mi^2 (D) 100π mi^2
   (E) 50π mi^2

4. If $5x + 3y = 23$, and x and y are positive integers, then y can equal which one of the following? D
   (A) 3   (B) 4   (C) 5
   (D) 6   (E) 7

5. $\frac{1}{4} \div \left(\frac{1}{4} \div \frac{1}{4} \right) = $? A
   (A) $\frac{1}{4}$ (B) 4   (C) 1
   (D) $\frac{1}{16}$ (E) 16

6. In $5p$ years, Bill will be $10q + 1$ times his current age. Find Bill's current age in terms of p and q. B
   (A) $\frac{5p - 1}{10q - 1}$ (B) $\frac{p}{2q}$
   (C) $\frac{5p}{10q + 1}$ (D) $10q + 1 - 5p$
   (E) It cannot be determined from the information given.

7. If $x + y = 4$ and $2x - y = 5$, then $x + 2y = $? A
   (A) 5   (B) 2   (C) 1
   (D) 4   (E) 6

8.

   The interior of a rectangle is shaded inside a regular hexagon as shown above. What is the ratio of the shaded area to the area of the hexagon? B
   (A) $\frac{1}{2}$ (B) $\frac{2}{3}$ (C) $\frac{3}{4}$
   (D) $\frac{4}{5}$ (E) $\frac{5}{6}$

9. $\sqrt{1 - \left(\frac{1}{2} + \frac{1}{4} + \frac{1}{8} + \frac{1}{16} + \frac{1}{32} \right)} = $ E
   (A) $\frac{3\sqrt{6}}{8}$ (B) $\frac{\sqrt{62}}{8}$ (C) $\frac{1}{2\sqrt{2}}$
   (D) $\frac{1}{4}$ (E) $\frac{\sqrt{2}}{8}$

10. $\left(-27x^6 y^{21} \right)^{\frac{2}{3}} = $? B
    (A) $-9x^4 y^{14}$ (B) $9x^4 y^{14}$
    (C) $18x^2 y^7$ (D) $-18x^2 y^7$
    (E) $-27x^4 y^{14}$

Additional Answers, page 278

Written Exercises

7. $-\sqrt{11}$
8. $11\sqrt{5}$
11. $30\sqrt{2} - 5\sqrt{6}$
12. $5\sqrt{5} + 12\sqrt{3}$
16. $4x - 4$
17. $8\sqrt{7}$
18. $-5\sqrt{21}$
19. $120\sqrt{2}$
20. $-48\sqrt{3}$
21. $42 - 4\sqrt{6} + 4\sqrt{15}$
27. $6x - \sqrt{15xy} - 5y$
30. $4x^2 y \sqrt{3xy}$

Brainteaser
They are the same. After the transfers, the total amount of liquid in each cup remains the same. Also, the total amounts of both cocoa and milk remain the same. Thus, if a certain amount of milk is now in the cocoa, then it must have displaced the same amount of cocoa, which is now in the milk cup.

For additional standardized test practice, see the SAT/ACT Practice Test for Cumulative Test, Chapters 5–8.

12. $\{x \mid x \le \frac{7}{2}\}$
13. $\{x \mid x \le -3 \text{ or } x > 4\}$
14. $\{x \mid 1 < x < 5\}$
15. $\{x \mid x \le -2 \text{ or } x \ge 8\}$
18.

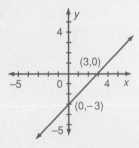

19.

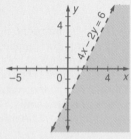

20. 200 volts
22.

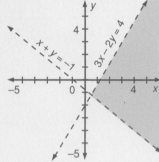

28. $(x + 2)(x - 2)(x^2 + 2)$
29. $(3x - 1)(y + 2)$ **30.** 1,2,3
33. $\frac{-4a}{5a - 25}$ **34.** $\frac{6 - x}{x^2 - 9}$
42. 7, 21, 23

▨ **Cumulative Review** *(Chapters 1–8)*

Match each equation with one of the properties (A-E) for operations with real numbers.

1. $0 + 8x = 8x$ A   *1.3*
2. $5x + 5y = 5(x + y)$ B
3. $-xy + xy = 0$ D
4. $x + (y + 7) = (y + 7) + x$ E
5. $7 + (y + 9) = (7 + y) + 9$ C

   (A) Add Identity
   (B) Distr
   (C) Associate for Add
   (D) Add Inverse
   (E) Commutative for Add

Solve the equation. (Exercises 6–10)

6. $2(3x + 8) - 3(x - 2) =$ *1.4*
   $9x - (4x - 6)$ 8
7. $|2x - 8| = 20$ −6, 14   *2.3*
8. $3^{2n-4} = 27$ $\frac{7}{2}$ *5.1*
9. $\dfrac{5}{2y + 6} = \dfrac{2y - 4}{y^2 - y - 12}$ 12   *7.6*
10. $x^2 - 6 = \frac{1}{4}$ $\pm\frac{5}{2}$ *8.1*
11. Solve $ax + b = c$ for x. $\frac{c - b}{a}$ *1.4*

Find the solution set.

12. $3(4 - 2x) \ge 4x - 23$ *2.1*
13. $2x + 9 \le 3 \text{ or } 3x - 7 > 5$ *2.2*
14. $|2x - 6| < 4$ *2.3*
15. $x^2 - 6x - 16 \ge 0$ *6.4*

Write an equation in standard form ($ax + by = c$) for the line described.

16. Passing through $P(2, -1)$ and   *3.4*
   $Q(12,3)$ $2x - 5y = 9$
17. Passing through $A(-5,2)$ and   *3.7*
   parallel to the line $y = -x$
   $x + y = -3$

Draw the graph on a coordinate plane. (Exercises 18–19)

18. $5x - 5y - 15 = 0$ *3.5*
19. $4x - 2y > 6$ *3.9*
20. The voltage V of a given   *3.8*
   electric circuit varies directly
   as the current I. If V is 120
   volts when I is 9 amperes,
   what is the voltage V when
   the current I is 15 amperes?
21. Solve the system: *4.3*
   $5x - 2y = 20$
   $7x + 4y = 11$ $(3, -\frac{5}{2})$
22. Solve the system by graphing: *4.7*
   $x + y > -1$
   $3x - 2y > 4$

For Exercises 23–25, $f(x) = 2x - 7$ and $g(x) = x^2 - 4x - 2$. Find the following values.

23. $f(-4) - g(-2)$ −25   *3.2*
24. $f(g(-1))$ −1   *5.9*
25. $\dfrac{f(a) - f(b)}{a - b}$ 2   *7.1*
26. Simplify $\dfrac{4a^{-3}b^5}{8a^2b^{-2}c^{-3}}$ *5.2*
   using positive exponents. $\frac{b^7c^3}{2a^5}$

Factor completely. (Exercises 27–29)

27. $8x^3 - 1$ $(2x - 1)(4x^2 + 2x + 1)$ *5.6*
28. $x^4 - 2x^2 - 8$ *5.7*
29. $3xy + 6x - y - 2$
30. Use synthetic division and the   *6.6*
   Factor Theorem to solve
   $2x^3 - 12x^2 + 22x - 12 = 0$.
31. Find the domain of the   *7.1*
   function $f(x) = \dfrac{x - 7}{x^2 - 5x + 4}$.
   $\{x \mid x \ne 1 \text{ and } x \ne 4\}$

Simplify. (Exercises 32–41)

32. $3x(4x - 2)^2$ $48x^3 - 48x^2 + 12x$ *5.5*

33. $\dfrac{4a + 8}{5a - 20} \div \dfrac{10 + 3a - a^2}{a^2 - 4a}$ *7.2*

34. $\dfrac{5x}{x^2 - 9} - \dfrac{4}{x + 3} + \dfrac{2}{3 - x}$ *7.4*

35. $\dfrac{1 - \dfrac{6}{x} + \dfrac{5}{x^2}}{1 - \dfrac{3}{x} - \dfrac{10}{x^2}}$ $\dfrac{x - 1}{x + 2}$ *7.5*

36. $3y\sqrt{27y^{11}}, \; y \geq 0$ $9y^6\sqrt{3y}$ *8.2*

37. $7\sqrt{27} - 4\sqrt{3} - 8\sqrt{12}$ $\sqrt{3}$ *8.3*

38. $(2\sqrt{3} - \sqrt{2})(4\sqrt{3} + 3\sqrt{2})$ $18 + 2\sqrt{6}$

39. $\dfrac{28}{4 - \sqrt{2}}$ $8 + 2\sqrt{2}$ *8.4*

40. $27^{-\frac{2}{3}}$ $\frac{1}{9}$ *8.6*

41. $(16a^{16})^{\frac{3}{4}}$ $8a^{12}$ *8.7*

42. The second of three numbers is 3 times the first. The third number is 2 more than the second number. Seven less than twice the second is the same as 12 more than the third. Find the three numbers.   *1.6*

43. The length of a rectangle is 4 m longer than twice its width. The perimeter is 26 m. Find the length and the width.   *l:* 10 m, *w:* 3 m   *1.7*

44. Carol is 6 times as old as her nephew, Juan. Betty is 22 years younger than her aunt, Carol. In 4 years, Carol's age will be twice the sum of Juan's and Betty's ages now. How old is each person now?   *2.5*
Carol: 36 yr; Betty: 14 yr; Juan: 6 yr

45. If 3 less than a positive integer is divided by 2, the result is less than 1. Find the set of all such positive integers.   *2.4*
{1, 2, 3, 4}

46. A passenger train and a freight train start toward each other at the same time from stations 870 km apart. The passenger train travels at 80 km/h and the freight train at 65 km/h. In how many hours will they meet?   6 h   *2.6*

47. Bill has twice as many quarters as dimes. The total face value of the money is $9.00. How many coins are there of each type?   30 quarters, 15 dimes   *4.4*

48. Find three consecutive even integers such that the first integer times the second integer is 80.   *6.2*
8, 10, 12 and $-10, -8, -6$

49. A square and a rectangle have the same width. The length of the rectangle is 4 times its width. If the sum of the two areas is 45 cm^2, what is the area of each figure?   *6.3*
Square: 9 cm^2, rectangle: 36 cm^2

50. The denominator of a fraction is 3 more than twice the numerator. If the numerator is decreased by 1 and the denominator is increased by 1, the resulting fraction is equal to $\frac{1}{4}$. Find the original fraction. $\frac{4}{11}$ *7.6*

51. Work Crew *A* takes 15 h to do a job that Crew *B* can do in 10 h. If Crew *B* joins Crew *A* 3 h after Crew *A* has begun, what is the total time for the job?   7 h 48 min   *7.7*

9 COMPLEX NUMBERS

OVERVIEW

In this chapter, students are introduced to complex numbers and operations with complex numbers. Both an algebraic and a geometric view are presented. The algebraic approach emphasizes complex solutions to quadratic equations. The geometric view develops the fact that complex numbers are vectors. Lesson 9.8 focuses on complex numbers and their vector properties.

OBJECTIVES

- To perform operations with complex numbers
- To solve a quadratic equation for its complex roots
- To find the maximum value of a quadratic function
- To draw sums, differences, and scalar multiples of vectors
- To solve word problems using the quadratic formula

PROBLEM SOLVING

There are two lessons, 9.4 and 9.6, that are specifically devoted to problem solving based on quadratic relationships.

In addition to these lessons, an *Application* feature, Stopping Distance on page 312, enables students to practice the problem-solving strategy *Using a Formula.* A *Mixed Problem Solving* page appears immediately following the last lesson, enabling students to practice a variety of problem-solving methods.

READING AND WRITING MATH

This chapter includes one *Focus on Reading* (on page 318) and one question in which students are asked to explain a mathematical concept in their own words (Exercise 38 on page 309).

TECHNOLOGY

Calculator: Students may wish to use a calculator to find the roots of a quadratic equation by evaluating the Quadratic Formula of Lesson 9.5 for appropriate values of a, b, and c. They will find it best first to evaluate the quantity $b^2 - 4ac$ (this is the *discriminant*; see Lesson 10.2). Then they may use the square-root function followed in succession by the addition of $-b$ and the division of the entire numerator by $2a$.

Students that use calculators will be quickly alerted to those roots that have imaginary parts by the letter E (for "error") that appears on the display screen when they attempt to find the square root of a negative number.

SPECIAL FEATURES

Mixed Review pp. 306, 309, 312, 319, 322, 327, 330
Application: Stopping Distance p. 312
Midchapter Review p. 316
Focus on Reading p. 318
Extension: The Field of Complex Numbers pp. 331–332
Mixed Problem Solving p. 333
Key Terms p. 334
Key Ideas and Review Exercises pp. 334–335
Chapter 9 Test p. 336
College Prep Test p. 337

PLANNING GUIDE

Lesson	Basic	Average	Above Average	Resources
9.1 pp. 305–306	CE all WE 1–37 odd	CE all WE 9–45 odd	CE all WE 11–47 odd	Reteaching 60 Practice 60
9.2 pp. 308–309	CE all WE 1–33 odd	CE all WE 9–41 odd	CE all WE 15–47 odd	Reteaching 61 Practice 61
9.3 pp. 311–312	CE all WE 1–17 odd Application	CE all WE 7–23 odd Application	CE all WE 13–29 odd Application	Reteaching 62 Practice 62
9.4 pp. 315–316	CE all WE 1–11 odd Midchapter Review	CE all WE 3–13 odd Midchapter Review	CE all WE 7–17 odd Midchapter Review	Reteaching 63 Practice 63
9.5 pp. 318–319	FR all CE all WE 1–23 odd	FR all CE all WE 3–25 odd	FR all CE all WE 7–29 odd	Reteaching 64 Practice 64
9.6 p. 322	CE all WE 1–5	CE all WE 2–6	CE all WE 3–7	Reteaching 65 Practice 65
9.7 pp. 326–327	CE all WE 1–15 odd	CE all WE 7–21 odd	CE all WE 11–25 odd	Reteaching 66 Practice 66
9.8 pp. 330–333	CE all WE 1–9 odd Extension Mixed Problem Solving	CE all WE 3–11 odd Extension Mixed Problem Solving	CE all WE 7–15 odd Extension Mixed Problem Solving	Reteaching 67 Practice 67
Chapter 9 Review pp. 334–335	all	all	all	
Chapter 9 Test p. 336	all	all	all	
College Prep Test p. 337	all	all	all	

CE = Classroom Exercises   WE = Written Exercises   FR = Focus on Reading

NOTE: For each level, all students should be assigned all Mixed Review exercises.

■ INVESTIGATION

Project: This investigation is designed to extend the idea of *iteration* from the real number line (see Investigation, page 266) to the complex plane.

Materials Each student will need a copy of Project Worksheet 9. The students should work in groups.

The students will work with the following iteration:

Take a *complex* number
$z = a + bi$. Square it.
Let the result be the new number.
Then square the new number.

The object of this investigation is to lead students to the discovery that numbers that lie inside a circle of radius 1 centered at the origin go to zero under the iteration process, while numbers outside the circle get larger and larger. Numbers that lie on the circle *stay on the circle* under the iteration process.

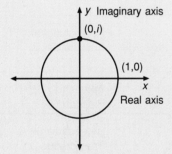

Each group should report its findings to the class, a part at a time. Students will need time to complete and discuss their calculations.

The unit circle of the complex plane is the simplest example of a **Julia set**, a type of set which is studied in connection with fractals. Another Julia set, which has been specially colored by a computer, appears on this page. Julia sets are the basis of the **Mandelbrot set** (pages xvi and 198).

Point out to students that the iteration on the number line (Investigation, page 266) is a *special case* of the iteration on the complex plane.

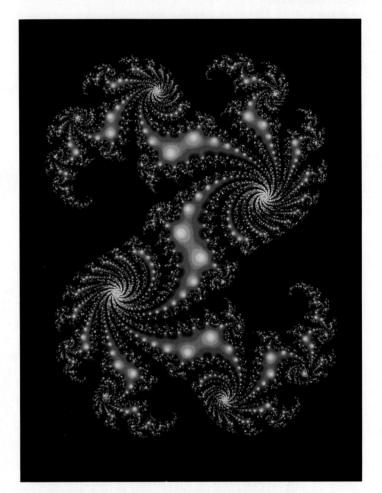

The image shown here is an example of a type of fractal known as a Julia set. Such images are produced by repeating a simple procedure over and over. With Julia sets, the process involves squaring *complex numbers*, numbers with both real and "imaginary" components.

More About the Image

Julia sets are produced by the iteration of $z_{n+1} = z_n^2 + c$ on the complex plane, where c is a constant. All points *inside* the set are attracted to some point or points inside the set as they are iterated indefinitely. All points *outside* the set increase without bound. The points that lie *on the boundary* of the set remain on the boundary during iteration. The value of c for the Julia set shown here is $0.32 + 0.043i$. For $c = 0$, the boundary of the set is the unit circle centered at the origin. (See Project Worksheet 9.)

9.1 Complex Numbers: Addition and Absolute Value

Teaching Resources

Quick Quizzes 60
Reteaching and Practice
   Worksheets 60
Transparency 20

Objectives

To solve equations of the form $x^2 = k$, where $k < 0$
To add and subtract complex numbers
To find the absolute value of complex numbers

The equation $x^2 = -1$ has no *real-number* roots because there is no real number whose square is -1. To solve this equation, mathematicians have defined the numbers i and $-i$ as follows:

$$i = \sqrt{-1} \qquad i^2 = -1 \qquad -i = -\sqrt{-1} \qquad (-i)^2 = -1$$

Having defined i as $\sqrt{-1}$, we can extend the idea to other radicals of the form $\sqrt{-k}$. First observe that $(2i)^2 = 2^2 \cdot i^2 = 4(-1) = -4$. Then

$$(2i)^2 = -4$$
$$2i = \sqrt{-4}$$

Also, $3i = \sqrt{-9}$, and so on.

In general, $\sqrt{-y} = i\sqrt{y}$ if $y > 0$, that is, if $-y < 0$.

EXAMPLE 1 Simplify each expression below.

  a. $-2 + \sqrt{-16}$ **b.** $6 - 4\sqrt{-5}$

Solutions

  $= -2 + i\sqrt{16}$ $= 6 - 4i\sqrt{5}$
  $= -2 + 4i$

Definition

> A **complex number** is a number that can be written in the **standard form** $a + bi$, where a and b are real numbers.

The numbers in Example 1 are *complex numbers*. If $a = 0$, then the complex number $a + bi$, or bi, is called an **imaginary number**. If $b = 0$, then the complex number $a + bi$, or a, is a *real number*.

Equations such as $x^2 + 16 = 0$ and $3y^2 = -24$ have roots in the set of complex numbers. Thus, the general solution of $x^2 = k$ (see Lesson 8.1) can be extended from $k \geq 0$ to include $k < 0$ as follows:

    If $x^2 = k$, then $x = \pm\sqrt{k}$, for all real numbers k.

9.1 Complex Numbers: Addition and Absolute Value   **303**

■■■■**GETTING STARTED**

Prerequisite Quiz

Describe each number. Is it rational? Irrational? Real?

1. $\frac{3}{4}$ Rational, real

2. $\sqrt{5}$ Irrational, real

3. 5.04   Rational, real

Solve.

4. $x^2 = 16$ ± 4

5. $y^2 = 3$ $\pm\sqrt{3}$

Simplify.

6. $5\sqrt{3} - \sqrt{27}$ $2\sqrt{3}$

7. $\sqrt{3^2 + 5^2}$ $\sqrt{34}$

Motivator

Have students attempt to find $\sqrt{-9}$ using scientific calculators:

$$\boxed{9}\ \boxed{+/-}\ \boxed{\sqrt{x}}.$$

Have students discuss reasons why the calculator gives an error message.
If $\sqrt{-9}$ exists, it is *not* a *real* number.

Additional Example 1

Simplify each expression.

a. $5 - \sqrt{-12}$ $5 - 2i\sqrt{3}$

b. $4\sqrt{-25}$ $20i$

c. $-3 + 2\sqrt{-10}$ $-3 + 2i\sqrt{10}$

Highlighting the Standards

Standard 4b: This lesson exemplifies a theme of the Standards that new concepts should be introduced as natural extensions of familiar concepts.

TEACHING SUGGESTIONS

Lesson Note

The symbol i is written before rather than after a radical, as in $i\sqrt{5}$ instead of $\sqrt{5}i$, to avoid confusion with $\sqrt{5i}$.

Beginning with this lesson, the replacement set when solving equations is extended from the real numbers to the complex numbers. Each complex number $a + bi$ is related to an ordered pair of real numbers (a,b). In Lesson 9.8, each complex number $a + bi$ will be paired with a unique *vector* (a,b) in standard form.

Math Connections

Electrical Impedance: Engineers find it convenient to express some properties of electric circuits in terms of complex numbers. For example, the formula for the complex impedance Z of a circuit is $Z = R + i(\Omega L - \frac{1}{\Omega C})$ where R, L, and C are the resistance, inductance, and capacitance of the circuit and Ω is 2π times f, the frequency of the alternating current (usually equal to 60).

Critical Thinking Questions

Analysis: Ask students to find a subset of the set of complex numbers that is *closed* (see page 8) under either addition or multiplication. The set $\{i, -i, 1, -1\}$ is closed under multiplication.

EXAMPLE 2 Solve each equation below.

a. $x^2 + 16 = 0$ **b.** $3y^2 = -24$

Solutions

$$x^2 = -16 \qquad\qquad y^2 = -8$$
$$x = \pm\sqrt{-16} = \pm 4i \qquad y = \pm\sqrt{-8} = \pm 2i\sqrt{2}$$

The roots are $4i$ and $-4i$. The roots are $2i\sqrt{2}$ and $-2i\sqrt{2}$.

The equation $(2x - 1)(2x + 1)(x^2 + 9) = 0$ has *four complex* roots. Two $(\pm\frac{1}{2})$ are real numbers obtained from $2x - 1 = 0$ and $2x + 1 = 0$. Two $(\pm 3i)$ are imaginary numbers obtained from $x^2 + 9 = 0$.

EXAMPLE 3 Solve $y^4 + 21y^2 - 100 = 0$.

Plan Factor completely. Then use the Zero-Product Property.

Solution
$$y^4 + 21y^2 - 100 = 0$$
$$(y^2 - 4)(y^2 + 25) = 0$$
$$(y + 2)(y - 2)(y^2 + 25) = 0$$
$$y + 2 = 0 \quad or \quad y - 2 = 0 \quad or \quad y^2 + 25 = 0$$
$$y = -2 \qquad\qquad y = 2 \qquad\qquad y^2 = -25 \text{ and } y = \pm 5i$$

The solutions are $-2, 2, 5i,$ and $-5i$.

In Example 3, the equation $y^4 + 21y^2 - 100 = 0$ is in **quadratic form** because, by substituting z for y^2, it is possible to obtain the equation $z^2 + 21z - 100 = 0$, which is a quadratic equation in z.

The eleven properties of a field listed in Lesson 1.3 also apply to the set of complex numbers. Therefore, you can add and subtract complex numbers just as you would real-number binomials.

EXAMPLE 4 Write in the form $a + bi$, where a and b are real numbers.

a. $(7 - 8i) - (5 - 3i)$ **b.** $(-7 + \sqrt{-3}) + (2 - \sqrt{-48})$

Solutions

$$= (7 - 8i) + (-5 + 3i) \qquad = (-7 + i\sqrt{3}) + (2 - 4i\sqrt{3})$$
$$= (7 - 5) + (-8 + 3)i \qquad = (-7 + 2) + (\sqrt{3} - 4\sqrt{3})i$$
$$= 2 - 5i \qquad\qquad = -5 - 3i\sqrt{3}$$

Each complex number $a + bi$ can be associated with a unique ordered pair of real numbers (a,b), which in turn can be associated with a unique point on a coordinate plane. In this coordinate plane, the x- and y-axes are called the **real axis** and the **imaginary axis**, respectively.

304 Chapter 9 Complex Numbers

Additional Example 2

a. Solve $x^2 + 49 = 0$ $\pm 7i$
b. Solve $3y^2 + 40 = y^2 + 4$ $\pm 3i\sqrt{2}$

Additional Example 3

Solve $y^4 - 8y^2 - 9 = 0$. $\pm 3, \pm i$

Additional Example 4

Write in the form $a + bi$, where a and b are real numbers.

a. $(-3 + 7i) - (-8 + 2i)$ $5 + 5i$

b. $(5 - \sqrt{-8}) + (-7 + \sqrt{-18})$ $-2 + i\sqrt{2}$

Complex number	Ordered pair	Point
$-5 + 3i$	$(-5,3)$	A
$-2i$	$(0,-2)$	B
-7	$(-7,0)$	C

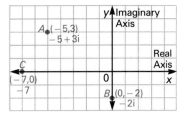

The distance PO between $P(a,b)$ and the origin is the length of the hypotenuse of the right triangle OPM (see diagram). Therefore, PO can be determined by the Pythagorean relation.

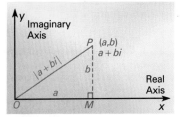

$$a^2 + b^2 = (PO)^2, \text{ or}$$
$$\sqrt{a^2 + b^2} = PO$$

The distance between $A(-5, 3)$ and the origin is $\sqrt{(-5)^2 + 3^2}$, or $\sqrt{34}$. This distance is called the *absolute value* of $-5 + 3i$ and is written as $|-5 + 3i| = \sqrt{34}$.

For the imaginary number $-2i$, $B(0,-2)$ is 2 units from the origin and $|0 - 2i| = \sqrt{0^2 + (-2)^2} = \sqrt{4} = 2$. For the real number -7, $C(-7,0)$ is 7 units from the origin and $|-7 + 0i| = \sqrt{(-7)^2 + 0^2} = \sqrt{49} = 7$, which confirms the real-number definition of $|-7|$ as 7. In general, the **absolute value of a complex number** is given by $|a + bi| = \sqrt{a^2 + b^2}$.

EXAMPLE 5 Find the absolute value of each complex number. Simplify the answer.

  a. $3\sqrt{2} - 2i$ **b.** $3i\sqrt{2}$ **c.** -6

Solutions

a. $|3\sqrt{2} + (-2)i|$ **b.** $|0 + (3\sqrt{2})i|$ **c.** $|-6 + 0i|$
$= \sqrt{(3\sqrt{2})^2 + (-2)^2}$ $= \sqrt{0^2 + (3\sqrt{2})^2}$ $= \sqrt{(-6)^2 + 0^2}$
$= \sqrt{18 + 4}$, or $\sqrt{22}$ $= \sqrt{18}$, or $3\sqrt{2}$ $= \sqrt{36}$, or 6

Classroom Exercises

Simplify.

1. $\sqrt{-4}$ $2i$ **2.** $\sqrt{-5}$ $i\sqrt{5}$ **3.** $(4 + 2i) - (2 - i)$ $2 + 3i$ **4.** $|3 + 4i|$ 5

Solve each equation.

5. $x^2 = -1$ $\pm i$ **6.** $y^2 = -4$ $\pm 2i$ **7.** $2n^2 = -10$ $\pm i\sqrt{5}$ **8.** $4x^2 + 36 = 0$ $\pm 3i$

9.1 Complex Numbers: Addition and Absolute Value **305**

Checkpoint

Simplify.

1. $-5 - 2\sqrt{-24}$ $-5 - 4i\sqrt{6}$
2. $(-3 - 4i) - (7 + 2i)$ $-10 - 6i$
3. $|-6 + 2i|$ $2\sqrt{10}$

Solve.

4. $6x^2 + 30 = 0$ $\pm i\sqrt{5}$
5. $y^4 - 5y^2 - 36 = 0$ $\pm 3, \pm 2i$

Closure

Have students define i and $-i$.
$i = \sqrt{-1}; -i = -\sqrt{-1}$ Ask students if the expression $3i$ is a complex number. Yes; It is also an imaginary number. Ask them if $-6 + 3i$ is a complex number. Yes
Ask students to give the formula for finding the absolute value of $3 + 4i$. $|a + bi| = \sqrt{a^2 + b^2}$; 5

◼◼◼FOLLOW UP

Guided Practice

Classroom Exercises 1–8

Independent Practice

Ⓐ Ex. 1–34, Ⓑ Ex. 35–46, Ⓒ Ex. 47,48

Basic: WE 1–37 odd
Average: WE 9–45 odd
Above Average: WE 11–47 odd

Additional Answers

Written Exercises

47. $|a + bi| = \sqrt{a^2 + b^2} = \sqrt{a^2 + (-b)^2} = |a - bi|$
48. $|a - bi| = \sqrt{a^2 + (-b)^2} = \sqrt{(-a)^2 + b^2} = |-a + bi|$

Additional Example 5

Find the absolute value of each complex number. Simplify the answer.

a. $7 + i\sqrt{2}$ $\sqrt{51}$
b. $-4i\sqrt{3}$ $4\sqrt{3}$
c. -0.5 0.5

305

Written Exercises

Write in simplest radical form using i.

1. $\sqrt{-64}$ $\quad 8i$ **2.** $\sqrt{-81}$ $\quad 9i$ **3.** $-\sqrt{-36}$ $\quad -6i$ **4.** $-\sqrt{-14}$ $\quad -i\sqrt{14}$

5. $\sqrt{-21}$ $\quad i\sqrt{21}$ **6.** $3\sqrt{-4}$ $\quad 6i$ **7.** $-\sqrt{-3}$ $\quad -i\sqrt{3}$ **8.** $-5\sqrt{-16}$ $\quad -20i$

9. $\sqrt{-50}$ $\quad 5i\sqrt{2}$ **10.** $-3\sqrt{-20}$ $\quad -6i\sqrt{5}$ **11.** $-4\sqrt{-40}$ $\quad -8i\sqrt{10}$ **12.** $10\sqrt{-80}$ $\quad 40i\sqrt{5}$

13. $8 - \sqrt{-36}$ **14.** $-18 - \sqrt{-30}$ **15.** $-7 + 3\sqrt{-50}$ **16.** $3 - \sqrt{-45}$
$\qquad 8 - 6i$ $\qquad\qquad -18 - i\sqrt{30}$ $\qquad\quad -7 + 15i\sqrt{2}$ $\qquad\quad 3 - 3i\sqrt{5}$

Solve each equation.

17. $x^2 = -25$ $\quad \pm 5i$ **18.** $2y^2 = -20$ $\quad \pm i\sqrt{10}$ **19.** $3a^2 + 150 = 0$ $\quad \pm 5i\sqrt{2}$

20. $a^4 + 7a^2 - 18 = 0$ **21.** $x^4 - 16 = 0$ $\quad \pm 2, \pm 2i$ **22.** $y^4 + 13y^2 + 36 = 0$
$\quad \pm\sqrt{2}, \pm 3i$ $\qquad\qquad\qquad\qquad\qquad\qquad\qquad\qquad\qquad\qquad \pm 2i, \pm 3i$

Write in the form $a + bi$, where a and b are real numbers.
$\qquad\qquad\qquad\qquad\qquad 8 + 8i \qquad\qquad\qquad\qquad 4 + 3i \qquad\qquad\qquad\qquad\qquad 4 + 4i$

23. $(2 + 5i) + (6 + 3i)$ **24.** $(6 + 7i) - (2 + 4i)$ **25.** $(-3 - 6i) + (7 + 10i)$

26. $(6 - 8i) - (-3 + 4i)$ **27.** $(-3 - i) - (5 - 2i)$ **28.** $(-10 + i) + (8 - 4i)$
$\quad 9 - 12i$ $\qquad\qquad\qquad\qquad -8 + i$ $\qquad\qquad\qquad\qquad -2 - 3i$

Find the absolute value of each complex number. Simplify the answer.

29. $-3 - 4i$ $\quad 5$ **30.** $6 - 2i$ $\quad 2\sqrt{10}$ **31.** $-4i$ $\quad 4$

32. $5\sqrt{2} + 3i$ $\quad \sqrt{59}$ **33.** $4i\sqrt{2}$ $\quad 4\sqrt{2}$ **34.** $-2 + 3i\sqrt{5}$ $\quad 7$

Solve each equation.
$\qquad\qquad\qquad\qquad \pm 4i \qquad\qquad\qquad\qquad\qquad \pm i\sqrt{3}$

35. $7t^2 + 20 = 4t^2 - 28$ **36.** $5x^2 + 21 = 3 - x^2$ **37.** $7y^2 + 12 = 8 - 2y^2$ $\quad \pm \frac{2}{3}i$

38. $y^4 = 7y^2 + 8$ $\quad \pm 2\sqrt{2}, \pm i$ **39.** $n^4 + 2n^2 = 15$ **40.** $x^4 + 10x^2 + 16 = 0$
$\qquad\qquad\qquad\qquad\qquad\qquad\qquad\qquad \pm\sqrt{3}, \pm i\sqrt{5} \qquad\qquad \pm i\sqrt{2}, \pm 2i\sqrt{2}$

Write in the form $a + bi$, where a and b are real numbers.

41. $15 - 10\sqrt{-12}$ $\quad 15 - 20i\sqrt{3}$ **42.** $(5 + \sqrt{-36}) + (-9 - \sqrt{-4})$ $\quad -4 + 4i$

Simplify.

43. $|6 + 8i| + |6 - 8i|$ $\quad 20$ **44.** $|-2 + 4i| + |-2 - 4i|$ $\quad 4\sqrt{5}$

45. $|5 - 3i| - |-5 + 3i|$ $\quad 0$ **46.** $|3 - i| - |3 + i|$ $\quad 0$

Prove that each theorem is true for all complex numbers.

47. $|a + bi| = |a - bi|$ **48.** $|a - bi| = |-a + bi|$

Mixed Review

Simplify. *8.3, 8.4*
$\qquad\qquad\qquad\qquad\qquad\qquad\qquad 18 - 5\sqrt{15} \qquad\qquad\qquad \frac{3\sqrt{2}}{7} \qquad\quad \frac{6\sqrt{3} + 2\sqrt{5}}{11}$

1. $2\sqrt{3}(3\sqrt{6} - 4\sqrt{3})$ **2.** $(3\sqrt{5} - 4\sqrt{3})(2\sqrt{5} + \sqrt{3})$ **3.** $\frac{6}{7\sqrt{2}}$ **4.** $\frac{4}{3\sqrt{3} - \sqrt{5}}$
$\quad 18\sqrt{2} - 24$

Enrichment

Discuss with students the different sets of numbers that they have learned about over the years. As you do, make a list on the chalkboard. The list should include the following.

N: Natural (counting) numbers
W: Whole numbers
I: Integers
Rt: Rational numbers

Ir: Irrational numbers
R: Real numbers
Im: Imaginary numbers
Cx: Complex numbers

Now have the students list all the sets that contain each of the following numbers.

1. 0   *W, I, Rt, R, Cx*
2. −7   *I, Rt, R, Cx*

3. $\sqrt{2}$ *Ir, R, Cx*
4. $9i$ *Im, Cx*
5. $2 + i$ *Cx*
6. $\sqrt[3]{7}$ *Ir, R, Cx*
7. $-1 - i$ *Cx*
8. $\frac{1}{2}$ *Rt, R, Cx*
9. 1   *N, W, I, Rt, R, Cx*
10. i *Im, Cx*

9.2 Products, Quotients, Conjugates

Teaching Resources

Project Worksheet 9
Quick Quizzes 61
Reteaching and Practice
 Worksheets 61

Objectives   To simplify products, powers, and quotients of complex numbers
To factor $x^2 + y^2$

The fact that i^2 is defined as -1 can be used to simplify the products of complex numbers. Begin by writing each factor in the form $a + bi$.

EXAMPLE 1   Simplify each product.

 a. $5i \cdot 4i$ b. $-\sqrt{-10} \cdot \sqrt{-5}$ c. $5\sqrt{6} \cdot 4\sqrt{-3}$

Solutions

a. $= 20i^2$
$= 20(-1)$
$= -20$

b. $= -i\sqrt{10} \cdot i\sqrt{5}$
$= -\sqrt{50} \cdot i^2$
$= -5\sqrt{2} \cdot (-1)$
$= 5\sqrt{2}$

c. $= 5\sqrt{6} \cdot 4i\sqrt{3}$
$= 20i\sqrt{18}$
$= 20i \cdot 3\sqrt{2} = 60i\sqrt{2}$

The real-number property for multiplying square-root radicals,

$$\sqrt{x} \cdot \sqrt{y} = \sqrt{x \cdot y}, \text{ where } x \geq 0 \text{ and } y \geq 0,$$

does *not* extend to the case where $x < 0$ and $y < 0$. As a counter-example, $\sqrt{-4} \cdot \sqrt{-9}$ is *not* equal to $\sqrt{-4(-9)}$ or $\sqrt{36}$, which is 6. Instead, $\sqrt{-4} \cdot \sqrt{-9} = 2i \cdot 3i = 6i^2 = 6(-1) = -6$, and in general, $\sqrt{-x} \cdot \sqrt{-y} = i\sqrt{x} \cdot i\sqrt{y} = i^2\sqrt{xy} = -\sqrt{xy}$, where $-x < 0$ and $-y < 0$.

Simplifying a quotient such as $\frac{-3}{8i}$ is similar to rationalizing the denominator in the real-number system, as shown below.

$$\frac{-3}{8i} = \frac{-3}{8i} \cdot \frac{i}{i} = \frac{-3i}{8i^2} = \frac{-3i}{-8} = \frac{3i}{8} \quad \begin{array}{l}\text{(no imaginary} \\ \text{number in the} \\ \text{denominator)}\end{array}$$

To simplify a product of the form $(a + bi)(c + di)$, use the FOIL method.

EXAMPLE 2   Simplify $(2 + 3i)(4 - 5i)$.

Solution
$$\begin{array}{cccc} \text{F} & \text{O} & \text{I} & \text{L} \end{array}$$
$$(2 + 3i)(4 - 5i) = 8 - 10i + 12i - 15i^2$$
$$= 8 + 2i - 15(-1) = 23 + 2i$$

Thus, $(2 + 3i)(4 - 5i) = 23 + 2i$.

GETTING STARTED

Prerequisite Quiz

Simplify.

1. $\sqrt{6} \cdot \sqrt{3}$ $3\sqrt{2}$
2. $\frac{-5}{3\sqrt{5}}$ $\frac{-\sqrt{5}}{3}$
3. $\sqrt{2}(3\sqrt{2} - 5\sqrt{6})$ $6 - 10\sqrt{3}$
4. $(8 + 2\sqrt{5})(8 - 2\sqrt{5})$ 44

Motivator

Ask students to recall the definition of i^2.
-1 Have them simplify the following product: $2i \cdot 2i$. -4 Ask students to give a real-number value to the expression i^4.   1

TEACHING SUGGESTIONS

Lesson Note

Dividing by a complex number is analogous to dividing by an irrational number and rationalizing the denominator, as shown in Example 3.

The concept of a pair of conjugates is extended from the irrationals to the complex numbers.

Factoring a binomial such as $4x^2 + 25$, a sum of squares, cannot be done with respect to the integers. However, if complex numbers are allowed, then $4x^2 + 25$ can be factored as shown in Example 4.

Additional Example 1

Simplify each product.

a. $-3i \cdot 6i$ 18
b. $\sqrt{-20} \cdot \sqrt{-2}$ $-2\sqrt{10}$
c. $-3\sqrt{5} \cdot 2\sqrt{-8}$ $-12i\sqrt{10}$

Additional Example 2

Simplify $(4 - 2i)(1 + 4i)$. $12 + 14i$

Highlighting the Standards

Standard 14d: Work with products, quotients, and conjugates is an important step toward the full development of operational skills with the complex number system.

Math Connections

Geometric Patterns: The powers of i discussed on page 308 have a simple geometric representation, namely, as a set of four points that are symmetrically positioned on the real and imaginary axes of a coordinate plane at a distance of one unit from the origin. Viewed in another way, the points are a graph of the four fourth roots of 1: i, $-i$, 1, and -1. In Lesson 19.10, this geometric viewpoint is generalized to cover the n different nth roots of a complex number $a + bi$.

Critical Thinking Questions

Analysis: After discussing the real-number property for multiplying square-root radicals that appears under Example 1, ask students to find the restrictions that must be placed on x and y for the Division Property $\sqrt{\dfrac{x}{y}} = \dfrac{\sqrt{x}}{\sqrt{y}}$ to be true.   Students may be surprised to discover that, except for the obvious restriction $y \neq 0$, the only condition on x and y that does not work is the conjunction "$x > 0$ and $y < 0$." All other combinations of inequalities are allowed.

Checkpoint

Simplify each product or power.

1. $(6 - 2i)^2$ $32 - 24i$
2. $(-2i)^5$ $-32i$
3. $(-2 + 4i)(\overline{-2 + 4i})$ 20

Simplify each quotient by rationalizing the denominator.

4. $-7 \div (-2i)$ $\dfrac{-7i}{2}$
5. $\dfrac{-5}{-2 + i}$ $2 + i$

Recall that $5 + \frac{1}{2}\sqrt{2}$ and $5 - \frac{1}{2}\sqrt{2}$ are a pair of *irrational* conjugates, and that their product $24\frac{1}{2}$ is a *rational* number. Similarly, $5 + \pi i$ and $5 - \pi i$ are *complex* conjugates, and their product $25 + \pi^2$ is a *real* number. In general, two complex numbers of the form $a + bi$ and $a - bi$ are **complex conjugates**. Their product, $(a + bi)(a - bi)$, is a real number, $a^2 + b^2$.

The conjugate of $a + bi$ is sometimes written as $\overline{a + bi}$. For example, $(5 + 3i)(\overline{5 + 3i}) = (5 + 3i)(5 - 3i) = 25 - 9i^2 = 34$.

A quotient with a complex-number divisor can be simplified by using a conjugate.

EXAMPLE 3   Simplify $\dfrac{8}{5 - 3i}$ by rationalizing the denominator.

Solution

$$\dfrac{8}{5 - 3i} \cdot \dfrac{5 + 3i}{5 + 3i} = \dfrac{8(5 + 3i)}{25 - 9i^2} = \dfrac{8(5 + 3i)}{34}$$

$$= \dfrac{4(5 + 3i)}{17}, \text{ or } \dfrac{20 + 12i}{17}$$

Powers of i can be simplified using (1) $i^2 = -1$ and (2) $i^4 = i^2 \cdot i^2 = -1(-1) = 1$, or $i^4 = 1$.

For example, $i^7 = i^4 \cdot i^2 \cdot i = 1(-1)i = -i$, and
$(2i)^{10} = 2^{10} \cdot i^{10} = 2^{10} \cdot (i^4)^2 \cdot i^2 = 1{,}024 \cdot 1^2 \cdot (-1) = -1{,}024$.

EXAMPLE 4   Factor $4x^2 + 25$. Use complex numbers.

Plan   Write 25 as $-(-25)$, or $-25i^2$.

Solution $4x^2 + 25 = 4x^2 - (-25) = 4x^2 - 25i^2 = (2x - 5i)(2x + 5i)$

Thus, $4x^2 + 25 = (2x - 5i)(2x + 5i)$.

Classroom Exercises

Determine whether each statement is true or false.

1. $\sqrt{-2} \cdot \sqrt{-8} = \sqrt{16}$ F   2. $\overline{5 - 2i} = -5 + 2i$ F   3. $i^{15} = i^3 = i^2 \cdot i = -i$ T

Simplify.

4. $-i\sqrt{8} \cdot i\sqrt{2}$ 4   5. $(1 - i)^2$ $-2i$ 6. $(6 - 2i) \div i$ $-2 - 6i$

Additional Example 3

Simplify $\dfrac{-7}{3 + 4i}$ by rationalizing the denominator. $\dfrac{-21 + 28i}{25}$

Additional Example 4

Factor $16x^2 + 9$. $(4x + 3i)(4x - 3i)$

Written Exercises

Simplify each product or power.

1. $4i \cdot 7i$ -28

2. $-i\sqrt{5} \cdot i\sqrt{2}$ $\sqrt{10}$

3. $3\sqrt{2} \cdot 2\sqrt{-3}$ $6i\sqrt{6}$

4. $i\sqrt{3} \cdot i\sqrt{6} \cdot i\sqrt{2}$ $-6i$

5. $\sqrt{-10} \cdot \sqrt{-2} \cdot \sqrt{-5}$

6. $-2\sqrt{-3} \cdot 5\sqrt{-12}$ 60

7. $\sqrt{-3}(2\sqrt{3} - \sqrt{-12})$

8. $2\sqrt{-9}(5 + \sqrt{-16})$

9. $3\sqrt{2}(5 - 2\sqrt{-24})$

10. $(3 + 4i)(6 + 5i)$

11. $(2 - 2i)(3 - 4i)$

12. $(5 - 6i)(-2 + i)$ $-4 + 17i$

13. $(4 + 3i)^2$ $7 + 24i$

14. $(5 - 2i)^2$ $21 - 20i$

15. $(-3 - i)^2$ $8 + 6i$

16. $(5 + 8i)(\overline{5 + 8i})$ 89

17. $(-3 - 5i)(\overline{-3 - 5i})$ 34

18. $(6 - 2i)(6 - 2i)$ 40

19. $(-5i)^3$ $125i$

20. $(i\sqrt{5})^4$ 25

21. $(2i)^5$ $32i$

Simplify each quotient by rationalizing the denominator.

22. $29 \div i$ $-29i$

23. $5 \div (4i)$ $-\dfrac{5i}{4}$

24. $-4 \div (3i)$ $\dfrac{4i}{3}$

25. $7 \div (-2i)$ $\dfrac{7i}{2}$

26. $\dfrac{3 - 5i}{4i}$ $\dfrac{5 + 3i}{-4}$

27. $\dfrac{2 + 5i}{-6i}$ $\dfrac{-5 + 2i}{6}$

28. $\dfrac{10}{4 - 3i}$ $\dfrac{8 + 6i}{5}$

29. $\dfrac{6i}{-1 - i}$ $-3 - 3i$

30. $\dfrac{1 - 5i}{2 + 4i}$ $\dfrac{-9 - 7i}{10}$

31. $\dfrac{1 + i}{1 - i}$ i

32. $\dfrac{1 - 2i}{-2 - 6i}$ $\dfrac{1 + i}{4}$

33. $\dfrac{4\sqrt{3}}{2\sqrt{3} + i\sqrt{3}}$ $\dfrac{8 - 4i}{5}$

Factor each binomial.

34. $x^2 + 16$ $(x - 4i)(x + 4i)$

35. $9y^2 + 4$ $(3y + 2i)(3y - 2i)$

36. $36x^2 + y^2$ $(6x + yi)(6x - yi)$

37. $5a^2 + 3$

38. Write in your own words how you would simplify an odd power of i greater than i^4.

Simplify each power.

39. $(-10i)^6$

40. i^{14} -1

41. $(4 - 2i\sqrt{3})^2$ $4 - 16i\sqrt{3}$

42. $(2i\sqrt{3} + i\sqrt{2})^2$ $-14 - 4\sqrt{6}$

43. $(2i)^{-3}$ $\dfrac{i}{8}$

44. $(3 + i)^{-1}$ $\dfrac{3 - i}{10}$

45. $i^{-2} + i^{-4}$ 0

46. $(1 - 2i)^{-2}$ $\dfrac{-3 + 4i}{25}$

In Exercises 47–50, w and z are defined as the complex numbers $a + bi$ and $c + di$, respectively. Show each of the following.

47. $\overline{w + z} = \overline{w} + \overline{z}$

48. $\overline{w - z} = \overline{w} - \overline{z}$

49. $\overline{wz} = \overline{w} \cdot \overline{z}$

50. $\overline{\overline{w}} = w$

Mixed Review 5.5, 5.6, 8.1

1. Simplify $(x - 6)^2$.
$x^2 - 12x + 36$

2. Factor $t^2 - 8t + 16$.
$(t - 4)^2$ $(t - 4)^2$

3. Solve $y^2 = 18$.
$\pm 3\sqrt{2}$ $\pm 3\sqrt{2}$

Closure

Have students give the conjugate of $a + bi$.
$a - bi$ or $\overline{a + bi}$ Ask students how they
rationalize the denominator of $\dfrac{4}{2 - 3i}$.

Multiply by $\dfrac{2 + 3i}{2 + 3i}$

■■■■FOLLOW UP

Guided Practice

Classroom Exercises 1–6

Independent Practice

A Ex. 1–29, **B** Ex. 30–46, **C** Ex. 47–50

Basic: WE 1–33 odd
Average: WE 9–41 odd
Above Average: WE 15–47 odd

Additional Answers

Written Exercises

5. $-10i$
7. $6 + 6i$
8. $-24 + 30i$
9. $15\sqrt{2} - 24i\sqrt{3}$
10. $-2 + 39i$
11. $-2 - 14i$
37. $(a\sqrt{5} + i\sqrt{3})(a\sqrt{5} - i\sqrt{3})$
38. Answers will vary. One possible answer: Factor out the greatest power of i that is a multiple of 4. Then the remainder will give the answer: $i^3 = -i$, $i^2 = -1$, and $i^1 = i$. A zero remainder means the answer is 1.

See page 337 for the answers to Ex. 39, 47–50.

Enrichment

Offer students this explanation of points located on the i axis. Multiplying numbers such as 1, 2, or 3 by -1 can be viewed as a 180° or $\dfrac{2}{4}$ counterclockwise rotation of the graphs of these numbers.

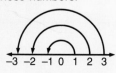

Since $i^2 = -1$, multiplication of a number by i can be viewed as a 90° or $\dfrac{1}{4}$ rotation. Thus, the graphs of i, $2i$, and $3i$ are points on the i axis.

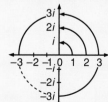

Multiplication by i^3 results in a $\dfrac{3}{4}$ rotation, producing the graphs of i, $-2i$, $-3i$. Thus, i^4 results in $\dfrac{4}{4}$ rotation, i^5 in a $\dfrac{5}{4}$ rotation, and so on.

Challenge the students to use this interpretation of powers of i to evaluate the following.

1. i^4 1

2. i^5 i

3. i^6 -1

4. i^7 $-i$

5. i^8 1

6. i^{11} $-i$

7. i^{16} 1

8. i^{42} -1

GETTING STARTED

Prerequisite Quiz

Solve.

1. $x^2 = 12$ $\pm 2\sqrt{3}$
2. $y^2 = -8$ $\pm 2i\sqrt{2}$

Write as the square of a binomial.

3. $y^2 - 10y + 25$ $(y - 5)^2$
4. $x^2 + 3x + \frac{9}{4}$ $(x + \frac{3}{2})^2$

Motivator

Ask students what needs to be added to the equation $x^2 + 4x + ? = 0$ to complete the square on the left side.   4   Have students divide the coefficient of the x-term ($4x$) by 2 and square the result. Have them record their result. Then have them perform the same operations on the x-terms of the equations $x^2 + 6x + 9$ and $x^2 + 8x + 16$. Ask them if they've discovered a pattern.

TEACHING SUGGESTIONS

Lesson Note

Most students should recall the method of *completing the square* from their work in first-year algebra. If necessary, use an example with real roots—for instance, $y^2 - 6y = 11$—before discussing Example 1.

Objective   To solve quadratic equations for their complex roots by completing the square

You have solved quadratic equations such as $y^2 = 28$ and $n^2 = -9$ by using the following property: If $x^2 = k$, then $x = \pm\sqrt{k}$.

If $y^2 = 28$, then $y = \pm\sqrt{28}$, or $y = \pm 2\sqrt{7}$; the two roots are real numbers.

If $n^2 = -9$, then $n = \pm\sqrt{-9}$, or $n = \pm 3i$; the two roots are imaginary numbers.

This property can be extended to equations of the form $(x - t)^2 = k$, where $x - t = \pm\sqrt{k}$, or $x = t \pm \sqrt{k}$. For example:

If $(y + 3)^2 = 8$,
then $y + 3 = \pm\sqrt{8} = \pm 2\sqrt{2}$,
or $y = -3 \pm 2\sqrt{2}$
The roots are $-3 + 2\sqrt{2}$ and $-3 - 2\sqrt{2}$.

If $(n - 2)^2 = -25$,
then $n - 2 = \pm\sqrt{-25} = \pm 5i$,
or $n = 2 \pm 5i$
The roots are $2 + 5i$ and $2 - 5i$.

The equation $x^2 - 5x = -7$ can be solved in a similar way after **completing the square** on the left side. Notice that the coefficient of x in $x^2 - 5x = -7$ is -5, and the square of $\frac{1}{2}$ of -5 is $-(\frac{5}{2})^2$, or $\frac{25}{4}$.

EXAMPLE 1   Solve $x^2 - 5x = -7$ by completing the square.

Plan   Add $(-\frac{5}{2})^2$, or $\frac{25}{4}$, to each side of the equation to complete the square.

Solution

$$x^2 - 5x = -7$$
$$x^2 - 5x + \frac{25}{4} = -7 + \frac{25}{4}$$
$$(x - \frac{5}{2})^2 = \frac{-3}{4}$$

Use the property:
If $y^2 = k$, then $y = \pm\sqrt{k}$.

$$x - \frac{5}{2} = \pm\sqrt{\frac{-3}{4}} = \pm\frac{\sqrt{-3}}{2}$$
$$x - \frac{5}{2} = \pm\frac{i\sqrt{3}}{2}$$
$$x = \frac{5}{2} \pm \frac{i\sqrt{3}}{2}, \text{ or } \frac{5 \pm i\sqrt{3}}{2}$$

The roots are $\dfrac{5 + i\sqrt{3}}{2}$ and $\dfrac{5 - i\sqrt{3}}{2}$.

Highlighting the Standards

Standard 3c: Completing the square, as introduced here, is an excellent way to expand the role of mathematical reasoning.

Additional Example 1

Solve $x^2 + 3x + 4 = 0$ by completing the square. $\dfrac{-3 \pm i\sqrt{7}}{2}$

Solving a Quadratic Equation by Completing the Square
In general, to solve a quadratic equation of the form $x^2 + bx = c$, divide b by 2, square the result, and add $(\frac{b}{2})^2$ to each side. Then factor and use the property: If $y^2 = k$, then $y = \pm\sqrt{k}$.

To solve $6y^2 + 7y - 5 = 0$ by completing the square, begin by writing the equation in the form $y^2 + by = c$, as shown in Example 2.

EXAMPLE 2   Solve $6y^2 + 7y - 5 = 0$ by completing the square.

Plan   Add 5 to each side; then divide each side by 6 to obtain the form $y^2 + by = c$.
Then $b = \frac{7}{6}$, so $\frac{b}{2} = \frac{7}{6} \cdot \frac{1}{2} = \frac{7}{12}$ and $(\frac{b}{2})^2 = (\frac{7}{12})^2 = \frac{49}{144}$.

Solution
$$6y^2 + 7y = 5$$
$$y^2 + \frac{7}{6}y = \frac{5}{6}$$
$$y^2 + \frac{7}{6}y + \frac{49}{144} = \frac{5}{6} + \frac{49}{144}$$
$$(y + \frac{7}{12})^2 = \frac{169}{144}$$
$$y + \frac{7}{12} = \pm\frac{13}{12}$$
$$y = \frac{-7 \pm 13}{12}$$
$$y = \frac{6}{12} \quad or \quad y = -\frac{20}{12}$$

The roots are $\frac{1}{2}$ and $-\frac{5}{3}$.

Classroom Exercises

Find the number to be added to each binomial to complete the square.

1. $a^2 + 12a$ 36
2. $n^2 - 4n$ 4
3. $x^2 + 3x$ $\frac{9}{4}$
4. $y^2 - \frac{5}{3}y$ $\frac{25}{36}$
5. $x^2 + 2x\sqrt{5}$ 5
6. $y^2 - 10y\sqrt{3}$ 75

Solve each equation.

7. $(x - 3)^2 = 25$ $-2, 8$
8. $y^2 + 4y = 3$ $-2 \pm \sqrt{7}$
9. $t^2 - 8t + 25 = 0$ $4 \pm 3i$

Math Connections
Geometric Models: Students might benefit by seeing a simple picture that geometrically represents the algebraic process of "completing the square." The figure below is an "incomplete square" of area $x^2 + 6x$ ($= x^2 + 3x + 3x$). The "missing" region is a square of dimensions 3 by 3. When this region is restored, a square of dimensions $x + 3$ by $x + 3$ is obtained. Its area is $x^2 + 6x + 9$.

	x	3
x	x^2	3x
3	3x	9

Critical Thinking Questions
Analysis: Ask students whether the technique of completing the square can be easily extended to that of "completing the cube" in order to solve a cubic equation such as $x^3 + 6x^2 - x = 7$. They will find that the process quickly becomes unmanageable since much more is involved than merely adding a properly chosen constant, which is all that is required when completing the square. Have them multiply out $(x + 1)^3$ and compare the result with $(x + 1)^2$. The complexity of $x^3 + 3x^2 + 3x + 1 (= (x + 1)^3)$ compared to $x^2 + 2x + 1 (= (x + 1)^2)$ should make it clear why "completing the cube" is an impractical technique.

Checkpoint
Solve by completing the square.
1. $x^2 + 5x = 3$ $\frac{-5 \pm \sqrt{37}}{2}$
2. $y^2 - 8y + 20 = 0$ $4 \pm 2i$
3. $2x^2 + 3x - 20 = 0$ $\frac{5}{2}, -4$
4. $y^2 - 4y\sqrt{2} - 10 = 0$ $-\sqrt{2}, 5\sqrt{2}$

Additional Example 2
Solve $6y^2 - 7y - 3 = 0$ by completing the square. $\frac{3}{2}, -\frac{1}{3}$

Closure

Have students give a formula for solving a quadratic equation by completing the square. See page 311. Have students solve $x^2 - 2\pi x = 3\pi^2$ by completing the square. $3\pi, -\pi$

See page 311.

◤◤◤FOLLOW UP

Guided Practice

Classroom Exercises 1–9

Independent Practice

Ⓐ Ex. 1–15, Ⓑ Ex. 16–24, Ⓒ Ex. 25–30

Basic: WE 1–17 odd, Application

Average: WE 7–23 odd, Application

Above Average: WE 13–29 odd, Application

Additional Answers

Written Exercises

4. $-5 \pm 2\sqrt{5}$
5. $4 \pm 3\sqrt{2}$
6. $-3 \pm 5\sqrt{2}$
10. $\dfrac{3 \pm \sqrt{33}}{2}$
11. $\dfrac{-7 \pm 3\sqrt{5}}{2}$
12. $\dfrac{-5 \pm \sqrt{17}}{2}$
13. $3 \pm 4i$
14. $-5 \pm 3i$
15. $-4 \pm 2i$
16. $2\sqrt{5}, 4\sqrt{5}$
17. $-\sqrt{2}, 5\sqrt{2}$
18. $-9\sqrt{6}, \sqrt{6}$
19. $\dfrac{-7 \pm \sqrt{33}}{2}$
20. $\dfrac{5 \pm 3\sqrt{5}}{2}$
22. $\dfrac{5 \pm \sqrt{65}}{4}$
23. $\dfrac{7 \pm \sqrt{29}}{10}$
24. $\dfrac{7 \pm i\sqrt{11}}{2}$
26. $2i, 8i$
28. $\dfrac{-3\sqrt{2} \pm 2\sqrt{6}}{2}$

Written Exercises

Solve each equation by completing the square.

1. $x^2 - 4x = 21$ $-3, 7$
2. $n^2 + 12n = -20$ $-10, -2$
3. $y^2 - 7y = -10$ $2, 5$
4. $a^2 + 10a + 5 = 0$
5. $c^2 - 8c - 2 = 0$
6. $x^2 + 6x - 41 = 0$
7. $2x^2 + 5x = 3$ $-3, \frac{1}{2}$
8. $3y^2 - 8y + 4 = 0$ $\frac{2}{3}, 2$
9. $6x^2 + 5x + 1 = 0$ $-\frac{1}{3}, -\frac{1}{2}$
10. $y^2 - 3y - 6 = 0$
11. $t^2 + 7t = -1$
12. $a^2 + 5a + 2 = 0$
13. $x^2 - 6x + 25 = 0$
14. $a^2 + 10a + 34 = 0$
15. $x^2 + 8x + 20 = 0$
16. $y^2 - 6y\sqrt{5} + 40 = 0$
17. $n^2 - 4n\sqrt{2} = 10$
18. $a^2 + 8a\sqrt{6} = 54$
19. $x^2 + 7x + 4 = 0$
20. $y^2 - 5y - 5 = 0$
21. $16x^2 + 24x = -9$ $-\frac{3}{4}$
22. $2y^2 = 5y + 5$
23. $7x = 5x^2 + 1$
24. $y^2 = 7y - 15$
25. $z^2 - 4iz = -12$ $-2i, 6i$
26. $a^2 - 10ia - 16 = 0$
27. $x^2 + 6ix - 5 = 0$ $-5i, -i$
28. $2x^2 + 6x\sqrt{2} = 3$
29. $a^2 - 4a\sqrt{2} = -8$ $2\sqrt{2}$
30. $n^2 + 6n\sqrt{5} = -45$ $-3\sqrt{5}$

Mixed Review

Simplify. Use positive exponents. 5.1, 5.2

1. $5x^2(-2x^3y)^3$ $-40x^{11}y^3$
2. $5x^{-2}y \cdot (-3)x^{-4}y^5 \cdot xy^{-1}$ $-\dfrac{15y^5}{x^5}$
3. $\dfrac{-10a^{-3}b^4c^2}{15a^{-5}b^{-1}c^{-3}}$ $-\dfrac{2a^2b^5c^5}{3}$
4. $(5m^{-2}n^4)^{-3}$ $\dfrac{m^6}{125n^{12}}$

◤◤◤ Application: Stopping Distance

The distance d (in ft) that a car needs to stop on a particular road surface is given by the formula $d = \dfrac{s^2}{30F}$, where s is the car's speed (in mi/h) and F is the friction coefficient of the road.

Friction Coefficients		
	Concrete	**Tar**
Wet	0.4	0.5
Dry	0.8	1.0

Police often measure skid marks to find the speed at which the car that made them was traveling. Use the friction coefficients above to find the speed, to the nearest tenth, at which the car that made each of these marks was traveling. (HINT: First solve the formula for s.)

1. 132 ft on a dry tar road 62.9 mi/h
2. 280 ft on a wet concrete road 58.0 mi/h

Enrichment

Challenge the students to use the method of completing the square to solve these quadratic equations with imaginary coefficients.

1. $x^2 + 6ix + 1 = 0$ $-3i \pm i\sqrt{10}$
   or $i(-3 \pm \sqrt{10})$
2. $x^2 - 4ix + i = 0$ $2i \pm i\sqrt{4 + i}$

Have students check their solutions. This will provide valuable practice in operating with the imaginary unit, i.

9.4 Problem Solving: Maximum Values

Objectives

To find the maximum value of given quadratic expressions
To solve word problems involving maximum value

The quadratic polynomial $30t - 5t^2$ represents the height h in meters at the end of t seconds of an object sent upward with an initial velocity of 30 m/s. When $h = 30t - 5t^2$, the value of h increases from 0 to some *maximum* value and then decreases again to 0 as the value of t increases from 0 to 6.

t	0	2	3	4	6
$30t - 5t^2$	$0 - 0$	$60 - 20$	$90 - 45$	$120 - 80$	$180 - 180$
h	0	40	45	40	0

It seems from the table that the maximum height is 45 m, and that this occurs at the end of 3 s. The technique of completing the square can be used to prove that this is true.

Let $h = 30t - 5t^2$.

Then $h = -5t^2 + 30t$.

Factor out -5, the coefficient of t^2.

$$h = -5(t^2 - 6t)$$

Complete the square *inside* the parentheses by adding $(\frac{-6}{2})^2$, or 9. Add $-5 \cdot 9$ to the left side to preserve the equality.

$$h - 5 \cdot 9 = -5(t^2 - 6t + 9)$$
$$h - 45 = -5(t - 3)^2$$

Now, either $t = 3$ or $t \neq 3$. Consider these two cases.

If $t = 3$, then: $\quad t - 3 = 0$ $\qquad$ If $t \neq 3$, then: $\quad t - 3 \neq 0$
$$(t - 3)^2 = 0 \qquad\qquad\qquad\qquad (t - 3)^2 > 0$$
$$-5(t - 3)^2 = 0 \qquad\qquad\qquad\quad -5(t - 3)^2 < 0$$

Substitute. $\quad h - 45 = 0$ $\qquad$ Substitute. $\quad h - 45 < 0$
$$h = 45 \qquad\qquad\qquad\qquad\qquad h < 45$$

Thus, if $h - 45 = -5(t - 3)^2$, then $h = 45$ when $t = 3$, and $h < 45$ when $t \neq 3$. The maximum height h is 45 m, and this occurs when the time t is 3 s. In general:

If $y - y_1 = a(x - x_1)^2$ and $a < 0$, then the maximum value of y is y_1 and this occurs when $x = x_1$.

Teaching Resources

Quick Quizzes 63
Reteaching and Practice Worksheets 63

▬ GETTING STARTED

Prerequisite Quiz

Factor out the coefficient of x^2.

1. $-7x^2 + 28x$ $\quad -7(x^2 - 4x)$

2. $70x - 5x^2$ $\quad -5(x^2 - 14x)$

Complete the square and write the result as the square of a binomial.

3. $x^2 - 18x$ $\quad x^2 - 18x + 81; (x - 9)^2$

4. $x^2 - 40x$ $\quad x^2 - 40x + 400; (x - 20)^2$

Find the value of P when $x = 4$.

5. $P - 80 = -10(x - 4)^2$ $\quad 80$

6. $P - 200 - 10 \cdot 16 = -10(x - 4)^2$ $\quad 360$

Motivator

Have students find the maximum value of h and the corresponding value of t given that $h = 20t - 5t^2$ by substituting 0, 1, 2, 3, and 4 for t to find values of h. Maximum value is 20 at $t = 2$.

Highlighting the Standards

Standard 1b: This lesson follows the recommendation that problems and applications be used to introduce new mathematical content.

Lesson Note

The skill of completing the square, introduced in the previous lesson, is applied to solving word problems that involve finding maximum values.

The table of increases $(0, 1, 2, 3, \cdots, x)$ in Example 2 helps the student write the product when the number of increases is represented by x.

Math Connections

Minimum Values: Problems can be devised that involve finding a *minimum* value using the techniques of this lesson. For example, if a 40-cm wire is to be cut into two pieces and the pieces are each to be bent into squares, what is the minimum value of the sum of the areas of the two squares?   50 cm²

Critical Thinking Questions

Analysis: The problems of this lesson involve quadratic polynomials, all of which assume either a maximum value or, in Exercises 14–18, a minimum value for some value of x. Ask students how they can tell whether a maximum or a minimum is involved.   As a hint, give them the polynomial for the problem of *Math Connections* above $P = 2x^2 - 80x + 1600$. They should see that if the coefficient of x^2 is positive, the polynomial will have a minimum; if the coefficient is negative, the polynomial will have a maximum.

EXAMPLE 1 Find the maximum value of P and the corresponding value of x given that $P = 500 + 80x - 10x^2$.

Solution

Subtract the constant term, 500. $P - 500 = -10x^2 + 80x$

Factor out -10, the $P - 500 = -10(x^2 - 8x + \underline{\ ?\ })$
coefficient of x^2.

Complete the square and $P - 500 - 10 \cdot 16 = -10(x^2 - 8x + 16)$
preserve the equality. $P - 660 = -10(x - 4)^2$

The maximum value is 660, and it occurs when $x = 4$.

One objective in the business world is to sell your product or service at a rate that will yield the maximum income.

Suppose you can sell 80 items at $6 each, and for every 50¢ decrease in price, you can sell 10 more items. Sales will keep increasing, but there is a "best price" that will yield the maximum income, as shown in the table at the right. Two 50¢ decreases from $6, or $5, is that best price, and the maximum income earned is $500 from selling $80 + 20$, or 100 items.

No. of 50¢ decreases	No. of items	× Item Price	= Income
0	80	$6	$480
1	80 + 10	6 − 0.50	495
2	80 + 20	6 − 1.00	500
3	80 + 30	6 − 1.50	495
4	80 + 40	6 − 2.00	480

EXAMPLE 2 If tickets cost $4 each, 800 people will attend a certain concert. For each 25¢ increase in the ticket price, attendance will decrease by 20 people. What ticket price will yield the maximum income? What is the maximum income from the concert? How many tickets need to be sold to yield the maximum income?

Solution

No. of 25¢ increases	No. of tickets	×	Ticket price
0	800	·	$4
1	$(800 - 20 \cdot 1)$	·	$[4 + 0.25(1)]$
2	$(800 - 20 \cdot 2)$	·	$[4 + 0.25(2)]$
3	$(800 - 20 \cdot 3)$	·	$[4 + 0.25(3)]$
.	.		.
.	.		.
.	.		.
x	$(800 - 20x)$	·	$(4 + 0.25x) =$ income I

for x 25¢ increases

Additional Example 1

Find the maximum value of P and the corresponding value of x if $P = 600 + 60x - 5x^2$. $P = 780$ when $x = 6$

Additional Example 2

A publisher will sell 1,000 copies of a magazine if the price is $6 per copy. Research shows that an additional 50 copies will be sold for each 10-cent decrease in price. For the maximum income I, find (a) the number of 10-cent decreases, (b) the price of one copy, (c) the maximum income, and (d) the number of copies sold.
(a) 20, (b) $4, (c) $8,000, (d) 2,000

Now,
$$I = (800 - 20x)(4 + 0.25x)$$
$$I = -5x^2 + 120x + 3{,}200$$
$$I - 3{,}200 = -5x^2 + 120x$$
$$I - 3{,}200 = -5(x^2 - 24x + \underline{\ ?\ })$$
$$I - 3{,}200 - 5 \cdot 144 = -5(x^2 - 24x + 144)$$
$$I - 3{,}920 = -5(x - 12)^2$$
So, $I = 3{,}920$ when $x = 12$.

The maximum income I occurs when the number x of 25¢ increases is 12. The best ticket price, $(4 + 0.25x)$ dollars, is $4 + 0.25(12)$, or $7. The maximum income of $3,920 occurs when the number of tickets sold, $(800 - 20x)$, is $800 - 20 \cdot 12$, or 560.

Classroom Exercises

Find the maximum value of P and the corresponding value of x.

1. $P - 18 = -2(x - 3)^2$ **2.** $P - 16 - 34 = -1(x - 4)^2$

3. $P - 2{,}000 - 20 \cdot 25 = -20(x - 5)^2$ **4.** $P + 10 - 4 \cdot 4 = -4(x - 2)^2$
$P_{max} = 2{,}500,\ x = 5$ $P_{max} = 6,\ x = 2$

Fill in the blanks to complete the square in parentheses and to preserve the equality.

5. $P - \underline{\ ?\ } = -4(x^2 - 6x + \underline{\ ?\ })$ 36, 9   **6.** $A - \underline{\ ?\ } = -3(w^2 - 8w + \underline{\ ?\ })$ 48, 16

7. $y - \underline{\ ?\ } = -(x^2 - 4x + \underline{\ ?\ })$ 4, 4   **8.** $I - \underline{\ ?\ } = -20(t^2 - 10t + \underline{\ ?\ })$ 500, 25

Written Exercises

Find the maximum value of each polynomial P and the corresponding value of x.

1. $P = -2x^2 + 20x$ **2.** $P = 64x - 16x^2$ **3.** $P = 25 + 6x - x^2$

4. $P = 50 + 12x - 3x^2$ **5.** $P = 675 + 200x - 20x^2$ **6.** $P = 750 + 500x - 25x^2$

7. An arrow is shot upward at 40 m/s. Find its maximum height and the corresponding time. Use the formula $h = vt - 5t^2$. $h_{max} = 80$ m, $t = 4$ s

8. If there are 20 passengers, a chartered bus ride costs $30 per ticket. The ticket price is reduced $1 for each additional passenger beyond 20. How many passengers will produce the maximum income? What is the maximum income? What is the best ticket price? 25 passengers; $625; $25

9. A second number is 12 decreased by a first number. Among all such pairs of numbers, find the pair with the greatest product. 6,6

10. One number is 24 decreased by 3 times another number. Among all such pairs, find the pair with the greatest product. 4,12

Checkpoint

1. Find the maximum value of h and the corresponding value of t if $h = 50 + 160t - 16t^2$.
$h = 450$ when $t = 5$

2. A second number is 16 decreased by twice the first number. Among all such pairs of numbers, find the pair with the greatest product. 4, 8

3. If $I = (100 + 10x)(4 - 0.2x)$, find the maximum value of I and the corresponding values of x, $4 - 0.2x$, and $100 + 10x$.
$I = 450$; $x = 5$; $4 - 0.2x = 3$; $100 + 10x = 150$

Closure

Have students find the maximum value of P and the corresponding value of x given that $P = 2(x - 4)^2 + 300$. No maximum value can be found. Ask students if they can find a minimum value, and if so, have them give the minimum value and the value of x.
300; $x = 4$

▮▮▮▮FOLLOW UP

Guided Practice

Classroom Exercises 1–8

Independent Practice

Ⓐ Ex. 1–10, Ⓑ Ex. 11–13, Ⓒ Ex. 14–18

Basic: WE 1–11 odd, Midchapter Review

Average: WE 3–13 odd, Midchapter Review

Above Average: WE 7–17 odd, Midchapter Review

Enrichment

Explain that a shortcut for finding the maximum (or minimum) value of a quadratic function can be borrowed from calculus, an advanced course.

At every point on the graph of $y = ax^2 + bx + c$, the slope of the tangent to the curve is given by $2ax + b$. At a maximum (or minimum) point of the curve, the slope is 0.

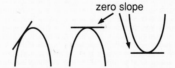
zero slope

To find the value of x at which the maximum (or minimum) point occurs, set $2ax + b$ equal to 0. For the function of Example 1, $P = -10x^2 + 80x + 500$, in which $a = -10$ and $b = 80$. Therefore, $2ax + b =$

$-20x + 80$. Set $-20x + 80$ equal to 0.

$-20x + 80 = 0$; $x = \dfrac{-80}{-20} = 4$

Since $P(4) = 660$, the maximum value of the function is 660.

Have students try this method with some of the Written Exercises.

11. A rectangular field next to the straight bank of a river is to be fenced in with no fencing along the riverbank. If 240 yd of fencing are available, what is the maximum area that can be enclosed? What will the dimensions of the field be? 7,200 yd²; 60 yd by 120 yd

12. Six hundred people will buy tickets at $4 each for a district championship game. For each 20¢ decrease in price, 50 more people will buy tickets. Find the best ticket price, the maximum income, and the corresponding number of tickets. $3.20; $2,560; 800 tickets

13. An orange grove has 20 trees per acre, and the average yield is 300 oranges per tree. For each additional tree per acre, the average yield will be reduced by 10 oranges per tree. How many trees per acre will yield the maximum number of oranges per acre? What is the maximum number of oranges per acre? 25 trees per acre; 6,250 oranges per acre

14. Prove: If $y - y_1 = a(x - x_1)^2$ and $a > 0$, then the *minimum* value of y is y_1, and this occurs when $x = x_1$. (HINT: Either $x = x_1$ or $x \neq x_1$.)

Find the minimum value of P and the corresponding value of x.

15. $P = 10x + x^2$ $P_{min} = -25$, $x = -5$
16. $P = 3x^2 - 18x$ $P_{min} = -27$, $x = 3$
17. $P = 5x^2 + 40x - 20$ $P_{min} = -100$, $x = -4$
18. One number is 12 more than twice another number. Among all such pairs of numbers, find the pair with the minimum product. What is the product? −3,6; −18

Midchapter Review

Simplify. *9.1, 9.2*

1. $5 - 2\sqrt{-18}$ $5 - 6i\sqrt{2}$
2. $(8 - 2i) - (-3 + 4i)$ $11 - 6i$
3. $4\sqrt{-6} \cdot \sqrt{-3}$ $-12\sqrt{2}$
4. $(-3i)^3$ $27i$
5. $(3 - 4i)(2 + 5i)$ $26 + 7i$
6. $\dfrac{-7}{3 + 5i}$ $\dfrac{-21 + 35i}{34}$

7. Write $9x^2 + 4$ as the difference of two squares and factor the result. *9.2* $(3x)^2 - (2i)^2 = (3x + 2i)(3x - 2i)$

Solve by completing the square. *9.3*

8. $x^2 + 10x = 11$ $-11, 1$
9. $x^2 - 4x = 4$ $2 \pm 2\sqrt{2}$
10. $y^2 + 6y + 14 = 0$ $-3 \pm i\sqrt{5}$
11. Find the maximum value of P and the corresponding value of x given that $P = 100 + 48x - 6x^2$. *9.4* $P_{max} = 196$, $x = 4$

9.5 The Quadratic Formula

Objectives
To solve quadratic equations for their complex roots using the quadratic formula
To find the three cube roots of nonzero integers that are perfect cubes

Every quadratic equation can be written in the form $ax^2 + bx + c = 0$, where $a > 0$. This general equation can be solved by completing the square. Begin by adding $-c$ to each side of the equation.

$$ax^2 + bx = -c$$

Divide by a and complete the square.

$$x^2 + \frac{b}{a}x + \left(\frac{b}{2a}\right)^2 = \left(\frac{b}{2a}\right)^2 + \frac{-c}{a}$$

$$x^2 + \frac{b}{a}x + \left(\frac{b}{2a}\right)^2 = \frac{b^2}{4a^2} + \frac{-c}{a} \cdot \frac{4a}{4a}$$

Factor the left side.

$$\left(x + \frac{b}{2a}\right)^2 = \frac{b^2 - 4ac}{4a^2}$$

Use: If $y = k^2$, then $y = \pm \sqrt{k}$.

$$x + \frac{b}{2a} = \frac{\pm\sqrt{b^2 - 4ac}}{2a}$$

$$x = \frac{-b \pm \sqrt{b^2 - 4ac}}{2a}$$

The Quadratic Formula
The solutions of a quadratic equation of the form $ax^2 + bx + c = 0$, where $a > 0$, are given by the formula

$$x = \frac{-b \pm \sqrt{b^2 - 4ac}}{2a}.$$

A polynomial equation with coefficients that are integers is an **integral polynomial equation**. The roots of a second-degree integral polynomial equation can be a pair of rational numbers, a pair of irrational conjugates, or, as in Example 1 below, a pair of complex conjugates.

EXAMPLE 1   Solve $3x^2 + 4x = -2$ using the quadratic formula.

Solution   Write the general form. $3x^2 + 4x + 2 = 0 \leftarrow a = 3, b = 4, c = 2$

Write the quadratic formula. $x = \frac{-b \pm \sqrt{b^2 - 4ac}}{2a}$

9.5 The Quadratic Formula   **317**

Additional Example 1

Solve $3x^2 = 2 - 6x$ using the quadratic formula. $\frac{-3 \pm \sqrt{15}}{3}$

Teaching Resources

Quick Quizzes 64
Reteaching and Practice Worksheets 64

▰▰▰GETTING STARTED

Prerequisite Quiz

Simplify, if possible.

1. $\frac{6x + 8}{4}$ $\frac{3x + 4}{2}$
2. $\sqrt{16 - 8}$ $2\sqrt{2}$
3. $\sqrt{6 - 18}$ $2i\sqrt{3}$
4. $\sqrt{2 - 4 \cdot 3(-2)}$ $\sqrt{26}$
5. $\frac{-4 - \sqrt{12}}{2}$ $-2 - \sqrt{3}$

Motivator

Have students solve for x in $x^2 + 6x + 5 = 0$. $x = -5, x = -1$ Write the following formula on the chalkboard.

$x = \frac{-b \pm \sqrt{b^2 - 4ac}}{2a}$

Have students use $a = 1$, $b = 6$, and $c = 5$ to solve for x. $x = -5, x = -1$ Have students compare their answers to that of the trinomial above and discuss their findings.

▰▰▰TEACHING SUGGESTIONS

Lesson Note

To solve a quadratic equation, students should write the equation in the form $ax^2 + bx + c = 0$, where $a > 0$. If $ax^2 + bx + c$ is not easily factored, they should identify the values of a, b, and c and use the quadratic formula.

Highlighting the Standards

Standard 2d: The Focus on Reading for this lesson provides a good exercise for students to test their ability to read and understand mathematics.

Math Connections

Solvability of Polynomial Equations:
Students may be interested in knowing that formulas solving third- and fourth-degree equations do exist. However, they are too complicated to be of practical use. No such formulas exist for polynomial equations of higher degree.

Critical Thinking Questions

Analysis: Ask students whether the nonexistence of a formula for finding the roots of a polynomial equation of degree five or more (see *Math Connections* above) means that such an equation has no roots.
A polynomial equation of degree five does have roots; however, there is no assured method of finding the roots.

Common Error Analysis

Error: Students may simplify $\dfrac{4 \pm 6\sqrt{3}}{4}$ as $\pm 6\sqrt{3}$.

Remind students that they can *factor* the numerator of such a fraction before dividing out a common factor, as shown below.

$$\frac{4 \pm 6\sqrt{3}}{4} = \frac{2(2 \pm 3\sqrt{3})}{2 \cdot 2} = \frac{2 \pm 3\sqrt{3}}{2}$$

Checkpoint

1. Use the quadratic formula to solve $4y^2 + 2y + 1 = 0$. $\dfrac{-1 \pm i\sqrt{3}}{4}$

2. Write an equation, without fractions, equivalent to $\frac{1}{6}x^2 = \frac{1}{2}x + \frac{1}{3}$. Solve.
$x^2 - 3x - 2 = 0$, $\dfrac{3 \pm \sqrt{17}}{2}$

3. Find the three cube roots of -27.
$-3, \dfrac{3 + 3i\sqrt{3}}{2}, \dfrac{3 - 3i\sqrt{3}}{2}$

Substitute. $x = \dfrac{-4 \pm \sqrt{16 - 4 \cdot 3 \cdot 2}}{2 \cdot 3}$

$= \dfrac{-4 \pm \sqrt{-8}}{6} = \dfrac{-4 \pm 2i\sqrt{2}}{6}$, or $\dfrac{-2 \pm i\sqrt{2}}{3}$

The roots are $\dfrac{-2 + i\sqrt{2}}{3}$ and $\dfrac{-2 - i\sqrt{2}}{3}$.

EXAMPLE 2 Find the three cube roots of -8.

Plan Solve a third-degree equation. The 3 cube roots of -8 are the 3 roots of the equation $x^3 = -8$. Solve $x^3 = -8$.

First rewrite the equation as $x^3 + 8 = 0$. Then write $x^3 + 8$ as the sum of two cubes, $x^3 + 2^3$. Use $a^3 + b^3 = (a + b)(a^2 - ab + b^2)$ to factor.

Solution
$$x^3 + 2^3 = 0$$
$$(x + 2)(x^2 - 2x + 4) = 0$$

Use the Zero-Product Property to solve the last equation.

$x + 2 = 0$ *or* $x^2 - 2x + 4 = 0$

$x = -2$ $x = \dfrac{2 \pm \sqrt{4 - 16}}{2} = \dfrac{2 \pm \sqrt{-12}}{2}$

$= \dfrac{2 \pm 2i\sqrt{3}}{2}$, or $1 \pm i\sqrt{3}$

The three cube roots of -8 are -2, $1 + i\sqrt{3}$, and $1 - i\sqrt{3}$.

▰ Focus on Reading

Place the following steps in the best sequence for solving a quadratic equation using the quadratic formula. e, c, f, b, a, d

a. Simplify the radical, if possible.

b. Simplify the radicand.

c. Identify the values of a, b, and c.

d. Simplify the quotient, if possible.

e. Write the equation in the form $ax^2 + bx + c = 0$, where $a > 0$.

f. Substitute for a, b, and c in the quadratic formula.

318 Chapter 9 Complex Numbers

Classroom Exercises

Identify a, b, and c in the general form of the quadratic equation.

1. $5x^2 + 3x - 2 = 0$ $a = 5, b = 3, c = -2$
2. $0 = x^2 - 4x + 3$ $a = 1, b = -4, c = 3$
3. $x^2 - x - 4 = 0$ $a = 1, b = -1, c = 4$

Simplify.

4–6. Solve the equations of Classroom Exercises 1–3 using the quadratic formula.   4. $-1, \frac{2}{5}$ 5. 1, 3   6. $\frac{1 \pm \sqrt{17}}{2}$

Written Exercises

Solve each equation using the quadratic formula.

1. $x^2 - 6x + 8 = 0$ 2, 4
2. $n^2 + 3n - 10 = 0$ $-5, 2$
3. $y^2 - 5y - 2 = 0$ $\frac{5 \pm \sqrt{33}}{2}$
4. $5x^2 + x - 2 = 0$
5. $2n^2 - n - 1 = 0$ $-\frac{1}{2}, 1$
6. $2y^2 - 5y + 4 = 0$
7. $4y^2 = 2y - 1$
8. $2x^2 = 6x - 3$
9. $4t^2 + 13 = 12t$
10. $3n^2 + 2 = 2n$ $\frac{1 \pm i\sqrt{5}}{3}$
11. $8y^2 = 4y - 5$ $\frac{1 \pm 3i}{4}$
12. $5x^2 = 1 - 2x$ $\frac{-1 \pm \sqrt{6}}{5}$
13. $t^2 + 4t = 1$ $-2 \pm \sqrt{5}$
14. $y^2 + 2y + 2 = 0$ $-1 \pm i$
15. $n^2 = 6n + 11$ $3 \pm 2\sqrt{5}$

Find the three cube roots of each number. Simplify the results.

16. -1 17. 27   18. 8   19. 64   20. 125   21. $-1,000$

Solve each equation using the quadratic formula. Simplify each answer.

22. $\frac{1}{4}x^2 = \frac{1}{2}x + \frac{3}{8}$ $\frac{2 \pm \sqrt{10}}{2}$
23. $\frac{1}{3}y^2 + \frac{3}{2} = \frac{1}{3} - y$ $\frac{-3 \pm i\sqrt{5}}{2}$
24. $2x^2\sqrt{2} + 3x - \sqrt{2} = 0$ $-\sqrt{2}, \frac{\sqrt{2}}{4}$
25. $x^2\sqrt{3} - 2x - \sqrt{3} = 0$ $-\frac{\sqrt{3}}{3}, \sqrt{3}$
26. $x^2 - 2ix + 3 = 0$ $-i, 3i$
27. $2ix^2 - 5x - 2i = 0$ $-2i, -\frac{i}{2}$
28. Find the 4 fourth roots of 16. $\pm 2i, \pm 2$
29. Find the 3 cube roots of $\frac{1}{8}$. $\frac{1}{2}, \frac{-1 \pm i\sqrt{3}}{4}$

Mixed Review

Simplify each expression.   *7.1, 7.4, 8.2, 8.5*

1. $\frac{f(a) - f(b)}{a - b}$, given that $f(x) = 5x + 4$ 5
2. $\frac{9x}{x^2 - 9} - \frac{2}{x + 3} + \frac{5}{3 - x}$ $\frac{2x - 9}{x^2 - 9}$
3. $\sqrt{48x^{10}y^9}$, $x \geq 0$, $y \geq 0$ $4x^5y^4\sqrt{3y}$
4. $\sqrt[3]{-16x^4y^6}$ $-2xy^2\sqrt[3]{2x}$

9.5 The Quadratic Formula   **319**

Have students give the Quadratic Formula. See page 317. Ask them what restriction is placed on the coefficient of a in the Quadratic Formula. $a > 0$ Ask students to give the procedure for finding the cube roots of $x^3 = -64$. See page 318, Example 2.

▰▰ FOLLOW UP

Guided Practice

Classroom Exercises 1–6

Independent Practice

Ⓐ Ex. 1–21, Ⓑ Ex. 22–23, Ⓒ Ex. 24–29

Basic: FR, WE 1–23 odd
Average: FR, WE 3–25 odd
Above Average: FR, WE 7–29 odd

Additional Answers

Written Exercises

4. $\frac{-1 \pm \sqrt{41}}{10}$
6. $\frac{5 \pm i\sqrt{7}}{4}$
7. $\frac{1 \pm i\sqrt{3}}{4}$
8. $\frac{3 \pm \sqrt{3}}{2}$
9. $\frac{3 \pm 2i}{2}$
16. $-1, \frac{1 \pm i\sqrt{3}}{2}$
17. $3, \frac{-3 \pm 3i\sqrt{3}}{2}$
18. $2, -1 \pm i\sqrt{3}$
19. $4, -2 \pm 2i\sqrt{3}$
20. $5, \frac{-5 \pm 5i\sqrt{3}}{2}$
21. $-10, 5 \pm 5i\sqrt{3}$

Enrichment

Explain that the quadratic formula can sometimes be used in factoring quadratic trinomials. For example, to factor $6x^2 - 7x + 2$, write the equation $6x^2 - 7x + 2 = 0$. Then solve the equation by using the quadratic formula. The roots are $\frac{2}{3}$ and $\frac{1}{2}$. Thus, an equation in factored form with these roots is $(x - \frac{2}{3})(x - \frac{1}{2}) = 0$.

Multiplying both sides of this equation by 3 and 2 results in $(3x - 2)(2x - 1) = 0$.

The factors of $6x^2 - 7x + 2$ are $(3x - 2)$ and $(2x - 1)$.

Have the students use this technique to factor these trinomials.

1. $2x^2 - 11x + 5$ $(x - 5)(2x - 1)$
2. $5x^2 + 16x + 3$ $(x + 3)(5x + 1)$
3. $16x^2 - 14x + 3$ $(2x - 1)(8x - 3)$

 **GETTING STARTED**

Prerequisite Quiz

Solve without using the quadratic formula.

1. One leg of a right triangle is 1 ft longer than the other leg and 1 ft shorter than the hypotenuse. Find the area of the triangle.   6 ft^2
2. Use the formula $h = vt - 16t^2$ to find t in seconds when $h = 240$ ft and $v = 128$ ft/s.
   3 s, 5 s

Motivator

Have students find the area, to the nearest tenth, of a rectangle with length and width $2 + \sqrt{7}$ and $5 + 2\sqrt{7}$, respectively, using a scientific calculator.   47.8

TEACHING SUGGESTIONS

Lesson Note

Some answers to the exercises are to be given in simplest radical form and the others are to be approximated to the nearest tenth of a unit. Calculators will be an aid in many of these exercises. In some exercises, one of the roots of an equation cannot be used since it is a negative number. This is illustrated in Examples 1 and 3.

9.6 Problem Solving: Irrational Answers

Objective   To solve word problems using the quadratic formula

The Pythagorean relation, $a^2 + b^2 = c^2$, and the quadratic formula can be used to solve some problems involving right triangles.

EXAMPLE 1   The length of a rectangle is 1 ft more than twice its width, and 1 ft less than the length of a diagonal. Find the length and the area of the rectangle. State the answers in simplest radical form.

What are you to find?   the length and the area

Choose a variable.   Let $w =$ the width.
What does it represent?   Then $2w + 1 =$ the length of the rectangle, and $2w + 2 =$ the length of the diagonal.

What do you know? $\triangle ABC$ is a right triangle. Use the Pythagorean relation, $a^2 + b^2 = c^2$.

Write an equation. $w^2 + (2w + 1)^2 = (2w + 2)^2$

Solve it. $w^2 + 4w^2 + 4w + 1 = 4w^2 + 8w + 4$

$$w^2 - 4w - 3 = 0$$

$$w = \frac{4 \pm \sqrt{16 - 4 \cdot 1(-3)}}{2} \leftarrow \text{quadratic formula}$$

$$= \frac{4 \pm \sqrt{28}}{2}, \text{ or } 2 \pm \sqrt{7}$$

$$w = 2 + \sqrt{7} \leftarrow w \neq 2 - \sqrt{7} \text{ because } 2 - \sqrt{7} < 0.$$

$$2w + 1 = 2(2 + \sqrt{7}) + 1 = 5 + 2\sqrt{7}$$

$$\text{Area} = w(2w + 1)$$

$$= (2 + \sqrt{7})(5 + 2\sqrt{7})$$

$$= 10 + 9\sqrt{7} + 14, \text{ or } 24 + 9\sqrt{7}$$

Check in the original problem.   The check is left for you.

State your answer.   The length is $5 + 2\sqrt{7}$ ft, and the area is $24 + 9\sqrt{7}$ ft^2.

Highlighting the Standards

Standard 1b: Through problem solving, students see that the complex numbers they have learned can occur as answers to real-world problems.

Additional Example 1

The length of a rectangle is 2 m more than the width and 1 m less than the length of a diagonal. Find the length of the diagonal and the area of the rectangle in simplest radical form. $4 + \sqrt{6}$ m; $9 + 4\sqrt{6}$ m^2

EXAMPLE 2 An object is sent upward from the earth's surface with an initial velocity of 96 ft/s. When will the height of the object be 136 ft? State the answer to the nearest tenth of a second.

Solution Use the formula $h = vt - 16t^2$. Substitute 136 for h and 96 for v.

$$h = vt - 16t^2$$
$$136 = 96t - 16t^2$$
$$17 = 12t - 2t^2 \leftarrow \text{Each side is divided by 8, the GCF of each term.}$$
$$2t^2 - 12t + 17 = 0$$
$$t = \frac{12 \pm \sqrt{144 - 136}}{4} = \frac{12 \pm 2\sqrt{2}}{4} = \frac{6 \pm \sqrt{2}}{2} \approx \frac{6 \pm 1.414}{2}$$

$$\frac{6 + 1.414}{2} = 3.707 \quad or \quad \frac{6 - 1.414}{2} = 2.293$$

The height to the nearest 0.1 s will be 136 ft at the end of 2.3 s (on the way up) and again at the end of 3.7 s (on the way down).

EXAMPLE 3 A rectangular lawn is 3 times as long as it is wide. It is surrounded by a sidewalk with a uniform width of 5.00 ft. The total area of the lawn and sidewalk is 2,600 ft². Find the dimensions of the lawn to the nearest 0.1 ft.

Solution Use the given data to draw and label a figure such as that at the right.

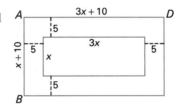

$$AB \cdot AD = \text{total area}$$
$$(x + 10)(3x + 10) = 2,600$$
$$3x^2 + 40x + 100 = 2,600$$
$$3x^2 + 40x - 2,500 = 0$$

Now use the quadratic formula to solve the equation above.

$$x = \frac{-40 \pm \sqrt{1,600 + 30,000}}{6} = \frac{-40 \pm \sqrt{31,600}}{6} = \frac{-40 \pm 20\sqrt{79}}{6}$$

$$x = \frac{-20 + 10\sqrt{79}}{3} \approx \frac{-20 + 10(8.888)}{3} \approx 23.0 \leftarrow \text{The negative value of } x \text{ is discarded.}$$

$$3x = -20 + 10\sqrt{79} \approx -20 + 10(8.888) \approx 68.9$$

The lawn measures 23.0 ft by 68.9 ft, to the nearest 0.1 ft.

Math Connections

Geometry: The Quadratic Formula often arises in geometry. For example, consider the theorem that if two chords intersect within a circle, then the product of the lengths of the segments of one chord is equal to the product of the lengths of the segments of the other. Thus, if the length of one chord is 10 and the chord divides another chord into two segments of lengths 2 and 4, the lengths of the segments of the first chord are found by solving the quadratic equation $2 \cdot 4 = x(10 - x)$.

Critical Thinking Questions

Analysis: In Example 1, the value $2 - \sqrt{7}$ is rejected, "because $2 - \sqrt{7} < 0$." Ask students if it is necessary to calculate $\sqrt{7}$ in order to see that "$2 - \sqrt{7} < 0$" is true. Students should see that since $4 < 7$, $\sqrt{4} < \sqrt{7}$. Also, since $2 = \sqrt{4}$, 2 must be less than $\sqrt{7}$. That is, "$2 - \sqrt{7} < 0$" must be true.

Checkpoint

1. The length of a rectangle is twice the width and 3 in. less than the length of a diagonal. Find the area of the rectangle in simplest radical form. $162 + 72\sqrt{5}$ in.²
2. To the nearest tenth, find two values of t if $h = vt - 5t^2$, $h = 40$, and $v = 60$. 0.7, 11.3
3. To the nearest tenth, find the positive value of x if $(x + 4)(3x + 6) = 180$. 4.8

Additional Example 2

Use the formula $h = vt - 16t^2$ to find t to the nearest tenth (0.1) of a second when $h = 72$ ft and $v = 80$ ft/s. 1.2 s, 3.8 s

Additional Example 3

Rectangle I is x units by $2x$ units. Rectangle II is 8 units wider and 8 units longer than rectangle I. The area of rectangle II is 140 square units. Find the length of rectangle I to the nearest tenth of a unit. 5.2 units

Have students give the two formulas they need to use to find the area of a rectangle with width, w, length $3w + 1$, and a diagonal of length $3w + 7$. Use the Pythagorean relation to write an equation. Then use the quadratic formula to solve for w.

◼◼◼ FOLLOW UP

Guided Practice

Classroom Exercises 1, 2

Independent Practice

A Ex. 1–4, **B** Ex. 5, 6, **C** Ex. 7

Basic: WE 1–5

Average: WE 2–6

Above Average: WE 3–7

Additional Answers

Written Exercises

3. $(2 + \sqrt{5}, -2 + \sqrt{5}), (2 - \sqrt{5}, -2 - \sqrt{5})$

4. $\left(\dfrac{1 + \sqrt{7}}{2}, -1 + \sqrt{7}\right), \left(\dfrac{1 - \sqrt{7}}{2}, -1 - \sqrt{7}\right)$

6. $\left(\dfrac{-3 + i}{2}, -2 + 2i\right), \left(\dfrac{-3 - i}{2}, -2 - 2i\right)$

Classroom Exercises

Solve the following problems.

1. A diagonal of a square is 5 cm longer than a side of the square. Find the length of the diagonal and the area of the square. Give your answer in simplest radical form. Diag: $10 + 5\sqrt{2}$ cm, A: $75 + 50\sqrt{2}$ cm^2

2. An object is sent upward with an initial velocity of 144 ft/s. To the nearest 0.1 s, when will the height of the object be 96 ft? Use the formula $h = vt - 16t^2$. 0.7 s, 8.3 s

Written Exercises

Solve. Write the answer in simplest form.

1. One leg of a right triangle is 2 ft longer than the other leg and 4 ft shorter than the hypotenuse. Find the length of the longer leg and the area of the triangle. l: $6 + 4\sqrt{3}$ ft, A: $36 + 20\sqrt{3}$ ft^2

2. If a pellet is launched straight upward with an initial velocity of 200 ft/s, when will its altitude be 568 ft? Answer to the nearest 0.1 s. Use the formula $h = vt - 16t^2$. 4.4 s, 8.1 s

3. The product of a pair of numbers is 1. The second number is 4 less than the first number. Find all such pairs of numbers.

4. A second number is 2 less than twice a first number. Find all such pairs of numbers whose product is 3.

5. A second complex number is 6 less than a first complex number. Find all such pairs of numbers given that their product is the real number -14. $(3 + i\sqrt{5}, -3 + i\sqrt{5}), (3 - i\sqrt{5}, -3 - i\sqrt{5})$

6. If a second number is multiplied by 2 more than a first number, the product is -2. Find all such pairs of numbers given that the second is 4 more than 4 times the first.

7. A rectangular lawn measures 20 yd by 40 yd and is surrounded by a sidewalk of uniform width. The sidewalk's outer edge is a rectangle with an area of 1,000 yd^2. Find the width of a strip of the sidewalk to the nearest tenth of a yard. 1.6 yd

Mixed Review

Solve each equation. *1.4, 6.1, 7.6*

1. $\dfrac{5x - 3}{10} = \dfrac{6x + 2}{15}$ $\dfrac{13}{3}$

2. $\dfrac{x}{6} + \dfrac{x}{3} + \dfrac{x}{4} = 1$ $\dfrac{4}{3}$

3. $2(x - 1)^2 + 3x = (x + 3)^2 - x$ $-1, 7$

4. $\dfrac{y + 8}{3y} = \dfrac{y + 3}{2y - 1}$ $2, 4$

Enrichment

Challenge the students to write the quadratic equation, given its solutions. Students may use the quadratic formula to identify the coefficients a, b, and c.

1. $x = \dfrac{3 \pm 1}{4}$ $2x^2 - 3x + 1 = 0$

2. $x = \dfrac{7 \pm 3\sqrt{5}}{2}$ $x^2 - 7x + 1 = 0$

3. $x = \dfrac{-1 \pm \sqrt{21}}{10}$ $5x^2 + x - 1 = 0$

4. $x = \dfrac{-1 \pm i\sqrt{3}}{2}$ $x^2 + x + 1 = 0$

9.7 Vectors: Addition and Subtraction

Objectives To add and subtract vectors by drawing
To draw scalar multiples of vectors
To draw vectors whose resultants are zero vectors

The displacement (movement) of an object through a distance of 8 mi to the southeast is represented by the directed line segment $\overrightarrow{AB}$ shown at the right. Such directed line segments are called **vectors**. The vector shown is written as $\overrightarrow{AB}$ (with a half-arrow $\rightharpoonup$) and read as "vector AB."

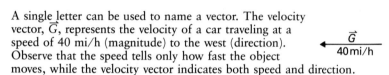

Notice:

(1) $\overrightarrow{AB}$ has a **magnitude** of 8 mi and its **direction** is southeast, as shown by the arrowhead in the drawing.

(2) The initial point (*tail*) of $\overrightarrow{AB}$ is A. The terminal point (*head*) of $\overrightarrow{AB}$ is B.

A second use of vectors is to represent forces. The force vector $\overrightarrow{CD}$ represents the force exerted by a 7-newton weight (magnitude) suspended on a coiled spring, stretching it downward (direction).

A single letter can be used to name a vector. The velocity vector, $\overrightarrow{G}$, represents the velocity of a car traveling at a speed of 40 mi/h (magnitude) to the west (direction). Observe that the speed tells only how fast the object moves, while the velocity vector indicates both speed and direction.

Four vectors are shown at the right in a coordinate plane. Vectors $\overrightarrow{R}$ and $\overrightarrow{R'}$ are **equivalent vectors** because they have the same magnitude and the same direction. Vectors $\overrightarrow{T}$ and $-\overrightarrow{T}$ are **opposite vectors** because they have the same magnitude but opposite directions.

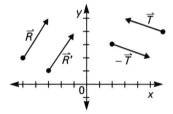

Vectors can be added, the sum of two vectors being the result of two consecutive displacements. If an object moves from A to B and then from B to C, the result is equivalent to one move directly from A to C. Thus, $\overrightarrow{AB} + \overrightarrow{BC} = \overrightarrow{AC}$. Notice that the tail of $\overrightarrow{BC}$ is the head of $\overrightarrow{AB}$ and that the sum, $\overrightarrow{AC}$, is drawn from the tail of $\overrightarrow{AB}$ to the head of $\overrightarrow{BC}$.

$$\overrightarrow{AB} + \overrightarrow{BC} = \overrightarrow{AC}$$

Teaching Resources

Application Worksheet 9
Quick Quizzes 66
**Reteaching and Practice
Worksheets** 66

▬▬ GETTING STARTED

Prerequisite Quiz

Identify the property illustrated.

1. $8.2 + \sqrt{2}$ is a real number.   Clos for Add
2. $-a^3b + a^3b = 0$ Add Inv Prop
3. $-2(x + y) = -2x - 2y$ Distr Prop

Motivator

Have students consider a baseball diamond. Let H = home plate, F = first base, S = second base, and T = third base. Tell students that if the vector between home plate and first base is $\overrightarrow{HF}$ and the vector between first base and second base is $\overrightarrow{FS}$, then the vector between home plate and second base can be represented as $\overrightarrow{HS} = \overrightarrow{HF} + \overrightarrow{FS}$. Now ask students to write an equation for the vector between home plate and third base. $\overrightarrow{HT} = \overrightarrow{HF} + \overrightarrow{FS} + \overrightarrow{ST}$

▬▬ TEACHING SUGGESTIONS

Lesson Note

Following Example 2, the concept of a scalar multiple is introduced. Note that $3 \cdot \overrightarrow{D}$ is a vector, and is a scalar multiple of the vector $\overrightarrow{D}$.

Highlighting the Standards

Standard 2d: This lesson offers another good opportunity for students to read mathematics and demonstrate their understanding of what they have read.

Math Connections

Physics: As mentioned on page 323, *force* (a push or pull) and *velocity* (the time rate of change of position) are vectors, since they have both magnitude and direction. *Acceleration* (the time rate of change of velocity) is also a vector.

Critical Thinking Questions

Analysis: The lesson implicitly assumes that addition of vectors is a commutative operation. Have students start with the drawings of two vectors $\vec{A}$ and $\vec{B}$ and demonstrate graphically that $\vec{A} + \vec{B} = \vec{B} + \vec{A}$. Then have them demonstrate graphically that addition of vectors is associative; that is, for any three vectors $\vec{A}$, $\vec{B}$, and $\vec{C}$, $(\vec{A} + \vec{B}) + \vec{C} = \vec{A} + (\vec{B} + \vec{C})$.

Checkpoint

Draw each of the following.

1. $\vec{B} + \vec{C}$
2. $\vec{A} - \vec{C}$
3. $\vec{A} + \vec{B} - \vec{C}$
4. $\vec{V}$ so that $\vec{B} + \vec{C} + \vec{V}$ is a zero vector

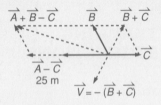

The method just shown is called the *triangle method* of addition for vectors. The sum, $\overrightarrow{AC}$, is the **resultant** of $\overrightarrow{AB}$ and $\overrightarrow{BC}$. Given the disjoint vectors $\vec{V}$ and $\vec{W}$ below, you can use the equivalent vector $\vec{W}'$ to draw $\vec{V} + \vec{W}$ as shown below at the right. Move the tail of $\vec{W}'$ to the head of $\vec{V}$, and use the triangle method to draw the resultant.

Three vectors can be added by extending the method above.

Because vectors $\vec{F}$ and $\vec{G}$ below have the same initial point, the *parallelogram method* can be used to draw $\vec{F} + \vec{G}$. Draw equivalent vectors $\vec{F}'$ and $\vec{G}'$ to form a parallelogram with $\vec{F}$ and $\vec{G}$ as shown. The diagonal from the common initial point is $\vec{F} + \vec{G}$.

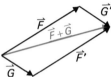

EXAMPLE 1 Use the figure at the right to find the following vectors.

a. $\overrightarrow{AG} + \overrightarrow{GH}$ b. $\overrightarrow{DB} + \overrightarrow{BF}$ c. $\overrightarrow{CH} + \overrightarrow{HD}$
d. $\overrightarrow{BE} + \overrightarrow{HI}$ e. $\overrightarrow{CE} + \overrightarrow{BA}$ f. $\overrightarrow{AI} + \overrightarrow{CB}$
g. $\overrightarrow{AB} + \overrightarrow{AD}$ h. $\overrightarrow{HB} + \overrightarrow{HG}$ i. $\overrightarrow{EF} + \overrightarrow{EH}$
j. $\overrightarrow{GB} + \overrightarrow{BC} + \overrightarrow{CF}$ k. $\overrightarrow{DB} + \overrightarrow{EF} + \overrightarrow{BH}$

Solutions Use the triangle method for parts a–c.

a. $\overrightarrow{AH}$ b. $\overrightarrow{DF}$ c. $\overrightarrow{CD}$

Use equivalent vectors for parts d–f.

d. $\overrightarrow{BE} + \overrightarrow{EF} = \overrightarrow{BF}$ e. $\overrightarrow{CE} + \overrightarrow{ED} = \overrightarrow{CD}$ f. $\overrightarrow{AI} + \overrightarrow{IH} = \overrightarrow{AH}$

Use the parallelogram method for parts g–i.

g. $\overrightarrow{AE}$ h. $\overrightarrow{HA}$ i. $\overrightarrow{EI}$

Use any method or a combination of methods for parts j and k.

j. $\overrightarrow{GF}$ k. $\overrightarrow{DB} + \overrightarrow{BC} + \overrightarrow{CI} = \overrightarrow{DI}$

Additional Example 1

Use the figure above to find the following.

a. $\overrightarrow{AF} + \overrightarrow{FE}$ $\overrightarrow{AE}$
b. $\overrightarrow{DB} + \overrightarrow{BF}$ $\overrightarrow{DF}$

c. $\overrightarrow{AE} + \overrightarrow{BC}$ $\overrightarrow{AD}$
d. $\overrightarrow{DB} + \overrightarrow{EF}$ $\overrightarrow{DA}$
e. $\overrightarrow{AB} + \overrightarrow{AF}$ $\overrightarrow{AE}$
f. $\overrightarrow{DC} + \overrightarrow{DF}$ $\overrightarrow{DA}$
g. $\overrightarrow{FE} + \overrightarrow{ED}$ $\overrightarrow{FD}$
h. $\overrightarrow{CA} + \overrightarrow{AB}$ $\overrightarrow{CB}$
i. $\overrightarrow{FA} + \overrightarrow{AB} + \overrightarrow{BD}$ $\overrightarrow{FD}$
j. $\overrightarrow{AB} + \overrightarrow{ED} + \overrightarrow{BE}$ $\overrightarrow{AD}$

The Rule of Subtraction for vectors is similar to the Rule of Subtraction for real numbers, as stated below.

$$a - b = a + (-b) \text{ for all real numbers } a \text{ and } b.$$
$$\vec{X} - \vec{Y} = \vec{X} + (-\vec{Y}) \text{ for all vectors } \vec{X} \text{ and } \vec{Y}.$$

EXAMPLE 2 Draw $\vec{P} - \vec{Q}$ for each of the cases shown.

a.

b.

Solutions **a.** Draw $-\vec{Q}$. Use the parallelogram method to draw $\vec{P} + (-\vec{Q})$.

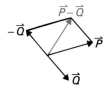

b. Draw $-\vec{Q}$ from the head of $\vec{P}$. Use the triangle method to draw $\vec{P} + (-\vec{Q})$.

A vector can also be multiplied by a real number. Consider, for example, a displacement of 4 mi to the east occurring every hour for 3 h. The result is a total displacement of 12 mi to the east, as shown below.

$\vec{D}$ above was multiplied by the real number 3 to give the vector $3 \cdot \vec{D}$. In this case, the real number 3 is called a **scalar**, and the vector $3 \cdot \vec{D}$ is called a **scalar multiple** of $\vec{D}$.

EXAMPLE 3 Given $\vec{A}$ as shown, draw $-2 \cdot \vec{A}$.

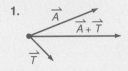

Solution $-2 \cdot \vec{A} = 2(-\vec{A})$

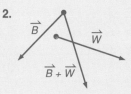

Additional Example 2

Draw $\vec{A} - \vec{B}$ for each case.

a.

b.

Additional Example 3

Given $\vec{A}$ as shown, draw $2 \cdot \vec{A}$ and $-\frac{3}{4} \cdot \vec{A}$.

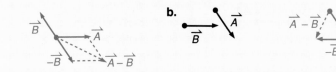

3.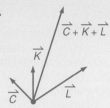

4.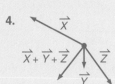

5.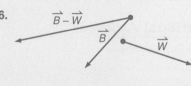

6.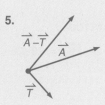

7.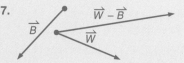

EXAMPLE 4 Draw $\frac{1}{2} \cdot \vec{P} - 2 \cdot \vec{Q}$ given $\vec{P}$ and $\vec{Q}$.

Plan Draw $\frac{1}{2} \cdot \vec{P}$ and $-2 \cdot \vec{Q}$.

Use the parallelogram method to draw $\frac{1}{2} \cdot \vec{P} + (-2 \cdot \vec{Q})$.

Solution

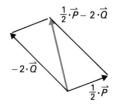

The set of vectors is closed for addition. That is, the sum of any two vectors is a vector. Consider two forces of 50 lb each acting in opposite directions on an object at C, as shown in the figure at the right. The object will remain in equilibrium because $\overrightarrow{CD} + \overrightarrow{CE} = \overrightarrow{CC}$.

$\overrightarrow{CC}$ is a *zero vector*. Its magnitude is 0 and it has no direction. In general, the sum of a pair of *opposite vectors* is a **zero vector**.

EXAMPLE 5 Using the figure at the right, draw $\vec{R}$ so that $\vec{P} + \vec{Q} + \vec{R}$ is a zero vector ($\overrightarrow{AA}$) that leaves the object at A in equilibrium.

Solution Draw $\vec{P} + \vec{Q}$.   Draw $\vec{R} = -(\vec{P} + \vec{Q})$.

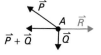

$\vec{P} + \vec{Q} + \vec{R} = \overrightarrow{AA}$

Classroom Exercises

Match the phrase at the left with a vector in the figure.

1. velocity vector $\vec{C}$
2. force vector $\vec{A}$
3. displacement vector $\vec{B}$
4. northeast direction $\vec{B}$
5. west direction $\vec{C}$
6. upward direction $\vec{A}$
7. magnitude of 10 km $\vec{B}$
8. eastward direction $\vec{D}$

326 Chapter 9 Complex Numbers

Additional Example 4

Draw $3 \cdot \vec{A} - \frac{1}{2} \cdot \vec{B}$, given $\vec{A}$ and $\vec{B}$.

Additional Example 5

Use the figure below to draw $\vec{V}$ so that $\vec{P} + \vec{Q} + \vec{V}$ is a zero vector $\overrightarrow{AA}$.

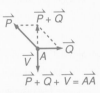

Use the figure at the right to find the following.

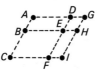

9. two vectors equivalent to $\overrightarrow{ED}$ $\overrightarrow{BA}, \overrightarrow{HG}$
10. three vectors that are opposites of $\overrightarrow{AD}$ $\overrightarrow{DA}, \overrightarrow{EB}, \overrightarrow{FC}$

11. $\overrightarrow{CE} + \overrightarrow{EA}$ $\overrightarrow{CA}$ 12. $\overrightarrow{CB} + \overrightarrow{CF}$ $\overrightarrow{CE}$ 13. $\overrightarrow{IE} + \overrightarrow{FC}$ $\overrightarrow{IB}$ 14. $\overrightarrow{GI} + \overrightarrow{GD}$ $\overrightarrow{GF}$
15. $\overrightarrow{AF} + \overrightarrow{DG}$ $\overrightarrow{AI}$ 16. $\overrightarrow{AB} + \overrightarrow{BF} + \overrightarrow{FI}$ 17. $\overrightarrow{CF} + \overrightarrow{IH} + \overrightarrow{BA}$ 18. $\overrightarrow{AD} - \overrightarrow{HG}$ $\overrightarrow{AE}$
19. $\overrightarrow{HI} - \overrightarrow{DG}$ $\overrightarrow{HF}$ 20. $\overrightarrow{FE} - \overrightarrow{AD}$ $\overrightarrow{FB}$ 21. $\overrightarrow{AE} - \overrightarrow{FE}$ $\overrightarrow{AF}$ 22. $\overrightarrow{FB} - \overrightarrow{DA}$ $\overrightarrow{FE}$

Written Exercises

Carefully copy the figures below. Then draw and label the resultant vectors given in Exercises 1–26.

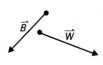

1. $\overrightarrow{A} + \overrightarrow{T}$ 2. $\overrightarrow{B} + \overrightarrow{W}$ 3. $\overrightarrow{C} + \overrightarrow{K} + \overrightarrow{L}$ 4. $\overrightarrow{X} + \overrightarrow{Y} + \overrightarrow{Z}$
5. $\overrightarrow{A} - \overrightarrow{T}$ 6. $\overrightarrow{B} - \overrightarrow{W}$ 7. $\overrightarrow{W} - \overrightarrow{B}$ 8. $\overrightarrow{T} - \overrightarrow{A}$
9. $3 \cdot \overrightarrow{Q}$ 10. $\frac{2}{3} \cdot \overrightarrow{R}$ 11. $-2 \cdot \overrightarrow{Q}$ 12. $-1.5 \cdot \overrightarrow{P}$
13. $2 \cdot \overrightarrow{A} + 3 \cdot \overrightarrow{T}$ 14. $3 \cdot \overrightarrow{A} - 2 \cdot \overrightarrow{T}$ 15. $3 \cdot \overrightarrow{B} - 2 \cdot \overrightarrow{W}$ 16. $\frac{1}{2} \cdot \overrightarrow{W} - 2 \cdot \overrightarrow{B}$
17. $\overrightarrow{C} + \overrightarrow{K} - \overrightarrow{L}$ 18. $\overrightarrow{K} + \overrightarrow{L} - \overrightarrow{C}$ 19. $\overrightarrow{X} + \overrightarrow{Y} - \overrightarrow{Z}$ 20. $\overrightarrow{X} - \overrightarrow{Y} + \overrightarrow{Z}$
21. $\overrightarrow{V}$, so that $\overrightarrow{A} + \overrightarrow{T} + \overrightarrow{V}$ is a zero vector   22. $\overrightarrow{V}$, so that $\overrightarrow{B} + \overrightarrow{W} + \overrightarrow{V}$ is a zero vector
23. $\overrightarrow{C} - (\overrightarrow{K} + \overrightarrow{L})$ 24. $2 \cdot (\overrightarrow{C} + \overrightarrow{K} + \overrightarrow{L})$
25. $\overrightarrow{V}$, so that $\overrightarrow{C} + \overrightarrow{L} + \overrightarrow{K} + \overrightarrow{V}$ is a zero vector
26. $\overrightarrow{V}$, so that $\overrightarrow{X} + \overrightarrow{Y} + \overrightarrow{Z} + \overrightarrow{V}$ is a zero vector

Mixed Review

Identify the property of operations with real numbers illustrated below. All variables represent real numbers. *1.3*

1. $-5 + 3.26$ is a real number.   2. $x + x^2 = x^2 + x$
3. $(-8 + 17) + 23 = -8 + (17 + 23)$ 4. $7.1 + (-7.1) = 0$
5. $8(c + 12) = 8 \cdot c + 8 \cdot 12$ 6. $1(c + d) = c + d$

8.
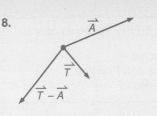

9.

$\overrightarrow{Q}$
7 cm

$3 \cdot \overrightarrow{Q}$
21 cm

10.
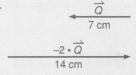
$\overrightarrow{R}$
9 cm

$\frac{2}{3} \cdot \overrightarrow{R}$
6 cm

11.
$\overrightarrow{Q}$
7 cm

$-2 \cdot \overrightarrow{Q}$
14 cm

12.
$\overrightarrow{P}$
12 cm

$-1.5\overrightarrow{P}$
18 cm

13.
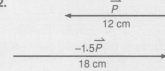

See page 810 for the answers to Written Ex. 14–26 and Mixed Review Ex. 1–6.

Enrichment

In air navigation, directions or bearings are measured clockwise from the north.

Vector $\overrightarrow{PQ}$ represents a direction of 40, $\overrightarrow{PR}$ represents a direction of 160, and $\overrightarrow{PT}$ represents a direction of 320.
Have the students solve the following problem by drawing vectors and making the necessary calculations.

A pilot flies 300 km in a direction of 60, and then 400 km in a direction of 330, both at the same altitude. How far is the pilot from the initial position?   500 km

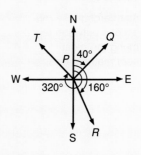

▐▬▬ GETTING STARTED

Prerequisite Quiz

Determine the quadrant or axis that contains each point $P(x,y)$.

1. $A(-6,10)$ II
2. $B(0,8)$ y-axis
3. $C(5,-2)$ IV
4. $D(-4,0)$ x-axis

Simplify.

5. $(6 + 4i) + (-4 - 7i)$ $2 - 3i$
6. $(-3 - 2i) - (5 - 6i)$ $-8 + 4i$

Motivator

Ask students which is the real axis and which is the imaginary axis in a complex coordinate plane. The x-axis is the real axis; the y-axis is the imaginary axis.

Have students give the complex number that corresponds to the ordered pair $(-5,3)$.
$-5 + 3i$

9.8 Vectors and Complex Numbers

Objectives   To match vectors in standard position with corresponding complex numbers

To draw sums and differences of two vectors corresponding to the sums and differences of two complex numbers.

To draw scalar multiples of vectors corresponding to the products of real numbers and complex numbers

Recall that each complex number $a + bi$ can be matched with a unique ordered pair of real numbers (a,b), and that each ordered pair can in turn be matched with a unique point in a coordinate plane.

Now, let $\vec{V}$ be a vector with its initial point at the origin of a coordinate plane and its terminal point matched with a unique ordered pair of real numbers (a,b). The table below illustrates this matching for the four vectors shown in the diagram. The corresponding complex numbers for these vectors are also shown.

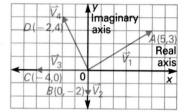

Vector	Terminal point	Ordered pair (a,b)	Complex number
$\vec{V_1}$	A	$(5,3)$	$5 + 3i$
$\vec{V_2}$	B	$(0,-2)$	$-2i$
$\vec{V_3}$	C	$(-4,0)$	-4
$\vec{V_4}$	D	$(-2,4)$	$-2 + 4i$

As suggested by the table, there is a one-to-one correspondence between all complex numbers $a + bi$ and all vectors $\vec{V}$ with their initial points at the origin. Such vectors are said to be in **standard position**. If $\vec{V}$ is in standard position with its terminal point at (a,b), then (a,b) can be used to represent $\vec{V}$. (a,b) is called the **rectangular form** of $\vec{V}$.

Addition and subtraction of complex numbers correspond to addition and subtraction of vectors in standard position, as shown below.

Addition of complex numbers:
$(5 + 3i) + (-3 + 2i) = 2 + 5i$

Addition of vectors:
Let $\vec{V_1} = (5,3)$ and $\vec{V_2} = (-3,2)$. Use the parallelogram method to draw $\vec{V_1} + \vec{V_2}$.

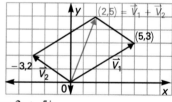

Notice that $(5 + 3i) + (-3 + 2i) = 2 + 5i$ corresponds to $(5,3) + (-3,2) = (2,5)$.

328   Chapter 9 Complex Numbers

Highlighting the Standards

Standard 4c: The Standards make the point that "the connections between algebra and geometry are among the most important in high school mathematics."

Subtraction of complex numbers:
$(3 + 4i) - (5 - 2i) =$
$= (3 + 4i) + (-5 + 2i)$
$= -2 + 6i$

Subtraction of vectors:
Let $\overrightarrow{V_1} = (3,4)$ and $\overrightarrow{V_2} = (5,-2)$.
Then $-\overrightarrow{V_2} = (-5,2)$.
Draw $\overrightarrow{V_1} - \overrightarrow{V_2} = \overrightarrow{V_1} + (-\overrightarrow{V_2})$.

Notice that $(3,4) - (5,-2)$
$= (3,4) + (-5,2)$
$= (-2,6)$

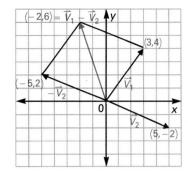

EXAMPLE   Draw the two vectors that correspond to $4 - 2i$ and $1 + 3i$. Then draw their sum and their difference.

Solutions

Draw $(4,-2)$ and $(1,3)$.

Sum: $(4,-2) + (1,3)$ $= (5,1)$

Difference: Draw $-(1,3) = (-1,-3)$. Then, $(4,-2) - (1,3)$ $= (3,-5)$

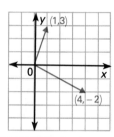
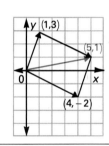
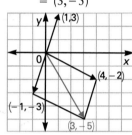

The product of a real number and a complex number corresponds to the product of a scalar and a vector. For $3(-2 + i) = -6 + 3i$, let $\overrightarrow{V} = (-2,1)$.
Draw $3 \cdot \overrightarrow{V}$ as shown above at the right. Notice that

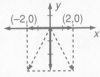

$$3 \cdot \overrightarrow{V} = 3 \cdot (-2,1) = (-6,3), \text{ and } 3(-2 + i) = -6 + 3i$$

■■■■**TEACHING SUGGESTIONS**

Lesson Note

On a coordinate plane, a vector is said to be in *standard position* if its initial point is at the origin. The terminal point $P(x,y)$ corresponds to the rectangular form of the vector (x,y).

Correspondences among vectors in rectangular form, points on a coordinate plane, ordered pairs of real numbers, and complex numbers are illustrated below.

$(-7,3)$: vector in rectangular form
$P(-7,3)$: point
$(-7,3)$: ordered pair of numbers
$-7 + 3i$: complex number

Math Connections

Electrical Impedance: It is the vectorial nature of complex numbers that makes them a useful model for representing the impedance of an electric circuit (see *Math Connections* for Lesson 9.1). In the vector diagram of these three components of impedance, the resistance vector runs along the positive real axis starting from the origin. The capacitance and inductance each run vertically from the origin along the imaginary axis. The capacitance vector points up; the inductance vector points down. The vector sum of the three vectors is the resultant vector for the impedance of the circuit.

Critical Thinking Questions

Analysis: Ask students how they think that the *zero vector* should be represented using complex numbers. $0 + 0i$

Additional Example

Draw the two vectors that correspond to the two given complex numbers. Then draw their sum and difference.

a. $-1 + 2i$ and $-2 - 3i$
Sum: $(-3,-1)$; difference: $(1,5)$

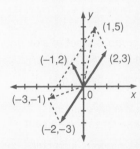

b. $-3i$ and 2

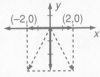

Sum: $(2,-3)$; difference: $(-2,-3)$

Checkpoint

Draw the vector (a,b) in standard position that is paired with the given complex number a + bi.

1. $4 + 2i$ (4,2)
2. $-1 + 3i$ (−1,3)
3. $(4 + 2i) + (-1 + 3i)$ (3,5)
4. $(4 + 2i) - (-1 + 3i)$ (5,−1)
5. conjugate of $-1 + 3i$ (−1,−3)
6. $2(-1 + 3i)$ (−2,6)

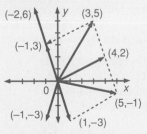

Closure

Have students produce three complex numbers whose sum is $0 + 0i$. Then have students draw the corresponding vectors and their sum.

■■■FOLLOW UP

Guided Practice

Classroom Exercises 1–9

Independent Practice

A Ex. 1–8, B Ex. 9–12, C Ex. 13–15

Basic: WE 1–9 odd, Extension

Average: WE 3–11 odd, Extension

Above Average: WE 7–15 odd, Extension

See pages 811 and 812 for answers to Classroom Ex. 9, Written Ex. 1–15, and Mixed Review Ex. 1–4.

Classroom Exercises

Match each complex number at the left with the rectangular form of the corresponding vector at the right.

1. 5 g
2. $2 + 3i$ c
3. $5i$ f
4. $-2 - 3i$ d
5. $\overline{2 + 3i}$ e
6. $-2 + 3i$ b
7. $\overline{5 - 5i}$ a
8. $\overline{5 + 5i}$ h

a. (5,5)
b. (−2,3)
c. (2,3)
d. (−2,−3)
e. (2,−3)
f. (0,5)
g. (5,0)
h. (5,−5)

9. Draw the vectors for $5 - i$ and $6 + 2i$. Then draw their sum and the difference of the second from the first.

Written Exercises

Draw the two vectors that correspond to the two given complex numbers. Then draw their sum and the difference of the second from the first.

1. $4 + 2i, -2 + 3i$
2. $-4 - i, 3 - 2i$
3. $3 + 5i, 2 - 3i$
4. $-2i, 5$
5. $-4 - 2i, 5i$
6. $2 + 4i, -5$
7. $5i, -5i$
8. $2 - 4i, 2 + 4i$

In Exercises 9–12, each expression corresponds to the product $s\vec{V}$ of a scalar s and a vector $\vec{V}$, represented as a complex number. Draw both $\vec{V}$ and $s\vec{V}$.

9. $2(3 - 4i)$
10. $1.5(-4 + 2i)$
11. $-3(2 + 2i)$
12. $-0.5(6i)$

13. Use a vector diagram to prove that the sum of a pair of conjugate complex numbers is a real number. (HINT: Use the general complex number $a + bi$ and its conjugate $a - bi$, and recall that the x-axis is the axis of real numbers.)

14. Use a vector diagram to prove that the difference of two conjugate complex numbers is an imaginary number.

15. Draw the vectors, in standard position, that correspond to the consecutive integral powers of i: $i, i^2, i^3, i^4, \ldots, i^8$.

Mixed Review

Identify the property of operations with real numbers illustrated below. All variables represent real numbers.   *1.3*

1. $-8 \cdot 112$ is a real number.
2. $5(3c) = (5 \cdot 3)c$
3. $y(x + 2) = (x + 2)y$
4. $m \cdot \dfrac{1}{m} = 1, m \neq 0$

Enrichment

A mathematical *group* is a set of elements that satisfies these conditions.

1. It is closed under a binary operation.
2. It is associative.
3. It contains an identity element.
4. It has an inverse for every element.

If the commutative property also holds, then the set is called a *commutative group*.

Have the students draw vector diagrams to determine whether vectors form a commutative group under the operation of addition.

Vectors do form a commutative group under addition. Drawings will vary.

NOTE: This exploration indicates that the set of complex numbers also forms a commutative group under addition.

Extension: The Field of Complex Numbers

The eleven properties of a field that are true for real numbers can be used to prove that the complex numbers form a field under the operations of addition and multiplication. The following definitions will be used.

Definitions 1, 2, 3, and 4

1. $z = x + yi$ is a complex number, where x and y are real numbers.

If $z_1 = a + bi$ and $z_2 = c + di$, then:

2. $z_1 = z_2$ if and only if $a = c$ and $b = d$,

3. $z_1 + z_2 = (a + bi) + (c + di) = (a + c) + (b + d)i$, and

4. $z_1 \cdot z_2 = (a + bi)(c + di) = (ac - bd) + (ad + bc)i$.

Proofs of several theorems about complex numbers are outlined below with $z_1 = a + bi$ and $z_2 = c + di$, where a, b, c, and d are real numbers. Try to supply a reason for each step.

Theorem 9.1
Closure for addition

$z_1 + z_2$
$= (a + bi) + (c + di)$ Subst
$= (a + c) + (b + d)i$ Def 3
$a + c$ and $b + d$ are real numbers.
$z_1 + z_2$ is a complex number.
Closure for Add; Def 1

Theorem 9.2
Addition is commutative.

$z_1 + z_2$
$= (a + bi) + (c + di)$ Subst
$= (a + c) + (b + d)i$ Def 3
$= (c + a) + (d + b)i$ Comm Prop Add
$= (c + di) + (a + bi)$ Def 3
$= z_2 + z_1$ So, $z_1 + z_2 = z_2 + z_1$.
Subst; Trans Prop Eq

If $z_1 + z_2 = z_1$, then z_2 is an additive identity. For example, $0 + 0i$ is an additive identity since $(a + bi) + (0 + 0i) = (a + 0) + (b + 0)i = a + bi$. Is $0 + 0i$ the only additive identity?

Theorem 9.3
$0 + 0i$ is *the* additive identity.

$(a + bi) + (x + yi) = a + bi$ if $x + yi$ is an identity element.

So, $(a + x) + (b + y)i = a + bi$. By Definition 2: $a + x = a$ and $b + y = b$.

The only solutions are $x = 0$ and $y = 0$.

Thus, $x + yi$, or $0 + 0i$ is the *only* additive identity.

If $z_1 + z_2 = 0 + 0i$, then z_2 is an additive inverse of z_1. For example, $-a - bi$ is an additive inverse of $a + bi$ since $(a + bi) + (-a - bi) = [a + (-a)] + [b + (-b)]i = 0 + 0i$. You can show that $-a - bi$ is the *only* additive inverse of $a + bi$.

4. $z_1 \cdot z_2$
$= (a + bi)(c + di)$
$= (ac - bd) + (ad + bc)i$
$ac - bd$ and $ad + bc$ are real numbers. $z_1 \cdot z_2$ is a complex number.

5. $z_1 \cdot z_2$
$= (a + bi)(c + di)$
$= (ac - bd) + (ad + bc)i$
$= (ca - db) + (cb + da)i$
$= (c + di)(a + bi)$
$= z_2 \cdot z_1$

6. $(z_1 + z_2) + z_3$
$= [(a + bi) + (c + di)] + (e + fi)$
$= [(a + c) + (b + d)i] + (e + fi)$
$= [(a + c) + e] + [(b + d) + f]i$
$= [a + (c + e)] + [b + (d + f)]i$
$= (a + bi) + [(c + e) + (d + f)i]$
$= (a + bi) + [(c + di) + (e + fi)]$
$= z_1 + (z_2 + z_3)$

7. $(a + bi)(1 + 0i)$
$= (a \cdot 1 - b \cdot 0) + (a \cdot 0 + b \cdot 1)i$
$= a + bi$

8. $(a + bi)(x + yi) = a + bi$
$(ax - by) + (ay + bx)i = a + bi$
So, $ax - by = a$ and $ay + bx = b$

Solve the system.

$ax - by = a$
$bx + ay = b$
_____
$(-b^2 - a^2)y = 0$
$y = 0$
$ax = a$
$x = 1$
So, $x + yi = 1 + 0i$.

9. $(3 + 4i)\left(\dfrac{3}{3^2 + 4^2} + \dfrac{-4}{3^2 + 4^2}i\right)$

$= (3 + 4i)\left[\dfrac{3}{25} + \left(-\dfrac{4}{25}\right)i\right]$

$= \left(\dfrac{9}{25} + \dfrac{16}{25}\right) + \left(-\dfrac{12}{25} + \dfrac{12}{25}\right)i$

$= 1 + 0i$

10. $(a + bi)(x + yi) = 1 + 0i$

$(ax - by) + (ay + bx)i = 1 + 0i$

$ax - by = 1$ and $ay + bx = 0$

Solve the system.

$\quad ax - by = 1$
$\quad bx + ay = 0$
_____
$(-b^2 - a^2)y = b$

$\qquad y = \dfrac{-b}{a^2 + b^2};$

$bx + a\left(\dfrac{-b}{a^2 + b^2}\right) = 0$

$\qquad bx = \dfrac{ab}{a^2 + b^2}$

$\qquad x = \dfrac{a}{a^2 + b^2}$

So, $x + yi = \dfrac{a}{a^2 + b^2} + \dfrac{-b}{a^2 + b^2}i$

Theorem 9.4

$-a - bi$ is *the* additive inverse of $a + bi$.

$(a + bi) + (x + yi) = 0 + 0i$ if $x + yi$ is an additive inverse of $a + bi$.

So, $(a + x) + (b + y)i = 0 + 0i$, or $a + x = 0$ and $b + y = 0$.

The only solutions are $x = -a$ and $y = -b$.

Thus, $x + yi$, or $-a - bi$, is the *only* additive inverse of $a + bi$.

Exercises

Use Definition 2 to solve each equation for x and y.

1. $x + yi = 5 - 2i$ $x = 5, y = -2$
2. $4 - yi = x + 6i$ $x = 4, y = -6$
3. $3x - 4yi = -6 + 12i$ $x = -2, y = -3$

Outline the proofs of the following theorems for complex numbers. Use the outlines for Theorems 9.1–9.4 as models. Also, use Definition 4 for $z_1 \cdot z_2$.

4. $z_1 \cdot z_2$ is a complex number. (Closure for multiplication)
5. Multiplication is commutative: $z_1 \cdot z_2 = z_2 \cdot z_1$.
6. Addition is associative: $(z_1 + z_2) + z_3 = z_1 + (z_2 + z_3)$.

If $z_1 \cdot z_2 = z_1$, then z_2 is a **multiplicative identity**.

7. Show that $1 + 0i$ is a multiplicative identity by showing that $(a + bi)(1 + 0i) = a + bi$.

8. Prove that $1 + 0i$ is the *only* multiplicative identity by solving $(a + bi)(x + yi) = a + bi$ for $x + yi$.

If $z_1 \cdot z_2 = 1 + 0i$, then z_2 is a **multiplicative inverse** of z_1, $z_1 \neq 0 + 0i$.

9. Show that a multiplicative inverse of $z = 3 + 4i$ is

$\dfrac{3}{3^2 + 4^2} + \dfrac{-4}{3^2 + 4^2}i$, or that $(3 + 4i)\left(\dfrac{3}{3^2 + 4^2} + \dfrac{-4}{3^2 + 4^2}i\right) = 1 + 0i$.

10. Prove that $\dfrac{a}{a^2 + b^2} + \dfrac{-b}{a^2 + b^2}i$ is the *only* multiplicative inverse of $a + bi$, $a + bi \neq 0 + 0i$ by solving $(a + bi)(x + yi) = 1 + 0i$ for $x + yi$.

Solve each problem.

1. A rowing crew traveled 8.5 km downstream in the same time it took to travel 6.5 km upstream. Find the rate of the current given that the crew travels 30.0 km/h in still water. 4 km/h

2. Machine A can do a job in 18 h. If Machines A and B work together, the time is cut to 8 h. How long would it take Machine B to do the job working alone? 14 h 24 min

3. Find three consecutive multiples of 4 such that the product of the second and third numbers is 192. −20, −16, −12 and 8, 12, 16

4. Convert a pressure of 180 pounds per square foot to ounces per square inch. 20 oz/in^2

5. How many liters of water must be added to 6 liters of a 40% iodine solution to dilute it to a 10% iodine solution? 18 liters

6. How many milliliters of a 25% iodine solution should be added to 600 ml of a 65% iodine solution to obtain a 55% iodine solution? 200 ml

7. The height of a triangle is 6 cm more than 3 times the length of its base. Find the height, given that the area of the triangle is 36 cm^2. 18 cm

8. Ninety-six limes are packed in boxes so that the number of boxes is 2 less than 3 times the number of limes in a box. Find the number of boxes. 16 boxes

9. One leg of a right triangle is twice as long as the other leg. The hypotenuse is 5 in. longer than the longer leg. Find the length of the hypotenuse in simplest radical form. $10 + 5\sqrt{5}$ in

10. Ralph is 3 years older than Sonya, and Teresa is 3 times as old as Ralph. In 8 years, Teresa's age will be 6 years more than the combined ages of Ralph and Sonya then. Find their present ages. Sonya: 8 yr; Ralph: 11 yr; Teresa: 33 yr

11. The stretch S in a spring balance varies directly as the applied weight w. Given that S = 4.8 in. when w = 7.2 lb, find the weight needed to cause a stretch of 6.4 in. 9.6 lb

12. The denominator of a fraction is 2 less than the square of its numerator. If the numerator and denominator are both increased by 1, the fraction is equal to $\frac{1}{2}$. Find the original fraction. $\frac{3}{7}$

13. A patio floor is 2 yd longer than it is wide. It is then extended by a second floor 3 yd longer and 1 yd narrower than the first floor. Find the area of the original floor given that the new total area of the patio is 31 yd^2. 15 yd^2

14. Some red and some black pens are sold in packages of 20. The red pens are worth 50¢ each and the black pens are worth 30¢ each. The package of 20 pens is worth $8.20. How many black pens are in the package? 9

15. The units digit of a two-digit number is 1 more than 3 times the tens digit. If the digits are reversed, the new number is 9 less than 3 times the original number. Find the new number. 72

16. Six times the tens digit of a two-digit number is 4 more than the units digit. If the digits are reversed, the sum of the new number and twice the original one is 138. Find the original number. 28

19.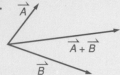

20.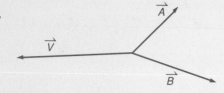

21.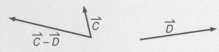

22.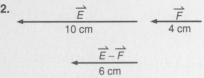

$\overrightarrow{E}$
10 cm $\overrightarrow{F}$
4 cm

$\overrightarrow{E - F}$
6 cm

23.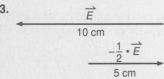

$\overrightarrow{E}$
10 cm

$-\frac{1}{2} \cdot \overrightarrow{E}$
5 cm

24.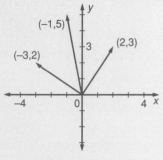

$(-3 + 2i) + (2 + 3i) = -1 + 5i$

Chapter 9 Review

Key Terms

absolute value of a complex number (p. 305)
completing the square (p. 310)
complex conjugates (p. 308)
complex number (p. 303)
direction of a vector (p. 323)
equivalent vectors (p. 323)
imaginary axis (p. 304)
imaginary number (p. 303)
integral polynomial equation (p. 317)
magnitude of a vector (p. 323)
opposite vectors (p. 323)

quadratic form (p. 304)
Quadratic Formula (p. 317)
real axis (p. 304)
rectangular form of a vector (p. 328)
resultant (p. 324)
scalar (p. 325)
scalar multiple of a vector (p. 325)
standard form of a complex number (p. 303)
standard position of a vector (p. 328)
vector (p. 323)
zero vector (p. 326)

Key Ideas and Review Exercises

9.1 To simplify the sum or difference of two complex numbers, use $\sqrt{-x} = i\sqrt{x}$, where $-x < 0$, and combine like terms. The absolute value of $a + bi$, denoted by $|a + bi|$, is $\sqrt{a^2 + b^2}$.

Write in the form $a + bi$, where a and b are real numbers.

1. $(6 - 4i) - (-3 + 2i)$ $9 - 6i$ **2.** $(5 - 3\sqrt{-12}) + (-2 - \sqrt{-27})$ $3 - 9i\sqrt{3}$

3. Simplify $|2 - 5i|$. $\sqrt{29}$ **4.** Simplify $|2 + 4i| + |2 - 4i|$. $4\sqrt{5}$

9.1 If $x^2 = k$, then $x = \pm\sqrt{k}$, for all real numbers k.

5. Solve $3y^2 + 33 = 0$. $\pm i\sqrt{11}$ **6.** Solve $x^4 + 2x^2 = 24$. $\pm 2, \pm i\sqrt{6}$

9.2 To simplify the product of two complex numbers, use $\sqrt{-1} = i$ and $i^2 = -1$. Also use the conjugate of $a + bi$, denoted by $\overline{a + bi}$, which is $a - bi$. To simplify the quotients $\dfrac{2}{3i}$ and $\dfrac{3}{4 + 2i}$, multiply by $\dfrac{i}{i}$ and $\dfrac{4 - 2i}{4 - 2i}$, respectively.

Simplify.

7. $3\sqrt{-2} \cdot 2\sqrt{-6}$ $-12\sqrt{3}$ **8.** i^7 $-i$ **9.** $4 \div (-5i)$ $\frac{4}{5}i$ **10.** $\dfrac{2}{3 - 4i}$ $\frac{6 + 8i}{25}$

9.2 To factor $x^2 + y^2$, write $x^2 - (-y^2) = x^2 - y^2i^2 = (x + yi)(x - yi)$.

11. Factor $16x^2 + 9$. $(4x + 3i)(4x - 3i)$ **12.** Factor $49a^2 + 121$. $(7a + 11i)(7a - 11i)$

9.3 To complete the square for $x^2 + bx$, write $x^2 + bx + (\frac{b}{2})^2 = (x + \frac{b}{2})^2$.

13. Solve $x^2 - 8x - 4 = 0$ by completing the square. $4 \pm 2\sqrt{5}$

9.4   If $y - y_1 = -a(x - x_1)^2$ and $-a < 0$, then the maximum value of y is y_1 and this occurs when $x = x_1$.

14. Find the maximum value of the polynomial P and the corresponding value of x given that $P = -5x^2 + 30x + 12$. $P_{max} = 57, x = 3$

15. Leon High School will sell 400 of its yearbooks if each book sells for \$18.00. For each 50¢ decrease in the price, 20 more books will be sold. What book price will produce the maximum income?   \$14

9.5   To solve a quadratic equation $ax^2 + bx + c = 0$ using the quadratic formula, use $x = \dfrac{-b \pm \sqrt{b^2 - 4ac}}{2a}$. To find the three cube roots of 125, solve the equation $x^3 = 125$, or $x^3 - 125 = 0$, by writing $x^3 - 125$ as $x^3 - 5^3$ and factoring.

16. Solve $2x^2 + 7 = 6x$ using the quadratic formula. Simplify. $\dfrac{3 \pm i\sqrt{5}}{2}$

17. Find the three cube roots of 216. Simplify. $6, -3 \pm 3i\sqrt{3}$

9.6   To solve a word problem involving geometric figures, first draw and label the figures.

18. The length of a rectangle is 1 ft more than twice its width, and 2 ft less than its diagonal's length. Find the length of the diagonal in simplest radical form. $11 + 4\sqrt{6}$ ft

9.7   If $\vec{X}$ and $\vec{Y}$ are noncollinear vectors, $\vec{X} + \vec{Y}$ is drawn using the triangle method or the parallelogram method. If $\vec{X}$ and $\vec{Y}$ are collinear, draw $\vec{X} + \vec{Y}$ using the tail-to-head method. To draw $\vec{X} - \vec{Y}$, draw $-\vec{Y}$ and then draw $\vec{X} + (-\vec{Y})$.

Carefully copy the figures at the right. Then draw and label the following.

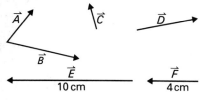

19. $\vec{A} + \vec{B}$

20. $\vec{V}$, so that $\vec{A} + \vec{B} + \vec{V}$ is a zero vector.

21. $\vec{C} - \vec{D}$ **22.** $\vec{E} - \vec{F}$ **23.** $-\dfrac{1}{2} \cdot \vec{E}$

9.8   The vector with its initial point at the origin of a coordinate plane and its terminal point at $P(a,b)$ corresponds to the complex number $a + bi$.

24. Draw the two vectors that correspond to $-3 + 2i$ and $2 + 3i$ and draw their sum.

25. Draw the vector $\vec{V}$ that corresponds to $5 + 2i$ and the scalar multiple $s\vec{V}$ that corresponds to $-2(5 + 2i)$.

25.

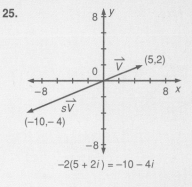

$-2(5 + 2i) = -10 - 4i$

Chapter Test

21.

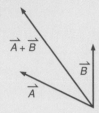

$\vec{A} + \vec{B}$

$\vec{B}$

$\vec{A}$

22.

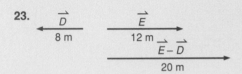

$\vec{B}$

$\vec{C}$

$\vec{C} - \vec{B}$

23.

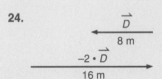

$\vec{D}$ 8 m

$\vec{E}$ 12 m

$\vec{E} - \vec{D}$ 20 m

24.

$\vec{D}$ 8 m

$-2 \cdot \vec{D}$ 16 m

25.

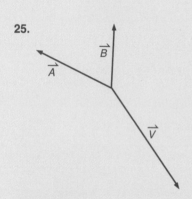

$\vec{B}$

$\vec{A}$

$\vec{V}$

■ **Chapter 9 Test**

A Exercises: 1, 3, 5–14, 16–19, 22–25, 27
B Exercises: 2, 4, 15, 20, 21, 26   C Exercises: 28, 29

Write in the form $a + bi$, where a and b are real numbers. $5 - 11i\sqrt{2}$

1. $(5 - 2i) - (-7 + 4i)$ $12 - 6i$

2. $(7 - 3\sqrt{-8}) + (-2 - \sqrt{-50})$

3. Simplify $|5 - 3i|$. $\sqrt{34}$

4. Simplify $|7 - i| - |7 + i|$. 0

5. Solve $4y^2 + 40 = 0$. $\pm i\sqrt{10}$

6. Solve $x^4 + 4x^2 - 32 = 0$. $\pm 2, \pm 2i\sqrt{2}$

Simplify.

7. $2\sqrt{-3} \cdot 3\sqrt{-6}$ $-18\sqrt{2}$

8. i^9 i

9. $(3 + 5i)(2 - 4i)$ $26 - 2i$

10. $5 \div (-6i)$ $\frac{5i}{6}$

11. $(3 + 4i)^2$ $-7 + 24i$

12. $(-i\sqrt{3})^5$ $-9i\sqrt{3}$

13. $(3 - 6i)(\overline{3 - 6i})$ 45

14. $\dfrac{3}{5 + 2i}$ $\frac{15 - 6i}{29}$

15. Factor $25x^2 + 36$. $(5x + 6i)(5x - 6i)$

16. Solve $x^2 - 8x - 5 = 0$ by completing the square. $4 \pm \sqrt{21}$

17. Three hundred people will enter a bowling tournament if the entry fee is \$10 each. The number of entries will decrease by 20 for each \$2 increase in the fee. What entry fee will yield the maximum income from all the fees? $20

18. Solve $2x^2 + 5 = 4x$ using the quadratic formula. Simplify. $\frac{2 \pm i\sqrt{6}}{2}$

19. Find the three cube roots of 64. Simplify. $4, -2 \pm 2i\sqrt{3}$

20. The length of a rectangle is 3 ft more than twice its width, and 2 ft less than a diagonal's length. Find the length of the diagonal in simplest radical form. $13 + 8\sqrt{2}$ ft

Copy the figures below. Then draw and label the following for Exercises 21–25.

$\vec{B}$ $\vec{A}$ $\vec{C}$ $\vec{D}$ 8 m $\vec{E}$ 12 m

21. $\vec{A} + \vec{B}$ **22.** $\vec{C} - \vec{B}$ **23.** $\vec{E} - \vec{D}$ **24.** $-2 \cdot \vec{D}$

25. $\vec{V}$, so that $\vec{A} + \vec{B} + \vec{V}$ is a zero vector.

26. Draw the two vectors that correspond to $-3 + 4i$ and $4 + 2i$ and draw their sum.

27. Draw the vector $\vec{V}$ that corresponds to $2 - i$ and the scalar multiple $s\vec{V}$ that corresponds to $-3(2 - i)$.

28. Prove that $|a + bi| = |a|\sqrt{2}$, if $a = b$.

29. Solve $x^2 + 10ix - 16 = 0$ by completing the square. $-8i, -2i$

26.

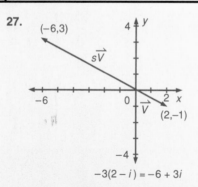

$(1,6)$ $(-3,4)$ $(4,2)$

$(-3 + 4i) + (4 + 2i) = 1 + 6i$

27.

$(-6,3)$ $s\vec{V}$ $\vec{V}$ $(2,-1)$

$-3(2 - i) = -6 + 3i$

28. $|a + bi|$
$= |a + ai|$
$= \sqrt{a^2 + a^2}$
$= \sqrt{2a^2}$
$= \sqrt{a^2} \cdot \sqrt{2}$
$= |a|\sqrt{2}$

College Prep Test

Choose the *one* best answer to each question or problem.

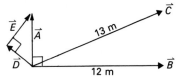

1. Use the drawing above to find the magnitude of $\vec{E}$, given that $\vec{A} + \vec{B} = \vec{C}$, $\vec{D} + \vec{E} = \vec{A}$, and the magnitude of $\vec{D}$ is 3 m. B
   (A) 5 m (B) 4 m (C) 3 m
   (D) 2 m (E) None of these

2. $a + bi = c + di$ if and only if $a = c$ and $b = d$. Find x and y given that $3x + y + 9i = 1 + (2x - y)i$. D
   (A) $x = \frac{8}{5}$ and $y = \frac{11}{5}$
   (B) $x = 4$ and $y = -1$
   (C) $x = 3$ and $y = -8$
   (D) $x = 2$ and $y = -5$
   (E) None of these

3. For every complex number z, z^* is defined to be z^{-2}. Find the value of $[(-3i)^*]^*$. B
   (A) $81i$ (B) 81 (C) $-81i$
   (D) -81 (E) None of these

4. The formula $h = 5t(40 - t)$ gives the height h of an object at t seconds. The greatest height is reached at which time? C
   (A) 0 s (B) 10 s (C) 20 s (D) 40 s
   (E) All heights are the same.

5. Choose the number that is not equal to the other three. E
   (A) $4 + 2\sqrt{5}$ (B) $\dfrac{8 + \sqrt{80}}{2}$
   (C) $6 + \sqrt{80} - 2 - \sqrt{20}$
   (D) $\sqrt{2}(\sqrt{8} + \sqrt{10})$
   (E) They are all equal.

6. The complex number $x + yi$ corresponds to the point $P(x,y)$. The sum of $5 - 4i$, the conjugate of $2 + 3i$, and the opposite of $6 - i$ corresponds to a point in which quadrant? D
   (A) Quadrant I
   (B) Quadrant II
   (C) Quadrant III
   (D) Quadrant IV
   (E) None of these

7. If $7(3 - 2i)^2 = n(2 + 3i)^2$, then $n = \underline{\ ?\ }$. B
   (A) 7 (B) -7 (C) $7i$
   (D) $-7i$ (E) None of these

8. For which equation is the sum of the roots the greatest? A
   (A) $(x - 6)^2 = 4$
   (B) $(x - 2)^2 = 9$
   (C) $(x + 5)^2 = 16$
   (D) $(x + 8)^2 = 25$
   (E) $x^2 = 36$

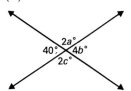

9. In the figure above, what is the value of $a + b + c$? C
   (A) 40 (B) 140 (C) 150
   (D) 160 (E) 320

10. Each of the numbers below is the product of two consecutive positive integers. For which of these is the greater of the two consecutive integers an even integer? D
   (A) 20 (B) 42 (C) 72
   (D) 90 (E) None of these

Additional Answers, page 309

39. $-1,000,000$

47. $\overline{w + z}$
   $= \overline{(a + bi) + (c + di)}$
   $= \overline{(a + c) + (b + d)i}$
   $= (a + c) - (b + d)i$
   $= (a - bi) + (c - di)$
   $= \overline{a + bi} + \overline{c + di}$
   $= \overline{w} + \overline{z}$

48. $\overline{w - z}$
   $= \overline{(a + bi) - (c + di)}$
   $= \overline{(a - c) + (b - d)i}$
   $= (a - c) - (b - d)i$
   $= (a - bi) - (c - di)$
   $= \overline{a + bi} - \overline{c + di}$
   $= \overline{w} - \overline{z}$

49. $\overline{wz}$
   $= \overline{(a + bi)(c + di)}$
   $= \overline{ac + adi + bci + bdi^2}$
   $= \overline{(ac - bd) + (ad + bc)i}$
   $= (ac - bd) - (ad + bc)i$
   $= ac - adi - bci - bd$
   $= (a - bi)(c - di)$
   $= \overline{(a + bi)}\ \overline{(c + di)}$
   $= \overline{w} \cdot \overline{z}$

50. $\overline{\overline{w}} = \overline{\overline{a + bi}}$
   $= \overline{a - bi}$
   $= a + bi = w$

337

10 POLYNOMIAL EQUATIONS AND FUNCTIONS

OVERVIEW

In Chapter 10, students continue their study of relationships between the roots of a polynomial equation and the coefficients of the polynomial. Second-degree equations are stressed in Lessons 10.1 and 10.2, while higher-degree equations are covered in Lessons 10.4 through 10.7. Lesson 10.3 considers algebraic equations in which at least one polynomial is under a radical symbol. The techniques of Lesson 10.3 will be needed when developing some of the formulas of Chapter 11.

OBJECTIVES

- To describe the nature of the roots of a quadratic equation
- To solve radical equations
- To find the complex zeros of an integral polynomial
- To graph polynomial functions

PROBLEM SOLVING

Lesson 10.6 asks students to use the problem solving strategy *Making a Graph* to locate the real zeros of functions. For problem solving of a less routine kind, students may try the *Brainteaser* on page 360.

READING AND WRITING MATH

There is a Focus on Reading on page 345 that asks students to interpret the discriminant of a quadratic equation. Also, there are two exercises in which the students are asked to explain an algebraic concept in their own words. The first is on page 359 in Lesson 10.5; the second is on page 371 in the Chapter Review.

TECHNOLOGY

Calculator: Both a scientific calculator and a graphing calculator can be used in this chapter. The graphing calculator can be used to help evaluate polynomials in Lesson 10.6 using the techniques mentioned in Lesson 6.5. Also in Lesson 10.6, a graphing calculator can be used to help graph some of the more difficult polynomial functions.

SPECIAL FEATURES

Mixed Review pp. 342, 346, 355, 360, 364, 369
Brainteaser pp. 342, 360
Focus on Reading p. 345
Midchapter Review p. 349
Application: Escape Velocity p. 350
Brainteaser p. 360
Statistics: Correlation p. 365
Application: Rational Powers In Music p. 369
Key Terms p. 370
Key Ideas and Review Exercises pp. 370–371
Chapter 10 Test p. 372
College Prep Test p. 373
Cumulative Review (Chapters 1–10) pp. 374–375

PLANNING GUIDE

Lesson	Basic	Average	Above Average	Resources
10.1 pp. 341–342	CE all WE 1–23 odd Brainteaser	CE all WE 7–29 odd Brainteaser	CE all WE 11–33 odd Brainteaser	Reteaching 68 Practice 68
10.2 pp. 345–346	FR all CE all WE 1–17 odd	FR all CE all WE 5–21 odd	FR all CE all WE 9–25 odd	Reteaching 69 Practice 69
10.3 pp. 348–350	CE all WE 1–19 odd Midchapter Review Application	CE all WE 7–25 odd Midchapter Review Application	CE all WE 9–27 odd Midchapter Review Application	Reteaching 70 Practice 70
10.4 p. 355	CE all WE 1–13 odd	CE all WE 5–17 odd	CE all WE 7–19 odd	Reteaching 71 Practice 71
10.5 pp. 359–360	CE all WE 1–9 odd Brainteaser	CE all WE 5–17 odd Brainteaser	CE all WE 9–19 odd Brainteaser	Reteaching 72 Practice 72
10.6 pp. 363–365	CE all WE 1–9 odd Statistics	CE all WE 5–13 odd Statistics	CE all WE 7–15 odd Statistics	Reteaching 73 Practice 73
10.7 pp. 368–369	CE all WE 1–7 odd Application	CE all WE 3–11 odd Application	CE all WE 7–13 odd Application	Reteaching 74 Practice 74
Chapter 10 Review pp. 370–371	all	all	all	
Chapter 10 Test p. 372	all	all	all	
College Prep Test p. 373	all	all	all	
Cumulative Review pp. 374–375	1–59 odd	5–65 odd	7–69 odd	

CE = Classroom Exercises   WE = Written Exercises   FR = Focus on Reading
NOTE: For each level, all students should be assigned all Mixed Review exercises.

■■■INVESTIGATION

Project: In this investigation, the students are guided in an exploration of Descartes' Rule of Signs for certain special cases.

Materials: Students may work individually or in groups. Each student or group of students will need a graphing calculator, and each student should have a copy of Project Worksheet 10.

There are four different questions on the worksheet. Allow time for each question individually. When most of the students have completed the work on each question, ask for different students to report on their findings.

The worksheet leads the students to certain insights about Descartes' Rule by asking them to observe the relationship of a graph of a polynomial factored into binomials to its graph.

Some of the students may be motivated to carry the investigation beyond the questions on the worksheet.

Mathematical equations are used to study population growth. The equations are often simple, but the repetition of them reveals fascinating patterns. The "science-fiction landscape" shown here is actually a computer study of what happens when different arrangements of numbers are used in a simple population-growth equation.

More About the Image

The population growth equation $x_{n+1} = rx_n(1 - x_n)$ is used to model successive populations. When iterated indefinitely, certain values of r produce *periodic cycles* for the values of the population, while others become "chaotic," with *no regularly recurring values*. In the illustration, the process has been carried a step further, by allowing the constant r to take on two different values A and B in a regularly recurring pattern throughout the iteration: $AAAAAABBBBBB, AAAAAABBBBBB, \ldots$. The illustration is a plotting of points (A,B) on the coordinate plane. The dark regions consist of the values of A and B for which population growth is chaotic.

10.1 Sums and Products of Roots

Objectives

To find the sums and products of roots of given quadratic equations without solving the equations

To write quadratic equations in the general form $ax^2 + bx + c = 0$ given their roots

Teaching Resources

Application Worksheet 10
Problem Solving Worksheet 10
Quick Quizzes 68
Reteaching and Practice
   Worksheets 68

The polynomial equation $6x^5 - 17x^4 - 29x^3 + 74x^2 + 20x - 24 = 0$ has roots, but cannot be solved directly for x. Even so, some information about its roots *can* be obtained by carefully examining the coefficients. You will study such higher-degree equations later in this chapter, but first, some simpler cases. One such case is the quadratic equation.

An interesting property of the general quadratic equation is that the sum and product of its solutions, or roots, can be found without solving the equation for the roots themselves. For example, if x_1 and x_2 are the roots of $3x^2 + 7x - 20 = 0$, where $a = 3$, $b = 7$, and $c = -20$, then $x_1 + x_2 = -\frac{7}{3} \left(= -\frac{b}{a}\right)$ and $x_1 \cdot x_2 = \frac{-20}{3} \left(= \frac{c}{a}\right)$. This can be shown using the quadratic formula, $x = \dfrac{-b \pm \sqrt{b^2 - 4ac}}{2a}$.

If $3x^2 + 7x - 20 = 0$,

then $x_1 = \dfrac{-7 + \sqrt{289}}{6}$ and $x_2 = \dfrac{-7 - \sqrt{289}}{6}$.

$$x_1 + x_2 = \frac{-7 + \sqrt{289}}{6} + \frac{-7 - \sqrt{289}}{6} = \frac{-14}{6} = -\frac{7}{3} \left(= -\frac{b}{a}\right).$$

$$x_1 \cdot x_2 = \frac{-7 + \sqrt{289}}{6} \cdot \frac{-7 - \sqrt{289}}{6}$$

$$= \frac{49 - 289}{36} = \frac{-240}{36} = \frac{-20}{3} \left(= \frac{c}{a}\right)$$

This suggests Theorem 10.1 below.

Theorem 10.1

If $ax^2 + bx + c = 0$, or $x^2 + \frac{b}{a}x + \frac{c}{a} = 0$, where $a \neq 0$ and x_1 and x_2 are the roots, then

$$x_1 + x_2 = -\frac{b}{a} \text{ and } x_1 \cdot x_2 = \frac{c}{a}.$$

■■■■ **GETTING STARTED**

Prerequisite Quiz

Solve each equation using the quadratic formula.

1. $2x^2 + 3x - 9 = 0$ $\frac{3}{2}, -3$
2. $y^2 - 6y + 7 = 0$
    $3 + \sqrt{2}, 3 - \sqrt{2}$
3. $4t^2 - 4t + 1 = 0$ $\frac{1}{2}$

Find the sum and the product of each pair of numbers.

4. $3\sqrt{5}$ and $-3\sqrt{5}$ Sum = 0; prod = -45
5. $-5i$ and $5i$ Sum = 0; prod = 25

Motivator

Ask students to write a quadratic equation given that its roots are -3 and $\frac{1}{2}$ by working backward from $x = -3$ or $x = \frac{1}{2}$.

$x + 3 = 0$ or $2x - 1 = 0$
$(x + 3)(2x - 1) = 0$
$2x^2 + 5x - 3 = 0$

Highlighting the Standards

Standard 2A: Constructing equations according to given conditions helps build facility with the language of mathematics, which "is an integral part of thinking mathematically."

Lesson Note

The converse of Theorem 10.1 can be used to check the roots of a given quadratic equation. For example, check that $-\frac{3}{2}$ and 1 are the roots of $2x^2 + x - 3 = 0$.

$-\frac{3}{2} + 1 = -\frac{1}{2} = -\frac{b}{a}; \; -\frac{3}{2} \cdot 1 = -\frac{3}{2} = \frac{c}{a}$

Math Connection

The Linear Equation: The lesson considers the relationship between the roots of a polynomial equation and the coefficients of the polynomial. For a simpler instance of this relationship, students may consider the polynomial equation of first degree, that is, the linear equation $ax + b = 0$. The solution $x = -\frac{b}{a}$ suggests the relationship $x_1 + x_2 = -\frac{b}{a}$ in the quadratic case.

Critical Thinking Questions

Analysis: The text points out that "a quadratic equation may have exactly one root." Ask students whether a quadratic equation may have *no* roots or *more than two* roots. By contemplating the quadratic formula, they should be able to answer *No* to both of these suggested possibilities. See also the *Fundamental Theorem of Algebra* in Lesson 10.4.

EXAMPLE 1 Find the sum and product of the roots of $5x^2 + 2 = 6x$ without solving the equation.

Solution

$5x^2 - 6x + 2 = 0 \qquad a = 5, b = -6, c = 2$

$x_1 + x_2 = -\frac{b}{a} = -\left(\frac{-6}{5}\right) = \frac{6}{5} \qquad x_1 \cdot x_2 = \frac{c}{a} = \frac{2}{5}$

The sum of the roots is $\frac{6}{5}$, and their product is $\frac{2}{5}$.

The converse of Theorem 10.1 is also true.

Theorem 10.2

If $x_1 + x_2 = -\frac{b}{a}$ and $x_1 \cdot x_2 = \frac{c}{a}$, then x_1 and x_2 are the roots of the equation $x^2 + \frac{b}{a}x + \frac{c}{a} = 0$, or $ax^2 + bx + c = 0$.

Therefore, if $\frac{6}{5}$ and $\frac{2}{5}$ are the sum and product, respectively, of the two roots of a quadratic equation, then

$$\frac{6}{5} = -\frac{b}{a}, \text{ or } -\frac{6}{5} = \frac{b}{a}, \text{ and } \frac{2}{5} = \frac{c}{a},$$

and the equation is

$$x^2 - \frac{6}{5}x + \frac{2}{5} = 0, \text{ or } 5x^2 - 6x + 2 = 0.$$

EXAMPLE 2 Write a quadratic equation, in general form, that has $\frac{4}{5}$ and $-\frac{1}{2}$ as its roots.

Plan Find the sum and product of the roots. Then use Theorem 10.2.

Solution

Let $x_1 = \frac{4}{5}$ and $x_2 = -\frac{1}{2}$.

$x_1 + x_2 = \frac{4}{5} + \frac{-1}{2} = \frac{8}{10} + \frac{-5}{10} = \frac{3}{10}$

So, $-\frac{b}{a} = \frac{3}{10}$ and $\frac{b}{a} = -\frac{3}{10}$

$x_1 \cdot x_2 = \frac{4}{5} \cdot \frac{-1}{2} = -\frac{4}{10}$

So, $\frac{c}{a} = -\frac{4}{10}$

$x^2 + \frac{b}{a}x + \frac{c}{a} = 0$ becomes $x^2 - \frac{3}{10}x - \frac{4}{10} = 0$.

The equation, in general form, is $10x^2 - 3x - 4 = 0$.

A quadratic equation may have exactly one root. For example, if $x^2 - 6x + 9 = 0$, then $(x - 3)(x - 3) = 0$ and $x - 3 = 0$ *or* $x - 3 = 0$. The only root is 3. For this case, let x_1 and x_2 represent the roots of the quadratic equation, and let $x_1 = x_2$.

340 Chapter 10 Polynomial Equations and Functions

Additional Example 1

Find the sum and product of the roots of $3x^2 = 12 - 4x$, without solving the equation. $-\frac{4}{3}, -4$

Additional Example 2

Write a quadratic equation, in general form, that has -2 and $\frac{4}{3}$ as the roots. $3x^2 + 2x - 8 = 0$

EXAMPLE 3 Write a quadratic equation, in general form, that has $\frac{3}{5}$ as its only root.

Solution Let $x_1 = x_2 = \frac{3}{5}$.

$x_1 + x_2 = \frac{3}{5} + \frac{3}{5} = \frac{6}{5} = -\frac{b}{a}$ and $x_1 \cdot x_2 = \frac{3}{5} \cdot \frac{3}{5} = \frac{9}{25} = \frac{c}{a}$

Because $-\frac{b}{a} = \frac{6}{5}, \frac{b}{a} = -\frac{6}{5}$.

Thus, $x^2 + \frac{b}{a}x + \frac{c}{a} = x^2 - \frac{6}{5}x + \frac{9}{25} = 0$, or $25x^2 - 30x + 9 = 0$.

The procedure shown above can also be used with roots that are pairs of conjugate irrational numbers or conjugate complex numbers, as illustrated in Example 4.

EXAMPLE 4 Write a quadratic equation, in general form, that has $-5 \pm 3\sqrt{2}$ as its roots.

Solution Let $x_1 = -5 + 3\sqrt{2}$ and $x_2 = -5 - 3\sqrt{2}$.

Then $x_1 + x_2 = (-5 + 3\sqrt{2}) + (-5 - 3\sqrt{2}) = -10 \leftarrow -\frac{b}{a} = -10$

$\frac{b}{a} = 10$

and $x_1 \cdot x_2 = (-5 + 3\sqrt{2})(-5 - 3\sqrt{2}) = 25 - 18 = 7 \leftarrow \frac{c}{a} = 7$

Thus, the equation with the given roots is $x^2 + 10x + 7 = 0$.

Classroom Exercises

Match each equation at the left with one of the items a–j in the column at the right.

1. $x^2 + 11x + 24 = 0$ d
2. $x^2 + 5x - 24 = 0$ c
3. $x^2 - 11x + 24 = 0$ a
4. $x^2 - 5x - 24 = 0$ b
5. $x^2 + 6x - 10 = 0$ g
6. $x^2 - 6x + 10 = 0$ i
7. $x^2 - 6x - 10 = 0$ e
8. $2y^2 - 5y + 8 = 0$ j
9. $2y^2 + 5y - 8 = 0$ h
10. $2y^2 + 5y + 8 = 0$ f

a. roots: 3, 8
b. roots: -3, 8
c. roots: 3, -8
d. roots: -3, -8
e. sum and product of roots: 6, -10
f. sum and product of roots: $-\frac{5}{2}$, 4
g. sum and product of roots: -6, -10
h. sum and product of roots: $-\frac{5}{2}$, -4
i. sum and product of roots: 6, 10
j. sum and product of roots: $\frac{5}{2}$, 4

Checkpoint

1. Without solving the equation, find the sum and the product of the roots of $6x^2 + 3x - 2 = 0$.
Sum $= -\frac{1}{2}$; prod $= -\frac{1}{3}$

Write a quadratic equation, in general form, that has the given root or roots.

2. -3 and 7 $x^2 - 4x - 21 = 0$
3. $\frac{3}{4}$ and $-\frac{5}{8}$ $32x^2 - 4x - 15 = 0$
4. $-3\sqrt{2}$ and $3\sqrt{2}$ $x^2 - 18 = 0$

Closure

Ask students to explain how they can find both the sum and the product of the roots of a quadratic equation without solving the equation. See Theorem 10.1. Have students write a quadratic equation if one of its roots is $3 + \sqrt{2}$.
$x^2 - 6x + 7 = 0$

Additional Example 3

Write a quadratic equation, in general form, that has $-\frac{3}{2}$ as its only root.
$4x^2 + 12x + 9 = 0$

Additional Example 4

Write a quadratic equation, in general form, that has $2 \pm \sqrt{6}$ as its roots.
$x^2 - 4x - 2 = 0$

Guided Practice

Classroom Exercises 1–10

Independent Practice

A Ex. 1–17, **B** Ex. 18–31, **C** Ex. 32, 33

Basic: WE 1–23 odd, Brainteaser

Average: WE 7–29 odd, Brainteaser

Above Average: WE 11–33 odd, Brainteaser

Additional Answers

Written Exercises

14. $32x^2 + 20x + 3 = 0$
15. $5x^2 - 11x - 12 = 0$
16. $x^2 + 10x + 25 = 0$
17. $9x^2 - 12x + 4 = 0$
18. $x^2 - 12 = 0$
19. $x^2 + 16 = 0$
20. $x^2 - 6x + 2 = 0$
21. $x^2 + 12x + 16 = 0$

Written Exercises

Without solving each equation, find the sum and product of its roots.

1. $x^2 + 9x - 22 = 0$ $-9, -22$
2. $y^2 - 7y - 18 = 0$ $7, -18$
3. $3n^2 - 12n = 0$ $4, 0$
4. $5y^2 - 6y - 1 = 0$ $\frac{6}{5}, -\frac{1}{5}$
5. $6n^2 - 12n + 3 = 0$ $2, \frac{1}{2}$
6. $t^2 = 6 - 5t$ $-5, -6$
7. $-2y^2 + y + 6 = 0$ $\frac{1}{2}, -3$
8. $8x^2 + 6x = 0$ $-\frac{3}{4}, 0$
9. $2y^2 - 5 = 0$ $0, -\frac{5}{2}$

Write a quadratic equation, in general form, that has the given root or roots.

10. 4 and 6 $x^2 - 10x + 24 = 0$
11. -5 and 3 $x^2 + 2x - 15 = 0$
12. 0 and 2 $x^2 - 2x = 0$
13. $\frac{1}{6}$ and $-\frac{1}{3}$ $18x^2 + 3x - 1 = 0$

14. $-\frac{3}{8}$ and $-\frac{1}{4}$
15. $-\frac{4}{5}$ and 3
16. -5
17. $\frac{2}{3}$

18. $\pm 2\sqrt{3}$
19. $\pm 4i$
20. $3 \pm \sqrt{7}$
21. $-6 \pm 2\sqrt{5}$

22. $4 \pm 2i$ $x^2 - 8x + 20 = 0$
23. $\frac{5 \pm \sqrt{10}}{2}$ $4x^2 - 20x + 15 = 0$
24. $\frac{-1 \pm 2\sqrt{6}}{3}$ $9x^2 + 6x - 23 = 0$
25. $\frac{-6 \pm 5i\sqrt{2}}{4}$ $8x^2 + 24x + 43 = 0$

Without solving each equation, find the sum and product of its roots.

26. $3x^2 + \frac{1}{2}x - \frac{2}{3} = 0$ $-\frac{1}{6}, -\frac{2}{9}$
27. $\frac{1}{2}y^2 - 6y + 3 = 0$ $12, 6$
28. $\frac{2}{3}t^2 + 4 = \frac{1}{6}t$ $\frac{1}{4}, 6$

29. $0.4x^2 + 1.2x = 4.8$ $-3, -12$
30. $12y^2 = 1.44$ $0, -0.12$
31. $y^2 - 3y\sqrt{2} = 20$ $3\sqrt{2}, -20$

32. For what values of k will the sum of the roots of $x^2 - (k^2 - 3k)x + 21 = 0$ be 10? $-2, 5$

33. For what values of k will the product of the roots of $x^2 - 8x + (k^2 - 10k + 21) = 0$ be 12? $1, 9$

Mixed Review

Find the solution set.   *6.1, 6.4*

1. $x^2 - 4x - 12 = 0$ $-2, 6$
2. $-12k^2 + 48k + 144 = 0$ $-2, 6$
3. $x^2 - 4x - 12 > 0$ $\{x \mid x < -2 \text{ or } x > 6\}$
4. $x^2 - 4x - 12 < 0$ $\{x \mid -2 < x < 6\}$

■/ Brainteaser

Let $f(x) = 10x^2 - 29x + 10$. For what real values of x does $f\left(x + \frac{1}{x}\right) = 0$? $\frac{1}{2}, 2$

Enrichment

Have the students prove the following theorem, which provides a method for factoring $ax^2 + bx + c$. Tell the students that they will have to use Theorem 10.1.

If r and s are roots of the equation $ax^2 + bx + c = 0$, then $ax^2 + bx + c = a(x - r)(x - s)$.

Proof:

$ax^2 + bx + c = a\left(x^2 + \frac{b}{a}x + \frac{c}{a}\right)$
$= a[x^2 - (r + s)x + rs]$
$= a(x^2 - rx - sx + rs)$
$= a[x(x - r) - s(x - r)]$
$= a(x - r)(x - s)$

10.2 The Discriminant

Objectives

To describe the nature of the roots of quadratic equations by finding their discriminants

To find coefficients of quadratic equations, such that the equations will have exactly one real root, two real roots, or two nonreal roots

It is not necessary to solve a quadratic equation in order to know the number of its roots and whether they are rational, irrational, or complex. This information can be determined instead by using the value of $b^2 - 4ac$ in $x = \dfrac{-b \pm \sqrt{b^2 - 4ac}}{2a}$.

Equation	$b^2 - 4ac$	Nature of roots	Roots
$8x^2 - 2x - 3 = 0$	100 (positive, a perfect square)	2 rational roots	$\dfrac{3}{4}, -\dfrac{1}{2}$
$x^2 + 5x - 2 = 0$	33 (positive, not a perfect square)	2 irrational roots	$\dfrac{-5 \pm \sqrt{33}}{2}$
$4x^2 - 12x + 9 = 0$	0 (perfect square)	1 rational root	$\dfrac{3}{2}$
$2x^2 + 5x + 4 = 0$	-7 (negative)	2 nonreal roots	$\dfrac{-5 \pm i\sqrt{7}}{4}$

Nature of the Roots of a Quadratic Equation

If $ax^2 + bx + c = 0$, where a, b, and c are real numbers ($a \neq 0$),

then $x = \dfrac{-b \pm \sqrt{b^2 - 4ac}}{2a}$, and $b^2 - 4ac$ is the **discriminant** of the equation.

(1) If $b^2 - 4ac > 0$, there are two real roots.
(2) If $b^2 - 4ac = 0$, there is exactly one real root.
(3) If $b^2 - 4ac < 0$, there are two nonreal roots.

Also, if a, b, and c are rational and if $b^2 - 4ac > 0$, then the roots are *rational* when $b^2 - 4ac$ is a perfect square, and the roots are *irrational* when $b^2 - 4ac$ is not a perfect square.

10.2 The Discriminant **343**

Teaching Resources

Quick Quizzes 69
Reteaching and Practice
   Worksheets 69
Transparency 22

■■■GETTING STARTED

Prerequisite Quiz

Solve the equations and find the solution sets of the inequalities.

1. $x^2 - 3x - 10 = 0$ $-2, 5$
2. $x^2 - 3x - 10 < 0$ $\{x \mid -2 < x < 5\}$
3. $x^2 - 3x - 10 > 0$ $\{x \mid x < -2 \text{ or } x > 5\}$
4. $64 - 4(k - 1)^2 = 0$ $5, -3$

Motivator

Ask students to determine the value of $b^2 - 4ac$ for the perfect-square trinomial $x^2 + 4x + 4$. $b^2 - 4ac = 0$ Have them factor the trinomial and give the number of roots. One Then have students perform the same operations for $x^2 + 6x + 9$. $b^2 - 4ac = 0$; one root Ask them if they think $b^2 - 4ac = 0$ will hold true for any perfect-square trinomial that factors as $(ax + b)^2$. Yes

Highlighting the Standards

Standard 5d: Properties of the discriminant will lead students to a sense of strength of mathematical procedures and notation.

343

Lesson Note

Rather than memorizing the four cases for the roots of a quadratic equation shown on page 343, the students can determine the nature of the roots of a quadratic equation by writing $\dfrac{-b \pm \sqrt{\square}}{2a}$ and replacing only the $\square$ (not the $-b$ and $2a$) with the discriminant of the equation.

Math Connections

The Discriminant of a Cubic Equation:
The cubic equation $z^3 + pz + q = 0$, like the quadratic equation $ax^2 + bx + c = 0$, has a discriminant, namely the expression $-4p^3 - 27q^2$. If this expression is nonnegative, then the corresponding cubic equation has only real roots. If it is negative, then the cubic equation has two nonreal roots (and one real root).

EXAMPLE 1 Use the discriminant to determine the nature of the roots.

a. $16x^2 + 24x + 9 = 0$ b. $x^2 = 6x - 25$
c. $2x^2 + 3x - 7 = 0$ d. $2x^2 = 9x + 5$

Plan Write each equation in general form in order to identify a, b, and c.

Solutions

a. $16x^2 + 24x + 9 = 0$
$a = 16, b = 24, c = 9$
$b^2 - 4ac$
$= 24^2 - 4 \cdot 16 \cdot 9$
$= 576 - 576$, or 0

Because $b^2 - 4ac = 0$, there is exactly one root, and it is real and rational.

b. $x^2 = 6x - 25$
$x^2 - 6x + 25 = 0$
$a = 1, b = -6, c = 25$
$b^2 - 4ac$
$= (-6)^2 - 4 \cdot 1 \cdot 25$
$= 36 - 100$, or -64

Because $b^2 - 4ac < 0$, there are two nonreal roots.

c. $2x^2 + 3x - 7 = 0$
$b^2 - 4ac$
$= 9 - 4 \cdot 2(-7)$
$= 9 + 56$, or 65

Because 65 is positive, there are two real roots. Because 65 is not a perfect square, the roots are irrational.

d. $2x^2 = 9x + 5$
$2x^2 - 9x - 5 = 0$
$b^2 - 4ac$
$= (-9)^2 - 4 \cdot 2(-5)$
$= 81 + 40$, or 121

Because 121 is a perfect square, there are two real, rational roots.

In an equation such as $4x^2 + 2kx + 9 = 0$, the letter k can be treated as a temporary variable. If k is replaced by 6, the equation formed is $4x^2 + 12x + 9 = 0$. Its discriminant, $b^2 - 4ac$, is $12^2 - 4 \cdot 4 \cdot 9$, or 0, and $4x^2 + 12x + 9 = 0$ has exactly one root.

If $k = 7$, then $4x^2 + 2kx + 9 = 0$ becomes $4x^2 + 14x + 9 = 0$, where $b^2 - 4ac = 14^2 - 4 \cdot 4 \cdot 9 = 52$.

If $k = 5$, then $4x^2 + 2kx + 9 = 0$ becomes $4x^2 + 10x + 9 = 0$, where $b^2 - 4ac = 10^2 - 4 \cdot 4 \cdot 9 = -44$.

Thus, $b^2 - 4ac > 0$ and $4x^2 + 14x + 9 = 0$ has two real roots.

Thus, $b^2 - 4ac < 0$ and $4x^2 + 10x + 9 = 0$ has two nonreal roots.

344   Chapter 10 Polynomial Equations and Functions

Additional Example 1

From the discriminant, $b^2 - 4ac$, determine the nature of the roots of equations a–d.

a. $4x^2 - 20x + 25 = 0$ 0; exactly one real, rational root
b. $x^2 = 2x - 10$ -36; two nonreal roots
c. $3x^2 - 6x + 2 = 0$ 12; two irrational roots
d. $6x^2 = x + 2$ 49; two real, rational roots

EXAMPLE 2   For what values of k will $(k - 4)x^2 - 12x + 3k = 0$ have exactly one root?

Plan   Express the discriminant $b^2 - 4ac$ in terms of k.

Solution   The equation $(k - 4)x^2 - 12x + 3k = 0$ is in the general form $ax^2 + bx + c = 0$. Thus, $a = k - 4$, $b = -12$, and $c = 3k$.

$$b^2 - 4ac = 144 - 4(k - 4) \cdot 3k$$
$$= 144 - 12k^2 + 48k = -12k^2 + 48k + 144$$

To have exactly one root, $b^2 - 4ac = 0$.
$$-12k^2 + 48k + 144 = 0$$
Divide by -12. $k^2 - 4k - 12 = 0$
$$(k + 2)(k - 6) = 0$$
$$k = -2 \quad or \quad k = 6$$

The equation will have exactly one root if $k = -2$ or $k = 6$.

Focus on Reading

Given the quadratic equation $ax^2 + bx + c = 0$, where a, b, and c are integers, match each expression or statement at the left with exactly one phrase at the right.

1. $b^2 - 4ac$ d
2. $b^2 - 4ac < 0$ b
3. $b^2 - 4ac = 0$ e
4. $b^2 - 4ac > 0$ c
5. $b^2 - 4ac$ is a positive perfect square.

a. two rational roots
b. two nonreal roots
c. two real roots
d. the discriminant
e. exactly one root
a

Classroom Exercises

Match each value of the discriminant of a quadratic equation with the nature of the roots of the equation. The numbers a, b, and c are rational.

1. $b^2 - 4ac = 0$ c
2. $b^2 - 4ac = 39$ d
3. $b^2 - 4ac = -16$ a
4. $b^2 - 4ac = 25$ b

a. two nonreal
b. two rational
c. one rational
d. two irrational

Critical Thinking Questions

Analysis: The third line from the bottom of page 343 begins "Also, if a, b, and c are rational . . . ." Ask students whether the word *rational* can be replaced with another word without changing the truth of the entire sentence.   With a little thought, students will probably realize that the word *integral* (or "integers") could be used instead of *rational*. Indeed, any quadratic equation with rational coefficients can be transformed to one with integral coefficients by multiplying both sides of the equation by the LCD of the coefficients.

Checkpoint

Find the discriminant $b^2 - 4ac$ and determine the nature of the roots, without solving the equation.

1. $2n^2 = 4n - 3$ -8; two nonreal roots
2. $3x^2 = 4 - x$ 49; two rational roots
3. $5t^2 + 2t = 1$ 24; two irrational roots
4. $9y^2 + 12y + 4 = 0$ 0; exactly one rational root

Additional Example 2

For what values of k will $x^2 + (k + 3)x + 16 = 0$ have exactly one root? $-11, 5$

Have students summarize the role for the discriminant in determining the nature of the roots of a quadratic equation.   See page 343.   Have students find the values of m and n for which $x^2 + (m + n)x + mn = 0$ has exactly one root. $m = n$

▬▬FOLLOW UP

Guided Practice

Classroom Exercises 1–10

Independent Practice

A Ex 1–13, **B** Ex. 14–22, **C** Ex. 23–26

Basic: FR, WE 1–17 odd

Average: FR, WE 5–21 odd

Above Average: FR, WE 9–25 odd

Additional Answers

Written Exercises

4. -108, 2 nonreal
5. 1, 2 rational
6. 0, 1 rational

Determine whether the quadratic equation has exactly one rational root, two rational roots, two irrational roots, or two nonreal roots.

5. $(x - 4)(x - 4) = 0$ One rational
6. $(x - 2i)(x + 2i) = 0$ Two nonreal
7. $(2x + 5)(3x - 2) = 0$ Two rational
8. $(x + 3\sqrt{2})(x - 3\sqrt{2}) = 0$ Two irrational
9. $(x + 3)^2 = 0$ One rational
10. $x^2 - 5x - 1 = 0$ Two irrational

Written Exercises

Find the discriminant, $b^2 - 4ac$, and determine the nature of the roots without solving the equation.

0; one rational   72; two irrational   64; two rational
1. $4x^2 - 20x + 25 = 0$ **2.** $9y^2 + 12y + 2 = 0$ **3.** $4n^2 + 4n - 3 = 0$
4. $9t^2 - 6t + 4 = 0$ **5.** $x^2 = 21x - 110$ **6.** $49y^2 + 25 = 70y$
7. $1 = 8n - 4n^2$ **8.** $-t^2 + 6t - 10 = 0$ **9.** $7x^2 + 10x = 0$
48; two irrational -4; two nonreal   100; two rational

For what values of k will the quadratic equation have exactly one root?

10. $kx^2 - 12x + k = 0$ ± 6 **11.** $y^2 + 8y + k^2 = 0$ ± 4
12. $2y^2 + 5ky + 50 = 0$ ± 4 **13.** $x^2 - (k + 6)x + 16 = 0$ $-14, 2$

For what values of k will the quadratic equation have two *real* roots?

14. $kx^2 + 10x + k = 0$ **15.** $y^2 - 6y + k = 0$ **16.** $y^2 - 2ky + 16 = 0$
$-5 < k < 5, k \neq 0$ $-3 < k < 3$ $k < -4$ or $k > 4$

For what values of k will the quadratic equation have two *nonreal* roots?

17. $x^2 + 5kx + 25 = 0$ $-2 < k < 2$ **18.** $y^2 - (k + 6)y + 64 = 0$ $-22 < k < 10$

Determine the nature of the roots without solving the equation.

19. $1.5x^2 - 2.5x - 2 = 0$ Two irrational   **20.** $5y^2 + 3.5y + 2.2 = 0$ Two nonreal
21. $1.8x^2 - 7.2x + 7.2 = 0$ One rational   **22.** $(2x - 7)(3x + 8) = -50$ Two rational

Determine the nature of the roots without solving the equation. Note that at least one coefficient in each equation is an irrational number.

23. $x^2 - 4x\sqrt{3} + 12 = 0$ One irrational   **24.** $y^2\sqrt{2} - 6y - 20\sqrt{2} = 0$ Two irrational
25. $4z^2 + 12iz - 9 = 0$ One nonreal   **26.** $6ix^2 + 11x - 3i = 0$ Two nonreal

Mixed Review

Simplify each expression.   *8.3, 8.5*

1. $(\sqrt{x + 5})^2$ **2.** $(\sqrt[3]{2x - 1})^3$ **3.** $(3\sqrt{x - 2})^2$ **4.** $(3 + \sqrt{x})^2$
$x + 5$ $2x - 1$ $9x - 18$ $x + 6\sqrt{x} + 9$

Enrichment

Remind the students that the roots of $ax^2 + bx + c = 0$ are nonreal complex roots if $b^2 - 4ac$ is negative. Then show this equation on the chalkboard.

$ix^2 - 3ix + 2i = 0$

Have the students evaluate $b^2 - 4ac$, and then solve for the roots. $b^2 - 4ac = -1$ and the roots are 1 and 2.

Ask the students to explain why the roots can be real numbers even though the discriminant is negative.   The discriminant test applies only if a, b, and c are real numbers. In the given equation, a, b, and c are imaginary numbers.

Point out to students that when the given equation is divided through by i, the equation is equivalent to $x^2 - 3x - 2 = 0$.

Have the students find other equations in which the discriminant is negative but the roots are real numbers.

10.3 Radical Equations

Teaching Resources

Quick Quizzes 70
Reteaching and Practice
  Worksheets 70

Objective   To solve radical equations

Equations such as $x - \sqrt{2x + 1} = 7$, $\sqrt[3]{y^2 - 6y} = -2$, and $2\sqrt[4]{3x} = \sqrt[4]{3x + 15}$ are called **radical equations** because variables appear in the radicands. Such equations can be solved by (1) isolating the radicals each on one side, (2) raising each side of the equation to the power indicated by the index of the radicals, (3) solving for the variable, and (4) checking the apparent solutions. Although radical equations are not polynomial equations, polynomial equations often arise in the steps of the solution.

In Example 1 below, each radical can be isolated on one side of the equation.

EXAMPLE 1   Solve $2\sqrt[4]{3x} - \sqrt[4]{3x + 15} = 0$. Check.

Plan   The indices are both 4, so isolate the radicals and raise each side to the fourth power.

Solution

$$2\sqrt[4]{3x} - \sqrt[4]{3x + 15} = 0$$
$$2\sqrt[4]{3x} = \sqrt[4]{3x + 15}$$
$$(2\sqrt[4]{3x})^4 = (\sqrt[4]{3x + 15})^4$$
$$16 \cdot 3x = 3x + 15$$
$$45x = 15$$
$$x = \frac{1}{3}$$

$$2\sqrt[4]{3x} - \sqrt[4]{3x + 15} = 0$$
$$2\sqrt[4]{3 \cdot \frac{1}{3}} - \sqrt[4]{3 \cdot \frac{1}{3} + 15} \stackrel{?}{=} 0$$
$$2\sqrt[4]{1} - \sqrt[4]{16} \stackrel{?}{=} 0$$
$$0 = 0$$
          True

The solution is $\frac{1}{3}$.

The apparent solutions of a radical equation must always be checked in the original equation because raising each side to the same power may introduce an *extraneous* solution, one that does not satisfy the original equation (see Lesson 7.6). Notice that it is possible to square each side of an equation that is false for all values of x, such as $3 = -\sqrt{x}$, and obtain an equation, such as $3^2 = x$, that is true for some values of x ($x = 9$). In this case, 9 is an extraneous solution because it does not satisfy the original equation, $3 = -\sqrt{x}$, even though it satisfies the squared equation, $9 = x$.

GETTING STARTED

Prerequisite Quiz

Simplify.

1. $(\sqrt{x - 3})^2$ $x - 3$
2. $(\sqrt[3]{2y + 5})^3$ $2y + 5$
3. $(3\sqrt{3t - 1})^2$ $27t - 9$
4. $(2\sqrt[3]{4x + 3})^3$ $32x + 24$
5. $(4 - \sqrt{y})^2$ $16 - 8\sqrt{y} + y$

Motivator

Ask students how they would solve for x in the equation $\sqrt[4]{x} = 2$. Raise each side of the equation to the power indicated by the index of the radical. Have students give a definition of the term "extraneous solution." See page 253.

Additional Example 1

Solve $\sqrt[3]{7x + 25} - 2\sqrt[3]{4x} = 0$. Check.   1

Highlighting the Standards

Standard 1b: This lesson illustrates that problem situations often spring from within mathematics itself.

Lesson Note

The concept of extraneous solutions appeared in Chapter 7 in solving fractional equations. For example, if $\frac{2n-9}{n-7} + \frac{n}{2} = \frac{5}{n-7}$ and each side is multiplied by $2(n-7)$, then $4n - 18 + n^2 - 7n = 10$, $n^2 - 3n - 28 = 0$, $(n-7)(n+4) = 0$, and $n = 7$ or $n = -4$. But 7 is not a root since the fraction $\frac{2n-9}{n-7}$, or $\frac{14-9}{7-7}$, is undefined for $n = 7$. (The denominator of a fraction cannot be 0.)

Extraneous solutions may appear while solving a *radical* equation by squaring each side, or raising each side to a higher power. Thus, students must check "apparent" solutions in the original equation.

Math Connections

String Instruments: The pitch of a vibrating string depends on the fundamental frequency f of the vibrations of the string. This frequency in turn is given by the following radical equation.

$$f = \frac{1}{2L}\sqrt{\frac{T}{\mu}}$$

In this formula, L is the length of the string, T is the tension in the string and μ is the mass per unit length of string. The formula expresses the fact that the frequency of a string varies inversely as its length. This is seen, for example, in the case of the relatively long strings of a double bass compared with the strings of a violin.

EXAMPLE 2 Solve $x - \sqrt{2x + 1} = 7$. Check for extraneous solutions.

Plan Isolate the radical and square each side.

Solution
$$x - 7 = \sqrt{2x + 1}$$
$$(x - 7)^2 = (\sqrt{2x + 1})^2$$
$$x^2 - 14x + 49 = 2x + 1$$
$$x^2 - 16x + 48 = 0$$
$$(x - 4)(x - 12) = 0$$
$$x = 4 \quad or \quad x = 12$$

Checks
$x = 4$:
$$x - \sqrt{2x + 1} = 7$$
$$4 - \sqrt{9} \overset{?}{=} 7$$
$$1 = 7$$
$$\text{False}$$

$x = 12$:
$$x - \sqrt{2x + 1} = 7$$
$$12 - \sqrt{25} \overset{?}{=} 7$$
$$7 = 7$$
$$\text{True}$$

The only solution is 12. (4 is extraneous.)

An equation such as $\sqrt{3x + 1} - \sqrt{x - 1} = 2$ has two radicals and a third term. To solve it, it is necessary to isolate radicals twice.

EXAMPLE 3 Solve $\sqrt{3x + 1} - \sqrt{x - 1} = 2$. Check.

Solution Isolate one of the radicals and square each side.

$$\sqrt{3x + 1} = 2 + \sqrt{x - 1}$$

Square each side. $\quad 3x + 1 = 4 + 4\sqrt{x - 1} + (x - 1)$
Isolate the radical. $\quad 2x - 2 = 4\sqrt{x - 1}$
Divide each side by 2. $\quad x - 1 = 2\sqrt{x - 1}$
Square each side.
$$x^2 - 2x + 1 = 4(x - 1)$$
$$x^2 - 6x + 5 = 0$$
$$(x - 1)(x - 5) = 0$$
$$x = 1 \quad or \quad x = 5$$

The solutions are 1 and 5. The check is left for you.

Classroom Exercises

For each equation, the apparent solutions are -2 and/or 5. Check -2 and 5 in each equation to find the correct root, or roots, of the equation. If there are no roots, write *no roots*.

1. $\sqrt{x + 11} = 3$ -2 **2.** $\sqrt{x + 4} = -3$ No roots   **3.** $\sqrt{x + 11} = x - 1$ 5

Additional Example 2

Solve $x - \sqrt{3x + 3} = 5$. Check for extraneous solutions. The only solution is 11. (2 is extraneous.)

Additional Example 3

Solve $\sqrt{5x + 1} - \sqrt{3x - 5} = 2$. Check.
3, 7

Solve the equation. Check.

4. $\sqrt{x + 7} = 2$ -3

5. $\sqrt{x + 12} = x$ 4

6. $\sqrt[3]{y - 5} = -2$ -3

Written Exercises

Solve the equation. Check.

1. $\sqrt{3x - 5} = 4$ 7

2. $\sqrt[3]{4a} = 4$ 16

3. $\sqrt{6n - 2} = \sqrt{4n + 4}$ 3

4. $y + 1 = \sqrt{y + 7}$ 2

5. $x = 4 + \sqrt{2x}$ 8

6. $\sqrt{2w - 1} = w - 2$ 5

7. $n = 5 + \sqrt{2n + 5}$ 10

8. $\sqrt[3]{c^2 - 8} = 2$ ±4

9. $\sqrt[3]{2a^2 - 9} = 3$ ±3$\sqrt{2}$

10. $\sqrt[4]{x^3 + 8} = 2$ 2

11. $\sqrt[4]{x^2 + 9} = 3$ ±6$\sqrt{2}$

12. $2\sqrt{y} = \sqrt{y} + 2$ 4

13. $\sqrt{3x} - \sqrt{x - 2} = 2$ 3

14. $\sqrt{x + 11} = \sqrt{5x} - 1$ 5

15. $\sqrt{n + 7} - \sqrt{n} = 1$ 9

16. $4 + \sqrt{5t + 1} = 0$ No roots

17. $-3 - \sqrt{4t + 1} = 0$ No roots

18. $2 + \sqrt{3x + 7} = 2x$ 3

19. $2\sqrt{3c - 2} - c - 2 = 0$ 2, 6

20. $\sqrt{x^2 - x + 5} = 5$ -4, 5

21. $\sqrt{y^2 + 8y} = 4\sqrt{3}$ -12, 4

22. $\sqrt[4]{x^2 + 7} = \sqrt[4]{5x + 1}$ 2, 3

23. $\sqrt{4x - 3} - \sqrt{2x - 5} = 2$ 3, 7

24. The radius r of a circle with area A is given by the formula

$r = \sqrt{\dfrac{A}{\pi}}$. Solve this formula for A in terms of r. $A = \pi r^2$

25. The radius of the base of a cylinder with height h and volume V is

given by $r = \sqrt{\dfrac{V}{\pi h}}$. Solve for V in terms of r and h. $V = \pi r^2 h$

Solve each equation. Check.

26. $\sqrt{x + 1} = \sqrt[4]{10x + 1}$ 0, 8

27. $\sqrt[3]{x + 2} = \sqrt[6]{9x + 10}$ -1, 6

Midchapter Review

1. Without solving $4x^2 - 3x - 8 = 0$, find the sum and the product of its roots. **10.1** $\frac{3}{4}$, -2

Find the discriminant and determine the nature of the roots without solving the equation. **10.2**

2. $2x^2 - 3x - 4 = 0$ 41, two irrational

3. $9y^2 = 12y - 4$ 0, one rational

Solve the equation. Check. **10.3**

4. $n = 3 + \sqrt{2n + 2}$ 7

5. $\sqrt[4]{2x^2 - 19} = 3$ ±5$\sqrt{2}$

Analysis: Ask students what they think will happen in the examples of the lesson if the radical is not "isolated" before squaring both sides of the original equation. Will the equations be impossible to solve? They will discover that if a radical is not isolated in the first step, then it will be necessary to do so in a later step with a good deal more expenditure of time and effort.

Checkpoint

Solve each equation. Check.

1. $\sqrt[3]{6x + 10} = -2$ -3

2. $3\sqrt{2y + 2} = 2\sqrt{5y - 1}$ 11

3. $3 + \sqrt{3x + 1} = x$ 8 (1 is extraneous.)

4. $\sqrt{2x + 7} - \sqrt{x} = 2$ 9, 1

Closure

Have students give the four-step process for solving radical equations. See page 347. Ask students to explain why it is important to check each solution. Raising each side of the equation to the same power may introduce an extraneous solution.

◤◤◤◤ FOLLOW UP

Guided Practice

Classroom Exercises 1–6

Independent Practice

A Ex. 1–12, **B** Ex. 13–25, **C** Ex. 26, 27

Basic: WE 1–19 odd, Midchapter Review, Application

Average: WE 7–25 odd, Midchapter Review, Application

Above Average: WE 9–27 odd, Midchapter Review, Application

Enrichment

Challenge the students to solve this equation.

$\sqrt{\sqrt{\sqrt{x - 1} - 1} - 1} = 2$ 677

Application

3. After being launched at the escape velocity, the object keeps moving away from the planet with some positive velocity forever. (If it ever completely stopped, it would eventually fall back.) So, we really can say an object has "escaped" only when it will always keep getting higher and higher above the planet. Now in the formula, as h gets bigger and bigger the radius of the planet becomes insignificant compared to the great height, and we can effectively ignore it. Thus, we can treat the ratio $\dfrac{h}{h+r}$ as $\dfrac{h}{h}$.

So we have:

$$v \approx \sqrt{2gr}\left(\sqrt{\dfrac{h}{h}}\right)$$

$$= \sqrt{2gr}(\sqrt{1})$$

$$= \sqrt{2gr}, \text{ the escape velocity}$$

When an object is launched upward from the earth, the force of gravity acts to pull it back down. Whether or not it actually falls back, however, depends on the velocity with which it is launched. With a strong enough boost, even a 6,200,000-lb rocket like the Saturn V can overcome the earth's gravitational field, never to return.

An object launched vertically at an initial speed of 11.2 km/s, for instance, will never come down because 11.2 km/s is the earth's escape velocity. The **escape velocity** for a particular planet is the minimum speed at which an object must be launched in order to escape that planet's gravitational attraction permanently.

Escape velocity can be calculated using the following formula,

$$v_e = \sqrt{2gr},$$

where g is the surface gravity of the planet and r is its radius.

Example   The radius of the earth is 6,380,000 m. Its surface gravity is 9.8 m/s². Find its escape velocity.

Solution
$$v_e = \sqrt{2gr}$$
$$= \sqrt{2(9.8)(6,380,000)}$$
$$\approx 11,200 \text{ m/s, or } 11.2 \text{ km/s}$$

The earth's escape velocity is about 11.2 km/s.

1. The surface gravity of the moon is 1.6 m/s². Its radius is 1,740,000 m. Find its escape velocity. About 2.4 km/s

2. If the surface gravity of Mars is 3.6 m/s², and its escape velocity is 5.1 km/s, what is the radius of Mars? About 3,600,000 m

3. The formula for escape velocity can be derived from another formula, shown below, for the velocity needed to reach height h above a planet.

$$v = \sqrt{2gr}\left(\sqrt{\dfrac{h}{h+r}}\right)$$

Explain how the derivation is done. (HINT: Let $h =$ an infinite distance, or infinity.)

10.4 Higher-Degree Equations; Complex Roots

Objectives

To find all the roots of integral polynomial equations

To find upper and lower bounds for the real zeros of integral polynomials

Let $P(x) = x^4 - 2x^3 - 11x^2 + 16x + 24$. Recall that according to the Integral Zero Theorem (see Lesson 6.6), the integral roots, *if any*, of the equation $P(x) = 0$ must be among the integral factors of the constant term—in this case, 24.

$$\pm 1, \pm 2, \pm 3, \pm 4, \pm 6, \pm 8, \pm 12, \pm 24$$

After finding the integral zeros, if any, of $P(x)$, you can find its *irrational* zeros as shown below. Synthetic substitution is used.

$$x^4 - 2x^3 - 11x^2 + 16x + 24 \leftarrow P(x)$$

	1	−2	−11	16	24	
		(−1)	(3)	(8)	(−24)	
−1	1	−3	−8	24	⓪	$\leftarrow (x+1)(x^3 - 3x^2 - 8x + 24)$
			(3)	(0)	(−24)	
3	1	0	−8	⓪		$\leftarrow (x+1)(x-3)(x^2 - 8)$

Solve $x^2 - 8 = 0$.

$$x^2 = 8$$

$$x = 2\sqrt{2} \text{ or } x = -2\sqrt{2} \leftarrow (x+1)(x-3)(x - 2\sqrt{2})(x + 2\sqrt{2})$$

So, the zeros of $P(x)$ and the roots of $P(x) = 0$ are $-1, 3, 2\sqrt{2}$, and $-2\sqrt{2}$.

The *Fundamental Theorem of Algebra* states that each polynomial of degree n, where $n > 0$, has *at least one* zero.

Theorem 10.3

The Fundamental Theorem of Algebra
If $P(x)$ is a polynomial of degree n, where $n > 0$, then there is a zero r among the complex numbers for which $P(r) = 0$.

EXAMPLE 1   Let $Q(x) = x^3 + 2x^2 + 4x + 8$. Find all the roots of $Q(x) = 0$.

Solution   Find the zeros of $Q(x)$. Try the integral factors of 8: $\pm 1, \pm 2, \pm 4, \pm 8$.

Additional Example 1

Let $Q(x) = x^4 - 2x^3 + 6x^2 - 18x - 27 = 0$.
Find all the roots of the equation $Q(x) = 0$.
$-1, 3, 3i, -3i$

Teaching Resources

Quick Quizzes 71
Reteaching and Practice
   Worksheets 71

GETTING STARTED

Prerequisite Quiz

For Exercises 1–2, use $P(x) = 6x^4 + 13x^3 - 18x^2 - 7x + 6$.

1. List the potential integral zeros of $P(x)$.
   $\pm 1, \pm 2, \pm 3, \pm 6$
2. Find the zeros of $P(x)$. $1, -3, \frac{1}{2}, -\frac{2}{3}$

Solve each equation.

3. $x^2 - 8 = 0$ $2\sqrt{2}, -2\sqrt{2}$
4. $x^2 + 9 = 0$ $3i, -3i$

Motivator

Have students recall the Integral Zero Theorem (see page 224).   Then, have students determine if $x - 2$ is a factor of the polynomial $13x^2 - 4x + 6$. $x - 2$ is not a factor because the remainder is not 0.

Highlighting the Standards

Standard 14a: Students gain a sense of the structure of mathematics through the general accumulation of experience rather than by learning particular rules.

This lesson continues the strand of using synthetic substitution and the Factor Theorem to find the roots of an integral polynomial equation. Here, irrational roots and complex roots are found. The Rational Zero Theorem will be introduced in the next lesson. Remind students that the procedure of using synthetic substitution is continued until a second-degree factor is found. At that point, a second-degree equation is solved in the usual way.

Math Connections

Historical Note: The Fundamental Theorem of Algebra was proved four times by the great mathematician Carl Friedrich Gauss (1777–1855). His first proof, which was largely geometric in nature, was published as his doctoral thesis in 1799. He published two entirely different proofs of the theorem 17 years later. In 1850, he also published an algebraic proof.

The *compact form* of synthetic substitution is used.

$$
\begin{array}{c|cccc}
 & 1 & 2 & 4 & 8 \quad \leftarrow Q(x) = x^3 + 2x^2 + 4x + 8 \\
\hline
1 & 1 & 3 & 7 & 15 \quad \leftarrow \text{1 is not a root. Try another.} \\
-2 & 1 & 0 & 4 & \boxed{0} \quad \leftarrow -2 \text{ is a root.}
\end{array}
$$

$Q(x) = (x + 2)(x^2 + 4)$ Solve. The roots are -2, $2i$, and $-2i$.

Notice in Example 1 that synthetic substitution is not used after a second-degree factor is found. Instead, a quadratic equation is solved.

The potential integral zeros of $P(x) = 12x^4 + x^3 - 42x^2 - 3x + 18$ are ± 1, ± 2, ± 3, ± 6, ± 9, ± 18. It is possible to reduce this list to ± 1, ± 2 by using the **Upper- and Lower-Bound Theorem** below.

Theorem 10.4

Let $P(x)$ be an integral polynomial, and U and L be real numbers such that $U \geq 0$ and $L \leq 0$.

(1) Let $P(U)$ be found by synthetic substitution. If the bottom row of numbers is *all nonnegative* or *all nonpositive*, then U is an **upper bound** for the real zeros of $P(x)$.

(2) Let $P(L)$ be found by synthetic substitution. If the bottom row of numbers *alternates in sign*, then L is a **lower bound** for the real zeros of $P(x)$.

This theorem can be used to find upper and lower bounds for the real zeros of $P(x) = 12x^4 + x^3 - 42x^2 - 3x + 18$, as shown below.

1. Find $P(x)$ for $x = 0, 1, 2, \ldots$ by synthetic substitution.

x	12	1	-42	-3	18	$P(x)$
0	12	1	-42	-3	18	18
1	12	13	-29	-32	-14	-14
2	12	25	8	13	44	44

All nonnegative

3	12	37	69	204	630	630

All nonnegative

These numbers and $P(x)$ will continue to increase. So, $P(x)$ cannot be zero for $x > 2$. Thus, 2 is an upper bound for the real zeros of $P(x)$.

2. Find $P(x)$ for $x = -1, -2, -3, \ldots$ by synthetic substitution.

x	12	1	-42	-3	18	$P(x)$
-1	12	-11	-31	28	-10	-10
-2	12	-23	4	-11	40	40

Signs alternate

-3	12	-35	63	-192	594	594

Signs alternate

These signs will continue to alternate, and $P(x)$ will continue to be positive. So, $P(x)$ cannot be zero for $x < -2$. Thus, -2 is a lower bound for the real zeros of $P(x)$.

Because 2 and -2 are bounds for the real zeros of $P(x)$, the list of potential integral zeros has been reduced from $\pm 1, \pm 2, \pm 3, \pm 6, \pm 9, \pm 18$ to $\pm 1, \pm 2$.

The zeros of $P(x) = 12x^4 + x^3 - 42x^2 - 3x + 18$, or
$$P(x) = (4x + 3)(3x - 2)(x - \sqrt{3})(x + \sqrt{3}),$$
are $-\frac{3}{4}, \frac{2}{3}, \sqrt{3}$, and $-\sqrt{3}$. Each zero z is in the interval $-2 \le z \le 2$, where 2 is an upper bound and -2 is a lower bound for the four real zeros.

EXAMPLE 2   Find the least integral upper bound U and the greatest integral lower bound L for the real zeros of $P(x) = -2x^4 + 3x^3 + 7x^2 + 3x + 9$.

Solution   **1.** For U, try $x = 0, 1, 2, \ldots$ **2.** For L, try $x = -1, -2, -3, \ldots$

x	-2	3	7	3	9
0	-2	3	7	3	9
1	-2	1	8	11	20
2	-2	-1	5	13	35
3	-2	-3	-2	-3	0

All nonpositive: $U = 3$

x	-2	3	7	3	9
-1	-2	5	2	1	8
-2	-2	7	-7	17	-25

Signs alternate: $L = -2$

Thus, 3 is the least integral upper bound and -2 is the greatest integral lower bound.

The Integral Zero Theorem and the Upper- and Lower-Bound Theorem can often be used together to solve higher-degree equations.

Critical Thinking Questions

Analysis: It might be well to anticipate the Rational Zero Theorem of Lesson 10.5 to explain the appearance of the two zeros $-\frac{3}{4}$ and $\frac{2}{3}$ of $12x^4 + x^3 - 42x^2 - 3x + 18$ in the middle of page 353. Ask students whether they see how the numerators of the two roots might be related to the constant term, 18, and how the two denominators, 4 and 3, might be related to the leading coefficient, 12.   As a hint, ask them to review the Integral Zero Theorem.

Checkpoint

1. Find an upper bound U and a lower bound L for the real zeros of $x^3 + 2x^2 - 5x - 10$. U: 3, L: -4

2. Solve $x^3 + 2x^2 - 5x - 10 = 0$. (HINT: Use the work from Exercise 1 above.) $-2, \sqrt{5}, -\sqrt{5}$

3. Find all the roots of $x^5 - 2x^4 + 11x^3 - 22x^2 - 12x + 24 = 0$. $1, -1, 2, 2i\sqrt{3}, -2i\sqrt{3}$

Closure

Have students explain in their own words the Upper- and Lower-Bound Theorem. Answers will vary.   Ask students to give the Fundamental Theorem of Algebra.   See page 351.

Additional Example 2

Find an upper bound U and a lower bound L for the real zeros of $P(x) = 2x^4 - 5x^3 + 5x^2 - 20x - 12$. U: 3, L: -1

Guided Practice

Classroom Exercises 1–7

Independent Practice

A Ex. 1–14, **B** Ex. 15–18, **C** Ex. 19, 20

Basic: WE 1–13 odd
Average: WE 5–17 odd
Above Average: WE 7–19 odd

Additional Answers

Written Exercises

11. $U = 3, L = -4$
15. $-4, -1, \pm 2i\sqrt{3}$
16. $-2, 3, \pm 2\sqrt{2}$
17. $-1, -\frac{1}{2}, \frac{2}{3}, 1, 2$

EXAMPLE 3 The polynomial $P(x) = 6x^5 - 17x^4 - 29x^3 + 74x^2 + 20x - 24$ has five distinct zeros. Solve $P(x) = 0$.

Plan The integral roots of $P(x) = 0$ are among the integral factors of -24: $\pm 1, \pm 2, \pm 3, \pm 4, \pm 6, \pm 8, \pm 12, \pm 24$. Use synthetic substitution.

Solution Use $x = 0, 1, 2, \ldots$ in synthetic substitution. Look for zeros and an upper bound.

x	6	-17	-29	74	20	-24
0	6	-17	-29	74	20	-24
1	6	-11	-40	34	54	30
2	6	-5	-39	-4	12	$\boxed{0}$

Thus, 2 is a zero of $P(x)$.

$P(x) = (x - 2)(6x^4 - 5x^3 - 39x^2 - 4x + 12)$

Next, find the zeros of the *depressed* polynomial $Q(x) = 6x^4 - 5x^3 - 39x^2 - 4x + 12$.

x	6	-5	-39	-4	12
3	6	13	0	-4	$\boxed{0}$

Thus, 3 is a zero of $Q(x)$ and also of $P(x)$.

$P(x) = (x - 2)(x - 3)(6x^3 + 13x^2 - 4)$

Next, find the zeros of the depressed polynomial $R(x) = 6x^3 + 13x^2 - 4$.

x	6	13	0	-4
4	6	37	148	588

All nonnegative: 4 is not a zero, but it *is* an upper bound.

Use $x = -1, -2, -3, \cdots$ in synthetic substitution with $R(x) = 6x^3 + 13x^2 - 4$. Look for negative zeros or a lower bound.

x	6	13	0	-4
-1	6	7	-7	3
-2	6	1	-2	$\boxed{0}$

$P(x) = (x - 2)(x - 3)(x + 2)(6x^2 + x - 2)$

Factor. $6x^2 + x - 2 = (3x + 2)(2x - 1)$

Thus, $P(x) = (x - 2)(x - 3)(x + 2)(3x + 2)(2x - 1)$, and the roots of $P(x) = 0$ are $2, 3, -2, -\frac{2}{3}$, and $\frac{1}{2}$.

Additional Example 3

The polynomial $P(x) = 4x^4 + 4x^3 - 25x^2 - x + 6$ has four distinct zeros. Solve $P(x) = 0$.
$2, -3, \frac{1}{2}, -\frac{1}{2}$

Classroom Exercises

Find all the zeros of each polynomial.

1. $5x + 2$ $-\frac{2}{5}$ **2.** $x^2 + 25$ $\pm 5i$ **3.** $x^2 - 18$ $\pm 3\sqrt{2}$ **4.** $(x - 12)(x^2 - 12)$ $12, \pm 2\sqrt{3}$

5. $(x + 1)(x - 4)(x^2 - 7)$ **6.** $(x - 1)(x + 2)(x^2 + 9)$ **7.** $x^3 - 5x^2 + x - 5$
$-1, 4, \pm\sqrt{7}$ $1, -2, \pm 3i$ $5, \pm i$

Written Exercises

Find all the roots of each equation.

1. $x^3 - 2x^2 - x + 2 = 0$ $-1, 1, 2$ **2.** $x^3 - x^2 - 14x + 24 = 0$ $-4, 2, 3$

3. $a^3 + 3a^2 + a + 3 = 0$ $-3, \pm i$ **4.** $x^3 - 5x^2 + 9x - 45 = 0$ $5, \pm 3i$

5. $y^3 - 2y^2 - 18y + 36 = 0$ $2, \pm 3\sqrt{2}$ **6.** $c^3 + 4c^2 - 20c - 80 = 0$ $-4, \pm 2\sqrt{5}$

7. $x^3 + 5x^2 + 2x + 10 = 0$ $-5, \pm i\sqrt{2}$ **8.** $a^3 - 3a^2 + 12a - 36 = 0$ $3, \pm 2i\sqrt{3}$

Find the least integral upper bound U and the greatest integral lower
bound L for the real zeros of each polynomial.

                                                         $U = 1, L = -1$

9. $y^3 - 2y^2 + 6y - 12$ $U = 2, L = 0$ **10.** $6x^4 - x^3 + 58x^2 - 10x - 20$

11. $x^4 + x^3 - 10x^2 - 6x + 36$ **12.** $t^4 - 7t^2 - 15$ $U = 3, L = -3$

13. $4y^4 - 8y^3 - 16y^2 + 40y - 20$ **14.** $x^3 + 2x^2 - 10x - 20$ $U = 4, L = -5$
                            $U = 4, L = -3$

Find all the roots of each equation.

15. $x^4 + 5x^3 + 16x^2 + 60x + 48 = 0$ **16.** $a^4 - a^3 - 14a^2 + 8a + 48 = 0$

17. $6y^5 - 13y^4 - 6y^3 + 17y^2 - 4 = 0$ (HINT: Insert a missing term.)

18. $9y^4 - 146y^2 + 32 = 0$ $\pm 4, \pm\dfrac{\sqrt{2}}{3}$

19. $x^4 - x^3 - 21x^2 + 43x - 10 = 0$ **20.** $x^4 + 6x^3 + 27x^2 + 32x - 66 = 0$
                $-5, 2, 2 \pm \sqrt{3}$ $-3, 1, -2 \pm 3i\sqrt{2}$

Mixed Review

1. Write an equation, in standard form, of the line through $P(-3, 2)$
and $Q(7, 4)$. **3.4** $x - 5y = -13$

2. Factor $5x^3 - 40y^3$ completely. **5.7** $5(x - 2y)(x^2 + 2xy + 4y^2)$

3. If $f(x) = 2x^2$ and $g(x) = -3x$, what is $f(g(2))$? **5.9** 72

4. Solve $\dfrac{3x}{x^2 - 5x - 14} - \dfrac{5}{x - 7} = \dfrac{4}{x + 2}$. **7.6** 3

Enrichment

Challenge students to find the three cube
roots of -1. Point out that this is equivalent
to solving the equation, $\sqrt[3]{-1} = x$.

Note that students may remember similar
problems in Lesson 9.5. At that time they
solved third-degree equations by factoring
the sum or difference of two cubes. Now
they are able to use synthetic substitution.

$\sqrt[3]{-1} = x$
$-1 = x^3$
$x^3 + 1 = 0$

$$\begin{array}{r|rrrr} & 1 & 0 & 0 & 1 \\ -1 & & -1 & 1 & -1 \\ \hline & 1 & -1 & 1 & 0 \end{array}$$

-1 is a root.

$x^2 - x + 1 = 0$
$x = \dfrac{1 \pm i\sqrt{3}}{2}$

The three cube roots of -1 are -1,
$\dfrac{1 + i\sqrt{3}}{2}$, and $\dfrac{1 - i\sqrt{3}}{2}$.

■■■GETTING STARTED

Prerequisite Quiz

1. List the potential integral zeros of $x^4 - 3x^3 - 6x^2 + 6x + 8$. $\pm 1, \pm 2, \pm 4, \pm 8$
2. Solve $x^4 - 3x^3 - 6x^2 + 6x + 8 = 0$.
   $-1, 4, \sqrt{2}, -\sqrt{2}$

Motivator

Have students use the Integral Zero Theorem to attempt to solve $6x^4 - x^3 + 5x^2 - x - 1 = 0$.   No zeros can be found. Then have students offer explanations of why the Integral Zero Theorem is inadequate.   The zeros are not integers.

■■■TEACHING SUGGESTIONS

Lesson Note

This lesson continues the strand of solving polynomial equations using synthetic substitution and the Factor Theorem. The Rational Zero Theorem is introduced so that all potential zeros can be listed when necessary.

10.5 The Rational Zero Theorem

Objective To solve integral polynomial equations using the Rational Zero Theorem

The polynomial $P(x) = 12x^3 - 4x^2 - 3x + 1$ has no integral zeros. Instead, it has three *rational zeros*, $\frac{1}{3}$, $\frac{1}{2}$, and $-\frac{1}{2}$, which cannot be found by the Integral Zero Theorem. It can be proved, however, that if $\frac{1}{r}$ is a zero of $P(x)$, and r is an integer, then r is a factor of the first coefficient of $P(x)$, which in this instance is 12.

Let $\frac{1}{r}$ be a zero of $12x^3 - 4x^2 - 3x + 1$, where r is an integer.

$$12 \cdot \frac{1}{r^3} - 4 \cdot \frac{1}{r^2} - 3 \cdot \frac{1}{r} + 1 = 0$$

Next, multiply each side by r^3. $12 - 4r - 3r^2 + r^3 = 0$

$$r^3 - 3r^2 - 4r = -12$$

Factor. $r(r^2 - 3r - 4) = -12$

So, since r is a factor of -12, it is also a factor of 12, and is thus among the integers

$$\pm 1, \ \pm 2, \ \pm 3, \ \pm 4, \ \pm 6, \ \pm 12.$$

Thus, $\frac{1}{r}$ is among $\pm \frac{1}{1}$, $\pm \frac{1}{2}$, $\pm \frac{1}{3}$, $\pm \frac{1}{4}$, $\pm \frac{1}{6}$, $\pm \frac{1}{12}$.

This line of reasoning can be used to extend the Integral Zero Theorem to rational numbers in the *Rational Zero Theorem* below.

Theorem 10.5

Rational Zero Theorem

If $\frac{c}{d}$, where c and d are relatively prime integers, is a zero of an integral polynomial $P(x)$, then c is a factor of the constant term in $P(x)$ and d is a factor of the coefficient of the term of highest degree. That is, if $\frac{c}{d}$ is a zero of

$$a_0 x^n + a_1 x^{n-1} + a_2 x^{n-2} + \cdots + a_{n-1} x^1 + a_n,$$

then c is a factor of a_n and d is a factor of a_0.

NOTE: The Rational Zero Theorem does not assure the existence of rational zeros. In fact, an integral polynomial may have no rational zeros.

Highlighting the Standards

Standard 5a: This lesson exhibits the Standards' point that "algebra in grades 9–12 will focus on its own logical framework and consistency."

For $P(x) = 6x^4 + 7x^3 - 21x^2 - 21x + 9$, you can list the *potential* rational zeros that are not integers as shown below.

c is a factor of 9: 1, 3, 9. $\qquad$ d is a factor of 6: 1, 2, 3, 6.

Thus, any nonintegral rational zeros $\frac{c}{d}$, if they exist, are among $\pm\frac{1}{2}$, $\pm\frac{1}{3}$, $\pm\frac{1}{6}$, $\pm\frac{3}{2}$, $\pm\frac{9}{2}$.

EXAMPLE 1 $\quad$ $P(x) = 6x^4 + 7x^3 - 21x^2 - 21x + 9$ has four distinct zeros. Find the zeros.

Solution $\qquad$ **1.** Test the potential *integral* zeros first: ±1, ±3, ±9.

$$
\begin{array}{r|rrrrr}
 & 6 & 7 & -21 & -21 & 9 \\
\hline
1 & 6 & 13 & -8 & -29 & -20 \\
3 & 6 & 25 & 54 & 141 & 432 \\
\end{array}
\qquad
\begin{array}{r|rrrrr}
 & 6 & 7 & -21 & -21 & 9 \\
\hline
-1 & 6 & 1 & -22 & 1 & 8 \\
-3 & 6 & -11 & 12 & -57 & 180 \\
\end{array}
$$

All nonnegative: $\qquad\qquad\qquad$ Signs alternate:
3 is an upper bound. $\qquad\qquad\quad$ -3 is a lower bound.

It is unnecessary to test ±9. $P(x)$ has no integral zeros.

2. Test the remaining potential rational zeros between the bounds -3 and 3: $\pm\frac{1}{2}$, $\pm\frac{1}{3}$, $\pm\frac{1}{6}$, $\pm\frac{3}{2}$.

$$
\begin{array}{r|rrrrr}
 & 6 & 7 & -21 & -21 & 9 \\
\hline
\frac{1}{3} & 6 & 9 & -18 & -27 & 0 \\
-\frac{3}{2} & 6 & 0 & -18 & & 0 \\
\end{array}
$$

Successive Factorizations:

$(x - \frac{1}{3})(6x^3 + 9x^2 - 18x - 27)$

$(x - \frac{1}{3})(x + \frac{3}{2})(6x^2 - 18)$

After two rational zeros are found, identify a second-degree factor.

3. Solve $6x^2 - 18 = 0$.

$\qquad 6(x^2 - 3) = 0 \qquad\qquad \leftarrow \quad (x - \frac{1}{3})(x + \frac{3}{2}) \cdot 6(x^2 - 3)$

$\qquad\qquad x^2 = 3$

$\qquad\qquad x = \pm\sqrt{3} \qquad \leftarrow \quad 6(x - \frac{1}{3})(x + \frac{3}{2})(x - \sqrt{3})(x + \sqrt{3})$

The zeros of $P(x)$ are $\frac{1}{3}$, $-\frac{3}{2}$, $\sqrt{3}$, and $-\sqrt{3}$.

The fourth-degree integral polynomial in Example 1 has two rational (but nonintegral) zeros, $\frac{1}{3}$ and $-\frac{3}{2}$, and two irrational zeros that are conjugates, $\sqrt{3}$ and $-\sqrt{3}$. A fourth-degree integral polynomial may also have one integral zero, one rational zero, and two complex zeros that are conjugates, as in $P(x) = 2x^4 - x^3 + 3x^2 - 2x - 2$.

Math Connections

Prime Numbers: The Factorization Theorem on page 358, which applies to the set of polynomials of degree n, suggests a somewhat similar theorem that applies to the set of counting numbers: Any counting number can be expressed uniquely (except for the order of the factors) as a product of positive prime numbers. This theorem is called the *Fundamental Theorem of Arithmetic.*

Critical Thinking Questions

Analysis: In Example 2, the potential rational zero $\frac{1}{2}$ is tested twice and found to be a double root of the polynomial equation $P(x) = 0$. Ask students whether there is any way of telling how many times a potential zero should be tested. As a general matter, students should continue testing until the potential zero fails the test. For some polynomials, this could mean testing the potential zero many times. However, it will often happen, as in Example 2, that the remaining zeros can be discovered directly by solving a linear or quadratic equation.

Checkpoint

Solve the equation.
$12x^4 + 8x^3 - 9x^2 - 3x + 2 = 0$
State the multiplicity m of a root when $m > 1$.
-1, $-\frac{2}{3}$, $\frac{1}{2}$ with $m = 2$

Additional Example 1

$P(x) = 6x^4 + x^3 - 31x^2 - 5x + 5$ has four distinct zeros. Find the zeros.
$-\frac{1}{2}$, $\frac{1}{3}$, $\sqrt{5}$, $-\sqrt{5}$

Ask students to explain in their own words the Factorization Theorem. Answers will vary. Ask students if it is beneficial to test a potential rational zero more than once. Yes, it could have a multiplicity greater than 1. Ask them if it is possible for a fourth-degree polynomial to have an integral zero, a rational zero, and two complex zeros. Yes

▰▰▰ FOLLOW UP

Guided Practice

Classroom Exercises 1–6

Independent Practice

A Ex. 1–4, **B** Ex. 5–17, **C** Ex. 16–20

Basic: WE 1–9 odd, Brainteaser

Average: WE 5–17 odd, Brainteaser

Above Average: WE 9–19 odd, Brainteaser

Additional Answers

Classroom Exercises

3. $\pm 1, \pm 2, \pm 3, \pm 6, \pm\frac{1}{3}, \pm\frac{2}{3}$

4. $\pm 1, \pm 2, \pm 4, \pm\frac{1}{2}, \pm\frac{1}{3}, \pm\frac{1}{6}, \pm\frac{2}{3}, \pm\frac{4}{3}$

Written Exercises

5. $2, -\frac{2}{3}, \pm\sqrt{5}$

6. $-1, \frac{3}{4}, \pm 2i\sqrt{2}$

7. $5, -\frac{2}{3}, \pm i\sqrt{2}$

8. $-\frac{1}{2}\ (m = 2), \pm\sqrt{3}$

9. $\frac{1}{2}\ (m = 3), -\frac{2}{3}$

10. $\frac{3}{2}\ (m = 2), \pm 2i$

Recall that a polynomial may have zeros with multiplicities greater than 1. If $P(x) = (x - 5)(x - 5)(x - 5)(2x + 3)$, then the zero 5 has a multiplicity of 3. The zero $-\frac{3}{2}$ has a multiplicity of 1. Notice that the sum of the multiplicities is $3 + 1$, or 4, which is the same as the degree of $P(x)$.

The *Factorization Theorem* stated below guarantees that a polynomial $P(x)$ of degree n, where $n > 0$, has exactly n first-degree factors, and thus that the sum of the multiplicities of the zeros of $P(x)$ is equal to n.

Theorem 10.6

Factorization Theorem
Each polynomial $P(x)$ of degree n, where $n > 0$, can be factored into exactly n first-degree factors and a constant factor.

EXAMPLE 2 $P(x) = 16x^4 - 16x^3 - 32x^2 + 36x - 9$. Solve $P(x) = 0$. State the multiplicity m of a root when $m > 1$.

Solution   1. Test the potential integral zeros: $\pm 1, \pm 3, \pm 9$.

	16	−16	−32	36	−9	
1	16	0	−32	4	−5	
3	16	32	64	228	675	← 3 is an upper bound.
−1	16	−32	0	36	−45	
−3	16	−64	160	−444	1,323	← −3 is a lower bound.

Thus, there are no integral roots.

2. Try to "narrow" the bounds.

	16	−16	−32	36	−9	
2	16	16	0	36	63	← 2 is an upper bound.
−2	16	−48	64	−92	175	← −2 is a lower bound.

3. List the remaining potential rational zeros between the bounds -2 and 2:

$$\pm\frac{1}{2}, \pm\frac{1}{4}, \pm\frac{1}{8}, \pm\frac{1}{16}, \pm\frac{3}{2}, \pm\frac{3}{4}, \pm\frac{3}{8}, \pm\frac{3}{16}, \pm\frac{9}{8}, \pm\frac{9}{16}$$

Additional Example 2

$P(x) = 18x^3 - 21x^2 + 8x - 1$. Solve $P(x) = 0$. State the multiplicity m of a root when $m > 1$. $\frac{1}{2}; \frac{1}{3}$ with $m = 2$

4. Test these zeros until a second-degree factor is found.

$$\begin{array}{r|rrrrr} & 16 & -16 & -32 & 36 & -9 \\ \frac{1}{2} & & 16 & -8 & -36 & 18 & 0 \end{array}$$

$$16-8-36180$$

Try $\frac{1}{2}$ again. $\quad \begin{array}{r|rrrr} \frac{1}{2} & 16 & 0 & -36 & 0 \end{array}$

$$P(x) = (x - \tfrac{1}{2})(x - \tfrac{1}{2})(16x^2 - 36)$$

5. Solve $16x^2 - 36 = 0$.
$$4(4x^2 - 9) = 0$$
$$4(2x - 3)(2x + 3) = 0 \quad P(x) = 4(x - \tfrac{1}{2})(x - \tfrac{1}{2})(2x - 3)(2x + 3)$$

The roots are $\frac{1}{2}, \frac{3}{2}$, and $-\frac{3}{2}$. The multiplicity of $\frac{1}{2}$ is 2.

Classroom Exercises

State the potential integral zeros of each polynomial. Then give the potential rational, but nonintegral, zeros of the polynomial. $\quad \pm 1, \ \pm 2, \ \pm 5, \ \pm 10;$ none

1. $8x^3 + 6x^2 - 5x + 1$ $\quad \pm 1, \ \pm\frac{1}{2}, \ \pm\frac{1}{4}, \ \pm\frac{1}{8}$ **2.** $y^4 - 8y^3 + y - 10$

3. $3z^4 - z^3 + 2z - 6$ **4.** $6x^3 + x^2 - x + 4$

Solve each equation. State the multiplicity m of a root when $m > 1$.

5. $(x - 3)(x + 4)(x - 3) = 0$ $\quad 3(m = 2), \ -4$ **6.** $5(n + 2)(n + 2)(n - 4)(n + 2) = 0$ $\quad -2(m = 3), \ 4$

Written Exercises

Solve each equation. State the multiplicity m of a root when $m > 1$.

1. $3y^3 - y^2 - 24y + 8 = 0$ $\quad \frac{1}{3}, \ \pm 2\sqrt{2}$ **2.** $4x^3 + 3x^2 + 36x + 27 = 0$ $\quad -\frac{3}{4}, \ \pm 3i$

3. $18c^3 + 9c^2 - 23c + 6 = 0$ $\quad -\frac{3}{2}, \frac{1}{3}, \frac{2}{3}$ **4.** $18y^3 + 3y^2 - 4y - 1 = 0$ $\quad -\frac{1}{3}(m = 2), \frac{1}{2}$

5. $3x^4 - 4x^3 - 19x^2 + 20x + 20 = 0$ **6.** $4n^4 + n^3 + 29n^2 + 8n - 24 = 0$

7. $3y^4 - 13y^3 - 4y^2 - 26y - 20 = 0$ **8.** $4x^4 + 4x^3 - 11x^2 - 12x - 3 = 0$

9. $24x^4 - 20x^3 - 6x^2 + 9x - 2 = 0$ **10.** $4c^4 - 12c^3 + 25c^2 - 48c + 36 = 0$

11. $9c^4 + 23c^2 - 12 = 0$ $\quad \pm\frac{2}{3}, \ \pm i\sqrt{3}$ **12.** $4y^4 - 17y^2 + 18 = 0$ $\quad \pm\frac{3}{2}, \ \pm\sqrt{2}$

13. $16x^5 - 12x^4 - 68x^3 + 51x^2 + 16x - 12 = 0$ $\quad \pm\frac{1}{2}, \frac{3}{4}, \ \pm 2$

14. $2y^5 + y^4 - 16y^3 + 7y^2 + 24y - 18 = 0$ $\quad -3, \ \pm\sqrt{2}, \frac{3}{2}, 1$

15. Write in your own words how you would form a list of the potential rational zeros of an integral polynomial $P(x)$.

---

15. Answers will vary. One possible answer: Make a list of all the integral factors of the constant term and a list of all the integral factors of the coefficient of the term of highest degree. Then list all the distinct rational numbers formed by using a number from the first list for the numerator and a number from the second list for the denominator.

16. Case I: Suppose x is even.
$$O \cdot E \cdot E + O \cdot E + O$$
$$= E + E + O$$
$$= O, \text{ which contradicts}$$
statement c

Case II: Suppose x is odd.
$$O \cdot O \cdot O + O \cdot O + O$$
$$= O + O + O$$
$$= O, \text{ which is again a}$$
contradiction

Thus, x cannot be an integer, since it is neither even nor odd.

Mixed Review

1.

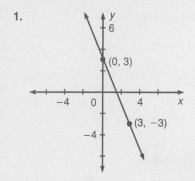

Enrichment

Have the students answer these questions.

1. If $f(x) = ax^8 + x$ is an eighth-degree polynomial, can $a = 0$? Explain your answer. No, because $f(x)$ would then be a first-degree polynomial.

2. If the constant term of a polynomial with degree $n > 1$ is 0, what is a *certain* rational zero of the function, that is, a number that must be a zero? 0

3. Name all the certain and potential zeros of $f(x) = 3x^4 - 2x^3 + x$. 0 is a certain zero. ± 1 and $\pm\frac{1}{3}$ are potential zeros.

4. One zero of $g(x) = x^3$ is 0. Name all the other potential zeros of $g(x)$. There are none.

5. If $h(x) = ax^3 + 3x^2 - x + b$ ($a \neq 0$, $b \neq 0$) has the least possible number of potential zeros, what can be said of a and b? Each equals 1 or -1.

2.

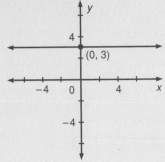

3.

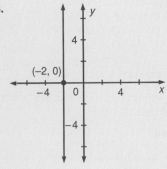

4.

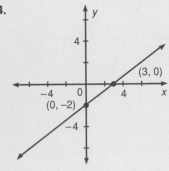

16. If A, B, and C are odd integers, then $Ax^2 + Bx + C = 0$ has no integral roots. Prove that this is true using the following analysis:

a. A, B, and C are odd and zero is even.   **b.** Let O mean odd and E mean even.

c. If $Ax^2 + Bx + C = 0$, then $O \cdot x \cdot x + O \cdot x + O = E$.

d. Show that x can be neither odd nor even, and thus, not an integer.
Consider two cases. (Case I) x is even. (Case II) x is odd.

Irrational zeros of integral polynomials occur in conjugate pairs. Thus, if $4 - 3\sqrt{2}$ is a zero of $P(x)$, then $4 + 3\sqrt{2}$ is another zero of $P(x)$.

Example: Solve $x^4 - 8x^3 - 5x^2 + 24x + 6 = 0$, given that $x_1 = 4 - 3\sqrt{2}$ is one of its roots.

$$
\begin{array}{r|rrrrr}
 & 1 & -8 & -5 & 24 & 6 \\
4 - 3\sqrt{2} & 1 & -4 - 3\sqrt{2} & -3 & 12 + 9\sqrt{2} & \boxed{0} \\
\hline
4 + 3\sqrt{2} & 1 & 0 & -3 & \boxed{0} \\
\end{array}
$$

Solve $x^2 - 3 = 0$. $x^2 = 3$, or $x = \pm\sqrt{3}$

The roots are $4 - 3\sqrt{2}$, $4 + 3\sqrt{2}$, $\sqrt{3}$, and $-\sqrt{3}$.

Solve each equation, given one of its roots.

17. $x^4 - 14x^2 + 24 = 0$, $x_1 = 2\sqrt{3}$ $\pm 2\sqrt{3},\ \pm\sqrt{2}$

18. $n^4 - 2n^3 - 19n^2 + 36n + 18 = 0$, $n_1 = -3\sqrt{2}$ $\pm 3\sqrt{2},\ 1 \pm \sqrt{2}$

19. $x^4 - 4x^3 - 9x^2 + 32x + 8 = 0$, $x_1 = 2 + \sqrt{5}$ $2 \pm \sqrt{5},\ \pm 2\sqrt{2}$

20. $c^4 + 2c^3 - 20c^2 - 36c + 36 = 0$, $c_1 = -1 - \sqrt{3}$ $-1 \pm \sqrt{3},\ \pm 3\sqrt{2}$

Mixed Review

Graph each equation in a coordinate plane.   *3.5*

1. $y = -2x + 3$ **2.** $2y - 6 = 0$ **3.** $5x + 10 = 0$ **4.** $2x - 3y = 6$

Brainteaser

The mathematics department of Kenyon High School is planning to hold a departmental meeting next month.

- The date of the meeting is an even number.
- The date is sometime after the 12th.
- The date is sometime before the 19th.
- The date is not a perfect cube.
- The date is a perfect square.

Only one of these statements is true. What is the date of the meeting?   27th

10.6 Graphing Polynomial Functions

Objectives To graph integral polynomial functions of degree n with n distinct real zeros
To bound (locate) between consecutive integers the real zeros of given polynomials

The graph of $y = x^3 - x^2 - 14x + 14$ can be sketched using a table of ordered pairs, as demonstrated below. Note that all the integral values of x fall between upper and lower bounds for the real zeros.

x	1	-1	-14	14	(x,y)
-4	1	-5	6	-10	$(-4,-10)$

Signs alternate:
-4 *is a lower bound.*

-3	1	-4	-2	20	$(-3,20)$
-2	1	-3	-8	30	$(-2,30)$
-1	1	-2	-12	26	$(-1,26)$
0	1	-1	-14	14	$(0,14)$
1	1	0	-14	0	$(1,0)$
2	1	1	-12	-10	$(2,-10)$
3	1	2	-8	-10	$(3,-10)$
4	1	3	-2	6	$(4,6)$
5	1	4	6	44	$(5,44)$

All nonnegative:
5 is an upper bound.

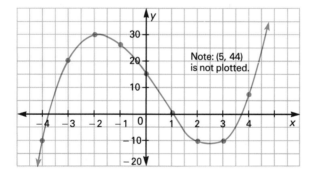

Note: (5, 44) is not plotted.

Teaching Resources

Quick Quizzes 73
**Reteaching and Practice
    Worksheets** 73
Transparency 23

GETTING STARTED

Prerequisite Quiz

Use synthetic substitution and the Remainder Theorem to evaluate $P(x) = x^3 + 4x^2 - 7x - 10$ for the given value of x.

1. $x = -6$ -40
2. $x = -2$ 12
3. $x = 1$ -12
4. $x = 3$ 32

Motivator

Have students find the real zeros of
$y = 2x^2 + 3x - 20$
$= (x + 4)(2x - 5)$
and locate each zero as an integer or between consecutive integers. -4 is a zero; $2\frac{1}{2}$ is a zero between 2 and 3.   Then, have students graph $y = 2x^2 + 3x - 20$ on a coordinate plane.   Check students' graphs.

Highlighting the Standards

Standard 4a: Here the equation, table, and graph present equivalent representations of the same concept and thus reinforce student learning.

Lesson Note

Point out the following facts to the students:

1. The tables of ordered pairs (x,y) on page 361 and in the Example are presented in sequence from top-to-bottom beginning with the x-value that is the lower bound and ending with the x-value that is the upper bound. While working the exercises, students will find it more efficient to use x-values beginning with 0 $(0, -1, -2, \ldots)$ and to continue until a lower bound is found. Then positive values of x can be used until an upper bound is found.
2. Convenient scales for the axes are chosen to stretch the graph horizontally and shrink the graph vertically.

Math Connections

Interpolation: The main emphasis of the lesson is on locating the real zeros of a polynomial between successive integers. The graph of the corresponding polynomial function is a visual aid in locating the zeros. If students wish to find the zeros to a greater degree of precision, they may use the graph together with the technique of *linear interpolation*, which is discussed on pages 490–492 of Chapter 13. A convenient way to proceed is to combine the methods of linear interpolation with computer techniques.

The graph of $y = x^3 - x^2 - 14x + 14$ reveals the following facts.

(1) The curve crosses the x-axis between -4 and -3, at 1, and between 3 and 4. So, this third-degree function has three real zeros: 1 is a zero. There is a zero between -4 and -3, and a third zero between 3 and 4.

(2) The "*turning points*" (where the graph changes direction) in Quadrants II and IV cannot be determined precisely at this time. However, approximate turning points can be located.

(3) The graph of this third-degree function has this characteristic shape.

The table can be used to locate each real zero as follows.

(1) The ordered pair $(1,0)$ indicates that 1 is a zero.

(2) The ordered pairs $(-4,-10)$ and $(-3,20)$ have y-values that differ in sign $(-10$ and 20$)$. So, there is a zero between the x-values -4 and -3.

(3) The pairs $(3,-10)$ and $(4,6)$ have y-values with different signs. So, there is a zero between 3 and 4.

If we were to use a table of ordered pairs to graph the second-degree equation $y = 4x^2 - 20x - 20$, and locate each real zero, we could find that:

(1) there are no integral zeros for the function;

(2) there is a zero between -1 and 0, because the y-values of $(-1,4)$ and $(0,-20)$ change sign; and

(3) there is a zero between 5 and 6 for the same reason.

The graph would show that:

(1) this second-degree function with two real zeros has this characteristic shape, and

(2) the turning point cannot be determined precisely at this time. However, an approximate turning point can be located.

EXAMPLE Use a table of ordered pairs to locate each real zero of the function described by $y = 6x^4 - 5x^3 - 58x^2 + 45x + 36$. Graph the function.

Solution Construct a table of ordered pairs (x,y) for all integral values of x between an upper bound and a lower bound.

362   Chapter 10 Polynomial Equations and Functions

Additional Example

Use a table of ordered pairs to locate each real zero for the function described by $y = 4x^4 - 8x^3 - 23x^2 + 27x + 18$. Graph the function.

(x,y): $(-3,270)$, $(-2,0)$, $(-1,-20)$, $(0,18)$, $(1,18)$, $(2,-20)$, $(3,0)$, $(4,270)$
Zeros: -2, 3, between -1 and 0, between 1 and 2

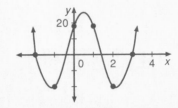

x	6	−5	−58	45	36	(x,y)	Zeros
−4	6	−29	58	−187	784	(−4,784)	There are no zeros for x ≤ −4.
				−4 is a lower bound.			
−3	6	−23	11	12	0	(−3,0)	⟵ −3 is a zero.
−2	6	−17	−24	93	−150	(−2,−150)	
−1	6	−11	−47	92	−56	(−1,−56)	There is a zero between
0	6	−5	−58	45	36	(0,36)	−1 and 0: −56 < 0 < 36.
1	6	1	−57	−12	24	(1,24)	There is a zero between
2	6	7	−44	−43	−50	(2,−50)	1 and 2: −50 < 0 < 24.
3	6	13	−19	−12	0	(3,0)	⟵ 3 is a zero.
4	6	19	18	117	504	(4,504)	There are no zeros for x ≥ 4.
				4 is an upper bound.			

A scientific calculator can be quite helpful in constructing these tables of ordered pairs. Review the *Using the Calculator* feature in Lesson 6.5. Remember to record the display each time you press $=$.

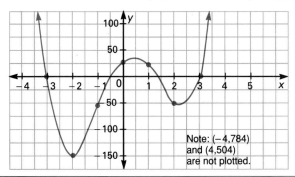

Note: (−4,784) and (4,504) are not plotted.

In the example, observe that (1) the graph of this fourth-degree function with four real zeros has this characteristic shape, and that (2) the turning points cannot be determined precisely at this time.

Classroom Exercises

Match the degree of a polynomial function at the left with its characteristic graph at the right.

1. degree 4 c
2. degree 3 e
3. degree 2 b
4. degree 1 a
5. degree 0 d

 a
 b
 c
 d
 e

Critical Thinking Questions

Analysis: Ask students how they can use the "characteristic shape" of certain polynomial graphs (see pages 362 and 363) to show that a polynomial might have no real zeros. They should see that the "characteristic shape" of the graph of a second-degree or fourth-degree polynomial function need not necessarily cross the x-axis, and thus the polynomial need not be zero for any real value of x.

Checkpoint

Graph the function described by $y = x^3 - 3x^2 - 3x + 9$. Locate each real zero as an integer or locate it between consecutive integers.

(x,y): (−2,−5), (−1,8), (0,9), (1,4), (2,−1), (3,0), (4,13) Zeros: 3, between −2 and −1, between 1 and 2

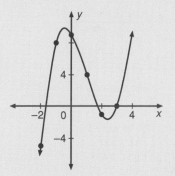

Closure

Have students use a *graphing* calculator to graph $y = x^3 - x^2 - 12x + 12$ and verify that the three zeros are 1, between −4 and −3, and between 3 and 4.
(Set the x-axis from −6 to 6 with scale factor: 1, and the y-axis from −20 to 40 with scale factor: 10.)

FOLLOW UP

Guided Practice

Classroom Exercises 1–8

Independent Practice

A Ex. 1–6, **B** Ex. 7–12, **C** Ex. 13–15

Basic: WE 1–9 odd, Statistics
Average: WE 5–13 odd, Statistics
Above Average: WE 7–15 odd, Statistics

Additional Answers

Written Exercises

1. 3

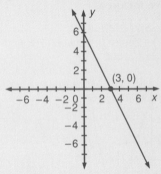

2. Between 1 and 2

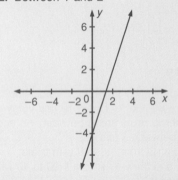

See page 371 for the answers to Written Ex. 3–7 and page 373 for Ex. 8–12.

Use the graph of each function to locate its real zeros, either as integers or between consecutive integers.

6.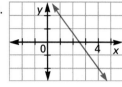
Between 2 and 3

7.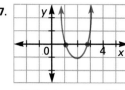
At 1, between 2 and 3

8.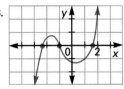
Between −3 and −2, at −1, between 1 and 2

Written Exercises

Graph each function. Locate each real zero as an integer or between consecutive integers.

1. $y = -2x + 6$
2. $y = 3x - 4$
3. $y = x^2 + 4x - 4$
4. $y = x^3 - 4x^2 - 2x + 8$
5. $y = x^4 + x^3 - 14x^2 - 2x + 24$
6. $y = x^4 - 10x^2 + 9$
7. $y = 4x^2 + 20x + 19$
8. $y = -3x^2 - 6x + 4$
9. $y = 4x^3 - 4x^2 - 19x + 10$
10. $y = -2x^3 + x^2 + 18x - 9$
11. $y = 2x^4 - x^3 - 35x^2 + 16x + 48$
12. $y = -x^4 + 10x^2 - 9$

The turning points for the third-degree function $y = x^3 + bx^2 + cx + d$ occur where the values of x are the solutions of the equation

$$3x^2 + 2bx + c = 0, \text{ or at } x = \frac{-2b \pm \sqrt{4b^2 - 12c}}{6}.$$

Find the values of x, both in simplest radical form and to the nearest tenth, for the two turning points of each graph described below.

13. $y = x^3 - x^2 - 14x + 14$ (the first graph of this lesson) $\frac{1 \pm \sqrt{43}}{3}$; −1.9, 2.5
14. $y = x^3 - 4x^2 - 2x + 8$ (the graph in Exercise 4) $\frac{4 \pm \sqrt{22}}{3}$; −0.2, 2.9

15. Determine the range of values of c for which the function $f(x) = 2x^3 - x^2 - 6x + c$ has a zero between $x = 0$ and $x = 1$. (HINT: Find values of c so that $f(x)$ is positive at one of the points 0 and 1 and negative at the other.) $0 < c < 5$

Mixed Review

Determine whether the lines are parallel, perpendicular, or neither. *3.7*

1. $3x - 2y = 7$
   $4y = 6x + 9$
   Parallel

2. $5x - 2y = 10$
   $2y + 5x = 10$
   Neither

3. $4x - 3y = 12$
   $4y + 3x = 12$
   Perpendicular

4. $2 - 3x = 14$
   $4y - 5 = 23$
   Perpendicular

Enrichment

Refer the students to the Example, in which irrational zeros are known to lie between −1 and 0 and between 1 and 2. Explain that the approximate zeros can be found by interpolation. Consider the zero that lies between 1 and 2.

x	1	?	2
y	24	0	−50

Note that the y-value 0 lies $\frac{24}{74}$ of the way from 24 to −50. Thus, the value of the zero must be *about* $\frac{24}{74}$ of the way from $x = 1$ to $x = 2$. An approximation of the zero is

$$x = 1\frac{24}{74}, \text{ or } 1.\overline{324}.$$

To check, have the students substitute $1.\overline{324}$ for x in the function.

$$y \approx 0.717$$

This is close enough to 0 to use $1.\overline{324}$ as an approximate irrational zero of the function. Have the students use this strategy to approximate other irrational zeros in some of the examples and exercises.

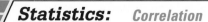

Statistics: *Correlation*

Researchers often assert that there is a "high correlation" between two things, such as smoking and cancer or certain pollutants and ozone depletion. That is, they seem to be *connected* or *related*.

Statisticians may use a line to show the relation between 2 sets of data plotted on a *scatter diagram* (see page 123). A number between −1 and 1 called the **correlation coefficient,** denoted by the letter r, tells how closely the points cluster about a line. If r is positive, there is a **positive correlation** between the data sets, and a line drawn to fit the points has positive slope. If r is negative, there is a **negative correlation,** and a line drawn to fit the points has negative slope.

If r is −1 or 1, the points all lie on a line, as shown at the right. If r is zero, they appear to be randomly distributed, and show no relationship. The closer r is to −1 or 1, the *stronger* the correlation, and the flatter an ellipse drawn to contain the points becomes. Some illustrations of values of r are shown below.

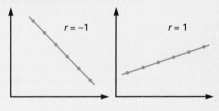

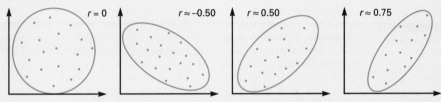

The Dow Jones Industrial Average and Standard and Poor's 500 Index are two of the primary measures of the stock market. The table at the right gives each measure for 15 consecutive trading days in a recent summer, with the Dow Average listed first.

(2895.30, 359.10)	(2862.38, 358.71)
(2893.56, 358.47)	(2897.33, 363.15)
(2882.18, 356.88)	(2911.65, 364.96)
(2935.89, 362.91)	(2925.00, 366.64)
(2928.22, 362.90)	(2935.19, 367.40)
(2929.95, 364.90)	(2900.97, 363.16)
(2933.42, 366.25)	(2876.66, 361.23)
(2892.57, 361.63)	

1. Construct a scatter diagram of the points. Is the correlation positive or negative? Does one measure cause the other to go up or down or might other factors, called *hidden variables*, control both?

2. These ordered pairs group the average monthly rainfall in inches with the average monthly percent of sunshine blocked by clouds (listed first) in Amarillo, Texas. Construct a scatter diagram. Describe the correlation. Are the results surprising?
   (32, 0.49), (32, 0.46), (31, 0.57), (29, 0.87), (27, 1.08), (28, 2.79)
   (23, 3.50), (22, 2.70), (23, 2.95), (26, 1.72), (25, 1.39), (28, 0.58)

Statistics **365**

Statistics

1.
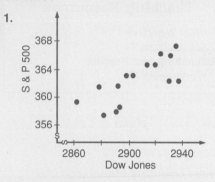

Positive; Answers will vary. One possible answer: Since each measure reflects the state of the stock market, it is likely that large-scale economic factors influence each in a similar way.

2.
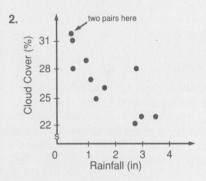

There is a negative correlation. Answers will vary.

365

■■■GETTING STARTED

Prerequisite Quiz

Find the conjugate of each complex number.

1. $3 - 2i$ $3 + 2i$
2. $-4 + i\sqrt{3}$ $-4 - i\sqrt{3}$
3. $5i$ $-5i$

Simplify.

4. $(3 + 2i)(-2 + 2i) + 1$ $-9 + 2i$
5. $(3 + 2i)(-9 + 2i) + 49$ $18 - 12i$

Motivator

Have students determine the nature of the roots of the equation $9x^2 - 6x + 4$ without solving.   Two nonreal roots   Ask students what they know about the two nonreal roots. They are conjugates.   Have students find the multiplicities of the zeros of $(x + 5)^2(x^2 - 25)$. 5 has $m = 1$, -5 has $m = 3$

10.7 The Conjugate Zero Theorem and Descartes' Rule of Signs

Objectives

To find the remaining zeros of integral polynomials given one nonreal zero

To predict the sums of the multiplicities of the positive and negative real zeros of integral polynomials

The zeros of $P(x) = x^2 - 8x + 20$ are a pair of conjugates, as shown below.

Let $P(x) = 0$. Then by the quadratic formula,

$$x = \frac{8 \pm \sqrt{(-8)^2 - 80}}{2} = \frac{8 \pm \sqrt{-16}}{2} = \frac{8 \pm 4i}{2}.$$

The zeros of $P(x)$ are $4 + 2i$ and $4 - 2i$, a pair of complex conjugates. In general, for any integral polynomial $P(x)$, its nonreal zeros, if any, appear in *conjugate pairs*. This is stated in the *Conjugate Zero Theorem* below.

Theorem 10.7

Conjugate Zero Theorem
If $a + bi$ is a zero of an integral polynomial $P(x)$, then $a - bi$ is also a zero of $P(x)$.

EXAMPLE 1 Given that $P(x) = x^4 - 5x^3 + x^2 + 49x - 78$ and that $3 - 2i$ is a zero of $P(x)$, find the remaining zeros of $P(x)$.

Plan If $3 - 2i$ is a zero, then $3 + 2i$ is also a zero. Use synthetic substitution to find a second-degree factor of $P(x)$.

Solution

	1	-5	1	49	-78
		$(3 - 2i)$	$(-10 - 2i)$	$(-31 + 12i)$	(78)
$3 - 2i$	1	$-2 - 2i$	$-9 - 2i$	$18 + 12i$	$\boxed{0}$

		$(3 + 2i)$	$(3 + 2i)$	$(-18 - 12i)$
$3 + 2i$	1	1	-6	$\boxed{0}$

$x^2 + x - 6 = (x + 3)(x - 2)$

The zeros of $P(x)$ are $3 - 2i$, $3 + 2i$, -3, and 2.

Highlighting the Standards

Standard 14d: This lesson takes students well into the realm of complex numbers. They will be able to see many similarities with the set of real numbers.

Additional Example 1

Given: $P(x) = 6x^4 + x^3 + 52x^2 + 9x - 18$ and $3i$ is a zero of $P(x)$. Find the remaining zeros of $P(x)$. $3i, -3i, -\frac{2}{3}, \frac{1}{2}$

The polynomial $P(x)$ in Example 1 has one positive real zero, one negative real zero and two nonreal zeros, each with a multiplicity of 1. The sum of these multiplicities, or the total number of first-degree factors, is equal to 4, which is also the degree of $P(x)$.

Consider $P(x) = 2x^5 - 11x^4 + x^3 + 65x^2 - 39x - 90$.

$$= (x - 3)^2(2x - 5)(x + 1)(x + 2)$$

$P(x)$ has two positive real zeros: 3 (with a multiplicity of 2) and $\frac{5}{2}$ (with a multiplicity of 1). The sum of these multiplicities is $2 + 1$, or 3.

$P(x)$ has two negative real zeros, -1 and -2, each with a multiplicity of 1. The sum of these multiplicities is $1 + 1$, or 2.

There are three *changes in sign* in the coefficients of $P(x)$, as shown below.

$$P(x) = +2x^5 \underset{1}{\overset{}{\frown}} -11x^4 \underset{2}{\overset{}{\frown}} +1x^3 \quad +65x^2 \underset{3}{\overset{}{\frown}} -39x \quad -90$$

Notice that the number of sign changes in $P(x)$ is the same as the sum of the multiplicities of the positive real zeros.

There are two sign changes in $P(-x)$, as shown below.

$$P(-x) = 2(-x)^5 - 11(-x)^4 + 1(-x)^3 + 65(-x)^2 - 39(-x) - 90$$
$$= -2x^5 - 11x^4 \underset{1}{\overset{}{\frown}} -1x^3 + 65x^2 \underset{2}{\overset{}{\frown}} +39x - 90$$

Notice that the number of sign changes in $P(-x)$ is the same as the sum of the multiplicities of the negative real zeros. Also notice that the sum of all the multiplicities is 5, which is the degree of $P(x)$.

This example suggests *Descartes' Rule of Signs*, stated below. It allows us to predict both the number of first-degree factors of $P(x)$ that give *positive* real roots and the number that give *negative* real roots, that is, the sums of the multiplicities of the positive and negative real zeros.

Descartes' Rule of Signs

The sum of the multiplicities of the positive real zeros of an integral polynomial $P(x)$ is either equal to the number of sign changes in $P(x)$ or is less than that number by a multiple of two.

The sum of the multiplicities of the negative real zeros of $P(x)$ is either equal to the number of sign changes in $P(-x)$ or is less than that number by a multiple of two.

Lesson Note

In the answer to Example 2, the sum of the multiplicities of the complex nonreal zeros can be included in the table, as shown below.

positive real	2	2	0	0
negative real	3	1	3	1
complex	0	2	2	4

Notice that the sum for each column above is 5, the same as the degree of $P(x)$.

Math Connections

Radical Conjugates: The Conjugate Zero Theorem has a counterpart in the set of numbers of the form $a + b\sqrt{c}$, where a and b are integers and c is a counting number that is not a perfect square. If $a + b\sqrt{c}$ is a zero of an integral polynomial, then its conjugate $a - b\sqrt{c}$ is a zero. (See also page 281). For example, by direct substitution it can be shown that $2 + 3\sqrt{2}$ is a zero of $x^3 - 4x^2 - 14x$. By the "Conjugate Zero Theorem for radical conjugates" it is the case that $2 - 3\sqrt{2}$ is a zero also.

Critical Thinking Questions

Analysis: Ask students to show, without solving, that $x^6 - x^2 - 7$ has exactly two real zeros. By the first part of Descartes' Rule of Signs, there is one positive zero. By the second part of the Rule, there is one negative zero. The possibility that 0 is a zero is eliminated by direct substitution. Thus, there are exactly two real zeros.

Common Error Analysis

Error: Some students have difficulty in finding the number of sign changes in $P(-x)$, given a polynomial $P(x)$. Have these students determine whether each of the following is a positive or a negative number.

$$(-2)^2, (-2)^4, (-2)^6, \cdots$$
$$(-2)^1, (-2)^3, (-2)^5, \cdots$$

Elicit the conclusion that $(-2)^n$ is positive if n is even and $(-2)^n$ is negative if n is odd. The students can then be asked for the number of sign changes in $P(-x)$, if $P(x) = x^5 + 3x^4 - 2x^3 - 4x^2 + 6x - 8$.

$$P(-x) = (-x)^5 + 3(-x)^4 - 2(-x)^3$$
$$- 4(-x)^2 + 6(-x) - 8$$
$$= -x^5 + 3x^4 + 2x^3 - 4x^2 - 6x - 8$$

There are two sign changes in the sequence $-\ +\ +\ -\ -\ -$.

$$1\qquad 2$$

Checkpoint

1. Use the Conjugate Zero Theorem to find all the zeros of $P(x) = x^3 + x^2 - 4x + 6$, if $1 - i$ is a zero of $P(x)$. $1 - i, 1 + i, -3$
2. Find the possible combinations of the sums of the multiplicities of the positive real zeros and the negative real zeros of $x^5 - x^4 + x^2 + x + 2$.

Positive real	2	2	0	0
Negative real	3	1	3	1

Closure

Have students summarize Descartes' Rule of Signs in their own words. Then, ask them to explain how a fourth-degree polynomial with four distinct zeros can have two positive zeros and no negative zeros.   It can have two complex nonreal zeros.

EXAMPLE 2 Find the possible combinations of the sums of the multiplicities of the positive and negative real zeros of $P(x)$, where
$$P(x) = x^5 - 9x^3 - x^2 + 9.$$

Solution Find the number of sign changes in $P(x)$ and in $P(-x)$.

$$P(x) \quad = \quad +x^5\underset{1}{} -9x^3 \quad -x^2\underset{2}{} +9 \quad \leftarrow 2 \text{ sign changes}$$

So, the sum of the multiplicities of the positive real zeros of $P(x)$ is either 2 or 0.

$$P(-x) = (-x)^5 - 9(-x)^3 - (-x)^2 + 9$$
$$= -x^5\underset{1}{} +9x^3\underset{2}{} -x^2\underset{3}{} +9 \leftarrow 3 \text{ sign changes}$$

So, the sum of the multiplicities of the negative real zeros of $P(x)$ is either 3 or 1.

Thus, there are four possible combinations for the sums of the multiplicities of the real zeros, as shown in the table at the right.

positive	2	2	0	0
negative	3	1	3	1

Classroom Exercises

Determine another zero of the integral polynomial $P(x)$ given the non-real zero of $P(x)$ below.

1. $-3i$ $3i$
2. $5i$ $-5i$
3. $2 + 4i$ $2 - 4i$
4. $-1 - 6i$ $-1 + 6i$

Find the number of sign changes in the coefficients of each polynomial.

5. $P(x) = 7x^4 + 6x^3 - 5x^2 + 4x - 8$ 3
6. $Q(x) = 6x^5 + 3x^4 - 4x^2 - 1$ 1
7. $P(-x) = 7(-x)^4 + 6(-x)^3 - 5(-x)^2 + 4(-x) - 8$ 1
8. Find the remaining zeros of $P(x) = 2x^3 - x^2 + 32x - 16$ given that $x_1 = 4i$. $-4i, \frac{1}{2}$

Written Exercises

Find the remaining zeros of $P(x)$ given that x_1 is a zero of $P(x)$.

1. $P(x) = 3x^3 - x^2 + 75x - 25, x_1 = 5i$ $-5i, \frac{1}{3}$
2. $P(x) = x^3 - 4x^2 + 4x - 16, x_1 = -2i$ $2i, 4$
3. $P(x) = x^4 - 5x^3 + 15x^2 - 5x - 26, x_1 = 2 - 3i$ $2 + 3i, -1, 2$
4. $P(x) = 3x^4 - 22x^3 + 50x^2 - 16x - 40, x_1 = 3 + i$ $3 - i, -\frac{2}{3}, 2$

Additional Example 2

Find the possible combinations of the sums of the multiplicities of the positive real zeros and the negative real zeros of $P(x)$, where $P(x) = 2x^5 - 3x^4 + 7x^2 + 4x - 2$.

Positive real	3	1	3	1
Negative real	2	2	0	0

Enrichment

Refer the students to this general form of an integral polynomial.
$$P(x) = a_0x^n + a_1x^{n-1} + a_2x^{n-2}$$
$$+ \cdots + a_{n-1}x + a_n\ (a_0 \geq 1)$$

Ask what conditions must exist for the given statement to be true.

1. All the potential zeros must be integers.
   $a_0 = \pm 1$

2. The function must have at least one real zero. n is odd.
3. The maximum number of potential rational zeros is 2. $a_0 = \pm 1; a_n = \pm 1$
4. The existence of a complex zero is impossible. $n = 1$
5. The only potential rational zeros are $\pm \frac{1}{2}$ and ± 1. $a_0 = \pm 2; a_n = \pm 1$
6. A certain zero of the function is 0. $a_n = 0$

Find the possible combinations of the sums of the multiplicities of the positive and negative real zeros of each polynomial.

5. $x^5 - x^4 + 12$ **6.** $8x^3 + 4x^2 + x + 2$ **7.** $x^4 - 4x^3 + 6x^2 - 5$

8. $x^6 - x^5 + x^2 + x + 4$ **9.** $2x^5 - x^3 + 3x^2 - x - 8$

10. $x^7 - x^6 + x^4 + x^2 + 1$ **11.** $2x^8 + x^7 - 3x^6 + x + 2$

12. Find all the zeros of $P(x) = 2x^5 - 7x^4 + 8x^3 - 2x^2 + 6x + 5$ given that i and $2 - i$ are two of the zeros. $\pm i, 2 \pm i, -\frac{1}{2}$

13. Find all the zeros of $P(x) = x^5 - 5x^4 + 8x^3 - 40x^2 + 16x - 80$ given that $2i$ is one of the zeros and that it has a multiplicity of two. $\pm 2i, 5$

Mixed Review *5.4, 7.3*

1. Solve $a(bx - c) - (3x + ac) = d$ for x. $\dfrac{d + 2ac}{ab - 3}$

2. Convert 75 mi/h to feet per second.   110 ft/s

▰ Application: *Rational Powers in Music*

On a piano, a **half-step** is the *musical interval* from one key to the next. For example, the interval from B to C is a half-step. The interval from C to C-sharp (a black key) is also a half-step. The interval from one C to the next C, which is called an octave, has 12 half-steps.

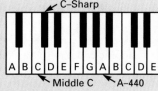

In the **well-tempered system of tuning**, invented by J. S. Bach, there is a constant ratio of *musical frequencies* of any note to the next higher note. That ratio is $1 : 2^{\frac{1}{12}}$. If the frequency of a note is f, then the frequency f' of the note n half-steps *higher* is

$$f' = f \times 2^{\frac{n}{12}}, \text{ or } f \times \sqrt[12]{2^n}.$$

The frequency f'' of a note n half-steps *lower* is

$$f'' = f \times 2^{-\frac{n}{12}}, \text{ or } f \div \sqrt[12]{2^n}.$$

The standard tuning note is the A above Middle C. This note, which is called A-440, has a frequency of 440 cycles per second.

1. From A-440 to the C above there are 3 half-steps. Find the frequency of the C.   523.25

2. From A-440 to the C below (Middle C) there are 9 half-steps. Find the frequency of Middle C.   261.63

3. What is the ratio of Middle C to the C above? How could you show this using a frequency formula?

Guided Practice

Classroom Exercises 1–8

Independent Practice

A Ex. 1–4, **B** Ex. 5–11, **C** Ex. 12, 13

Basic: WE 1–7 odd, Application

Average: WE 3–11 odd, Application

Above Average: WE 7–13 odd, Application

Additional Answers

Written Exercises

5. 2 pos, 1 neg; 0 pos, 1 neg

6. 0 pos, 3 neg; 0 pos, 1 neg

7. 3 pos, 1 neg; 1 pos, 1 neg

8. 2 pos, 2 neg; 2 pos, 0 neg; 0 pos, 2 neg; 0 pos, 0 neg

9. 3 pos, 2 neg; 3 pos, 0 neg; 1 pos, 2 neg; 1 pos, 0 neg

10. 2 pos, 1 neg; 0 pos, 1 neg

11. 2 pos, 2 neg; 2 pos, 0 neg; 0 pos, 2 neg; 0 pos, 0 neg

Application

3. 1:2; The ratio of middle C to the C above can be written as follows.

$$\frac{440 \times 2^{-\frac{9}{12}}}{440 \times 2^{\frac{3}{12}}} = \frac{2^{-\frac{9}{12}}}{2^{\frac{3}{12}}}$$

$$= 2^{-\frac{12}{12}} = 2^{-1} = \frac{1}{2}, \text{ or } 1:2$$

7. −7, two nonreal

8. 121, two rational

9. 0, one rational

21. Between −3 and −2, between 0 and 1, between 2 and 3

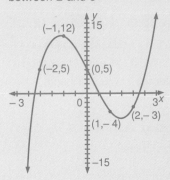

22. −3, −1, 2, 4

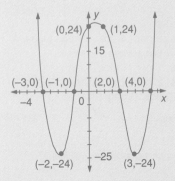

23. Answers will vary. One possible answer: The conjugate of the given nonreal zero will also be a zero of the polynomial. Use synthetic substitution with these two roots to find a second-degree factor of the polynomial, which can then be solved for the remaining roots.

Chapter 10 Review

Key Terms

Conjugate Zero Theorem (p. 366)
Descartes' Rule of Signs (p. 367)
discriminant (p. 343)
Factorization Theorem (p. 358)
Fundamental Theorem of Algebra (p. 351)
lower bound (p. 352)

radical equation (p. 347)
rational zero (p. 356)
Rational Zero Theorem (p. 356)
Upper- and Lower-Bound Theorem (p. 352)
upper bound (p. 352)

Key Ideas and Review Exercises

10.1 If $ax^2 + bx + c = 0$, or $x^2 + \frac{b}{a}x + \frac{c}{a} = 0$, has roots x_1 and x_2, then

$x_1 + x_2 = -\frac{b}{a}$ and $x_1 \cdot x_2 = \frac{c}{a}$.

If $x_1 + x_2 = -\frac{b}{a}$ and $x_1 \cdot x_2 = \frac{c}{a}$, then x_1 and x_2 are the roots of $x^2 + \frac{b}{a}x + \frac{c}{a} = 0$, or $ax^2 + bx + c = 0$.

Without solving the equation, find the sum and product of its roots.

1. $3w^2 - 5w + 6 = 0$
Sum: $\frac{5}{3}$; product: 2

2. $\frac{1}{2}y^2 + 4y = 3$
Sum: −8, product: −6

3. $-x^2 + 6 = 4x$
Sum: −4; product: −6

Write a quadratic equation, in general form, that has the given roots.

4. $-\frac{1}{3}$ and 4
$3x^2 - 11x - 4 = 0$

5. $2 \pm 3\sqrt{2}$
$x^2 - 4x - 14 = 0$

6. 7 $x^2 - 14x + 49 = 0$

10.2 The discriminant $b^2 - 4ac$ determines the nature of the roots of any quadratic equation $ax^2 + bx + c = 0$. It determines the number of roots (1 or 2), and whether they are real or nonreal. If the roots are real, and a, b, and c are rational, then the discriminant also determines whether the roots are rational or irrational.

Find the discriminant and determine the nature of the roots without solving the equation.

7. $4x^2 - 5x + 2 = 0$

8. $6y^2 = 4 - 5y$

9. $20x = 25x^2 + 4$

10. For what values of k will $y^2 + (k + 2)y + 9 = 0$ have exactly one root? (HINT: $b^2 - 4ac = 0$ when there is exactly one root) −8, 4

11. For what values of k will $kx^2 - 8x + (k + 6) = 0$ have two real roots? (HINT: $b^2 - 4ac > 0$ when there are two real roots) $-8 < k < 2, k \neq 0$

10.3 To solve a radical equation such as $\sqrt{5y - 16} - y = 2$, (1) isolate the radical, (2) raise each side to the appropriate power, (3) solve for the variable, and (4) check the apparent solutions.

Solve the equation. Check.

12. $3\sqrt{y-5} = \sqrt{3y-3}$ 7

13. $\sqrt{2x+7} - \sqrt{x+4} = 2$ 21

14. Solve $y = \sqrt{\dfrac{3x}{z}}$ for x, in terms of y and z. $\dfrac{y^2z}{3}$

10.4   The Upper- and Lower-Bound Theorem is used to identify intervals within which all the real zeros of a polynomial lie. It can be used to reduce the number of potential zeros that must be tested.

15. Find the least integral upper bound U and the greatest integral lower bound L for the real zeros of $3x^4 - 5x^3 - 25x^2 - 5x - 28$. $U = 4, L = -3$

Use the Integral Zero Theorem to solve the equation.

16. $x^3 - 3x^2 + 4x - 12 = 0$ $3, \pm 2i$

17. $x^4 - 2x^3 - 8x^2 + 10x + 15 = 0$ $-1, 3, \pm\sqrt{5}$

18. $4y^5 + 8y^4 - 29y^3 - 42y^2 + 45y + 54 = 0$ $-3, 2, -1, \pm\frac{3}{2}$

10.5   The Rational Zero Theorem can be used to list the potential rational zeros of a polynomial.

Use the Rational Zero Theorem to solve the equation.

19. $3x^4 - 2x^3 + 8x^2 - 6x - 3 = 0$ $-\frac{1}{3}, 1, \pm i\sqrt{3}$

20. $8y^4 - 6y^3 + 17y^2 - 12y + 2 = 0$ $\frac{1}{4}, \frac{1}{2}, \pm i\sqrt{2}$

10.6   To graph a function such as $y = 2x^3 - x^2 - 10x + 5$, use synthetic substitution to form a table of ordered pairs (x,y) for all integral values of x between an upper bound and a lower bound for the real zeros of the function.

Graph each function. Locate each real zero as an integer or between consecutive integers.

21. $y = 2x^3 - x^2 - 10x + 5$

22. $y = x^4 - 2x^3 - 13x^2 + 14x + 24$

10.7   Review the Conjugate Zero Theorem and Descartes' Rule of Signs.

23. Write in your own words how you would find all the zeros of a fourth-degree integral polynomial given one of its nonreal zeros.

24. Find all the remaining zeros of $P(x) = x^4 - 5x^3 - 3x^2 + 43x - 60$ given that $2 + i$ is one of the zeros. $2 \pm i, 4, -3$

25. Use Descartes' Rule of Signs to find all the possible combinations of the sums of the multiplicities of the positive and negative real zeros of $3x^5 - 2x^4 + 5x^3 - 4x - 2$.
   3 pos, 2 neg; 3 pos, 0 neg; 1 pos, 2 neg; 1 pos, 0 neg

3. Between -5 and -4, between 0 and 1

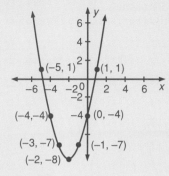

4. Between -2 and -1, between 1 and 2, 4

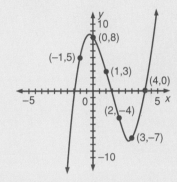

5. -4, between -2 and -1, between 1 and 2, 3

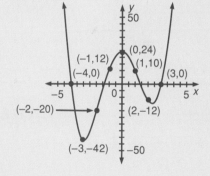

6. $-3, -1, 1, 3$

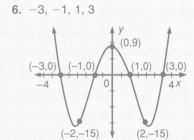

7. Between -4 and -3, between -2 and -1

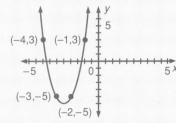

7. -44, two nonreal
8. 0, one rational
9. 84, two irrational
17. -1, 1, between -3 and -2, between 2 and 3

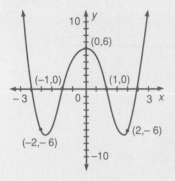

Chapter 10 Test

Without solving the equation, find the sum and product of its roots.

1. $2x^2 + 7x - 8 = 0$
   Sum: $-\frac{7}{2}$, product: -4

2. $\frac{1}{3}y^2 = 5y - 2$ Sum: 15,
   product: 6

3. $14 = -\frac{1}{2}y^2$ Sum: 0,
   product: 28

Write a quadratic equation, in general form, that has the given root or roots.

4. $\frac{1}{2}$ and -3
   $2x^2 + 5x - 3 = 0$

5. $3 \pm i$ $x^2 - 6x + 10 = 0$

6. -6 $x^2 + 12x + 36 = 0$

Find the discriminant, $b^2 - 4ac$, and determine the nature of the roots without solving the equation.

7. $3x^2 - 2x + 4 = 0$

8. $16y^2 = 8y - 1$

9. $5z^2 + 2z - 4 = 0$

10. For what values of k will $2x^2 + (k + 2)x + 18 = 0$ have exactly one solution? $-14, 10$

Solve the equation. Check.

11. $2\sqrt[4]{x} = \sqrt[4]{14x + 32}$ 16

12. $3 + \sqrt{3x + 1} = x$ 8

13. $\sqrt{x + 16} - \sqrt{x} = 2$ 9

14. Find the least integral upper bound U and the greatest integral lower bound L for the real zeros of $2x^4 + x^3 - 11x^2 - 5x + 5$. $U = 3, L = -3$

Find all the roots of the equation.

15. $2x^4 + x^3 - 11x^2 - 5x + 5 = 0$ $-1, \frac{1}{2}, \pm\sqrt{5}$

16. $x^4 + 4x^3 - x^2 + 16x - 20 = 0$ $-5, 1, \pm 2i$

17. Graph $y = x^4 - 7x^2 + 6$. Locate each real zero as an integer or between consecutive integers.

18. Find all the remaining zeros of $2x^4 - 7x^3 + 14x^2 - 63x - 36$ given that $3i$ is one of the zeros. $-3i, -\frac{1}{2}, 4$

19. Use Descartes' Rule of Signs to find all the possible combinations of the sums of the multiplicities of the positive and negative real zeros of $5x^4 - 2x^3 + 7x^2 - 10x - 4$. 3 pos, 1 neg; 1 pos, 1 neg

20. For what values of k will the sum of the roots of $x^2 - (k^2 - 2k)x + 12 = 0$ be 8? $-2, 4$

21. Solve $\sqrt{x + 2} = \sqrt[4]{10x + 11}$. $-1, 7$

22. Solve $x^4 - 6x^3 + 5x^2 + 12x - 14 = 0$ given that $3 - \sqrt{2}$ is one of the roots. $3 \pm \sqrt{2}, \pm\sqrt{2}$

College Prep Test

Choose the *one* best answer to each question or problem.

1. A polynomial with *exactly* one integral zero is __?__. D
- (A) $x^2 - 1$ (B) $4z^2 + 1$
- (C) $a^3 + a^2 + a - 1$
- (D) $15y^3 - 10y^2 - 4y - 1$
- (E) None of these

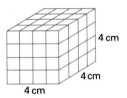

4 cm

4 cm

4 cm

2. A solid cube is 4 cm long on each edge. How many straight cuts, *through* the cube, are needed to produce 64 smaller cubes that are 1 cm long on each edge? C
- (A) 3 (B) 6 (C) 9
- (D) 12 (E) 16

3. If $\dfrac{2(\sqrt{2x} - \sqrt{2x - 5})}{2\sqrt{x + 5} - 2\sqrt{x}} = 1$,

then $x = $ __?__
- (A) 3 (B) 4 (C) 5
- (D) 6 (E) None of these C

4. If $y = \sqrt[3]{\dfrac{x}{z}}$, and y is tripled while z is doubled, by what number is x multiplied? E
- (A) 2 (B) 3 (C) 8
- (D) 27 (E) 54

5. One diagonal of a rectangle is $\sqrt{15}$ cm long. The rectangle's length is $\sqrt{12}$ cm. Find the rectangle's area. B
- (A) $3\sqrt{2}$ cm^2 (B) 6 cm^2
- (C) 9 cm^2 (D) $6\sqrt{5}$ cm^2
- (E) None of these

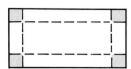

6. A rectangular piece of cardboard is 80 cm by 70 cm. A 10-cm square is cut from each corner, and the four flaps are folded up to form an open box. Find the volume of the box. B
- (A) 3,000 cm^3 (B) 30,000 cm^3
- (C) 42,000 cm^3 (D) 56,000 cm^3
- (E) None of these

7. If -2, with a multiplicity of 3, is a zero of $P(x) = x^4 + 4x^3 - 16x - 16$, what is another zero of $P(x)$? C
- (A) -1 (B) 1 (C) 2
- (D) 3 (E) 8

8. $\sqrt[4]{(\sqrt{8})^3} = $ C
- (A) $\sqrt[4]{8}$ (B) $\sqrt[7]{8}$ (C) $2\sqrt[8]{2}$
- (D) $4\sqrt[4]{4}$ (E) $\sqrt[12]{8}$

9. If $(a,b)^* = \sqrt{a^2 + b^2}$, which of the following is the greatest? D
- (A) $(3,4)^*$ (B) $(-4,-3)^*$
- (C) $(5,0)^*$ (D) $(-1,-5)^*$
- (E) $(\sqrt{7},\sqrt{10})^*$

10. $4 + 3\sqrt{2}$ is a root of which equation?
- (A) $x^2 - 8x - 2 = 0$ A
- (B) $x^2 - 8x - 18 = 0$
- (C) $x^2 - 16x - 2 = 0$
- (D) $x^2 - 16x + 18 = 0$
- (E) None of these

11. If $kx^2 = k^2$, for what values of k will there be exactly two real values of x?
- (A) All values of k C
- (B) All values of $k \neq 0$
- (C) All values of $k > 0$
- (D) All values of $k < 0$
- (E) None of these

College Prep Test **373**

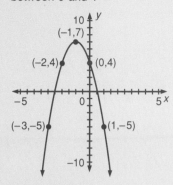
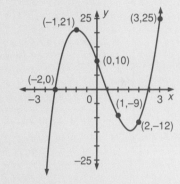
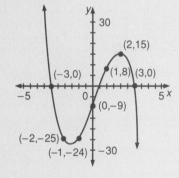
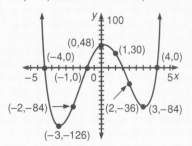
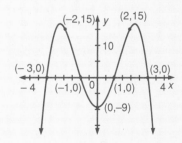

20. $x = \dfrac{7a}{2a + b}$; 3

25. $\{a \mid a < -3 \text{ or } a > 4\}$

34.

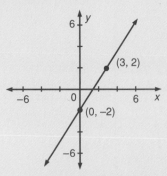

35.

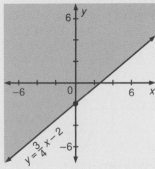

38.

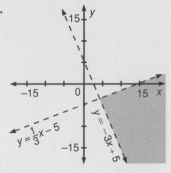

41. $\dfrac{a^4(x + 1)}{40x^4}$

Cumulative Review (Chapters 1–10)

Identify the property of operations   *1.3*
with real numbers. All variables
represent real numbers.

1. $3a \cdot \dfrac{1}{3a} = 1$ Mult Inverse Prop

2. $3a \cdot 1 = 3a$ Mult Identity Prop

3. $(c + 4)d = cd + 4d$ Distr Prop

4. $cd + 4d = 4d + cd$ Comm Prcp Add

5. $(c + 4)d = d(c + 4)$ Comm Prop Mult

6. $(c + 4) + d = c + (4 + d)$
     Assoc Prop Add

Solve.

7. $5a - (6 - 3a) - 2(a - 4) = 14$ 2   *1.4*

8. $\dfrac{x}{6} + \dfrac{x}{4} + \dfrac{x}{3} = 1$ $\dfrac{4}{3}$

9. $a + 2.65 = 7.45 - 2.2a$ 1.5

10. $|4y - 6| = 14$ $-2, 5$ *2.3*

11. $2^{3n-1} = 32$ 2   *5.1*

12. $\dfrac{2x - 1}{x - 1} = \dfrac{3}{x + 3}$ $-1, 0$ *7.6*

13. $\dfrac{24}{x + 2} - \dfrac{x - 4}{x^2 - 5x - 14} = \dfrac{2}{x - 7}$ 8

14. $\dfrac{5}{n} - \dfrac{3}{n + 8} = 1$ $-10, 4$

15. $x^2 + 14 = 0$ $\pm i\sqrt{14}$ *9.1*

16. $2t^2 - 5t + 4 = 0$ $\dfrac{5 \pm i\sqrt{7}}{4}$ *9.5*

17. $3\sqrt{2x - 1} = \sqrt{x + 25}$ 2   *10.3*

18. $\sqrt[3]{3x^2 - 11} = 4$ ± 5

19. $y + \sqrt{5y + 1} = 7$ 3

20. Solve $a(2x - 4) = 3a - bx$ *7.6*
for x. Use the result to find x
given that $a = 1.5$ and $b = 0.5$.

Find the solution set. $\{x \mid x \le 5\}$

21. $5x - 4(2x - 3) \ge x - 8$ *2.1*

22. $3 - 5x < x - 9$ *and*   *2.2*
$8 > 2x - 6$ $\{x \mid 2 < x < 7\}$

23. $-14 < 3t + 4 < 22$ $\{t \mid -6 < t < 6\}$

24. $|4y - 6| < 10$ $\{y \mid -1 < y < 4\}$ *2.3*

25. $a^2 - a - 12 > 0$ *6.4*

26. $y^3 - 36y > 0$
    $\{y \mid -6 < y < 0 \text{ or } y > 6\}$

Find the domain of each function.
                     $\{3, 4, 5, 6\}$

27. $\{(3, -1), (4, 2), (5, 0), (6, 7)\}$ *3.1*

28. $y = 4$ $\{$real numbers$\}$ *3.5*

29. $y = \dfrac{x + 5}{x^2 - 3x - 10}$ *7.1*
    $\{x \mid x \ne -2 \text{ and } x \ne 5\}$

Given $f(x) = 4x - 3$ and $g(x) =$
$x^2 + 3$, find the following values.

30. $f(5) - g(-2)$ 10   *3.2*

31. $f(g(c + 4))$ $4c^2 + 32c + 73$ *5.9*

32. $\dfrac{f(a) - f(b)}{a - b}$ 4   *7.1*

33. Use the slope formula and   *3.3*
the point-slope form to write   *3.4*
an equation, in standard
form, for the line $\overleftrightarrow{PQ}$
through $P(2, -4)$ and
$Q(-1, 5)$. $3x + y = 2$

Draw the graph in a coordinate plane.

34. $4x - 3y - 6 = 0$ *3.5*

35. $3x - 4y \le 8$ *3.9*

36. If p varies directly as v, and   *3.8*
$p = 314$ when $v = 50$, what
is v when $p = 785$?   125

37. Solve the system.   *4.3*
$2x - 5y = 16$
$3x + 4y = 1$ $(3, -2)$

38. Solve the system by graphing.   *4.7*
$y + 3x > 5$
$y < \dfrac{1}{3}x - 5$

Simplify.

39. $\left(\dfrac{2a^{-2}}{b^{-1}}\right)^3$ $\dfrac{8b^3}{a^6}$ *5.2*

40. $2a(3a - 4)^2$ $18a^3 - 48a^2 + 32a$ *5.3*

41. $\dfrac{a^5x^2}{5x + 20} \div \dfrac{8ax^6}{x^2 + 5x + 4}$ *7.2*

42. $\dfrac{3y}{y^2 - 2y - 8} + \dfrac{4}{y - 4} - \dfrac{5}{y + 2}$ *7.4*

43. $2c\sqrt{12c^7}$, $c \ge 0$ $4c^4\sqrt{3c}$ *8.2*

44. $5\sqrt{8} + 6\sqrt{32} - 4\sqrt{2}$ $30\sqrt{2}$ *8.3*

45. $(5\sqrt{3} - \sqrt{5})(5\sqrt{3} + \sqrt{5})$ 70

46. $\dfrac{5}{4 - 2\sqrt{2}}$ $\dfrac{10 + 5\sqrt{2}}{4}$ *8.4*

47. $5a\sqrt[3]{16a^6b^4}$ $10a^3b\sqrt[3]{2b}$ *8.5*

48. $(7 + \sqrt{-8}) - (2 - 3\sqrt{-2})$ *9.1*

49. $3\sqrt{-2} \cdot 5\sqrt{-12}$ $-30\sqrt{6}$ *9.2*

50. $(6 + 2i)^2$ $32 + 24i$

51. $\dfrac{2}{4 - 3i}$ $\dfrac{8 + 6i}{25}$

Factor completely. (Exercises 52–53)

52. $12y^3 - 27y$ $3y(2y + 3)(2y - 3)$

53. $4x^3 + 32$ $4(x + 2)(x^2 - 2x + 4)$

54. Use synthetic substitution to solve. $-1, -\dfrac{1}{2}, 1, 3$ *6.6*
$2x^4 - 5x^3 - 5x^2 + 5x + 3 = 0.$

55. Convert 8 oz/in² to pounds per square foot.   72 lb/ft²   *7.3*

Evaluate. (Exercises 56–59)

56. $\dfrac{3 + 5^2 - (-3)^2}{(4 - 10)^2 + 9 \cdot 4}$ $\dfrac{19}{72}$ *1.2*

57. $(1.5 \times 10^{-11}) \times (4 \times 10^9)$ 0.06   *5.2*

58. $2 \cdot 81^{\frac{3}{4}}$ 54   *8.6*

59. $16^{-\frac{1}{4}}$ $\dfrac{1}{2}$

60. Find the maximum value of P given that $P = -3x^2 + 12x + 10.$ 22   *9.4*

Copy the vectors above and draw the following. (Exercises 61–64)

61. $\vec{A} + \vec{B}$ *9.7*

62. $\vec{B} - \vec{A}$

63. $-2 \cdot \vec{B}$

64. $\vec{V}$, so that $\vec{A} + \vec{B} + \vec{V}$ is a zero vector.

65. Use the Rational Zero Theorem to find all the roots of $4x^4 + 3x^2 - 1 = 0.$ $\pm\dfrac{1}{2}, \pm i$ *10.5*

66. A van left a terminal at 8:00 A.M. and traveled east at 25 mi/h. At 10:00 A.M., an auto left the same terminal and averaged 55 mi/h along the same route. At what time did the auto overtake the van?   11:40 A.M.   *2.6*

67. Raul has three more dimes than quarters. The total face value of his coins is $1.35. How many coins of each kind does Raul have?   3 q, 6 d   *4.4*

68. An airplane traveled 840 mi in 3 h aided by a tailwind. It would have taken 4 h to travel the same distance against the wind. Find the wind speed.   35 mi/h   *4.5*

69. Machine A can produce 300 bolts in 6 h, while Machine B produces 300 bolts in 4 h. How long will it take Machines A and B, operating together, to make 1,500 bolts?   12 h   *7.7*

70. The length of a rectangle is 2 ft more than its width, and 1 ft less than the length of a diagonal. Find the length of the diagonal in simplest radical form. $4 + \sqrt{6}$ ft   *9.6*

---

42. $\dfrac{2y + 28}{(y - 4)(y + 2)}$

48. $5 + 5i\sqrt{2}$

61.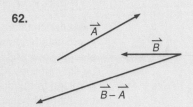

62.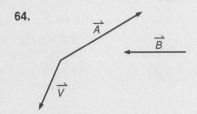

63.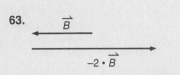

64.

11 COORDINATE GEOMETRY AND QUADRATIC FUNCTIONS

OVERVIEW

In earlier chapters, graphing in the coordinate plane was directed mainly to straight lines and to linear regions bounded by straight lines. Chapter 11 extends graphing to quadratic functions. The treatment of basic geometric ideas, such as the distance between two points and the midpoint of a line segment, is also approached through coordinate geometry. Finally, the concept of symmetry is illustrated using coordinate geometry.

OBJECTIVES

- To apply the distance and midpoint formulas
- To graph absolute value functions
- To graph parabolas
- To solve maximum and minimum problems

PROBLEM SOLVING

In Lesson 11.1, students are introduced to the Distance Formula and the Midpoint Formula to explore the strategy *Using a Formula* to solve problems. The *Problem Solving Strategies* activity on page 390 teaches the strategy *Using a Graph* to plot geometric figures in space. Lesson 11.6 presents problem solving in the context of finding maxima and minima. The *Mixed Problem Solving* on page 403 allows students to choose from a variety of strategies to solve problems.

READING AND WRITING MATH

The *Focus on Reading* on page 399 asks students to interpret the truth or falsity of mathematical statements. Exercise 22 in Lesson 11.4 and Exercise 20 in the *Chapter Review* ask students to explain a mathematical procedure or concept in their own words.

TECHNOLOGY

Calculator: The graphs of this chapter may be done using a graphing calculator. Students should be encouraged to use the graphing calculator to explore what happens when they change the value of the coefficients in equations of the form $y = ax^2 + bx + c$.

SPECIAL FEATURES

Mixed Review pp. 381, 385, 389, 396, 400, 402
Brainteaser p. 381
Application: Highway Curves p. 385
Problem Solving Strategies: Using a Graph p. 390
Midchapter Review p. 394
Focus on Reading p. 399
Mixed Problem Solving p. 403
Key Terms p. 404
Key Ideas and Review Exercises pp. 404–405
Chapter 11 Test p. 406
College Prep Test p. 407

PLANNING GUIDE

Lesson	Basic	Average	Above Average	Resources
11.1 pp. 380–381	CE all WE 1–31 odd Brainteaser	CE all WE 3–33 odd Brainteaser	CE all WE 7–37 odd Brainteaser	Reteaching 75 Practice 75
11.2 pp. 384–385	CE all WE 1–13 odd Application	CE all WE 3–17 odd Application	CE all WE 7–21 odd Application	Reteaching 76 Practice 76
11.3 pp. 389–390	CE all WE 1–12 Problem Solving	CE all WE 3–14 Problem Solving	CE all WE 7–18 Problem Solving	Reteaching 77 Practice 77
11.4 pp. 393–394	CE all WE 1–17 odd Midchapter Review	CE all WE 3–19 odd Midchapter Review	CE all WE 5–23 odd Midchapter Review	Reteaching 78 Practice 78
11.5 p. 396	CE all WE 1–15 odd	CE all WE 7–21 odd	CE all WE 9–23 odd	Reteaching 79 Practice 79
11.6 pp. 399–400	FR all CE all WE 1–12	FR all CE all WE 3–14	FR all CE all WE 6–17	Reteaching 80 Practice 80
11.7 pp. 402–403	CE all WE 1–10 Mixed Problem Solving	CE all WE 3–12 Mixed Problem Solving	CE all WE 5–14 Mixed Problem Solving	Reteaching 81 Practice 81
Chapter 11 Review pp. 404–405	all	all	all	
Chapter 11 Test p. 406	all	all	all	
College Prep Test p. 407	all	all	all	

CE = Classroom Exercises   WE = Written Exercises   FR = Focus on Reading

NOTE: For each level, all students should be assigned all Mixed Review exercises.

▰ INVESTIGATION

Project: This investigation is designed to lead students to draw connections between properties of quadratic equations studied earlier and the equation of the parabola in standard form.

Materials: Each student should have a copy of Project Worksheet 11. Students may work individually or in groups.

Allow sufficient time for each question on the worksheet. Then ask different individuals or groups to report their results.

On page 397 of the textbook, the students are shown how to convert a quadratic equation in general form, $y = ax^2 + bx + c$, to the standard form of a parabola,

$$y - h = a(x - k)^2.$$

The students should recognize the expression in the numerator of the k term as the opposite of the *discriminant* of a quadratic equation. Remind the students that the discriminant of a quadratic equation gives information about the number of real roots that it has.

The remaining questions on the worksheets lead the student to see that when the discriminant of a parabola is negative, the graph of the parabola will not cross the x-axis at any point and will thus have no roots.

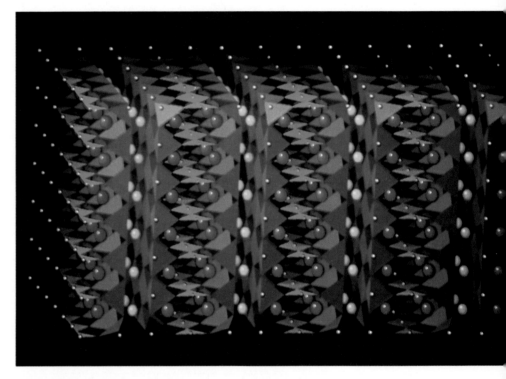

Scientists use computer graphics to construct images of objects that they cannot see. The pattern above is a computer-generated picture showing the crystal structure of a high temperature superconductor.

More About the Image

The crystalline structure of a superconductor forms planes through which electric current passes. Since these planes provide unobstructed pathways, the electrons that make up the flow of electric current are not slowed by collisions with other atoms. Thus, electricity flows without depletion of any kind. The superconductor represented in the image above is considered a "high temperature" superconductor because it superconducts above the temperature of liquid nitrogen ($-319°$ F).

11.1 Applying Coordinate Geometry

Objectives

To find distances between points
To find midpoints of segments
To apply the distance and midpoint formulas

The distance between points A and B on a number line is the absolute value of the difference between their coordinates. AB is read as "the *distance* between A and B."

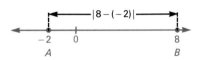

$AB = |8 - (-2)| = |10| = 10$, or $AB = |-2 - 8| = |-10| = 10$

The concept of distance between points on a number line can be extended to the distance between points on a horizontal or vertical line in a plane.

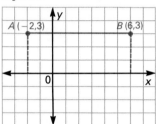
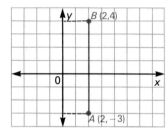

$AB = |6 - (-2)| = 8$ $AB = |4 - (-3)| = 7$

In general, $AB = |x_2 - x_1|$ or $AB = |y_2 - y_1|$.

The Pythagorean relation, $a^2 + b^2 = c^2$, can be used to find the distance between points on a nonvertical, nonhorizontal line. To find the distance between $A(7,5)$ and $B(3,2)$, first draw a right triangle with $\overline{AB}$ (read "segment AB") as its hypotenuse. Then apply the Pythagorean relation to find c, the length of the hypotenuse.

$$c^2 = a^2 + b^2$$
$$c = \sqrt{a^2 + b^2}$$
$$= \sqrt{|7 - 3|^2 + |5 - 2|^2}$$
$$= \sqrt{4^2 + 3^2}$$
$$= \sqrt{25} = 5$$

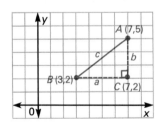

Prerequisite Quiz

Find the missing length for the given data in the right triangle shown. Express answers that are not rational in simplest radical form.

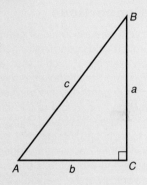

1. $b = 4$, $a = 3$, $c = $ ___?___ 5
2. $b = 5$, $c = 13$, $a = $ ___?___ 12
3. $a = 4$, $b = 4$, $c = $ ___?___ $4\sqrt{2}$
4. $c = 6$, $b = 3$, $a = $ ___?___ $3\sqrt{3}$

Motivator

Have students recall the Pythagorean relation, $a^2 + b^2 = c^2$. Ask students if the segment joining $B(3,2)$ and $A(7,5)$ is on a horizontal or a vertical line. No Ask them how they would find the distance between A and B. They must use $\overline{AB}$ as the hypotenuse of a right triangle. Then use the Pythagorean relation to find the length of $\overline{AB}$.

Highlighting the Standards

Standard 4c: This chapter returns to the most important connection in high school mathematics—the relationship between algebra and geometry.

TEACHING SUGGESTIONS

Lesson Note

It is often easier to begin finding distances and midpoints by first using horizontal and vertical line segments. Emphasize that the distance formula is merely an application of the Pythagorean relation.

Some students are less intimidated by the following verbalized form of the distance formula than by the formula using subscripts.

$$d = \sqrt{(\text{diff of } x\text{'s})^2 + (\text{diff. of } y\text{'s})^2}$$

Math Connections

Geometry: Some theorems of geometry can be proved using coordinate methods. Here is an example of such a theorem: *The segment joining the midpoints of two sides of a triangle is parallel to the third side, and its length is half the length of the third side.* The theorem is easily proved using the distance formula and the fact that two lines are parallel if they have equal slopes. For convenience and without loss of generality, the three vertices of the triangle can be taken as (0,0), (2a,2b), and (2c,0).

In the triangle at the right, the lengths of the legs are $|x_2 - x_1|$ and $|y_2 - y_1|$.

Use the Pythagorean relation.

$$c^2 = a^2 + b^2$$
$$c = \sqrt{a^2 + b^2}$$
$$ = \sqrt{|x_2 - x_1|^2 + |y_2 - y_1|^2}$$
$$ = \sqrt{(x_2 - x_1)^2 + (y_2 - y_1)^2} \leftarrow \text{ for all real numbers } a, |a|^2 = a^2$$

This generalized result is known as the **distance formula**.

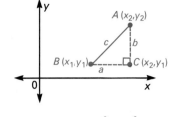

Theorem 11.1

The distance between any two points $A(x_1,y_1)$ and $B(x_2,y_2)$ is
$$d = \sqrt{(x_2 - x_1)^2 + (y_2 - y_1)^2}.$$

EXAMPLE 1   Find the distance between $R(3,-5)$ and $S(5,-9)$.

Solution
$$d = \sqrt{(x_2 - x_1)^2 + (y_2 - y_1)^2}$$
$$= \sqrt{(5 - 3)^2 + [-9 - (-5)]^2} = \sqrt{2^2 + (-4)^2} = \sqrt{20} = 2\sqrt{5}$$

Calculator check:

(5 − 3) x^2 + (9 +/− − 5 +/−) x^2 = √

Also, 2 × 5 √ =   Display: 4.472136

Thus, the distance between R and S is $2\sqrt{5}$.

EXAMPLE 2   Use the distance formula to determine whether $\triangle ABC$ has at least two congruent sides (is *isosceles*).

Solution
$$AB = \sqrt{(8 - 2)^2 + (2 - 4)^2}$$
$$= \sqrt{36 + 4} = \sqrt{40} = 2\sqrt{10}$$
$$AC = \sqrt{(7 - 2)^2 + (9 - 4)^2}$$
$$= \sqrt{25 + 25} = \sqrt{50} = 5\sqrt{2}$$
$$BC = \sqrt{(7 - 8)^2 + (9 - 2)^2}$$
$$= \sqrt{1 + 49} = \sqrt{50} = 5\sqrt{2}$$

Thus, $\triangle ABC$ is isosceles, because $AC = BC = 5\sqrt{2}$.

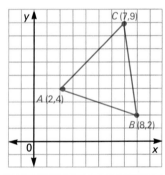

378   Chapter 11 Coordinate Geometry and Quadratic Functions

Additional Example 1

Find the distance between $A(2,-3)$ and $B(6,-1)$. $2\sqrt{5}$

Additional Example 2

Use the distance formula to determine whether points $A(2,4)$, $B(8,5)$, and $C(2,6)$ are vertices of an isosceles triangle.
Yes, $AB = BC = \sqrt{37}$

A point is the **midpoint of a segment** if it divides the segment into two segments of equal length. In the figure at the right, the distance formula can be used to show that M is the midpoint of $\overline{AB}$ by demonstrating that $AM = MB$.

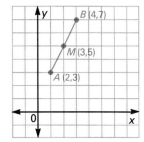

$$AM = \sqrt{(3-2)^2 + (5-3)^2}$$
$$= \sqrt{1+4} = \sqrt{5}$$

$$MB = \sqrt{(4-3)^2 + (7-5)^2}$$
$$= \sqrt{1+4} = \sqrt{5}$$

Thus, M is the midpoint of $\overline{AB}$ because $AM = MB$.

Notice that the coordinates of M, the midpoint of $\overline{AB}$, are the arithmetic means (averages) of the coordinates of A and B.

$$\text{Coordinates of } M = \left(\frac{2+4}{2}, \frac{3+7}{2}\right) = (3,5)$$

This suggests the *Midpoint Formula*.

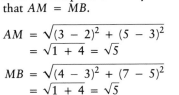

Theorem 11.2

The Midpoint Formula: The coordinates of the midpoint M of the segment joining $A(x_1, y_1)$ and $B(x_2, y_2)$ are $M\left(\dfrac{x_1 + x_2}{2}, \dfrac{y_1 + y_2}{2}\right)$.

EXAMPLE 3   Determine the coordinates of Q, an endpoint of $\overline{PQ}$, given that the other endpoint is $P(-6,-4)$ and the midpoint is $M(-2,1)$.

Solution

$P(-6,-4)$, $M(-2,1)$, and $Q(x_2,y_2)$.
Use the midpoint formula.

$$x_M = \frac{x_1 + x_2}{2} \qquad y_M = \frac{y_1 + y_2}{2}$$

$$-2 = \frac{-6 + x_2}{2} \qquad 1 = \frac{-4 + y_2}{2}$$

$$-4 = -6 + x_2 \qquad 2 = -4 + y_2$$

$$x_2 = 2 \qquad y_2 = 6$$

Thus, the coordinates of Q are $(2,6)$.

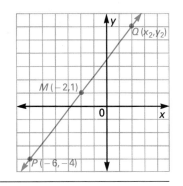

Closure

Ask students how they can find the distance between two points which do not lie on a line parallel to either axis if they are given the coordinates of the points. Use the Distance Formula. Then, ask students how they would find the coordinates of the midpoint of a segment if they are given the coordinates of the endpoints of the segment. Use the Midpoint Formula.

◼◼◼FOLLOW UP

Guided Practice

Classroom Exercises 1–8

Independent Practice

Ⓐ Ex. 1–31, Ⓑ Ex. 32–34, Ⓒ Ex. 35–37

Basic: WE 1–31 odd, Brainteaser
Average: WE 3–33 odd, Brainteaser
Above Average: WE 7–37 odd, Brainteaser

Additional Answers

Written Exercises

24. $\left(\dfrac{-5}{2}, \dfrac{-11}{2}\right)$

36. Coordinates of P: (c,d); Coordinates of Q: $(a + c, b + d)$.
$PQ = \sqrt{[(a + c) - c]^2 + [(b + d) - d]^2}$
$= \sqrt{a^2 + b^2}$
$AB = \sqrt{(2a - 0)^2 + (2b - 0)^2}$
$= \sqrt{4a^2 + 4b^2}$
$= 2\sqrt{a^2 + b^2}$
So, $PQ = \dfrac{1}{2}AB$.

37. $\sqrt{a^2 + b^2} + \sqrt{c^2 + d^2} +$
$\sqrt{(c - a)^2 + (d - b)^2}$

Classroom Exercises

Determine the length of $\overline{AB}$.

1.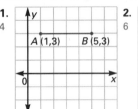
2.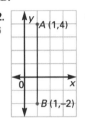
3. $\sqrt{74}$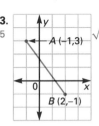
4.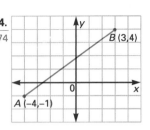

5–8. Using the diagrams for Exercises 1–4, find the coordinates of M, the midpoint of $\overline{AB}$. **5.** (3,3) **6.** (1,1) **7.** $\left(\frac{1}{2},1\right)$ **8.** $\left(-\frac{1}{2},\frac{3}{2}\right)$

Written Exercises

Find the distance between the given points. Give the result in simplest radical form.

1. $A(3,0)$, $B(7,0)$ 4
2. $C(0,5)$, $D(0,-2)$ 7
3. $P(2,4)$, $Q(2,-1)$ 5

4. $A(4,7)$, $B(3,10)$ $\sqrt{10}$
5. $P(2,4)$, $Q(6,6)$ $2\sqrt{5}$
6. $M(2,1)$, $N(6,5)$ $4\sqrt{2}$

7. $S(5,6)$, $T(9,10)$ $4\sqrt{2}$
8. $G(3,6)$, $H(5,12)$ $2\sqrt{10}$
9. $F(2,9)$, $G(8,13)$ $2\sqrt{13}$

10. $A(-2,5)$, $B(6,7)$ $2\sqrt{17}$
11. $T(-4,-3)$, $U(0,2)$ $\sqrt{41}$
12. $Z(-3,-6)$, $J(4,-4)$ $\sqrt{53}$

13. $A(2,-4)$, $B(4,6)$ $2\sqrt{26}$
14. $K(1,5)$, $L(-3,7)$ $2\sqrt{5}$
15. $U\left(-\frac{3}{2},2\right)$, $V\left(\frac{1}{2},\frac{1}{4}\right)$ $\dfrac{\sqrt{113}}{4}$

Use the distance formula to determine whether the points with the given coordinates are the vertices of an isosceles triangle.

16. $A(8,4)$, $B(4,2)$, $C(6,0)$ Yes
17. $A(0,0)$, $B(2,4)$, $C(4,2)$ Yes

18. $P(-6,0)$, $Q(1,-3)$, $R(0,4)$ No
19. $G(0,4)$, $H(10,10)$, $K(4,0)$ Yes

Determine the coordinates of the midpoint M of the segment joining points P and Q.

20. $P(8,6)$, $Q(4,10)$ (6,8)
21. $P(6,5)$, $Q(8,3)$ (7,4)
22. $P(5,4)$, $Q(7,-10)$ (6,-3)

23. $P(-5,-7)$, $Q(-1,-3)$
(-3,-5)
24. $P(-1,-5)$, $Q(-4,-6)$
25. $P(-8,-6)$, $Q(-3,1)$
$\left(-\frac{11}{2},-\frac{5}{2}\right)$

Determine the coordinates of Q, an endpoint of $\overline{PQ}$, given that the other endpoint is P and the midpoint is M.
(-13,-3)   (1,16)

26. $P(2,3)$, $M(7,-7)$ (12,-17) **27.** $P(7,-1)$, $M(-3,-2)$ **28.** $P(9,-2)$, $M(5,7)$

29. $P(-3,-2)$, $M(-5,-4)$ **30.** $P(8,-2)$, $M(-4,-3)$ **31.** $P(0,-5)$, $M(-2,-6)$
(-7,-6)   (-16,-4)   (-4,-7)

Use the figure at the right for Exercises 32–34.

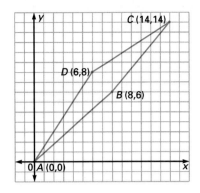

32. A *rhombus* is a four-sided figure with all sides equal in length. Is *ABCD* a rhombus? Yes

33. Determine whether the diagonals $\overline{AC}$ and $\overline{BD}$ bisect each other. (HINT: Determine whether they have the same midpoint.) They do.

34. Find the perimeter of the quadrilateral formed by joining the midpoints of the four sides of *ABCD*.
$16\sqrt{2}$

Use the figure at the right for Exercises 35–37.

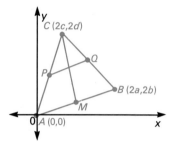

35. Find the length of $\overline{CM}$, where *M* is the midpoint of $\overline{AB}$. $\sqrt{(2c - a)^2 + (2d - b)^2}$

36. Show that the length of $\overline{PQ}$, the segment joining the midpoints of $\overline{AC}$ and $\overline{BC}$, is half the length of $\overline{AB}$.

37. Find the perimeter of the triangle *PQM* formed by joining the midpoints of the three sides.

Mixed Review

Write an equation of the line containing the given points. *3.4*

1. $A(5,7)$ and $B(0,-3)$ $y = 2x - 3$

2. $P(-4,7)$ and $Q(5,7)$ $y = 7$

Find the slope of the line described. *3.7*

3. perpendicular to the line containing $P(3,7)$, and $Q(8,9)$ $-\frac{5}{2}$

4. parallel to the line $3x - 4y = 12$ $\frac{3}{4}$

Brainteaser

In the figure at the right, point *C* is the center of the square and also a vertex of the triangle. The lengths of the sides of the square and the triangle are as shown. Find the area of the shaded region. 16 sq units

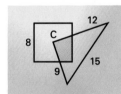

Enrichment

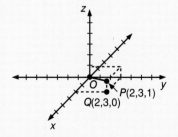

In a 3-dimensional coordinate space, each point has coordinates (x,y,z). The coordinates of *O* are $(0,0,0)$, the coordinates of *P* are $(2,3,1)$, and the coordinates of *Q* are $(2,3,0)$.

Challenge the students to verify that the 3-dimensional distance formula,

$$d = \sqrt{(x_2 - x_1)^2 + (y_2 - y_1)^2 + (z_2 - z_1)^2}$$

is valid in determining *OP*. (Have the students confirm the result of using the formula by working with right triangle *OQP* in the plane

determined by that triangle.)
Applying the formula,

$OP = \sqrt{2^2 + 3^2 + 1^2} = \sqrt{14}$.

In the plane determined by the *x*-axis and *y*-axis,

$OQ = \sqrt{2^2 + 3^2} = \sqrt{13}$. Then, in the plane determined by right triangle *OQP*,

$OQ = \sqrt{13}$, $PQ = 1$, and

$OP = \sqrt{(\sqrt{13})^2 + 1^2} = \sqrt{14}$.

▬▬GETTING STARTED

Prerequisite Quiz

Use the points $A(2,8)$ and $B(6,16)$.

1. Find the coordinates of M, the midpoint of $\overline{AB}$. (4,12)
2. Find the length of $\overline{AB}$. $4\sqrt{5}$
3. Find the slope of $\overline{AB}$. 2
4. Find an equation of $\overleftrightarrow{AB}$ in standard form. $2x - y = -4$

Motivator

Ask students the following questions. Given points $A(3,1)$ and $B(9,5)$, how do you find each of the following?

1. The slope of $\overline{AB}$? See page 84.
2. The slope of a line parallel to $\overline{AB}$? See page 100.
3. The slope of a line perpendicular to $\overline{AB}$? See page 101.
4. The midpoint of $\overline{AB}$? See page 379.
5. The length of $\overline{AB}$? See page 378.
6. An equation of $\overline{AB}$? See page 88.

11.2 Other Coordinate Geometry Applications

Objective To apply the formulas of coordinate geometry

Recall from Chapter 3 how to find the slope of a line given two of its points, and how to write the equation of a line using the point-slope form. Consider segment $\overline{AB}$ with endpoints $A(3,1)$ and $B(9,5)$.

(1) Slope of $\overline{AB}$: $m = \dfrac{y_2 - y_1}{x_2 - x_1} = \dfrac{5 - 1}{9 - 3} = \dfrac{4}{6}$, or $\dfrac{2}{3}$

(2) Equation of $\overleftrightarrow{AB}$:

Point-slope form: $y - y_1 = m(x - x_1)$

$$y - 1 = \tfrac{2}{3}(x - 3) \qquad \text{or} \qquad y - 5 = \tfrac{2}{3}(x - 9)$$
$$3y - 3 = 2x - 6 \qquad\qquad\quad 3y - 15 = 2x - 18$$
$$2x - 3y = 3 \qquad\qquad\qquad 2x - 3y = 3$$

Standard form: $ax + by = c$

A **median** of a triangle is the segment joining a vertex to the midpoint of the opposite side.

EXAMPLE 1 Determine whether the median $\overline{CM}$ of $\triangle ABC$ is also an altitude of the triangle.

Plan Because $\overline{CM}$ is a median, M is the midpoint of $\overline{AB}$. Find the coordinates of M.
Then determine whether $\overline{CM}$ is perpendicular to $\overline{AB}$. Remember that the product of the slopes of two nonvertical, nonhorizontal perpendicular lines is -1.

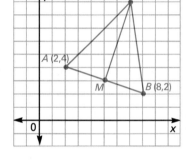

Solution Coordinates of M: $\left(\dfrac{2 + 8}{2}, \dfrac{4 + 2}{2}\right)$
$= (5,3)$

Slope of $\overline{AB}$: $\dfrac{2 - 4}{8 - 2} = \dfrac{-2}{6} = -\dfrac{1}{3}$ Slope of $\overline{CM}$: $\dfrac{9 - 3}{7 - 5} = \dfrac{6}{2} = 3$

$\overline{CM}$ is perpendicular to $\overline{AB}$ because the product of their slopes is -1.

Thus, $\overline{CM}$ is an altitude of $\triangle ABC$.

Highlighting the Standards

Standard 8b: Examples 1 and 2 demonstrate how coordinate geometry can be used to determine properties of geometric figures.

Additional Example 1

Determine whether the median to $\overline{BC}$ of the isosceles triangle of Example 1 is also an altitude. No; the slope of the median is $\dfrac{3}{11}$ and the slope of $\overline{BC}$ is -7. The product of the slopes is not -1.

In the figure at the right, M is the mid-point of $\overline{AB}$. This makes $\overleftrightarrow{CM}$ a *bisector* of $\overline{AB}$. $\overleftrightarrow{CM}$ is also *perpendicular* to $\overline{AB}$. Therefore, $\overleftrightarrow{CM}$ is the **perpendicular bisector** of $\overline{AB}$.

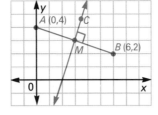

To find an equation of $\overleftrightarrow{CM}$, it suffices to know the coordinates of a point on the line and the slope of the line. Both can be found as follows.

(1) Point M is a point on both $\overline{AB}$ and $\overleftrightarrow{CM}$. The midpoint formula can be used to find the coordinates of M.

(2) Because $\overleftrightarrow{CM}$ is perpendicular to $\overline{AB}$, its slope is the opposite of the reciprocal of the slope of $\overline{AB}$.

EXAMPLE 2 Write an equation in standard form of the perpendicular bisector $\overleftrightarrow{CM}$ of $\overline{AB}$ for $A(0,4)$ and $B(6,2)$.

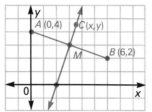

Plan Use the two given facts about $\overleftrightarrow{CM}$: (1) that it bisects $\overline{AB}$, and (2) that it is perpendicular to $\overline{AB}$. Then use the point-slope form to find an equation of $\overleftrightarrow{CM}$.

Solution

Find the coordinates of M, the midpoint of $\overline{AB}$.

$$\left(\frac{0+6}{2}, \frac{4+2}{2}\right) = (3,3)$$

Find the slope of $\overleftrightarrow{CM}$, the opposite of the reciprocal of the slope of $\overline{AB}$.

Slope of $\overline{AB} = \dfrac{2-4}{6-0} = -\dfrac{1}{3}$

Slope of $\overleftrightarrow{CM} = 3$

Write an equation of $\overleftrightarrow{CM}$ using the point-slope form.

$$y - y_1 = m(x - x_1)$$
$$y - 3 = 3(x - 3)$$
$$y - 3 = 3x - 9$$

Write an equation in standard form.

$$6 = 3x - y, \text{ or } 3x - y = 6$$

Thus, $3x - y = 6$ is an equation in standard form of the perpendicular bisector of $\overline{AB}$.

■TEACHING SUGGESTIONS

Lesson Note

It may be necessary to review the concepts of coordinate geometry needed to answer the questions in the Prerequisite Quiz. Outline the procedures for finding the following, given the coordinates of two points, A and B: the coordinates of the midpoint of $\overline{AB}$, the length of $\overline{AB}$, the slope of $\overline{AB}$, the slope of a line parallel or perpendicular to $\overline{AB}$, and the equation of $\overleftrightarrow{AB}$.

Math Connections

Geometry: The following theorem can be proved using the methods of coordinate geometry: *If the midpoints of the four sides of a quadrilateral are connected in order, then the resulting figure is a parallelogram.* Without loss of generality, the four vertices of the quadrilateral can be taken as $(0,0)$, $(2a,0)$, $(2b,2c)$, and $(2d,2e)$. The proof relies on the definition of *parallelogram* in Exercise 17 and on the fact that two line segments are parallel if they have equal slopes.

Additional Example 2

Write an equation in standard form for the perpendicular bisector of $\overline{AB}$ for $A(2,-4)$ and $B(2,6)$. $y = 1$

Critical Thinking Questions

Analysis: Ask students whether they can find an alternate proof of the theorem stated in the *Math Connections* on page 383 by using properties of a parallelogram other than those provided in its definition.   For example, they might use the fact that if two sides of a quadrilateral are both parallel and congruent, then the quadrilateral is a parallelogram. Then the coordinate proof will use the distance formula as well as the concept of slope.

Common Error Analysis

Error: Students frequently forget to find the *negative* of the reciprocal of the slope of a line when they write an equation of the perpendicular bisector of the line.

Have students review Lesson 3.7 on parallel and perpendicular lines.

Checkpoint

Use A(0,0), B(8,2), and C(2,9) as the vertices of triangle ABC.

1. Write an equation of the perpendicular bisector of $\overline{AB}$. $4x + y = 17$
2. Write an equation of the median from A. $11x - 10y = 0$

Determine whether quadrilateral ABCD is a rectangle.

3. $A(0,0)$, $B(4,-2)$, $C(6,2)$, $D(2,4)$ Yes
4. $A(2,1)$, $B(7,6)$, $C(8,13)$, $D(3,8)$ No

Classroom Exercises

Tell whether $\overline{CM}$ is a median of $\triangle ABC$.

1.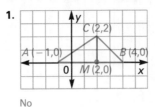

No

2.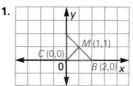

Yes

3.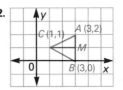

No

Written Exercises

Determine whether the median $\overline{CM}$ is also an altitude of the given triangle. M is the midpoint of $\overline{AB}$.

1. 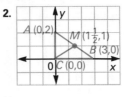   Yes

2. 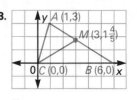   Yes

3. $A(2,-4)$, $B(8,-6)$, $C(6,-8)$ No   4. $A(4,-3)$, $B(6,3)$, $C(-4,3)$ Yes

Write an equation in standard form of the perpendicular bisector of $\overline{PQ}$.

5. $P(2,4)$, $Q(6,6)$ $2x + y = 13$
6. $P(0,4)$, $Q(-4,0)$ $x + y = 0$
7. $P(2,-1)$, $Q(3,7)$ $2x + 16y = 53$
8. $P(5,8)$, $Q(-8,-5)$ $x + y = 0$

Use the coordinates of the following triangles for Exercises 9–14.

$\triangle ABC$: $A(0,0)$, $B(8,4)$, $C(1,8)$ $\triangle PQR$: $P(0,0)$, $Q(8,4)$, $R(6,8)$

9. Is $\triangle ABC$ isosceles? Yes   10. Is $\triangle PQR$ isosceles? No

Find an equation of the line containing the following.

11. median from C to $\overline{AB}$ $2x + y = 10$ 12. perpendicular bisector of $\overline{AB}$ $2x + y = 10$
13. median from R to $\overline{PQ}$ $3x - y = 10$ 14. perpendicular bisector of $\overline{PQ}$ $2x + y = 10$

A *rectangle* is a quadrilateral in which adjacent sides form right angles. Determine whether quadrilateral $ABCD$ is a rectangle.

15. $A(0,0)$, $B(2,8)$, $C(-2,9)$, $D(-4,1)$ Yes 16. $A(-2,0)$, $B(8,-1)$, $C(13,3)$, $D(3,4)$ No

17. The vertices of quadrilateral $ABCD$ are $A(1,-1)$, $B(9,5)$, $C(15,13)$, and $D(7,7)$. Determine whether $ABCD$ is a parallelogram (a quadrilateral with two pairs of parallel sides); a rectangle (see Exercises 15–16); a rhombus (a parallelogram with four sides of equal length). *ABCD is a parallelogram and a rhombus.*

18. Using quadrilateral $ABCD$ of Exercise 17, determine whether the diagonals are perpendicular bisectors of each other. *Yes*

19. Any point on the perpendicular bisector of a segment is equidistant from the endpoints of that segment. Verify this statement for the segment with endpoints $A(2,4)$ and $B(8,6)$. (HINT: Find a random point whose coordinates satisfy the equation of the perpendicular bisector of $\overline{AB}$, and show that this point is the same distance from A as it is from B.)

20. The coordinates of the vertices of $\triangle ABC$ are $A(0,0)$, $B(4a,0)$, and $C(2a, 2a\sqrt{3})$. Show that $\triangle ABC$ is equilateral.

21. Use coordinate geometry to prove that any median of an equilateral triangle is also an altitude.

22. The coordinates of the vertices of $\triangle ABC$ are $A(2,1)$, $B(5,5)$, and $C(6,3)$. Show that $\triangle ABC$ is a right triangle.

Mixed Review

1. Graph the equation $y = x^2 - 5$. Use the following values for x to make a table of ordered pairs: $-3, -2, -1, 0, 1, 2, 3$. *3.2*

2. Solve $x^2 - 4 = 0$. *6.1* ± 2 3. Solve $\sqrt{2x - 1} = 2$. *10.3* $\frac{5}{2}$

 Application: *Highway Curves*

When designing curves, highway planners take the following into account: the degree D of the curve, which measures the number of degrees turned while traveling 100 feet; the superelevation e, which indicates how steeply the curve is banked; the coefficient of side friction f, which describes the friction between tire and pavement; and the speed v (in miles per hour) for which the road was designed. They use the following formula.

$$D_{max} = \frac{85,950(e + f)}{v^2}$$

Find the maximum degree of curve for a road with a superelevation of 0.06, a coefficient of side friction of 0.13, and a design speed of 60 mi/h. Give your answer to the nearest tenth. *4.5 degrees*

Ask students to give a definition of median of a triangle. *A median of a triangle is the segment joining a vertex to the midpoint of the opposite side.* Ask them how they would determine if a median of a triangle is also an altitude. *See Example 1.* Have students give the steps in writing an equation of a perpendicular bisector of a segment given the coordinates of the segment's endpoints. *See Example 2.*

◼◼◼ FOLLOW UP

Guided Practice

Classroom Exercises 1–3

Independent Practice

A Ex. 1–8, **B** Ex. 9–18, **C** Ex. 19–22

Basic: WE 1–13 odd, Application

Average: WE 3–17 odd, Application

Above Average: WE 7–21 odd, Application

Additional Answers

Written Exercises

19. Slope of $\overline{AB}$: $\frac{1}{3}$; midpoint of $\overline{AB}$: (5,5); perpendicular bisector of $\overline{AB}$:
$y = -3x + 20$
For the point $P(0,20)$:
$PA = \sqrt{(2 - 0)^2 + (4 - 20)^2} = \sqrt{260}$
$PB = \sqrt{(8 - 0)^2 + (6 - 20)^2} = \sqrt{260}$
Thus, $PA = PB$.

See page 403 for the answers to Written Ex. 20–22 and Mixed Review Ex. 1.

Enrichment

Challenge students to prove by methods of coordinate geometry that an angle inscribed in a semicircle is a right angle.

Suggest that the students use a drawing such as the one given, where the circle has its center at the origin, a semicircle has endpoints $A(-k,0)$ and $B(k,0)$, and $C(a,b)$ is any other point on the circle so that $\angle C$ is inscribed in a semicircle.

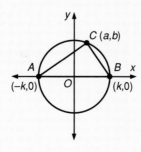

Slope of $\overline{BC} = \frac{b}{a - k}$; slope of $\overline{AC} = \frac{b}{a + k}$; and the product of the slopes is $\frac{b^2}{a^2 - k^2}$.

Since $OC = \sqrt{(a - 0)^2 + (b - 0)^2} = k$, it follows that $a^2 + b^2 = k^2$, or $a^2 - k^2 = -b^2$. Then $\frac{b^2}{a^2 - k^2} = \frac{b^2}{-b^2} = -1$, so $\overline{BC}$ is perpendicular to $\overline{AC}$, and $\angle C$ is a right angle.

Prerequisite Quiz

Evaluate each expression for the given value of x.

1. $|x|; x = -4$ 4
2. $|2x - 6|; x = 2$ 2
3. $-4|3x - 2|; x = -5$ -68
4. $|3x + 4| - 7; x = -2$ -5
5. $-3|2x - 8|; x = -4$ -48

Motivator

Have students form small groups and graph each of the following on a graphing calculator.

1. $y = x, y = 2x, y = 3x, y = 4x$
2. $y = x, y = \frac{1}{2}x, y = \frac{1}{3}x, y = \frac{1}{4}x$
3. $y = x, y = -2x, y = -3x$

Ask students what effect changing the coefficient of x has on the graph of the basic shape $y = x$.   The slope of the line changes.

11.3 Absolute Value Functions

Objective	To graph equations of the form $y - k = a|x - h|$

Recall that $|x| = x$ when $x \geq 0$, and $|x| = -x$ when $x < 0$.

EXAMPLE 1   Graph the function whose equation is $y = |x|$.

Solution   First, prepare a table of values.

x	-6	-4	-2	0	2	4	6
y	6	4	2	0	2	4	6

The resulting graph is V-shaped.

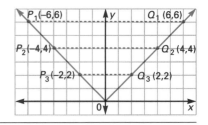

The function in Example 1 is called the basic **absolute value function**. Notice that the y-axis is the perpendicular bisector of $\overline{P_1Q_1}$. Also, if the graph of $y = |x|$ is folded along the y-axis, P_1 and Q_1 coincide. So, P_1 and Q_1 are a pair of **mirror-image points** with respect to the y-axis.

In fact, every point on the graph of $y = |x|$ (except the origin) can be paired with another point that is its mirror image. Thus, the graph is *symmetric* with respect to the y-axis, and the y-axis is the **axis of symmetry** of the graph. The point $(0,0)$ is the **vertex** of the graph.

Compare the graphs of $y = \frac{1}{2}|x|$ and $y = 2|x|$ with the graph of the basic absolute value function $y = |x|$.

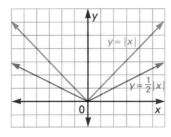

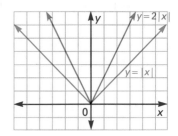

The graph of $y = \frac{1}{2}|x|$ is a *widening* of the graph of $y = |x|$. The slopes of the rays are $\frac{1}{2}$ and $-\frac{1}{2}$.

The graph of $y = 2|x|$ is a *narrowing* of the graph of $y = |x|$. The slopes of the rays are 2 and -2.

386   Chapter 11 Coordinate Geometry and Quadratic Functions

Highlighting the Standards

Standard 6e: The analysis of parameter changes on graphs of absolute value functions is a step toward a deeper understanding of the workings of equations in general.

Additional Example 1

Graph $y = 2|x|$.

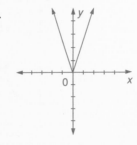

EXAMPLE 2 Graph each function below on the same set of axes as a graph of
$y = |x|$, and compare its graph with the graph of $y = |x|$.

a. $y = -|x|$

b. $y = -\frac{1}{2}|x|$

Solutions

x	-4	-2	0	2	4
y	-4	-2	0	-2	-4

x	-4	-2	0	2	4
y	-2	-1	0	-1	-2

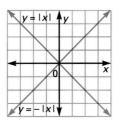

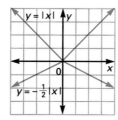

The graph of $y = -|x|$ has the *same shape* as that of $y = |x|$. The vertex is at the origin, and the graph opens *downward*.

The graph of $y = -\frac{1}{2}|x|$ is a *widening* of the graph of $y = |x|$. The vertex is at the origin, and the graph opens *downward*.

In Example 2, the graphs are inverted Vs because the coefficients of $|x|$ are negative. Another way in which the basic absolute value function $y = |x|$ can be modified is by adding a constant to $|x|$.

EXAMPLE 3 Graph the functions below on the same set of axes as a graph of
$y = |x|$, and compare each graph with the graph of $y = |x|$.

a. $y = |x| + 3$

b. $y = |x| - 2$

Solutions

x	-4	-2	0	2	4
y	7	5	3	5	7

x	-4	-2	0	2	4
y	2	0	-2	0	2

a. When $y = |x|$ is changed to $y = |x| + 3$, the graph of $y = |x|$ moves up 3 units. The shape remains unchanged.

b. When $y = |x|$ is changed to $y = |x| - 2$, the graph of $y = |x|$ moves down 2 units. The shape remains unchanged.

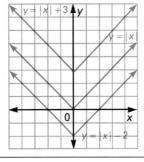

■■■**TEACHING SUGGESTIONS**

Lesson Note

Emphasize the characteristics of the absolute value functions that are listed at the end of this lesson.

Note that the standard form of the equation of an absolute value function is $y - k = a|x - h|$ rather than $y = a|x - h| + k$, which might seem easier to understand now. The form $y - k = a|x - h|$ provides a basis for understanding equations in standard form for conic sections with reference to translations of their graphs. For example, the equation of a circle with radius r and center (h,k) is $(x - h)^2 + (y - k)^2 = r^2$.

Math Connections

Optics: There is a simple physical model for the absolute-value equation $y = a|x|$. If the x-axis is interpreted as the cross-section of a mirror, then a ray of light will be reflected from the mirror's surface in such a manner that a "V" is formed. The y-axis is the normal (perpendicular line) to the mirror at the point of reflection. The angle formed by the normal and the incident ray is congruent to the angle formed by the normal and the reflected ray, so that the "V" that is formed may be wide or narrow, depending on whether the incident ray strikes the mirror relatively head-on or in a more glancing fashion. Thus, the figure formed by the combined incident and reflected rays can be represented as an absolute-value equation.

Additional Example 2

Graph the basic absolute value function $y = |x|$. Then graph each of the following functions on the same set of axes. Compare them with the graphs of $y = |x|$.

a. $y = -2|x|$ **b.** $y = \frac{1}{3}|x|$

The graph of $y = -2|x|$ opens downward and is narrower. The graph of $y = \frac{1}{3}|x|$ opens upward and is wider.

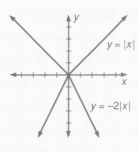

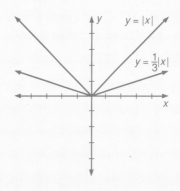

387

Critical Thinking Questions

Analysis: The text summarizes the characteristics for the absolute value function $y - k = a|x - h|$ for $a < 0$ and for $a > 0$. Ask students to describe the function for $a = 0$. They should see that the function becomes the horizontal line $y = k$.

Checkpoint

Graph the function using a table of values.

1. $y = -\frac{1}{3}|x|$

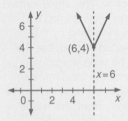

Graph the function without using a table of values. Give the vertex and the equation of the axis of symmetry.

2. $y - 4 = 2|x - 6|$

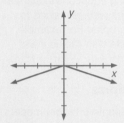

Closure

Have students describe the shape of the graph of $y = |x|$. **V-shaped** Ask them to give the standard form of the equation of the absolute value function. $y - k = a|x - h|$

A move, such as one of those in Example 3, that neither changes the shape of a graph nor rotates it, is called a **translation** of the graph. The graph of $y = |x|$ can also be translated horizontally, left or right.

Notice that when the equation $y = |x|$ is changed to $y = |x - 5|$, the vertex moves 5 units to the *right*. The coordinates of the new vertex are $(5,0)$. The equation of the new axis of symmetry is $x = 5$.

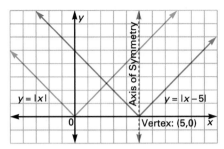

The shape of the graph, however, is unchanged, as is the *direction* of the graph, in the sense that it is not inverted.

A function such as $y = -\frac{1}{3}|x - 2| + 4$ involves a horizontal translation, a vertical translation, a widening, and an inversion. It can be graphed by first rewriting the equation in the **standard form** $y - k = a|x - h|$.

EXAMPLE 4 Graph $y = -\frac{1}{3}|x - 2| + 4$ without using a table of values.

Plan Rewrite the equation in standard form to find the vertex.

Solution

$y - 4 = -\frac{1}{3}|x - 2|$
The vertex is at (h,k), or $(2,4)$.

Draw the axis of symmetry, $x = h$, or $x = 2$.
Draw the V *downward* because the coefficient, $a = -\frac{1}{3}$, is negative.
The slopes of the two rays of the V are $\frac{1}{3}$ and $-\frac{1}{3}$ $(\pm a)$.

Thus, the V is wider in appearance.

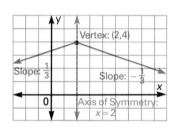

Here are the important characteristics of the absolute value function:

(1) Equation (standard form): $y - k = a|x - h|$
(2) Vertex: (h,k)
(3) Equation of the axis of symmetry: $x = h$
(4) Graph opens upward for $a > 0$. Graph opens downward for $a < 0$.
(5) V widens for $|a| < 1$. V narrows for $|a| > 1$.
(6) Slopes of the rays: a and $-a$

Additional Example 3

Graph $y = |x|$. Then graph each of the following functions on the same set of axes.

a. $y = |x| - 5$

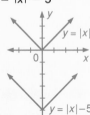

b. $y = |x| + 6$

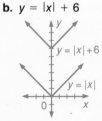

Additional Example 4

Graph $y = 2|x + 4| - 6$ without using a table of values.

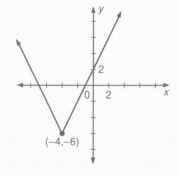

Classroom Exercises

Compare the shape and the direction of the graph of each function with the graph of the function $y = |x|$.

1. $y = 3|x|$
Narrower, upward

2. $y = -\frac{1}{4}|x|$
Wider, downward

3. $y = \frac{2}{3}|x|$
Wider, upward

4. $y = -5|x|$
Narrower, downward

Give the coordinates of the vertex of each graph and the equation of its axis of symmetry.

5. $y = \frac{1}{2}|x|$ (0,0); $x = 0$

6. $y - 3 = |x - 5|$ (5,3); $x = 5$

7. $y + 4 = |x + 6|$ (-6,-4); $x = -6$

8. $y = |x + 1| + 2$ (-1,2); $x = -1$

9-12. Graph each function in Classroom Exercises 5-8.

Written Exercises

Graph each function. Use a table of values for x and y. Give the coordinates of the vertex.

1. $y = 4|x|$

2. $y = \frac{1}{4}|x|$

3. $y = -3|x|$

4. $y = |x| + 6$

5. $y = |x| - 7$

6. $y = |x| - 5$

7. $y = |x - 8|$

8. $y = |x + 3|$

Graph each function without using a table of values. Give the coordinates of the vertex and the equation of the axis of symmetry.

9. $y - 1 = |x - 1|$

10. $y + 1 = 2|x + 1|$

11. $y - 2 = -|x + 2|$

12. $y = 2|x - 4| + 6$

13. $y = -|x + 5| - 4$

14. $y = \frac{1}{2}|x - 1| + 7$

15. $y = 4|x - 2| - 3$

16. $y = -\frac{1}{4}|x + 5| + 2$

17. $y = -2|x + 6| - 8$

Give the coordinates of the vertex of each graph and the equation of its axis of symmetry. (HINT: $|a(x + b)| = |a||x + b|$)

18. $y = |4x - 8| - 10$

19. $y + 5 = |3x - 12|$

20. $y = -\frac{1}{2}|7x + 21| - 10$

21. $y + 6 = |\frac{1}{2}x - 2|$

22. $y = 3|\frac{1}{3}x + 1| - 1$

23. $6 - y = 2|-2x + 3|$

Graph the following relations. (HINT: Consider separate cases, one for $y \geq 0$ and one for $y < 0$.)

24. $x = |y|$

25. $|y| = |x| + 4$

26. $|x| - 2|y| = 2$

Mixed Review

Simplify. Use positive exponents. *5.2, 8.2, 8.6, 9.2*

1. $(3x^{-2})^2 x^{-3}$ $\frac{9}{x^7}$

2. $\sqrt{28a^3b^4}$; $a \geq 0$ $2ab^2\sqrt{7a}$

3. $(-27)^{-\frac{2}{3}}$ $\frac{1}{9}$

4. $(3 - 4i)^2$ $-7 - 24i$

11.3 Absolute Value Functions **389**

FOLLOW UP

Guided Practice

Classroom Exercises 1–12

Independent Practice

A Ex. 1–17, **B** Ex. 18–23, **C** Ex. 24–26

Basic: WE 1–12, Problem Solving

Average: WE 3–14, Problem Solving

Above Average: WE 7–18, Problem Solving

Additional Answers

Classroom Exercises

9.

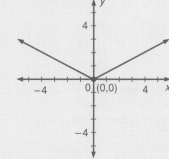

10.

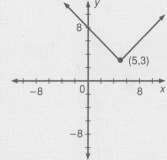

See page 390 for the answers to Classroom Ex. 11–12 and Written Ex. 1, 2. See pages 812–813 for answers to Written Ex. 3–26.

Enrichment

Have the students graph $y = |x + 2|$ and $y = |x - 2|$ on a single set of axes. Then, working from these graphs alone, have them graph the combined function, $y = |x + 2| + |x - 2|$. This can be done by selecting various values of x and combining the y-values from each of the graphs to obtain the y-value of the combined function.

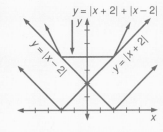

When the combined function has been graphed, it can be verified algebraically by substituting key values of x in the equation. Now have the students challenge each other to find graphs of similar combinations of absolute value functions.

389

11.

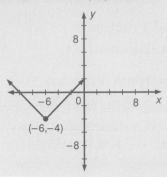

$(-6, -4)$

12.

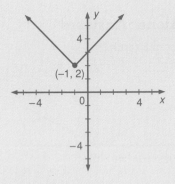

$(-1, 2)$

Written Exercises

1.

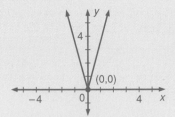

$(0,0)$

2.

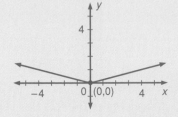

$(0,0)$

Using a Graph

Using a graph can provide the key to solving many new problems. For example, by examining a graph it is possible to derive the distance formula for points in space, much as was done for points in a plane.

In space, there are three mutually perpendicular axes: x, y, and z. So, three planes are represented: xy, yz, and xz. By convention, the xy-plane is represented as a horizontal plane. A point in space is represented by an **ordered triple**, (x, y, z).

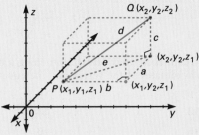

Plotting a point in space resembles doing so in a plane. For example, to plot the point $(4, -4, 3)$, move 4 units in the positive x-direction, 4 units in the negative y-direction, and 2 units in the positive z-direction, as shown at the right.

To find the distance between two points $P(x_1, y_1, z_1)$ and $Q(x_2, y_2, z_2)$, refer to the graph at the right. As in the plane, $e^2 = a^2 + b^2$. The Pythagorean relation also gives the following.

$$d^2 = e^2 + c^2$$
$$d^2 = a^2 + b^2 + c^2 \text{ (Subst for } e^2)$$
$$d = \sqrt{a^2 + b^2 + c^2}$$

Now since a, b, and c are the distances between P and Q along the x-, y-, and z-axes, the formula above becomes the following.

$$d = \sqrt{|x_2 - x_1|^2 + |y_2 - y_1|^2 + |z_2 - z_1|^2}, \text{ or}$$
$$d = \sqrt{(x_2 - x_1)^2 + (y_2 - y_1)^2 + (z_2 - z_1)^2}$$

Exercises

Use the graph at the right for Exercises 1–3.

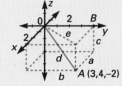

1. Find the distances a, b, c, and e. $a=3, b=4, c=2, e=5$
2. Find the distance d. $\sqrt{29}$
3. Find the distance between A and B. $\sqrt{13}$
4. Find the distance between $P(1,2,2)$ and $Q(4,6,14)$. 13
5. Which is farther from the origin, $(0,0,15)$ or $(9,9,9)$? (9,9,9)

390 Chapter 11 Coordinate Geometry and Quadratic Functions

11.4 Quadratic Functions

Objectives

Given parabolas of the form $y - k = a(x - h)^2$:
To determine the coordinates of their vertices
To write equations for their axes of symmetry
To graph the parabolas

The graph of the quadratic function $y = x^2$ behaves much like the graph of the absolute value function $y = |x|$.

$y = x^2$:

x	-3	-2	-1	0	1	2	3
y	9	4	1	0	1	4	9

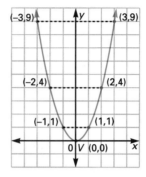

Notice the pairs of mirror-image points with respect to the y-axis: $(-3,9)$, $(3,9)$; $(-2,4)$, $(2,4)$, $(-1,1)$, $(1,1)$

This is the basic graph of a **parabola**. The point $V(0,0)$ is called the *turning point*, or *vertex*, of the parabola. The y-axis is the *axis of symmetry* of the parabola.

EXAMPLE 1

Graph each of the following on the same set of axes.

a. $y = x^2$ b. $y = 2x^2$ c. $y = -\frac{1}{2}x^2$

Solutions

Make a table of values for each graph.

a. The table of values for $y = x^2$ is given above.

b. $y = 2x^2$

x	-2	-1	0	1	2
y	8	2	0	2	8

c. $y = -\frac{1}{2}x^2$

x	-3	-2	-1	0	1	2	3
y	$-4\frac{1}{2}$	-2	$-\frac{1}{2}$	0	$-\frac{1}{2}$	-2	$-4\frac{1}{2}$

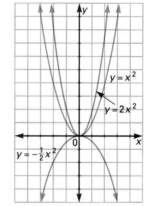

Teaching Resources

Application Worksheet 11
Quick Quizzes 78
Reteaching and Practice
   Worksheets 78
Transparency 26

■■■ GETTING STARTED

Prerequisite Quiz

Find the vertex and the axis of symmetry for each absolute value function.

1. $y = |x| - 7$ $(0,-7)$; $x = 0$
2. $y - 2 = 3|x - 4|$ $(4,2)$; $x = 4$
3. $y = 2|x + 7| - 4$ $(-7,-4)$; $x = -7$
4. $y = 4|x - 6| + 1$ $(6,1)$; $x = 6$

Motivator

Have students form a table of x- and y-values for $y = x^2$ using the following values of x; -5, -4, -3, -2, -1, 0, 1, 2, 3, 4, 5.

Have them plot the corresponding points and draw a smooth curve through them. Ask them to give the coordinates of the vertex. $(0,0)$

Have students guess how the graph of $y = 2x^2$ will compare with the graph of $y = x^2$.

Additional Example 1

Graph the function $y = x^2$. Then graph each of the following.

a. $y = 3x^2$
b. $y = -x^2$
c. $y = -\frac{1}{8}x^2$

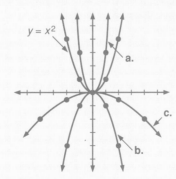

Highlighting the Standards

Standard 8b: The introductory material in the lesson and the examples shows how coordinate geometry is used to find the properties of the parabola.

Lesson Note

Emphasize similarities in graphing a quadratic function and an absolute value function. Compare the characteristics of the quadratic function listed after Example 2 with the summary given for the absolute value function in the preceding lesson.

Math Connection

Trajectories: The shape of a projectile's trajectory is a parabola with a *maximum*, not a *minimum*. Students may relate this fact to the statement on page 392 that "the coefficient a of x^2 in $y = ax^2$ determines the shape of the parabola and the direction in which it opens," since a parabola with a maximum opens downward, not upward.

Critical Thinking Questions

Analysis: The summary at the top of page 393 for the quadratic function does not cover the case $a = 0$. Ask students why this is so. They should see that if $a = 0$, then the function is not quadratic at all, since the equation reduces to the equation of the straight line $y = k$. See also the *Critical Thinking Question* for Lesson 11.3.

Common Error Analysis

Error: Given an equation of a parabola such as $y + 2 = 3(x + 4)^2$, students tend to identify the vertex as (4,2). Emphasize that the vertex for $y - k = a(x - h)^2$ is (h,k). ← *Opposite signs*

Compare the graphs of $y = 2x^2$ and $y = -\frac{1}{2}x^2$ with that of $y = x^2$.

(1) $y = 2x^2$: The graph is *narrower* and opens *upward*.

(2) $y = -\frac{1}{2}x^2$: The graph is *wider* and opens *downward*.

So, the coefficient a of x^2 in $y = ax^2$ determines the shape of the parabola and the direction in which it opens.

Next, look at the graph of $y = (x + 3)^2 - 6$ below.

x	-6	-5	-4	-3	-2	-1	0
y	3	-2	-5	-6	-5	-2	3

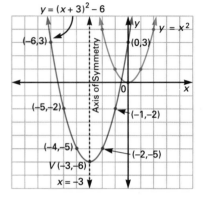

Rewrite $y = (x + 3)^2 - 6$ in *standard form:*
$y - k = a(x - h)^2$

Notice the pattern below.

$y - (-6) = [x - (-3)]^2$

Vertex: $(-3, -6)$

Axis of symmetry: $x = -3$

Thus, the equation $y + 6 = (x + 3)^2$ is a *translation* of the basic parabola $y = x^2$ to a new position 3 units to the left and 6 units down.

EXAMPLE 2   Graph $y = -2(x - 7)^2 + 4$.

Solution   Rewrite the equation in standard form: $y - 4 = -2(x - 7)^2$

Find and plot the vertex, (7,4).

Draw the axis of symmetry from its equation, $x = 7$.

Find and plot four more points. Let $x = 7 \pm 1$ (that is, 6 or 8). Let $x = 7 \pm 2$ (that is, 5 or 9).

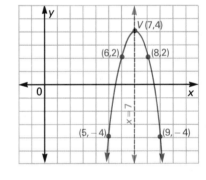

x	5	6	7	8	9
y	-4	2	4	2	-4

Sketch the graph.

The graph is a parabola with vertex (7,4), opening downward.

Additional Example 2

Graph $y = 3(x + 4)^2 + 2$.

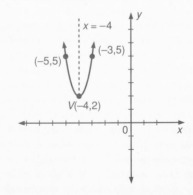

In summary, the graph of a quadratic function is a parabola with the following characteristics:

(1) Equation (standard form):
$y - k = a(x - h)^2$

(2) Vertex: (h,k)

(3) Equation of the axis of symmetry: $x = h$

(4) Graph opens upward for $a > 0$.
Graph opens downward for $a < 0$.

(5) Graph widens for $|a| < 1$.
Graph narrows for $|a| > 1$.

The standard form $y - k = a(x - h)^2$ can also be used to find the equation of a quadratic function when one point of its graph is given along with the vertex of the parabola.

EXAMPLE 3 Find an equation of the function whose graph is given below.

Plan Because the graph is a parabola, the function is quadratic. Use the vertex V to find the values of h and k in $y - k = a(x - h)^2$. Then find the value of a using point $P(0,1)$ of the parabola.

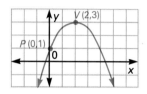

Solution

Substitute 2 for h and 3 for k.
Substitute 0 for x and 1 for y.
Solve for a.

$$y - k = a(x - h)^2$$
$$y - 3 = a(x - 2)^2$$
$$1 - 3 = a(0 - 2)^2$$
$$-2 = a(-2)^2$$
$$-2 = 4a, \text{ or } a = -\frac{1}{2}$$

The equation of the quadratic function is $y - 3 = -\frac{1}{2}(x - 2)^2$.

Classroom Exercises

Give the vertex of each parabola.

1. $y - 1 = 2(x - 2)^2$ (2,1) **2.** $y = (x + 3)^2 - 9$ $(-3, -9)$ **3.** $y = (x - 3)^2 + 1$ (3,1)

4–6. Give the equation of the axis of symmetry of each parabola in Exercises 1–3. **4.** $x = 2$ **5.** $x = -3$ **6.** $x = 3$

7. Graph the parabola of Classroom Exercise 1.

8. What is the equation of the quadratic function which passes through $(4,10)$ and whose vertex is $(2,2)$? $y - 2 = 2(x - 2)^2$

9. How do the shape and direction of the graph of $y = -\frac{1}{3}(x - 2)^2 + 7$ compare with those of the graph of $y = x^2$? Wider, downward

Graph each quadratic function. Give the vertex of the parabola and the equation of the axis of symmetry.

1. $y = -2x^2$

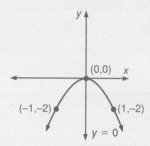

2. $y = 2(x - 3)^2 - 4$

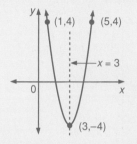

Closure

Have students give the standard form of an equation of a parabola. $y - k = a(x - h)^2$ Have them give the coordinates of the vertex. (h,k) Ask them how they would determine which way the parabola opens—upward or downward. The coefficient a in x^2 determines the direction it opens. For $a > 0$, it opens upward. For $a < 0$, it opens downward.

Additional Example 3

Write the equation of the function whose graph is given at the right.
$y - 6 = -3(x - 2)^2$

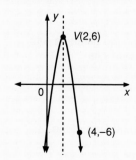

393

Guided Practice

Classroom Exercises 1–9

Independent Practice

A Ex. 1–18, **B** Ex. 19–22, **C** Ex. 23–25

Basic: WE 1–17 odd, Midchapter Review

Average: WE 3–19 odd, Midchapter Review

Above Average: WE 5–23 odd, Midchapter Review

Additional Answers

Classroom Exercises

7.

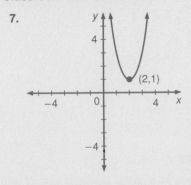

See pages 813–815 for answers to Written Ex. 1–22 and Midchapter Review Ex. 3–5.

Written Exercises

Graph each function and $y = x^2$ in the same coordinate plane. Use a table of values.

1. $y = 3x^2$ 2. $y = \frac{1}{3}x^2$ 3. $y = -3x^2$ 4. $y = 2x^2$ 5. $y = -\frac{1}{4}x^2$ 6. $y = -4x^2$

Graph each quadratic function. Give the vertex of the parabola and write the equation of the axis of symmetry.

7. $y = (x - 1)^2$
8. $y - 2 = x^2$
9. $y - 1 = (x - 2)^2$
10. $y = -(x + 4)^2 + 2$
11. $y = 3(x - 2)^2 - 5$
12. $y = -2(x + 4)^2 + 3$
13. $y = (x - 8)^2$
14. $y = -(x + 5)^2 - 1$
15. $y = x^2 - 7$
16. $y = \frac{1}{2}(x - 7)^2 - 4$
17. $y = -\frac{1}{2}(x + 2)^2 + 3$
18. $y = \frac{2}{3}(x + 1)^2 - 5$

Find an equation of the quadratic function whose graph is given.

19.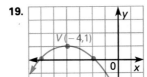

20.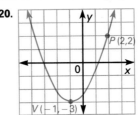

21.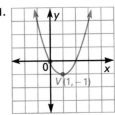

22. Write, in your own words, the steps in graphing $y - 1 = 2(x - 3)^2$.

23. For what values of p will the graph of the parabola $y = (-2p + 5)(x - 5)^2 + 4$ open downward? $\{p \mid p > \frac{5}{2}\}$

24. Determine the values of a and b for which the vertex of the parabola $y = -3(x + 2a + b)^2 + a - b$ will be $(-5,8)$. $a = \frac{13}{3}, b = \frac{-11}{3}$

25. For what values of h and k will the vertex of the graph of $y - 6 = a(x + h)^2 + k$ be in Quadrant I? $\{(h,k) \mid h < 0 \text{ and } k > -6\}$

Midchapter Review

Determine the coordinates of the midpoint M of the segment $\overline{AB}$. Then determine whether the median $\overline{CM}$ is also an altitude of $\triangle ABC$. *11.1, 11.2*

1. $A(-3,-4)$, $B(2,5)$, $C(-5,3)$ $\left(-\frac{1}{2},\frac{1}{2}\right)$; Yes 2. $A(2,2)$, $B(4,-1)$, $C(6,1)$ $\left(3,\frac{1}{2}\right)$; No

Graph each function. Give the vertex and the equation of the axis of symmetry. *11.3, 11.4*

3. $y - 1 = |x + 3|$ 4. $y = -|x - 5| - 2$ 5. $y - 1 = (x + 3)^2$

Enrichment

Have the students graph the parabola with equation $x = y^2$ and discuss its characteristics. The graph opens to the right. It is not the graph of a function. The vertex is (0,0). The x-axis, with equation $y = 0$, is the axis of symmetry.

Point out that the standard form of the equation of a horizontal parabola is $x - h = a(y - k)^2$. Then have the students graph a horizontal parabola, label the vertex and another point on the parabola, and give the graph to another student. Have the students determine the equation of the graph they received.

11.5 Changing to Standard Form

Objectives
To change equations of parabolas from general form to standard form
To determine the vertices and axes of symmetry of parabolas described by equations in general form

Sometimes the equation of a parabola is given in the **general form**, $y = ax^2 + bx + c$, rather than the **standard form**, $y - k = a(x - h)^2$. Because certain characteristics of the parabola are not apparent in the general form of the equation, it can be helpful to change the equation to standard form by completing the square.

EXAMPLE 1   Rewrite $y = x^2 - 8x + 15$ in standard form. Identify the vertex of the parabola and write the equation of its axis of symmetry.

Solution
$$y = x^2 - 8x + 15$$
Isolate the x- and x^2-terms.
$$y - 15 = x^2 - 8x$$
Complete the square.
$$y - 15 + 16 = x^2 - 8x + 16$$
Factor the right side.
$$y + 1 = (x - 4)^2$$
Write in the form $y - k = a(x - h)^2$.
$$y - (-1) = 1(x - 4)^2$$
Vertex: $(4, -1)$ Equation of the axis of symmetry: $x = 4$

EXAMPLE 2   Graph the parabola described by $y = -2x^2 + 12x - 13$.

Solution   Write the equation in standard form by completing the square.
$$y = -2x^2 + 12x - 13$$
$$y + 13 = -2(x^2 - 6x)$$
$$y + 13 - 2 \cdot 9 = -2(x^2 - 6x + 9)$$
$$y - 5 = -2(x - 3)^2$$

Graph the parabola $y - 5 = -2(x - 3)^2$.

Vertex: $(3, 5)$

Equation of the axis of symmetry: $x = 3$

Make a table of values.

x	1	2	3	4	5
y	-3	3	5	3	-3

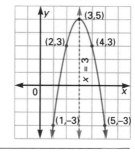

Teaching Resources

Project Worksheet 11
Quick Quizzes 79
Reteaching and Practice
   Worksheets 79

▄▄▄ GETTING STARTED

Prerequisite Quiz

Solve by completing the square.

1. $x^2 - 8x + 12 = 0$ 6,2
2. $x^2 - 10x + 25 = 0$ 5
3. $x^2 - 4x - 12 = 0$ 6,−2

Motivator

Have students define a perfect square trinomial.   The square of a binomial   Ask them what they need to add to each side of the equation $y - 15 = x^2 - 8x$ so that the right side of the equation will be a perfect square trinomial.   16   What is the resulting equation in standard form?
$y - (-1) = 1(x - 4)^2$

▄▄▄ TEACHING SUGGESTIONS

Lesson Note

Review the standard form of the equation of a parabola developed in the preceding lesson.

Math Connections

Completing the Square: In this lesson students are reintroduced to *completing the square*, a useful technique that they have used in Lessons 9.3–9.5.

Additional Example 1

Rewrite $y = x^2 + 4x - 3$ in standard form. Identify the vertex of the parabola and write the equation of its axis of symmetry.
$y + 7 = (x + 2)^2; (-2, -7); x = -2$

Additional Example 2

Graph the parabola described by $y = 2x^2 - 12x + 18$.

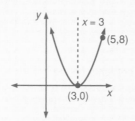

Highlighting the Standards

Standard 14b: In learning how other different forms of an equation reveal different facts about the graph, students are also learning about the structure of algebra.

Critical Thinking Questions

Analysis: Ask students how the coefficients a, b, and c in the general form of the equation of a parabola are related to a, h, and k in the standard form. By working it out directly, they will find that the simplest relationship is the obvious one, $a = a$. The remaining relationships are $b = -2ah$ and $c = ah^2 + k$.

Checkpoint

Rewrite each equation in standard form. Identify the vertex of the parabola and write the equation of the axis of symmetry.

1. $y = x^2 + 4x + 2$
   $y + 2 = (x + 2)^2$; $(-2, -2)$; $x = -2$
2. $y = 2x^2 + 12x + 13$
   $y + 5 = 2(x + 3)^2$; $(-3, -5)$, $x = -3$

Closure

Ask students to give the steps in changing an equation like $y = -5x^2 - 10x - 8$ to the standard form of an equation of a parabola so it can be graphed. See page 395.

■■■FOLLOW UP

Guided Practice

Classroom Exercises 1–12

Independent Practice

A Ex. 1–12, **B** Ex. 13–21, **C** Ex. 22–25

Basic: WE 1–15 odd

Average: WE 7–21 odd

Above Average: WE 9–23 odd

Classroom Exercises

Tell what number must be added to each side of the equation to put it in the standard form of the equation of a parabola.

1. $y - 2 = x^2 - 10x$ 25   2. $y - 1 = x^2 + 8x$ 16   3. $y = x^2 - 4x$ 4
4. $y + 3 = x^2 - 12x$ 36   5. $y = x^2 - 6x$ 9   6. $y + 2 = x^2 + 14x$ 49
7. $y - 7 = x^2 + 3x$ $\frac{9}{4}$ 8. $y + 4 = x^2 + x$ $\frac{1}{4}$ 9. $y - 2 = x^2 + 5x$ $\frac{25}{4}$

10–12. Rewrite the equations of Classroom Exercises 1–3 in standard form. Give the vertex of the parabola and the equation of its axis of symmetry. Then graph the parabola.

Written Exercises

Rewrite each equation in standard form. Give the vertex and the equation of the axis of symmetry. Then graph the parabola.

1. $y = x^2 + 4x - 1$ 2. $y = x^2 - 6x + 2$ 3. $y = x^2 - 10x + 15$
4. $y = x^2 - 6x + 12$ 5. $y = x^2 + 8x + 20$ 6. $y = x^2 + 4x - 5$
7. $y = x^2 + 10x + 27$ 8. $y = x^2 - 6x$ 9. $y = x^2 - 2x$
10. $y = x^2 - 4x$ 11. $y = x^2 - 8x$ 12. $y = x^2 + 6x$
13. $y = 2x^2 - 8x + 6$ 14. $y = 3x^2 - 6x - 5$ 15. $y = 4x^2 - 8x + 1$
16. $y = -5x^2 - 10x + 7$ 17. $y = -2x^2 - 4x + 6$ 18. $y = -2x^2 - 16x + 3$
19. $y = -x^2 - 6x - 4$ 20. $y = -3x^2 + 18x$ 21. $y = -x^2 - 2x$
22. Graph $y = \left| \frac{1}{4}x^2 + \frac{3}{2}x \right|$ for $-9 \leq x \leq 3$.
23. Graph $y = \frac{1}{4}x^2 + \frac{1}{2}|x| + 2$ for $-4 \leq x \leq 4$.
24. The graph of a parabola has the same shape as the graph of $y = -2x^2 - 8x + 3$, but is shifted (translated) 2 units to the right and 4 units up. Also, this parabola opens in the opposite direction. Write an equation of this parabola in standard form. $y - 15 = 2x^2$
25. Find b and c so that the graph of $y = x^2 + bx + c$ has its vertex at $(3, 4)$. $b = -6$, $c = 13$

Mixed Review

For the points $A(-2, 4)$ and $B(6, 8)$, determine each of the following.
3.4, 11.1, 11.2

1. the equation in standard form of the line containing the points A and B $x - 2y = -10$
2. the distance between A and B $4\sqrt{5}$
3. the coordinates of the midpoint of $\overline{AB}$ $(2, 6)$
4. the equation of the perpendicular bisector of $\overline{AB}$ $2x + y = 10$

Additional Answers

See pages 815–817 for the answers to Classroom Ex. 10–12 and Written Ex. 1–23.

Enrichment

Challenge students to graph the following equations on one set of axes in the intervals $-4 < x < 4$ and $-4 < y < 4$.

$y = |x^2 - 1| + 1$ $x = |y^2 - 1| + 1$
$y = -|x^2 - 1| - 1$ $x = -|y^2 - 1| - 1$

Now have the students create other symmetric designs by combining absolute value graphs with graphs of parabolas.

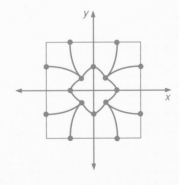

11.6 Problem Solving: Maximum and Minimum Problems

Teaching Resources

Quick Quizzes 80
Reteaching and Practice
   Worksheets 80

Objective   To solve maximum and minimum problems

In the last two lessons, it was shown that the graph of a quadratic (second-degree) function is a parabola.

The parabola $y = -x^2 + 6x + 1$ can be expressed in standard form as $y - 10 = -(x - 3)^2$. Its graph is shown at the right.

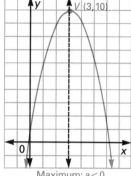

Maximum: $a < 0$

Because $a < 0$, the parabola opens *downward*.

The coordinates of the vertex are $(3,10)$.

Because the vertex is the highest point of the parabola, the y-coordinate of the vertex, 10, is the greatest value, or **maximum** value, of the function.

Similarly, for a parabola that opens *upward*, (with $a > 0$), the vertex is the lowest point of the parabola. In the graph of $y - 5 = (x - 4)^2$ at the right, the y-coordinate of the vertex, 5, is the least, or **minimum**, value of the function. The general form of its equation is $y = x^2 - 8x + 21$.

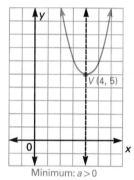

Minimum: $a > 0$

The general equation $y = ax^2 + bx + c$ can be written in standard form by completing the square. The result follows.

$$y - \left(\frac{4ac - b^2}{4a} \right) = a \left(x + \frac{b}{2a} \right)^2$$

The coordinates of the vertex follow from the equation above.

$$(h,k) = \left(-\frac{b}{2a}, \frac{4ac - b^2}{4a} \right)$$

This result is expressed in the theorem at the top of the next page.

GETTING STARTED

Prerequisite Quiz

Write each equation in standard form and give the vertex of the parabola.

1. $y = x^2 - 6x + 1$
   $y + 8 = (x - 3)^2$; $(3,-8)$
2. $y = x^2 - 8x + 21$
   $y - 5 = (x - 4)^2$; $(4,5)$
3. $y = x^2 + 4x$
   $y + 4 = (x + 2)^2$; $(-2,-4)$
4. $y = -2x^2 + 12x - 13$
   $y - 5 = -2(x - 3)^2$; $(3,5)$

Motivator

Ask students to describe in their own words what is meant by the *maximum/minimum* value of a function.

1. $y = -x^2 + 6x + 1$
2. $y = x^2 - 8x + 21$

Without graphing, have students tell which of the two equations above attains a maximum y value.   Equation 1   Ask them if Equation 2 attains a minimum y value.
Yes

Highlighting the Standards

Standard 13a: Determining maximum and minimum values is a significant step along the road leading toward calculus.

Lesson Note

It is possible to solve maximum and minimum problems involving quadratic functions by completing the square. Problems involving minimum values were solved this way in Lesson 9.4.

A simpler method of finding a maximum or minimum value of y, when $y = ax^2 + bx + c$, is to use Theorem 11.3. Find the x-coordinate of the vertex using $x = -\frac{b}{2a}$. Then substitute in the equation to find the maximum or minimum value of y. There is no need to memorize the formula for the y value, $\frac{4ac - b^2}{4a}$.

Math Connections

Calculus: Some students may benefit from the knowledge that there are advanced methods for determining whether a function has any extreme values, at least in some cases, and determining what those values are. Without being given too many confusing details, students might be told that the location and value of the extreme values of a function, if such exist, can often be determined using the methods of *differential calculus*.

Theorem 11.3

The x-coordinate of the vertex of the parabola defined by $y = ax^2 + bx + c$ is $x = -\frac{b}{2a}$.

EXAMPLE 1 Find the maximum value of the function $y = -x^2 + 8x - 4$.

Plan Since the coefficient of x^2 is negative, the function has a *maximum* at the vertex of the parabola. Use Theorem 11.3 to find the x-coordinate of the vertex.

Solution From the general equation $y = ax^2 + bx + c$, identify a and b.

$$a = -1, b = 8$$

Next, use Theorem 11.3 to find the x-coordinate of the vertex.

$$x = -\frac{b}{2a} = -\frac{8}{2(-1)}, \text{ or } 4$$

Now substitute 4 for x to find the y-coordinate of the vertex, the function's maximum value.

$$y = -x^2 + 8x - 4$$
$$y = -(4)^2 + 8(4) - 4 = 12$$

The maximum value of the function $y = -x^2 + 8x - 4$ is 12.

EXAMPLE 2 Determine whether $y = 2x^2 + 10x - 4$ has a maximum or a minimum value. Then compute that value.

Solution Because 2, the coefficient of x^2, is positive, y has a *minimum*. Use $x = -\frac{b}{2a}$ to find the x-coordinate of the vertex of the parabola.

Substitute 2 for a and 10 for b.

$$x = -\frac{b}{2a}$$
$$= -\frac{10}{2(2)} = -\frac{10}{4}, \text{ or } -\frac{5}{2}$$

Substitute $-\frac{5}{2}$ for x.

$$y = 2x^2 + 10x - 4$$
$$y = 2\left(-\frac{5}{2}\right)^2 + 10\left(-\frac{5}{2}\right) - 4$$
$$y = 2\left(\frac{25}{4}\right) - 25 - 4$$
$$y = 12\frac{1}{2} - 25 - 4 = -16\frac{1}{2}$$

Thus, the minimum value of $y = 2x^2 + 10x - 4$ is $-16\frac{1}{2}$.

Additional Example 1

Find the maximum value of the quadratic function.

$y = -x^2 + 10x + 14$ 39

Additional Example 2

Determine whether $y = 4x^2 - 8x + 5$ has a maximum or minimum. Then compute that value.   Minimum; 1

EXAMPLE 3 Plans call for a rectangular berry patch next to a summer cottage to be enclosed on three sides by a total of 100 ft of fencing and on the fourth side by a cottage wall. Find the maximum area that can be enclosed in this way. Also, find the corresponding dimensions of the berry patch.

Plan Distribute the 100 ft of fencing among the three sides as shown in the diagram. Let x = the length of each of the two shorter sides. Then the length of the third side is $100 - x - x$, or $100 - 2x$.

Solution Use the formula for the area of a rectangle.

$A = lw$

$A = (100 - 2x)x$

$\quad = 100x - 2x^2$, or $-2x^2 + 100x$

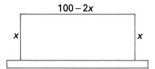

Use Theorem 11.3 to find the value of x for which A is a maximum.

Substitute -2 for a and 100 for b. $x = -\dfrac{b}{2a} = -\dfrac{100}{2(-2)} = 25$

When $x = 25$, $A = -2(25)^2 + 100(25)$, or 1,250. The maximum value of A is 1,250 and this occurs when $x = 25$.

Thus, the maximum area of the berry patch is $1{,}250 \text{ ft}^2$ when the width x is 25 ft and the length, $100 - 2x$, is 50 ft.

Focus on Reading

Indicate whether each of the following is true or false. If false, tell why.

1. The equation $y = x^2 - 5x + 2$ is in standard form. False; general form
2. The function $y = 4x^2 - 7x + 8$ has a minimum value. True
3. The minimum value of the function $y = 4x^2 - 8x - 10$ occurs when $x = 1$. True
4. The function $y = x^2 + 1$ has a maximum value when $x = 0$.
   False; it has a minimum value there.

Classroom Exercises

Determine whether each function has a maximum value or a minimum value.

1. $y = 3x^2 - 12x + 2$ Min   2. $y = -x^2 - 8x - 10$ Max   3. $y = 4x^2 + 8x - 4$ Min

4. $y = -3x^2 - 6x - 12$ Max   5. $y = -6x^2 - x - 7$ Max   6. $y = x^2 - x + 1$ Min

7. Determine whether $y = -2x^2 - 16x + 5$ has a maximum or a minimum value. Then compute that value. Max; 37

Critical Thinking Questions

Analysis: Ask students whether it is the case that the graph of every function must have either a maximum or a minimum. Many students will quickly realize that a linear function, such as $y = 2x + 5$, has no maximum or minimum if its domain is the set of real numbers. If the domain is restricted, the question of whether a function must have an extreme value may not be so easy for students to answer, since they have not yet encountered a function such as $y = \frac{1}{x}$, $x > 0$, which has no extreme value. (See Lesson 12.5.)

Common Error Analysis

Error: When students are given the perimeter of a rectangle, say 24, and are asked to represent the lengths of the sides in terms of a single variable x, they often write x and $24 - x$ for the length and width. Have them draw a diagram of the rectangle.

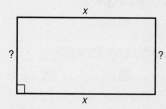

If x represents each length, then the sum of the two widths is $24 - 2x$, so that each width is $\frac{24 - 2x}{2}$, or $12 - x$.

Additional Example 3

Find two numbers whose sum is 24 and whose product is a maximum. 12, 12

Checkpoint

Determine whether each function has a maximum or a minimum value. Then compute that value.

1. $y = x^2 + 4x - 3$ Minimum; -7
2. $y = x^2 - 4x$ Minimum; -4
3. $y = -2x^2 + 8x - 1$ Maximum; 7

Closure

Have students give the formula for determining the x-coordinate of the vertex of a parabola. $-\dfrac{b}{2a}$ If given an algebraic relationship between the length and width of a rectangle, ask students to give the steps in determining the length and width to produce a rectangle with maximum area. See Example 3.

◼◼◼◼FOLLOW UP

Guided Practice

Classroom Exercises 1–7

Independent Practice

A Ex. 1–10, **B** Ex. 11–14, **C** Ex. 15–17

Basic: FR all, WE 1–12
Average: FR all, WE 3–14
Above Average: FR all, WE 6–17

Additional Answers

Written Exercises

17.
$$y = ax^2 + bx + c$$
$$y - c = ax^2 + bx$$
$$y - c = a\left(x^2 + \frac{b}{a}x\right)$$
$$y - c + \frac{b^2}{4a} = a\left(x^2 + \frac{b}{a}x + \frac{b^2}{4a^2}\right)$$
$$y - \frac{(4ac - b^2)}{4a} = a\left[x - \left(\frac{-b}{2a}\right)\right]^2$$

Written Exercises

1–6. For Classroom Exercises 1–6, find the maximum or minimum value of each function. **1.** −10 **2.** 6 **3.** −8 **4.** −9 **5.** $-\frac{167}{24}$ **6.** $\frac{3}{4}$

7. For a typical day, the net income I, in hundreds of dollars, for a small pizza stand is given by the formula $I = -\frac{1}{2}p^2 + 7p - \frac{45}{2}$, where p is the asking price in dollars per pizza. What price maximizes net income? What is this income? $7, \$200

8. The formula for the height h reached by a rocket fired straight up from a 50-ft platform with an initial velocity of 96 ft/s is $h = -16t^2 + 96t + 50$. Find the number of seconds t required to reach the maximum height. Find the maximum height. 3 s, 194 ft

9. A biologist's formula for predicting the number of impulses fired after stimulation of a nerve is $i = -x^2 + 40x - 45$, where i is the number of impulses per millisecond after stimulation, and x is a measure of the strength of the stimulus. Find the maximum possible number of impulses. 355

10. In a 220-volt electric circuit with a resistance of 15 ohms, the available power in watts W is given by the formula $W = 220I - 15I^2$, where I is the amount of current in amperes. Find the maximum power that the electric circuit can deliver. $806\frac{2}{3}$ W

11. A rectangular piece of ground is to be enclosed on three sides by 160 ft of fencing. The fourth side is the side of a barn. Find the dimensions of the enclosure that maximizes the area. 40 ft by 80 ft

12. A rectangular field is to be enclosed by 300 ft of fencing. Find the dimensions of the enclosure that maximizes the area. 75 ft by 75 ft

13. Find two numbers whose sum is 20 and the sum of whose squares is a minimum. 10, 10

14. Find two numbers summing to 16.8 whose product is maximal. 8.4, 8.4

15. A rain gutter is made by bending up the edges of a piece of metal of width 36 in. so that the cross-sectional area is a maximum. Find the maximal cross-sectional area. 162 in^2

16. Find the dimensions of the rectangle of maximum area that has a perimeter of p units. $\frac{p}{4}$ by $\frac{p}{4}$

17. From the general form of the equation of a parabola, $y = ax^2 + bx + c$, derive the standard form, $y - k = a(x - h)^2$.

Mixed Review

Solve for x.   2.3, 6.1, 6.6, 9.5

1. $|2x - 8| = 12$ $-2, 10$

2. $x^2 - 8x + 12 = 0$ 2, 6

3. $x^3 - 7x^2 + 14x - 8 = 0$ 1, 2, 4

4. $x^3 - 8 = 0$ 2, $-1 \pm i\sqrt{3}$

400   Chapter 11 Coordinate Geometry and Quadratic Functions

Enrichment

Challenge students to find an equation of the form $y = ax^2 + bx + c$ for a parabola that contains the points (1,26), (2,11), and (3,2).

Point out to the students that they can substitute the ordered pairs in $y = ax^2 + bx + c$ and then solve a system of three linear equations in a, b, and c to find values for these constants. $y = 3x^2 - 24x + 47$

The students can make up similar problems for each other. Each student can choose an equation for another student to discover using three of its ordered pairs.

Note: If points such as (1,5), (2,7), and (3,9) are given, the students will find that $a = 0$, $b = 2$, and $c = 3$. Thus, the function is $y = 2x + 3$, which is linear, and the three points lie on a line.

11.7 Odd and Even Functions

Objective   To determine whether a function is odd or even

Recall that the graph of $y = x^2$, or $f(x) = x^2$, is symmetric with respect to the y-axis. Notice that $f(-2) = f(2) = 4$, and $f(-3) = f(3) = 9$. In general, $f(-x) = f(x)$. Such a function is called an *even function*.

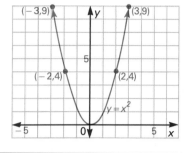

Definition   A function is an **even function** if $f(-x) = f(x)$ for all x in the domain of f.

EXAMPLE 1   Determine whether or not each function is even.

a. $f(x) = x^5$ b. $f(x) = x^4$

Solution   Replace x by $-x$.

a. $f(-x) = (-x)^5 = -x^5$
$$x^5 \neq -x^5$$
$$f(-x) \neq f(x) \leftarrow f \text{ is } not \text{ even.}$$

b. $f(-x) = (-x)^4 = x^4$
$$x^4 = x^4$$
$$f(-x) = f(x) \leftarrow f \text{ is even.}$$

Consider the following definition.

Definition   A curve is said to be **symmetric with respect to the origin** if for any segment through the origin with its endpoints on the curve, the segment is bisected by the origin.

The graph of $f(x) = \frac{1}{3}x^3$ is symmetric with respect to the origin, as can be seen at the right. The origin bisects the segment $\overline{PP'}$. So, $P(3,9)$ and $P'(-3,-9)$ are a pair of symmetric points on the curve. Notice that $f(-x) = -f(x)$; that is, $f(-x)$ and $f(x)$ are *opposites*. So, $f(-3) = -9$, while $f(3) = 9$, the opposite of -9.

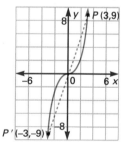

Teaching Resources

Quick Quizzes 81
Reteaching and Practice
Worksheets 81

▬▬GETTING STARTED

Prerequisite Quiz

For $f(x) = x^2$, find each value.

1. $f(3)$ 9
2. $f(-3)$ 9
3. $f(-x)$ x^2

Motivator

Have students graph the equation $y = x^2$. Have them compare the values of the y-coordinate when the x-coordinates are 1 and -1, 2 and -2, 3 and -3. Ask them to summarize their findings.

▬▬TEACHING SUGGESTIONS

Lesson Note

Emphasize the graphic interpretation of the odd and even functions. For even functions, the graphs are symmetric with respect to the y-axis. For odd functions, the graphs are symmetric with respect to the origin.

Math Connections

Trigonometry: There are examples of odd and even functions which students will encounter when they study trigonometric functions. For example, the sine function is an odd function and the cosine functions is an even function.

Additional Example 1

Determine whether each function is even.

a. $f(x) = -2x^2$ yes
b. $f(x) = x^2 + x^3$ no

Highlighting the Standards

Standard 8a: Students will make conjectures as they work at determining the different properties of functions.

Analysis: Ask students to devise a credible explanation for the use of the words odd and even in describing functions. Polynomial functions with only even-powered terms are symmetrical about the y-axis while those with only odd-powered terms are symmetrical about the origin.

Checkpoint

Determine whether each function is odd or even.

1. $2x^9$ odd   **2.** x^4 even

Closure

Have students give definitions of odd and even functions. See page 401.

▰▰▰FOLLOW UP

Guided Practice

Classroom Exercises 1–8

Independent Practice

A Ex. 1–8, **B** Ex. 9–12, **C** Ex. 13, 14

Basic: WE 1–10, Mixed Problem Solving

Average: WE 3–12, Mixed Problem Solving

Above Average: WE 5–14, Mixed Problem Solving

Additional Answers

Written Exercises

  3. Even   7. Odd   8. Even
 12. Even   14. 2, $-2 \pm \sqrt{3}$

See page 403 for the answer to Mixed Review Ex. 3.

When the graph of f is symmetric with respect to the origin, then f is called an *odd* function.

Definition | A function f is an **odd function** if $f(-x) = -f(x)$ for all x in the domain of f.

EXAMPLE 2 Determine whether or not $f(x) = -2x^3$ is odd.

Solution

$$f(x) = -2x^3$$
Replace x by $-x$. $f(-x) = -2(-x)^3$ ⎫
$$= -2(-x^3)$$ ⎬ opposites
$$= 2x^3$$ ⎭

Thus, $f(-x) = -f(x)$, and $f(x) = -2x^3$ is an odd function.

Classroom Exercises

Determine whether or not each function is symmetric with respect to the origin, and state the reason for your answer.

1. $f(x) = x^4$ No;   **2.** $f(x) = -x^5$ **3.** $f(x) = -3x^2$ **4.** $f(x) = x^3 + 8$
    $f(-x) \neq -f(x)$ Yes; $f(-x) = -f(x)$ No; $f(-x) \neq -f(x)$ No; $f(-x) \neq -f(x)$

Determine whether each function is odd, even, or neither.

5. $f(x) = -4x^2$ **6.** $f(x) = 6x^3$ Odd   **7.** $f(x) = 3x^2$ Even   **8.** $f(x) = x^3 + 8$
   Even   Neither

Written Exercises

Determine whether each function is odd, even, or neither.

1. $f(x) = 6x^2$ Even   **2.** $f(x) = 2x^5$ Odd   **3.** $f(x) = 8 - x^2$ **4.** $f(x) = -6x$ Odd
5. $f(x) = -3x$ Odd   **6.** $y = 3x$ Odd   **7.** $f(x) = -2x^3$ **8.** $f(x) = 8 - x^4$
9. $f(x) = x^4 + x^2$ **10.** $y - 4 = x^3 + x^2$ **11.** $y + 2 = x^3$ **12.** $y - x^6 = 4x^2$
   Even   Neither   Neither

Points A and B are symmetric with respect to the origin. Find the possible values of x.

13. $A(x^2 - 2x, 5)$, $B(2x - 9, -5)$ ± 3 **14.** $A(x^3 + 2x^2, 1)$, $B(-7x - 2, -1)$

Mixed Review   5.6, 2.1, 3.5, 4.2, 8.2

  $(3x - 5)(3x + 5)$ $\{x \mid x > -6\}$
1. Factor $9x^2 - 25$.   **2.** Solve $3x - 4 < 5x + 8$.   **3.** Graph $3x - 2y = 8$.

4. Solve the system $\begin{array}{l} 2x - y = 4 \\ 3x + 4y = 17. \end{array}$ (3,2)   **5.** Simplify $\sqrt{28x^7}$ for $x \geq 0$. $2x^3\sqrt{7x}$

402   Chapter 11 Coordinate Geometry and Quadratic Functions

Additional Example 2

Determine whether $f(x) = 6 - 2x^3$ is odd.

no

Enrichment

Have students verify that the graph of the sine function is an odd function (see *Math Connections*). Using a scientific calculator, have them make a table of values of sin θ in 30-degree increments ranging from −360 to 360. Ask them if the values of sin θ, if graphed, are symmetrical with respect to the origin. Yes

Mixed Problem Solving

1. A number y varies directly as the square of x. If $y = 100$ when $x = 5$, what is y when $x = 3$? 36

2. Find the set of positive integers for which 3 more than 8 times the sum of a number and 9 is less than 115.

3. Convert 30 mi/h to ft/s. 44 ft/s

4. Find the positive number whose square exceeds 3 times the number by 88. 11

5. Find the maximum height reached by an object shot upward from the earth's surface with an initial velocity of 160 ft/s. Use the formula $h = vt - 16t^2$. 400 ft

6. One diagonal of a square is 2 cm longer than the length of a side. Find the length of the side in simplest radical form. $2 \pm 2\sqrt{2}$ cm

7. The perimeter of a rectangle is 24 ft, and its area is 32 ft^2. Find its dimensions. 8 ft by 4 ft

8. Find two numbers whose product is 20 and the sum of whose reciprocals is $\frac{9}{20}$. 4, 5

9. How many liters of a 20% alcohol solution must be added to 12 liters of a 60% alcohol solution to obtain a 35% solution? 20 liters

10. The product of two numbers is 2, and the second number is 4 less than the first number. Find two pairs of such numbers.

11. Working together, Terry and Kim can sand a set of chairs in 6 h. Working alone, Terry takes 5 h longer than Kim. How long does Kim take to sand the chairs working alone? 10 h

12. Ted drove a distance of 80 mi in 2 h. Because of construction work, his average speed for the first 20 mi was half of his speed for the rest of his trip. What was his average speed for the first 20 mi? 25 mi/h

13. A racing crew rowed 1,120 yd downstream in 3.5 min, and rowing at the same pace, returned the same distance upstream in 4 min. Find the rate of the current in yd/min. 20 yd/min

14. Teresa mixed Brand A dog food worth 55¢/lb with Brand B dog food worth 77¢/lb to make a 20-lb package worth $12.76. Find the amount of Brand A that was used. 12 lb

15. If the numerator of a fraction is increased by 3 and the denominator is decreased by 9, the result is a fraction equal to $\frac{5}{3}$. Find the original fraction, given that the numerator is 8 less than the denominator. $\frac{7}{15}$

16. A rectangular garden is 3 yd longer than it is wide. A second rectangular garden is planned so that it will be 6 yd wider and twice as long as the first garden. Find the area of the first garden if the sum of the areas of both gardens will be 216 yd^2. 40 yd^2

17. The sum of the digits of a two-digit number is 15. If the digits are reversed, the new number is 27 less than the original number. Find the original number. 96

18. The tens digit of a two-digit number is twice the units digit. If the digits are reversed, the new number is 36 less than the original number. Find the original number. 84

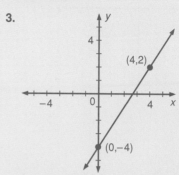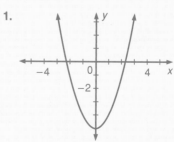

11.

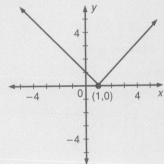

12.

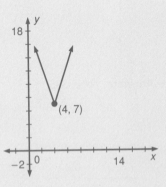

13.

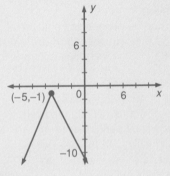

Chapter 11 Review

Key Terms

absolute value function (p. 386)
axis of symmetry (p. 386)
distance formula (p. 378)
even function (p. 401)
general form of a parabola (p. 395)
maximum (p. 397)
median (p. 382)
midpoint formula (p. 379)
midpoint of a segment (p. 379)
minimum (p. 397)

mirror-image points (p. 386)
odd function (p. 402)
ordered triple (p. 390)
parabola (p. 391)
perpendicular bisector (p. 383)
standard form (pp. 388, 395)
symmetry with respect to
the origin (p. 401)
translation (p. 388)
vertex (p. 386)

Key Ideas and Review Exercises

11.1   Given two points $P_1(x_1,y_1)$ and $P_2(x_2,y_2)$, the following are true.
   (1) The distance d between P_1 and P_2 is $\sqrt{(x_2 - x_1)^2 + (y_2 - y_1)^2}$.
   (2) The midpoint M of $\overline{P_1P_2}$ is $M\left(\dfrac{x_1 + x_2}{2}, \dfrac{y_1 + y_2}{2}\right)$.

Give the distance between the following points in simplest radical form.

1. $A(2,-4)$; $B(4,6)$ $2\sqrt{26}$ **2.** $Z(-3,-6)$; $J(4,-4)$ $\sqrt{53}$

3. $D(-2,0)$; $E(10,\sqrt{5})$ $\sqrt{149}$ **4.** $W\left(-\dfrac{2}{3},-\dfrac{1}{3}\right)$; $Q\left(\dfrac{1}{6},-1\right)$ $\dfrac{\sqrt{41}}{6}$

Use $\triangle ABC$ with vertices $A(-3,2)$, $B(1,-2)$, and $C(3,4)$ for Exercises 5–7.

5. Find the distance between A and B. $4\sqrt{2}$ **6.** Find the midpoint of $\overline{AB}$. $(-1,0)$

7. Determine whether $\triangle ABC$ is isosceles.   Yes

8. Determine the coordinates of Q, an endpoint of $\overline{PQ}$, given that the other endpoint is $P(-2,4)$ and the midpoint is $M(1,5)$.   (4,6)

11.2   To illustrate geometric relationships involving either
   (1) lengths and midpoints of line segments, or
   (2) parallelism and perpendicularity, use the properties in Lesson 11.2.

For Exercises 9 and 10, use $\triangle ABC$ of Exercises 5–7.

9. Determine whether the median from C to $\overline{AB}$ is also an altitude of the triangle.   Yes

10. Write an equation in standard form of the perpendicular bisector of $\overline{BC}$.
   $x + 3y = 5$

14.

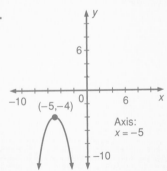

Axis:
$x = -5$

15.

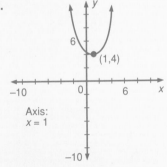

Axis:
$x = 1$

16.

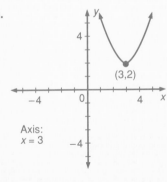

Axis:
$x = 3$

11.3 To graph a function of the form $y - k = a|x - h|$, use the properties summarized in Lesson 11.3.

Graph each function without using a table of values.

11. $y = |x - 1|$ **12.** $y = 3|x - 4| + 7$ **13.** $y = -2|x + 5| - 1$

11.4 To graph a quadratic function of the form $y - k = a(x - h)^2$, use the properties summarized in Lesson 11.4.

Graph each quadratic function. Give the vertex of the parabola and the equation of its axis of symmetry.

14. $y = -(x + 5)^2 - 4$ **15.** $y = \frac{1}{2}(x - 1)^2 + 4$ **16.** $y = (x - 3)^2 + 2$

11.5 To graph a parabola of the form $y = ax^2 + bx + c$, first use the technique of completing the square to change the equation into standard form.

For each parabola, give the vertex and the equation of its axis of symmetry. Then graph the parabola.

17. $y = x^2 - 2x + 5$ **18.** $y = -4x^2 - 8x - 1$ **19.** $y = 2 + 2x - 2x^2$

20. Write, in your own words, the effect of the coefficient a on the shape and the direction of the parabola $y = ax^2 + bx + c$.

11.6 To find the maximum or minimum value of the quadratic function $y = ax^2 + bx + c$, use $x = -\frac{b}{2a}$ to find the x-coordinate of the vertex of the parabola. Then use this x-value to find the corresponding y-value.

Determine whether the given function has a maximum value or a minimum value. Then compute that value.

21. $y = -8x^2 + 16x - 9$ **22.** $y = 3x^2 - 9x + 1$ **23.** $y = 2x - x^2 + 3$

24. The sum of two numbers is 18. Their product is a maximum. Find the two numbers. 9, 9

25. Sarita intends to use 180 yards of fencing to build two pens of equal size to separate two incompatible pets. To make the best use of the fencing, she plans to build the pens adjacent to each other, as shown at the right. What are the dimensions for each pen that will maximize the area enclosed?
30 yd by $22\frac{1}{2}$ yd.

$\frac{1}{2}(180 - 3x)$

11.7 A function $y = f(x)$ is even if $f(-x) = f(x)$. A function $y = f(x)$ is odd if $f(-x) = -f(x)$.

Determine whether each of the following functions is *even, odd,* or *neither*.

26. $f(x) = -2x^2$ Even   **27.** $f(x) = \frac{1}{2}x^3$ Odd   **28.** $f(x) = -|x + 1|$
Neither

17.
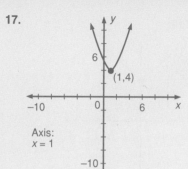
Axis:
$x = 1$

18.
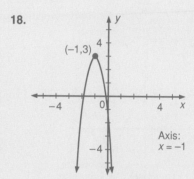
Axis:
$x = -1$

19.
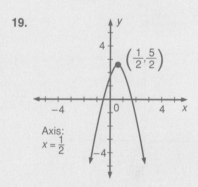
Axis:
$x = \frac{1}{2}$

20. Answers will vary. One possible answer: For $|a| < 1$, the basic graph of the parabola widens. For $|a| > 1$, the basic graph of the parabola narrows. For $a < 0$, the parabola opens downward. For $a > 0$, the parabola opens upward.

21. Max; -1 **22.** Min; $-\frac{23}{4}$ **23.** Max; 4

Chapter Test

9.

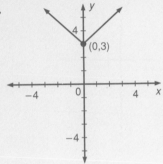

10.

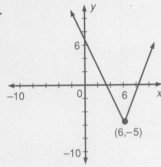

11.

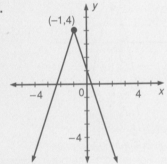

See page 407 for the answers to Ex. 12–14, 22, 23.

Use $\triangle ABC$ with vertices $A(2,4)$, $B(6,0)$, and $C(10,8)$ for Exercises 1–5.

1. Determine whether or not $\triangle ABC$ is isosceles. Yes
2. Write an equation in standard form of the perpendicular bisector of $\overline{AC}$. $2x + y = 18$
3. Determine whether or not the median from A to $\overline{BC}$ is also an altitude of the triangle. No
4. Find the length of the median from C to $\overline{AB}$. $6\sqrt{2}$
5. Find the length of $\overline{AB}$. $4\sqrt{2}$

Use the following points for Exercises 6 and 7: $A(0,0)$, $B(5,12)$, $C(-19,22)$, $D(-24,10)$.

6. Determine whether or not $ABCD$ is a rectangle. Yes
7. Determine whether or not the diagonals of $ABCD$ are perpendicular bisectors of each other. No
8. Determine the coordinates of Q, an endpoint of $\overline{PQ}$, given that the other endpoint is $P(-6,3)$ and the midpoint is $M(4,5)$. (14,7)

Graph each function without using a table of values.

9. $y = |x| + 3$ 10. $y + 5 = 2|x - 6|$ 11. $y = -3|x + 1| + 4$

For each parabola, give the vertex and the equation of its axis of symmetry. Then graph the parabola.

12. $y = -\frac{1}{8}(x + 5)^2 + 2$ 13. $y = -(x - 1)^2 + 6$ 14. $y = x^2 - 12x + 11$

Determine whether the given function has a maximum value or a minimum value. Then compute that value.   Min; $\frac{159}{40}$ Max; 29

15. $y = -x^2$ Max; 0   16. $y = 10x^2 - x + 4$ 17. $y = 10x - x^2 + 4$
18. Find two numbers whose sum is 30 and the sum of whose squares is a minimum. 15, 15

Determine whether the functions in Exercises 19–21 are *even*, *odd*, or *neither*.

19. $f(x) = 55x - 3$ Neither   20. $f(x) = 4$ Even   21. $f(x) = \sqrt[3]{x^4}$ Even
22. Graph $y = |36 - x^2|$.   23. Graph $x = -|y - 6| + 5$.
24. What is the equation in standard form of the parabola formed by translating the graph of the parabola $y = x^2 - 4x + 2$ four units to the left and four units upward and then inverting it so that it opens downward? $y - 2 = -(x + 2)^2$

Choose the *one* best answer to each question or problem.

1. If $y - 3 = 7$, then $y + 6 =$ __?__ . B
(A) 10   (B) 16   (C) 24
(D) 4   (E) 27

2. If $k = \frac{n}{18} + \frac{n}{18} + \frac{n}{18}$, then the least positive integer n for which k is an integer is __?__ . D
(A) 18   (B) 4   (C) 12
(D) 6   (E) 18

3.
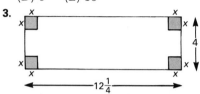

In the figure above, the area of the unshaded region is __?__ . A
(A) $(7 - 2x)(7 + 2x)$
(B) $(12\frac{1}{4} - x)(4 - x)$
(C) $(12\frac{1}{4} - 4x)(4 - 4x)$
(D) $16\frac{1}{4}x$ (E) $49 - x^2$

4. Which of the following values for N will maximize $\left(-\frac{1}{3}\right)^N$?   E
(A) 2   (B) 3   (C) 7
(D) 5   (E) 0

5. $P = 2 - \frac{9}{10}$, $Q = 2 - 0.099$,
$R = 2 \div 9$
In which list below are P, Q, and R in order from greatest to least? B
(A) P, Q, R (B) Q, P, R
(C) R, P, Q (D) P, R, Q
(E) R, Q, P

6. Fifty percent of 50% of 2 is __?__ . B
(A) 1   (B) $\frac{1}{2}$ (C) $\frac{1}{4}$
(D) 4   (E) 2

7.

A 9-hour clock is shown. If at 12 noon today the pointer is at 0, where will the pointer be at 12 noon tomorrow? C
(A) 0   (B) 2   (C) 6
(D) 7   (E) 8

8. After giving her $5, Ian had $9 more than Gail. How much more money than Gail did Ian have originally? C
(A) $14   (B) $4   (C) $19
(D) $9   (E) None of these

9. If $a^3b^2c < 0$, which of the following must be true? A
I. $abc < 0$ II. $ac < 0$ III. $bc < 0$
(A) II only   (B) I only
(C) II and III   (D) I and III
(E) None of these

10. If $y = \frac{1}{xz}$, what is the value of x when $y = 12^{-1}$ and $z = 3$? D
(A) -36 (B) $-\frac{1}{36}$ (C) -4
(D) 4   (E) None of these

11.

If the area of the triangle above is 30, then the length of $\overline{DB}$ is __?__ . B
(A) 2   (B) 8   (C) 10
(D) 3   (E) 5

12.
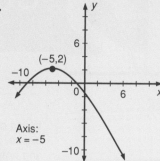
Axis:
$x = -5$

13.
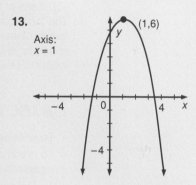
Axis:
$x = 1$

14.
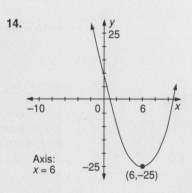
Axis:
$x = 6$

22.
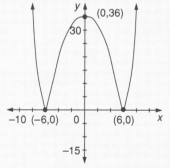

23.
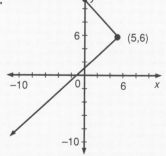

12 CONIC SECTIONS

OVERVIEW

In this chapter, equations of the circle, ellipse, parabola, and hyperbola are written and graphed. Vertices and foci of ellipses and hyperbolas are found, as well as equations of the asymptotes of the latter. The focus and directrix of a parabola are found. Linear-quadratic and quadratic-quadratic systems are solved and quadratic inequalities are graphed.

OBJECTIVES

- To write equations of conic sections
- To graph conic sections given their equations
- To find missing values in direct, inverse, joint, and combined variations
- To solve quadratic-quadratic systems of equations

PROBLEM SOLVING

In Lessons 12.5 and 12.6, several instances are provided that illustrate how inverse, joint, and combined variation are reflected in the physical world. In Lesson 12.8, the strategy *Using a Diagram* is employed to illustrate the conic sections introduced in this chapter. The problem solving strategies of *Using a Formula* and *Making a Graph* are used to solve problems in many of the lessons of this chapter.

READING AND WRITING MATH

A *Focus on Reading* appearing on page 438 asks students to fill in missing words to make a sentence true. Exercise 21 on page 442 asks students to explain a mathematical concept in their own words.

TECHNOLOGY

Calculator: Graphing calculators are now available that can accurately and rapidly display every graph shown in this chapter, including systems of quadratic equalities and quadratic inequalities.

SPECIAL FEATURES

Mixed Review pp. 412, 417, 422, 425, 434, 439, 442, 446, 451
Midchapter Review p. 429
Application: Eccentricity of an Ellipse p. 430
Focus on Reading p. 438
Application: LORAN p. 442
Brainteaser p. 446
Key Terms p. 452
Key Ideas and Review Exercises pp. 452–453
Chapter 12 Test p. 454
College Prep Test p. 455
Cumulative Review (Chapters 1–12) pp. 456–457

PLANNING GUIDE

Lesson	Basic	Average	Above Average	Resources
12.1 pp. 411–412	CE all WE 1–25 odd	CE all WE 7–31 odd	CE all WE 13–37 odd	Reteaching 82 Practice 82
12.2 pp. 416–417	CE all WE 1–19 odd	CE all WE 5–23 odd	CE all WE 9–29 odd	Reteaching 83 Practice 83
12.3 pp. 421–422	CE all WE 1–17 odd	CE all WE 5–21 odd	CE all WE 9–23 odd	Reteaching 84 Practice 84
12.4 p. 425	CE all WE 1–8	CE all WE 5–13	CE all WE 8–16	Reteaching 85 Practice 85
12.5 pp. 428–430	CE all WE 1–19 odd Midchapter Review Application	CE all WE 3–21 odd Midchapter Review Application	CE all WE 5–25 odd Midchapter Review Application	Reteaching 86 Practice 86
12.6 pp. 433–434	CE all WE 1–8	CE all WE 3–10	CE all WE 6–16	Reteaching 87 Practice 87
12.7 pp. 438–439	FR all, CE all WE 1–17 odd	FR all, CE all WE 5–21 odd	FR all, CE all WE 7–23 odd	Reteaching 88 Practice 88
12.8 pp. 441–442	CE all WE 1–14 Application	CE all WE 6–19 Application	CE all WE 8–24 Application	Reteaching 89 Practice 89
12.9 pp. 445–446	CE all WE 1–13 odd Brainteaser	CE all WE 7–19 odd Brainteaser	CE all WE 11–23 odd Brainteaser	Reteaching 90 Practice 90
12.10 pp. 450–451	CE all WE 1–15 odd	CE all WE 7–21 odd	CE all WE 11–27 odd	Reteaching 91 Practice 91
Chapter 12 Review pp. 452–453	all	all	all	
Chapter 12 Test p. 454	all	all	all	
College Prep Test p. 455	all	all	all	
Cumulative Review pp. 456–457	1–47 odd	9–49 odd	15–51 odd	

CE = Classroom Exercises   WE = Written Exercises   FR = Focus on Reading

NOTE: For each level, all students should be assigned all Mixed Review exercises.

■ INVESTIGATION

The students are invited to explore the elliptical orbits of major and minor planets.

A planet orbiting about the Sun follows an elliptical path. The Sun will be at one of the points F_1 and F_2 in the figure.

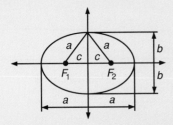

The *eccentricity* e of an orbit is defined as $e = \frac{c}{a}$, which is a ratio that varies from 0 to 1. Given the eccentricity e and the semi-major axis a, it is easy to calculate c.

The values of a and e for the Earth and the asteroid Apollo are shown below. The value of the Earth's semi-major axis is *defined* as 1, which is called 1 astronomical unit (abbreviated 1 a.u.). Have the students calculate the value of c for the Earth and Apollo.

	Earth	Apollo
a	1.000	1.471
e	0.017	0.560

Earth: $c = 0.017$ Apollo: $c = 0.824$

The value of b, the semi-minor axis, can be calculated using the Pythagorean relation. In the above figure,

$$b^2 = a^2 - c^2.$$

Have the students calculate the value of b for the Earth and Apollo.
Earth: $b = 0.99986$, Apollo: $b = 1.219$

Have the students sketch the orbits of the Earth and the asteroid Apollo using the values for a and b. They should show the Sun at one focus of each ellipse.

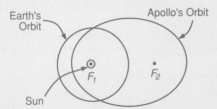

Apollo is known as an *Earth-crossing asteroid*, because its orbit crosses inside the Earth's orbit.

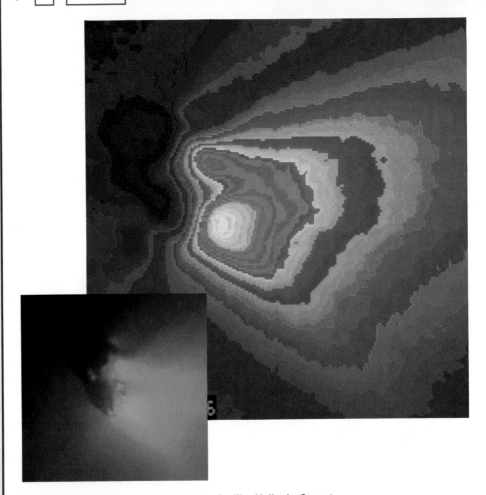

Comets that reappear on a regular basis, like Halley's Comet, have elliptical orbits. These images of Halley's Comet were produced from data gathered from the European Giotto space mission. Giotto is the name of an Italian pre-Renaissance artist who represented the Christmas star as a comet.

More About the Image

In 1986, a flotilla consisting of five separate spacecraft launched earlier by the USSR, Japan, and the European Space Agency (ESA) sent information back to the Earth about Halley's Comet. ESA's satellite, called Giotto, flew closer to the comet than any of the other four spacecraft. It sent back valuable information, including the photo and the computer-enhanced image above. The data from the mission revealed that the nucleus of the comet is a velvety-black elongated object shaped like a peanut or a potato about the size of Manhattan Island.

12.1 The Circle

Objectives   To write equations of circles given certain conditions
To graph circles given their equations

People have been fascinated by circles for thousands of years. Around 2000 B.C., an ancient people erected huge standing stones in a large circular ditch at Stonehenge in southern England. Most archaeologists believe that Stonehenge served a religious function, but some scientists suspect that it also served as an astronomical instrument for tracing the movements of the sun and the moon and for observing eclipses.

Definitions

A **circle** is the set of all points in a plane that are the same distance from a given point, called the **center**. This distance is equal to the length of any segment, called a **radius**, which joins the center to a point on the circle.

The word *radius* can also denote the *length* of the radius. Thus, *radius 5* means "radius of length 5." In a coordinate plane, the distance formula, $d = \sqrt{(x_2 - x_1)^2 + (y_2 - y_1)^2}$, can be used to write an equation of the circle of radius 5 shown below, which has its center at the origin. Let $P(x,y)$ be any point on the circle. Then use the distance formula to express the distance between $C(0,0)$ and $P(x,y)$, which is already known to be 5.

$$\sqrt{(x - 0)^2 + (y - 0)^2} = 5$$
$$\sqrt{x^2 + y^2} = 5$$
$$(\sqrt{x^2 + y^2})^2 = 5^2$$
$$x^2 + y^2 = 25$$

Thus, an equation of the circle of radius 5 having its center at the origin is $x^2 + y^2 = 5^2$, or $x^2 + y^2 = 25$.

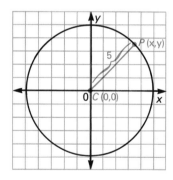

12.1 The Circle   **409**

Prerequisite Quiz

Use the distance formula to find the distance between the two given points.

1. $P(0,0)$, $Q(3,4)$ 5
2. $P(0,0)$, $Q(5,-12)$ 13
3. $P(1,1)$, $Q(5,5)$ $4\sqrt{2}$

Simplify.

4. $(\sqrt{5})^2$ 5
5. $\sqrt{(x^2 + y^2)^2}$ $x^2 + y^2$

Motivator

Ask students why the equation $x^2 + y^2 = 25$ yields *two* values of y for $x = 3$.   Positive and negative values will solve the equation when squared.   What are they?   4, −4

Have students make a table of x- and y-values for the equation above using the following values of x: 5, 4, 3, 0, −3, −4, −5. Plot the points corresponding to these ordered pairs. Draw a smooth curve through them. What does the shape appear to be? A circle.

Highlighting the Standards

Standard 8a: The circle has been studied from various perspectives through different levels of mathematics. This study can help students grow in mathematical understanding.

Lesson Note

For students who have already studied geometry, this lesson will be a good review of the properties of circles. Correlate the idea of locus, studied in geometry, with the set of points, $P(x,y)$, that satisfy an equation.

You can have students verify that the perpendicular bisector of a chord of a circle passes through the center of the circle. The students can find the equation of the perpendicular bisector of chord $\overline{AB}$, and then show that $(0,0)$ satisfies this equation.

Math Connections

Circular Motion: Objects in circular motion often produce strange effects. In an experiment, a man stood at the center of a turntable holding his arms extended straight out to his side. When he folded his arms close to his body, his rotational speed more than doubled. The phenomenon illustrates the physical principle called the *conservation of angular momentum*.

Critical Thinking Questions

Analysis: Ask students to describe the set of points described by the equation of Theorem 12.2 for the case $r = 0$. Since the radius of a circle cannot be zero, some students may think that the set is empty. However, they should understand that the set of points in question consists of the single point (h,k). Thus, a point may be thought of as a "degenerate" circle, a circle with a radius of zero.

Theorem 12.1 | The standard form of the equation of a circle of radius r with its center at the origin is $x^2 + y^2 = r^2$.

EXAMPLE 1 Write the equation of the circle centered at the origin and passing through the point $A(12,5)$. Graph the circle.

Plan Substitute 12 for x and 5 for y in the equation above to find r^2.

Solution
$$x^2 + y^2 = r^2$$
$$12^2 + 5^2 = r^2$$
$$144 + 25 = 169 = r^2; \; r = \sqrt{169}, \text{ or } 13$$

Thus, the equation is $x^2 + y^2 = 169$. Plot convenient points 13 units from the origin and draw the circle.

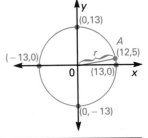

Circles can be *translated* (moved without rotation) so that their centers are not at the origin. In the figure at the right, circle S of radius 3 is centered at the origin. Circle T is a translation of circle S to $T(4, -2)$. The length of the radius, however, remains the same.

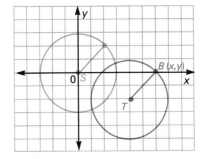

An equation of circle T can be found using the distance formula. Let $B(x,y)$ be any point on circle T. The distance between $T(4, -2)$ and $B(x,y)$ is 3. Therefore,

$$\sqrt{(x - 4)^2 + [y - (-2)]^2} = 3$$
$$\sqrt{(x - 4)^2 + (y + 2)^2} = 3$$
$$(x - 4)^2 + (y + 2)^2 = 9 \quad \text{Square each side.}$$

Notice the following relationship: $\qquad (x - 4)^2 + (y + 2)^2 = 9$

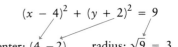

center: $(4, -2)$ radius: $\sqrt{9} = 3$

Theorem 12.2 | The standard form of the equation of a circle of radius r with its center at $C(h,k)$ is $(x - h)^2 + (y - k)^2 = r^2$.

Additional Example 1

Write an equation of a circle in standard form with its center at the origin and passing through the point $A(-6,-8)$. $x^2 + y^2 = 100$

EXAMPLE 2   Write, in standard form, the equation of the circle centered at $C(-4,5)$ and passing through the point $P(2,10)$. Graph the circle.

Solution   Substitute 2 for x, 10 for y, -4 for h, and 5 for k in the standard form.

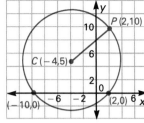

$$(x - h)^2 + (y - k)^2 = r^2$$
$$[2 - (-4)]^2 + (10 - 5)^2 = r^2$$
$$6^2 + 5^2 = r^2$$
$$61 = r^2$$
$$r = \sqrt{61} \approx 7.8$$

Thus, the standard form of the equation of this circle is $(x + 4)^2 + (y - 5)^2 = 61$.

Plot several points 7.8 units from $C(-4,5)$. Draw the graph as shown.

EXAMPLE 3   Rewrite $x^2 - 8x + y^2 + 4y + 16 = 0$ in standard form. Identify the circle's center and radius.

Plan   Isolate the x- and y-terms. Then complete the squares in both x and y.

Solution
$$x^2 - 8x + \underline{\quad\quad} + y^2 + 4y + \underline{\quad\quad} = -16$$
$$x^2 - 8x + \left(\tfrac{1}{2} \cdot 8\right)^2 + y^2 + 4y + \left(\tfrac{1}{2} \cdot 4\right)^2 = -16 + 16 + 4$$
$$(x - 4)^2 + (y + 2)^2 = 4 \leftarrow r^2 = 4; r = 2$$

Standard form: $(x - 4)^2 + (y + 2)^2 = 4$; center: $(4, -2)$; radius: 2

Classroom Exercises

Give the radius and the coordinates of the center of each circle. Graph each circle.

1. $x^2 + y^2 = 49$ **2.** $x^2 + y^2 = 19$
3. $(x - 4)^2 + (y - 1)^2 = 100$ **4.** $(x - 1)^2 + y^2 = 144$

Written Exercises

Write, in standard form, the equation of the circle with center C and radius r given, or with center C and a point P on the circle given.

1. $C(0,0)$, $r = 13$ **2.** $C(0,0)$, $r = 3.1$ **3.** $C(2,3)$, $P(8,0)$ **4.** $C(4,2)$, $P(0,8)$
5. $C(-2,8)$, $r = 9$ **6.** $C(1,1)$, $r = \sqrt{3}$ **7.** $C(1,-1)$, $P\left(\tfrac{1}{2},1\right)$ **8.** $C\left(\tfrac{1}{4},\tfrac{1}{4}\right)$, $P\left(1,-1\right)$
9–16. Graph each circle in Written Exercises 1–8.

Common Error Analysis

Error: When writing the equation of a circle in standard form, students tend to forget to square the radius. Thus, for the circle with center $(4,-2)$ and radius 3, they may incorrectly write the equation as $(x - 4)^2 + (y + 2)^2 = 3$.

Another typical error is to write the square of $\sqrt{5}$ as 25 rather than as 5.

Emphasize that $(\sqrt{a})^2$ is a, not a^2.

Checkpoint

Write an equation, in standard form, of each circle described by the given conditions.

1. center: $(0,0)$, radius: 12 $x^2 + y^2 = 144$
2. center: $(0,0)$; radius: $\sqrt{7}$ $x^2 + y^2 = 7$
3. center: $(0,0)$; passing through $P(5,-2)$ $x^2 + y^2 = 29$
4. center: $(-2,-4)$; radius: 5 $(x + 2)^2 + (y + 4)^2 = 25$

Closure

Ask students to give the standard form of the equation of a circle, the coordinates of the center, and the radius. $(x - h)^2 + (y - k)^2 = r^2$; (h,k); r Have students give the steps in writing an equation like $x^2 - 6x + y^2 - 14y - 42 = 0$ so that the circle can be graphed.   See Example 3.

Additional Example 2

Write an equation of the circle in standard form with center $C(2,-5)$ and passing through the point $P(7,7)$. Graph the circle.
$(x - 2)^2 + (y + 5)^2 = 169$

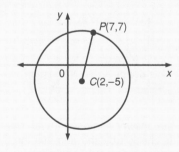

Additional Example 3

Rewrite $x^2 - 4x + y^2 + 8y - 16 = 0$ in standard form. Identify the center and radius.
$(x - 2)^2 + (y + 4)^2 = 36$; $C(2,-4)$; $r = 6$

Guided Practice

Classroom Exercises 1–4

Independent Practice

A Ex. 1–24, B Ex. 25–31, C Ex. 32–38

Basic: WE 1–25 odd

Average: WE 7–31 odd

Above Average: WE 13–37 odd

Additional Answers

Classroom Exercises

1.

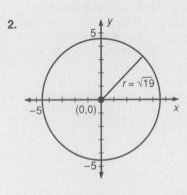

2.

y
5
$r = \sqrt{19}$
–5   (0,0)   x
–5

See page 455 for the answers to Classroom Ex. 3 and 4 and pages 817–819 for Written Ex. 1–28, 30, 31, 33, 36, 38, and Mixed Rev. Ex. 1 and 2.

Graph, rewriting each equation in standard form if necessary. Identify the circle's center and radius.

17. $x^2 + y^2 = 4$ 18. $x^2 + y^2 = 25$ 19. $x^2 + y^2 = 64$ 20. $x^2 + y^2 = 81$

21. $x^2 + y^2 = 8$ 22. $x^2 + y^2 = 12$ 23. $x^2 + y^2 = 22$ 24. $x^2 + y^2 = 15$

25. $x^2 - 6x + y^2 - 10y - 2 = 0$ 26. $x^2 + 8x + y^2 - 4y - 5 = 0$

27. $x^2 + 6x + y^2 - 14y - 42 = 0$ 28. $x^2 - 12x + y^2 - 2y - 8 = 0$

29. A point $P(x,y)$ moves in a coordinate plane about a fixed point $C(-1,2)$ so that its distance from C is always $2\sqrt{3}$ units. Use the distance formula, $d = \sqrt{(x_2 - x_1)^2 + (y_2 - y_1)^2}$, to write an equation describing the path of point P. $(x + 1)^2 + (y - 2)^2 = 12$

Use the distance formula for Exercises 30 and 31.

30. Prove Theorem 12.1.   31. Prove Theorem 12.2.

32. The circle with equation $(x - 1)^2 + (y + 1)^2 = 9$ is translated 4 units to the right and 2 units downward. Find the equation of the resulting circle. $(x - 5)^2 + (y + 3)^2 = 9$

33. The circle with equation $x^2 - 8x + y^2 + 4y + 9 = 0$ is translated 2 units to the left and 5 units upward. Find the equation of the resulting circle. $(x - 2)^2 + (y - 3)^2 = 11$

Write, in standard form, the equation of the circle having the given properties.

34. center on the line $x = 5$, tangent to the y-axis at $P(0,9)$ $(x - 5)^2 + (y - 9)^2 = 25$

35. tangent to both rays of the graph of $y = |x|$, radius 4
(HINT: The slope of $y = x$ is one, so the triangle formed by the origin, the circle's center, and the point of tangency is isosceles.) $x^2 + (y - 4\sqrt{2})^2 = 16$

Use the circle $x^2 + y^2 = 100$ and triangle ABC for Exercises 36–38.

36. Show that the vertices $A(-6,8)$, $B(0,10)$, and $C(6,-8)$ of triangle ABC are points on the circle, and that the triangle is a right triangle.

37. Write an equation of the line tangent to the circle at $A(-6,8)$. $3x - 4y = -50$

38. Show that the perpendicular bisector of $\overline{AB}$ passes through the center of the circle.

Mixed Review

1. Graph $y = |x - 4| + 5$. **11.3**   2. Graph $y = x^2 - 8x + 2$. **11.5**

Simplify. **5.2, 8.2, 8.6, 9.2**

3. $(2x - 4)^0, x \neq 2$ 1   4. $\sqrt{32x^5}$ $4x^2\sqrt{2x}$ 5. $27^{-\frac{2}{3}}$ $\frac{1}{9}$ 6. $(3 + 4i)^2$
$-7 + 24i$

Enrichment

Challenge students to find the coordinates of the center of the circle that contains the points (1,2), (7,8), and (13,0).

You may wish to draw the figure in the center column on the chalkboard and give the hint that the perpendicular bisector of every chord of a circle passes through the center of the circle.

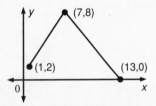

When the answer has been found, have the students check it by showing that the three points are equidistant from the center.

The center is $\left(\frac{50}{7}, \frac{13}{7}\right)$. The distance of each point from the center is $\frac{\sqrt{1850}}{7}$, or $\frac{5\sqrt{74}}{7}$ units.

12.2 The Ellipse

Objectives
To graph ellipses given their equations
To write equations of ellipses given certain conditions

The figure at the right shows a 20-cm piece of string fastened to two points, F_1 and F_2, 16 cm apart. A pencil holds the string taut. As it moves, it generates the points P_1, P_2, P_3, and so on. These points form an "egg-shaped" curve called an *ellipse*. At each point on this curve, the sum of the distance to F_1 and the distance to F_2 is always 20 cm, the length of the string. For example, $P_3F_1 + P_3F_2 = 20$. This experiment suggests the following definitions.

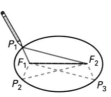

Definitions

An **ellipse** is the set of all points in a plane such that for each point, the sum of the distances from two fixed points is constant. Each fixed point is called a **focus** (plural: *foci*). The distance from any point on the ellipse to a focus is called a **focal radius**.

In the ellipse at the right, the foci are $F_1(-8,0)$ and $F_2(8,0)$. The midpoint of $\overline{F_1F_2}$ is the *center*, $C(0,0)$, of the ellipse. The sum of the distances from any point $P(x,y)$ on the ellipse to the foci is 20. The equation of the ellipse can be written using the definition above and the distance formula.

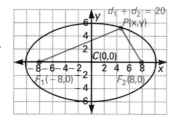

Use the definition of ellipse. $d_1 + d_2 = 20$

Substitute. $\sqrt{(x+8)^2 + y^2} + \sqrt{(x-8)^2 + y^2} = 20$

Isolate a radical. $\sqrt{x^2 + 16x + 64 + y^2} = 20 - \sqrt{x^2 - 16x + 64 + y^2}$

Square each side. $x^2 + 16x + 64 + y^2 = 400 - 40\sqrt{x^2 - 16x + 64 + y^2} + x^2 - 16x + 64 + y^2$

Isolate a radical. $32x - 400 = -40\sqrt{x^2 - 16x + 64 + y^2}$

Divide each side by 8. $4x - 50 = -5\sqrt{x^2 - 16x + 64 + y^2}$

Square each side. $16x^2 - 400x + 2{,}500 = 25x^2 - 400x + 1{,}600 + 25y^2$

12.2 The Ellipse   **413**

Teaching Resources

Application Worksheet 12
Project Worksheet 12
Quick Quizzes 83
Reteaching and Practice Worksheets 83
Transparencies 27A, 27B, 27C, 27D, 27E

GETTING STARTED

Prerequisite Quiz

Find the indicated lengths. Refer to the figure below.

1. $b = 6$, $f = 8$, $a = \underline{\ ?\ }$ 10
2. $a = 13$, $b = 12$, $f = \underline{\ ?\ }$ 5
3. $b = 2$, $f = 4$, $a = \underline{\ ?\ }$ $2\sqrt{5}$
4. Simplify $(4 - \sqrt{2x-5})^2$.
   $11 - 8\sqrt{2x-5} + 2x$

Motivator

Ask students if they think that the graph of $16x^2 + 25y^2 = 400$ will be a circle. Have them graph some points. HINT: Find the two y-values for $x = 0$ and x-values for $y = 0$. Then have them find y-values for $x = 4$ and $x = -4$ and draw a smooth curve through the points. Now, ask students to name its shape.   Oval

Highlighting the Standards

Standard 2d: This lesson on the ellipse, as an extension of the work on the circle, can be used to see how well students read mathematics with understanding.

Lesson Note

Emphasize that in the general equation of the ellipse with center at the origin, a is the length of the semi-major axis and a is always greater than b.

The students should memorize the following patterns.

$$\frac{x^2}{a^2} + \frac{y^2}{b^2} = 1 \qquad \frac{y^2}{a^2} + \frac{x^2}{b^2} = 1$$

Notice that in both cases the length of the hypotenuse is equal to a.

Math Connections

Planetary Orbits: In the early seventeenth century, Johannes Kepler published his three laws of planetary motion. (Sir Isaac Newton relied heavily on these laws while formulating his own discoveries concerning gravitation late in the century.) Two of Kepler's laws involve the ellipse. The first law states that the orbit of a planet is an ellipse with the sun at one of the two foci. The second law says that the focal radius connecting a planet to one focus of its orbit sweeps through equal areas in equal times.

Critical Thinking Questions

Analysis: Ask students why it is important to "isolate a radical" when deriving the equation of an ellipse. (See, for example, page 413.) They should realize that if this is not done, then the algebraic manipulations become completely unmanageable.

The resulting equation, $9x^2 + 25y^2 = 900$, can be analyzed to reveal facts about the ellipse it describes.

Divide each side by 900 and simplify. $\dfrac{x^2}{100} + \dfrac{y^2}{36} = 1$

Rewrite the denominators as squares. $\dfrac{x^2}{10^2} + \dfrac{y^2}{6^2} = 1$

If $y = 0$, then $x = \pm 10$. The x-intercepts of the ellipse are ± 10.

If $x = 0$, then $y = \pm 6$. The y-intercepts are ± 6.

In the diagram at the right, $\overline{V_1V_2}$ is the **major axis** of the ellipse, and $\overline{B_1B_2}$ is the **minor axis**. The points B_1, B_2, V_1, and V_2 are the **vertices** of the ellipse. Notice that the numbers 10 and 6 in the equation of the ellipse are, respectively, the lengths of *half* the major axis $\overline{V_1V_2}$ and *half* the minor axis $\overline{B_1B_2}$.

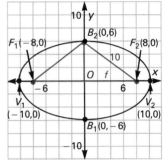

Another relationship can be found by calculating B_2F_1 and B_2F_2.

By the distance formula,
$$B_2F_1 = \sqrt{[0 - (-8)]^2 + (6 - 0)^2} = \sqrt{100} = 10,$$
and $B_2F_2 = \sqrt{(0 - 8)^2 + (6 - 0)^2} = \sqrt{100} = 10.$

Therefore, both B_2F_1 and B_2F_2 are equal to OV_2, half the length of the major axis.

In the figure, triangle OB_2F_2 is a right triangle. The Pythagorean relation can be used to relate the lengths 6, 10, and the distance f from the origin to the focus F_2. That relationship is $f^2 + 6^2 = 10^2$, so $f = 8$.

■■■ *Theorem 12.3*

The equation, in standard form, of an ellipse with its center at the origin and a horizontal major axis is

$$\frac{x^2}{a^2} + \frac{y^2}{b^2} = 1, \text{ where } 0 < b < a.$$

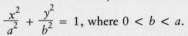

The length of the *semimajor axis* is a.

The length of the *semiminor axis* is b.

The coordinates of the vertices are $(\pm a, 0)$ and $(0, \pm b)$.

The coordinates of the foci are $(\pm f, 0)$, where $f^2 + b^2 = a^2$, or $f = \sqrt{a^2 - b^2}$.

414 Chapter 12 Conic Sections

EXAMPLE 1

Graph $16x^2 + 25y^2 = 400$. Find the coordinates of the ellipse's vertices and foci.

Solution

Write in standard form.

$$\frac{x^2}{25} + \frac{y^2}{16} = 1$$

Plot the vertices: $(\pm 5,0)$, $(0,\pm 4)$
Draw the graph.

$$f^2 + 4^2 = 5^2$$

$$f = \sqrt{25 - 16} = \sqrt{9} = 3$$

The coordinates of the foci are $(\pm 3,0)$.

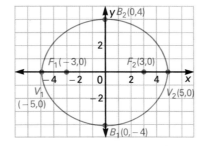

The major axis of an origin-centered ellipse can also lie along the y-axis. In this case, the foci will be points on the y-axis. The resulting changes in the standard form of the equation of an ellipse are outlined in the following theorem.

Theorem 12.4

The equation of an ellipse with its center at the origin and a vertical major axis is

$$\frac{y^2}{a^2} + \frac{x^2}{b^2} = 1, \text{ where } 0 < b < a.$$

The length of the semimajor axis is a.

The length of the semiminor axis is b.

The coordinates of the vertices are $(\pm b,0)$ and $(0,\pm a)$.

The coordinates of the foci are $(0,\pm f)$, where $f = \sqrt{a^2 - b^2}$.

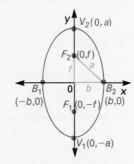

Both Theorems 12.3 and 12.4 can be used to write equations for origin-centered ellipses given the coordinates of their vertices, or given those of a vertex and a focus. Sketching each ellipse can help determine whether its major axis (length a) is horizontal or vertical, and thus whether a^2 is the denominator of x^2 or y^2 in the equation.

EXAMPLE 2

Given the following information about each ellipse, write its equation in standard form.

a. Vertices: $(\pm 8,0)$, $(0,\pm 5)$

b. Vertices: $(0,\pm 4)$; Foci: $(0,\pm 2)$

Common Error Analysis

Error: Students tend to think that for an equation like $25x^2 + 16y^2 = 400$, the major axis is horizontal since $25 > 16$ and 25 is the coefficient of the x-term.

Point out that the above equation is *not* in standard form. To put the equation in standard form, each side must be divided by 400.

$$\frac{x^2}{16} + \frac{y^2}{25} = 1$$

Now 25, the larger number, is the denominator of the y-term. Therefore, the major axis is vertical, or along the y-axis.

Checkpoint

Graph the ellipse. Find the coordinates of the vertices and foci.

1. $\frac{x^2}{100} + \frac{y^2}{49} = 1$

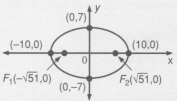

Write an equation, in standard form, of each ellipse.

2. Vertices: $(\pm 2,0)$, $(0,\pm 1)$

$$\frac{x^2}{4} + \frac{y^2}{1} = 1$$

3. Vertices: $(\pm 5,0)$
   foci: $(0,\pm 4)$

$$\frac{y^2}{41} + \frac{x^2}{25} = 1$$

Additional Example 1

Graph $4x^2 + 9y^2 = 36$. Find the coordinates of the vertices and foci.

Vertices: $(\pm 3,0)$, $(0,\pm 2)$
Foci: $(\pm\sqrt{5},0)$

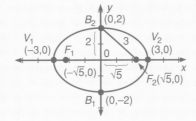

Additional Example 2

Information about an ellipse is given. In each case write, in standard form, an equation of the ellipse.

a. Vertices: $(\pm 6,0)$, $(0,\pm\sqrt{5})$

$$\frac{x^2}{36} + \frac{y^2}{5} = 1$$

b. Vertices: $(0,\pm 6)$; Foci: $(0,\pm 3)$

$$\frac{y^2}{36} + \frac{x^2}{27} = 1$$

Closure

Ask students to give a general equation of an ellipse with its center at the origin and a horizontal axis.   See Theorem 12.3.
a vertical axis.   See Theorem 12.4.   Ask them how they would find the coordinates of the foci.   See Example 1.

▰▰▰▰FOLLOW UP

Guided Practice

Classroom Exercises 1–8

Independent Practice

A Ex. 1–18, **B** Ex. 19–24, **C** Ex. 25–29

Basic: WE 1–19 odd
Average: WE 5–23 odd
Above Average: WE 9–29 odd

Additional Answers

Classroom Exercises

1. Vert; $a = 4$, $b = 2$
2. Horiz; $a = 7$, $b = 5$
3. Vert; $a = 10$, $b = 4$
4. Horiz; $a = 9$, $b = 3$

Written Exercises

1.

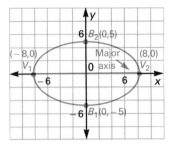

Solutions   First, sketch each ellipse.

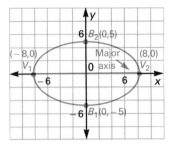

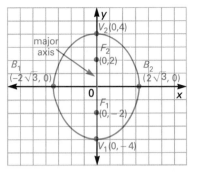

Major axis is horizontal: $8 > 5$

Use $\dfrac{x^2}{a^2} + \dfrac{y^2}{b^2} = 1$, with $a = 8$ and $b = 5$.

$\dfrac{x^2}{8^2} + \dfrac{y^2}{5^2} = 1$, or $\dfrac{x^2}{64} + \dfrac{y^2}{25} = 1$

The equation is $\dfrac{x^2}{64} + \dfrac{y^2}{25} = 1$.

Foci lie on the y-axis so major axis is vertical.

Use $\dfrac{y^2}{a^2} + \dfrac{x^2}{b^2} = 1$, with $a = 4$.

$f = \sqrt{a^2 - b^2}$

$2 = \sqrt{4^2 - b^2}$, so $4 = 16 - b^2$, or $b^2 = 12$

The equation is $\dfrac{y^2}{16} + \dfrac{x^2}{12} = 1$.

Classroom Exercises

Indicate whether the major axis is horizontal or vertical. Find the lengths of the semimajor axis *a* and the semiminor axis *b*.

1. $\dfrac{y^2}{16} + \dfrac{x^2}{4} = 1$ **2.** $\dfrac{x^2}{49} + \dfrac{y^2}{25} = 1$ **3.** $\dfrac{y^2}{100} + \dfrac{x^2}{16} = 1$ **4.** $\dfrac{x^2}{81} + \dfrac{y^2}{9} = 1$

5. $\dfrac{x^2}{49} + \dfrac{y^2}{4} = 1$ **6.** $\dfrac{y^2}{144} + \dfrac{x^2}{64} = 1$ **7.** $\dfrac{x^2}{25} + \dfrac{y^2}{4} = 1$ **8.** $\dfrac{x^2}{5} + \dfrac{y^2}{3} = 1$
    Horiz; $a = 7$, $b = 2$ Vert; $a = 12$, $b = 8$ Horiz; $a = 5$, $b = 2$ Horiz; $a = \sqrt{5}$, $b = \sqrt{3}$

Written Exercises

Graph each ellipse. Give the coordinates of the vertices and the foci.

1. $\dfrac{x^2}{25} + \dfrac{y^2}{9} = 1$ **2.** $\dfrac{y^2}{25} + \dfrac{x^2}{16} = 1$ **3.** $\dfrac{y^2}{169} + \dfrac{x^2}{25} = 1$
4. $4x^2 + 25y^2 = 100$ **5.** $4x^2 + 49y^2 = 196$ **6.** $4x^2 + y^2 = 4$
7. $x^2 + 25y^2 = 25$ **8.** $9x^2 + y^2 = 36$ **9.** $16x^2 + 9y^2 = 144$
10. $x^2 + 4y^2 = 4$ **11.** $25x^2 + 4y^2 = 100$ **12.** $100x^2 + 36y^2 = 3{,}600$

416   Chapter 12 Conic Sections

Given the following information, write an equation in standard form of each ellipse.

13. 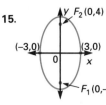 14. 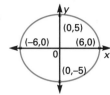 15.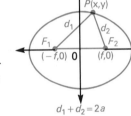

16. vertices: $(\pm 5, 0)$, $(0, \pm 8)$ $\frac{y^2}{64} + \frac{x^2}{25} = 1$

17. vertices: $(\pm\sqrt{5}, 0)$, $(0, \pm\sqrt{3})$ $\frac{x^2}{5} + \frac{y^2}{3} = 1$

18. vertices: $(\pm\sqrt{7}, 0)$, $(0, \pm\sqrt{6})$ $\frac{x^2}{7} + \frac{y^2}{6} = 1$

19. vertices: $(\pm 8, 0)$ foci: $(\pm 3, 0)$ $\frac{x^2}{64} + \frac{y^2}{55} = 1$

20. vertices: $(\pm 6, 0)$ foci: $(\pm 4, 0)$ $\frac{x^2}{36} + \frac{y^2}{20} = 1$

21. vertices: $(0, \pm 8)$ foci: $(0, \pm 3)$ $\frac{y^2}{64} + \frac{x^2}{55} = 1$

22. vertices: $(0, \pm 5)$ foci: $(0, \pm 2)$ $\frac{y^2}{25} + \frac{x^2}{21} = 1$

23. vertices: $(\pm 4, 0)$ foci: $(0, \pm 3)$ $\frac{y^2}{25} + \frac{x^2}{16} = 1$

24. vertices: $(0, \pm 8)$ foci: $(\pm 5, 0)$ $\frac{x^2}{89} + \frac{y^2}{64} = 1$

25. Find the standard form of the equation of the ellipse with foci $(0, \pm 3)$ and the sum of whose focal radii is 10. $\frac{y^2}{25} + \frac{x^2}{16} = 1$

26. Find the standard form of the equation of the ellipse with vertices $(\pm 2, 0)$ and that contains the point $\left(1, -\frac{\sqrt{3}}{2}\right)$. $\frac{x^2}{4} + y^2 = 1$

27. Given the ellipse with equation $9x^2 + 36y^2 = 324$ and the circle with equation $x^2 - 2x + y^2 - 2y + 1 = 0$ in the same coordinate plane, find the area of the region inside the ellipse, but not inside the circle. (HINT: Area of the ellipse $= \pi ab$) 17π

28. Prove Theorem 12.3 using the definition of an ellipse and the figure at the right. (HINT: Let the two fixed points, the foci, be $(\pm f, 0)$. Let $P(x, y)$ be any point on the ellipse. Let the sum of the distances d_1 and d_2 be $2a$, $a > 0$. At the final stage of development, group the terms that contain $a^2 - f^2$ as a factor. Then substitute b^2 for $a^2 - f^2$ so that b becomes the length of the semiminor axis.)

29. Prove Theorem 12.4. (HINT: Let the foci be $(0, \pm f)$. Then refer to the hint for Exercise 28.)

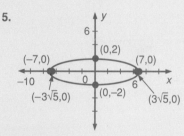

Mixed Review

Graph each equation. *3.5, 11.3, 11.5, 12.1*

1. $4x - 3y + 12 = 0$
2. $y = 2|x - 5| - 4$
3. $x^2 + 6x - 8y + 25 = 0$
4. $x^2 - 6x + y^2 + 4y - 23 = 0$

12.2 The Ellipse **417**

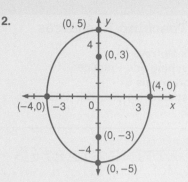

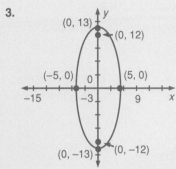

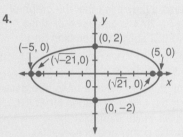

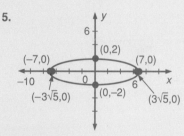

See pages 819–820 for the answers to Written Ex. 6–15, 28–29, and Mixed Review Ex. 1–4.

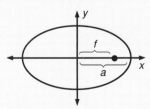
417

GETTING STARTED

Prerequisite Quiz

1. Find algebraically the x-intercepts of
$\dfrac{x^2}{9} - \dfrac{y^2}{16} = 1$. ± 3

2. Solve $\sqrt{x + 12} - \sqrt{2x + 1} = 1$. 4

3. Solve $16x^2 - 9y^2 = 144$ for y in terms of x.

$$y = \pm \frac{4}{3}\sqrt{x^2 - 9}$$

Motivator

Ask students if they think that the graph of $16x^2 - 9y^2 = 144$ is an ellipse or circle. Have them first solve the equation for y.

$$y = \pm \frac{4}{3}\sqrt{x^2 - 9}$$

Then have students make a table of ordered pairs using the following values for x: ± 3, ± 5, ± 8, ± 10. Have them plot the corresponding points and observe the graph.

12.3 The Hyperbola

Objectives

To graph hyperbolas given their equations
To write equations of hyperbolas given certain conditions
To find the vertices, foci, and equations of the asymptotes of hyperbolas given their equations

Consider the following definitions.

Definition

A **hyperbola** is the set of all points in a plane such that from each point, the difference of the distances to two fixed points is constant. The fixed points are called the **foci**, and the distances are called **focal radii**.

The hyperbola at the right has foci $F_1(-5,0)$ and $F_2(5,0)$. Let $P(x,y)$ represent any point on the hyperbola. Given that the difference of the distances from $P(x,y)$ to the foci is 6, an equation for the hyperbola can be written using the definition above and the distance formula.

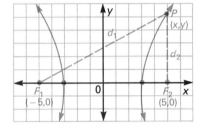

$$d_1 = \sqrt{[x - (-5)]^2 + (y - 0)^2} \qquad d_2 = \sqrt{(x - 5)^2 + (y - 0)^2}$$

$$d_1 - d_2 = \sqrt{x^2 + 10x + 25 + y^2} \quad - \quad \sqrt{x^2 - 10x + 25 + y^2} = 6$$

Simplify this equation in the same manner as that of the ellipse in the previous lesson. The result is $16x^2 - 9y^2 = 144$.

The equation above can be solved for y to give the equation $y = \pm \frac{4}{3}\sqrt{x^2 - 9}$. Because $x^2 - 9$ must be nonnegative for y to be real, x cannot assume values between -3 and 3. Thus, the domain of the relation $16x^2 - 9y^2 = 144$ is $\{x \mid x \le -3 \ or \ x \ge 3\}$. Note that the two extreme values, $x = \pm 3$, occur when the two *branches* of the hyperbola intersect the x-axis at the two *vertices*, $V_1(-3,0)$ and $V_2(3,0)$.

The equation $y = \pm \frac{4}{3}\sqrt{x^2 - 9}$ can also be rewritten as follows.

$$y = \pm \frac{4}{3}\sqrt{x^2 - 9} = \pm \frac{4}{3}\sqrt{x^2\left(1 - \frac{9}{x^2}\right)} = \pm \frac{4x}{3}\sqrt{1 - \frac{9}{x^2}}$$

Highlighting the Standards

Standard 14b: This third use of the Distance Formula in the development of conic sections can lead to a deeper understanding of algebraic procedures.

When $|x|$ is very large, $\dfrac{9}{x^2}$ is very close to zero and $\sqrt{1 - \dfrac{9}{x^2}}$ is very close to 1. Therefore, $y = \pm\dfrac{4x}{3}\sqrt{1 - \dfrac{9}{x^2}} \approx \pm\dfrac{4x}{3}$.

In the figure below, as $|x|$ gets very large, each branch of the hyperbola gets very close to the two lines $y = \pm\dfrac{4}{3}x$. These lines are called the **asymptotes**. Their slopes can be found as follows.

Original equation: $16x^2 - 9y^2 = 144$

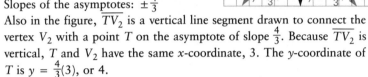

Divide each side by 144.

$$\dfrac{16x^2}{144} - \dfrac{9y^2}{144} = \dfrac{144}{144}$$

$$\dfrac{x^2}{9} - \dfrac{y^2}{16} = 1$$

$$\dfrac{x^2}{3^2} - \dfrac{y^2}{4^2} = 1$$

Slopes of the asymptotes: $\pm\dfrac{4}{3}$

Also in the figure, $\overline{TV_2}$ is a vertical line segment drawn to connect the vertex V_2 with a point T on the asymptote of slope $\dfrac{4}{3}$. Because $\overline{TV_2}$ is vertical, T and V_2 have the same x-coordinate, 3. The y-coordinate of T is $y = \dfrac{4}{3}(3)$, or 4.

Note that, by the distance formula, $OT = \sqrt{3^2 + 4^2} = 5$. Thus, OT is equal to f, the distance from the center of the hyperbola, $(0,0)$, to either of the foci, F_1 or F_2.

Theorem 12.5

The equation, in standard form, of a hyperbola centered at the origin with its foci on the x-axis is

$\dfrac{x^2}{a^2} - \dfrac{y^2}{b^2} = 1$, where $a > 0$ and $b > 0$.

The coordinates of the vertices are $(\pm a, 0)$.

The coordinates of the foci are $(\pm f, 0)$, where $f = \sqrt{a^2 + b^2}$.

The slopes of the asymptotes are $\pm\dfrac{b}{a}$.

The equations of the asymptotes are $y = \pm\dfrac{b}{a}x$.

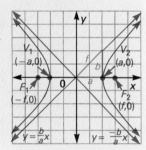

Lesson Note

The meaning of asymptote is taught using an intuitive limit concept. As shown on page 419, $1 - \dfrac{9}{x^2}$ gets very close to 1 as x becomes very large. This happens because $\dfrac{9}{x^2}$ gets very close to zero as x increases without bound.

In drawing the graph of a hyperbola, emphasize that as $|x|$ gets very large, the branches of the hyperbola get very close to the asymptotes. Thus, for very large numerical values of x, the graph of each branch of the hyperbola is almost the same as the graph of the asymptotes.

Math Connections

Hyperbolic Functions: Ordinary trigonometric functions (Chapters 17–19) are also called *circular functions* because they can be defined in terms of the unit circle $x^2 + y^2 = 1$. A similar set of functions, the *hyperbolic functions*, can be defined based on the hyperbola $x^2 - y^2 = 1$. The relationships among the hyperbolic functions are similar, but not identical, to the relationships among the trigonometric functions. For example, the hyperbolic sine (sinh) and hyperbolic cosine (cosh) are related by the identity $\sinh^2 x - \cosh^2 x = 1$, which is analogous to the trigonometric identity $\sin^2 x + \cos^2 x = 1$.

Critical Thinking Questions

Analysis: Many students will know that a point can be considered as a "degenerate circle." For example, the equation $x^2 + y^2 = 0$ can be thought of as a "circle" of radius 0 centered at the origin, that is, the origin itself. Ask students what the "degenerate hyperbola" $x^2 - y^2 = 0$ represents. Students should see that this is the equation of two intersecting lines, $y = \pm x$.

Common Error Analysis

Error: In the ellipse, f is the length of a leg of a right triangle along the semi-major axis. For the hyperbola, f is the length of the hypotenuse. Students tend to confuse these two ideas.

It may help to draw contrasting diagrams of the two conic sections to emphasize the difference in the role played by f in each case.

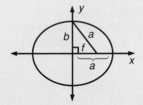

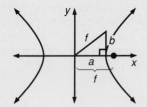

EXAMPLE 1   Graph $25x^2 - 4y^2 = 100$. Find the coordinates of the vertices and the foci. Write the equations of the asymptotes.

Solution   Divide each side by 100 to obtain the standard form.

$$\frac{25x^2}{100} - \frac{4y^2}{100} = \frac{100}{100}$$

$$\frac{x^2}{4} - \frac{y^2}{25} = 1 \leftarrow a = 2 \text{ and } b = 5$$

The vertices are $V(\pm 2, 0)$.

Because $f = \sqrt{a^2 + b^2} = \sqrt{4 + 25} = \sqrt{29}$, the foci are $F(\pm\sqrt{29}, 0)$.

The equations of the asymptotes are $y = \pm\frac{5}{2}x$.

Draw the asymptotes and plot the vertices and foci. It is possible to find a few other points on the hyperbola by solving for y and making a table as shown below.

x	$y = \pm\frac{5}{2}\sqrt{x^2 - 4}$
± 2	0
± 2.5	± 3.8
± 3	± 5.6

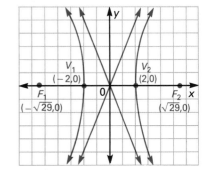

Draw the graph so that as $|x|$ increases, the hyperbola approaches the asymptotes without touching them.

If the foci of a hyperbola lie on the y-axis, the following is true.

Theorem 12.6   The equation, in standard form, of a hyperbola centered at the origin with its foci on the y-axis is $\frac{y^2}{a^2} - \frac{x^2}{b^2} = 1$, where $a > 0$ and $b > 0$.

The coordinates of the vertices are $(0, \pm a)$. The coordinates of the foci are $(0, \pm f)$, where $f = \sqrt{a^2 + b^2}$. The slopes of the asymptotes are $\pm\frac{a}{b}$. The equations of the asymptotes are $y = \pm\frac{a}{b}x$.

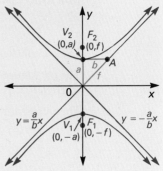

420 Chapter 12 Conic Sections

Additional Example 1

Write the equation of the hyperbola $4x^2 - 9y^2 = 36$ in standard form. Graph the hyperbola. Find the coordinates of the vertices and foci. Write the equations of the asymptotes.

$$\frac{x^2}{9} - \frac{y^2}{4} = 1$$

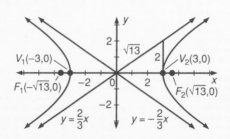

Vertices: $(\pm 3, 0)$

Foci: $(\pm\sqrt{13}, 0)$

Asymptotes: $y = \pm\frac{2}{3}x$

420

EXAMPLE 2   A hyperbola has vertices $(0, \pm 4)$ and foci $(0, \pm 5)$.

Write the equation of the hyperbola in standard form. Write the equations of the asymptotes.

Solution   Plot the vertices and foci. Sketch the hyperbola. The foci are points on the y-axis. So, $a = 4$ and $a^2 = 16$. $OF_2 = OA = f = 5$

Use $f = \sqrt{a^2 + b^2}$ to find b^2 and b.

$$f = \sqrt{a^2 + b^2}$$
$$5 = \sqrt{4^2 + b^2}$$
$$25 = 16 + b^2$$
$$9 = b^2, \text{ or } b^2 = 9 \quad \text{Thus, } b = 3.$$

Write the equation using the standard form. The equation of the hyperbola is
$$\frac{y^2}{16} - \frac{x^2}{9} = 1.$$

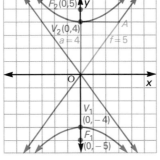

The equations of the asymptotes are $y = \pm\frac{4}{3}x$.

Classroom Exercises

For each hyperbola, indicate whether the foci are points on the x-axis or the y-axis.

1. $\frac{x^2}{4} - \frac{y^2}{9} = 1$ *x-axis*   **2.** $\frac{y^2}{36} - \frac{x^2}{16} = 1$ *y-axis*   **3.** $\frac{x^2}{25} - \frac{y^2}{1} = 1$ *x-axis*   **4.** $y^2 - \frac{x^2}{64} = 1$ *y-axis*

5–8. Give the slopes of the asymptotes of each hyperbola in Classroom Exercises 1–4. **5.** $\pm\frac{3}{2}$ **6.** $\pm\frac{3}{2}$ **7.** $\pm\frac{1}{5}$ **8.** $\pm\frac{1}{8}$

9–12. Graph each hyperbola in Classroom Exercises 1–4. Give the coordinates of the vertices and the foci. Write the equations of the asymptotes.

Written Exercises

Graph each hyperbola using the properties of a hyperbola stated in Theorems 12.5 and 12.6. Give the coordinates of the vertices and the foci. Write the equations of the asymptotes.

1. $\frac{x^2}{4} - \frac{y^2}{16} = 1$ **2.** $\frac{x^2}{49} - \frac{y^2}{9} = 1$ **3.** $\frac{x^2}{36} - \frac{y^2}{16} = 1$

4. $\frac{x^2}{25} - \frac{y^2}{100} = 1$ **5.** $\frac{y^2}{64} - \frac{x^2}{49} = 1$ **6.** $\frac{y^2}{121} - \frac{x^2}{16} = 1$

Checkpoint

Graph the hyperbola. Find the coordinates of the vertices and foci. Write the equations of the asymptotes.

1. $\frac{x^2}{4} - \frac{y^2}{1} = 1$

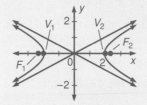

Vertices: $(\pm 2, 0)$
Foci: $(\pm\sqrt{5}, 0)$
Asymptotes: $y = \pm\frac{1}{2}x$

Given the following data, write an equation of each hyperbola in standard form.

2. Vertices: $(\pm 8, 0)$ $\frac{x^2}{64} - \frac{y^2}{36} = 1$
    Foci: $(\pm 10, 0)$

3. Vertices: $(0, \pm 4)$ $\frac{y^2}{16} - \frac{x^2}{20} = 1$
    Foci: $(0, \pm 6)$

Closure

Ask students to give the general equation of a hyperbola with its center at the origin and foci on the x-axis.   See Theorem 12.5. Repeat for the y-axis.   See Theorem 12.6. Have students give a definition of asymptotes.   See page 419.

Additional Example 2

A hyperbola has vertices with coordinates $(0, \pm 2)$ and foci $(0, \pm 4)$. Write the standard form of the equation of the hyperbola. Write the equations of the asymptotes.
$$\frac{y^2}{4} - \frac{x^2}{12} = 1$$
Asymptotes: $y = \pm\frac{\sqrt{3}}{3}x$

Guided Practice

Classroom Exercises 1–12

Independent Practice

A Ex. 1–12, **B** Ex. 13–20, **C** Ex. 21–23

Basic: WE 1–17 odd

Average: WE 5–21 odd

Above Average: WE 19–23 odd

Additional Answers

Classroom Exercises

9.

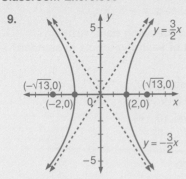

10.

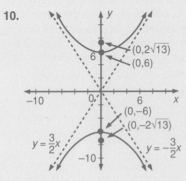

See pages 820–821 for the answers to Classroom Ex. 11–12, Written Ex. 1–12, 19, 20, 22, 23, and Mixed Review Ex. 1–2.

7. $x^2 - 16y^2 = 144$ 8. $9x^2 - 25y^2 = 225$ 9. $4x^2 - 16y^2 = 64$

10. $y^2 - x^2 = 144$ 11. $25x^2 - 36y^2 = 900$ 12. $36x^2 - y^2 = 144$

Use the information in Exercises 13–20 to write the equation of each hyperbola in standard form.

13.

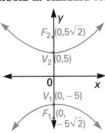

$\dfrac{y^2}{25} - \dfrac{x^2}{25} = 1$

14.

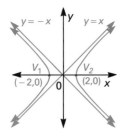

$\dfrac{x^2}{4} - \dfrac{y^2}{4} = 1$

15. vertices: $(\pm 4, 0)$
    foci: $(\pm 5, 0)$ $\dfrac{x^2}{16} - \dfrac{y^2}{9} = 1$

16. vertices: $(0, \pm 2)$
    foci: $(0, \pm 8)$ $\dfrac{y^2}{4} - \dfrac{x^2}{60} = 1$

17. vertices: $(0, \pm 5)$
    foci: $(0, \pm 7)$ $\dfrac{y^2}{25} - \dfrac{x^2}{24} = 1$

18. vertices: $(0, \pm 3)$
    foci: $(0, \pm 5)$ $\dfrac{y^2}{9} - \dfrac{x^2}{16} = 1$

19. equations of asymptotes: $5y + 2x = 0$, $5y - 2x = 0$; vertices: $(\pm 10, 0)$

20. equations of asymptotes: $2y + 7x = 0$, $2y - 7x = 0$; vertices: $(0, \pm 14)$

21. The foci of a hyperbola are $F_1(13, 0)$ and $F_2(-13, 0)$. Write its equation in standard form given that the difference between its focal radii is 10. $\dfrac{x^2}{25} - \dfrac{y^2}{144} = 1$

22. Use the definition of hyperbola to prove Theorem 12.5. (HINT: Let the coordinates of the foci be $F(\pm f, 0)$, and let the difference between the focal radii be $2a$. At the final stage of development, group terms that contain $f^2 - a^2$ as a factor. Then substitute b^2 for $f^2 - a^2$.)

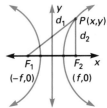

23. Prove Theorem 12.6. (HINT: Let the foci be $F(0, \pm f)$.) Then refer to the hint for Exercise 22.)

Mixed Review

1. Graph $4x^2 + 9y^2 = 36$. *12.2*

2. Graph $25x^2 + 4y^2 = 100$. *12.2*

3. Solve by completing the square.
   $x^2 - 4x = -3$ *9.3* 1, 3

4. Write an equation of the ellipse with vertices $(\pm 10, 0)$ and foci $(\pm 8, 0)$. *12.2*
   $\dfrac{x^2}{100} + \dfrac{y^2}{36} = 1$

Enrichment

Discuss the results of Theorems 12.5 and 12.6 that identify the asymptotes of a hyperbola as the lines
$y = \pm\dfrac{b}{a}x$ or $y = \pm\dfrac{a}{b}x$.

Explain that the asymptotes can be drawn easily by forming a rectangle, centered at the origin, with sides of lengths $2a$ and $2b$. In Example 1, where $a = 2$ and $b = 5$, the rectangle looks like this.

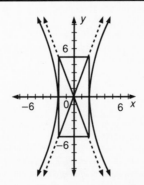

If the students extend the diagonals of this rectangle, the lines are the asymptotes of the hyperbola. Have the students use this strategy in several of the exercises.

12.4 Translations of Ellipses and Hyperbolas

Objectives To find the centers, vertices, and foci of ellipses and hyperbolas given their equations

To graph ellipses and hyperbolas given their equations

Just as a circle, an ellipse can be translated so that its center is no longer at the origin. Shown below are the standard forms of the equation of an ellipse with center (h,k). For both forms, the distance from the center to either focus is $f = \sqrt{a^2 - b^2}$.

Horizontal major axis
$$\frac{(x-h)^2}{a^2} + \frac{(y-k)^2}{b^2} = 1,$$
$$0 < b < a$$

Vertical major axis
$$\frac{(y-k)^2}{a^2} + \frac{(x-h)^2}{b^2} = 1,$$
$$0 < b < a$$

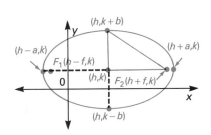

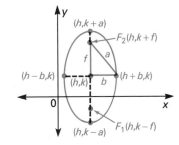

EXAMPLE 1 For the ellipse $\dfrac{(y-2)^2}{9} + \dfrac{(x+1)^2}{4} = 1,$ find the coordinates of the center, the vertices, and the foci. Graph the ellipse.

Solution Center: $(-1,2)$

Because $9 > 4$, and 9 is the denominator for $(y-2)^2$, $a = \sqrt{9} = 3$, $b = \sqrt{4} = 2$, and the major axis is vertical.

Vertices $(h, k \pm a)$, $(h \pm b, k)$:

$(-1,5)$, $(-1,-1)$, $(1,2)$, $(-3,2)$

Use $f = \sqrt{a^2 - b^2}$ to find the foci.
$$f = \sqrt{9 - 4} = \sqrt{5}$$

Foci $(h, k \pm f)$: $F_1(-1, 2 - \sqrt{5})$, $F_2(-1, 2 + \sqrt{5})$

Use the vertices to graph the ellipse as shown.

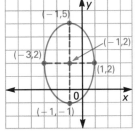

Additional Example 1

For the ellipse $\dfrac{(x+3)^2}{25} + \dfrac{(y-1)^2}{4} = 1,$ find the coordinates of the center, vertices, and foci. Graph the ellipse. Center: $(-3,1)$; vertices: $(2,1)$, $(-8,1)$, $(-3,3)$, $(-3,-1)$; foci: $(-3 - \sqrt{21}, 1)$ $(-3 + \sqrt{21}, 1)$

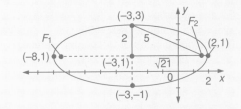

Teaching Resources

Quick Quizzes 85
Reteaching and Practice
   Worksheets 85

GETTING STARTED

Prerequisite Quiz

Tell what constant must be added to each binomial to produce a perfect square trinomial.

1. $x^2 + 4x$ 4   **2.** $x^2 - 10x$ 25
3. $y^2 - 12y$ 36   **4.** $x^2 - x$ $\frac{1}{4}$

Motivator

Ask students to describe the graph of $x^2 + y^2 = 25$.   A circle   Ask them how changing the equation to $(x - 3)^2 + (y + 4)^2 = 25$ changes the first graph.   The center of the circle is moved from the origin to $(3, -4)$. Ask students what this is called.
A translation

TEACHING SUGGESTIONS

Lesson Note

Ask the students to copy in their notes the graphs of the hyperbola and ellipse with centers at (h,k). Have them label the vertices and foci. Equations of asymptotes are not generally required for hyperbolas that are not centered at the origin.

Math Connections

Transformations: In this lesson, students are reintroduced to translations. Students may be interested in knowing that translations are an example of a transformation.

Highlighting the Standards

Standard 3c: The properties of ellipses and hyperbolas are logical, but complex enough to require careful attention to the mathematical development and discussion.

Critical Thinking Questions

Analysis: In the *Critical Thinking Questions* for Lesson 12.3, students were shown that the equation for the degenerate circle $x^2 + y^2 = 0$ is an equation of the origin. Ask students whether they can devise an equation of the point $(-1, 2)$. The equation of Example 1 can be modified as follows:

$$\frac{(y-2)^2}{9} + \frac{(x+1)^2}{4} = 0$$

It is satisfied only if each term on the left is zero, that is, only if $x = -1$ and $y = 2$.

Checkpoint

For each ellipse or hyperbola, find the center, vertices, and foci. Then graph the ellipse or hyperbola.

1. $4x^2 + y^2 + 8x - 6y + 9 = 0$ Center: $(-1,3)$; vertices: $(-1,5)$, $(-1,1)$, $(0,3)$, $(-2,3)$; foci: $(-1, 3 - \sqrt{3})$, $(-1, 3 + \sqrt{3})$

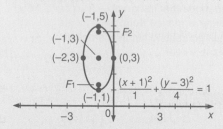

$$\frac{(x+1)^2}{1} + \frac{(y-3)^2}{4} = 1$$

2. $9x^2 - 16y^2 - 18x + 32y - 151 = 0$ Center: $(1,1)$; vertices: $(5,1)$, $(-3,1)$; foci: $(-4,1)$, $(6,1)$

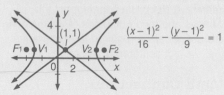

$$\frac{(x-1)^2}{16} - \frac{(y-1)^2}{9} = 1$$

By translating a hyperbola so that its center moves from the origin to $C(h,k)$, the standard forms change as follows. The distance from the center to the foci is $f = \sqrt{a^2 + b^2}$, where $a > 0$ and $b > 0$. The foci and vertices lie on lines parallel to the x- and y-axes respectively.

$$\frac{(x-h)^2}{a^2} - \frac{(y-k)^2}{b^2} = 1 \qquad \frac{(y-k)^2}{a^2} - \frac{(x-h)^2}{b^2} = 1$$

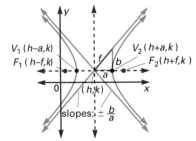 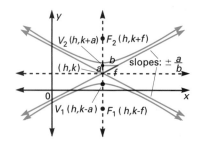

EXAMPLE 2 For the hyperbola $9x^2 + 54x - 4y^2 + 8y + 41 = 0$, find the coordinates of the center, the vertices, and the foci. Write the equations of the asymptotes. Sketch the hyperbola.

Plan Complete the squares of both the x- and y-terms.

Solution
$$9(x^2 + 6x + \underline{\quad}) - 4(y^2 - 2y + \underline{\quad}) = -41 + \underline{\quad}$$
$$9(x^2 + 6x + 9) - 4(y^2 - 2y + 1) = -41 + 9 \cdot 9 - 4 \cdot 1$$
$$9(x+3)^2 - 4(y-1)^2 = 36$$
$$\frac{9(x+3)^2}{36} - \frac{4(y-1)}{36} = 1$$
$$\frac{(x+3)^2}{4} - \frac{(y-1)^2}{9} = 1 \leftarrow a^2 = 4, b^2 = 9$$

Center: $(-3, 1)$

Use $a = 2$ to find the vertices.

Vertices: $(-3 \pm 2, 1)$, or $(-1, 1)$, $(-5, 1)$

Use $f = \sqrt{3^2 + 2^2} = \sqrt{13}$ to find the foci.

Foci: $F_1(-3 - \sqrt{13}, 1)$, $F_2(-3 + \sqrt{13}, 1)$

Use $C(-3, 1)$ and $m = \pm\dfrac{b}{a} = \pm\dfrac{3}{2}$ to find equations of the asymptotes.

$$y - y_1 = m(x - x_1)$$
$$y - 1 = \pm\frac{3}{2}[x - (-3)], \text{ or } y = \frac{3}{2}x + \frac{11}{2} \text{ and } y = -\frac{3}{2}x - \frac{7}{2}$$

Use the vertices and the asymptotes to sketch the hyperbola.

Additional Example 2

For the hyperbola
$4y^2 - 9x^2 - 48y - 18x + 99 = 0$,
find the coordinates of the center, vertices, and foci. Graph the hyperbola.
Center: $(-1,6)$; vertices: $(-1,9)$, $(-1,3)$; foci: $(-1, 6 - \sqrt{13})$, $(-1, 6 + \sqrt{13})$

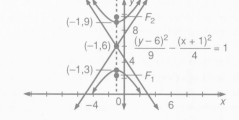

$$\frac{(y-6)^2}{9} - \frac{(x+1)^2}{4} = 1$$

Classroom Exercises

Give the coordinates of the center of each ellipse or hyperbola.

1. $\dfrac{(x-4)^2}{25} + \dfrac{(y+1)^2}{9} = 1$ $(4,-1)$

2. $\dfrac{(y-5)^2}{16} + \dfrac{(x+7)^2}{4} = 1$ $(-7,5)$

3. $\dfrac{(x-8)^2}{9} - \dfrac{(y-5)^2}{4} = 1$ $(8,5)$

4. $\dfrac{(y+1)^2}{9} + \dfrac{(x-6)^2}{49} = 1$ $(6,-1)$

Written Exercises

For each ellipse or hyperbola, find the coordinates of the center, the vertices, and the foci. Then graph the ellipse or hyperbola.

1. $\dfrac{(x-3)^2}{100} + \dfrac{(y-2)^2}{36} = 1$

2. $\dfrac{(x+6)^2}{25} + \dfrac{(y+5)^2}{16} = 1$

3. $\dfrac{(x-3)^2}{16} - \dfrac{(y-2)^2}{9} = 1$

4. $\dfrac{(x-6)^2}{64} + \dfrac{(y+1)^2}{100} = 1$

5. $\dfrac{(x+5)^2}{36} - \dfrac{(y-1)^2}{64} = 1$

6. $\dfrac{(y+6)^2}{25} - \dfrac{(x-1)^2}{144} = 1$

7. $16x^2 + 4y^2 - 96x + 8y + 84 = 0$

8. $25x^2 - 4y^2 - 150x - 16y + 109 = 0$

9. $x^2 - y^2 - 2x - 4y - 4 = 0$

10. $x^2 + 4y^2 - 2x + 40y + 100 = 0$

11. $4x^2 - y^2 + 24x + 4y + 28 = 0$

12. $4x^2 + 25y^2 + 16x + 50y - 59 = 0$

13. $4x^2 + 9y^2 - 24x + 18y + 9 = 0$

14. $9x^2 - 16y^2 - 90x + 32y + 65 = 0$

15. Find the equation of the ellipse with foci $(2,-3)$ and $(2,5)$, and the sum of whose focal radii equals 10. $\dfrac{(y-1)^2}{25} + \dfrac{(x-2)^2}{9} = 1$

16. The hyperbola with equation $4x^2 - 9y^2 + 32x + 18y + 19 = 0$ is translated 4 units to the right and 1 unit downward. Give the standard form of the equation of the hyperbola in its new position. $\dfrac{x^2}{9} - \dfrac{y^2}{4} = 1$

17. Write an equation of the hyperbola with foci at $F_1(-5,2)$ and $F_2(3,2)$ and asymptotes $y - 2 = \pm\sqrt{3}(x + 1)$. $\dfrac{(x+1)^2}{4} - \dfrac{(y-2)^2}{12} = 1$

18. Write an equation for the set of all points, the sum of whose distances from $A(2,3)$ and $B(-4,3)$ is always 12. Draw the resulting graph.

Mixed Review

Solve.   *7.6, 9.5, 6.6*

1. $\dfrac{3}{x-5} = \dfrac{x}{2}$ $-1, 6$

2. $x^2 - 2x + 4 = 0$ $1 \pm i\sqrt{3}$

3. $x^3 - 7x^2 + 14x - 8 = 0$ $1, 2, 4$

12.4 Translations of Ellipses and Hyperbolas   **425**

Enrichment

Discuss the relationship that involves a, b, and f in the ellipse and the hyperbola. In the ellipse, $a^2 - b^2 = f^2$. In the hyperbola, $a^2 + b^2 = f^2$.

Now, draw this figure on the board.

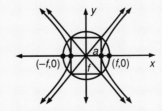

Explain that f is the radius of the circle that circumscribes the rectangle with dimensions

$2a$ and $2b$. Thus, the foci of the hyperbola occur at the points where the circle intersects the axis containing the vertices. This is called the *transverse* axis. As with the ellipse, the ratio $\frac{f}{a}$ defines the eccentricity of the hyperbola, but, for the hyperbola, $f > a$. Hyperbolas with the same eccentricity can be defined as similar. Have students sketch several similar hyperbolas with different values of a and b.

GETTING STARTED

Prerequisite Quiz

1. Which equations describe direct
   variation?   a, c
   a. $y = 4x$
   b. $xy = 4$
   c. $\dfrac{y}{x} = 6$
2. A number y varies directly as x. If $y = 16$
   when $x = 4$, find y when $x = 12$.   48

Motivator

Have students solve the equation $xy = 4$ for
y. $y = \dfrac{4}{x}$ Ask them what happens to y as
x gets larger. y approaches zero.   Ask
students if there is any value of x for which
y actually attains this value.   No

TEACHING SUGGESTIONS

Lesson Note

Compare the two types of variation, direct
and inverse.

Direct: x and y have a *constant quotient*.
Inverse: x and y have a *constant product*.

For both types of variation, problems are
easier to solve if the following formulas are
used.

Direct: $\dfrac{y_1}{x_1} = \dfrac{y_2}{x_2}$

Inverse: $x_1 y_1 = x_2 y_2$

12.5 The Rectangular Hyperbola and Inverse Variation

Objectives   To graph rectangular hyperbolas
To determine whether equations express inverse variations
To solve word problems involving inverse variation

In 1662, Robert Boyle noted that if a fixed amount of gas is held at a
constant temperature, then its volume goes *down* when it pressure goes
up. This observation, known as *Boyle's Law*, is represented more precisely by the equation $pv = k$, where k is a real nonzero constant. A
similar equation is shown in Example 1.

EXAMPLE 1   Graph $xy = 4$.

Plan   Solve for y in terms of x.

Then make a table of values.

Solution $xy = 4$

$y = \dfrac{4}{x}$

x	-4	-2	-1	1	2	4
y	-1	-2	-4	4	2	1

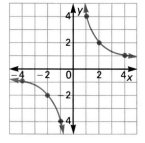

Plot the points. Draw the two branches of the resulting hyperbola.

Example 1 suggests that the graph of an equation of the form $xy = k$,
where k is a real nonzero constant, is a hyperbola with the x- and
y-axes as its asymptotes. Such a hyperbola is called a **rectangular
hyperbola**.

If $k > 0$, then the branches of the   If $k < 0$, then the branches lie in
hyperbola lie in quadrants I and   quadrants II and IV.
III.

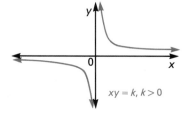

$xy = k, k > 0$

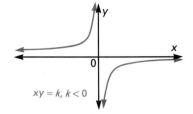

$xy = k, k < 0$

Additional Example 1

Graph $xy = -48$.

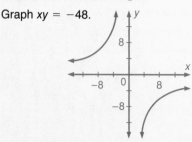

If the product xy is constant, then as one factor increases, the other factor decreases, and y is said to "vary inversely as x."

Definitions

> A function defined by an equation of the form $xy = k$, where k is a real nonzero constant, is called an **inverse variation**. The constant k is called the **constant of variation**.

EXAMPLE 2 Determine whether the equation or table represents an inverse variation. If so, give the constant of variation.

a. $\frac{4}{r} = t$

b. $y = \frac{x}{4}$

c.

x	1	3	4	12
y	12	4	3	1

Solutions

a. Rewrite as $rt = 4$, which is of the form $xy = k$.

Inverse variation; constant of variation: 4

b. Rewrite as $\frac{y}{x} = \frac{1}{4}$, which is not of the form $xy = k$.

Not an inverse variation

c. This is a table for $xy = 12$, which is of the form $xy = k$.

Inverse variation; constant of variation: 12

Note that the equation in Example 2b is an example of *direct* variation.

Suppose (x_1, y_1) and (x_2, y_2) represent two ordered pairs of an inverse variation. By definition, $x_1 y_1 = k$ and $x_2 y_2 = k$, since the product of x and y is *constant*.

By substitution, $x_1 y_1 = x_2 y_2$. This equation is applied below.

EXAMPLE 3 A number y varies inversely as x. If $y = 27$ when $x = 2$, what is y when $x = 9$?

Plan Use $x_1 y_1 = x_2 y_2$.

Solution Let $(x_1, y_1) = (2, 27)$ and $(x_2, y_2) = (9, y)$.

$2 \cdot 27 = 9 \cdot y$

$54 = 9y$

$y = 6$

Thus, $y = 6$ when $x = 9$.

EXAMPLE 4 The illumination I from a lamp varies inversely as the square of the distance d between the lamp and the illuminated object. If $I = 8$ ft-candles when $d = 3$ ft, what is d when $I = 2$ ft-candles? (THINK: Will the distance be less than 3 ft or greater than 3 ft? How do you know?)

12.5 The Rectangular Hyperbola and Inverse Variation   **427**

Math Connections

Rotations: In Lesson 12.8 it will be pointed out that the general equation of a conic is $Ax^2 + Bxy + Cy^2 + Dx + Ey + F = 0$. However, the rectangular hyperbola $xy = k$ is the only conic section in this book that has the term xy in its equation. This might be a good time to mention that the presence of an xy term in the equation of a conic indicates that the conic is "tilted," that is, it is not symmetric with respect to a line of symmetry that is parallel to either coordinate axis.

Critical Thinking Questions

Analysis: Ask students to describe the "degenerate" hyperbola $xy = 0$. They should see that this equation can be true only if $x = 0$ or $y = 0$; that is, only for a point on the x- or y-axis. Thus, the "hyperbola" is the two coordinate axes.

Checkpoint

1. Determine whether the equation $-\frac{4}{k} = h$ expresses an inverse variation. If so, name the constant of variation.   Yes; -4

In Exercises 2-3, a number y varies inversely as x.

2. If $y = 18$ when $x = 3$, find y when $x = -6$. -9
3. If $y = -8$ when $x = 12$, find x when $y = -24$.   4

Closure

Have students give the basic formula for inverse variation and the constant of variation. $xy = k$; k

Ask students to write a formula for inverse variation if y varies inversely as the square of x. $x^2 y = k$

Additional Example 2

Determine whether the equation or table represents an inverse variation. If so, name the constant of variation.

a. $\frac{x}{5} = y$ Not inverse variation

b. $m = \frac{7}{n}$ Inverse variation; constant of variation: 7

c.

x	3	-12	-8	9
y	-24	6	9	-8

Inverse variation; const of var: -72

Additional Example 3

A number y varies inversely as x. If $x = -42$ when $y = 4$, find x when $y = 8$. -21

Additional Example 4

A number y varies inversely as the square root of x. If $y = 4$ when $x = 25$, find x when $y = 10$.   4

Guided Practice

Classroom Exercises 1–8

Independent Practice

A Ex. 1–18, **B** Ex. 19–22, **C** Ex. 23–26

Basic: WE 1–19 odd, Midchapter Review, Application

Average: WE 3–21 odd, Midchapter Review, Application

Above Average: WE 5–25 odd, Midchapter Review, Application

Additional Answers

Written Exercises

1.

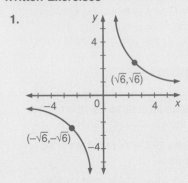

$(\sqrt{6}, \sqrt{6})$
$(-\sqrt{6}, -\sqrt{6})$

2.

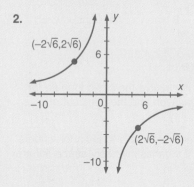

$(-2\sqrt{6}, 2\sqrt{6})$
$(2\sqrt{6}, -2\sqrt{6})$

Plan Because I varies inversely as the square of d, use $x_1 y_1 = x_2 y_2$.

Solution

$I_1 \cdot d_1^2 = I_2 \cdot d_2^2$ ← Let $x_1 = I_1$, $y_1 = d_1^2$, $x_2 = I_2$, and $y_2 = d_2^2$.

$8 \cdot 3^2 = 2 \cdot d_2^2$

$72 = 2d_2^2$

$36 = d_2^2$

$d_2 = \sqrt{36} = 6$

Thus, $d = 6$ ft when $I = 2$ ft-candles.

Classroom Exercises

Complete each table so that an inverse variation will be represented.

1.

x	2	?	16
y	24	8	3

6

2.

a	−8	40	?
b	5	−1	−20

2

3.

x	?	25	10
y	50	4	10

2

Determine whether each equation expresses an inverse variation (c is a nonzero constant).

4. $4x = y$ No

5. $x = \dfrac{c}{y}$ Yes

6. $\dfrac{c}{x} = \dfrac{y}{1}$ Yes

7. $1 = \dfrac{-x}{y}$ No

8. $y = \dfrac{1}{cx + x}, c \neq -1$ Yes

Written Exercises

Graph each rectangular hyperbola.

1. $xy = 6$ **2.** $xy = -24$ **3.** $y = -\dfrac{1}{x}$ **4.** $y = \dfrac{20}{x}$ **5.** $4xy = -48$

Determine whether the equation or table expresses an inverse variation. If so, find the constant of variation.

6. $y = 8x - 7$ No

7. $\dfrac{a}{3} = b$ No

8. $A = 7.3M$ No

9. $p = \dfrac{5}{q}$ Yes; 5

10. $7b = \dfrac{3}{c}$ Yes; $\frac{3}{7}$

11.

x	2	3	5	6	12
y	1.5	1	0.6	0.5	0.25

Yes; 3

12.

x	1	2	3	6	10
y	5	10	15	30	50

No

In Exercises 13–16, y varies inversely as x.

13. If $y = 24$ when $x = 3$, what is y when $x = 8$? 9

14. If $y = -32$ when $x = 2$, what is x when $y = 4$? −16

15. If $y = 4$ when $x = 9$, what is x when $y = 72$? $\frac{1}{2}$

16. If $y = -3$ when $x = 25$, what is y when $x = 15$? −5

17. The current in an electric circuit of constant voltage varies inversely as the resistance. When the current is 30 amps, the resistance is 20 ohms. Find the current when the resistance is 25 ohms. 24 amps

18. The length of a rectangle with constant area varies inversely as the width. When the length is 18, the width is 8. Find the length when the width is 9. 16

19. A number y varies inversely as the square of x. If $y = 2$ when $x = 6$, what is y when $x = 3$? 8

20. A number y varies inversely as $\sqrt{x}$. If $y = 3$ when $x = 64$, what is x when $y = 6$? 16

21. The illumination I from a light varies inversely as the square of its distance d from the illuminated object. If $I = 8.00$ ft-candles when $d = 5.00$ ft, what is I when $d = 4.00$ ft? 12.5 ft-candles

22. The height of a cylinder with constant volume is inversely proportional to the square of its radius. If $h = 8$ when $r = 4$, what is r when $h = 2$? 8

23. The volume of a fixed amount of an ideal gas at constant temperature is inversely proportional to the pressure. What happens to the volume when the pressure is doubled? It is halved.

24. The graph of $xy - 3y + 4x - 12 = 8$ represents a translation of the rectangular hyperbola $xy = 8$. Find the center of the translated hyperbola. $(3, -4)$

25. The weight of an object varies inversely as the square of its distance from the center of the earth. If Daniel weighs 170.0 lb on the surface of the earth 3960 mi from its center, what will he weigh on top of a nearby mountain rising 3 mi above him? 169.7 lb

26. Prove that if y is inversely proportional to x, then $\dfrac{1}{y}$ is directly proportional to x.

Midchapter Review

Write an equation of each circle, ellipse, or hyperbola given the following conditions. *12.1, 12.2, 12.3*

1. Circle: center $(-4, 2)$, radius 5

2. Ellipse: vertices $(\pm 6, 0)$, foci $(0, \pm 8)$

3. Hyperbola: vertices $(\pm 5, 0)$, foci $(\pm 13, 0)$

Graph each circle, ellipse, or hyperbola. If the graph is a circle, label the center and radius. If the graph is an ellipse, label the vertices, center, and foci. If the graph is a hyperbola, label the vertices, center, foci, and equations of the asymptotes. *12.1, 12.2, 12.3, 12.4*

4. $x^2 + y^2 = 100$

5. $4x^2 - 25y^2 = 100$

6. $16x^2 + 25y^2 = 400$

7. $9x^2 + 4y^2 - 18x + 16y - 11 = 0$

8. $y^2 - 4x^2 - 6y - 16x - 23 = 0$

9. A number y varies inversely as the square root of x. If $y = 8$ when $x = 9$, what is x when $y = 12$? *12.5* 4

3.

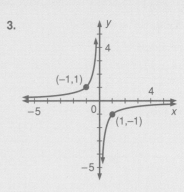

4.

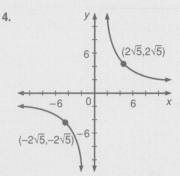

5.

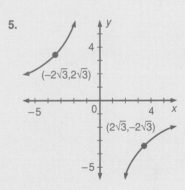

26. $xy = k_1;\ x = k_1\left(\dfrac{1}{y}\right);\ \dfrac{1}{y} = \dfrac{1}{k_1}x = k_2 x_1,$

since the reciprocal of a constant is still a constant.

See page 430 for the answers to Midchapter Review Ex. 1–8.

Enrichment

Refer the students to the graph of $xy = 4$ in Example 1. Challenge the students to find the coordinates of the two foci of this hyperbola.

If you wish, give the hint that they should draw in the two lines, $y = x$ and $y = -x$, and label these lines x' and y', the axes of a new rectangular coordinate framework. In this framework, the vertices of the hyperbola lie on the x'-axis and the values of a, b, and f can be calculated.

$a = b = \sqrt{2^2 + 2^2} = 2\sqrt{2}$

$f = \sqrt{a^2 + b^2} = \sqrt{16} = 4$

$F_1 = (2\sqrt{2},\ 2\sqrt{2});\ F_2 = (-2\sqrt{2},\ -2\sqrt{2})$

Midchapter Review

1. $(x + 4)^2 + (y - 2)^2 = 25$

2. $\dfrac{y^2}{100} + \dfrac{x^2}{36} = 1$

3. $\dfrac{x^2}{25} - \dfrac{y^2}{144} = 1$

4.
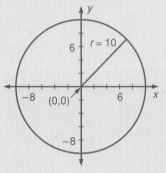

5.
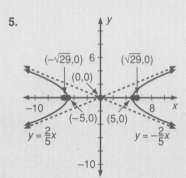

6.
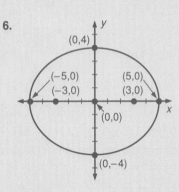

7.
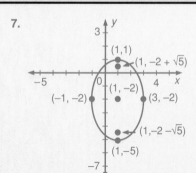

8.
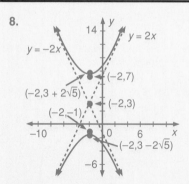

Application: *Eccentricity of an Ellipse*

When undisturbed by outside forces, a satellite of the earth—whether a communications satellite, a space station, or an astronaut—follows an elliptical orbit with the center of the earth at one focus. The **eccentricity**, denoted by e, tells how round or flat the ellipse is. The eccentricity is the ratio of the distance between the foci of any ellipse to the length of its major axis. So, $e = \dfrac{2f}{2a} = \dfrac{f}{a}$.

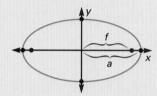

If $e = 0$, then $f = 0$ and the ellipse is actually a circle. For values of e closer to 1, the ellipse becomes flatter.

The eccentricity of a satellite can be found from its maximum and minimum distances from the surface of the earth, often called its **apogee** and **perigee**. The satellite OV$_{1\ 13}$ had an apogee of 9,320 km and a perigee of 560 km. Using 6,370 km as the radius of the earth,

$$2a = 9{,}320 + 2(6{,}370) + 560$$
$$= 22{,}620, \text{ so } a = 11{,}310.$$

$$f = a - (6370 + 560)$$
$$= 11{,}310 - 6{,}930 = 4{,}380$$

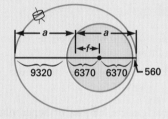

So, $e = \dfrac{f}{a} = \dfrac{4{,}380}{11{,}310} \approx 0.387$.

1. Vanguard I, launched in 1958, was the second U.S. satellite. Its apogee was 3,950 km and its perigee 660 km. Find its eccentricity. About 0.190

2. 10th Molniya$_1$, a Soviet satellite launched in 1968, had an apogee of 39,600 km and a perigee of 430 km. Find its eccentricity. About 0.742

3. Explorer 41, launched in 1969 to explore interplanetary fields, had an apogee of 213,850 km (over half the distance to the moon) and an eccentricity of 0.9405. Find its closest pass to the earth. About 380 km

430 Chapter 12 Conic Sections

12.6 Problem Solving: Joint and Combined Variation

Teaching Resources

Quick Quizzes 87
Reteaching and Practice
   Worksheets 87

Objective   To solve word problems involving joint or combined variation

Recall the two types of variation studied so far: direct variation (see Lesson 3.8), in which the *quotient* of y and x is constant ($\frac{y}{x} = k$, $k \neq 0$); and inverse variation, in which the *product* of y and x is constant ($yx = k$, $k \neq 0$).

Direct variation can be extended to several variables. For example, if y varies directly as v, w, and x, then the equation of variation is $\frac{y}{vwx} = k$.

The statement "y varies directly as v, w, and x" can also be written as "y varies *jointly* as v, w, and x."

EXAMPLE 1   A number y varies jointly as x and z. If $y = 2$ when $x = 4$ and $z = 6$, what is y when $x = 3$ and $z = 8$?

Plan   Write "y varies jointly as x and z" as an equation with subscripted variables.

Solution

$$\frac{y_1}{x_1 z_1} = \frac{y_2}{x_2 z_2} \quad \leftarrow \quad \frac{y_1}{x_1 z_1} = k = \frac{y_2}{x_2 z_2}$$

$$\frac{2}{4 \cdot 6} = \frac{y_2}{3 \cdot 8}$$

$$\frac{1}{12} = \frac{y_2}{24}$$

$$12 y_2 = 24$$

$$y_2 = 2$$

You can use a calculator to find $y_2 = \dfrac{2 \cdot 3 \cdot 8}{4 \cdot 6}$ directly.
2 $\boxed{\times}$ 3 $\boxed{\times}$ 8 $\boxed{\div}$ 4 $\boxed{\div}$ 6 $\boxed{=}$ 2.

Thus, $y = 2$ when $x = 3$ and $z = 8$.

EXAMPLE 2   The safe load for a beam varies directly as the width of the beam and the square of its depth. A particular beam 1.6 cm wide and 3.5 cm deep can safely support 6,300 lb. Find the safe load for a similar beam 1.4 cm wide and 4.0 cm deep.

Prerequisite Quiz

Write an equation that describes each variation. Use *k* as the constant of variation.

1. y varies directly as x. $\frac{y}{x} = k$
2. y varies inversely as x. $yx = k$
3. y varies inversely as the square of x. $yx^2 = k$
4. a is directly proportional to b. $\frac{a}{b} = k$, or $\frac{a_1}{b_1} = \frac{a_2}{b_2}$

Motivator

Have students give the equation representing "y varies directly as the product of x, w, and z." $\frac{y}{wxz} = k$ Introduce students to the statement "y varies *jointly* as x, w, and z," and have them write an equation for "y varies jointly as m and the square of t." $\frac{y}{mt^2} = k$

Additional Example 1

A number a varies jointly as b and c. If $a = 48$ when $b = 6$ and $c = 4$, find a when $b = 6$ and $c = 7$.   84

Additional Example 2

A number y varies directly as x and the square of z. If $y = 32$ when $x = 4$ and $z = 2$, find y when $x = 6$ and $z = 3$.   108

Highlighting the Standards

Standard 1d: Gravitational attraction and electrical resistance are examples of combined variation. They are real-world applications in that they have a profound effect on our lives.

Lesson Note

Example 3 illustrates combined variation, where F varies directly as $m \cdot M$ and inversely as r^2, so that $\frac{Fr^2}{m \cdot M}$ is a constant. To show the direct variation as a quotient, $\frac{F}{m \cdot M}$ can be circled, while Fr^2 can be boxed to show the inverse variation as a product.

Math Connections

Mass and Force: In countries that use the metric system of measurement in everyday life, the "kilo" (kilogram) is used as if it were a unit of weight (a particular kind of force). As Example 3 illustrates, the kilogram is a unit of mass, not force. Physicists customarily operate in two related metric systems, the centimeter/gram/second (cgs) system and the meter/kilogram/second (mks) system. In the cgs system the unit of force is the *dyne*; in the mks system the unit of force is the *newton*. Units from both systems appear in Example 3 in order to avoid decimals, but in actual practice, physical calculations would ordinarily be carried out in one system or the other.

Solution   Let s = the safe load, w = the width, and d = the depth. So, s varies jointly as w and the *square* of d.

$$\frac{s_1}{w_1 d_1^{\,2}} = \frac{s_2}{w_2 d_2^{\,2}} \quad \leftarrow \frac{s}{wd^2} \text{ is constant.}$$

$$\frac{6{,}300}{(1.6)(3.5)^2} = \frac{s_2}{(1.4)(4.0)^2}$$

$$\frac{6{,}300}{19.6} = \frac{s_2}{22.4} \quad \leftarrow \text{Use a calculator to find } s_2 = \frac{6{,}300(1.4)(4.0)^2}{(1.6)(3.5)^2}.$$

$$6{,}300 \; \boxed{\times} \; 1.4 \; \boxed{\times} \; 4.0 \; \boxed{x^2} \; \boxed{\div} \; 1.6 \; \boxed{\div} \; 3.5 \; \boxed{x^2} \; \boxed{=}$$
$$7{,}200$$

$$19.6 s_2 = 141{,}120$$

$$s_2 = 7{,}200$$

Thus, the safe load for a similar beam 1.4 cm wide and 4.0 cm deep is 7,200 lb.

Both direct and inverse variation may occur at the same time. Such instances are examples of **combined variation**. It is helpful to keep in mind that the *quotient* is constant for direct variation, and the *product* is constant for inverse variation.

EXAMPLE 3   The force of gravitational attraction, F, between two objects varies directly as the product of their masses m and M and inversely as the square of the distance r between them. The force F_1 is 134 dynes when $m_1 = 600$ kg, $M_1 = 750$ kg, and $r_1 = 15$ cm. Find the value of F_2 when $m_2 = 1{,}080$ kg, $M_2 = 6{,}700$ kg, and $r_2 = 201$ cm.

Plan   First, write an equation that defines the variation.
F *varies directly as* $m \cdot M$ *and inversely as* r^2.

This means that $\dfrac{Fr^2}{m \cdot M}$ is constant.

Solution

$$\frac{F_1 r_1^{\,2}}{m_1 M_1} = \frac{F_2 r_2^{\,2}}{m_2 M_2}$$

$$\frac{134 \cdot 15^2}{600 \cdot 750} = \frac{F_2 \cdot 201^2}{1{,}080 \cdot 6{,}700}$$

Simplify the fractions. $\dfrac{67}{1{,}000} = \dfrac{67 F_2}{12{,}000}$ Solve for F_2. $F_2 = 12$

Thus, the value of F_2 is 12 dynes.

Additional Example 3

Heat loss, H, per hour through a glass window varies jointly as the difference d between the inside and outside temperatures and the area A of the window, and inversely as the thickness t of the pane of glass. If the temperature difference is 30°, there is a heat loss of 9,000 calories in one hour through a windowpane of area 1,500 cm² and thickness of 0.25 cm. Find the heat loss through a glass of the same area and 0.2 cm thickness when the temperature difference is 15°.   5,625 calories

Classroom Exercises

Write an equation to describe each variation.

1. a varies jointly as b and c. $\dfrac{a}{bc} = k$
2. t varies directly as m and g. $\dfrac{t}{mg} = k$
3. y varies directly as x and inversely as z.
4. p varies directly as q and the square of r.
5. y varies directly as x and the square root of z, and inversely as m. $\dfrac{ym}{x\sqrt{z}} = k$
6. y varies jointly as the square root of x and the cube of r. $\dfrac{y}{r^3\sqrt{x}} = k$
7. A number y varies directly as x and z. If $y = 6$ when $x = 4$ and $z = 5$, what is x when $y = 21$ and $z = 7$? 10

Written Exercises

1. A number y varies directly as x and z. If $y = 6$ when $x = 8$ and $z = 3$, what is y when $x = 12$ and $z = 4$? 12
2. A number y varies jointly as x and z. If $y = 5$ when $x = 3$ and $z = 10$, what is y when $x = 15$ and $z = 4$? 10
3. A number a varies jointly as b and the square of c. If $a = 8$ when $b = 9$ and $c = 2$, what is c when $a = 4$ and $b = 2$? ±3
4. A number y varies directly as x and the square root of z, and inversely as t. If $y = 4$ when $x = 2$, $z = 9$, and $t = 5$, what is t when $y = 5$, $x = 12$, and $z = 4$? 16
5. A number z varies directly as m and n, and inversely as the square of p. If $z = 30$ when $m = 5$, $n = 4$, and $p = 2$, what is p when $z = 15$, $m = 5$, and $n = 32$? ±8
6. A number y varies directly as the cube root of x and inversely as the square of z. If $y = 3$ when $x = 8$ and $z = 4$, what is y when $x = 27$ and $z = 6$? 2
7. The volume V of a given mass of gas varies directly as the absolute temperature T and inversely as the pressure P. If $V = 462$ cm^3 when $T = 42°$ and $P = 40$ kg/cm^2, what is the volume when $T = 30°$ and $P = 30$ kg/cm^2? 440 cm^3
8. The electrical resistance in a wire varies directly as its length and inversely as the area of its cross section. A silver wire 100 m long and 0.0125 cm in cross-sectional radius has a resistance of 30 ohms. Find the length of silver wire of radius 0.075 cm needed to achieve a resistance of 20 ohms. (Assume the wires are circular in cross section, and use $A = \pi r^2$.) 2,400 m
9. If y varies directly as x and the cube of z, what is the effect on y of doubling z? Multiplies y by 8
10. If a varies jointly as b and the square root of c, what is the effect on a of multiplying c by 4? Doubles a
11. If y varies directly as x and inversely as the square of z, what is the effect on y of doubling x and doubling z? Halves y
12. If t varies directly as s and inversely as the square of r, what is the effect on t when the value of r is increased by 50%? Divides t by 2.25

Critical Thinking Questions

Analysis: Ask students whether they can think of an example other than gravitation that illustrates a force that varies inversely as the square of the distance between two objects. One example is the magnetic force that exists between two bar magnets.

Common Error Analysis

Error: Students sometimes mix the roles of direct and inverse variation when writing the equation for a combined variation. Offer the following remedial practice.

Write an equation that describes each variation. Use k as the constant.

1. A number y varies directly as x and inversely as z. $\dfrac{yz}{x} = k$
2. A number y varies jointly as p and q and inversely as the square of t. $\dfrac{yt^2}{pq} = k$

Checkpoint

1. A number y varies jointly as x and z. If $y = 12$ when $x = 4$ and $z = 3$, find y when $x = 9$ and $z = 8$. 72
2. A number y varies directly as the square of x and inversely as z. If $y = 12$ when $x = 2$ and $z = 7$, find y when $x = 3$ and $z = 9$. 21

Closure

Ask students how they would write an equation expressing the variation of y jointly as several variables. $\dfrac{y}{abc} = k$ Ask them how they would write an equation expressing the variation of y directly as one variable and inversely as another variable. For y varies directly as z and inversely as w, the equation is $\dfrac{yw}{z} = k$.

Guided Practice

Classroom Exercises 1–7

Independent Practice

A Ex. 1–6, **B** Ex. 7–10, **C** Ex. 11–16

Basic: WE 1–8
Average: WE 3–10
Above Average: WE 6–16

Additional Answers

Classroom Exercises

3. $\frac{yz}{x} = k$

4. $\frac{p}{qr^2} = k$

Written Exercises

15. $F = \frac{km_1m_2}{d^2}$; 4.19×10^8 m from the earth, or 4.10×10^7 m from the moon

13. Under certain conditions of artificial light, the exposure time to photograph an object is directly proportional to the square of its distance from the source of light and inversely proportional to the candlepower of the light. If the exposure time is 0.01 s when the light is 6.00 ft from the object, how far from the object should the light be placed when both the candlepower and the exposure time are doubled? 12 ft

14. The lifting force exerted by the atmosphere on the wings of an airplane in flight varies directly as the surface area of the wings and the square of the plane's airspeed. A small private plane has a cruising airspeed of 250 mi/h. In order to obtain three times the lifting force, a new plane is designed with a wing surface area twice that of the older model. What cruising speed (to the nearest mile per hour) is planned for the new model? 306 mi/h

15. As a consequence of Newton's Law of Universal Gravitation, the magnitude of the attraction between two spheres varies jointly as the product of their masses and inversely as the square of the distance between their centers. Write this relationship as a formula and use it to find the point between the centers of the earth and the moon at which a space vehicle would experience equal attraction from both the earth and the moon. Use the following data: earth's mass: 5.98×10^{24} kg; moon's mass: 7.35×10^{22} kg; mean distance between the centers of the earth and the moon: 4.60×10^8 m.

16. The maximum safe load (in pounds) for a horizontal beam varies jointly as its width and the square of its depth, and inversely as the distance between its supports. For the construction of a certain house, a beam is positioned between two supports 16 ft apart. The width and depth of the beam are 2.0 in. and 6.0 in., respectively. The beam can safely support 1,200 lbs. A second beam of similar material has a width of 4.0 in. and a depth of 8.0 in. The supports for this beam are 18 ft apart. Find the safe load for the second beam to the nearest hundred pounds. 3,400 lb

Mixed Review

Simplify. *7.1, 7.4, 8.4, 9.2*

1. $\frac{x^3 - 8}{x^2 - 4x + 4} \cdot \frac{x^2 + 2x + 4}{x - 2}$

2. $\frac{5}{x^2 - 7x + 12} - \frac{2x - 1}{x - 4} \cdot \frac{-2x^2 + 7x + 2}{x^2 - 7x + 12}$

3. $\frac{28}{4 - \sqrt{2}}$ $8 + 2\sqrt{2}$

4. $\frac{13}{3 - 2i}$ $3 + 2i$

5. Find all the roots of $x^4 + 3x^3 - 3x^2 + 3x - 4 = 0$. *10.4* $-4, 1, \pm i$

Enrichment

Discuss a few of the many practical applications of direct and inverse variation.

Hooke's law, which states that the stretch of a spring is directly proportional to the force applied to it, provides a good example of direct variation.

An example of inverse variation is that the volume of a gas at constant temperature is inversely proportional to the pressure on the gas.

Have students use the library and any available resource centers to find other significant examples of direct, inverse, and joint variation. Often, a physics book is a good source of examples.

12.7 The Parabola: Focus and Directrix

Objectives

To write equations of parabolas given certain conditions

To write an equation for the directrix and to find the coordinates of the focus of a parabola given its equation

To graph parabolas

In the graph at the right, the distance from $P_1(4,4)$ to $F(0,1)$ is $\sqrt{(4-0)^2 + (4-1)^2} = \sqrt{16+9} = \sqrt{25} = 5$.

The distance from P_1 to H on the line d is also 5. Thus, P_1 is equidistant from the point F and the line d.

The graph also shows several other points that are equidistant from the point F and the line d. These points form a *parabola*.

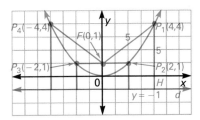

Definitions

A **parabola** is the set of all points in a coordinate plane equidistant from a given point, the **focus**, and a given line, the **directrix**. The **vertex** of the parabola is the midpoint of the perpendicular segment from the focus to the directrix.

Theorem 12.7

If a vertical parabola has focus $(0,p)$, directrix $y = -p$, and its vertex at the origin, then its equation in standard form is $x^2 = 4py$.

There are two alternate forms of the equation of a parabola with its vertex at the origin and the y-axis as its axis of symmetry.

$$x^2 = 4py, \text{ or } y = \frac{1}{4p}x^2, \text{ and } y = ax^2 \text{ (from Lesson 11.4)}$$

For both forms, a *positive* coefficient indicates that the parabola opens *upward*, and a *negative* coefficient indicates that the parabola opens *downward*. For example, the parabola

$y = \frac{1}{4}x^2$ opens *upward* because $\frac{1}{4} > 0$.

Or, $x^2 = 4y$ opens *upward* because the coefficient of y is *positive*.

$y = -5x^2$ opens *downward* because $-5 < 0$.

Or, $x^2 = -\frac{1}{5}y$ opens *downward* because the coefficient of y is *negative*.

Teaching Resources

Quick Quizzes 88
Reteaching and Practice
  Worksheets 88
Transparency 29

GETTING STARTED

Prerequisite Quiz

Graph each parabola.

1. $(y - 4) = (x + 5)^2$
2. $(x + 2) = (y - 1)^2$

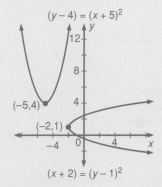

Motivator

Remind students that the graph of $y = x^2$ is a parabola with vertex at the origin and opening upward. Then have them guess what the graph of $x = y^2$ will be.   A parabola with vertex at the origin opening to the right.   Have students make a table of values of x and y for the following values of x: 0, 1, 4, 9, 16, 25. To check their guesses, have them graph the points.

Highlighting the Standards

Standard 4a: Students have studied properties and solved quadratic equations. They should be reminded that the parabola is simply another expression of the quadratic function.

Lesson Note

This lesson introduces a new approach to graphing the parabola. The definition of a parabola in terms of a locus ties in with the definitions used to derive equations for the circle, ellipse, and hyperbola. Students will now have to identify two additional properties when graphing the parabola, the focus and the directrix. Point out that the students will still begin by completing the square. The only additional skill is writing the coefficient of the nonsquare term in the form $4 \cdot p$.

Math Connections

Parabolic Mirrors: If a parabola is rotated about its axis of symmetry, a "three-dimensional parabola," or *paraboloid of revolution*, is formed. This shape is very useful in the design of parabolic mirrors for search lights and similar optical instruments. If a light source is placed at the focus of a parabolic mirror, then all rays emanating from that source will be parallel to one another after they are reflected from the mirror's surface. The result is a beam of light that is concentrated in one direction.

The figures below show that p represents the *directed* (signed) distance from the vertex to the focus.

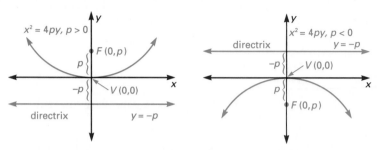

These relationships can also be extended to parabolas whose vertices are not at the origin.

The standard form of the equation of a vertical parabola with vertex (h,k) (see Lesson 11.4), $y - k = a(x - h)^2$, can be written as $y - k = \frac{1}{4p}(x - h)^2$, where $a = \frac{1}{4}p$. Then $(x - h)^2 = 4p(y - k)$, where p is the directed distance from the vertex to the focus.

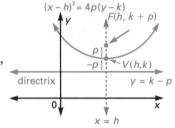

EXAMPLE 1 Write an equation of the parabola with focus $F(3, -6)$ and directrix $y = 4$. Graph the parabola.

Plan Find the vertex $V(h,k)$. Then use points F and V to find p, the directed distance from V to F. Finally, substitute in $(x - h)^2 = 4p(y - k)$.

Solution (1) Plot the focus $F(3, -6)$ and draw the directrix $y = 4$.

(2) Draw a line through F perpendicular to the directrix at D. This is the axis of symmetry. Points D and F have the same x-coordinate, 3.

(3) Find V, the midpoint of $\overline{FD}$.
$$V = \left(\frac{3 + 3}{2}, \frac{-6 + 4}{2} \right) = (3, -1)$$

(4) Find p, the directed distance from V to F.
$$p = -|-1 - (-6)| = -5$$

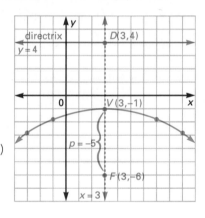

Additional Example 1

Write an equation of the parabola with focus $F(-4,2)$ and directrix $y = -6$. Graph the parabola.
$$(x + 4)^2 = 16(y + 2)$$

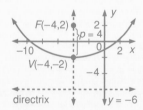

(5) Write the equation using $p = -5$ and $V(3, -1)$.

$$(x - h)^2 = 4p(y - k)$$
$$(x - 3)^2 = 4(-5)[y - (-1)]$$
$$(x - 3)^2 = -20(y + 1)$$

Thus, an equation of the parabola is $(x - 3)^2 = -20(y + 1)$.

Plot the vertex. Also, find a few other points on the parabola by solving for y and making a table. Draw the graph as shown.

Each figure below shows a horizontal parabola with vertex $V(0,0)$, focus $(p,0)$, and directrix $x = -p$.

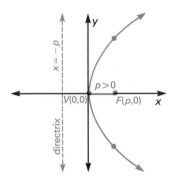

If $p > 0$, the parabola opens to the *right*.

If $p < 0$, the parabola opens to the *left*.

The general equation for a horizontal parabola with its vertex at the origin is given in Theorem 12.8.

Theorem 12.8

If a horizontal parabola has focus $(p,0)$, directrix $x = -p$, and its vertex at the origin, then its equation is $y^2 = 4px$.

The standard form of the equation of a horizontal parabola with vertex (h, k), $x - h = a(y - k)^2$, can be written as $(y - k)^2 = 4p(x - h)$, where p is the directed distance from the vertex to the focus.

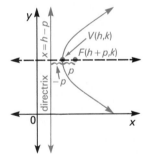

Critical Thinking Questions

Analysis: Ask students whether they can think of an application of the focal properties of a parabola other than that mentioned before in the *Math Connections*. Perhaps some students will realize that the process of having light emanate from the focus can be reversed. Light rays that strike a parabolic mirror parallel to the mirror's axis of symmetry will be concentrated at the focus, creating an intense concentration of energy that may be strong enough to ignite any combustible material that might be positioned there.

Common Error Analysis

Error: When putting the equation in the form $(x - h)^2 = 4 \cdot p \cdot (y - k)$ or $(y - k)^2 = 4 \cdot p \cdot (x - h)$, students make mistakes in factoring out a 4 when the coefficient of the nonsquare term does not contain 4 as a factor. Oral practice such as the following may be helpful.

Determine the value of p.

1. $(x + 1)^2 = 3(y - 6)$ $4p = 3; p = \frac{3}{4}$
2. $(y - 5)^2 = -2(x + 8)$
   $4p = -2; p = -\frac{1}{2}$
3. $(x + 8)^2 = (y - 5)$ $4p = 1; p = \frac{1}{4}$

Checkpoint

Use the information provided to write an equation of each parabola.

1. Focus: $F(3,2)$
   Directrix: $y = 6$ $(x - 3)^2 = -8(y - 4)$
2. Vertex: $V(5,2)$
   Directrix: $x = 1$ $(y - 2)^2 = 16(x - 5)$
3. Focus: $F(-1,8)$
   Vertex: $V(-1,6)$ $(x + 1)^2 = 8(y - 6)$

438

Closure

Have students find the vertex, focus, and equation of the directrix for an equation of the indicated form.

$(y - k)^2 = 4p(x - h)$
V: (h,k); F: $(h + p,k)$; Dir: $x = h - p$
$(x - h)^2 = 4p(y - k)$
V: (h,k); F: $(h,k + p)$; Dir: $y = k - p$

▰▰FOLLOW UP

Guided Practice

Classroom Exercises 1–6

Independent Practice

A Ex. 1–14, **B** Ex. 15–21, **C** Ex. 22, 23

Basic: FR, WE 1–17 odd
Average: FR, WE 5–21 odd
Above Average: FR, WE 7–23 odd

Additional Answers

Classroom Exercises

4.

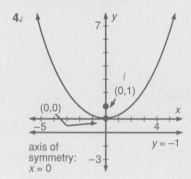

axis of symmetry: $x = 0$

EXAMPLE 2 For the parabola $y^2 - 2y - 2x + 5 = 0$, find the coordinates of the vertex and the equation of the axis of symmetry. Then find the coordinates of the focus and the equation of the directrix. Graph the parabola.

Plan The graph is a horizontal parabola because there is a y^2-term.

Solution

$$y^2 - 2y - 2x + 5 = 0$$

Complete the square.

$$y^2 - 2y + \underline{\quad} = 2x - 5 + \underline{\quad}$$
$$y^2 - 2y + 1 = 2x - 5 + 1$$
$$(y - 1)^2 = 2x - 4$$

Factor out 2 and then rewrite it as $4(\frac{1}{2})$ to obtain standard form.

$$(y - 1)^2 = 2(x - 2)$$
$$(y - 1)^2 = 4 \cdot \frac{1}{2} \cdot (x - 2)$$

Vertex: $(2,1)$; $p = \frac{1}{2}$

The axis of symmetry is *horizontal*.

Axis of symmetry: $y = 1$

Use $p = \frac{1}{2}$ to find the focus.

Focus: $(2 + p,1)$, or $(2\frac{1}{2},1)$

The directrix is *vertical*.

Directrix: $x = 2 - p$, or $x = 1\frac{1}{2}$

Plot the vertex and draw the axis of symmetry. Because $p > 0$, the parabola opens to the right.

Find several other points on the parabola by solving the equation for y and making a table of values.

Draw the horizontal parabola.

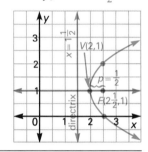

▰▰ *Focus on Reading*

Complete each of the following to make a true statement.

1. The graph of $y^2 = 4px$ is a _____ parabola with its vertex at V(_____). horizontal; (0,0)
2. The graph of $x^2 = 4py$ is a _____ parabola opening _____ when $p > 0$. vertical; upward
3. The graph of $(x - h)^2 = 4p(y - k)$ is a _____ parabola with its vertex at V(_____). vertical; (h,k)

Additional Example 2

For the parabola $y^2 - 8y - 12x - 8 = 0$, find the coordinates of the vertex and the equation of the axis of symmetry. Then find the coordinates of the focus and the equation of the directrix. Graph the parabola.
Vertex: $(-2,4)$
Axis of symmetry: $y = 4$
Focus: $(1,4)$
Directrix: $x = -5$

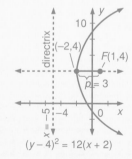

Classroom Exercises

Determine whether each parabola is vertical or horizontal.

1. $x^2 = 4y$ Vert   **2.** $y^2 = -8x$ Horiz   **3.** $x^2 = -20y$ Vert

4–6. For each parabola in Classroom Exercises 1–3, find the coordinates of the vertex and the focus, and the equations of the directrix and the axis of symmetry. Graph the parabola.

Written Exercises

Use the information provided to write an equation of each parabola.

1. Focus: $(-8,0)$ **2.** Focus: $(3,1)$ **3.** Focus: $(3,8)$ **4.** Focus: $(5,-1)$
Directrix: $x = 8$ Directrix: $y = 10$ Directrix: $x = 6$ Directrix: $x = 6$

5. Vertex: $(-2,3)$ **6.** Vertex: $(-4,3)$ **7.** Focus: $(3,8)$ **8.** Focus: $(-8,-5)$
Directrix: $y = -3$ Directrix: $x = 5$ Vertex: $(3,2)$ Vertex: $(2,-5)$

For each parabola, find the coordinates of the vertex and the focus, and the equations of the directrix and the axis of symmetry. Graph the parabola.

9. $(x - 7)^2 = 8(y - 2)$ **10.** $(x - 4)^2 = -12(y + 5)$ **11.** $(x - 6)^2 = 10(y + 1)$

12. $(x + 3)^2 = -2(y - 4)$ **13.** $(y - 6)^2 = -8(x - 5)$ **14.** $(y - 1)^2 = 12(x + 7)$

15. $x^2 - 6x - 8y + 25 = 0$ **16.** $x^2 + 10x + 12y + 31 = 0$

17. $x^2 - 2x - 2y - 9 = 0$ **18.** $y^2 - 8y - 4x + 28 = 0$

19. $y^2 - 10y + 2x + 27 = 0$ **20.** $x^2 - 4x + 8y + 12 = 0$

21. A point P moves in a coordinate plane in such a way that it is always equidistant from the line $x = -2$ and the point $Q(2,0)$. Use the distance formula, $d = \sqrt{(x_2 - x_1)^2 + (y_2 - y_1)^2}$, to write an equation describing the path of point P. $y^2 = 8x$

22. Prove Theorem 12.8.

23. Write an equation for the set of all points equidistant from the point $F(h, p + k)$ and the line $y = k - p$. Verify that this equation is equivalent to the standard form of the equation of a vertical parabola. $(x - h)^2 = 4p(y - k); y - k = a(x - h)^2$, or $(x - h)^2 = \frac{1}{a}(y - k)$, and letting $a = \frac{1}{4p}$, $(x - h)^2 = 4p(y - k)$.

Mixed Review

Graph each equation.   *3.5, 11.4, 12.1, 12.3*

1. $3x - 4y = 8$ **2.** $x^2 + y^2 = 25$ **3.** $4x^2 - 9y^2 = 36$ **4.** $y = -x^2$

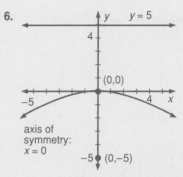

5. axis of symmetry: $y = 0$ $(-2,0)$, $(0,0)$, $x = 2$

6. $y = 5$, axis of symmetry: $x = 0$, $(0,0)$, $(0,-5)$

Written Exercises

1. $y^2 = -32x$

2. $(x - 3)^2 = -18\left(y - \frac{11}{2}\right)$

3. $(y - 8)^2 = -6\left(x - \frac{9}{2}\right)$

4. $(y + 1)^2 = -2\left(x - \frac{11}{2}\right)$

5. $(x + 2)^2 = 24(y - 3)$

6. $(y - 3)^2 = -36(x + 4)$

7. $(x - 3)^2 = 24(y - 2)$

8. $(y + 5)^2 = -40(x - 2)$

See pages 822–823 for the answers to Written Ex. 9–20, 22, and Mixed Review Ex. 1–4.

Enrichment

Point out that any ray emanating from the focus of a parabola will reflect off the parabolic surface in a line parallel to the axis of the parabola.

This is the principle behind the sealed-beam lights of an automobile. The 3-dimensional equivalent of the parabola is called a *paraboloid*. Have the students do research to find other practical examples of the use of parabolas and paraboloids. Two such examples are solar ovens and dish antennae.

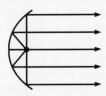

Teaching Resources

Quick Quizzes 89
Reteaching and Practice
Worksheets 89

▰▰GETTING STARTED

Prerequisite Quiz

Graph each equation.

1. $8y = x^2$

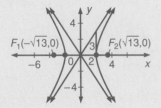

2. $9x^2 - 4y^2 = 36$ $\dfrac{x^2}{4} - \dfrac{y^2}{9} = 1$

Motivator

Have students give a value of B that will result in the graph of $9x^2 + By^2 = 36$ being the indicated one below.

1. circle   9
2. ellipse   4
3. hyperbola -4

Objective   To identify conic sections

The graphs of circles, parabolas, ellipses, and hyperbolas can be represented by passing a plane through a hollow double cone. Therefore, each of these curves is called a **conic section**.

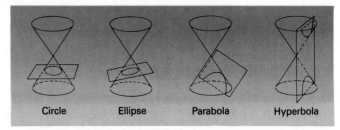

| Circle | Ellipse | Parabola | Hyperbola |

The general equation of a conic section can be written in the form
$$Ax^2 + Bxy + Cy^2 + Dx + Ey + F = 0,$$
where A, B, C, D, E, and F are not all zero.

To identify conic sections described by equations of the form above, change the equations to one of the standard forms shown below.

Conic Section	Center at (0,0)	Center at (h,k)	
Circle	$x^2 + y^2 = r^2$	$(x-h)^2 + (y-k)^2 = r^2$	$r > 0$
Ellipse	$\dfrac{x^2}{a^2} + \dfrac{y^2}{b^2} = 1$	$\dfrac{(x-h)^2}{a^2} + \dfrac{(y-k)^2}{b^2} = 1$	$0 < b < a$
	$\dfrac{y^2}{a^2} + \dfrac{x^2}{b^2} = 1$	$\dfrac{(y-k)^2}{a^2} + \dfrac{(x-h)^2}{b^2} = 1$	
Hyperbola	$\dfrac{x^2}{a^2} - \dfrac{y^2}{b^2} = 1$	$\dfrac{(x-h)^2}{a^2} - \dfrac{(y-k)^2}{b^2} = 1$	$a > 0,$ $b > 0$
	$\dfrac{y^2}{a^2} - \dfrac{x^2}{b^2} = 1$	$\dfrac{(y-k)^2}{a^2} - \dfrac{(x-h)^2}{b^2} = 1$	
Rectangular Hyperbola	$xy = k$		$k \neq 0$
Parabola	$x^2 = 4py$ $y^2 = 4px$	$(x-h)^2 = 4p(y-k)$ $(y-k)^2 = 4p(x-h)$	$p \neq 0$

Highlighting the Standards

Standard 4c: Conic sections and their properties have been thoroughly covered. Connections among the conics and their equations are now explored.

EXAMPLE 1 Identify each conic section whose equation is given.

　　a. $4x^2 + 4y^2 = 36$　　b. $4x^2 + 9y^2 = 36$　　c. $4x^2 - 9y^2 = 36$
　　d. $-3xy = 24$　　e. $12 + 2x^2 = y + 7x$

Plan If there is only one squared term, and no other second-degree variable, then the conic section is a parabola. Otherwise, change the equation to a standard form.

Solutions

Equation	Standard form	Conic section
a. $4x^2 + 4y^2 = 36$	$x^2 + y^2 = 3^2$	circle
b. $4x^2 + 9y^2 = 36$	$\dfrac{x^2}{3^2} + \dfrac{y^2}{2^2} = 1$	ellipse
c. $4x^2 - 9y^2 = 36$	$\dfrac{x^2}{3^2} - \dfrac{y^2}{2^2} = 1$	hyperbola
d. $-3xy = 24$	$xy = -8$	rectangular hyperbola
e. $12 + 2x^2 = y + 7x$		parabola

EXAMPLE 2 Identify the conic section defined by
$4x^2 + 9y^2 + 32x - 90y + 253 = 0$.

Plan Complete both squares. Then write in a standard form.

Solution
$4x^2 + 32x + \underline{\quad} + 9y^2 - 90y + \underline{\quad} = -253 + \underline{\quad} + \underline{\quad}$

$4(x^2 + 8x + \underline{\quad}) + 9(y^2 - 10y + \underline{\quad}) = -253 + \underline{\quad} + \underline{\quad}$

$4(x^2 + 8x + 16) + 9(y^2 - 10y + 25) = -253 + 4 \cdot 16 + 9 \cdot 25$

$4(x + 4)^2 + 9(y - 5)^2 = 36$

$\dfrac{(x + 4)^2}{9} + \dfrac{(y - 5)^2}{4} = 1$　　← standard form of an ellipse

Thus, the conic section is an ellipse.

Classroom Exercises

Identify the conic section whose equation is given.

1. $\dfrac{x^2}{25} - \dfrac{y^2}{9} = 1$
Hyperbola

2. $\dfrac{x^2}{25} + \dfrac{y^2}{9} = 1$
Ellipse

3. $\dfrac{x^2}{9} + \dfrac{y^2}{9} = 1$
Circle

4. $y = x^2$
Parabola

Lesson Note

Point out that it is not necessary to complete the square in order to identify the conic section, providing that the restrictions on r, a, and b are satisfied. The students may be interested in equations that do not satisfy these restrictions, such as the following.

$x^2 + y^2 = 0 \longrightarrow$　　Graph: 1 point
$3x^2 + y^2 = -2 \longrightarrow$　　No Graph

See also Exercise 24.

Math Connections

History of Mathematics: Students may be interested in knowing that conic sections have a long and rich history, one that dates from antiquity. Archimedes (287?−212 B.C.), who worked extensively with conic sections, knew that the area of an ellipse is πab, where a and b are the semimajor and semiminor axes.

Critical Thinking Questions

Analysis: The general equation of a conic on page 440 has as its coefficients, A, B, C, D, E, and F, not all of which are zero. Ask students to assume that $B = 0$ (in order to avoid "tilted" conics) and then to specify which other coefficients can be chosen to be zero in order to have a parabola.　One answer: C　Then ask the same question for each of the other conic sections.

Checkpoint

Identify each conic section whose equation is given.

1. $3x^2 - 12x + y + 7 = 0$　Parabola
2. $x^2 + 10x + y^2 + 12y = 60$　Circle
3. $(5y - x)(5y + x) = 100$　Hyperbola
4. $5y^2 + 50y + 275 = 100x - 2x^2$　Ellipse

Additional Example 1

Identify each conic section whose equation is given.

a. $4x^2 - 16y^2 = 16$　Hyperbola
b. $8x - 2y - y^2 - 25 = 0$　Parabola
c. $16x^2 + 16y^2 = 32$　Circle
d. $16x^2 + 4y^2 = 16$　Ellipse
e. $2xy - 10 = 0$　Rectangular hyperbola

Additional Example 2

Identify the conic section defined by
$(x - 6)(x + 4) - (y - 2)(y + 4) = 0$.
$\dfrac{(x - 1)^2}{16} - \dfrac{(y + 1)^2}{16} = 1$; hyperbola

Closure

Have students explain how they can use the values of A, B, and C to identify the conic section defined by the equation given below.

$$Ax^2 + By^2 = C$$

Change the equation to standard form. Ask students how they can quickly identify whether the graph of an equation is a parabola. The parabola has only one second-degree term.

▰▰FOLLOW UP

Guided Practice

Classroom Exercises 1–4

Independent Practice

A Ex. 1–12, **B** Ex. 13–21, **C** Ex. 22–24

Basic: WE 1–14, Application

Average: WE 6–19, Application

Above Average: WE 8–24, Application

Additional Answers

Written Exercises

1. Hyperbola   2. Parabola
3. Parabola   4. Circle
5. Ellipse   6. Rect hyp
7. Hyperbola   8. Ellipse
13. Circle   14. Hyperbola
21. Answers will vary. One possible answer: The equation of a parabola has one first-degree and one second-degree term. No combination of real-number coefficients can make either second-degree term a first-degree term, so the equation cannot represent a parabola.

Written Exercises

Identify each conic section whose equation is given.

1. $4x^2 - 8y^2 = 32$ 2. $x^2 + 4 = y$ 3. $5x^2 - y = 20$ 4. $5x^2 + 5y^2 = 20$

5. $6x^2 + 4y^2 = 24$ 6. $7xy = -28$ 7. $4x^2 - y^2 = 4$ 8. $8x^2 + 5y^2 = 40$

9. $\dfrac{(x-4)^2}{16} + \dfrac{(y+3)^2}{4} = 1$ Ellipse   10. $\dfrac{(x+8)^2}{25} - \dfrac{(y-1)^2}{4} = 1$ Hyperbola

11. $\dfrac{(x+3)^2}{4} + (y-5)^2 = 1$ Ellipse   12. $4(x-2)^2 + 4(y+6)^2 = 100$ Circle

13. $6x^2 + 6y^2 + 36y + 18x - 2 = 0$ 14. $25x^2 - 9y^2 + 200x - 175 = 0$

15. $y^2 + x^2 - 8y + 2x - 8 = 0$ Circle   16. $16x^2 - 9y^2 - 72y - 288 = 0$ Hyperbola

17. $x^2 - 12x - y + 30 = 0$ Parabola   18. $x^2 + y^2 - 4x + 6y - 1 = 0$ Circle

19. $(3x - 2y)(3x + 2y) = 36$ Hyperbola   20. $(x-6)(x+8) = (y-2)^2$ Hyperbola

21. Write, in your own words, why an equation of the form $Ax^2 + By^2 = C$, where A, B, and C are real numbers, cannot describe a parabola.

22. For what values of k will $(k^2 - 2k)x^2 - 4x + 8y^2 + 2y + 3 = 0$ define a circle? $-2, 4$

23. For what values of k will $x^2 + (k^2 - 7k + 12)y^2 - 5x - y + 6 = 0$ define a parabola? $3, 4$

24. The graph of $4x^2 - 9y^2 = F$ is a hyperbola if $F = 36$. What is the graph if $F = 0$? It is the graph of the hyperbola's asymptotes.

Mixed Review

For Exercises 1–4, use $f(x) = x^2 - 5x + 4$ and $g(x) = 2x - 6$.   3.2, 5.9

1. Find $f(3)$.   2. Find $f(-3) + g(-3)$.   3. Give the domain   4. Simplify $f(g(x))$.
-2 16   of f. {Real numbers} $4x^2 - 34x + 70$

▰▰ Application: *LORAN*

LORAN is a long-range navigation system that uses a pair of radio transmitters placed several hundred miles apart. LORAN receivers measure the difference in microseconds between the reception of the two signals. Each pair of signals is synchronized so that, knowing the speed of the signal, a navigator can calculate the difference between the distances from the ship to the transmitters.

The locus of all possible ship locations is what conic section? What are its foci? Hyperbola; the transmitters

Enrichment

Explain that there are many interesting graphs other than the conic sections. One such graph is the cycloid. Challenge the students to use their imaginations to graph the path traveled by a point on the circumference of a circle as the circle rolls along a straight line.

When the graph is determined, ask the students to calculate the height, h, of the curves if the distance between two successive cusps (points on the line) is one linear unit.

$$h = \frac{1}{\pi}$$

12.9 Linear-Quadratic Systems

Objectives

To solve systems consisting of a linear and a quadratic equation
To solve word problems involving linear-quadratic systems

The graphs of the line $y = 2x + 1$ and the parabola $y = x^2 - 2$ are shown at the right. They have two points in common, $A(-1, -1)$ and $B(3, 7)$. The coordinates of these points satisfy both equations, which can be shown by substituting these values for x and y in each equation.

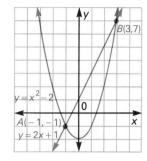

Thus, $(-1, -1)$ and $(3, 7)$ are the solutions of the linear-quadratic system consisting of $y = 2x + 1$ and $y = x^2 - 2$. As illustrated below, a system of one quadratic and one linear equation may have 0, 1, or 2 real solutions.

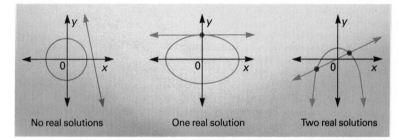

| No real solutions | One real solution | Two real solutions |

EXAMPLE 1 Solve this system graphically: $\dfrac{x^2}{4} + \dfrac{y^2}{25} = 1$
$5x + 2y = 10$

Solution Ellipse: $\dfrac{x^2}{4} + \dfrac{y^2}{25} = 1$

x-intercepts: ± 2; y-intercepts: ± 5

Line: $5x + 2y = 10$

x-intercept: 2; y-intercept: 5

The solutions are $(2, 0)$ and $(0, 5)$ because the graphs intersect as shown at the right.

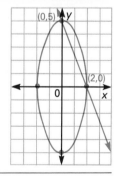

Teaching Resources

Quick Quizzes 90
**Reteaching and Practice
  Worksheets** 90

▬ GETTING STARTED

Prerequisite Quiz

Solve algebraically by the substitution method.

1. $3x - 2y = 2$
   $y = 2x - 4$ 6, 8
2. $y = 2x$
   $4x + 3y = 60$ 6, 12

Motivator

Ask students if they think a system like $x = 2y - 4$ and $x^2 + 4y^2 = 16$ can be solved by substitution. Ask them to explain what happens if they substitute $2y - 4$ for x in x^2. $(2y - 4)^2$ Then ask them to give the resulting equation. $(2y - 4)^2 + 4y^2 = 16$

▬ TEACHING SUGGESTIONS

Lesson Note

Emphasize that an algebraic method for solving a linear-quadratic system of equations is far more reliable than the graphing method.

To illustrate a system with no solution, consider Example 1 with $x + y = 10$ for the second equation.

Additional Example 1

Solve this system graphically.

$x^2 + y^2 = 25$
$y + 3x = 5$
$(0, 5), (3, -4)$

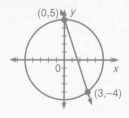

Highlighting the Standards

Standard 1c: These equations are another example of how problem situations now spring from within mathematics itself.

Earth Satellites: Communications satellites travel about the earth in elliptical orbits that have the center of the earth as one focus. A radio signal transmitted to or from the satellite may be thought of as a straight line. Thus, the system consisting of the satellite's orbit and the path of the radio signal may be thought of as the graph of a linear-quadratic system.

Critical Thinking Questions

Analysis: Ask students to use their imaginations to extend the idea of linear-quadratic systems into three dimensions. They may already know that an equation such as $x + y + z = 0$ is the graph of a plane in a coordinate space with three mutually perpendicular reference axes. (See Lesson 4.6)   Ask them to guess the shape of the graph of the equation $x^2 + y^2 + z^2 = 25$.   By analogy with $x^2 + y^2 = 25$, they should guess that it is a sphere with its center at the origin (0, 0, 0) and with radius 5.   What are the possible shapes of the intersection of the plane and sphere? Students should see that there are three possible answers: no points at all (in case the plane and sphere do not intersect), a point (in case the plane is tangent to the sphere), and a circle (in case the plane and sphere intersect). Some students will see that for the plane and sphere mentioned here, the intersection is a circle.

Algebraic methods are usually more accurate than graphical methods for solving linear-quadratic systems. An algebraic solution by substitution is illustrated in Example 2.

EXAMPLE 2   Solve $\begin{aligned} 2x - 3y &= 3 \\ x^2 - 2y^2 &= 7 \end{aligned}$ algebraically.

Solution

Solve the linear equation for one of its variables.

$$2x - 3y = 3$$
$$x = \frac{3y + 3}{2}$$

Substitute this expression for x in the quadratic equation.

$$x^2 - 2y^2 = 7$$
$$\left(\frac{3y + 3}{2}\right)^2 - 2y^2 = 7$$
$$\frac{9y^2 + 18y + 9}{4} - 2y^2 = 7$$

Multiply each side by 4.

$$9y^2 + 18y + 9 - 8y^2 = 28$$

Solve the quadratic equation.

$$y^2 + 18y - 19 = 0$$
$$(y + 19)(y - 1) = 0$$
$$y = -19 \qquad or \qquad y = 1$$

Find the corresponding values of x by substituting for y in the linear equation already solved for x.

$$x = \frac{3y + 3}{2} \qquad\qquad x = \frac{3y + 3}{2}$$
$$x = \frac{3(-19) + 3}{2} \qquad x = \frac{3(1) + 3}{2}$$
$$x = -27 \qquad\qquad x = 3$$

Check by substituting the ordered pairs $(-27, -19)$ and $(3,1)$ in both equations. This is left for you.

Thus, $(-27, -19)$ and $(3,1)$ are the solutions of the system.

EXAMPLE 3   Find the length of a side of each of two squares given that the sum of their perimeters is 48 cm and the sum of their areas is 74 cm^2.

Plan

Let x = the length of a side of the larger square and y = the length of a side of the smaller square. Then the respective perimeters are $4x$ and $4y$, and the respective areas are x^2 and y^2.

Additional Example 2

Solve algebraically.

$$\frac{x^2}{4} + \frac{y^2}{25} = 1$$
$$5x + 2y = 10$$

(2,0), (0,5)

Additional Example 3

A two-digit number is represented by $10t + u$. The sum of the squares of the digits is 13. If the number is decreased by the number with its digits reversed, the result is 9. Find the two-digit number.   32

Write a system and solve by substitution.

The sum of the perimeters is 48 cm.   (1) $4x + 4y = 48$

The sum of the areas is 74 cm².   (2) $x^2 + y^2 = 74$

Solve (1) for x.

$$4x = 48 - 4y$$
$$x = 12 - y$$

Substitute for x in (2).

$$(12 - y)^2 + y^2 = 74$$
$$144 - 24y + y^2 + y^2 = 74$$
$$2y^2 - 24y + 70 = 0$$
$$y^2 - 12y + 35 = 0$$
$$(y - 5)(y - 7) = 0$$
$$y = 5 \text{ or } y = 7$$

Use $x = 12 - y$ to find x.   If $y = 5$, then $x = 12 - 5 = 7$.

If $y = 7$, then $x = 12 - 7 = 5$.

Because $x > y$, discard $x = 5$, $y = 7$.

Check $x = 7$ and $y = 5$.   Sum of perimeters = $28 + 20 = 48$

Sum of areas = $49 + 25 = 74$

Thus, 7 cm and 5 cm are the lengths of the sides of the two squares.

Classroom Exercises

Tell what substitution might be made as a first step in the solution of each system.

1. $x^2 + 4y^2 = 16$
$x = 2y - 4$

2. $x^2 + y^2 = 13$
$y = 2x - 1$

3. $4x^2 + 9y^2 = 36$
$x = y - 3$

4. $x^2 + 4y^2 = 25$
$x + 2y = 1$

5–8. Solve the systems of Classroom Exercises 1–4 algebraically.

     5. $(-4,0)$, $(0,2)$ **6.** $\left(-\frac{6}{5}, -\frac{17}{5}\right)$, $(2,3)$ **7.** $(-3,0)$, $\left(-\frac{15}{13}, \frac{24}{13}\right)$ **8.** $(-3,2)$, $\left(4, -\frac{3}{2}\right)$

Written Exercises

Solve each system of equations graphically.

1. $x^2 + y^2 = 4$
$x + y = 2$

2. $y = x^2 - 5$
$2x - y = -3$

3. $xy = 9$
$y = x$

4. $\dfrac{x^2}{16} + \dfrac{y^2}{9} = 1$
$3x - 4y = 12$

Solve each system of equations algebraically.

5. $y = x^2 + 3x - 1$
$4x - y = -1$
$(2,9)$, $(-1,-3)$

6. $x^2 + y^2 = 25$
$2y = x + 5$
$(-5,0)$, $(3,4)$

7. $x + y = 4$
$y = x^2 + 4x - 20$
$(3,1)$, $(-8,12)$

8. $2a + b = 7$
$a^2 - b^2 = 8$
(a,b): $(3,1)$, $\left(\frac{19}{3}, -\frac{17}{3}\right)$

Common Error Analysis

Error: When substituting an expression such as $\dfrac{3y + 3}{2}$ for x in $x^2 - 2y^2 = 7$, as in Example 2, students frequently square $\dfrac{3y + 3}{2}$ incorrectly. They may forget to square the denominator or they may omit the middle term in squaring $3y + 3$.

Emphasize that $(3y + 3)^2$ is *not* $9y^2 + 9$, but rather $9y^2 + 18y + 9$.

Checkpoint

Solve each system of equations algebraically. Give irrational solutions in simplest radical form.

1. $y = x^2 - 2x + 6$
$y = -x + 6$ $(0,6)$, $(1,5)$

2. $xy = 8$
$y = 4x$ $(\sqrt{2}, 4\sqrt{2})$, $(-\sqrt{2}, -4\sqrt{2})$

3. $2a - b = 4$
$a^2 - b^2 = 4$ $(2,0)$, $\left(\frac{10}{3}, \frac{8}{3}\right)$

Closure

Ask students to explain how to solve a linear/quadratic system graphically. See Example 1. Ask them to give the steps in solving the system $y = x + 3$ and $y = x^2 + 3x - 1$ algebraically. See Example 2 for the steps to an algebraic solution.

Guided Practice

Classroom Exercises 1–8

Independent Practice

A Ex. 1–8, **B** Ex. 9–20, **C** Ex. 21–24

Basic: WE 1–13 odd, Brainteaser
Average: WE 7–19 odd, Brainteaser
Above Average: WE 11–23 odd, Brainteaser

Additional Answers

Classroom Exercises

Answers may vary for Exercises 1–4. Possible answers are given.

1. $(2y - 4)^2$ for x^2
2. $(2x - 1)^2$ for y^2
3. $4(y - 3)^2$ for $4x^2$
4. $(1 - 2y)^2$ for x^2

Written Exercises

1.

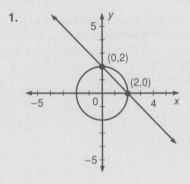

See page 455 for the answers to Written Ex. 2–4 and 9–16.

---

9. $4x^2 + 9y^2 = 36$
$3y = 12 - 4x$

10. $c^2 + d^2 = 25$
$3c - 4d = 25$

11. $y^2 - x^2 = 16$
$3x + 5y = 16$

12. $x^2 + y^2 = 50$
$9x + 7y = 70$

13. $y = x^2 + 4$
$y = x + 1$

14. $x^2 - y^2 = 1$
$x - y = 3$

15. $x^2 + 4y^2 = 4$
$2x - 2y = -6$

16. $l = w + 6$
$l = -w^2 + 6w$

Use a linear-quadratic system in two variables to solve each problem.

17. Find the length of a side of each of two squares given that the sum of their perimeters is 44 ft and the sum of their areas is 73 ft². 8 ft, 3 ft

18. The perimeter of a rectangular lot is 88 m, and its area is 480 m². Find the width and the length. 20 m × 24 m

19. If the numerator of a fraction $\frac{n}{d}$ is increased by 3, and the denominator is decreased by 3, the resulting fraction is the reciprocal of the original fraction. The numerator of the original fraction is 1 more than one-half its denominator. What is the original fraction? $\frac{5}{8}$

20. A positive two-digit number is represented by $10t + u$. The sum of the squares of the digits is 26. If the number is decreased by the number with its digits reversed, the result is 36. Find the two-digit number. 51

Solve each system of equations. Give irrational solutions in simplest radical form. $\left(\frac{4}{5}, \frac{9}{5}\right), \left(\frac{16}{5}, \frac{1}{5}\right)$ $(2,0), \left(-\frac{10}{3}, -\frac{8}{3}\right)$ $\left(\frac{1}{3}, \frac{1}{2}\right), (2, -2)$

21. $2x + 3y = 7$
$\frac{x^2}{2y} + 2y = \frac{5}{y} + 1$

22. $x - 2y = 2$
$\frac{y^2}{x + 2} = x - 2$

23. $\frac{1}{x} - \frac{1}{y} = 1$
$3x + 2y = 2$

24. For what values of m will the line $y = mx + 8$ and the circle $x^2 + y^2 = 25$ have exactly one point in common? $\pm\frac{\sqrt{39}}{5}$

Mixed Review

Simplify. 7.1, 8.3, 8.4, 8.6, 9.2

1. $\frac{x^3 - 27}{x - 3}$ $x^2 + 3x + 9$

2. $\sqrt{32} - 6\sqrt{8}$ $-8\sqrt{2}$

3. $\frac{26}{4 - \sqrt{3}}$ $8 + 2\sqrt{3}$

4. $(-27)^{-\frac{2}{3}}$ $\frac{1}{9}$

5. $\frac{-4}{1 - i}$ $-2 - 2i$

Brainteaser

A mule and a donkey were carrying bundles of wheat to a market. The mule said, "If you give me one bundle of wheat, then I will have twice as many as you. However, if I give you one bundle of wheat, then we will each be carrying the same number." How many bundles of wheat was each carrying? Mule: 7 bundles, donkey: 5 bundles

---

Enrichment

Draw this figure on the board.

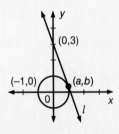

Line *l* passes through (0,3) and is tangent to the circle $x^2 + y^2 = 1$ at the point having coordinates (a,b). Challenge the students to find the values of a and b.

If the students have trouble deciding on a strategy to find a solution, you can offer this hint: A radius drawn to (a,b) will be ⊥ to line *l*. $a = \frac{2\sqrt{2}}{3}, b = \frac{1}{3}$

12.10 Quadratic-Quadratic Systems

Objectives

To solve systems of two quadratic equations
To solve word problems involving systems of two quadratic equations
To graph quadratic inequalities

A system of two quadratic equations may have 0, 1, 2, 3, or 4 real solutions, as illustrated by the systems of a parabola and a circle shown below.

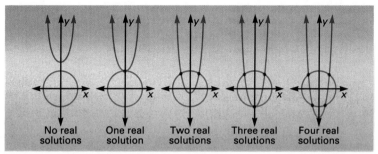

| No real solutions | One real solution | Two real solutions | Three real solutions | Four real solutions |

The graphical solution of a system of quadratic equations usually yields only approximate results, as illustrated in Example 1.

EXAMPLE 1

Solve the system $\begin{array}{c} x^2 + y^2 = 25 \\ y = x^2 + 1 \end{array}$ graphically.

Estimate the solutions to the nearest half unit.

Solution

$x^2 + y^2 = 25$: circle of radius 5 with its center at $(0,0)$

Write $y = x^2 + 1$ in the standard form of a parabola.

$(x - 0)^2 = (y - 1)$: parabola opening upward with its vertex at $(0,1)$

The graphs intersect at two points whose approximate coordinates are $(2,4.5)$ and $(-2,4.5)$.

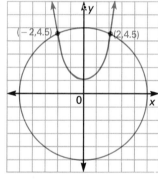

Checks

Check by substituting $(2,4.5)$ and $(-2,4.5)$ in both equations.

The approximate solutions are $(2,4.5)$ and $(-2,4.5)$, or $(\pm 2,4.5)$.

Additional Example 1

Solve the system by graphing.

$4x^2 - 9y^2 = 36$
$xy = 5$

Estimate the solutions to the nearest half-unit. $(3.5,1.5), (-3.5,-1.5)$

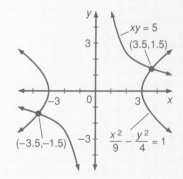

Teaching Resources

Quick Quizzes 91
Reteaching and Practice
  Worksheets 91

▮▮GETTING STARTED

Prerequisite Quiz

Solve each system.

1. $x + y = 8$ (7,1)
   $x - y = 6$
2. $3x + 4y = 18$ (2,3)
   $2x + 5y = 19$
3. $2x + 3y = 4$ (2,0)
   $7x = 4y + 14$
4. $5x - 3y = -2$ (2,4)
   $2x = 4y - 12$

Motivator

Recall that to solve a linear system like $3x + 2y = 5$, $5x + 7y = 9$, substitution is not very practical since *no* variable coefficient is 1. Ask students to give an alternate method for solving the system above. Linear combination method To solve a system such as $3x^2 + 2y^2 = 14$
    $4x^2 + 5y^2 = 21$,
ask students which method they would use. Linear combination method Ask them what multipliers they would use to eliminate the y-terms. Answers will vary. One possible answer: (1) = 5; (2) = −2

Highlighting the Standards

Standard 4c: Solutions of systems of quadratics can be easier when students have a mental picture of the graphs and the possible points of intersection.

Lesson Note

The graphical method of solving a system of equations allows students to visualize the number of possible solutions. However, the more reliable algebraic methods should be emphasized.

Math Connections

Interplanetary Orbits: An example of the graph of a quadratic-quadratic system is the collision of a planet with another heavenly body, such as a comet. In both cases, the orbits are elliptical.

Critical Thinking Questions

Analysis: After students have become familiar with Example 4, ask them to consider the following compound inequality.

$$x^2 + y^2 \geq 25 \text{ and } x^2 + y^2 \leq 49$$

What is the shape of its graph? Some students will, without graphing the inequality, see that the graph is a ring framed by two origin-centered circles, one with radius 5, the other with radius 7.

Common Error Analysis

Error: When the process of solving a system leads to irrational answers, as in Example 3, some students tend to forget that there are two potential solutions.

Emphasize that for an equation such as $x^2 = 12$, there are two possible solutions, $\pm 2\sqrt{3}$, even though one of them may not check in the original system of equations.

Systems of two quadratic equations can also be solved by the substitution and linear-combination methods. The system in the next example is solved algebraically by the linear-combination method.

EXAMPLE 2   Solve $\dfrac{x^2}{4} + \dfrac{y^2}{10} = 5$ algebraically.
$$2x^2 + 3y^2 = 84$$

Give irrational solutions in simplest radical form, and then find each to the nearest tenth using a square root table or a calculator.

Solution

Multiply each side of the first equation by the LCD, 20, to eliminate the fractions.	$\dfrac{x^2}{4} + \dfrac{y^2}{10} = 5$
	$5x^2 + 2y^2 = 100$

Solve the system.
     (1) $5x^2 + 2y^2 = 100$
     (2) $2x^2 + 3y^2 = 84$

Multiply each side of (1) by 3. $15x^2 + 6y^2 = 300$
Multiply each side of (2) by -2. $\underline{-4x^2 - 6y^2 = -168}$

Add the resulting equations. $11x^2 \quad\quad = 132$
$$x^2 = 12$$
$$x = \pm\sqrt{12} = \pm 2\sqrt{3}$$

Find y by substituting $\pm 2\sqrt{3}$ for x in (1).
$$5x^2 + 2y^2 = 100$$
$$5(\pm 2\sqrt{3})^2 + 2y^2 = 100$$
$$5(12) + 2y^2 = 100$$
$$2y^2 = 40$$
$$y^2 = 20$$
$$y = \pm\sqrt{20} = \pm 2\sqrt{5}$$

The four solutions are $(\pm 2\sqrt{3}, \pm 2\sqrt{5})$.

$2\sqrt{3}$ to the nearest tenth is 3.5. $2\sqrt{5}$ to the nearest tenth is 4.5.

Thus, the solutions to the nearest tenth are $(\pm 3.5, \pm 4.5)$.

The check is left for you.

EXAMPLE 3   Find the dimensions of the rectangle whose area is 12 ft^2 and whose diagonal is 5 ft long.

Plan

What are you to find?	the length and width
Choose two variables.	Let l = the length.
What do they represent?	Let w = the width.

Additional Example 2

Solve algebraically.

$$x^2 + 4y^2 = 16$$
$$3x^2 - 4y^2 = 12$$

$(\sqrt{7}, \tfrac{3}{2}), (\sqrt{7}, -\tfrac{3}{2}), (-\sqrt{7}, \tfrac{3}{2}), (-\sqrt{7}, -\tfrac{3}{2})$

Additional Example 3

The area of a rectangle is 15 cm^2. The length of a diagonal is $6\tfrac{1}{2}$ cm. Find the perimeter of the rectangle.   17 cm

| What is given? | The area is 12, so $lw = 12$. |
| | The diagonal is the hypotenuse of a right triangle, so $l^2 + w^2 = 25$. |

Solution

Write a system.	(1) $lw = 12$ (2) $l^2 + w^2 = 25$
Solve the system.	Solve (1) for either variable. (3) $l = \frac{12}{w}$
Substitute (3) in (2).	(2) $\left(\frac{12}{w}\right)^2 + w^2 = 25$

Solve for w.

$$\frac{144}{w^2} + w^2 = 25$$

$$144 + w^4 = 25w^2$$

$$w^4 - 25w^2 + 144 = 0$$

$$(w^2 - 16)(w^2 - 9) = 0$$

$w^2 - 16 = 0$	or	$w^2 - 9 = 0$
$w^2 = 16$	or	$w^2 = 9$
$w = \pm 4$	or	$w = \pm 3$
$w = 4$	or	$w = 3$

Lengths are positive; discard the negative solutions.

Substitute 4 and 3 for w in (3).　$l = \frac{12}{w}$　　$l = \frac{12}{w}$

$$l = \frac{12}{4} = 3 \qquad l = \frac{12}{3} = 4$$

So, $l = 3$ and $w = 4$, or $l = 4$ and $w = 3$.

The two solutions describe the same rectangle, so choose $l = 4$ and $w = 3$.

Check in the original problem.	The area is 12: $3 \cdot 4 = 12$ True
	The diagonal's length is 5: $\sqrt{3^2 + 4^2} = 5$ True
State the answer.	The dimensions are 3 ft by 4 ft.

The graph of $x^2 + y^2 = 25$ at the right separates the plane into three sets of points: a circle, points outside the circle, and points inside the circle. When graphing such inequalities as $x^2 + y^2 < 25$ or $x^2 + y^2 > 25$, you can decide whether to shade the *interior* or the *exterior* by choosing sample points.

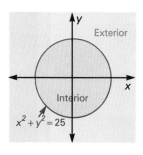

Additional Answers

1. Multiply first row by −2 or 3.
2. Multiply first row by −1.
3. Multiply first row by 2 or second row by −2.
4. Multiply second row by −1.
5. Multiply first row by 2 or second row by −2.
6. Multiply first row by 4 and second row by −3.
7. (±5, ±3)
8. (0,−5), (±3,4)
9. (±2, ±3)

Written Exercises

1.

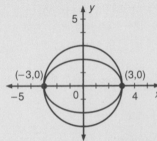

2.

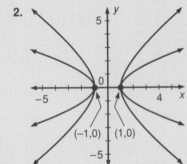

EXAMPLE 4   Graph $x^2 + y^2 > 25$.

Solution   First, graph the circle $x^2 + y^2 = 25$. Use a dashed curve because the inequality symbol $>$ indicates that the circle is not part of the graph. Test sample points.

Checks   Choose (0,0) in the interior.

$$x^2 + y^2 > 25$$
$$0^2 + 0^2 \overset{?}{>} 25$$
$$0 + 0 \overset{?}{>} 25$$
$$0 > 25 \text{ False}$$

Choose (7,6) in the exterior.

$$x^2 + y^2 > 25$$
$$7^2 + 6^2 \overset{?}{>} 25$$
$$49 + 36 \overset{?}{>} 25$$
$$85 > 25 \text{ True}$$

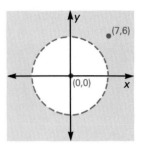

So, the graph consists of all points exterior to the circle.

Classroom Exercises

Tell what multipliers can be used to solve each system algebraically.

1. $x^2 + y^2 = 34$
   $2x^2 − 3y^2 = 23$

2. $x^2 − y = 5$
   $x^2 + y^2 = 25$

3. $2x^2 − y^2 = −1$
   $x^2 + 2y^2 = 22$

4. $5x^2 − y^2 = 1$
   $x^2 − y^2 = −3$

5. $2x^2 + y^2 = 8$
   $x^2 − 2y^2 = 4$

6. $3x^2 + 2y^2 = 13$
   $4x^2 + 5y^2 = 22$

7–9. Solve the systems of Classroom Exercises 1–3 algebraically.

10. Determine whether $\dfrac{x^2}{9} − \dfrac{y^2}{16} > 1$ describes the points in the interior or the exterior of the branches of the hyperbola $\dfrac{x^2}{9} − \dfrac{y^2}{16} = 1$. Is the point (6,8) in the interior or the exterior?   Interior; exterior

Written Exercises

Solve each system of equations graphically.

1. $x^2 + y^2 = 9$
   $4x^2 + 9y^2 = 36$

2. $x^2 − 4y^2 = 1$
   $x^2 − 1 = y^2$

3. $x^2 + y^2 = 4$
   $4x^2 − 9y^2 = 36$

Additional Example 4

Graph $4x^2 + 9y^2 < 36$.

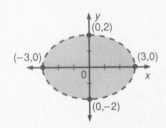

Enrichment

Explain that the hyperbolas $\dfrac{x^2}{a^2} − \dfrac{y^2}{b^2} = 1$ and $\dfrac{x^2}{a^2} − \dfrac{y^2}{b^2} = −1$ are called *conjugate* hyperbolas. Have the students choose suitable values of a and b, such as 3 and 2, and graph a pair of conjugate hyperbolas on the same rectangular axes. Then have them draw a conclusion about conjugate hyperbolas.   Conjugate hyperbolas have the same asymptotes.

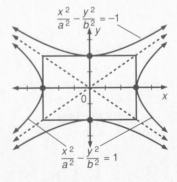

Solve each system algebraically. Give irrational solutions in simplest
radical form.

4–6. Solve the systems of Classroom Exercises 4–6. **4.** $(\pm 1, \pm 2)$ **5.** $(\pm 2, 0)$ **6.** $(\pm\sqrt{3}, \pm\sqrt{2})$

7. $x^2 - y^2 = 5$ $(-3, -2)$,   **8.** $xy = 6$ $(-\sqrt{6}, -\sqrt{6})$,   **9.** $9x^2 - 4y^2 = 32$

   $xy - 6 = 0$ $(3,2)$ $x^2 + y^2 = 12$ $(\sqrt{6}, \sqrt{6})$ $xy = 2$ $(-2, -1)$, $(2,1)$

10–12. Algebraically solve the systems of Written Exercises 1–3. **10.** $(\pm 3, 0)$ **11.** $(\pm 1, 0)$
                                                       12. No real solutions

Solve each system algebraically. Use a calculator to approximate
irrational solutions to the nearest tenth.

13. $y^2 - x^2 = -10$ **14.** $\dfrac{y^2}{6} - \dfrac{x^2}{3} = 1$ **15.** $\dfrac{x^2}{5} - \dfrac{4y^2}{5} = 1$ **16.** $\dfrac{y^2 - x^2}{5} = 1$

$\dfrac{y^2}{6} - \dfrac{x^2}{9} = 1$ $xy = 6$ $y = \dfrac{3}{x}$ $(-3, -1)$, $(3,1)$ $xy + 6 = 0$

$(\pm 6.9, \pm 6.2)$ $(-1.7, -3.5)$, $(1.7, 3.5)$ $(-2,3)$, $(2, -3)$

Write a system of two quadratic equations and solve the problem.

17. Find the dimensions of a rectangle with an area of 60 ft^2 and a
diagonal of 13 ft. 12 ft × 5 ft

18. The area of a rectangle is 48 m^2. The length of a diagonal is 10 m.
Find the perimeter of the rectangle. 28 m

19. The area of a rectangle is 12 in^2. The perimeter is 24 in. Find the
dimensions of the rectangle. $6 + 2\sqrt{6}$ in by $6 - 2\sqrt{6}$ in

20. Find two negative numbers such that the sum of their squares is
170, and twice the square of the first minus three times the square
of the second is 95. $-11, -7$

Graph each inequality.

21. $x^2 + y^2 > 64$ **22.** $\dfrac{x^2}{100} - \dfrac{y^2}{25} \leq 1$ **23.** $x^2 + y^2 \leq 1$ **24.** $\dfrac{x^2}{36} + \dfrac{y^2}{4} > 1$

Solve each system algebraically. Answer in simplest radical form.

25. $x^2 - 3xy + 2y^2 = 0$ $(2,1)$, $(-2, -1)$, **26.** $x^2 + 3xy = 28$ $(4,1)$, $(-4, -1)$,

$x^2 - xy + y^2 = 3$ $(\sqrt{3}, \sqrt{3})$, $(-\sqrt{3}, -\sqrt{3})$ $xy + 4y^2 = 8$ $(14, -4)$, $(-14, 4)$

27. Find an equation of the ellipse centered at the origin, with its foci
on the x-axis, containing the points $(\sqrt{2}, \sqrt{3})$, $(\sqrt{6}, 1)$. $\dfrac{x^2}{8} + \dfrac{y^2}{4} = 1$

Mixed Review

Simplify. **6.5, 7.1, 8.6, 9.2**

1. $\dfrac{x^3 - 27}{x^2 - 9}$ **2.** $32^{-\frac{4}{5}}$ $\dfrac{1}{16}$ **3.** $\dfrac{30}{2 + i}$ $12 - 6i$ **4.** $(x^3 + x^2 - 2x + 8)$

                                                                   $\div (x + 3)$

                                                                $x^2 - 2x + 4$, R: -4

3.

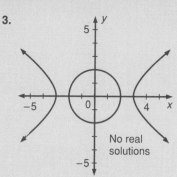

No real
solutions

21.

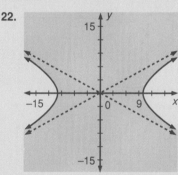

22.

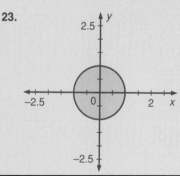

23.

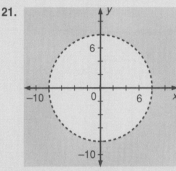

24.

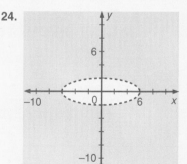

Mixed Review

1. $\dfrac{x^2 + 3x + 9}{x + 3}$

1.

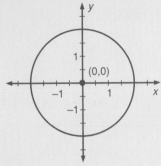

2.

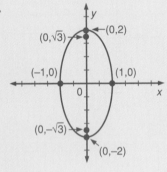

5.

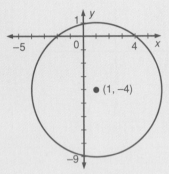

Chapter 12 Review

Key Terms

asymptote (p. 419)
center (p. 409)
circle (p. 409)
combined variation (p. 432)
constant of variation (p. 427)
directrix (p. 435)
ellipse (p. 413)
focal radius (pp. 413, 418)

focus (pp. 413, 418, 435)
hyperbola (p. 418)
inverse variation (p. 427)
joint variation (p. 431)
parabola (p. 435)
radius (p. 409)
rectangular hyperbola (p. 426)
vertex (pp. 414, 418, 435)

Key Ideas and Review Exercises

12.1 To identify the center (h,k) and radius r of a circle, write the equation in standard form: $(x - h)^2 + (y - k)^2 = r^2$, where $r > 0$.

Graph each circle whose equation is given.

1. $x^2 + y^2 = 4$

2. $x^2 - 2x + y^2 + 8y - 8 = 0$

Write the equation in standard form of the circle with given center C and given radius r, or with given center C and a given point P on the circle.

3. $C(0,0)$, $r = \sqrt{2}$ $x^2 + y^2 = 2$

4. $C(-3,1)$, $P(3,9)$ $(x + 3)^2 + (y - 1)^2 = 100$

12.2, To graph an origin-centered ellipse or hyperbola with foci on the x- or
12.3 y-axis, first write its equation in standard form.
          For an ellipse, use Theorems 12.3 and 12.4 in Lesson 12.2.
          For a hyperbola, use Theorems 12.5 and 12.6 in Lesson 12.3.

Graph each ellipse. Find the coordinates of its vertices and foci.

5. $4x^2 + y^2 = 4$

6. $36x^2 + 9y^2 = 144$

Given the following, write an equation in standard form of each ellipse.

7. Vertices: $(\pm 6, 0)$, $(0, \pm 2)$ $\dfrac{x^2}{36} + \dfrac{y^2}{4} = 1$

8. Vertices: $(0, \pm 7)$, Foci: $(0, \pm 4)$ $\dfrac{y^2}{49} + \dfrac{x^2}{33} = 1$

Graph each hyperbola. Find the coordinates of its vertices and foci. Write the equations of its asymptotes.

9. $x^2 - 25y^2 = 25$

10. $100y^2 - 25x^2 = 2{,}500$

Given the following, write an equation in standard form of each hyperbola.

11. Vertices: $(0, \pm 8)$; foci: $(0, \pm 10)$ $\dfrac{y^2}{64} - \dfrac{x^2}{36} = 1$

12. Vertices: $(\pm 4, 0)$; foci: $(\pm 5, 0)$ $\dfrac{x^2}{16} - \dfrac{y^2}{9} = 1$

452 Chapter 12 Review

6.

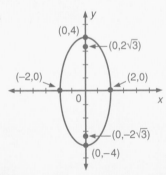

9.

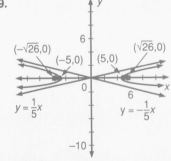

10.

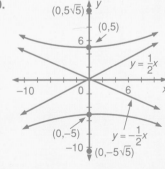

12.4 To graph an ellipse or hyperbola not centered at the origin, use the appropriate standard form (see Lesson 12.4).

Graph each ellipse or hyperbola. Label the vertices, center, and foci.

13. $\dfrac{(x + 7)^2}{36} + \dfrac{(y - 6)^2}{9} = 1$

14. $4x^2 - 24x - 9y^2 - 18y - 9 = 0$

12.5, To graph a rectangular hyperbola, $xy = k$, $k \neq 0$, first solve for y.
12.6 To solve problems involving variation, use the following: (1) $xy = k$, when y varies inversely as x, (2) $\dfrac{y}{ab} = k$, when y varies directly (jointly) as a and b, or (3) $\dfrac{yd}{bc} = k$, when y varies directly (jointly) as b and c and inversely as d.

15. Graph $xy = -8$.

16. Graph $xy = 48$.

17. If y varies inversely as x, and $y = 12$ when $x = 5$, what is y when $x = 18$? $\frac{10}{3}$

18. A number y varies jointly as x and z. If $y = -24$ when $x = 4$ and $z = 3$, what is y when $x = -6$ and $z = -2$? -24

19. A number y varies jointly as x and z and inversely as $\sqrt{w}$. If $y = 12$ when $x = 2$, $z = 6$, and $w = 9$, what is y when $x = 5$, $z = 7$, and $w = 25$? 21

12.7 To graph a parabola, or to write its equation, use Theorems 12.7 and 12.8 in Lesson 12.7.

Use the information provided to write an equation of each parabola.

20. focus: $F(6,8)$; directrix: $y = 12$
$(x - 6)^2 = -8(y - 10)$

21. focus: $F(-2,4)$; vertex: $V(6,4)$
$(y - 4)^2 = -32(x - 6)$

Graph each parabola. Label the vertex, focus, directrix, and axis of symmetry.

22. $x^2 - 6x + 8y - 7 = 0$

23. $y^2 - 8y + 4x - 8 = 0$

12.8 To identify a conic section, change its equation to a standard form.

Identify each conic section whose equation is given.

24. $4(x - 3)^2 + 9(y + 1)^2 = 36$ Ellipse

25. $4x^2 + 8x - 9y^2 + 36y - 68 = 0$ Hyperbola

12.9, Solve linear-quadratic and quadratic-quadratic systems by graphing, or
12.10 use one of the algebraic methods: substitution or linear-combination.

26. Solve by graphing. Estimate answers to the nearest half unit.
$2x - 3y = 6$
$9x^2 + 16y^2 = 144$

27. Solve algebraically. Give irrational answers to the nearest tenth.
$4x^2 + 2y^2 = 20$
$3x^2 - 4y^2 = 4$ $(\pm 2, \pm 1.4)$

13.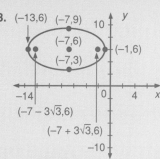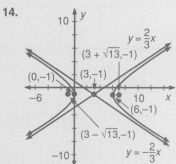

14.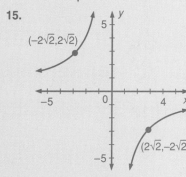

15.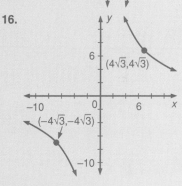

16.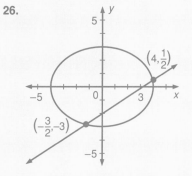

22.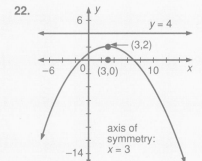

23.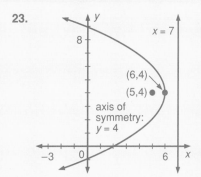

26.

7.

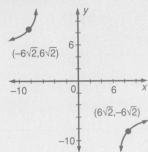

$(-6\sqrt{2},6\sqrt{2})$

$(6\sqrt{2},-6\sqrt{2})$

8.

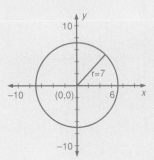

r=7

(0,0)

9.

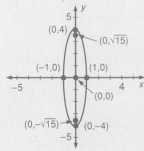

(0,4)

$(0,\sqrt{15})$

(−1,0) (1,0)

(0,0)

$(0,-\sqrt{15})$ (0,−4)

10.

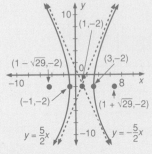

(1,−2)

$(1-\sqrt{29},-2)$ (3,−2)

(−1,−2)

$(1+\sqrt{29},-2)$

$y=\frac{5}{2}x$ $y=-\frac{5}{2}x$

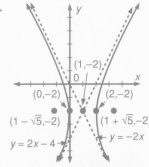

Given the following data, write the equation of each conic section in
standard form.

$\dfrac{x^2}{36}+\dfrac{y^2}{16}=1$

1. circle; center: $(-6,4)$
radius: 7 $(x+6)^2+(y-4)^2=49$

2. ellipse; vertices: $(\pm 6,0)$, $(0,\pm 4)$

3. ellipse; vertices: $(\pm 3,0)$ $\dfrac{y^2}{25}+\dfrac{x^2}{9}=1$
foci: $(0,\pm 4)$

4. hyperbola; vertices: $(0,\pm 5)$ $\dfrac{y^2}{25}-\dfrac{x^2}{39}=1$
foci: $(0,\pm 8)$

5. parabola; focus: $(4,6)$ $(x-4)^2=24y$
directrix: $y=-6$

6. parabola; focus: $(8,2)$ $(y-2)^2=20(x-3)$
vertex: $(3,2)$

7. Graph the rectangular hyperbola $xy=-72$.

Graph each equation. For a circle, label the center and the radius.
For a parabola, label the vertex, the focus, and the directrix. For an
ellipse, label the vertices, the center, and the foci. For a hyperbola,
label the vertices, the center, and the foci, and write the equations
of the asymptotes.

8. $x^2+y^2=49$

9. $16x^2+y^2=16$

10. $\dfrac{(x-1)^2}{4}-\dfrac{(y+2)^2}{25}=1$

11. $\dfrac{(x-4)^2}{4}+\dfrac{(y+1)^2}{9}=1$

12. $y^2-4y-6x-14=0$

13. $4x^2-y^2-4y-8x-4=0$

14. A number y varies directly as x and z and inversely as the square
of r. If $y=6$ when $x=3$, $z=4$, and $r=7$, what is y when
$x=6$, $z=8$, and $r=4$? 73.5

15. A number y varies jointly as x and the cube of z. If $y=160$ when
$x=4$ and $z=2$, what is y when $x=-5$ and $z=3$? −675

16. The natural frequency of a string under constant tension varies inversely as its length. If a string 40 cm long vibrates 680 times per
second, what length must the string be to vibrate 850 times per
second under the same tension? 32 cm

17. Solve $\begin{aligned}x^2+y^2&=25\\x^2-y^2&=25\end{aligned}$ graphically.

Solve each system algebraically. Give irrational solutions in simplest
radical form.
$(2\sqrt{3},\sqrt{3})$, $(-2\sqrt{3},-\sqrt{3})$, $(\sqrt{3},2\sqrt{3})$, $(-\sqrt{3},-2\sqrt{3})$

18. $\begin{aligned}2x^2+y^2&=22\\x^2+3y^2&=21\end{aligned}$ $(\pm 3,\pm 2)$

19. $\begin{aligned}x^2+y^2&=15\\xy&=6\end{aligned}$

20. Solve algebraically.
$\begin{aligned}x^2-5xy+6y^2&=0\\x+y^2-7y&=0\end{aligned}$ (0,0), (10,5), (12,4)

21. Find the square roots of $3-4i$. (HINT:
Let $x+yi$ be a square root. Then
$(x+yi)^2=3-4i$.) $-2+i$, $2-i$

11.

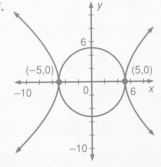

(4,2)

$(4,-1+\sqrt{5})$

(2,−1)

(4,−1)
(6,−1)

$(4,-1-\sqrt{5})$ (4,−4)

12.

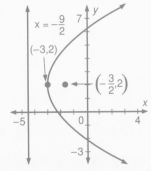

$x=-\frac{9}{2}$

(−3,2)

$\left(-\frac{3}{2},2\right)$

13.

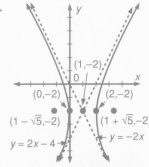

(1,−2)

(0,−2)   (2,−2)

$(1-\sqrt{5},-2)$ $(1+\sqrt{5},-2)$

$y=2x-4$ $y=-2x$

17.

(−5,0)   (5,0)

College Prep Test

Choose the *one* best answer to each question or problem.

Sample Question

*(x, y, z) is defined as follows:

$$*(x, y, z) = \dfrac{\frac{x}{y}}{z}$$

Then *(6,10,12) = ___?___

(A) $\frac{1}{20}$ (B) $\frac{1}{5}$ (C) $\frac{1}{6}$

(D) 5 (E) $\frac{36}{5}$

The answer is A: $\frac{6}{10} \div 12 = \frac{1}{20}$.

1. Use the definition in the sample question.

*(3,9,3) = ___?___ E

(A) *(1,3,1) (B) *(9,3,9)
(C) *(3,1,3) (D) *(6,18,6)
(E) *(12,9,12)

2. If ★p means p(p + 1)(p + 2), then $\frac{★8}{★2}$ = ___?___ D

(A) 4 (B) 10
(C) 20 (D) 30
(E) 36

3. (a,b) * (c,d) is defined as (ac + bd, ad + bc). If (a,b) * (x,y) = (a,b), then (x,y) = ___?___ B

(A) (0,0) (B) (1,0)
(C) (0,1) (D) (1,1)
(E) None of these

4. The lines with equations x = 2, x = 5, y = 7, and y = 3 form a rectangle. What is the area of this rectangle? A

(A) 12 (B) 14 (C) 15
(D) 17 (E) 28

5. The following are the dimensions of five rectangular solids. All have the same volume *except* ___?___ B

(A) 10 by 15 by 2
(B) 6 by 50 by 3
(C) 4 by 75 by 1
(D) 8 by 75 by $\frac{1}{2}$
(E) $\frac{1}{4}$ by 25 by 48

6.

PQRS is a square, and the two shaded regions are semicircles. What is the ratio of the total area of the shaded regions to the area of the square? A

(A) $\frac{\pi}{4}$ (B) $\frac{\pi}{2}$ (C) π

(D) $\frac{2\pi}{3}$ (E) $\frac{3\pi}{4}$

7. What is one-third of the perimeter of the square whose area is 36 square units? C

(A) 4 (B) 6 (C) 8
(D) 12 (E) 48

8. Which of the following conditions guarantee that a − b is a negative number? C

(A) b > 0 (B) a > 0
(C) b > a (D) b = a
(E) a > b

9. How many points do the graphs of y = x² and xy = 27 have in common? B

(A) 0 (B) 1 (C) 2
(D) 3 (E) 4

2.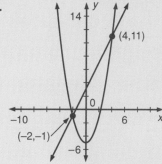

3.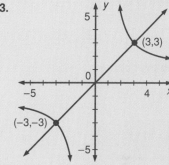

4.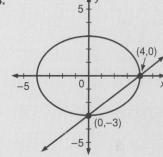

9. (3,0), $\left(\frac{9}{5},\frac{8}{5}\right)$ **10.** (c,d): (3,−4)

11. (−3,5) **12.** (7,1), $\left(\frac{35}{13},\frac{85}{13}\right)$

13. No real solutions **14.** $\left(\frac{5}{3},-\frac{4}{3}\right)$

15. No real solutions **16.** (w,l): (2,8), (3,9)

3.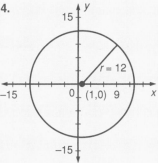

4.

For additional standardized test practice, see the SAT/ACT test booklet for Cumulative Tests (Chapters 9–12).

Cumulative Review

4. $-2, \frac{1}{2}, 2$ (mult = 2)

8.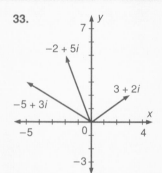
$-8 -6 -4 -2\ 0\ 2$
$\{x \mid x \ge -6\}$

9.
$-8 -6 -4 -2\ 0\ 2$
$\{n \mid -6 < n < 2\}$

10.
$-9 -6 -3\ 0\ 3\ 5\ 6$
$\{x \mid x < -6 \text{ and } x > 5\}$

14. $\frac{4}{3}$; up to the right

15. $4x - 3y = -12$

21.

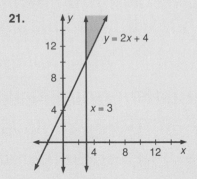

$y = 2x + 4$
$x = 3$

29. $16x^4y^3\sqrt{7x}$

---

1. What property of operations with real numbers is illustrated by $(3a + 2b) + c = 3a + (2b + c)$? Assoc Prop Add   **1.3**

Solve for x.

2. $2 - 3[4 - 2(3 - x)] = -16$ 4   **1.4**

3. $6x^3 - 27x^2 + 30x = 0$ 0, 2, $\frac{5}{2}$ **6.1**

4. $2x^4 - 5x^3 - 6x^2 + 20x = 8$ **6.6**

5. $M = \frac{P}{3}(x + y)$ $\frac{3M}{P} - y$ **7.6**

6. $x^2 + 2x + 4 = 0$ $-1 \pm i\sqrt{3}$ **9.5**

7. $\sqrt{2x - 5} + 4 = x$ 7   **10.3**

Find the solution set of each inequality. Graph the solutions.

8. $2x - 4 \le 4 - (2 - 3x)$ **2.1**

9. $|2n + 4| < 8$ **2.3**

10. $x^2 + x - 30 > 0$ **6.4**

For Exercises 11–13, use the relations described by $f(x) = 3x + 1$ and $g(x) = x^2 - 2$.

11. Determine the range of $f(x)$ for $D = \{-6, 0, 3, 9\}$. Is $f(x)$ a function? $\{-17, 1, 10, 28\}$; yes   **3.1**

12. Find $f(g(x))$. $3x^2 - 5$ **5.9**

13. Find $g(f(-2))$. 23

Use points $A(3,8)$ and $B(9,16)$ for Exercises 14–17.

14. Find the slope of $\overleftrightarrow{AB}$, and describe its slant.   **3.3**

15. Write an equation of $\overleftrightarrow{AB}$.   **3.4**

16. Find AB. 10   **11.1**

17. Write an equation of the perpendicular bisector of $\overline{AB}$.   **11.2**
$3x + 4y = 66$

Solve each system algebraically.

18. $2x + y = 8$ $\left(\frac{4}{3}, \frac{16}{3}\right)$ **4.2**
$x = y - 4$

19. $3x - 2y = -2$ (2,4)   **4.3**
$5x + 3y = 22$

20. $2x + y - 3z = -5$ (1,2,3)   **4.6**
$3x - y + z = 4$
$x + 3y - z = 4$

Solve by graphing.

21. $2x - y \le -4$ and $x \ge 3$ **4.7**

Simplify. Use positive exponents.

22. $(4a^{-3})^2$ $\frac{16}{a^6}$ **5.2**

23. $\frac{28x^{-4}y^3}{21x^{-2}y^{-1}}$ $\frac{4y^4}{3x^2}$

Factor completely.

24. $27y^3 - 125$ $(3y - 5)(9y^2 + 15y + 25)$ **5.6**

25. $xa - ay - 2x + 2y$ **5.7**
$(a - 2)(x - y)$

Divide.

26. $(3x^3 - 18x + 12) \div (x - 3)$ **5.8**
$3x^2 + 9x + 9$, R: 39

Simplify.

27. $\frac{x^2 - 5x}{3x^3} \div \frac{25 - x^2}{9x^2}$ $\frac{-3}{x + 5}$ **7.2**

28. $\frac{1 - 5x^{-1} - 14x^{-2}}{2 - 3x^{-1} - 14x^{-2}}$ $\frac{x - 7}{2x - 7}$ **7.5**

29. $8x^2y\sqrt{28x^5y^4}$, $x > 0$, $y > 0$ **8.2**

30. $\frac{6}{\sqrt{5} - \sqrt{3}}$ $3\sqrt{5} + 3\sqrt{3}$ **8.4**

31. $\frac{26}{3 + 2i}$ $6 - 4i$ **9.2**

33.

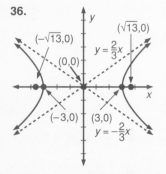

$-2 + 5i$
$-5 + 3i$
$3 + 2i$

35.

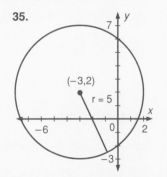

$(-3,2)$
$r = 5$

36.

$(-\sqrt{13},0)$
$(0,0)$
$(\sqrt{13},0)$
$y = \frac{2}{3}x$
$(-3,0)$ $(3,0)$
$y = -\frac{2}{3}x$

37.

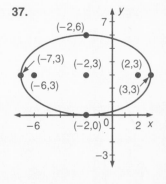

$(-2,6)$
$(-7,3)$ $(-2,3)$ $(2,3)$
$(-6,3)$ $(3,3)$
$(-2,0)$

32. Find the maximum value of P given that $P = -2x^2 + 8x + 4$. 12   *9.4*

33. Draw the vectors corresponding to $-5 + 3i$ and $3 + 2i$. Then draw their sum.   *9.8*

34. Determine the nature of the roots of $3x^2 + 5x - 2 = 0$ without solving the equation.   *10.2*
Two, rational

Graph each conic section. Label whatever applies: center, radius, vertices, foci, directrix, or asymptotes.

35. $x^2 + 6x + y^2 - 4y - 12 = 0$ *12.1*

36. $4x^2 - 9y^2 = 36$ *12.3*

37. $9x^2 + 25y^2 + 36x - 150y + 36 = 0$ *12.4*

38. $y = -\dfrac{24}{x}$ *12.5*

39. $y^2 + 8x - 6y + 1 = 0$ *12.7*

Write the equation in standard form of the given conic section. Graph the equation. (Exercises 40–41)

40. Circle; center: $(-5,4)$ *12.1*
radius: 6

41. Parabola; directrix: $y = -4$ *12.7*
focus: $(2,6)$

42. A number y varies jointly as r and s^2, and inversely as t. If $y = 120$ when $r = 5$, $s = 3$, and $t = 2$, what is r when $y = 80$, $s = 1$, and $t = 6$? 90   *12.6*

Solve each system. Give irrational answers to the nearest tenth. (Exercises 43–44)

43. $y = x^2 - 7x - 3$ *12.9*
$2x - y = 3$ (0,−3), (9,15)

44. $4x^2 + 5y^2 = 21$ *12.10*
$xy = 1$ $\left(-\frac{1}{2},-2\right), \left(\frac{1}{2},2\right), (-2.2, -0.4), (2.2, 0.4)$

45. A rectangle is 3 cm longer than it is wide. Find the set of all possible widths given that the perimeter is less than 34 cm and greater than 22 cm.   *2.4*

46. A passenger train leaves a station 2 h after a freight train, and, traveling at an average speed of 60 mi/h, overtakes it in 7 h. What is the average speed of the freight train?   *2.6*

47. How many ounces of water must be added to 30 oz of a 20% salt solution to dilute it to a 10% salt solution? 30 oz   *4.4*

48. Find three consecutive multiples of 4 such that the first times the third is 240.   *6.2*

49. A painter working alone can paint a room in 6 h, while her helper needs 9 h to do it. The painter works alone for 2 h, and then they finish the room together. What is the total time needed to paint the room? 4.4 h   *7.7*

50. One leg of a right triangle is 3 cm longer than twice the length of the other leg, and 1 cm shorter than the hypotenuse. Find the length of the hypotenuse in simplest radical form. $8 + 2\sqrt{11}$ cm   *9.6*

51. The mass of a metal cylinder varies jointly as its height and the square of the radius of its base. One cylinder has a mass of 120 g. Find the mass of a second cylinder made of the same metal, 3 times as high, and having one-half the base radius of the first. 90 g   *12.6*

38.
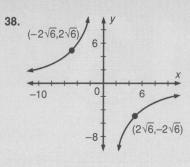

39.
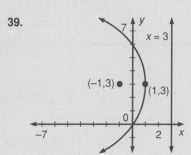

40. $(x + 5)^2 + (y - 4)^2 = 36$
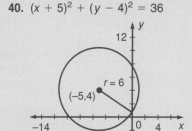

41. $(x - 2)^2 = 20(y - 1)$
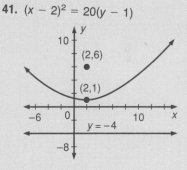

45. $\{w \mid 4\text{ cm} < w < 7\text{ cm}\}$

46. $46\frac{2}{3}$ mi/h

48. $(-20, -16, -12), (12, 16, 20)$

13 EXPONENTIAL AND LOGARITHMIC FUNCTIONS

OVERVIEW

Exponential functions, together with their inverses, the logarithmic functions, are the core of the chapter. After an introductory lesson on inverses, several lessons are devoted to these two functions and to their applications. The last lesson of the chapter extends the brief introduction to exponential equations presented in Lesson 8.7.

OBJECTIVES

- To graph exponential functions
- To graph logarithmic functions
- To solve exponential equations
- To solve logarithmic equations
- To solve word problems involving exponential or logarithmic functions

PROBLEM SOLVING

The problem-solving technique known as *linear interpolation* is thoroughly discussed in the *Extension* on pages 490–492. The *Mixed Problem Solving* page for this chapter offers students an opportunity to practice various problem-solving techniques.

READING AND WRITING MATH

This chapter has two sets of *Focus on Reading* exercises, on pages 462 and 479. The *Application* on page 463 asks students to explain the properties of parabolic reflectors. An exercise in the *Chapter Review* (Exercise 19) asks students to explain, in their own words, how to solve logarithmic equations.

TECHNOLOGY

Calculator: Scientific calculators, with or without graphing capability, can be used throughout this chapter. In particular, if a graphing calculator is available, then it can be used to graph all exponential and logarithmic functions with speed and accuracy. Throughout the chapter, suggestions and hints are provided for using a calculator either to facilitate the work or to illuminate understanding of a concept.

SPECIAL FEATURES

Mixed Review pp. 463, 466, 468, 476, 480, 485, 489
Focus on Reading pp. 462, 479
Application: Paraboloidal Reflectors p. 463
Midchapter Review p. 472
Brainteaser p. 480
Application: Carbon Dating p. 485
Extension: Linear Interpolation pp. 490–492
Mixed Problem Solving p. 493
Key Terms p. 494
Key Ideas and Review Exercises pp. 494–495
Chapter 13 Test p. 496
College Prep Test p. 497

PLANNING GUIDE

Lesson	Basic	Average	Above Average	Resources
13.1 pp. 462–463	FR all CE all WE 1–21 odd Application	FR all CE all WE 9–31 odd Application	FR all CE all WE 13–33 odd Application	Reteaching 92 Practice 92
13.2 pp. 465–466	CE all WE 1–19 odd	CE all WE 7–23 odd	CE all WE 11–27 odd	Reteaching 93 Practice 93
13.3 p. 468	CE all WE 1–17 odd	CE all WE 5–23 odd	CE all WE 9–25 odd	Reteaching 94 Practice 94
13.4 pp. 471–472	CE all WE 1–11 odd Midchapter Review	CE all WE 5–17 odd Midchapter Review	CE all WE 9–19 odd Midchapter Review	Reteaching 95 Practice 95
13.5 pp. 475–476	CE all WE 1–21 odd	CE all WE 7–27 odd, 28	CE all WE 9–29 odd	Reteaching 96 Practice 96
13.6 pp. 479–480	FR all CE all WE 1–21 odd Brainteaser	FR all CE all WE 9–31 odd Brainteaser	FR all CE all WE 13–33 odd Brainteaser	Reteaching 97 Practice 97
13.7 pp. 484–485	CE all WE 1–21 odd Application	CE all WE 7–29 odd Application	CE all WE 11–31 odd Application	Reteaching 98 Practice 98
13.8 pp. 488–493	CE all WE 1–13 odd Extension Mixed Problem Solving	CE all WE 7–21 odd Extension Mixed Problem Solving	CE all WE 11–25 odd Extension Mixed Problem Solving	Reteaching 99 Practice 99
Chapter 13 Review pp. 494–495	all	all	all	
Chapter 13 Test p. 496	all	all	all	
College Prep Test p. 497	all	all	all	

CE = Classroom Exercises   WE = Written Exercises   FR = Focus on Reading
NOTE: For each level, all students should be assigned all Mixed Review exercises.

▬ INVESTIGATION

In this investigation, the students explore the reflection property of inverse relations.

Ask the students to consider what happens to a point (a,b) if the coordinates are reversed to become (b,a).

Have the students calculate the midpoint between the two points $M(a,b)$ and $N(b,a)$. They can use the midpoint formula from plane geometry.

$$(x_m, y_m) = \left(\frac{x_1 + x_2}{2}, \frac{y_1 + y_2}{2} \right)$$

The midpoint is $\left(\frac{a + b}{2}, \frac{b + a}{2} \right)$.

Ask the students to describe the relation between the two points and to explain the relation to the line $y = x$.

The x and y coordinates are equal, so they must lie on the line whose graph is $y = x$.

Next, ask the students to find the slope of the segment MN connecting the two points and describe the relation of this line to the line $y = x$.

Using the slope formula,

$$m_{\overline{MN}} = \frac{b - a}{a - b} = -1.$$

The slope of the line $y = x$ is 1, so the slopes of the lines are negative reciprocals. Therefore, the lines are perpendicular.

In geometry, point A is said to be the *reflection* of the point B about the line $y = x$. The students should experiment with the cases $a > b$, $a < b$, and $a = b$ to convince themselves that this is true in all cases.

Ask the students what happens if they fold the graph of the two points along the line $y = x$. The two points overlap each other.

The subtle patterns in this computer graphic of the Monterey coast were generated by repeated functions. The rock formations, the spray of the water, and other features have fractal-like structures that readily lend themselves to computer representation.

More About the Image

Fractal images can be "made to order," as the computer graphic of the Monterrey coast demonstrates. Mathematical investigators discovered that fascinating fractal images are produced by a method called the **chaos game,** in which certain mathematical transformations are randomly chosen from a list and applied to points on a plane. By using a mathematical result known as the **collage theorem,** mathematicians can determine a list of transformations which will generate a desired image.

13.1 Inverse Relations and Functions

Objectives
To describe the inverses of relations
To determine whether the inverses of functions are functions

In earlier chapters, you solved exponential equations such as $125 = 5^{2x}$, where the exponent either was a variable or contained a variable. In this chapter, you will study *exponential functions*, such as $y = 5^{2x}$, and related functions known as *logarithmic functions*. These functions have applications both in science and everyday life. In order to study them properly, however, you must first understand what is meant by the *inverse* of a function.

Adding 2 and *subtracting 2* are *inverse operations* because one "reverses" the effect of the other. For example, if you begin with 7 and use both operations in succession, you will return to 7, as shown below.

$$7 + 2 = 9 \text{ and } 9 - 2 = 7 \qquad 7 - 2 = 5 \text{ and } 5 + 2 = 7$$

These two operations can be represented by the relations $A: y = x + 2$ and $S: y = x - 2$. If their domain (see Lesson 3.1) is the set of integers, then the following are true.

$$A = \{\ldots, (-2,0), (-1,1), (0,2), (1,3), (2,4), \ldots\}$$
$$S = \{\ldots, (0,-2), (1,-1), (2,0), (3,1), (4,2), \ldots\}$$

Notice that reversing each ordered pair (a,b) in A gives the ordered pairs (b,a) in S. Such relations are called *inverse relations*.

Definition

> The inverse of a relation A is the relation, denoted A^{-1}, obtained by reversing the order of the elements of each ordered pair in A. A and A^{-1} are called **inverse relations.**

Note that here the symbol -1 is *not* an exponent. For the inverse relations A and S above, S is the inverse of A, and A is the inverse of S.

$$S = A^{-1} \quad \text{and} \quad A = S^{-1}$$

If the domain of A is the set of all real numbers, then for each real number x, A is described by the equation $y = x + 2$. You can *trade x and y* as follows to obtain the equation for A^{-1}, or S.

	$A: y = x + 2$
Trade x and y.	$S \text{ (or } A^{-1}): x = y + 2$
Solve for y.	$S: y = x - 2$

Teaching Resources

Problem Solving Worksheet 13
Quick Quizzes 92
Reteaching and Practice
   Worksheets 92

GETTING STARTED

Prerequisite Quiz

Graph each relation. Determine its domain and range. Is it (a) a function, (b) a linear function, (c) a constant function?

1. $y = \frac{1}{2}x - 1$
   D: the set of real numbers,
   R: the set of real numbers; yes, yes, no
2. $x = 3$
   D: {3}, *R*: the set of real numbers; no, no, no
3. $y = 4$
   D: the set of real numbers,
   R: {4}; yes, yes, yes
4. $y = x^2 + 1$
   D: the set of real numbers,
   R: {$y \mid y \geq 1$}; yes, no, no

1–4

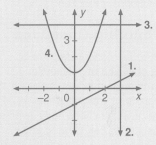

Motivator

Ask students to define the term *inverse*. Then have students give the *operation* that "reverses" the effect of *multiplying by 3 and then subtracting by 4.*
Adding 4 and then dividing by 3.

Additional Example 1

Describe the inverse of each function. Is the inverse a function?

a. Function *f*, described by the equation
   $y = -\frac{3}{4}x + 12$. $y = -\frac{4}{3}x + 16$; yes
b. Constant function *g*, with equation $2y + 8 = 0$. $x = -4$; no: $(-4,1)$ and $(-4,2)$ have the same first element.

Highlighting the Standards

Standard 6b: This lesson uses equations, tables, and graphs to reinforce the concept of inverse functions.

Lesson Note

A relation can be described by a list of ordered pairs, an equation, or a graph. The *inverse* of a given relation R is found by reversing each ordered pair (a,b) in R to (b,a) in R^{-1}, the inverse of R. Emphasize that the graphs of a relation and its inverse are symmetric with respect to the line $y = x$.

Remind students that a relation is not a function if two of its ordered pairs have the same first element.

Math Connections

Inverses of Nonlinear Functions: Theorem 13.1 assures that linear, nonconstant functions have inverses that are functions. In addition, the other polynomial functions of this lesson illustrate that the inverse of a nonlinear polynomial function may not be a function. (It may not even be a polynomial.) In general, only if a polynomial function is known to be linear (and nonconstant) can it be definitely stated that the inverse of a polynomial function is a function.

Critical Thinking Questions

Analysis: Since the nonconstant linear function is the only kind of polynomial function guaranteed to have an inverse that is a function, students may think that *all* nonlinear polynomial functions have inverses that are not functions. Ask them to show that this is not so by giving a counterexample, such as $y = x^3$, which has $y = \sqrt[3]{x}$ as its inverse function.

The relation A is a function because no two ordered pairs have the same first coordinate. Its inverse S is also a function. However, the inverse of a function is not always a function.

EXAMPLE 1   Describe the inverse of each function. Is the inverse a function?

a. Function f, described by the equation $y = \frac{3}{2}x - 6$.

$$f: y = \frac{3}{2}x - 6$$

Trade x and y. $f^{-1}: x = \frac{3}{2}y - 6$

Solve for y. $f^{-1}: y = \frac{2}{3}x + 4$

f^{-1} is described by $y = \frac{2}{3}x + 4$. There is exactly one value of y for each value of x. Thus, f^{-1} is a function.

b. Constant function G, with equation $y = 3$.
Replace y with x in $y = 3$.
G^{-1}, with equation $x = 3$, is *not* a function because it contains many pairs, such as $(3,5)$ and $(3,6)$, with the same first coordinate of 3.

The diagram below shows the graphs of the inverse relations C and C^{-1}. The vertical line test (see Lesson 3.1) will determine whether they are functions.

$$C = \{(1,3), (0,2), (-2,2), (-3,1)\}$$
$$C^{-1} = \{(3,1), (2,0), (2,-2), (1,-3)\}$$

The graphs of C and C^{-1} are mirror images about the line $y = x$; that is, they are symmetric to each other with respect to the line $y = x$.

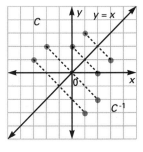

C is a function, but C^{-1} is not a function because a vertical line intersects its graph at two points when $x = 2$.

EXAMPLE 2   The function H is described by its graph at the right. Draw the graph of H^{-1}, the inverse of H. Is H^{-1} a function?

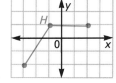

Plan   The graph of H consists of two segments with endpoints $(2,1)$, $(-1,1)$ and $(-1,1)$, $(-3,-2)$. Plot $(1,2)$, $(1,-1)$, and $(-3,-2)$. Then draw the two segments that form the mirror image of H with respect to the line $y = x$.

Additional Example 2

The function R is described by the graph at the right. Draw the graph of R^{-1}, the inverse of R. Is R^{-1} a function?   No

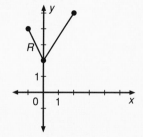

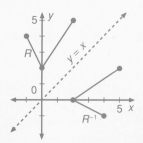

Solution

H^{-1} is represented by the graph at the right.

Although H is a function, H^{-1} is *not* a function because its graph does not pass the vertical line test at $x = 1$.

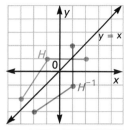

EXAMPLE 3

Write the slope-intercept form of the equation that describes g^{-1} if g is described by $3x - 5y = 15$.

Solution

g: $3x - 5y = 15$

g^{-1}: $3y - 5x = 15$

$3y = 5x + 15$; $y = \frac{5}{3}x + 5$

Thus, $y = \frac{5}{3}x + 5$ describes g^{-1}.

In Example 3, both g and g^{-1} are *linear nonconstant functions* because their equations are of the form $ax + by = c$, where $a \neq 0$ and $b \neq 0$.

Theorem 13.1 | The inverse of any linear nonconstant function is also a linear nonconstant function.

The inverse of a quadratic function is described in Example 4.

EXAMPLE 4

Function Q is described by $y = 2x^2 - 6$. Write the equation describing Q^{-1}, solving for y. Are Q and Q^{-1} a pair of inverse functions?

Solution

Q: $y = 2x^2 - 6$

Q^{-1}: $x = 2y^2 - 6$

$2y^2 = x + 6$

$y = \pm\sqrt{\dfrac{x + 6}{2}}$

Q^{-1} is described by $y = \pm\sqrt{\dfrac{x + 6}{2}}$. Choose a value for x, such as 8. Then Q^{-1} contains $(8, \sqrt{7})$ and $(8, -\sqrt{7})$, so Q^{-1} is not a function.

So, Q and Q^{-1} are *not* a pair of inverse functions.

Checkpoint

Describe the inverse of the function by writing its equation, solving for *y*, or by drawing its graph. Is the inverse a function?

1. $y = -4x + 8$ $y = -\frac{1}{4}x + 2$; yes
2. $5x - 2y = 10$ $y = \frac{2}{5}x + 2$; yes
3. $y = \frac{1}{2}x^2 - 5$ $y = \pm\sqrt{2x + 10}$; no: $(-3, 2)$ and $(-3, -2)$ have the same first element.

Closure

Ask students if the inverse of a linear nonconstant function is always a function. Yes   Ask them if the same is true of a quadratic function. No   Have students create a function whose inverse is not a function.   Answers will vary.

◾◾◾FOLLOW UP

Guided Practice

Classroom Exercises 1–8

Independent Practice

▪A Ex. 1–18, ▪B Ex. 19–32, ▪C Ex. 33

Basic: FR all, WE 1–21 odd, Application

Average: FR all, WE 9–31 odd, Application

Above Average: FR all, WE 13–33 odd, Application

Additional Answers

Written Exercises

4. $y = -\frac{1}{3}x$; yes

Additional Example 3

Write the equation that describes g^{-1}, solving for y, if g is described by $4x + 3y + 8 = 0$.

g^{-1}: $y = -\frac{3}{4}x - 2$

Additional Example 4

Function Q is described by $y = x^2 + 4$. Write the equation describing Q^{-1}, solving for y. Are Q and Q^{-1} a pair of inverse functions? Q^{-1}; $y = \pm\sqrt{x - 4}$; no: $(8, 2)$ and $(8, -2)$ have the same first element.

11.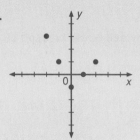

12.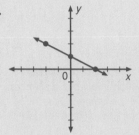

13.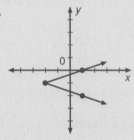

14.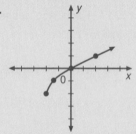

Is the statement always, sometimes, or never true?

1. The inverse of a function is a function. Sometimes
2. The inverse of a relation is a relation. Always
3. If A is a relation, then $(A^{-1})^{-1} = A$. Always
4. The inverse of a constant function $y = c$ is a function. Never
5. The inverse of a linear nonconstant function is a function. Always
6. If f is a quadratic function whose domain is the set of real numbers, then the inverse of f is a function. Never

Classroom Exercises

Describe the inverse of each.

1. $\{(6,3), (4,4), (2,5)\}$ $\{(3,6), (4,4), (5,2)\}$
2. subtracting 5 Adding 5
3. dividing by 2 Multiplying by 2
4. multiplying by -4 Dividing by -4
5. $y = -2$ $x = -2$
6. $y = x$ $y = x$
7. $y = x + 6$ $y = x - 6$
8. $y = 2x$ $y = \frac{1}{2}x$

Written Exercises

Describe the inverse of each function. Is the inverse a function?

1. $\{(1,4), (-3,4), (2,0)\}$ $\{(4,1), (4,-3), (0,2)\}$; no
2. $\{(7,8), (6,4), (5,0)\}$ $\{(8,7), (4,6), (0,5)\}$; yes
3. $y = 4x$ $y = \frac{1}{4}x$; yes
4. $y = -3x$
5. $y = 5$ $x = 5$; no
6. $y = x$ $y = x$; yes
7. $3y = 2x$ $y = \frac{3}{2}x$; yes
8. $y = 2x - 4$ $y = \frac{1}{2}x + 2$; yes
9. $y = 2x - 6$ $y = \frac{1}{2}x + 3$; yes
10. $y = -3x + 6$ $y = -\frac{1}{3}x + 2$; yes

Graph the inverse of each relation.

11. 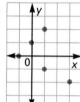 **12.** 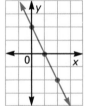 **13.** 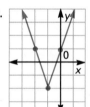 **14.**

15–18. For the graphs of Exercises 11–14, is the inverse a function? Are the relation and its inverse a pair of inverse functions?

15. Yes, no 16. Yes, yes 17. No, no 18. Yes, yes

Enrichment

Explain that if the graph of a function is drawn, there is an easy way to determine whether its inverse is a function. Show students that they need only look at the graph of the original function to determine whether it passes a *horizontal line test*. If no horizontal line passes through the graph in more than one point, then the inverse is a function.

Have the students verify this by applying the horizontal line test to graphs for Examples 3 and 4 as well as to the graphs of a few of the functions in Exercises 19–30.

Write the equation of the inverse of each function, solving for *y*.
Are the function and its inverse a pair of inverse functions?

19. $y = \frac{3}{4}x - 3$ **20.** $y = -\frac{5}{3}x + 10$ **21.** $y = -\frac{2}{3}x + 2$ **22.** $y = \frac{1}{2}x^2$

23. $y = 2x^2$ **24.** $y = -x^2$ **25.** $2x - 3y = 6$ **26.** $3x + 4y = 12$

27. $2x + 5y = 10$ **28.** $y = x^2 + 2$ **29.** $y = -2x^2 + 3$ **30.** $y = \frac{1}{2}x^2 - 3$

For each function *f*, find f^{-1} and the values $f(12)$, $f^{-1}(12)$, $f(f^{-1}(12))$,
and $f^{-1}(f(12))$.

31. $f(x) = 3x - 3$ **32.** $f(x) = \frac{3}{4}x + 6$

33. Given $g(x) = \frac{1}{2}x + 5$, choose the function *t* such that $g(t(x)) = x$
and $t(g(x)) = x$. b

    a. $t(x) = 2x + 10$ **b.** $t(x) = 2x - 10$ **c.** $t(x) = 2x - 5$

Mixed Review

1. Find the solution set of $5 - 3x < 17$ *and* $9 > 2x - 3$.   **2.2** $\{x \mid -4 < x < 6\}$

2. Factor $x^3 - 8$.   **5.6** $(x - 2)(x^2 + 2x + 4)$ **3.** Simplify $\dfrac{x^2 - x - 2}{x^2 - 1}$.   **7.1** $\dfrac{x - 2}{x - 1}$

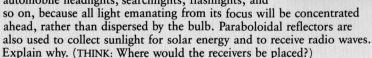

Application: *Paraboloidal Reflectors*

Parabolas have a very interesting property. Take any parabola, where *F* is the focus, *P* is any point on the curve, and *l* is the line passing through *P* parallel to the parabola's axis of symmetry. It can be proved that any beam of light from a source at *F* striking a parabolic mirror at *P* will reflect along the line *l* parallel to the parabola's axis of symmetry.

Now imagine a parabola spun around its axis of symmetry until it becomes three-dimensional, a shape much like a cup. This shape, called a **paraboloid**, is used to make the reflectors for automobile headlights, searchlights, flashlights, and so on, because all light emanating from its focus will be concentrated ahead, rather than dispersed by the bulb. Paraboloidal reflectors are also used to collect sunlight for solar energy and to receive radio waves. Explain why. (THINK: Where would the receivers be placed?)

19. $y = \frac{4}{3}x + 4$; yes

20. $y = -\frac{3}{5}x + 6$; yes

21. $y = -\frac{3}{2}x + 3$; yes

22. $y = \pm\sqrt{2x}$; no

23. $y = \pm\sqrt{\frac{x}{2}}$; no

24. $y = \pm i\sqrt{x}$; no

25. $y = \frac{3}{2}x + 3$; yes

26. $y = -\frac{4}{3}x + 4$; yes

27. $y = -\frac{5}{2}x + 5$; yes

28. $y = \pm\sqrt{x - 2}$; no

29. $y = \pm\sqrt{\frac{3 - x}{2}}$; no

30. $y = \pm\sqrt{2x + 6}$; no

31. $f^{-1}(x) = \dfrac{x + 3}{3}$, or $\frac{1}{3}x + 1$

     $f(12) = 33$
     $f^{-1}(12) = 5$
     $f(f^{-1}(12)) = 12$
     $f^{-1}(f(12)) = 12$

32. $f^{-1}(x) = \dfrac{4(x - 6)}{3}$, or $\frac{4}{3}x - 8$

     $f(12) = 15$
     $f^{-1}(12) = 8$
     $f(f^{-1}(12)) = 12$
     $f^{-1}(f(12)) = 12$

Application

Paraboloidal reflectors are used to collect light and radio waves because they concentrate the incoming signals.

Given the reflective properties of the paraboloid, all waves reaching the dish from a given source will be reflected to, and therefore concentrated at, a receiver located at the focus of the paraboloid.

GETTING STARTED

Prerequisite Quiz

Simplify.

1. x^{-3} $\quad \frac{1}{x^3}$ 2. 3^{-x} $\quad \frac{1}{3^x}$

3. 2^{-3} $\quad \frac{1}{8}$ 4. $\left(\frac{1}{2}\right)^{-3}$ $\quad 8$

Evaluate the expression using 2, 1, 0, −1, and −2 for x.

5. 3^x $\quad 9, 3, 1, \frac{1}{3}, \frac{1}{9}$

6. $2 \cdot 3^x$ $\quad 18, 6, 2, \frac{2}{3}, \frac{2}{9}$

Motivator

Ask students to evaluate 4^x for the following values without a calculator.

1. $x = 2$ 2. $x = \frac{1}{2}$ 3. $x = 2.5$
   16   2   32

Have students use a scientific calculator to approximate the following to one decimal place. $5^3, 5^{3.1}, 5^{\sqrt{10}}$ 125; 146.8; 162.3

TEACHING SUGGESTIONS

Lesson Note

Exponential functions have applications in business and science, such as compound interest, growth of bacteria, and radioactive decay. Encourage students to use a scientific calculator for this lesson. It will also be useful in Lessons 13.7 and 13.8, and in later chapters.

13.2 Exponential Functions

Objectives

To graph exponential functions
To interpret graphs of exponential functions
To approximate irrational powers of positive numbers

The equation $y = 5 \cdot 2^x$ describes an **exponential function** with a base of 2. Notice that both x and y can have rational and irrational values.

x	-2	-1	0	1	1.5	2	3	$\sqrt{10}$
2^x	0.25	0.5	1	2	2.83	4	8	8.95
$y = 5 \cdot 2^x$	1.25	2.5	5	10	14.1	20	40	44.8

1. If $x = -2$, then $2^x = 2^{-2} = \frac{1}{4}$. So, $y = 5 \cdot \frac{1}{4} = 5(0.25) = 1.25$.

2. If $x = 1.5$, then $2^x = 2^{1.5} = 2^{\frac{3}{2}} = \sqrt{2^3} = \sqrt{8}$.
   $\sqrt{8} = 2.828427\ldots$, an irrational number, so $y \approx 5(2.828) \approx 14.1$.

3. If $x = \sqrt{10}$, an *irrational exponent*, then $2^x = 2^{\sqrt{10}} = 2^{3.1622776}\cdots$.

Since the rational numbers 3, 3.1, 3.16, 3.162, 3.1622, 3.16227, . . . are getting closer and closer to $\sqrt{10}$, it follows that the rational powers $2^3, 2^{3.1}, 2^{3.16}, 2^{3.162}, 2^{3.1622}, 2^{3.16227}, \ldots$ are getting closer and closer to $2^{\sqrt{10}}$. These powers can be found using a calculator's $\boxed{x^y}$ key:
8, 8.5741 . . . , 8.9382 . . . , 8.9506 . . . , 8.9519 . . . , 8.9523 . . .

Notice that the list appears to be rounding off near 8.95.

So, $2^{\sqrt{10}} \approx 8.95$, and $y = 5 \cdot 2^{\sqrt{10}} \approx 44.8$.

Also, a scientific calculator can be used to approximate $5 \cdot 2^{\sqrt{10}}$ directly: $5 \; \boxed{\times} \; 2 \; \boxed{x^y} \; 10 \; \boxed{\sqrt{}} \; \boxed{=} \; 44.762098 \approx 44.8$.

The graph of $y = 5 \cdot 2^x$ is drawn using the table. Note the scales on the axes. The graph reveals the following properties of $f: y = 5 \cdot 2^x$.

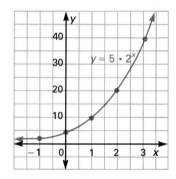

(1) The domain is the set of real numbers.
(2) The range is $\{y \mid y > 0\}$.
(3) The function is *always increasing*; that is, if $x_2 > x_1$, then $y_2 > y_1$.
(4) The ordered pair $(0,5)$ belongs to the function because the graph has 5 as its y-intercept.

The base of an exponential function must be a positive number not equal to 1. It may, however, be a fraction between 0 and 1, as in $y = \left(\frac{1}{3}\right)^x$. Recall that $\left(\frac{1}{3}\right)^x = \frac{1}{3^x} = 3^{-x}$ (see Lesson 5.2).

EXAMPLE

Draw the graphs of $A: y = 3^x$ and $B: y = \left(\frac{1}{3}\right)^x$ in the same coordinate plane. Choose convenient scales for the axes.

Plan

Construct tables of ordered pairs for A and for B.
Then plot the points and draw two smooth curves.

Solution

$A: y = 3^x$

x	y
-2	$\frac{1}{9}$
-1	$\frac{1}{3}$
0	1
1	3
2	9
2.5	15.6

$B: y = \left(\frac{1}{3}\right)^x$

x	y
2	$\frac{1}{9}$
1	$\frac{1}{3}$
0	1
-1	3
-2	9
-2.5	15.6

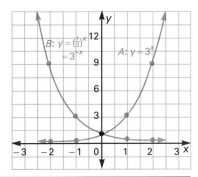

The graphs above show the following four properties of A and B. These properties hold for all exponential functions of the form $y = b^x$, where $b > 0$, $b \neq 1$.

(1) The domains of A and B are the set of real numbers.

(2) The ranges of A and B are $\{y \mid y > 0\}$.

(3) A is always increasing: $y_2 > y_1$ if $x_2 > x_1$.
    B is always decreasing: $y_2 < y_1$ if $x_2 > x_1$.

(4) The ordered pair $(0,1)$ belongs to both A and B.

Note also that the combined graph of A and B is symmetric with respect to the y-axis. But A and B are *not* inverses of each other because their graphs are not symmetric to each other with respect to the line $y = x$.

Classroom Exercises

For each equation, give another equation such that the combined graph of the equations is symmetric with respect to the y-axis.

1. $y = 2^x$ $y = \left(\frac{1}{2}\right)^x$ **2.** $y = \left(\frac{1}{5}\right)^x$ $y = 5^x$ **3.** $y = \left(\frac{1}{4}\right)^x$ $y = 4^x$ **4.** $y = 6^x$ $y = \left(\frac{1}{6}\right)^x$

Limit of a Sequence: The concept of *limit* is introduced informally in this lesson and will reappear in connection with infinite geometric series in Lesson 15.7. A formal definition of the *limit of a sequence* is given in more advanced texts. Briefly, this definition states that if it happens that there is a number L with the property that $|L - a_n|$ is arbitrarily small for all sufficiently large values of n, then a_n *converges* to L as a limit.

Critical Thinking Questions

Analysis: The exponential function $y = b^x$ has as a restriction that $b > 0$ and $b \neq 1$. Ask students to think of reasons why these two restrictions are necessary.   They should see that if, for example, $b = -4$, then the function will have an imaginary value for $x = \frac{1}{2}$. If $b = -1$, then the graph of the function will be a horizontal line, that is, the function will be the constant function, $y = 1$.

Common Error Analysis

Error: Many students have difficulty in evaluating an expression such as $\left(\frac{1}{4}\right)^x$ for negative values of x. Students will find it easier to evaluate $\left(\frac{1}{4}\right)^x$ in the form 4^{-x} when x is replaced by -1, -2, and -3.

Additional Example

Draw the graphs of $A: y = 5 \cdot 4^x$ and $B: y = 5\left(\frac{1}{4}\right)^x$ on the same coordinate plane. Choose convenient scales for the axes. List four properties of each function.

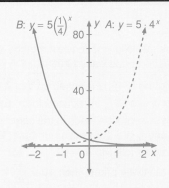

For $A: y = 5 \cdot 4^x$ and $B: y = 5\left(\frac{1}{4}\right)^x$:
D: set of real numbers
R: $\{y \mid y > 0\}$
The ordered pair $(0,5)$ belongs to each function, so that each graph has a y-intercept of 5.
$A: y = 5 \cdot 4^x$ is always increasing.
$B: y = 5\left(\frac{1}{4}\right)^x$ is always decreasing.

Checkpoint

Complete the table for the function $f\colon y = 2\left(\frac{1}{5}\right)^x$. List four properties of the function.

x	$\left(\frac{1}{5}\right)^x = 5^{-x}$	$2\left(\frac{1}{5}\right)^x$
2	$\frac{1}{25}$	? $\frac{2}{25}$
1	$\frac{1}{5}$	? $\frac{2}{5}$
0	1	? 2
−1	? 5	? 10
−2	? 25	? 50

D: set of real numbers; R: $\{y \mid y > 0\}$; The function is always decreasing: ($y_1 > y_2$ if $x_1 < x_2$). The ordered pair (0,2) belongs to the function, so that its graph has a y-intercept of 2.

Closure

Have students use a graphing calculator to draw and interpret the graph of $y = 3 + 5^x$. Set the x-axis from −2 to 2 with scale 1, and the y-axis from −1 to 30 with scale 4. Check students' graphs.

▬▬▬FOLLOW UP

Guided Practice

Classroom Exercises 1–12

Independent Practice

A Ex. 1–16, **B** Ex. 17–21, **C** Ex. 23–27

Basic: WE 1–19 odd

Average: WE 7–23 odd

Above Average: WE 11–27 odd

Additional Answers

See page 472 for the answers to Classroom Ex. 5–12, page 480 for Written Ex. 1–7, and pages 824–825 for Ex. 8–12, 17–27.

Determine whether the function is always increasing or always decreasing.

5. $y = 4^x$ **6.** $y = \left(\frac{1}{2}\right)^x$ **7.** $y = \left(\frac{1}{6}\right)^x$ **8.** $y = 5^x$

9. Graph $y = 4^x$. Choose convenient scales for the axes.

10. Graph $y = \left(\frac{1}{4}\right)^x$. Choose convenient scales for the axes.

11–12. Give the domain and range for Classroom Exercises 9 and 10.

Written Exercises

Graph each function. Choose convenient scales for the axes. Find the domain and range. Determine whether the function is always increasing or always decreasing. Find the y-intercept of the graph.

1. $y = 2^x$ **2.** $y = 5^x$ **3.** $y = \left(\frac{1}{2}\right)^x$ **4.** $y = \left(\frac{1}{5}\right)^x$

5. $y = 3 \cdot 2^x$ **6.** $y = 3\left(\frac{1}{2}\right)^x$ **7.** $y = 3 \cdot 4^x$ **8.** $y = 2\left(\frac{1}{3}\right)^x$

9. $y = -2 \cdot 3^x$ **10.** $y = -\left(\frac{1}{4}\right)^x$ **11.** $y = 2^x + 6$ **12.** $y = 3^x - 7$

Use the $\boxed{x^y}$ key on a scientific calculator to approximate each power to the nearest tenth.

13. $3^{4.2}$ 100.9   **14.** $6.5^{3.4}$ 580.6   **15.** $5^{\sqrt{2}}$ 9.7   **16.** $14^{\sqrt{3}}$ 96.6

Sketch the graph of $y = 2^x$ and the given function in the same coordinate plane. Describe the relationship between the two graphs.

17. $y = \left(\frac{1}{2}\right)^x$ **18.** $y = 4 + 2^x$ **19.** $y = 4 \cdot 2^x$

20. $y = 2^{x+3}$ **21.** $y = 2^{x-1}$ **22.** $y = -2^x$

Graph each function for the given domain.

23. $y = 3 \cdot 2^{x+1} - 4$ for $-3 \le x \le 3$ **24.** $y = 9^{\sqrt{x}}$ for $0 \le x \le 4$

25. Why is $y = a^x$ not an exponential function when $a = 1$?

26. Will any exponential function of the form $y = a^x$, $a > 0$, $a \ne 1$ intersect the x-axis? Why or why not?

27. Find the asymptote for $y = a^x - c$, where $a > 0$, $a \ne 1$, and c is a constant.

Mixed Review

Find the value of each expression.   *5.2, 8.6*

1. 8^0 1   **2.** 3^{-2} $\frac{1}{9}$ **3.** $25^{\frac{1}{2}}$ 5   **4.** $16^{\frac{3}{4}}$ 8

5. Write $\sqrt[3]{x^2}$ in exponential form.   *8.6* $x^{\frac{2}{3}}$

466   Chapter 13 Exponential and Logarithmic Functions

Enrichment

The introduction to this lesson refers to $\sqrt{10}$ as an irrational number. Challenge the students to prove that $\sqrt{10}$ is irrational.

Some students may use the Rational Zero Theorem (Lesson 10.5) by letting $x = \sqrt{10}$ and then showing that $x^2 - 10 = 0$ has no rational solution (or that $x^2 - 10$ has no rational zeros).

Discuss the following indirect method of proof used by the early Greeks.

Assume the $\sqrt{10}$ is rational, so that it can be expressed as $\frac{a}{b}$ in reduced form. Explain that this assumption leads to the contradiction that a and b are both even numbers. Begin the proof as follows. If $\sqrt{10} = \frac{a}{b}$, with a and b relatively prime, then $10 = \frac{a^2}{b^2}$, or

$10b^2 = a^2$. Then a^2 and a must be even. Now let $a = 2c$ so that $\sqrt{10} = \frac{2c}{b}$. Have students complete the proof by showing that b must also be even. If $\sqrt{10} = \frac{2c}{b}$, then $10b^2 = 4c^2$, or $5b^2 = 2c^2$. Therefore, b^2 and b must be even. If both a and b are even, this contradicts the assumption that a and b are relatively prime. Therefore, $\sqrt{10}$ is not a rational number; that is, $\sqrt{10}$ is irrational.

13.3 Base-*b* Logarithms

Objectives
To find base-*b* logarithms of positive numbers
To write base-*b* logarithmic equations as exponential equations
To solve base-*b* logarithmic equations

The base-2 exponential function f: $y = 2^x$ is a mapping from the exponent x to the power 2^x.

The inverse of f is the mapping from the power 2^x to x. The exponent x is called a **logarithm**.

Exponent	Power
4	2^4, or 16
3	2^3, or 8
0	2^0, or 1
-3	2^{-3}, or $\frac{1}{8}$

Power	Exponent
16, or 2^4	4
8, or 2^3	3
1, or 2^0	0
$\frac{1}{8}$, or 2^{-3}	-3

The *base-2 logarithm* of a positive number, denoted by $\log_2$, is found by changing an exponential equation into a **logarithmic equation** as shown below. $\log_2 8$ is read as "the base-two log of eight."

Base-2 exponential equation

$$16 = 2^4$$
$$8 = 2^3$$
$$1 = 2^0$$
$$\frac{1}{8} = 2^{-3}$$
$$\vdots$$
$$y = 2^x$$

Base-2 logarithmic equation

$$\log_2 16 = \log_2 2^4 = 4$$
$$\log_2 8 = \log_2 2^3 = 3$$
$$\log_2 1 = \log_2 2^0 = 0$$
$$\log_2 \frac{1}{8} = \log_2 2^{-3} = -3$$
$$\vdots$$
$$\log_2 y = \log_2 2^x = x$$

In general, if $y = 2^x$, then $\log_2 y = \log_2 2^x = x$. Any positive number b not equal to 1 can be the base of the **base-*b* logarithm**.
For example, $\log_5 125$ is a base-5 logarithm: $\log_5 125 = \log_5 5^3 = 3$.

Definition

For each positive number y and each positive number b where $b \neq 1$, if $y = b^x$, then $\log_b y = \log_b b^x = x$.

EXAMPLE 1 Find each logarithm.
a. $\log_5 625$ b. $\log_3 \frac{1}{9}$ c. $\log_6 \sqrt[3]{6}$ d. $\log_{10} \sqrt[4]{1,000}$

Solutions a. $\log_5 5^4 = 4$ b. $\log_3 3^{-2} = -2$ c. $\log_6 6^{\frac{1}{3}} = \frac{1}{3}$ d. $\log_{10} 10^{\frac{3}{4}} = \frac{3}{4}$

13.3 Base-*b* Logarithms **467**

Additional Example 1

Find each logarithm.

a. $\log_7 49$ 2
b. $\log_5 \frac{1}{25}$ -2
c. $\log_7 \sqrt{7}$ $\frac{1}{2}$
d. $\log_5 \sqrt[3]{25}$ $\frac{2}{3}$

Teaching Resources

Quick Quizzes 94
Reteaching and Practice
  Worksheets 94

GETTING STARTED

Prerequisite Quiz

1. 2^{-4} $\frac{1}{16}$ 2. 5^0 1

Solve for the variable.

3. $2^x = 64$ 6   4. $b^{-3} = \frac{1}{8}$ 2

Motivator

Have students write the following numbers in the form 5^x and determine the value of x.

1. 125 5^3 2. $\frac{1}{25}$ 5^{-2}
3. 1 5^0 4. $\sqrt[3]{5}$ $5^{\frac{1}{3}}$

TEACHING SUGGESTIONS

Lesson Note

Emphasize that $\log_b b^x$ is simply the exponent x. To find $\log_2 8$, write 8 as a power of the base 2. Thus, $\log_2 8 = \log_2 2^3 = 3$. The skill of changing a logarithmic equation, $\log_b y = x$, to an exponential equation, $b^x = y$, requires practice.

Math Connections

Mappings: Students have studied mappings earlier in the course, but the two mappings shown at the beginning of this lesson are the first significant *one-to-one* mappings of the book. When such a two-way correspondence exists between a function and its inverse, then the inverse relation is the inverse *function* of the original function.

Highlighting the Standards

Standard 4c: The fact that a logarithm is an exponent cannot be repeated too often to be sure that this fundamental connection is understood and used.

467

Analysis: Ask students what is wrong with the idea of a "base-1 logarithm." To begin with, you would be able to take the logarithm of only one number, 1, and the value of that logarithm could be any number.

Common Error Analysis

Error: Some students are undecided as to whether $\log_b a = c$ is equivalent to $b^c = a$ or $b^a = c$. Use the Classroom Exercises to drill on this.

Checkpoint

Find each logarithm.

1. $\log_9 81$ 2
2. $\log_9 1$ 0
3. $\log_9 9$
4. $\log_9 \frac{1}{81}$ -2
5. $\log_9 \frac{1}{9}$ -1 1

Closure

Have students create a base-5 logarithmic equation and then change it to an exponential equation.   Answers will vary.

◼◼◼◼FOLLOW UP

Guided Practice

Classroom Exercises 1–8

Independent Practice

Ⓐ Ex. 1–14, Ⓑ Ex. 15–22, Ⓒ Ex. 23–26

Basic: WE 1–17 odd
Average: WE 5–23 odd
Above Average: WE 9–25 odd

Additional Answers

Classroom Exercises

1. $4^c = 16$
2. $2^d = 16$
3. $2^4 = c$
4. $3^x = 9$

EXAMPLE 2   Find the value of the variable in each logarithmic equation.
a. $\log_b 25 = 2$ b. $\log_2 128 = x$ c. $\log_3 y = 4$ d. $\log_b \frac{1}{16} = -4$

Plan   Rewrite each logarithmic equation as an exponential equation.

Solutions
a. $b^2 = 25$
$b = 5$
($b > 0$, so
$b \neq -5$)

b. $2^x = 128$
$2^x = 2^7$
$x = 7$

c. $3^4 = y$
$y = 81$

d. $b^{-4} = \frac{1}{16}$
$b^{-4} = 2^{-4}$
$b = 2$
($b > 0$, so
$b \neq -2$)

Classroom Exercises

Write each logarithmic equation as an exponential equation.

1. $\log_4 16 = c$
2. $\log_2 16 = d$
3. $\log_2 c = 4$
4. $\log_3 9 = x$
5. $\log_5 p = -2$
$5^{-2} = p$
6. $\log_4 m = 0$
$4^0 = m$
7. $\log_b \sqrt[5]{3} = \frac{1}{5}$
$b^{\frac{1}{5}} = \sqrt[5]{3}$
8. $\log_n p = m$
$n^m = p$

Written Exercises

Find each logarithm.

1. $\log_8 64$ 2
2. $\log_5 1$ 0
3. $\log_6 6$ 1
4. $\log_6 \frac{1}{6}$ -1
5. $\log_2 \frac{1}{16}$ -4
6. $\log_3 81$ 4
7. $\log_4 1$ 0
8. $\log_2 \sqrt[5]{2}$ $\frac{1}{5}$
9. $\log_5 \sqrt{5}$ $\frac{1}{2}$
10. $\log_3 \sqrt[4]{3}$ $\frac{1}{4}$

Find the value of the variable in each logarithmic equation.

11. $\log_b 100 = 2$ 10
12. $\log_2 y = 6$ 64
13. $\log_3 243 = x$ 5
14. $\log_4 y = 3$ 64
15. $\log_5 \sqrt[3]{5} = x$ $\frac{1}{3}$
16. $\log_{\frac{1}{3}} y = 2$ $\frac{1}{9}$
17. $\log_b \frac{1}{8} = -3$ 2
18. $\log_{\frac{2}{3}} \frac{8}{27} = x$ 3

Find each logarithm.

19. $\log_2 \sqrt[5]{8}$ $\frac{3}{5}$
20. $\log_{10} \sqrt[3]{100}$ $\frac{2}{3}$
21. $\log_5 \sqrt[4]{125}$ $\frac{3}{4}$
22. $\log_{\frac{1}{4}} 16$ -2
23. $\log_{\frac{1}{2}} 4^3$ -6
24. $\log_c (c^2)^d$ $2d$
25. $\log_{4n} 16n^2$ 2
26. $\log_{2n} \frac{1}{8n^3}$ -3

Mixed Review

1. Solve $3^{4x-2} = 81$.   **5.1** $\frac{3}{2}$
2. Find $f(2)$ for $f(x) = 8x^{-3}$.   **5.2**   1
3. Factor $8a^3 - b^3$.   **5.6**
$(2a - b)(4a^2 + 2ab + b^2)$
4. Solve $x^2 + 8x + 25 = 0$.   **9.5**
$-4 \pm 3i$

Additional Example 2

Find the value of the variable in each logarithmic equation.

a. $\log_b 100 = 2$ 10
b. $\log_3 81 = x$ 4
c. $\log_5 y = 3$ 125
d. $\log_b \frac{1}{9} = -2$ 3

Enrichment

Discuss reasons why the base of an exponential or logarithmic function cannot be negative. For example, using integral values of x, have the students find y-values if $y = (-3)^x$, and note that the signs of y-alternate.

x	-2	-1	0	1	2	3
y	$\frac{1}{9}$	$-\frac{1}{3}$	1	-3	9	-27

Discuss the graph of $y = (-3)^x$ between $(0,1)$ and $(1,-3)$. Are there any points for which $y = 0$?   No   What if $x = \frac{1}{2}$?
y would not be a real number, since $(-3)^{\frac{1}{2}}$ is an even root of a negative number.

Thus, the graph of $y = (-3)^x$ would have many "holes." Corresponding discontinuities would occur for $y = \log_{-3} x$.

13.4 Graphing $y = \log_b x$

Teaching Resources

Project Worksheet 13
Quick Quizzes 95
Reteaching and Practice
   Worksheets 95
Transparencies 30A, 30B, 30C, 30D,
   30E

Objectives

To describe the inverses of exponential functions as logarithmic
functions

To graph logarithmic functions

To interpret the graphs of logarithmic functions

To describe the inverse of an exponential function such as $A: y = 3^x$,
trade x and y in $y = 3^x$. Then write the new equation, solving for y.

$$A: \qquad y = 3^x$$
$$A^{-1}: \qquad x = 3^y \quad \leftarrow \text{ Trade } x \text{ and } y.$$
$$\log_3 x = y, \quad \leftarrow \text{ Change the exponential equation}$$
$$\qquad\qquad\qquad \text{ to a logarithmic equation.}$$
$$\text{or } y = \log_3 x$$

Thus, the inverse of $A: y = 3^x$ is described by $y = \log_3 x$. In general,
the inverse of an exponential function is a **logarithmic function**.

Theorem 13.2

> The inverse of the exponential function
> $$y = b^x \, (y > 0, b > 0, b \neq 1)$$
> is the logarithmic function described by
> $$y = \log_b x \, (x > 0, b > 0, b \neq 1).$$
> The converse is also true.

You will be asked to prove Theorem 13.2 in Exercises 16 and 17.

EXAMPLE 1

Write the equation for the inverse of each function, solving for y.

a. $B: y = \left(\frac{1}{3}\right)^x$

b. $C: y = \log_5 x$

Solutions

a. The inverse of $B: y = \left(\frac{1}{3}\right)^x$ is described by $y = \log_{\frac{1}{3}} x$.

b. The inverse of $C: y = \log_5 x$ is described by $y = 5^x$.

Because $y = \log_3 x$ and $y = 3^x$ describe inverse functions, their graphs
are symmetric to each other with respect to the line $y = x$. To draw the
graph of $y = \log_3 x$, make a table of ordered pairs (x,y) for $y = 3^x$.
Then reverse the order in each pair to form a table for $y = \log_3 x$.
Plot these points and draw a smooth curve through them, as shown at
the top of the next page.

Prerequisite Quiz

Write the equation for the inverse of the
function, solving for y.

1. $y = x - 2 \quad y = x + 2$
2. $y = \frac{1}{2}x \quad y = 2x$

Complete the table.

3.
x	2	1	$\frac{1}{2}$	0	-1	-2
$y = 4^x$	16	4	2	1	$\frac{1}{4}$	$\frac{1}{16}$

4.
x		2	1	0	-1	-2
$y = \frac{1}{3^x} = 3^{-x}$		$\frac{1}{9}$	$\frac{1}{3}$	1	3	9

Motivator

Have students graph (plot) the following
sets of points in the same coordinate plane.
Then have them explain how A^{-1} was
determined and the symmetry shown.

1. $A = \{(-1,\frac{1}{2}), (0,1), (1,2), (2,4), (3,8)\}$,
2. A^{-1}, and
3. $y = x$

Check students' graphs.

Additional Example 1

Write the equation for the inverse of each
function, solving for y.

$A: y = 8^x \quad y = \log_8 x$
$B: y = \log_{\frac{1}{2}} x \quad y = \left(\frac{1}{2}\right)^x$

Highlighting the Standards

Standard 6c: Once again, the threefold
presentation—table, symbols, and graphs—
is used to insure understanding of a basic
function concept.

Lesson Note

The base-b logarithmic function, $y = \log_b x$, is defined as the *inverse* of the base-b exponential function, $y = b^x$. To graph $y = \log_3 x$, a table of ordered pairs (n,m) is made for the inverse, $y = 3^x$. Then the order is reversed to find ordered pairs (m,n) for $y = \log_3 x$.

Math Connections

Asymptotes—Actual and Supposed:
Students who recall the asymptotes of the rectangular hyperbola of Lesson 12.5 will notice that the y-axis is an asymptote of the graph of $y = \log_3 x$; that is, the curve approaches the y-axis from the right but never crosses it.

Critical Thinking Questions

Analysis: Ask students if a vertical asymptote exists for the exponential curve $y = 3^x$.   Since it is rising with extraordinary rapidity, it appears that there must be a vertical asymptote which the curve is approaching from the left. However, this is not the case, since the x-coordinate of such an "asymptote" would be greater than the greatest value of x for which 3^x is defined. There is no such greatest value.

$y = 3^x$ $y = \log_3 x$

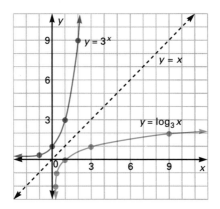

x	y	x	y
-2	$\frac{1}{9}$	$\frac{1}{9}$	-2
-1	$\frac{1}{3}$	$\frac{1}{3}$	-1
0	1	1	0
1	3	3	1
2	9	9	2

The graph of $y = \log_3 x$ shows four properties of the $\log_3$ function.

(1) The domain is $\{x \mid x > 0\}$.

(2) The range is the set of real numbers.

(3) The function is always increasing: $y_2 > y_1$ if $x_2 > x_1$.

(4) The ordered pair $(1,0)$ belongs to the function because 1 is the x-intercept of the graph. That is, $\log_3 1 = 0$.

EXAMPLE 2 Draw the graphs of C: $y = \log_5 x$ and D: $y = \log_{\frac{1}{5}} x$ in the same coordinate plane. Choose convenient scales for the axes.

Plan Because it is not yet convenient to use a calculator to find such values as $\log_5 11.2$ directly, make tables of ordered pairs for C^{-1} and D^{-1}. Then reverse the order to make tables for C and D.

Solution

C^{-1}:   C: D^{-1}:   D:

$y = 5^x$ $y = \log_5 x$ $y = \left(\frac{1}{5}\right)^x = 5^{-x}$ $y = \log_{\frac{1}{5}} x$

x	y	x	y	x	y	x	y
-2	$\frac{1}{25}$	$\frac{1}{25}$	-2	-2	25	25	-2
-1	$\frac{1}{5}$	$\frac{1}{5}$	-1	-1.5	11.2	11.2	-1.5
0	1	1	0	-1	5	5	-1
1	5	5	1	0	1	1	0
1.5	11.2	11.2	1.5	1	$\frac{1}{5}$	$\frac{1}{5}$	1
2	25	25	2	2	$\frac{1}{25}$	$\frac{1}{25}$	2

Additional Example 2

Draw the graphs of A: $y = \log_8 x$ and B: $y = \log_{\frac{1}{8}} x$ on the same coordinate plane. Choose convenient scales for the axes. List four properties of each function.

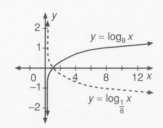

Domain: $\{x \mid x > 0\}$
Range: set of real numbers
The ordered pair $(1,0)$ belongs to each function, so that each graph has an x-intercept of 1.
A: $y = \log_8 x$ is always increasing.
B: $y = \log_{\frac{1}{8}} x$ is always decreasing.

Using the tables on the previous page yields the graphs of $C: y = \log_5 x$ and $D: y = \log_{\frac{1}{5}} x$ as shown at the right.

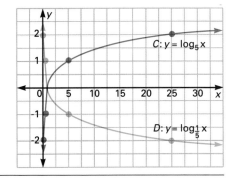

These graphs show the following for $C: y = \log_5 x$ and $D: y = \log_{\frac{1}{5}} x$.

(1) The domains of C and D are $\{x \mid x > 0\}$.

(2) The ranges of C and D are the set of real numbers.

(3) The function $C: y = \log_5 x$ is always increasing.

The function $D: y = \log_{\frac{1}{5}} x$ is always decreasing.

(4) The ordered pair (1,0) belongs to both C and D.

Notice that the graphs of the logarithmic functions $C: y = \log_5 x$ and $D: y = \log_{\frac{1}{5}} x$ are symmetric to each other with respect to the x-axis.

Pairs of Functions ($b > 1$)	Inverses?	Line of symmetry
$y = b^x$ and $y = \log_b x$	yes	$y = x$
$y = \left(\frac{1}{b}\right)^x$ and $y = \log_{\frac{1}{b}} x$	yes	$y = x$
$y = \log_b x$ and $y = \log_{\frac{1}{b}} x$	no	x-axis

Classroom Exercises

What is the line of symmetry for the graph of each pair of functions? Are the functions a pair of inverse functions?

1. $y = \log_4 x$, $y = \log_{\frac{1}{4}} x$ x-axis; no

2. $y = 4^x$, $y = \log_4 x$ $y = x$; yes

3. $y = \log_{\frac{1}{6}} x$, $y = \left(\frac{1}{6}\right)^x$ $y = x$; yes

4. $y = \log_{\frac{1}{6}} x$, $y = \log_6 x$ x-axis; no

5. Use the first table to complete the second.

x	4	0	1	-3	2
$y = 2^x$	16	1	2	$\frac{1}{8}$	4

x	16	1	2	$\frac{1}{8}$	4
$y = \log_2 x$	4	0	1	-3	2

Complete the tables.

1.

x	2	1	0	-1	-2
$y = \frac{1}{3^x} = 3^{-x}$	$\frac{1}{9}$	$\frac{1}{3}$	1	3	9

2.

x	$\frac{1}{9}$	$\frac{1}{3}$	1	3	9
$y = \log_{\frac{1}{3}} x$	2	1	0	-1	-2

3. List four properties of the function $y = \log_{\frac{1}{3}} x$.

D: $\{x \mid x > 0\}$; R: set of real numbers; The function is always decreasing. The function contains (1,0).

Closure

Ask students if the functions $y = b^x$ and $y = \log_b x$ are inverses.   Yes   Then ask if the functions $y = \log_8 x$ and $y = \log_{\frac{1}{8}} x$ are inverses.   No   Why or why not?   The line of symmetry is the x-axis.

◼◼◼FOLLOW UP

Guided Practice

Classroom Exercises 1–9

Independent Practice

A Ex. 1–9, **B** Ex. 10–15, **C** Ex. 16–20

Basic: WE 1–11 odd, Midchapter Review

Average: WE 5–17 odd, Midchapter Review

Above Average: WE 9–19 odd, Midchapter Review

Additional Answers

See pages 491–493 for the answers to Written Ex. 4–20 and Midchapter Review Ex. 5–6.

Classroom Exercises

5. Increasing
6. Decreasing
7. Decreasing
8. Increasing

9.

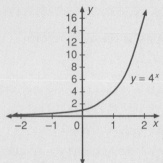

10.

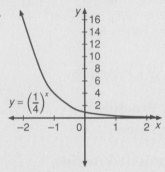

11–12. Domain: {real numbers};
Range: $\{y \mid y > 0\}$

Determine whether the function is always increasing or always
decreasing. What two quadrants contain the graph of the function?

6. $y = \log_4 x$
Increasing,
I and IV

7. $y = 4^x$
Increasing,
I and II

8. $y = \log_{\frac{1}{4}} x$
Decreasing,
I and IV

9. $y = \left(\frac{1}{4}\right)^x$, or
$y = 4^{-x}$
Decreasing,
I and II

Written Exercises

Write the equation of the inverse of each function, solving for y.

1. $y = 2^x$ $y = \log_2 x$

2. $y = \log_{\frac{1}{2}} x$ $y = \left(\frac{1}{2}\right)^x$

3. $y = \log_8 x$ $y = 8^x$

Graph each function. Choose convenient scales for the axes. Give the
domain and range. Determine whether the function is always increasing
or always decreasing. Find the x-intercept of the graph.

4. $y = \log_2 x$

5. $y = \log_4 x$

6. $y = \log_{\frac{1}{2}} x$

7. $y = \log_{\frac{1}{4}} x$

8. $y = \log_6 x$

9. $y = \log_{\frac{1}{6}} x$

10. $y = 3 \cdot \log_2 x$

11. $y = 2 \cdot \log_3 x$

12. $y = 4 \cdot \log_{\frac{1}{2}} x$

13. $y = 3 + \log_2 x$

14. $y = 2 + \log_3 x$

15. $y = 4 + \log_{\frac{1}{2}} x$

16. Prove that the inverse of $f: y = b^x$ is $y = \log_b x$.

17. Prove that the inverse of $g: y = \log_b x$ is $y = b^x$.

Graph each function. Choose convenient scales for the axes.

18. $y = \log_2 (x - 4)$

19. $y = \log_3 (x + 2)$

20. $y = \log_4 (x - 4)$

Midchapter Review

Write the equation that describes the inverse of the function, solving
for y.   *13.1, 13.4*   1. $y = \frac{5}{2}x + 5$

1. $2x - 5y = 10$

2. $y = x^2 + 4$
$y = \pm\sqrt{x - 4}$

3. $y = 6^x$ $y = \log_6 x$

4. $y = \log_4 x$
$y = 4^x$

Graph each function. Choose convenient scales for the axes.   *13.2, 13.4*

5. $y = \left(\frac{1}{2}\right)^x$

6. $y = \log_2 x$

Find each logarithm.   *13.3*

7. $\log_6 36$ 2

8. $\log_2 \frac{1}{8}$ -3

9. $\log_5 \sqrt[3]{5}$ $\frac{1}{3}$

10. $\log_3 1$ 0

Find the value of the variable.   *13.3*

11. $\log_b 81 = 4$ 3

12. $\log_2 y = -3$ $\frac{1}{8}$

Enrichment

Have the students complete the tables of
values for each of the following functions.

1. $y = 2^{\log_2 x}$

x	$\log_2 x$	$2^{\log_2 x} = y$
4	?	? 2, 4
8	?	? 3, 8
2	?	? 1, 2
1	?	? 0, 1

2. $y = \log_2 2^x$

x	2^x	$\log_2 2^x = y$
3	?	? 8, 3
2	?	? 4, 2
1	?	? 2, 1
0	?	? 1, 0

$\frac{1}{2}$	?	? $-1, \frac{1}{2}$
$\frac{1}{4}$	?	? $-2, \frac{1}{4}$
2^n	?	? $n, 2^n$

-1	?	? $\frac{1}{2}, -1$
-2	?	? $\frac{1}{4}, -2$
n	?	? $2^n, n$

Now have the students name another
function with exactly the same graph as the
given function.

3. $y = 2^{\log_2 x}$ $y = x$ for every $x > 0$

4. $y = \log_2 2^x$ $y = x$ for every x

13.5 $\text{Log}_b (x \cdot y)$ and $\text{Log}_b \frac{x}{y}$

Objectives
To expand logarithms of products and quotients
To simplify sums and differences of logarithms
To solve logarithmic equations

In the late sixteenth-century, Scottish nobleman John Napier devised tables of numbers he called *logarithms* that could be used to convert multiplications and divisions into additions and subtractions, respectively. Contemporaries soon adopted Napier's tables to help reduce the extraordinary drudgery of their arithmetic calculations. The methods for calculating with logarithms follow.

The logarithm of the *product* of two positive numbers can be written as the *sum* of two logarithms. For example, consider $\log_2 (16 \cdot 8)$.

$$\log_2 (16 \cdot 8) = \log_2 (2^4 \cdot 2^3) = \log_2 (2^7) = 7$$
$$\log_2 16 + \log_2 8 = \log_2 2^4 + \log_2 2^3 = 4 + 3 = 7$$

So, $\log_2 (16 \cdot 8) = \log_2 16 + \log_2 8$. This suggests Theorem 13.3.

Theorem 13.3

$\log_b xy = \log_b x + \log_b y$, where $x > 0$, $y > 0$, $b > 0$, and $b \neq 1$.

Proof

(1) Let $\qquad \log_b x = p \qquad$ and $\qquad \log_b y = q$.

(2) Then $\qquad x = b^p \qquad$ and $\qquad y = b^q$.

(3) So, $\qquad xy = b^p \cdot b^q = b^{p+q}$

(4) and $\qquad \log_b xy = \log_b b^{p+q} = p + q$.

(5) Thus, $\qquad \log_b xy = \log_b x + \log_b y$ (Steps 1 and 4).

The logarithm of the *quotient* of two positive numbers can be written as the *difference* of two logarithms. Consider $\log_2 \frac{128}{16}$.

$$\log_2 \frac{128}{16} = \log_2 \frac{2^7}{2^4} = \log_2 2^3 = 3$$
$$\log_2 128 - \log_2 16 = \log_2 2^7 - \log_2 2^4 = 7 - 4 = 3$$

So, $\log_2 \frac{128}{16} = \log_2 128 - \log_2 16$.

This suggests Theorem 13.4, which you will be asked to prove in Exercise 30.

13.5 $\text{Log}_b (x \cdot y)$ and $\text{Log}_b \frac{x}{y}$ **473**

Teaching Resources

Quick Quizzes 96
Reteaching and Practice
Worksheets 96

GETTING STARTED

Prerequisite Quiz

Solve.

1. $c(c + 2) = \frac{6c + 40}{2}$ 5, −4
2. $\frac{32}{x} = 2x$ −4, 4
3. $t(t - 6) = 16$ 8, −2
4. $\frac{3x + 6}{x - 2} = x + 2$ 5, −2
5. $4t(t + 4) = 84$ −7, 3

Motivator

Have students determine which one of the following is true.

1. $2^4 \cdot 2^6 = 2^{24}$ False
2. $2^4 \cdot 2^6 = 2^{10}$ True

TEACHING SUGGESTIONS

Lesson Note

Two laws of logarithms, $\log_b (xy) = \log_b x + \log_b y$ and $\log_b \frac{x}{y} = \log_b x - \log_b y$, are used to expand and simplify logarithmic expressions. Logarithmic equations are solved by simplifying each side and then using the property that if $\log_b x = \log_b y$, then $x = y$. *Apparent* solutions of logarithmic equations must be checked, since $\log_b x$ is defined only for $x > 0$.

Highlighting the Standards

Standard 5d: The interplay between logarithms and exponents demonstrates the power and importance of fundamental notation.

Critical Thinking Questions

Analysis: It is pointed out at the bottom of page 474 that the logarithm of a negative number is undefined. Ask students to explain why this is so. If they rewrite $y = \log_3 (-2)$ as $-2 = 3^y$, they will see that there is no real-number exponent y such that 3^y is a negative number. Students should generalize this example to conclude that since the base is restricted to positive values, there is no exponent (logarithm) to which the base can be raised to produce a negative number.

Checkpoint

1. Expand $\log_b \frac{4t}{5v}$.

   $\log_b 4 + \log_b t - \log_b 5 - \log_b v$

2. Simplify.

   $\log_3 (c^2 - 4) - \log_3 (c + 2) + \log_3 5$
   $\log_3 (5c - 10)$

3. Solve for t.

   $\log_2 t + \log_2 (t - 5) = \log_2 14$ 7

Theorem 13.4 $\log_b \frac{x}{y} = \log_b x - \log_b y$, where $x > 0$, $y > 0$, $b > 0$, and $b \neq 1$.

Theorems 13.3 and 13.4 can be used to *expand the logarithm* of a product or a quotient.

EXAMPLE 1   Write each expression in expanded form.

   a. $\log_3 7rt$ **b.** $\log_5 \frac{4c}{d}$ **c.** $\log_b \frac{mn}{5p}$

Solutions   **a.** $\log_3 7rt = \log_3 (7r \cdot t) = \log_3 7r + \log_3 t$
                     $= \log_3 7 + \log_3 r + \log_3 t$

   b. $\log_5 \frac{4c}{d} = \log_5 4c - \log_5 d = \log_5 4 + \log_5 c - \log_5 d$

   c. $\log_b \frac{mn}{5p} = \log_b mn - \log_b 5p$
                     $= \log_b m + \log_b n - (\log_b 5 + \log_b p)$
                     $= \log_b m + \log_b n - \log_b 5 - \log_b p$

Similarly, the sum or difference of two logarithms can be written as the logarithm of one term. To do this, use the *converse* (see page 103) of Theorems 13.3 and 13.4 as shown in Example 2.

EXAMPLE 2   Simplify $\log_3 8 - \log_3 m + \log_3 t^2 - \log_3 2t$.

Solution

	$\log_3 8 - \log_3 m + \log_3 t^2 - \log_3 2t$
Rearrange the terms.	$= \log_3 8 + \log_3 t^2 - \log_3 m - \log_3 2t$
Factor out -1.	$= \log_3 8 + \log_3 t^2 - (\log_3 m + \log_3 2t)$
$\log_b x + \log_b y = \log_b xy$	$= \log_3 8t^2 - \log_3 2mt$
$\log_b x - \log_b y = \log_b \frac{x}{y}$	$= \log_3 \frac{8t^2}{2mt}$
Simplify.	$= \log_3 \frac{4t}{m}$

Logarithmic equations can be solved by using the following property.

Property of Equality for Logarithms: If $\log_b x = \log_b y$, then $x = y$.

Thus, if $\log_2 5c = \log_2 (c + 12)$, then $5c = c + 12$, and $c = 3$.

Each apparent solution of a logarithmic equation must be checked, however, because $\log_b x$ is defined only for $x > 0$. For example, if $\log_3 2y = \log_3 (y - 5)$, then $2y = y - 5$, and $y = -5$. However, if $y = -5$, then $\log_3 2y = \log_3 (-10)$, but $\log_3 (-10)$ is undefined because $-10 < 0$. Thus, $\log_3 2y = \log_3 (y - 5)$ has no solution.

Additional Example 1

Write each expression in expanded form.

a. $\log_2 6cd$ $\log_2 6 + \log_2 c + \log_2 d$

b. $\log_b \frac{mn}{3}$ $\log_b m + \log_b n - \log_b 3$

c. $\log_3 \frac{at}{6r}$ $\log_3 a + \log_3 t - \log_3 6 - \log_3 r$

Additional Example 2

Simplify.

$\log_5 6 - \log_5 c^2 - \log_5 3d + \log_5 c$

$\log_5 \frac{2}{cd}$

EXAMPLE 3 Solve $\log_b c + \log_b (c + 2) = \log_b (6c + 40) - \log_b 2$ for c. Check.

Solution Simplify each side.

$$\log_b c(c + 2) = \log_b \frac{6c + 40}{2}$$

$$c(c + 2) = \frac{6c + 40}{2} \leftarrow \text{If } \log_b x = \log_b y, \text{ then } x = y.$$

$$c^2 + 2c = 3c + 20$$

$$c^2 - c - 20 = 0$$

$$(c - 5)(c + 4) = 0$$

$$c = 5 \quad \text{or} \quad c = -4$$

Check $c = 5$: $\log_b c + \log_b (c + 2) = \log_b 5 + \log_b 7$

$$= \log_b (5 \cdot 7) = \log_b 35, \text{ and } \log_b (6c + 40) -$$

$$\log_b 2 = \log_b 70 - \log_b 2 = \log_b \frac{70}{2} = \log_b 35$$

Check $c = -4$: $\log_b c = \log_b (-4)$, and $\log_b (-4)$ is undefined

because $-4 < 0$.

Thus, 5 is the only solution.

Classroom Exercises

Determine whether each statement is true or false.

1. $\log_2 (5 \cdot 6) = \log_2 5 + \log_2 6$ T
2. $\log_5 18 = \log_5 6 + \log_5 3$ T
3. $\log_b 10 = \log_b 4 + \log_b 6$ F
4. $\log_6 7 + \log_6 4 = \log_6 28$ T
5. $\log_b \frac{23}{5} = \log_b 23 - \log_b 5$ T
6. $\log_2 \frac{27}{3} = \log_2 27 \div \log_2 3$ F
7. $\log_4 \left(\frac{5 \cdot 9}{11} \right) = \log_4 5 + \log_4 9 - \log_4 11$ T
8. $\log_2 3 - \log_2 5 + \log_2 6 - \log_2 7 = \log_2 3 + \log_2 6 - (\log_2 5 + \log_2 7)$ T
9. If $\log_3 2x + \log_3 4y = \log_3 z$, then $2x \cdot 4y = z$. T
10. If $\log_2 (c + 7) - \log_2 5 = \log_2 3c$, then $\frac{c + 7}{5} = 3c$. T

Match each expression at the left with one expression at the right.

11. $\log_3 2 + \log_3 a - \log_3 c - \log_3 d$ d
a. $\log_3 \frac{2c}{ad}$

12. $\log_3 2 - \log_3 a + \log_3 c - \log_3 d$ a
b. $\log_3 \frac{ad}{2c}$

13. $\log_3 a + \log_3 c - (\log_3 2 + \log_3 d)$ c
c. $\log_3 \frac{ac}{2d}$

14. $\log_3 a - \log_3 2 + \log_3 d - \log_3 c$ b
d. $\log_3 \frac{2a}{cd}$

Closure

Have students give the restrictions to x, y and b for the equation $\log_b xy = \log_b x + \log_b y$. $x > 0, y > 0, b > 0$, and $b \neq 1$.

Have students complete the following sentence:

If $\log_b x = \log_b y$, then $\underline{\ ?\ }$. $x = y$

◼◼◼FOLLOW UP

Guided Practice

Classroom Exercises 1–14

Independent Practice

Ⓐ Ex. 1–20, Ⓑ Ex. 21–27, Ⓒ Ex. 28–30

Basic: WE 1–21 odd
Average: WE 7–27 odd, 28
Above Average: WE 9–29 odd

Additional Answers

Written Exercises

1. $\log_2 6 + \log_2 c$
2. $\log_3 5 + \log_3 t + \log_3 v$
3. $\log_4 n - \log_4 9$
4. $\log_5 7 - \log_5 m$
5. $\log_b 3 + \log_b t + \log_b w + \log_b z$
6. $\log_b 4 + \log_b k - \log_b p$
7. $\log_2 5 + \log_2 m - \log_2 c - \log_2 d$
8. $\log_3 10 + \log_3 a - \log_3 \pi - \log_3 r$
30. Let $\log_b x = p$ and $\log_b y = q$.
Then $x = b^p$ and $y = b^q$.
So, $\frac{x}{y} = \frac{b^p}{b^q} = b^{p-q}$, and $\log_b \frac{x}{y} = \log_b b^{p-q} = p - q$.
Thus, $\log_b \frac{x}{y} = \log_b x - \log_b y$.

Additional Example 3

Solve for c.

$\log_3 2 + \log_3 c = \log_3 (4c + 36) - \log_3 (c - 1)$ 6

Written Exercises

Write each expression in expanded form.

1. $\log_2 6c$

2. $\log_3 5tv$

3. $\log_4 \frac{n}{9}$

4. $\log_5 \frac{7}{m}$

5. $\log_b 3twz$

6. $\log_b \frac{4k}{p}$

7. $\log_2 \frac{5m}{cd}$

8. $\log_3 \frac{10a}{\pi r}$

Simplify each expression.

9. $\log_2 10 + \log_2 3c$ $\log_2 30c$

10. $\log_3 12a - \log_3 4$ $\log_3 3a$

11. $\log_4 9 + \log_4 5 - \log_4 3a$ $\log_4 \frac{15}{a}$

12. $\log_b 8t^2 - \log_b 2 - \log_b t$ $\log_b 4t$

13. $\log_5 2c - \log_5 3d + \log_5 6c$ $\log_5 \frac{4c^2}{d}$

14. $\log_b (c^2 - 9) - \log_b (c + 3)$ $\log_b (c - 3)$

Solve each equation. Check.

15. $\log_2 3k + \log_2 5 = \log_2 45$ 3

16. $\log_b (6c - 12) - \log_b 3 = \log_b 6$ 5

17. $\log_b 5t + \log_b 2t = \log_b 40$ 2

18. $\log_5 t + \log_5 (t - 6) = \log_5 16$ 8

19. $\log_b 2c + \log_b 4 = \log_b (3c + 10)$ 2

20. $\log_3 32 - \log_3 x = \log_3 2x$ 4

21. $\log_4 (3x + 6) - \log_4 (x + 2) = \log_4 (x - 2)$ 5

22. $\log_b (2x + 5) - \log_b (x + 6) = \log_b (2x - 2) - \log_b x$ $\frac{12}{5}$

23. $\log_3 y^2 + \log_3 2y^2 = \log_3 10 + \log_3 5y^2$ ± 5

24. $\log_2 (x + 4) + \log_2 (x - 1) = \log_2 (x - 2) + \log_2 (2x + 2)$ 5

25. $\log_5 4 + \log_5 t + \log_5 (t + 4) = \log_5 84$ 3

Simplify each expression.

26. $\log_b 3t - \log_b 2v - \log_b 9t^3 + \log_b 12v^2$ $\log_b \frac{2v}{t^2}$

27. $\log_4 5c^2 - \log_4 (c^2 - 4c) + \log_4 (c^2 - 9) - \log_4 (2c + 6)$ $\log_4 \frac{5c^2 - 15c}{2c - 8}$

Solve each equation. Check. (HINT: If $\log_b x = y$, then $x = b^y$.)

28. $\log_2 t + \log_2 (t - 4) = 5$ 8

29. $\log_8 (3c + 1) + \log_8 (c - 1) = 2$ 5

30. Prove Theorem 13.4.

Mixed Review

Write each expression in exponential form.   *8.6*

1. $\sqrt{5}$ $5^{\frac{1}{2}}$

2. $\sqrt[5]{4}$ $4^{\frac{1}{5}}$

3. $\sqrt[3]{x}$ $x^{\frac{1}{3}}$

4. $\sqrt[4]{c^3}$ $c^{\frac{3}{4}}$

Write each expression in radical form.   *8.6*

5. $x^{\frac{1}{4}}$ $\sqrt[4]{x}$

6. $10^{\frac{2}{3}}$ $\sqrt[3]{100}$

7. $5x^{\frac{1}{2}}$ $5\sqrt{x}$

8. $3^{-\frac{1}{2}}$ $\frac{1}{\sqrt{3}}$

Enrichment

Write the following information on the chalkboard.

$$\log_a 5 \approx 0.7$$
$$\log_a 2 \approx 0.3$$

Have the students use this information to determine the decimal value of each of the following logarithms.

1. $\log_a \frac{5}{2}$ 0.4

2. $\log_a \frac{2}{5}$ −0.4

3. $\log_a 10$ 1.0

4. $\log_a 20$ 1.3

5. $\frac{\log_a 5}{\log_a 2}$ 2.3

6. $\log_a 100$ 2.0

Ask the students if they can determine the value of *a* from the answers to any of these exercises.   *a* = 10, from Ex. 3 and 6

13.6 $\log_b x^r$ and $\log_b \sqrt[n]{x}$

Objectives

To expand logarithms of powers and radicals
To simplify multiples of logarithms
To solve logarithmic equations

Because $\log_2 7^3 = \log_2 (7 \cdot 7 \cdot 7) = \log_2 7 + \log_2 7 + \log_2 7$, it follows that $\log_2 7^3 = 3 \log_2 7$. That is, the logarithm of a *power* is a *multiple* of a logarithm. This suggests Theorem 13.5.

Theorem 13.5

$\log_b x^r = r \cdot \log_b x$, where $x > 0$, $b > 0$, $b \neq 1$, and r is a rational number.

Proof

1. Let $\log_b x = p$.
2. Then $x = b^p$.
3. So, $x^r = (b^p)^r = b^{r \cdot p}$
4. and $\log_b x^r = \log_b b^{r \cdot p} = r \cdot p$.
5. Thus, $\log_b x^r = r \cdot \log_b x$ (Steps 1 and 4).

If $b = 2$ and $r = \frac{1}{3}$, then $\log_b x^r = \log_2 x^{\frac{1}{3}} = \frac{1}{3} \cdot \log_2 x$, or $\log_2 \sqrt[3]{x} = \frac{1}{3} \cdot \log_2 x$. This illustrates a *corollary* to Theorem 13.5.

A **corollary** is a statement that can be immediately inferred from another statement that has already been proved.

Corollary

For each positive number x and each positive integer n,
$\log_b \sqrt[n]{x} = \frac{1}{n} \cdot \log_b x$, where $b > 0$ and $b \neq 1$.

This corollary is used to expand the logarithms of *radicals*, while Theorem 13.5 is used to expand the logarithms of *powers*.

EXAMPLE 1

Write in expanded form.

a. $\log_4 (tv^2)^3$

b. $\log_b \sqrt[4]{\dfrac{c^3}{d}}$

Plan

a. Use $\log_b x^r = r \cdot \log_b x$.

b. Use $\log_b \sqrt[n]{x} = \frac{1}{n} \cdot \log_b x$.

Solutions

a. $\log_4 (tv^2)^3 = 3 \log_4 tv^2$
$= 3(\log_4 t + \log_4 v^2)$
$= 3(\log_4 t + 2 \log_4 v)$

b. $\log_b \sqrt[4]{\dfrac{c^3}{d}} = \frac{1}{4} \log_b \dfrac{c^3}{d}$
$= \frac{1}{4}(\log_b c^3 - \log_b d)$
$= \frac{1}{4}(3 \log_b c - \log_b d)$

Additional Example 1

Write in expanded form.

a. $\log_5 (7c^3)^4$ $4(\log_5 7 + 3 \log_5 c)$

b. $\log_5 \sqrt[3]{\dfrac{m}{n^2}}$ $\frac{1}{3}(\log_5 m - 2 \log_5 n)$

Teaching Resources

Quick Quizzes 97
Reteaching and Practice Worksheets 97

▮▮ GETTING STARTED

Prerequisite Quiz

Write in exponential form.

1. $\sqrt[3]{x}$ $x^{\frac{1}{3}}$

2. $\sqrt[n]{x}$ $x^{\frac{1}{n}}$

Find the value of each expression.

3. $25^{\frac{1}{2}}$ 5

4. $8^{\frac{1}{3}}$ 2

Motivator

Ask students to determine which of the following is true.

1. $(x^5)^2 = x^{(5)^2} = x^{25}$ False
2. $(x^5)^2 = x^{5 \cdot 2} = x^{10}$ True
3. $\log_2 7^3 = \log_2 (7 \cdot 7 \cdot 7) =$
$\log_2 7 + \log_2 7 + \log_2 7$ True

▮▮ TEACHING SUGGESTIONS

Lesson Note

Two more laws of logarithms, $\log_b x^r = r \cdot \log_b x$ and $\log_b \sqrt[n]{x} = \frac{1}{n} \cdot \log_b x$, are used to expand and to simplify logarithmic expressions and to solve equations. Apparent solutions of logarithmic equations must be checked, since $\log_b x$ is undefined if $x \leq 0$.

Highlighting the Standards

Standard 4a: The exercises emphasize the recognition of equivalent forms. This skill will be of continuing importance in theoretical and applied mathematics.

Corollaries and Lemmas: The text introduces the term *corollary*, a statement that follows from a proven theorem. Students may be interested in knowing another term, *lemma*, that also plays an auxiliary role with respect to a theorem. A **lemma** is a preliminary "helping theorem" that is presented in order to simplify the statement or proof of the main theorem.

Critical Thinking Questions

Analysis: Ask students whether a logarithm equation such as those presented in Examples 3 and 4 might have *no* roots. The answer is "Yes." All that is required is for the resulting apparent roots all to be negative.

Checkpoint

Expand.

1. $\log_b (c^2 d)^3$ $3(2 \log_b c + \log_b d)$

Simplify each expression.

2. $\frac{1}{4} \log_5 6 + \frac{1}{4} \log_5 c - 3 \log_5 d$

$\log_5 \frac{\sqrt[4]{6c}}{d^3}$

Solve each equation.

3. $2 \log_5 x - \log_5 x = \log_5 (20 - 3x)$ 5

Closure

Have students complete the following statement: The logarithm of a power is a __?__ of a logarithm.   Multiple   Ask students to give the restiction on r for the equation $\log_b x^r = r \cdot \log_b x$. r is a rational number

---

In Examples 2–4, multiples of logarithms are simplified using the converse forms of Theorem 13.5 and its corollary.

EXAMPLE 2   Simplify each expression.

a. $4 \log_2 m + \frac{1}{3} \log_2 n - 5 \log_2 p$ **b.** $\frac{1}{4}(\log_5 c + 3 \log_5 d - \log_5 f)$

Solutions   **a.** $\log_2 m^4 + \log_2 \sqrt[3]{n} - \log_2 p^5$ **b.** $\frac{1}{4}(\log_5 c + \log_5 d^3 - \log_5 f)$

$= \log_2 (m^4 \sqrt[3]{n}) - \log_2 p^5$ $= \frac{1}{4} \log_5 \frac{cd^3}{f}$

$= \log_2 \frac{m^4 \sqrt[3]{n}}{p^5}$ $= \log_5 \sqrt[4]{\frac{cd^3}{f}}$

EXAMPLE 3   Solve $2 \log_b t = \log_b 3 + \log_b (t + 6)$ for t. Check.

Solution

$$2 \log_b t = \log_b 3 + \log_b (t + 6)$$
$$\log_b t^2 = \log_b 3(t + 6)$$
$$t^2 = 3(t + 6)$$
$$t^2 - 3t - 18 = 0$$
$$(t - 6)(t + 3) = 0$$
$$t = 6 \quad \text{or} \quad t = -3$$

Check $t = 6$:
$$2 \log_b t = \log_b 3 + \log_b (t + 6)$$
$$2 \log_b 6 \stackrel{?}{=} \log_b 3 + \log_b 12$$
$$\log_b 6^2 \stackrel{?}{=} \log_b (3 \cdot 12)$$
$$\log_b 36 = \log_b 36 \quad \text{True}$$
If $t = -3$, $\log_b t$ is undefined.

Thus, the only root is 6.

EXAMPLE 4   Solve $\frac{1}{2} \log_3 t + \frac{1}{2} \log_3 (t - 5) = \log_3 6$ for t. Check.

Solution

$$\frac{1}{2} \log_3 t + \frac{1}{2} \log_3 (t - 5) = \log_3 6$$
$$\frac{1}{2} \log_3 (t^2 - 5t) = \log_3 6$$
$$\log_3 \sqrt{t^2 - 5t} = \log_3 6$$
$$\sqrt{t^2 - 5t} = 6$$
$$t^2 - 5t = 36$$
$$t^2 - 5t - 36 = 0$$
$$(t - 9)(t + 4) = 0$$
$$t = 9 \quad \text{or} \quad t = -4$$

Check $t = 9$: $\frac{1}{2} \log_3 t + \frac{1}{2} \log_3 (t - 5) = \frac{1}{2} \log_3 9 + \frac{1}{2} \log_3 4 =$

$\frac{1}{2} \log_3 (9 \cdot 4) = \log_3 36^{\frac{1}{2}} = \log_3 6 \quad \text{True}$

If $t = -4$, $\log_3 t$ is undefined. Thus, the only root is 9.

Additional Example 2

Simplify each expression.

a. $3 \log_4 c - 2 \log_4 d + \frac{1}{2} \log_4 7$

$\log_4 \frac{c^3 \sqrt{7}}{d^2}$

b. $\frac{1}{3}(\log_6 m - \log_6 5 + 2 \log_6 d)$

$\log_6 \sqrt[3]{\frac{md^2}{5}}$

Additional Example 3

Solve for t.

$2 \log_5 t = \log_5 4 + \log_5 (t + 8)$
8

Additional Example 4

Solve for t.

$\frac{1}{2} \log_5 t + \frac{1}{2} \log_5 (t - 6) = \log_5 4$ 8

Focus on Reading

Find the expression that does *not* belong in each list.

1. a. $3 \log_5 2x$ b

  b. $\log_5 2x^3$

  c. $\log_5 8x^3$

  d. $\log_5 (2x)^3$

2. a. $\log_3 \sqrt[4]{2x}$ c

  b. $\frac{1}{4} \log_3 2x$

  c. $4 \log_3 2x$

  d. $\log_3 (2x)^{\frac{1}{4}}$

3. a. $\frac{2}{3} \log_b 5$ d

  b. $\frac{1}{3} \log_b 25$

  c. $\log_b \sqrt[3]{5^2}$

  d. $2 \log_b \frac{5}{3}$

Classroom Exercises

Match each expression at the left with one expression at the right.

1. $\log_5 \sqrt[3]{x}$ i

2. $\log_5 x^3$ c

3. $\log_5 x^2$ h

4. $\log_5 \sqrt{x}$ b

5. $4 \log_5 x$ d

6. $\frac{1}{4} \log_5 x$ g

7. $5 \log_4 x$ j

8. $\frac{1}{5} \log_4 x$ a

9. $4(\log_3 c - \log_3 d)$ k

10. $\frac{1}{4}(\log_3 c - \log_3 d)$ f

11. $4(\log_3 c + \log_3 d)$ l

12. $\log_3 c + \frac{1}{4} \log_3 d$ e

a. $\log_4 \sqrt[5]{x}$

b. $\frac{1}{2} \log_5 x$

c. $3 \log_5 x$

d. $\log_5 x^4$

e. $\log_3 c \sqrt[4]{d}$

f. $\log_3 \sqrt[4]{\frac{c}{d}}$

g. $\log_5 \sqrt[4]{x}$

h. $2 \log_5 x$

i. $\frac{1}{3} \log_5 x$

j. $\log_4 x^5$

k. $\log_3 \left(\frac{c}{d}\right)^4$

l. $\log_3 (cd)^4$

Written Exercises

Write each expression in expanded form.

1. $\log_3 (mn)^2$
  $2(\log_3 m + \log_3 n)$

2. $\log_b \sqrt{5g}$
  $\frac{1}{2}(\log_b 5 + \log_b g)$

3. $\log_5 \sqrt[3]{\frac{4}{h}}$
  $\frac{1}{3}(\log_5 4 - \log_5 h)$

4. $\log_b \left(\frac{t}{y}\right)^3$
  $3(\log_b t - \log_b y)$

Simplify each expression.

5. $3 \log_2 x + 4 \log_2 x$ $\log_2 x^7$

6. $\frac{1}{2} \log_5 3c + \frac{1}{2} \log_5 2d$ $\log_5 \sqrt{6cd}$

7. $\frac{1}{3}(\log_4 7 + \log_4 t)$ $\log_4 \sqrt[3]{7t}$

8. $4(\log_b 2 + \log_b c)$ $\log_b 16c^4$

9. $5 \log_b m - 2 \log_b n$ $\log_b \frac{m^5}{n^2}$

10. $\frac{1}{4} \log_3 ax - \frac{1}{4} \log_3 cy$ $\log_3 \sqrt[4]{\frac{ax}{cy}}$

11. $\frac{1}{2} \log_5 c + \frac{1}{2} \log_5 (c - 2)$
  $\log_5 \sqrt{c^2 - 2c}$

12. $2 \log_b x - 2 \log_b (x + 3)$
  $\log_b \left(\frac{x}{x+3}\right)^2$

Solve each equation. Check.

13. $2 \log_3 t = \log_3 2 + \log_3 (t + 12)$ 6

14. $2 \log_b x - \log_b 2 = \log_b (3x + 8)$ 8

15. $\frac{1}{2} \log_4 (3c - 5) = \log_4 5$ 10

16. $\log_5 2 + \frac{1}{3} \log_5 (x - 1) = \frac{1}{3} \log_5 4x$ 2

17. $\log_b 3 + 2 \log_b y = \log_b (8y + 3)$ 3

18. $\log_5 2 + 2 \log_5 t = \log_5 (3 - t)$ 1

19. $\frac{1}{2} \log_3 a + \frac{1}{2} \log_3 (a - 6) = \log_3 4$ 8

20. $\frac{1}{3} \log_b y + \frac{1}{3} \log_b (y - 2) = \log_b 2$ 4

Guided Practice

Classroom Exercises 1–12

Independent Practice

A Ex. 1–20, **B** Ex. 21–30, **C** Ex. 31–34

Basic: FR, WE 1–21 odd, Brainteaser

Average: FR, WE 9–31 odd, Brainteaser

Above Average: FR, WE 13–33 odd, Brainteaser

Additional Answers

Written Exercises

25. $2(\log_5 x + 3 \log_5 y - 2 \log_5 z)$

26. $\frac{1}{4}(\log_b 6 + \log_b m - \log_b n - 3 \log_b p)$

27. $\log_3 \frac{6\sqrt{m}}{n^2}$

28. $\log_4 \frac{\sqrt[3]{10ab}}{a^3 b}$

29. $\log_b \frac{8m^3 n^6}{p^3}$

30. $\log_b \sqrt[4]{\frac{4x^3}{15y}}$

Mixed Review

4.

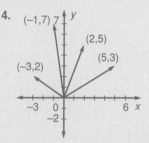

$(2,5) + (-3,2) = (-1,7)$
$(2,5) - (-3,2) = (5,3)$

Brainteaser

$\log_8 \left(\frac{1}{2}\right)$ is negative, so the direction of the inequality in the second step must be reversed.

Enrichment

Write the following information on the chalkboard.

$$\log_b 5 \approx 2.32$$
$$\log_b 10 \approx 3.32$$

Have the students use this information to determine the decimal value of each of the following logarithms.

1. $\log_b 125$ 6.96

2. $\log_b \sqrt{10}$ 1.66

3. $\log_b 5{,}000$ 12.28

4. $\log_b \sqrt{100}$ 3.32

5. $\log_b 2$ 1.00

6. $\log_b \sqrt{50}$ 2.82

Ask the students if they can determine the value of b from the answers to any of these exercises. $b = 2$, from Ex. 5

Written Exercises

1.

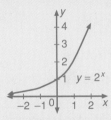

Domain: {real numbers};
Range: {y | y > 0};
Increasing; y-intercept: (0,1)

2.

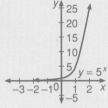

Domain: {real numbers};
Range: {y | y > 0};
Increasing; y-intercept: (0,1)

3.

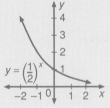

Domain: {real numbers};
Range: {y | y > 0};
Decreasing; y-intercept: (0,1)

---

21. $2 \log_b (2y + 4) = \log_b 9 + 2 \log_b (y - 1)$ 7

22. $2 \log_5 (3y - 1) = \log_5 4 + 2 \log_5 (y + 2)$ 5

23. $\frac{1}{2} \log_3 (c + 2) + \frac{1}{2} \log_3 (c - 3) = \log_3 6$ 7

24. $\frac{1}{2} \log_b (2x - 1) - \frac{1}{2} \log_b (3x + 1) = \log_b 3 - \log_b 4$ 5

Write each expression in expanded form.

25. $\log_5 \left(\dfrac{xy^3}{z^2} \right)^2$

26. $\log_b \sqrt[4]{\dfrac{6m}{np^3}}$

Simplify each expression.

27. $\log_3 6 + \frac{1}{2} \log_3 m - 2 \log_3 n$

28. $\frac{1}{3} \log_4 5a + \frac{1}{3} \log_4 2b - 3 \log_4 a - \log_4 b$

29. $3(\log_b 2m + 2 \log_b n - \log_b p)$

30. $\frac{1}{4}(\log_b 4 + 3 \log_b x - \log_b 3 - \log_b 5y)$

Solve each equation. Check. (HINT: If $\log_b x = y$, then $x = b^y$.)

31. $\log_5 (a^2 + 2a + 5) - \log_5 (a - 5) = 2$ **32.** $2 \log_3 (y - 3) - \log_3 (y + 1) = 2$

  13, 10   15

Simplify.

33. $\log_b \sqrt[3]{y^2} - \log_b \sqrt[4]{y^3} + 4 \log_b \sqrt[3]{y} - 5 \log_b \sqrt[4]{y}$ 0

34. $\frac{2}{5} \log_b 3 - \frac{1}{5} \log_b 2 - \frac{2}{5} \log_b 4 + \frac{1}{5} \log_b 27$ $\log_b \frac{3}{2}$

Mixed Review

Rationalize the denominator.   *8.4, 9.2*

1. $\dfrac{4}{2\sqrt{7} - 5}$ $\dfrac{8\sqrt{7} + 20}{3}$

2. $\dfrac{11}{5 + 3i}$ $\dfrac{55 - 33i}{34}$

3. Solve $\sqrt{6x - 2} = x + 1$. Check.   *10.3*   1, 3

4. Draw the vectors (2,5) and (−3,2) on a coordinate plane. Then draw their sum and difference.   *9.8*

▰▰▱ *Brainteaser*

Find the error in the sequence of steps below.

$$3 > 2$$
$$3 \log_8 \left(\tfrac{1}{2}\right) > 2 \log_8 \left(\tfrac{1}{2}\right)$$
$$\log_8 \left(\tfrac{1}{2}\right)^3 > \log_8 \left(\tfrac{1}{2}\right)^2$$
$$\left(\tfrac{1}{2}\right)^3 > \left(\tfrac{1}{2}\right)^2$$
$$\tfrac{1}{8} > \tfrac{1}{4}$$

480   Chapter 13 Exponential and Logarithmic Functions

4.

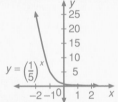

Domain: {real numbers};
Range: {y | y > 0};
Decreasing; y-intercept: (0,1)

5.

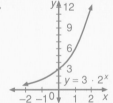

Domain: {real numbers};
Range: {y | y > 0};
Increasing; y-intercept: (0,3)

6.

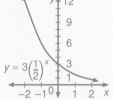

Domain: {real numbers};
Range: {y | y > 0};
Decreasing; y-intercept: (0,3)

7.

Domain: {real numbers};
Range: {y | y > 0};
Increasing; y-intercept: (0,3)

13.7 Common Logarithms and Antilogarithms

Objectives   To find base-10 logarithms of positive rational numbers
To find base-10 antilogarithms of rational numbers
To solve word problems involving pH

Three rows in a table of **common logarithms** (base-10) are displayed below. From the table, $\log_{10} 5.26 = 0.7210$ to four decimal places.

n	0	1	2	3	4	5	6	7	8	9
5.1	7076	7084	7093	7101	7110	7118	7126	7135	7143	7152
5.2	7160	7168	7177	7185	7193	7202	7210	7218	7226	7235
5.3	7243	7251	7259	7267	7275	7284	7292	7300	7308	7316

If $\log_b q = r$, then q is the base-b **antilogarithm** (antilog) of r. To find antilog$_{10}$ 0.7210, reverse the steps above. Look for 7210 in the table. It is in row 5.2, column 6. So, antilog$_{10}$ 0.7210 = 5.26.

When approximating common logarithms, it is customary to use the *equal* sign (=) rather than the *approximately equal* sign (≈).

EXAMPLE 1   **a.** Find $\log_{10} 5.31$.   **b.** Find antilog$_{10}$ 0.7093.

Solutions   **a.** $\log_{10} 5.31 = 0.7251$ **b.** antilog$_{10}$ 0.7093 = 5.12

The table on pages 790–791 lists the common logarithms of numbers between 1 and 10. However, using scientific notation (see Lesson 5.2), you can also use the table to estimate the common logarithms of positive numbers less than 1 or greater than 10.

Using $\log_{10} 5.29 = 0.7235$ and scientific notation, for example, you can estimate $\log_{10} 5{,}290$ as shown below.

$$\log_{10} 5{,}290 = \log_{10} (5.29 \times 10^3)$$
$$= \log_{10} 5.29 + \log_{10} 10^3$$
$$= 0.7235 + 3$$

So, $\log_{10} 5{,}290 = 3 + 0.7235$, or 3.7235.

The integer 3 is the *characteristic* of 3.7235, and the decimal 0.7235 is the *mantissa*.

GETTING STARTED

Prerequisite Quiz

Write in scientific notation.

1. 804,000 8.04×10^5

2. 0.00752 7.52×10^{-3}

3. 6.93 6.93×10^0

Find the base-10 logarithm.

4. $\log_{10} 1{,}000$ 3

5. $\log_{10} 10$ 1

6. $\log_{10} \frac{1}{100}$ -2

7. $\log_{10} \frac{1}{10}$ -1

Motivator

Have students express the number 10,000 in scientific notation. 1.0×10^4 Now have students use a scientific calculator to find $\log_{10} 10{,}000$.   4

Additional Example 1

a. Find $\log_{10} 5.18$ 0.7143

b. Find antilog$_{10}$ 0.7185   5.23

Highlighting the Standards

Standard 6b: The use of log tables is not merely an exercise in finding numbers but an opportunity to "represent and analyze relationships using tables."

Lesson Note

Students should learn to use a scientific calculator to find

1. the base-10 logarithm of a number using the $\boxed{\log}$ key, and
2. the base-10 antilogarithm of a decimal using the $\boxed{\text{INV}}$ and $\boxed{\log}$ keys (or using the $\boxed{10^x}$ key).

The pH of a solution is defined in terms of common logarithms, and problems from chemistry are solved in Examples 4 and 5. Examples 4b and 5 can be solved using a scientific calculator as follows.
Example 4b: pH $= -\log 0.000375$
Press: .000375, $\boxed{\log}$, $\boxed{+/-}$
Display: 3.4259687
Example 5: 8.2 $= -\log [H^+]$
  $\log [H^+] = -8.2$
Press: 8.2, $\boxed{+/-}$, $\boxed{\text{INV}}$, $\boxed{\log}$
  (or press: 8.2, $\boxed{+/-}$ $\boxed{10^x}$)
Display: 6.3096 $- 09$, which means
  6.3096×10^{-9}

Definitions

If $y = n \cdot 10^c$, where $1 \leq n < 10$ and c is an integer, then $\log_{10} y = c + \log_{10} n$. The integer c is the **characteristic** and $\log_{10} n$ is the **mantissa** of $\log_{10} y$.

Notice that the mantissa, $\log_{10} n$, is a *positive* decimal less than 1. The subscript 10 can be omitted from expressions such as $\log_{10} 5{,}290$ when working with common logarithms. Thus, log 5,290 (without a written base) means $\log_{10} 5{,}290$.

EXAMPLE 2 **a.** Find log 534.   **b.** Find log 0.00534

Solutions

a. $\log 534 = \log (5.34 \times 10^2)$
      $= 2 + 0.7275$
      $= 2.7275$

b. $\log 0.00534$
      $= \log (5.34 \times 10^{-3})$
      $= -3 + 0.7275$
      $= -2.2725$

The log key on a calculator can be used to find the logarithms in Example 2 as shown below.

a. log 534: 534 $\boxed{\log}$ $= 2.7275413 \approx 2.7275$

b. log 0.00534: 0.00534 $\boxed{\log}$ $= -2.2724587 \approx -2.2725$

If you are using a table of logarithms to find the antilogarithm of a *negative* number, remember to write the number as the sum of an integer (the characteristic) and a *positive* decimal (the mantissa).

Thus, from Example 2b above comes the following.

  antilog (-2.2725)
  $=$ antilog $(-3 + 0.7275)$ ←— The mantissa must be between 0 and 1.
  $= 5.34 \times 10^{-3}$ ←— Take the antilog of 0.7275.
  $= 0.00534$

EXAMPLE 3 **a.** Find antilog $(6.7193 - 10)$.   **b.** Find antilog 4.7135.

Solutions

a. antilog $(6.7193 - 10)$
    characteristic: $6 - 10$, or -4
    mantissa: 0.7193
    antilog $0.7193 = 5.24$
  antilog $(6.7193 - 10)$
    $=$ antilog $(-4 + 0.7193)$
    $= 5.24 \times 10^{-4} = 0.000524$

b. antilog 4.7135
    characteristic: 4
    mantissa: 0.7135
    antilog $0.7135 = 5.17$
  antilog 4.7135
    $=$ antilog $(4 + 0.7135)$
    $= 5.17 \times 10^4 = 51{,}700$

Additional Example 2

a. Find log 5,300. 3.7243
b. Find log 0.000530. -3.2757

Additional Example 3

a. Find antilog $(7.7235 - 10)$. 0.00529
b. Find antilog 5.7101. 513,000

The antilogarithms in Example 3 can be found using the inverse and log keys on a calculator, as shown below.

a. antilog $(6.7193 - 10)$: 6.7193 $\boxed{-}$ 10 $\boxed{=}$ $\boxed{INV}$ $\boxed{log}$ $=$
   0.000523962

b. antilog 4.7135: 4.7135 $\boxed{INV}$ $\boxed{log}$ $= 51701.126$

On some calculators, the inverse key is shown as $\boxed{2nd}$ or $\boxed{2nd\ f}$.

Also, a calculator display of $5.23962 \quad -04$ in Example a above means 5.23962×10^{-4}.

In the study of chemistry, common logs are used to calculate the pH of a solution. pH is a measure of the solution's relative acidity or alkalinity, and it is defined as follows: $pH = -\log [H^+]$, where $[H^+]$ is the concentration of hydrogen ions in moles per liter.

EXAMPLE 4   Calculate the pH of a solution containing the given concentration of H^+ per liter.

   a. 0.001 mole   b. 0.000375 mole

Plan   Use the formula $pH = -\log [H^+]$.

Solutions   a. $[H^+] = 0.001 = 10^{-3}$
              $pH = -\log [H^+] = -\log_{10} 10^{-3} = -(-3) = 3$

              The pH of the solution is 3.

              b. $[H^+] = 0.000375 = 3.75 \times 10^{-4}$
              $pH = -\log [H^+] = -\log (3.75 \times 10^{-4}) = -(-4 + 0.5740)$
              $\qquad\qquad\qquad\qquad\qquad\qquad\qquad = -(-3.4260) \approx 3.4$

              Calculator steps: 0.000375 $\boxed{log}$ $\boxed{+/-}$ $= 3.4259687$

              The pH of the solution is 3.4.

EXAMPLE 5   Calculate the concentration of H^+ in a solution with a pH of 8.2.

Plan   Use the formula $pH = -\log [H^+]$, or $\log [H^+] = -pH$.

Solution $pH = 8.2$

              $8.2 = -\log [H^+]$, or $\log [H^+] = -8.2$
              antilog $-8.2 =$ antilog $[-8 + (-0.2)]$ ← -0.2 is negative.
              $\qquad\qquad\qquad\qquad\qquad\qquad$ The mantissas in the
              $\qquad\qquad\qquad\qquad\qquad\qquad$ table are positive.
              $\qquad\qquad\qquad\qquad\qquad\qquad$ Rewrite -8.2 as $-9 + 0.8$.
              $\qquad\qquad = $ antilog $[-9 + (0.8)]$
              $\qquad\qquad = 6.3 \times 10^{-9}$

              The concentration is 6.3×10^{-9} moles of H^+ per liter.

Math Connections

Historical Note: We take it for granted that logarithms must have a base, since we regard them as inverses of expressions of the form b^x. It is interesting to note that neither Napier nor Burgi, the two men who independently invented logarithms near the end of the sixteenth century, developed their ideas using the concept of a base. Rather, their ideas were suggested by a calculating technique of the time known as prosthaphaeresis (a Greek word meaning addition and subtraction). This procedure uses addition formulas from trigonometry to convert multiplication into addition and subtraction.

Critical Thinking Questions

Analysis: Ask students which of the solutions of Examples 4 and 5 is more acidic (contains a heavier concentration of hydrogen ions).   Although this appears to be a chemistry question, it really tests how well students understand exponential notation. In Example 4, the concentrations are 10^{-3} and $10^{-3.4}$ moles per liter; in Example 5 the concentration is $10^{-8.2}$ moles per liter. The concentrations of hydrogen ions in Example 4 are significantly higher than those of Example 5. (A neutral solution has a pH of 7, i.e., a hydrogen ion concentration of 10^{-7} moles per liter.)

Additional Example 4

Find the pH for the given concentration of H^+ ions per liter.

a. 0.00001 mole   5
b. 0.000058 mole   4.2

Additional Example 5

Find the concentration of H^+ ions for the given pH.

a. pH = 5   0.00001 mole/liter
b. pH = 10.6 2.5×10^{-11} mole/liter

Common Error Analysis

Error: When using calculators, students may have difficulty in finding the mantissa and characteristic of a negative logarithm. For example, $\log_{10} 0.00358$ is found by pressing .00358 and log. This displays -2.4461, which is equal to $-2 + (-0.4461)$.

To find the mantissa, which is positive, add 10 and read 7.5539 as $7.5539 - 10$. The mantissa is 0.5539 and the characteristic is -3.

Checkpoint

Find the logarithm or antilogarithm.

1. log 761,000   5.8814
2. log 0.000761 $6.8814 - 10$
3. log 0.0001 -4
4. antilog 4.3729   23,600
5. antilog $(8.3729 - 10)$ 0.0236
6. Use pH = $-\log$ [H$^+$] to find the pH if [H$^+$] = 0.00078.   3.1

Closure

Have students give definitions for the terms *mantissa* and *characteristic*.   See page 482.   Ask students to estimate the value of the following:

1. antilog$_{10}$ 2.7958   about 625
2. log$_{10}$ 2790   about 3.4

Classroom Exercises

Given log 536 = 2.7292, match each expression at the left with its equivalent at the right.

1. log 536 f
2. log 5.36 d
3. log 0.0536 b
4. characteristic of log 536 e
5. characteristic of log 53,600 g
6. characteristic of log 0.00536 a
7. mantissa of log 536 d
8. mantissa of log 5.36 d
9. antilog 2.7292 i
10. antilog 0.7292 h
11. antilog $(9.7292 - 10)$ c

a. -3
b. $8.7292 - 10$
c. 0.536
d. 0.7292
e. 2
f. 2.7292
g. 4
h. 5.36
i. 536

Express each number in scientific notation.

12. 234,000 2.34×10^5
13. 78.9 7.89×10^1
14. 0.000321 3.21×10^{-4}

Express each number in ordinary decimal notation.

15. 5.43×10^{-1} 0.543
16. 4.56×10^0 4.56
17. 1.25×10^3 1,250

Find the characteristic of each logarithm.

18. log 3,450   3
19. log 6.54   0
20. log 0.000321 -4

Written Exercises

Find each logarithm to four decimal places or antilogarithm to three significant digits. Where possible, give an exact answer.

1. log 18,700   4.2718
2. log 0.305 -0.5157
3. antilog 2.4548   285
4. antilog $(7.5490 - 10)$
5. log 0.00594 -2.2262
6. antilog $(-2 + 0.9320)$
7. log 23,500,000   7.3711
8. antilog $(6.7582 - 10)$
9. antilog 4.8457   70,100
10. log 1,000   3
11. log 0.001 -3
12. antilog 3   1,000
13. antilog -1 0.1
14. log 1   0
15. log 0.1 -1
16. antilog 1   10
17. log 10   1
18. antilog -4 0.0001
19. log $\sqrt{10}$ $\frac{1}{2}$
20. antilog $\frac{1}{3}$ $\sqrt[3]{10}$
21. log $\sqrt[3]{100}$ $\frac{2}{3}$

Calculate the pH of the solution containing the given concentration of H$^+$ per liter.

22. 0.0001 mole   4
23. 0.0029 mole   2.5
24. $7.5 \cdot 10^{-12}$ mole   11.1

Enrichment

Inform the students that in addition to the common logarithms, which use base 10, there is a system of natural logarithms, abbreviated ln N, which uses as its base the number e.

$$e \approx 2.71828$$

A definition of e found in advanced courses is the value that $\left(1 + \frac{1}{n}\right)^n$ approaches as n increases without limit.

Have the students use scientific calculators to evaluate this expression for $n = 5$, $n = 10$, $n = 20$, $n = 100$, and $n = 100,000$.
2.48832, 2.59374, 2.65330, 2.70481, 2.71825

Point out that e can be found on a scientific calculator by clearing the calculator and then pressing 1, INV, and then ln x (or by pressing 1 and then e^x). This also reveals that ln $e = 1$, just as log 10 = 1.

Calculate the concentration of H^+ in a solution with the given pH.

25. pH = 5   **26.** pH = 10.2   **27.** pH = 4.8

28. Find log (antilog (log (antilog 3.9350))).   3.9350

29. Simplify antilog 3 + antilog 2 + antilog 1 + antilog 0 + antilog (−1).   1,111.1

Use the information below for Exercises 30 and 31.

$$6^3 = 216 \qquad\qquad 6^4 = 1{,}296 \qquad\qquad 6^8 = 1{,}679{,}616$$
$$3 \cdot \log 6 \approx 2 + 0.33 \qquad 4 \cdot \log 6 \approx 3 + 0.11 \qquad 8 \cdot \log 6 \approx 6 + 0.23$$

30. Find the number of digits in the numeral for 6^{20}, the 20th power of 6.   16

31. For two positive integers x, the numeral for 4^x has exactly 17 digits. Find the greater value of x.   28

Mixed Review

Solve each equation.   *5.1, 7.6, 8.7*

1. $5^{2x-1} = 125$ 2

2. $\dfrac{8}{x+5} + \dfrac{3}{x} = 2$ $\dfrac{-5}{2}, 3$

3. $6^{x+2} = \dfrac{1}{36}$ −4

4. $8^{2x} = 16$ $\dfrac{2}{3}$

Application: *Carbon Dating*

Radioactive elements decay at fixed exponential rates, called half-lives. The half-life of radioactive carbon-14, for example, is 5,730 years. In that time, half of any particular mass of carbon-14 will decay to form stable carbon-12 atoms. Because carbon-14 is found in all organic matter, it is often used to date fossils and other ancient artifacts.

The formula for radioactive decay used in carbon dating is $\dfrac{r}{r_0} = 2^{-\frac{t}{h}}$, where r is the amount of the radioactive element currently in the sample, r_0 is the amount in the sample when it was alive, t is the age of the sample, and h is the half-life of the radioactive element. The amount of the radioactive element present during life is generally assumed to be approximately that present in living matter today.

Find the age of a fossil having the given ratio of r to r_0 and $h = 5{,}730$ yr.

1. $\dfrac{1}{2}$ 5,730 yr   **2.** $\dfrac{1}{4}$ 11,460 yr   **3.** $\dfrac{1}{\sqrt{2}}$ 2,865 yr   **4.** $\dfrac{\sqrt{2}}{32}$ 25,785 yr

13.7 Common Logarithms and Antilogarithms   **485**

Guided Practice

Classroom Exercises 1–20

Independent Practice

Ⓐ Ex. 1–18, Ⓑ Ex. 19–27, Ⓒ Ex. 28–31

Basic: WE 1–21 odd, Application

Average: WE 7–29 odd, Application

Above Average: WE 11–31 odd, Application

Additional Answers

Written Exercises

4. 0.00354
6. 0.0855
8. 0.000573
25. 1×10^{-5} mole/l
26. 6.31×10^{-11} mole/l
27. 1.6×10^{-5} mole/l

▬▬GETTING STARTED

Prerequisite Quiz

Solve for x.

1. $3^x = 81$ 4
2. $5^{-x} = 125$ -3

Change the logarithmic equation to an exponential equation.

3. $\log_4 64 = x$ $4^x = 64$
4. $\log_b y = x$ $b^x = y$

Motivator

Have students locate the solution of $5^x = 80$ between consecutive integers by placing 80 between consecutive integral powers of 5.

$25 < 80 < 125$
$5^2 < 5^x < 5^3$
$2 < x < 3$

Objectives   To solve exponential equations using common logarithms
To find base-b logarithms of positive numbers using common logarithms
To solve word problems using exponential formulas

In the *exponential equation* $3^{2x+1} = 27$, 27 is an integral power of three. So, $3^{2x+1} = 3^3$, $2x + 1 = 3$, and $x = 1$. In the equation $3^{2x+1} = 212$, however, 212 is not an integral power of three. Instead, it lies in an interval between two integral powers of three. Because $81 < 212 < 243$, it follows that $3^4 < 3^{2x+1} < 3^5$. So, $4 < 2x + 1 < 5$, or $3 < 2x < 4$, and thus $1.5 < x < 2$.

To find x to three significant digits, use base-10 logarithms.

EXAMPLE 1   Solve $3^{2x+1} = 212$ for x to three significant digits.

Solution
$$3^{2x+1} = 212, \text{ so } \log 3^{2x+1} = \log 212.$$
$$(2x + 1)\log 3 = \log 212 \leftarrow \log a^r = r \log a$$
$$2x + 1 = \frac{\log 212}{\log 3} = \frac{2.3263}{0.4771}$$
$$2x + 1 = 4.876 \leftarrow \text{Divide to four significant digits.}$$
$$2x = 3.876$$
$$x = 1.938$$

Thus, $x = 1.94$ to three significant digits.

A base-b logarithm such as $\log_4 6$ can be found to three significant digits by solving an exponential equation.

EXAMPLE 2   Find $\log_4 6$ to three significant digits.

Solution   Let $\log_4 6 = x$.   Then $4^x = 6$. ← Change to an exponential equation.
$$\log 4^x = \log 6$$
$$x \cdot \log 4 = \log 6$$
$$x = \frac{\log 6}{\log 4} = \frac{0.7782}{0.6021} = 1.292$$

Calculator steps: 6 (log) (÷) 4 (log) (=) 1.2924813
So, $\log_4 6 = 1.29$ to three significant digits.

Highlighting the Standards

Standard 4d: Examples 2, 3, and 4 show how exponential equations are used in several areas of business and science.

Additional Example 1

Solve $14^{3x-2} = 2.5$ for x to three significant digits.   0.782

Additional Example 2

Find $\log_2 45$ to three significant digits.   5.49

Notice in Example 2 that $\log_4 6 = \dfrac{\log 6}{\log 4}$. In general, $\log_b y = \dfrac{\log y}{\log b}$, where $b > 0$, $b \neq 1$, and $y > 0$.

Exponential formulas are used in business and science, and may lead to large powers of rational numbers.

If \$8,000 is invested at 14% per year and compounded semiannually, then every 6 months the interest earned for the half-year is added to the previous principal to form a new principal. The total value of this investment over several 6-month periods is found and generalized below. Observe that 14% per year is 7% per 6 months.

Number of 6-month periods	Total value (principal plus interest)
0	8,000
1	$8,000 + 8,000(0.07)$ $= 8,000(1 + 0.07) = 8,000(1.07)^1$
2	$8,000(1.07) + 8,000(1.07)(0.07)$ $= 8,000(1.07)(1 + 0.07) = 8,000(1.07)^2$
3	$8,000(1.07)^2 + 8,000(1.07)^2(0.07)$ $= 8,000(1.07)^2(1 + 0.07) = 8,000(1.07)^3$
. . .	. . .
x	$8,000(1.07)^x$

This suggests the formula for *compound interest*: $A = p\left(1 + \dfrac{r}{n}\right)^{nt}$, where A is the total amount at the end of t years when p dollars are invested at $r\%$ per year, compounded n times each year.

EXAMPLE 3 An investment of \$7,400 at 12% per year is compounded quarterly (4 times each year). How much will the investment be worth in 15 years? Use logarithms or a calculator with the x^y function.

Solution $A = p\left(1 + \dfrac{r}{n}\right)^{nt} = 7,400\left(1 + \dfrac{0.12}{4}\right)^{4 \cdot 15} = 7,400(1.03)^{60}$.

$\log A = \log 7,400(1.03)^{60} = \log 7,400 + 60 \cdot \log 1.03$
$= 3.8692 + 60(0.0128)$
$= 4.6372$

$A = \text{antilog}\,(4 + 0.6372) = 4.34 \times 10^4 = \$43,400$

Using logarithms as shown above, you will find that the \$7,400 investment will be worth \$43,400 in 15 years. Using a scientific calculator, you will obtain a value of \$43,597.86.

■■■■**TEACHING SUGGESTIONS**

Lesson Note

Common (base-10) logarithms are used to approximate the solution of an exponential equation and to approximate base-b logarithms, where $b > 0$ and $b \neq 1$. A calculator is essential for computations in this lesson, where quotients such as $\dfrac{\log 212}{\log 3}$, or $\dfrac{2.3263}{0.4771}$, must be found to three significant digits. Students should know that the following numbers contain exactly three significant digits.

0.543   530   304
0.780   0.0602   8.60

Math Connections

Exponential Growth and Decay: The lesson provides examples of exponential *growth*, that is, the successive doublings of a starting amount of material in equal time periods. The *Application* on page 485 is an example of exponential decay, the successive halvings of a starting amount of material in equal time periods.

Additional Example 3

Find the value of an investment of \$6,500, paying 8% compounded semiannually, at the end of 5 years.   \$9,620 by logs; \$9,621.59 by calculator

487

Analysis: The $7,400 of Example 3 will grow to well over $200,000 if it is invested at the same 12% rate of interest for a second 15 years. Ask students to discuss whether this means that they each can be guaranteed a retirement nest egg simply by placing $7,400 aside as an investment for 30 years and forgetting about it.   The basic approach is sound, but students should be aware of the factors that will weaken somewhat the returns from such an investment program. These include the effects of inflation on the purchasing power of money, the difficulty of sustaining a 12% annual return for 30 years, and the annual reduction in income created by tax payments.

Checkpoint

1. Find x to three significant digits.
   $8^{2x+2} = 500$ 0.494
2. Find $\log_5 400$ to three significant digits.
   3.72
3. Find the value of a $2,500 investment, paying 8% compounded quarterly, at the end of 12 years. Use $A = p\left(1 + \frac{r}{n}\right)^{nt}$.
   $6,470 by logs; $6,467.68 by calculator

Closure

Have students explain the steps in solving for x in $\log_5 9 = x$.   See Example 2.

By a process called *simple fission*, some bacteria reproduce by splitting themselves into two new bacteria at the end of each growth period. If 600 bacteria are present at the start of an experiment, and simple fission takes place every 20 minutes, the number of bacteria increases rapidly.

Periods	0	1	2	3	4	5
Number of bacteria	600	1,200	2,400	4,800	9,600	19,200
	600	$600 \cdot 2$	$600 \cdot 2^2$	$600 \cdot 2^3$	$600 \cdot 2^4$	$600 \cdot 2^5$

At the end of 5 hours, the bacteria have gone through 15 growth periods, and the number of bacteria is $600 \cdot 2^{15}$, or 19,660,800.

This suggests the formula $N = N_0 \cdot 2^k$, where N is the number of bacteria present at the end of k periods of simple fission and N_0 is the number of bacteria at the start.

EXAMPLE 4   Find the number of bacteria produced from 800 bacteria at the end of 6 h if simple fission occurs every 12 min.

Plan   There are 5 twelve-minute periods in 1 hour, and 30 such periods in 6 h. Use $N = N_0 \cdot 2^k$, where $N_0 = 800$ and $k = 30$.

Solution
$$N = N_0 \cdot 2^k = 800 \cdot 2^{30} \leftarrow \text{Calculator steps: } 800 \; \boxed{\times} \; 2 \; \boxed{x^y} \; 30 \; \boxed{=}$$
$$8.58993 \times 10^{11}$$

$$\log N = \log(800 \cdot 2^{30})$$
$$= \log 800 + 30 \cdot \log 2$$
$$= 2.9031 + 30(0.3010) = 11.9331$$
$$N = \text{antilog } 11.9331 = \text{antilog}(11 + 0.9331) = 8.57 \times 10^{11}$$

There will be 8.57×10^{11} bacteria at the end of 6 h.
If a scientific calculator is used, this value will be 8.59×10^{11} bacteria.

Classroom Exercises

The exponent x lies between what two consecutive integers?

1. $3^x = 22$ 2, 3   **2.** $2^x = 22$ 4, 5   **3.** $5^x = 22$ 1, 2   **4.** $10^x = 754$ 2, 3

Solve each equation for x to three significant digits.

5. $3^x = 22$ 2.81   **6.** $5^{x-1} = 45$ 3.37   **7.** $\log_2 7 = x$ 2.81   **8.** $x = \log_4 5.2$ 1.19

Additional Example 4

Find N, the number of bacteria present, if $N_0 = 15,000$ and simple fission occurs every 20 min for 6 h. 3.93×10^9 bacteria

Written Exercises

Solve each equation for x to three significant digits.

1. $4^{2x} = 30$ 1.23   **2.** $8^{2x+1} = 600$ 1.04 **3.** $7^{3x-2} = 750$ 1.80 **4.** $9^{2x-4} = 5$ 2.37

Find each logarithm to three significant digits.

5. $\log_7 4$ 0.712   **6.** $\log_5 35$ 2.21   **7.** $\log_3 45.3$ 3.47   **8.** $\log_6 850$ 3.76

Find the value to the nearest cent of each investment at the given annual interest rate. Use the formula $A = p\left(1 + \frac{r}{n}\right)^{nt}$ and a calculator.

9. $600, paying 10%, compounded semiannually for 4 years $886.47
10. $850, paying 8%, compounded quarterly for 3 years $1,078.01
11. $8,000, paying 12%, compounded quarterly for 11 years $29,371.62
12. $7,200, paying 14%, compounded semiannually for 20 years $107,816.10

Find N, the number of bacteria present, for the given data to three significant digits. Use $N = N_0 \cdot 2^k$ and a calculator.

13. $N_0 = 500$, and simple fission occurs every 30 min for 4 h. 1.28×10^5
14. $N_0 = 4,000$, and simple fission occurs every 15 min for 10 h. 4.40×10^{15}
15. $N_0 = 7,500$, and simple fission occurs every 20 min for 15 h. 2.64×10^{17}
16. $N_0 = 35,000$, and simple fission occurs every 90 min for 24 h. 2.29×10^9

Solve each equation for x to three significant digits.

17. $5^x = 3 \cdot 4^x$ 4.92   **18.** $3 \cdot 5^{x+1} = 5 \cdot 3^{x+2}$ 2.15   **19.** $2^{4x+3} = 5 \cdot 2^{3x}$ −0.678
20. $x = 2^{\sqrt{6}}$ 5.46   **21.** $3x = 4^{\sqrt{5}}$ 7.40   **22.** $2x - 1 = 12^{\sqrt{2}}$ 17.3

Graph $y = \log_2 x$ and the given function in the same coordinate plane. Describe the relationship between the two graphs.

23. $y = 3 + \log_2 x$ **24.** $y = \log_2 (x - 4)$ **25.** $y = \log_2 (x + 3)$

Mixed Review

1. Solve the system: $\begin{array}{l} 3x - 2y = 18 \\ 10x + 3y = 2 \end{array}$ *4.3* (2, −6)
2. Convert 18 lb/ft^2 to ounces per square inch. *7.3* 2 oz/in^2
3. Use synthetic substitution and the Integral Zero Theorem to solve $x^4 + x^3 + 7x^2 + 9x - 18 = 0$. *10.4* −2, 1, ±3i
4. Find the maximum value of y given that $y = -2x^2 - 12x + 35$. *9.4* 53
5. A number b varies inversely as the square of a. If $b = 4$ when $a = 3$, what is b when $a = 2$? *12.5* 9

FOLLOW UP

Guided Practice

Classroom Exercises 1–8

Independent Practice

Ⓐ Ex. 1–10, Ⓑ Ex. 11–16, Ⓒ Ex. 17–25

Basic: WE 1–13 odd, Extension, Mixed Problem Solving

Average: WE 7–21 odd, Extension, Mixed Problem Solving

Above Average: WE 11–25 odd, Extension, Mixed Problem Solving

Additional Answers

Written Exercises

23.

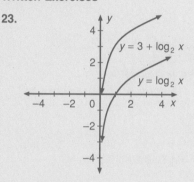

The graph of $y = 3 + \log_2 x$ is that of $y = \log_2 x$ shifted 3 units up.

See page 490 for the answers to Written Ex. 24–25.

Enrichment

The generalization following Example 2 provides an easy way to convert common logarithms to natural logarithms (ln). (See the Enrichment for Lesson 13.7.)

$\ln N = \dfrac{\log N}{\log e}$

Have the students use this formula and a scientific calculator to find decimal constants for converting from each form to the other.

$\ln N \approx 2.3 \log N$
$\log N \approx 0.43 \ln N$

489

24.

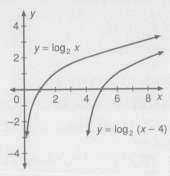

The graph of $y = \log_2 (x - 4)$ is that of $y = \log_2 x$ shifted 4 units right.

25.

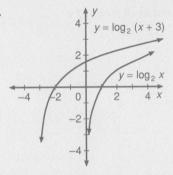

The graph of $y = \log_2 (x + 3)$ is that of $y = \log_2 x$ shifted 3 units left.

Extension: *Linear Interpolation*

The table below shows values of x to one decimal place and values of $f(x)$ to three significant digits.

x	2.0	2.1	2.2	2.3	2.4	2.5	2.6	2.7	2.8	2.9
$f(x)$	5.40	5.47	5.54	5.61	5.68	5.75	5.82	5.89	5.96	6.03

The equation for the function is $f(x) = 0.7x + 4$, but even when the equation is not known, the value of $f(x)$ for $x = 2.76$ (two decimal places) can still be found using only the table and a process called **linear interpolation**. Study the graph for the interval $2.7 < x < 2.8$.

In the graph, notice the following.
(1) $f(2.7) = 5.89$, $f(2.8) = 5.96$, and
$f(2.76) = 5.89 + BD$

(2) Triangle ABD is similar to triangle ACE.

Hence, $\dfrac{AD}{AE} = \dfrac{BD}{CE}$, or $\dfrac{0.06}{0.1} = \dfrac{BD}{0.07}$.

$BD = \dfrac{0.06 \times 0.07}{0.1} = 0.042 \approx 0.04$

Then $5.89 + BD = 5.89 + 0.04 = 5.93$.
So, $f(2.76) = 5.93$.

Linear interpolation can be performed as shown below.

Example 1   Find $f(2.53)$ to three significant digits. Use the table above.

Solution

$$
0.1 \left\{ 0.03 \left\{ \begin{array}{c|c} x & f(x) \\ \hline 2.5 & 5.75 \\ 2.53 & ? \\ 2.6 & 5.82 \end{array} \right\} n \right\} 0.07
$$

$$\frac{0.03}{0.1} = \frac{n}{0.07}$$

$$n = \frac{0.03 \times 0.07}{0.1}, \text{ or } n = 0.021 \approx 0.02$$

Thus, $f(2.53) = 5.75 + n = 5.75 + 0.02 = 5.77$.

Example 2 Find x to two decimal places given that $f(x) = 5.58$. Use the table on the previous page.

Solution

$$0.1 \left\{ n \left\{ \begin{array}{c|c} x & f(x) \\ \hline 2.2 & 5.54 \\ ? & 5.58 \\ 2.3 & 5.61 \end{array} \right\} 0.04 \right\} 0.07 \right.$$

$$\frac{n}{0.1} = \frac{0.04}{0.07}$$

$$n = \frac{0.1 \times 0.04}{0.07}$$

$$n \approx 0.06$$

Thus, $x = 2.2 + n = 2.2 + 0.06 = 2.26$.

Over narrow intervals, the graphs of logarithmic functions are close to straight lines. Hence, linear interpolation can be used to find $\log x$ when x is given to four significant digits. This is shown in Example 3.

Example 3 Find $\log 2{,}346$ to four decimal places. Use linear interpolation and the logarithm table.

Plan 2,346 is between 2,340 and 2,350. From the table, $\log 2340 = 3.3692$ and $\log 2350 = 3.3711$. Omit all decimal points and commas while interpolating.

Solution

$$10 \left\{ 6 \left\{ \begin{array}{c|c} x & \log x \\ \hline 2340 & 33692 \\ 2346 & ? \\ 2350 & 33711 \end{array} \right\} n \right\} 19 \right.$$

$$\frac{6}{10} = \frac{n}{19}, \text{ or } n = \frac{6 \times 19}{10} \approx 11$$

So, $33692 + n = 33692 + 11 = 33703$.

Thus, $\log 2{,}346 = 3.3703$.

Extension: Linear Interpolation **491**

Written Exercises

4.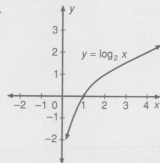

Domain: $x > 0$; Range: reals; Increasing; x-intercept: $(1,0)$

5.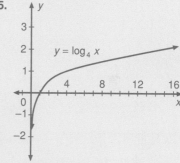

Domain: $x > 0$; Range: reals; Increasing; x-intercept: $(1,0)$

6.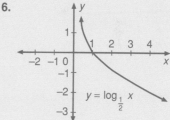

Domain: $x > 0$; Range: reals; Decreasing; x-intercept: $(1,0)$

7.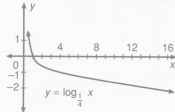

Domain: $x > 0$; Range: reals; Decreasing; x-intercept: $(1,0)$

8.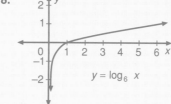

Domain: $x > 0$; Range: reals; Increasing; x-intercept: $(1,0)$

9.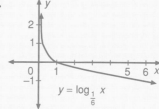

Domain: $x > 0$; Range: reals; Decreasing; x-intercept: $(1,0)$

491

10.

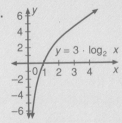

Domain: $x > 0$; Range: reals;
Increasing; x-intercept: (1,0)

11.

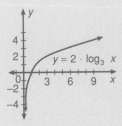

Domain: $x > 0$; Range: reals;
Increasing; x-intercept: (1,0)

12.

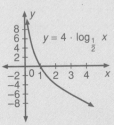

Domain: $x > 0$; Range: reals;
Decreasing; x-intercept: (1,0)

13.

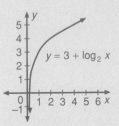

Domain: $x > 0$; Range: reals;
Increasing; x-intercept: $\left(\frac{1}{8}, 0\right)$

Example 4 Find antilog 1.5123 to four significant digits. Use linear interpolation and the logarithm table.

Plan If antilog 1.5123 = x, then log x = 1.5123. From the table, 5123 is between 5119 and 5132. So, antilog 1.5123 is between antilog 1.5119 = 32.5 and antilog 1.5132 = 32.6. While interpolating, omit all decimal points and commas.

Solution

$$10 \left\{ n \left\{ \begin{array}{c|c} x & \log x \\ 3250 & 15119 \\ ? & 15123 \\ 3260 & 15132 \end{array} \right\} 4 \right\} 13$$

$$\frac{n}{10} = \frac{4}{13}$$

$$n = \frac{10 \times 4}{13} \approx 3$$

So, $3250 + n = 3250 + 3 = 3253$.

Thus, antilog 1.5123 = 32.53.

Exercises

Use linear interpolation and the table below for Exercises 1–8.

x	3.0	3.1	3.2	3.3	3.4	3.5	3.6	3.7
$f(x)$	10.52	10.77	11.02	11.27	11.52	11.77	12.02	12.27

Find $f(x)$ to two decimal places.

1. $x = 3.42$ 11.57 **2.** $x = 3.04$ 10.62 **3.** $x = 3.52$ 11.82 **4.** $x = 3.36$ 11.42

Find x to two decimal places.

5. $f(x) = 11.08$ 3.22 **6.** $f(x) = 10.67$ 3.06 **7.** $f(x) = 10.93$ 3.16 **8.** $f(x) = 11.09$ 3.23

Find each logarithm to four decimal places and each antilogarithm to four significant digits. Use linear interpolation.

9. log 3,582
3.5541

10. log 23.74
1.3755

11. antilog 2.5735
374.5

12. antilog 4.3454
22,150

492 Extension: Linear Interpolation

14.

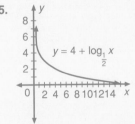

Domain: $x > 0$; Range: reals;
Increasing; x-intercept: $\left(\frac{1}{9}, 0\right)$

15.

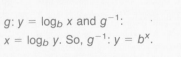

Domain: $x > 0$; Range: reals;
Decreasing; x-intercept: (16,0)

16. $f: y = b^x$ and $f^{-1}: x = b^y$.
So, $f^{-1}: y = \log_b x$.

17. $g: y = \log_b x$ and g^{-1}:
$x = \log_b y$. So, $g^{-1}: y = b^x$.

18.

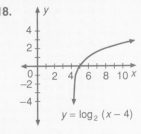

Mixed Problem Solving

1. Find the value after 5 years of $7,500 earning 12% per year, compounded quarterly, to the nearest cent. Use the formula $A = p\left(1 + \frac{r}{n}\right)^{nt}$ and a calculator. $13,545.83

2. The pressure P of a gas at constant temperature varies inversely as the volume V. If $V = 600$ in^3 when $P = 45$ lb/in^2, what is P when $V = 54$ in^3? 500 lb/in^2

3. If y varies jointly as x and $\sqrt{z}$, and $y = 14$ when $x = 4$ and $z = 36$, what is z when $x = 12$ and $y = 63$? 81

4. Find the maximum area A of a rectangle whose dimensions are x and $800 - 2x$ units. 80,000 sq units

5. Corn worth $2.10/kg and oats worth $2.80/kg are mixed to make 100 kg of feed worth a total of $227.50. How many kilograms of oats are used? 25 kg

6. How many milliliters of a 40% acid solution must be added to 150 milliliters of a 65% acid solution to obtain a 50% acid solution? 225 milliliters

7. Two cars took the same trip. One averaged 40 mi/h and the other averaged 45 mi/h. If the slower car took 30 min longer, what distance did the cars drive? 180 mi

8. Alice is 4 years younger than Betty and Carl is 3 times as old as Alice. Five years ago, Carl's age was 3 years less than twice the sum of the girls' ages then. How old is Carl now? 30 yr

9. The length of a rectangle is 1 ft more than its width and 3 ft less than the length of a diagonal. Find the length of the diagonal in simplest radical form. $7 + 2\sqrt{6}$ ft

10. An object is shot upward from the earth's surface with an initial velocity of 80 m/s. To the nearest 0.1 second, when will its height be 200 m? Use the formula $h = vt - 5t^2$. 3.1 s, 12.9 s

11. The Drama Workshop raised $615 by selling 280 tickets to its annual play. Adult tickets cost $3.00 each and children's tickets cost $1.50 each. How many $3.00 tickets were sold? 130

12. The perimeter of a rectangle is 72 in. If the width and length are each increased by 6 inches, their ratio will be 7:9. What are the dimensions of the original rectangle? 15 in. × 21 in.

13. A certain boat can travel 50 km/h in still water. One day it travels 10.4 km downstream in the same time that it travels 9.6 km upstream. Find the rate of the current. 2 km/h

14. Machine A can do a certain job in 18 hours. Machines B and C can do the same job in 6 h and 4.5 h, respectively. How long will the job take if all three machines operate at the same time? 2 h 15 min

15. Find three consecutive odd integers such that the sum of the squares of the second and third integers is 130.

16. The product of two numbers is 8 and the sum of their squares is 20. Find all such pairs of numbers. 4, 2; −4, −2

17. The tens digit of a two-digit number is one more than 4 times the units digit. The sum of the number and the number with its digits reversed is 121. Find the number. 92

18. Working together, Jake and Felicia can clean a house in 3 hours. It takes Jake 4 times longer than Felicia to do it alone. How long would it take each alone? Jake: 15 h; Felicia: $3\frac{3}{4}$ h

Mixed Problem Solving

15. 5, 7, 9, or −11, −9, −7

Additional Answers, page 471

19.
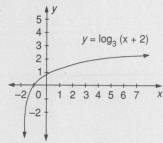

20.
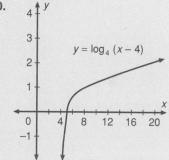

Midchapter Review

5.
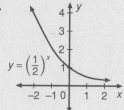

Domain: {real nos.}; Range: {$y \mid y > 0$}
Decreasing; y-intercept: (0, 1)

6.
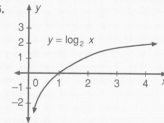

Domain: {$x \mid x > 0$}; Range: {real nos.}
Increasing; x-intercept: (1, 0)

1.

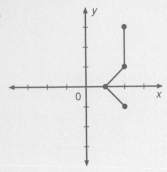

5.

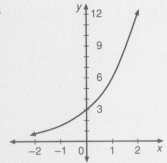

Domain: {all real nos.} Range: $\{y \mid y > 0\}$
Increasing; y-intercept: (0,3)

12.

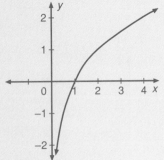

Domain: $\{x \mid x > 0\}$ Range: {all real nos.}
Increasing; x-intercept: (1,0)

Chapter 13 Review

Key Terms

antilogarithm (p. 481)
base (p. 467)
base-b logarithm (p. 467)
characteristic (p. 482)
common logarithm (p. 481)
corollary (p. 477)
exponential function (p. 464)
inverse relations (p. 459)

linear interpolation (p. 490)
logarithm (p. 467)
logarithmic equation (p. 467)
logarithmic function (p. 469)
mantissa (p. 482)
Property of Equality for
  Logarithms (p. 474)

Key Ideas and Review Exercises

13.1 For the inverse relations A and A^{-1}, each ordered pair (a,b) in A is reversed to (b,a) in A^{-1}. The graphs of A and A^{-1} are symmetric with respect to the line $y = x$. The equation for A^{-1} can be obtained from the equation for A by trading x and y.

The graph of relation A is shown at the right.

1. Draw the graph of A^{-1}.

2. Is A^{-1} a function? No

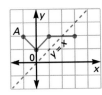

The function B is described by $y = x^2 + 6$.

3. Write the equation for B^{-1}, solving for y. $y = \pm\sqrt{x - 6}$

4. Is B^{-1} a function? No

13.2 Exponential functions are described by the equations $y = b^x$ and $y = \left(\frac{1}{b}\right)^x$ (or $y = b^{-x}$), where $b > 1$. Their domains are the set of real numbers. Their ranges are $\{y \mid y > 0\}$. $y = b^x$ is always increasing; $y = \left(\frac{1}{b}\right)^x$ is always decreasing. Their y-intercepts are 1, and their combined graph is symmetric with respect to the y-axis.

5. Draw the graph of g: $y = 3 \cdot 2^x$. Find the domain, range, and y-intercept. Is the function always increasing or always decreasing?

13.3 The equation $\log_b y = x$ can be written as $y = b^x$. Then $\log_b y = \log_b b^x = x$.

Find each logarithm.

6. $\log_4 64$ 3   **7.** $\log_8 1$ 0   **8.** $\log_7 \sqrt{7}$ $\frac{1}{2}$ **9.** $\log_2 2^6$ 6

10. Solve $\log_3 y = 4$ for y. 81   **11.** Solve $\log_b 16 = 4$ for b. 2

494   Chapter 13 Review

13.4 Logarithmic functions are described by the equations $y = \log_b x$ and $y = \log_{\frac{1}{b}} x$, where $b > 1$.

Their domains are $\{x \mid x > 0\}$. Their ranges are the set of real numbers. $y = \log_b x$ is always increasing; $y = \log_{\frac{1}{b}} x$ is always decreasing.

Their x-intercepts are 1, and their combined graph is symmetric with respect to the x-axis.

Note that $f: y = b^x$ where $y > 0$, $b > 0$, $b \neq 1$, and $f^{-1}: y = \log_b x$ where $x > 0$, $b > 0$, $b \neq 1$ are a pair of inverse functions.

12. Draw the graph of $f: y = \log_2 x$. Find the domain, range, and x-intercept. Determine whether the function is always increasing or always decreasing.

13.5, 13.6 To expand or simplify a logarithmic expression, use the following.

$$\log_b xy = \log_b x + \log_b y \qquad \log_b \frac{x}{y} = \log_b x - \log_b y$$

$$\log_b x^r = r \cdot \log_b x \qquad \log_b \sqrt[n]{x} = \frac{1}{n} \cdot \log_b x$$

To solve a logarithmic equation, use: If $\log_b x = \log_b y$, then $x = y$.

Write each expression in expanded form.

13. $\log_3 \frac{6a}{m}$ **14.** $\log_b c^2 d$ **15.** $\log_5 \sqrt[4]{tv}$ **16.** $\log_4 \sqrt{\frac{x}{y}}$

Simplify each expression.

17. $\log_b 7 + \log_b a - \log_b d \quad \log_b \frac{7a}{d}$ **18.** $4 \log_2 c + \frac{1}{3} \log_2 d \quad \log_2 c^4 \sqrt[3]{d}$

19. Write in your own words how to solve $\frac{1}{4} \log_6 (x - 3) = \log_6 3$ for x.

20. Solve $2 \log_5 y = \log_5 2 + \log_5 (3y + 8)$ for y. Check. 8

13.7 Common logarithms ($\log x$) are base-10 logarithms. Log x means $\log_{10} x$.

21. Find $\log 0.00453$. -2.3439

22. Find antilog 4.8007. $63{,}200$

23. Find the pH of a solution that has 0.00060 mole of H^+ per liter. Use $pH = -\log [H^+]$. 3.2

24. Find the concentration of H^+ in a solution with a pH of 3.8. Use $\log [H^+] = -pH$.
$1.6 \cdot 10^{-4}$ mole/liter

13.8 To solve $a^x = b$ for x, use the fact that $\log a^x = \log b$.

Then $x \log a = \log b$, and $x = \dfrac{\log b}{\log a}$.

To find $\log_b y$, let $\log_b y = x$, rewrite as $b^x = y$, and solve for x.

25. Solve $6^{3x-2} = 476$ for x to three significant digits. 1.81

26. Find $\log_3 78$ to three significant digits. 3.97

13. $\log_3 6 + \log_3 a - \log_3 m$

14. $2 \log_b c + \log_b d$

15. $\frac{1}{4}(\log_5 t + \log_5 v)$

16. $\frac{1}{2}(\log_4 x - \log_4 y)$

19. $\frac{1}{4}\log_6 (x - 3) = \log_6 3$.

Use $\frac{1}{n} \cdot \log_b x = \log_b \sqrt[n]{x}$:

$\log_6 \sqrt[4]{x - 3} = \log_6 3$.

Use the fact that if $\log_b c = \log_b d$, then $c = d$: $\sqrt[4]{x - 3} = 3$. Raise each side to the 4th power: $x - 3 = 81$, or $x = 84$. Check 84 in the original equation:
$\frac{1}{4} \log_6 (x - 3) = \frac{1}{4} \log_6 81 = \log_6 \sqrt[4]{81} = \log_6 3$.

Chapter Test

1.

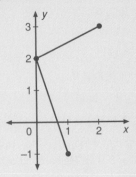

5.

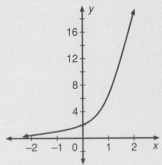

Domain: {all real nos.}
Range: {y | y > 0}
y-intercept: (0,2)

6.

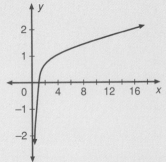

Domain: {x | x > 0}
Range: {all real nos.}
x-intercept: (1,0)

The graph of relation A is shown at the right.

1. Draw the graph of A^{-1}.   No
2. Is A^{-1} a function?   No
3. Are A and A^{-1} a pair of inverse functions? No
4. Write the equation for the inverse of
   $B: y = x^2 - 3$, solving for y. $y = \pm \sqrt{x + 3}$

Graph each function. Find the domain, range, and x- or y-intercept.

5. $y = 2 \cdot 3^x$
6. $y = \log_4 x$

Find each logarithm.

7. $\log_2 64$ 6
8. $\log_2 \sqrt{32}$ $\frac{5}{2}$

Find the value of the variable in each logarithmic equation.

9. $\log_5 y = 4$ 625
10. $\log_b 64 = 2$ 8
11. Write the equation for the inverse of $f: y = 4^x$, solving for y. $y = \log_4 x$

Write each expression in expanded form.

12. $\log_2 \frac{7a}{c}$ $\log_2 7 + \log_2 a - \log_2 c$
13. $\log_5 \sqrt[3]{pc}$ $\frac{1}{3}(\log_5 p + \log_5 c)$

Simplify each expression.

14. $\log_b 5 + \log_b 3v - \log_b c$ $\log_b \frac{15v}{c}$
15. $3 \log_b c + \frac{1}{5} \log_b d$ $\log_b c^3 \sqrt[5]{d}$

Solve each equation. Check.

16. $\log_4 a + \log_4 (a - 5) = \log_4 24$ 8
17. $\frac{1}{3} \log_b (x - 2) = \log_b 2$ 10
18. Find log 0.0532. −1.2741
19. Find antilog 3.4942. 3,120
20. Calculate the pH of a solution containing 0.000025 mole of H^+ per liter. Use pH = $-\log [H^+]$. 4.6
21. Solve $3^x = 6 \cdot 2^x$ for x to three significant digits. 4.42
22. Find $\log_4 25$ to three significant digits. 2.32
23. Use $A = p\left(1 + \frac{r}{n}\right)^{nt}$ and a calculator to find the value of a $750 investment, paying 10% per year, compounded semiannually for 8 years, to the nearest cent. $1,637.16
24. Draw the graph of $y = \log_2 (x - 3)$. Find the domain, range, and x-intercept.
25. Solve $\log_2 (3c - 4) + \log_2 c = 5$. Check. 4

24.

Domain: {x | x > 3}
Range: {real numbers}
x-intercept: (4,0)

Choose the *one* best answer to each question or problem.

1. Let A be any constant function $y = c$. Which statement(s) is (are) true? D
 I. A^{-1} is a function.
 II. A^{-1} is a relation.
 III. A and A^{-1} are a pair of inverse relations.

 (A) I only   (B) II only
 (C) I and II only
 (D) II and III only
 (E) I, II, and III

2.

$\overline{QR}$ is a diameter of the circle above. Point P is on the circle so that m $\angle PQR = 2x$. Find m $\angle PRQ$. D
 (A) $6x$
 (B) $\dfrac{180 - 2x}{2}$
 (C) $2x - 90$
 (D) $2(45 - x)$
 (E) There is not enough information.

3. $\log \dfrac{1}{mn} = $ ___?___ C
 (A) $1 - \log m - \log n$
 (B) $\log m + \log n$
 (C) $-\log m - \log n$
 (D) $\log m - \log n$
 (E) None of these

4. If x and y are real numbers, and $y^2 = 6 - 2x$, then ___?___ D
 (A) $x \geq 6$ (B) $x \geq 3$
 (C) $x \leq 6$ (D) $x \leq 3$
 (E) None of these

5. The only solution to the equation $\log_2 x - \log_3 (x + 1) = \log_5 (x - 3)$ is ___?___ A
 (A) 8   (B) 6
 (C) 4   (D) 2
 (E) None of these

6. If $\log_2 y = x$, then ___?___ B
 (A) $y < x$ (B) $y > x$
 (C) $y = x$ (D) $y = 2$
 (E) None of these

7. If $\log_b (4x - 8) < \log_b (2x + 6)$, where $b > 1$, then ___?___ D
 (A) $-3 < x < 7$ (B) $0 < x < 7$
 (C) $-3 < x < 2$ (D) $2 < x < 7$
 (E) None of these

8. If $2^y = 50$ and $y = 2x - 1$, then ___?___ D
 (A) $x = 13$
 (B) $16.5 < x < 32.5$
 (C) $2.0 < x < 2.5$
 (D) $3.0 < x < 3.5$
 (E) None of these

9. Find the false statement, if any.
 (A) If $y = \log_{\frac{1}{2}} x$, then $x = 2^{-y}$. C
 (B) $\log_b (x^2 - 9) = \log_b (x + 3) + \log_b (x - 3)$
 (C) $\log_5 (x^2 + y^2) = 2 \log_5 x + 2 \log_5 y$
 (D) $\log_{\frac{1}{2}} 8 = \log_2 \dfrac{1}{8}$
 (E) None of these

10. The graphs of which pair of functions do *not* intersect? C
 (A) $y = \log_2 x, \; y = \log_{\frac{1}{2}} x$
 (B) $y = 3x - 2, \; y = -3x + 2$
 (C) $y = 2^x, \; y = \log_2 x$
 (D) $y = 3^x, \; y = \left(\dfrac{1}{3}\right)^x$
 (E) Each pair of graphs intersects.

14 PERMUTATIONS, COMBINATIONS, PROBABILITY

OVERVIEW

The chapter opens with a lesson on the Fundamental Counting Principle. This is the basis for the following three lessons on permutations and the two lessons on combinations. In turn, these lessons provide the groundwork for three lessons on probability. The last two lessons of the chapter provide an introduction to statistics.

OBJECTIVES

- To find the number of permutations of a set of objects
- To find the number of combinations of a set of objects
- To find the probability of an event
- To find the probability of selecting at random more than one object from a set of objects
- To solve problems involving permutations, combinations, and probability
- To find the mean, median, mode(s), variance, and standard deviation for a set of data

PROBLEM SOLVING

Students should be encouraged to use the strategy *Drawing a Diagram* to help them solve the problems of this chapter. A diagram enables students to see permutations, combinations, and outcomes of events (see page 518). Lesson 14.11 allows students the opportunity to use graphs to solve problems.

READING AND WRITING MATH

The *Focus on Reading* in Lesson 14.3 gives students practice in recognizing and using a new mathematical symbol. The *Brainteaser* on page 528 and Exercise 21 in the *Chapter Review* ask students to explain mathematical procedure in their own words.

TECHNOLOGY

Calculator: A scientific calculator has a factorial key that can be used with problems involving factorials, such as the permutations of Lessons 14.2–14.4 and the combinations of Lessons 14.5 and 14.6 (see *Using the Calculator* on pages 506 and 513). Many scientific calculators have a statistics mode that allows the user to calculate the mean and standard deviation of a set of numbers. This feature can be used with statistics lessons at the end of the chapter, especially with Lesson 14.10.

SPECIAL FEATURES

Mixed Review pp. 502, 506, 508, 510, 513, 520, 524, 528, 532, 535
Using the Calculator pp. 506, 513
Brainteaser pp. 506, 528
Application: The Decibel Scale p. 506
Focus on Reading p. 508
Midchapter Review p. 516
Extension: Sampling a Population pp. 536–537
Key Terms p. 538
Key Ideas and Review Exercises pp. 538–539
Chapter 14 Test p. 540
College Prep Test p. 541
Cumulative Review (Chapters 1–14) pp. 542–543

PLANNING GUIDE

Lesson	Basic	Average	Above Average	Resources
14.1 pp. 501–502	CE all WE 1–13 odd	CE all WE 5–17 odd, 18	CE all WE 7–19 odd	Reteaching 100 Practice 100
14.2 pp. 504–506	CE all, WE 1–11 odd Using the Calculator Brainteaser, Application	CE all, WE 5–15 odd Using the Calculator Brainteaser, Application	CE all, WE 9–21 odd Using the Calculator Brainteaser, Application	Reteaching 101 Practice 101
14.3 p. 508	FR all, CE all WE 1–13 odd	FR all, CE all WE 3–17 odd	FR all, CE all WE 7–19 odd	Reteaching 102 Practice 102
14.4 p. 510	CE all WE 1–6	CE all WE 3–9	CE all WE 6–11	Reteaching 103 Practice 103
14.5 pp. 512–513	CE all, WE 1–13 odd Using the Calculator	CE all, WE 5–19 odd Using the Calculator	CE all, WE 11–23 odd Using the Calculator	Reteaching 104 Practice 104
14.6 pp. 515–516	CE all, WE 1–15 odd Midchapter Review	CE all, WE 5–23 odd Midchapter Review	CE all, WE 11–27 odd Midchapter Review	Reteaching 105 Practice 105
14.7 pp. 519–520	CE all WE 1–17 odd	CE all WE 11–27 odd, 28	CE all WE 13–29 odd	Reteaching 106 Practice 106
14.8 pp. 523–524	CE all WE 1–21 odd	CE all WE 9–31 odd	CE all WE 13–33 odd	Reteaching 107 Practice 107
14.9 pp. 527–528	CE all WE 1–15 odd Brainteaser	CE all WE 11–29 odd Brainteaser	CE all WE 11–31 odd Brainteaser	Reteaching 108 Practice 108
14.10 pp. 531–532	CE all WE 1–9 odd	CE all WE 3–13 odd	CE all WE 7–15 odd	Reteaching 109 Practice 109
14.11 pp. 535–537	CE all WE 1–8, Extension	CE all WE 5–13, Extension	CE all WE 6–13, Extension	Reteaching 110 Practice 110
Chapter 14 Review pp. 538–539	all	all	all	
Chapter 14 Test p. 540	all	all	all	
College Prep Test p. 541	all	all	all	
Cumulative Review pp. 542–543	1–45 odd	7–49 odd	16–53 odd	

CE = Classroom Exercises   WE = Written Exercises   FR = Focus on Reading

NOTE: For each level, all students should be assigned all Mixed Review exercises.

◤ INVESTIGATION

In this investigation, students explore the use of random numbers to simulate random events.

Lists of random numbers have been used by experimenters to insure that their choices have no particular bias that would effect the results. Now, computers and even calculators have the ability to produce lists of random numbers.

Random numbers are typically numbers greater than 0 and less than 1. They have no particular order. This is their most important characteristic.

For example, a modern calculator might produce the following sequence:

.054
.632
.992
.697

The experimenter can use such a list for a large variety of purposes.

Challenge the students to convert a list like this to a list of integers between 1 and 10, inclusive.

Multiply each number by 10, add 1, and take the integer part.

$10 \times .054 + 1 = 1.54$
$10 \times .632 + 1 = 7.32$
$10 \times .992 + 1 = 10.92$
$10 \times .397 + 1 = 4.97$

The integer parts of the new numbers are 1, 7, 10, and 4.

Ask the students how they could use the list of random numbers to imitate the throwing of a die (values 1 through 6).

Multiply each number by 6, add 1, and take the integer part of the result. The above list gives the values 1, 4, 6, and 3.

Invite discussion about other uses of a list of random numbers to simulate real world events.

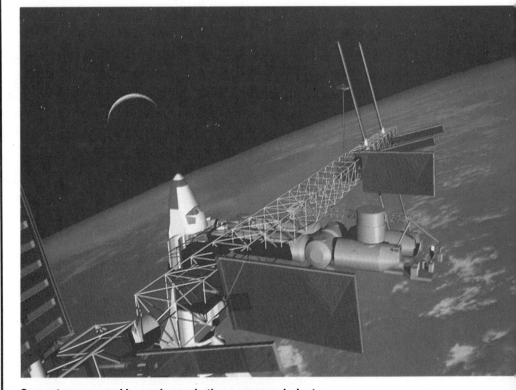

Computers are used by engineers in the aerospace industry to design and test complex machinery. The image above is a computer-generated model of the space station Freedom—a project that is currently being developed at NASA's Marshall Space Flight Center.

More About the Image

Sophisticated CAD programs offer NASA engineers a powerful tool for designing models of a working space station. The drafting capability of the computer can be linked to a data base that calculates physical properties such as mass, volume, center of gravity, etc. Animated models that contain programmed motion constraints simulate real mechanical movement. Thus, for a design as complicated as a space station, computer models provide vital information on how the entire system will work.

14.1 Fundamental Counting Principle

Objective

To find the number of possible arrangements of objects

There are 3 different paths that joggers can choose to run from the west side of Hudson Park to the center. From the center of the park, a jogger has a choice of 4 paths leading to the east side of the park.

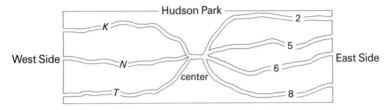

One example of a west-to-east route is to take Path *K* and then take Path 2. This route can be called *K2*. All of the possible routes are listed below. By counting you find that there are 12.

K2   K5   K6   K8   N2   N5   N6   N8   T2   T5   T6   T8

A west-to-east jogger has to make 2 *decisions*.

1st decision: Choose a west-to-center path.
There are 3 *choices*: *K*, *N*, or *T*.

2nd decision: Choose a center-to-east path.
There are 4 *choices*: 2, 5, 6, or 8.

$$3 \times 4 = 12 \text{ possible routes}$$
(letter-choices)   (number-choices)

This illustrates a short way for computing the total number of possible choices. It is called the *Fundamental Counting Principle*.

Fundamental Counting Principle
If a choice can be made in *a* ways, and for each of these a second choice can be made in *b* ways, then the choices can be made in *a* × *b* ways.

The Fundamental Counting Principle can be extended to 3 or more choices. If the jogger above wants to make a round trip, west-center-east-center-west, that jogger has to make 4 decisions. The number of possible routes for the round trip is as follows.

$$\underline{\ 3\ } \times \underline{\ 4\ } \times \underline{\ 4\ } \times \underline{\ 3\ } = 144$$

Teaching Resources

Application Worksheet 14
Quick Quizzes 100
**Reteaching and Practice
    Worksheets** 100
Transparency 31

GETTING STARTED

Prerequisite Quiz

Find the value of each expression.

1. $1 \times 2 \times 3 \times 4 \times 5 \times 6$ 720
2. $7 \times 6 \times 5 \times 4 \times 3 \times 2 \times 1$ 5,040
3. Write a 4-digit number in which no digit is repeated.   Answers will vary.
4. Write a 5-digit number in which a digit is repeated.   Answers will vary.

Motivator

Have students use a scientific calculator to find $5 \cdot 4 \cdot 3 \cdot 2 \cdot 1$ and 5! and compare the results. Introduce factorial notation, explaining to students that $5! = 5 \cdot 4 \cdot 3 \cdot 2 \cdot 1$.

Highlighting the Standards

Standard 2a: The clarity of the counting principle makes it an ideal starting point for the study of probability. It will help students clarify their thinking.

Lesson Note

To use the FCP, students must distinguish between *decisions* and *choices*. Have them draw a blank for each decision to be made, and then write the number of choices for each decision in the corresponding blank. This procedure is useful in determining the factors to be multiplied.

Encourage the students to use a scientific calculator with the *x*! function to find factorials such as 8!.

Math Connections

Tree Diagrams: An alternate way of visualizing the Fundamental Counting Principle is to make a *tree diagram* such as the one below for the jogging-path example in the text.

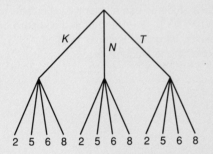

EXAMPLE 1   How many 3-digit numbers can be formed from the 5 digits 1, 2, 4, 7, 9, if no digit may be repeated in a number?

Plan   Three decisions must be made, one for each digit. Form a number from the given digits, say 427, to study the available choices.

Solution   There are 5 choices for a 100s digit: 1, 2, 4, 7, 9. Choose the 4.

This leaves 4 choices for a 10s digit: 1, 2, 7, 9. Choose the 2.

Three choices remain for a 1s digit: 1, 7, 9. Choose the 7.

Now use the Fundamental Counting Principle.

$$\underline{\ 5\ } \times \underline{\ 4\ } \times \underline{\ 3\ } = 60$$

Thus, 60 3-digit numbers can be formed.

At times, digits may be repeated in a number. For example, at least one digit is repeated in each of the numbers 5,383, 5,355, and 3,838.

EXAMPLE 2   How many 4-digit numbers can be formed from the digits 2, 4, 6, 7, 8, if a digit may be repeated in a number?

Solution   There are 5 choices for each of the 4 digits. $\underline{\ 5\ } \times \underline{\ 5\ } \times \underline{\ 5\ } \times \underline{\ 5\ } = 625$

Thus, 625 4-digit numbers can be formed.

EXAMPLE 3   A manufacturer makes school jackets in 4 different colors. Each jacket is available in 5 fabrics, 3 kinds of collars, and a choice of buttons or a zipper. If the jackets are also available with or without the school name, how many different types of jackets does the manufacturer make?

Solution   There are 5 decisions to be made.

$$\underset{\text{color}}{\underline{\ 4\ }} \times \underset{\text{fabric}}{\underline{\ 5\ }} \times \underset{\text{collar}}{\underline{\ 3\ }} \times \underset{\substack{\text{button/}\\\text{zipper}}}{\underline{\ 2\ }} \times \underset{\substack{\text{name/}\\\text{no name}}}{\underline{\ 2\ }} = 240$$

Thus, 240 types of jackets are made.

There is a short way to indicate a product of consecutive positive integers starting with 1. For example, 1 × 2 × 3 × 4 × 5 is abbreviated as 5!, which is read "*5 factorial*." Other examples using **factorial** notation are shown below.

$$3! = 1 \times 2 \times 3 = 6 \qquad 6! = 1 \times 2 \times 3 \times 4 \times 5 \times 6 = 720$$

500   Chapter 14 Permutations, Combinations, Probability

Additional Example 1

How many 4-digit numbers can be formed from the digits 2, 3, 5, 6, 8, 9, if no digit may be repeated in a number?   360

Additional Example 2

How many 3-digit numbers can be formed from the digits 2, 4, 6, 8, if a digit may be repeated in a number?   64

Additional Example 3

A truck can be ordered with choices of 6 exterior colors, 3 interior colors, and with or without a CB radio. How many different types of trucks are available?   36

For each positive integer n, $n!$ is defined as follows.

$$n! = 1 \cdot 2 \cdot 3 \cdot \ldots \cdot n$$

EXAMPLE 4   In how many different ways can 7 waiters be assigned to 7 tables if each waiter serves one table?

Plan   The first waiter has a choice of 7 tables, the second waiter then has a choice of 6 tables, and so on.

Solution $\underline{7} \times \underline{6} \times \underline{5} \times \underline{4} \times \underline{3} \times \underline{2} \times \underline{1} = 7!$, or 5,040

A calculator with an $\boxed{x!}$ key can be used to find 7!.
Press 7 $\boxed{x!}$ and read 5,040.
The waiters can be assigned in 5,040 different ways.

Classroom Exercises

In Exercises 1–6, match each item at the left with one item at the right.

1. 3! f   **2.** 5! c   **3.** 9! g
4. the number of digits in our numer-
   ation system b
5. a 4-digit number in which a digit is repeated d
6. a 4-digit number in which no digit is repeated e
7. How many 2-digit numbers can be formed from the 4 digits 2, 3, 5, 8, if no digit may be repeated in a number? 12

a. 9   e. 1,236
b. 10   f. 6
c. 120   g. $9 \times 8!$
d. 2,880   h. Infinitely many

Written Exercises

1. How many 5-digit numbers can be formed from the digits 1, 2, 3, 4, 7, 8, if no digit may be repeated in a number? 720
2. If a digit may be repeated in a number, how many 5-digit numbers can be formed from the digits 2, 3, 4, 7, 8, 9? 7,776
3. Find the number of ways that 8 classes can be assigned to 8 rooms for the first hour of school. 40,320
4. In how many ways can 6 different books be placed side by side on a shelf? 720
5. Find the number of 3-letter arrangements of the letters in the word MONTH if no letter may be repeated in an arrangement. 60
6. How many 4-letter code words can be formed from the letters in NUMBER if each letter may be repeated in a word? 1,296

14.1 Fundamental Counting Principle   **501**

Critical Thinking Questions

Analysis: In the jogging path example, ask students to calculate the east-side to west-side routes.   Intuition suggests that the answer should be the same as for the west-side to east-side route discussed in the text, and this is indeed the case.
Ask students which *field postulate* is illustrated by the fact that the two answers are the same.   They should see that since $3 \times 4 = 4 \times 3$, the Commutative Property of Multiplication is illustrated.

Checkpoint

1. How many 4-letter code words can be formed from the letters in NUMERAL under each condition?
   a. A letter may not be repeated in a code word. 840
   b. A letter may be repeated in a code word. 2,401
2. How many different signals can be shown by arranging 3 flags in a row if 7 different flags are available? 210
3. How many different automobile license plate numbers can be formed by 2 different letters followed by any 4 digits? 6,500,000

Additional Example 4

In how many ways can a family of 5 be arranged in a line? 120

Closure

A school building having 4 floors has 4 different stairways from the first to the second floor, 3 different stairways from the second to the third floor, and 2 different stairways from the third to the fourth floor. Have the students find the number of different ways to go from the first floor to the fourth floor. $4 \times 3 \times 2 = 24$ ways

◾◾FOLLOW UP

Guided Practice

Classroom Exercises 1–7

Independent Practice

A Ex. 1–10, **B** Ex. 11–17, **C** Ex. 18–19

Basic: WE 1–13 odd
Average: WE 5–17 odd, 18
Above Average: WE 7–19 odd

Additional Answers

Mixed Review

1. $3x - 4y = -8$
2. $2x + 5y = -1$

7. How many different automobile license plates can be made if each plate has a letter followed by three digits? 26,000

8. Find the number of 7-digit phone numbers that can be dialed if the first digit cannot be zero. 9,000,000

9. A school cafeteria offers each student 3 choices of meat, 4 choices of vegetable, 5 choices of drink, and 2 choices of fruit. How many different meat-vegetable-drink-fruit lunch trays are available? 120

10. A car-rental service offers 2-door and 4-door models. Each model is available in 5 different exterior colors, 2 different interior colors, 3 different types of interior upholstery, with or without a radio, and with or without air-conditioning. How many different types of cars are available for rent? 240

A sports arena has 4 west gates and 6 east gates. Solve Exercises 11–12 using this information.

11. In how many ways can you enter the arena through a west gate and later leave through an east gate? 24

12. In how many different ways can a person enter and then leave the arena? 100

13. How many different signals can be shown by arranging 5 flags in a row if 8 different flags are available? 6,720

14. In how many different ways can 6 multiple-choice questions be answered if each question has 4 choices for the answers? 4,096

15. A baseball coach must present a batting order for his 9 starting hitters before a game can begin. How many different batting orders are possible? 362,880

16. In 1988, Jefferson High School had 4 valedictorians. In how many different orders could they have given their graduation speeches? 24

17. A secretary typed 4 letters and addressed 4 envelopes. The letters were placed in the envelopes at random with one letter per envelope. In how many ways could this be done? 24

Write each of the following in factorial notation.

18. $1 \times 2 \times 3 \times \cdots (n - 4)(n - 3)(n - 2)$ $(n - 2)!$
19. $(n + 3)(n + 2)(n + 1) \cdots \times 3 \times 2 \times 1$ $(n + 3)!$

Mixed Review

Write an equation in standard form for the line described. 3.4

1. slope of $\frac{3}{4}$ and y-intercept at 2
2. through $P(2, -1)$ and $Q(-3, 1)$
3. vertical and through $P(2, -1)$
   $x = 2$
4. horizontal and through $Q(-3, 1)$
   $y = 1$

502 Chapter 14 Permutations, Combinations, Probability

Enrichment

Challenge the students with these exercises involving factorial notation.

1. Write 12 as the product of two factorials.
   $3! \cdot 2!$
2. Write 90 as the quotient of two factorials.
   $10! \div 8!$
3. Write $10 \cdot 8 \cdot 6 \cdot 4 \cdot 2$ as a multiple of a factorial. $32 \cdot (5!)$

4. Write $5! + 4!$ as a multiple of a single factorial. $6(4!)$
5. Write $5(5!)$ as the difference of two factorials. $6! - 5!$

14.2 Permutations

Objective   To find the number of permutations of objects under given conditions

A **permutation** is an arrangement of objects in a definite order.
For example, HDNA and ANDH are two permutations of letters in the
word HAND. Altogether, there are $4 \times 3 \times 2 \times 1$, or 24 permutations. If two letters are taken at a time, there are only 12 permutations:
HA, HN, HD, AN, AD, AH, NA, NH, ND, DH, DA, DN.

EXAMPLE 1   Find the number of permutations of letters in the word UNTIL under
the following conditions.
   a. The permutation ends with the letter U or the letter N.
   b. The permutation begins with the prefix UN-.

Solutions   **a.** The 1st decision consists of 2 choices for the last letter: U or N.
The number of choices for the remaining decisions follow.

$$\underset{5th}{\underline{1}} \times \underset{4th}{\underline{2}} \times \underset{3rd}{\underline{3}} \times \underset{2nd}{\underline{4}} \times \underset{1st}{\underline{2}} = 48$$

Thus, 48 permutations of U, N, T, I, L end with U or N.
   b. There is 1 choice for each of the first two letters.

$$\underline{1} \times \underline{1} \times \underline{3} \times \underline{2} \times \underline{1} = 6$$

Thus, 6 permutations of U, N, T, I, L begin with the prefix UN-.

EXAMPLE 2   How many even numbers with one, two, *or* three digits can be formed
from the digits 6, 7, 8, if no digit is repeated in a number?

Plan   Since the units digit must be even for the number to be even, there are
2 choices (6 and 8) for the units digit. Add the numbers of permutations for each type of number.

Solution

	Number of permutations	The even numbers
one-digit numbers	2	6 8
two-digit numbers	$2 \times 2 = 4$	76 78 68 86
three-digit numbers	$1 \times 2 \times 2 = 4$	876 678 768 786

Thus, there are $2 + 4 + 4$, or 10 even numbers.

Example 2 demonstrates that in permutations involving the connector
or, numbers of permutations are *added* to find the total number.

Prerequisite Quiz

**Use the digits 5, 6, 7 to write the list of
numbers described.**

1. 2-digit *odd* numbers if a digit may be
repeated in a number   55, 65, 75, 57,
67, 77
2. 2-digit *even* numbers if a digit may not
be repeated in a number   56, 76
3. numbers with *one or two* digits if a digit
may not be repeated in a number   5, 6,
7, 56, 57, 65, 67, 75, 76
4. numbers *less than 70* if a digit may be
repeated in a number   5, 6, 7, 55, 56,
57, 65, 66, 67

Motivator

Ask students the following questions.

1. Using the digits 5, 6, 7 only once, how
many choices are possible for the first
digit of a three-digit number?   Three
choices
2. How many choices are possible for the
second digit?   2
3. How many choices are possible for the
third digit?   1
4. How many arrangements are possible?
$3 \times 2 \times 1 = 6$

Additional Example 1

Find the number of permutations of the 6
letters in the word SINGLE under the
following conditions.

a. The permutation begins with E, L, or S.
360
b. The permutation ends with the suffix
-ING.   6

Additional Example 2

How many odd numbers of one or more
digits can be formed from the digits 5, 6, 7,
8, if no digit is repeated in a number?   32

Highlighting the Standards

Standard 3c: The logic of permutations, as
presented in the examples and explored in
the Classroom Exercises, continues the
solid foundation on which the study of
probability will rest.

Lesson Note

The word *permutation* is introduced and defined. When conditions are attached to an arrangement, these determine the first decision(s) to be made, as illustrated in Examples 1–3.

Students should associate addition with the connector *or* and multiplication with the connector *and*.

The notation $_nP_r$ is introduced and applied in the C Exercises.

The Calculator Activities will help students to develop skills that will be needed in lessons that follow.

Math Connections

The Genetic Code: The four nucleotides of RNA (symbolized as A, C, G, and U) are the coding material for the approximately 22 amino acids that are the building blocks of the human body. Early investigators theorized that the nucleotides must form a coding unit of fixed length for each amino acid. The possible lengths were 2, 3, or 4 nucleotides per unit. The possibility of a unit of four nucleotides in each coding unit (for example, AUGG) was rejected as improbable since the number of permutations of four nucleotides, with repetitions allowed, is $4 \times 4 \times 4 \times 4$, or 256, which is far more than needed to make 22 amino acids.

The conjunction *and* suggests the use of *multiplication*. For example, the list below shows 12 permutations of the letters A, B, C *and* the digits 1 and 2, with the letters together and at the left of the digits.

ABC12	ACB12	BAC12	BCA12	CAB12	CBA12
ABC21	ACB21	BAC21	BCA21	CAB21	CBA21

There are 6 ways to arrange the letters and 2 ways to arrange the digits. The letters *and* digits are arranged in 6×2, or 12 ways.

EXAMPLE 3 Four different biology books and 3 different chemistry books are to be placed on a shelf with the biology books together and at the right of the chemistry books. In how many ways can this be done?

Solution

Chemistry books (left) **Biology books (right)**

$$\underbrace{(\underset{1st}{\ 3\ } \times \underset{2nd}{\ 2\ } \times \underset{3rd}{\ 1\ })} \times \underbrace{(\underset{1st}{\ 4\ } \times \underset{2nd}{\ 3\ } \times \underset{3rd}{\ 2\ } \times \underset{4th}{\ 1\ })} = 144$$

There are $3! \times 4!$, or 144 possible arrangements.

Classroom Exercises

Match each description with one or more examples at the right.

1. three even numbers b   2. three odd numbers c   **a.** 80, 4, 56, 770
3. numbers with two or more digits b   **b.** 324, 16, 4,000
4. numbers less than 1,000 a, c   **c.** 7, 59, 995

Match the given condition with the number of permutations of all the letters in the word NUMERAL indicated at the right.

5. end with the suffix -ER g   **d.** $1 \cdot 2 \cdot 3 \cdot 4 \cdot 5 \cdot 6 \cdot 6$
6. begin with E or R e   **e.** $2 \cdot 6 \cdot 5 \cdot 4 \cdot 3 \cdot 2 \cdot 1$
7. do not end with R d   **f.** $6 \cdot 5 \cdot 4 \cdot 1 \cdot 3 \cdot 2 \cdot 1$
8. have N for the middle letter f   **g.** $1 \cdot 2 \cdot 3 \cdot 4 \cdot 5 \cdot 1 \cdot 1$

Find the number of permutations of all the letters in the word PINCER under the following conditions. (Exercises 9 and 10)

9. The permutation begins with the letter N, C, or E. 360
10. The permutation ends with the suffix -NCE. 6
11. How many even 4-digit numbers can be formed from the digits 1, 4, 8, 9, if a digit may be repeated? 128

Additional Example 3

In how many ways can 4 males and 6 females be arranged in a line with the females together and to the left of the males? 17,280

Written Exercises

1. Find the number of permutations of the 6 letters in the word JUNIOR that begin with I and end with O. 24

2. How many odd 4-digit numbers can be formed from the digits 1, 2, 3, 5, 6, 7, if a digit may be repeated in a number? 864

3. How many even 5-digit numbers can be formed from the digits 1, 2, 4, 7, 8, if no digit may be repeated in a number? 72

4. How many numbers of one or more digits can be formed from the digits 1, 2, 3, 4, if no digit may be repeated in a number? 64

5. If digits may be repeated in a number, how many numbers less than 500 can be formed from the digits 2, 3, 4, 5, 6? 105

6. In how many ways can 6 different algebra books be placed together on a shelf at the right of 4 different geometry books? 17,280

7. Five female cheerleaders and 4 male cheerleaders want to line up for a cheer with the males at the left of the females. In how many ways can this be done? 2,880

8. How many 3-digit numbers less than 600 can be formed from 0, 1, 5, 7, 9, without repeating digits? Consider 075 as a 2-digit number. 24

9. How many numbers less than 350 can be formed from 1, 2, 3, 4, 5, 6, if a digit may be repeated in a number? 138

10. Find the number of odd numbers less than 1,600 that can be formed from 1, 2, 5, 7, 8, if repetition of digits is allowed. 138

11. How many 2-, 3-, or 4-digit numbers can be formed from 0, 3, 6, 9, if no digit may be repeated in a number? Consider 093 the same as 93. 45

12. Five novels and 3 short stories are displayed on a shelf with the novels together and the short stories together. In how many ways can this be done? (NOTE: The novels can be at the left *or* the right.) 1,440

13. Find the number of ways that a family of 5 can stand in a line, if the mother and father are not to be separated. 48

14. In how many ways can 7 students be seated in a row of 7 chairs if 2 of the students are a brother and sister who do not want to sit together? 3,600

15. A 2-volume dictionary, a 4-volume atlas, and a 5-volume collection of plays are placed on a shelf with volumes of the same type together. In how many ways can this be done? 34,560

16. In how many ways can a tennis game of mixed doubles (male and female versus male and female) be arranged from a group of 6 males and 4 females? 180

17. The number of 4-flag signals in a row that can be formed when 10 different flags are available is the number of permutations of 10 things taken 4 at a time, or $_{10}P_4$, where $_{10}P_4 = 10 \cdot 9 \cdot 8 \cdot 7 = 5,040$.

    In general, $_nP_r = n(n - 1)(n - 2) \cdots (n - r + 1)$.

    Prove that $_nP_r = \dfrac{n!}{(n - r)!}$.

Critical Thinking Questions

Analysis: Ask students which of the two remaining possibilities (see Math Connections) is the more likely. They should reason that the number of possible nucleotide pairs, such as AG, GC, CG, UU, and so on, is 4 × 4, or 16. This is less than the number of amino acids required. The number of nucleotide triplets (trinucleotides), such as AAG, ACU, and so on, is 4 × 4 × 4, which is more than enough to make 22 amino acids. Because of this calculation, it was concluded that each amino acid is encoded by one or more trinucleotides.

Checkpoint

How many numbers of the kind described can be formed from the digits 2, 3, 5, 6, 7, if no digit is repeated in a number? (Ex. 1–4)

1. even 3-digit numbers   24
2. odd 4-digit numbers   72
3. numbers of one or more digits   325
4. numbers less than 600   61
5. In how many ways can 4 marines and 5 sailors be arranged in a line with all the marines to the left of the sailors?   2,880

Closure

Have students give a definition of the term *permutation.*   See page 503.   Ask students to find the number of permutations of all the letters in the word *numeral* that begin with the prefix *re-* and end with *l* or *m.*
1 × 1 × 4 × 3 × 2 × 1 × 2 = 48

Guided Practice

Classroom Exercises 1–11

Independent Practice

A Ex. 1–7, **B** Ex. 8–12, **C** Ex. 13–21

Basic: WE 1–11 odd, Using the Calculator, Brainteaser, Application

Average: WE 5–15 odd, Using the Calculator, Brainteaser, Application

Above Average: WE 9–21 odd, Using the Calculator, Brainteaser, Application

Additional Answers

Written Exercises

17. $\dfrac{n!}{(n-r)!}$

$= \dfrac{n(n-1)(n-2)\cdots(n-r+1)(n-r)!}{(n-r)!}$

$= n(n-1)(n-2)\cdots(n-r+1) = {}_nP_r$

Use the formula ${}_nP_r = \dfrac{n!}{(n-r)!}$ from Exercise 17 for Exercises 18–21.

Find the value of the expression, or the factored form in terms of n.

18. ${}_8P_3$ 336

19. ${}_{10}P_6$ 151,200

20. ${}_nP_2$ $n(n-1)$

21. ${}_nP_4$ $n(n-1)(n-2)(n-3)$

Mixed Review

Given $f(x) = 3x - 4$ and $g(x) = x^2 + 2$, find the following.   *3.2, 5.5, 5.9, 7.1*

1. $f(-2)$ -10

2. $g(5c)$ $25c^2 + 2$

3. $g(f(a+1))$ $9a^2 - 6a + 3$

4. $\dfrac{f(x+h) - f(x)}{h}$ 3

◼◼ *Using the Calculator*

Use a calculator with the $\boxed{x!}$ function to compute the following. Answers greater than 1,000,000 may be left in scientific notation.

1. $3! \cdot 4!$ 144

2. $6! \cdot 5! \cdot 3!$ 518,400

3. $4! + 7!$ 5,064

4. $8! + 7! + 6! + 5!$ 46,200

5. $15! \cdot 14!$ 1.14×10^{23}

6. $15! + 14!$ 1.3949×10^{12}

◼◼ *Brainteaser*

The first fifty natural numbers are multiplied together. How many zeros appear in the right part of the product?   12

◼◼ *Application:*   *The Decibel Scale*

The decibel (dB) scale is used to measure sound intensity. Since the human ear responds to changes in sound pressure, the decibel scale is based on the ratio of a measured sound pressure P to a reference value P_0. The most commonly used value for P_0 is 0.0002 dynes/cm^2, the lowest sound pressure that can be detected through the air by the average human ear. Decibel values can be computed using the formula

$$dB = 20 \log_{10} \frac{P}{P_0},$$ where P and P_0 are measured in dynes/cm^2.

1. A rock band produces a sound pressure P of 1,000 dynes/cm^2. To the nearest decibel, what is the sound intensity? Use $P_0 = 0.0002$ dynes/cm^2.
   134 dB

2. In underwater acoustics, a different reference value P_0 is used. Suppose an underwater sound has a sound pressure of 10,000 dynes/cm^2. The decibel level is calculated to be 80 dB. What reference value was used?
   $P_0 = 1$ dyne/cm^2

Challenge the students to solve this puzzle.

Three discs, small, medium and large, are positioned on post 3 in the order shown.

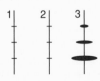

If you move the discs one at a time from one post to another, never placing a larger disc over a smaller disc, can you position the discs in the same order on post 2? What is the minimum number of moves required?

Yes: 7 moves are required. Here is one solution. Move small disc to post 2; medium disc to post 1; small disc to post 1; large disc to post 2; small disc to post 3; medium disc to post 2; small disc to post 2.

14.3 Distinguishable Permutations

Objectives   To find the value of an expression with the factorial notation
To find the number of distinguishable permutations of objects

The five letters in the word DAILY can be arranged in 5!, or 120 different ways. In the word DADDY, there are fewer than 5! *distinguishable* arrangements because the 3 Ds are alike. Consider the arrangement ADDYD. The 3 Ds can be arranged in 3! different ways.

$$AD_1D_2YD_3 \quad AD_1D_3YD_2 \quad AD_2D_1YD_3 \quad AD_2D_3YD_1 \quad AD_3D_1YD_2 \quad AD_3D_2YD_1$$

Without the subscripts, these 6 arrangements are not distinguishable. Thus, there are $\frac{5!}{3!}$ *distinguishable permutations* of the letters in the word DADDY. This example illustrates the following rule.

The number of **distinguishable permutations** of n objects of which r of the objects are alike is $\frac{n!}{r!}$.

You can extend this rule to find the number of distinguishable permutations of n objects of which more than one set of the objects are alike.

EXAMPLE 1   How many distinguishable 7-digit numbers can be formed from the digits of 6,563,656?

Plan   Find the number of permutations of 7 digits of which four are 6s and two are 5s. Use the fact that $7! = 7 \cdot 6 \cdot 5 \cdot (4!)$.

Solution

$$\frac{7!}{4!2!} = \frac{7 \cdot \overset{3}{6} \cdot \overset{1}{5} \cdot (4!)}{(4!) \cdot \underset{1}{2} \cdot \underset{1}{1}} = 7 \cdot 3 \cdot 5 = 105$$

There are 105 distinguishable 7-digit numbers.

EXAMPLE 2   Find the number of distinguishable signals that can be formed by displaying 12 flags in a row if 4 flags are green, 5 are red, 2 are blue, and 1 is yellow.

Solution

$$\frac{12!}{4!5!2!} = \frac{\overset{1}{12} \cdot 11 \cdot 10 \cdot 9 \cdot \overset{4}{8} \cdot 7 \cdot \overset{3}{6} \cdot (5!)}{\underset{1}{4} \cdot \underset{1}{3} \cdot \underset{1}{2} \cdot 1 \cdot (5!) \cdot \underset{1}{2} \cdot 1} = 83{,}160$$

Thus, 83,160 distinguishable signals can be formed.

Teaching Resources

Quick Quizzes 102
Reteaching and Practice Worksheets 102

GETTING STARTED

Prerequisite Quiz

True or false?

1. $8! = 8 \cdot 7 \cdot 6 \cdot (5!)$ True

2. $\frac{7 \cdot 6 \cdot 5 \cdot (4!)}{3 \cdot 2 \cdot (4!)} = 35$ True

Motivator

Have students list and count the distinct permutations of all the letters in the word TOT.   TOT, TTO, OTT; 3 distinct permutations

TEACHING SUGGESTIONS

Lesson Note

Encourage students to use a scientific calculator with the $x!$ function to evaluate expressions such as $\frac{12!}{3!4!2!}$.

Math Connections

Combinations: It may be noticed that in the case of distinguishable permutations of n objects of which r are alike and of which also $n - r$ are alike, the formula is identical to the formula for the combination of n objects taken r at a time. (See Lesson 14.5.)

Additional Example 1

How many distinguishable 8-digit numbers can be formed from the digits of 77,337,722?   420

Additional Example 2

Find the number of distinguishable signals that can be formed by displaying 11 flags in a row if 5 flags are yellow, 4 are blue, and the remainder are black.   6,930

Highlighting the Standard

Standard 3d: This would be an ideal lesson on which to have a classroom discussion about the meaning of permutations and how the solutions are developed.

Analysis: Ask students which is greater, 12! or 12^{12}. Although both are very large, they should see that the second number has 12 factors all of which are 12. The second number is greater.

Checkpoint

1. Find the value of $\frac{18!}{15!3!3!2!}$. 68
2. Find the number of distinguishable permutations of all the letters in LOLLIPOP. 1,680

Closure

Have students explain how the number of distinguishable permutations can differ from the number of permutations.

 **FOLLOW UP**

Guided Practice

Classroom Exercises 1–5

Independent Practice

A Ex. 1–9, **B** Ex. 10–15, **C** Ex. 16–19

Basic: FR, WE 1–13 odd

Average: FR, WE 3–17 odd

Above Average: FR, WE 7–19 odd

▰ Focus on Reading

Which number does not belong in the list?

1. $\frac{12!}{8!}$; $99(5!)$; $\frac{8!4!}{8!}$; $\frac{12 \cdot 11 \cdot 10 \cdot 9 \cdot (8!)}{8!}$ $\frac{8!4!}{8!}$

2. 9!; 5! + 4!; 12^2; 4! · 3! 9!

3. $\frac{6!}{3!2!}$; $\frac{5!}{2!}$; $\frac{5 \cdot 4!}{2}$; $\frac{4 \cdot 5!}{3!}$ $\frac{4 \cdot 5!}{3!}$

Classroom Exercises

Find the number of distinguishable permutations of all the letters in the word.

1. ace 6
2. add 3
3. toot 6
4. poor 12
5. deemed 60

Written Exercises

Find the value of each expression.

1. $\frac{8!}{4!}$ 1,680
2. $\frac{11!}{8!}$ 990
3. $\frac{14!}{11!3!}$ 364
4. $\frac{10!}{3!6!2!}$ 420
5. $\frac{12!}{7!5!3!}$ 132

Find the number of distinguishable permutations of all the letters in the word.

6. beef 12
7. puppy 20
8. pepper 60
9. scissors 1,680
10. parallel 3,360
11. murmur 90
12. nonsense 1,680
13. Tennessee 3,780

14. How many distinguishable signals can be formed with 10 flags in a line if 3 flags are black, 4 are orange, and the remainder are green? 4,200

15. In how many ways can 4 nickels, 2 dimes, 3 quarters, 1 half-dollar, and 1 penny be distributed among 11 children if each child is to receive 1 coin? 138,600

Simplify.

16. $\frac{n!}{(n-2)!}$ $n^2 - n$
17. $\frac{(n+3)!}{(n+1)!}$ $n^2 + 5n + 6$
18. $\frac{(n+1)!}{(n-1)!}$ $n^2 + n$
19. $\frac{(n-1)!}{(n-3)!}$ $n^2 - 3n + 2$

Mixed Review

1. Simplify $\frac{3}{5-i}$. **9.2** $\frac{15 + 3i}{26}$

2. Solve $x^4 - 2x^3 + 6x^2 - 18x = 27$ **10.4** $-1, 3, \pm 3i$

3. Solve $\log_5 t + \log_5 (t - 2) = \log_5 24$. Check. **13.5** 6

4. Simplify $\log_b 4c + 2 \log_b 3 - \log_b 6d$. **13.6** $\log_b \frac{6c}{d}$

Enrichment

Challenge the students with this problem.

In how many ways can $1.14 in pennies, nickels, dimes, quarters, and half-dollars be distributed among 12 children if each child is to receive one coin?

There are 12 coins, 1 fifty-cent piece, 1 quarter, 1 dime, 5 nickels, and 4 pennies. $\frac{12!}{5!4!} = 166,320$

14.4 Circular Permutations

Objective

To find the number of circular permutations

Three objects A, B, and C can be arranged in a row in 3!, or 6 ways. However, they can be placed in a circle in only 2!, or 2 ways.

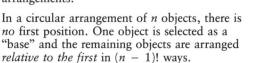

Consider a lazy Susan (revolving platter) with trays A, B, and C. As shown at the right, the two arrangements on top are different. However, each arrangement below the first row can be found from the one above it by rotating the tray. Therefore, they are not different arrangements.

In a circular arrangement of n objects, there is *no* first position. One object is selected as a "base" and the remaining objects are arranged *relative to the first* in $(n - 1)!$ ways.

The number of **circular permutations** of n objects is $(n - 1)!$.

EXAMPLE 1

In how many ways can 8 different types of bushes be planted in a circle around a flagpole?

Plan

Arrange 7 of the bushes in relation to 1 "base" bush.

Solution

$(n - 1)! = (8 - 1)! = 7! = 5,040$

There are 5,040 different arrangements.

EXAMPLE 2

In how many ways can 3 males and 3 females be seated around a circular table if no 2 females may be seated next to each other?

Plan

Seat one person as a base. The remaining two of the *same sex* can be seated in 2! ways. The 3 members of the opposite sex can be seated in 3! ways.

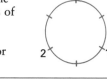

Solution

The 6 people can be seated alternately in 2! · 3!, or 12 ways.

If keys are placed on a ring, the ring can be turned around so that the two views, front and back, represent just one arrangement. So, the number of permutations of n objects on a *ring* is $\dfrac{(n - 1)!}{2}$.

Teaching Resources

Quick Quizzes 103
Reteaching and Practice Worksheets 103

▬ GETTING STARTED

Prerequisite Quiz

Evaluate the expression for $n = 6$.

1. $n!$ 720   2. $(n - 1)!$ 120

3. $\dfrac{(n - 1)!}{2}$ 60   4. $(n - 3)! \cdot (n - 2)!$ 144

Motivator

Have students list and count the ways in which an apple, a banana, and a carrot can be arranged in a row. Then have the students discuss whether they would get a different arrangement if the items were placed on a revolving platter.

▬ TEACHING SUGGESTIONS

Lesson Note

The illustration of a revolving platter can be supplemented in the classroom by having 3 (or 4) students sit in a circle in 2 (or 6) different ways relative to one designated person.

Math Connections

Permutation Groups: As a part of the study of abstract algebra, permutations themselves become objects of study. From an abstract approach, a permutation is defined as a "one-to-one transformation of a finite set into itself."

Additional Example 1

In how many ways can 9 flags be placed around a circular platform?   40,320

Additional Example 2

In how many ways can 6 brothers and 6 sisters be seated around a circular table if no two males may be seated next to each other?   86,400

Highlighting the Standard

Standard 1a: Each exercise in circular permutations is essentially a problem-solving situation through which students can grow in understanding.

509

Critical Thinking Questions

Analysis: Ask students which is greater, $(n + 1)! - n!$ or $n! - (n - 1)!$. By working actual examples or by working algebraically, they will find that the first expression, which can be written as $n \cdot n!$, is greater than the second, which can be written as $(n - 1)[(n - 1)!]$.

Checkpoint

1. In how many ways can a scoutmaster and 11 scouts be seated in a circle around a campfire? If possible, use a calculator.   39,916,800
2. In how many ways can 5 keys be arranged on a key ring?   12

Closure

Have the students explain why 5 people can be seated in a circle in 24 ways but 5 keys can be arranged on a key ring in only 12 ways.   See page 509 for explanation.

 **FOLLOW UP**

Guided Practice

Classroom Exercises 1–8

Independent Practice

A Ex. 1–4, **B** Ex. 5–8, **C** Ex. 9–11

Basic: WE 1–6
Average: WE 3–9
Above Average: WE 6–11

Classroom Exercises

Describe each arrangement as linear, circular, or either of these.

1. books on a table Either
2. keys on a key ring Circular
3. digits in a 3-digit number Linear
4. charms on a bracelet Circular
5. trees in a park Either
6. letters in the word LETTERS Linear

Tell which number of permutations is the greater for each pair below.

7. six people in a *line* or 6 people in a *circle* 6 in a line
8. all the letters in POLO or all the letters in POLE POLE

Written Exercises

3,628,800

1. Find the number of ways that a football team of 11 players can form a circular huddle.
2. Find the number of different ways that a landscaper can plant 10 different types of bushes around a circular flower bed. 362,880
3. In how many ways can the principal and 7 department heads at Jackson High School be seated around a circular conference table? 5,040
4. A mother and daughter each invite two friends to lunch. In how many ways can all of them be seated around a circular table? 120

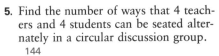

5. Find the number of ways that 4 teachers and 4 students can be seated alternately in a circular discussion group. 144
6. In how many ways can 5 couples be seated around a circular banquet table if no 2 men may be seated by each other? 2,880
7. In how many ways can 4 keys be arranged on a key ring? 3
8. In how many ways can 6 charms be arranged on a bracelet? 60
9. In how many ways can a family of 6 be seated around a circular table if the mother and father must not be seated next to each other? 72

Solve each equation for n.

10. $(n - 1)! = 720$ 7
11. $\dfrac{(n - 1)!}{2} = 60$ 6

Mixed Review

Find the slope and describe the slant of $\overleftrightarrow{AB}$.   3.3

1. $A(-2,3)$, $B(4,-2)$
   $-\dfrac{5}{6}$; down to right
2. $A(1,3)$, $B(7,3)$
   0; horizontal
3. $A(-1,-2)$, $B(3,4)$
   $\dfrac{3}{2}$; up to right
4. $A(8,2)$, $B(8,7)$
   Undefined; vertical

510   Chapter 14 Permutations, Combinations, Probability

Enrichment

Have the students solve this problem.

Six chairs are arranged around a circular dinner table. One position is uncomfortable because the chair is in front of a drafty window and nobody wants to sit there. In how many different ways can 5 persons be seated comfortably around this table? Explain your answer.

5! or 120. The $(n - 1)!$ rule does not apply here. Because of the drafty-window position, there is a first position in the circle of chairs. As a result, the "circular" situation yields a straight-line permutation.

14.5 Combinations

Objectives

To find the number of possible selections of n objects taken r at a time without regard to order

To find the value of $\binom{n}{r}$ for $n \geq r$

A selection of objects *without regard to order* is called a **combination**.

How many combinations of 4 letters can be selected from the 6 letters A, B, C, D, E, F? The list below shows all 15 combinations.

ABCD	ABDE	ACDE	BCDE	DEFA
ABCE	ABDF	ACDF	BCDF	DEFB
ABCF	ABEF	ACEF	BCEF	DEFC

The symbol $\binom{6}{4}$ is used to denote *the number of combinations of 6 things taken 4 at a time*. As shown above, $\binom{6}{4} = 15$.

To evaluate $\binom{6}{4}$ without listing combinations, you can reason as follows. There are $6 \cdot 5 \cdot 4 \cdot 3$, or 360, different *arrangements* of 4 letters from the 6. Each *combination* of 4 letters has 4! arrangements. Thus,

$$\binom{6}{4} = \frac{6 \cdot 5 \cdot 4 \cdot 3}{4 \cdot 3 \cdot 2 \cdot 1} = 15.$$

This fraction can also be expressed using factorial notation.

$$\binom{6}{4} = \frac{6 \cdot 5 \cdot 4 \cdot 3}{4 \cdot 3 \cdot 2 \cdot 1} = \frac{6 \cdot 5 \cdot 4 \cdot 3 \cdot 2 \cdot 1}{4!2!} = \frac{6!}{4!2!}$$

In general, $\binom{n}{r}$ is the number of combinations of n things taken r at a time. For positive integers n and r, where $n > r$,

$$\binom{n}{r} = \frac{n!}{r!(n - r)!}.$$

EXAMPLE 1

How many different 4-member committees can be formed from 9 people.

Plan

Find $\binom{9}{4}$, the number of combinations of 9 people taken 4 at a time.

Solution

$$\binom{9}{4} = \frac{9!}{4!(9 - 4)!} = \frac{9!}{4!5!} = \frac{9 \cdot 8 \cdot 7 \cdot 6}{4 \cdot 3 \cdot 2 \cdot 1} = 126$$

Thus, 126 different 4-member committees can be formed.

Additional Example 1

In how many ways can a homeroom of 30 students select 3 of its members?   4,060

Prerequisite Quiz

Find the value of each expression.

1. $\frac{8!}{5!3!}$ 56
2. $\frac{12!}{4!8!}$ 495
3. Give one 3-letter name for each triangle determined by the vertices of a rectangle *ABCD*. $\triangle ABC$, $\triangle ABD$, $\triangle BCD$, $\triangle ACD$ (The triangles may be named in other ways.)

Motivator

Have the students list and count the ways that *A*, *B*, *C*, and *D* can be arranged in order.   4! or 24   Ask students how many triangles can be formed from the 4 noncollinear points *A*, *B*, *C*, and *D*, where any 3 points may be chosen as vertices.   4 For each of the triangles formed, in how many ways can the letters be arranged.   3! or 6 ways.

TEACHING SUGGESTIONS

Lesson Note

Emphasize the difference between permutations and combinations. A permutation is an arrangement *in a definite order*. A combination is a selection *without regard to order*. In Example 2, note that no three points are collinear since the points are located on a circle.

Highlighting the Standards

Standard 5d: The meaning of the combination symbol is developed concretely on the first page of the lesson. In time, students will appreciate its breadth and power.

Math Connections

Poker: The total number of possible 5-card hands in a game of cards such as poker is 2,598,960 (see Exercise 11). Thus, a serious player of the game needs to know the number of possible combinations for the many kinds of hands. For example, there are $9(4\times1\times1\times1\times1)$, or 36 ways of drawing a straight flush but only $(4\times1\times1\times1\times1)$, or 4 ways of drawing a royal flush. The easiest poker hand that can be drawn (other than "nothing") is one pair, which can be accomplished in 1,098,240 ways.

Critical Thinking Questions

Analysis: Have students review the definition of $_nP_r$ at the bottom of page 505 and then find how it is related to the combination of n objects taken r at a time. They should discover that $_nP_r = r! \cdot \binom{n}{r}$.

Checkpoint

Find the value of each expression.

1. $\binom{12}{5}$ 792
2. $\binom{12}{7}$ 792
3. $\binom{15}{15}$ 1
4. $\binom{10}{0}$ 1

5. How many 6-member panels of experts can be formed if 10 experts are available?   210

In Example 1, notice that for each group of 4 people selected, a group of 5 people is *not* selected. Therefore $\binom{9}{4} = \binom{9}{5}$, or $\frac{9!}{4!5!} = \frac{9!}{5!4!}$. This suggests that $\binom{n}{r} = \binom{n}{n-r}$.

EXAMPLE 2   Eight points, A through H, are on a circle.
a. How many lines are determined by the 8 points?
b. How many triangles are determined by the 8 points?

Solutions   a. A line is determined by any 2 points.

$$\binom{8}{2} = \frac{8!}{2!6!} = \frac{8 \cdot 7}{2 \cdot 1} = 28$$

Thus, 28 lines are determined by the 8 points.

b. A triangle is determined by any 3 noncollinear points.

$$\binom{8}{3} = \frac{8!}{3!5!} = \frac{8 \cdot 7 \cdot 6}{3 \cdot 2 \cdot 1} = 56$$

Calculator steps: 8 $\boxed{x!}$ $\boxed{\div}$ 3 $\boxed{x!}$ $\boxed{\div}$ 5 $\boxed{x!}$ $\boxed{=}$ 56

Thus, 56 triangles are determined by the 8 points.

With the *definition*, $0! = 1$, the meaning of $\binom{n}{r} = \binom{n}{n-r}$ can be extended to include $r = n$ and $r = 0$.

For example, consider the number of combinations of 9 objects taken 9 at a time: $\binom{9}{9} = \frac{9!}{9!(9-9)!} = \frac{9!}{9!0!} = \frac{1}{0!} = 1$.

Similarly, $\binom{9}{0} = \frac{9!}{0!(9-0)!} = \frac{9!}{0!9!} = \frac{1}{0!} = 1$.

Classroom Exercises

Tell whether a permutation or a combination is indicated.
(Exercises 1–6)

1. a 3-letter name of a triangle   Combination
2. a 5-digit number   Permutation
3. a 6-person committee   Combination
4. a 5-card hand in a card game   Combination
5. a 4-letter code word   Permutation
6. a 5-flag signal   Permutation

7. Find the 4 different amounts of money determined by selecting 3 coins from 1 penny, 1 nickel, 1 dime, and 1 quarter.   16¢, 31¢, 36¢, 40¢

8. How many 3-member committees can be formed from 7 people?   35
9. How many 5-member committees can be formed from 7 people?   21

Additional Example 2

Sixteen points, A through P, are located on a circle.

a. How many lines are determined by the 16 points?   120
b. How many triangles are determined by the 16 points?   560

Written Exercises

Find the value of each expression.

1. $\binom{9}{2}$ 36 **2.** $\binom{9}{7}$ 36 **3.** $\binom{15}{3}$ 455 **4.** $\binom{8}{8}$ 1 **5.** $\binom{7}{0}$ 1 **6.** $\binom{50}{1}$ 50

7. Find the number of lines determined by 12 points, A through L, on a circle. 66

8. Find the number of triangles determined by 14 points, A through N, on a circle. 364

9. How many different 12-member juries can be formed if 18 people are available for jury duty? 18,564

10. In how many ways can a team of 15 soccer players select 3 captains for its next game? 455

11. Find the number of different 5-card hands that can be drawn from a deck of 52 cards. 2,598,960

12. A student must select 8 of 10 functions to be graphed. How many selections are possible? 45

13. How many baseball games will be played by 12 teams if each team plays each other team once? 66

14. Each of 6 teams in a league will play each other team 3 times. How many games will be played? 45

Solve each equation for n.

15. $\binom{n}{4} = \binom{12}{8}$ 12 **16.** $\binom{n}{3} = \binom{n}{7}$ 10 **17.** $\binom{n}{5} = \binom{8}{0}$ 5 **18.** $\binom{n}{n-1} = \binom{9}{1}$ 9

Find 4 pairs of integers (n, r) for each equation.

19. $\binom{n}{6} = \binom{9}{r}$ **20.** $\binom{n}{20} = \binom{30}{r}$ **21.** $\binom{8}{r} = \binom{n}{2}$ **22.** $\binom{n}{10} = \binom{12}{r}$

23. Prove: $\binom{n}{r} = \binom{n}{n-r}$ for all integers r and n, where $0 \le r \le n$.

Mixed Review 1.3, 14.2

1. Evaluate $5(3x - 2y) - (12x + 10y)$ for $x = -3$ and $y = -4$. 71

2. Evaluate $3x^2y - xy - 5xy^2$ for $x = 5$ and $y = -2$. −240

3. In how many ways can 5 girls and 3 boys stand in a line with the girls at the left and the boys at the right? 720

4. How many numbers of 1 or more digits can be formed from the digits 4, 5, 6, if no digit is repeated in a number? 15

Using the Calculator

Use a scientific calculator to find the value of each expression.

1. $\binom{3}{1} + \binom{3}{2} + \binom{3}{3}$ 7 **2.** $\binom{4}{1} + \binom{4}{2} + \binom{4}{3} + \binom{4}{4}$ 15 **3.** $\binom{12}{4} \cdot \binom{8}{3} \cdot \binom{4}{2}$ 166,320

4. $\binom{8}{5} \cdot \binom{10}{4} + \binom{8}{4} \cdot \binom{10}{5}$ 29,400 **5.** $\binom{7}{3} \cdot \binom{5}{2} + \binom{7}{4} \cdot \binom{5}{1} + \binom{7}{5}$ 546

14.5 Combinations **513**

Closure

Have students do each of the following.

1. Use a scientific calculator to find 0!. 1
2. Find the number of games played in an 8-school conference if each school plays the other teams 4 times each. 112

▬▬FOLLOW UP

Guided Practice

Classroom Exercises 1–9

Independent Practice

A Ex. 1–12, **B** Ex. 13–18, **C** Ex. 19–23

Basic: WE 1–13 odd, Using the Calculator

Average: WE 5–19 odd, Using the Calculator

Above Average: WE 11–23 odd, Using the Calculator

Additional Answers

Written Exercises

19. (9,6), (6,9), (9,3), (6,0)
20. (30,20), (20,30), (30,10), (20,0)
21. (8,2), (8,6), (2,8), (2,0)
22. (12,10), (12,2), (10,12), (10,0)

23. $\binom{n}{n-r} = \dfrac{n!}{(n-r)![n-(n-r)]!}$
$= \dfrac{n!}{(n-r)!r!}$
$= \dfrac{m!}{r!(n-r)!} = \binom{n}{r}$

Enrichment

Have the students try this problem.
The dial of a combination lock is shown at the right. The lock is opened by

1. setting one of the letters at the pointer,
2. turning the dial to a second letter,
3. turning the dial in the opposite direction to a third letter, and

4. turning the dial again in the original direction to a fourth letter. No turn can be more than a full turn. Repetitions of letters are allowed.

How many different possible combinations are there for a lock of this type?

$8 \cdot 16 \cdot 8 \cdot 8$ or 8,192
The turn to the second letter offers 16 choices, since the turn can be either to the left or to the right.

513

▖▖▖GETTING STARTED

Prerequisite Quiz

Find the value of each expression. Use a calculator if possible.

1. $\binom{8}{2} \cdot \binom{6}{3}$ 560   2. $\binom{8}{2} + \binom{6}{3}$ 48

3. $\binom{4}{1} + \binom{4}{2} + \binom{4}{3} + \binom{4}{4}$ 15

4. $\binom{4}{2} \cdot \binom{6}{4} + \binom{4}{3} \cdot \binom{6}{3}$ 170

Motivator

Have the students use a scientific calculator to find the value of $\binom{8}{6} \cdot \binom{6}{3} + \binom{8}{5} \cdot \binom{6}{4}$.
1,400

▖▖▖TEACHING SUGGESTIONS

Lesson Note

Emphasize the following points.

1. Multiplication is used with the connector *and*.
2. Addition is used with the connector *or*.
3. Selecting *at least 2* cards from a group of 4 cards means 2 cards *or* 3 cards *or* 4 cards.
4. *Exactly 3* boys on a team of 5 means 3 boys *and* 2 girls.
5. *At least 3* girls on a team of 5 means 3 girls *and* 2 boys *or* 4 girls *and* 1 boy *or* 5 girls and no boys, if at least 5 girls and 2 boys are available.

14.6 Conditional Combinations

Objective To find the number of combinations of objects given certain conditions.

Recall that in Lesson 14.2, multiplication is used with the conjunction *and*, while addition is used with the disjunction *or*.

If a box contains 7 yellow cards and 5 white cards, you can select 2 yellow cards in $\binom{7}{2}$ ways and 3 white cards in $\binom{5}{3}$ ways. So, 2 yellow cards *and* 3 white cards can be selected in

$$\binom{7}{2} \times \binom{5}{3} \text{ ways (multiplication),}$$

2 yellow cards *or* 3 white cards can be selected in

$$\binom{7}{2} + \binom{5}{3} \text{ ways (addition),}$$

and 1 *or* more white cards can be selected in

$$\binom{5}{1} + \binom{5}{2} + \binom{5}{3} + \binom{5}{4} + \binom{5}{5} \text{ ways.}$$

EXAMPLE 1 In how many ways can at least 2 girls be selected from among 4 girls?

Solution "At least 2 girls" means 2 girls, *or* 3 girls, *or* 4 girls.

$$\binom{4}{2} + \binom{4}{3} + \binom{4}{4} = \frac{4!}{2!2!} + \frac{4!}{3!1!} + \frac{4!}{4!0!} = 6 + 4 + 1 = 11$$

There are 11 ways to select at least 2 girls from among 4 girls.

EXAMPLE 2 Six boys and 3 girls are eligible for a 5-member team.
  a. In how many ways can the team be formed with exactly 3 boys?
  b. In how many ways can the team be formed with at least 2 girls?

Solutions **a.** "Exactly 3 boys" implies 3 boys *and* 2 girls. Use multiplication.

$$\binom{6}{3} \cdot \binom{3}{2} = \frac{6!}{3!3!} \cdot \frac{3!}{2!1!} = 20 \cdot 3 = 60$$

There are 60 ways that the team of 5 can have exactly 3 boys.

b. "At least 2 girls" on a team of 5 implies either 2 girls *and* 3 boys *or* 3 girls *and* 2 boys. Use multiplication coupled with addition.

$$\binom{3}{2} \cdot \binom{6}{3} + \binom{3}{3} \cdot \binom{6}{2} = \frac{3!}{2!1!} \cdot \frac{6!}{3!3!} + \frac{3!}{3!0!} \cdot \frac{6!}{2!4!} = 75$$

There are 75 ways that the team can have at least 2 girls.

Highlighting the Standards

Standard 2d: This is not an easy lesson. But it has only one page of presentation and it would be helpful for students to read and study it carefully before the classroom explanation.

Additional Example 1

In how many ways can at least 5 cards be selected from a group of 7 different cards?
29

Additional Example 2

Box A has 5 green tags and box B has 7 yellow tags. Six of the 12 tags are to be selected.

a. In how many ways can the 6 tags be selected with exactly 4 yellow tags?
350

b. In how many ways can the 6 tags be selected with at least 4 green tags?
112

Classroom Exercises

Five males and 8 females are eligible to begin training as astronauts. Match each selection described with one expression at the right.

1. 3 males or 4 females f
2. 3 males and 4 females c
3. 3 or more males d
4. at least 6 females a

a. $\binom{8}{6} + \binom{8}{7} + \binom{8}{8}$ **d.** $\binom{5}{3} + \binom{5}{4} + \binom{5}{5}$

b. $\binom{8}{6}$ **e.** $\binom{5}{3}$

c. $\binom{5}{3} \cdot \binom{8}{4}$ **f.** $\binom{5}{3} + \binom{8}{4}$

A 4-member shuttle crew will be selected from the 5 males and 8 females. Match the crew description with one expression at the right. (Exercises 5–8)

5. exactly 2 males k
6. exactly 3 females j
7. at least 3 females l
8. no females i

g. $\binom{5}{2}$ **j.** $\binom{8}{3} \cdot \binom{5}{1}$

h. $\binom{8}{3} + \binom{8}{4}$ **k.** $\binom{5}{2} \cdot \binom{8}{2}$

i. $\binom{5}{4} \cdot \binom{8}{0}$ **l.** $\binom{8}{3} \cdot \binom{5}{1} + \binom{8}{4} \cdot \binom{5}{0}$

9. Find the number of different ways to select exactly 3 boys from 6 boys and 4 girls for a scholarship if just 4 scholarships are given. 80

Written Exercises

A box contains 8 blue chips and 10 yellow chips. Find the number of different ways to select the following.

1. 5 blue chips and 3 yellow chips 6,720
2. exactly 3 blue chips 56
3. 2 blue chips or 8 yellow chips 73
4. 6 or more blue chips 37
5. exactly 9 yellow chips 10
6. at least 8 yellow chips 56

From a group of 15 females and 10 males, a 6-member team will be selected to represent their school at a science fair.

7. In how many ways can the team of 6 be selected if it is to have exactly 4 females? 61,425

8. Find the number of different 6-member teams that have at least 4 males. 26,040

9. How many teams of 6 could have at least 5 females? 35,035

10. In how many ways can a team of 6 have exactly 3 males? 54,600

11. How many different amounts of money can be obtained by selecting 2 or 3 coins from 1 penny, 1 nickel, 1 dime, and 1 quarter? 10

12. A box contains 1 silver dollar, 1 half-dollar, 1 quarter, 1 dime, 1 nickel, and 1 penny. How many different amounts of money can be obtained by selecting one or more coins from the box? 63

Math Connections

Compound Sentences: Example 2b involves the multiple occurrence of the connectors *and* and *or*. Compound sentences involving such multiple occurrences were considered in Lesson 6.4 on Inequalities of Degree 2 and Degree 3.

Critical Thinking Questions

Analysis: Ask students if they can think of an alternate wording to describe in how many ways the team can be formed in Examples 2a and 2b, without changing the meaning of the problem. A possible alternate wording for 2a is "with exactly 2 girls." A possible alternate wording for 2b is "with at most 3 boys."

Checkpoint

Some people are to be selected from a group of 5 males and 9 females. Find the number of ways to select the following.

1. exactly 5 females 126
2. at least 3 males 16
3. 3 males and 4 females 1,260
4. 3 males or 4 females 136
5. a group of 7 with exactly 2 males 1,260
6. a group of 7 with at least 5 females 1,716

Closure

Have students define the words *exactly* and *at least*. Ask students to discuss what is implied by the following statements. If 4 men and 3 women are eligible for a 4-member committee, what is the implication of (1) exactly 2 men on a committee? Two men and two women. (2) at least 2 women on a committee? Two women and two men, or three women and one man.

Guided Practice

Classroom Exercises 1–9

Independent Practice

A Ex. 1–10, **B** Ex. 11–19, **C** Ex. 20–27

Basic: WE 1–15 odd, Midchapter Review

Average: WE 5–23 odd, Midchapter Review

Above Average: WE 11–27 odd, Midchapter Review

Additional Answers

Written Exercises

25. $1 + 5 + 10 + 10 + 5 + 1 = 32 = 2^5$

26. $1 + 6 + 15 + 20 + 15 + 6 = 63 = 64 - 1 = 2^6 - 1$

27. $\binom{n}{r} + \binom{n}{r+1}$

$= \dfrac{n!}{r!(n-r)!} + \dfrac{n!}{(r+1)!(n-r-1)!}$

$= \dfrac{(r+1)n!}{(r+1)r!(n-r)!} +$

$\dfrac{n!(n-r)}{(r+1)!(n-r)(n-r-1)!}$

$= \dfrac{n!(r+1+n-r)}{(r+1)!(n-r)!}$

$= \dfrac{(n+1)n!}{(r+1)!(n-r)!}$

$= \dfrac{(n+1)!}{(r+1)!(n-r)!}$

$= \dfrac{(n+1)!}{(r+1)![n+1-(r+1)]!}$

$= \binom{n+1}{r+1}$

Three red marbles, 4 white marbles, and 5 blue marbles are numbered 1 through 12. Use this information in Exercises 13–16.

13. In how many ways can 2 red, or 2 white, or 3 blue marbles be selected? 19

14. Find the number of ways to select 3 red, 3 white, and 2 blue marbles. 40

15. If 5 marbles are selected, find the number of ways that exactly 3 of them can be blue. 210

16. If 6 marbles are selected, in how many ways can at least 4 of them be blue? 112

Five sophomores, 8 juniors, and 10 seniors are eligible for selection to a 5-member group that will tour Europe during the summer.

17. In how many ways can the group of 5 be selected and have 1 sophomore, 2 juniors, and 2 seniors? 6,300

18. Find the number of ways to select the 5 members so that there are exactly 2 juniors in the group. 12,740

19. How many 5-member groups can have at least 4 seniors? 2,982

The number of combinations of n things taken r at a time, or $\binom{n}{r}$, can be denoted by $_nC_r$ where $_nC_r = \dfrac{n!}{r!(n-r)!}$.

Use the formula above to evaluate the following. (Exercises 20–24)

20. $_{20}C_{17}$ 1,140 **21.** $_{53}C_{53}$ 1 **22.** $_{16}C_1$ 16 **23.** $_8C_3 \cdot _7C_4$ 1,960

24. $_4C_0 + _4C_1 + _4C_2 + _4C_3 + _4C_4$ 16

25. Show that $_5C_0 + _5C_1 + _5C_2 + _5C_3 + _5C_4 + _5C_5 = 2^5$.

26. Show that $_6C_6 + _6C_5 + _6C_4 + _6C_3 + _6C_2 + _6C_1 = 2^6 - 1$.

27. Prove: $\binom{n}{r} + \binom{n}{r+1} = \binom{n+1}{r+1}$ for all integers r and n, $0 \le r \le n$.

Midchapter Review

For Exercises 1 and 2, the digits 1, 2, 3, 5, 6, 7 are available. How many numbers, described below, can be formed from these 6 digits? *14.1, 14.2*

1. four-digit numbers if no digit is repeated in a number 360

2. three-digit odd numbers if a digit may be repeated in a number 144

3. Find the number of distinguishable permutations of all the letters in the word NINETEEN. *14.3* 1,120

4. In how many ways can 6 people be seated around a circular table? *14.4* 120

5. Each of 8 teams in a baseball league will play the other teams 5 times each. How many games will be played? *14.5* 140

6. From a group of 6 seniors and 4 juniors, in how many ways can a 5-member committee be selected with at least 4 seniors on the committee? *14.6* 66

Enrichment

Ask the students to find a pattern in the following sums.

$\binom{1}{1} + \binom{1}{0}$ 2

$\binom{2}{2} + \binom{2}{1} + \binom{2}{0}$ $4 = 2^2$

$\binom{3}{3} + \binom{3}{2} + \binom{3}{1} + \binom{3}{0}$ $8 = 2^3$

$\binom{4}{4} + \binom{4}{3} + \binom{4}{2} + \binom{4}{1} + \binom{4}{0}$ $16 = 2^4$

Then ask the students to use the pattern to predict, and then verify, the following sum.

$\binom{8}{8} + \binom{8}{7} + \binom{8}{6} + \binom{8}{5} + \binom{8}{4} + \binom{8}{3} +$

$\binom{8}{2} + \binom{8}{1} + \binom{8}{0}$ $2^8 = 256$

$1 + 8 + 28 + 56 + 70 + 56 + 28 + 8 + 1$
$= 256$

Now ask the students to use the pattern to predict the sum below where n is a positive integer.

$\binom{n}{n} + \binom{n}{n-1} + \binom{n}{n-2} + \cdots + \binom{n}{0}$ 2^n

14.7 Probability

Objective   To find the probability of an event

When a die (plural: dice) is tossed, the upper face will show 1, 2, 3, 4, 5, or 6 dots. These are equally likely outcomes if the die is "fair." If the die is tossed many times, the 4-dot face should appear on top about 1 out of 6 times, or $\frac{1}{6}$ of the time. Thus, the *probability* of rolling a 4 in one toss is $\frac{1}{6}$. In the ratio $\frac{1}{6}$, 1 is the number of successful outcomes s (ways the desired event can occur), and 6 is the total number of possible outcomes t. This probability is written as follows.

$$P(4) = \frac{s}{t} = \frac{1}{6}$$

The set of outcomes {1, 2, 3, 4, 5, 6} is called a *sample space* for rolling a die. A **sample space** is the set of *all* possible outcomes; a *subset* of the sample space is called an **event**.

If a sample space consists of t equally likely outcomes, then the **probability** of an event E is given by the following formula.

$$P(E) = \frac{s}{t} = \frac{\text{number of successful outcomes}}{\text{total number of outcomes}}$$

EXAMPLE 1   In one toss of a die, what is the probability of each event?
     a. rolling a 5   **b.** rolling an odd number
     c. rolling a number less than 7   **d.** rolling a 7

Plan   The sample space is {1, 2, 3, 4, 5, 6}. There are 6 possible outcomes in this set, so $t = 6$.

Solutions
a. rolling a 5
     1 successful outcome: $s = 1$
     $P(5) = \frac{s}{t} = \frac{1}{6}$
c. rolling a number less than 7
     6 successful outcomes: $s = 6$
     $P(\text{number} < 7) = \frac{s}{t} = \frac{6}{6} = 1$
     This event *is certain to occur.*

b. rolling an odd number: 1, 3, 5
     3 successful outcomes: $s = 3$
     $P(\text{odd number}) = \frac{s}{t} = \frac{3}{6}$, or $\frac{1}{2}$
d. rolling a 7
     no successful outcome: $s = 0$
     $P(7) = \frac{s}{t} = \frac{0}{6}$, or 0
     This event *cannot occur.*

Example 1 illustrates the *range* for the probability of an event E. If E cannot occur, $P(E) = 0$. If E is certain to occur, $P(E) = 1$. Thus:

$$0 \leq P(E) \leq 1$$

14.7 Probability   **517**

517

Math Connections

Life Skills: Students are familiar with various applications of probability to daily life situations. For example, the U.S. Weather Bureau may issue forecasts for rain or snow in terms of probabilities. Insurance companies may charge higher or lower auto insurance premiums depending on the probability that drivers in different age groups will be involved in an accident. Have the students give other examples of probability applied to real life situations.

Critical Thinking Questions

Analysis: Have students use the data on poker hands given in *Math Connections* for Lesson 14.5 to find the probabilities for a straight flush and a royal flush. For the straight flush, the probability is $\frac{36}{2,598,960}$, or 0.000014. For the royal flush, the probability is $\frac{4}{2,598,960}$, or 0.0000015.

Checkpoint

Find the odds (a) for the event and (b) against the event.

1. rolling a 3 with a die
   a. 1 to 5   b. 5 to 1
2. exactly 1 girl in a 3-child family
   a. 3 to 5   b. 5 to 3
3. drawing an 8 from a deck of 52 playing cards
   a. 1 to 12   b. 12 to 1

In Exercises 4–6, $P(E) = \frac{5}{12}$. Find the following.

4. the odds for E 5 to 7
5. the odds against E 7 to 5
6. $P(\text{not } E)$ $\frac{7}{12}$

For a family with 3 children, there are 8 possibilities for the order in which boys B and girls G can be born. These outcomes are shown in the sample space below, where BGG means the first is a boy and the next two are girls.

| Sample | BBB | BGB | GBB | GGB |
| space | BBG | BGG | GBG | GGG |

Use this sample space for Example 2. Assume that the 8 outcomes are equally likely.

EXAMPLE 2   For a family of 3 children, find the probability that the children can be described as follows.

   a. boy, girl, boy   **b.** exactly 2 girls   **c.** at least 1 boy

Solutions   **a.** boy, girl, boy   There is 1 successful outcome: *BGB*.
$$P(BGB) = \frac{s}{t} = \frac{1}{8}$$

   b. exactly 2 girls   There are 3 successful outcomes: *BGG, GBG, GGB.*
$$P(\text{exactly 2 girls}) = \frac{s}{t} = \frac{3}{8}$$

   c. at least 1 boy   There are 7 successful outcomes: all except *GGG.*
$$P(\text{at least 1 boy}) = \frac{s}{t} = \frac{7}{8}$$

In Example 3, cards are drawn at random from a deck of 52 playing cards. "At random" means that each card has an equal chance of being drawn. The 52 cards provide the sample space. They appear in 2 colors (red, black), 4 suits (clubs, diamonds, hearts, spades), 3 types of face cards (kings, queens, jacks), aces, and 9 numbers (2–10), as shown below.

faces

clubs	–	2 3 4 5 6 7 8 9 10 J Q K A	–	black
diamonds	–	2 3 4 5 6 7 8 9 10 J Q K A	–	red
hearts	–	2 3 4 5 6 7 8 9 10 J Q K A	–	red
spades	–	2 3 4 5 6 7 8 9 10 J Q K A	–	black

Additional Example 2

For a family of 3 children, find the probability that the children can be described as follows.

a. girl, girl, boy $\frac{1}{8}$
b. at least 2 girls $\frac{1}{2}$
c. The middle child is a boy. $\frac{1}{2}$

EXAMPLE 3 One card is drawn at random from an ordinary deck of 52 cards. Find the probability of each event.

  a. a red face card   **b.** the 6 of clubs   **c.** a king   **d.** not a king

Solutions

a. red face card
6 face cards are red: $s = 6$

$P(\text{red face card}) = \frac{6}{52}$, or $\frac{3}{26}$

b. 6 of clubs
$s = 1$

$P(\text{6 of clubs}) = \frac{1}{52}$

c. a king
4 cards are kings: $s = 4$

$P(\text{king}) = \frac{4}{52}$, or $\frac{1}{13}$

d. not a king
$s = 52 - 4 = 48$

$P(\text{not a king}) = \frac{48}{52}$, or $\frac{12}{13}$

In Example 3, notice that $P(\text{king}) = \frac{1}{13}$, $P(\text{not a king}) = \frac{12}{13}$, and $\frac{1}{13} + \frac{12}{13} = 1$. This fact can be generalized as follows.

If $P(E)$ is the probability of an event E occurring, and $P(\text{not } E)$ is the probability of E not occurring, then the following is true.

$$P(E) + P(\text{not } E) = 1 \quad \text{and} \quad P(\text{not } E) = 1 - P(E)$$

To understand the meaning of the word *odds* in an expression such as "The odds are 3 to 2" that an event will occur, consider drawing a red marble from a bag filled with 8 red and 6 blue marbles.

In this case, $t = 14$, the total number of possible outcomes,
$\quad\quad s = 8$, the number of successful outcomes, and
$\quad\quad u = 6$, the number of unsuccessful outcomes.

The odds *for* drawing a red marble are 8 to 6, or 4 to 3, the *ratio of s to u*. The odds *against* drawing a red marble are 3 to 4, the *ratio of u to s*.

In general, if an event E has s successful outcomes and u unsuccessful outcomes, the **odds** for E occurring are s to u, and the odds against E occurring are u to s.

Classroom Exercises

One die is tossed. Find the probability that the top face will show the following number of dots. $\frac{1}{2}$ (1 is not prime.)

1. an even number $\frac{1}{2}$ **2.** a prime number   **3.** a number greater than 6 0

Use the sample space for the 3-child family, shown on the previous page, to find the number of successful outcomes for each event.

4. two girls and a boy 3   **5.** boy, boy, girl (in that order) 1

6. exactly 1 girl 3   **7.** at least 1 girl 7

8–11. Find the probability of each event for Classroom Exercises 4–7.

  8. $\frac{3}{8}$ **9.** $\frac{1}{8}$ **10.** $\frac{3}{8}$ **11.** $\frac{7}{8}$

14.7 Probability **519**

Closure

Have students give definitions of probability and sample space. See page 517. If the odds for drawing a face card from an ordinary deck of cards are 3:13, have students give the odds against drawing a face card. 13:3

■■■■**FOLLOW UP**

Guided Practice

Classroom Exercises 1–11

Independent Practice

A Ex. 1–13, **B** Ex. 14–27, **C** Ex. 28, 29

Basic: WE 1–17 odd
Average: WE 11–27 odd, 28
Above Average: WE 13–29 odd

Additional Example 3

One card is drawn from a deck of 52 playing cards. Find the probability of each event.

a. card with an odd number $\frac{4}{13}$

b. an 8 $\frac{1}{13}$

c. not an 8 $\frac{12}{13}$

d. black ace $\frac{1}{26}$

Written Exercises

For a 3-child family, find the indicated probability of each event.

1. P(girl, then 2 boys) $\frac{1}{8}$ **2.** P(exactly 2 boys) $\frac{3}{8}$ **3.** P(at least 2 boys) $\frac{1}{2}$

4. P(oldest child is a boy) $\frac{1}{2}$ **5.** P(at least 1 boy and at least 1 girl) $\frac{3}{4}$

One card is drawn at random from an ordinary deck of 52 playing cards. Find the indicated probability of each event.

6. P(black) $\frac{1}{2}$ **7.** P(face card) $\frac{3}{13}$ **8.** P(not a 2) $\frac{12}{13}$

9. P(heart) $\frac{1}{4}$ **10.** P(7 of spades) $\frac{1}{52}$ **11.** P(not a diamond) $\frac{3}{4}$

12. What are the odds for drawing a red marble from a bag containing 12 red and 18 white marbles? 2 to 3

13. What are the odds against drawing a club at random from a deck of 52 playing cards? 3 to 1

In Exercises 14–16, $P(E) = \frac{3}{4}$. Find each of the following.

14. P(not E) $\frac{1}{4}$ **15.** the odds for E 3 to 1   **16.** the odds against E 1 to 3

Prepare a sample space showing all outcomes for tossing 3 coins. Use T for tails and H for heads. Find the probability of each event.

17. 3 tails $\frac{1}{8}$ **18.** exactly 1 tail $\frac{3}{8}$ **19.** 2 heads and a tail $\frac{3}{8}$

20. at least 2 heads $\frac{1}{2}$ **21.** THT $\frac{1}{8}$ **22.** exactly 2 heads $\frac{3}{8}$

Prepare a sample space showing all possible outcomes for boys and girls in a 4-child family. Find the probability of each event.

23. 3 girls and a boy $\frac{1}{4}$ **24.** 2 girls and 2 boys $\frac{3}{8}$ **25.** exactly 2 girls $\frac{3}{8}$

26. at least 2 boys $\frac{11}{16}$ **27.** at least 1 girl and at least 1 boy $\frac{7}{8}$

28. The probability that it will rain on any given day in April in Pensacola is $\frac{7}{10}$. About how many days in April is it likely not to rain? 9

29. If the probability that it will rain on a given day in Denver is $\frac{2}{9}$, what are the odds that it will not rain on a given day? 7 to 2

Mixed Review

Find the solution set. *2.2, 2.3, 6.4*

1. $5 - x < 7 \text{ or } 9 < 2x - 3$ $\{x \mid x > -2\}$ **2.** $8 - x < 2 \text{ and } 4 > 2x - 14$ $\{x \mid 6 < x < 9\}$

3. $|2x - 7| < 5$ $\{x \mid 1 < x < 6\}$ **4.** $x^2 - 5x - 14 > 0$ $\{x \mid x < -2 \text{ or } x > 7\}$

Enrichment

Point out that the probability of some events must be determined by experiment or from collected data, as in weather forecasts. Such probability is called *empirical* probability.

The students can calculate an empirical probability as follows. When a thumbtack is tossed, it will land in one of two positions.

UP   DOWN

Divide the class into teams of 2, and have each team conduct an experiment to find the probability that a tossed thumbtack will land point upward. Each team should toss the thumbtack at least 20 times. Now collect all the totals and calculate the empirical probability of the thumbtack landing point upwards.

Similar experiments have yielded a probability of about 0.606.

14.8 Adding and Multiplying Probabilities

Objectives

To find the probability of inclusive and mutually exclusive events
To find the probability of independent events

To determine $P(A \ or \ B)$, begin by deciding whether the two events A and B are *inclusive* or *mutually exclusive events*. If two events can occur at the same time, they are called **inclusive events**. If they cannot occur at the same time, they are called **mutually exclusive events**.

Both types of events can be illustrated by tossing a pair of dice. Suppose that the first die is red and the second die is green. Then the ordered pair (3,5) represents "3 on the red and 5 on the green." The 36-pair sample space for all such outcomes is shown at the right.

Green						
6	(1,6)	(2,6)	(3,6)	(4,6)	(5,6)	(6,6)
5	(1,5)	(2,5)	(3,5)	(4,5)	(5,5)	(6,5)
4	(1,4)	(2,4)	(3,4)	(4,4)	(5,4)	(6,4)
3	(1,3)	(2,3)	(3,3)	(4,3)	(5,3)	(6,3)
2	(1,2)	(2,2)	(3,2)	(4,2)	(5,2)	(6,2)
1	(1,1)	(2,1)	(3,1)	(4,1)	(5,1)	(6,1)
	1	2	3	4	5	6

Red

Let r and g represent outcomes on the red die and green die, respectively. There are 12 pairs where $r \geq 5$ and 18 pairs where $g \leq 3$. Since $(r \geq 5)$ and $(g \leq 3)$ have 6 pairs in common, they are *inclusive events*. The 6 pairs cannot be counted twice for $(r \geq 5 \ or \ g \leq 3)$.

Green						
6					(5,6)	(6,6)
5					(5,5)	(6,5)
4					(5,4)	(6,4)
3	(1,3)	(2,3)	(3,3)	(4,3)	(5,3)	(6,3)
2	(1,2)	(2,2)	(3,2)	(4,2)	(5,2)	(6,2)
1	(1,1)	(2,1)	(3,1)	(4,1)	(5,1)	(6,1)
	1	2	3	4	5	6

Red

The number of pairs (outcomes) for this event is $12 + 18 - 6$, or 24.

Therefore, $P(r \geq 5 \ or \ g \leq 3) = \frac{s}{t} = \frac{12 + 18 - 6}{36} = \frac{24}{36} = \frac{2}{3}$.

This result can also be found by the alternate method below.

$$P(r \geq 5) = \frac{12}{36} = \frac{1}{3} \qquad P(g \leq 3) = \frac{18}{36} = \frac{1}{2}$$
$$P(r \geq 5 \ and \ g \leq 3) = \frac{6}{36} = \frac{1}{6}$$

Then, in terms of the above probabilities,

$$P(r \geq 5 \ or \ g \leq 3) = P(r \geq 5) + P(g \leq 3) - P(r \geq 5 \ and \ g \leq 3)$$
$$= \frac{1}{3} + \frac{1}{2} - \frac{1}{6} = \frac{2}{3}$$

Teaching Resources

Problem Solving Worksheet 14
Quick Quizzes 107
**Reteaching and Practice
   Worksheets** 107

■■■ GETTING STARTED

Prerequisite Quiz

Simplify.

1. ratio of 24 to 60   ratio of 2 to 5
2. If $P(E) = \frac{3}{8}$, find the odds for E.   3 to 5

Find the probability of the event.

3. rolling an even number with a die $\frac{1}{2}$
4. rolling a number greater than or equal to 3 with a die $\frac{2}{3}$
5. drawing a red face card from a deck of 52 playing cards $\frac{3}{26}$

Motivator

Ask students if it is possible for both the Boston Red Sox and the Chicago White Sox to win the baseball game if they are playing each other.   No   If the Red Sox play the Detroit Tigers and the White Sox play the Texas Rangers, is it possible for both the Red Sox and the White Sox to win?   Yes   In this case, both teams can win at the same time.   Why?   They are independent events.

Highlighting the Standards

Standard 11b: This lesson lends itself to the kind of simulations that will best demonstrate how probabilities work.

521

Lesson Note

This lesson covers the probability of two events that may or may not occur at the same time. The two events are related by either the connector *or* or the connector *and*. Students can find probabilities by either (1) counting in the sample, or (2) adding and multiplying probabilities.

Note that when events *A* and *B* are dependent, the rule for finding $P(A \text{ and } B)$ is given in Lesson 14.9.

Math Connections

The Birthday Problem: What is the probability that two or more people in the same room have the same birthday? The answer depends upon the number of people in the room. It can be calculated that if the number of people is 23, then the probability of this event is around 0.5. To develop a general formula, it is best to find the probability that no two people in the room have the same birthday and to subtract this probability from 1.

Now consider the events $(g \leq 2)$ and $(g \geq 6)$. These are mutually exclusive events because they cannot occur at the same time (see the sample space). There are 12 pairs where $g \leq 2$ and 6 pairs where $g \geq 6$, with *no* pairs common to both sets. Therefore:

$$P(g \leq 2 \text{ or } g \geq 6) = \frac{s}{t} = \frac{12 + 6}{36} = \frac{1}{2}$$

This result can also be computed as follows.

$$P(g \leq 2) = \frac{12}{36} = \frac{1}{3} \qquad P(g \geq 6) = \frac{6}{36} = \frac{1}{6}$$

Then $P(g \leq 2 \text{ or } g \geq 6) = P(g \leq 2) + P(g \geq 6) = \frac{1}{3} + \frac{1}{6} = \frac{1}{2}$.

Probability of (A or B)
If A and B are inclusive events, then
$$P(A \text{ or } B) = P(A) + P(B) - P(A \text{ and } B).$$
If A and B are mutually exclusive events, then
$$P(A \text{ or } B) = P(A) + P(B).$$

EXAMPLE 1 Use the 36-pair sample space on the previous page for tossing a red die and a green die to find the probabilities or odds below.

Solutions

a. $P(r \leq 4 \text{ or } g = 1)$

The two events are inclusive. There are 24 pairs where $r \leq 4$, 6 pairs where $g = 1$, and 4 pairs where $r \leq 4$ *and* $g = 1$.
$$P(r \leq 4 \text{ or } g = 1) = \frac{24 + 6 - 4}{36} = \frac{13}{18}$$

b. $P(r = 2 \text{ or } r > 4)$

The two events are mutually exclusive. There are 6 pairs for $r = 2$ and 12 pairs for $r > 4$.
$$P(r = 2 \text{ or } r > 4) = \frac{6 + 12}{36} = \frac{1}{2}$$

c. $P(\text{sum of } 9)$

Find $P(r + g = 9)$. The sum 9 occurs in 4 ways: (3,6), (4,5), (5,4), (6,3).
$$P(\text{sum of } 9) = \frac{4}{36} = \frac{1}{9}$$

d. odds for a sum of 9

There are 4 successful ways and $36 - 4$, or 32, unsuccessful ways to have a sum of 9. The odds for a sum of 9 are 4 to 32, or 1 to 8.

Additional Example 1

Use the 36-pair sample space for tossing a red die and a green die to find the probabilities or odds.

a. $P(r = 5 \text{ or } g \geq 2)$ $\frac{31}{36}$

b. $P(r < 4 \text{ or } r = 5)$ $\frac{2}{3}$

c. $P(\text{sum of } 6)$ $\frac{5}{36}$

d. odds against a sum of 6   31 to 5

For the event (A and B), described by the conjunction *and*, count the outcomes that are in *both* Event A and Event B. Consider again the tossing of a red die and a green die. As noted before, there are 6 outcomes out of the 36 for which $r \geq 5$ and $g \leq 3$. Thus:

$$P(r \geq 5 \text{ and } g \leq 3) = \frac{6}{36} = \frac{1}{6}$$

You can also multiply probabilities to find $P(r \geq 5 \text{ and } g \leq 3)$. Recall that $P(r \geq 5) = \frac{12}{36} = \frac{1}{3}$ and $P(g \leq 3) = \frac{18}{36} = \frac{1}{2}$.

$$P(r \geq 5 \text{ and } g \leq 3) = P(r \geq 5) \cdot P(g \leq 3) = \frac{1}{3} \cdot \frac{1}{2} = \frac{1}{6}$$

In this case, events $(r \geq 5)$ and $(g \leq 3)$ are called **independent events** because the outcome of throwing a red die does not affect the outcome of throwing a green die, and vice versa. This example illustrates the following rule.

Probability of (A and B)
If A and B are independent events, then $P(A \text{ and } B) = P(A) \cdot P(B)$.

The rule for finding $P(A \text{ and } B)$ is modified when A and B are *dependent* events. This will be shown in the next lesson.

EXAMPLE 2   Use the 36-pair sample space for tossing a red die and a green die to find each probability.

    a. $P(r = 3 \text{ and } g > 4)$ **b.** $P(r < 3 \text{ and } g \geq 4)$

Solutions   **a.** $P(r = 3 \text{ and } g > 4) = P(r = 3) \cdot P(g > 4) = \frac{6}{36} \cdot \frac{12}{36} = \frac{1}{18}$
            b. $P(r < 3 \text{ and } g \geq 4) = P(r < 3) \cdot P(g \geq 4) = \frac{12}{36} \cdot \frac{18}{36} = \frac{1}{6}$

Classroom Exercises

Use the 36-pair sample space for tossing a red die and a green die to match each event at the left with the number of its successful outcomes at the right.

1. $r = 3$ or $g = 4$ g   **2.** $r = 3$ and $g = 4$ a   **a.** 1   **e.** 2
3. $r \leq 2$ or $g > 5$ h   **4.** $r \leq 2$ and $g > 5$ e   **b.** 3   **f.** 4
5. $r + g = 11$ e   **6.** $r + g \leq 5$ c   **c.** 10   **g.** 11
7. $r + g = 3$ or $r + g = 12$ b   **8.** $r \cdot g = 6$ f   **d.** 12   **h.** 16

9–16. Find the probability of each event for Classroom Exercises 1–8.

     9. $\frac{11}{36}$ **10.** $\frac{1}{36}$ **11.** $\frac{4}{9}$ **12.** $\frac{1}{18}$ **13.** $\frac{1}{18}$ **14.** $\frac{5}{18}$ **15.** $\frac{1}{12}$ **16.** $\frac{1}{9}$

Critical Thinking Questions

Analysis: Have students calculate the probability that at least two people in a room of 23 people (or of some other number of people) have the same birthday (see *Math Connections*). In order to help them find the appropriate pattern, have them start with a room of just two people, then three people, and so on. The probability that two people do not have the same birthday is $\frac{364}{365}$, so the probability that they do is $1 - \frac{364}{365}$. The probability that three people do not have the same birthday is $\frac{364}{365} \cdot \frac{363}{365}$, so the probability that the three people do have the same birthday is $1 - \frac{364}{365} \cdot \frac{363}{365}$, and so on.

Checkpoint

Use the 36-pair sample space for tossing a red die and a green die to find the probability or odds.

1. $P(r = 2 \text{ or } g = 4)$ $\frac{11}{36}$
2. $P(r = 2 \text{ and } g = 4)$ $\frac{1}{36}$
3. $P(\text{sum of } 4)$ $\frac{1}{12}$
4. odds for $r = 3$ 1 to 5
5. odds against $g = 6$ 5 to 1
6. $P(r \leq 2 \text{ or } g \geq 5)$ $\frac{5}{9}$
7. $P(r \leq 2 \text{ and } g \geq 5)$ $\frac{1}{9}$

Additional Example 2

Use the 36-pair sample space for tossing a red die and green die to find each probability.

a. $P(r < 5 \text{ and } g = 6)$ $\frac{1}{9}$

b. $P(r \geq 2 \text{ and } g < 3)$ $\frac{5}{18}$

Closure

Have students complete the following statements.

1. If A and B are inclusive events, then $P(A \text{ or } B) = \underline{}$.
2. If A and B are mutually exclusive events, then $P(A \text{ or } B) = \underline{}$. For Ex. 1–2, see page 522.
3. If A and B are independent events, then $P(A \text{ and } B) = \underline{}$. See page 523.

■■■FOLLOW UP

Guided Practice

Classroom Exercises 1–16

Independent Practice

A Ex. 1–16, B Ex. 17–30, C Ex. 31–34

Basic: WE 1–21 odd

Average: WE 9–31 odd

Above Average: WE 13–33 odd

Written Exercises

Use the 36-pair sample space for tossing a red die and a green die to find the following probabilities or odds.

1. $P(r = 3 \text{ or } r = 5)$ $\frac{1}{3}$
2. $P(r \le 2 \text{ or } r > 3)$ $\frac{5}{6}$
3. $P(g = 2 \text{ and } g = 4)$ 0
4. $P(r \ge 1 \text{ and } g \le 6)$ 1
5. $P(r < 3 \text{ or } g > 3)$ $\frac{2}{3}$
6. $P(r \ge 3 \text{ and } g \ge 5)$ $\frac{2}{9}$
7. $P(r + g = 7)$ $\frac{1}{6}$
8. $P(r + g \ge 8)$ $\frac{5}{12}$
9. $P(r + g = 6 \text{ or } r + g = 8)$ $\frac{5}{18}$
10. $P(r + g = 5 \text{ or } r + g = 9)$ $\frac{2}{9}$
11. $P(r + g < 6)$ $\frac{5}{18}$
12. $P(r \cdot g = 12)$ $\frac{1}{9}$
13. odds for $r = 2$ 1 to 5
14. odds against $g = 5$ 5 to 1
15. odds for sum of 7 1 to 5
16. odds against sum of 11 17 to 1

One card is drawn at random from an ordinary deck of 52 playing cards. Find the probability of drawing the indicated card. The sample space appears after Example 2 of Lesson 14.7.

17. a red card *or* an ace $\frac{7}{13}$
18. a black card *or* a face card $\frac{8}{13}$
19. a club *or* a 10 $\frac{4}{13}$
20. an odd number *or* a diamond $\frac{25}{52}$
21. a red queen *or* a heart $\frac{7}{26}$
22. a black 7 *or* a spade $\frac{7}{26}$
23. a black card *or* a heart $\frac{3}{4}$
24. a 2 *or* a face card $\frac{4}{13}$
25. an ace *or* a 2 $\frac{2}{13}$
26. a club *or* a red card $\frac{3}{4}$
27. a red card *and* a face card $\frac{3}{26}$
28. an even number *and* a black card $\frac{5}{26}$
29. an odd number *and* a club $\frac{1}{13}$
30. a diamond *and* a face card $\frac{3}{52}$

Fifteen slips of paper are numbered from 1 through 15 and placed in a box. One slip is drawn at random from the box. Find the probability that the slip drawn has the number described.

31. divisible by 3 *or* 5 $\frac{7}{15}$
32. divisible by 2 *or* 3 $\frac{2}{3}$
33. prime *or* divisible by 7 $\frac{7}{15}$
34. factor of 15 *or* 24 $\frac{3}{5}$

Mixed Review

Simplify. 5.2, 8.5

1. $\dfrac{8.8 \times 10^{-12}}{4 \times 10^{-9}}$ 0.0022
2. $\sqrt[3]{16x^6 y^8}$ $2x^2 y^2 \sqrt[3]{2y^2}$

Rationalize the denominator. 8.4, 9.2

3. $\dfrac{5}{3\sqrt{3} - 5}$ $\dfrac{15\sqrt{3} + 25}{2}$
4. $\dfrac{-2}{4 + 3i}$ $\dfrac{-8 + 6i}{25}$

Enrichment

Have the students try this problem.

A square, 10 units on a side, is inscribed in a circle. A dart is thrown by a person with no special skill other than the ability always to get the dart to land inside the circle. What is the probability of the dart landing inside the square in one throw? $\frac{2}{\pi}$

14.9 Selecting More than One Object at Random

Objective

To find the probability of selecting at random more than one object from a set of objects

Given a box containing 6 white marbles and 4 red marbles, two marbles can be drawn at random in either of the following two ways.

(1) Draw one marble, record its color, and replace it before making the second draw. The sample space is the same for each draw.

(2) Draw one marble, record its color, do *not* replace it, and draw a second marble. The sample space for the second draw is reduced by 1 marble from the sample space for the first draw. In this case, the two events are **dependent** because the first event affects the outcome of the second event.

If events A and B are dependent and B follows A, then the probability of B is written as $P(B$ given $A)$, or $P(B|A)$. So, the probability of A occurring, *followed* by B, is $P(A,$ then $B) = P(A) \cdot P(B|A)$.

EXAMPLE 1

A box contains 6 white marbles and 4 red marbles. Find the probability of each event when two marbles are drawn at random.
a. a white and then a red *without* replacement after the first draw
b. a red and then a white *with* replacement after the first draw
c. a white and a red *without* replacement after the first draw

Plan

The sample space for the first draw is the set of 10 marbles. For the second draw, the sample space has 10 marbles if there is replacement and 9 marbles if there is no replacement. Let W represent a *white* marble and R represent a *red* marble.

Solutions

a. $P(W,$ then $R) = P(W) \cdot P(R|W) = \frac{6}{10} \cdot \frac{4}{9} = \frac{4}{15}$ ← For R, the sample space is reduced by 1, from 10 to 9.
   $P(W,$ then $R)$ without replacement is $\frac{4}{15}$.

b. $P(R,$ then $W) = P(R) \cdot P(W) = \frac{4}{10} \cdot \frac{6}{10} = \frac{6}{25}$ ← The second sample space is unchanged.
   $P(R,$ then $W)$ with replacement is $\frac{6}{25}$.

c. $(W$ and $R)$ is equivalent to $[(W,$ then $R)$ *or* $(R,$ then $W)]$.
   $P(W$ and $R) = P[(W,$ then $R)$ *or* $(R,$ then $W)]$
   $= P(W) \cdot P(R|W) + P(R) \cdot P(W|R)$
   $= \frac{6}{10} \cdot \frac{4}{9} + \frac{4}{10} \cdot \frac{6}{9} = \frac{4}{15} + \frac{4}{15} = \frac{8}{15}$
   Therefore, $P(W$ and $R)$ without replacement is $\frac{8}{15}$.

Teaching Resources

Quick Quizzes 108
Reteaching and Practice
Worksheets 108

GETTING STARTED

Prerequisite Quiz

Simplify. Write answers as fractions in lowest terms. Use a calculator, if possible.

1. $\dfrac{\binom{8}{6}}{\binom{9}{6}}$ $\dfrac{1}{3}$

2. $\dfrac{\binom{6}{4} \cdot \binom{3}{2}}{\binom{9}{6}}$ $\dfrac{15}{28}$

3. $\dfrac{\binom{3}{3} \cdot \binom{6}{3} + \binom{3}{2} \cdot \binom{6}{4} + \binom{3}{1} \cdot \binom{6}{5}}{\binom{9}{6}}$ $\dfrac{83}{84}$

Motivator

Students will find the probability of selecting more than one object from a set of objects. Have students consider drawing two cards from an ordinary deck in one of two ways. Draw a first card and (1) replace it before drawing the second card, or (2) do not replace it before drawing the second card. Have students give the sample spaces for both draws in both cases.   Case 1: For both draws, the sample space is 52 cards. Case 2: The sample space for the first draw is 52 cards; the sample space for the second is 52 cards minus the first card drawn.

Additional Example 1

Two tags are drawn from a box containing 3 green tags and 7 black tags. Find the probability of the event described.

a. black, then green, with replacement $\dfrac{21}{100}$

b. green, then black, without replacement $\dfrac{7}{30}$

c. green and black, with replacement $\dfrac{21}{50}$

Highlighting the Standards

Standard 11c: The exercises are carefully sequenced to reinforce the concept of randomness and illustrate how it operates in different situations.

Lesson Note

Remind students again that (1) multiplication is used with the connector *and*, (2) addition is used with the connector *or*, and (3) a scientific calculator with $x!$ function is a helpful tool in computing with $\binom{n}{r}$ notation, as in the Prerequisite Quiz on the previous page.

Math Connections

Random Selections with a Computer:
The BASIC programming language has a function, the RND function, that allows the user to simulate random events, such as tossing a coin. The computer randomly selects a value from the interval zero to one. It is up to the user to decide how the randomly selected numbers are to apply to his or her particular experiment. For example, if a coin tossing experiment is being simulated, then the user might program the experiment so that any selected value less than 0.5 corresponds to "heads."

Critical Thinking Question

Analysis: Ask students to think of uses for the RND function (see *Math Connections*) other than the tossing of a coin. They could mention, for example, the dealing of cards or the tossing of a die.

When you select objects *without replacement and without regard to order*, you can use combinations to compute s and t and then $P(E)$. For example, a box of 10 microchips has 7 functional (good) chips and 3 defective (bad) chips. If 4 chips are drawn at random, what is the probability that all 4 are functional?

Number of successful outcomes: $\binom{7}{4}$ Total number of outcomes: $\binom{10}{4}$

$$P(E) = \frac{s}{t} = \frac{\binom{7}{4}}{\binom{10}{4}} = \frac{\frac{7!}{4!3!}}{\frac{10!}{4!6!}} = \frac{\frac{7 \cdot 6 \cdot 5}{3 \cdot 2 \cdot 1}}{\frac{10 \cdot 9 \cdot 8 \cdot 7}{4 \cdot 3 \cdot 2 \cdot 1}} = \frac{35}{210} = \frac{1}{6}$$

To find $P(E)$ above by calculator, calculate the numerator and denominator separately. Then simplify the fraction.

EXAMPLE 2   A 5-member committee is to be formed by selecting 5 people at random from a group of 4 males and 6 females. Find the probability that the committee has the following.

a. all females   **b.** exactly 3 males   **c.** at least 2 males

Plan   The number t of all possible outcomes is $\binom{10}{5} = \frac{10!}{5!5!}$, or 252. Let M represent *male* and F *female*.

Solutions   **a.** $P(\text{all } F) = P(5F \text{ and no } M)$

To find s, select 5 from among 6 and 0 from among 4. Multiply.

$$P(\text{all } F) = \frac{s}{t} = \frac{\binom{6}{5} \cdot \binom{4}{0}}{\binom{10}{5}} = \frac{6 \cdot 1}{252} = \frac{1}{42}$$

b. $P(\text{exactly } 3M) = P(3M \text{ and } 2F)$

To find s, select 3 from among 4 and 2 from among 6. Multiply.

$$P(3M \text{ and } 2M) = \frac{s}{t} = \frac{\binom{4}{3} \cdot \binom{6}{2}}{\binom{10}{5}} = \frac{4 \cdot 15}{252} = \frac{60}{252} = \frac{5}{21}$$

c. $P(\text{at least } 2M) = P[(2M \text{ and } 3F) \text{ or } (3M \text{ and } 2F) \text{ or } (4M \text{ and } 1F)]$

Find s as follows: $(2M,3F)$ or $(3M,2F)$ or $(4M,1F)$

$$s = \binom{4}{2} \cdot \binom{6}{3} + \binom{4}{3} \cdot \binom{6}{2} + \binom{4}{4} \cdot \binom{6}{1}$$
$$= 6 \cdot 20 + 4 \cdot 15 + 1 \cdot 6 = 186$$

$$P(\text{at least } 2M) = \frac{s}{t} = \frac{186}{252} = \frac{31}{42}$$

Additional Example 2

The 4 members of a school's Brain Bowl team are selected at random from a group of 6 males and 3 females. Find the probability of the indicated team membership.

a. all males $\frac{5}{42}$

b. exactly 2 females $\frac{5}{14}$

c. at least 1 female $\frac{37}{42}$

Classroom Exercises

A bag contains 5 white balls and 3 black balls. Two balls are drawn at random *without* replacement. Tell whether the given expression represents the probability of (white, then black), (both black), (white *and* black), or (at least one white).

1. $\frac{3}{8} \cdot \frac{2}{7}$ 2. $\frac{5}{8} \cdot \frac{3}{7} + \frac{3}{8} \cdot \frac{5}{7}$ 3. $\frac{\binom{5}{1} \cdot \binom{3}{1} + \binom{5}{2} \cdot \binom{3}{0}}{\binom{8}{2}}$ 4. $\frac{5}{8} \cdot \frac{3}{7}$
   Both black   White *and* black   White, then black

At least one white

Two cards are drawn *with* replacement from a deck of 52 cards. Match the probability of the given event with one expression at the right.

5. First is a 7 and second   a. $\frac{4}{52} \cdot \frac{4}{52}$ c. $\frac{13}{52} \cdot \frac{39}{52} + \frac{39}{52} \cdot \frac{13}{52}$
   is a club. b

6. One is a 7 and the   b. $\frac{4}{52} \cdot \frac{13}{52}$ d. $\frac{4}{52} \cdot \frac{13}{52} + \frac{13}{52} \cdot \frac{4}{52}$
   other is a club. d

7. Both are 7s. a   8. Exactly one is a club. c

A bag contains 8 green tokens and 9 red tokens. Two tokens are drawn. Find the probability of each event.

9. red, then green *with* replacement $\frac{72}{289}$ 10. red, then green *without* replacement $\frac{9}{34}$

Written Exercises

A bag contains 3 green marbles and 6 blue marbles. Two marbles are drawn at random *with* replacement. Find the probability of each event.

1. green, then blue $\frac{2}{9}$ 2. green and blue $\frac{4}{9}$ 3. neither one green $\frac{4}{9}$

A bag has 5 yellow buttons, 3 red buttons, and 1 white button. Two buttons are drawn *without* replacement. Find the probability of each event.

4. red, then yellow $\frac{5}{24}$ 5. yellow, then white $\frac{5}{72}$ 6. red *and* white $\frac{1}{12}$

From a group of 4 boys and 6 girls, three piano students are to be selected at random to represent their school at a music recital. Determine the probability of each selection.

7. all girls $\frac{1}{6}$ 8. exactly 1 boy $\frac{1}{2}$ 9. at least 2 girls $\frac{2}{3}$

Two cards are drawn at random *with* replacement from a deck of 52 playing cards. Find the probability that the cards are as follows.

10. club, then heart $\frac{1}{16}$ 11. black *and* red $\frac{1}{2}$ 12. exactly 1 face card $\frac{60}{169}$

527 **14.9 Selecting More than One Object at Random**

Common Error Analysis

Error: When selecting more than one object in a set, students may not recognize dependent events.

Emphasize that they should notice whether the first object is, or is not, replaced before the second draw is made. For the tags in Additional Example 1, consider the probability of drawing two green tags. If the first tag is replaced before the second tag is drawn, the events are independent and $P(2$ green tags$) = \frac{3}{10} \cdot \frac{3}{10} = \frac{9}{100}$. If the first tag is not replaced, the events are dependent, and $P(2$ green tags$) = \frac{3}{10} \cdot \frac{2}{9} = \frac{1}{15}$.

Checkpoint

Two cards are drawn at random from a deck of 52 playing cards without replacing the first card. Find the probability that the cards are as described.

1. an even number, then a face card $\frac{20}{221}$
2. an even number and an ace $\frac{40}{663}$
3. both even numbers $\frac{95}{663}$

4–6. Find the probabilities in Exercises 1–3 if the first card is replaced before the second card is drawn.

4. $\frac{15}{169}$ 5. $\frac{10}{169}$ 6. $\frac{25}{169}$

Closure

Given that 5 coins are to be selected from a box of 6 dimes and 3 nickels, have the students show how they would use combinations to find the probability that at least 2 of the 5 coins are nickels.

$$\frac{\binom{3}{2} \cdot \binom{6}{3} + \binom{3}{3} \cdot \binom{6}{2}}{\binom{9}{5}}$$

Enrichment

Remind students that to select an item at random means that each item has an equal chance of being selected. This is usually done by assigning numbers to the individual items and then selecting numbers at random.

To be sure that the numbers selected are truly random, tables of random numbers are often used. Write this small section taken from such a table on the chalkboard.

10	09	73	25	33	76	52	01	35	86
37	54	20	48	05	64	89	47	42	96
08	42	26	89	53	19	64	50	93	03

By going from left to right, row by row, have the students select 5 students at random from a group of students numbered 01–20. Students numbered 10, 09, 01, 20, and 05

Point out that a scientific calculator with a statistical mode will also generate random numbers.

Have the students find textbooks with tables of random numbers, and have them use these tables to make various random selections.

Guided Practice

Classroom Exercises 1–10

Independent Practice

A Ex. 1–9, **B** Ex. 10–27, **C** Ex. 28–31

Basic: WE 1–15 odd, Brainteaser
Average: WE 11–29 odd, Brainteaser
Above Average: WE 11–31 odd, Brainteaser

Additional Answers

Brainteaser

Yes; The proof is based on the following principle: If n objects are placed in $n - 1$ slots, at least one slot must contain 2 objects. It is possible to bring any of the 24 people to his or her place card by rotating the table. There are 23 positions of the table to account for these 24 correct seatings, so one of the possible rotations must place at least two of the people in front of the proper place cards.

Two cards are drawn at random from a deck of 52 playing cards *without* replacing the first card. Find the probability that the cards are as follows.

13. both red $\frac{25}{102}$ **14.** both clubs $\frac{1}{17}$ **15.** spade, then heart $\frac{13}{204}$

16. red, then black $\frac{13}{51}$ **17.** club *and* diamond $\frac{13}{102}$ **18.** face card *and* ace $\frac{8}{221}$

19. exactly 1 ace $\frac{32}{221}$ **20.** exactly 1 red $\frac{26}{51}$ **21.** no face card $\frac{10}{17}$

Six members of a jury are selected at random from a group of 8 males and 7 females. Find the probability of each indicated jury selection.

22. all males $\frac{4}{715}$ **23.** all females $\frac{1}{715}$ **24.** exactly 3 females $\frac{56}{143}$

25. exactly 2 males $\frac{28}{143}$ **26.** at least 4 males $\frac{54}{143}$ **27.** at least 4 females $\frac{3}{13}$

28. A box contains 5 white marbles and 10 black marbles. Two marbles are drawn *without* replacement. What is the probability that both are the same color? $\frac{11}{21}$

29. If 5 cards are drawn at random from an ordinary deck of 52 cards, what is the probability that all 5 cards are hearts? $\frac{33}{66,640}$

30. The letters of the alphabet are printed on separate cards and placed in a box. Four cards are drawn at random. Find the probability that the cards in the order drawn spell the word FOUR. $\frac{1}{358,800}$

31. In a carton of 12 eggs, exactly two are cracked. If 2 eggs are selected at random, what is the probability that (**a**) both eggs are not cracked and (**b**) both eggs are cracked. a: $\frac{15}{22}$; b: $\frac{1}{66}$

Mixed Review

1. y varies directly as x and $y = 15$ when $x = 2.7$. Find y when $x = 1.8$. **3.8** 10

2. Solve the system for (x,y).
$3x + 2y = 5$
$5x = 7 - 4y$ **4.3** $(3, -2)$

3. Factor completely.
$4x^3 - 32$ **5.7** $4(x - 2)(x^2 + 2x + 4)$

4. Solve and check.
$2 \log_b x = \log_b 3 + \log_b (x + 6)$
13.6 6

■■■ *Brainteaser*

A conference is held at a circular table that seats 24 people, equally spaced. Place cards that have been put on the table list 24 different names. Suppose the conference participants arrive and take seats at random. They then find that no one is in the correct seat. Regardless of the seats taken by the participants, is it always possible to rotate the table until at least 2 people are seated in front of their place cards? Explain.

14.10 Frequency Distributions

Objective

To find the mean, median, mode(s), variance, and standard deviation for a set of data

The table at the right shows the scores and their frequency for 20 students who took an algebra test. The same data can be displayed in a bar graph called a **histogram** as shown below.

Test score	Number of students
100	3
97	4
95	2
92	1
86	5
85	2
82	1
76	1
72	1

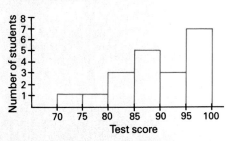

In the histogram, scores have been assigned to intervals of 5 units, with each boundary score, 75, 80, 85, 90, 95, 100, placed in the interval at the left. For example, 85 is assigned to the interval 80–85.

The table and the histogram provide information about the scores and their frequency. Such displays are called **frequency distributions**. Three numbers that measure the central tendency of a frequency distribution are the *mean*, the *median*, and the *mode*. The mean is often referred to as the "average."

Definitions

The **mean** is the number found by adding the scores and dividing the sum by the number of scores.

When the scores are arranged in order, the **median** is the middle score in an odd number of scores, or the mean (average) of the two middle scores in an even number of scores.

The **mode** is the score that occurs most frequently. There may be no mode if no score occurs more frequently than another, or there may be more than one mode if two or more scores have the same highest frequency.

For the 20 test scores above, the mean is 90, the median is $(92 + 86) \div 2$, or 89, and the mode is 86.

14.10 Frequency Distributions **529**

Teaching Resources

Project Worksheet 14
Quick Quizzes 109
Reteaching and Practice
  Worksheets 109

GETTING STARTED

Prerequisite Quiz

1. Which person is in the middle of a line of 17 people?   The ninth
2. Which people are the two middle ones in a line of 40 people?   The 20th and 21st

Simplify.

3. $(-11)^2$ 121   4. $\frac{506}{11}$ 46

Motivator

Given that a student has test scores of 81, 94, 82, and 78, have your students determine a fifth test score so that the average of the 5 scores is 85.   90

TEACHING SUGGESTIONS

Lesson Note

The *mean*, *median*, and *mode* are reviewed as measures of the central tendency of a frequency distribution. Then the *variance* and *standard deviation* are introduced as measures of dispersion (spread) from the mean for a set of scores. The significance of a standard deviation is shown in the next lesson. Before undertaking Example 2, use the Classroom Exercises to check that the students have the prerequisite skills and understanding for finding a standard deviation.

Highlighting the Standards

Standard 10c: The introduction shows students the vital connection between central tendency and frequency distribution.

Math Connections

Quality Control: One application of statistical measures such as the mean or median is in the area of *quality control*. To take a simple example, a quality-control team in a vegetable canning factory might test several cans of corn to determine the mean weight of kernels per can. If the weight falls below a company standard, then corrective action is taken.

Critical Thinking Questions

Analysis: Ask students whether the mean or the median is a more appropriate measure of central tendency in the following situation.

In a certain neighborhood, a survey of the family incomes of the residents produced the following results: $15,800, $21,000, $13,200, $18,000, $35,000, $34,000, $27,000, $46,000, and $375,000. Students should see that the choice of a measure depends upon the purpose of the survey. If the purpose is to identify a *typical* salary earner, then the median would be more appropriate than the mean, since the latter gives a misleading impression due to the presence of one unusually high income ($375,000).

Checkpoint

Find the mean, median, mode(s), variance, and standard deviation for the following frequency distribution.

score	69	76	81	84	86
frequency	1	2	3	6	4

Mean = 82; median = 84; mode = 84; variance = 20.75; standard deviation ≈ 4.6

EXAMPLE 1 Find the mean, median, and mode(s) for the following scores.
2, 7, 5, 6, 5, 7, 2, 8, 5, 9, 2   Make a frequency table.

Solution

			score	frequency	score × frequency
2	2	2	2	3	6
5	5	5	5	3	15
6			6	1	6
7	7		7	2	14
8			8	1	8
9			9	1	9
				11 ←Totals→	58

The mean (M) is $\frac{58}{11}$, or 5.27 to the nearest 0.01.

The median is the 6th score (5) in the ordered list of 11 scores. There are two modes, 2 and 5, each with a frequency of 3. So, the mean is 5.27, the median is 5, and the modes are 2 and 5.

The scores in a list may cluster around the mean or may be spread from the mean. The simplest measure of the spread, or dispersion, is the **range**, which is the difference between the highest and lowest scores.

The two most common measures of dispersion are the *variance* and the *standard deviation*. Both of these measures involve the deviation from the mean for each score in a list of scores. Consider the list with scores of 92, 89, 83, 77, 77, 68 and a mean (M) of 81.

score	92	89	83	77	77	68
score minus 81	92–81	89–81	83–81	77–81	77–81	68–81
deviation from M	+11	+8	+2	−4	−4	−13

The **variance** of a set of scores is the sum of the squares of the deviations, divided by the number of entries. For the 6 entries above, the variance is 65, as shown below.

$$\text{variance} = \frac{11^2 + 8^2 + 2^2 + 2(-4)^2 + (-13)^2}{6} = \frac{390}{6} = 65$$

The **standard deviation** of a set of scores is the positive square root of the variance for the set. For the 6 entries above with a variance of 65, the standard deviation is $\sqrt{65}$, or 8.06 to 2 decimal places. The Greek letter sigma (σ) is used to denote the standard deviation. For the 6 scores above,

$$\sigma = \sqrt{\text{variance}} = \sqrt{65} \approx 8.06$$

Additional Example 1

Find the mean, median, and mode(s) for the following scores.

3, 8, 6, 7, 6, 8, 3, 9, 6, 12
Mean = 6.8; median = 6.5; mode = 6

In general, the variance and standard deviation can be found as follows.

Formulas for Variance and Standard Deviation

For a set of n scores $(x_1, x_2, x_3, \ldots, x_n)$ with a mean of M, the variance is equal to the following expression.

$$\frac{(x_1 - M)^2 + (x_2 - M)^2 + (x_3 - M)^2 + \cdots + (x_n - M)^2}{n}$$

Standard deviation, σ: $\sigma = \sqrt{\text{variance}}$

EXAMPLE 2   Find the variance and standard deviation for the following list of scores: 35, 56, 23, 47, 68, 47, 23, 50, 56, 23, 78.

Plan   Organize the data in a frequency table and find the mean (M).

Solution   No. = score   Freq. = frequency   Dev. = deviation from mean (M)

No.	Freq.	No. × Freq.	Dev. (No. − 46)	Dev.^{2}	Freq. × Dev.^{2}
78	1	78	+32	1,024	1,024
68	1	68	+22	484	484
56	2	112	+10	100	200
50	1	50	+4	16	16
47	2	94	+1	1	2
35	1	35	−11	121	121
23	3	69	−23	529	1,587
	11	506			3,434

$$\text{mean}(M) = \frac{506}{11} = 46 \qquad \text{variance} = \frac{3,434}{11} \approx 312.18$$

$$\sigma = \sqrt{\frac{3,434}{11}} \approx 17.7$$

The variance is 312.2 and the standard deviation is 17.7.

Classroom Exercises

Find the mean, median, and mode(s) for each set of data.

1. 2, 3, 5, 5, 5   4, 5, 5   **2.** 1, 3, 3, 7, 7, 9   5, 5, 3 and 7   **3.** 2, 2, 2, 2, 4, 6   3, 2, 2

The mean for each list below is 20. Determine the deviation from the mean for each score in the list.

4. 28, 23, 15, 14   **5.** 17, 18, 20, 22, 23   **6.** 30, 25, 20, 20, 15, 10
8, 3, −5, −6   −3, −2, 0, 2, 3   10, 5, 0, 0, −5, −10

Additional Example 2

Find the variance and standard deviation for the following list of scores. 50, 60, 55, 44, 60, 55   Variance ≈ 31.67; σ ≈ 5.6

Written Exercises

Find the mean, median, and mode(s) for each set of data.

1. 37, 44, 54, 64, 23, 33, 43, 53, 54 45, 44, 54

2. 75, 85, 45, 75, 35, 85, 25, 65 61.25, 70, 75 and 85

3. 16, 22, 19, 31, 15, 22, 16, 18, 40, 16 21.5, 18.5, 16

4. 70, 80, 40, 10, 20, 40, 80, 70, 10, 30, 80 48.2, 40, 80

5. score	frequency	**6.** score	frequency	**7.** score	frequency
95	2	8	7	80	2
91	3	11	3	70	2
87	4	15	4	60	3
85	6	20	4	50	1
82	3	23	1	40	1
77	2	35	1	30	3
65	1				

5: 85.05, 85, 85   6: 14.35, 13, 8   7: 55, 60, 60 and 30

Find the variance and standard deviation for the following data. First organize the data in a frequency table and find the mean (see Example 2).

8. 80, 80, 75, 70, 70, 65, 65, 65, 60 44.44, 6.67

9. 14, 9, 14, 16, 20, 7, 10, 14, 16, 10 14, 3.74

10. 100, 97, 94, 93, 95, 93, 88, 76, 95, 89 39.4, 6.28

11. 35, 65, 42, 35, 55, 25, 65, 30, 35 199.78, 14.13

12. The mean for 8 scores is 94. Seven of the scores are 99, 98, 90, 95, 98, 92, 96. Find the missing score. 84

13. The mean for 10 scores is 48. Nine of the scores are 53, 53, 41, 41, 47, 47, 47, 45, 45. What is the missing score? 61

14. Fifty seniors from Northside High School had a mean score of 980 on the SAT. Thirty seniors from Southside High School had a mean score of 940 on the same test. What was the mean score for the 80 seniors? 965

15. A herd of 27 dairy cows produces a monthly mean of 70 lb of milk per cow. Another herd of 36 dairy cows produces a monthly mean of 84 lb of milk per cow. Find the monthly mean of milk per cow produced by both herds. 78 lb

Mixed Review

1. Find the slope and describe the slant of the line whose equation is $3x - 4y = 8$. **3.3** $\frac{3}{4}$, up to right

2. Convert 21.6 lb/ft^3 to ounces per cubic inch (oz/in^3). **7.3** 0.2 oz/in^3

3. Simplify $\dfrac{8x}{x^2 - 25} + \dfrac{2}{5 - x} - \dfrac{4}{x + 5}$ **7.4** $\dfrac{2}{x - 5}$

4. Solve $\dfrac{2x}{3} + \dfrac{5x}{8} - \dfrac{x}{4} = 2$. **7.6** $\dfrac{48}{25}$

Enrichment

Explain that many national tests are graded with a *standard score* as well as with a *raw score*. The standard score indicates how many standard deviations a raw score is above or below the mean score. For example, if the mean is 70 and the standard deviation is 10, then a raw score of 80 would have a standard score of 1, and a raw score of 65 would have a standard score of −0.5.

Using the same mean and standard deviation, have the students calculate the standard scores for the following raw scores.

Raw Score	Standard Score
68	−0.2
87	1.7
70	0.0
55	−1.5

14.11 Normal Distribution

Objective To solve problems involving a normal distribution

If a large number of families with 4 children were surveyed to find the sexes and order of birth for their 4 children, a "representative" sample of 16 families would yield the following results.

		BBGG GGBB		
	BBBG BBGB BGBB	BGBG GBGB GBGB GBGB	GGGB GGBG GBGG	
BBBB	GBBB	GBBG	BGGG	GGGG
4B	3B, 1G	2B, 2G	3G, 1B	4G
$n = 1$	$n = 4$	$n = 6$	$n = 4$	$n = 1$

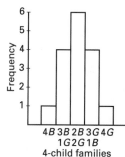

4-child families

The histogram at the right shows the frequency distribution for the data without regard to order of birth.

EXAMPLE 1 Mr. and Mrs. Mason want to have 4 children. Use the histogram above to find the probability of each event for a 4-child family.

  a. 4 girls **b.** 2 boys and 2 girls **c.** at least 2 boys

Plan Let the area of the bar for 4 boys be one square unit. Then the total area of the 5 bars is $1 + 4 + 6 + 4 + 1$, or 16 square units.
$$P(E) = \frac{s}{t} = \frac{s}{16}$$

Solutions **a.** $\frac{1}{16}$ **b.** $\frac{6}{16}$, or $\frac{3}{8}$ **c.** $\frac{6 + 4 + 1}{16}$, or $\frac{11}{16}$

The bell-shaped curve, drawn through the midpoints of the tops of the bars in this histogram, is called a **normal curve**. It represents a **normal distribution** of data, such as the weights of many children, the foot sizes of many females, and so on, all selected at random.

A normal curve reflects the mean M and the dispersion from the mean in terms of the standard deviation.

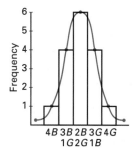

Teaching Resources

Quick Quizzes 110
Reteaching and Practice Worksheets 110

▬▬ GETTING STARTED

Prerequisite Quiz

Evaluate the following expressions if $a = 34.1\%$ and $b = 13.6\%$.

1. $a + b$ 47.7%
2. $a + 50\%$ 84.1%
3. $50\% - a$ 15.9%
4. $2(a + b)$ 95.4%

Motivator

Have students informally collect data such as (1) the weight of classmates, (2) the shoe size of female classmates. Then have them analyze how the data is distributed. More scores should be near the mean than at the highest and lowest scores.

▬▬ TEACHING SUGGESTIONS

Lesson Note

Emphasize that a *normal* distribution of measurements is approached only after a *large* number of measurements is taken at *random*.

As a rough approximation, $\frac{2}{3}$ of the scores are in the range $M \pm \sigma$ in a normal distribution.

Additional Example 1

Use the histogram above Example 1 to find the probability of the event for a 4-child family.

a. exactly 3 girls $\frac{1}{4}$
b. at least 1 girl $\frac{15}{16}$
c. no girls $\frac{1}{16}$

Highlighting the Standards

Standard 12a: The Focus on Reading and Classroom Exercises build skills and concepts related to discrete mathematics.

Math Connections

The Poisson Curve: There are distribution curves other than the normal distribution. For example, the *Poisson*, or *J-shaped* curve, is used to show the distribution of rare events. An example is the appointment of a Supreme Court judge by the president of the United States. A classic example from the nineteenth century is the frequency of deaths from kicks of horses per year in the Russian Army.

Critical Thinking Questions

Analysis: Ask students whether *exactly* 100% of the normal curve is confined between some two values of *x* (as it clearly is between two values of *y*). Students may be able to reason that if two such points existed, they would have the effect of placing sharp demarcation points between events that are merely rare and those that are impossible. For practical purposes, this cannot be done. The curve must be wide enough to embrace even the most unlikely events and thus cannot really be bounded at all.

Checkpoint

The life expectancy of a certain type of tractor approximates a normal distribution under ordinary use. The mean life of this tractor is 12 years with σ = 1 yr 6 mo.

1. Find the probability that a tractor's life is between 10 yr 6 mo and 13 yr 6 mo. 0.682
2. Find the probability that a tractor will be in use for more than 10 yr 6 mo. 0.841
3. If a company buys 300 of these tractors, how many should be in use between 9 yr and 15 yr? 286 tractors

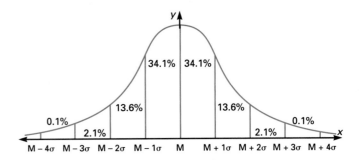

The normal curve above has the following properties.
(1) The curve is symmetric with respect to the *y*-axis.
(2) The area between the curve and the *x*-axis is 1.
(3) 50% of the area is at the right (or left) of $x = M$.
(4) 68.2% of the area is between $M - 1\sigma$ and $M + 1\sigma$.
(5) 95.4% of the area is between $M - 2\sigma$ and $M + 2\sigma$.
(6) 99.6% of the area is between $M - 3\sigma$ and $M + 3\sigma$.

EXAMPLE 2   The life expectancy of a certain type of automobile tire approximates a normal distribution. The mean life of a tire is 35,000 mi, with one standard deviation equal to 4,000 mi.
a. What is the probability of a tire lasting from 31,000 to 39,000 mi?
b. What is the probability that a tire will last more than 31,000 mi?
c. If a car rental agency buys 500 tires, how many of them can be expected to last between 27,000 and 43,000 mi?

Plan   Use the normal curve with $M = 35,000$ and $\sigma = 4,000$.

$M - 2\sigma$	$M - \sigma$	M	$M + \sigma$	$M + 2\sigma$
27,000	31,000	35,000	39,000	43,000

Solutions   a. The interval from 31,000 to 39,000 is between $M - \sigma$ and $M + \sigma$. 34.1% + 34.1%, or 68.2%, of the area is in this interval. The probability is 0.682.
b. $31,000 = M - \sigma$
The area at the right of $M - \sigma$ is 34.1% + 50%, or 84.1%. The probability is 0.841.
c. The interval from 27,000 to 43,000 is between $M - 2\sigma$ and $M + 2\sigma$. 2(13.6% + 34.1%), or 95.4%, of the area is in this interval. 95.4% of 500 = 0.954(500) = 477

So, 477 tires can be expected to last between 27,000 and 43,000 mi.

Additional Example 2

Refer to Example 2 where the life expectancy of a tire is approximately 35,000 miles, with a standard deviation of 4,000 miles.

a. What is the probability that a tire will last between 31,000 and 43,000 miles? 0.818
b. What is the probability that a tire will last more than 27,000 miles? 0.977

c. How many of 400 tires should last between 35,000 and 43,000 miles? 190.8, or about 190 tires

Classroom Exercises

The histogram shows the frequency distribution of the children in one type of family.

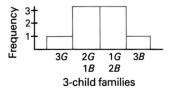

3-child families

1. How many children are in this type of family? 3
2. The area of the bar for 3 girls is 1 square unit. What is the total area of the 4 bars? 8 square units

Find the probability of each event for a 3-child family.

3. 3 boys $\frac{1}{8}$ 4. 2 girls *and* 1 boy $\frac{3}{8}$ 5. at least 2 girls $\frac{1}{2}$ 6. at least 1 boy $\frac{7}{8}$
7. fewer than 3 girls $\frac{7}{8}$ 8. exactly 2 boys *or* exactly 2 girls $\frac{3}{4}$

Written Exercises

The histogram at the right shows the frequency distribution for tossing 5 coins. Use the histogram to find the probability of each event.

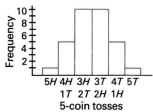

5-coin tosses

1. 4 tails *and* 1 head $\frac{5}{32}$ 2. 3 heads *and* 2 tails $\frac{5}{16}$
3. 2 heads *or* 2 tails $\frac{5}{8}$ 4. 4 tails *or* 1 tail $\frac{5}{16}$
5. at least 3 heads $\frac{1}{2}$ 6. more than 2 tails $\frac{1}{2}$

The life expectancy of a certain tire approximates a normal distribution. The mean life of a tire is 45,000 mi, with one standard deviation equal to 5,000 mi. Find the probability that a tire will last the given mileage.

7. more than 45,000 mi $\frac{1}{2}$
8. between 40,000 and 50,000 mi 0.682
9. more than 40,000 mi 0.841
10. between 40,000 and 55,000 mi 0.818
11. Out of 25 tires, how many should last between 40,000 and 50,000 mi? About 17
12. Out of 200 tires, how many should last between 35,000 and 55,000 mi? About 190
13. The heights, to the nearest inch, for 200 randomly selected adults approximate a normal distribution and are tallied in the table below. Which heights are between $M - \sigma$ and $M + \sigma$?

Height (in.)	78	76	73	71	68	66	63	61	59
Frequency	1	2	27	32	69	34	30	3	2

66 in; 68 in; 71 in.

Mixed Review

Find the following, given $f(x) = 2x - 3$ and $g(x) = x^2 + 2$.   3.2, 5.9

1. $f(4.5)$ 6   2. $g(-5)$ 27   3. $f(g(a))$ $2a^2 + 1$ 4. $g(f(a))$ $4a^2 - 12a + 11$

14.11 Normal Distribution   **535**

Closure

Have students give an informal definition of the term *normal distribution*. Have students find the number of students out of 600 that should score between 400 and 700 on a college entrance test that has a mean of 500 with a standard deviation of 100.
About 490

FOLLOW UP

Guided Practice

Classroom Exercises 1–8

Independent Practice

A Ex. 1–6, **B** Ex. 7–12, **C** Ex. 13

Basic: WE 1–8, Extension
Average: WE 5–13, Extension
Above Average: WE 6–13, Extension

Enrichment

Explain that not all distributions are *normal* distributions. When a population is made up of two disparate groups, a bimodal distribution often results.

Such a curve might result from an experiment involving push-up capability in a group made up of pre-teenagers and teenagers.

Another distribution might be skewed to the left, as shown below.

This curve might show daily rainfall amounts in a year in which there was a drought during much of the year, with the mean being the daily average for the year.

Have the students provide additional examples which might yield the various types of graphs.

Extension

For Exercises 1–7, answers will vary. Possible answers are given.

1. Poor; males at the Naval Academy might be in better shape than most males.
2. Poor; some people do not read the newspaper. Also, people on one side of the issue might be more likely to vote than those on the other.
3. The students in the English classes should be fairly representative since most students take English. Students in the trigonometry classes and drama club might not be representative, however, since these are much smaller groups and since people with different interests might have somewhat different habits.
4. Poor; this method samples only those people viewing a particular program. Also, people on one side of the issue might be more likely to vote than those on the other.
5. Reasonable; this is a fairly large sample. Also, it seems reasonable that errors will be scattered fairly uniformly throughout the book as a whole.
6. Reasonable; the sample size is a significant percentage of the total population. Also, it seems unlikely that the number of chocolate chips will vary greatly depending upon a cookie's location in the bag.
7. Poor; it is quite possible that the quality of automobiles built immediately before the weekend might be significantly different from that of those built at other times.

Extension

Sampling a Population

Several weeks prior to an election, a straw poll can be taken to predict the percent of the votes that each candidate for mayor will receive on election day. Since contacting every eligible voter would be impractical, a subset, or *sample*, of all eligible voters is polled. This sample should represent the total population of voters as accurately as possible.

Example

In a certain city with 80,000 eligible voters, the population lives in four areas that roughly correspond to the annual family incomes shown below.

Area	No. of voters	Family income
Rolling Hills	400	over $80,000
South Park	40,600	$40,000–$60,000
West Side	27,000	$15,000–$35,000
North End	12,000	under $12,000

Explain why each of the following samples is not a good representation of the total population of voters in this city.

a. every home in Rolling Hills

b. 100 homes in each of the four areas

c. every 10th person entering the terminal building at the local commercial airport that serves this city

Solutions

a. The Rolling Hills area has families with the highest income level. Voters from this area might tend to vote for a candidate who will represent their interests by promising low taxes, wanting to provide services based on property and home values, and so on.

b. The 80,000 voters are not equally divided among the four areas. South Park and West Side would be *under*represented while Rolling Hills and North End would be *over*represented.

c. The frequency of airline trips is not a characteristic that is equally distributed among the total population. Businesspeople tend to fly more often than many other people.

In contrast to the faulty sampling procedures shown above, there are efficient procedures that provide a more representative sample of a given population. Two examples follow.

(1) Telephone each nth person listed in a telephone book.
(2) Mail a questionnaire with a stamped return envelope to each nth person in the telephone book.

Notice that even these methods are not truly representative. For example, some people don't have telephones, some have unlisted numbers, some might not return their questionnaires, and so on. Still, for a specific city or area, they often give a fairly representative sample.

Exercises

Tell whether each sampling procedure below provides a poor or a reasonable representation of the total population. Justify your answer.

1. To determine the mean (average) weight of all 20-year-old males, find the weights of all 20-year-old males at the U.S. Naval Academy.

2. To determine the majority opinion on whether a city should fluoridate its water supply, publish a ballot in the city's newspaper and ask the public to vote and mail it to the newspaper's office.

3. To determine what percent of the students in a high school drink milk at breakfast, survey all the students in the following classes or organizations.
   a. trigonometry classes
   b. English classes
   c. drama club

4. To determine the majority opinion on a very controversial topic, have television viewers telephone a specific number for a "yes" response or another specific number for a "no" response.

5. To determine the overall accuracy of a 700-page novel, randomly choose 70 of the pages to examine for errors.

6. To determine the average number of chocolate chips per cookie in a bag of cookies, find the average number in every fifth cookie.

7. To determine the quality of a given model of automobile, test the last one assembled every Friday of the model year.

21. Answers will vary. One possible
    answer: To find the mean of a set of
    scores, multiply each score times the
    number of times that it occurs, add
    these results, and divide by the total
    number of scores.

Chapter 14 Review

Key Terms

circular permutation (p. 509)
combination (p. 511)
dependent events (p. 525)
distinguishable permutation (p. 507)
event (p. 517)
factorial (p. 500)
frequency distribution (p. 529)
Fundamental Counting Principle (p. 499)
histogram (p. 529)
inclusive events (p. 521)
independent events (p. 523)
mean (p. 529)

median (p. 529)
mode (p. 529)
mutually exclusive events (p. 521)
normal curve (p. 533)
normal distribution (p. 533)
odds (p. 519)
permutation (p. 503)
probability (p. 517)
range (p. 530)
sample space (p. 517)
standard deviation (p. 530)
variance (p. 530)

Key Ideas and Review Exercises

14.1   The Fundamental Counting Principle is stated on page 499.

1. How many different types of jackets can be made using 6 colors, 3
   choices of collar, 2 choices of cuffs, and 4 choices of fabric? 144

14.2   In finding the number of permutations of objects, determine the decisions
to be made and the number of choices for each decision.

2. How many odd 5-digit numbers can be formed from the digits 1, 2,
   3, 4, 6, 8, 9, if a digit may be repeated in a number? 7,203

3. How many numbers of one or more digits can be formed from the
   digits 2, 4, 6, 8, if no digit may be repeated in a number? 64

4. Find the number of ways that 6 males and 4 females can be arranged in a line with the females together and at the left of the
   males. 17,280

14.3   The number of distinguishable permutations of n objects with r of them
alike is $\frac{n!}{r!}$.

5. How many distinguishable permutations can be made using all the
   letters in the word SCISSORS? 1,680

14.4   There are $(n - 1)!$ ways to arrange n objects in a circle, and $\dfrac{(n - 1)!}{2}$
ways to arrange n objects on a ring with two views.

6. In how many ways can 8 different desserts be arranged in a circle
   around a floral centerpiece? 5,040

7. In how many ways can 8 different charms be arranged on a bracelet? 2,520

14.5, The number of combinations of n things taken r at a time is given by
14.6 $\binom{n}{r} = \dfrac{n!}{r!(n-r)!}$.

8. Find the number of lines determined by 7 points, A–G, on a circle. 21

9. From a group of 7 females and 6 males, in how many ways can you select a 5-member committee having exactly 3 females? 525

14.7 $P(E) = \dfrac{s}{t} = \dfrac{\text{number of successful outcomes}}{\text{total number of outcomes}}$; $P(\text{not } E) = 1 - P(E)$

The odds for E are s to u and the odds against E are u to s, where $u =$ the number of unsuccessful outcomes.

A playing card is drawn from a regular 52-card deck. Find the probability or odds.

10. $P(\text{red})$ $\frac{1}{2}$ **11.** $P(\text{not a club})$ $\frac{3}{4}$ **12.** odds for an ace 1 to 12   **13.** odds against a 10 12 to 1

14.8 $P(A \text{ or } B) = P(A) + P(B) - P(A \text{ and } B)$ for inclusive events A and B.
$P(A \text{ or } B) = P(A) + P(B)$ for mutually exclusive events A and B.
$P(A \text{ and } B) = P(A) \cdot P(B)$ for independent events A and B.

If a red die and a green die are tossed, find the probability of each event.

14. $g = 4 \text{ or } g \le 2$ $\frac{1}{2}$ **15.** $r > 4 \text{ or } g > 3$ $\frac{2}{3}$ **16.** $r \ge 4 \text{ and } g \le 3$ $\frac{1}{4}$

14.9 $P(A, \text{then } B) = P(A) \cdot P(B \mid A)$, where the sample space for B is reduced if A and B are dependent events.

A box contains 3 red tags and 6 green tags. A tag is drawn, replaced, and a second tag is drawn. Find the probability that the tags are as follows.

17. green, then red $\frac{2}{9}$ **18.** both red $\frac{1}{9}$ **19.** red and green $\frac{4}{9}$

14.10 The definitions of mean, median, mode, variance, and standard deviation are given on pages 529–531.

20. Find the mean, median, mode, variance, and standard deviation for the frequency distribution shown in the table at the right.

score	32	26	24	20	18
frequency	4	3	2	3	2

25, 25, 32, 26.71, 5.17

21. Describe in your own words how to find the mean of a set of scores.

14.11 For a normal distribution, 68.2% of the scores are between $M - \sigma$ and $M + \sigma$, and 95.4% of the scores are between $M - 2\sigma$ and $M + 2\sigma$.

22. The scores on a college entrance test are normally distributed with a mean M of 500 and a standard deviation σ of 100. What is the probability that a student selected at random will score from 500 to 700? 0.477

1. How many different automobile license plates can be made if each plate has two different letters followed by any four digits? 6,500,000

2. How many odd 4-digit numbers can be formed from the digits 1, 3, 4, 7, 8, if a digit may be repeated in a number? 375

3. How many numbers of one or more digits can be formed from the digits 4, 6, 7, if a digit may not be repeated in a number? 15

4. In how many ways can 3 different novels be placed together on a shelf at the right of 5 different biographies? 720

5. How many distinguishable 6-digit numbers can be formed by using all the digits in 747,475? 60

6. In how many ways can 6 people be seated around a circular table? 120

7. In how many ways can 5 different keys be arranged on a key ring? 12

8. Find the number of triangles determined by 6 points, A–F, on a circle. 20

9. How many tennis matches will be played by 6 people if each person plays each other person 3 times? 45

10. From a group of 9 males and 7 females, in how many ways can you select a 6-member committee having exactly 3 males? 2,940

11. How many different amounts of money are determined by selecting one or more bills from a box that contains one $1 bill, one $5 bill, one $10 bill, one $20 bill, and one $50 bill? 31

A playing card is drawn from a regular 52-card deck. Find the probability or odds.

12. P(black face card) $\frac{3}{26}$ 13. P(not a diamond) $\frac{3}{4}$ 14. odds for a club 1 to 3

A red die and a green die are tossed. Find the probability of each event.

15. $r > 4$ *or* $g < 3$ $\frac{5}{9}$ 16. $r > 4$ *and* $g < 3$ $\frac{1}{9}$ 17. $r + g = 11$ $\frac{1}{18}$

A bag contains 5 red tags and 6 green tags. A tag is drawn and replaced before a second tag is drawn. Find the probability of each event.

18. first is red, second is green $\frac{30}{121}$

19. one is green, the other is red $\frac{60}{121}$

20. Two cards are drawn *without* replacement from a deck of 52 playing cards. Find the probability that the two cards are a 9 and a 10. $\frac{8}{663}$

21. A 5-member committee is to be selected at random from a group of 3 boys and 7 girls. Find the probability that at least 4 of the 5 will be girls. $\frac{1}{2}$

In each item, compare a quantity in Column 1 with a quantity in Column 2. Write the letter of the correct answer from these choices.

A—The quantity in Column 1 is greater than the quantity in Column 2.
B—The quantity in Column 2 is greater than the quantity in Column 1.
C—The quantity in Column 1 is equal to the quantity in Column 2.
D—The relationship cannot be determined from the given information.

NOTE: Information centered over both columns refers to one or both of the quantities to be compared.

Sample Items and Answers

	Column 1	Column 2
S1.	$(7 - 7)!$	$7! - 7!$
	$n = 12$ and $r = 6$	
S2.	$\binom{n}{r-1}$	$\binom{n}{r+1}$

S1: The answer is A: $(7 - 7)! = 0!$, or 1; $7! - 7! = 0$, and $1 > 0$.

S2: The answer is C: If $n = 12$ and $r = 6$, then $\binom{n}{r-1} = \dfrac{12!}{5!7!}$, $\binom{n}{r+1} = \dfrac{12!}{7!5!}$, and $\dfrac{12!}{5!7!} = \dfrac{12!}{7!5!}$.

	Column 1	Column 2	
1.	$5! + 4!$	$5! \cdot 4!$	B
2.	$\binom{50}{0}$	$\binom{30}{30}$	C
3.	$\binom{7}{0} + \binom{7}{1} + \binom{7}{2}$	$\binom{7}{7} + \binom{7}{6} + \binom{7}{5}$	C

$n = 8$ and $r = 3$

	Column 1	Column 2	
4.	$\binom{n}{r-1}$	$\binom{n-1}{r}$	B
5.	number of triangles determined by 6 points on a circle	number of lines determined by 6 points on a circle	A
6.	$50! \times 30!$	$51! \times 29!$	B

n is an integer, $n > 4$

	Column 1	Column 2	
7.	$\binom{n}{2}$	$\binom{n}{5}$	D
8.	the number of ways to select 2 males or 2 females from a group of 6 males and 4 females	the number of ways to select 1 male and 1 female from a group of 6 males and 4 females	B

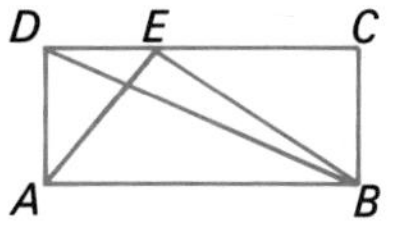

ABCD is a rectangle.

	Column 1	Column 2	
9.	area of $\triangle DBC$	area of $\triangle ABE$	C

10. $\dfrac{2y - 6}{y^2 - 36}$

22. $\dfrac{4a}{3a + 2b}$

27. $-1, \dfrac{1}{2}, \pm\sqrt{3}$

30.

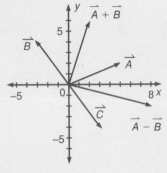

Cumulative Review (Chapters 1–14)

Solve the equation.

1. $3^{2n+2} = 243$ $\quad \frac{3}{2}$ **5.1**

2. $\dfrac{2x - 3}{x + 1} = \dfrac{x}{6}$ $\quad 2, 9$ **7.6**

3. $3t^2 - 2t + 2 = 0$ $\quad \dfrac{1 \pm i\sqrt{5}}{3}$ **9.5**

4. $3\sqrt{5x + 4} = 2\sqrt{10x + 14}$ $\quad 4$ **10.3**

5. $\log_b 64 = 3$ $\quad 4$ **13.3**

6. $\log_5 x + \log_5 (x - 1) = \log_5 2$ **13.5**
$\quad 2$

Find the solution set.

7. $-4x + 5 < x - 10$ or $\quad$ **2.2**
$7 < 2x - 3$ $\quad \{x \mid x > 3\}$

8. $a^2 + 2a - 15 < 0$ $\quad$ **6.4**
$\{a \mid -5 < a < 3\}$

Simplify.

9. $\left(\dfrac{5a^{-3}}{-2b^{-2}}\right)^2$ $\quad \dfrac{25b^4}{4a^6}$ **5.2**

10. $\dfrac{7y}{y^2 - 36} - \dfrac{2}{y + 6} + \dfrac{3}{6 - y}$ **7.4**

11. $2a\sqrt[3]{24a^{12}b^5}$, $b \geq 0$ $\quad 4a^5b\sqrt[3]{3b^2}$ **8.5**

12. $2\sqrt{-3} \cdot 4\sqrt{-8}$ $\quad -16\sqrt{6}$ **9.2**

13. $\dfrac{4}{3 + 4i}$ $\quad \dfrac{12 - 16i}{25}$

14. $\log_2 32$ $\quad 5$ **13.3**

15. $\log_2 x + 3 \log_2 y$ $\quad \log_2 xy^3$ **13.6**

For Exercises 16–20, use the relations
$C = \{(2,6), (3,5), (4,4), (5,3)\}$ **and**
$D = \{(-2,6), (6,-2), (3,3), (6,6)\}$.

16. Find the domain of C. $\{2, 3, 4, 5\}$ **3.1**

17. What is the range of D? $\{-2, 3, 6\}$

18. Is C a function? Yes

19. Is D a function? No

20. Are C and D a pair of **13.1**
inverse functions? No

21. Solve the system for (x,y). **4.3**
$5x - 2y = 24$
$2x + 3y = 2$ $(4, -2)$

22. Solve $a(3x + 1) = 5a - 2bx$ **5.4**
for x.

23. Factor $ax^3 - 27a$ completely. **5.7**
$a(x - 3)(x^2 + 3x + 9)$

For Exercises 24–26, $f(x) = 3x + 2$ and
$g(x) = x^2 - 4$. **Find the following.**

24. $g(f(2))$ $\quad 60$ **5.9**

25. $f(g(c + 1))$ $\quad 3c^2 + 6c - 7$

26. $\dfrac{f(a) - f(b)}{a - b}$ $\quad 3$

27. Solve $2x^4 + x^3 - 7x^2 - 3x + 3 = 0$, using the Integral Zero **10.5**
and Rational Zero theorems.

28. Convert 4 oz/in^2 to pounds per **7.3**
square foot. 36 lb/ft^2

29. Find the maximum value of P **11.6**
if $P = -5x^2 + 20x + 7$. $\quad 27$

30. On a coordinate plane, draw the **9.7**
vectors $\vec{A} = (5,2)$, $\vec{B} = (-3,4)$,
and $\vec{C} = -\vec{B}$. Use the
parallelogram method to draw
$\vec{A} + \vec{B}$ and $\vec{A} - \vec{B}$.

For Exercises 31–35, use $A(-2,3)$,
$B(6,-1)$, and $C(4,5)$. **Find the**
following.

31. distance between A and B, in **11.1**
simplest radical form $\quad 4\sqrt{5}$

32. coordinates of the midpoint
of $\overline{BC}$ $\quad (5,2)$

33. slope and slant of $\overleftrightarrow{AB}$ **3.3**
$\quad -\frac{1}{2}$; down to right

34. equation of the horizontal line   *3.4*
through A $y = 3$

35. equation of the line parallel to   *3.7*
$\overleftrightarrow{AB}$ and passing through C $x + 2y = 14$

For points A, B, and C in Exercises 31–35, prove the following.

36. $\triangle ABC$ is an isosceles   *11.2*
triangle. $AC = BC = 2\sqrt{10}$

37. $\overline{AC}$ is perpendicular to $\overline{BC}$.   *3.7*
$\frac{2}{6}\left(-\frac{6}{2}\right) = -1$

Name the conic section.   *12.8*

38. $4x^2 - 9y^2 = 36$ Hyperbola

39. $9x^2 + 4y^2 = 36$ Ellipse

40. $4x^2 + 4y^2 = 36$ Circle

41. $9y = 9x^2 + 36$ Parabola

42. $4xy = 36$ Rectangular hyperbola

43. If y varies directly as the square   *3.8*
of x, and $y = 36$ when $x = 8$,
find y when $x = 4$.   9

44. M varies inversely as d. Find   *12.5*
M when $d = 1.8$, if $M = 42$
when $d = 2.4$.   56

45. If y varies jointly as x and $\sqrt{z}$,   *12.6*
and $y = 44$ when $x = 6$ and
$z = 6.25$, find z when
$y = 22$ and $x = 5$.   2.25

For Exercises 46–48, write the equation for A^{-1}, the inverse of A, solving for y, and determine whether A^{-1} is a function.

46. $A{:}y = 4x - 8$ $y = \frac{1}{4}x + 2$, yes   *13.1*

47. $A{:}y = x^2 + 5$ $y = \pm\sqrt{x - 5}$, no

48. $A{:}y = 3^x$ $y = \log_3 x$, yes   *13.3*

49. Rico and José entered a non-   *2.6*
stop, cross-country cycling
race. Rico began the race at
6:30 A.M. and averaged 34 mi/h.
José began the race at 7:00
A.M. and averaged 36 mi/h.
At what time did José pass
Rico on the race course?   3:30 P.M.

50. Find three consecutive multiples   *6.2*
of 4 such that the square of the
second number is 32 more than
7 times the third number.   8, 12, 16

51. A triangle's height is 4 cm less   *6.3*
than 3 times the length of the
base. Find the height if the
area of the triangle is 16 cm^2.   8 cm

52. The product of two numbers is   *9.6*
4, and the second number is 4 less
than the first number. Find two such
pairs of numbers.
$(2 + 2\sqrt{2}, -2 + 2\sqrt{2})$, $(2 - 2\sqrt{2}, -2 - 2\sqrt{2})$

53. The perimeter of a rectangle is   *12.9*
28 ft and the area is 48 ft^2. Find
the length and the width of the
rectangle.   8 ft, 6 ft

54. How many odd numbers of one   *14.2*
or more digits can be formed from
the digits 1, 2, 3, 4, 5, if no digit
may be repeated in a number?   195

55. How many different 7-digit   *14.3*
numbers can be formed using all the
digits in 3,838,835?   140

56. How many different 4-member   *14.5*
committees can be selected if 12
people are available for selection?   495

57. Find the probability of drawing   *14.7*
a red face card in one draw from a
deck of 52 playing cards. $\frac{3}{26}$

15 SEQUENCES, SERIES, BINOMIAL THEOREM

OVERVIEW

Chapter 15 introduces students to sequences and series, especially those that are either arithmetic or geometric. A special variety of arithmetic sequence, covered in Lesson 15.2, is called the *harmonic sequence*.

In the first six lessons, attention is confined to sequences and series that are finite. Lesson 15.7 considers infinite geometric series. This lesson distinguishes between those series that converge and those that do not. The chapter ends with a lesson on the binomial theorem.

OBJECTIVES

- To find a specified term in an arithmetic, geometric, or harmonic sequence
- To find means between two terms in an arithmetic, geometric, or harmonic sequence
- To find the sum of an arithmetic, geometric, or infinite geometric series
- To solve word problems that involve arithmetic or geometric series
- To expand a positive integral power of a binomial

PROBLEM SOLVING

Although no lesson is specifically identified as a problem-solving lesson, the strategy *Making a Table to Find a Pattern* should be used by students to solve many of the problems in this chapter. The *Mixed Problem Solving* exercises on page 579 give students an opportunity to practice various problem-solving strategies.

READING AND WRITING MATH

The introduction of sigma-notation occurs on page 555. Students are encouraged to verbalize how the symbol should be read. Two exercises in this chapter ask students to discuss a mathematical concept in their own words (Exercise 26 on page 565 and Exercise 6 on page 580).

TECHNOLOGY

Calculator: Some calculators enable students to use a constant difference or constant ratio by the simple procedure of pressing the appropriate operation key *twice* rather than once. For example, to calculate five doublings of one dollar (as in a compound interest problem), one would execute the following keystrokes:
$2 \times \times = = = =$. (2 represents the first doubling. The second $\times$ and the four equality symbols produce the last four doublings.)

SPECIAL FEATURES

Mixed Review pp. 548, 552, 557, 565, 569, 574, 577
Focus on Reading p. 551
Midchapter Review p. 561
Brainteaser p. 565
Extension: Mathematical Induction p. 570
Statistics: Expected Value p. 571
Application: Quality-Control Charts p. 578
Mixed Problem Solving p. 579
Key Terms p. 580
Key Ideas and Review Exercises pp. 580–581
Chapter 15 Test p. 582
College Prep Test p. 583

PLANNING GUIDE

Lesson		Basic	Average	Above Average	Resources
15.1	pp. 547–548	CE all WE 1–25 odd	CE all WE 7–31 odd	CE all WE 9–33 odd	Reteaching 111 Practice 111
15.2	pp. 551–552	FR all CE all WE 1–21 odd	FR all CE all WE 5–25 odd	FR all CE all WE 7–27 odd	Reteaching 112 Practice 112
15.3	pp. 556–557	CE all WE 1–25 odd	CE all WE 9–31 odd	CE all WE 11–35 odd	Reteaching 113 Practice 113
15.4	pp. 560–561	CE all WE 1–21 odd Midchapter Review	CE all WE 7–29 odd Midchapter Review	CE all WE 9–31 odd Midchapter Review	Reteaching 114 Practice 114
15.5	pp. 564–565	CE all WE 1–21 odd Brainteaser	CE all WE 9–31 odd Brainteaser	CE all WE 11–37 odd Brainteaser	Reteaching 115 Practice 115
15.6	pp. 568–571	CE all WE 1–25 odd Extension Statistics	CE all WE 5–31 odd Extension Statistics	CE all WE 11–35 odd Extension Statistics	Reteaching 116 Practice 116
15.7	pp. 573–574	CE all WE 1–17 odd	CE all WE 5–23 odd	CE all WE 7–23 odd, 24	Reteaching 117 Practice 117
15.8	pp. 577–579	CE all WE 1–15 odd Application Mixed Problem Solving	CE all WE 7–25 odd Application Mixed Problem Solving	CE all WE 11–27 odd Application Mixed Problem Solving	Reteaching 118 Practice 118
Chapter 15 Review pp. 580–581		all	all	all	
Chapter 13 Test p. 582		all	all	all	
College Prep Test p. 583		all	all	all	

CE = Classroom Exercises   WE = Written Exercises   FR = Focus on Reading

NOTE: For each level, all students should be assigned all Mixed Review exercises.

■ INVESTIGATION

Using their calculators, the students explore a property of the Fibonacci sequence.

The first few terms of the Fibonacci sequence are 1, 1, 2, 3, 5, 8. Ask students whether this is (a) an arithmetic sequence (Lesson 15.1), (b) a geometric sequence (Lesson 15.4), or (c) neither.
(c) Neither. There is neither a constant difference nor a constant ratio between terms.

Ask students to extend the sequence to higher values by continuing to add successive terms to produce each new term.

1, 1, 2, 3, 5, 8, 13, 21, 34, 55, 89, 144, 233, 377, 610, 987, 1,597, 2,584, 4,181, 6,765, 10,946, 17,711, 28,657, 46,368, 75,025 are the first 25 terms.

Ask students to calculate the ratio between successive terms of the sequence for the first few terms of the sequence. Ask them if they observe a pattern.

The first few ratios of successive terms are 1, 2, 1.5, 1.667, 1.6, 1.625, 1.615, 1.619, 1.618, 1.618, 1.618, · · ·.

Ask students if they recognize the value 1.618. Some students may recognize this value as the golden ratio.

Ask the students to compute the ratio of the 25th to the 24th term of the sequence and compare this with the value of the golden ratio, $(1 + \sqrt{5}) \div 2$. $75,025 \div 46,368 = 1.618033989$. This number agrees with the value of the golden ratio to all 9 decimal places.

Have students discuss the nature of the Fibonacci sequence in light of this result. The Fibonacci sequence behaves very much like a geometric sequence for larger values of its terms.

This "surreal" image of a flower was created using the drawing capability of the computer. Artists can sketch their designs, then *paint* the figure with any color imaginable. Today's computers have color capacities of more than a million hues.

More About the Image

Computers give artists immeasurable versatility for creating fine-art images. Software programs provide millions of color gradients as well as the texturing effects of various types of paint or coloring tools. For example, the artist can choose to apply color with chalk, oil paint, colored pencils, or any other tool—and the special physical characteristics of each one is true to life. These paint tools are pressure sensitive so that the artist can control the saturation of color. Some programs provide the artist with effects for background. Textured paper can be selected that reacts to the paint tools like the actual paper.

15.1 Arithmetic Sequences

Teaching Resources

Quick Quizzes 111
Reteaching and Practice
   Worksheets 111

Objectives   To write several consecutive terms in an arithmetic sequence
To find a specified term in an arithmetic sequence

The list of numbers 1, 1, 2, 3, 5, 8, 13, 21, . . . is called a *Fibonacci sequence*. This sequence can be used to describe certain structures in nature, such as the distribution of leaves around a stem. A **sequence** (or *progression*) is an ordered list of numbers. The numbers in the list are called *terms* of the sequence. Often, as with the Fibonacci sequence, a pattern can be observed in the sequence. Add any two consecutive terms of the Fibonacci sequence. What do you observe?

Notice the pattern in the infinite sequences below.

(1) 23, 28, 33, 38, . . .
   Add 5 to any term to obtain
   the next term.

(2) 4, -2, -8, -14, . . .
   Add -6 to any term to obtain
   the next term.

Each sequence above is an *arithmetic sequence*. Notice that the same number is added to each term to obtain the next term. The number added is called a *common difference*. To find the common difference, subtract any term from the one that follows it.

$$\text{In (1), } 5 = 28 - 23 = 33 - 28 = 38 - 33 = \ldots$$

$$\text{In (2), } -6 = -2 - 4 = -8 - (-2) = -14 - (-8) = \ldots$$

A sequence such as $\frac{1}{2}$, 1, 2, 4, 8, . . . is *not* an arithmetic sequence because there is no common difference. For example, $1 - \frac{1}{2} \neq 2 - 1$.

Definitions

The sequence a_1, $a_1 + d$, $a_1 + 2d$, $a_1 + 3d$, $a_1 + 4d$, . . .
is an **arithmetic sequence** with first term a_1 and **common difference** d.
If $x_1, x_2, x_3, x_4, \ldots$ is an arithmetic sequence with common difference d, then $d = x_2 - x_1 = x_3 - x_2 = \ldots$.

EXAMPLE 1   For each sequence that is arithmetic, find the common difference d.

   a. 6, 8, 11, 15, 20, . . .

   b. -7, -2, 3, 8, 13, . . .

Solutions   **a.** Since $8 - 6 \neq 11 - 8$, the sequence is not arithmetic.

   b. Since $-2 - (-7) = 3 - (-2) = 8 - 3 = 13 - 8 = 5$, the sequence is arithmetic, with $d = 5$.

GETTING STARTED

Prerequisite Quiz

Find the value of $a + (n - 1)d$ for the given values of *a*, *n*, and *d*.

1. $a = -5$, $n = 36$, $d = 2$ 65
2. $a = 30$, $n = 26$, $d = -10$ -220
3. $a = 2.5$, $n = 51$, $d = 1.2$ 62.5

Solve for *n*.

4. $135 = 5 + (n - 1)10$ 14
5. $-68 = -4 + (n - 1)(-2)$ 33

Motivator

Have students write a sequence of numbers in which a constant term is added to each term to obtain the next term. Then have students exchange papers with a partner. Ask students to find the common difference.

Additional Example 1

If the sequence is arithmetic, find the common difference *d*.

a. 3, 6, 12, 24, · · ·   Not arithmetic
b. 8.1, 7.6, 7.1, 6.6, · · ·   arithmetic, $d = -0.5$

Highlighting the Standards

Standard 2c: The mathematics of arithmetic sequences provides another good opportunity for class discussion. The basic ideas are simple, but the applications are varied.

Lesson Note

This lesson on arithmetic sequences provides the foundation for lessons on arithmetic means, harmonic sequences and means, and arithmetic series. Notice that the first term in a sequence can be represented either by a_1 or by a, the 6th term by a_6, and the nth term by a_n.

Encourage students to use a calculator, preferably a scientific calculator, to avoid time-consuming calculations for the exercises in this chapter. When the word *arithmetic* is an adjective, as in *arithmetic sequence*, it is pronounced arith-MEH-tik.

Math Connections

Fibonnaci Sequence: The problem that originally gave rise to the Fibonnaci sequence, which is given in the text, is to determine how many pairs of rabbits will be produced in a year beginning with a single pair, if, in every month, each pair bears a new pair that reproduces beginning in the second month. In 1202, the problem appeared in the classic *Liber Abaci* (book of the abacus) written by Leonardo of Pisa ("Fibonnaci").

Critical Thinking Questions

Analysis: Ask students to think of the simplest arithmetic sequence that they can. The simplest sequence is the set of natural, or counting numbers. The set of whole numbers (the set consisting of the natural numbers and zero) could also be chosen.

EXAMPLE 2   Write the next three terms of the arithmetic sequence 11, 3, −5, −13, ...

Solution   The common difference $d = 3 - 11$, or -8. When -8 is added to each term to obtain the next term, the next three terms are found to be $-21, -29, -37$.

EXAMPLE 3   Write the first four terms of the arithmetic sequence whose first term a_1 is 4 and whose common difference d is -5.

Solution

2nd term:	3rd term:	4th term:
$4 + (-5) = -1$	$-1 + (-5) = -6$	$-6 + (-5) = -11$

Thus, the first four terms are $4, -1, -6, -11$.

In an arithmetic sequence, you may need to find a specific term. For example, find the 71st term in the sequence 23, 28, 33, 38, ... To find the term, it is helpful to rewrite the sequence as follows.

23	28	33	38	. . .	?
$23 + 0 \cdot 5$	$23 + 1 \cdot 5$	$23 + 2 \cdot 5$	$23 + 3 \cdot 5$	. . .	$23 + 70 \cdot 5$
1st term	2nd term	3rd term	4th term	. . .	71st term

The 71st term is $23 + (71 - 1) \cdot 5$, or 373. This suggests the following formula for the nth term of an arithmetic sequence.

---

nth term of an arithmetic sequence

The nth term of an arithmetic sequence is given by

$$a_n = a_1 + (n - 1)d,$$

where a_n is the nth term ($n \geq 1$), a_1 is the first term, and d is the common difference.

EXAMPLE 4   Find the specified term of each arithmetic sequence.

a. 26th term of $-5, -2, 1, 4, \ldots$

b. 41st term of $7, 4.4, 1.8, -0.8, \ldots$

Solutions

a. $a_1 = -5, n = 26, d = 3$
$a_n = a_1 + (n - 1)d$
$\quad = -5 + (26 - 1)3$
$\quad = 70$
The 26th term is 70.

b. $a_1 = 7, n = 41, d = -2.6$
$a_n = a_1 + (n - 1)d$
$\quad = 7 + (41 - 1)(-2.6)$
$\quad = -97$
The 41st term is -97.

546   Chapter 15 Sequences, Series, Binomial Theorem

Additional Example 2

Write the next three terms of the arithmetic sequence.

$5\frac{1}{2}, 5\frac{1}{3}, 5\frac{1}{6}, 5, \cdots$ $4\frac{5}{6}, 4\frac{2}{3}, 4\frac{1}{2}$

Additional Example 3

Write the first four terms of the arithmetic sequence whose first term a is $-3\frac{1}{2}$ and whose common difference d is $2\frac{1}{2}$.

$-3\frac{1}{2}, -1, 1\frac{1}{2}, 4$

Additional Example 4

Find the specified term of each arithmetic sequence.

a. 21st term of $7, 3, -1, -5, \cdots$ -73
b. 46th term of $-6, -4.6, -3.2, -1.8, \cdots$ 57

You can also use the formula $a_n = a_1 + (n - 1)d$ to find which term a given number is in an arithmetic sequence. For example, which term is 189 in the sequence 9, 15, 21, 27, . . . ? In this case, $a_n = 189$, $a_1 = 9$, and $d = 6$.

Solve for n in the formula $a_n = a_1 + (n - 1)d$.

$189 = 9 + (n - 1)6; \quad 180 = (n - 1)6; \quad 30 = n - 1; \quad n = 31$

So, 189 is the 31st term of the sequence.

EXAMPLE 5 A computer programmer's starting salary is $25,500 with a guaranteed minimum annual increase of $1,100. What minimum annual salary should be expected for the twelfth year?

Plan Note that the salaries form an arithmetic sequence, 25,500, 26,600, 27,700, . . . , with 1,100 as the common difference. Use the formula $a_n = a_1 + (n - 1)d$ to find the 12th term.

Solution $a_1 = 25,500 \qquad n = 12 \qquad d = 1,100$

$a_n = a_1 + (n - 1)d = 25,500 + (12 - 1)1,100 = 37,600$

The minimum salary is $37,600 for the twelfth year.

Classroom Exercises

Match each arithmetic sequence with exactly one statement at the right.

1. 6, 4, 2, . . . b   2. 4, 5, 6, . . . d   a. $d = 6$ c. Sixth term is 12.
3. 3, 9, 15, . . . a   4. $-8, -4, 0, . . .$ c   b. First term is 6.   d. Third term is 6.

For each sequence that is arithmetic, find the common difference and the next two terms.

5. 3, 8, 13, 18, . . .   6. $-4, -2, 0, 2, . . .$ 7. 1, 5, 25, 125, . . .
   $d = 5$; 23, 28 $d = 2$; 4, 6   Not arithmetic

Written Exercises

For each sequence that is arithmetic, find the common difference d.

1. 33, 45, 57, 69, . . . 12   2. $35, 22, 9, -4, . . . \; -13$ 3. $-11, -9, -7, -5, . . . \; 2$
4. 5, 10, 20, 40, . . .   5. $8\frac{1}{2}, 10\frac{1}{4}, 12, . . . \; 1\frac{3}{4}$ 6. $-3, 6, -12, . . .$
   Not arithmetic   Not arithmetic

Find the next three terms of each arithmetic sequence.

7. $8, 5, 2, -1, . . .$ 8. 21, 22.3, 23.6, . . .   9. $\frac{1}{4}, \frac{1}{3}, \frac{5}{12}, \frac{1}{2}, . . . \; \frac{7}{12}, \frac{2}{3}, \frac{3}{4}$
   $-4, -7, -10$ 24.9, 26.2, 27.5

15.1 Arithmetic Sequences **547**

Checkpoint

1. Find the next three terms of the arithmetic sequence 7.2, 8.4, 9.6, · · ·. 10.8, 12, 13.2
2. Find the 53rd term of the arithmetic sequence $-17, -15.5, -14, · · ·$. 61
3. An object in a free fall travels 16 ft in the 1st second, 48 ft in the 2nd second, 80 ft in the 3rd second, and so on. How far will it fall in the 8th second? 240 ft

Closure

Ask students to give the formula for finding the nth term of an arithmetic sequence. $a_n = a_1 + (n - 1)d$ Have students write several terms of arithmetic sequences and find specified terms in such sequences. Have students create an arithmetic sequence by writing its first three terms. Then have them find a_1, d, and a_{21} for the sequence.

■■■FOLLOW UP

Guided Practice

Classroom Exercises 1–7

Independent Practice

Ⓐ Ex. 1–21, Ⓑ Ex. 22–31, Ⓒ Ex. 32, 33

Basic: WE 1–25 odd
Average: WE 7–31 odd
Above Average: WE 9–33 odd

Additional Example 5

For the first three months of 1990, the beginning balances in Paula's savings account were $1,500, $1,575, and $1,650. If her balance increased by the same amount each month, what was her balance on December 1, 1991? $2,325

Additional Answers

Written Exercises

28. $-5 + 10\sqrt{2}$
29. $25 + 18\sqrt{3}$

Mixed Review

4.

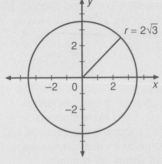

$r = 2\sqrt{3}$

Write the first four terms of the arithmetic sequence, given its first term a_1 and common difference d.

10. $a_1 = 23, d = 7$ 23, 30, 37, 44   **11.** $a_1 = 11, d = -5$ 11, 6, 1, -4 **12.** $a_1 = -8, d = -3$ $-8, -11, -14, -17$

13. $a_1 = 4, d = 2.6$ 4, 6.6, 9.2, 11.8   **14.** $a_1 = 7°C, d = 5°C$ 7°C, 12°C, 17°C, 22°C   **15.** $a_1 = 14$ oz, $d = -3\frac{1}{2}$ oz   14 oz, $10\frac{1}{2}$ oz, 7 oz, $3\frac{1}{2}$ oz

Find the specified term of each arithmetic sequence.

16. 26th term of 12, 16, 20, . . .   112   **17.** 21st term of 9, 3, -3, -9, . . . -111

18. 31st term of 9.3, 9, 8.7, . . .   0.3   **19.** 36th term of -8, -6.6, -5.2, . . .   41

20. 83rd term of 6, $6\frac{1}{2}$, 7, . . .   47   **21.** 77th term of $11\frac{4}{5}$, $10\frac{2}{5}$, 9, . . . $-94\frac{3}{5}$

22. Some cartons are stacked with 4 in the top row, 7 in the 2nd row, 10 in the 3rd row, and so on. How many cartons are in the 16th row?   49

23. A parachutist in a free fall travels 5 m in the 1st second, 15 m in the 2nd second, 25 m in the 3rd second, and so on. How far will he travel in the 9th second?   85 m

24. Mrs. Cooper deposited $300 in a bank on January 2 and each month after that she deposited $45 more than the preceding month. What was her deposit on December 2? $795

25. A landscaper's starting salary is $17,500. If he is guaranteed a minimum annual increase of $800, what minimum salary should he expect for his 10th year? $24,700

For each sequence that is arithmetic, find the indicated term.
(Exercises 26–31)

26. $2 + i$, 5, $8 - i$, . . . (fourth) $11 - 2i$ **27.** $\sqrt{11}$, $\sqrt{13}$, $\sqrt{15}$, $\sqrt{17}$, . . . (fifth)   Not arithmetic

28. 5, $4 + \sqrt{2}$, $3 + 2\sqrt{2}$, . . . (eleventh)   **29.** $6 - \sqrt{3}$, 7, $8 + \sqrt{3}$, . . . (twentieth)

30. Find n if 128 is the nth term of 4, 8, 12, 16, ...   32   **31.** Which term of -5, -2, 1, 4, . . . is 61?   23rd

32. Find the value of c so that $c + 4$, $4c + 1$, $8c - 4$ is an arithmetic sequence.   2

33. Find the first term of the arithmetic sequence whose 8th term is 29 and 20th term is 65.   8

Mixed Review

1. Simplify $5a^2(3a^2bc^3)^4$.   *5.1* $405a^{10}b^4c^{12}$

2. Express the value of $\dfrac{2.3 \times 10^{-12}}{4.6 \times 10^{-14}}$ in ordinary notation.   *5.2*   50

3. Find the maximum value of y if $y = -2x^2 + 12x + 9$.   *9.4, 11.6*   27

4. Graph the circle described by $4x^2 + 4y^2 = 48$.   *12.1*

Enrichment

Write these sequences of prime numbers on the chalkboard.

2, 3, 5 · · ·
2, 7, 13 · · ·
2, 11, 19 · · ·

Have the students determine that they are *not* arithmetic sequences. Then ask if they can find an arithmetic sequence of three numbers such that the first is the number 2, and the second and third are prime numbers. If they decide that this is not possible, have them explain why.

Not possible. The third number would have to be $2 + 2d$, or $2(1 + d)$. Thus, it would contain the factor 2 and would not be prime.

15.2 Arithmetic Means and Harmonic Sequences

Objectives

To find arithmetic means between two given numbers
To find *the* arithmetic mean of two numbers
To find a given term and harmonic means in a harmonic sequence

A sequence with a specific number of terms is a *finite* sequence. Thus, the arithmetic sequence 2, 5, 8, 11, 14, is a *finite* sequence of five terms. The terms between the first and last terms of an arithmetic sequence are called the **arithmetic means** between those two terms, as illustrated below.

2, 5, 8, 11, 14

arithmetic means

You can find any number of arithmetic means between two numbers.

EXAMPLE 1   Find the four arithmetic means between 36 and 32.

Plan   Draw 4 blanks for the means: 36, ____, ____, ____, ____, 32.

The 4 means and the given numbers will form a 6-term arithmetic sequence with $a_1 = 36$, $n = 6$, and $a_6 = 32$.

Use the formula $a_n = a_1 + (n - 1)d$ to find d, the common difference.

Solution

$$a_n = a_1 + (n - 1)d$$
$$32 = 36 + (6 - 1)d$$
$$-4 = 5d$$
$$-0.8 = d$$

Add -0.8 to each term to obtain the next term in the sequence.

36, 35.2, 34.4, 33.6, 32.8, 32

The four means are 35.2, 34.4, 33.6, and 32.8.

The procedure above can be used to find the *one* arithmetic mean between two numbers. This expression is the mean, or the average, of two numbers x and y as defined in Lesson 14.10.

The **arithmetic mean** *of* x *and* y *is* $\dfrac{x + y}{2}$.

Teaching Resources

Quick Quizzes 112
**Reteaching and Practice
Worksheets** 112

GETTING STARTED

Prerequisite Quiz

1. Write the next four terms in the arithmetic sequence with $a_1 = 28$ and $d = -0.6$.
   27.4, 26.8, 26.2, 25.6

Find the average (mean) for each pair of numbers.

2. 18 and 6   12
3. −7 and 15   4
4. 23 and 26   24.5, or $24\frac{1}{2}$
5. Find the 21st term in the sequence 2, 6, 10, 14, · · ·.   82

Motivator

Have students use the formula $a_n = a_1 + (n - 1)d$ to find test scores x and y in the list 83%, x, y, 95% so that the four test scores form an arithmetic sequence.
87%, 91%

Additional Example 1

Find three arithmetic means between 17 and 27.   19.5, 22, 24.5

Highlighting the Standards

Standard 12a: The Focus on Reading and Classroom Exercises build skills and concepts related to discrete mathematics.

Lesson Note

Encourage students to draw an appropriate number of blanks when inserting two or more means between two given numbers, as shown in Examples 1 and 4. This procedure helps in counting the number of terms n and in writing the terms after finding d.

The *one* arithmetic mean between two numbers is *the* arithmetic mean (average) of the two numbers. To find the arithmetic mean between two numbers x and y, students should find $\frac{x + y}{2}$ rather than use the method of Example 1.

Math Connections

Harmonic Series: Arithmetic and geometric series are studied in Lessons 15.3, 15.6, and 15.7. Harmonic series also are of some interest. For example, it is known that the harmonic series $\frac{1}{1} + \frac{1}{2} + \frac{1}{3} + \frac{1}{4} + \cdots$ diverges (does not exist as a real number) but that the same series with alternating signs converges as shown below.

$$\frac{1}{1} - \frac{1}{2} + \frac{1}{3} - \frac{1}{4} + \frac{1}{5} - \cdots = 0.693 \ldots$$

EXAMPLE 2 The arithmetic mean of two numbers is 15. If the greater number is decreased by 6 times the smaller number, the result is 2. Find the two numbers.

Plan Write and solve a system of two equations.

Solution Let x = the smaller number and y = the greater number.

The arithmetic mean of x and y is 15.   (1) $\frac{x + y}{2} = 15$

If y is decreased by 6 times x, the result is 2.   (2) $y - 6x = 2$

(1) $x + y = 30$ (1) $x + (6x + 2) = 30$
(2) $y = 6x + 2$ $7x = 28$
                            $x = 4$

(2) $y = 6 \cdot 4 + 2$
    $y = 26$

So, the two numbers are 4 and 26. The check is left for you.

A sequence is *harmonic* if the reciprocals of its terms form an arithmetic sequence. Harmonic sequences have applications in music theory. Two examples are shown below.

Harmonic sequence	Corresponding arithmetic sequence
(1) $\frac{1}{2}, \frac{1}{3}, \frac{1}{4}, \frac{1}{5}$	2, 3, 4, 5
(2) $4, \frac{8}{3}, 2$	$\frac{1}{4}, \frac{3}{8}, \frac{1}{2}$, or $\frac{2}{8}, \frac{3}{8}, \frac{4}{8}$

In the second example above, $\frac{8}{3}$ is the **harmonic mean** of 4 and 2, while its reciprocal, $\frac{3}{8}$, is the arithmetic mean of $\frac{1}{4}$ and $\frac{1}{2}$.

Definition   If $x_1, x_2, x_3, \ldots$ is an arithmetic sequence with no term equal to zero, then $\frac{1}{x_1}, \frac{1}{x_2}, \frac{1}{x_3}, \ldots$ is a **harmonic sequence**.

EXAMPLE 3 Find the 12th term in the harmonic sequence $\frac{1}{3}, \frac{1}{9}, \frac{1}{15}, \ldots$.

Plan First, find the 12th term in the corresponding arithmetic sequence 3, 9, 15, ... Then find the reciprocal of that term.

Solution For the 12th term in 3, 9, 15, . . . , $a_1 = 3$, $n = 12$, and $d = 6$.
So, $a_n = a_1 + (n - 1)d = 3 + (12 - 1)6$, or 69.

Thus, the 12th term in the harmonic sequence is $\frac{1}{69}$.

550   Chapter 15 Sequences, Series, Binomial Theorem

Additional Example 2

The arithmetic mean of two numbers is 2. If the greater number is increased by twice the smaller number, the result is −4. Find both numbers.   −8 and 12

Additional Example 3

Find the 16th term in the harmonic sequence $4, \frac{8}{3}, 2, \cdots$. $\frac{8}{17}$

EXAMPLE 4   Find the three harmonic means between 2 and $\frac{1}{3}$.

Plan   Find three arithmetic means between $\frac{1}{2}$ and 3: $\frac{1}{2}$, ____, ____, ____, 3.

Solution   Use $a_n = a_1 + (n-1)d$: $3 = \frac{1}{2} + (5-1)d$, $\frac{5}{2} = 4d$, and $d = \frac{5}{8}$.

Arithmetic sequence
$$\underbrace{\frac{4}{8}, \frac{9}{8}, \frac{14}{8}, \frac{19}{8}, \frac{24}{8}}_{} \quad \text{or} \quad \frac{1}{2}, \underbrace{\frac{9}{8}, \frac{7}{4}, \frac{19}{8}}_{\text{means}}, 3$$

Harmonic Sequence
$$2, \underbrace{\frac{8}{9}, \frac{4}{7}, \frac{8}{19}}_{\text{means}}, \frac{1}{3}$$

The three harmonic means are $\frac{8}{9}$, $\frac{4}{7}$, and $\frac{8}{19}$.

EXAMPLE 5   Find *the* harmonic mean of 4 and $\frac{3}{2}$.

Plan   Find the arithmetic mean of $\frac{1}{4}$ and $\frac{2}{3}$.

Solution $\dfrac{x + y}{2} = \dfrac{\frac{1}{4} + \frac{2}{3}}{2} = \dfrac{\frac{11}{12}}{2} = \dfrac{11}{24}$

Thus, the harmonic mean of 4 and $\frac{3}{2}$ is $\frac{24}{11}$.

Focus on Reading

Identify each sequence as arithmetic or harmonic.

1. 5, 8, 11, 14 Arithmetic
2. $\frac{1}{5}, \frac{1}{8}, \frac{1}{11}, \frac{1}{14}$ Harmonic
3. $-6, 2, 10$ Arithmetic

4. $\frac{2}{3}, 1, \frac{4}{3}, \frac{5}{3}$ Arithmetic
5. $\frac{1}{4}, \frac{1}{7}, \frac{1}{10}$ Harmonic
6. $\frac{1}{2}, \frac{5}{16}, \frac{1}{8}$ Arithmetic

7. $\sqrt{1}, \sqrt{4}, \sqrt{9}, \sqrt{16}$ Arithmetic
8. 0.2, 0.9, 1.6 Arithmetic
9. 0.2, 0.5, -1, -0.25 Harmonic

Classroom Exercises

In each arithmetic sequence, find the missing arithmetic means between the first and last terms.

1. 11, _8_, 5, 2
2. 4, _6_, 8, _10_, _12_, 14
3. 5, _1_, _−3_, −7

Find *the* arithmetic mean of each pair of numbers.

4. 5 and 9 7
5. 30 and 20 25
6. 9 and −5 2

Find the arithmetic sequence related to the given harmonic sequence.

7. $\frac{1}{8}, \frac{1}{10}, \frac{1}{12}, \ldots$
8, 10, 12, . . .

8. $2, \frac{12}{5}, 3, \ldots$
$\frac{6}{12}, \frac{5}{12}, \frac{4}{12}, \ldots$

9. $\frac{2}{5}, \frac{4}{11}, \frac{1}{3}, \ldots$
$\frac{10}{4}, \frac{11}{4}, \frac{12}{4}, \ldots$

15.2 Arithmetic Means and Harmonic Sequences   **551**

Guided Practice

Classroom Exercises 1–9

Independent Practice

A Ex. 1–20, **B** Ex. 21–25, **C** Ex. 26, 27

Basic: FR, WE 1–21 odd
Average: FR, WE 5–25 odd
Above Average: FR, WE 7–27 odd

Additional Answers

Written Exercises

24. $2 - 4i$, $4 - 3i$, $6 - 2i$, $8 - i$

26. $x, \_\_\_, y$: $y = x + 2d$, $d = \dfrac{y - x}{2}$,

$x + \dfrac{y - x}{2} = \dfrac{2x + y - x}{2} = \dfrac{x + y}{2}$

Written Exercises

Find the indicated number of arithmetic means between each pair of numbers.

1. two, between 14 and 35   21, 28

2. three, between 40 and 12   33, 26, 19

3. three, between 9 and 7   8.5, 8, 7.5

4. four, between 8 and 16   9.6, 11.2, 12.8, 14.4

5. four, between -3 and 14.5   0.5, 4, 7.5, 11

6. five, between 24 and -7.2
18.8, 13.6, 8.4, 3.2, -2

Find *the* arithmetic mean of each pair of numbers.

7. 36 and 82   59

8. 57 and -18 19.5

9. 14.8 and 6.2   10.5

10. $\sqrt{8}$ and $\sqrt{32}$ $3\sqrt{2}$

11. The arithmetic mean of two numbers is 16. Three more than the greater number is 6 times the smaller number. Find the two numbers.   5, 27

12. The arithmetic mean of two numbers is 25. If one number is increased by twice the other number, the result is 68. Find both numbers.   18, 32

Find the 16th term in each harmonic sequence.

13. $\frac{1}{6}, \frac{1}{7}, \frac{1}{8}, \frac{1}{9}, \ldots$ $\frac{1}{21}$

14. $\frac{2}{7}, \frac{1}{3}, \frac{2}{5}, \ldots$ $-\frac{1}{4}$

15. $12, 6, 4, 3, \ldots$ $\frac{3}{4}$

16. $16, 8, \frac{16}{3}, \ldots$ 1

17. Find two harmonic means between $\frac{4}{21}$ and $\frac{4}{3}$. $\frac{4}{15}, \frac{4}{9}$

18. Find three harmonic means between 2 and 1. $\frac{8}{5}, \frac{4}{3}, \frac{8}{7}$

19. Find the harmonic mean of $\frac{1}{8}$ and $\frac{1}{2}$. $\frac{1}{5}$

20. Find the harmonic mean of 5 and $2\frac{1}{2}$. $\frac{10}{3}$

21. A chemist's annual salary for 8 consecutive years was in an arithmetic sequence from \$18,600 to \$24,900. Find the salary for each year.   \$18,600; \$19,500; \$20,400; \$21,300; \$22,200; \$23,100; \$24,000; \$24,900

Find the indicated number of arithmetic means between each pair of numbers.

$8 + \sqrt{2}$, $12 + 2\sqrt{2}$, $16 + 3\sqrt{2}$

22. two, between $3\sqrt{6}$ and $15\sqrt{6}$ $7\sqrt{6}$, $11\sqrt{6}$

23. three, between 4 and $20 + 4\sqrt{2}$

24. four, between $-5i$ and 10

25. two, between $\sqrt{5}$ and $\sqrt{80}$ $2\sqrt{5}$, $3\sqrt{5}$

26. Prove: The arithmetic mean of x and y is $\dfrac{x + y}{2}$. (HINT: Use $a_n = a_1 + (n - 1)d$.)

27. Find the harmonic mean of x and y $(x \neq 0, y \neq 0, x \neq -y)$. $\dfrac{2xy}{x + y}$

Mixed Review

Factor completely.   5.7

1. $2x^3 - 250$ $2(x - 5)(x^2 + 5x + 25)$

2. $ax - ab - cx + bc$ $(a - c)(x - b)$

3. $4x^4 - 17x^2 + 4$
$(2x + 1)(2x - 1)(x + 2)(x - 2)$

4. $mx^2 - my^2$ $m(x + y)(x - y)$

Enrichment

Show the students that it is easy to generate an arithmetic sequence on some scientific calculators. For example, if the first term is 7.25 and the common difference is 5.25, you simply add 5.25 to 7.25. Then 12.5, the second term of the sequence, will appear on the display. To get successive terms, it is merely necessary to keep pressing the equal button.

Have the students use this simple method to write several terms of a few arithmetic sequences.

Note: A similar method will also work in generating geometric sequences (the topic of Lesson 15.4).

15.3 Arithmetic Series and Sigma-Notation

Objectives

To find the sum of the first n terms of an arithmetic series
To write an arithmetic series using sigma-notation

When the terms of the arithmetic sequence 7, 2, -3, -8, . . . are also the terms of the indicated sum $7 + 2 + (-3) + (-8) + \ldots$, or $7 + 2 - 3 - 8 - \ldots$, the result is called an *arithmetic series*.

Definition

> An **arithmetic series** is the indicated sum of the terms of an arithmetic sequence.

The sum S_{20} of the first 20 terms of the arithmetic series $6 + 11 + 16 + 21 + \ldots$ can be found by writing the series in its given order and then in reverse order. The series is added to itself to give twice the sum of the first 20 terms.

$$
\begin{aligned}
S_{20} &= 6 + 11 + 16 + \ldots + 96 + 101 \quad \leftarrow \text{20th term} \\
S_{20} &= 101 + 96 + 91 + \ldots + 11 + 6 \quad \text{is } 6 + 19 \cdot 5, \\
\hline
2S_{20} &= 107 + 107 + 107 + \ldots + 107 + 107 \quad \text{or } 101.
\end{aligned}
$$

$$2S_{20} = 20 \cdot 107 \text{ and } S_{20} = 1{,}070$$

Notice that $S_{20} = \frac{20}{2}(6 + 101)$, or $1{,}070$.

The steps above suggest a method for deriving a formula for the sum S_n of the first n terms of an arithmetic series with a_1 as the first term and a_n as the nth term.

$$
\begin{aligned}
S_n &= a_1 + (a_1 + d) + (a_1 + 2d) + \ldots + (a_n - 2d) + (a_n - d) + a_n \\
S_n &= a_n + (a_n - d) + (a_n - 2d) + \ldots + (a_1 + 2d) + (a_1 + d) + a_1 \\
\hline
2S_n &= (a_1 + a_n) + (a_1 + a_n) + (a_1 + a_n) + \ldots + (a_1 + a_n) + (a_1 + a_n) + (a_1 + a_n)
\end{aligned}
$$

$$2S_n = n(a_1 + a_n)$$

$$S_n = \frac{n}{2}(a_1 + a_n)$$

Teaching Resources

Quick Quizzes 113
Reteaching and Practice Worksheets 113

GETTING STARTED

Prerequisite Quiz

1. Evaluate $\frac{n}{2}(a + b)$ for $n = 40$, $a = -2.6$, and $b = 53.2$.   1,012

2. Evaluate $\frac{n}{2}[2a + (n - 1)d]$ for $n = 31$, $a = 4.5$, and $d = 0.5$.   372

3. Find the sum $1 + 2 + 3 \cdots + 8 + 9 + 10$ using the fact that $1 + 10 = 2 + 9 = 3 + 8 = \cdots = 11$. $11 \cdot 5 = 55$

Motivator

Have students find (a) the sum of the first 20 natural numbers $1 + 2 + 3 + \cdots + 18 + 19 + 20$ by computing $(1 + 20) + (2 + 19) + (3 + 18) + \cdots + (10 + 11) = 10 \cdot 21 = 210$ and (b) the sum of the first 100 natural numbers.   (b) $50 \cdot 101 = 5050$

Highlighting the Standards

Standard 13b: The limit and sigma-notation as explained here are ideas that will be used again in the beginnings of calculus.

Lesson Note

Emphasize that the sum formula $S_n = \frac{n}{2}(a_1 + a_n)$ is used when a_n is given and $S_n = \frac{n}{2}[2a_1 + (n - 1)d]$ is used when a_n is *not* given.

For Example 4, show students the following pattern.

Since $d = 7$, write the terms in the form $7j + c$, where

$j = 1, 2, 3, \cdots$.

$$7 \cdot j + c$$

$18 = 7 \cdot 1 + 11$, so $c = 11$
$25 = 7 \cdot 2 + 11$
$32 = 7 \cdot 3 + 11$

.
.
.

$158 = 7 \cdot j + 11 = 7 \cdot 21 + 11$,

  so 158 is the 21st term.

Thus, $\displaystyle\sum_{j=1}^{21} (7j + 11)$ represents the series.

Math Connections

Pi-Notation: The sigma-notation introduced in this lesson allows one to symbolize an indicated sum in a compact way. Here, the capital Greek letter sigma suggests the first letter of the word "sum." Students may be interested in knowing that an analogous notation exists for symbolizing an indicated product, using the capital Greek letter Π. As an example, the indicated product $1 \cdot 3 \cdot 5 \cdot 7$ can be expressed as

$$\prod_{n=1}^{4} (2n - 1).$$

EXAMPLE 1 Find the sum of the arithmetic series in which the number of terms n is 45, the first term a_1 is 14.3, and the last term a_{45} is 80.3.

Plan Use the formula $S_n = \frac{n}{2}(a_1 + a_n)$ and the given values of n, a_1, and a_n.

Solution $S_n = \frac{n}{2}(a_1 + a_n) = \frac{45}{2}(14.3 + 80.3) = \frac{45}{2}(94.6) = 2{,}128.5$

The sum of the series is $2{,}128.5$.

The two formulas $S_n = \frac{n}{2}(a_1 + a_n)$ and $a_n = a_1 + (n - 1)d$ can be combined to provide a formula for the sum S_n that does not involve the last term a_n.

If $S_n = \frac{n}{2}(a_1 + a_n)$ and $a_n = a_1 + (n - 1)d$, then

$S_n = \frac{n}{2}[a_1 + a_1 + (n - 1)d]$, or $S_n = \frac{n}{2}[2a_1 + (n - 1)d]$.

Sum of the first n terms of an arithmetic series

The sum S_n of the first n terms of an arithmetic series is given by

(1) $S_n = \frac{n}{2}(a_1 + a_n)$ or

(2) $S_n = \frac{n}{2}[2a_1 + (n - 1)d]$,

where a_1 is the first term, a_n is the nth term, and d is the common difference.

EXAMPLE 2 Find the sum of the first 35 terms of the arithmetic series $27.5 + 26 + 24.5 + 23 + \ldots$.

Plan The 35th term is not given. Use the formula $S_n = \frac{n}{2}[2a_1 + (n - 1)d]$.

Solution $d = 26 - 27.5$, or -1.5 $\qquad n = 35 \qquad a_1 = 27.5$

$S_n = \frac{n}{2}[2a_1 + (n - 1)d]$

$\quad = \frac{35}{2}[2(27.5) + (35 - 1)(-1.5)]$

$\quad = \frac{35}{2}[55 + (-51)]$, or 70

The sum of the first 35 terms is 70.

The indicated sum $12 + 15 + 18 + 21$ can be written in the compact form $\displaystyle\sum_{k=4}^{7} 3k$, where the terms of the series have the general form $3k$, and k takes on all integral values from 4 to 7.

Additional Example 1

Find the sum of the arithmetic series in which the number of terms n is 40, the first term a_1 is 59.4, and the last term a_{40} is -10.2. 984

Additional Example 2

Find the sum of the first 21 terms of the arithmetic series $80 + 78.4 + 76.8 + 75.2 \cdots$. 1,344

To see how this compact form works, see the example below.

$$\overbrace{\sum_{k=4}^{7} 3k = 3 \cdot 4 + 3 \cdot 5 + 3 \cdot 6 + 3 \cdot 7}^{\text{expanded form}}$$

$$= \underbrace{12 + 15 + 18 + 21}_{\text{series form}} = 66$$

The Greek letter Σ (read: *sigma*) is a **summation symbol**, and k is called the **index of summation**. $\sum_{k=4}^{7} 3k$ is read "the summation of $3k$ from 4 to 7." Notice that the number of terms of the series is one more than the difference between the upper and lower values of the index. That is, the series has $(7 - 4) + 1$, or 4 terms.

To write $\sum_{j=1}^{3} (5j - 11)$ in expanded form, replace j by 1, 2, and 3.

$$\sum_{j=1}^{3} (5j - 11) = (5 \cdot 1 - 11) + (5 \cdot 2 - 11) + (5 \cdot 3 - 11)$$

$$= -6 - 1 + 4 = -3$$

Notice that 5, the coefficient of j in $5j - 11$, is the common difference of the series.

EXAMPLE 3 Write the expanded form of $\sum_{p=5}^{40} (70 - 3p)$.

Then find the sum of the series.

Solution $\sum_{p=5}^{40} (70 - 3p) = (70 - 3 \cdot 5) + (70 - 3 \cdot 6) + (70 - 3 \cdot 7) +$

$$\dots + (70 - 3 \cdot 40) \leftarrow \text{expanded form}$$

$$= 55 + 52 + 49 + \dots - 50 \leftarrow \text{series form}$$

Next, use $S_n = \frac{n}{2}(a_1 + a_n)$, where $n = (40 - 5) + 1$, or 36, $a_1 = 55$, and $a_{36} = -50$.

$$55 + 52 + 49 + \dots - 50 = \frac{36}{2}(55 - 50) = 90$$

The sum of the series is 90.

Critical Thinking Questions

Analysis: To see if students understand the Pi-notation mentioned in the *Math Connections* on page 554, ask them to write $\prod_{n=0}^{3} (2^n)$ in expanded form. $2^0 \cdot 2^1 \cdot 2^2 \cdot 2^3$
Then have them write $4 \cdot 6 \cdot 8 \cdot 10$ in compact form. $\prod_{n=2}^{5} (2n)$, or some equivalent expression

Checkpoint

Find the sum of each arithmetic series for the given data.

1. $n = 25$, $a_1 = -23$, $a_n = 57$ 425
2. $n = 35$; $-8 - 5.5 - 3 - 0.5 + \dots$
   1,207.5
3. $\sum_{j=4}^{7} (10 - 2j)$ -4
4. $\sum_{k=1}^{30} (3k + 2)$ 1,455

Write each arithmetic series using Σ notation.

5. $4 + 8 + 12 + 16 + 20$ $\sum_{j=1}^{5} 4j$
6. $5 + 8 + 11 + 14 + \dots + 35$
   $\sum_{k=1}^{11} (3k + 2)$

Additional Example 3

Write the expanded form of $\sum_{k=4}^{20} (30 - 2k)$.
Then find the sum of the series.   102

Ask students to translate the expression

$$\sum_{y=2}^{9} 4y$$ into an English phrase.

The summation of $4y$ from 2 to 9

Have students find the sum of the terms of

$$\sum_{k=0}^{4} k!.$$

$0! + 1! + 2! + 3! + 4! = 1 + 1 + 2 + 6 + 24 = 34$

■■■FOLLOW UP

Guided Practice

Classroom Exercises 1–8

Independent Practice

A Ex. 1–24, **B** Ex. 25–29, **C** Ex. 30–36

Basic: WE 1–25 odd
Average: WE 9–31 odd
Above Average: WE 11–35 odd

Additional Answers

Classroom Exercises

1. The summation of $10k$ from 1 to 5; $n = 5$; $10 + 20 + 30 + 40 + 50$
2. The summation of $-j$ from 7 to 10; $n = 4$; $-7 - 8 - 9 - 10$
3. The summation of $p + 7$ from 2 to 6; $n = 5$; $9 + 10 + 11 + 12 + 13$
4. The summation of $-2j$ from 3 to 8; $n = 6$; $-6 - 8 - 10 - 12 - 14 - 16$
5. The summation of $5 - k$ from 0 to 4; $n = 5$; $5 + 4 + 3 + 2 + 1$

EXAMPLE 4 Write $18 + 25 + 32 + 39 + \ldots + 158$ using sigma-notation.

Plan Since $d = 7$, let the general term be $7j + c$, where $j = 1, 2, 3, \ldots$. Then find the constant c and the greatest value of j.

Solution To find c, use the first term 18 and solve $7 \cdot 1 + c = 18$: $c = 11$. To find j when $a_n = 158$, solve the equation $7j + 11 = 158$: $j = 21$.

So, the required notation is $\displaystyle\sum_{j=1}^{21}(7j + 11)$

Note that the answer in Example 4 is not unique. The series can also be expressed with the notation $\displaystyle\sum_{j=2}^{22}(7j + 4)$.

Classroom Exercises

State each summation in words. Predict the number of terms in the series. Then write the series form.

1. $\displaystyle\sum_{k=1}^{5} 10k$ 2. $\displaystyle\sum_{j=7}^{10} -j$ 3. $\displaystyle\sum_{p=2}^{6} (p + 7)$ 4. $\displaystyle\sum_{j=3}^{8} -2j$ 5. $\displaystyle\sum_{k=0}^{4} (5 - k)$

Find the sum of each arithmetic series for the given data.

6. $n = 4$, $a_1 = 2$, $a_4 = -16$ -28

7. $n = 25$, $a_1 = -1$, $a_{25} = 10$ 112.5

8. $\displaystyle\sum_{k=1}^{10} (8 + 3k)$ 245

Written Exercises

Find the sum of each arithmetic series for the given data.

1. $n = 30$, $a_1 = 1$, $a_{30} = 134$ 2,025   2. $n = 20$, $a_1 = 4$, $a_{20} = -115$ $-1,110$
3. $n = 35$, $a_1 = -7.2$, $a_{35} = 58$ 889   4. $n = 44$, $a_1 = -2.5$, $a_{44} = -56.7$
5. $n = 40$; $6 + 11 + 16 + 21 + \ldots$ 4,140 6. $n = 30$; $10 + 6 + 2 - 2 - \ldots$
7. $n = 55$; $6.7 + 7 + 7.3 + 7.6 + \ldots$ 8. $n = 65$; $-3 - 1.9 - 0.8 + 0.3 + \ldots$
9. $\displaystyle\sum_{k=1}^{5} 6k$ 90   10. $\displaystyle\sum_{p=0}^{39} (p + 2)$ 860   11. $\displaystyle\sum_{j=3}^{7} (2j - 5)$ 25   12. $\displaystyle\sum_{p=30}^{33} (7 - p)$ -98

814 (for 7)   2,093 (for 12)

Additional Example 4

Write $54 + 50 + 46 + 42 + \cdots + 14$ using Σ notation.

$$\sum_{k=1}^{11} (-4k + 58)$$

Write each arithmetic series using sigma-notation.

13. $8 + 16 + 24 + 32$

14. $-6 - 12 - 18 - 24 - 30$

15. $14 + 16 + 18 + 20 + 22$

16. $15 + 12 + 9 + 6$

17. $7 + 9 + 11 + 13 + 15$

18. $14 + 11 + 8 + 5$

19. $15 + 11 + 7 + 3 - \ldots - 17$

20. $-7 - 5 - 3 - 1 + \ldots + 23$

21. $19 + 27 + 35 + 43 + \ldots + 107$

22. $75 + 66 + 57 + 48 + \ldots - 15$

Find the sum of each arithmetic series for the given data.

23. $n = 31$, $a_1 = 3\sqrt{2}$, $a_{31} = 153\sqrt{2}$
$2{,}418\sqrt{2}$

24. $n = 18$, $a_1 = 5x - 2y$,
$a_{18} = 32y - 12x$ $-63x + 270y$

25. $n = 36$; $-0.2 + 0.3 + 0.8 + 1.3 + \ldots$ 307.8

26. $n = 27$; $35.5 + 34.3 + 33.1 + 31.9 + \ldots$ 537.3

27. Cartons are stacked in 16 rows with 4 in the top row, 7 in the second row, 10 in the third row, and so on. Find the total number of cartons in the stack. 424

28. In the month of June, Kevin saved 1 quarter the 1st day, 2 quarters the 2nd day, 3 quarters the 3rd day, and so on. How much money did he save in June? $116.25

29. The cost for printing 1000 copies of a 128-page book is $2500. The cost for printing each additional eight pages is $72. How much will it cost to print 1000 copies of a 208-page book? $3220

30. A traveling salesperson receives a 5% commission on each of the first 5 vacuum cleaners sold, a 6% commission on each of the next 3 sold, a 7% commission on each of the next 3, and so on. If the vacuum cleaners are selling for $400 each and Mary Ings sells 23, what is her commission on the last three sold? $132

31. Solve $\displaystyle\sum_{k=5}^{8}(kx - 4) = 36$ for x. 2

32. Solve $\displaystyle\sum_{j=1}^{4}\frac{j}{x} = 75$ for x. $\frac{2}{15}$

33. Prove $\displaystyle\sum_{k=1}^{20}5kx = 5x \cdot \sum_{k=1}^{20}k$.

34. Prove $\displaystyle\sum_{j=1}^{15}(j + c) = 15c + \sum_{j=1}^{15}j$.

35. Prove $\displaystyle\sum_{n=1}^{k}(2n - 1) = k^2$.

36. Prove $\displaystyle\sum_{n=1}^{k}2n = k(k + 1)$.

Mixed Review

1. Factor $x^2 - 2xy + y^2 - 16$ as the difference of two squares. $(x - y + 4)(x - y - 4)$ *5.6*

2. Find $f(g(3a))$ if $f(x) = x^2 + 4$ and $g(x) = x - 2$. *5.9* $9a^2 - 12a + 8$

3. Find the solution set for $x^2 - 2x - 8 < 0$. *6.4* $\{x \mid -2 < x < 4\}$

4. What is the harmonic mean of $\frac{1}{2}$ and $\frac{1}{8}$? *15.2* $\frac{1}{5}$

4. $-1{,}302.4$

6. $-1{,}440$

13. $\displaystyle\sum_{k=1}^{4}8k$

14. $\displaystyle\sum_{k=1}^{5}-6k$

15. $\displaystyle\sum_{k=1}^{5}(2k + 12)$

16. $\displaystyle\sum_{k=1}^{4}(18 - 3k)$

17. $\displaystyle\sum_{k=1}^{5}(2k + 5)$

18. $\displaystyle\sum_{k=1}^{4}(17 - 3k)$

19. $\displaystyle\sum_{k=1}^{9}(19 - 4k)$

20. $\displaystyle\sum_{k=1}^{16}(2k - 9)$

21. $\displaystyle\sum_{k=1}^{12}(8k + 11)$

22. $\displaystyle\sum_{k=1}^{11}(84 - 9k)$

33. $\displaystyle\sum_{k=1}^{20}5kx = 5x + 10x + 15x + \cdots + 100x$
$= 5x \cdot (1 + 2 + 3 + \cdots + 20)$
$= 5x \cdot \displaystyle\sum_{k=1}^{20}k$

34. $\displaystyle\sum_{j=1}^{15}(j + c) = 1 + c + 2 + c + 3 + c + \cdots + 15 + c$
$= 15c + (1 + 2 + 3 + \cdots + 15)$
$= 15c + \displaystyle\sum_{j=1}^{15}j$

35. $\displaystyle\sum_{n=1}^{k}(2n - 1) = 1 + 3 + 5 + \cdots + (2k - 1)$
$= \frac{k}{2}(1 + 2k - 1) = \frac{k}{2} \cdot 2k$
$= k^2$

36. $\displaystyle\sum_{n=1}^{k}2n = 2 + 4 + 6 + \cdots + 2k$
$= \frac{k}{2}(2 + 2k) = \frac{k}{2} \cdot 2(k + 1)$
$= k(k + 1)$

Enrichment

Have the students use the summation formula $S_n = \frac{n}{2}(a_1 + a_n)$ to derive a formula for $\displaystyle\sum_{i=1}^{n}i$, the sum of the first n counting numbers. If necessary, point out that $a_1 = 1$ and $a_n = n$.

$$\sum_{i=1}^{n}i = \frac{n}{2}(1 + n) \text{ or } \sum_{i=1}^{n} = \frac{n(n + 1)}{2}$$

Let the students verify this formula with a particular case, say $n = 5$.

$$\sum_{i=1}^{5}i = 1 + 2 + 3 + 4 + 5 = 15$$
and $\frac{5(5 + 1)}{2} = 15$

Now have the students use the same method to derive these two formulas.

1. The sum of the first n even counting numbers $\displaystyle\sum_{i=1}^{n}2i = n(n + 1)$

2. The sum of the first n odd counting numbers $\displaystyle\sum_{i=1}^{n}(2i - 1) = n^2$

Next, have the students use the above formulas to find the sum of each of the following.

1. The first 85 counting numbers 3,655

2. The first 61 odd numbers 3,721

3. The first 100 even numbers 10,100

Teaching Resources

Problem Solving Worksheet 15
Quick Quizzes 114
Reteaching and Practice
   Worksheets 114
Transparency 33

■■■■GETTING STARTED

Prerequisite Quiz

Find the value of each expression. Use a scientific calculator and the y^x key, if possible.

1. 2^{10} 1,024
2. $5 \cdot 2^9$ 2,560
3. $-4 \cdot 3^6$ −2,916
4. $\left(\frac{1}{5}\right)^6$ $\frac{1}{15,625}$ or 0.000064

Motivator

Have students write a sequence of numbers in which each consecutive term is multiplied by a constant. Then have students exchange sequences and find the common ratio of the sequence.

15.4 Geometric Sequences

Objectives   To write several consecutive terms in a geometric sequence
To find a specified term in a geometric sequence

In the sequence below, each consecutive term is obtained by multiplying the preceding term by 2.

$$3, 6, 12, 24, \ldots$$

A sequence formed in this manner is called a *geometric sequence*. Therefore, the ratio of any two successive terms is the same. In the sequence above, the *common ratio* is 2, since $2 = \frac{6}{3} = \frac{12}{6} = \frac{24}{12}$.

Definitions $a_1, a_1r, a_1r^2, a_1r^3, \ldots$ is a **geometric sequence** with first term a_1 and **common ratio** r ($a_1 \neq 0, r \neq 0$).

If $x_1, x_2, x_3, x_4, \ldots$ is a geometric sequence with common ratio r, then $r = \frac{x_2}{x_1} = \frac{x_3}{x_2} = \frac{x_4}{x_3} = \ldots$.

EXAMPLE 1   For each sequence that is geometric, find the common ratio r.

a. $-2, -10, -50, -250, \ldots$ b. $2, 4, 6, 8, \ldots$

Solutions   a. Since $\frac{-10}{-2} = \frac{-50}{-10} = \frac{-250}{-50} = 5$,
the sequence is geometric with
$r = 5$.

b. Since $\frac{4}{2} \neq \frac{6}{4}$, the sequence
is not geometric.

EXAMPLE 2   Write the next three terms of the geometric sequence $64, -32, 16, -8, \ldots$.

Solution   The common ratio $r = \frac{-32}{64}$, or $-\frac{1}{2}$. Multiply each term by $-\frac{1}{2}$ to obtain the next three terms.

Thus, the next three terms are $4, -2$, and 1.

EXAMPLE 3   Write the first four terms of the geometric sequence whose first term a_1 is -3 and whose common ratio r is -4.

Solution $-3(-4) = 12; \quad 12(-4) = -48; \quad -48(-4) = 192$

Thus, the first four terms are $-3, 12, -48$, and 192.

Highlighting the Standards

Standard 3c: The examples demonstrate the logic of geometric sequences. Students will gain confidence as they practice the related skills.

Additional Example 1

If the sequence is geometric, find the common ratio r.

a. $-3, 6, -12, 24, \cdots$ -2
b. $7, 14, 21, 28, \cdots$ Not geometric

Additional Example 2

Write the next three terms of the following geometric sequence.

$243, -81, 27, -9, \cdots$ $3, -1, \frac{1}{3}$

A geometric sequence in the form $a_1, a_1r, a_1r^2, \ldots$ can be used to find a specific term, such as the 20th term of $3, 6, 12, 24, \ldots$.

$$
\begin{array}{ccccccc}
3 & 6 & 12 & 24 & 48 & \cdots & ? \\
\downarrow & \downarrow & \downarrow & \downarrow & \downarrow & & \downarrow \\
3 & 3 \cdot 2^1 & 3 \cdot 2^2 & 3 \cdot 2^3 & 3 \cdot 2^4 & \cdots & 3 \cdot 2^{19} \\
\text{1st} & \text{2nd} & \text{3rd} & \text{4th} & \text{5th} & & \text{20th} \\
\text{term} & \text{term} & \text{term} & \text{term} & \text{term} & & \text{term}
\end{array}
$$

The 20th term is $3 \cdot 2^{20-1}$, or 1,572,864. This suggests the following formula for the nth term in a geometric sequence.

---

nth term of a geometric sequence

The nth term of a geometric sequence is given by
$a_n = a_1 \cdot r^{n-1}$, where a_n is the nth term ($n \geq 1$), a_1 is the first term, and r is the common ratio.

EXAMPLE 4 Find the specified term of each geometric sequence.

  a. 5th term: $a_1 = 7$ and $r = 0.2$ **b.** 10th term of $-8, 4, -2, \ldots$

Solutions **a.** $a_1 = 7, n = 5, r = 0.2$

  $a_n = a_1 \cdot r^{n-1} = 7(0.2)^{5-1} = 7(0.2)^4 = 7(0.0016) = 0.0112$

  Calculator steps: 7 $\boxed{\times}$ 0.2 $\boxed{x^y}$ 4 $\boxed{=}$ 0.0112

  The 5th term is 0.0112.

  b. $a_1 = -8, n = 10, r = -\dfrac{1}{2}$

  $a_n = a_1 \cdot r^{n-1} = -8\left(-\dfrac{1}{2}\right)^{10-1} = \dfrac{-8}{1}\left(\dfrac{1}{-2}\right)^9 = \dfrac{(-2)^3}{(-2)^9} = \dfrac{1}{(-2)^6}$

  $= \dfrac{1}{64}$

  The 10th term is $\dfrac{1}{64}$.

EXAMPLE 5 At the end of each hour, one-fourth of the gas in a tank is released. If the tank contains 2,048 ft^3 of gas at the beginning, how much gas will be in the tank at the end of 7 h?

TEACHING SUGGESTIONS

Lesson Note

This lesson on geometric sequences leads to lessons on geometric means, geometric series, and infinite geometric series. A scientific calculator with the y^x function is very useful in this lesson.

For Exercise 27 and 28, point out that the bouncing ball travels through a sequence of descents and rebounds (upward), that each sequence is geometric, but that the two sequences have different first terms. The word *geometric* is pronounced gee-oh-MEH-trik.

Math Connection

Population Studies: Thomas Malthus (1766–1834) contended that human poverty is unavoidable. His reasoning was that human populations tend to grow in a geometric way, increasing by a fixed ratio, while the total food supply grows only in an arithmetic way, increasing only by a constant amount. Malthus concluded that since, in the long run, geometric growth always must overwhelm arithmetic growth, efforts to improve the human condition are futile. His ideas influenced many other thinkers, including Ricardo and Darwin. The current belief is that Malthus underestimated the increase in the productivity of agriculture.

Additional Example 3

Write the first four terms of the geometric sequence whose first term a is 2 and whose common ratio r is -5. $2, -10, 50, -250$

Additional Example 4

Find the specified term of each geometric sequence.

  a. 6th term: $a = 9$ and $r = \dfrac{1}{3}$ $\dfrac{1}{27}$

  b. 9th term of $\dfrac{1}{27}, \dfrac{1}{9}, \dfrac{1}{3}, \cdots$ 243

Additional Example 5

A ball, dropped from a height of 729 cm, rebounds on each bounce one-third of the distance from which it fell. How high does it go on the 7th rebound? $\dfrac{1}{3}$ cm

Critical Thinking Questions

Analysis: Students will better understand the ideas of Malthus (see p. 559) by considering the two functions $y = a + (n - 1)d$ and $y = ar^{n-1}$, which represent, respectively, the nth term of an arithmetic and a geometric sequence. Ask them to show graphically in the ny-plane that the nth term of the geometric sequence must eventually overtake the nth term of the arithmetic sequence. Have them choose a simple value, such as 2, for a, d, and r. Students should be able to see that the exponential curve (representing geometric growth) will eventually cross the straight, nonvertical line (arithmetic growth), even if the line is steep.

Checkpoint

1. Write the next three terms of the geometric sequence 40, 4, 0.4, $\cdots$.
   0.04, 0.004, 0.0004

2. Find the 9th term of the geometric sequence $\frac{1}{16}$, $\frac{1}{4}$, 1, $\cdots$. 4,096

3. At the end of each hour, one-third of the gas in a tank is removed. If the tank contains 2,430 ft^3 of gas at noon, how much gas is in the tank at 5 P.M.?
   320 ft^3

Plan Find the amount remaining at the end of each of the first 3 h. 1st hour: $\frac{3}{4}(2,048)$, or 1,536 ft^3 remain; 2nd hour: $\frac{3}{4}(1,536)$, or 1,152 ft^3 remain; 3rd hour: $\frac{3}{4}(1,152)$, or 864 ft^3 remain. Then find the 7th term of the geometric sequence 1,536; 1,152; 864; . . . .

Solution $a_1 = 1,536$, $r = \frac{3}{4}$, $n = 7$

$a_7 = 1,536\left(\frac{3}{4}\right)^6 = \dfrac{1,536 \cdot 729}{4,096}$

$\qquad = 273.375$

At the end of 7 h, 273.375, or $273\frac{3}{8}$ ft^3 of gas will be in the tank.

Classroom Exercises

For each sequence that is geometric, find the common ratio and the next two terms. $-\frac{1}{3}$; 1, $-\frac{1}{3}$ Not geometric -1; 8, -8

1. 81, -27, 9, -3, . . .   **2.** 4, 8, 12, 16, . . .   **3.** 8, -8, 8, -8, . . .

4. $\frac{1}{12}$, $\frac{1}{4}$, $\frac{3}{4}$, $\frac{9}{4}$, . . . 3; $\frac{27}{4}$, $\frac{81}{4}$ **5.** 18, 12, 8, $\frac{16}{3}$, . . .   **6.** 0.3, 0.06, 0.012, . . .

7. $\sqrt{2}$, 2, $2\sqrt{2}$, 4, . . .   **8.** $\sqrt{3}$, 3, $2\sqrt{3}$, 6, . . .   **9.** -4, 8, -4, 8, . . .

10. Write the first four terms of a geometric sequence in which $a_1 = -3$ and $r = -2$. -3, 6, -12, 24

11. Find the 10th term of a geometric sequence in which $a_1 = 15$ and $r = 3$. 295,245

Written Exercises

Write the next three terms of each geometric sequence. 2. -1, $-\frac{1}{4}$, $-\frac{1}{16}$

1. 15, 30, 60, . . .   **2.** -64, -16, -4, . . .   **3.** $-\frac{2}{9}$, $\frac{2}{3}$, -2, . . .
   120, 240, 480   6, -18, 54

Write the first four terms of each geometric sequence given its first term a_1 and common ratio r. $-\frac{1}{8}$, $-\frac{1}{2}$,   0.75, 1.5, 3, 6   64, 32, 16, 8

4. $a_1 = -\frac{1}{8}$, $r = 4$ -2, -8 **5.** $a_1 = 0.75$, $r = 2$ **6.** $a_1 = 64$, $r = \frac{1}{2}$

7. $a_1 = -8$, $r = -\frac{1}{4}$ **8.** $a_1 = 7$, $r = 0.1$ **9.** $a_1 = 128$, $r = -\frac{3}{8}$
                                       7, 0.7, 0.07, 0.007   128, -48, 18, $-6\frac{3}{4}$

Find the specified term of each geometric sequence.

10. 6th term: $a_1 = 23$, $r = 10$ 2,300,000   **11.** 5th term: $a_1 = -10$, $r = 4$ $-2,560$

12. 7th term: $a_1 = -500$, $r = -0.1$ -0.0005 **13.** 6th term: $a_1 = 15$, $r = -0.2$ -0.0048

14. 11th term: $-\frac{1}{8}$, $-\frac{1}{4}$, $-\frac{1}{2}$, . . . -128 **15.** 10th term: $\frac{1}{4}$, $\frac{1}{2}$, 1, . . . 128

16. 11th term: -96, 48, -24, . . . $-\frac{3}{32}$ **17.** 8th term: -0.36, -3.6, -36, . . .
                                                              $-3,600,000$

560   Chapter 15 Sequences, Series, Binomial Theorem

18. 6th term: $a_1 = 25$, $r = \frac{2}{5}$ $\frac{32}{125}$ **19.** 7th term: $a_1 = -27$, $r = \frac{1}{3}$ $-\frac{1}{27}$
20. 5th term: $a_1 = \sqrt{2}$, $r = 3\sqrt{2}$ $324\sqrt{2}$ **21.** 6th term: $a_1 = -4$, $r = \sqrt{3}$ $-36\sqrt{3}$

Find the next two terms of each geometric sequence.

22. $1, i, -1, -i, \ldots$ $1, i$ **23.** $-2x^2, 8x^5, -32x^8, \ldots$ $128x^{11}, -512x^{14}$
24. $\sqrt{3}y^3, -3y^5, 3\sqrt{3}y^7, \ldots$ $-9y^9, 9\sqrt{3}y^{11}$ **25.** $2i, 2, -2i, -2, \ldots$ $2i, 2$

26. A tank contains 729 liters of oil. One-third of the oil is released each time that a valve is operated. How much oil will be in the tank after the valve is operated 7 times? $42\frac{2}{3}$ liters

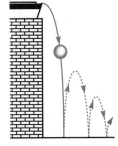

27. On each bounce, a certain golf ball rebounds one-half of the distance from which it falls. If it is dropped from a height of 128 ft, how high does it go on its 10th rebound (upward)?

28. If a rubber ball dropped from a height of 81 yd rebounds on each bounce two-thirds of the distance from which it fell, how far does it fall on its 7th descent (downward)? $7\frac{1}{9}$ yd

29. Find two values of y so that $\frac{1}{4}$, $y - 2$, $4y$ is a geometric sequence. $1, 4$
30. Find two values of t for which $4t$, $t - 2$, $\frac{2}{3}$ is a geometric sequence. $\frac{2}{3}, 6$

31. The 7th term of a geometric sequence is 512 and the 8th term is 1,024. Find the first term. 8
32. A geometric sequence has a 9th term of $\frac{1}{32}$ and a 10th term of $\frac{1}{64}$. What is the first term? 8

Midchapter Review

For Exercises 1–3, use the arithmetic sequence 8, 6.5, 5, 3.5, ... Find each of the following. *15.1*

2, 0.5, −1

1. the common difference d **2.** the next three terms   **3.** a_{31}, the 31st term -37
4. Find the four arithmetic means between 22 and 26. *15.2* 22.8, 23.6, 24.4, 25.2
5. Find the arithmetic mean of -3.6 and 7.2. *15.2* 1.8
6. Find the two harmonic means between 4 and $\frac{8}{5}$. *15.2* $\frac{8}{3}, 2$

For Exercises 7 and 8, find the sum of the arithmetic series for the given number of terms. *15.3*

553.5

7. $n = 40$; $a_1 = 5$, $a_n = -144$ $-2{,}780$ **8.** $n = 45$; $5.7 + 6 + 6.3 + 6.6 + \ldots$
9. Write $9 + 14 + 19 + 24 + 29 + 34$ using sigma-notation. *15.3* $\sum\limits_{k=1}^{6}(5k + 4)$

For Exercises 10–12, use the geometric sequence $-\frac{1}{4}, \frac{1}{2}, -1, \ldots$. Find each of the following. *15.4*

10. the common ratio r -2 **11.** the next three terms   **12.** a_{10}, the tenth term 128
2, −4, 8

Closure

Have students give the formula for finding the nth term of a geometric sequence.
$a_n = a_1 \cdot r^{n-1}$ Have students create a geometric sequence by writing its first three terms. Then have them find a_1, r, and a_{11} for the sequence.

◤◤◤ FOLLOW UP

Guided Practice

Classroom Exercises 1–11

Independent Practice

A Ex. 1–17, **B** Ex. 18–28, **C** Ex. 29–32

Basic: WE 1–21 odd, Midchapter Review
Average: WE 7–29 odd, Midchapter Review
Above Average: WE 9–31 odd, Midchapter Review

Additional Answers

Classroom Exercises

5. $\frac{2}{3}, \frac{32}{9}, \frac{64}{27}$
6. 0.2; 0.0024, 0.00048
7. $\sqrt{2}$; $4\sqrt{2}$, 8
8. Not geometric
9. Not geometric

Written Exercises

7. $-8, 2, -\frac{1}{2}, \frac{1}{8}$
27. $\frac{1}{8}$ ft

Midchapter Review

1. -1.5

Enrichment

Challenge the students to write a geometric sequence that meets these conditions.

1. It uses only two of the nonzero digits of our number system.
2. The sum of 100 terms equals zero.   Any nonzero digit can be the first term, and then choose $r = -1$. For example, one such sequence is $3, -3, 3, -3, \cdots$.

Prerequisite Quiz

1. Write the first four terms in the geometric
   sequence with $a = \frac{1}{4}$ and $r = -2$.
   $\frac{1}{4}, -\frac{1}{2}, 1, -2$

Solve each equation for its real roots.

2. $r^4 = 81$ $\quad -3, 3$
3. $r^3 = 64$ $\quad 4$
4. $r^3 = -27$ $\quad -3$
5. $r^3 = 5\sqrt{5}$ $\quad \sqrt{5}$

Motivator

Have students use the formula $a_n = a_1 \cdot r^{n-1}$
to find x and y in the list below to form a
geometric sequence.

$$5, x, 20, y, 80$$

$x = 10, y = 40$

15.5 Geometric Means

Objectives To find geometric means between two given numbers
To find *the* geometric mean of two numbers

The terms between two given terms in a geometric sequence are called
the **geometric means** between the two terms. For example, 10, 20, 40,
and 80 are the four geometric means between 5 and 160 in the geometric sequence below.

$$5, \underline{10, 20, 40, 80}, 160$$

geometric means

EXAMPLE 1 Find three real geometric means between $-\frac{2}{3}$ and -54.

Plan The three means and the given numbers will form a 5-term geometric
sequence with $a_1 = -\frac{2}{3}$, $a_5 = -54$, and $n = 5$:
$$-\frac{2}{3}, \underline{\quad}, \underline{\quad}, \underline{\quad}, -54.$$
To find r, use the formula $a_n = a_1 \cdot r^{n-1}$.

Solution
$$-54 = -\frac{2}{3} \cdot r^{5-1} \leftarrow \text{Substitute for } a_1, a_n, \text{ and } n.$$
$$81 = r^4$$
$$r = \sqrt[4]{81} \quad \text{or} \quad r = -\sqrt[4]{81}$$
$$r = 3 \quad\quad \text{or} \quad r = -3 \leftarrow \text{two real numbers for } r$$

Multiply each term by r to obtain the next term in the sequence.

$$r = 3: -\frac{2}{3}, \underline{-2}, \underline{-6}, \underline{-18}, -54 \quad\quad r = -3: -\frac{2}{3}, \underline{2}, \underline{-6}, \underline{18}, -54$$

The three real means are $-2, -6, -18$ or $2, -6, 18$.

In Example 1, $x^4 = 81$ has two other roots, $3i$ and $-3i$. These values
of r were not used since the problem asked for *real* number means.

When finding n real geometric means between two numbers, you will
obtain two sets of means if n is odd, as in Example 1, and one set of
means if n is even, as in Example 2.

Highlighting the Standards

Standard 12a: The work with geometric
means offers another taste of discrete
mathematics. Most students like the limited
scope of the exercises.

Additional Example 1

Find three real geometric means between
32 and 2.
16, 8, 4 or −16, 8, −4

EXAMPLE 2 Find the two real geometric means between 18 and $-\frac{16}{3}$.

Solution To find r, use $a_n = a_1 \cdot r^{n-1}$ with $a_1 = 18$, $n = 4$, and $a_4 = -\frac{16}{3}$.

Substitute for a_1, a_n, $\rightarrow$ $-\frac{16}{3} = 18 \cdot r^{4-1}$
and n.

$$-\frac{16}{3} \cdot \frac{1}{18} = \frac{1}{18} \cdot 18 \cdot r^3$$

$$-\frac{8}{27} = r^3$$

$$r = \sqrt[3]{-\frac{8}{27}} = -\frac{2}{3}$$

Multiply each term by $-\frac{2}{3}$. $\quad 18, \underline{-12}, \underline{8}, -\frac{16}{3}$

The two real geometric means are -12 and 8.

EXAMPLE 3 Find the geometric mean between 2 and 6.

Solution $2, \underline{\quad}, 6; \quad a_n = a_1 \cdot r^{n-1} \qquad 6 = 2 \cdot r^{3-1} \qquad 3 = r^2 \qquad r = \pm\sqrt{3}$

Use $r = \sqrt{3}$. The mean is $2\sqrt{3}$: $2, \underline{2\sqrt{3}}, 6$.

Use $r = -\sqrt{3}$. The mean is $-2\sqrt{3}$: $2, \underline{-2\sqrt{3}}, 6$.

The geometric mean is $2\sqrt{3}$ or $-2\sqrt{3}$.

The one geometric mean between two numbers is called *the geometric mean* of the two numbers. In Example 3, notice that the geometric mean between 2 and 6 is $\sqrt{2 \cdot 6}$, which equals $2\sqrt{3}$, or it is $-\sqrt{2 \cdot 6}$, which equals $-2\sqrt{3}$. This suggests the following formula.

The **geometric mean** of x and y is $\sqrt{xy}$ or $-\sqrt{xy}$ $(x \neq 0, y \neq 0)$.

EXAMPLE 4 The positive geometric mean of two numbers is 6 and one number is 9 more than the other number. Find the two numbers.

Plan Let x and y be the two numbers. Write and solve a system of equations.

Solution
(1) $\sqrt{xy} = 6$ (2) $y = x + 9$

(1) $\sqrt{x(x + 9)} = 6 \qquad x(x + 9) = 36 \qquad x^2 + 9x - 36 = 0$

$$(x - 3)(x + 12) = 0$$

$$x = 3 \ or \ x = -12$$

(2) $y = x + 9$ If $x = 3$, then $y = 12$.
If $x = -12$, then $y = -3$.

The numbers are 3 and 12 or -12 and -3.

564

Checkpoint

1. Find three real geometric means between 8 and $\frac{81}{2}$.   12, 18, 27 or −12, 18, −27
2. Solve the system.
$$\sqrt{xy} = 6$$
$$x - 2y = 1$$
   $x = 9, y = 4$ or $x = -8, y = -4\frac{1}{2}$

Closure

Asks students to give the formula for finding the geometric mean of x and y. $\sqrt{xy}$ or $-\sqrt{xy}$, $(x \neq 0, y \neq 0)$ Have students use a calculator to find the positive geometric mean of $\sqrt{6.48}$ and $\sqrt{12.5}$.   3

◾FOLLOW UP

Guided Practice

Classroom Exercises 1–8

Independent Practice

A Ex. 1–18, **B** Ex. 19–30, **C** Ex. 31–38

Basic: WE 1–21 odd, Brainteaser
Average: WE 9–31 odd, Brainteaser
Above Average: WE 11–37 odd, Brainteaser

Classroom Exercises

In each geometric sequence, find the missing geometric means between the first and last terms.

1. 3, __12__, 48, 192   2. −2, __−1__, −$\frac{1}{2}$, __−$\frac{1}{4}$__, −$\frac{1}{8}$ 3. 32, __16__, __8__, __4__, 2

Find *the* positive geometric mean of each pair of numbers.

4. 20 and 5 10   5. 1 and 4 2   6. 4 and $\frac{1}{4}$ 1   7. −2 and −8 4   8. 5 and 3 $\sqrt{15}$

Written Exercises

Find the indicated number of real geometric means between the two numbers.

1. three, between 2 and 162   6, 18, 54 or −6, 18, −54
2. three, between 64 and 4   32, 16, 8 or −32, 16, −8
3. two, between 9 and $\frac{1}{3}$ 3, 1
4. two, between $\frac{3}{4}$ and 48   3, 12
5. three, between −4 and −324
6. three, between −$\frac{3}{8}$ and −6
7. two, between −$\frac{1}{4}$ and 16   1, −4
8. two, between 75 and −$\frac{3}{5}$ −15, 3
9. four, between 9 and 288   18, 36, 72, 144
10. four, between 300,000 and 3
11. three, between $\sqrt{2}$ and $4\sqrt{2}$ 2, $2\sqrt{2}$, 4 or −2, $2\sqrt{2}$, −4
12. three, between 6 and 54 $6\sqrt{3}$, 18, $18\sqrt{3}$ or −$6\sqrt{3}$, 18, −$18\sqrt{3}$

Find *the* positive geometric mean of each pair of numbers.

13. 3 and 12 6   14. 36 and 4 12   15. −5 and −80 20   16. −49 and −2 $7\sqrt{2}$

17. One number is 9 times another number. The positive geometric mean of the two numbers is 6. Find the two numbers.   2, 18, or −2, −18

18. Find two numbers whose positive geometric mean is 4, if one number is 6 less than the other number.   8, 2 or −8, −2

19. Find two numbers whose difference is 12 and whose negative geometric mean is −8.   16, 4 or −4, −16

20. Nine different weights are in a geometric sequence. Find each weight if the lightest is 0.25 g and the median (middle) weight is 4 g.   0.25 g, 0.5 g, 1 g, 2 g, 4 g, 8 g, 16 g, 32 g, 64 g

21. Wheat is removed from a storage silo so that the amounts remaining at the end of successive hours are in a geometric sequence. At the end of 6 h, 270 tons remain. How much wheat was in the silo at the end of 4 h if there were 7,290 tons at the end of 3 h?   2,430 tons

Find *the* negative geometric mean of each pair of numbers.

22. 6, 3   −$3\sqrt{2}$ 23. $2\sqrt{3}$, $8\sqrt{3}$ −$4\sqrt{3}$ 24. −$\sqrt{2}$, −$\sqrt{8}$ −2   25. −5, −10   −$5\sqrt{2}$

564   Chapter 15 Sequences, Series, Binomial Theorem

Enrichment

Have the students solve this problem.

The first three terms of an arithmetic sequence have a sum of 90. If 5 is added to the third term, the three terms form a geometric sequence. Find the three numbers.

Solve the system.
$$a + (a + d) + (a + 2d) = 90$$
$$\frac{a}{a + d} = \frac{a + d}{a + 2d + 5}$$

The three numbers are 45, 30, 15 or 20, 30, 40.

26. Write in your own words how the positive geometric mean of two positive numbers is found.

Find the positive geometric mean of each pair of numbers.

27. $0.2, 0.008$ 0.04   **28.** $-1.8, -0.2$ 0.6   **29.** $-2.7, -0.3$ 0.9   **30.** $6, 0.24$ 1.2

For the right triangle PQR at the right, the altitude to the hypotenuse separates the hypotenuse into two segments. It is proved in geometry that $\frac{a}{m} = \frac{m}{b}$, where m is called the *mean proportional* between a and b.

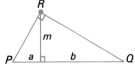

Observe that m is the geometric mean of a and b since

if $\frac{a}{m} = \frac{m}{b}$, then $m^2 = ab$ and $m = \sqrt{ab}$.

Use the information and the figure above with Exercises 31–36.

31. Find m if $a = 10$ and $b = 6.4$. 8   **32.** Find m if $a = 2$ and $b = 6$. $2\sqrt{3}$

33. Find b if $a = 4$ and $m = 2\sqrt{6}$. 6   **34.** Find m if $PR = 18$ and $a = 6$. $12\sqrt{2}$

35. Find PR if $a = 3$ and $b = 6$. $3\sqrt{3}$ **36.** Find PQ if $a = 3$ and $m = 6$. 15

37. Prove: The one positive geometric mean between x and y is $\sqrt{xy}$. (HINT: Use the formula $a_n = a_1 \cdot r^{n-1}$ to find r and fill the blank in $x, \underline{\hspace{1cm}}, y$ to form a geometric sequence.)

38. Prove: If a, g, and h are the arithmetic, geometric, and harmonic means, respectively, of two numbers x and y, then $g^2 = ah$.

Mixed Review

Simplify. *5.2, 5.3, 7.4, 8.3*

1. $\dfrac{-10x^{-3}y^{-2}z^3}{15x^5y^{-6}z^{-1}}$ $\dfrac{-2y^4z^4}{3x^8}$

2. $8x - (7 - 2x) + 3x(4x - 5)$ $12x^2 - 5x - 7$

3. $\dfrac{5x}{x^2 - 7x + 12} - \dfrac{4}{6 - 2x}$ $\dfrac{7x - 8}{(x - 3)(x - 4)}$

4. $(3\sqrt{2} + \sqrt{6})(4\sqrt{2} - \sqrt{6})$ $18 + 2\sqrt{3}$

▨ *Brainteaser*

Using an 11-minute hourglass and a 7-minute hourglass, what is the easiest way to time the boiling of an egg for 15 minutes?

Additional Answers

Written Exercises

5. $-12, -36, -108$ or $12, -36, 108$

6. $-\frac{3}{4}, -\frac{3}{2}, -3$ or $\frac{3}{4}, -\frac{3}{2}, 3$

10. $30,000, 3,000, 300, 30$

26. Answers will vary. One possible answer: Multiply the two numbers and find the positive square root of the product.

37. $x, \underline{\hspace{0.5cm}}, y$: $y = xr^2$, $r^2 = \frac{y}{x}$,

$r = \frac{\sqrt{y}}{\sqrt{x}} \cdot \frac{\sqrt{x}}{\sqrt{x}} = \frac{\sqrt{xy}}{x}$,

$x \cdot \frac{\sqrt{xy}}{x} = \sqrt{xy}$

38. $a = \frac{x + y}{2}$, $g = \pm\sqrt{xy}$,

$h = \dfrac{\frac{1}{x} + \frac{1}{y}}{2} = \dfrac{2xy}{x + y}$

$a \cdot h = \frac{x + y}{2} \cdot \frac{2xy}{x + y} = xy$ and

$g^2 = (\pm\sqrt{xy})^2 = xy$, so $g^2 = ah$.

Brainteaser

When the egg is dropped into the boiling water, start both the 7- and 11-minute hourglasses. When the sand stops running in the 7-minute glass, turn it over. When the sand stops in the 11-minute glass, the sand in the 7-minute glass will have been running for 4 minutes, so turn the 7-minute glass again. When the sand stops again in the 7-minute glass, 15 minutes will have elapsed.

▮▮▮GETTING STARTED

Prerequisite Quiz

Find the value of each expression. Use a scientific calculator if available.

1. $\dfrac{2{,}187 - \left(-\frac{1}{3}\right)(-9)}{1 - \left(-\frac{1}{3}\right)}$ 1,638

2. $\dfrac{8 - \left(-\frac{1}{2}\right)\left(-\frac{1}{16}\right)}{1 - \left(-\frac{1}{2}\right)}$ $5\frac{5}{16}$ or 5.3125

3. $\dfrac{12(1 - 3^6)}{1 - 3}$ 4,368

4. $\dfrac{-\frac{1}{4}[1 - (-2)^7]}{1 - (-2)}$ $-10\frac{3}{4}$ or -10.75

5. $\displaystyle\sum_{k=1}^{5} (2k - 3)$ 15

Motivator

Have students draw four squares with sides of 5 mm, 10 mm, 20 mm, and 40 mm and ask them to find the following.

a. The sum of the perimeters
b. The sum of the areas
   (a) 300 mm, (b) 2125 mm²

Ask students if the four widths, four perimeters, and four areas are in three geometric sequences.   Yes

15.6 Geometric Series

Objective

To find the sum of the first n terms of a geometric series
To solve problems involving the sum of a geometric series

The terms of the geometric sequence 3, -6, 12, -24, . . . can be written as an indicated sum. The result is the *geometric series* $3 - 6 + 12 - 24 + \ldots$ with a common ratio of -2.

Definition

A geometric series is the indicated sum of the terms of a geometric sequence.

The sum S of the geometric series $2 + 6 + 18 + 54 + \ldots + 39{,}366$, with $r = 3$, can be found as shown below. The terms are multiplied by $-r$, or -3, and then added to the original terms.

$$
\begin{aligned}
S &= 2 + 6 + 18 + 54 + \ldots + 39{,}366 \\
-3 \cdot S &= \quad -6 - 18 - 54 - \ldots - 39{,}366 - 118{,}098 \\
\hline
-2 \cdot S &= 2 + 0 + \ 0 + \ 0 + \ldots + \quad\quad 0 - 118{,}098 \\
S &= \frac{2 - 118{,}098}{-2} \\
&= \frac{-118{,}096}{-2} = 59{,}048
\end{aligned}
$$

The steps above suggest a formula for the sum S_n of the terms of the geometric series with first term a_1, common ratio r, and last term a_n.

$$(1 - r)S_n = a_1 - ra_n, \quad \text{or} \quad S_n = \frac{a_1 - ra_n}{1 - r}$$

EXAMPLE 1   Find the sum of the terms in the geometric series with first term $a_1 = -9$, common ratio $r = -4$, and last term $a_n = 9{,}216$.

Solution $S_n = \dfrac{a_1 - ra_n}{1 - r} = \dfrac{-9 - (-4)(9{,}216)}{1 - (-4)} = \dfrac{36{,}855}{5} = 7{,}371$

The sum of the series is 7,371.

The two formulas $S_n = \dfrac{a_1 - ra_n}{1 - r}$ and $a_n = a_1 r^{n-1}$ can be combined to provide a formula for the sum S_n that does not involve the last term a_n.

Highlighting the Standards

Standard 4c: The connection between sequences and series is most important and one requiring frequent clarification.

Additional Example 1

Find the sum of the terms in the geometric series with first term $a_1 = -7$, common ratio $r = -3$, and last term $a_n = 15{,}309$.   11,480

> **Sum of the terms of a finite geometric series**
> The sum S_n of the terms of a finite geometric series is given by
>
> (1) $S_n = \dfrac{a_1 - ra_n}{1 - r}$ or (2) $S_n = \dfrac{a_1(1 - r^n)}{1 - r}$,
>
> where a_1 is the first term, a_n is the last term, r is the common ratio, and n is the number of terms.

EXAMPLE 2 Find the sum of the terms of the geometric series
$16 - 8 + 4 - 2 + \ldots - \frac{1}{8}$.

Plan The last term, $-\frac{1}{8}$, is given. Use the formula $S_n = \dfrac{a_1 - ra_n}{1 - r}$.

Solution $a_1 = 16,\; r = \dfrac{-8}{16} = -\dfrac{1}{2},\; a_n = -\dfrac{1}{8}$

$$S_n = \frac{a_1 - ra_n}{1 - r} = \frac{16 - \left(-\frac{1}{2}\right)\left(-\frac{1}{8}\right)}{1 - \left(-\frac{1}{2}\right)} = \frac{16 - \frac{1}{16}}{\frac{3}{2}} = \frac{255}{16} \cdot \frac{2}{3} = \frac{85}{8}$$

The sum is $10\frac{5}{8}$, or 10.625.

EXAMPLE 3 Find the sum of the first 9 terms of the geometric series
$-\frac{1}{8} + \frac{1}{4} - \frac{1}{2} + 1 - \ldots$.

Plan The last term is not given. Use the formula $S_n = \dfrac{a_1(1 - r^n)}{1 - r}$.

Solution $a_1 = -\dfrac{1}{8},\qquad r = \dfrac{1}{4} \div \left(-\dfrac{1}{8}\right) = -2,\qquad n = 9$

$$S_n = \frac{a_1(1 - r^n)}{1 - r} = \frac{-\frac{1}{8}[1 - (-2)^9]}{1 - (-2)} = \frac{-\frac{1}{8}(1 + 512)}{3} = -21\frac{3}{8}$$

The sum is $-21\frac{3}{8}$, or -21.375.

Recall that $\displaystyle\sum_{k=2}^{5} 3k$ symbolizes $3 \cdot 2 + 3 \cdot 3 + 3 \cdot 4 + 3 \cdot 5$, or the arithmetic series $6 + 9 + 12 + 15$ with a common difference of 3.

Similarly, $\displaystyle\sum_{k=2}^{5} 3^k$ represents $3^2 + 3^3 + 3^4 + 3^5$, or the geometric series $9 + 27 + 81 + 243$ with a common ratio of 3.

Lesson Note

Emphasize that the sum formula, $S_n = \dfrac{a_1 - ra_n}{1 - r}$, is used when a_n is given and $S_n = \dfrac{a_1(1 - r^n)}{1 - r}$ is used when a_n is not given.

Math Connections

Historical Note: A formula for finding the sum of a geometric series has been known for well over 2,000 years. Book IX of Euclid's *Elements* is devoted to the theory of numbers. In it, Euclid includes the following statement as Proposition 35: "If as many numbers as we please be in continued proportion, and there be subtracted from the second and the last numbers equal to the first, then as the excess of the second is to the first, so will the excess of the last be to all those before it." When this proposition is expressed in modern algebraic notation, a proportion is obtained that can be shown to be equivalent to the first of the two formulas at the top of page 567.

Critical Thinking Questions

Analysis: Ask students to show how Formula (2) at the top of page 567 can be derived from Formula (1). Have them begin by substituting $a_1 r^{n-1}$ for a_n in Formula (1). The second formula follows directly.

Additional Example 2

Find the sum of the terms of the geometric series $-\frac{1}{4} + 1 - 4 + \cdots - 1{,}024$. $\quad -819\frac{1}{4}$

Additional Example 3

Find the sum of the first 10 terms of the geometric series.

$128 - 64 + 32 - 16 + \cdots$. $\quad 85\frac{1}{4}$

Checkpoint

Find the sum of the terms for each geometric series described below.

1. $a_1 = -4$, $r = 10$, $a_n = -40{,}000$
   $-44{,}444$

2. $\frac{1}{4} + 1 + 4 + \cdots + 256$ $341\frac{1}{4}$

3. $n = 9$; $-3 + 6 - 12 + \cdots$ -513

4. $\sum\limits_{j=3}^{6} (-2)^j$ 40

Closure

Have students give the formula for finding the sum of the terms of a finite geometric series. See page 567 for the two forms of the formula. Have students find the number of terms n and the common ratio r for the geometric series represented by

$\sum\limits_{k=5}^{15} -2(3^{k-4})$. $n = 11$, $r = 3$

 FOLLOW UP

Guided Practice

Classroom Exercises 1–10

Independent Practice

A Ex. 1–22, B Ex. 23–29, C Ex. 30–35

Basic: WE 1–25 odd, Extension, Statistics

Average: WE 5–31 odd, Extension, Statistics

Above Average: WE 11–35 odd, Extension, Statistics

EXAMPLE 4   Write the expanded form for $\sum\limits_{j=1}^{4} 6(-3)^j$. Find the sum of the terms.

Solution
$$\sum\limits_{j=1}^{4} 6(-3)^j = 6(-3)^1 + 6(-3)^2 + 6(-3)^3 + 6(-3)^4 \leftarrow \text{expanded form}$$
$$= -18 + 54 - 162 + 486 = 360$$

The sum of the terms is 360.

EXAMPLE 5   On each bounce, a ball dropped from a height of 800 ft rebounds one-half of the distance from which it falls. How far will it have traveled when it hits the ground the 8th time? See the figure on page 561.

Plan   The ball descends 8 times and rebounds 7 times.

The descent and rebound distances form a series, $800 + 400 + 400 + 200 + 200 + \ldots$, for 15 terms.

The series of descents is the geometric series given by $D = 800 + 400 + 200 + \ldots$, for 8 terms with $r = \frac{1}{2}$.

The series of rebounds is the geometric series given by $U = 400 + 200 + 100 + \ldots$, for 7 terms with $r = \frac{1}{2}$.

Solution
$$D = \frac{a_1(1 - r^n)}{1 - r} = \frac{800\left[1 - \left(\frac{1}{2}\right)^8\right]}{1 - \frac{1}{2}} = \frac{800\left(1 - \frac{1}{256}\right)}{\frac{1}{2}} = 1{,}593.75$$

$$U = \frac{a_1(1 - r^n)}{1 - r} = \frac{400\left[1 - \left(\frac{1}{2}\right)^7\right]}{1 - \frac{1}{2}} = \frac{400\left(1 - \frac{1}{128}\right)}{\frac{1}{2}} = 793.75$$

$$D + U = 1{,}593.75 + 793.75 = 2{,}387.5$$

The ball will have traveled $2{,}387.5$ ft when it hits the ground the 8th time.

Classroom Exercises

Find the common ratio r for each geometric series.

1. $18 + 0.18 + 0.0018 + \ldots$ 0.01

2. $0.007 + 0.07 + 0.7 + 7 + \ldots$ 10

3. $\sum\limits_{k=1}^{6} (-2)^k$ -2

4. $\sum\limits_{j=1}^{6} 3(5^j)$ 5

5. $\sum\limits_{j=0}^{2} \left(\frac{1}{3}\right)^{j-1}$ $\frac{1}{3}$

6. $\sum\limits_{p=3}^{10} 5(2^{p-3})$ 2

7–10. Find the sum of the terms of each series in Exercises 3–6 above.
   7. 42   **8.** 58,590   **9.** $\frac{13}{3}$ **10.** 1,275

Additional Example 4

Write the expanded form for $\sum\limits_{k=1}^{4} 2\left(\frac{1}{4}\right)^{k-1}$.

Find the sum of the terms.
$2 + \frac{1}{2} + \frac{1}{8} + \frac{1}{32}; \frac{85}{32}$

Additional Example 5

Find the sum of the first 17 terms of the series $400 + 200 + 200 + 100 + 100 + 50 + 50 + \cdots$.   1,196.875 or $1{,}196\frac{7}{8}$

Written Exercises

Find the sum of the terms for each geometric series described below.

1. $a_1 = 5$, $r = 10$, $a_n = 50,000$ 55,555   **2.** $a_1 = -8$, $r = -4$, $a_n = -32,768$ −26,216

3. $a_1 = -486$, $r = \frac{1}{3}$, $a_n = -6$ −726   **4.** $a_1 = 2,187$, $r = -\frac{1}{3}$, $a_n = 27$ 1,647

5. $32 + 16 + 8 + \ldots + \frac{1}{4}$ $63\frac{3}{4}$ **6.** $\frac{1}{3} + 1 + 3 + \ldots + 243$ $364\frac{1}{3}$

7. $0.4 + 4 + 40 + \ldots + 40,000$ 44,444.4   **8.** $800 - 80 + 8 - \ldots + 0.0008$ 727.2728

9. $n = 10$; $3 - 6 + 12 - 24 + \ldots$ **10.** $n = 9$; $-5 - 10 - 20 - 40 - \ldots$

11. $n = 7$; $0.03 + 0.3 + 3 + 30 + \ldots$ **12.** $n = 7$; $-1 + 10 - 100 + 1,000 - \ldots$

13. $n = 8$; $-4 + 8 - 16 + 32 - \ldots$ 340   **14.** $n = 8$; $64 - 32 + 16 - 8 + \ldots$ 42.5

15. $a_1 = 11$, $r = 3$, $n = 6$ 4,004   **16.** $a_1 = -6$, $r = -3$, $n = 5$ −366

17. $a_1 = \frac{1}{5}$, $r = 10$, $n = 6$ 22,222.2   **18.** $a_1 = 0.6$, $r = -10$, $n = 7$ 545,454.6

19. $\sum\limits_{k=1}^{7} 2^k$ 254   **20.** $\sum\limits_{j=4}^{7} (-3)^j$ −1,620   **21.** $\sum\limits_{p=0}^{6} \left(\frac{1}{2}\right)^p$ $\frac{127}{64}$ **22.** $\sum\limits_{k=5}^{8} (0.2)^{k-4}$ 0.2496

23. $\sum\limits_{j=1}^{4} 5(-2)^{j-1}$ −25   **24.** $\sum\limits_{p=1}^{4} \frac{1}{6}(3^p)$ 20   **25.** $\sum\limits_{k=1}^{5} -32\left(\frac{1}{4}\right)^{k-1}$ $-42\frac{5}{8}$ **26.** $\sum\limits_{j=2}^{7} 27\left(-\frac{1}{3}\right)^{j-2}$ $20\frac{2}{9}$

27. Smaller and smaller squares are formed consecutively, as shown at the right. Find the sum of the perimeters of the first 9 squares if the first square is 40 in. wide. $319\frac{3}{8}$ in

28. Find the sum of the areas of the first 6 squares at the right if the first square is 40 in. wide. $2,132\frac{13}{16}$ in^2

29. On each bounce, a certain golf ball rebounds two-thirds of the distance from which it falls. If the ball falls from a height of 81 yd, how far will it have traveled when it hits the ground the 7th time? $376\frac{5}{9}$ yd

Find the sum of the terms of each series.

30. $\sum\limits_{k=1}^{5} (k - 1)!$ 34   **31.** $\sum\limits_{n=1}^{5} \frac{1}{(n - 1)!}$ $2\frac{17}{24}$ **32.** $\sum\limits_{j=1}^{5} \binom{4}{j - 1}$ 16

33. $\sum\limits_{p=1}^{5} \frac{p}{(p - 1)!}$ $5\frac{3}{8}$ **34.** $\sum\limits_{k=1}^{5} \frac{2^{k-1}}{(k - 1)!}$ 7   **35.** $\sum\limits_{j=1}^{3} \frac{(-1)^{j-1}(0.2)^{2j-2}}{(2j - 2)!}$

   $\frac{14,701}{15,000}$

Mixed Review

Let $f(x) = 3x - 2$ and $g(x) = x^2 + 4$. Find the following.   *3.2, 5.5, 5.9*

1. $f(2.8)$ 6.4   **2.** $g(c - 5)$ $c^2 - 10c + 29$ **3.** $f(g(-4))$ 58   **4.** $g(f(-1))$ 29

Additional Answers

Written Exercises

9. −1,023
10. −2,555
11. 33,333.33
12. −909,091

Enrichment

Have the students evaluate these expressions by inspection alone.

1. $\sum\limits_{n=1}^{10} n - \sum\limits_{n=1}^{9} n$ 10   **4.** $\sum\limits_{n=1}^{20} 2n + \sum\limits_{n=1}^{10} 3n - \sum\limits_{n=1}^{19} 2n - \sum\limits_{n=1}^{9} 3n$ 70

2. $\sum\limits_{n=1}^{3} 5n - \sum\limits_{n=1}^{3} n$ 24

3. $\sum\limits_{n=1}^{100} n - \sum\limits_{n=1}^{99} n$ 100

1. (1) Verify for $n = 1$:

$2 + 4 + 6 + \cdots + 2n = n(n + 1)$

$2(1) \stackrel{?}{=} 1(1 + 1)$

$\quad 2 = 2$ True

(2) Assume true for $n = k$:

$2 + 4 + 6 + \cdots + 2k = k(k + 1)$

Adding the $(k + 1)$st term to both sides:

$2 + 4 + 6 + \cdots + 2k + (2k + 2)$

$= k(k + 1) + 2k + 2$

$= k^2 + 3k + 2$

$= (k + 1)(k + 2)$

If $n = k + 1$:

$2 + 4 + 6 + \cdots + 2(k + 1)$

$= (k + 1)[(k + 1) + 1]$

$= (k + 1)(k + 2)$

(1) and (2) are both true. So, by mathematical induction, the statement is true.

2. (1) Verify for $n = 1$:

$7 + 9 + 11 + \cdots + (2n + 5) = n(n + 6)$

$(2 \cdot 1 + 5) \stackrel{?}{=} 1(1 + 6)$

$\quad\quad\quad 7 = 7$ True

(2) Assume true for $n = k$:

$7 + 9 + 11 + \cdots + (2k + 5) = k(k + 6)$

Adding the $(k + 1)$st term to both sides:

$7 + 9 + 11 + \cdots + (2k + 5) + (2k + 7)$

$= k(k + 6) + 2k + 7$

$= k^2 + 8k + 7$

$= (k + 1)(k + 7)$

If $n = k + 1$:

$7 + 9 + 11 + \cdots + [2(k + 1) + 5]$

$= (k + 1)(k + 1 + 6)$

$= (k + 1)(k + 7)$

(1) and (2) are both true. So, by mathematical induction, the statement is true.

See page 571 for the answer to Ex. 3.

Extension

Mathematical Induction

Suppose you want to prove the following statement, S_1: For all natural numbers n, $5 + 11 + 17 + \ldots + (6n - 1) = n(3n + 2)$. Trials using values such as $n = 1, 2, 3$ may prove S_1 true for these values, but a finite number of trials will not show that S_1 is true for *all* natural numbers n. To prove the statement, the following principle must be used.

Principle of Mathematical Induction

A statement S is true for all natural numbers n if both (1) and (2) are true: (1) S is true for $n = 1$.

(2) If S is true for $n = k$, *then* S is true for $n = k + 1$.

Since (1) guarantees that S is true for $n = 1$ and (2) guarantees that S is true for $n = 1 + 1$, or 2, and so on forever, the principle is reasonable.

Example

Prove S_1: For all natural numbers n, $5 + 11 + 17 + \ldots + (6n - 1)$
$= n(3n + 2)$.

Solution

(1) Show that S_1 is true for $n = 1$. $5 = 1(3 \cdot 1 + 2)$, or 5

(2) Show that if S_1 is true for $n = k$, then S_1 is true for $n = k + 1$, or
$5 + 11 + 17 + \ldots + [6(k + 1) - 1] = (k + 1)[3(k + 1) + 2]$.

Now if we assume that S_1 is true for $n = k$, then
$5 + 11 + 17 + \ldots + (6k - 1) = k(3k + 2)$. Adding the
$(k + 1)$st term, $6(k + 1) - 1$, to each side gives the following.

$5 + 11 + 17 + \ldots + (6k - 1) + [6(k + 1) - 1]$

$= k(3k + 2) + [6(k + 1) - 1]$

$= 3k^2 + 8k + 5 = (k + 1)(3k + 5) = (k + 1)[3(k + 1) + 2]$

So, if S_1 is true for $n = k$, then S_1 is true for $n = k + 1$. Thus, (1) and (2) are true, and by mathematical induction, S_1 is true.

Exercises

Use mathematical induction to prove that each statement is true for all natural numbers n.

1. $2 + 4 + 6 + \ldots + 2n = n(n + 1)$

2. $7 + 9 + 11 + \ldots + (2n + 5) = n(n + 6)$

3. $1 + 3 + 5 + \ldots + (2n - 1) = n^2$

Statistics: *Expected Value*

An insurance company must calculate the expected value of claims in order to cover their costs and set rates. A sociologist may want to know the average time-sentence for felons convicted of a particular crime. A person thinking about entering a contest may wish to compare the expected return with the contest's entry fee.

Suppose a random variable x assumes values $x_1, x_2, \ldots, x_n$, and that these values have equal probabilities of occurring. Then, the expected value is simply the numerical average of the values. However, in many instances, some values are more likely to occur than others. To reflect this, expected value is calculated by multiplying each possible value of the random variable by its probability and adding the results.

For a random variable x that assumes values $x_1, x_2, \ldots, x_i, \ldots, x_n$ with probabilities $P(x = x_1), P(x = x_2), \ldots, P(x = x_i), \ldots,$ $P(x = x_n)$, the expected value of x, or $E(x)$, is defined by

$$E(x) = \sum_{i=1}^{n} x_i \cdot P(x = x_i).$$

You can use the above summation to solve the following problem.

Data is collected by a company that writes hospitalization-insurance policies. They estimate the average claim for a hospital stay is $600 per year, with a probability of 0.42 for this type of claim. The average out-patient claim is $45 per year, with a probability of 0.03. Special-care claims average $520 per year, with a 0.15 probability of occurring. The probability of no claim is 0.4. What must the insurance company charge per policy, just to cover expected claims?

The random variable x is the amount paid on a claim, with the probabilities as given. So, using the formula above gives the following.

$$E(x) = 600(0.42) + 45(0.03) + 520(0.15) + 0(0.4) = 331.35$$

The company needs to charge $331.35 per year to cover costs.

Exercises

1. If a first offender is convicted of a certain crime, the probabilities of various jail sentences are as follows: $P(3 \text{ years}) = 0.1$, $P(2 \text{ years}) = 0.2$, $P(1 \text{ year}) = 0.3$, $P(\text{probation}) = 0.4$. What is the expected length of time a first offender will spend in jail? 1 year

2. One million entries are accepted for a state contest that offers the following prizes: one $50,000 prize, 2 prizes of $25,000, 6 prizes of $5,000, 100 prizes of $100, and 1,000 prizes of $10. What is the expected value of an entry? If the entry fee is $1 per entry, is it to your advantage to enter? 15¢, no

3. (1) Verify for $n = 1$:
$1 + 3 + 5 + \cdots + (2n - 1) = n^2$
$(2 \cdot 1 - 1) \stackrel{?}{=} 1^2$
$\qquad 1 = 1$ True
(2) Assume true for $n = k$:
$1 + 3 + 5 + \cdots + (2k - 1) = k^2$
Adding the $(k + 1)$st term to both sides:
$1 + 3 + 5 + \cdots + (2k - 1) + (2k + 1)$
$= k^2 + 2k + 1$
$= (k + 1)^2$
If $n = k + 1$:
$1 + 3 + 5 + \cdots + [2(k + 1) - 1]$
$= (k + 1)^2$
(1) and (2) are both true. So, by mathematical induction, the statement is true.

Teaching Resources

Project Worksheet 15
Quick Quizzes 117
Reteaching and Practice
   Worksheets 117

▬▬GETTING STARTED

Prerequisite Quiz

Find the value of each sum or expression.

1. $3 + 1 + \frac{1}{3} + \frac{1}{9} + \frac{1}{27}$ $4\frac{13}{27}$

2. $\dfrac{16}{1 - \left(-\frac{1}{4}\right)}$ $\frac{64}{5}$, or $12\frac{4}{5}$

3. $\dfrac{5}{1 - 0.1}$ $\frac{50}{9}$, or $5\frac{5}{9}$

4. $\dfrac{0.37}{1 - 0.01}$ $\frac{37}{99}$

5. $50 + 5 + 0.5 + 0.05 + 0.005$ 55.555

Motivator

Draw the figure from Written Exercise 18 on the chalkboard. Tell students that the midpoints of the sides of the square are joined to create a new square. If this process is carried out infinitely many times, ask students if the sum of the areas of the squares will be infinitely large. Have them discuss this question.

▬▬TEACHING SUGGESTIONS

Lesson Note

Emphasize that an infinite geometric series such as $3 + 6 + 12 + 24 + \cdots$, with $r = 2$, does not have a sum.

Highlighting the Standards

Standard 13b: The concept that an infinite series can have a sum is one of the great discoveries of mathematics. For some students, it will lead to the study of calculus.

Objectives
To find the sum, if it exists, of an infinite geometric series
To solve problems involving the sum of an infinite geometric series
To write a repeating decimal as an infinite geometric series
To write a repeating decimal in the form $\frac{x}{y}$, where x and y are integers

---

An **infinite geometric series**, such as $2 + 1 + \frac{1}{2} + \frac{1}{4} + \ldots$, has no last term. However, the "sum" of its terms appears to approach 4 when you add the first n terms for $n = 1, 2, 3, 4, \ldots$ as follows.

$$2, \quad 2 + 1 = 3, \quad 2 + 1 + \tfrac{1}{2} = 3\tfrac{1}{2}, \quad 2 + 1 + \tfrac{1}{2} + \tfrac{1}{4} = 3\tfrac{3}{4}$$

The sequence $2, 3, 3\frac{1}{2}, 3\frac{3}{4}, \ldots$ is called the sequence of **partial sums** for the given series. The table below shows how the partial sums get closer and closer to 4, or *converge* to 4.

n	1	2	3	4	5	6	7	8	. . .
Partial sums	2	3	$3\frac{1}{2}$	$3\frac{3}{4}$	$3\frac{7}{8}$	$3\frac{15}{16}$	$3\frac{31}{32}$	$3\frac{63}{64}$	. . .

If $r = \frac{1}{2}$ in the sum formula $S_n = \dfrac{a_1(1 - r^n)}{1 - r}$, notice how rapidly r^n decreases as n increases.

$$\left(\tfrac{1}{2}\right)^1 = \tfrac{1}{2}, \ \left(\tfrac{1}{2}\right)^5 = \tfrac{1}{32}, \ \left(\tfrac{1}{2}\right)^{10} = \tfrac{1}{1,024}, \ \left(\tfrac{1}{2}\right)^{15} = \tfrac{1}{32,768}, \ \cdots$$

As n increases without end, $\left(\frac{1}{2}\right)^n$ approaches 0 as a *limit*. In general, when n increases without end and $-1 < r < 1$, $r \neq 0$, then

$$S_n = \frac{a_1(1 - r^n)}{1 - r} \text{ approaches } \frac{a_1(1 - 0)}{1 - r}, \text{ or } \frac{a_1}{1 - r}, \text{ as a } \textit{limit.}$$

This limit is called the *sum* of the infinite geometric series
$$a_1 + a_1 r + a_1 r^2 + a_1 r^3 + \ldots.$$

Sum of an infinite geometric series
The sum S of an infinite geometric series is given by
$$S = \frac{a_1}{1 - r} \quad (-1 < r < 1, r \neq 0),$$
where a_1 is the first term and r is the common ratio.

EXAMPLE 1 Find the sum, if it exists, of each infinite geometric series.

a. $2 + 1 + \frac{1}{2} + \frac{1}{4} + \ldots$ b. $16 - 4 + 1 - \frac{1}{4} + \ldots$

c. $\frac{1}{8} + \frac{1}{4} + \frac{1}{2} + \ldots$

Solutions

a. $r = \frac{1}{2}$; $S = \dfrac{a_1}{1-r} = \dfrac{2}{1 - \frac{1}{2}} = 2 \div \frac{1}{2} = 4$

b. $r = -\frac{1}{4}$; $S = \dfrac{a_1}{1-r} = \dfrac{16}{1 - \left(-\frac{1}{4}\right)} = 16 \div \frac{5}{4} = \frac{64}{5} = 12\frac{4}{5}$

c. $r = 2$; Since 2 is not between -1 and 1, the sum in **c** does not exist.

The repeating decimal $7.77\ldots$ can be written as the infinite geometric series $7 + 0.7 + 0.07 + \ldots$, where $a_1 = 7$ and $r = 0.1$.

The sum $S = \dfrac{a_1}{1-r} = \dfrac{7}{1 - 0.1} = \dfrac{7(10)}{0.9(10)} = \dfrac{70}{9}$. Thus, $7.77\ldots = \dfrac{70}{9}$.

EXAMPLE 2 Write the repeating decimal $0.535353\ldots$, or $0.53\overline{53}$, as indicated.

a. as an infinite geometric series
b. in the form $\frac{x}{y}$, where x and y are integers

Solutions

a. $0.53\overline{53} = 0.53 + 0.0053 + 0.000053 + \ldots$

b. $a_1 = 0.53$ $r = 0.01$ $S = \dfrac{0.53}{1 - 0.01} = \dfrac{0.53}{0.99} \cdot \dfrac{100}{100} = \dfrac{53}{99}$

So, $0.53\overline{53} = \dfrac{53}{99}$. ← THINK: Can you use a calculator to check the answer?

Classroom Exercises

For each sequence of partial sums that converges, find the limiting value.

1. $\dfrac{1}{2}, \dfrac{3}{4}, \dfrac{7}{8}, \dfrac{15}{16}, \dfrac{31}{32}, \dfrac{63}{64}, \ldots$ 1 2. $\dfrac{1}{8}, \dfrac{3}{16}, \dfrac{9}{32}, \dfrac{27}{64}, \dfrac{81}{128}, \dfrac{243}{256}, \dfrac{729}{512}, \dfrac{2{,}187}{1{,}024}, \ldots$

Does not converge

Find the sum, if it exists, of each infinite geometric series.

3. $\dfrac{2}{3} + \dfrac{1}{3} + \dfrac{1}{6} + \dfrac{1}{12} + \ldots$ $1\frac{1}{3}$ 4. $0.1 + 0.01 + 0.001 + 0.0001 + \ldots$ $\frac{1}{9}$

Find the first three terms when the repeating decimal is written as an infinite geometric series.

5. $0.222\ldots$ 6. $4.4\overline{4}$ 7. $0.333\ldots$ 8. $0.34\overline{34}$ 9. $37.\overline{37}$ 10. $3.9\overline{39}$

15.7 Infinite Geometric Series **573**

Math Connections

Infinite Products: Students who have been introduced to Pi-notation for expressing products (see *Critical Thinking Questions* for Lesson 15.3) may be interested in the following example of an infinite product.

$$\frac{2}{\pi} = \prod_{n=1}^{\infty} \left(1 - \frac{1}{4n^2}\right)$$

This formula allows one to calculate π to any desired degree of accuracy. However, it can be written in a more accessible form as the following infinite product discovered by the English mathematician John Wallis (1616–1703).

$$\frac{\pi}{2} = \frac{2}{1} \cdot \frac{2}{3} \cdot \frac{4}{3} \cdot \frac{4}{5} \cdot \frac{6}{5} \cdot \frac{6}{7} \cdots$$

Critical Thinking Questions

Analysis: Ask students to derive Wallis' Formula from the infinite product shown in *Math Connections* above. As a hint, have them begin by showing that the expression in parentheses is equivalent to $\dfrac{(2n-1)(2n+1)}{2n \cdot 2n}$.

Checkpoint

1. Find the sum of the infinite geometric series $24 - 12 + 6 - 3 \cdots$. 16
2. Write the repeating decimal $0.37\overline{37}$ as an infinite geometric series and in the form $\frac{x}{y}$, where x and y are integers.
$0.37 + 0.0037 + 0.000037 + \cdots$; $\frac{37}{99}$

Closure

Ask students to give the formula for finding the sum of an infinite geometric series.

$$S_n = \frac{a_1}{1-r} \quad (-1 < r < 1, r = 0)$$

Have students determine the conditions under which $\dfrac{a(1-r^n)}{1-r}$ will be equal to $\dfrac{a}{1-r}$.
$-1 < r < 1, r \neq 0, n$ increases without end.

Additional Example 1

Find the sum, if it exists, of each infinite geometric series.

a. $18 + 6 + 2 + \frac{2}{3} + \cdots$ 27

b. $\frac{1}{12} + \frac{1}{2} + 3 + \cdots$ Sum does not exist.

c. $50 - 10 + 2 - \frac{2}{5} + \cdots$ $\frac{125}{3}$, or $41\frac{2}{3}$

Additional Example 2

Write the repeating decimal $4.4\overline{4}$ as indicated.

a. As an infinite geometric series
$4 + 0.4 + 0.04 + \cdots$

b. In the form $\frac{x}{y}$, where x and y are integers
$\frac{40}{9}$

■ FOLLOW UP

Guided Practice

Classroom Exercises 1–10

Independent Practice

A Ex. 1–12, **B** Ex. 13–22, **C** Ex. 23, 24

Basic: WE 1–17 odd
Average: WE 5–23 odd
Above Average: WE 7–23 odd, 24

Additional Answers

Classroom Exercises

5. $0.2 + 0.02 + 0.002 + \cdots$
6. $4 + 0.4 + 0.04 + \cdots$
7. $0.3 + 0.03 + 0.003 + \cdots$
8. $0.34 + 0.0034 + 0.000034 + \cdots$
9. $37 + 0.37 + 0.0037 + \cdots$
10. $3.9 + 0.039 + 0.00039 + \cdots$

Written Exercises

1. Does not exist; $r = 2$
10. $0.8 + 0.08 + 0.008 + \cdots$; $\frac{8}{9}$
11. $5 + 0.5 + 0.05 + \cdots$; $\frac{50}{9}$
12. $0.75 + 0.0075 + 0.000075 + \cdots$; $\frac{25}{33}$
13. $8.1 + 0.081 + 0.00081 + \cdots$; $\frac{90}{11}$
14. $0.064 + 0.00064 + 0.0000064 + \cdots$; $\frac{32}{495}$
15. $3.12 + 0.00312 + 0.00000312 + \cdots$; $\frac{1,040}{333}$
19. $4, 6, 7, 7\frac{1}{2}, 7\frac{3}{4}, 7\frac{7}{8}, 7\frac{15}{16}, 7\frac{31}{32}$; converges to 8.
20. $27, 36, 39, 40, 40\frac{1}{3}, 40\frac{4}{9}, 40\frac{13}{27}, 40\frac{40}{81}$; converges to $40\frac{1}{2}$.
21. $27, 18, 21, 20, 20\frac{1}{3}, 20\frac{2}{9}, 20\frac{7}{27}, 20\frac{20}{81}$; converges to $20\frac{1}{4}$.

Written Exercises

Find the sum, if it exists, of each infinite geometric series.

1. $\frac{1}{8} + \frac{1}{4} + \frac{1}{2} + \dots$
2. $-1 + \frac{1}{3} - \frac{1}{9} + \dots$ $-\frac{3}{4}$
3. $-\frac{1}{4} - \frac{1}{8} - \frac{1}{16} - \dots$ $-\frac{1}{2}$
4. $27 - 9 + 3 - \dots$ $20\frac{1}{4}$
5. $4 + 0.4 + 0.04 + \dots$ $4\frac{4}{9}$
6. $60 + 6 + 0.6 + \dots$ $66\frac{2}{3}$
7. $16 + 8 + 4 + \dots$ 32
8. $\frac{1}{2} - \frac{1}{4} + \frac{1}{8} - \dots$ $\frac{1}{3}$
9. $12 + 8 + \frac{16}{3} + \dots$ 36

Write each repeating decimal as an infinite geometric series and then in the form $\frac{x}{y}$, where x and y are integers.

10. $0.888 \ldots$
11. $5.55 \ldots$
12. $0.75\overline{75}$
13. $8.1\overline{81}$
14. $0.064\overline{64}$
15. $3.12\overline{312}$

16. A golf ball, dropped from a height of 81 yd, rebounds on each bounce two-thirds of the distance from which it falls. How far does it travel before coming to rest? 405 yd

17. The midpoints of the sides of an equilateral triangle are connected to create a new triangle. This procedure is repeated on each new triangle. Find the sum of the perimeters of an infinite sequence of the triangles if one side of the original triangle is 80 cm long. 480 cm

80 cm

18. The midpoints of the sides of a square are joined to create a new square. This process is performed on each new square. Find the sum of the areas of an infinite sequence of such squares if one side of the first square is 10 ft long. 200 ft²

10 ft

Find the sequence of partial sums of each series for $n = 1, 2, 3, \ldots, 8$. To what value does the sequence of partial sums converge?

19. $4 + 2 + 1 + \dots$
20. $27 + 9 + 3 + \dots$
21. $27 - 9 + 3 - \dots$

22. Find the sum of the infinite geometric series
$6\sqrt{2} + 6 + 3\sqrt{2} + 3 + \dots$ $12\sqrt{2} + 12$

For what values of x does each infinite geometric series have a sum?

23. $(x - 2)^1 + (x - 2)^2 + (x - 2)^3 + \dots$ $1 < x < 3$
24. $(x - 1)^0 + (x - 1)^1 + (x - 1)^2 + \dots$ $0 < x < 2, x \neq 1$

Mixed Review

Evaluate each expression using $\binom{n}{r} = \dfrac{n!}{r!(n - r)!}$. *14.5*

1. $\binom{7}{4}$ 35
2. $\binom{5}{0}$ 1
3. $\binom{6}{6}$ 1
4. $\binom{3}{0} + \binom{3}{1} + \binom{3}{2} + \binom{3}{3}$ 8

Enrichment

Write this expression on the chalkboard.

$$\lim_{x\to\infty} \frac{2x^2 - x + 1}{5x^2 + x - 1}$$

Explain that this refers to the value of the rational expression as x becomes larger and larger indefinitely, or to "the limit of the expression as x approaches positive infinity."

After the students try to evaluate this expression, explain that the value is easy to find if each term is divided by the highest power of x, in this case, x^2.

$$\lim_{x\to\infty} \frac{2 - \frac{1}{x} + \frac{1}{x^2}}{5 + \frac{1}{x} - \frac{1}{x^2}}$$

In this form, it is clear that as x gets larger,

all the terms except 2 and 5 tend to equal zero, making the limit $\frac{2}{5}$.

Have the students use this strategy to evaluate these expressions, if a limit exists.

$$\lim_{x\to\infty} \frac{x^4 - x^3}{x^5}$$ 0

$$\lim_{x\to\infty} \frac{x^5}{x^4 - x^3}$$ The limit does not exist.

15.8 The Binomial Theorem

Objectives

To expand a positive integral power of a binomial
To find a specified term of a binomial expansion

Lesson 5.5 presented the special product $(a + b)^2 = a^2 + 2ab + b^2$. Is there also a "special product" for $(a + b)^3$ or $(a + b)^4$? In this lesson you will learn how to expand such products.

The expansions of $(a + b)^n$ for $n = 0, 1, 2, 3, 4$ show patterns that can be used to expand $(a + b)^n$ for any positive integer n.

$$(a + b)^0 = 1$$
$$(a + b)^1 = 1a^1 + 1b^1$$
$$(a + b)^2 = 1a^2 + 2a^1b^1 + 1b^2$$
$$(a + b)^3 = 1a^3 + 3a^2b^1 + 3a^1b^2 + 1b^3$$
$$(a + b)^4 = 1a^4 + 4a^3b^1 + 6a^2b^2 + 4a^1b^3 + 1b^4$$

Notice the following in the expansions of $(a + b)^n$ above.

(1) The number of terms is always $n + 1$.
(2) The first term is a^n and the last term is b^n.
(3) The exponent of a decreases by 1 from term to term.
(4) The exponent of b increases by 1 from term to term.
(5) For each term, the sum of the exponents of a and b is n.
(6) The coefficients are symmetrical. That is, they read the same from left to right as from right to left.
(7) There is a pattern for finding the coefficients in any expansion.

Note that $(a + b)^4 = 1 \cdot a^4 + 4 \cdot a^3b + 6 \cdot a^2b^2 + 4 \cdot ab^3 + 1 \cdot b^4$

$$= \binom{4}{0}a^4 + \binom{4}{1}a^3b + \binom{4}{2}a^2b^2 + \binom{4}{3}ab^3 + \binom{4}{4}b^4$$

To expand $(a + b)^5$: (1) The coefficients will be $\binom{5}{k}$ for $k = 0, 1, 2, 3, 4, 5$.

(2) The exponents of a will decrease from 5 to 0.
(3) The exponents of b will increase from 0 to 5.

So, $(a + b)^5 = \binom{5}{0}a^5 + \binom{5}{1}a^4b^1 + \binom{5}{2}a^3b^2 + \binom{5}{3}a^2b^3 + \binom{5}{4}ab^4 + \binom{5}{5}b^5$

$$= 1a^5 + 5a^4b + 10a^3b^2 + 10a^2b^3 + 5ab^4 + 1b^5$$

15.8 The Binomial Theorem **575**

Teaching Resources

Quick Quizzes 118
Reteaching and Practice
   Worksheets 118
Transparency 34

GETTING STARTED

Prerequisite Quiz

Simplify. Recall that $\binom{n}{r} = \dfrac{n!}{r!(n - r)!}$.

1. $\binom{6}{0}$ 1   **2.** $\binom{8}{8}$ 1   **3.** $\binom{7}{3}$ 35

4. $\binom{4}{2}(2x)^2(-3)^2$ $216x^2$

Motivator

Have students expand $(a + b)^4$ as follows.

$(a + b)^2 \cdot (a + b)^2 =$
$(a^2 + 2ab + b^2)(a^2 + 2ab + b^2) =$
$a^4 + 4a^3b + 6a^2b^2 + 4ab^3 + b^4$

Have them estimate the time it would take to expand $(a + b)^7$ by this technique. Tell them that they will expand $(a + b)^n$, where n is an integer, by a simpler procedure in this lesson.

TEACHING SUGGESTIONS

Lesson Note

Before considering Example 1, write the expansion of $(a + b)^3$ using some of the patterns 1–7 listed on page 575.

Highlighting the Standards

Standard 3d: The inductive approach to the binomial theorem stretches and challenges students to follow closely a mathematical argument and to judge for themselves whether it seems convincing.

Math Connections

Pascal's Triangle: An alternate method of obtaining the coefficients of an expanded binomial is to use Pascal's Triangle, as illustrated below. The numbers of each row are the coefficients of an expanded binomial with the nth row representing the coefficients of the binomial $(a + b)^n$, $n = 0, 1, 2, 3, \cdots$

```
            1
          1   1
        1   2   1
      1   3   3   1
    1   4   6   4   1
  1        · · ·
```

Each pair of successive numbers in each row is added to give the number that lies between the pair in the row directly below.

Critical Thinking Questions

Analysis: Ask students whether they can write the rule for generating the numbers of Pascal's Triangle (see *Math Connections* above) using the notaton for combinations. The relationship in question is

$$\binom{n}{r-1} + \binom{n}{r} = \binom{n+1}{r}.$$

Common Error Analysis

When students expand a binomial such as $(2x - y^2)^3$ in Example 1, they may forget to place parentheses about $2x$ and $-y^2$ within the terms of the expansion.

Emphasize that these grouping symbols must be used in order to achieve the proper expansion.

Theorem 15.1

Binomial Theorem: If n is a positive integer, then
$$(a + b)^n = \binom{n}{0}a^n + \binom{n}{1}a^{n-1}b + \binom{n}{2}a^{n-2}b^2 + \binom{n}{3}a^{n-3}b^3 + \ldots + \binom{n}{n}b^n.$$

EXAMPLE 1 Expand $(2x - y^2)^3$. Simplify each term.

Plan Rewrite $(2x - y^2)^3$ as $[2x + (-y^2)]^3$ and use the binomial theorem.

Solution $a = 2x$, $b = -y^2$, $n = 3$

$(a + b)^n = [2x + (-y^2)]^3$

$= \binom{3}{0}(2x)^3 + \binom{3}{1}(2x)^2(-y^2)^1 + \binom{3}{2}(2x)^1(-y^2)^2 + \binom{3}{3}(-y^2)^3$

$= 1 \cdot 8x^3 + 3 \cdot 4x^2(-y^2) + 3 \cdot 2x \cdot y^4 + 1(-y^6)$

$= 8x^3 - 12x^2y^2 + 6xy^4 - y^6$

Notice the pattern of signs in a binomial expansion.
(1) In the expansion of $(a + b)^n$, the signs are all $+$.
(2) In the expansion of $(a - b)^n$, the signs alternate $+$ and $-$.

The rth term in the expansion of $(a + b)^n$
The rth term in the expansion of $(a + b)^n$ is
$$\binom{n}{r-1}a^{n-(r-1)}b^{r-1}, \quad \text{or} \quad \binom{n}{r-1}a^{n-r+1}b^{r-1}.$$

EXAMPLE 2 Find the 7th term in the expansion of $(a + b)^9$.

Solution **Method 1:** The factors are $\binom{9}{6}$, b^6, and a^3, since the sum of the exponents is 9. Thus, the 7th term is $\binom{9}{6}a^3b^6$, or $84a^3b^6$.

Method 2: Use the formula for the rth term with $n = 9$ and $r = 7$. The 7th term is $\binom{9}{7-1}a^{9-7+1}b^{7-1} = \binom{9}{6}a^3b^6$, or $84a^3b^6$.

EXAMPLE 3 Find the 6th term in the expansion of $(3x - y^2)^7$.

Solution $(3x - y^2)^7 = [3x + (-y^2)]^7$ Use $\binom{7}{5}a^2b^5$ with $a = 3x$ and $b = -y^2$.

The 6th term is $\binom{7}{5}(3x)^2(-y^2)^5 = 21 \cdot 9x^2(-y^{10}) = -189x^2y^{10}$.

Additional Example 1

Expand $(x^2 - 3y)^4$. Simplify each term.
$x^8 - 12x^6y + 54x^4y^2 - 108x^2y^3 + 81y^4$

Additional Example 2

Find the 4th term in the expansion of $(a + b)^7$. $35a^4b^3$

Additional Example 3

Find the middle term in the expansion of $(x^2 - 2y)^8$. $1,120x^8y^4$

Classroom Exercises

For each term of a binomial expansion, find the value of k.

1. $\binom{k}{3}y^3$ is the 4th term. 3   **2.** $\binom{2}{0}c^k$ is the first term. 2   **3.** $\binom{4}{k}x^{4-k}y^k$ is the middle term. 2

Simplify.

4. $\binom{7}{0}c^7$ c^7 **5.** $\binom{8}{7}c(-d)^7$ $-8cd^7$ **6.** $\binom{8}{8}(-d)^8$ d^8 **7.** $\binom{6}{5}(2c)(-1)^5$ $-12c$

8. Expand and simplify the terms. $(x+1)^8$ $x^8 + 8x^7 + 28x^6 + 56x^5 + 70x^4 + 56x^3 + 28x^2 + 8x + 1$

Written Exercises

Expand each power of a binomial. Simplify the terms.

1. $(a+b)^6$ **2.** $(a-b)^7$ **3.** $(x-3)^4$ **4.** $(2m+1)^5$

5. $(3c+2d)^5$ **6.** $(2t-3v)^4$ **7.** $(x^2-2y)^6$ **8.** $(3p+r^3)^4$

Find the specified term of each expansion. Simplify the term.

9. 4th term of $(3c+d)^5$ $90c^2d^3$ **10.** 4th term of $(x-y^2)^6$ $-20x^3y^6$

11. middle term of $(x^2-2y)^4$ $24x^4y^2$ **12.** 5th term of $(4x+2y)^7$ $35{,}840x^3y^4$

13. 3rd term of $(2c^3-3d)^5$ $720c^9d^2$ **14.** 5th term of $(3x^2+\sqrt{y})^6$ $135x^4y^2$

15. middle term of $\left(4x^2+\dfrac{x}{2}\right)^6$ $160x^9$ **16.** 6th term of $\left(27y^4-\dfrac{y^2}{3}\right)^7$ $-63y^{18}$

Expand each power of a binomial. Simplify the terms.

17. $(3c^3+2d^2)^5$ **18.** $(p-\sqrt{3})^6$ **19.** $(2p-\sqrt{p})^4$ **20.** $(1+\sqrt[5]{4})^5$

Find the first four terms of each expansion. Simplify the terms.

21. $(x+1)^{14}$ **22.** $(x-3y)^{11}$ **23.** $(x^2-y^3)^{10}$ **24.** $(1+0.02)^8$

Write each power in sigma-notation. Then find the sum of the terms.

25. $(2+0.1)^4$ **26.** $(1+0.1)^4$ **27.** $(1.5)^3$ **28.** $(2.2)^3$

Mixed Review

Solve.   *1.4, 2.3, 8.1, 9.1*

1. $\dfrac{x}{3}=\dfrac{4}{x-4}$ $-2, 6$ **2.** $|3x+2|=17$ $5, -\dfrac{19}{3}$ **3.** $2x^2=20$ $\pm\sqrt{10}$ **4.** $x^2+4=0$ $\pm 2i$

Checkpoint

1. Expand $(3x-y^2)^3$.
Simplify each term.
$27x^3 - 27x^2y^2 + 9xy^4 - y^6$

2. Find the 5th term of $(c^2-10d)^6$.
Simplify the term.
$150{,}000c^4d^4$

Closure

Ask students to give the Binomial Theorem. See page 576.   Have students give the formula for finding the rth term in the expansion of $(a+b)^n$.   See page 576.

▬▬FOLLOW UP

Guided Practice

Classroom Exercises 1–8

Independent Practice

A Ex. 1–12, **B** Ex. 13–24, **C** Ex. 25–28

Basic: WE 1–15 odd, Application, Mixed Problem Solving

Average: WE 7–25 odd, Application, Mixed Problem Solving

Above Average: WE 11–27 odd, Application, Mixed Problem Solving

Additional Answers

Written Exercises

1. $a^6 + 6a^5b + 15a^4b^2 + 20a^3b^3 + 15a^2b^4 + 6ab^5 + b^6$

2. $a^7 - 7a^6b + 21a^5b^2 - 35a^4b^3 + 35a^3b^4 - 21a^2b^5 + 7ab^6 - b^7$

See page 578 for the answers to Written Ex. 3–8 and 17–28.

Enrichment

Introduce the students to Pascal's Triangle, an interesting device for finding the numerical coefficients of a binomial expansion.

```
              1
           1     1
        1     2     1
     1     3     3     1
  1     4     6     4     1
```

Point out that each entry, other than 1, is the sum of the two numbers that flank it from above. The triangle can be continued to find the coefficients of $(a+b)^n$ for any positive integral value of n.

Have the students find the coefficients when $n=5$ and $n=6$.

$n=5$: 1, 5, 10, 10, 5, 1

$n=6$: 1, 6, 15, 20, 15, 6, 1

Pascal's Triangle has other interesting properties. Have students investigate these properties as a research project.

3. $x^4 - 12x^3 + 54x^2 - 108x + 81$

4. $32m^5 + 80m^4 + 80m^3 + 40m^2 + 10m + 1$

5. $243c^5 + 810c^4d + 1{,}080c^3d^2 + 720c^2d^3 + 240cd^4 + 32d^5$

6. $16t^4 - 96t^3v + 216t^2v^2 - 216tv^3 + 81v^4$

7. $x^{12} - 12x^{10}y + 60x^8y^2 - 160x^6y^3 + 240x^4y^4 - 192x^2y^5 + 64y^6$

8. $81p^4 + 108p^3r^3 + 54p^2r^6 + 12pr^9 + r^{12}$

17. $243c^{15} + 810c^{12}d^2 + 1{,}080c^9d^4 + 720c^6d^6 + 240c^3d^8 + 32d^{10}$

18. $p^6 - 6\sqrt{3}p^5 + 45p^4 - 60\sqrt{3}p^3 + 135p^2 - 54\sqrt{3}p + 27$

19. $16p^4 - 32\sqrt{p}p^3 + 24p^3 - 8\sqrt{p}p^2 + p^2$

20. $1 + 5\sqrt[5]{4} + 10\sqrt[5]{16} + 10\sqrt[5]{64} + 5\sqrt[5]{256} + 4$

21. $x^{14} + 14x^{13} + 91x^{12} + 364x^{11} + \cdots$

22. $x^{11} - 33x^{10}y + 495x^9y^2 - 4{,}455x^8y^3 + \cdots$

23. $x^{20} - 10x^{18}y^3 + 45x^{16}y^6 - 120x^{14}y^9 + \cdots$

24. $1 + 0.16 + 0.0112 + 0.000448 + \cdots$

25. $\displaystyle\sum_{k=0}^{4} \binom{4}{k} 2^{4-k}(0.1)^k$, 19.4481

26. $\displaystyle\sum_{k=0}^{4} \binom{4}{k}(0.1)^k$, 1.4641

27. $\displaystyle\sum_{k=0}^{3} \binom{3}{k}(0.5)^k$, 3.375

28. $\displaystyle\sum_{k=0}^{3} \binom{3}{k} 2^{3-k}(0.2)^k$, 10.648

Application: *Quality-Control Charts*

The concepts of probability, standard deviation, and the normal distribution can be used to create a quality-control chart that monitors the number of defective parts produced by a factory in a day. Such a chart includes the number of defective parts produced each day, the mean number of defective parts M, and upper and lower limits that represent three standard deviations from that mean, $M \pm 3\sigma$. Recall that in a normal distribution (see Lesson 14.11), 99.6% of the values will fall between $M - 3\sigma$ and $M + 3\sigma$. So, on days when the number of defective parts falls above or below the upper or lower limits, the production process should be checked to determine the cause of the unusually poor or excellent results.

Consider a process that produces 2,000 parts per day. Data shows that the number of defective parts is normally distributed and that the probability p of a defective part is 0.05. Thus, the mean number of defective parts per day is 0.05(2,000), or 100. Standard deviation from the mean can be calculated using the formula $\sigma = \sqrt{npq}$, where n = number of parts, p = probability of a defective part, and $q = 1 - p$. Thus, upper and lower limits are found as follows.

$$\sigma = \sqrt{npq}$$
$$\sigma = \sqrt{2{,}000(0.05)(0.95)}$$
$$\sigma = \sqrt{95}$$
$$\sigma \approx 9.75$$
$$3\sigma \approx 29.25$$

$M = 100$

Upper limit: $M + 3\sigma = 100 + 29.25$
$$= 129.25$$

Lower limit: $M - 3\sigma = 100 - 29.25$
$$= 70.75$$

Suppose, over a seven-day period, the plant described above produces the following numbers of defective parts: 100, 130, 115, 95, 105, 110, 65. This data is shown on the quality control chart at the right.

1. In the example above, on which day(s) should the process be checked to find the cause of the unusual performance? Days 2 and 7

2. Suppose a plant produces 5,000 parts per day. The probability of defective parts is 0.04. Over a five-day period, the plant produces the following daily totals of defective parts: 240, 155, 225, 220, 230. Use the data to make a quality-control chart.

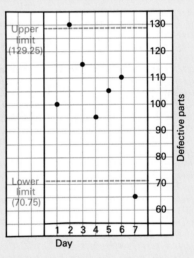

Application

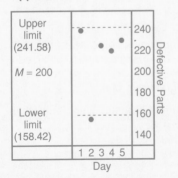

Mixed Problem Solving

Solve each problem.

1. How many games will be played if each of 8 baseball teams plays the other 7 teams 4 times each? 112 games

2. How many signals can be formed by displaying 5 flags in a row if 10 different flags are available? 30,240 signals

3. Book A has 125 fewer pages than Book B, and Book C has twice as many pages as Book A. Find the number of pages in Book C if the three books have a total of 1,225 pages. 550 pages

4. The length of a rectangle is 3 m less than 4 times the width. The perimeter is 54 m. Find the following.
   a. the width 6 m   b. the length 21 m
   c. the area 126 m^2
   d. the length of a diagonal in simplest radical form $3\sqrt{53}$ m

5. Convert 18 oz/pt to pounds per gallon. 9 lb/gallon

6. If x is inversely proportional to $\sqrt[3]{y}$, and $x = 3.6$ when $y = 64$, find x when $y = 27$. 4.8

7. How many even numbers of one or more digits can be formed from the digits 1, 2, 4, 5, 8, if no digit may be repeated in a number? 195

8. How many liters of water must be evaporated from 75 liters of a 12% salt solution to obtain an 18% salt solution? 25 liters

9. Find the number of quarters in a coin collection with 8 more dimes than nickels and twice as many quarters as dimes if the collection is worth $9.35. 30

10. Corn worth $2/kg is mixed with oats worth $3/kg. There are 40 kg more oats than corn. How many kilograms of oats are used if the mixture is worth $320? 80 kg

11. A triangle's height is 7 cm less than 5 times the length of its base. Find the height if the area is 12 cm^2. 8 cm

12. If 7 times a number is decreased by 9, the square root of the result is the same as twice the square root of the number. Find the number. 3

13. Ellen drove 4 h at an average speed of 42 mi/h. Her car averages 28 mi/gal. If gas costs $1.30/gal, how much did she pay for gas? $7.80

14. In a shipment of 20 lamps, two are broken. If 2 of the lamps are selected at random, what is the probability that exactly 1 of them is broken? $\frac{18}{95}$

15. Marie bought a new car for $9,600. Each year its value will depreciate by 30% of its value at the start of the year. What will its value be at the end of 5 years? Give your answer to the nearest dollar. $1,613

16. The numerator of a fraction is one-half the sum of the denominator and 1. If the numerator is increased by 2 and the denominator is decreased by 2, the resulting fraction is the reciprocal of the original fraction. What is the original fraction? $\frac{3}{5}$

3. $12 + \sqrt{6}$, $16 + 2\sqrt{6}$, $20 + 3\sqrt{6}$, $24 + 4\sqrt{6}$

6. Answers will vary. One possible answer: Take the reciprocals of both numbers. Next, find the arithmetic mean of the reciprocals. Finally, take the reciprocal of this arithmetic mean.

Chapter 15 Review

Key Terms

arithmetic means (p. 549)
arithmetic sequence (p. 545)
arithmetic series (p. 553)
Binomial Theorem (p. 576)
common difference (p. 545)
common ratio (p. 558)
geometric means (p. 562)
geometric sequence (p. 558)

geometric series (pp. 566, 572)
harmonic means (p. 550)
harmonic sequence (p. 550)
index of summation (p. 555)
infinite geometric series (p. 572)
partial sums (p. 572)
sequence (p. 545)
summation symbol (p. 555)

Key Ideas and Review Exercises

15.1 For the arithmetic sequence $a_1, a_1 + d, a_1 + 2d, \ldots, a_1 + (n - 1)d$, the nth term is given by $a_n = a_1 + (n - 1)d$.

1. Write the next 3 terms of the arithmetic sequence -3.8, -1.5, $0.8, \ldots$. 3.1, 5.4, 7.7

2. Find the 26th term of the arithmetic sequence -4, -2.5, $-1, \ldots$. 33.5

15.2 For the arithmetic sequence x_1, x_2, x_3, x_4, x_5, the following are true.
(1) x_2, x_3, and x_4 are the three arithmetic means between x_1 and x_5.
(2) $\dfrac{1}{x_1}, \dfrac{1}{x_2}, \dfrac{1}{x_3}, \dfrac{1}{x_4}, \dfrac{1}{x_5}$, is a harmonic sequence.
(3) The arithmetic mean of x and y is $\dfrac{x + y}{2}$.

3. Find the four arithmetic means between 8 and $28 + 5\sqrt{6}$.

4. The arithmetic mean of two numbers is 16 and one number is 4 more than 3 times the other number. Find both numbers. 7, 25

5. Find the 31st term in the harmonic sequence $\dfrac{1}{3}, \dfrac{2}{9}, \dfrac{1}{6}, \dfrac{2}{15}, \ldots$. $\dfrac{1}{48}$

6. Write in your own words how to find the harmonic mean of two nonzero numbers.

15.3 The sum of the first n terms of an arithmetic series is given by
(1) $S_n = \dfrac{n}{2}(a_1 + a_n)$ or   (2) $S_n = \dfrac{n}{2}[2a_1 + (n - 1)d]$.

Find the sum of each arithmetic series for the given data.

7. $n = 34$; $a_1 = -2.5$, $a_n = 33.8$ 532.1

8. $n = 41$; $14 + 9 + 4 - 1 - 6 - \ldots$ $-3{,}526$

9. $\displaystyle\sum_{k=1}^{31} (2k - 7)$ 775

10. $\displaystyle\sum_{j=8}^{10} \dfrac{j + 2}{3}$ 11

11. Use sigma-notation to represent the series $6 + 10 + 14 + \ldots + 42$.

Answers will vary. One possible answer: $\displaystyle\sum_{k=1}^{10}(4k + 2)$

15.4 The nth term, a_n, of the geometric sequence $a_1, a_1r, a_1r^2, \ldots$ is a_1r^{n-1}.

Find the specified term of each geometric sequence.

12. 7th term: $a_1 = 0.03$, $r = 10$ 30,000   **13.** 9th term: 256, 128, 64, . . . 1

14. A ball rebounds one-half of the distance that it falls. If the ball is dropped from a height of 256 ft, how high does it go on its 8th rebound? 1 ft

15.5 For the geometric sequence x_1, x_2, x_3, x_4, x_5, the three geometric means between x_1 and x_5 are x_2, x_3, and x_4.

The geometric mean of x and y is $\sqrt{xy}$ or $-\sqrt{xy}$.

15. Find three positive geometric means between $\frac{2}{3}$ and 54. 2, 6, 18

16. What is the negative geometric mean of 0.4 and 250? -10

17. The positive geometric mean of two numbers is 8. Find the two numbers if one number is 12 less than the other number. $-4, -16$ or 16,4

15.6 The sum of a finite geometric series of n terms is given by

$$(1)\ S_n = \frac{a_1 - ra_n}{1 - r} \quad \text{or} \quad (2)\ S_n = \frac{a_1(1 - r^n)}{1 - r}.$$

Find the sum of the terms of the geometric series described.

18. $a_1 = -2$, $r = 5$, $a_n = -6{,}250$ $-7{,}812$ **19.** $\sum_{k=4}^{6} \frac{1}{6}(3^k - 2)$ $19\frac{1}{2}$

20. $n = 10$; $3 + 6 + 12 + 24 + \ldots$ 3,069

21. Smaller and smaller squares are formed, one after the other, as shown at the right. Find the sum of the perimeters of the first 7 squares if the first square is 12 ft wide. $95\frac{1}{4}$ ft

15.7 The sum of an infinite geometric series is

$$\frac{a_1}{1 - r},\ -1 < r < 1,\ r \neq 0.$$

22. Find the sum, if it exists, of the infinite series $64 + 48 + 36 + 27 + \ldots$. 256

23. Write $0.17\overline{17}$ as an infinite geometric series and then in the form $\frac{x}{y}$, where x and y are integers. $0.17 + 0.0017 + 0.000017 + \ldots$, $\frac{17}{99}$

24. Refer to the squares in Exercise 21. Find the sum of the areas of an infinite sequence of the squares if the first square is 12 ft wide. 192 ft^2

15.8 $(a + b)^n = \binom{n}{0}a^n + \binom{n}{1}a^{n-1}b + \binom{n}{2}a^{n-2}b^2 + \binom{n}{3}a^{n-3}b^3 + \ldots + \binom{n}{n}b^n$

In the expansion of $(a + b)^n$, the rth term is $\binom{n}{r-1}a^{n-r+1}b^{r-1}$.

25. Expand $(x^2 - 3y)^4$. Simplify the terms. $x^8 - 12x^6y + 54x^4y^2 - 108x^2y^3 + 81y^4$

26. Find the 6th term in the expansion of $(c^2 + 2d)^8$. Simplify the term. $1{,}792c^6d^5$

Chapter Test

11. Answers will vary. One possible answer:

$$\sum_{k=1}^{13} (4k + 3)$$

22. $243x^5 - 405x^4y^2 + 270x^3y^4 - 90x^2y^6 + 15xy^8 - y^{10}$

1.3, 2.7, 4.1

1. Write the next three terms of the arithmetic sequence $-2.9, -1.5, -0.1, \ldots$.

2. Find the 36th term of the arithmetic sequence $6, 2, -2, -6, \ldots$. -134

3. Find the three arithmetic means between 5 and $9 + 12\sqrt{5}$. $6 + 3\sqrt{5}, 7 + 6\sqrt{5}, 8 + 9\sqrt{5}$

4. What is the arithmetic mean of -5.6 and 8? 1.2

5. The arithmetic mean of two numbers is 45 and one number is 6 more than 5 times the other number. Find both numbers. 14, 76

6. Find the 21st term in the harmonic sequence $\frac{1}{2}, \frac{4}{9}, \frac{2}{5}, \frac{4}{11}, \ldots$. $\frac{1}{7}$

Find the sum of each arithmetic series for the given data. $-1,953$

7. $n = 42; a_1 = -3.4, a_{42} = 29.4$ 546

8. $n = 31; 12 + 7 + 2 - 3 - \ldots$

9. $\displaystyle\sum_{k=1}^{21} (3k - 5)$ 588

10. $\displaystyle\sum_{j=7}^{10} \frac{j + 1}{2}$ 19

11. Use sigma-notation to represent the series $7 + 11 + 15 + \ldots + 55$.

12. Boxes are stacked in 21 rows with 2 in the top row, 5 in the 2nd row, 8 in the 3rd row, and so on. How many boxes are in the stack? 672

13. Find the 10th term of the geometric sequence $128, -64, 32, \ldots$. $-\frac{1}{4}$

14. A cylinder contains $32,000$ cm^3 of oxygen. One-half of the oxygen is released each time that a valve is opened. How much oxygen will be in the tank after the valve is opened 8 times? $125\ cm^3$

15. A certain steel ball rebounds on each bounce one-third of the distance from which it fell. If it is dropped from a height of 729 cm, how high does it go on its 7th rebound (upward)? $\frac{1}{3}$ cm

16. Find the three real, positive geometric means between $\frac{3}{4}$ and 12. $\frac{3}{2}, 3, 6$

17. What is the positive geometric mean of 0.9 and 40? 6

Find the sum of the terms of the geometric series described. 2,555

18. $a_1 = 3, r = 4, a_n = 12,288$ 16,383

19. $n = 9; 5 + 10 + 20 + 40 + \ldots$

20. Find the sum, if it exists, of the infinite series $27 + 18 + 12 + 8 + \ldots$. 81

21. Write $0.07\overline{3}$ as an infinite geometric series, and then in the form $\frac{x}{y}$, where x and y are integers. $0.073 + 0.00073 + 0.0000073 + \ldots, \frac{73}{990}$

22. Expand $(3x - y^2)^5$. Simplify the terms.

23. Find the 5th term in the expansion of $(2c + d^2)^7$. Simplify the term. $280c^3d^8$

24. Find the 1st term of the arithmetic sequence whose 6th term is 31 and 20th term is 52. 23.5

25. Find two values of c so that $\frac{1}{5}, c - 3, 4c$ is a geometric sequence. $5, 1\frac{4}{5}$

26. Find the sum of the first n terms of the geometric series $1 + 2x^2 + 4x^4 + 8x^6 \cdots + 2^{(n-1)}x^{(2n-2)}$. $\frac{1 - 2^n x^{2n}}{1 - 2x^2}$

In each item, compare a quantity in Column 1 with a quantity in
Column 2. Write the letter of the correct answer from these choices.

  A—The quantity in Column 1 is greater than the quantity in Column 2.
  B—The quantity in Column 2 is greater than the quantity in Column 1.
  C—The quantity in Column 1 is equal to the quantity in Column 2.
  D—The relationship cannot be determined from the given information.

  NOTE: Information centered over both columns refers to one or both of the
  quantities to be compared.

Sample Items and Answers

Column 1	Column 2	
$x = 3$ and $y = -3$		
S1. Next term of the sequence: $2x, 3x, 4x, \ldots$	Next term of the sequence: $y^2, y^3, y^4, \ldots$	S1: The answer is A: $5x > y^5$, since $5(3) > (-3)^5$, or $15 > -243$.
S2. $\displaystyle\sum_{k=1}^{6} 3^{k-1}$	$\displaystyle\sum_{j=4}^{9} 3^{j-4}$	S2: The answer is C: Each symbol represents the same sum, $3^0 + 3^1 + 3^2 + 3^3 + 3^4 + 3^5$.

Column 1	Column 2
$x = 4$ and $y = -5$	
1. Next term of the sequence: (C)	Next term of the sequence:
$\dfrac{1}{x}\sqrt{\dfrac{1}{x}}, \dfrac{1}{x}, \sqrt{\dfrac{1}{x}} \ldots$	$y^3, y^2, y^1, \ldots$
$x = y$ and $x > 0$	
2. Arithmetic mean of x and y (C)	Positive geometric mean of x and y
3. Arithmetic mean of $\frac{1}{3}$ and $\frac{1}{6}$ (A)	Harmonic mean of $\frac{1}{3}$ and $\frac{1}{6}$
4. Coefficient of the 5th term in the expansion of $(x - 4)^{10}$ (C)	Coefficient of the 5th term in the expansion of $(y + 4)^{10}$

Column 1	Column 2
5. $(x + 2)^4$ (D)	$(x + 2)^5$

Midpoints of sides of a triangle are connected to form a new triangle. This procedure is continued in each new triangle.

Column 1	Column 2
6. Sum of the perimeters of an infinite sequence of the triangles (C)	Twice the perimeter of the first triangle

16 LINEAR SYSTEMS, MATRICES, DETERMINANTS

OVERVIEW

This chapter introduces students to matrices and operations on matrices. These are approached from both a theoretical and a practical point of view.

On the theoretical level, properties of real-number operations do not always carry over into the realm of matrices. For this reason, after an operation with matrices has been defined in a lesson, it is examined to determine whether it obeys the properties with which students are familiar, such as commutativity and associativity. These are the central concerns of Lessons 16.1 and 16.2.

The practical aspect of matrices, such as applications, are stressed in Lessons 16.5, and 16.6. In Lesson 16.5, for example, students are shown how to solve a linear system using matrices. Since determinants can also be useful in solving linear systems, two-by-two and three-by-three determinants are introduced in the chapter, in Lessons 16.3 and 16.4.

OBJECTIVES

- To find the sum of two matrices
- To find the product of two matrices
- To solve systems of two or of three linear equations using matrices
- To solve problems using matrices

PROBLEM SOLVING

Determinants give students the opportunity to employ the problem-solving strategy *Use an Alternate Approach* to solve systems of three linear equations. Lesson 16.6 is a problem-solving lesson that shows how matrices can be used to solve problems involving a large amount of data. The data is first organized in a table and then translated into matrix form. The problem is solved using matrix operations.

READING AND WRITING MATH

A *Focus on Reading* on page 592 familiarizes students with the symbolic representation of matrices by asking them to translate an algebraic sentence into matrix notation. Exercise 16 on page 605 asks students to explain, in their own words, the process of a matrix operation.

TECHNOLOGY

Calculator: Graphing calculators can simplify calculations. A good graphing calculator can display 6×6 matrices and allow manipulations to be performed on them, including finding the inverse and transpose and performing elementary row operations.

SPECIAL FEATURES

PLANNING GUIDE

Lesson	Basic	Average	Above Average	Resources
16.1 p. 588	CE all WE 1–5	CE all WE 3–8	CE all WE 4–10	Reteaching 119 Practice 119
16.2 pp. 592–593	FR all CE all WE 1–13 odd Brainteaser	FR all CE all WE 3–15 odd Brainteaser	FR all CE all WE 5–21 odd Brainteaser	Reteaching 120 Practice 120
16.3 pp. 595–596	CE all WE 1–17 odd Midchapter Review	CE all WE 5–23 odd Midchapter Review	CE all WE 11–27 odd Midchapter Review	Reteaching 121 Practice 121
16.4 pp. 600–601	CE all WE 1–13 odd Brainteaser	CE all WE 5–15 odd Brainteaser	CE all WE 5–17 odd Brainteaser	Reteaching 122 Practice 122
16.5 pp. 605–606	CE all WE 1–11 odd Application	CE all WE 3–17 odd Application	CE all WE 5–21 odd Application	Reteaching 123 Practice 123
16.6 pp. 609–611	CE all WE 1–18 Application	CE all WE 7–22 Application	CE all WE 8–26 Application	Reteaching 124 Practice 124
Chapter 16 Review pp. 612–613	all	all	all	
Chapter 16 Test p. 614	all	all	all	
College Prep Test p. 615	all	all	all	
Cumulative Review pp. 616–617	1–51 odd	1–51 odd	1–51 odd	

CE = Classroom Exercises   WE = Written Exercises   FR = Focus on Reading
NOTE: For each level, all students should be assigned all Mixed Review exercises.

■ INVESTIGATION

The students explore properties of determinants (Lessons 16.3 and 16.4).

Ask students to consider the effect of interchanging any two rows or columns of a matrix upon its determinant.

Have them first consider the determinant of a 2×2 matrix, such as

$$\begin{vmatrix} 1 & 2 \\ 3 & 4 \end{vmatrix} = (1)(4) - (2)(3) = -6.$$

Interchange the two rows of the original matrix.

$$\begin{vmatrix} 3 & 4 \\ 1 & 2 \end{vmatrix} = (3)(2) - (4)(1) = 6.$$

Interchange the two columns of the original matrix.

$$\begin{vmatrix} 2 & 1 \\ 4 & 3 \end{vmatrix} = (2)(3) - (1)(4) = 6.$$

In either case, the resulting determinant is the negative of the determinant of the matrix before the interchange of the rows or the columns. This will be true of any 2×2 matrix.

Ask the student what happens if they interchange any two rows or columns of a 3×3 matrix.

They should consider the determinant of the following 3×3 matrix.

$$\begin{vmatrix} 1 & 2 & 3 \\ 2 & 1 & 3 \\ 3 & 1 & 2 \end{vmatrix} = \begin{matrix} (1)(1)(2) + (2)(3)(3) + (3)(2)(1) \\ - (1)(3)(1) - (2)(2)(2) - (3)(1)(3) \\ = 2 + 18 + 6 - 3 - 8 - 9 = 6 \end{matrix}$$

Interchange the first two rows.

$$\begin{vmatrix} 2 & 1 & 3 \\ 1 & 2 & 3 \\ 3 & 1 & 2 \end{vmatrix} = \begin{matrix} (2)(2)(2) + (1)(3)(3) + (3)(1)(1) \\ - (2)(3)(1) - (1)(1)(2) - (3)(2)(3) \\ = 8 + 9 + 3 - 6 - 2 - 18 = -6 \end{matrix}$$

And so the result is the negative of the original determinant. The result will be the same for interchanging any two rows or columns.

Invite the students to explore what happens in the following cases:

1. Two rows (or columns) are identical.
   The determinant is 0.
2. One row is composed entirely of zeros.
   The determinant is 0.

Artists use the graphic and color capabilities of computers as a medium of artistic expression. The image above is called "After Picasso." It is a still print taken from an animated film created by computer artist Lillian Schwartz.

More About the Image

Artists have discovered that the capabilities of the computer expand the realm of artistic expression. Computer artists can go beyond still images to create animated works such as Lillian Schwartz's animated film mentioned in the Student text. The computer is instrumental in creating **virtual realities**—works of art that interact with the viewer. Virtual realities enable the viewer to become an active participant in the creative process.

16.1 Addition of Matrices

Objectives
To determine whether two matrices are equal
To add matrices
To prove properties of matrix addition and to use these properties

A **matrix** (plural: **matrices**) is a rectangular array of numbers arranged in rows and columns and enclosed by brackets. A matrix that contains only one row is called a *row* matrix. A matrix that contains only one column is called a *column* matrix. A matrix that has the same number of rows and columns is called a *square* matrix.

Following are examples of matrices and their classifications.

Row Matrix	Column Matrix	Square Matrix	General Rectangular Matrix (3 by 2)
$\begin{bmatrix} 1 & 4 & -3 \end{bmatrix}$	$\begin{bmatrix} 8 \\ \sqrt{3} \\ 12 \end{bmatrix}$	$\begin{bmatrix} 1 & -2 \\ 0 & 3 \end{bmatrix}$	$\begin{bmatrix} a_{11} & a_{12} \\ a_{21} & a_{22} \\ a_{31} & a_{32} \end{bmatrix}$
1 row 3 columns	3 rows 1 column	2 rows 2 columns	3 rows 2 columns

The individual numbers, such as a_{31}, are called **elements** of the matrix and are real numbers. The first number of the subscript indicates the row of the element, and the second number indicates the column of the element. So, a_{31} means the element in the third row and first column. The number of rows and columns determines the **dimensions** of a matrix. The dimensions of the matrices above in order are 1 by 3 (1×3), 3 by 1 (3×1), 2 by 2 (2×2), and 3 by 2 (3×2).

Two matrices are *equal* if and only if they have the same dimensions and the elements in corresponding positions are equal.

Two matrices can be added if they have the same dimensions. The **sum of two matrices** with the same dimensions is the matrix obtained by adding the corresponding elements of the two matrices.

For example: $\begin{bmatrix} a_{11} & a_{12} \\ a_{21} & a_{22} \\ a_{31} & a_{32} \end{bmatrix} + \begin{bmatrix} b_{11} & b_{12} \\ b_{21} & b_{22} \\ b_{31} & b_{32} \end{bmatrix} = \begin{bmatrix} a_{11} + b_{11} & a_{12} + b_{12} \\ a_{21} + b_{21} & a_{22} + b_{22} \\ a_{31} + b_{31} & a_{32} + b_{32} \end{bmatrix}$

Notice that the sum of two matrices is a unique matrix with elements that are real numbers. Thus, the **Closure Property** holds for the sum of two matrices of the same dimensions.

16.1 Addition of Matrices **585**

Lesson Note

For matrices that can be added, the following properties hold.

Closure: $A + B$ is a matrix.
Identity: $A + I = A$
Associative: $(A + B) + C = A + (B + C)$
Inverse: $A + (-A) = I$
Commutative: $A + B = B + A$

Thus, matrices form a **commutative** or **Abelian group** under addition.

Math Connections

Historical Note: Matrices have been studied as objects of mathematical interest since the middle of the nineteenth century, when Arthur Cayley (1821–1895) was working on the theory of transformations. Out of his research came the realization that a mathematical operation (in this case the following of one transformation by another) need not be commutative. This investigation led naturally to a broader view of matrices as a class of objects in themselves and ultimately to the construction of an algebra of matrices.

Critical Thinking Questions

Analysis: Example 3 illustrates the commutative and associative properties for the addition of two matrices of the same dimensions. Ask students to show that the commutative property of addition of two matrices of the same dimensions is true generally. If the notation $|a_{ij}|$ is used in the place of A, then the proof is straightforward:
$$A + B = |a_{ij}| + |b_{ij}| = |a_{ij} + b_{ij}| =$$
$$|b_{ij} + a_{ij}| = |b_{ij}| + |a_{ij}| = B + A.$$

EXAMPLE 1   Matrices A, B, and C are given as follows.

$$A = \begin{bmatrix} 3 & -1 & 2 \\ 4 & 5 & -4 \end{bmatrix}, B = \begin{bmatrix} 1 & -2 & -2 \\ -3 & 4 & -1 \end{bmatrix}, C = \begin{bmatrix} -1 & 5 \\ 3 & -2 \\ 5 & -4 \end{bmatrix}$$

Find the following sums if they exist. If a sum does not exist, give a reason.

a. $A + B$

b. $A + C$

Solutions   **a.** $A + B =$

$$\begin{bmatrix} 3 + 1 & -1 + (-2) & 2 + (-2) \\ 4 + (-3) & 5 + 4 & -4 + (-1) \end{bmatrix} = \begin{bmatrix} 4 & -3 & 0 \\ 1 & 9 & -5 \end{bmatrix}$$

b. Since A and C do not have the same dimensions, the sum does not exist.

Recall that in the set of real numbers there exists a unique additive identity, the number 0, such that for each real number a, $a + 0 = a$. Also, each real number a has an opposite, or additive inverse, $-a$, such that $a + (-a) = 0$. Similarly, additive identities and inverses exist for matrix addition.

Definitions   If matrices A and O have the same dimensions and $A + O = A$, then O is called an **additive identity matrix** (zero matrix) for A. If matrices A and $-A$ have the same dimensions and $A + (-A) = O$, then $-A$ is called an **additive inverse matrix** (opposite matrix) of A.

EXAMPLE 2   Matrices A, $-A$, and O are given as follows.

$$A = \begin{bmatrix} 3 & -2 \\ -4 & 5 \end{bmatrix}, \quad -A = \begin{bmatrix} -3 & 2 \\ 4 & -5 \end{bmatrix}, \quad \text{and} \quad O = \begin{bmatrix} 0 & 0 \\ 0 & 0 \end{bmatrix}$$

a. Prove that $A + O = A$.

b. Prove that $A + (-A) = O$.

Solutions   **a.** $A + O = \begin{bmatrix} 3 + 0 & -2 + 0 \\ -4 + 0 & 5 + 0 \end{bmatrix} = \begin{bmatrix} 3 & -2 \\ -4 & 5 \end{bmatrix} = A$

Thus, $A + O = A$.

b. $A + (-A) = \begin{bmatrix} 3 + (-3) & -2 + 2 \\ -4 + 4 & 5 + (-5) \end{bmatrix} = \begin{bmatrix} 0 & 0 \\ 0 & 0 \end{bmatrix} = O$

Thus, $A + (-A) = O$.

Additional Example 1

Matrices A, B, and C are given as follows.

$$A = \begin{bmatrix} 3 & 2 & 5 \\ -1 & -3 & -4 \end{bmatrix},$$

$$B = \begin{bmatrix} 4 & -1 \\ -3 & 2 \\ -1 & 3 \end{bmatrix}, C = \begin{bmatrix} 1 & 3 \\ -1 & 5 \\ -2 & -3 \end{bmatrix}$$

Find the following sums, if they exist. If a sum does not exist, give a reason.

a. $A + B$ Since A and B do not have the same dimensions, the sum does not exist.

b. $B + C$
$$B + C = \begin{bmatrix} 5 & 2 \\ -4 & 7 \\ -3 & 0 \end{bmatrix}$$

Additional Example 2

Matrices B, $-B$, and O are given as follows.

$$B = \begin{bmatrix} -1 \\ 7 \end{bmatrix}, -B = \begin{bmatrix} 1 \\ -7 \end{bmatrix}, O = \begin{bmatrix} 0 \\ 0 \end{bmatrix}$$

a. Prove that $B + O = B$.
$$B + O = \begin{bmatrix} -1 + 0 \\ 7 + 0 \end{bmatrix} = \begin{bmatrix} -1 \\ 7 \end{bmatrix} = B$$

b. Prove that $B + (-B) = O$.
$$B + (-B) = \begin{bmatrix} -1 + 1 \\ 7 + (-7) \end{bmatrix} = \begin{bmatrix} 0 \\ 0 \end{bmatrix} = 0$$

The results of Example 2 can be generalized in two ways. First, all matrices of the same dimensions have the same additive identity matrix, with elements that consist entirely of zeros. Second, each element of an additive inverse matrix, $-A$, is equal to the opposite of its corresponding element in matrix A. These two points are further illustrated below.

$$A = \begin{bmatrix} a_{11} & a_{12} & a_{13} \\ a_{21} & a_{22} & a_{23} \end{bmatrix}$$

The additive identity (zero) matrix for A and the additive inverse (opposite) matrix for A are, respectively,

$$O_{2\times3} = \begin{bmatrix} 0 & 0 & 0 \\ 0 & 0 & 0 \end{bmatrix} \quad \text{and} \quad -A_{2\times3} = \begin{bmatrix} -a_{11} & -a_{12} & -a_{13} \\ -a_{21} & -a_{22} & -a_{23} \end{bmatrix},$$

where the subscript 2×3 is used to indicate the dimensions of the matrix. In general, an m-by-n matrix A may be written as $A_{m\times n}$ whenever it is desired to emphasize the dimensions of the matrix.

EXAMPLE 3   Matrices A, B, and C are given as follows.

$$A = \begin{bmatrix} 5 & -3 \\ -4 & 6 \end{bmatrix}, B = \begin{bmatrix} 2 & 4 \\ 5 & -2 \end{bmatrix}, \text{ and } C = \begin{bmatrix} -8 & -3 \\ 4 & -1 \end{bmatrix}$$

a. Prove that $A + B = B + A$.   **b.** Prove that $(A + B) + C = A + (B + C)$.

Solutions   **a.** $A + B = \begin{bmatrix} 5 + 2 & -3 + 4 \\ -4 + 5 & 6 + (-2) \end{bmatrix} = \begin{bmatrix} 7 & 1 \\ 1 & 4 \end{bmatrix}$

$$B + A = \begin{bmatrix} 2 + 5 & 4 + (-3) \\ 5 + (-4) & -2 + 6 \end{bmatrix} = \begin{bmatrix} 7 & 1 \\ 1 & 4 \end{bmatrix}$$

So, $A + B = B + A$. ← Matrix addition is commutative.

b. $(A + B) + C = \begin{bmatrix} 5 + 2 & -3 + 4 \\ -4 + 5 & 6 + (-2) \end{bmatrix} + \begin{bmatrix} -8 & -3 \\ 4 & -1 \end{bmatrix}$

$$= \begin{bmatrix} 7 + (-8) & 1 + (-3) \\ 1 + 4 & 4 + (-1) \end{bmatrix} = \begin{bmatrix} -1 & -2 \\ 5 & 3 \end{bmatrix}$$

$$A + (B + C) = \begin{bmatrix} 5 & -3 \\ -4 & 6 \end{bmatrix} + \begin{bmatrix} 2 + (-8) & 4 + (-3) \\ 5 + 4 & -2 + (-1) \end{bmatrix}$$

$$= \begin{bmatrix} 5 + (-6) & -3 + 1 \\ -4 + 9 & 6 + (-3) \end{bmatrix} = \begin{bmatrix} -1 & -2 \\ 5 & 3 \end{bmatrix}$$

So, $(A + B) + C = A + (B + C)$. ← Matrix addition is associative.

Common Error Analysis

Error: Students may try to add or subtract any two matrices.

Emphasize that matrices can be added only if they have the same dimensions. Use examples to show the students why they cannot add or subtract two matrices if their dimensions are not the same.

Checkpoint

Give the dimensions of each matrix. Find each sum, if possible.

1. $\begin{bmatrix} 5 & 3 \\ 2 & -1 \end{bmatrix} + \begin{bmatrix} -9 & 6 \\ 7 & 4 \end{bmatrix}$

2 by 2, 2 by 2; $\begin{bmatrix} -4 & 9 \\ 9 & 3 \end{bmatrix}$

2. $\begin{bmatrix} -1 & 7 & -4 \\ 3 & 5 & 6 \\ 4 & -1 & -2 \end{bmatrix} + \begin{bmatrix} 9 & -2 & 5 \\ -3 & 2 & -7 \\ 6 & -3 & -2 \end{bmatrix}$

3 by 3, 3 by 3; $\begin{bmatrix} 8 & 5 & 1 \\ 0 & 7 & -1 \\ 10 & -4 & -4 \end{bmatrix}$

3. $\begin{bmatrix} -2 & -9 & 7 \\ 5 & -1 & 9 \end{bmatrix} + \begin{bmatrix} 2 & -9 & -6 \\ -1 & 3 & 10 \end{bmatrix}$

2 by 3, 2 by 3; $\begin{bmatrix} 0 & -18 & 1 \\ 4 & 2 & 19 \end{bmatrix}$

4. $\begin{bmatrix} -4 & 3 & -1 \end{bmatrix} + \begin{bmatrix} 5 \\ 2 \\ 3 \end{bmatrix}$

1 by 3, 3 by 1; sum doesn't exist.

Additional Example 3

Matrices A, B, and C are given as follows.

$$A = \begin{bmatrix} -1 \\ 3 \end{bmatrix}, B = \begin{bmatrix} 6 \\ -1 \end{bmatrix}, C = \begin{bmatrix} -5 \\ 2 \end{bmatrix}$$

a. Prove that $A + B = B + A$.

$$A + B = \begin{bmatrix} -1 + 6 \\ 3 + (-1) \end{bmatrix} = \begin{bmatrix} 5 \\ 2 \end{bmatrix}$$

$$B + A = \begin{bmatrix} 6 + (-1) \\ -1 + 3 \end{bmatrix} = \begin{bmatrix} 5 \\ 2 \end{bmatrix}$$

Thus, $A + B = B + A$.

b. Prove that $(A + B) + C = A + (B + C)$.

$$(A + B) + C = \begin{bmatrix} -1 + 6 \\ 3 + (-1) \end{bmatrix} + \begin{bmatrix} -5 \\ 2 \end{bmatrix}$$

$$= \begin{bmatrix} 5 + (-5) \\ 2 + 2 \end{bmatrix}$$

$$= \begin{bmatrix} 0 \\ 4 \end{bmatrix}$$

$$A + (B + C) = \begin{bmatrix} -1 \\ 3 \end{bmatrix} + \begin{bmatrix} 6 + (-5) \\ -1 + 2 \end{bmatrix}$$

$$= \begin{bmatrix} -1 + 1 \\ 3 + 1 \end{bmatrix}$$

$$= \begin{bmatrix} 0 \\ 4 \end{bmatrix}$$

Thus, $A + (B + C) = (A + B) + C$.

Closure

Remind students that matrices can be added only if they have the same *dimensions*. Ask students why they think this is so. By definition, there must be a ⌐-to-one correspondence between elements in the matrices. The sum is obtained by adding the corresponding elements of the two matrices. Ask them if they think matrices are multiplicable by other matrices or by constants.

▰▰▰FOLLOW UP

Guided Practice

Classroom Exercises 1–8

Independent Practice

A Ex. 1–3, **B** Ex. 4–7, **C** Ex. 8–10

Basic: WE 1–5
Average: WE 3–8
Above Average: WE 4–10

Additional Answers

Written Exercises

4. $A + O_{2\times2} = \begin{bmatrix} 5+0 & -3+0 \\ 7+0 & 2+0 \end{bmatrix}$
$= \begin{bmatrix} 5 & -3 \\ 7 & 2 \end{bmatrix} = A$

5. $A + (-A) = \begin{bmatrix} 5+(-5) & -3+3 \\ 7+(-7) & 2+(-2) \end{bmatrix}$
$= \begin{bmatrix} 0 & 0 \\ 0 & 0 \end{bmatrix} = O_{2\times2}$

See page 606 for the answers to Written Ex. 6–10.

Classroom Exercises

Matrices $A_{2\times2}$ and $B_{2\times2}$ are equal, as shown at the right. Find the value of each of the following.

$$A \qquad\qquad B$$
$$\begin{bmatrix} a_{11} & a_{12} \\ -3 & 2p \end{bmatrix} = \begin{bmatrix} 1 & 6p \\ b_{21} & 0 \end{bmatrix}$$

1. a_{11} 1
2. b_{21} -3
3. b_{11} 1
4. a_{21} -3
5. b_{22} 0
6. p 0
7. a_{22} 0
8. a_{12} 0

Written Exercises

Find each sum, if it exists. If it does not exist, give a reason.

1. $\begin{bmatrix} -3 & 5 & -6 \\ -2 & 1 & 3 \\ 5 & -2 & -9 \end{bmatrix} + \begin{bmatrix} 9 & 6 & -3 \\ 8 & -7 & 10 \\ -1 & 8 & -1 \end{bmatrix}$ $\begin{bmatrix} 6 & 11 & -9 \\ 6 & -6 & 13 \\ 4 & 6 & -10 \end{bmatrix}$

2. $\begin{bmatrix} -10 & 6 \\ -12 & -18 \\ 14 & 9 \end{bmatrix} + \begin{bmatrix} 1 & 3 & -8 \\ 4 & -6 & 5 \end{bmatrix}$ Does not exist; dimensions unequal

3. $\begin{bmatrix} 12 & -6 \\ -14 & 9 \end{bmatrix} + \begin{bmatrix} -14 & 9 \\ -8 & 4 \end{bmatrix}$ $\begin{bmatrix} -2 & 3 \\ -22 & 13 \end{bmatrix}$

In Exercises 4–7, $A = \begin{bmatrix} 5 & -3 \\ 7 & 2 \end{bmatrix}$, $B = \begin{bmatrix} -9 & 8 \\ -6 & 4 \end{bmatrix}$, and $C = \begin{bmatrix} -1 & -5 \\ -7 & -3 \end{bmatrix}$.

Use the definition of matrix addition to prove each of the following.

4. $A + O_{2\times2} = A$
5. $A + (-A) = O_{2\times2}$
6. $-(A + B) = -A + (-B)$
7. $A + (C + B) = (A + C) + B$

In Exercises 8–10, $A = \begin{bmatrix} a_{11} & a_{12} \\ a_{21} & a_{22} \end{bmatrix}$, $B = \begin{bmatrix} b_{11} & b_{12} \\ b_{21} & b_{22} \end{bmatrix}$, and $C = \begin{bmatrix} c_{11} & c_{12} \\ c_{21} & c_{22} \end{bmatrix}$,

where all elements are real numbers. Prove each of the following.

8. $A + O_{2\times2} = A$
9. $B + (-B) = O_{2\times2}$
10. $C = -(-C)$

Mixed Review

Solve by the linear combination method. *4.3*

1. $3a + b = 6$
   $4a - b = 8$
   $a = 2, b = 0$

2. $5a + 6b = 4$
   $-2a - 3b = -5$
   $a = -6, b = \frac{17}{3}$

3. $8a - 3b = 27$
   $-4a + 5b = -17$
   $a = 3, b = -1$

4. $6a - 5b = 20$
   $4a - 2b = 8$
   $a = 0, b = -4$

Enrichment

$$\begin{bmatrix} (-3)^2 & 5^0 & 8^{\frac{2}{3}} \\ 3! & |-2| & \sqrt[5]{-1} \\ -2^2 & \sqrt{\left(\frac{1}{9}\right)^{-1}} & \sqrt{100} \end{bmatrix}$$

Have students use the letters *A–I* to form a matrix equal to the matrix above. The numerical values of the letters are described as follows.

A: the number of sides in a hexagon
B: the largest digit in a decimal number system
C: the additive inverse of 2^2
D: the reciprocal of $\frac{1}{3}$
E: the largest digit in a binary number system
F: a necessary factor of an even number

G: the number of millimeters in one centimeter
H: the nonzero root of $x^2 = 4x$
I: 2 raised to this exponent equals $\frac{1}{2}$

$$\begin{bmatrix} B & E & H \\ A & F & I \\ C & D & G \end{bmatrix}$$

16.2 Matrices and Multiplication

Objectives
To multiply a scalar and a matrix
To multiply matrices
To prove properties of matrix multiplication and to use these properties

Any matrix can be multiplied by a real number. This real number is called a **scalar**.

The product of a scalar k and a matrix A is the matrix that is represented by kA. It is obtained by multiplying each element of A by k.

EXAMPLE 1 Find kA if $k = -5$ and $A = \begin{bmatrix} 3 & -1 & 5 \\ 2 & 3 & -4 \\ 5 & -2 & -7 \end{bmatrix}$.

Plan Multiply each element of A by -5.

Solution
$$-5A = \begin{bmatrix} -5(3) & -5(-1) & -5(5) \\ -5(2) & -5(3) & -5(-4) \\ -5(5) & -5(-2) & -5(-7) \end{bmatrix} = \begin{bmatrix} -15 & 5 & -25 \\ -10 & -15 & 20 \\ -25 & 10 & 35 \end{bmatrix}$$

One matrix can be multiplied by another if the number of *columns* of the first matrix equals the number of *rows* of the second matrix. The method is illustrated below for matrices C and D, whose product can be represented as either $C \cdot D$ or CD.

$$C = \begin{bmatrix} -3 & 5 & 2 \\ 1 & -2 & 4 \end{bmatrix}, D = \begin{bmatrix} 5 & -6 \\ -4 & 3 \\ 1 & -2 \end{bmatrix}$$

1. Multiply the elements of the *first row* of C by the corresponding elements of the *first column* of D. Then add the three products.

$$-3(5) + 5(-4) + 2(1)$$
$$= -15 - 20 + 2$$
$$= -33$$

2. The result is the element of the *first row* and *first column* of CD.

$$\begin{bmatrix} -33 & \underline{\quad} \\ \underline{\quad} & \underline{\quad} \end{bmatrix}$$

3. Multiply the elements of the *first row* of C by the corresponding elements of the *second column* of D. Then add the three products.

$$-3(-6) + 5(3) + 2(-2)$$
$$= 18 + 15 - 4$$
$$= 29$$

4. The result is the element of the *first row* and *second column* of CD.

$$\begin{bmatrix} -33 & 29 \\ \underline{\quad} & \underline{\quad} \end{bmatrix}$$

Teaching Resources

Application Worksheet 16
Problem Solving Worksheet 16
Project Worksheet 16
Quick Quizzes 120
Reteaching and Practice
   Worksheets 120

GETTING STARTED

Prerequisite Quiz

Given $A = \begin{bmatrix} 3 & 0 \\ 1 & -4 \end{bmatrix}$ and $B = \begin{bmatrix} -2 & 5 \\ 7 & 0 \end{bmatrix}$, find each sum.

1. $A + B$ $\begin{bmatrix} 1 & 5 \\ 8 & -4 \end{bmatrix}$

2. $B + B$ $\begin{bmatrix} -4 & 10 \\ 14 & 0 \end{bmatrix}$

3. $B + (-A)$ $\begin{bmatrix} -5 & 5 \\ 6 & 4 \end{bmatrix}$

Motivator

Write the following matrix on the chalkboard.

$$\begin{bmatrix} a & b \\ c & d \end{bmatrix}$$

Ask students to multiply the matrix by x.

$$\begin{bmatrix} ax & bx \\ cx & dx \end{bmatrix}$$

Ask students to consider how the matrix above might be multiplied by another 2 × 2 matrix. Answers will vary.

Additional Example 1

Find kA if $k = -3$ and

$$A = \begin{bmatrix} 5 & -3 & -2 \\ 7 & 12 & 6 \\ -10 & -1 & -4 \end{bmatrix}.$$

$$-3A = \begin{bmatrix} -15 & 9 & 6 \\ -21 & -36 & -18 \\ 30 & 3 & 12 \end{bmatrix}$$

Highlighting the Standards

Standard 14c: Students will soon see that matrices, although they appear new and different, have many of the same rules and elements as the arithmetic and algebraic systems with which they are familiar.

Compare the properties of matrices with the properties of real numbers. For example, for all real numbers a and b, $a \cdot b = b \cdot a$. But $A \cdot B \neq B \cdot A$ for all matrices A and B.

The following properties hold for matrices that can be multiplied.

Closure: $A \cdot B$ is a matrix.
Identity: $A \cdot I = A$
Associative: $(A \cdot B) \cdot C = A \cdot (B \cdot C)$
Inverse: $A \cdot A^{-1} = I$

Thus, matrices form a group under multiplication, but **not** a commutative group.

A square matrix $\begin{bmatrix} a & b \\ c & d \end{bmatrix}$ has an inverse if and only if $ad - bc \neq 0$.

Thus, the square matrix $\begin{bmatrix} 5 & -5 \\ 0 & 0 \end{bmatrix}$ does not have an inverse.

---

5. Use the elements of the second row of C to calculate the remaining two elements of CD in a similar manner. The entire process is summarized below.

$$CD = \begin{bmatrix} -3(5) + 5(-4) + 2(1) & -3(-6) + 5(3) + 2(-2) \\ 1(5) + (-2)(-4) + 4(1) & 1(-6) + (-2)(3) + 4(-2) \end{bmatrix}$$

$$= \begin{bmatrix} -33 & 29 \\ 17 & -20 \end{bmatrix}$$

The above example is generalized below for the case of a matrix A with 2 rows and 3 columns and a matrix B with 3 rows and 2 columns.

If $A = \begin{bmatrix} a_{11} & a_{12} & a_{13} \\ a_{21} & a_{22} & a_{23} \end{bmatrix}$ and $B = \begin{bmatrix} b_{11} & b_{12} \\ b_{21} & b_{22} \\ b_{31} & b_{32} \end{bmatrix}$, then

$$AB = \begin{bmatrix} a_{11}b_{11} + a_{12}b_{21} + a_{13}b_{31} & a_{11}b_{12} + a_{12}b_{22} + a_{13}b_{32} \\ a_{21}b_{11} + a_{22}b_{21} + a_{23}b_{31} & a_{21}b_{12} + a_{22}b_{22} + a_{23}b_{32} \end{bmatrix}.$$

In general, $A_{m \times n} \cdot B_{n \times r} = (AB)_{m \times r}$, where $m \times n$, $n \times r$, and $m \times r$ are the dimensions of the matrices.

Note that $A_{3 \times 1} \cdot B_{1 \times 2}$ exists and equals the 3-by-2 matrix $(AB)_{3 \times 2}$. However, $B_{1 \times 2} \cdot A_{3 \times 1}$ does not exist since the number of columns of B does not equal the number of rows of A.

EXAMPLE 2 Use the matrices C and D given earlier to find the matrix DC.

Plan Since the number of columns of D is 2 and the number of rows of C is 2, DC exists; that is, $D_{3 \times 2} \cdot C_{2 \times 3} = (DC)_{3 \times 3}$.

Solution
$$DC = \begin{bmatrix} 5(-3) + (-6)(1) & 5(5) - 6(-2) & 5(2) - 6(4) \\ -4(-3) + 3(1) & -4(5) + 3(-2) & -4(2) + 3(4) \\ 1(-3) - 2(1) & 1(5) - 2(-2) & 1(2) - 2(4) \end{bmatrix}$$

$$= \begin{bmatrix} -21 & 37 & -14 \\ 15 & -26 & 4 \\ -5 & 9 & -6 \end{bmatrix}$$

Compare matrix DC of Example 2 with matrix CD that was determined earlier. Notice that $CD \neq DC$. Thus, the multiplication of two matrices *is not* a commutative operation. However, multiplication of matrices *is* an associative operation. That is, for matrices A, B, and C, if all products exist, then $(A \cdot B) \cdot C = A \cdot (B \cdot C)$.

Recall that in the set of real numbers, there exists a unique multiplicative identity, the number 1, such that for each real number a, $a \cdot 1 = a$. Also, for each nonzero real number a, there exists a multiplicative inverse $\frac{1}{a}$ such that $a \cdot \frac{1}{a} = 1$.

590 Chapter 16 Linear Systems, Matrices, Determinants

Additional Example 2

Find the matrix AB if

$$A = \begin{bmatrix} 3 & -1 \\ -2 & 2 \\ 5 & 1 \end{bmatrix} \text{ and } B = \begin{bmatrix} 2 & -1 & 3 \\ -3 & -2 & 1 \end{bmatrix}.$$

$$AB = \begin{bmatrix} 9 & -1 & 8 \\ -10 & -2 & -4 \\ 7 & -7 & 16 \end{bmatrix}$$

For *square* matrices and the multiplication of two square matrices, the situation is similar.

For each square matrix A, if $A \cdot I = I \cdot A = A$, then I is called a **multiplicative identity matrix** for A.

For each square matrix A, if $A \cdot A^{-1} = A^{-1} \cdot A = I$, then A^{-1} is called a **multiplicative inverse matrix** of A.

All square matrices of the same dimensions have the same unique multiplicative identity matrix. For example, the unique multiplicative identity matrices for 2-by-2 and 3-by-3 matrices are as follows.

$$I_{2 \times 2} = \begin{bmatrix} 1 & 0 \\ 0 & 1 \end{bmatrix} \text{ and } I_{3 \times 3} = \begin{bmatrix} 1 & 0 & 0 \\ 0 & 1 & 0 \\ 0 & 0 & 1 \end{bmatrix}$$

EXAMPLE 3 If $A = \begin{bmatrix} -2 & 1 & -3 \\ 4 & 2 & -1 \\ 3 & -2 & 4 \end{bmatrix}$ and $I = \begin{bmatrix} 1 & 0 & 0 \\ 0 & 1 & 0 \\ 0 & 0 & 1 \end{bmatrix}$, prove that $A \cdot I = A$.

Solution $A \cdot I =$

$$\begin{bmatrix} -2(1) + 1(0) + (-3)(0) & -2(0) + 1(1) + (-3)(0) & -2(0) + 1(0) + (-3)(1) \\ 4(1) + 2(0) + (-1)(0) & 4(0) + 2(1) + (-1)(0) & 4(0) + 2(0) + (-1)(1) \\ 3(1) + (-2)(0) + 4(0) & 3(0) + (-2)(1) + 4(0) & 3(0) + (-2)(0) + 4(1) \end{bmatrix}$$

$$= \begin{bmatrix} -2 & 1 & -3 \\ 4 & 2 & -1 \\ 3 & -2 & 4 \end{bmatrix} = A \quad \text{Thus, } A \cdot I = A.$$

The next example illustrates how to find the multiplicative inverse matrix of a given square matrix.

EXAMPLE 4 If $A = \begin{bmatrix} -2 & 1 \\ 3 & -1 \end{bmatrix}$, find A^{-1}.

Plan Let $A^{-1} = \begin{bmatrix} a_1 & b_1 \\ a_2 & b_2 \end{bmatrix}$. Then use $A \cdot A^{-1} = I$.

Solution $\begin{bmatrix} -2 & 1 \\ 3 & -1 \end{bmatrix} \cdot \begin{bmatrix} a_1 & b_1 \\ a_2 & b_2 \end{bmatrix} = \begin{bmatrix} 1 & 0 \\ 0 & 1 \end{bmatrix}$

$$\begin{bmatrix} -2a_1 + a_2 & -2b_1 + b_2 \\ 3a_1 + (-1)a_2 & 3b_1 + (-1)b_2 \end{bmatrix} = \begin{bmatrix} 1 & 0 \\ 0 & 1 \end{bmatrix}$$

Math Connections

Communications Networks: A matrix can be used to represent the number of ways that a group of people can communicate with one another. In the matrix below, 1 represents the number of ways that x can communicate with y in one step and also the number of ways y can communicate with z in one step and z can communicate with x in one step. Zero represents the number of ways that x can communicate with z in one step and also the number of ways y can communicate with x in one step and z can communicate with y in one step.

$$\begin{array}{c} \\ \\ \\ \end{array} \quad \begin{array}{c} \; x \; \; y \; \; z \\ x \\ y \\ z \end{array} \begin{bmatrix} 0 & 1 & 0 \\ 0 & 0 & 1 \\ 1 & 0 & 0 \end{bmatrix}$$

The square of the matrix represents the number of ways that one person can communicate with another in two steps. The result of squaring the matrix is shown below.

$$\begin{array}{c} \; x \; \; y \; \; z \\ x \\ y \\ z \end{array} \begin{bmatrix} 0 & 0 & 1 \\ 1 & 0 & 0 \\ 0 & 1 & 0 \end{bmatrix}$$

Critical Thinking Questions

Analysis: In the communications diagram above, suppose that z can communicate with y as well as with x. Have students devise the matrices for communication between the three people in one step and in two steps.

The results are, respectively,

$$\begin{bmatrix} 0 & 1 & 0 \\ 0 & 0 & 1 \\ 1 & 1 & 0 \end{bmatrix} \text{ and } \begin{bmatrix} 0 & 0 & 1 \\ 1 & 1 & 0 \\ 0 & 1 & 1 \end{bmatrix}$$

Additional Example 3

If $A = \begin{bmatrix} -1 & 2 \\ 3 & 1 \end{bmatrix}$ and $I = \begin{bmatrix} 1 & 0 \\ 0 & 1 \end{bmatrix}$, prove that $A \cdot I = A$.

$$A \cdot I = \begin{bmatrix} -1(1) + 2(0) & -1(0) + 2(1) \\ 3(1) + 1(0) & 3(0) + 1(1) \end{bmatrix}$$

$$= \begin{bmatrix} -1 & 2 \\ 3 & 1 \end{bmatrix} = A$$

Additional Example 4

If $A = \begin{bmatrix} 5 & -1 \\ 2 & 3 \end{bmatrix}$, find A^{-1}.

$$A^{-1} = \begin{bmatrix} \dfrac{3}{17} & \dfrac{1}{17} \\ -\dfrac{2}{17} & \dfrac{5}{17} \end{bmatrix}$$

Common Error Analysis

Error: Students will have to be reminded of which elements to multiply. Do sufficient board work to avoid this problem.

In demonstrating multiplication of matrices, the phrase "row into column" can be a guide. This means that you multiply elements of row m by corresponding elements of column n, and then add to find the element x_{mn} of the product matrix.

Checkpoint

Can the following pairs of matrices be multiplied? Justify your answer.

1. $[-3\ 5\ -2]$, $\begin{bmatrix} 1 \\ -2 \\ 3 \end{bmatrix}$

Yes; $A_{1 \times 3} \cdot B_{3 \times 1} = (AB)_{1 \times 1}$

2. $\begin{bmatrix} 9 & 6 & -3 \\ -7 & -5 & 10 \end{bmatrix}$, $\begin{bmatrix} -1 & 12 \\ 9 & 6 \\ -8 & -1 \end{bmatrix}$

Yes; $A_{2 \times 3} \cdot B_{3 \times 2} = (AB)_{2 \times 2}$

Find each product if possible.

3. $-3 \begin{bmatrix} 7 & 1 & 8 \\ -6 & -2 & 10 \\ 5 & 9 & -3 \end{bmatrix}$

$\begin{bmatrix} -21 & -3 & -24 \\ 18 & 6 & -30 \\ -15 & -27 & 9 \end{bmatrix}$

4. $\begin{bmatrix} 5 & -2 \\ -3 & 4 \end{bmatrix} \cdot \begin{bmatrix} 7 & -1 & 3 \\ 3 & 2 & -1 \end{bmatrix}$

$\begin{bmatrix} 29 & -9 & 17 \\ -9 & 11 & -13 \end{bmatrix}$

5. Find the multiplicative identity and the multiplicative inverse for $\begin{bmatrix} -3 & -1 \\ 2 & 4 \end{bmatrix}$.

Identity: $\begin{bmatrix} 1 & 0 \\ 0 & 1 \end{bmatrix}$

Inverse: $\begin{bmatrix} -\frac{2}{5} & -\frac{1}{10} \\ \frac{1}{5} & \frac{3}{10} \end{bmatrix}$

Since the two matrices are equal, their corresponding elements are equal.

$$-2a_1 + a_2 = 1 \qquad -2b_1 + b_2 = 0$$
$$3a_1 - a_2 = 0 \qquad 3b_1 - b_2 = 1$$

Solve each system of equations by the linear combination method (see Lesson 4.3). The result is $a_1 = 1$, $a_2 = 3$, $b_1 = 1$, and $b_2 = 2$.

Thus, $A^{-1} = \begin{bmatrix} 1 & 1 \\ 3 & 2 \end{bmatrix}$.

Focus on Reading

Let $X = \begin{bmatrix} 3 & 5 \\ 4 & -1 \end{bmatrix}$, $Y = \begin{bmatrix} -2 & -7 \\ 6 & 5 \end{bmatrix}$, and $k = 3$. Match each product in the left column with exactly one matrix in the two right columns.

1. $X \cdot Y$ c
2. kX e
3. kY a
4. $Y \cdot X$ d
5. $X \cdot X^{-1}$ b

a. $\begin{bmatrix} -6 & -21 \\ 18 & 15 \end{bmatrix}$

b. $\begin{bmatrix} 1 & 0 \\ 0 & 1 \end{bmatrix}$

c. $\begin{bmatrix} 24 & 4 \\ -14 & -33 \end{bmatrix}$

d. $\begin{bmatrix} -34 & -3 \\ 38 & 25 \end{bmatrix}$

e. $\begin{bmatrix} 9 & 15 \\ 12 & -3 \end{bmatrix}$

Classroom Exercises

Find each product, if it exists. If it does not, give a reason.

1. $-2 \begin{bmatrix} 1 \\ -3 \end{bmatrix}$ $\begin{bmatrix} -2 \\ 6 \end{bmatrix}$

2. $\begin{bmatrix} 3 & 5 \end{bmatrix} \cdot \begin{bmatrix} 2 \\ 3 \end{bmatrix}$ $\begin{bmatrix} 21 \end{bmatrix}$

3. $\begin{bmatrix} 1 \\ 2 \end{bmatrix} \cdot \begin{bmatrix} -1 & 5 \end{bmatrix}$

4. $\begin{bmatrix} -3 \\ -2 \\ 1 \end{bmatrix} \cdot \begin{bmatrix} -1 & 2 & 5 \end{bmatrix}$

5. $\begin{bmatrix} 8 & 5 \\ -1 & 6 \end{bmatrix} \cdot \begin{bmatrix} -1 & 3 \\ 2 & 4 \end{bmatrix}$

6. $\begin{bmatrix} -3 & 1 & 4 \end{bmatrix} \cdot \begin{bmatrix} 1 & 2 \\ -4 & 5 \end{bmatrix}$

Written Exercises

Find each product, if it exists. If it does not exist, give a reason.

1. $6 \begin{bmatrix} 3 \\ 0 \\ 1 \end{bmatrix}$

2. $4 \begin{bmatrix} 5 & -3 \\ 7 & -8 \\ 4 & 6 \end{bmatrix}$

3. $\begin{bmatrix} -1 & 4 & -3 \\ 5 & -2 & 2 \end{bmatrix} \cdot \begin{bmatrix} 3 & 1 & -2 \\ 4 & 5 & 3 \\ -1 & 2 & 4 \end{bmatrix}$

4. $\begin{bmatrix} 4 & -6 & -3 \\ 5 & -1 & 7 \\ 8 & 6 & -5 \end{bmatrix} \cdot \begin{bmatrix} 3 & -1 & 2 \\ 5 & 6 & -3 \end{bmatrix}$ **5.** $\begin{bmatrix} 1 & 2 & -1 \\ 3 & 4 & -2 \\ -1 & 5 & 6 \end{bmatrix} \cdot \begin{bmatrix} 1 & -1 & 3 \\ 4 & 8 & 6 \\ -5 & 7 & -3 \end{bmatrix}$

Find the multiplicative inverse of each matrix if possible. If not possible, indicate this.

6. $\begin{bmatrix} 1 & 4 \\ 2 & 9 \end{bmatrix}$ **7.** $\begin{bmatrix} -2 & \frac{1}{2} \\ 1 & -1 \end{bmatrix}$ **8.** $\begin{bmatrix} 3 & -5 \\ -2 & 6 \end{bmatrix}$ **9.** $\begin{bmatrix} -5 & 7 \\ 8 & -2 \end{bmatrix}$

10. $\begin{bmatrix} 1 & 1 \\ 1 & 1 \end{bmatrix}$ **11.** $\begin{bmatrix} 1 & 2 & -3 \\ 4 & 6 & 4 \\ -5 & -1 & 2 \end{bmatrix}$ **12.** $\begin{bmatrix} 7 & -3 & 2 \\ -1 & -2 & 5 \\ 6 & -4 & 8 \end{bmatrix}$ **13.** $\begin{bmatrix} 3 & -2 \\ -6 & 4 \end{bmatrix}$

In Exercises 14–16, $A = \begin{bmatrix} 2 & 3 \\ -1 & 4 \end{bmatrix}$, $B = \begin{bmatrix} -1 & -3 \\ 2 & 1 \end{bmatrix}$, and $C = \begin{bmatrix} -2 & 3 \\ 4 & 2 \end{bmatrix}$.

Prove that each of the following statements is true.

14. $B \cdot I = I \cdot B$ **15.** $A \cdot C \neq C \cdot A$ **16.** $A \cdot (B \cdot C) = (A \cdot B) \cdot C$

For Exercises 17–22,

$A = \begin{bmatrix} a_{11} & a_{12} \\ a_{21} & a_{22} \end{bmatrix}$, $B = \begin{bmatrix} b_{11} & b_{12} \\ b_{21} & b_{22} \end{bmatrix}$, $C = \begin{bmatrix} c_{11} & c_{12} \\ c_{21} & c_{22} \end{bmatrix}$,

and each element represents a real number. Determine which of the following statements are true. Justify your answer by performing the computations.

17. $0B = O_{2 \times 2}$ T   **18.** $O_{2 \times 2} \cdot B = O_{2 \times 2}$ T **19.** $A \cdot B = B \cdot A$ F

20. $I \cdot C = C \cdot I = C$ T   **21.** $(A \cdot B)^2 = A^2 \cdot B^2$ F **22.** $A \cdot (B + C) = A \cdot B + A \cdot C$ T

Mixed Review

Simplify each expression. *8.5*

1. $\sqrt[3]{-8}$ -2 **2.** $\sqrt[3]{-125}$ -5 **3.** $\sqrt[3]{250a^9}$ $5a^3\sqrt[3]{2}$ **4.** $\sqrt[4]{81a^{16}}$ $3a^4$

███/ **Brainteaser**

A boat is carrying cobblestones on a small, shallow lake. The boat capsizes and the cobblestones sink. The boat, now empty, displaces less water than it did before it capsized. Does the lake's water level rise or drop because of the stones that sink to the lake's bottom?

Closure

Ask students to define the following.

1. A multiplicative identity matrix.
2. A multiplicative inverse matrix.

Have students describe the procedure of matrix multiplication.   See page 589.

███ **FOLLOW UP**

Guided Practice

Classroom Exercises 1–6

Independent Practice

Ⓐ Ex. 1–9, Ⓑ Ex. 10–16, Ⓒ Ex. 17–22

Basic: FR, WE 1–13 odd, Brainteaser
Average: FR, WE 3–15 odd, Brainteaser
Above Average: FR, WE 5–21 odd, Brainteaser

Additional Answers

Classroom Exercises

3. $\begin{bmatrix} -1 & 5 \\ -2 & 10 \end{bmatrix}$

4. $\begin{bmatrix} 3 & -6 & -15 \\ 2 & -4 & -10 \\ -1 & 2 & 5 \end{bmatrix}$

5. $\begin{bmatrix} 2 & 44 \\ 13 & 21 \end{bmatrix}$

6. Does not exist: dimensions do not fit.

Written Exercises

1. $\begin{bmatrix} 18 \\ 0 \\ 6 \end{bmatrix}$ **2.** $\begin{bmatrix} 20 & -12 \\ 28 & -32 \\ 16 & 24 \end{bmatrix}$

See page 825 for answers to Written Ex. 3–16 and Brainteaser.

Enrichment

Have the students find the values of a, b, c, and d in Matrix B given the information below.

$A = [1 \quad 2 \quad 1 \quad 4]$, $B = \begin{bmatrix} a \\ b \\ c \\ d \end{bmatrix}$

1. $A \cdot B = [15]$
2. $c = 2b$
3. $d = a + 1$
4. $a = b + 4$ $a = 3, b = -1, c = -2, d = 4$

▰▰▰GETTING STARTED

Prerequisite Quiz

Multiply and simplify.

1. $-3(2) - (4)(4)$ -22
2. $3(-4) - (-3)(2)$ -6
3. $-5(-3) - 4(-3)$ 27
4. $7(3) - (-5)(-3)$ 6
5. $(x + y)(-2) - (-3)(3x - y)$ $7x - 5y$
6. $-4(2x - y) - (x + 5)(-4)$
   $-4x + 4y + 20$

Motivator

Have students recall the methods used to solve systems of linear equations taught in Chapter 4. Graphing, substitution method, linear combination method. Have them describe each method. See Lessons 4.1, 4.2, and 4.3.

▰▰▰TEACHING SUGGESTIONS

Lesson Note

Distinguish between a determinant and a matrix. A matrix is an array of numbers or elements, while a determinant is a real number assigned to the matrix. For example, $\begin{bmatrix} -5 & 6 \\ 1 & -3 \end{bmatrix}$ is a matrix while $\begin{vmatrix} -5 & 6 \\ 1 & -3 \end{vmatrix}$ is the determinant assigned to the matrix. The determinant equals $(-5)(-3) - (1)(6)$, or 9.

Highlighting the Standards

Standard 2f: Determinants communicate a great deal because of the rules that govern them. The complexity of computation will lead students to an appreciation of the computer's importance in mathematics.

16.3 Two-by-Two Determinants

Objectives   To find the values of 2-by-2 determinants
To solve systems of two linear equations in two variables using determinants

For each square matrix there is a corresponding real number associated with it called a *determinant*. For the 2-by-2 square matrix $\begin{bmatrix} 3 & 4 \\ 2 & -1 \end{bmatrix}$, the value of the determinant is found by multiplying the elements of one diagonal, 3 and -1, and subtracting from this product the product of the elements of the other diagonal, 2 and 4. So, the determinant of the 2-by-2 square matrix is $3(-1) - 2(4) = -3 - 8$, or -11. The determinant of a matrix is written in the same form as the matrix, but with vertical line segments instead of brackets.

Definition

2-by-2 determinant   The determinant of the 2-by-2 square matrix $\begin{bmatrix} a & b \\ c & d \end{bmatrix}$ is symbolized by $\begin{vmatrix} a & b \\ c & d \end{vmatrix}$ and is equal to $ad - cb$.

EXAMPLE 1   Find the value of each determinant.

a. $\begin{vmatrix} -3 & 1 \\ 4 & 5 \end{vmatrix}$ b. $\begin{vmatrix} \sqrt{3} & \sqrt{6} \\ -1 & \sqrt{2} \end{vmatrix}$

Solutions

a. $ad - cb = -3(5) - 4(1)$
   $= -15 - 4 = -19$

b. $ad - ab = \sqrt{3} \cdot \sqrt{2} - (-1)(\sqrt{6})$
   $= \sqrt{6} + \sqrt{6} = 2\sqrt{6}$

In Chapter 4, systems of linear equations were solved using either the *substitution method* or *linear combination method*. Such systems can also be solved by using determinants in a method known as **Cramer's Rule**. This method is used to solve linear systems in the standard form $\begin{aligned} a_1x + b_1y &= c_1 \\ a_2x + b_2y &= c_2 \end{aligned}$, where a_1, a_2, b_1, b_2, c_1, and c_2 are real numbers.

To solve this linear system, use the following formulas.

$$x = \frac{\begin{vmatrix} c_1 & b_1 \\ c_2 & b_2 \end{vmatrix}}{\begin{vmatrix} a_1 & b_1 \\ a_2 & b_2 \end{vmatrix}} = \frac{c_1b_2 - c_2b_1}{a_1b_2 - a_2b_1} \text{ and } y = \frac{\begin{vmatrix} a_1 & c_1 \\ a_2 & c_2 \end{vmatrix}}{\begin{vmatrix} a_1 & b_1 \\ a_2 & b_2 \end{vmatrix}} = \frac{a_1c_2 - a_2c_1}{a_1b_2 - a_2b_1},$$

where $a_1b_2 - a_2b_1 \neq 0$.

Additional Example 1

Find the value of each determinant.

a. $\begin{vmatrix} 5 & 6 \\ -1 & -2 \end{vmatrix}$ -4

b. $\begin{vmatrix} 8 & -1 \\ -3 & -5 \end{vmatrix}$ -43

In the formulas for Cramer's Rule, the denominator is the determinant of the matrix of coefficients, $\begin{bmatrix} a_1 & b_1 \\ a_2 & b_2 \end{bmatrix}$. Notice that the determinant for each numerator differs in exactly one of its columns from the determinant of the denominator.

EXAMPLE 2   Solve the system $\begin{array}{l} 3x - 2y = 6 \\ y = 4x + 4 \end{array}$ using Cramer's Rule.

Solution   Write each equation in the standard form $ax + by = c$. Then determine the coefficients of the variables and find the constant terms.

$$3x + (-2)y = 6 \qquad\qquad 4x + (-1)y = -4$$
$$a_1x + b_1y = c_1 \qquad\qquad a_2x + b_2y = c_2$$
$$a_1 = 3,\ b_1 = -2,\ c_1 = 6 \qquad a_2 = 4,\ b_2 = -1,\ c_2 = -4$$

$$x = \frac{\begin{vmatrix} 6 & -2 \\ -4 & -1 \end{vmatrix}}{\begin{vmatrix} 3 & -2 \\ 4 & -1 \end{vmatrix}} = \frac{6(-1) - (-4)(-2)}{3(-1) - 4(-2)} \qquad y = \frac{\begin{vmatrix} 3 & 6 \\ 4 & -4 \end{vmatrix}}{\begin{vmatrix} 3 & -2 \\ 4 & -1 \end{vmatrix}} = \frac{3(-4) - 4(6)}{3(-1) - 4(-2)}$$

$$= \frac{-6 - 8}{-3 + 8}, \text{ or } -\frac{14}{5} \qquad\qquad = \frac{-12 - 24}{-3 + 8}, \text{ or } -\frac{36}{5}$$

Thus, $\left(-\frac{14}{5}, -\frac{36}{5}\right)$ is the solution of the system.

Classroom Exercises

State the value of each of the following determinants.

1. $\begin{vmatrix} 1 & 1 \\ 1 & 1 \end{vmatrix}$ 0   **2.** $\begin{vmatrix} 1 & -1 \\ 1 & 1 \end{vmatrix}$ 2   **3.** $\begin{vmatrix} 0 & -1 \\ -1 & 0 \end{vmatrix}$ -1   **4.** $\begin{vmatrix} -2 & 4 \\ 6 & 5 \end{vmatrix}$ -34

To solve $\begin{array}{l} -3x + 5y = 6 \\ 2x - 3y = 9 \end{array}$, what determinant would you use for each of the following?

5. the denominator for x **6.** the denominator for y **7.** the numerator for x **8.** the numerator for y

9. Solve the system of Classroom Exercises 5–8 using Cramer's Rule. (63,39)

Written Exercises

Find the value of each determinant.

1. $\begin{vmatrix} 3 & -1 \\ 2 & 5 \end{vmatrix}$ 17   **2.** $\begin{vmatrix} -4 & 3 \\ -2 & 6 \end{vmatrix}$ -18   **3.** $\begin{vmatrix} 9 & 4 \\ 8 & -6 \end{vmatrix}$ -86   **4.** $\begin{vmatrix} -10 & -3 \\ 4 & -8 \end{vmatrix}$ 92

Math Connections

Historical Note: One of the earliest references to determinants in the Western world was by Gottfried Wilhelm Leibniz (1646–1716), the codiscoverer of calculus. Cramer's Rule was published in 1750 by Gabriel Cramer (1704–1752), although not in the notation that is used today. Only in the nineteenth century was the subject developed by mathematicians such as Augustin-Louis Cauchy (1789–1857).

Critical Thinking Questions

Analysis: In the statement of Cramer's Rule, it is noted that the determinant formed from the coefficients of x and y must not be zero. Ask students to give the significance of a zero value for this determinant. They should be able to reason that if $a_1b_2 - a_2b_1 = 0$, then the slopes of the lines corresponding to the two equations will be equal. Thus, the two equations are either dependent or inconsistent.

Common Error Analysis

Error: Students often make errors in signs when finding the value of a determinant. To avoid such errors, have them use parentheses freely, as shown below.

$$\begin{vmatrix} -1 & -5 \\ -2 & -6 \end{vmatrix} = (-1)(-6) - (-2)(-5)$$
$$= 6 - 10 = -4$$

Additional Example 2

Solve the system using Cramer's Rule.

$$2x - 3y = 4$$
$$x + 2y = -1$$
$$\left(\frac{5}{7}, -\frac{6}{7}\right)$$

595

Checkpoint

Find the value of each determinant.

1. $\begin{vmatrix} -5 & -3 \\ 7 & -2 \end{vmatrix}$ 31

2. $\begin{vmatrix} \frac{1}{5} & -\frac{1}{2} \\ \frac{3}{4} & -\frac{1}{6} \end{vmatrix}$ $\frac{41}{120}$

Solve each system of linear equations using Cramer's Rule.

3. $-2x - 3y = 8$
$3x + y = -5$ $(-1,-2)$

4. $x + 4y = -3$
$-3x - y = -2$ $(1,-1)$

Closure

Ask students to define the term **determinant**. See page 594. Have students give the formulas for determining the *x*- and *y*-coordinates when solving a system of linear equations using Cramer's Rule. See page 594.

▰▰▰FOLLOW UP

Guided Practice

Classroom Exercises 1–9

Independent Practice

Ⓐ Ex. 1–15, Ⓑ Ex. 16–21, Ⓒ Ex. 22–27

Basic: WE 1–17 odd, Midchapter Review

Average: WE 5–23 odd, Midchapter Review

Above Average: WE 11–27 odd, Midchapter Review

Additional Answers

See page 825 for the answers to Classroom Ex. 5–8, Written Ex. 22, 24–27, and Midchapter Review Ex. 1–12.

---

5. $\begin{vmatrix} 1 & 0 \\ 0 & 1 \end{vmatrix}$ 1   **6.** $\begin{vmatrix} -1 & -6 \\ -7 & -10 \end{vmatrix}$ -32 **7.** $\begin{vmatrix} 0 & 5 \\ 10 & 0 \end{vmatrix}$ -50 **8.** $\begin{vmatrix} 7 & 9 \\ 11 & 12 \end{vmatrix}$ -15

9. $\begin{vmatrix} \frac{1}{2} & \frac{1}{3} \\ \frac{1}{5} & -\frac{2}{3} \end{vmatrix}$ $-\frac{6}{15}$ **10.** $\begin{vmatrix} \frac{3}{4} & \frac{1}{2} \\ -\frac{1}{4} & \frac{1}{3} \end{vmatrix}$ $\frac{3}{8}$ **11.** $\begin{vmatrix} -\frac{2}{3} & -\frac{1}{4} \\ \frac{3}{5} & -\frac{1}{2} \end{vmatrix}$ $\frac{29}{60}$ **12.** $\begin{vmatrix} 4\sqrt{3} & \sqrt{2} \\ -2\sqrt{3} & -3\sqrt{3} \end{vmatrix}$
$-36 + 2\sqrt{6}$

Solve each system of linear equations using Cramer's Rule.

13. $3x - 2y = -7$
$-x + y = 3$ $(-1,2)$

14. $-4x + 3y = 14$
$-2x + 2y = 8$ $(-2,2)$

15. $5x - 3y = 18$
$2x + 4y = 2$ $(3,-1)$

16. $-x + 4y = -8$
$6x - 25 = y$ $(4,-1)$

17. $6x - 5y = 15$
$2y = 3x - 6$ $(0,-3)$

18. $2x - 7 = y$
$3y = x - 16$ $(1,-5)$

19. $x + 2 = y$
$2y = 3x + 2$ $(2,4)$

20. $y = 3x - 1$
$y = 2x + 5$ $(6,17)$

21. $4x + 3y = 11.5$
$2x - y = 0.5$ $(1.3,2.1)$

22. $a_1x + b_1y = 3$
$a_2x + b_2y = -4$

23. $\sqrt{2}x + \sqrt{3}y = 10$
$\sqrt{3}x - \sqrt{2}y = 0$
$(2\sqrt{2},2\sqrt{3})$

24. $m_1x - n_1y = p_1$
$m_2x + n_2y = p_2$

Suppose that $A = \begin{bmatrix} a_1 & b_1 \\ c_1 & d_1 \end{bmatrix}$ and $B = \begin{bmatrix} a_2 & b_2 \\ c_2 & d_2 \end{bmatrix}$ are 2-by-2 matrices and k is a scalar. Define $|A|$, $|B|$, and $|A \cdot B|$ as the determinants formed from the corresponding elements of A, B, and $A \cdot B$; that is,

$$|A| = \begin{vmatrix} a_1 & b_1 \\ c_1 & d_1 \end{vmatrix}$$ and so on. Prove that each of the following is true.

25. $|-A| = |A|$ **26.** $|kA| = k^2|A|$ **27.** $|A \cdot B| = |A| \cdot |B|$

Midchapter Review

In Exercises 1–12, $A = \begin{bmatrix} 2 & 0 \\ -1 & 8 \end{bmatrix}$, $B = \begin{bmatrix} -4 & 4 \\ 2 & 2 \end{bmatrix}$,

$C = \begin{bmatrix} -6 & 3 \\ 0 & 8 \end{bmatrix}$, $D = \begin{bmatrix} -7 \\ 1 \end{bmatrix}$, $E = \begin{bmatrix} 1 & -1 \end{bmatrix}$, $O_{2 \times 2} = \begin{bmatrix} 0 & 0 \\ 0 & 0 \end{bmatrix}$,

and $I_{2 \times 2} = \begin{bmatrix} 1 & 0 \\ 0 & 1 \end{bmatrix}$. Find each of the following. If the indicated sum or product does not exist, give a reason.   *16.1, 16.2*

1. $A + B$ **2.** $B + D$ **3.** $-C$ **4.** $A + (-C)$

5. $A + O_{2 \times 2}$ **6.** $-2B$ **7.** AB **8.** BA

9. CE **10.** $CI_{2 \times 2}$ **11.** DE **12.** AD

13. Solve $\begin{matrix} 2x - 3y = 11 \\ 4x + 7y = -30 \end{matrix}$ using Cramer's Rule.   *16.3* $\left(-\frac{1}{2},-4\right)$

Enrichment

Have students attempt to find the solutions to these two systems of equations by using Cramer's Rule.

1. $2x - y = 3$
$-4x + 2y = -6$

2. $3x - y = -4$
$6x - 2y = -12$

After they have made the attempt, have them discuss the significance of their findings.

In system 1, computations for both *x* and *y* result in the indeterminate form $\frac{0}{0}$. This is a system of dependent equations.

In system 2, *x* is computed as $\frac{-4}{0}$ and *y* is computed as $\frac{-12}{0}$, both undefined. This is a system of inconsistent equations.

16.4 Three-by-Three Determinants

Objectives To find the values of 3-by-3 determinants
To solve systems of three linear equations in three variables using determinants

For the 3-by-3 square matrix $\begin{bmatrix} a_1 & b_1 & c_1 \\ a_2 & b_2 & c_2 \\ a_3 & b_3 & c_3 \end{bmatrix}$, the value of

the corresponding determinant can be found by the *method of repeated columns*. First, repeat columns 1 and 2 as shown below. Then multiply on the down diagonals, multiply on the up diagonals, and subtract the sum of the up-diagonal products from the sum of the down-diagonal products.

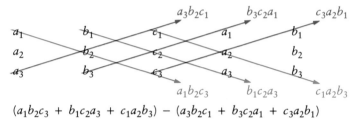

$$(a_1b_2c_3 + b_1c_2a_3 + c_1a_2b_3) - (a_3b_2c_1 + b_3c_2a_1 + c_3a_2b_1)$$

Definition

> **3-by-3 Determinant**   The determinant of a 3-by-3 square matrix is
> $$\begin{vmatrix} a_1 & b_1 & c_1 \\ a_2 & b_2 & c_2 \\ a_3 & b_3 & c_3 \end{vmatrix} = (a_1b_2c_3 + b_1c_2a_3 + c_1a_2b_3) - (a_3b_2c_1 + b_3c_2a_1 + c_3a_2b_1).$$

When evaluating 3-by-3 determinants, you will often find it easier to use the method of repeated columns than to refer to the definition above.

EXAMPLE 1 Find the value of $\begin{vmatrix} -3 & 1 & -2 \\ -4 & 3 & 2 \\ -1 & -2 & -1 \end{vmatrix}$.

Use the method of repeated columns to obtain the following.

Solution $[9 + (-2) + (-16)] - [6 + 12 + 4] = -9 - 22 = -31$

Thus, the value of the determinant is -31.

Additional Example 1

Find the value of $\begin{vmatrix} 5 & 1 & -2 \\ -3 & 2 & -1 \\ 3 & 6 & -4 \end{vmatrix}$.   23

Lesson Note

You may wish to work through the proofs of Exercises 16–18 with your class. The problems to prove can be written as follows.

16. $\begin{vmatrix} a_1 & a_2 & a_3 \\ b_1 & b_2 & b_3 \\ 0 & 0 & 0 \end{vmatrix} = 0$

17. $\begin{vmatrix} b_1 & b_2 & b_3 \\ a_1 & a_2 & a_3 \\ c_1 & c_2 & c_3 \end{vmatrix} = -\begin{vmatrix} a_1 & a_2 & a_3 \\ b_1 & b_2 & b_3 \\ c_1 & c_2 & c_3 \end{vmatrix}$

18. $\begin{vmatrix} a_1 & a_2 & a_3 \\ b_1 & b_2 & b_3 \\ b_1 & b_2 & b_3 \end{vmatrix} = 0$

Have the students create numerical examples for Exercises 16–18.

Math Connections

Geometry: There are some geometric formulas that have an interesting form when expressed using determinants. For example, the area of a triangle in a coordinate plane with vertices (x_1, y_1), (x_2, y_2), (x_3, y_3) is given by this expression.

$$\frac{1}{2}\begin{vmatrix} x_1 & y_1 & 1 \\ x_2 & y_2 & 1 \\ x_3 & y_3 & 1 \end{vmatrix}$$

The volume of a tetrahedron with its vertices located on a three-dimensional coordinate space is given by this expression.

$$\frac{1}{6}\begin{vmatrix} x_1 & y_1 & z_1 & 1 \\ x_2 & y_2 & z_2 & 1 \\ x_3 & y_3 & z_3 & 1 \\ x_4 & y_4 & z_4 & 1 \end{vmatrix}$$

Cramer's Rule can also be used to solve systems of three linear equations in three variables which are of the form

$$\begin{aligned} a_1x + b_1y + c_1z &= d_1 \\ a_2x + b_2y + c_2z &= d_2, \text{ where all coefficients are real numbers.} \\ a_3x + b_3y + c_3z &= d_3 \end{aligned}$$

To solve this linear system, use the following formulas.

$$x = \frac{\begin{vmatrix} d_1 & b_1 & c_1 \\ d_2 & b_2 & c_2 \\ d_3 & b_3 & c_3 \end{vmatrix}}{D}, \; y = \frac{\begin{vmatrix} a_1 & d_1 & c_1 \\ a_2 & d_2 & c_2 \\ a_3 & d_3 & c_3 \end{vmatrix}}{D}, \text{ and } z = \frac{\begin{vmatrix} a_1 & b_1 & d_1 \\ a_2 & b_2 & d_2 \\ a_3 & b_3 & d_3 \end{vmatrix}}{D},$$

where $D = \begin{vmatrix} a_1 & b_1 & c_1 \\ a_2 & b_2 & c_2 \\ a_3 & b_3 & c_3 \end{vmatrix}$ is the determinant of the matrix of coefficients of the variables x, y, and z.

EXAMPLE 2 Solve the system $\begin{aligned} -3x + 2y - z &= -8 \\ 4x + 3y + 2z &= 7 \\ -x - 4y + 3z &= 12 \end{aligned}$ using Cramer's Rule.

Solution

$$x = \frac{\begin{vmatrix} -8 & 2 & -1 \\ 7 & 3 & 2 \\ 12 & -4 & 3 \end{vmatrix}}{D}, \; y = \frac{\begin{vmatrix} -3 & -8 & -1 \\ 4 & 7 & 2 \\ -1 & 12 & 3 \end{vmatrix}}{D}, \; z = \frac{\begin{vmatrix} -3 & 2 & -8 \\ 4 & 3 & 7 \\ -1 & -4 & 12 \end{vmatrix}}{D}$$

$D:$

$D = [-27 + (-4) + 16] - [3 + 24 + 24] = -15 - 51 = -66$

$x = \dfrac{(-72 + 48 + 28) - (-36 + 64 + 42)}{-66} = \dfrac{4 - 70}{-66}$, or 1

$y = \dfrac{(-63 + 16 - 48) - (7 - 72 - 96)}{-66} = \dfrac{-95 + 161}{-66}$, or -1

$z = \dfrac{(-108 - 14 + 128) - (24 + 84 + 96)}{-66} = \dfrac{6 - 204}{-66}$, or 3

Thus, $(1, -1, 3)$ is the solution of the system. Check in the original equations.

Additional Example 2

Solve the system using Cramer's Rule.

$$\begin{aligned} 4x - 3y + z &= 5 \\ -x + y - 2z &= 2 \qquad (1, -1, -2) \\ 2x - y + 4z &= -5 \end{aligned}$$

Another way to find the determinant of a matrix and to solve a system of equations is to expand a determinant by *minors*. The **minor** of an element of a determinant is the determinant found by deleting the column and row in which the element lies.

Thus, for $\begin{vmatrix} a_1 & b_1 & c_1 \\ a_2 & b_2 & c_2 \\ a_3 & b_3 & c_3 \end{vmatrix}$, the minor of a_1 is $\begin{vmatrix} b_2 & c_2 \\ b_3 & c_3 \end{vmatrix}$, the minor of a_2 is $\begin{vmatrix} b_1 & c_1 \\ b_3 & c_3 \end{vmatrix}$, and the minor of a_3 is $\begin{vmatrix} b_1 & c_1 \\ b_2 & c_2 \end{vmatrix}$.

In general, the value of the determinant $\begin{vmatrix} a_1 & b_1 & c_1 \\ a_2 & b_2 & c_2 \\ a_3 & b_3 & c_3 \end{vmatrix}$ can be obtained by finding the minors of all the elements of any row or column, multiplying each minor by its element, and forming an *expansion by minors*, as shown below for column 1.

$$a_1\begin{vmatrix} b_2 & c_2 \\ b_3 & c_3 \end{vmatrix} - a_2\begin{vmatrix} b_1 & c_1 \\ b_3 & c_3 \end{vmatrix} + a_3\begin{vmatrix} b_1 & c_1 \\ b_2 & c_2 \end{vmatrix}$$

$$= a_1(b_2c_3 - b_3c_2) - a_2(b_1c_3 - b_3c_1) + a_3(b_1c_2 - b_2c_1)$$

$$= a_1b_2c_3 - a_1b_3c_2 - a_2b_1c_3 + a_2b_3c_1 + a_3b_1c_2 - a_3b_2c_1$$

Notice that in the second term in the original expansion above, the product of the element a_2 and its minor has been multiplied by -1. In general, if the sum of an element's row number and column number is an odd integer, multiply by -1; otherwise, do not.

a_1: 1st row, 1st column; $1 + 1 = 2$, an even integer
a_2: 2nd row, 1st column; $2 + 1 = 3$, an odd integer
a_3: 3rd row, 1st column; $3 + 1 = 4$, an even integer

Thus, only the product of a_2 and its minor is multiplied by -1.

EXAMPLE 3 Use minors to find the value of $D = \begin{vmatrix} -3 & 4 & 5 \\ 1 & -2 & -3 \\ 2 & 3 & -1 \end{vmatrix}$.

Plan Expand the determinant using the minor of the elements of column 2.

Solution $D = -4\begin{vmatrix} 1 & -3 \\ 2 & -1 \end{vmatrix} + (-2)\begin{vmatrix} -3 & 5 \\ 2 & -1 \end{vmatrix} - 3\begin{vmatrix} -3 & 5 \\ 1 & -3 \end{vmatrix}$

$= -4(-1 + 6) - 2(3 - 10) - 3(9 - 5)$

$= -20 + 14 - 12$, or -18

Thus, the value of the determinant is -18.

Critical Thinking Questions

Analysis: Ask students to devise a one-dimensional version of the two- and three-dimensional formulas shown in *Math Connections*. They should be able to reason backwards from the established pattern to this determinant, which has the value $x_1 - x_2$.

$$\begin{vmatrix} x_1 & 1 \\ x_2 & 1 \end{vmatrix}$$

The value, $|x_1 - x_2|$, is the length of a line segment connecting points x_1 and x_2 on the x-axis.

Common Error Analysis

Error: Trouble with signs may occur when students use either of the methods shown to find the value of a determinant.

Some board work may alleviate this source of error.

Checkpoint

1. Find the value of the determinant $\begin{vmatrix} 1 & -2 & 5 \\ 3 & -1 & 2 \\ -3 & -4 & -1 \end{vmatrix}$. -60

2. Solve the system using minors and Cramer's Rule.
$-3x + 2y - 3z = -6$
$4x - y + 2z = 2$
$-x + 3y - z = -2$
$(-1,0,3)$

3. Using minors, find the value of $\begin{vmatrix} 6 & 2 & -3 \\ -5 & -1 & 4 \\ -2 & 3 & -4 \end{vmatrix}$. -53

Additional Example 3

Use minors to find the value of this determinant.

$\begin{vmatrix} 3 & -1 & 2 \\ 1 & -2 & 5 \\ -3 & 2 & -1 \end{vmatrix}$. -18

Have students explain the repeated columns method. See page 597. Have students define the minor of an element of a determinant. See page 599. Have students explain expansion by minors. See page 599.

■■■**FOLLOW UP**

Guided Practice

Classroom Exercises 1–9

Independent Practice

A Ex. 1–12, **B** Ex. 13–15, **C** Ex. 16–18

Basic: WE 1–13 odd, Brainteaser
Average: WE 5–15 odd, Brainteaser
Advanced: WE 5–17 odd, Brainteaser

Additional Answers

Written Exercises

7. $(-1, 1, 2)$

EXAMPLE 4 Solve the system
$$-3x + 4y + 5z = 5$$
$$x - 2y - 3z = -5$$
$$2x + 3y - z = -2$$
using minors and Cramer's Rule.

Solution The determinant of the matrix of coefficients of the system was evaluated as -18 in Example 3. Use this value and Cramer's rule.

$$x = \frac{\begin{vmatrix} 5 & 4 & 5 \\ -5 & -2 & -3 \\ -2 & 3 & -1 \end{vmatrix}}{-18}, \quad y = \frac{\begin{vmatrix} -3 & 5 & 5 \\ 1 & -5 & -3 \\ 2 & -2 & -1 \end{vmatrix}}{-18}, \text{ and } z = \frac{\begin{vmatrix} -3 & 4 & 5 \\ 1 & -2 & -5 \\ 2 & 3 & -2 \end{vmatrix}}{-18}$$

$$x = \frac{5\begin{vmatrix} -2 & -3 \\ 3 & -1 \end{vmatrix} - (-5)\begin{vmatrix} 4 & 5 \\ 3 & -1 \end{vmatrix} + (-2)\begin{vmatrix} 4 & 5 \\ -2 & -3 \end{vmatrix}}{-18} \quad \leftarrow \text{Numerator is expanded using Column 1.}$$

$$= \frac{5(2 + 9) + 5(-4 - 15) - 2(-12 + 10)}{-18} = \frac{55 - 95 + 4}{-18} = \frac{-36}{-18}, \text{ or } 2$$

In a similar manner, y and z can be evaluated using an expansion by minors. The values obtained are $y = -1$ and $z = 3$. Thus, $(2, -1, 3)$ is the solution of the system.

Classroom Exercises

For the determinant shown, state the value of the minor for the indicated entries.
$$\begin{vmatrix} 0 & 0 & 2 \\ 1 & 1.6 & 4 \\ 0 & 8 & 5 \end{vmatrix}$$

1. 1.6 0 **2.** 2 8 **3.** 1 −16 **4.** 4 0 **5.** 5 0 **6.** 8 −2

Find the value of the determinant of Exercises 1–6 using the minors of the following.

7. the elements of row 1 16 **8.** the elements of column 1 16 **9.** the elements of row 3 16

Written Exercises

Use the definition of a 3-by-3 determinant or the method of repeated columns to find the value of each determinant.

1. $\begin{vmatrix} -2 & -4 & 2 \\ 3 & 5 & 1 \\ 4 & -1 & -3 \end{vmatrix}$ −70   **2.** $\begin{vmatrix} 5 & -1 & -3 \\ 2 & -4 & 2 \\ 1 & 4 & -2 \end{vmatrix}$ −42   **3.** $\begin{vmatrix} 3 & -2 & 4 \\ 1 & -3 & 5 \\ -3 & 2 & -1 \end{vmatrix}$ −21

600   Chapter 16 Linear Systems, Matrices, Determinants

Additional Example 4

Solve the system using minors and Cramer's Rule.

$$2x - y + z = 7$$
$$x + 3y + 2z = -5$$
$$x + y - 2z = 3$$

$(3, -2, -1)$

Use minors to find the value of each determinant.

4. $\begin{vmatrix} -3 & 2 & -1 \\ 4 & 3 & 6 \\ -1 & 5 & 1 \end{vmatrix}$ 38   **5.** $\begin{vmatrix} 4 & -2 & -4 \\ -2 & 1 & -3 \\ -3 & 3 & 2 \end{vmatrix}$ 30   **6.** $\begin{vmatrix} 8 & -6 & -1 \\ 4 & 3 & -2 \\ -5 & -2 & 1 \end{vmatrix}$ −51

Solve each system of equations using Cramer's Rule. $(3, -1, 0)$

$(-2, 2, 1)$

7. $-3x + 2y - z = 3$
$2x + 4y + 3z = 8$
$-x - 3y + 2z = 2$

8. $2x - y + 3z = 7$
$-x + 3y - 2z = -6$
$3x - 2y + 4z = 11$

9. $-x + 3y - 2z = 6$
$4x - 2y + 3z = -9$
$3x + y - z = -5$

10. $-3x + 2y - 2z = -10$
$2x + 3y + z = -1$
$-x - y + 2z = -8$
$(4, -2, -3)$

11. $-4x - y + 2z = -4$
$-3x + 2y - z = 10$
$5x - 3y + 3z = -20$
$(-1, 2, -3)$

12. $3x - 2y - 4z = 10$
$4x + 3y - z = 16$
$-x - 2y + 3z = -12$
$(2, 2, -2)$

Solve each system of equations using minors and Cramer's Rule.

13. $-4x - 3y + z = 7$
$5x + y - 3z = -16$
$-x - 2y + 4z = 3$
$(-3, 2, 1)$

14. $2x + 3y - 2z = 3$
$-3x + 4y - z = 12$
$-2x - y + 4z = -5$
$(-2, 1, -2)$

15. $-x - 2y + 3z = 8$
$-4x - y - 2z = -22$
$2x - 3y - z = 6$
$\left(\frac{244}{63}, -\frac{38}{63}, \frac{224}{63}\right)$

Prove that the following are true for the determinant below.

$$\begin{vmatrix} a_1 & a_2 & a_3 \\ b_1 & b_2 & b_3 \\ c_1 & c_2 & c_3 \end{vmatrix}$$

16. If the elements of a row are zero, the value of the determinant is 0.

17. If any two rows of a determinant are interchanged, the sign of the value of the determinant is changed.

18. If two rows in a determinant are the same, the value of the determinant is 0.

Mixed Review

Simplify and write each expression using only positive exponents. *5.1, 5.2*

1. $(-2a^4b^2c)^2$
$4a^8b^4c^2$

2. $\dfrac{-18x^6y^4z}{-6x^4yz^6}$
$\dfrac{3x^2y^3}{z^5}$

3. $-3(-2x^{-3}y^2)^4$
$-\dfrac{48y^8}{x^{12}}$

4. $(2a^4b)(-3a^{-3}b^2)$
$-6ab^3$

▰▰▰ Brainteaser

Can an irrational number raised to an irrational power be a rational number? (HINT: Begin with $\sqrt{2}^{\sqrt{2}}$.)

16. Given the three-by-three determinant:

$$\begin{vmatrix} a_1 & a_2 & a_3 \\ b_1 & b_2 & b_3 \\ c_1 & c_2 & c_3 \end{vmatrix}$$

To prove that the value of the given determinant is zero, expand about the row composed of zero elements. Since each element of the row is zero, all the minors will be multiplied by zero, yielding a net value of zero. For example, if $a_1 = 0$, $a_2 = 0$, and $a_3 = 0$:

$$D = a_1\begin{vmatrix} b_2 & b_3 \\ c_2 & c_3 \end{vmatrix} - a_2\begin{vmatrix} b_1 & b_3 \\ c_1 & c_3 \end{vmatrix}$$
$$+ a_3\begin{vmatrix} b_1 & b_2 \\ c_1 & c_2 \end{vmatrix}$$

$$D = 0\begin{vmatrix} b_2 & b_3 \\ c_2 & c_3 \end{vmatrix} + 0\begin{vmatrix} b_1 & b_3 \\ c_1 & c_3 \end{vmatrix}$$
$$+ 0\begin{vmatrix} b_1 & b_2 \\ c_1 & c_2 \end{vmatrix} = 0$$

See pages 615–616 for the answers to Written Ex. 17–18 and Brainteaser.

Enrichment

Have the students use Cramer's Rule or matrix transformations to solve the following system of equations.

$5a - 3b + 3c - 5d = -2$
$-a + 8b - 3c + 8d = 7$
$4a - 2b + 2c - 4d = -2$
$2a + 6b - c + 4d = 4$

$a = 1, b = -1, c = 0, d = 2$

16.5 Solving Linear Systems with Matrices

Objectives
To solve systems of two linear equations in two variables using the inverse of a matrix
To solve systems of three linear equations in three variables using matrix transformations

Prerequisite Quiz

1. Find the inverse of $\begin{bmatrix} 3 & 2 \\ -1 & 5 \end{bmatrix}$.

$\begin{bmatrix} \frac{5}{17} & -\frac{2}{17} \\ \frac{1}{17} & \frac{3}{17} \end{bmatrix}$

2. Multiply. $\begin{bmatrix} 4 & -1 \\ 2 & -3 \end{bmatrix}\begin{bmatrix} -5 \\ -6 \end{bmatrix}$ $\begin{bmatrix} -14 \\ 8 \end{bmatrix}$

3. Find the value of $\begin{vmatrix} 5 & -6 \\ 7 & -8 \end{vmatrix}$.   2

Motivator

Have students recall from Lesson 16.2 the Multiplicative Identity matrix for any 2-by-2 matrix.

$\begin{bmatrix} 1 & 0 \\ 0 & 1 \end{bmatrix}$

Ask them to give the procedure for finding the Multiplicative Inverse matrix of any 2-by-2 matrix. $A^{-1} \cdot A = I$, so multiply and then set each element equal to the corresponding element in the inverse matrix. Solve each system by the linear combination method.

The system of two linear equations $\begin{array}{l} a_1x + b_1y = c_1 \\ a_2x + b_2y = c_2 \end{array}$

can be written in matrix form as $\begin{bmatrix} a_1 & b_1 \\ a_2 & b_2 \end{bmatrix}\begin{bmatrix} x \\ y \end{bmatrix} = \begin{bmatrix} c_1 \\ c_2 \end{bmatrix}$.

To see this, multiply the two matrices on the left side of the equation above.

Thus, the system $\begin{array}{l} 3x - 2y = 9 \\ -5x + 9y = -6 \end{array}$ can be written in matrix form as

$\begin{bmatrix} 3 & -2 \\ -5 & 9 \end{bmatrix}\begin{bmatrix} x \\ y \end{bmatrix} = \begin{bmatrix} 9 \\ -6 \end{bmatrix}$.

A system of two linear equations can be solved by writing it in matrix form and then using $\begin{bmatrix} x \\ y \end{bmatrix} = \begin{bmatrix} a_1 & b_1 \\ a_2 & b_2 \end{bmatrix}^{-1}\begin{bmatrix} c_1 \\ c_2 \end{bmatrix}$, where

$\begin{bmatrix} a_1 & b_1 \\ a_2 & b_2 \end{bmatrix}^{-1}$ is the multiplicative inverse of $\begin{bmatrix} a_1 & b_1 \\ a_2 & b_2 \end{bmatrix}$.

The above statement can be verified by first writing the following.

$\begin{bmatrix} a_1 & b_1 \\ a_2 & b_2 \end{bmatrix} \cdot \begin{bmatrix} x \\ y \end{bmatrix} = \begin{bmatrix} c_1 \\ c_2 \end{bmatrix}$

Then, substitute A, X, and C, respectively, for the above matrices to obtain this equation:

$A \cdot X = C$

Next, multiply each side on its left by A^{-1}. $A^{-1}(A \cdot X) = A^{-1}C$
Now, use the Associative Property. $(A^{-1} \cdot A) \cdot X = A^{-1}C$
Use $A^{-1} \cdot A = I$. $I \cdot X = A^{-1}C$
Use $I \cdot X = X$. $X = A^{-1}C$

Thus, $\begin{bmatrix} x \\ y \end{bmatrix} = \begin{bmatrix} a_1 & b_1 \\ a_2 & b_2 \end{bmatrix}^{-1}\begin{bmatrix} c_1 \\ c_2 \end{bmatrix}$.

Highlighting the Standards

Standard 5e: Solving systems of linear equations is usually considered the primary goal for working with matrices. This is particularly true with systems of three equations.

EXAMPLE 1 Solve the system $\begin{array}{r} 3x - 2y = 12 \\ -2x + 5y = -19 \end{array}$ using the inverse of a matrix.

Plan Use $X = A^{-1}C$, where $X = \begin{bmatrix} x \\ y \end{bmatrix}$, $A = \begin{bmatrix} 3 & -2 \\ -2 & 5 \end{bmatrix}$, and $C = \begin{bmatrix} 12 \\ -19 \end{bmatrix}$.

Solution $X = \begin{bmatrix} 3 & -2 \\ -2 & 5 \end{bmatrix}^{-1} \begin{bmatrix} 12 \\ -19 \end{bmatrix}$

First find the inverse of $A = \begin{bmatrix} 3 & -2 \\ -2 & 5 \end{bmatrix}$.

Use $A \cdot A^{-1} = I$. $\begin{bmatrix} 3 & -2 \\ -2 & 5 \end{bmatrix} \begin{bmatrix} a_1 & b_1 \\ a_2 & b_2 \end{bmatrix} = \begin{bmatrix} 1 & 0 \\ 0 & 1 \end{bmatrix}$

Find the matrix product. $\begin{bmatrix} 3a_1 - 2a_2 & 3b_1 - 2b_2 \\ -2a_1 + 5a_2 & -2b_1 + 5b_2 \end{bmatrix} = \begin{bmatrix} 1 & 0 \\ 0 & 1 \end{bmatrix}$

Use the definition of $\begin{array}{cc} 3a_1 - 2a_2 = 1 & 3b_1 - 2b_2 = 0 \\ -2a_1 + 5a_2 = 0 & -2b_1 + 5b_2 = 1 \end{array}$
matrix equality.

When the above two systems are solved, the results are as follows.

$a_1 = \dfrac{5}{11} \qquad a_2 = \dfrac{2}{11} \qquad b_1 = \dfrac{2}{11} \qquad b_2 = \dfrac{3}{11}$

So, $A^{-1} = \begin{bmatrix} a_1 & b_1 \\ a_2 & b_2 \end{bmatrix} = \begin{bmatrix} \frac{5}{11} & \frac{2}{11} \\ \frac{2}{11} & \frac{3}{11} \end{bmatrix}$. Next, use A^{-1} to find X.

$X = A^{-1}C$,

so $X = \begin{bmatrix} x \\ y \end{bmatrix} = \begin{bmatrix} \frac{5}{11} & \frac{2}{11} \\ \frac{2}{11} & \frac{3}{11} \end{bmatrix} \begin{bmatrix} 12 \\ -19 \end{bmatrix} = \begin{bmatrix} \frac{60}{11} + \left(-\frac{38}{11}\right) \\ \frac{24}{11} + \left(-\frac{57}{11}\right) \end{bmatrix}$, or $\begin{bmatrix} 2 \\ -3 \end{bmatrix}$.

Thus, $(2, -3)$ is the solution of the system.

Another way to solve a system of linear equations is to apply a series of steps called *transformations*. The coefficients and constants of a system can be modified until the system is in **triangular form**, as shown below for a system of three linear equations.

Once the system is in this form, the solution is easily obtained from $z = -2$.

$5x + 2y - z = 3$
$3y + 2z = 5$
$z = -2$

$3y + 2(-2) = 5$, or $y = 3$

$5x + 2(3) - (-2) = 3$, or $x = -1$

The solution is $(-1, 3, -2)$.

To transform a system to triangular form, it is convenient first to write the **augmented matrix** of the system, as illustrated on the next page.

16.5 Solving Linear Systems with Matrices **603**

■TEACHING SUGGESTIONS

Lesson Note

You may wish to discuss the formula given for Exercises 17–20:

$\begin{bmatrix} a_1 & b_1 \\ a_2 & b_2 \end{bmatrix}^{-1} = \dfrac{1}{\begin{vmatrix} a_1 & b_1 \\ a_2 & b_2 \end{vmatrix}} \cdot \begin{bmatrix} b_2 & -b_1 \\ -a_2 & a_1 \end{bmatrix}$

Show how the formula can be applied to the determinant in Example 1 as follows.

$\begin{bmatrix} 3 & -2 \\ -2 & 5 \end{bmatrix}^{-1} = \dfrac{1}{\begin{vmatrix} 3 & -2 \\ -2 & 5 \end{vmatrix}} \cdot \begin{bmatrix} 5 & -(-2) \\ -(-2) & 3 \end{bmatrix}$

$= \dfrac{1}{11} \cdot \begin{bmatrix} 5 & 2 \\ 2 & 3 \end{bmatrix}$

$= \begin{bmatrix} \frac{5}{11} & \frac{2}{11} \\ \frac{2}{11} & \frac{3}{11} \end{bmatrix}$

Additional Example 1

Solve the system using the inverse of a matrix.

$-x + 2y = -4$
$2x + 3y = 1$

$(2, -1)$

Math Connections

Transformations: The matrix form of the linear system near the top of page 602 can be interpreted in another way. In this interpretation the matrix $\begin{bmatrix} x \\ y \end{bmatrix}$ is thought of as a vector (x, y) with its endpoint at the origin. Because of its vertical appearance, it is also called a column vector. The 2×2 matrix $\begin{bmatrix} a_1 & b_1 \\ a_2 & b_2 \end{bmatrix}$, by which the column vector is multiplied, is a **transformation matrix** under this new point of view, since the result of its multiplication with $\begin{bmatrix} x \\ y \end{bmatrix}$ is a new column vector.

For example, the matrix multiplication of Example 1 can be interpreted as a transformation of the column vector $\begin{bmatrix} 12 \\ -19 \end{bmatrix}$ to a new column vector, $\begin{bmatrix} 2 \\ -3 \end{bmatrix}$, as a result of the multiplication by the transformation matrix $\begin{bmatrix} 3 & -2 \\ -2 & 5 \end{bmatrix}$.

Critical Thinking Questions

Analysis: In the solution of the matrix equation at the bottom of page 602, each side of the equation $A \cdot X = C$ is multiplied by A^{-1} "on the left." Ask students why this qualification is necessary. They should recall that, in general, multiplication of matrices is not a commutative operation. Therefore, multiplication on the left must be distinguished from multiplication on the right.

Linear System

$$4x - 3y + 5z = -6$$
$$-5x + 4y - 3z = 8$$
$$9x - 7y + 6z = 5$$

Augmented Matrix

$$\begin{bmatrix} 4 & -3 & 5 & -6 \\ -5 & 4 & -3 & 8 \\ 9 & -7 & 6 & 5 \end{bmatrix}$$

To solve a system of three linear equations using matrices, transform its augmented matrix into an *equivalent matrix* that is in triangular form and that has a main diagonal that consists only of 1s, with 0s below the diagonal.

$$\begin{bmatrix} 1 & b_1 & c_1 & d_1 \\ 0 & 1 & c_2 & d_2 \\ 0 & 0 & 1 & d_3 \end{bmatrix}$$

Definitions

Two augmented matrices are said to be **equivalent** if one matrix can be obtained from the other by performing one or more of the following steps.
(1) Interchange two rows.
(2) Multiply each element of a row by a nonzero number.
(3) Add the elements of one row to the corresponding elements of another row.

The above three steps are called **elementary matrix transformations.**

EXAMPLE 2 Solve the system $\begin{array}{l} -3x + 2y - 3z = 1 \\ 2x - 3y + z = 3 \\ -x + 4y + 2z = -9 \end{array}$ using matrix transformations.

Plan Write the augmented matrix. Use elementary matrix transformations.

Solution

$$\begin{bmatrix} -3 & 2 & -3 & 1 \\ 2 & -3 & 1 & 3 \\ -1 & 4 & 2 & -9 \end{bmatrix}$$ Add $(2 \cdot \text{row } 2)$ to row 1. $\rightarrow$ $$\begin{bmatrix} 1 & -4 & -1 & 7 \\ 2 & -3 & 1 & 3 \\ -1 & 4 & 2 & -9 \end{bmatrix}$$

Add $(2 \cdot \text{row } 3)$ to row 2. $\rightarrow$ $$\begin{bmatrix} 1 & -4 & -1 & 7 \\ 0 & 5 & 5 & -15 \\ -1 & 4 & 2 & -9 \end{bmatrix}$$ Multiply row 2 by $\frac{1}{5}$. $\rightarrow$ $$\begin{bmatrix} 1 & -4 & -1 & 7 \\ 0 & 1 & 1 & -3 \\ -1 & 4 & 2 & -9 \end{bmatrix}$$

Add row 1 to row 3. $\rightarrow$ $$\begin{bmatrix} 1 & -4 & -1 & 7 \\ 0 & 1 & 1 & -3 \\ 0 & 0 & 1 & -2 \end{bmatrix}$$ Write the equivalent $\rightarrow$ system.
$$\begin{array}{l} x - 4y - z = 7 \\ y + z = -3 \\ z = -2 \end{array}$$

Solve the system. $\rightarrow$

$$\begin{array}{l} y + z = -3 \\ y - 2 = -3 \\ y = -1 \end{array} \rightarrow \begin{array}{l} x - 4y - z = 7 \\ x - 4(-1) - (-2) = 7 \\ x + 4 + 2 = 7 \\ x = 1 \end{array} \rightarrow \begin{array}{l} x = 1 \\ y = -1 \\ z = -2 \end{array}$$

Thus, $(1, -1, -2)$ is the solution of the system.

Chapter 16 Linear Systems, Matrices, Determinants

Additional Example 2

Solve the system using matrix transformations.

$$2x - y + 3z = 5$$
$$-x + 2y - z = -3$$
$$x + 3y + 2z = 0$$

$(-1, -1, 2)$

Classroom Exercises

State the matrix equation for each of the following systems.

1. $3x + 4y = 6$
$-2x - y = 7$

2. $x + 2y = 3$
$x - y = 1$

3. $2x + 3y = 0$
$x = 1$

4. $x = 1$
$y = 2$

5–8. Give the augmented matrix for the four systems of Classroom Exercises 1–4.

9. Give the steps in transforming the augmented matrix $\begin{bmatrix} 2 & 4 & 6 \\ 3 & 2 & 1 \end{bmatrix}$

into the equivalent augmented matrix $\begin{bmatrix} 1 & 2 & 3 \\ 0 & 1 & 2 \end{bmatrix}$.

Written Exercises

Solve each system using the inverse of a matrix.

1. $3x - 2y = 8$
$x + 3y = -1$ $(2, -1)$

2. $-2x + 4y = 2$
$3x - y = 7$ $(3, 2)$

3. $5x - 2y = -8$
$-x + 3y = -1$ $(-2, -1)$

4. $-3x + 5y = -1$
$4x - 2y = -8$ $(-3, -2)$

5. $6x - 2y = 4$
$-5x + 3y = -2$ $(1, 1)$

6. $-4x - 3y = 8$
$2x + y = -2$ $(1, -4)$

Solve each system using matrix transformations.

7. $2x + 3y = 11$
$-x + y = 12$ $(-5, 7)$

8. $y - 2z = 2$ $(y, z) = \left(\frac{1}{2}, -\frac{3}{4}\right)$
$-6y + 8z = -9$

$(x, z) = (-10, 5)$

9. $2x + 4z = 0$
$5x - 9z = -95$

10. $2x - 3y + 2z = 1$
$4x - 3y + 2z = 5$
$3x - 2y + z = -5$

11. $2x + 3y - z = -7$
$x + y - z = -4$
$3x - 2y - 3z = -7$

12. $3x - 2y + z = -5$
$-2x + 3y - 3z = 12$
$3x - 2y - 2z = 4$

13. $x + 3y - 2z = 5$
$-2x - y - 3z = 4$
$4x - 2y + z = 2$

14. $3x - y + 2z = 9$
$x - 2y - 3z = -1$
$2x - 3y + z = 10$

15. $-3x + 2y + 2z = 9$
$2x - 5y - 3z = -2$
$-6x + 3y - 4z = 4$

16. Explain in writing how to solve a system of two linear equations using the inverse of a matrix.

If $A = \begin{bmatrix} a_1 & b_1 \\ a_2 & b_2 \end{bmatrix}$ and A^{-1} exists, it can be shown that

$A^{-1} = \dfrac{1}{\text{Det } A} \begin{bmatrix} b_2 & -b_1 \\ -a_2 & a_1 \end{bmatrix}$, where $\text{Det } A = \begin{vmatrix} a_1 & b_1 \\ -a_2 & b_2 \end{vmatrix}$.

Use this formula to find A^{-1} in Exercises 17–19.

17. $A = \begin{bmatrix} 4 & -1 \\ 2 & 0 \end{bmatrix}$

18. $A = \begin{bmatrix} 6 & 3 \\ -2 & 1 \end{bmatrix}$

19. $A = \begin{bmatrix} a & -b \\ b & a \end{bmatrix}$

Checkpoint

1. Write $-4x + y = 8$ $\begin{bmatrix} -4 & 1 \\ 5 & 4 \end{bmatrix}\begin{bmatrix} x \\ y \end{bmatrix} = \begin{bmatrix} 8 \\ -6 \end{bmatrix}$
$5x + 4y = -6$
in matrix form.

2. Solve the system using the inverse of a matrix.
$3x - 2y = -7$
$-5x + y = 21$
$(-5, -4)$

3. Solve the system using matrix transformations.
$-4x + 2y - z = -3$
$2x - y + 3z = -6$
$-2x + 3y - 2z = 1$
$(1, -1, -3)$

Closure

Ask students to give an informal definition of the process of *transformation*. Remind students that a system of equations that has been transformed assumes a triangular shape. Ask them to explain why a system in this form is beneficial. The system can be solved easily. Ask students to describe how they would write an augmented matrix and how they could transform that matrix into an equivalent matrix that is in triangular form. See the definition and Example 2 on page 604.

▬▬FOLLOW UP

Guided Practice

Classroom Exercises 1–9

Independent Practice

A Ex. 1–9, **B** Ex. 10–16, **C** Ex. 17–21

Basic: WE 1–11 odd, Application
Average: WE 3–17 odd, Application
Above Average: WE 5–21 odd, Application

Enrichment

Point out that the triangular form of an augmented matrix, $\begin{bmatrix} 1 & b_1 & c_1 & d_1 \\ 0 & 1 & c_2 & d_2 \\ 0 & 0 & 1 & d_3 \end{bmatrix}$, makes the value of the third variable explicit, and the values of the first two variables implicit. If row transformations are used to change b_1, c_1, and c_2 into zeros, the matrix is said to be in **reduced form,**

namely $\begin{bmatrix} 1 & 0 & 0 & d_1 \\ 0 & 1 & 0 & d_2 \\ 0 & 0 & 1 & d_3 \end{bmatrix}$.

In this form, the values of all three variables become explicit.

Refer to the following triangular form of the matrix.

$\begin{bmatrix} 1 & -4 & -1 & 7 \\ 0 & 1 & 1 & -3 \\ 0 & 0 & 1 & -2 \end{bmatrix}$

Have the students use row transformations to transform the matrix to reduced form.

$\begin{bmatrix} 1 & 0 & 0 & 1 \\ 0 & 1 & 0 & -1 \\ 0 & 0 & 1 & -2 \end{bmatrix}$

See pages 612 and 613 for the answers to Classroom Ex. 1–9 and Written Ex. 10–21.

Application

1. 4–door: $[120 \; 210] \begin{bmatrix} 12{,}000 \\ 16{,}000 \end{bmatrix}$

   $= [4{,}800{,}000]$

   2–door: $[200 \; 150] \begin{bmatrix} 11{,}000 \\ 15{,}000 \end{bmatrix}$

   $= [4{,}450{,}000]$

   Sports car: $[70 \; 90] \begin{bmatrix} 18{,}000 \\ 22{,}000 \end{bmatrix}$

   $= [3{,}240{,}000]$

2. $12,490,000

Additional Answers, page 588

6. $-(A + B) = -\begin{bmatrix} 5 + (-9) & -3 + 8 \\ 7 + (-6) & 2 + 4 \end{bmatrix}$

   $= -\begin{bmatrix} -4 & 5 \\ 1 & 6 \end{bmatrix} = \begin{bmatrix} 4 & -5 \\ -1 & -6 \end{bmatrix}$;

   $-A + -B = \begin{bmatrix} -5 + 9 & 3 + (-8) \\ -7 + 6 & -2 + (-4) \end{bmatrix}$

   $= \begin{bmatrix} 4 & -5 \\ -1 & -6 \end{bmatrix}$

   Thus, $-(A + B) = -A + (-B)$.

7. $A + (C + B) = \begin{bmatrix} 5 & -3 \\ 7 & 2 \end{bmatrix} +$

   $\begin{bmatrix} -1 + (-9) & (-5) + 8 \\ -7 + (-6) & (-3) + 4 \end{bmatrix}$

   $= \begin{bmatrix} 5 + (-10) & -3 + 3 \\ 7 + (-13) & 2 + 1 \end{bmatrix}$

   $= \begin{bmatrix} -5 & 0 \\ -6 & 3 \end{bmatrix}$;

   $(A + C) + B = \begin{bmatrix} 5 + (-1) & -3 + (-5) \\ 7 + (-7) & 2 + (-3) \end{bmatrix}$

   $+ \begin{bmatrix} -9 & 8 \\ -6 & 4 \end{bmatrix}$

   $= \begin{bmatrix} 4 + (-9) & -8 + 8 \\ 0 + (-6) & -1 + 4 \end{bmatrix}$

   $= \begin{bmatrix} -5 & 0 \\ -6 & 3 \end{bmatrix}$;

   Thus, $A + (C + B) = (A + C) + B$.

8. $A + O_{2 \times 2} = \begin{bmatrix} a_{11} + 0 & a_{12} + 0 \\ a_{21} + 0 & a_{22} + 0 \end{bmatrix}$

   $= \begin{bmatrix} a_{11} & a_{12} \\ a_{21} & a_{22} \end{bmatrix} = A$

9. $B + (-B)$

   $= \begin{bmatrix} b_{11} + (-b_{11}) & b_{12} + (-b_{12}) \\ b_{21} + (-b_{21}) & b_{22} + (-b_{22}) \end{bmatrix}$

   $= \begin{bmatrix} 0 & 0 \\ 0 & 0 \end{bmatrix} = O_{2 \times 2}$

10. $-(-C) = -\begin{bmatrix} -c_{11} & -c_{12} \\ -c_{21} & -c_{22} \end{bmatrix}$

    $= \begin{bmatrix} c_{11} & c_{12} \\ c_{21} & c_{22} \end{bmatrix} = C$

---

20. Prove the formula used in Exercises 17–19.

21. Two square matrices X and Y have the same dimensions. The matrices and their product, $X \cdot Y$, all have multiplicative inverses. Prove that $(X \cdot Y)^{-1} = Y^{-1} \cdot X^{-1}$. (HINT: Use reversible steps starting with this equality to obtain $I = I$. Then reverse the steps.)

Mixed Review

Factor each of the following.   *5.4, 5.7*

1. $x^2 - 2x - 3$ $(x - 3)(x + 1)$
2. $6x^2 - 5x - 4$ $(3x - 4)(2x + 1)$
3. $16x^2 - 1$ $(4x - 1)(4x + 1)$
4. $-27x^3 + 12x$ $-3x(3x - 2)(3x + 2)$
5. $x^4 + 5x^2 + 4$ $(x^2 + 4)(x^2 + 1)$
6. $x^4 - 13x^2 + 36$
   $(x - 2)(x + 2)(x - 3)(x + 3)$

Application: *Car Sales*

A foreign car company sells three different models: a 4-door car, a 2-door car, and a sports car. Each model is produced in two types: a normally equipped car and a fully equipped car. Table 1 shows the number of cars sold by model and by equipment. Table 2 shows the cost of each model based on the cost of equipment.

TABLE 1: Cars Sold

	Normal equipment	Full equipment
4-door	120	210
2-door	200	150
Sports car	70	90

TABLE 2: Cost in Dollars

	4-door	2-door	Sports car
Normal equipment	12,000	11,000	18,000
Full equipment	16,000	15,000	22,000

Exercises

1. Use matrices to show total dollar sales for each model.
2. Calculate total sales in dollars.

606   Chapter 16 Linear Systems, Matrices, Determinants

16.6 Problem Solving: Using Matrices

Objective   To apply matrices

Matrices can be used to represent a variety of data. They can also be used to solve a variety of practical problems.

EXAMPLE 1   A company manufactures three models of television sets and three models of videocassette recorders (VCRs) in two plants. The following tables represent the production in each plant.

<table>
<tr><td colspan="4" align="center">PLANT 1</td></tr>
<tr><td></td><td colspan="3" align="center">Model</td></tr>
<tr><td></td><td>X</td><td>Y</td><td>Z</td></tr>
<tr><td>TV sets</td><td>120</td><td>72</td><td>97</td></tr>
<tr><td>VCRs</td><td>67</td><td>46</td><td>71</td></tr>
</table>

<table>
<tr><td colspan="4" align="center">PLANT 2</td></tr>
<tr><td></td><td colspan="3" align="center">Model</td></tr>
<tr><td></td><td>X</td><td>Y</td><td>Z</td></tr>
<tr><td>TV sets</td><td>79</td><td>86</td><td>105</td></tr>
<tr><td>VCRs</td><td>75</td><td>68</td><td>53</td></tr>
</table>

a. Represent the data in matrix form.
b. Find the sum of the two matrices.
c. Find the difference of the two matrices (Plant 1 − Plant 2).
d. Interpret the results of Steps *b* and *c*.

Solutions   a.
$$\text{Plant 1} \qquad \text{Plant 2}$$
$$\begin{bmatrix} 120 & 72 & 97 \\ 67 & 46 & 71 \end{bmatrix} \qquad \begin{bmatrix} 79 & 86 & 105 \\ 75 & 68 & 53 \end{bmatrix}$$

b.
$$\text{Plant 1 + Plant 2}$$
$$\begin{bmatrix} 120+79 & 72+86 & 97+105 \\ 67+75 & 46+68 & 71+53 \end{bmatrix} = \begin{bmatrix} 199 & 158 & 202 \\ 142 & 114 & 124 \end{bmatrix}$$

c.
$$\text{Plant 1 − Plant 2}$$
$$\begin{bmatrix} 120-79 & 72-86 & 97-105 \\ 67-75 & 46-68 & 71-53 \end{bmatrix} = \begin{bmatrix} 41 & -14 & -8 \\ -8 & -22 & 18 \end{bmatrix}$$

d. The company produces 199 Model-X TVs, 158 Model-Y TVs, 202 Model-Z TVs, 142 Model-X VCRs, 114 Model-Y VCRs, and 124 Model-Z VCRs. They also produce 41 more Model-X TV sets in Plant 1 than in Plant 2.

Complete the interpretation on your own.

The cost of manufacturing products is determined by the type of materials used, the quantity of materials used, and the costs, including labor, involved in processing these materials. The next example illustrates the total cost of manufacturing 100 boats.

Teaching Resources

Quick Quizzes 124
Reteaching and Practice
   Worksheets 124

▰▰▰ GETTING STARTED

Prerequisite Quiz

Refer to matrices *A*, *B*, *C*, and *D*.

$$A = \begin{bmatrix} 3 & 1 & 2 \\ 4 & 5 & 3 \end{bmatrix} \qquad B = \begin{bmatrix} 7 & 3 & 1 \\ 4 & 1 & 0 \end{bmatrix}$$

$$C = [10 \; 20] \qquad D = \begin{bmatrix} 8 \\ 4 \\ 2 \end{bmatrix}$$

Find the following matrices.

1. $(A + B) \cdot D$ $\begin{bmatrix} 102 \\ 94 \end{bmatrix}$

2. $(A \cdot D) \cdot C$ $\begin{bmatrix} 320 & 640 \\ 580 & 1{,}160 \end{bmatrix}$

Motivator

Have students write tables of data like those in the Application on page 606. Then have them represent the data in matrix form.

▰▰▰ TEACHING SUGGESTIONS

Lesson Note

Point out that data used in business, economics, science, and social statistics is often stored, and operated on, in matrix form.

Highlighting the Standards

Standard 1d: Beyond the solutions to systems of equations, matrices are useful in modeling a variety of real-world situations.

Computer Programming: When the dimensions of a matrix are very large, it becomes impractical to perform the calculations by hand. Some graphing calculators can manipulate 6×6 matrices, but a computer is required for larger matrices. In the BASIC program language, there are several MAT (matrix) commands that are useful. For example, MAT READ A, B, C, $\cdots$ reads all elements of each of the matrices A, B, C, $\cdots$. First, all elements of A are read from the DATA in row-wise sequence followed by readings from B, C, and so on. Several other MAT commands are also available, each designed to carry out a particular phase of the matrix manipulation process.

Critical Thinking Questions

Analysis: Ask students to draw upon their own interests to devise a problem that illustrates an application of matrices.

EXAMPLE 2   A boat manufacturer builds three models of boats: Model X, Model Y, and Model Z. Each model is produced in two styles: a leisure style and a racing style. Each leisure boat is constructed with the same relative amounts of wood, plastic, fiberglass, and other materials as any other leisure boat, regardless of model. Similarly, each racing boat is constructed with the same relative amounts of materials as any other racing boat, regardless of model. The company will produce 100 boats during the month of April.

The matrix at the right represents the number of boats produced for each model and style.

$$\text{Models} \begin{array}{c} X \\ Y \\ Z \end{array} \begin{array}{cc} \text{Leisure} & \text{Racing} \\ \left[\begin{array}{cc} 30 & 10 \\ 10 & 20 \\ 0 & 30 \end{array} \right] \end{array} = A$$

(Styles: Leisure, Racing)

The next matrix represents the number of units of materials used for each style of boat.

$$\text{Styles} \begin{array}{c} \text{Leisure} \\ \text{Racing} \end{array} \begin{array}{cccc} \text{Wood} & \text{Plastic} & \text{Fiberglass} & \text{Other} \\ \left[\begin{array}{cccc} 15 & 12 & 3 & 6 \\ 1 & 4 & 25 & 15 \end{array} \right] \end{array} = B$$

(Materials: Wood, Plastic, Fiberglass, Other)

The next matrix represents the unit costs of the materials in dollars per unit, with the costs of the labor and processing included.

$$\text{Materials} \begin{array}{l} \text{Wood (unit: board feet, or bd ft)} \\ \text{Plastic (unit: square feet, or ft}^2) \\ \text{Fiberglass (unit: square feet, or ft}^2) \\ \text{Other (unit: square feet, or ft}^2) \end{array} \begin{array}{c} \text{Costs} \\ \left[\begin{array}{c} 65 \\ 70 \\ 90 \\ 35 \end{array} \right] \end{array} = C$$

a. Find the cost of production of each boat.
b. Find the cost of production of each model.

Solutions

a. Find a matrix, each element of which is the production cost of a boat.

From Lesson 7.3, (units) $\times \dfrac{\text{dollars}}{\text{unit}} = $ dollars. Therefore, find $B \cdot C$.

$$B \cdot C = \begin{bmatrix} 15 & 12 & 3 & 6 \\ 1 & 4 & 25 & 15 \end{bmatrix} \begin{bmatrix} 65 \\ 70 \\ 90 \\ 35 \end{bmatrix} \quad \leftarrow 15 \text{ bd ft of wood} \times \dfrac{65 \text{ dollars}}{\text{bd ft}}$$

$$= \begin{bmatrix} 15(65) + 12(70) + 3(90) + 6(35) \\ 1(65) + 4(70) + 25(90) + 15(35) \end{bmatrix}, \text{ or } \begin{bmatrix} 2{,}295 \\ 3{,}120 \end{bmatrix}$$

So, the cost of production is \$2,295 for each leisure boat and \$3,120 for each racing boat.

Additional Example 1

The tables below show enrollments in Jordan and Parker High Schools.

Jordan	Fr.	Soph.	Jr.	Sr.
1989	420	400	380	370
1990	510	450	400	375

Parker	Fr.	Soph.	Jr.	Sr.
1989	380	340	345	305
1990	400	385	350	340

a. Show the data in matrix form.

$$\text{Jordan} \begin{bmatrix} 420 & 400 & 380 & 370 \\ 510 & 450 & 400 & 375 \end{bmatrix}$$

$$\text{Parker} \begin{bmatrix} 380 & 340 & 345 & 305 \\ 400 & 385 & 350 & 340 \end{bmatrix}$$

b. Find the sum of the two matrices.

$$\begin{bmatrix} 800 & 740 & 725 & 675 \\ 910 & 835 & 750 & 715 \end{bmatrix}$$

c. Find the difference of the two matrices (Jordan − Parker).

$$\begin{bmatrix} 40 & 60 & 35 & 65 \\ 110 & 65 & 50 & 35 \end{bmatrix}$$

d. Interpret the results of parts **b** and **c**.
   b. The matrix shows the total enrollment for the two schools by class and year.
   c. The matrix shows the amounts by which enrollment figures at Jordan H.S. exceeded those at Parker H.S.

b. To determine the cost of production for each model, find $A \cdot (B \cdot C)$.

$$A \cdot (B \cdot C) = \begin{bmatrix} 30 & 10 \\ 10 & 20 \\ 0 & 30 \end{bmatrix} \begin{bmatrix} 2{,}295 \\ 3{,}120 \end{bmatrix} \quad \begin{matrix} \leftarrow & 30 \text{ Model-}X & \times & \dfrac{2{,}295 \text{ dollars}}{\text{leisure boat}} \\ & \text{leisure boats} & & \end{matrix}$$

$$= \begin{bmatrix} 30(2{,}295) + 10(3{,}120) \\ 10(2{,}295) + 20(3{,}120) \\ 0(2{,}295) + 30(3{,}120) \end{bmatrix}, \text{ or } \begin{bmatrix} 100{,}050 \\ 85{,}350 \\ 93{,}600 \end{bmatrix}$$

So, the cost of production is $100,050 for 40 Model-X boats, $85,350 for 30 Model-Y boats, and $93,600 for 30 Model-Z boats.

Classroom Exercises

Refer to Example 2. Then find each of the following.

1. total production cost for all 100 boats $279,000
2. production cost for one Model-X leisure boat $2,295
3. production cost for one Model-X racing boat $3,120
4. production cost for one Model-Y leisure boat $2,295
5. production cost for one Model-Y racing boat $3,120
6. production cost for one Model-Z leisure boat $2,295
7. production cost for one Model-Z racing boat $3,120

Written Exercises

Each year each member of the Clemente family invests money in bonds, stocks, and savings certificates. The following tables represent their investments (in dollars) for two consecutive years (Exercises 1–16).

YEAR 1

	Bonds	Stocks	Certificates
Mother	400	1,200	1,000
Father	500	900	600
Paul	100	200	500
Rosa	200	300	400

YEAR 2

	Bonds	Stocks	Certificates
Mother	800	700	1,500
Father	300	900	2,000
Paul	0	800	200
Rosa	500	300	0

Use the tables above for Exercises 1–8.

1. Represent the data of Year 1 in matrix form.
2. Represent the data of Year 2 in matrix form.
3. Find the sum of the two matrices, Year 1 + Year 2.
4. Find the difference of the two matrices, Year 2 − Year 1.

Checkpoint

The matrices below represent the numbers of vehicles driving north and south over a toll bridge one Saturday and Sunday.

Saturday

	Cars	Trucks	Buses	
North	180	100	10	= A
South	200	125	10	

Sunday

	Cars	Trucks	Buses	
North	150	35	12	= B
South	120	20	12	

The next matrix represents the toll fees (in dollars) collected for each type of vehicle.

Toll Fees

$$\begin{matrix} \text{Cars} \\ \text{Trucks} \\ \text{Buses} \end{matrix} \begin{bmatrix} \$1 \\ \$2 \\ \$5 \end{bmatrix} = C$$

1. How many cars traveled north on the weekend? 330
2. Find the matrix $(A + B) \cdot C$.
$$\begin{bmatrix} 710 \\ 720 \end{bmatrix}$$
3. What were the total receipts over the weekend for northbound traffic? $710 for southbound traffic? $720

Additional Example 2

A company makes three types of board games, each in a deluxe and a regular model. Production figures for May are shown below.

Model

	De.	Reg.	
X	20	70	
Games Y	25	45	= A
Z	15	25	

The next table represents the numbers of units of material and labor (in hours) that are needed to produce each type of game.

	Mat.	Lab.	
Deluxe	3	1.5	= B
Regular	2	1	

The third table shows the unit costs of materials and labor (dollars per unit).

Costs

$$\begin{matrix} \text{Materials} \\ \text{Labor} \end{matrix} \begin{bmatrix} 1 \\ 6 \end{bmatrix} = C$$

Use matrices to find the following.

a. The cost of producing each model $12 for deluxe; $8 for regular
b. The cost of producing each type of game in May $800 for type X; $660 for type Y; $380 for type Z

609

Closure

Give students the following information: A book-of the-month club offers three packages of books (A, B, and C). Each package contains a variety of books listed in the table below.

	Novels	Poems	Plays
A	3	1	1
B	2	2	1
C	2	1	2

Each package is available in hardcover or paperback with costs listed in the table below.

	Hardcover	Paperback
Novels	$12	$4
Poems	$8	$5
Plays	$10	$6

Using the tables, have students think of questions that involve matrix multiplication. Solve them as a group.

◾◾◾FOLLOW UP

Guided Practice

Classroom Exercises 1–7

Independent Practice

A Ex. 1–16, **B** Ex. 17–26

Basic: WE 1–18, Application

Average: WE 7–22, Application

Above Average: WE 8–26, Application

Additional Answers

Written Exercises

1.
$$\begin{bmatrix} 400 & 1{,}200 & 1{,}000 \\ 500 & 900 & 600 \\ 100 & 200 & 500 \\ 200 & 300 & 400 \end{bmatrix}$$

5. How much did Mrs. Clemente invest in savings certificates in two years? $2,500

6. How much did Mr. Clemente invest in bonds in two years? $800

7. How much did son Paul invest in stock in two years? $1,000

8. How much more did daughter Rosa invest in bonds in the second year compared to the first year? $300

In the third year each member of the family decided to double the investments of the second year.

9. Represent the third-year investments as a product of a scalar and a matrix.

10. Find the product of the scalar and matrix.

In the fourth year, after inheriting some money, each family member was able to multiply the investments of the second year as follows: Mrs. Clemente by 3, Mr. Clemente by 2, Paul by 4, and Rosa by 5.

11. Represent the product of the above multiples and the second-year investments.

12. Find the matrix representing the fourth-year investments.

13. How much did Mr. Clemente invest in stocks in Year 4? $1,800

14. How much did Mrs. Clemente invest in bonds in Year 4? $2,400

15. How much did Rosa invest in savings certificates in Year 4? $0

16. How much did Paul invest in savings certificates in Year 4? $800

A builder builds two models of houses in three styles. Each year she builds exactly 65 houses. Table 1 represents the number of houses produced by model and style. Table 2 represents the units of materials used for each house by style and by kinds of materials used. Table 3 represents the costs of materials per unit, with labor costs included.

TABLE 1

	Colonial	Split-Entry	Ranch
Model X	20	15	10
Model Y	2	6	12

TABLE 2

	Stone	Brick	Wood	Concrete
Colonial	200	200	100	50
Split-Entry	20	120	150	300
Ranch	30	250	300	200

TABLE 3

	Costs
Stone	160
Brick	140
Wood	180
Concrete	122

Enrichment

Remind students of the methods given in Lesson 16.5 for finding the inverse of a square matrix.

1. By solving systems of linear equations
2. By using a formula that involves the determinant of the matrix (C Exercises)

Show students the following simple method using matrix transformations. Consider

finding the inverse of $\begin{bmatrix} 2 & -1 \\ 3 & 1 \end{bmatrix}$.

Set up the matrix shown below, with the identity matrix on the right.

$$\begin{bmatrix} 2 & -1 \\ 3 & 1 \end{bmatrix} \begin{bmatrix} 1 & 0 \\ 0 & 1 \end{bmatrix}$$

The strategy is to use row transformations to change the matrix on the left into the identity matrix. For each transformation on the left, make corresponding tranformations on the right matrix. When you succeed in

obtaining the identity matrix on the left, the matrix on the right will be the inverse of the original matrix. Challenge students to use this strategy to find the inverse of the matrix at the left.

$$\begin{bmatrix} 1 & 0 \\ 0 & 1 \end{bmatrix} \begin{bmatrix} \frac{1}{5} & \frac{1}{5} \\ -\frac{3}{5} & \frac{2}{5} \end{bmatrix}$$

Note that the same method can be used to find the inverse of a 3-by-3 matrix, if it exists.

17. Write Table 1 in matrix form.
18. Write Table 2 in matrix form.
19. Write Table 3 in matrix form.
20. Find the cost to build each colonial house. $84,100
21. Find the cost to build each split-entry house. $83,600
22. Find the cost to build each ranch house. $118,200
23. Find the cost to build all of the Model-*Y* ranch houses. $1,418,400
24. Find the cost to build all of the Model-*X* houses. $4,118,000
25. Find the cost to build all of the Model-*Y* houses. $2,088,200
26. Find the cost to build all of the houses. $6,206,200

Mixed Review

Let $f(x) = -3x^2 + 1$. Find each of the following. *3.2, 5.5*

1. $f(-1)$ -2
2. $f(5)$ -74
3. $f(a)$ $-3a^2 + 1$
4. $f(a + t)$ $-3a^2 - 6at - 3t^2 + 1$

Simplify and write each expression using positive exponents. *5.2*

5. $(8p^0 q)^{-1}$ $\frac{1}{8q}$
6. $(-3a^{-2}b^3)^2$ $\frac{9b^6}{a^4}$
7. $(-2x^4 y^{-3})^3$ $\frac{-8x^{12}}{y^9}$
8. $(-4a^3 b^{-4})(-2a^{-2}b^6)$ $8ab^2$

Application: *Networks*

A **network** is a set of points connected by arcs. Networks can represent roadways, electrical systems, and ecological relationships. A **directed network** shows the possible directions of travel or influence along each arc.

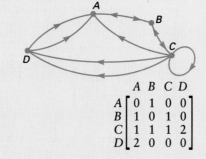

A matrix conveniently represents a network. The 0 in the first row, third column of the matrix at the right means that there are *no* paths from *A* to *C*. Likewise, there is *one* path from *C* to *C*, and there are *two* possible paths from *C* to *D*.

$$
\begin{array}{c}
 \\
A \\
B \\
C \\
D
\end{array}
\begin{array}{cccc}
A & B & C & D \\
\left[\begin{array}{cccc}
0 & 1 & 0 & 0 \\
1 & 0 & 1 & 0 \\
1 & 1 & 1 & 2 \\
2 & 0 & 0 & 0
\end{array}\right]
\end{array}
$$

1–2. Construct matrices for the directed networks at the right.

1.
2.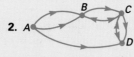

3–4. Construct directed networks for the matrices at the right.

3. $\begin{bmatrix} 2 & 1 & 1 \\ 1 & 0 & 1 \\ 1 & 1 & 0 \end{bmatrix}$

4. $\begin{bmatrix} 1 & 0 & 2 & 0 \\ 1 & 0 & 2 & 1 \\ 0 & 1 & 0 & 0 \\ 0 & 1 & 0 & 0 \end{bmatrix}$

16.6 Problem Solving: Using Matrices **611**

2. $\begin{bmatrix} 800 & 700 & 1,500 \\ 300 & 900 & 2,000 \\ 0 & 800 & 200 \\ 500 & 300 & 0 \end{bmatrix}$

3. $\begin{bmatrix} 1,200 & 1,900 & 2,500 \\ 800 & 1,800 & 2,600 \\ 100 & 1,000 & 700 \\ 700 & 600 & 400 \end{bmatrix}$

4. $\begin{bmatrix} 400 & -500 & 500 \\ -200 & 0 & 1,400 \\ -100 & 600 & -300 \\ 300 & 0 & -400 \end{bmatrix}$

9. $2 \cdot \begin{bmatrix} 800 & 700 & 1,500 \\ 300 & 900 & 2,000 \\ 0 & 800 & 200 \\ 500 & 300 & 0 \end{bmatrix}$

10. $\begin{bmatrix} 1,600 & 1,400 & 3,000 \\ 600 & 1,800 & 4,000 \\ 0 & 1,600 & 400 \\ 1,000 & 600 & 0 \end{bmatrix}$

11. $\begin{bmatrix} 3(800) & 3(700) & 3(1,500) \\ 2(300) & 2(900) & 2(2,000) \\ 4(0) & 4(800) & 4(200) \\ 5(500) & 5(300) & 5(0) \end{bmatrix}$

12. $\begin{bmatrix} 2,400 & 2,100 & 4,500 \\ 600 & 1,800 & 4,000 \\ 0 & 3,200 & 800 \\ 2,500 & 1,500 & 0 \end{bmatrix}$

17. $\begin{bmatrix} 20 & 15 & 10 \\ 2 & 6 & 12 \end{bmatrix}$

18. $\begin{bmatrix} 200 & 200 & 100 & 50 \\ 20 & 120 & 150 & 300 \\ 30 & 250 & 300 & 200 \end{bmatrix}$

19. $\begin{bmatrix} 160 \\ 140 \\ 180 \\ 122 \end{bmatrix}$

Application

1. $\begin{bmatrix} 1 & 1 & 1 \\ 0 & 0 & 2 \\ 1 & 1 & 0 \end{bmatrix}$

2. $\begin{bmatrix} 0 & 2 & 0 & 1 \\ 0 & 0 & 2 & 0 \\ 0 & 1 & 0 & 2 \\ 0 & 0 & 1 & 0 \end{bmatrix}$

3. Answers will vary. One possible answer is given.

4. Answers will vary. One possible answer is given.

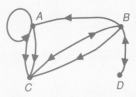

1. $\begin{bmatrix} 13 & 9 & 8 \\ 3 & 4 & 1 \\ -8 & 14 & -8 \end{bmatrix}$

3. $\begin{bmatrix} 13 & 5 & 28 \\ 30 & -16 & -1 \end{bmatrix}$

Additional Answers, page 606

Classroom Exercises

1. $\begin{bmatrix} 3 & 4 \\ -2 & -1 \end{bmatrix} \begin{bmatrix} x \\ y \end{bmatrix} = \begin{bmatrix} 6 \\ 7 \end{bmatrix}$

2. $\begin{bmatrix} 1 & 2 \\ 1 & -1 \end{bmatrix} \begin{bmatrix} x \\ y \end{bmatrix} = \begin{bmatrix} 3 \\ 1 \end{bmatrix}$

3. $\begin{bmatrix} 2 & 3 \\ 1 & 0 \end{bmatrix} \begin{bmatrix} x \\ y \end{bmatrix} = \begin{bmatrix} 0 \\ 1 \end{bmatrix}$

4. $\begin{bmatrix} 1 & 0 \\ 0 & 1 \end{bmatrix} \begin{bmatrix} x \\ y \end{bmatrix} = \begin{bmatrix} 1 \\ 2 \end{bmatrix}$

5. $\begin{bmatrix} 3 & 4 & 6 \\ -2 & -1 & 7 \end{bmatrix}$ **6.** $\begin{bmatrix} 1 & 2 & 3 \\ 1 & -1 & 1 \end{bmatrix}$

7. $\begin{bmatrix} 2 & 3 & 0 \\ 1 & 0 & 1 \end{bmatrix}$ **8.** $\begin{bmatrix} 1 & 0 & 1 \\ 0 & 1 & 2 \end{bmatrix}$

9. Answers will vary. One possible answer:

Multiply the first row of $\begin{bmatrix} 2 & 4 & 6 \\ 3 & 2 & 1 \end{bmatrix}$ by $\frac{1}{2}$

to obtain $\begin{bmatrix} 1 & 2 & 3 \\ 3 & 2 & 1 \end{bmatrix}$.

Add -3 times the first row of this matrix to the second row to obtain

$\begin{bmatrix} 1 & 2 & 3 \\ 0 & -4 & -8 \end{bmatrix}$.

Finally, multiply the second row of the

new matrix by $-\frac{1}{4}$ to obtain $\begin{bmatrix} 1 & 2 & 3 \\ 0 & 1 & 2 \end{bmatrix}$.

Written Exercises

10. $(2, 19, 27)$ **11.** $(-1, -1, 2)$

12. $(0, 1, -3)$ **13.** $(1, 0, -2)$

14. $(1, -2, 2)$ **15.** $(-3, -2, 2)$

Chapter 16 Review

Key Terms

additive identity matrix (p. 586)
additive inverse matrix (p. 586)
augmented matrix (p. 603)
Cramer's Rule (p. 594)
determinant (2-by-2) (p. 594)
determinant (3-by-3) (p. 597)
dimensions of a matrix (p. 585)

equivalent matrices (p. 604)
matrix (p. 585)
matrix transformations (p. 604)
minor (p. 599)
multiplicative identity matrix (p. 591)
multiplicative inverse matrix (p. 591)
triangular form (p. 603)

Key Ideas and Review Exercises

16.1 To add two matrices, check that their dimensions are the same. Then add corresponding elements.

Find each sum, if it exists.

1. $\begin{bmatrix} 4 & 3 & 9 \\ -5 & -1 & 4 \\ 2 & 6 & -7 \end{bmatrix} + \begin{bmatrix} 9 & 6 & -1 \\ 8 & 5 & -3 \\ -10 & 8 & -1 \end{bmatrix}$ **2.** $\begin{bmatrix} 6 & -3 \\ 9 & -7 \\ -8 & 4 \end{bmatrix} + \begin{bmatrix} 8 & 3 & -5 \\ 6 & -2 & 4 \end{bmatrix}$

Does not exist

16.2 To multiply two matrices, check that the number of columns of the first matrix equals the number of rows of the second matrix; then multiply the row elements of the first by the column elements of the second, and add.

Find each product, if it exists.

3. $\begin{bmatrix} 8 & -2 & 3 \\ 2 & 1 & -4 \end{bmatrix} \cdot \begin{bmatrix} 4 & -1 & 3 \\ 2 & -2 & 1 \\ -5 & 3 & 2 \end{bmatrix}$ **4.** $\begin{bmatrix} -4 & -2 & 3 \\ 5 & -3 & -4 \\ 6 & 9 & 3 \end{bmatrix} \cdot \begin{bmatrix} 8 & -3 & 2 \\ -4 & 9 & 8 \end{bmatrix}$

Does not exist

16.3 The value of a 2-by-2 determinant $\begin{vmatrix} a & b \\ c & d \end{vmatrix}$ is $ad - cb$.

To use Cramer's Rule to solve the system of two equations

$\begin{aligned} a_1 x + b_1 y &= c_1 \\ a_2 x + b_2 y &= c_2 \end{aligned}$, write $x = \dfrac{\begin{vmatrix} c_1 & b_1 \\ c_2 & b_2 \end{vmatrix}}{\begin{vmatrix} a_1 & b_1 \\ a_2 & b_2 \end{vmatrix}}$ and $y = \dfrac{\begin{vmatrix} a_1 & c_1 \\ a_2 & c_2 \end{vmatrix}}{\begin{vmatrix} a_1 & b_1 \\ a_2 & b_2 \end{vmatrix}}$.

Solve each system of equations using Cramer's Rule.

5. $2x - 5y = 16$
$\quad -3x + 2y = -13$ $(3, -2)$

6. $5x - 4y = -1$
$\quad 3y + 8 = 2x$ $(-5, -6)$

16.4 To find the value of a 3-by-3 determinant:
1. use the method of repeated columns, as shown on page 597, or
2. use expansion by minors, as shown on page 599.
To solve a system of three linear equations by Cramer's Rule, use 3-by-3 determinants, as shown on page 600.

Find the value of each determinant.

7. $\begin{vmatrix} 4 & -1 & 2 \\ 3 & -2 & 1 \\ -3 & 2 & 3 \end{vmatrix}$ -20

8. $\begin{vmatrix} -3 & -2 & 3 \\ 4 & -1 & 2 \\ -1 & 3 & -2 \end{vmatrix}$ 33

Solve each system of equations using Cramer's Rule.

9. $2x - 3y + z = -7$
$-3x + 4y + 2z = 3$
$3x - 2y - 2z = -1$ $(-1,1,-2)$

10. $4x + y - 2z = 3$
$-x + 3y + 5z = 5$
$2x - 3y + z = 9$ $(2,-1,2)$

16.5 To solve a system like $\begin{array}{l} -3x + 2y = 2 \\ 4x - 3y = -4 \end{array}$ using the inverse of a matrix:

1. find A^{-1} where $A = \begin{bmatrix} -3 & 2 \\ 4 & -3 \end{bmatrix}$; then

2. use $\begin{bmatrix} x \\ y \end{bmatrix} = A^{-1}C$, where $C = \begin{bmatrix} 2 \\ -4 \end{bmatrix}$.

Solve each system using the inverse of a matrix.

11. $2x - 3y = -5$
$-3x + 5y = 9$ $(2,3)$

12. $3x + 4y = 6$
$-2x - y = 7$ $\left(-\frac{34}{5}, \frac{33}{5}\right)$

16.6 To solve problems using matrices:
1. represent the data in matrix form,
2. determine which matrices should be multiplied,
3. multiply matrices, and
4. interpret data.

A company produced 65 Tudor-style houses and 35 Western-style houses. Table I represents the units of materials used for each style of house. Table II represents the dollar cost of materials by unit.

TABLE I

	Brick	Concrete	Lumber	Stone
Tudor	300	100	50	100
Western	100	300	200	100

TABLE II

	Cost
Brick	130
Concrete	110
Lumber	150
Stone	200

13. Find the cost of the building materials for all 100 houses. $8,397,500

Chapter 16 Review **613**

16. Answers will vary. One possible answer:
First, put the equations in the form
$a_1x + b_1y = c_1$
$a_2x + b_2y = c_2$.

Next, write the system in matrix form.
Find the inverse matrix $\begin{bmatrix} a_1 & b_1 \\ a_2 & b_2 \end{bmatrix}^{-1}$.

Use $\begin{bmatrix} x \\ y \end{bmatrix} = \begin{bmatrix} a_1 & b_1 \\ a_2 & b_2 \end{bmatrix}^{-1} \cdot \begin{bmatrix} c_1 \\ c_2 \end{bmatrix}$

to find $\begin{bmatrix} x \\ y \end{bmatrix}$.

17. $\begin{bmatrix} 0 & \frac{1}{2} \\ -1 & 2 \end{bmatrix}$

18. $\begin{bmatrix} \frac{1}{12} & -\frac{1}{4} \\ \frac{1}{6} & \frac{1}{2} \end{bmatrix}$

19. $\begin{bmatrix} \frac{a}{a^2 + b^2} & \frac{b}{a^2 + b^2} \\ \frac{-b}{a^2 + b^2} & \frac{a}{a^2 + b^2} \end{bmatrix}$

20. Let $A = \begin{bmatrix} a_1 & b_1 \\ a_2 & b_2 \end{bmatrix}$

and $A^{-1} = \begin{bmatrix} x_1 & y_1 \\ x_2 & y_2 \end{bmatrix}$.

By definition, $A(A)^{-1} = I$.
Then $\begin{bmatrix} a_1 & b_1 \\ a_2 & b_2 \end{bmatrix} \cdot \begin{bmatrix} x_1 & y_1 \\ x_2 & y_2 \end{bmatrix} = \begin{bmatrix} 1 & 0 \\ 0 & 1 \end{bmatrix}$,
which yields the systems
$a_1x_1 + b_1x_2 = 1$
$a_2x_1 + b_2x_2 = 0$
and
$a_1y_1 + b_1y_2 = 0$
$a_2y_1 + b_2y_2 = 1$.

$x_1 = \dfrac{b_2}{a_1b_2 - a_2b_1}$, $x_2 = \dfrac{-a_2}{a_1b_2 - a_2b_1}$,

$y_1 = \dfrac{-b_1}{a_1b_2 - a_2b_1}$, $y_2 = \dfrac{a_1}{a_1b_2 - a_2b_1}$;

$a_1b_2 - a_2b_1 = \begin{vmatrix} a_1 & b_1 \\ a_2 & b_2 \end{vmatrix}$,

the determinant of A.

Thus, $A^{-1} = \dfrac{1}{\text{Det } A} \begin{bmatrix} b_2 & -b_1 \\ -a_2 & a_1 \end{bmatrix}$.

21. 1. $I = I$: **(Reflexive Prop Eq)**
2. $(X \cdot Y)(X \cdot Y)^{-1} = I$: **($A \cdot A^{-1} = I$)**
3. $X^{-1} \cdot X \cdot Y \cdot (X \cdot Y)^{-1}$
$= X^{-1} \cdot I$: **(Mult Prop Eq)**
4. $I \cdot Y \cdot (X \cdot Y)^{-1} = X^{-1}$: **($A \cdot I = A$)**
5. $Y^{-1} \cdot Y \cdot (X \cdot Y)^{-1}$
$= Y^{-1} \cdot X^{-1}$: **(Mult Prop Eq)**
6. $(X \cdot Y)^{-1} = Y^{-1} \cdot X^{-1}$: **($A \cdot A^{-1} = I$,
$A \cdot I = A$)**

Chapter Test

6. $\begin{bmatrix} -\frac{1}{2} & 1 \\ -\frac{3}{2} & 2 \end{bmatrix}$

7. $A + C = \begin{bmatrix} 4+5 & -2+(-1) \\ 3+(-3) & -1+2 \end{bmatrix}$

$= \begin{bmatrix} 9 & -3 \\ 0 & 1 \end{bmatrix}$

$C + A = \begin{bmatrix} 5+4 & -1+(-2) \\ -3+3 & 2+(-1) \end{bmatrix}$

$= \begin{bmatrix} 9 & -3 \\ 0 & 1 \end{bmatrix}$

Thus, $A + C = C + A$.

8. $B + (-B) = \begin{bmatrix} -3+3 & 2+(-2) \\ -2+2 & 4+(-4) \end{bmatrix}$

$= \begin{bmatrix} 0 & 0 \\ 0 & 0 \end{bmatrix} = O_{2 \times 2}$

9. $AC = \begin{bmatrix} 4(5)+(-2)(-3) & 4(-1)+(-2)(2) \\ 3(5)+(-1)(-3) & 3(-1)+(-1)(2) \end{bmatrix}$

$= \begin{bmatrix} 26 & -8 \\ 18 & -5 \end{bmatrix}$

$CA = \begin{bmatrix} 5(4)+(-1)(3) & 5(-2)+(-1)(-1) \\ (-3)(4)+2(3) & (-3)(-2)+2(-1) \end{bmatrix}$

$= \begin{bmatrix} 17 & -9 \\ -6 & 4 \end{bmatrix}$

Thus, $AC \neq CA$.

Chapter 16 Test

A Exercises: 1–5, 10–18; B Exercises: 6–9; C Exercise: 19

Find each sum, if it exists.

1. $\begin{bmatrix} -5 & 8 & 7 \\ -1 & -3 & -2 \\ 6 & 9 & 10 \end{bmatrix} + \begin{bmatrix} 5 & -3 & 9 \\ -8 & 6 & 5 \\ 2 & -7 & -3 \end{bmatrix}$ $\begin{bmatrix} 0 & 5 & 16 \\ -9 & 3 & 3 \\ 8 & 2 & 7 \end{bmatrix}$

2. $\begin{bmatrix} 1 \\ -3 \\ 5 \end{bmatrix} + \begin{bmatrix} 3 & 0 & -4 \end{bmatrix}$ Does not exist

3. $\begin{bmatrix} -8 & -3 & -9 \\ -6 & 5 & 8 \end{bmatrix} + \begin{bmatrix} -10 & 7 & 1 \\ 9 & 6 & 2 \end{bmatrix}$ $\begin{bmatrix} -18 & 4 & -8 \\ 3 & 11 & 10 \end{bmatrix}$

Find each product, if it exists.

4. $-2 \begin{bmatrix} 7 & -6 \\ -3 & 5 \end{bmatrix}$ $\begin{bmatrix} -14 & 12 \\ 6 & -10 \end{bmatrix}$

5. $\begin{bmatrix} -1 & 2 & 3 \\ 4 & -2 & 3 \end{bmatrix} \cdot \begin{bmatrix} 2 & 1 & 3 \\ -1 & -2 & -3 \\ 4 & 1 & -2 \end{bmatrix}$ $\begin{bmatrix} 8 & -2 & -15 \\ 22 & 11 & 12 \end{bmatrix}$

In Exercises 6–9, $A = \begin{bmatrix} 4 & -2 \\ 3 & -1 \end{bmatrix}$, $B = \begin{bmatrix} -3 & 2 \\ -2 & 4 \end{bmatrix}$, and $C = \begin{bmatrix} 5 & -1 \\ -3 & 2 \end{bmatrix}$.

6. Find A^{-1}.

7. Prove that $A + C = C + A$.

8. Prove that $B + (-B) = O_{2 \times 2}$.

9. Prove that $AC \neq CA$.

10. Find the value of the determinant. $\begin{vmatrix} -3 & 5 \\ -1 & 6 \end{vmatrix}$ -13

11. Use minors to find the value of the determinant. $\begin{vmatrix} 2 & -3 & 1 \\ 4 & -2 & 3 \\ -2 & 1 & 5 \end{vmatrix}$ 52

Solve using Cramer's Rule.

12. $4x - 3y = 2$
$-2x + 4y = -6$ $(-1, -2)$

13. $2x - y + 3z = 5$
$-3x + 2y - z = -9$
$-x + 3y - 2z = -6$ $\left(\frac{19}{9}, -\frac{13}{9}, -\frac{2}{9} \right)$

14. $3y = x + 12$
$2x = y - 9$ $(-3, 3)$

15. $3x - 2y = -0.4$
$-5x - y = -1.5$ $(0.2, 0.5)$

Solve each system using the inverse of a matrix.

16. $4x - 3y = 2$
$-2x + 4y = -6$ $(-1, -2)$

17. $2x - 5y = 15$
$3x - 7y = 22$ $(5, -1)$

Solve using matrix transformations.

18. $3x - 2y + 3z = 3$
$-2x + 4y - z = -9$
$x + 3y + 2z = -9$ $(-2, -3, 1)$

19. $2a - 3b + c - 2d = 1$
$-a + 2b - 3c + d = 5$
$-3a - 2b + 2c + 2d = -14$
$a + b - 2c + 3d = 4$ $(2, 1, -2, -1)$

Choose the *one* best answer to each question or problem.

1. If $R = \begin{bmatrix} 3 & 5 \\ -2 & 0 \end{bmatrix}$ and $N = \begin{bmatrix} 1 & -1 \\ 3 & 1 \end{bmatrix}$,

then $2R - N =$ ___?___ B

(A) $\begin{bmatrix} -2 & 6 \\ -5 & -1 \end{bmatrix}$ (B) $\begin{bmatrix} 5 & 11 \\ -7 & -1 \end{bmatrix}$

(C) $\begin{bmatrix} 2 & 4 \\ 1 & -1 \end{bmatrix}$ (D) $\begin{bmatrix} 2 & 6 \\ -5 & -1 \end{bmatrix}$

(E) $\begin{bmatrix} 4 & 12 \\ -10 & -2 \end{bmatrix}$

2. If $5 \begin{bmatrix} 1 & x \\ x - y & 5 \end{bmatrix} = \begin{bmatrix} 5 & 15 \\ 20 & 25 \end{bmatrix}$,

find x and y.
(A) $x = 15, y = -5$ C
(B) $x = 3, y = 1$
(C) $x = 3, y = -1$
(D) $x = 3, y = -17$
(E) $x = 15, y = 5$

3. $[a \ b \ c] \cdot \begin{bmatrix} a \\ b \\ c \end{bmatrix} =$ ___?___ B

(A) $[a^2 \ b^2 \ c^2]$
(B) $[a^2 + b^2 + c^2]$
(C) $[2a + 2b + 2c]$
(D) $[2a \ 2b \ 2c]$
(E) none of these

4. If $\begin{bmatrix} x \\ y \end{bmatrix} = \begin{bmatrix} 2 & 1 \\ -1 & -3 \end{bmatrix} \cdot \begin{bmatrix} 3 \\ -2 \end{bmatrix}$,

find x and y. A
(A) $x = 4, y = 3$
(B) $x = 6, y = 2$
(C) $x = 9, y = 8$
(D) $x = 6, y = -6$
(E) $x = 4, y = -9$

5. If $3^{2x+1} = 27^{\frac{4}{3}}$, then $x =$ ___?___ D
(A) $\frac{1}{6}$ (B) $\frac{1}{2}$ (C) $\frac{2}{3}$ (D) $\frac{3}{2}$
(E) 3

6. If $A \cdot B = 36$, which of the following cannot be true? E
(A) $A + B > 12$
(B) $A + B < 12$
(C) $A + B = 12$
(D) $|A + B| > 12$
(E) $|A + B| < 12$

7. If $2 \begin{vmatrix} x & 4 & -1 \\ 0 & 2 & 3 \\ 0 & -2 & 2 \end{vmatrix} = 60$, then

$x =$ ___?___ B
(A) -15 (B) 3 (C) 5
(D) 10 (E) 30

8. If $ad - cb \neq 0$, then

$$\frac{\begin{vmatrix} a & b \\ c & d \end{vmatrix}}{\begin{vmatrix} -3b & a \\ -3d & c \end{vmatrix}} =$$ ___?___ D

(A) $-\frac{1}{3}$ (B) $-\frac{1}{9}$ (C) $\frac{1}{9}$
(D) $\frac{1}{3}$ (E) none of these

9. Given the figure below, with $\overline{AB}$ parallel to $\overline{CD}$, find x. D

(A) 20 (B) 40 (C) 60
(D) 80 (E) 100

Additional Answers, page 601

17. Suppose that consecutive rows are interchanged. If we expand about the rows with the same elements before and after the interchange, the same minors are produced. However, since the rows that are being expanded about have been shifted up or down one step, 1 is added to or subtracted from the row and column for each element of the row. So, the sign of the coefficient of each minor is reversed. For example, consider

the determinant $\begin{vmatrix} a_1 & a_2 & a_3 \\ b_1 & b_2 & b_3 \\ c_1 & c_2 & c_3 \end{vmatrix}$.

Expanding about the first row gives the following.

$$D = a_1 \begin{vmatrix} b_2 & b_3 \\ c_2 & c_3 \end{vmatrix} - a_2 \begin{vmatrix} b_1 & b_3 \\ c_1 & c_3 \end{vmatrix}$$
$$+ a_3 \begin{vmatrix} b_1 & b_2 \\ c_1 & c_2 \end{vmatrix}$$

However, if the first two rows of the determinant are interchanged, producing

the determinant $\begin{vmatrix} b_1 & b_2 & b_3 \\ a_1 & a_2 & a_3 \\ c_1 & c_2 & c_3 \end{vmatrix}$,

expanding about the row containing the "a" elements gives the following.

$$D = -a_1 \begin{vmatrix} b_2 & b_3 \\ c_2 & c_3 \end{vmatrix} + a_2 \begin{vmatrix} b_1 & b_3 \\ c_1 & c_3 \end{vmatrix}$$
$$- a_3 \begin{vmatrix} b_1 & b_2 \\ c_1 & c_2 \end{vmatrix}$$

Thus, the sign of the value of the determinant is changed. If the first and third rows are interchanged, the sign of each minor's coefficient remains the same. However, the minors are inverted; that is, the top row becomes the bottom row, and vice versa. Thus, the sign of the value of the determinant changes.

For additional standardized test practice, see the SAT/ACT Practice Tests for Cumulative Tests, Chapters 13–16.

Cumulative Review (Chapters 1–16)

1. Supply the reasons to justify Steps 1–5 in the following proof. **1.8**
   $x + xy = (y + 1)x$
   1. $x + xy = 1x + xy$ Ident Prop Mult
   2. $= 1x + yx$ Comm Prop Mult
   3. $= (1 + y)x$ Distr Prop
   4. $= (y + 1)x$ Comm Prop Add
   5. $x + xy = (y + 1)x$ Trans Prop Eq

Solve each equation for x.

2. $2x - (6x + 8) =$ **1.4**
   $-3(6x - 5) + 5$ 2

3. $|4x - 3| = 17$ 5 or $-\frac{7}{2}$ **2.3**

4. $\frac{3}{x} + \frac{5}{25} = 1$ $\frac{15}{4}$ **7.6**

5. $\frac{x - a}{b} = \frac{x + c}{d}$ $\frac{ad + bc}{d - b}$

6. $x^2 - 6x + 25 = 0$ $3 \pm 4i$ **9.5**

7. $x = 1 + \sqrt{2x + 6}$ 5 **10.3**

8. Find the solution set. **2.2**
   $-4x < 12$ *and*
   $x + 5 < 2x + 3$ $\{x \mid x > 2\}$

Find the following if $f(x) = 3x + 2$ **and** $g(x) = x^2 - 2.$

9. $f(-2.2)$ -4.6 **3.2**

10. $g(f(a))$ $9a^2 + 12a + 2$ **5.9**

11. $\frac{f(h + 4) - f(4)}{h}$ 3 **7.1**

12. Find the slope of $\overline{PQ}$ given $P(-3,2)$ and $Q(1,-1)$. $-\frac{3}{4}$ **3.3**

13. Write an equation in standard form, $ax + by = c$, of the line through $A(2, -4)$ and *perpendicular* to the line $y = \frac{2}{3}x + 1$. $3x + 2y = -2$ **3.7**

14. If y varies directly as x and $y = 48$ when $x = 72$, find x when $y = 60$. 90 **3.8**

15. Solve the system. **4.3**
    $2x + 3y + 4 = 0$
    $4y - 33 = 5x$ $(-5,2)$

16. Simplify $\left(\frac{2a^{-2}b}{3}\right)^3$ without using negative exponents. $\frac{8b^3}{27a^6}$ **5.2**

17. Factor completely. **5.7**
    $5x^4 - 25x^2 - 180$

18. Find the solution set. **6.4**
    $x^2 - 6x - 16 < 0$
    $\{x \mid -2 < x < 8\}$

Simplify. $15n^3 - 11n^2 + 22n - 8$

19. $(5n - 2)(3n^2 - n + 4)$ **5.3**

20. $\frac{x^2 - 49}{x^2 - 7x + 12} \div \frac{5x - 35}{4x - 12}$ **7.2**

21. $2c\sqrt{8c} + \sqrt{18c^3} - c\sqrt{32c},$ $c > 0$ $3c\sqrt{2c}$ **8.3**

22. $\frac{5}{4\sqrt{2} - \sqrt{3}}$ $\frac{20\sqrt{2} + 5\sqrt{3}}{29}$ **8.4**

23. $\sqrt[3]{4x^2} \cdot \sqrt[3]{16x}$ $4x$ **8.5**

24. $2\sqrt{-3} \cdot \sqrt{-12}$ -12 **9.2**

25. $\frac{7}{5 - 2i}$ $\frac{35 + 14i}{29}$

26. Copy the figure below and draw $\vec{A} + \vec{B}$. **9.7**

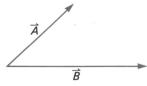

27. Use the Rational Zero Theorem to find the four zeros of $2x^4 - 5x^3 - 2x^2 + 10x - 4$. **10.5**

28. Given $A(3,7)$, $B(-2,2)$, and $C(5,-1)$, find the distance from B to the midpoint of $\overline{AC}$. $\sqrt{37}$ **11.1**

29. Graph the circle described by $x^2 + y^2 - 6x + 4y + 8 = 0$. *12.1*

30. Write an equation in standard form of the ellipse with vertices $(0, \pm 2)$ and $(\pm 2\sqrt{2}, 0)$. Then graph the ellipse. *12.2*

31. A number y varies inversely as the cube of x, and $y = 9$ when $x = 4$. Find y when $x = 6$. $y = \frac{8}{3}$ *12.5*

32. Solve the system: $\begin{array}{l} x + y = 9 \\ xy = 18 \end{array}$ *12.9*

33. If $y = \log_2 x$, find the values of y when $x = \frac{1}{4}, \frac{1}{2}, 1, 2, 4, 8$. *13.4*
$-2, -1, 0, 1, 2, 3$

Solve. Check if necessary.

34. $3^{x+1} = 243$ 4 *5.1*

35. $\log_b 125 = 3$ 5 *13.3*

36. $2 \log_3 t = \log_3 4 + \log_3 (t + 3)$ 6 *13.6*

37. Solve $5^{2x-1} = 800$ for x to 3 significant digits. 2.57 *13.8*

38. How many odd 4-digit numbers can be formed from the digits 3, 4, 5, 6, 7, 8, if no digit is repeated in a number? 180 *14.2*

39. In how many ways can a 5-member committee be selected if 8 people are available? 56 *14.5*

40. One card is drawn from a deck of 52 playing cards. Find the probability that it is a red card or an ace. $\frac{7}{13}$ *14.8*

A box contains 4 white and 8 blue marbles. Two marbles are drawn, with the first replaced before the second is drawn. Find the probability of each event.

41. a blue, then a white $\frac{2}{9}$ *14.9*

42. a blue and a white $\frac{4}{9}$

43. Find the two arithmetic means between 5 and 11. 7,9 *15.2*

44. Find the harmonic mean of $\frac{1}{2}$ and $\frac{1}{10}$. $\frac{1}{6}$

45. Find the sum of the first 18 terms of the arithmetic series $-7 - 2 + 3 + 8 + \ldots$. 639 *15.3*

46. Find the 10th term in the geometric sequence: 3, 6, 12 . . . . 1,536 *15.4*

47. Find the 3rd term in the expansion of $(2x - y^2)^5$. Simplify. $80x^3y^4$ *15.8*

48. Find the product. *16.2*

$$\begin{bmatrix} 2 & -2 & 4 \\ -1 & 3 & 5 \end{bmatrix} \cdot \begin{bmatrix} 3 & 5 \\ 2 & -2 \\ -1 & 4 \end{bmatrix}$$

49. Find three consecutive even integers such that the square of the second integer, increased by the product of the first two, is the same as 12 more than 3 times the square of the first integer. 2, 4, 6; and 4, 6, 8 *6.2*

50. Crew A can complete a job in 8 h and Crew B can do it in 12 h. If Crew C helps, all three together can do the job in 3 h. How long would it take Crew C working alone? 8 h *7.7*

51. Find the value of $7,500 invested at 8% compounded quarterly for 10 years. Use the formula $A = p\left(1 + \frac{r}{n}\right)^{nt}$. $16,560.30 *13.8*

52. The arithmetic mean of two numbers is 20. One number increased by 5 is 4 times the other number. Find both numbers. 9, 31 *15.2*

17. $5(x^2 + 4)(x - 3)(x + 3)$

20. $\frac{4(x + 7)}{5(x - 4)}$

26.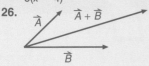

27. $\frac{1}{2}, 2, \pm\sqrt{2}$

29.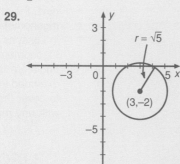

30. $\frac{x^2}{8} + \frac{y^2}{4} = 1$

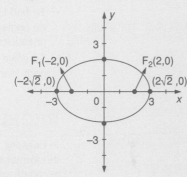

32. (6,3), (3,6)

48. $\begin{bmatrix} -2 & 30 \\ -2 & 9 \end{bmatrix}$

17 TRIGONOMETRIC FUNCTIONS

OVERVIEW

Chapter 17 is the first of three chapters on trigonometry. In this chapter and part of the next, the emphasis is on triangles. The trigonometry of right triangles is developed in Lessons 17.1–17.3. Here, the basic trigonometric functions of a positive acute angle are taught. The ideas are then extended in the chapters' remaining lessons to cover the general case, in which angles with any real measure are allowed.

OBJECTIVES

- To find the six trigonometric functions of a positive acute angle in degree or radian measure
- To find the six trigonometric functions of an angle of any real measure in degree or radian measure
- To solve trigonometric problems involving right triangles

PROBLEM SOLVING

In Lessons 17.1 and 17.2, students learn the formulas for the trigonometric ratios of an acute angle of a right triangle. Lesson 17.3 enables students to employ the strategy *Using a Formula* to solve problems involving right-triangle trigonometry.

READING AND WRITING MATH

The *Focus on Reading* on page 641 asks students to determine the truth value of English statements about angles of rotation. The *Focus on Reading* on page 650 asks students to complete English sentences with the correct mathematical expression. Exercise 35 on page 627 and Exercise 34 in the *Chapter Review* ask students to explain mathematical concepts in their own words.

TECHNOLOGY

Calculator: Students who study trigonometry are immeasurably aided by calculators, especially scientific calculators. There are frequent references to this tool throughout all three trigonometry chapters. (See the feature *Using the Calculator* on page 627.)

SPECIAL FEATURES

Mixed Review pp. 622, 627, 631, 642, 647, 651, 655
Brainteaser pp. 622, 642, 651
Using the Calculator p. 627
Application: Horizon Distance p. 632
Midchapter Review p. 637
Focus on Reading pp. 641, 650
Application: Tunneling p. 655
Key Terms p. 656
Key Ideas and Review Exercises pp. 656–657
Chapter 17 Test p.658
College Prep Test p. 659

PLANNING GUIDE

Lesson	Basic	Average	Above Average	Resources
17.1 pp. 621–622	CE all WE 1–11 odd Brainteaser	CE all WE 3–15 odd Brainteaser	CE all WE 7–17 odd Brainteaser	Reteaching 125 Practice 125
17.2 pp. 626–627	CE all WE 1–21 odd Using the Calculator	CE all WE 9–31 odd Using the Calculator	CE all WE 11–37 odd Using the Calculator	Reteaching 126 Practice 126
17.3 pp. 630–632	CE all WE 1–10 Application	CE all WE 2–12 Application	CE all WE 3–13 Application	Reteaching 127 Practice 127
17.4 pp. 636–637	CE all WE 1–27 odd Midchapter Review	CE all WE 3–29 odd Midchapter Review	CE all WE 5–31 odd Midchapter Review	Reteaching 128 Practice 128
17.5 pp. 641–642	FR all CE all WE 1–27 odd Brainteaser	FR all CE all WE 9–35 odd Brainteaser	FR all CE all WE 11–37 odd Brainteaser	Reteaching 129 Practice 129
17.6 pp. 646–647	CE all WE 1–35 odd	CE all WE 13–47 odd	CE all WE 17–53 odd	Reteaching 130 Practice 130
17.7 pp. 650–651	FR all CE all WE 1–19 odd Brainteaser	FR all CE all WE 3–23 odd Brainteaser	FR all CE all WE 5–25 odd Brainteaser	Reteaching 131 Practice 131
17.8 pp. 654–655	CE all WE 1–15 odd Application	CE all WE 5–19 odd Application	CE all WE 7–21 odd Application	Reteaching 132 Practice 132
Chapter 17 Review pp. 656–657	all	all	all	
Chapter 17 Test p. 658	all	all	all	
College Prep Test p. 659	all	all	all	

CE = Classroom Exercises   WE = Written Exercises   FR = Focus on Reading
NOTE: For each level, all students should be assigned all Mixed Review exercises.

■ INVESTIGATION

In this investigation, students explore the basic trigonometric functions experimentally.

Have the students draw the graph of a line with a negative slope on a piece of notebook paper. If the horizontal and vertical rules are thought of as right angles, a number of similar right triangles are formed.

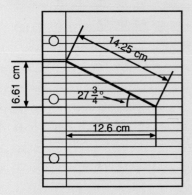

Have the students measure the sides of a triangle in centimeters as accurately as possible. Also have them measure the angle at the lower right with a protractor. The illustration shows some possible measurements.

Have the students compute the sine, cosine, and tangent of the angle based on their measurements. Then have them compare their results with the actual values of the functions to 3 decimal places. Using the measurements in the example, a table of values can be made.

	Estimate	Calculator value
$\sin 27.75 = \frac{6.61}{14.25} = 0.464$		0.466
$\cos 27.75 = \frac{12.6}{14.25} = 0.884$		0.885
$\tan 27.75 = \frac{6.61}{12.6} = 0.525$		0.526

Have students select another triangle using the same sloping line but a different rule for the base. Have them verify that the triangles are similar by checking the ratios of the first measurements to the second measurements. For example, if the new measurements were 4.86 cm, 9.3 cm, and 10.5 cm, the ratios would be

$$\frac{4.86}{6.61} = 0.735 \quad \frac{9.3}{12.6} = 0.738 \quad \frac{10.5}{14.25} = 0.737.$$

The new estimates for the sine, cosine and tangent of the angle (27.75) are 0.463, 0.886, and 0.523.

Flight simulation programs use computer graphics to model the conditions of flying. Each image is generated in response to the actions of the pilot—creating a realistic panorama of an aircraft in flight.

More About the Image

Computer simulation creates a realistic environment so that pilots gain experience in all types of flying conditions without endangering themselves or their aircraft. The flight simulation data base often contains actual terrain mappings of the Earth to create a realistic horizon. Images outside the cockpit are synchronized to the movement of an aircraft in space. The entire scene is calculated and redrawn instantly in response to the pilot's use of the controls.

17.1 Introduction to Right-Triangle Trigonometry

Teaching Resources

Project Worksheet 17
Quick Quizzes 125
Reteaching and Practice
   Worksheets 125

Objective

To compute the sine, cosine, and tangent of an acute angle of a right triangle

In an earlier mathematics course, you may have used properties of *similar triangles* (triangles that have the same shape) to measure indirectly the distance between two objects, such as two trees on opposite sides of a pond. It is possible to use indirect measurement in a more effective way by employing similar *right* triangles. This is one of the many purposes of trigonometry.

A second purpose of trigonometry is unrelated to indirect measurement. Trigonometry can be used to describe many phenomena that are repetitive or cyclic in nature. One example is the air-pressure vibrations that create sound, such as middle C on a piano. This tone can be described by the trigonometric formula $y = \sin 528\pi t$ and corresponds to a frequency of 264 cycles per second.

Recall that in a right triangle, the side opposite the right angle is called the *hypotenuse*. The other two sides are called *legs* of the right triangle. Each leg is opposite one of the two acute angles and adjacent to the other, as illustrated below for $\angle A$ (read "angle A") and $\angle B$.

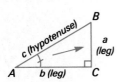

Leg a is *opposite* $\angle A$, and *adjacent* to $\angle B$.

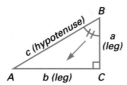

Leg b is *opposite* $\angle B$, and *adjacent* to $\angle A$.

The two right triangles shown below are *similar* triangles.

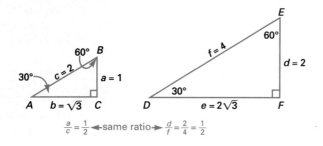

$\dfrac{a}{c} = \dfrac{1}{2}$ ◀same ratio▶ $\dfrac{d}{f} = \dfrac{2}{4} = \dfrac{1}{2}$

■■■ GETTING STARTED

Prerequisite Quiz

Simplify.

1. $(\sqrt{7})^2$ 7
2. $(\sqrt{6})^2$ 6
3. $(\sqrt{5})^2$ 5

Given $\triangle ABC$, **with** $\angle C$ **a right angle. Use the given data to find the indicated length.**

4. $b = 6, a = 8, c = $ _?_   10
5. $b = \sqrt{5}, a = 2, c = $ _?_   3
6. $c = \sqrt{6}, b = 2, a = $ _?_ $\sqrt{2}$

Motivator

Ask the class to graph a triangle on a coordinate plane with vertices $A(0,0)$, $B(3,0)$, $C(3,4)$. Ask them to classify the triangle. Right triangle   Have students use Thm 11.1 to find AC.   5   Have them measure $\angle A$ and find $\dfrac{BC}{AB}$ $\left(\dfrac{\text{opp}}{\text{adj}}\right)$. $\angle A = 53; \dfrac{BC}{AB} = 1.3$

Have students repeat the steps above for a new right $\triangle ABC$ with vertices $A(5, -2)$, $B(11, -2)$, $C(11,6)$. Then have them complete the following generalization. For any right triangle with an acute angle of measure 53, the ratio $\dfrac{\text{opp}}{\text{adj}}$ is _?_.   1.3

Highlighting the Standards

Standard 2a: The indirect techniques, in this case for measurement, are key mathematical concepts that require timely review and reinforcement.

Lesson Note

Students who have already studied geometry will be familiar with the properties of similar triangles. Even those who have not studied geometry should recognize these properties intuitively.

Math Connections

Historical Note: Although knowledge of trigonometry dates back to ancient times, hundreds of years elapsed before the modern view of trigonometric ratios was adopted. Rather, the ancients viewed trigonometry as the study of certain line segments, usually chords in a circle.

Interest in the subject increased during the age of exploration. However, it was not until the late sixteenth century that the name "trigonometry" began to be commonly used.

Critical Thinking Questions

Analysis: Ask students whether there seems to be any limitation on the possible values that the sine, cosine, and tangent may assume. They will immediately notice that negative values are not possible (although this restriction will change in later lessons). By drawing several right triangles, they may also notice that for an acute angle P, $0 < \sin P < 1$, $0 < \cos P < 1$, and $\tan P > 0$. In a later lesson, the values 0 and 1 will also be allowed as values for the sine and cosine ratios and the value 0 will be allowed for the tangent ratio.

Definition

Two triangles are said to be **similar** if their corresponding angle measures are equal and the lengths of their corresponding sides are proportional.

By the definition of similar triangles, $\frac{a}{d} = \frac{c}{f} = \frac{1}{2}$, and $m \angle A = m \angle D = 30$. Read $m \angle A$ as "measure of angle A."

Then, by the Property of Proportions, $af = dc$, or $af = cd$, which is equivalent to the following.

$$\frac{a}{c} = \frac{d}{f} \begin{array}{l} \leftarrow \text{length of leg opposite angle of 30} \\ \leftarrow \text{length of hypotenuse} \end{array}$$

This suggests that for all right triangles with a given acute angle of the same measure, the following ratio is always the same.

$$\frac{\text{length of the leg opposite the acute angle}}{\text{length of the hypotenuse}}$$

This *trigonometric ratio* is called the *sine ratio,* or *sine* of the angle.

For any given acute angle A of a right triangle, there are six trigonometric ratios. The first three of these are defined below. The remaining ratios will be defined in the next lesson.

Definitions

For all right triangles with given acute angle A:

Ratio of Lengths		Abbreviation
sine of $\angle A$	$= \dfrac{\text{opposite leg}}{\text{hypotenuse}}$	$\sin A = \dfrac{a}{c}$
cosine of $\angle A$	$= \dfrac{\text{adjacent leg}}{\text{hypotenuse}}$	$\cos A = \dfrac{b}{c}$
tangent of $\angle A$	$= \dfrac{\text{opposite leg}}{\text{adjacent leg}}$	$\tan A = \dfrac{a}{b}$

EXAMPLE 1   Find each trigonometric ratio to four decimal places.

a. $\sin P$
b. $\cos P$
c. $\tan P$

Additional Example 1

Find to four decimal places.

a. $\sin M$ $\sin M = 0.9231$
b. $\cos M$ $\cos M = 0.3846$
c. $\tan M$ $\tan M = 2.4000$

Plan	Use the definitions of the trigonometric ratios.

a. $\sin P = \dfrac{\text{opp}}{\text{hyp}}$ **b.** $\cos P = \dfrac{\text{adj}}{\text{hyp}}$ **c.** $\tan P = \dfrac{\text{opp}}{\text{adj}}$

$\quad\ = \dfrac{15}{17}$ $\qquad\quad = \dfrac{8}{17}$ $\qquad\quad = \dfrac{15}{8}$

$\quad\ \approx 0.8824$ $\qquad\quad \approx 0.4706$ $\qquad\quad = 1.8750$

Thus, $\sin P = 0.8824$, $\cos P = 0.4706$, and $\tan P = 1.8750$ to four decimal places.

EXAMPLE 2   Find cos A to four decimal places.

Solution   First, find c by using the Pythagorean relation.

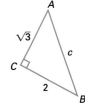

$c^2 = 2^2 + (\sqrt{3})^2$
$c^2 = 4 + 3$
$c^2 = 7$
$c = \sqrt{7}$

$\cos A = \dfrac{\text{adj}}{\text{hyp}} = \dfrac{\sqrt{3}}{\sqrt{7}} = \dfrac{\sqrt{3} \cdot \sqrt{7}}{\sqrt{7} \cdot \sqrt{7}} = \dfrac{\sqrt{21}}{7} \approx \dfrac{4.5825757}{7} \approx 0.6546536$

Thus, $\cos A = 0.6547$ to four decimal places.

In Example 2, a calculator was used to approximate $\sqrt{21}$ as 4.5825757. However, you should follow the instructions of your teacher in deciding when to use a calculator to solve the examples and exercises of Chapters 17, 18, and 19. The final answer to an example or exercise should be equivalent no matter which method is used.

Classroom Exercises

In Exercises 1–10 refer to the figure at the right to indicate the following.

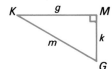

1. the leg adjacent to $\angle K$ g
2. the leg opposite $\angle G$ g
3. the leg adjacent to $\angle G$ k
4. the leg opposite $\angle K$ k
5. the hypotenuse m

Give each trigonometric ratio as a ratio of the lengths of a pair of sides.

6. $\sin G$ $\frac{g}{m}$ 7. $\cos K$ $\frac{g}{m}$ 8. $\tan G$ $\frac{g}{k}$ 9. $\cos G$ $\frac{k}{m}$ 10. $\sin K$ $\frac{k}{m}$

Common Error Analysis

Error: Students tend to think of trigonometric ratios in terms of letters associated with a particular triangle. This may lead to errors such as writing $\sin B = \frac{a}{c}$. Emphasize memorization of trigonometric ratios in terms of "opposite," "adjacent," and "hypotenuse."

Checkpoint

Find sin A, cos A, tan A, sin B, cos B, and tan B to four decimal places.

1.

$\sin A = 0.5547$
$\cos A = 0.8321$
$\tan A = 0.6667$
$\sin B = 0.8321$
$\cos B = 0.5547$
$\tan B = 1.5000$

Find each indicated ratio to four decimal places.

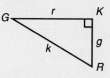

2. $\cos G$ if $k = 13$, $g = 12$ 0.3846
3. $\tan R$ if $k = 4$, $r = 1$ 0.2582
4. $\sin G$ if $r = 2$, $g = 2$ 0.7071

Additional Example 2

Find cos B to four decimal places.   0.8452

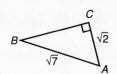

622

Closure

Ask students the following questions. For all right triangles with given acute angle A, in terms of opp, adj, and hyp, what is

1. sin A?
2. cos A?
3. tan A?

For Exercises 1–3, see definitions on page 620.

Have students give the first step in finding the sine of an acute angle of a right triangle if you are given the lengths of the two legs. Use the Pythagorean Relation to find the length of the hypotenuse.

◼◼◼FOLLOW UP

Guided Practice

Classroom Exercises 1–10

Independent Practice

A Ex. 1–8, **B** Ex. 9–14, **C** Ex. 15–18

Basic: WE 1–11 odd, Brainteaser

Average: WE 3–15 odd, Brainteaser

Above Average: WE 7–17 odd, Brainteaser

Additional Answers

Written Exercises

1. 0.6000; 0.8000; 0.7500; 0.8000; 0.6000; 1.3333
2. 0.3846; 0.9231; 0.4167; 0.9231; 0.3846; 2.4000
3. 0.7071; 0.7071; 1.0000; 0.7071; 0.7071; 1.0000

See page 627 for the answers to Written Ex. 4–15 and Mixed Review Ex. 3.

Written Exercises

Find sin A, cos A, tan A, sin B, cos B, and tan B to four decimal places.

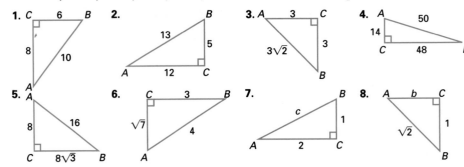

Find each trigonometric ratio to four decimal places. Use the figure at the right.

9. cos W if $t = 26$, $g = 10$
10. sin G if $t = 20$, $w = 16$
11. tan W if $w = 7$, $g = 24$
12. cos G if $g = 1$, $w = 1$
13. sin W if $w = 4$, $g = 6$
14. cos W if $w = \sqrt{5}$, $g = 3$

15. The lengths of the legs of a right triangle are p and f. Find the cosine of the angle opposite the leg of length p.

16. The length of a leg of a right triangle is half the length of the hypotenuse. Find the sine of the angle between that leg and the hypotenuse. 0.8660

17. The lengths of the legs of a right triangle are a and b. Simplify $(\sin A)^2 + (\cos B)^2$. $\frac{2a^2}{a^2 + b^2}$

18. The length of the hypotenuse of a right triangle is 1 less than twice the length of a leg. Find the sine of the angle between this leg and the hypotenuse if the other leg measures 4 units. 0.8000

Mixed Review 3.5, 8.2, 9.3, 9.5

1. Simplify $\sqrt{28x^3}$. $2|x|\sqrt{7x}$
2. Solve $x^2 + 4x + 6 = 0$. $-2 \pm \sqrt{2}\,i$
3. Graph $3x - 4y = 8$.

◼◼◼ Brainteaser

A triangle has a right angle C, two other angles A and B, and sin A = cos A. What are the measures of ∠A and ∠B? Classified by its side lengths, what type of triangle is the triangle? Expressed in terms of the hypotenuse c, what are the lengths of sides a and b? 45; 45; isos; $\frac{c\sqrt{2}}{2}$

Enrichment

Draw these right triangles on the chalkboard.

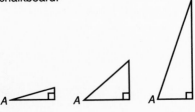

Have the students draw several triangles like these on paper, with the measure of angle A increasing from a little more than 0 to a little less than 90. Then have the students analyze the drawings to determine the behavior of sin A, cos A, and tan A, in the interval 0 < A < 90.

Sin A increases from a little more than 0 to a little less than 1. Cos A decreases from a little less than 1 to a little more than 0.

Tan A increases without bound from values a little more than 0.

17.2 Using Trigonometric Tables

Objectives

To compute the cosecant, secant, and cotangent of an acute angle of a right triangle

To use a trigonometric table to find a trigonometric ratio of an acute angle of a right triangle

The three trigonometric ratios, sine, cosine, and tangent, were defined in Lesson 17.1. The *reciprocals* of these ratios are defined below.

Definitions

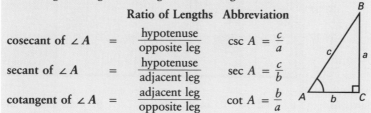

For all right triangles with a given acute angle A:

	Ratio of Lengths	Abbreviation
cosecant of $\angle A$ =	$\dfrac{\text{hypotenuse}}{\text{opposite leg}}$	$\csc A = \dfrac{c}{a}$
secant of $\angle A$ =	$\dfrac{\text{hypotenuse}}{\text{adjacent leg}}$	$\sec A = \dfrac{c}{b}$
cotangent of $\angle A$ =	$\dfrac{\text{adjacent leg}}{\text{opposite leg}}$	$\cot A = \dfrac{b}{a}$

EXAMPLE 1

Find $\csc T$, $\sec T$, and $\cot T$ to four decimal places.

Solution

Use the Pythagorean relation to find u.

$u^2 = 4^2 + 2^2$ ← THINK: Is u greater than 4? greater than 6? How do you know?

$u^2 = 16 + 4$

$u = \sqrt{20} = \sqrt{4 \cdot 5} = 2\sqrt{5}$

$$\csc T = \frac{\text{hyp}}{\text{opp}} \qquad \sec T = \frac{\text{hyp}}{\text{adj}} \qquad \cot T = \frac{\text{adj}}{\text{opp}}$$

$$= \frac{2\sqrt{5}}{4} \qquad\qquad = \frac{2\sqrt{5}}{2} \qquad\qquad = \frac{2}{4}$$

$$\approx \frac{2.2361}{2} \approx 1.1180 \qquad \approx 2.2361 \qquad\qquad = 0.5000$$

The tables on pages 792–796 give decimal approximations of the six basic trigonometric ratios for angles measuring from 0 to 90. Angle measure is given in multiples of ten minutes. One degree equals 60 minutes ($1° = 60'$). The values given by the table are approximate, but it is customary to use $=$ (is equal to) rather than $\approx$ (is approximately equal to).

Teaching Resources

Quick Quizzes 126
Reteaching and Practice
Worksheets 126

▬▬ GETTING STARTED

Prerequisite Quiz

Refer to a right triangle *ABC* with $\angle C$ a right angle. Use the given data to find each trigonometric ratio to four decimal places.

1. $a = 2$, $b = 4$, $\sin A = $ _?_ 0.4772
2. $a = 3$, $b = \sqrt{2}$, $\cos B = $ _?_ 0.9045
3. $c = \sqrt{13}$, $a = 2$, $\tan B = $ _?_ 1.5000
4. $c = \sqrt{3}$, $b = 1$, $\cos A = $ _?_ 0.5774

Motivator

Recall from the previous lesson that for an acute angle of measure approximately 53, the sine ratio is $\frac{4}{5}$ or 0.8000. This can be verified by calculator where trigonometric ratios have been built in.

Using a calculator, have students enter 53, then press the [sin] key and give the decimal that appears on the screen. 0.79863551

Now, have them press either the 2nd function or [INV] key followed by pressing the [sin] key. What appears in the calculator window? 53

Additional Example 1

Find csc *W*, sec *W*, and cot *W* to four decimal places.

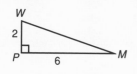

csc W = 1.0541
sec W = 3.1623
cot W = 0.3333

Highlighting the Standards

Standard 6b: Facility in the use of tables is an intermediate skill that, along with equations and graphs, will contribute to a more thorough understanding of trigonometric functions.

TEACHING SUGGESTIONS

Lesson Note

In the table of trigonometric ratios, emphasize that the top row of ratio labels is used with angle measures in the left column, while the bottom row is used with angle measures in the right column.

Although interpolation is not introduced at this stage, you may wish to point out that interpolation with trigonometric ratios is possible just as it was with logarithms. With the use of calculators, however, the practical value of interpolation is questionable.

Math Connections

Historical Note: It is believed that Hipparchus of Nicaea (180 B.C.?–125 B.C.) created the first trigonometric table, and is known as the "father of trigonometry." It seems probable that he drew up these tables for use in astronomy.

Critical Thinking Questions

Analysis: Have students find the restrictions on the possible values of the cosecant, secant, and cotangent ratios in the same way they did for the *Critical Thinking Questions* of Lesson 17.1. They will find that for an acute angle P, $\csc P > 1$, $\sec P > 1$, and $\cot P > 0$. In a later lesson, 0 will be allowed as a value for the cotangent and 1 will be allowed as a value for the secant and cosecant.

EXAMPLE 2 Find each trigonometric ratio using the portion of a trigonometric table shown below.

a. cos 38°50′ **b.** cot 39°20′

Plan The ratio names in the top row of the page must be used with degree measures in the left-hand column (0 through 45).

x	sin x	cos x	tan x	cot x	sec x	csc x	
30'	.6088	.7934	.7673	1.3032	1.2605	1.6427	30'
40'	.6111	.7916	.7720	1.2954	1.2633	1.6365	20'
50'	.6134	.7898	.7766	1.2876	1.2662	1.6304	10'
38° 0'	.6157	.7880	.7813	1.2799	1.2690	1.6243	52° 0'
10'	.6180	.7862	.7860	1.2723	1.2719	1.6183	50'
20'	.6202	.7844	.7907	1.2647	1.2748	1.6123	40'
30'	.6225	.7826	.7954	1.2572	1.2779	1.6064	30'
40'	.6248	.7808	.8002	1.2497	1.2808	1.6005	20'
38°50' → 50'	.6271	.7790	.8050	1.2423	1.2837	1.5948	10'
39° 0'	.6293	.7771	.8098	1.2349	1.2868	1.5890	51° 0'
10'	.6316	.7753	.8146	1.2276	1.2898	1.5833	50'
39°20' → 20'	.6338	.7735	.8195	1.2203	1.2929	1.5777	40'
30'	.6361	.7716	.8243	1.2131	1.2960	1.5721	30'
40'	.6383	.7698	.8292	1.2059	1.2991	1.5666	20'
50'	.6406	.7679	.8342	1.1988	1.3022	1.5611	10'

Solutions

a. Find 38°50′ in the left column. Read across to the cos column. Thus, cos 38°50′ = 0.7790.

b. Find 39°20′ in the left column. Read across to the cot column. Thus, cot 39°20′ = 1.2203.

The ratio names at the bottom of the page must be used with degree measures in the right-hand column (45 through 90).

EXAMPLE 3 Find each trigonometric ratio.

a. cos 45°40′ **b.** sec 46°50′

Solutions

43° 0'	.6820	.7314	.9325	1.0724	1.3673	1.4663	47° 0'
10'	.6841	.7294	.9380	1.0661	1.3711	1.4617	50' ← 46°50'
20'	.6862	.7274	.9435	1.0599	1.3748	1.4572	40'
30'	.6884	.7254	.9490	1.0538	1.3786	1.4527	30'
40'	.6905	.7234	.9545	1.0477	1.3824	1.4483	20'
50'	.6926	.7214	.9601	1.0416	1.3863	1.4439	10'
44° 0'	.6947	.7193	.9657	1.0355	1.3902	1.4396	46° 0'
10'	.6967	.7173	.9713	1.0295	1.3941	1.4352	50'
20'	.6988	.7153	.9770	1.0235	1.3980	1.4310	40' ← 45°40'
30'	.7009	.7133	.9827	1.0176	1.4020	1.4267	30'
40'	.7030	.7112	.9884	1.0117	1.4061	1.4225	20'
50'	.7050	.7092	.9942	1.0058	1.4101	1.4184	10'
45° 0'	.7071	.7071	1.0000	1.0000	1.4142	1.4142	45° 0'
	cos x	sin x	cot x	tan x	csc x	sec x	x

a. Thus, cos 45°40′ = 0.6988. **b.** Thus, sec 46°50′ = 1.4617.

Additional Example 2

Find each of the following trigonometric ratios using the portion of a trigonometric table shown on this page of the textbook.

a. sin 38° 10′ 0.6180
b. csc 39° 50′ 1.5611

Additional Example 3

Find each trigonometric ratio.

a. tan 46° 20′ 1.0477
b. sin 45° 20′ 0.7112

EXAMPLE 4 Find m ∠A if sin A = 0.8774 and ∠A is an acute angle of a right triangle. Reverse the process used in Example 2.

Solution

		cos x	sin x	cot x	tan x	csc x	sec x		
	10'	.4720	.8816	.5354	1.8676	1.1343	2.1185	50'	
	20'	.4746	.8802	.5392	1.8546	1.1361	2.1070	40'	
	30'	.4772	.8788	.5430	1.8418	1.1379	2.0957	30'	
	40'	.4797	.8774	.5467	1.8291	1.1397	2.0846	20'	61°20'
	50'	.4823	.8760	.5505	1.8165	1.1415	2.0736	10'	
29°	0'	.4848	.8746	.5543	1.8040	1.1434	2.0627	61° 0'	
	10'	.4874	.8732	.5581	1.7917	1.1452	2.0519	50'	
	20'	.4899	.8718	.5619	1.7796	1.1471	2.0413	40'	
	30'	.4924	.8704	.5658	1.7675	1.1490	2.0308	30'	
	40'	.4950	.8689	.5696	1.7556	1.1509	2.0204	20'	
	50'	.4975	.8675	.5735	1.7437	1.1528	2.0101	10'	
30°	0'	.5000	.8660	.5774	1.7321	1.1547	2.0000	60° 0'	
		cos x	sin x	cot x	tan x	csc x	sec x	x	

Thus, m ∠A = 61°20' if sin A = 0.8774.

EXAMPLE 5 Find m ∠B to the nearest ten minutes if ∠B is an acute angle of a right triangle and cot B = 2.2815.

Plan Find the number in the cot column that is closest to 2.2815. The number is 2.2817. Since "cot" is at the top of the page, read across to the angle column at the left portion of the page.

Solution

	x	sin x	cos x	tan x	cot x	sec x	csc x		
	30'	.3827	.9239	.4142	2.4142	1.0824	2.6131	30'	
	40'	.3854	.9228	.4176	2.3945	1.0837	2.5949	20'	
	50'	.3881	.9216	.4210	2.3750	1.0850	2.5770	10'	
23°40' → 23°	0'	.3907	.9205	.4245	2.3559	1.0864	2.5593	67° 0'	
	10'	.3934	.9194	.4279	2.3369	1.0877	2.5419	50'	
	20'	.3961	.9182	.4314	2.3183	1.0891	2.5247	40'	
	30'	.3987	.9171	.4348	2.2998	1.0904	2.5078	30'	
	40'	.4014	.9159	.4383	2.2817	1.0918	2.4912	20'	
	50'	.4041	.9147	.4417	2.2637	1.0932	2.4748	10'	

Thus, m ∠B = 23°40' to the nearest ten minutes.

Summary

Basic Trigonometric Ratios

$$\sin A = \frac{\text{opp}}{\text{hyp}} = \frac{a}{c} \qquad \cos A = \frac{\text{adj}}{\text{hyp}} = \frac{b}{c} \qquad \tan A = \frac{\text{opp}}{\text{adj}} = \frac{a}{b}$$

Reciprocals

$$\csc A = \frac{\text{hyp}}{\text{opp}} = \frac{c}{a} \qquad \sec A = \frac{\text{hyp}}{\text{adj}} = \frac{c}{b} \qquad \cot A = \frac{\text{adj}}{\text{opp}} = \frac{b}{a}$$

Checkpoint

Find csc A, sec A, cot A, csc B, sec B, and cot B to four decimal places.

1. A

csc A = 1.2500
sec A = 1.6667
cot A = 0.7500
csc B = 1.6667
sec B = 1.2500
cot B = 1.3333

Find each trigonometric ratio. Use the tables on pages 792–796.

2. sec 38   1.2690
3. cot 47° 10'   0.9271

Find m ∠A to the nearest ten minutes if ∠A is an acute angle of a right triangle.

4. sin A = 0.4950   29° 40'
5. csc A = 1.1530   60° 10'

Additional Example 4

Find m ∠A if cos A = 0.4924 and ∠A is an acute angle of a right triangle. 60° 30'

Additional Example 5

Find m ∠B to the nearest ten minutes if ∠B is an acute angle of a right triangle and sec B = 1.1006. 24° 40'

Closure

Have students answer the following questions.

1. What is the sine ratio?
2. What is its reciprocal called?
3. What is the cosine ratio?
4. What is its reciprocal called?
5. What is the tangent ratio?
6. What is its reciprocal called?

   See the Summary on page 625.

▰▰▰FOLLOW UP

Guided Practice

Classroom Exercises 1–16

Independent Practice

Ⓐ Ex. 1–24, Ⓑ Ex. 25–35, Ⓒ Ex. 36–38

Basic: WE 1–21 odd, Using the Calculator

Average: WE 9–31 odd, Using the Calculator

Above Average: WE 11–37 odd, Using the Calculator

Additional Answers

Written Exercises

1. 1.6667, 1.2500, 1.3333, 1.2500, 1.6667, 0.7500
2. 1.0833, 2.6000, 0.4167, 2.6000, 1.0833, 2.4000
3. 1.4142, 1.4142, 1.0000, 1.4142, 1.4142, 1.0000
4. 3.5714, 1.0417, 3.4286, 1.0417, 3.5714, 0.2917
35. Answers will vary. One possible answer: Find the length of the adjacent side by using the Pythagorean relation. Then divide that length by the length of the hypotenuse.

Classroom Exercises

Using the figure at the right, give each trigonometric ratio as the ratio of a pair of side lengths.

1. csc T $\frac{w}{t}$ 2. cot K $\frac{t}{k}$ 3. sec T $\frac{w}{k}$ 4. sin K $\frac{k}{w}$

5. cos T $\frac{k}{w}$ 6. tan T $\frac{t}{k}$ 7. sin T $\frac{t}{w}$ 8. cos K $\frac{t}{w}$

9–16. In the figure, suppose that $k = 3$, $t = 4$, and $w = 5$. Find the ratios of Classroom Exercises 1–8 to two decimal places.
9. 1.25 10. 1.33 11. 1.67 12. 0.60 13. 0.60 14. 1.33 15. 0.80 16. 0.80

Written Exercises

Find csc A, sec A, cot A, csc B, sec B, and cot B to four decimal places.

 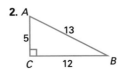 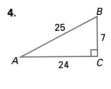

Find each trigonometric ratio. Use the tables on pages 792–796.

5. sin 28 0.4695   6. tan 83 8.1443   7. sec 32 1.1792   8. csc 53 1.2521

9. sin 60°20′ 0.8689 10. cot 37°20′ 1.3111 11. cos 39°30′ 0.7716 12. sin 80°40′ 0.9868

13. csc 19°30′ 2.9957 14. sec 47°10′ 1.4709 15. cot 87°50′ 0.0378 16. cos 11°40′ 0.9793

Find m ∠A to the nearest ten minutes if ∠A is an acute angle of a right triangle. Use the tables on pages 792–796.
17. sin A = 0.2079 18. csc A = 1.2361 19. tan A = 2.4751 20. sin A = 0.7934
   12   54   68   52°30′
21. sec A = 1.1471 22. cos A = 0.9492 23. sec A = 1.5721 24. cot A = 7.5958
 29°20′   18°20′   50°30′   7°30′

Find each indicated trigonometric ratio to four decimal places. Use the figure at the right.

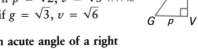

25. sec P, if $g = 6$, $p = 2$ 1.0541   26. cot P, if $p = 2$, $v = 4$ 1.7321

27. csc G, if $v = 6$, $p = 4$ 1.3416   28. cot G, if $p = \sqrt{2}$, $v = \sqrt{3}$ 1.4142

29. sec G, if $g = \sqrt{3}$, $p = \sqrt{2}$ 30. csc P, if $g = \sqrt{3}$, $v = \sqrt{6}$
  1.5811   1.4142

Find m ∠B to the nearest ten minutes if ∠B is an acute angle of a right triangle.

31. cot B = 0.7312 32. sin B = 0.1675 33. tan B = 2.5199 34. cos B = 0.8662
 53°50′   9°40′   68°20′   30°00′

Enrichment

Have the students use the table to compare the sine and tangent ratios for angle measures between 0 and 10. They should note that the values are almost the same. Now, draw this triangle on the chalkboard.

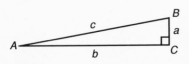

Have the students write a sentence or two explaining why sin A is almost equal to tan A in the interval $0 < A < 10$.

When angle A is small, the lengths of the leg adjacent to angle A and the hypotenuse are almost the same. Therefore, $b \approx c$ and $\frac{a}{b} \approx \frac{a}{c}$.

Next, have the students observe other ratios that have approximately equal values.

For $0 < A < 10$, csc $A \approx$ cot A.

For $80 < B < 90$, cos $B \approx$ cot B and sec $B \approx$ tan B.

35. Write how to find the cosine ratio of an acute angle A of a right triangle given the lengths of the hypotenuse and the leg opposite $\angle A$.

Simplify each expression (Exercises 36–37).

36. $(\sin 38°40')(\csc 38°40')$ 1

37. $(\sin 23°10')^2 + (\cos 23°10')^2$ 1

38. $\sin A = \sqrt{3} - 1$. Find $\csc A$ and simplify. $\dfrac{\sqrt{3}+1}{2}$

Mixed Review

Simplify.   *13.3, 8.7, 8.4, 5.2, 7.4* $8 + 4\sqrt{3}$ $\dfrac{5x+4}{x}$

1. $\log_3 27$ 3   **2.** $32^{-\frac{4}{5}}$ $\dfrac{1}{16}$ **3.** $\dfrac{4}{2-\sqrt{3}}$ **4.** $(3a - 7)^0$ 1   **5.** $5 + \dfrac{4}{x}$

Using the Calculator

Scientific calculators use angle measures in decimals, such as 27.4 degrees, rather than in degrees and minutes. Thus, $\cos 38°50'$ is written as $\cos\left(38 + \dfrac{50}{60}\right)$. Enter the angle measure as 50 ÷ 60 + 38 and press = to obtain 38.833333. Then press the $\boxed{\cos}$ key. The calculator will display 0.7789733. This is 0.7790 to four decimal places.

NOTE: Most calculators allow the use of three units of angle measure, the *degree,* the *radian* (see Lesson 18.5), and the *gradient* (a European unit of angle measure). A calculator is in the degree mode when DEG appears in the calculator display window.

EXAMPLE   Find $\cot 39°20'$ to four decimal places using a calculator.

   1. Find $\tan 39°20'$.

     $\tan 39°20' = \tan\left(39 + \dfrac{20}{60}\right)$

     Calculator steps:
     20 ÷ 60 + 39 = $\boxed{\tan}$ ⇒ 0.8194625

   2. Since tan and cot are reciprocals, press the reciprocal key $\boxed{1/x}$. To four decimal places, $\cot 39°20' = 1.2203$.

Exercises

Find the following ratios to four decimal places using a calculator.
   1.7362

1. $\sin 58.36$ 0.8514   **2.** $\tan 73°40'$ 3.4124   **3.** $\cot 42°20'$ 1.0977   **4.** $\csc 35°10'$

17.2 Using Trigonometric Tables   **627**

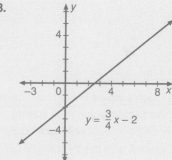
$y = \dfrac{3}{4}x - 2$

Teaching Resources

Application Worksheet 17
Problem Solving Worksheet 17
Quick Quizzes 127
Reteaching and Practice
   Worksheets 127

17.3 Solving Right Triangles

Objectives To find missing measures of sides and angles of right triangles using trigonometric ratios
To solve problems using trigonometric ratios

The trigonometric ratios can be used to form equations for finding indicated side lengths or angle measures in a right triangle.

EXAMPLE 1 If m $\angle A = 65$ and $c = 140$, find a to two significant digits.

Plan a is opp $\angle A$; $c = 140$ is the *hyp*.
Write an equation relating *opp*, *hyp*, and m $\angle A = 65$.

Solution Use $\sin 65 = \dfrac{opp}{hyp} = \dfrac{a}{140}$ or $\csc 65 = \dfrac{hyp}{opp} = \dfrac{140}{a}$.

$$\sin 65 = \frac{a}{140}$$
$$140 \sin 65 = a$$
$$a = 140(0.9063) = 126.882$$

Calculator Steps:
140 (×) 65 (sin) (=) 126.88309

Thus, $a = 130$ to two significant digits.

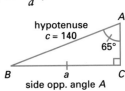

hypotenuse
$c = 140$

$65°$

side opp. angle A

Recall that the acute angles of a right triangle are *complementary*; that is, the sum of their measures is 90.

EXAMPLE 2 Right triangle ABC has a right angle at C, $a = 17$, and $b = 23$. Find m $\angle A$ and m $\angle B$ to the nearest degree.

Plan First, draw and label a diagram. Then, write an equation relating a, b, and one of the two angle measures required, such as m $\angle A$.

Solution $\dfrac{17}{23} = \dfrac{opp}{adj} = \tan A$ (or $\dfrac{23}{17} = \cot A$)

$\tan A = \dfrac{17}{23} = 0.7391$
Find m $\angle A$ using a table or calculator.

Calculator steps:
17 (÷) 23 (=) (INV) (tan) $\Rightarrow$ 36.469234

Thus, m $\angle A = 36$ to the nearest degree, and m $\angle B = 90 - 36 = 54$.

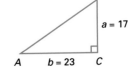

B

$a = 17$

A $b = 23$ C

GETTING STARTED

Prerequisite Quiz

Find each trigonometric ratio to four decimal places.

1. sin 27   0.4540
2. cos 87   0.0523
3. tan 68° 30'   2.5386

For the given trigonometric ratio, find the measure of $\angle A$ to the nearest degree.

4. sec A = 2.4563   66
5. csc A = 1.4556   43

Motivator

Have students describe how they would solve an equation such as $6 = \dfrac{x}{5}$. Now have them apply the same technique to solving the equation $\sin 65 = \dfrac{a}{140}$. Ask them how they intend to work with the term sin 65. Convert to decimal form.

Highlighting the Standards

Standard 9a: Solving applications of triangles is a basic trigonometric skill and one that is satisfying for students to learn and use.

Additional Example 1

Use the figure of Example 1 on this page. If m $\angle B = 74$ and $c = 180$, find b to two significant digits.   170

Additional Example 2

Right $\triangle ABC$ has a right angle at C, $a = 14$, and $b = 27$. Find m $\angle A$ and m $\angle B$ to the nearest degree.
m $\angle A = 27$, m $\angle B = 63$

In Examples 3 and 4 below, the **angle of elevation** or **angle of depression** is the angle between a horizontal line and the line of sight. The angles of elevation and depression are equal in measure.

EXAMPLE 3   A plane is flying at an altitude of 24,300 m. From the plane, the measure of the angle of depression of a ship is $55°30'$. How many meters, to three significant digits, is the plane from the ship?

Solution   Draw and label a diagram.
Let d = the required distance.
Write an equation with the variable in the numerator.

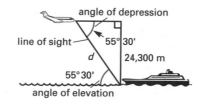

$$\frac{d}{24,300} = \csc 55°30'$$
$$d = 24,300 \csc 55°30'$$
$$= 24,300(1.2134)$$
$$= 29,486$$

Calculator steps:

55.5 (sin) (1/x) (×) $24,300$ (=) $29,485.776$ ← The sine is the *reciprocal* of the cosecant.

Thus, the distance is 29,500 m to three significant digits.

EXAMPLE 4   The captain of a ship wants to determine his distance from shore. Seeking a familiar landmark, he finds a 90-ft-high lighthouse on top of a cliff. He sights both the top and bottom of the lighthouse. The measures of the two angles of elevation are 46 and 39. How far, to two significant digits, is he from the base of the cliff?

Plan   Draw and label a sketch. Two angles of elevation are given. Therefore, a system of two equations in two variables may be used.

Solution   Let d = the captain's distance from shore.
Let h = the height of the cliff.

$$\tan 39 = \frac{h}{d}$$
$$d \tan 39 = h$$
$$(1)\ 0.8098d = h$$

$$\tan 46 = \frac{90 + h}{d}$$
$$d \tan 46 = 90 + h$$
$$(2)\ 1.0355d = 90 + h$$

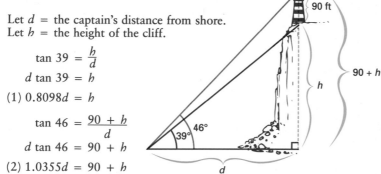

■■■**TEACHING SUGGESTIONS**

Lesson Note

From now on, the calculator will be used frequently. Determining the number of significant digits is important here and in science courses. Many students follow their study of trigonometry with a first course in physics, in which previous knowledge of the use of the calculator will be assumed.

Emphasize that for efficient use of the calculator it is helpful to write trigonometric equations in terms of sine, cosine, and tangent rather than in terms of the reciprocals of these ratios.

Math Connections

Historical Note: Although the ancient Egyptians and Babylonians were familiar with theorems on ratios of the sides of similar triangles, they were not able to develop a theory of trigonometry because they lacked a concept of angle measure. The Greeks began a systematic study of the relationships between the arcs (or angles) of a circle and their subtended chords. This study can be regarded as the precursor of trigonometry. It is believed that Eudoxus of Cnidus (408 B.C.–355 B.C.) used ratios and angle measures to determine the size of the earth as well as relative distances of the sun and the moon.

Additional Example 3

Show the solution of Example 3 using the sine ratio. This will involve an extra step in solving the original equation for d. The answer will be the same.

$$\frac{24,300}{d} = \sin 55° 30'$$
$$24,300 = d(\sin 55° 30')$$
$$\frac{24,300}{\sin 55° 30'} = d$$
$$d = 29,500 \text{ m}$$

Additional Example 4

Bill wants to find the height of a mountain. From some spot on the ground, he finds the measure of the angle of elevation to the top of the mountain is 36. After moving 1,500 m closer to the mountain, he now finds the measure of the angle of elevation to be 63. Find the height of the mountain to two significant digits.   1,700 m

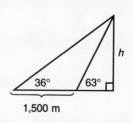

Critical Thinking Questions

Checkpoint

In right △ABC, ∠C is a right angle. Find the missing measures. Find lengths to two significant digits and angle measures to the nearest degree.

1. $a = 43$, $b = 26$, m $\angle A = \underline{?}$,
   m $\angle B = \underline{?}$ m $\angle A = 59$, m $\angle B = 31$
2. $a = 43$, m $\angle A = 47$, $b = \underline{?}$, $c = \underline{?}$
   $b = 40$, $c = 59$
3. m $\angle A = 17$, $b = 40$, $a = \underline{?}$, $c = \underline{?}$
   $a = 12$, $c = 42$
4. $c = 8.3$, $b = 4.6$, m $\angle A = \underline{?}$,
   m $\angle B = \underline{?}$ m $\angle A = 56$, m $\angle B = 34$
5. A lighthouse, standing at sea level, is 185 ft high. From the top of the lighthouse, the angle of depression of a ship at sea is 17° 40′. Find, to three significant digits, the distance of the ship from the base of the lighthouse.   581 ft

---

Solve the system.   (1) $h = 0.8098d$
                         (2) $1.0355d = 90 + h$

Substitute $0.8098d$ for h.   (3) $1.0355d = 90 + 0.8098d$
                                    $0.2257d = 90$
                                        $d = 398.7594$

Thus, the distance to the base of the vertical cliff is 400 ft to two significant digits.

The following relationships between lengths and angle measures were used to establish the number of significant digits in the answers to Examples 1–4.

Side Length		Angle measure
2 significant digits	$\longrightarrow$	nearest degree
3 significant digits	$\longrightarrow$	nearest 10 minutes or 0.1 degree

From now on, these relationships will be used in the examples and exercises.

Classroom Exercises

In right triangle *ABC*, give a trigonometric equation that relates the two given parts to the third part. Choose your equation so that the variable is not in a denominator.

1. m $\angle B = 40$, $c = 10$,
   $a = \underline{?}$
2. m $\angle A = 53$, $c = 25$,
   $b = \underline{?}$
3. m $\angle B = 32$, $a = 17$,
   $c = \underline{?}$
4. m $\angle B = 26$, $b = 19$,
   $c = \underline{?}$
5. m $\angle A = 39$, $c = 14$,
   $a = \underline{?}$
6. m $\angle B = 75$, $c = 43$,
   $b = \underline{?}$

Written Exercises

In Exercises 1–6, find lengths to two significant digits and angle measures to the nearest degree. Use the diagram above.

1. $a = 56$, $b = 21$, m $\angle A = \underline{?}$,
   m $\angle B = \underline{?}$ 69, 21
2. $c = 18$, $b = 13$, m $\angle B = \underline{?}$,
   m $\angle A = \underline{?}$ 46, 44
3. $a = 28$, m $\angle A = 39$, $b = \underline{?}$,
   $c = \underline{?}$ 35, 45
4. m $\angle A = 15$, $b = 30$, $a = \underline{?}$,
   $c = \underline{?}$ 8, 31
5. $c = 9.7$, $b = 5.3$, m $\angle A = \underline{?}$,
   m $\angle B = \underline{?}$ 57, 33
6. m $\angle A = 35$, $a = 19$, $b = \underline{?}$,
   $c = \underline{?}$ 27, 33

7. In the figure at the right, the tree casts a shadow on the ground because of the sun's rays. The length of the shadow is 34 ft. The measure of the angle of elevation is 37. Find the height of the tree to two significant digits. 26 ft

37°
◄34 ft►

8. From the top of a lighthouse 159 ft above sea level, the angle of depression of a ship at sea is 19°20′. Find, to three significant digits, the distance of the ship from the base of the lighthouse. 453 ft

9. At a point on the ground 27.6 m from the foot of a flagpole, the angle of elevation of the top of the pole is 29°50′. Find the height of the pole to three significant digits. 15.8 m

10. From a point 450 ft from the base of a building, the angles of elevation of the top and bottom of a flagpole on top of the building have measures of 60 and 55. Find the height of the flagpole to two significant digits. 140 ft

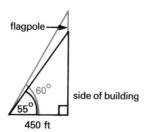
flagpole→
60°
side of building
55°
450 ft

11. Jane wants to find the height of a mountain. From some spot on the ground, she finds the angle of elevation to the top of the mountain to be 35°20′. After moving 1,000 m closer to the mountain, she now finds the angle of elevation to be 50°30′. Find the height of the mountain to three significant digits. 1,710 m

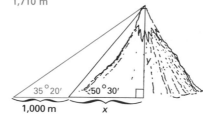
y
35°20′ 50°30′
1,000 m   x

12. Two tanks on a training mission are 1,800 m apart on a straight road. The drivers find the angles of elevation to a helicopter hovering over the road between them to be 33 and 52. Find the height of the helicopter to two significant digits. 780 m

13. The length of the longer leg of a right triangle is 2 more than twice the length of the shorter leg. The length of the hypotenuse is 1 more than the length of the longer leg. Find the measure, to the nearest degree, of the angle between the shorter leg and the hypotenuse. 67

Mixed Review

Solve. *6.4, 8.7, 9.5, 10.3*

1. $x^2 - 6x - 16 > 0$
$\{x | x < -2 \text{ or } x > 8\}$

2. $x^2 - 4x + 2 = 0$
$2 \pm \sqrt{2}$

3. $\sqrt{3x - 6} = 3$ 5

4. $x^{\frac{2}{3}} = 4$ 8

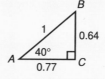
631

If you want to see farther, you try to find a higher vantage point, perhaps a rooftop, a mountaintop, or even a space shuttle. The added height lets you see over a greater part of the earth's surface, increasing the distance at which objects disappear behind the horizon.

To find the horizon distance d, let r be the earth's radius and h the viewer's height above the earth. The line of sight to the horizon is tangent to the earth's surface, and so is perpendicular to a radius drawn to the point of tangency. From the Pythagorean relation:

$$r^2 + d^2 = (r + h)^2$$
$$r^2 + d^2 = r^2 + 2rh + h^2$$
$$d^2 = 2rh + h^2$$
$$d = \sqrt{2rh + h^2}$$

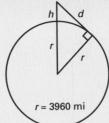

Since ordinarily $r \gg h$ ("r is much greater than h"), $rh \gg h^2$. So, $2rh \gg h^2$ and $2rh + h^2 \approx 2rh$. This gives the approximation

$$d \approx \sqrt{2rh} = \sqrt{2r}\sqrt{h} = \sqrt{7920}\sqrt{h} \approx 89.0\sqrt{h},$$

with d, r, and h in miles. If the height is entered as h *feet*, then

$$d \approx \sqrt{2rh} = \sqrt{(7920 \text{ mi})(h \text{ ft})\left(\frac{1 \text{ mi}}{5280 \text{ ft}}\right)} = \sqrt{\frac{7920}{5280} h \text{ mi}^2} \approx 1.22\sqrt{h} \text{ mi}.$$

The Rock of Gibralter on Spain's south coast looms 1400 ft above the Mediterannean Sea. By the formula above, a person on its peak can see about $1.22\sqrt{1400} \approx 46$ miles out to sea.

1. Leif perches 100 ft above the water in a ship's crow's nest, looking for a raft. At what distance will the raft enter his view? 12 mi

2. An astronaut is orbiting 200 miles above the earth in a space shuttle. Give the horizon distance to 3 significant digits, using both the original formula and the approximation. 1270 mi; 1260 mi

3. While riding a train across the Colorado plains at an elevation of 5000 ft, Trinh spots a mountain she believes to be Pike's Peak, elevation 14,110 ft. If the peak is 75 miles away, is this possible? Yes

17.4 Angles of Rotation

Objectives

To determine the quadrant containing the terminal side of an angle of rotation

To find measures of angles coterminal with a given angle

In geometry, the measure of an angle is restricted to values between 0 and 180, including 180. However, in many practical applications, such as navigation, it is necessary to extend the concept of *angle* to *angle of rotation*.

In the figure at the right, $\angle AOB$ is an **angle of rotation** in standard position. It is formed by holding $\overrightarrow{OA}$ (read "ray *OA*") stationary along the *x*-axis, and then rotating $\overrightarrow{OB}$ from the *x*-axis to some position, such as 70 degrees, in a counterclockwise direction.

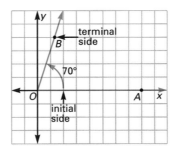

With $\overrightarrow{OA}$ (the **initial side**) fixed along the *x*-axis, allow $\overrightarrow{OB}$ (the **terminal side**) to continue to rotate counterclockwise. New angles are formed depending upon where $\overrightarrow{OB}$ terminates.

Some special angles of rotation, whose terminal sides coincide with a coordinate axis, are illustrated in the figures below.

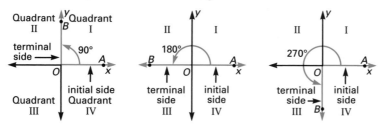

Notice that for each angle above, the initial and terminal sides lie along the coordinate axes. Such angles are called **quadrantal angles**.

It is also possible to generate angles of rotation by rotating the terminal side $\overrightarrow{OB}$ clockwise. *Clockwise* rotations correspond to *negative* angle measures.

Teaching Resources

Quick Quizzes 128
Reteaching and Practice Worksheets 128

▬ GETTING STARTED

Prerequisite Quiz

Find the least positive value of *k* if *n* is an integer.

1. $500 - 360n = k$ 140
2. $1,000 - 360n = k$ 280
3. $860 - 360n = k$ 140
4. $-1,000 + 360n = k$ 80

Motivator

Draw a set of axes on the chalkboard. Ask students to identify the four quadrants.

Draw $\overrightarrow{OA}$ on the positive side of the x-axis. Draw a second ray, $\overrightarrow{OB}$, rotated counterclockwise so that m $\angle BOA$ is approximately 45.

Ask students in which quadrant $\overrightarrow{OB}$ terminates or ends.   First quadrant
Repeat the above instruction for these angles.

1. 135   Second quadrant
2. 225   Third quadrant
3. 315   Fourth quadrant

Highlighting the Standards

Standard 8c: Angles of rotation are essentially transformations made by moving an angle of less than 90 degrees into various positions.

Lesson Note

A visual aid for illustrating angles of rotation is provided by either of the following devices.

1. A chalkboard compass, to show an angle opening that is greater than 90;
2. A grid for the overhead projector containing the x- and y-axes, with two cardboard arrows tacked to the origin so that one can be held fixed while the other rotates.

Emphasize that quadrantal angles do not have their terminal sides in any quadrant, just as the border between two states on a map is not part of either state.

Math Connections

Historical Note: The division of a full rotation (or circle) into 360 equal parts dates back to antiquity. It has been speculated that the practice may have begun with Hipparchus (see *Math Connections* for Lesson 17.2). He may have picked up the idea from contemporaries, who in turn may have been influenced by Babylonian astronomy.

EXAMPLE 1 Draw the angle of rotation in standard position that corresponds to each angle measure.

  a. -270 b. -360 c. -450

Solutions a. Rotate clockwise.   b. Rotate clockwise until the initial and terminal sides coincide.   c. Rotate clockwise 360° and then 90° more.

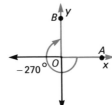
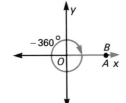
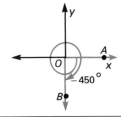

The next example involves angles of rotation that are not quadrantal angles. Note that it is customary to use the Greek letter θ (theta) to label angles of rotation.

EXAMPLE 2 Draw the angle of rotation in standard position that corresponds to each angle measure. Identify the quadrant containing the terminal side.

  a. $\theta = 140$ b. $\theta = -160$

Solutions a. 140 is between 90 and 180. Rotate the terminal side *counterclockwise* to a position between 90 and 180.   b. For -160, rotate the terminal side *clockwise* to a position between -90 and -180.

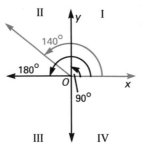
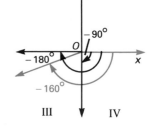

Thus, the angle with measure 140 terminates in Quadrant II.

Thus, the angle with measure -160 terminates in Quadrant III.

Additional Example 1

Draw the angle of rotation that corresponds to each angle measure.

  a. -180
  b. 0
  c. -90

Additional Example 2

Draw the angle of rotation in standard position that corresponds to each angle measure. Identify the quadrant containing the terminal side.

  a. $\theta = 225$ Quadrant III

  b. $\theta = -205$ Quadrant II

EXAMPLE 3   On the same set of coordinate axes, draw the angles of rotation with measures −30 and 330. State a relationship between the angles.

Solution   The two angles have the same initial and terminal sides. However, their angle measures are not equal since −30 ≠ 330. These two angles are called *coterminal* angles.

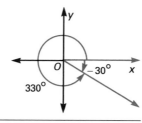

Definition   **Coterminal angles** are angles that share initial and terminal sides.

The pattern that follows suggests a formula for finding the measures of all angles coterminal with a given angle.

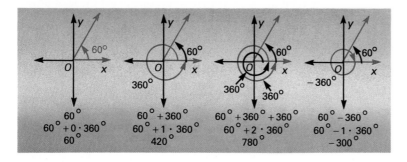

The angles with measures 60, 420, 780, and −300 share initial and terminal sides and thus are coterminal angles. Note the following.

$60 + 0 \cdot 360 = 60$
$60 + 1 \cdot 360 = 420$
$60 + 2 \cdot 360 = 780$
$60 + (-1) \cdot 360 = -300$

Notice the pattern:
$60 + n \cdot 360$,
where n is an integer.

This suggests the following formula for finding the measures of all angles coterminal with a given angle.

Coterminal Angle Formula
If angles θ_1 and θ_2 are coterminal, then $\theta_1 = \theta_2 + n \cdot 360$, where n is an integer.

Critical Thinking Questions

Analysis: The text indicates that the word "angle" may be used for the more precise "angle of rotation" when there is no danger of misunderstanding. Ask students whether they can think of other instances in mathematics of imprecise terminology that is accepted for the sake of convenience. Two examples with which they should be familiar are "line" and "function." One may refer informally to the "line" $y = 2x + 3$ (rather than to the line with equation $y = 2x + 3$) or to the "function" $y = \log x$. This is acceptable to the extent that the speaker's meaning is clearly understood.

Common Error Analysis

Error: When finding the least positive measure of an angle coterminal with a given angle, students tend to ignore the word "least" and merely subtract 360 from the measure of the given angle.

Emphasize that the required measure will be between 0 and 360.

Additional Example 3

On the same set of coordinate axes, draw the angles of rotation with measures of −160 and 200. Are the angles equal in measure? No Do the angles have the same terminal side? Yes

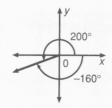

Checkpoint

Draw the angle in standard position that corresponds to each angle measure. If the angle is quadrantal, identify the axis along which the terminal side lies. If not, identify the quadrant that contains the terminal side.

1. 205   III

2. 180   Quadrantal; neg. *x*-axis

Find the least positive measure of an angle that is coterminal with the angle of given measure.

3. 950   230
4. −215   145

Closure

Ask students to define the following. What are coterminal angles?   See page 635.   What are quadrantal angles? See page 633.   In what quadrant does the terminal side of an angle lie if the angle has a measure

1. between 0 and 90?   I
2. between 270 and 360?   IV
3. between 180 and 270?   III
4. between 90 and 180?   II

EXAMPLE 4   Find two angles, one with a positive measure and the other with a negative measure, that are coterminal with an angle of measure 40.

Solution   Use the formula $\theta_1 = \theta_2 + n \cdot 360$. Assign values to *n*.

positive: $\theta_1 = 40 + 2 \cdot 360$ negative: $\theta_1 = 40 + (-1) \cdot 360$
        $= 40 + 720 = 760$ $= 40 - 360 = -320$

Thus, an angle of measure 760 and an angle of measure −320 are coterminal with an angle of measure 40. Many other answers are also possible, depending on the choice of values for *n*.

EXAMPLE 5   Find the least positive measure of an angle that is coterminal with each angle. Then draw the angle.

a. angle with measure 855   **b.** angle with measure −320

Plan   Add integral multiples of 360, such as ±360, ±720, ±1,080, and ±1,440, to the given angle measure to get the least positive measure.

Solutions   **a.** 855 − 720 = 135   **b.** −320 + 360 = 40

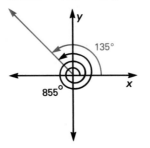 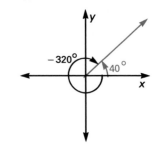

Classroom Exercises

For each angle measure, identify the quadrant in which each angle in standard position terminates, or, if the angle is quadrantal, describe the location of the terminal side.

1. 90   **2.** 160 II   **3.** −80 IV   **4.** 200 III   **5.** −180   **6.** −320 I
    Positive *y*-axis   Negative *x*-axis

For each angle measure, find two angle measures, one positive and one negative, so that all three angles are coterminal. Answers will vary. Possible answers are given.

7. 35 395,   **8.** −315   **9.** 180   **10.** −135   **11.** 160   **12.** −240
    −325   45, −675   540, −180   225, −495   520, −200   120, −600

636   Chapter 17 Trigonometric Functions

Additional Example 4

Find two angles, one with a positive measure and the other with a negative measure, that are coterminal with an angle measure of 70.   Answers may vary. One answer is 430 and −290.

Additional Example 5

Find the least positive measure of an angle that is coterminal with each angle. Then draw the angle.

a. angle with measure 920

b. angle with measure −210

Written Exercises

Draw the angle in standard position that corresponds to each angle measure. If the angle is quadrantal, identify the axis along which the terminal side lies. If not, identify the quadrant containing the terminal side.

1. 115   **2.** -220 **3.** -90 **4.** -306 **5.** 180   **6.** 280

7. 270   **8.** -130 **9.** 350   **10.** 220   **11.** 190   **12.** -25

For each angle measure, find two angle measures, one positive and one negative, so that all three angles are coterminal.

13. 30   **14.** 205   **15.** -130 **16.** -200 **17.** -310 **18.** 90

19. -50 **20.** -270 **21.** 320   **22.** -210 **23.** -44 **24.** 290

Find the least positive measure of an angle that is coterminal with the angle of given measure. Draw the angle.

25. 405   **26.** -275 **27.** 540   **28.** 810   **29.** 1,200   **30.** $-1,260$

31. Find a general formula for the measure of any angle θ such that its terminal side will lie in the second quadrant.

32. Find a general formula for the measure of any angle θ such that an angle with measure one-fourth as large will have its terminal side in the third quadrant. $\theta = x + 1{,}440n$, where $720 < x < 1{,}080$ and n is an integer.

Midchapter Review

In Exercises 1–8, refer to the figure at the right. Find each trigonometric ratio to four decimal places. *17.1, 17.2*

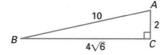

1. $\sin B$ 0.2000   **2.** $\cos B$ 0.9798   **3.** $\tan A$ 4.8990   **4.** $\cot B$ 4.8990

5. $\sin A$ 0.9798   **6.** $\csc B$ 5.0000   **7.** $\sec A$ 5.0000   **8.** $\cot A$ 0.2041

Find each trigonometric ratio. Use the tables on pages 792–796. *17.2*

9. $\sin 51$ 0.7771   **10.** $\cot 14$ 4.0108   **11.** $\csc 10$ 5.7588   **12.** $\cos 38°50'$ 0.7790

13. From a point in the middle of a park, a blimp is sighted in the distance. The measure of the angle of elevation to the blimp is $16°30'$. A range finder indicates that the blimp is 6,200 ft away. Find the altitude of the blimp to three significant digits. *17.3*   1,760 ft

14. Find the measures of two angles, one positive and one negative, that are coterminal with an angle that has a measure of -45. *17.4*   315, -405

17.4 Angles of Rotation   **637**

▮▮FOLLOW UP

Guided Practice

Classroom Exercises 1–12

Independent Practice

Ⓐ Ex. 1–24, Ⓑ Ex. 25–30, Ⓒ Ex. 31, 32

Basic: WE 1–27 odd, Midchapter Review

Average: WE 3–29 odd, Midchapter Review

Above Average: WE 5–31 odd, Midchapter Review

Additional Answers

Written Exercises

1. II

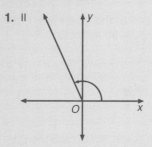

2. II

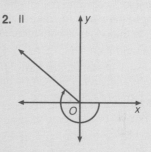

See page 642 for the answers to Written Ex. 3–10. See pages 658–659 for the answers to Written Ex. 11–31.

Enrichment

On the chalkboard, draw this picture of two wheels driven by a single belt.

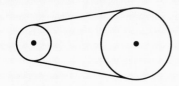

Discuss the concept of angular velocity. For example, if a wheel rotates 10 times per second, the angular velocity is 600/min. Now challenge the students with this question.

If the larger wheel has a radius of 1 m and the smaller wheel has a radius of 0.5 m, how does the angular velocity of the smaller wheel compare with the angular velocity of the larger wheel? The angular velocity of the smaller wheel will be twice the angular velocity of the larger wheel.

The students may note that in general the angular velocities of the two wheels vary inversely with the radii.

637

▰▰GETTING STARTED

Prerequisite Quiz

Give the signs of x and y for a point in the given quadrant.

1. first $(+,+)$
2. second $(-,+)$
3. third $(-,-)$
4. fourth $(+,-)$

Find the distance of the given point from the origin. Simplify all radicals.

5. $P(3,4)$ 5
6. $P(2,-4)$ $2\sqrt{5}$
7. $P(-6,6)$ $6\sqrt{2}$

Motivator

Ask students to graph $P(3,4)$, and then to form a right triangle by drawing a line from P perpendicular to the x-axis at point A, and another from P to the origin. Have students find PO and PA, and then sin $\angle POA$. $5,4;\frac{4}{5}$
If OP is called r, ask students how the sine of an angle can be defined in terms of y and r rather than in terms of opposite and hypotenuse. $\sin\theta = \frac{y}{r}$

17.5 Sines of Angles of Any Measure

Objectives   To find the sine of an angle given the coordinates of a point on the terminal side of the angle
To find the sines of angles of any measure

In the study of electricity it is necessary to work with formulas such as $E = E_{max} \sin\theta$. Here, E is the changing electromotive force (emf) created by a generator, and E_{max} is the maximum emf. However, θ is *not* the measure of an acute angle of a right triangle. In this lesson, you will learn how to define the sine ratio for angle measures that are not related to right triangles.

The sine of an angle has been defined in terms of an acute angle of a right triangle as follows.

$$\sin A = \frac{\text{length of leg opposite } \angle A}{\text{length of hypotenuse}}, \text{ or } \frac{\text{opp}}{\text{hyp}}$$

This definition can be altered to apply to an angle terminating in any quadrant.

EXAMPLE 1   Find the sine of an angle whose terminal side contains the point $P(3,4)$.

Plan   Form a right triangle by drawing both $\overline{OP}$, of length r, and $\overline{PQ}$, perpendicular to the x-axis. Use the triangle to find the sine.

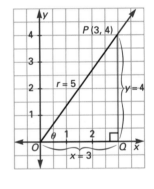

Solution $OQ = 3$ (x-coordinate of $P(3,4)$)

$QP = 4$ (y-coordinate of $P(3,4)$)

Find r, the length of the hypotenuse of the right triangle. The Pythagorean relation gives the following.

$$r^2 = 3^2 + 4^2 = 9 + 16 = 25$$
$$r = \sqrt{25} = 5$$

Thus, $\sin\theta = \frac{\text{opp}}{\text{hyp}} = \frac{4}{5}$.

Highlighting the Standards

Standard 8b: With an understanding of how angles of rotation are defined, students now deduce the sine values of these angles.

Additional Example 1

Find the sine of an angle whose terminal side contains the point $P(-8,-6)$.
$-\frac{3}{5}$

Notice in Example 1 that $\sin \theta = \frac{4}{5} = \frac{\text{opp}}{\text{hyp}}$, or $\frac{y\text{-coordinate of } P}{r}$.

Definition

Sine of an Angle of Any Measure

For any angle θ in standard position, with terminal side containing the point $P(x,y)$ and $r =$ distance from P to the origin, the following is true.

$$\sin \theta = \frac{y}{r} = \frac{y}{\sqrt{x^2 + y^2}}$$

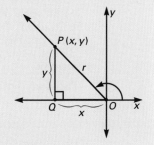

The right triangle formed by drawing a perpendicular segment from point P on a ray $\overrightarrow{OP}$ to the x-axis is called a **reference triangle**. In the figure, triangle OPQ is a reference triangle.

The sine of an angle is now no longer restricted to acute angles of right triangles. Note also that while the distance r is always positive, y and x may be positive, negative, or zero.

EXAMPLE 2 Find the sine of the angle whose terminal side contains the given point. Simplify all radicals.

a. $P(-3,4)$ **b.** $P(-1,-2)$ **c.** $P(3,-3)$

Solutions

a.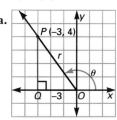

$r = \sqrt{x^2 + y^2}$
$r = \sqrt{(-3)^2 + 4^2}$
$r = \sqrt{25} = 5$
$\sin \theta = \frac{y}{r}$
$\quad = \frac{4}{5}$

b.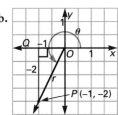

$r = \sqrt{x^2 + y^2}$
$r = \sqrt{(-1)^2 + (-2)^2}$
$r = \sqrt{5}$
$\sin \theta = \frac{y}{r}$
$\quad = \frac{-2}{\sqrt{5}} = -\frac{2\sqrt{5}}{5}$

c.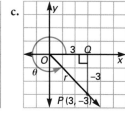

$r = \sqrt{x^2 + y^2}$
$r = \sqrt{3^2 + (-3)^2}$
$r = \sqrt{18} = 3\sqrt{2}$
$\sin \theta = \frac{y}{r}$
$\quad = \frac{-3}{3\sqrt{2}} = -\frac{\sqrt{2}}{2}$

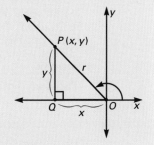

Lesson Note

Since students will be concerned for the first time with the sign of a trigonometric ratio, only the sine ratio is developed here. In the next lesson, the reference triangle will be used to define the remaining five ratios.

The calculator is not used in this lesson. Students must first understand the concept of the reference angle, so that later on they can find angle measures that correspond to a given ratio. For example, given $\sin \theta = 0.7070$, the calculator will display only one value for θ. The concept of reference angle is needed in order to find *both* angles between 0 and 360.

Math Connections

A Formula for the Sine: Students who are familiar with the Pi-notation introduced in the *Math Connections* for Lessons 15.3 and 15.7 may be interested in the following infinite product, which represents the sine of any number πx, where πx is the *radian* measure of an angle (see Lesson 18.5).

$$\sin \pi x = \pi x \prod_{n=1}^{\infty} \left(1 - \frac{x^2}{n^2}\right)$$

For example, if $\pi x = \frac{\pi}{3}$ (that is, 60°) then

$$\sin \frac{\pi}{3} = \frac{\pi}{3}\left(1 - \frac{1}{9}\right)\left(1 - \frac{1}{36}\right)\left(1 - \frac{1}{81}\right)\cdots,$$

which, however, will be found to converge quite slowly when checked against the table value of 0.8660.

Additional Example 2

Find the sine of the angle whose terminal side contains the given point. Simplify all radicals.

a. $P(-12,-5)$ $-\frac{5}{13}$

b. $P(-6,2)$ $\frac{\sqrt{10}}{10}$

c. $P(4,-4)$ $-\frac{\sqrt{2}}{2}$

Analysis: Have students consider the four reference triangles corresponding to the four angles in standard position that terminate at the points (3,4), (−3,4), (−3,−4), and (3,−4). These triangles are not useful in relating the four angles to a single reference angle unless the four triangles happen to be congruent to one another. Ask students what principle or theorem from geometry justifies concluding that the triangles are congruent.   They should see that it is the SAS Postulate.

Common Error Analysis

Error: When finding the sine of an angle, students frequently become so preoccupied with finding the reference angle that they forget the sign of the ratio.

Have the students label the signs of the legs in the reference triangle *before* computing the measure of the reference angle.

Checkpoint

Find the sine of the angle whose terminal side contains the given point. Simplify all radicals.

1. $P(-12,9)$ $\dfrac{3}{5}$
2. $P(-4,-4)$ $-\dfrac{\sqrt{2}}{2}$
3. $P(-2,-6)$ $-\dfrac{3\sqrt{10}}{10}$

Find the sine of each angle with the given degree measure.

4. 155   0.4226
5. 318   −0.6691
6. −100° 20′   −0.9838

In the figure at the right below, acute angle POQ is called the **reference angle** of reference triangle POQ. Note that trigonometric ratios of angles with degree measures greater than 90 do not appear in the table of trigonometric ratios. However, the sine of an angle such as 240 can be found in terms of its reference angle as follows.

(1) An angle of measure 240 terminates in Quadrant III. (2) The reference angle has degree measure 240 − 180 = 60. (3) The y-coordinate in Quadrant III is *negative*.

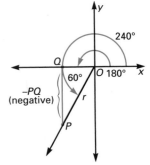

Therefore, $\sin 240 = \dfrac{-PQ}{r}$

$= -\sin 60$

$= -0.8660.$

EXAMPLE 3   Find $\sin 110°20'$.

Solution   Form the reference triangle.

Find the measure of the reference angle.

$180° - 110°20':$
$\qquad\quad 179°\ 60'$
$\qquad\underline{-110°\ 20'}$
$\qquad\quad\ \ 69°\ 40'$

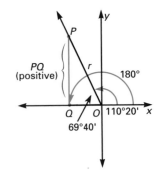

Determine the correct sign.
Since y is positive in Quadrant II,
$\sin 110°20' = +\sin 69°40' = 0.9377.$

EXAMPLE 4   Find $\sin 310°\ 40'$.

Solution   Find the measure of the reference angle.

$\qquad\quad 359°\ 60'$
$\qquad\underline{-310°\ 40'}$
$\qquad\quad\ \ 49°\ 20'$

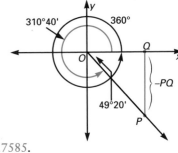

Since y is negative in Quadrant IV,
$\sin 310°40' = -\sin 49°20' = -0.7585.$

Additional Example 3

Find sin 232° 50′.   −0.7969

Additional Example 4

Find sin 319° 40′.   −0.6472

Use the procedure just shown to find the sine of a negative angle measure or an angle measure that is larger than 360.

EXAMPLE 5 **a.** Find sin (−132).　　　　　**b.** Find sin 415.

Solutions　**a.**

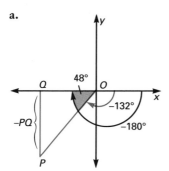

reference angle:
$|-180| - |-132|$
$= 180 - 132$, or 48
The y-coordinate is *negative*.
Thus, $\sin(-132) = -\sin 48$
$= -0.7431$

b.

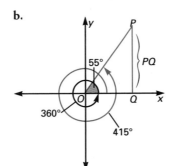

reference angle:
$415 - 360 = 55$

The y-coordinate is *positive*.
Thus, $\sin 415 = \sin 55$
$= 0.8192$

Focus on Reading

Indicate whether each of the following is *sometimes* true, *always* true, or *never* true.

1. The sine of an angle in Quadrant I is positive. Always
2. The sine of a negative angle of rotation is negative. Sometimes
3. The reference angle of any nonquadrantal angle of rotation is acute. Always
4. The sine of an angle of rotation with terminal side in Quadrant II is negative. Never

Classroom Exercises

Find the sine of the angle whose terminal side contains the given point. Simplify all radicals.

1. $P(-4,3)$ $\frac{3}{5}$　　　**2.** $P(6,8)$ $\frac{4}{5}$　　　**3.** $P(1,-1)$ $\frac{-\sqrt{2}}{2}$

3.

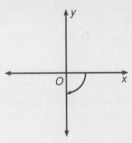

quadrantal: negative y-axis

4.

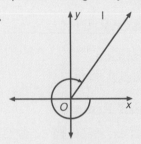

5.

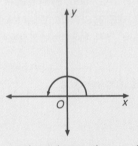

quadrantal: negative x-axis

6.

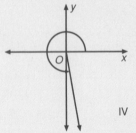

Give the measure of the reference angle for each angle measure.

4. 130 50   **5.** 200 20   **6.** 350 10   **7.** 400 40   **8.** −40 40   **9.** −120 60
10–15. Find the sine of each angle with the measures given in Exercises 4–9 above.
     10. 0.7660   **11.** −0.3420   **12.** −0.1736   **13.** 0.6428   **14.** −0.6428   **15.** −0.8660

Written Exercises

Find the sine of the angle whose terminal side contains the given point.
Simplify all radicals.

1. $P(4,3)$ $\frac{3}{5}$ **2.** $P(-6,8)$ $\frac{4}{5}$ **3.** $P(8,6)$ $\frac{3}{5}$ **4.** $P(5,-12)$ $\frac{-12}{13}$

5. $P(-12,-5)$ $\frac{-5}{13}$ **6.** $P(-1,3)$ **7.** $P(-2,-1)$ **8** $P(5,-2)$

9. $P(-1,-1)$ $\frac{-\sqrt{2}}{2}$ **10.** $P(6,-2)$ $\frac{-\sqrt{10}}{10}$ **11.** $P(4,-4)$ $\frac{-\sqrt{2}}{2}$ **12.** $P(-2,6)$ $\frac{3\sqrt{10}}{10}$

Find the sine of each angle with the given degree measure.

13. 175 0.8716	**14.** 230 −0.7660	**15.** 79 0.9816	**16.** 329 −0.5150
17. 100 0.9848	**18.** 129 0.7771	**19.** −45 −0.7071	**20.** −155 −0.4226
21. −205 0.4226	**22.** 370 0.1736	**23.** −86 −0.9976	**24.** −350 0.1736
25. 130°40′ 0.7585	**26.** −100°50′	**27.** 210°40′ −0.5100	**28.** 320°20′ −0.6383
29. 95°30′ 0.9954	**30.** −170°20′	**31.** 254°50′ −0.9652	**32.** −205°10′ 0.4253
33. 480 0.8660	**34.** −705 0.2588	**35.** −1,130 −0.7660	**36.** 1,015 −0.9063

37. Find the sine of an angle whose terminal side contains $P(a-1,a+1)$. $\dfrac{(a+1)\sqrt{2a^2+2}}{2a^2+2}$

38. For the angle described in Exercise 37, give the restrictions on a if the angle is to terminate in the second quadrant. $-1 < a < 1$

Mixed Review

Find each of the following. Use the tables on pages 792–796. *17.2*

1. sec 75 3.8637   **2.** cos 49 0.6561   **3.** csc 82 1.0098   **4.** cot 41 1.1504   **5.** tan 73°40′
                                                               3.4124

▰▰ / *Brainteaser*

A ship is sailing on a straight course *RST*. When the ship is at point *R*, the captain sights a lighthouse *L* and notes that m ∠*LRT* is 36°30′. After sailing for 8 mi to point *S*, the captain then observes that m ∠*LST* = 73. What is the distance from *S* to the lighthouse *L*? 8 mi

7.

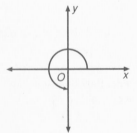

quadrantal: negative y-axis

8.

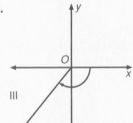

9.

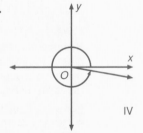

10.

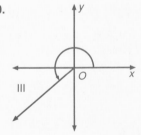

17.6 Trigonometric Ratios of Angles of Any Measure

Objectives

To find the six trigonometric ratios of an angle given the coordinates of a point on the terminal side of the angle

To find the six trigonometric ratios of angles of any measure

In Lesson 17.5, the sine ratio was defined for an angle of rotation terminating in any quadrant. The remaining five trigonometric ratios can also be defined given the coordinates of a point of the terminal side of an angle of rotation.

EXAMPLE 1 Find the six trigonometric ratios for the angle whose terminal side contains the point $P(8,6)$.

Plan Form the reference triangle and find the lengths of its sides. Then use the definitions of trigonometric ratios.

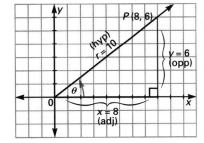

Solution

$adj = x = 8$ and $opp = y = 6$

$hyp = r = \sqrt{x^2 + y^2}$

$\quad = \sqrt{8^2 + 6^2}$

$\quad = \sqrt{100} = 10$

$\sin \theta = \dfrac{opp}{hyp} = \dfrac{y}{r} = \dfrac{6}{10} = \dfrac{3}{5}$ $\csc \theta = \dfrac{hyp}{opp} = \dfrac{r}{y} = \dfrac{10}{6} = \dfrac{5}{3}$

$\cos \theta = \dfrac{adj}{hyp} = \dfrac{x}{r} = \dfrac{8}{10} = \dfrac{4}{5}$ $\sec \theta = \dfrac{hyp}{adj} = \dfrac{r}{x} = \dfrac{10}{8} = \dfrac{5}{4}$

$\tan \theta = \dfrac{opp}{adj} = \dfrac{y}{x} = \dfrac{6}{8} = \dfrac{3}{4}$ $\cot \theta = \dfrac{adj}{opp} = \dfrac{x}{y} = \dfrac{8}{6} = \dfrac{4}{3}$

Definition

Trigonometric Ratios of an Angle of Any Measure

For any angle of rotation θ, with point $P(x,y)$ on the terminal side and located at a distance $r > 0$ from the origin, the trigonometric ratios of angle θ are defined as follows.

$\sin \theta = \dfrac{y}{r}$ $\cos \theta = \dfrac{x}{r}$ $\tan \theta = \dfrac{y}{x}, x \neq 0$

$\csc \theta = \dfrac{r}{y}, y \neq 0$ $\sec \theta = \dfrac{r}{x}, x \neq 0$ $\cot \theta = \dfrac{x}{y}, y \neq 0$

Teaching Resources

Quick Quizzes 130
**Reteaching and Practice
   Worksheets** 130
Transparencies 37A, 37B

▬▬GETTING STARTED

Prerequisite Quiz

1. Find the sine of the angle whose terminal side contains the point $P(8,6)$. $\dfrac{3}{5}$

Find the sine of each angle.

2. 160   0.3420
3. 313 -0.7314
4. $-170°\ 20'$ -0.1679

Motivator

In the last lesson, students learned to define $\sin A$ as $\dfrac{y}{r}$ rather than $\dfrac{opp}{hyp}$. This enabled them to find $\sin A$ even if A were not merely an acute angle. Now have students guess how to extend the definitions of the other five basic trigonometric ratios.

▬▬TEACHING SUGGESTIONS

Lesson Note

Emphasize again the importance of labeling the signs of the legs in the reference triangle *before* computing the measure of the reference angle. Otherwise students tend to disregard the sign of the trigonometric ratio.

Additional Example 1

Find the six trigonometric ratios for the angle θ whose terminal side contains the point $P(2,4)$. Simplify all radicals.

$\sin \theta = \dfrac{2\sqrt{5}}{5}$ $\csc \theta = \dfrac{\sqrt{5}}{2}$

$\cos \theta = \dfrac{\sqrt{5}}{5}$ $\sec \theta = \dfrac{\sqrt{5}}{1}$

$\tan \theta = \dfrac{2}{1}$ $\cot \theta = \dfrac{1}{2}$

Highlighting the Standards

Standard 8b: This is a natural extension of the preceding lesson. The examples lead students to new understandings for the values of all trigonometric functions for any angle.

Math Connections

Historical Note: The six trigonometric ratios were formulated in the way that we know them only in modern times; their origins, however, date back to antiquity. Ptolemy, who lived in the second century, created a table of chords of arcs that ranged in measure from $\frac{1}{2}$ to 180 in increments of $\frac{1}{2}$. This is virtually the same as a table of sines with measures from $\frac{1}{4}$ to 90 in increments of $\frac{1}{4}$. It has not been possible to attribute specific times to the tangent, secant, and cosecant ratios, although some evidence suggests that the concepts of secant and cosecant were known in Asia by the time of the Arabian scholar Thabit Ibn-Qurra (826–901).

Critical Thinking Questions

Analysis: Ask students to experiment with reference triangles of various sizes to write inequalities that describe the range of possible values for the six trigonometric ratios. (This means revising and extending the inequalities created in the *Critical Thinking Question* of Lesson 17.1.) For the sine, tangent, and secant, the inequalities are: $-1 \le \sin \theta \le 1$, $-\infty < \tan \theta < \infty$, $-\infty < \sec \theta \le -1$ or $1 \le \sec \theta < \infty$.

The cosine, cotangent, and cosecant have the same ranges as their respective cofunctions.

Common Error Analysis

Error: When finding the reference angle of a third quadrant angle, students tend to subtract from 270. Emphasize that in the reference triangle, one side of the reference angle is always on the *x*-axis. The measure of the reference angle is found as a difference from 180 or 360, never from 270 or 90.

EXAMPLE 2 Find the six trigonometric ratios for the angle whose terminal side contains the given point. Simplify all radicals.

a. $P(-4,3)$ **b.** $P(6,-2)$

Plan Form the reference triangle. Then, use the given values of x and y and the distance formula to find r.

Solutions **a.** 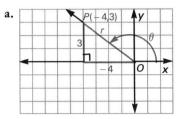 **b.**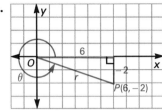

a.
$$r = \sqrt{x^2 + y^2} = \sqrt{(-4)^2 + 3^2}$$
$$= \sqrt{25} = 5$$

$\sin \theta = \dfrac{y}{r} = \dfrac{3}{5}$

$\cos \theta = \dfrac{x}{r} = \dfrac{-4}{5} = -\dfrac{4}{5}$

$\tan \theta = \dfrac{y}{x} = \dfrac{3}{-4} = -\dfrac{3}{4}$

$\cot \theta = \dfrac{x}{y} = \dfrac{-4}{3} = -\dfrac{4}{3}$

$\sec \theta = \dfrac{r}{x} = \dfrac{5}{-4} = -\dfrac{5}{4}$

$\csc \theta = \dfrac{r}{y} = \dfrac{5}{3}$

b.
$$r = \sqrt{x^2 + y^2} = \sqrt{6^2 + (-2)^2}$$
$$= \sqrt{40} = \sqrt{4 \cdot 10} = 2\sqrt{10}$$

$\sin \theta = \dfrac{-2}{2\sqrt{10}} = \dfrac{-1 \cdot \sqrt{10}}{\sqrt{10} \cdot \sqrt{10}}$

$\qquad = -\dfrac{\sqrt{10}}{10}$

$\cos \theta = \dfrac{6}{2\sqrt{10}} = \dfrac{3\sqrt{10}}{10}$

$\tan \theta = \dfrac{-2}{6} = -\dfrac{1}{3}$

$\cot \theta = \dfrac{6}{-2} = -\dfrac{3}{1}$, or -3

$\sec \theta = \dfrac{2\sqrt{10}}{6} = \dfrac{\sqrt{10}}{3}$

$\csc \theta = \dfrac{2\sqrt{10}}{-2} = \dfrac{\sqrt{10}}{-1}$, or $-\sqrt{10}$

As with the sine ratio, all trigonometric ratios of angles with measures greater than 90 or less than 0 can be found in terms of acute angles of reference triangles. Thus, to find a trigonometric ratio of an angle terminating in any quadrant, perform the following steps.

1. Find the corresponding trigonometric ratio of the reference angle using the tables on pages 792–796.
2. Assign the correct sign to the result; this will depend upon the signs of x and y for the given quadrant.

Additional Example 2

Find the six trigonometric ratios for the angle θ whose terminal side contains the given point. Simplify all radicals.

a. $P(-6,-8)$

$\sin \theta = -\dfrac{4}{5}$ $\cot \theta = \dfrac{3}{4}$

$\cos \theta = -\dfrac{3}{5}$ $\sec \theta = -\dfrac{5}{3}$

$\tan \theta = \dfrac{4}{3}$ $\csc \theta = -\dfrac{5}{4}$

b. $P(-4,2)$

$\sin \theta = \dfrac{\sqrt{5}}{5}$

$\cos \theta = -\dfrac{2\sqrt{5}}{5}$

$\tan \theta = -\dfrac{1}{2}$

$\cot \theta = -2$

$\sec \theta = -\dfrac{\sqrt{5}}{2}$

$\csc \theta = \sqrt{5}$

EXAMPLE 3 Find the six trigonometric ratios of 143°20′.

Solution Form the reference triangle. Find the measure of the reference angle.

$$180° - 143°20': \begin{array}{r} 179°\ 60' \\ -143°\ 20' \\ \hline 36°\ 40' \end{array}$$

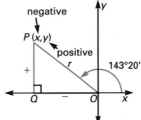

Determine the correct sign for each ratio.
In Quadrant II, x is negative and y is positive.

$\sin 143°20' = \dfrac{PQ}{r} = \sin 36°40' = 0.5972$

$\cos 143°20' = \dfrac{-OQ}{r} = -\cos 36°40' = -0.8021$

$\tan 143°20' = \dfrac{PQ}{-OQ} = -\tan 36°40' = -0.7445$

$\csc 143°20' = \dfrac{r}{PQ} = \csc 36°40' = 1.6746$

$\sec 143°20' = \dfrac{r}{-OQ} = -\sec 36°40' = -1.2467$

$\cot 143°20' = \dfrac{-OQ}{PQ} = -\cot 36°40' = -1.3432$

When the trigonometric ratios for angles of rotation are computed, it is essential first to establish the correct sign of x and y (NOTE: r is always positive.)

Sign Combinations of x and y for Angles Terminating in any Quadrant

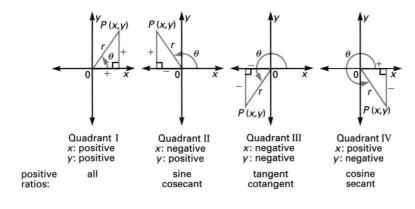

	Quadrant I	Quadrant II	Quadrant III	Quadrant IV
	x: positive	x: negative	x: negative	x: positive
	y: positive	y: positive	y: negative	y: negative
positive ratios:	all	sine cosecant	tangent cotangent	cosine secant

---

Checkpoint

Find the indicated trigonometric ratio for the angle whose terminal side contains the given point. Simplify all radicals.

1. sine; $P(-12,-5)$ $-\dfrac{5}{13}$

2. cosine; $P(-2,8)$ $-\dfrac{\sqrt{17}}{17}$

3. secant; $P(0,-6)$ Undef

Find each trigonometric ratio.

4. cos 132 -0.6691
5. tan (-270) Undef
6. sec 203° 20′ -1.0891

Closure

Have students define the 6 trigonometric ratios for an angle of rotation terminating in any quadrant.   See page 643 for definitions. Ask students how they determine the 6 trigonometric ratios for an angle whose terminal side contains a given point. Have them complete each statement.

1. You must first determine the ⌐?⌐.
   reference triangle.
2. You must then determine the correct ⌐?⌐.
   sign for each trigonometric ratio.

◼◼◼FOLLOW UP

Guided Practice

Classroom Exercises 1–20

Independent Practice

Ⓐ Ex. 1–28, Ⓑ Ex. 29–48, Ⓒ Ex. 49–53

Basic: WE 1–35 odd
Average: WE 13–47 odd
Above Average: WE 17–53 odd

Additional Example 3

Find the six trigonometric ratios of 207° 40′.

sin 207° 40′ = −0.4643
cos 207° 40′ = −0.8857
tan 207° 40′ = 0.5243
csc 207° 40′ = −2.1537
sec 207° 40′ = −1.1291
cot 207° 40′ = 1.9074

Classroom Exercises

Exercises 9–20 are in the following order: sin; cos; tan; csc; sec; cot.

9. 0.6428; −0.7660; −0.8391; 1.5557; −1.3054; −1.1918
10. −1; 0; undefined; −1; undefined; 0
11. −0.8480; −0.5299; 1.6003; −1.1792; −1.8871; 0.6249
12. −0.7660; 0.6428; −1.1918; −1.3054; 1.5557; −0.8391
13. 1; 0; undefined; 1; undefined; 0
14. 0.9063; 0.4226; 2.1445; 1.1034; 2.3662; 0.4663
15. −0.5000; −0.8660; 0.5774; −2.0000; −1.1547; 1.7321
16. 1; 0; undefined; 1; undefined; 0
17. $\frac{3}{5}; \frac{-4}{5}; \frac{-3}{4}; \frac{5}{3}; \frac{-5}{4}; \frac{-4}{3}$
18. $\frac{12}{13}; \frac{-5}{13}; \frac{-12}{5}; \frac{13}{12}; \frac{-13}{5}; \frac{-5}{12}$
19. $\frac{\sqrt{2}}{2}; \frac{\sqrt{2}}{2}; 1; \sqrt{2}; \sqrt{2}; 1$
20. $\frac{-3\sqrt{10}}{10}; -\frac{\sqrt{10}}{10}; 3; -\frac{\sqrt{10}}{3}; -\sqrt{10}; \frac{1}{3}$

Written Exercises

Exercises 1–40 are in the following order: sin; cos; tan; csc; sec; cot.

1. $\frac{4}{5}; \frac{3}{5}; \frac{4}{3}; \frac{5}{4}; \frac{5}{3}; \frac{3}{4}$
2. $\frac{-3}{5}; \frac{4}{5}; \frac{-3}{4}; \frac{-5}{3}; \frac{5}{4}; \frac{-4}{3}$
3. $\frac{-12}{13}; \frac{-5}{13}; \frac{12}{5}; \frac{-13}{12}; \frac{-13}{5}; \frac{5}{12}$
4. $\frac{-3}{5}; \frac{-4}{5}; \frac{3}{4}; \frac{-5}{3}; \frac{-5}{4}; \frac{4}{3}$
5. $\frac{5}{13}; \frac{-12}{13}; \frac{-5}{12}; \frac{13}{5}; \frac{-13}{12}; \frac{-12}{5}$
6. $\frac{\sqrt{2}}{2}; \frac{-\sqrt{2}}{2}; -1; \sqrt{2}; -\sqrt{2}; -1$
7. $\frac{-\sqrt{5}}{5}; \frac{2\sqrt{5}}{5}; \frac{-1}{2}; -\sqrt{5}; \frac{\sqrt{5}}{2}; -2$
8. $\frac{\sqrt{10}}{10}; \frac{-3\sqrt{10}}{10}; \frac{-1}{3}; \sqrt{10}; \frac{-\sqrt{10}}{3}; -3$

EXAMPLE 4 Express csc 325 in terms of the cosecant of a reference angle.

Solution Reference angle: 360 − 325 = 35

csc 325 = $\frac{r}{y}$ ← y is negative.

Therefore, csc 325 = − csc 35.

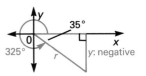

The six trigonometric ratios for quadrantal angles are defined in the same way as for nonquadrantal angles. However, as shown below, no reference triangles are needed for quadrantal angles.

To find sin 270 and tan 270, perform the following steps.
1. Draw the angle in standard position.
2. Choose any point on the terminal side, such as $P(0, -1)$.
3. Determine the x- and y-coordinates of P and the distance of P from the origin.
   x-coordinate of $P(0, -1)$: 0
   y-coordinate of $P(0, -1)$: −1
   Distance of $P(0, -1)$ from the origin: 1
4. sin 270 = $\frac{y}{r} = \frac{-1}{1} = -1$

   tan 270 = $\frac{y}{x} = \frac{-1}{0}$, which is *undefined*, since division by 0 is undefined.

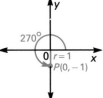

Therefore, sin 270 = −1 and tan 270 is undefined.

Classroom Exercises

Identify the quadrant in which each of the following angles terminates. For quadrantal angles, identify the two quadrants that are separated by the terminal side. Also, give the signs of x and y for the angle.

1. 140 II; −, +
2. −90 III, IV; 0, −
3. 238 III; −, −
4. −50 IV; +, −
5. 450 I, II; 0, +
6. 65 I; +, +
7. −150 III; −, −
8. −270 I, II; 0, +

9–16. Find the six trigonometric ratios for the angle measures of Exercises 1–8 above.

Find the six trigonometric ratios for each angle whose terminal side contains the given point. Simplify all radicals.

17. $P(-8, 6)$
18. $P(-5, 12)$
19. $P(3, 3)$
20. $P(-2, -6)$

Additional Example 4

Express tan 175 in terms of the tangent of a reference angle. tan 175 = −tan 5

Written Exercises

Find the six trigonometric ratios for the angle whose terminal side contains the given point. If the ratio is undefined, so indicate. Simplify all radicals.

1. $P(6,8)$ **2.** $P(4,-3)$ **3.** $P(-5,-12)$ **4.** $P(-8,-6)$

5. $P(-12,5)$ **6.** $P(-2,2)$ **7.** $P(4,-2)$ **8.** $P(-3,1)$

9. $P(-1,-3)$ **10.** $P(2,0)$ **11.** $P(4,-1)$ **12.** $P(0,-6)$

Find the six trigonometric ratios for each angle measure. If the ratio is undefined, so indicate.

13. 105 **14.** 255 **15.** 329 **16.** -49

17. -112 **18.** 55 **19.** -235 **20.** 186

21. 313 **22.** 420 **23.** -130 **24.** -223

25. 0 **26.** 180 **27.** -90 **28.** -270

29. $112°40'$ **30.** $312°20'$ **31.** $-120°50'$ **32.** $115°30'$

Find the six trigonometric ratios for the angle whose terminal side contains the given point. Simplify all radicals.

33. $P(2,-2\sqrt{3})$ **34.** $P(-\sqrt{2},-\sqrt{2})$ **35.** $P(-\sqrt{7},-3)$ **36.** $P(\sqrt{5},-\sqrt{5})$

37. $P(\sqrt{2},\sqrt{6})$ **38.** $P(-2\sqrt{5},\sqrt{5})$ **39.** $P(-4,4\sqrt{3})$ **40.** $P(6\sqrt{2},-2\sqrt{7})$

Express each trigonometric ratio in terms of the same trigonometric ratio of a reference angle.

41. $\sin 140$ sin 40   **42.** $\csc(-126)$ $-\csc 54$ **43.** $\tan 336$ $-\tan 24$ **44.** $\cot(-220)$ $-\cot 40$

45. $\sec(-50)$ sec 50   **46.** $\tan 185$ tan 5   **47.** $\sin 115°40'$ sin 64°20'   **48.** $\csc 290°20'$ $-\csc 69°40'$

If $0 < \theta < 90$, express each of the six trigonometric ratios for the angles below in terms of θ.

49. $180 - \theta$ **50.** $360 - \theta$ **51.** $180 + \theta$ **52.** $90 + \theta$

53. Is the expression $2n \cdot 90$ the measure of a quadrantal angle for all positive integral values of n? Which angles? Give an expression that will define all remaining measures of quadrantal angles for n, a positive integer.   Yes; 180, 360; $(2n - 1) \cdot 90$

Mixed Review

Solve each equation.   *8.7, 9.5, 10.4, 13.3*

1. $x^2 - 4x = -6$ $2 \pm \sqrt{2}i$ **2.** $x^3 - 27 = 0$ $3, \dfrac{-3 \pm 3\sqrt{3}i}{2}$ **3.** $4^{2x-3} = 64$ 3   **4.** $\log_8 x = \dfrac{2}{3}$ 4

17.6 Trigonometric Ratios of Angles of Any Measure   **647**

9. $\dfrac{-3\sqrt{10}}{10}$; $\dfrac{-\sqrt{10}}{10}$; 3; $\dfrac{-\sqrt{10}}{3}$; $-\sqrt{10}$; $\dfrac{1}{3}$

10. 0; 1; 0; undefined; 1; undefined

11. $\dfrac{-\sqrt{17}}{17}$; $\dfrac{4\sqrt{17}}{17}$; $\dfrac{-1}{4}$; $-\sqrt{17}$; $\dfrac{\sqrt{17}}{4}$; -4

12. -1; 0; undefined; -1; undefined; 0

13. 0.9659; -0.2588; -3.7321; 1.0353; -3.8637; -0.2679

14. -0.9659; -0.2588; 3.7321; -1.0353; -3.8637; 0.2679

15. -0.5150; 0.8572; -0.6009; -1.9416; 1.1666; -1.6643

16. -0.7547; 0.6561; -1.1504; -1.3250; 1.5243; -0.8693

17. -0.9272; -0.3746; 2.4751; -1.0785; -2.6695; 0.4040

18. 0.8192; 0.5736; 1.4281; 1.2208; 1.7434; 0.7002

19. 0.8192; -0.5736; -1.4281; 1.2208; -1.7434; -0.7002

20. -0.1045; -0.9945; 0.1051; -9.5668; -1.0055; 9.5144

21. -0.7314; 0.6820; -1.0724; -1.3673; 1.4663; -0.9325

22. 0.8660; 0.5000; 1.7321; 1.1547; 2.0000; 0.5774

23. -0.7660; -0.6428; 1.1918; -1.3054; -1.5557; 0.8391

24. 0.6820; -0.7314; -0.9325; 1.4663; -1.3673; -1.0724

25. 0; 1; 0; undefined; 1; undefined

26. 0; -1; 0; undefined; -1; undefined

27. -1; 0; undefined; -1; undefined; 0

28. 1; 0; undefined; 1; undefined; 0

29. 0.9228; -0.3854; -2.3945; 1.0837; -2.5949; -0.4176

30. -0.7392; 0.6734; -1.0977; -1.3527; 1.4849; -0.9110

31. -0.8587; -0.5125; 1.6753; -1.1646; -1.9511; 0.5969

32. 0.9026; -0.4305; -2.0965; 1.1079; -2.3228; -0.4770

See page 657 for the answers to Written Ex. 33–40, 49–52.

Enrichment

Write this equation and its solution on the chalkboard for $0 < x < 360$.

$2 \sin x - 1 = 0$
$\quad 2 \sin x = 1$
$\quad\quad \sin x = \dfrac{1}{2}$
$\quad\quad\quad x = 30$ or 150

The final step will require the use of the table or a calculator.

Have the students solve these equations for x to the nearest degree, where $0 < x < 360$.

1. $3 \sin x - 3 = 0$ 90

2. $\tan x - 2 = 0$ 63 or 243

3. $\sin x = \cos x$ 45 or 225

4. $\sin^2 x - 1 = 0$ 90 or 270

5. $3 \cos x + 4 = 0$ No solution

Note that this activity provides informal work with trigonometric equations, which will be studied in Chapter 19.

647

▨▨▨ GETTING STARTED

Prerequisite Quiz

Refer to triangle *ABC* with a right angle at *C*. Find the missing lengths.

1. $a = 4$, $b = 4$, $c = \underline{\ ?\ }$ $4\sqrt{2}$
2. $c = 2$, $a = 1$, $b = \underline{\ ?\ }$ $\sqrt{3}$
3. $b = 2\sqrt{3}$, $a = 2$, $c = \underline{\ ?\ }$ 4
4. $c = 2\sqrt{2}$, $b = 2$, $a = \underline{\ ?\ }$ 2
5. $a = 6$, $b = 6$, $c = \underline{\ ?\ }$ $6\sqrt{2}$

Motivator

Ask the students to draw an isosceles right triangle with legs of length 4. Have them give the measure of each acute angle. Each angle has a measure of 45. Have them use the Pythagorean relation to find the length of the hypotenuse. $4\sqrt{2}$ Ask them to give the sine of an acute angle. $\frac{\sqrt{2}}{2}$

Have them repeat the above steps for two more isosceles right triangles with legs of lengths 6; 10. Ask them to discuss what they now know about sin 45.

Objectives

To find the six trigonometric ratios of angles of rotation whose reference angles measure 60, 30, or 45

To evaluate expressions involving trigonometric ratios

There are three special reference angles whose trigonometric ratios can be found without referring to a table or calculator.

Recall that in an equilateral triangle, the measure of each angle is 60. The altitude $\overline{AD}$ bisects both the base $\overline{BC}$ and the vertex angle A. The altitude divides the equilateral triangle into two 30-60-90 triangles. In right triangle ABD, the leg opposite the angle measuring 30, $\overline{BD}$, has half the length of the hypotenuse $\overline{AB}$.

To find AD, use the Pythagorean relation.

$$(AB)^2 = (BD)^2 + (AD)^2$$
$$2^2 = 1^2 + (AD)^2$$
$$3 = (AD)^2$$
$$AD = \sqrt{3}$$

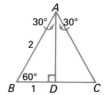

Thus, if an angle in standard position has its terminal side in the first quadrant and its reference triangle is a 30-60-90 triangle, it will be convenient to use one of the following figures, for which 2 is the distance of P from the origin.

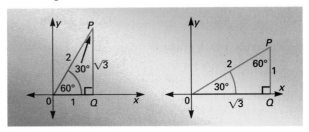

EXAMPLE 1   Find tan 30 and simplify.

Solution   Draw a reference triangle as at the right.

$$\tan 30 = \frac{y}{x} = \frac{1}{\sqrt{3}} = \frac{1 \cdot \sqrt{3}}{\sqrt{3} \cdot \sqrt{3}} = \frac{\sqrt{3}}{3}$$

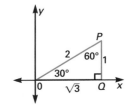

Highlighting the Standards

Standard 3a: After students work through the special angles in all quadrants, they are able to make conjectures in the Focus on Reading and Classroom Exercises.

Additional Example 1

Find csc 60 and simplify.

$\frac{2\sqrt{3}}{3}$

EXAMPLE 2   Find csc 120 and simplify.

Solution   Since the standard position of 120 has its terminal side in the second quadrant, draw a 30-60-90 reference triangle in the second quadrant.

$$\csc 120 = \frac{r}{y} = \frac{2}{\sqrt{3}} = \frac{2 \cdot \sqrt{3}}{\sqrt{3} \cdot \sqrt{3}} = \frac{2\sqrt{3}}{3}$$

If a right triangle is isosceles, then the legs have equal length and each of the acute angles has a measure of 45.

In the isosceles right triangle at the right, each leg has a length of 1. The Pythagorean relation can be used to find the length of the hypotenuse.

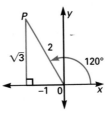

$$h^2 = 1^2 + 1^2 = 2, \text{ so } h = \sqrt{2}$$

Thus, if a reference angle has a measure of 45, it will be convenient to use the figure at the right for standard-position angles that terminate in the first quadrant.

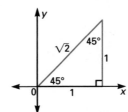

EXAMPLE 3   Find cos 45 and simplify.

Solution   The angle terminates in Quadrant I. Draw a reference triangle as shown, with legs each of length 1 and hypotenuse of length $\sqrt{2}$.

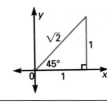

$$\cos 45 = \frac{x}{r} = \frac{1}{\sqrt{2}} = \frac{1 \cdot \sqrt{2}}{\sqrt{2} \cdot \sqrt{2}} = \frac{\sqrt{2}}{2}$$

EXAMPLE 4   Find csc (−135).

Solution   Draw and label the reference triangle in Quadrant III as shown at the right. Note that x and y are negative in Quadrant III.

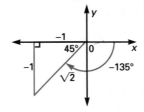

$$\csc (-135) = \frac{r}{y} = \frac{\sqrt{2}}{-1}, \text{ or } -\sqrt{2}$$

TEACHING SUGGESTIONS

Lesson Note

Emphasize the importance of memorizing the diagrams for the 45-45-90 and 30-60-90 triangles. Point out that the 45-45-90 triangle is easy to remember, since the legs opposite the two 45-degree angles have equal lengths. Using 1 for the length of each leg, it is easy to compute the hypotenuse by the Pythagorean relation.

Math Connections

Trigonometric Identities: Example 5 foreshadows one of the Pythagorean Identities of Chapter 19, namely, $\sin^2 \theta + \cos^2 \theta = 1$.

Critical Thinking Questions

Analysis: Ask students how they can show that it is not generally the case that sin θ is proportional to θ. It suffices just to show one counterexample.

For example, sin 60 ($= \frac{\sqrt{3}}{2} \approx 0.8660$) is not twice sin 30 ($= 0.5000$).

Common Error Analysis

Error: Students tend to confuse the locations of $\sqrt{3}$ and 1 in drawing the 30-60-90 triangle.

Remind the students that the leg opposite the angle measuring 30 is $\frac{1}{2}$ of the hypotenuse.

Additional Example 2

Find each of the following trigonometric ratios. Simplify.

a. sec 150 $-\frac{2\sqrt{3}}{3}$

b. sin 330 $-\frac{1}{2}$

Additional Example 3

Find csc 45 and simplify. $\sqrt{2}$

Additional Example 4

Find sin (−315) and simplify. $\frac{\sqrt{2}}{2}$

649

Find each trigonometric ratio. Simplify all radicals.

1. tan 30 $\frac{\sqrt{3}}{3}$
2. cot 45   1
3. csc 60 $\frac{2\sqrt{3}}{3}$
4. csc (−225) $\sqrt{2}$

Evaluate each expression for the indicated value of θ. Simplify all radicals.

5. cos θ; θ = 135 $-\frac{\sqrt{2}}{2}$
6. $\cos^2 \theta - \sin^2 \theta$;
   θ = 150 $\frac{1}{2}$

Closure

Ask students to give the lengths for the sides of a right triangle to find the trigonometric ratios for each of the following special angles.

1. 60 $r = 2, y = \sqrt{3}, x = 1$
2. 30 $r = 2, x = \sqrt{3}, y = 1$
3. 45 $r = \sqrt{2}, x = 1, y = 1$

▮▮▮FOLLOW UP

Guided Practice

Classroom Exercises 1–12

Independent Practice

A Ex. 1–16, **B** Ex. 17–22, **C** Ex. 23–26

Basic: FR, WE 1–19 odd, Brainteaser
Average: FR, WE 3–23 odd, Brainteaser
Above Average: FR, WE 5–25 odd, Brainteaser

In summary, note the Special Reference Angles: 60, 30, 45.

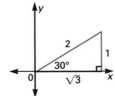 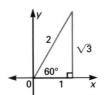 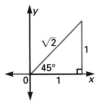

In the next example, note that it is accepted practice to write the squares of sin θ and cos θ as $\sin^2 \theta$ and $\cos^2 \theta$ instead of as $(\sin \theta)^2$ and $(\cos \theta)^2$. This is true for all of the trigonometric ratios.

EXAMPLE 5   Evaluate $\sin^2 \theta + \cos^2 \theta$ for θ = 210.

Solution
$$\sin^2 \theta + \cos^2 \theta = \sin^2 210 + \cos^2 210$$
$$= \left(-\frac{1}{2}\right)^2 + \left(-\frac{\sqrt{3}}{2}\right)^2$$
$$= \frac{1}{4} + \frac{3}{4}, \text{ or } 1$$

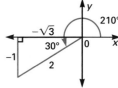

Therefore, 1 is the value of $\sin^2 \theta + \cos^2 \theta$ for θ = 210.

▮▮ *Focus on Reading*

Complete each of the following sentences.

1. In a 30-60-90 triangle, if the length of the hypotenuse is 2, then the length of the leg opposite the angle measuring 30 is __?__ . 1
2. In a 45-45-90 triangle, it is convenient to choose __?__ for the length of each leg and __?__ for the length of the hypotenuse. 1, $\sqrt{2}$

Classroom Exercises

Give the measure of the other acute angle and the missing values.

1.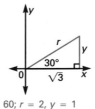
2.
3.
4.

45; $r = \sqrt{2}, y = 1$ 60; $r = 2, y = 1$ 30; $r = 2, x = -1$ 45; $r = \sqrt{2}, x = -1$

Additional Example 5

Evaluate sin θ · cot θ for θ = 240. $-\frac{1}{2}$

5.

6.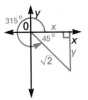

7.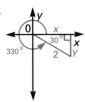

8.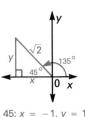

$60; r = 2, x = -\sqrt{3}$ $45; x = 1, y = -1$ $60; x = \sqrt{3}, y = -1$ $45; x = -1, y = 1$

Use the figure of Classroom Exercise 8 to find each trigonometric ratio.

9. sin 135 $\frac{\sqrt{2}}{2}$ **10.** cos 135 $\frac{-\sqrt{2}}{2}$ **11.** cot 135 -1 **12.** sec 135 $-\sqrt{2}$

Written Exercises

Find each trigonometric ratio. Simplify all radicals.

1. sin 30 $\frac{1}{2}$ **2.** csc 45 $\sqrt{2}$ **3.** cos 30 $\frac{\sqrt{3}}{2}$ **4.** sec 30 $\frac{2\sqrt{3}}{3}$

5. tan 60 $\sqrt{3}$ **6.** cot 45 1 **7.** sin 150 $\frac{1}{2}$ **8.** cos 240 $-\frac{1}{2}$

9. sec 315 $\sqrt{2}$ **10.** tan 120 $-\sqrt{3}$ **11.** csc 315 $-\sqrt{2}$ **12.** cot 225 1

13. sin(-225) $\frac{\sqrt{2}}{2}$ **14.** tan(-210) $\frac{-\sqrt{3}}{3}$ **15.** sec(-45) $\sqrt{2}$ **16.** cot(-330) $\sqrt{3}$

Evaluate each expression for the indicated value of θ. Simplify all radicals.

17. $\tan^2 \theta, \theta = 60$ 3

18. $2 \cos^2 \theta, \theta = 30$ $\frac{3}{2}$

19. $\cos^2 \theta - \sin^2 \theta, \theta = 240$ $-\frac{1}{2}$

20. $\sqrt{1 - \sin^2 \theta}, \theta = 135$ $\frac{\sqrt{2}}{2}$

21. $(2 \sin \theta)(\cos \theta), \theta = 150$ $\frac{-\sqrt{3}}{2}$

22. $\frac{\sin \theta}{\cos \theta}, \theta = 60$ $\sqrt{3}$

Simplify each expression. Rationalize all denominators.

23. $\cos 225 - \tan(-1,050) + \sin 135$ **24.** $(\sin 30 + \cos 30)^2$ $\frac{2 + \sqrt{3}}{2}$

25. $\frac{\cot 30 - \sec 45}{\tan 60 + \csc 45}$ $5 - 2\sqrt{6}$ $\frac{-\sqrt{3}}{3}$ **26.** $\frac{\csc 60}{\cot 60 + \tan 60}$ $\frac{1}{2}$

Mixed Review

Find θ to the nearest ten minutes if $0 < \theta < 90$.   *17.2*

1. $\sin \theta = 0.3746$ **2.** $\cos \theta = 0.1132$ **3.** $\tan \theta = 0.3959$ **4.** $\sec \theta = 2.430$
$22°$ $83°30'$ $21°40'$ $65°40'$

▰▰▰ Brainteaser

Find the value of tan 30 · cos 45 · sin 120 · sec 150 · cot 90 · csc 225. 0

Enrichment

Give the students the following problem.

Two sides of a triangle have lengths 8 cm and 10 cm. As the included angle increases through the measures 30, 45, 60, 90, 120, 135, and 150, find the changing values of the area of the triangle. Use the formula $A = \frac{1}{2}bh$, where h is the length of the altitude to either of the given sides.

30; area = 20 cm²
45; area = 20√2 cm²
60; area = 20√3 cm²
90; area = 40 cm²
120; area = 20√3 cm²
135; area = 20√2 cm²
150; area = 20 cm²

Note that in the next chapter, the students will be introduced to the formula $k = \frac{1}{2}bc \sin A$ for the area of a triangle.

GETTING STARTED

Prerequisite Quiz

Evaluate each expression.

1. sin 150 + sin 330   0
2. tan 30 · cot 150   −1
3. sin 60 − sin 120   0
4. sec 45 · sec 315   2
5. cos 135 − cos 315 $-\sqrt{2}$

Motivator

In each trigonometric ratio, it is assumed that $r > 0$. Suppose that $\sin A = -\frac{3}{5}$. Ignoring the negative sign, ask students what angle has a sine ratio of $\frac{3}{5}$. 37 Tell students that this is the reference angle. Now ask what the sign of y must be if the sine ratio is negative and r is positive. Negative In which quadrants is y negative? Quadrants 3 and 4. Ask students what two angles between 0 and 360 have a sine of $-\frac{3}{5}$. 217; 323

17.8 Finding Trigonometric Ratios and Angle Measures

Objectives   To find all angle measures between 0 and 360 corresponding to a given trigonometric ratio
To find all possible trigonometric ratios given two conditions

In the figure at the right, note that two angle measures between 0 and 360 have the same sine ratio ($\sin 210 = \sin 330 = -\frac{1}{2}$).

This happens because the sine ratio depends upon the values of y and r, which are the same for both angles.

Except for quadrantal angles, there are exactly *two* possible angles between 0 and 360 that correspond to any given trigonometric ratio. The procedure for finding such angles is illustrated in Example 1 below.

EXAMPLE 1   Find the two angle measures between 0 and 360 satisfying the equation $\cos \theta = -0.9397$.

Plan   Find the measure of a reference angle and use it together with two reference triangles to locate two angles of rotation.

Solution   When x is negative, $\cos \theta$ is negative. This occurs in Quadrants II and III. Draw the reference triangles in these two quadrants. Next, find the measure of a reference angle corresponding to $|\cos \theta| = 0.9397$. From the table, the measure is 20. Then, find the measures of the two angles of rotation using the reference triangles, as shown below.

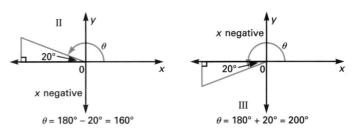

$\theta = 180° - 20° = 160°$ $\theta = 180° + 20° = 200°$

Therefore, the two angle measures satisfying the equation $\cos \theta = -0.9397$ are 160 and 200.

Highlighting the Standards

Standard 1a: The examples take students through the solution of equations under several simultaneous conditions—an important problem-solving concept.

Additional Example 1

Find the two angle measures between 0 and 360 satisfying the equation $\sin \theta = -0.2306$.   193° 20′, 346° 40′

EXAMPLE 2   Find all values of θ, $0 < \theta < 360$, for which $\tan \theta = -0.3574$. Find the results to the nearest ten minutes.

Solution   Tan θ is *negative*. The ratio $\frac{y}{x}$ is negative only if x and y have opposite signs ($\frac{+}{-}$ or $\frac{-}{+}$). This occurs in Quadrants II and IV (see chart at bottom of page 645). Draw the two reference triangles. Find the measure of a reference angle corresponding to $|\tan \theta| = 0.3574$. From the trigonometric table (or a calculator), the measure of each reference angle is $19°40'$ to the nearest 10 minutes.

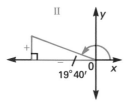   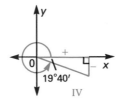

Tan θ is negative.

$\theta = 180° - 19° 40'$

$\quad = 160° 20'$

Tan θ is negative.

$\theta = 360° - 19° 40'$

$\quad = 340° 20'$

EXAMPLE 3   Find the value of each of the other five trigonometric ratios of θ if $\sin \theta = -\frac{1}{3}$ and $\tan \theta > 0$.

Plan   Locate the reference triangle and find values for x, y, and r. Do not find the measure of θ.

Solution   First, determine the quadrant of θ. The sine ratio $\frac{y}{r}$ is negative in Quadrants III and IV. Tan θ is positive in Quadrants I and III. To satisfy both conditions, the reference triangle must be in Quadrant III. Since $\sin \theta = -\frac{1}{3}$, or $\frac{-1}{3}$, choose $y = -1$ and $r = 3$. Use the distance formula to evaluate x.

$$r = \sqrt{x^2 + y^2}$$
$$3 = \sqrt{x^2 + (-1)^2}$$
$$9 = x^2 + 1$$
$$8 = x^2, \text{ or } x^2 = 8 \quad \text{Thus, } x = \pm\sqrt{8}.$$

Choose $x = -\sqrt{8} = -2\sqrt{2}$, since $x < 0$ in Quadrant III.

17.8 Finding Trigonometric Ratios and Angle Measures   **653**

▬▬TEACHING SUGGESTIONS

Lesson Note

Emphasize that a calculator will give only one of two possible solutions for an angle measure between 0 and 360 that corresponds to a trigonometric ratio. Suggestion: If a calculator is preferred to a table, teach the students to use the calculator to find the *reference* angle. Then use a diagram to find the two required angles, one of which might be the reference angle itself, if it occurs in Quadrant I.

Math Connections

Trigonometric Functions: This lesson provides a hint of the trigonometric functions of Chapter 18. Example 1 can be used as an illustration of the fact that if A is the set of all angle measures between 0 and 360, and C is the set of all cosine values, then the mapping of elements from A to C is such that each element of A corresponds to only one element of C. (Each angle has one and only one cosine value.) However, each element of C corresponds to one *or two* angle measures in the domain $0 \le \theta < 360$. Thus, the first mapping represents a function while the second mapping does not. (See Lesson 3.1.)

Critical Thinking Questions

Computer: Ask students what they think will be the result of solving Example 1 using a scientific calculator.   They should realize that only one of the two answers will appear. When they do use the calculator and find that only a measure of 160 is displayed, and not 200, you can indicate that the reason for this choice is somewhat arbitrary and is based on mathematical convenience. The matter is covered in more detail in Lesson 18.10.

Additional Example 2

Find all values of θ, $0 \le \theta < 360$, for which $\cos \theta = 0.8857$. Find the results to the nearest ten min. $27° 40'$, $332° 20'$

Additional Example 3

Find the value of each of the other five trigonometric ratios of θ if $\cos \theta = -\frac{3}{5}$ and $\cot \theta < 0$.

$\sin \theta = \frac{4}{5}$, $\tan \theta = -\frac{4}{3}$, $\sec \theta = -\frac{5}{3}$,

$\csc \theta = \frac{5}{4}$, $\cot \theta = -\frac{3}{4}$

Checkpoint

Find all values of θ, 0 < θ ≤ 360, to the nearest ten minutes for each given trigonometric ratio.

1. cos θ = 0.7765   39, 321
2. tan θ = −3.4567   106° 10′, 286° 10′
3. sin θ = −0.3567   200° 50′, 339° 10′
4. sec θ = 3.1000   71° 10′, 288° 50′
5. Given: sin θ = −$\frac{5}{13}$, cos θ > 0
   Find tan θ.   −$\frac{5}{12}$

Closure

Ask students how they determine the quadrant from the sign of the trigonometric ratio for each of the 6 trigonometric ratios. See discussion on page 652. Have students discuss how they would find the values of the five other trigonometric ratios if they are given the value of the sixth ratio. See Example 3 on page 653.

FOLLOW UP

Guided Practice

Classroom Exercises 1–10

Independent Practice

A Ex. 1–14, B Ex. 15–20, C Ex. 21, 22

Basic: WE 1–15 odd, Application

Average: WE 5–19 odd, Application

Above Average: WE 7–21 odd, Application

Additional Answers

Written Exercises

13. sin θ = $\frac{4}{5}$; tan θ = $\frac{4}{3}$; csc θ = $\frac{5}{4}$;
    sec θ = $\frac{5}{3}$; cot θ = $\frac{3}{4}$

So, cos θ = $\frac{-2\sqrt{2}}{3}$

tan θ = $\frac{-1}{-2\sqrt{2}} = \frac{1 \cdot \sqrt{2}}{2\sqrt{2} \cdot \sqrt{2}} = \frac{\sqrt{2}}{4}$

cot θ = $\frac{-2\sqrt{2}}{-1} = 2\sqrt{2}$

sec θ = $\frac{3}{-2\sqrt{2}} = \frac{3 \cdot \sqrt{2}}{-2\sqrt{2} \cdot \sqrt{2}} = -\frac{3\sqrt{2}}{4}$

csc θ = $\frac{3}{-1} = -3$

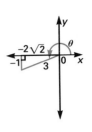

Classroom Exercises

Find the two quadrants which satisfy the given condition on θ.

1. cos θ > 0   I, IV
2. sin θ < 0   III, IV
3. tan θ > 0   I, III
4. sec θ < 0   II, III
5. cot θ < 0   II, IV

Find the one quadrant which satisfies the given conditions on θ.

6. cos θ < 0 and sin θ > 0   II
7. sec θ > 0 and tan θ > 0   I

Find, to the nearest ten minutes, the two angle measures between 0 and 360 satisfying the equation.

8. sin θ = 0.4723
   28°11′, 151°50′
9. cot θ = −3.1462
   162°20′, 342°20′
10. sec θ = 1.2385
    36°10′, 323°50′

Written Exercises

Find all values of θ to the nearest ten minutes, 0 < θ < 360, for each given trigonometric ratio. Use a calculator to find the measure of the reference angle.

1. cos θ = 0.9836   10°20′; 349°40′
2. tan θ = −2.7725   109°50′; 289°50′
3. cot θ = 1.2723   38°10′; 218°10′
4. sec θ = 1.2283   35°30′; 324°30′
5. csc θ = 1.4142   45°; 135°
6. sin θ = −0.4410   206°10′; 333°50′
7. cos θ = −0.3050   107°50′; 252°10′
8. sec θ = 2.1900   62°50′; 297°10′
9. cot θ = 3.6060   15°30′; 195°30′
10. tan θ = 3.3400   73°20′; 253°20′
11. cot θ = −9.5140   174°; 354°
12. sin θ = −0.8832   242°; 298°

Find the value of each of the other five trigonometric ratios of angle θ. Simplify all radicals.

13. cos θ = $\frac{3}{5}$ and sin θ > 0
14. tan θ = $\frac{4}{3}$ and sin θ < 0

Enrichment

Draw this figure on the chalkboard.

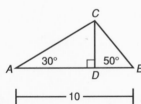

Challenge the students to find the measures of $\overline{AD}$ and $\overline{DB}$. If you wish, suggest that they let AD = x and DB = 10 − x.

$\frac{CD}{x}$ = tan 30; $\frac{CD}{10 - x}$ = tan 50

CD = x tan 30 and CD = (10 − x) tan 50

x tan 30 = (10 − x) tan 50

x tan 30 = 10 tan 50 − x tan 50

x tan 30 + x tan 50 = 10 tan 50

x(tan 30 + tan 50) = 10 tan 50

$x = \frac{10 \tan 50}{\tan 30 + \tan 50}$

$x = \frac{10(1.1918)}{0.5774 + 1.1918} \approx 6.7$

10 − x ≈ 3.3 So AD = 6.7 and DB = 3.3, to two significant digits.

15. $\sin \theta = \frac{1}{5}$ and $\cot \theta < 0$

16. $\tan \theta = -\frac{2}{3}$ and $\sin \theta > 0$

17. $\cos \theta = -\frac{1}{5}$ and $\cot \theta > 0$

18. $\cos \theta = \frac{2}{\sqrt{5}}$ and $\sin \theta < 0$

19. $\sec \theta = 3$ and $\tan \theta < 0$

20. $\cos \theta = -\frac{1}{3}$ and $\csc \theta > 0$

In Exercises 21–22, simplify all radicals.

21. If $\cot \theta = \frac{8}{15}$ and $\cos \theta < 0$, evaluate $\sqrt{\dfrac{1 - \cos \theta}{17}}$. $\frac{5}{17}$

22. If $\sec \theta = -\frac{5}{4}$ and $\tan \theta > 0$, evaluate

$$\frac{\tan 135 + \tan \theta}{\sin 30 \cos \theta + \sec 315 \sin 135} \cdot \frac{-5}{12}$$

Mixed Review

1. $f(x) = x^2 - 2x + 4, g(x) = 3x - 1.$ Find $f(g(x))$. **5.9** $9x^2 - 12x + 7$

2. Simplify. $27^{-\frac{2}{3}}$ **8.7** $\frac{1}{9}$

3. Solve for x. **1.4** $\frac{6c - ac - 2}{ab + 3}$
$$-3(x - 2c) = 2 + a(bx + c)$$

4. Solve the following system. $\begin{array}{l} 3(x - 2) = 2y - 7 \\ -2(y + 4) = -(x + 5) \end{array}$ **4.2, 4.3** $\left(-2, -\frac{5}{2}\right)$

Application: *Tunneling*

Tunnels can be constructed by starting at opposite sides of an obstruction and meeting in the center. To make sure that the tunnel crews will meet at the center, an engineer can choose points A and B on each side of the obstruction. Then, the engineer must find a point C for which angle ACB is 90. By measuring BC and AC, the engineer can compute the measures of $\angle A$ and $\angle B$ of $\triangle ACB$. Using this information, workers starting at point A follow a line which makes the calculated angle with $\overline{AC}$. Workers at point B follow a path that makes the calculated angle with $\overline{BC}$.

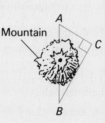

Mountain

Exercises

1. Explain how the engineer in the example above uses trigonometric information to make sure that the two tunnel crews will meet.

2. If AC is 120 m and BC is 90 m, what are m $\angle A$ and m $\angle B$? 37, 53

---

14. $\sin \theta = \frac{-4}{5}$; $\cos \theta = \frac{-3}{5}$; $\csc \theta = \frac{-5}{4}$; $\sec \theta = \frac{-5}{3}$; $\cot \theta = \frac{3}{4}$

15. $\cos \theta = \frac{-2\sqrt{6}}{5}$; $\tan \theta = \frac{-\sqrt{6}}{12}$; $\csc \theta = 5$; $\sec \theta = \frac{-5\sqrt{6}}{12}$; $\cot \theta = -2\sqrt{6}$

16. $\sin \theta = \frac{2\sqrt{13}}{13}$; $\cos \theta = \frac{-3\sqrt{13}}{13}$; $\csc \theta = \frac{\sqrt{13}}{2}$; $\sec \theta = \frac{-\sqrt{13}}{3}$; $\cot \theta = \frac{-3}{2}$

17. $\sin \theta = \frac{-2\sqrt{6}}{5}$; $\tan \theta = 2\sqrt{6}$; $\csc \theta = \frac{-5\sqrt{6}}{12}$; $\sec \theta = -5$; $\cot \theta = \frac{\sqrt{6}}{12}$

18. $\sin \theta = \frac{-\sqrt{5}}{5}$; $\tan \theta = \frac{-1}{2}$; $\csc \theta = -\sqrt{5}$; $\sec \theta = \frac{\sqrt{5}}{2}$; $\cot \theta = -2$

19. $\sin \theta = \frac{-2\sqrt{2}}{3}$; $\cos \theta = \frac{1}{3}$; $\tan \theta = -2\sqrt{2}$; $\csc \theta = \frac{-3\sqrt{2}}{4}$; $\cot \theta = \frac{-\sqrt{2}}{4}$

20. $\sin \theta = \frac{2\sqrt{2}}{3}$; $\tan \theta = -2\sqrt{2}$; $\csc \theta = \frac{3\sqrt{2}}{4}$; $\sec \theta = -3$; $\cot \theta = \frac{-\sqrt{2}}{4}$

Application

1. After measuring BC and AC, the engineer can calculate the angles that must be followed for the two teams to meet by using the tangent ratio. Thus, angle A is the angle whose tangent ratio is $\frac{BC}{AC}$, and angle B is the angle whose tangent ratio is $\frac{AC}{BC}$.

For Exercises 14–26, the answers are in the following order: sin; cos; tan; csc; sec; cot.

14. $\frac{3}{5}; \frac{-4}{5}; \frac{-3}{4}; \frac{5}{3}; \frac{-5}{4}; \frac{-4}{3}$

15. $\frac{-\sqrt{5}}{5}; \frac{2\sqrt{5}}{5}; \frac{-1}{2}; -\sqrt{5}; \frac{\sqrt{5}}{2}; -2$

16. $\frac{-1}{2}; \frac{\sqrt{3}}{2}; \frac{-\sqrt{3}}{3}; -2; \frac{2\sqrt{3}}{3}; -\sqrt{3}$

17. 0; −1; 0; undefined; −1; undefined

18. 0.9703; −0.2419; −4.0108; 1.0306; −4.1336; −0.2493

19. −0.7431; 0.6991; −1.1106; −1.3456; 1.4945; −0.9004

20. −0.8829; −0.4695; 1.8807; −1.1326; −2.1301; 0.5317

21. 0.9999; −0.0145; −68.7503; 1.0001; −68.7576; −0.0145

22. −1; 0; undefined; −1; undefined; 0

23. 0; −1; 0; undefined; −1; undefined

24. 0; 1; 0; undefined; 1; undefined

25. 1; 0; undefined; 1; undefined; 0

32. $\sin \theta = \frac{-3}{5}$; $\tan \theta = \frac{-3}{4}$; $\csc \theta = \frac{-5}{3}$; $\sec \theta = \frac{5}{4}$; $\cot \theta = \frac{-4}{3}$

33. $\sin \theta = \frac{-2\sqrt{13}}{13}$; $\cos \theta = \frac{-3\sqrt{13}}{13}$; $\cot \theta = \frac{3}{2}$; $\csc \theta = \frac{-\sqrt{13}}{2}$; $\sec \theta = \frac{-\sqrt{13}}{3}$

34. Answers will vary. One possible answer: Find the one quadrant which has the proper sign for both ratios. Then draw a reference triangle in that quadrant, labeling sides from the given ratio. Use the Pythagorean relation to determine the other side, and read off the remaining ratios.

Chapter 17 Review

Key Terms

angle of rotation (p. 633)
angles of depression and elevation (p. 629)
cosecant (p. 623)
cosine (p. 620)
cotangent (p. 623)
coterminal angles (p. 635)
initial side (p. 633)

quadrantal angles (p. 633)
reference angle (p. 640)
reference triangle (p. 639)
secant (p. 623)
similar triangles (p. 620)
sine (p. 620)
tangent (p. 620)
terminal side (p. 633)

Key Ideas and Review Exercises

17.1, 17.2 To find a trigonometric ratio for an acute angle of a right triangle, use one of the six basic trigonometric ratios.

$$\sin A = \frac{\text{opp}}{\text{hyp}} \leftarrow \text{reciprocals} \rightarrow \csc A = \frac{\text{hyp}}{\text{opp}}$$

$$\cos A = \frac{\text{adj}}{\text{hyp}} \leftarrow \text{reciprocals} \rightarrow \sec A = \frac{\text{hyp}}{\text{adj}}$$

$$\tan A = \frac{\text{opp}}{\text{adj}} \leftarrow \text{reciprocals} \rightarrow \cot A = \frac{\text{adj}}{\text{opp}}$$

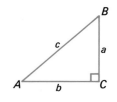

To find the measure of an acute angle of a right triangle, use one of the ratios above with a table or a calculator.

Find each trigonometric ratio to four decimal places.

1. sin 42 0.6691
2. tan 89 57.2900
3. sec 39°20′ 1.2929

Find θ to the nearest ten minutes if θ is an acute angle of a right triangle.

4. csc θ = 1.4530 43°30′
5. tan θ = 3.5199 74°10′
6. cos θ = 0.9765 12°30′

Find each indicated ratio to four decimal places.

7. csc R; $s = 4$, $r = 2$ 2.0000
8. sin T; $s = 6$, $r = 2$ 0.9428
9. cot R; $t = \sqrt{3}$, $r = \sqrt{2}$ 1.2247
10. cos T; $r = \sqrt{7}$, $t = 3$ 0.6614

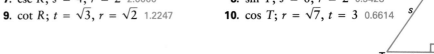

17.3 To find the measure of a side or acute angle of a right triangle, solve an equation involving a trigonometric ratio, the given data, and the missing part.

Find lengths to two significant digits and angle measures to the nearest degree. Use the figure for Exercises 7–10.

11. m $\angle T = 72$, $s = 53$,
$t = \underline{\ ?\ }$, $r = \underline{\ ?\ }$
50, 18

12. $s = 19$, $t = 13$,
m $\angle R = \underline{\ ?\ }$,
m $\angle T = \underline{\ ?\ }$ 47, 43

13. $r = 29$, m $\angle R = 42$,
$s = \underline{\ ?\ }$, $t = \underline{\ ?\ }$
43, 32

17.4, To find trigonometric ratios of angles of rotation, use the following
17.5, definitions, where $r = \sqrt{x^2 + y^2}$.
17.6 $\sin \theta = \frac{y}{r}$, $\cos \theta = \frac{x}{r}$, $\tan \theta = \frac{y}{x}$, $\cot \theta = \frac{x}{y}$, $\sec \theta = \frac{r}{x}$, $\csc \theta = \frac{r}{y}$

Find the six trigonometric ratios for the angle whose terminal side contains the given point. Simplify all radicals.

14. $P(-8,6)$ **15.** $P(4,-2)$ **16.** $P(3,-\sqrt{3})$ **17.** $P(-5,0)$

Find the six trigonometric ratios for each angle measure. If the ratio is undefined, so indicate.

18. 104 **19.** -48 **20.** -118 **21.** $450°50'$
22. 270 **23.** -180 **24.** 360 **25.** -270

17.7 To determine trigonometric ratios of the special angles 60, 30, and 45, refer to the top of p. 650.

Find each trigonometric ratio. Simplify all radicals.

26. $\sin 225$ $\frac{-\sqrt{2}}{2}$ **27.** $\tan 150$ $\frac{-\sqrt{3}}{3}$ **28.** $\sec 300$ 2 **29.** $\csc(-135)$ $-\sqrt{2}$

17.8 To find all values of θ for a given ratio, $0 < \theta < 360$ (where θ is not quadrantal), or to find the remaining trigonometric ratios given two conditions, locate the reference triangle(s) and use the reference angle and values of x, y, and r.

Find all values of θ to the nearest ten minutes, $0 < \theta < 360$.

30. $\sin \theta = -0.4413$ 206°10′; 333°50′ **31.** $\cot \theta = 3.6058$ 15°30′, 195°30′

Find the value of each of the other five trigonometric ratios of θ. Simplify all radicals.

32. $\cos \theta = \frac{4}{5}$ and $\tan \theta < 0$ **33.** $\tan \theta = \frac{2}{3}$ and $\sin \theta < 0$

34. Write in your own words how to find the five remaining trigonometric ratios given one ratio and the sign of a second ratio.

Chapter 17 Review **657**

Chapter Test

For Exercises 12–17, the answers are in the following order: sin; cos; tan; csc; sec; cot.

12. $\dfrac{3\sqrt{13}}{13}$; $\dfrac{-2\sqrt{13}}{13}$; $\dfrac{-3}{2}$; $\dfrac{\sqrt{13}}{3}$; $\dfrac{-\sqrt{13}}{2}$; $\dfrac{-2}{3}$

13. $\dfrac{-\sqrt{3}}{2}$; $\dfrac{1}{2}$; $-\sqrt{3}$; $\dfrac{-2\sqrt{3}}{3}$; 2; $\dfrac{-\sqrt{3}}{3}$

14. -1; 0; undefined; -1; undefined; 0

15. 0.1392; -0.9903; -0.1405; 7.1853; -1.0098; -7.1154

16. -0.6472; -0.7623; 0.8491; -1.5450; -1.3118; 1.1778

17. -0.9100; 0.4147; -2.1943; -1.0989; 2.4114; -0.4557

22. $\cos\theta = \dfrac{-12}{13}$; $\tan\theta = \dfrac{-5}{12}$; $\csc\theta = \dfrac{13}{5}$; $\sec\theta = \dfrac{-13}{12}$; $\cot\theta = \dfrac{-12}{5}$

23. $\sin\theta = \dfrac{-\sqrt{10}}{10}$; $\cos\theta = \dfrac{-3\sqrt{10}}{10}$; $\csc\theta = -\sqrt{10}$; $\sec\theta = \dfrac{-\sqrt{10}}{3}$; $\cot\theta = 3$

Additional Answers, page 637

11.
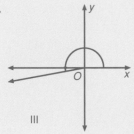

12.
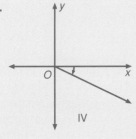

Find each trigonometric ratio to four decimal places.

1. cos 79   0.1908

2. csc 43°50′   1.4439

3. tan 23°30′   0.4348

Find m ∠ A to the nearest ten minutes if ∠ A is an acute angle of a right triangle.

4. sec A = 1.1575   30°10′

5. tan A = 0.6445   32°50′

6. sin A = 0.9686   75°40′

Find each indicated trigonometric ratio to four decimal places. Use the figure at the right.

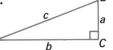

7. sin A; $c = 8$, $b = 4$ 0.8660

8. sec A; $a = 6$, $b = 4$ 1.8028

Find each length to two significant digits and each angle measure to the nearest degree. Use the figure above.

9. $a = 24$, m ∠ $A = 62$, $b = $ __?__, $c = $ __?__   13, 27

10. $c = 8$, $a = 6$, m ∠ $A = $ __?__, m ∠ $B = $ __?__   49, 41

11. The angle of elevation to the top of a mountain measures 38. From a point 1,500 ft closer, the angle of elevation to the top of the mountain measures 49. Find the height of the mountain to two significant digits.   3,700 ft

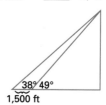

Find the six trigonometric ratios for the angle whose terminal side contains the given point. If the ratio is undefined, so indicate. Simplify all radicals.

12. $P(-4,6)$

13. $P(2,-2\sqrt{3})$

14. $(0,-8)$

Find the six trigonometric ratios for each angle measure.

15. 172

16. 220°20′

17. $-65°30′$

Find each trigonometric ratio. Simplify all radicals.

18. tan 90   Undef

19. sin(-180) 0

20. csc(-90) -1

21. cos 225 $\dfrac{-\sqrt{2}}{2}$

Find the other five trigonometric ratios of angle θ. Simplify all radicals.

22. $\sin\theta = \dfrac{5}{13}$ and $\cos\theta < 0$

23. $\tan\theta = \dfrac{1}{3}$ and $\sin\theta < 0$

24. If $\dfrac{\tan\theta_1 + \tan\theta_2}{1 - \tan\theta_1(\tan\theta_2)} = \tan 45$ and $\tan\theta_1 = \dfrac{2}{3}$, find $\tan\theta_2$. $\dfrac{1}{5}$

Exercises 13–24: Answers will vary. Possible answers are given.

13. 390, -330	20. 90, -630
14. 565, -155	21. 680, -40
15. 230, -490	22. 150, -570
16. 160, -560	23. 316, -404
17. 50, -670	24. 650, -70
18. 450, -270	
19. 310, -410	

25.
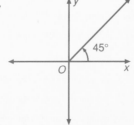

26.
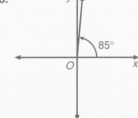

In each item, you are to compare a quantity in Column 1 with a quantity in Column 2. Write the letter of the correct answer from these choices.

A—The quantity in Column 1 is greater than the quantity in Column 2.
B—The quantity in Column 2 is greater than the quantity in Column 1.
C—The quantity in Column 1 is equal to the quantity in Column 2.
D—The relationship cannot be determined from the given information.

NOTE: Information centered over both columns refers to one or both of the quantities to be compared.

	Column 1	Column 2	
1.	$\frac{3}{5}$ of 5	$\frac{5}{3}$ of 3	B
2.	$\frac{2}{3}$ of 18	$\frac{3}{2}$ of 8	C
3.		$x > 1$	
	$\dfrac{x + x + x + x}{x \cdot x \cdot x}$	$\dfrac{4}{x^3}$	A
4.		$x \neq \pm 3$	
	$\dfrac{x^2 + 6x + 9}{x + 3}$	$\dfrac{x^2 - 9}{x - 3}$	C
5.	Ratio of $\frac{1}{3}$ to $\frac{1}{5}$	Ratio of $\frac{2}{5}$ to $\frac{1}{3}$	A
6.	Ratio of $\frac{1}{4}$ to $\frac{4}{1}$	Ratio of $\frac{3}{1}$ to $\frac{1}{3}$	B
7.		$a > 0$	
	$\dfrac{1}{a}$	a	D
8.	m	$\dfrac{3m + 1}{3}$	B
9.	$\sqrt{48} + \sqrt{80}$	$7 + 9$	B

	Column 1	Column 2	
10.	$\sin \theta$	1.2	B
11.	$\tan 45$	$\tan 315$	A
12.	$\sin 140$	$\cos 140$	A
13.	$\sin 150$	$\cos 240$	A
14.	$\csc 315$	$\sec(-45)$	B

15.

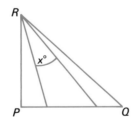

	Area of $\triangle ABC$	2	B

16. $PR = 4$, $PQ = 4$, and $QR = 4\sqrt{2}$

x	45	B

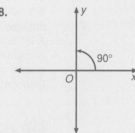

Additional Answers, page 637

27.
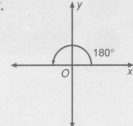
180°

28.
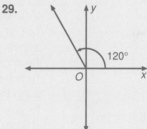
90°

29.
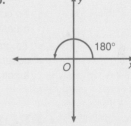
120°

30.
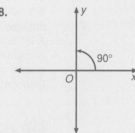
180°

31. $\theta = x + 360n$, where $90 < x < 180$ and n is an integer.

18 SOLVING TRIANGLES AND GRAPHING FUNCTIONS

OVERVIEW

In this chapter, the solving of triangles is extended to the case in which the triangles are oblique (nonright). This is done using the Law of Cosines and the Law of Sines. After Lesson 18.5 on radian measure, the remainder of the chapter is devoted to the graphs of trigonometric and inverse trigonometric functions.

OBJECTIVES

- To solve oblique triangles using the Law of Cosines
- To find areas of a triangle given the lengths of two sides and the measure of the angle between the sides
- To find the length of a side of a triangle using the Law of Sines
- To solve oblique triangles using the Law of Sines
- To solve word problems using the Law of Cosines or the Law of Sines
- To convert radian measure to degree measure and vice versa
- To graph the sine, cosine, tangent, cosecant, secant, and cotangent functions

PROBLEM SOLVING

The Law of Cosines and the Law of Sines give students an opportunity to use the problem-solving strategy *Using a Formula* to find the missing side or angle of a triangle. In Lesson 18.3, encourage students to use the problem-solving strategy *Drawing a Diagram* to solve Exercises 7–11 on page 671.

READING AND WRITING MATH

The *Focus on Reading* on page 703 asks students to read a mathematical statement and determine if it is always true, sometimes true, or never true.

Exercise 18 on page 689 asks students to explain why the sine function is an odd function. Exercise 19 on page 704 requires students to examine and explain a mathematical concept.

TECHNOLOGY

Many of the Examples in Chapter 18 give the appropriate calculator steps to simplify calculations and eliminate the need for trigonometric tables.

SPECIAL FEATURES

Mixed Review pp. 664, 667, 671, 676, 684, 689, 696, 700, 704
Midchapter Review p. 680
Problem Solving Strategies: Find an Estimate pp. 690–691
Focus on Reading p. 703
Application: Sine Curves p. 705
Key Terms p. 706
Key Ideas and Review Exercises pp. 706–707
Chapter 18 Test p. 708
College Prep Test p. 709

PLANNING GUIDE

Lesson	Basic	Average	Above Average	Resources
18.1 p. 664	CE all WE 1–5	CE all WE 1–9	CE all WE 3–11	Reteaching 133 Practice 133
18.2 p. 667	CE all WE 1–13 odd	CE all WE 3–15 odd	CE all WE 5–19 odd	Reteaching 134 Practice 134
18.3 pp. 670–671	CE all WE 1–11 odd	CE all WE 3–13 odd	CE all WE 5–15 odd	Reteaching 135 Practice 135
18.4 pp. 675–676	CE all WE 1–11 odd	CE all WE 5–15 odd	CE all WE 9–19 odd	Reteaching 136 Practice 136
18.5 pp. 679–680	CE all WE 1–33 odd Midchapter Review	CE all WE 3–35 odd Midchapter Review	CE all WE 7–39 odd Midchapter Review	Reteaching 137 Practice 137
18.6 pp. 683–684	CE all WE 1–5 odd	CE all WE 1–9	CE all WE 5–13	Reteaching 138 Practice 138
18.7 pp. 688–691	CE all WE 1–17 odd Problem Solving	CE all WE 5–19 odd Problem Solving	CE all WE 9–21 odd Problem Solving	Reteaching 139 Practice 139
18.8 p. 696	CE all WE 1–15 odd	CE all WE 5–19 odd	CE all WE 9–23 odd	Reteaching 140 Practice 140
18.9 p. 700	CE all WE 1–8	CE all WE 1–5, 11–15 odd	CE all WE 1–5, 13–17 odd	Reteaching 141 Practice 141
18.10 pp. 703–705	FR all CE all WE 1–13 odd Application	FR all CE all WE 3–15 odd Application	FR all CE all WE 7–19 odd Application	Reteaching 142 Practice 142
Chapter 18 Review pp. 706–707	all	all	all	
Chapter 18 Test p. 708	all	all	all	
College Prep Test p. 709	all	all	all	

CE = Classroom Exercises   WE = Written Exercises   FR = Focus on Reading
NOTE: For each level, all students should be assigned all Mixed Review exercises.

▬▬ INVESTIGATION

In this investigation, students use the graphing calculator to explore amplitudes and periods of trigonometric functions.

Have the students begin by graphing the sine function using the built-in calculator function. This will set the range of the graph to appropriate dimensions for the exercises that follow. The following keystrokes, for the Casio Fx-7000 or Fx-8000, should be used:

$$\boxed{\text{Graph}}\ \boxed{\text{sin}}\ \boxed{\text{EXE}}$$

Have the students graph $y = \sin 2x$ over the graph of the sine function and describe the resulting equation.

$$\boxed{\text{Graph}}\ \boxed{\text{sin}}\ 2\ \boxed{\text{alpha}}\ \boxed{+}\ \boxed{\text{EXE}}$$

The graph of sin 2x has a period that is one-half the period of the sine function. It oscillates twice as fast.

Have the students clear their screens by using the following keystrokes:

$$\boxed{\text{Shift}}\ \boxed{\text{G}\leftrightarrow\text{T}}\ \boxed{\text{EXE}}$$

Have the students graph the sine function as before and then graph $y = 2 \sin x$ over it. Ask them to describe the result.

The period is the same for the graph of $y = 2 \sin x$, but the amplitude is doubled.

Another set of graphs of instructional value is $y = \sin x$, $y = \sin (x + 90)$, $y = \sin x + 1$.

The calculators should be set in the degree mode (Deg) mode.

$$\boxed{\text{Graph}}\ \boxed{\text{sin}}\ \boxed{\text{EXE}}$$
$$\boxed{\text{Graph}}\ \boxed{\text{sin}}\ \boxed{(}\ \boxed{\text{Alpha}}\ \boxed{+}\ \boxed{+}\ 90\ \boxed{)}\ \boxed{\text{EXE}}$$
$$\boxed{\text{Graph}}\ \boxed{\text{sin}}\ \boxed{\text{Alpha}}\ \boxed{+}\ \boxed{+}\ 1\ \boxed{\text{EXE}}$$

The second function moves the sine curve to the *left* 90 degrees. The third function moves the sine curve *up* one unit.

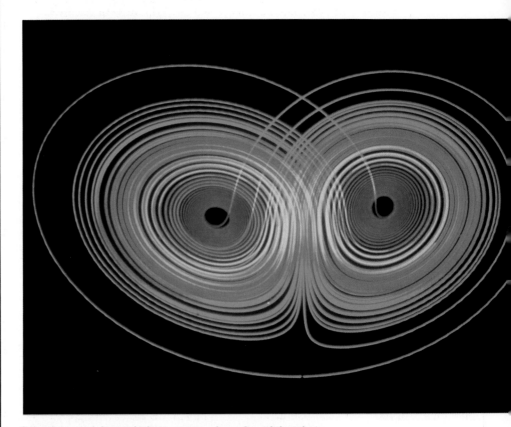

Scientists studying turbulent systems have found that there is often a surprisingly simple mathematical principle at work. The object pictured here is an example of a *strange attractor*—a graph which reveals the underlying mathematical structure of an otherwise unpredictable or "chaotic" system.

More About the Image

This mathematical object represents the motion of a fluid under certain convection conditions. The important thing to realize is that the curve is infinitely dense, and that a given position is *never* repeated. There can be no regularly recurring, predictable cycles. Also, *any* difference, no matter how small, between two starting points on the curve will be quickly magnified, so that the resulting motions will not bear any resemblance to each other. This famous curve is known as the Lorenz or "butterfly" attractor. It corresponds to the behavior of a simple waterwheel, whose motion turns out to be an exact mechanical analog of the convection equations under consideration.

18.1 The Law of Cosines

Objectives

To use the Law of Cosines to find the length of a side or the measure of an angle of a triangle

To solve problems using the Law of Cosines

It is sometimes possible to use trigonometric ratios to find the length of a side of a triangle, even if it is not a right triangle. For example, if the measures of two sides of a triangle and the included angle are known, it is possible to find the length of the third side.

In the nonright, or **oblique**, triangle at the right the measures of two sides and the included angle are 8, 10, and 58 respectively. Note (Figure 2) that h is the length of an altitude that divides triangle ABC into two right triangles, I and II. The Pythagorean relation can be used to find the length a.

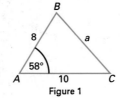
Figure 1

Triangle I: $8^2 = h^2 + x^2$, or $64 = h^2 + x^2$

Triangle II: $a^2 = h^2 + (10 - x)^2$, or

$$a^2 = h^2 + 100 - 20x + x^2$$

Subtract the equation for Triangle I from the equation for Triangle II.

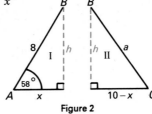
Figure 2

$$\begin{array}{l} a^2 = h^2 + 100 - 20x + x^2 \\ \underline{64 = h^2 \qquad\qquad\quad + x^2} \end{array}$$

$$a^2 - 64 = 100 - 20x$$

$$a^2 = 100 + 64 - 20x$$

$$a^2 = 100 + 64 - 20(8 \cos 58) \leftarrow \text{from the diagram, } \tfrac{x}{8} = \cos 58$$

$$a^2 = \underset{\text{sides}}{\underline{10^2}} + \underset{\text{sides}}{\underline{8^2}} - 2 \cdot \underset{\underset{\uparrow}{\text{included angle}}}{\underline{10 \cdot 8}} \cdot \cos 58$$

This pattern suggests the following theorem, called the *Law of Cosines*.

Theorem 18.1

Law of Cosines: For any triangle ABC,

$$a^2 = b^2 + c^2 - 2bc \cos A$$
$$b^2 = a^2 + c^2 - 2ac \cos B$$
$$c^2 = a^2 + b^2 - 2ab \cos C$$

Teaching Resources

Quick Quizzes 133
Reteaching and Practice Worksheets 133

GETTING STARTED

Prerequisite Quiz

Use a calculator to find each square root to two significant digits.

1. $\sqrt{82.8544}$ 9.1

2. $\sqrt{1,011,024}$ 1,000

3. $\sqrt{2,567,870}$ 1,600

Find m ∠A to the nearest degree for $0 \le$ m ∠A $\le$ 180.

4. $\cos A = -0.5750$ 125

5. $\cos A = 0.7789$ 39

Motivator

Try to get the students to discover the Law of Cosines. Draw Figure 1 on page 661 on the chalkboard. Ask the students to draw a perpendicular from B meeting $\overline{AC}$ at D. Let $AD = x$. Then have them find DC in terms of x. Draw the two triangles separately. Ask them to write equations for triangle 1 and triangle 2 using the Pythagorean relation. Follow the text at this point.

Highlighting the Standards

Standard 1a: The derivation of the Law of Cosines is a pleasing example of precise mathematical thinking. The Standards point out that problem solving essentially is "doing math."

Lesson Note

If students use a calculator to apply the Law of Cosines, they will not need to draw a reference triangle. The calculator will display the correct obtuse angle measure when the cosine of the angle is negative.

At this point, the students need not be concerned that a calculator will *not* display the angle measure that corresponds to a negative *tangent* value. For example, if tan $A = -0.3333$, the calculator will display -18.43322993 for A. This is because of the restricted range of the inverse tangent function, which is defined at the end of this chapter.

Math Connections

Historical Note: There are two propositions in Book II of Euclid's *Elements* that foreshadow the Law of Cosines. The first proposition is stated for obtuse triangles, the second for acute triangles. The statement of Proposition 13 for the acute case is as follows: In acute-angled triangles, the square on the side subtending the acute angle is less than the squares on the sides containing the acute angle by twice the rectangle contained by one of the sides about the acute angle, namely that on which the perpendicular falls, and the straight line cut off within by the perpendicular toward the acute angle.

Proposition 12 makes a similar statement of obtuse-angled triangles.

When the Law of Cosines is applied, it is important to use the correct sign for the cosine of the angle, as shown in Example 1.

EXAMPLE 1   In triangle ABC, $a = 4.0$, $b = 6.0$, and m $\angle C = 130$. Find c to two significant digits.

Plan   A calculator can be used to deter-mine the cosine of 130, including the correct sign. If a table is used, recall the following.

$$\cos 130 = -\cos (180 - 130)$$
$$= -\cos 50$$

This is illustrated in the reference triangle at the right.

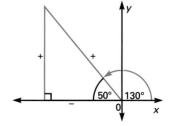

Solution $\cos 130 = -\cos 50 = -0.6428$ ← from the table on page 796
Use the appropriate form of the Law of Cosines.

$$c^2 = a^2 + b^2 - 2ab \cos C$$
$$c^2 = 4^2 + 6^2 - 2 \cdot 4 \cdot 6 \cdot \cos 130$$
$$c^2 = 16 + 36 - 48(-0.6428)$$
$$c^2 = 52 + 30.8544 = 82.8544$$
$$c = \sqrt{82.8544} = 9.1024392$$

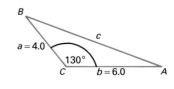

Therefore, $c = 9.1$ to two significant digits.

EXAMPLE 2   Two airplanes leave the same airport at the same time. After completing their take-offs, the planes assume flight paths that form an angle of $55°$. Their speeds are 450 mi/h and 600 mi/h. How far apart, to two significant digits, are the planes after 2 h?

Plan   Draw and label a diagram of the flight paths. After 2 h the plane flying at 450 mi/h had flown 900 mi; the other plane had flown 1,200 mi. Use the Law of Cosines to find d.

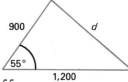

Solution $d^2 = 900^2 + 1,200^2 - 2 \cdot 900 \cdot 1,200 \cdot \cos 55$
$d^2 = 810,000 + 1,440,000 - 2,160,000(0.5736)$ ← from the table
$d^2 = 2,250,000 - 1,238,976 = 1,011,024$
$d = \sqrt{1,011,024} = 1005.4969$, or 1000 mi to 2 significant digits

Additional Example 1

In triangle ABC, $a = 1.8$, $b = 2.4$, and $m \angle C = 108$. Find c to two significant digits.   3.4

Additional Example 2

Two airplanes left the same airport and formed an angle with degree measure 65 in their flight paths. One flew at 600 km/h and the other at 800 km/h. Find the distance between them, to two significant digits, after two hours.   1,500 km

The Law of Cosines can also be used to find the measure of any angle of a triangle if the lengths of the three sides are known.

EXAMPLE 3   The lengths of the three sides of a triangular lot are 40 m, 50 m, and 80 m. Find, to the nearest degree, the measure of the largest angle of the triangular lot.

Plan   Draw and label a diagram. The largest angle is A, opposite the longest side, $a = 80$. Use the form of the Law of Cosines that involves cos A. Then solve for cos A.

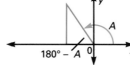

Solution

$$a^2 = b^2 + c^2 - 2bc \cos A$$
$$2bc \cos A = b^2 + c^2 - a^2$$
$$\cos A = \frac{b^2 + c^2 - a^2}{2bc}$$
$$= \frac{40^2 + 50^2 - 80^2}{2 \cdot 40 \cdot 50} = \frac{-2,300}{4,000} = -0.5750$$

Since cos $A < 0$, angle A must be obtuse. The reference triangle for angle A is illustrated at the right. Find the measure of the reference angle $180 - A$ as follows.

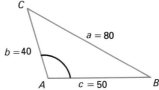

$$\cos (180 - A) = 0.5750$$
$$180 - A \approx 55 \leftarrow \text{from the table}$$
$$A \approx 125$$

The equation cos $A = -0.5750$ can also be solved using a calculator, as shown below.

Calculator Steps	Display
0.5750 (+/-) (INV) (cos)	125.09963

Thus, the measure of the largest angle is 125 to the nearest degree.

 Summary

The Law of Cosines can be used
(1) to find the measures of the angles of a triangle if the lengths of three sides are known (SSS), and
(2) to find the length of the third side of a triangle if the lengths of two sides and the measure of the included angle are known (SAS).

Checkpoint

In triangle *ABC*, find lengths to two significant digits and angle measures to the nearest degree.

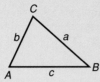

1. $a = 30$, $b = 60$, $c = 50$, m $\angle A = \underline{\ ?\ }$
   30
2. $b = 35$, $a = 50$, m $\angle C = 67$, $c = \underline{\ ?\ }$
   49
3. $a = 15$, $b = 30$, $c = 20$, m $\angle B = \underline{\ ?\ }$
   117
4. $a = 24$, $b = 20$, $c = 36$, measure of the largest angle $= \underline{\ ?\ }$ 109

Closure

Have students state the Law of Cosines in words without having to refer to letter labels for sides and angles. Given the lengths of two sides of a triangle, ask students which angle measure must be known to use the Law of Cosines.   The angle opposite the missing side.

◾️FOLLOW UP

Guided Practice

Classroom Exercises 1–4

Independent Practice

Ⓐ Ex. 1–4, Ⓑ Ex. 5–10, Ⓒ Ex. 11, 12

Basic: WE 1–5

Average: WE 1–9

Above Average: WE 3–11

Additional Answers

See pages 825–826 for the answers to Classroom Ex. 1–4 and Written Ex. 11–12.

Classroom Exercises

Use the Law of Cosines to find an equation involving *x* or *A*.

1.
2.
3.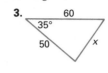
4.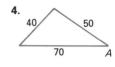

Written Exercises

In Exercises 1–4, refer to triangle *ABC* at the right. Find the lengths to two significant digits and the angle measures to the nearest degree.

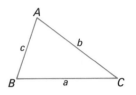

1. $a = 40$, $b = 50$, m $\angle C = 40$, $c = \underline{\ ?\ }$ 32
2. $a = 12$, $c = 14$, m $\angle B = 140$, $b = \underline{\ ?\ }$ 24
3. $a = 30$, $b = 70$, $c = 80$, m $\angle A = \underline{\ ?\ }$ 22
4. $a = 8.0$, $b = 4.1$, $c = 9.0$, m $\angle B = \underline{\ ?\ }$ 27

For Exercises 5–10, find lengths to two significant digits and angle measures to the nearest degree.

5. After two airplanes left the same airport at the same time, their flight paths formed an angle measuring 125. The first flew at 550 mi/h and the second flew at 620 mi/h. How far apart were they after 3 h? 3,100 mi

6. A baseball diamond forms a square 90 ft on a side. The pitcher's mound is 60 ft from home plate. How far is it from the mound to third base? 64 ft

7. The sides of an isosceles triangle have lengths 18, 18, and 10. Find the measure of the smallest angle of the triangle. 32

8. The length of the radius of a circle is 10. Two radii, $\overline{OA}$ and $\overline{OB}$, form an angle of measure $109°30'$. Find the length of chord $\overline{AB}$. 16

9. A triangular lot has side lengths of 16 m, 26 m, and 38 m. Find the measure of the largest angle of the lot. 128

10. The diagonals of a parallelogram bisect each other. If their lengths are 8.0 and 10 and they intersect at an angle of $20°40'$, how long are the sides? 1.9, 8.9

11. Prove the Law of Cosines (Theorem 18.1). Consider two cases. In the first case, no angle is obtuse. In the second case, one angle is obtuse.

12. Show that the Pythagorean relation, $a^2 + b^2 = c^2$, is a special case of the Law of Cosines.

Mixed Review

Write each trigonometric ratio in simplest radical form.   17.7, 17.8

1. sin 270 -1
2. tan 90 Undef
3. cos 135 $-\dfrac{\sqrt{2}}{2}$
4. csc 240 $-\dfrac{2\sqrt{3}}{3}$

Enrichment

Ask the students to use the Law of Cosines to prove that if a triangle is equilateral, it is also equiangular.

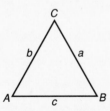

$c^2 = a^2 + b^2 - 2ab \cos C$
Since $a = b = c$,
$c^2 \quad = c^2 + c^2 - 2c^2 \cos C$
$c^2 \quad = 2c^2 - 2c^2 \cos C$
$c^2 \quad = 2c^2(1 - \cos C)$
$\dfrac{c^2}{2c^2} = 1 - \cos C$
$\dfrac{1}{2} \quad = 1 - \cos C$
$\cos C = \dfrac{1}{2}$

Therefore, m $\angle C = 60$.

In a similar way, it can be shown that m $\angle A = 60$ and m $\angle B = 60$.

18.2 The Area of a Triangle

Objectives
To find areas of triangles
To find areas of regular polygons

Recall that the area K of a triangle is half the product of the length of a side and the length of an altitude upon the side.

$$K = \tfrac{1}{2}bh$$

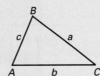

This formula can be used to derive another formula for the area of a triangle.

Write a trigonometric ratio. $\dfrac{h}{c} = \sin A$

Solve for h. $h = c \sin A$

Substitute for h in the area formula. $K = \tfrac{1}{2}b(c \sin A)$

Theorem 18.2

Area of a Triangle: The area K of a triangle is one-half the product of the lengths of any two sides and the sine of the included angle.

For any triangle ABC,

$K = \tfrac{1}{2}bc \sin A$,

$K = \tfrac{1}{2}ac \sin B$, and

$K = \tfrac{1}{2}ab \sin C$

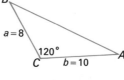

You will be asked to prove Theorem 18.2 for an obtuse included angle in Exercise 17.

EXAMPLE 1 In $\triangle ABC$, $a = 8$, $b = 10$, and $m \angle C = 120$.
Find the area of $\triangle ABC$ in simplest radical form.

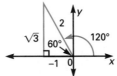

Plan Use $K = \tfrac{1}{2}ab \sin C$.

The reference angle for 120 is one of the special angles.

Solution $K = \tfrac{1}{2}ab \sin C$

$= \tfrac{1}{2} \cdot 8 \cdot 10 \sin 120 = 40 \cdot \dfrac{\sqrt{3}}{2}$, or $20\sqrt{3}$

Teaching Resources

Quick Quizzes 134
**Reteaching and Practice
Worksheets** 134

GETTING STARTED

Prerequisite Quiz

Use a calculator to find each value to four decimal places.

1. sin 72   0.9511
2. sin 112   0.9272
3. sin 115   0.9063

Simplify each radical.

4. $\sqrt{28}$ $2\sqrt{7}$
5. $\sqrt{12}$ $2\sqrt{3}$
6. $\sqrt{45}$ $3\sqrt{5}$

Motivator

Draw the first triangle shown on page 665 on the chalkboard. Have students give the formula for the area of a triangle from geometry. $K = \tfrac{1}{2}bh$ Ask students to write a trigonometric ratio involving h, c, and $\angle A$.
$\sin A = \dfrac{h}{c}$ Solve for h. $h = c \sin A$
Have students substitute this value of h in the area formula.
$K = \tfrac{1}{2}b(c \sin A)$

Additional Example 1

In $\triangle ABC$, $a = 16$, $b = 20$, and $m \angle C = 135$. Find the area of triangle ABC in simplest radical form.
$80\sqrt{2}$ square units

Highlighting the Standards

Standard 4a: The theorem and examples demonstrate different representations for the area of a triangle.

Lesson Note

The fundamental formula of this lesson is the one for SAS, $K = \frac{1}{2}bc \sin A$. Emphasize the verbalization of the formula so that the students will keep in mind that the angle is *between* the two sides: area = one-half the product of the lengths of two sides and the sine of the included angle.

Math Connections

Historical Note: The formula for the area of a triangle in terms of its three sides was proved by Heron of Alexandria in his *Metrica* sometime in the first century. The work was long lost but discovered in 1896 in a manuscript dating from about 1100. Heron's proof of this theorem is the first that is definitely known.

Critical Thinking Questions

Analysis: Some students may feel uncomfortable with the radical symbol in Heron's Formula. Ask them to use the methods of Lesson 7.3 (Dimensional Analysis) to show that the presence of the radical makes *dimensional* sense. They should see that if each side is measured in meters, then the dimensional form of the expression becomes $\sqrt{m(m)(m)(m)}$, or $\sqrt{m^4}$, or $(m^4)^{\frac{1}{2}}$, or m^2, which is the correct dimensional form for the measure of an area expressed in square meters.

EXAMPLE 2 Find, to two significant digits, the area of a regular pentagon inscribed in a circle of radius 6.0.

Plan Find the area K of one of the isosceles triangles in the figure at the right. Then find $5K$.

Since one complete circular rotation measures 360, each angle formed by a pair of consecutive radii has a measure of $\frac{360}{5}$, or 72.

Solution
$$K = \frac{1}{2} \cdot 6 \cdot 6 \cdot \sin 72$$
$$= 18(0.9511) = 17.1198 \leftarrow \text{from the table}$$
$$5K = 5 \cdot 17.1198 = 85.5990$$

The calculator steps for this example are shown below.

$0.5\ \boxed{\times}\ 6\ \boxed{\times}\ 6\ \boxed{\times}\ 72\ \boxed{\sin}\ \boxed{\times}\ 5\ \boxed{=}\ 85.5950867$

Therefore, the area of the pentagon is 86 to two significant digits.

If the lengths of the three sides of a triangle are known, then its area can be found by using *Heron's Formula*.

Heron's Formula for the Area of a Triangle

Let $s = \frac{1}{2}(a + b + c)$, half the perimeter, or *semiperimeter* of a triangle of area K. Then $K = \sqrt{s(s - a)(s - b)(s - c)}$.

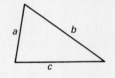

EXAMPLE 3 Find the area of a triangular lot with sides measuring 80 m, 50 m, and 70 m. Write the answer in simplest radical form.

Plan Find the semiperimeter. Then use Heron's Formula.

Solution
$$s = \frac{1}{2}(80 + 50 + 70) = 100$$
$$K = \sqrt{s(s - a)(s - b)(s - c)}$$
$$= \sqrt{100(100 - 80)(100 - 50)(100 - 70)}$$
$$= \sqrt{100 \cdot 20 \cdot 50 \cdot 30} = 1{,}000\sqrt{3}$$

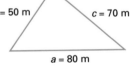

Therefore, the area is $1{,}000\sqrt{3}$ m^2 in simplest radical form.

Additional Example 2

Find, to two significant digits, the area of a regular decagon (10 sides) inscribed in a circle of radius 12.   420 square units

Additional Example 3

Find the area of a triangle with sides measuring 6 cm, 7 cm, and 9 cm. Write the answer in simplest radical form. $2\sqrt{110}$ cm^2

Classroom Exercises

2. $s = \frac{1}{2}(a + b + c)$, $\sqrt{s(s - a)(s - b)(s - c)}$

Choose an appropriate formula for finding the area of each triangle.

1. $a = 8$, $c = 10$, m $\angle B = 28$ $\frac{1}{2}ac \sin B$ 2. $a = 6$, $b = 7$, $c = 10$
3. $a = 9$, $b = 10$, m $\angle C = 35$ $\frac{1}{2}ab \sin C$ 4. $b = 9$, $a = 13$, $c = 6$
5–8. Find the area of each triangle of Classroom Exercises 1–4.

4. $s = \frac{1}{2}(a + b + c)$, $\sqrt{s(s - a)(s - b)(s - c)}$

5. 19 sq units
6. 21 sq units
7. 26 sq units
8. 24 sq units

Written Exercises

Find the area of triangle *ABC* to two significant digits.

1. $a = 18$, $c = 24$, m $\angle B = 42$ 2. $b = 10$, $c = 14$, m $\angle A = 75$
3. $c = 9.0$, $b = 4.0$, m $\angle A = 67$ 4. $a = 14$, $b = 16$, m $\angle C = 60$
5. $c = 20$, $b = 10$, m $\angle A = 110$ 6. $a = 50$, $c = 80$, m $\angle B = 125$

The lengths of the sides of a triangle are given. Find the area of the triangle in simplest radical form.

7. 4, 6, 8 $\frac{9\sqrt{119}}{4}$ sq units   8. 5, 12, 10   9. 5, 7, 10   10. 2, 4, 4
3√15 sq units   2√66 sq units   √15 sq units

Find, to two significant digits, the area of each regular polygon inscribed in a circle of the given radius.

11. octagon (8 sides), radius 12   410 sq units   12. decagon (10 sides), radius 4.0   47 sq units
13. hexagon (6 sides), radius 6.0   94 sq units   14. dodecagon (12 sides), radius 10   300 sq units
15. Find, in simplest radical form, the area of an equilateral triangle if each side has a length of 8. Use the formula $K = \frac{1}{2}bc \sin A$, and check your answer using Heron's Formula.   16√3 sq units
16. Find, in simplest radical form, the area of an isosceles triangle if each leg has a length of 6 and the angle formed by the legs measures 120.   9√3 sq units
17. Show that if angle *A* is an obtuse angle of triangle *ABC*, then the area of the triangle is given by $K = \frac{1}{2}bc \sin A$.
18. The length of each leg of an isosceles triangle is 8. Its area is 4√15 square units. Find the length of the third side.   4 units
19. For triangle *ABC*, $AB = 15$, $BC = 20$, and the area is 150 square units. Find the area of each of the two triangles formed by the altitude from *B* to $\overline{AC}$.   96 sq units, 54 sq units
20. The measure of the angle formed by two congruent sides of a triangle is 120. The area is 48√3 square units. Find the length of the base in simplest radical form.   24 units

Mixed Review

Graph each equation of a conic section.   12.1–12.3, 12.5

1. $x^2 + y^2 = 100$ 2. $4x^2 + 9y^2 = 36$ 3. $16x^2 - 25y^2 = 400$ 4. $xy = 48$

Checkpoint

Find the area of triangle *ABC* to two significant digits.

1. $b = 4.0$, $c = 6.0$, m $\angle A = 140$ 7.7
2. $b = 7$, $a = 10$, m $\angle C = 40$ 22

Find the area of triangle *ABC* in simplest radical form.

3. $a = 6$, $b = 8$, m $\angle C = 60$ 12√3
4. $a = 6$, $b = 12$, $c = 10$ 8√14

Closure

Have students give three formulas for the area of a triangle.   See Theorem 18.2 on page 665.   Ask students to explain the steps involved when using Heron's formula. See Example 3 on page 666.

◼◼◼ FOLLOW UP

Guided Practice

Classroom Exercises 1–8

Independent Practice

Ⓐ Ex. 1–10, Ⓑ Ex. 11–16, Ⓒ Ex. 17–20

Basic: WE 1–13 odd
Average: WE 3–15 odd
Above Average: WE 5–19 odd

Additional Answers

Written Exercises

1. 140 sq units   2. 68 sq units
3. 17 sq units   4. 97 sq units
5. 94 sq units   6. 1,600 sq units

See page 689 for the answers to Written Ex. 17 and Mixed Review Ex. 1–4.

Prerequisite Quiz

Solve each proportion.

1. $\frac{4}{7} = \frac{x}{21}$ 12   2. $\frac{12}{5} = \frac{x}{10}$ 24

Use a calculator to find c to two significant digits.

3. $c = \frac{12 \sin 44}{\sin 50}$ 11

4. $c = \frac{15 \sin 70}{\sin 45}$ 20

Motivator

Ask students if the Law of Cosines can be used to find the length of side *a* for $\triangle ABC$ given m $\angle A$ = 44, *b* = 14, and m $\angle B$ = 50.   No. The Law of Cosines can be used to find the measures of the angles if the sides are known (SSS), or the length of the third side of a triangle if two sides and the included angle are known (SAS).

■ TEACHING SUGGESTIONS

Lesson Note

In this lesson students are introduced to the Law of Sines to find lengths of sides only. Finding angle measures is deferred until the next lesson, when the ambiguous case will be introduced.

18.3 The Law of Sines

Objectives   To find a length of a side of a triangle using the Law of Sines
To solve problems using the Law of Sines

In this lesson, a law is developed that allows the missing measures of a triangle to be found if two angles and a side are known (AAS or ASA). The area of a triangle (Lesson 18.2) can be found using
$$K = \tfrac{1}{2}bc \sin A, \text{ or } K = \tfrac{1}{2}ac \sin B, \text{ or } K = \tfrac{1}{2}ab \sin C.$$

Therefore, $\qquad\qquad \tfrac{1}{2}bc \sin A = \tfrac{1}{2}ac \sin B = \tfrac{1}{2}ab \sin C.$

Multiply each part by 2. $bc \sin A = ac \sin B = ab \sin C$

Divide each part by abc. $\dfrac{bc \sin A}{abc} = \dfrac{ac \sin B}{abc} = \dfrac{ab \sin C}{abc}$

The result is $\dfrac{\sin A}{a} = \dfrac{\sin B}{b} = \dfrac{\sin C}{c}.$

Theorem 18.3

Law of Sines: For any triangle ABC,
$$\frac{\sin A}{a} = \frac{\sin B}{b} = \frac{\sin C}{c}.$$

EXAMPLE 1   In triangle ABC, $a = 12$, m $\angle A = 50$, and m $\angle C = 44$. Find c to two significant digits.

Plan   Use $\dfrac{\sin A}{a} = \dfrac{\sin C}{c}$, since m $\angle A$ and m $\angle C$ are given.

Solution
$$\frac{\sin 50}{12} = \frac{\sin 44}{c}$$
$$(\sin 50)c = 12 \sin 44$$
$$c = \frac{12 \sin 44}{\sin 50} = \frac{12(0.6947)}{0.7660} = 10.882$$

Therefore, the value of c is 11 to two significant digits.

In the Law of Sines each ratio in the proportion involves a side and the sine of the angle *opposite* that side. Therefore, when the measures of a side and two angles are given but neither angle is opposite the given side, it is necessary first to find the measure of the third angle.

Highlighting the Standards

Standard 9a: The solution of triangles is fundamental to the study of trigonometry. The applications in this lesson combine a variety of analytic skills.

Additional Example 1

In $\triangle ABC$, $a = 12$, m $\angle A = 36$, and m $\angle C = 48$. Find c to two significant digits.   15

EXAMPLE 2   In triangle ABC, $b = 15$, m $\angle A = 65$, and m $\angle C = 70$. Find c to two significant digits.

Plan   A proportion involving b must also include $\sin B$. To find m $\angle B$, use the fact that the sum of the measures of the angles of a triangle is 180.

Solution
$$m \angle B + 65 + 70 = 180$$
$$m \angle B = 45$$
Apply the Law of Sines to find c.

$$\frac{\sin 45}{15} = \frac{\sin 70}{c}$$
$$(\sin 45)c = 15 \sin 70$$
$$c = \frac{15 \sin 70}{\sin 45} = \frac{15(0.9397)}{0.7071} = 19.934$$

Therefore, $c = 20$ to two significant digits.

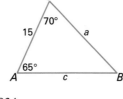

In ocean or air navigation, it is customary to measure angles of rotation clockwise from a north axis. In the diagram at the right, the plane is said to be flying at a **bearing** of 140.

EXAMPLE 3   A ship is sailing due north. The captain observes that the bearing of a lighthouse is 40. After sailing 60 km, the captain sees that the bearing of the lighthouse has become 135. How far, to two significant digits, is the ship from the lighthouse now?

Plan   Draw and label a figure.
Find m $\angle A$ and m $\angle B$ of the triangle.

Solution
$$m \angle A = 180 - 135 = 45$$
$$m \angle B = 180 - (45 + 40) = 95$$
Apply the Law of Sines to find c.

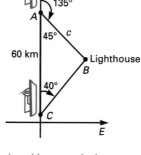

Lighthouse

$$\frac{\sin 95}{60} = \frac{\sin 40}{c}$$
$$(\sin 95)c = 60 \sin 40$$
$$c = \frac{60 \sin 40}{\sin 95}$$
$$= \frac{60(0.6428)}{0.9962} = 38.715 \leftarrow \text{Use the table or a calculator.}$$

So, the ship is 39 km from the lighthouse to two significant digits.

Math Connections

Geometry: The Law of Sines may be regarded as a refinement of the following theorem from plane geometry: *If one side of a triangle is longer than another side, then the measure of the angle opposite the longer side is greater than the measure of the angle opposite the shorter side.*

Critical Thinking Questions

Analysis: Ask students if the proportional relationships expressed in the Law of Sines are also true for the following relationships: $\frac{m \angle A}{a} \stackrel{?}{=} \frac{m \angle B}{b} \stackrel{?}{=} \frac{m \angle C}{c}$. Have them consider the 30-60-90 triangle with sides of measure 1, $\sqrt{3}$, and 2. They should see that although $60 = 2(30)$, it is not the case that $\sqrt{3} = 2(1)$.

Additional Example 2

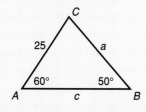

In $\triangle ABC$, $b = 25$, m $\angle A = 60$, and m $\angle B = 50$. Find c to two significant digits.   31

Additional Example 3

A ship is sailing due North. The captain notices that the bearing of a lighthouse 13 miles distant is 38 degrees. Later on, the captain observes that the bearing of the lighthouse has become 137 degrees. How far did the ship travel between the two observations of the lighthouse? Give the answer to two significant digits.   19 mi

Checkpoint

For triangle ABC, find the indicated lengths to two significant digits.

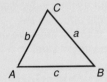

1. m ∠A = 75, a = 17, m ∠C = 38, c = ? 11
2. b = 29, m ∠B = 71, m ∠A = 72, c = ? 18
3.

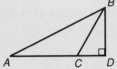

The angle of elevation from A to B is 24. The angle of elevation from C to B is 56. Find BC if AC is 80 m.   61 m

Closure

Have students give the Law of Sines and explain when it is most appropriate to use this law. $\frac{\sin A}{a} = \frac{\sin B}{b} = \frac{\sin C}{c}$. To find the missing measures of a triangle when two angles and a side are known.

EXAMPLE 4   The angle of elevation to the top of a mountain from a point P on the ground is 24°10′. The angle of elevation from a point Q directly in line with P and 1,350 ft closer is 61°40′. Find the height h of the mountain to three significant digits.

Plan   Draw and label a figure. First find m ∠1 and m ∠2.

Solution   m ∠1 = 180 − 61°40′ = 118°20′
m ∠2 = 180 − (24°10′ + 118°20′)
      = 37°30′

(1) Use the Law of Sines to find a.

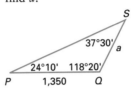

$$\frac{\sin 37°30′}{1,350} = \frac{\sin 24°10′}{a}$$

$(\sin 37°30′)a = 1,350 \sin 24°10′$

$$a = \frac{1,350(\sin 24°10′)}{\sin 37°30′}$$

$$= \frac{1,350(0.4094)}{0.6088}$$

$$= 907.8351$$

(2) Use right triangle QRS to find h.

$$\frac{h}{a} = \sin 61°40′$$

$$\frac{h}{907.8351} = \sin 61°40′$$

$$h = 907.8351(0.8802)$$

$$= 799.0765$$

The height of the mountain is 799 ft to three significant digits.

Classroom Exercises

Give a proportion for finding the indicated length. Do not find the length.

1. m ∠A = 23, m ∠B = 75, b = 14, a = ? $\frac{14 \sin 23}{\sin 75}$
2. m ∠B = 25, b = 24, m ∠C = 89, c = ? $\frac{24 \sin 89}{\sin 25}$
3. a = 18, m ∠C = 45, m ∠A = 89, c = ? $\frac{18 \sin 45}{\sin 89}$
4. b = 14, m ∠B = 62, m ∠A = 13, a = ? $\frac{14 \sin 13}{\sin 62}$
5. a = 110, m ∠A = 5, m ∠B = 44, b = ? $\frac{110 \sin 44}{\sin 5}$
6. m ∠C = 100, m ∠B = 46, b = 23, c = ? $\frac{23 \sin 100}{\sin 46}$

7-10. Find each indicated length to two significant digits in Classroom Exercises 1-4.   7. 5.7 8. 57 9. 13 10. 3.6

Additional Example 4

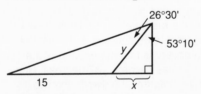

Refer to the figure above. Find x to three significant digits.   4.83

Written Exercises

For triangle ABC, find the indicated length to two significant digits.

1–2. Use the data for Classroom Exercises 5–6. **1.** 880 **2.** 31

3. m $\angle A$ = 59, m $\angle B$ = 63, b = 15, c = ___?___ 14

4. a = 26, m $\angle A$ = 78, m $\angle C$ = 18, b = ___?___ 26

5. m $\angle C$ = 6, m $\angle A$ = 38, c = 18, b = ___?___ 120

6. b = 14, m $\angle C$ = 110, m $\angle B$ = 54, a = ___?___ 4.8

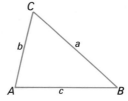

For Exercises 7–12, find the indicated measure to two significant digits, unless otherwise indicated.

7. The distance between Towns A and B is 56 mi. The angle formed by the road between Towns A and B and the road between Towns A and C mea-sures 46. The angle formed by $\overline{AB}$ and $\overline{BC}$ measures 115. Find the distance between Town B and Town C. 120 mi

8. On a ship sailing north, a woman no-tices that a hotel on shore has a bear-ing of 20. A little while later, after having sailed 40 km, she observes that the bearing of the hotel is now 100. How far is the ship from the hotel now? 14 km

9. A ship is steaming south. The naviga-tor notices that the bearing of a light-house is 120. After moving 8.0 mi/h for 2 h, he observes that the bearing of the lighthouse is 25. Find his dis-tance from the lighthouse at the time of the second sighting. 14 mi

10. Bill determines that the angle of eleva-tion to the top of a building measures 40°30'. If he walks 102 ft closer to the building, the measure of the new angle of elevation will be 50°20'. Find the height of the building to three sig-nificant digits. 299 ft

11. To determine the distance AB across a steep canyon, Megan walks 600 yd from B to another point, C. She then finds that m $\angle ACB$ = 35 and m $\angle CBA$ = 106. Find AB. 550 yd

12. Find the area of a regular pentagon if each side has a length of 12. 250 sq units

13. Show that if K is the area of triangle ABC, then $K = a^2 \left(\dfrac{\sin B \sin C}{2 \sin A} \right)$.

Use the figure at the right to answer Exercises 14 and 15.

14. Find AD and AB to three significant digits. 14.3, 13.5

15. Find the area of triangle ACD. 57.2 sq units

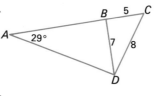

Mixed Review **1.** $\dfrac{x^2 + 3x + 9}{x + 3}$ **2.** $\dfrac{8x - 27}{x^2 - 7x + 12}$

Simplify. *7.1, 7.4, 8.4, 9.2*

1. $\dfrac{x^3 - 27}{x^2 - 9}$ **2.** $\dfrac{5}{x^2 - 7x + 12} - \dfrac{8}{3 - x}$ **3.** $\dfrac{6}{\sqrt{12}}$ $\sqrt{3}$ **4.** $\dfrac{3 + i}{4 - i}$ $\dfrac{11 + 7i}{17}$

Guided Practice

Classroom Exercises 1–10

Independent Practice

A Ex. 1–6, **B** Ex. 7–11, **C** Ex. 12–15

Basic: WE 1–11 odd
Average: WE 3–13 odd
Above Average: WE 5–15 odd

Additional Answers

Written Exercises

13. By Theorem 18.2, $K = \dfrac{1}{2}ab \sin C$.

By the Law of Sines, $\dfrac{\sin B}{b} = \dfrac{\sin A}{a}$, or $b = \dfrac{a \sin B}{\sin A}$.

Substituting for b, $K = \dfrac{1}{2}a\left(\dfrac{a \sin B}{\sin A}\right) \sin C$.

So, $K = a^2\left(\dfrac{\sin B \sin C}{2 \sin A}\right)$.

Enrichment

Tell the students that you can "prove" that the area of a square is zero. Then challenge the students to find the fallacy in the proof.

Remind students that a square can be inscribed in a circle. Draw this figure, where the circle has radius r.

Proof: Apply the area formula $K = \dfrac{1}{2}ab \sin C$ to each of the four triangles.

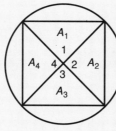

$A_1 = \dfrac{1}{2}r^2 \sin \angle 1$

$A_2 = \dfrac{1}{2}r^2 \sin \angle 2$

$A_3 = \dfrac{1}{2}r^2 \sin \angle 3$

$A_4 = \dfrac{1}{2}r^2 \sin \angle 4$

Then

$A_{total} = \dfrac{1}{2}r^2 (\sin \angle 1 + \sin \angle 2 + \sin \angle 3 + \sin \angle 4).$

Since $\angle 1 + \angle 2 + \angle 3 + \angle 4 = 2\pi$,

$A_{total} = \dfrac{1}{2}r^2 \sin 2\pi$

$A_{total} = \dfrac{1}{2}r^2 \cdot 0 = 0$

The fallacy is that $\sin \angle 1 + \sin \angle 2 + \sin \angle 3 + \sin \angle 4 \neq \sin(\angle 1 + \angle 2 + \angle 3 + \angle 4)$. In general, $\sin(A + B) \neq \sin A + \sin B$. Have the students demonstrate this, using different values of A and B.

▰▰▰GETTING STARTED

Prerequisite Quiz

1. What is the longest side of a right triangle? of an obtuse triangle?
   hypotenuse; the side opposite the obtuse angle

For the given right triangle, find each length to two significant digits and each angle measure to the nearest degree, if possible.

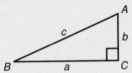

2. $c = 25$, $a = 12$, m $\angle A = $ _?_   29
3. $c = 13$, $a = 15$, m $\angle A = $ _?_   not possible
4. $c = 12$, m $\angle A = 40$, $a = $ _?_   7.7

Motivator

Have the class use their calculators and the Law of Sines to find m $\angle B$ to the nearest degree, given $a = 3$, $b = 28$, and m $\angle A = 80$.

Ask them to explain why there is no solution.   In cases involving two sides and a nonincluded angle, it is possible that no triangle can be formed.

18.4 The Ambiguous Case: Solving General Triangles

Objectives

To solve a triangle using the Law of Sines, given the measures of two sides and a nonincluded angle (SSA)

To determine whether one, two, or no triangles can be constructed when given the measures of two sides and a nonincluded angle

To use the Law of Sines and the Law of Cosines to solve triangles

In a triangle, if the measures of two sides and an angle opposite one of them are given (SSA), a value can be determined for h, the altitude to the unknown side of the triangle.

Suppose that m $\angle A$ and the lengths of sides a and b are known in the figure below. Using the sine ratio, the value of h can be found.

$$\sin A = \frac{h}{b}$$
$$h = b \sin A$$

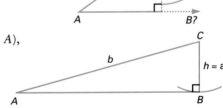

Notice that the length of side a must be greater than or equal to h. If it is not, no triangle can be formed using the given angle and sides.

The SSA case is often called the **ambiguous case**, since the number of possible triangles that can be constructed may be as few as zero and as many as two. All possibilities for constructing such triangles are summarized on this and the next page. In each possibility, the measure of $\angle A$ and the lengths of sides a and b are known.

Case 1

Angle A is an *acute* angle.

(1) If $a < h$ ($a < b \sin A$), then no triangle can be constructed.

(2) If $a = h$ ($a = b \sin A$), then one right triangle can be constructed.

Highlighting the Standards

Standard 9a: This lesson continues the work on solving triangles. The Law of Sines and Cosines together can be used to solve any triangle.

(3) If $a > h$ ($a > b \sin A$) and $a < b$, then two triangles can be constructed.

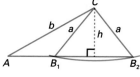

(4) If $a \geq b$, then one triangle can be constructed.

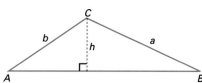

Case 2 Angle A is an *obtuse* or *right* angle.

(5) If $a \leq b$, then no triangle can be constructed.

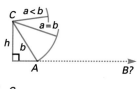

(6) If $a > b$, then one triangle can be constructed.

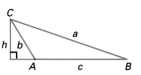

When the Law of Sines is used to find the measure of an angle for the ambiguous case, use the guidelines above to determine the number of solutions.

EXAMPLE 1 In triangle ABC, $a = 5$, $b = 20$, and m $\angle A = 40$. Find m $\angle B$ to the nearest degree.

Solution $h = b \sin A = 20(0.6428) = 12.856$
Since $a = 5$, $a < h$. Therefore, *no triangle* can be formed.

If we try to use the Law of Sines, we get an *impossible result*.

$$\frac{\sin A}{a} = \frac{\sin B}{b}$$
$$\frac{\sin 40}{5} = \frac{\sin B}{20}$$
$$5 \sin B = 20 \sin 40$$
$$\sin B = 4 \sin 40$$
$$= 4(0.6428) = 2.5712$$

There is *no solution*, since the sine ratio cannot be greater than 1.

■ **TEACHING SUGGESTIONS**

Lesson Note

Review the various cases summarized on pages 672–673. Use a board compass to illustrate the case with two possible solutions.

Have students go to the chalkboard and try to construct the cases where there is no solution or two solutions, using data such as that in Examples 1 and 2.

Math Connections

Historical Note: The ancient Greeks did not have a modern concept of trigonometry, but they did develop theorems about the relationships among chords of a circle that were equivalent to our Law of Sines. In the seventh century, the Indian mathematician Brahmagupta also touched on the Law of Sines while developing the formula for the radius of the circle circumscribed about a quadrilateral. The Law of Sines was more explicitly presented by the Persian scholar al-Biruni (973–1048).

Additional Example 1

In triangle ABC, $a = 3$, $b = 15$, and m $\angle A = 70$. Find m $\angle B$ to the nearest degree. No solution

Analysis: Students may recall a number of congruence theorems (or postulates) from plane geometry with the following abbreviations: SSS, SAS, ASA, and AAS. Some of these abbreviations have also been used in the lessons of this chapter. Ask students whether the abbreviations SSA can be associated with any of the theorems or postulates of geometry. Perhaps a few students will recall that *two right triangles are congruent if the hypotenuse and a leg of one are congruent, respectively, to the hypotenuse and corresponding leg of the other.* This theorem is often abbreviated HL, but could also be abbreviated HLA (or SSA), where A represents a right angle.

Common Error Analysis

Error: For the ambiguous case (SSA), students may solve only one triangle even when there are two possibilities.

Have students verbalize conditions for two possible triangles: Given that the side opposite an acute angle of a triangle is shorter than an adjacent side, if there is any solution other than a right triangle, then there are two solutions.

Notice that the expression for sin B in Example 1 is as follows.

$$\sin B = \frac{b \sin A}{a}$$

Substituting h for $b \sin A$ in the numerator gives the following.

$$\sin B = \frac{h}{a}$$

Therefore, if $\sin B \leq 1$, then $a \geq h$. So, it is not necessary to test if $a \geq h$ before applying the Law of Sines to find $\sin B$.

EXAMPLE 2 In triangle ABC, m $\angle A = 40$, $a = 9.0$, $b = 12$. *Solve* the triangle. (Find the measures of all the parts of the triangle.) Find lengths to two significant digits and angle measures to the nearest degree.

Solution

$$\frac{\sin 40}{9} = \frac{\sin B}{12}$$

$$9 \sin B = 12 \sin 40$$

$$\sin B = \frac{12(0.6428)}{9}$$

$$\sin B = 0.8571$$

Since $\sin B < 1$, $a > h$.
Therefore, two triangles can be constructed.
To the nearest degree, either m $\angle B = 59$, or
m $\angle B = 180 - 59 = 121$ (a Quadrant-II angle;
see the illustration at the right).

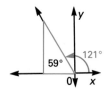

Next, find the values of m $\angle C$ and of c.

m $\angle C = 180 - 40 - 59$	m $\angle C = 180 - 40 - 121$
$= 81$	$= 19$
$c = \dfrac{a \sin C}{\sin A}$	$c = \dfrac{a \sin C}{\sin A}$
$= \dfrac{9 \sin 81}{\sin 40}$	$= \dfrac{9 \sin 19}{\sin 40}$
$= \dfrac{9(0.9877)}{0.6428}$	$= \dfrac{9(0.3256)}{0.6428}$
$= 13.8290 \approx 14$	$= 4.5584 \approx 4.6$

Thus, m $\angle B = 59$, m $\angle C = 81$, and $c = 14$.

Thus, m $\angle B = 121$, m $\angle C = 19$, and $c = 4.6$.

Additional Example 2

In triangle ABC, m $\angle A = 24$, $b = 14$, and $a = 10$. Solve the triangle. Find lengths to two significant digits and angle measures to the nearest degree.
m $\angle B = 35$, m $\angle C = 121$, $c = 21$, or m $\angle B = 145$, m $\angle C = 11$, $c = 4.7$

EXAMPLE 3

In triangle ABC, m $\angle A = 70$, $b = 2.0$, $a = 6.0$. Find the number of triangles that can be constructed.

Solution   Since $a > b$, there is only one solution.

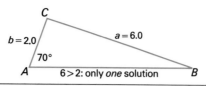

$b = 2.0$ $a = 6.0$ 70°

$6 > 2$: only *one* solution

Sometimes it is necessary to use both the Law of Sines and the Law of Cosines to solve a triangle.

For example, in triangle ABC at the right, $a = 6.0$, $b = 7.0$ and m $\angle C = 50$. How can you solve the triangle?

Note that the Law of Sines cannot be used directly, but the following sequence of steps will work.

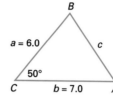

$a = 6.0$ c 50° $b = 7.0$

1. Since the pattern is SAS, use the Law of Cosines to find c.
2. Use the Law of Sines to find the smaller of the remaining angle measures.
3. Use the fact that the sum of the measures of the angles of a triangle is 180 to find the third angle measure.

Classroom Exercises

How many triangles can be constructed?

1. $a = 32$, $b = 21$, m $\angle A = 50$ 1
2. $b = 14$, m $\angle A = 55$, $a = 12$ and $a > b \sin A$ 2
3. $b = 8$, m $\angle A = 30$, $a = 4$ 1 right
4. m $\angle A = 160$, $a = 20$, $b = 60$ None
5. m $\angle B = 70$, $b = 12$, $a = 8$ 1

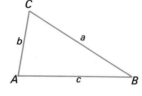

C b a A c B

Determine which of the two laws—the Law of Sines or the Law of Cosines—can be used to find the indicated measure. If both laws need to be used, indicate which must be used first.

6. $b = 15$, $c = 14$, m $\angle A = 40$, $a = \underline{\ ?\ }$ Law of Cosines
7. $b = 7$, $a = 10$, m $\angle A = 80$, m $\angle B = \underline{\ ?\ }$ Law of Sines
8. $a = 4$, $b = 5$, m $\angle C = 40$, m $\angle A = \underline{\ ?\ }$ Law of Sines; first use Law of Cosines

9–11. Find the indicated measure in each of Classroom Exercises 6–8 above.

9. 10   10. 44   11. 53

18.4 The Ambiguous Case: Solving General Triangles   **675**

Checkpoint

In each triangle *ABC*, find the number of triangles that can be constructed. If the answer is "one" or "two," then find the indicated angle measure(s) to the nearest degree.

1. m $\angle A = 80$, $b = 14$, $a = 3$, m $\angle B = \underline{\ ?\ }$ No solution
2. $a = 8$, m $\angle C = 30$, $c = 4$, m $\angle A = \underline{\ ?\ }$ One solution: 90
3. m $\angle B = 20$, $a = 10$, $b = 12$, m $\angle A = \underline{\ ?\ }$ One solution: 17
4. $a = 18$, $b = 16$, m $\angle B = 40$, m $\angle A = \underline{\ ?\ }$ Two solutions: 46, 134

Closure

Have students give the four possibilities for the number of triangles that can be constructed for SSA given $\angle A$ is acute. See pages 672–673. Then have them give the two possibilities if A is an obtuse or right angle. See page 673. Ask students to explain how they determine whether to use the Law of Sines or the Law of Cosines when solving a triangle. Answers may vary.

▬▬▬FOLLOW UP

Guided Practice

Classroom Exercises 1–11

Independent Practice

A Ex. 1–8, **B** Ex. 9–15, **C** Ex. 16–19

Basic: WE 1–11 odd
Average: WE 5–15 odd
Above Average: WE 9–19 odd

Additional Example 3

In triangle ABC, $a = 8$, $b = 12$, and m $\angle C = 40$. Solve the triangle.

m$\angle A = 41$, m$\angle B = 99$, $c = 7.8$

Additional Answers

Written Exercises

9. No solution
10. m $\angle B$ = 115, m $\angle C$ = 23, b = 26
11. m $\angle B$ = 53, m $\angle C$ = 47, a = 30
12. m $\angle A$ = 49, m $\angle B$ = 66, c = 12
13. m $\angle B$ = 124, m $\angle C$ = 21, b = 27
14. m $\angle A$ = 22, m $\angle B$ = 27, m $\angle C$ = 131

Written Exercises

In each triangle *ABC*, find the number of triangles that can be constructed. If the answer is "one" or "two," then find the indicated angle measure(s) to the nearest degree.

1. m $\angle A$ = 62, b = 14, a = 25,
   m $\angle B$ = ___?___ One; 30

2. m $\angle A$ = 65, b = 25, a = 20,
   m $\angle B$ = ___?___ None

3. m $\angle A$ = 85, b = 30, a = 6,
   m $\angle B$ = ___?___ None

4. m $\angle A$ = 100, b = 10, a = 25,
   m $\angle B$ = ___?___ One; 23

5. m $\angle C$ = 65, b = 14, c = 14,
   m $\angle B$ = ___?___ One; 65

6. m $\angle B$ = 57, a = 19, b = 3,
   m $\angle A$ = ___?___ None

7. m $\angle B$ = 66, a = 14, b = 11,
   m $\angle A$ = ___?___ None

8. m $\angle C$ = 43, b = 29, c = 26,
   m $\angle B$ = ___?___ Two; 50, 130

Solve triangle *ABC*, if possible. Find lengths to two significant digits and angle measures to the nearest degree.

9. a = 13, c = 110, m $\angle A$ = 45
10. m $\angle A$ = 42, a = 19, c = 11
11. b = 24, c = 22, m $\angle A$ = 80
12. a = 10, b = 12, m $\angle C$ = 65
13. a = 19, c = 12, m $\angle A$ = 35
14. a = 10, b = 12, c = 20

15. A triangular piece of property has sides with lengths of 40 m, 30 m, and 60 m. Find, to the nearest degree, the measure of each angle of the triangle. 36, 27, 117

16. The measures of two angles of a triangle are 50 and 70. The longest side is 6.0 ft longer than the shortest side. Find the lengths of the three sides to two significant digits. 26 ft, 30 ft, 32 ft

17. In triangle *ABC*, m $\angle A$ = 40, b = 8.0, and a = 10. Find the area of the triangle. 38 sq units

18. The lengths of two sides of a parallelogram are 16 and 20. One angle of the parallelogram measures 120. Find, to the nearest degree, the measure of the angle formed by the shorter side and the longer diagonal. 34

19. The lengths of the three sides of a triangle are 8.0, 10, and 12. Find, to the nearest degree, the measure of the angle formed by the shortest side and the median to the longest side. 47

Mixed Review

Solve each equation. 6.6, 9.5, 10.4, 13.3

1. $x^3 - 7x = 6$ 2. $x^2 - 2x + 6 = 0$ 3. $x^3 - 27 = 0$ 4. $\log_2 (x^2 - 2x) = 3$

   $-1, -2, 3$ $1 \pm i\sqrt{5}$ $3, \dfrac{-3 \pm 3i\sqrt{3}}{2}$ $-2, 4$

Enrichment

Ask the students to find a formula for the area of an isosceles triangle if the measures of the base and a base angle are given.

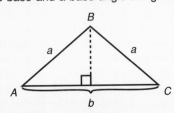

Let K = the area of the triangle.
Then $K = \dfrac{1}{2}ab \sin A$.

$\cos A = \dfrac{\frac{b}{2}}{a}$, $a \cos A = \dfrac{b}{2}$

So, $a = \dfrac{b}{2 \cos A}$

and $K = \dfrac{1}{2}\left(\dfrac{b}{2 \cos A}\right)b \sin A$.

$K = \dfrac{1}{4}b^2 \tan A$

Thus, the area of an isosceles triangle is one-fourth the product of the square of the length of the base and the tangent of a base angle.

18.5 Radian Measure

Objectives
To convert radian measure to degree measure and vice versa
To find trigonometric ratios for radian angle measures

Until now, angles of rotation have been measured in terms of degrees. Another way of measuring angles is suggested by the development below.

In the circle at the right, the center is located at O. The arc of the *central angle AOB* is equal in length to the radius r. Angle *AOB* is said to have a measure of one *radian*.

In this lesson, we will use degree symbols (°) to distinguish between *degree* measurements and *radian* measurements. After this lesson, you may again assume that an angle measure is given in degrees, unless it is in terms of π or unless otherwise indicated.

Definition

> If an angle is a central angle of a given circle and intercepts an arc of length equal to the length of a radius of the circle, then the measure of the angle is defined to be 1 **radian**.

The formula for the circumference of a circle is $C = 2\pi r$, or $C \approx 2(3.14)r$, or $C \approx 6.28r$.

In other words, there are approximately 6.28 *radii* in the circumference of a circle.

Therefore, the number of radians in a central angle that intercepts the entire circle is 2π, or approximately 6.28.

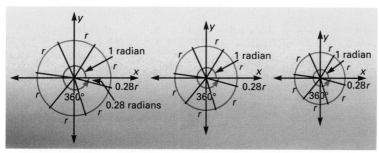

In each case above, regardless of the length of the radius, a rotation of 360° has a *radian* measure of 6.28, or 2π. When the measure of an angle is given in radians, it is customary to omit the word "radians."

18.5 Radian Measure **677**

Teaching Resources

Quick Quizzes 137
Reteaching and Practice
   Worksheets 137

GETTING STARTED

Prerequisite Quiz

Find the value of each trigonometric ratio in simplest radical form.

1. sin 240 $-\dfrac{\sqrt{3}}{2}$
2. tan 270   Undef
3. sec 150 $-\dfrac{2\sqrt{3}}{3}$
4. cos 180 -1
5. csc 135 $\sqrt{2}$
6. cot (-90) 0

Motivator

Remind students that the circumference of a circle is given by the formula $C = 2\pi r$. Ask them to give the circumference of a circle of radius 1. 2π Have them give the distance halfway around a circle of radius 1. π This would be equivalent to an *angular* rotation of __?__ degrees around the circle. 180

Highlighting the Standards

Standard 5d: Radian measure is a concept and symbol that gives increased breadth and power to the trigonometric functions.

TEACHING SUGGESTIONS

Lesson Note

Students need to remember the following ideas.

1. To convert degree measure to radian measure, multiply by $\frac{\pi}{180}$.
2. To convert radian measure to degree measure, replace π by 180, or use the conversion equation 1 radian $\approx$ 57.3 degrees.

Math Connections

A Formula for the Tangent: Radian measure lends itself more readily to work in higher mathematics than does degree measure. For example, it is possible to express the tangent ratio as the infinite series

$$\tan x = x + \frac{x^3}{3} + \frac{2x^5}{15} + \cdots,$$

but only if x is expressed in radians.

The relationship $360° = 2\pi$ radians is used to convert between radians and degrees.

EXAMPLE 1   Express $-200°$ in radian measure in terms of π.

Solution $360° = 2\pi$ radians.

Therefore, $1° = \frac{\pi}{180}$ radians. Multiply each side by -200.

$-200(1°) = -200 \cdot \frac{\pi}{180}$ radians $= \frac{-10\pi}{9}$ radians

Therefore, $-200°$ in radian measure is $-\frac{10\pi}{9}$.

EXAMPLE 2   Express $\frac{4\pi}{3}$ in degree measure.

Solution 2π radians $= 360°$

Therefore, π radians $= 180°$. Then $\frac{4\pi}{3}$ radians $= \frac{4(180°)}{3} = 240°$.

Therefore, $\frac{4\pi}{3}$ radians $= 240°$.

The degree measure of one radian can be found by using
$$\pi \text{ radians} = 180°.$$

Divide each side by π. $1 \text{ radian} = \frac{180°}{\pi}$ ←Some calculators have a key for entering π.

$1 \text{ radian} \approx \frac{180°}{3.1415927} \approx 57.29578$

$1 \text{ radian} = 57.3°$, to the nearest tenth of a degree.

EXAMPLE 3   Express the given radian measure in degrees, to the nearest degree.

     a. 3   **b.** -2.6

Solutions   **a.**   1 radian $\approx 57.3°$ **b.**   1 radian $\approx 57.3°$

           $3(1 \text{ radian}) \approx 3(57.3°)$ $-2.6(1 \text{ radian}) \approx -2.6(57.3°)$

           $3 \text{ radians} \approx 171.9°$ $-2.6 \text{ radians} \approx -148.98°$

Thus, 3 radians $\approx 172°$ and -2.6 radians $\approx -149°$.

Radian/Degree Equivalencies

When changing degrees to radians, use	When changing radians to degrees, use
$1° = \frac{\pi}{180}$ radians.	π radians $= 180°$, or 1 radian $\approx 57.3°$.

Additional Example 1

Express 240 in radian measure in terms of π. $\frac{4\pi}{3}$

Additional Example 2

Express $\frac{11\pi}{3}$ in degree measure.   660

Additional Example 3

Express the given radian measure in degrees, to the nearest degree.

a. 2   115   **b.** -5.4 -309

Some frequently used angle measures are represented in both degrees and radians at the right.

degrees	30	45	60	90	360
radians	$\frac{\pi}{6}$	$\frac{\pi}{4}$	$\frac{\pi}{3}$	$\frac{\pi}{2}$	2π

EXAMPLE 4 Find the value of each trigonometric ratio in simplest radical form.

a. $\sin \frac{4\pi}{3}$

b. $\tan \frac{3\pi}{2}$

Solutions

a. $\sin \frac{4\pi}{3}$

$\sin \left(\frac{4 \cdot 180°}{3}\right)$

$\sin 240° = -\frac{\sqrt{3}}{2}$

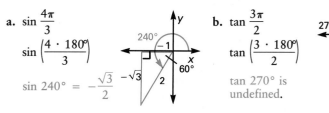

b. $\tan \frac{3\pi}{2}$

$\tan \left(\frac{3 \cdot 180°}{2}\right)$

tan 270° is undefined.

To check using a calculator, use the radian (RAD) mode and proceed as follows.

$4 \boxed{\times} \boxed{\pi} \boxed{\div} 3 \boxed{=} \boxed{\sin} \Rightarrow -0.8660254 \leftarrow$ Check: $3 \boxed{\sqrt{}} \boxed{\div} 2 \boxed{=}$

$\boxed{+/-} \Rightarrow -0.8660254$

$3 \boxed{\times} \boxed{\pi} \boxed{\div} 2 \boxed{=} \boxed{\tan} \Rightarrow$ E Since tan 270° is undefined, the calculator displays an error message.

Classroom Exercises

Tell how to perform each conversion.

1. $\frac{7\pi}{6}$ to degrees   2. 120° to radians   3. 4 to degrees   4. $\frac{3\pi}{4}$ to degrees

5. π to degrees   6. $-60°$ to radians   7. $-\frac{\pi}{3}$ to degrees   8. 8 to degrees

9–16. Perform the indicated conversion in each of Classroom Exercises 1–8 above. Express degree measures to the nearest degree.

9. 210 **10.** $\frac{2\pi}{3}$ **11.** 229 **12.** 135 **13.** 180 **14.** $-\frac{\pi}{3}$ **15.** -60 **16.** 458

Written Exercises

1. $\frac{5\pi}{6}$ 2. $\frac{5\pi}{4}$ 3. $\frac{5\pi}{3}$ 4. $\frac{-\pi}{2}$ 5. $\frac{11\pi}{6}$ 6. π 7. $\frac{7\pi}{6}$ 8. $\frac{7\pi}{4}$

Express as a radian measure in terms of π.

1. 150°   **2.** 225°   **3.** 300°   **4.** $-90°$

5. 330°   **6.** 180°   **7.** 210°   **8.** 315°

9. 160° $\frac{8\pi}{9}$ **10.** 80° $\frac{4\pi}{9}$ **11.** $-100°$ $\frac{-5\pi}{9}$ **12.** 30° $\frac{\pi}{6}$

Ask students to explain how to change degree measure to radian measure and vice versa. Students' answers should summarize the solutions to Examples 1 and 2. Have students give the number of degrees in 1 radian. 57.3 degrees

▰▰▰FOLLOW UP

Guided Practice

Classroom Exercises 1–16

Independent Practice

A Ex. 1–32, **B** Ex. 33–35, **C** Ex. 36–39

Basic: WE 1–33 odd, Midchapter Review
Average: WE 3–35 odd, Midchapter Review
Above Average: WE 7–39 odd, Midchapter Review

Additional Answers

Classroom Exercises

1. Mult by $\frac{180}{\pi}$ 2. Mult by $\frac{\pi}{180}$
3. Mult by 57.3   4. Mult by $\frac{180}{\pi}$
5. Mult by $\frac{180}{\pi}$ 6. Mult by $\frac{\pi}{180}$
7. Mult by $\frac{180}{\pi}$ 8. Mult by 57.3

Written Exercises

1. $\frac{5\pi}{6}$ 2. $\frac{5\pi}{4}$
3. $\frac{5\pi}{3}$ 4. $\frac{-\pi}{2}$
5. $\frac{11\pi}{6}$ 6. π
7. $\frac{7\pi}{6}$ 8. $\frac{7\pi}{4}$

Express in degree measure.

13. $\frac{2\pi}{3}$ 120
14. $-\frac{5\pi}{6}$ -150
15. $\frac{3\pi}{4}$ 135
16. $\frac{11\pi}{6}$ 330
17. $\frac{7\pi}{4}$ 315
18. $\frac{5\pi}{6}$ 150
19. $-\frac{7\pi}{6}$ -210
20. $\frac{\pi}{10}$ 18

Express the given radian measure in degree measure to the nearest degree.

21. 2   115
22. -4 -229
23. 2.3   132
24. -5.1 -292
25. -1 -57
26. 1.8   103
27. 6   344
28. -3 -172

Find the value of each trigonometric ratio. Give answers in simplest radical form.

29. $\sin \frac{\pi}{6}$ $\frac{1}{2}$
30. $\cos \frac{\pi}{4}$ $\frac{\sqrt{2}}{2}$
31. $\cos \frac{5\pi}{6}$ $\frac{-\sqrt{3}}{2}$
32. $\sin \frac{3\pi}{2}$ -1

33. $\frac{1}{3} \tan^2 \frac{2\pi}{3}$ 1
34. $2 \cos \left(-\frac{3\pi}{4}\right)$ $-\sqrt{2}$
35. $\cos^2 \left(\frac{3\pi}{4}\right) + \sin^2 \left(\frac{3\pi}{4}\right)$ 1

36. If $\theta = \frac{\pi}{6}$ and $r = 6$, find s, the length of $\overset{\frown}{AB}$. π
37. Find a general formula for s, the length of $\overset{\frown}{AB}$, in terms of r and θ. $s = r \cdot \theta$
38. If $\theta = \frac{\pi}{4}$ and $r = 4$, find the area of the sector of the circle formed by radii $\overline{OA}$ and $\overline{OB}$ and arc $\overset{\frown}{AB}$. 2π
39. Write a general formula for the area A of a sector of a circle in terms of r and θ, and then in terms of r and s. $A = \frac{\theta r^2}{2}; A = \frac{rs}{2}$

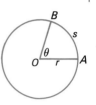

Midchapter Review

Exercises 1–4 refer to triangle *ABC*. Find lengths to two significant digits and angle measures to the nearest degree.   *18.1, 18.3*

1. $a = 13, b = 7, m \angle C = 47,$ $c = \underline{\ ?\ }$ 9.7
2. $b = 95, m \angle B = 39, m \angle C = 65,$ $c = \underline{\ ?\ }$ 140
3. $b = 12, c = 12, a = 20,$ $m \angle A = \underline{\ ?\ }$ 113
4. $c = 5, m \angle A = 43, m \angle C = 22,$ $b = \underline{\ ?\ }$ 12

Find the area of triangle *ABC*.   *18.2*

5. $a = 10, b = 14, m \angle C = 100$ (to two significant digits)   69
6. $a = 6, b = 7, c = 9$ (simplest radical form) $2\sqrt{110}$

Find the number of triangles *ABC* that can be constructed. If the answer is "one" or "two," then find the indicated angle measure(s) to the nearest degree.   *18.4*

7. $a = 6, b = 12, m \angle A = 30,$ $m \angle B = \underline{\ ?\ }$ One; 90
8. $a = 3, b = 12, m \angle A = 40,$ $m \angle B = \underline{\ ?\ }$ None
9. Express 15° as a radian measure in terms of π.   *18.5* $\frac{\pi}{12}$
10. Express $\frac{13\pi}{6}$ in degree measure.   *18.5*   390

Enrichment

Have the students solve the following problem.

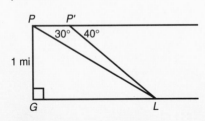

A plane flying at a constant altitude of one mile is approaching a landing strip. The pilot measures the angle of depression and finds it is 30. Exactly 10 sec later, the angle of depression becomes 40.

What is the speed of the plane?

Give your answer in mi/h to two significant digits.

The speed is $\frac{0.54 \text{ mi}}{10 \text{ s}} \cdot \frac{3,600 \text{ s}}{1 \text{ h}}$, or about 190 mi/h.

18.6 Periodic Functions

Objectives
To determine whether a periodic function is odd, even, or neither
To graph periodic functions
To determine the amplitude of a periodic function

A certain function f is defined for all real numbers x. Its graph is shown below. The red portion repeats itself every 8 units along the x-axis.

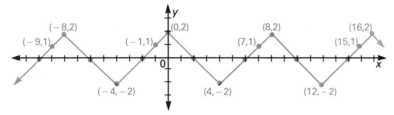

Note that the y-value for any value of x is the same as the y-value for x plus 8 or x plus an integral multiple of 8. For $x = -9$, for example,

$$f(x): \quad f(-9) = 1$$
$$f(x + 8): \quad f(-9 + 8) = f(-1) = 1$$
$$f(x + 2 \cdot 8): \quad f(-9 + 2 \cdot 8) = f(7) = 1$$
$$f(x + 3 \cdot 8): \quad f(-9 + 3 \cdot 8) = f(15) = 1$$

The basic shape of the graph of f is repeated every 8 units. Any horizontal interval of 8 units is said to be a **cycle** of the graph of function f. The function is said to be *periodic* with a *period* of 8.

Definitions

A function f with domain D is a **periodic function** if and only if for some positive constant p, $f(x) = f(x + p)$ for every x in D. The smallest such value of p is the **period** of f.

Note that in the graph of f above, the *maximum* y-value is $M = 2$. The minimum y-value is $m = -2$. Half the difference of the maximum and minimum y-values of the function is referred to as the *amplitude* of the periodic function f. The amplitude is $\dfrac{M - m}{2} = \dfrac{2 - (-2)}{2} = \dfrac{4}{2}$, or 2.

Definition

If a periodic function has a maximum value M and a minimum value m, then its **amplitude** is $\dfrac{M - m}{2}$.

18.6 Periodic Functions   **681**

◤▬▬▬**GETTING STARTED**

Prerequisite Quiz

Given that $f(x) = x^2$, find each of the following.

1. $f(3)$ 9
2. $f(-3)$ 9

Determine whether each function is odd or even.

3. $f(x) = x^4$ Even
4. $f(x) = x^3$ Odd
5. $f(x) = |x|$ Even

Motivator

Have students graph the following points. $A(0,0)$, $B(1,1)$, $C(2,0)$, $D(3,3)$, $E(4,0)$, $F(5,5)$, $G(6,0)$. Now have them connect, beginning with $A(0,0)$, each pair of consecutive points with a line segment. Ask students to discuss how this graph differs from typical straight line graphs such as $y = 4x$.

Highlighting the Standards

Standard 6b: The examples use equations, values, and graphs to demonstrate the workings of periodic functions.

TEACHING SUGGESTIONS

Lesson Note

The definition of amplitude may appear contrived to the students. They are more inclined to think of amplitude as the maximum value of the function. Point out that this is true only if the maximum and minimum values are additive inverses of each other. For example, if the function displayed on this page were translated up 4 units, the maximum value would then be 6 and the minimum, 2. The amplitude would still be 2; using the formula,

$$\frac{M - m}{2} = \frac{6 - 2}{2} = 2.$$

Math Connections

AM and FM radio: Both AM and FM radio depend on varying, or modulating, a carrier wave that oscillates several times per second. The shape of the carrier wave is that of a periodic function. The sounds that are transmitted from a radio station modify the shape of the function's graph. This is done by modulating the wave in one of two ways. The first way is to modulate the amplitude of the periodic function; that is, to magnify or diminish the amplitude several times each second in a manner that corresponds to the highs and lows of the transmitted sound. The second form of modulation, frequency modulation, maintains a constant amplitude, but allows the frequency of the oscillations to magnify or diminish in a way that matches the transmitted sound.

Recall that for a function f with x in the domain D of f,

(1) f is an *even* function if $f(-x) = f(x)$ for every x in D, and

(2) f is an *odd* function if $f(-x) = -f(x)$ for every x in D.

The graph of an even function is symmetric with respect to the y-axis. The graph of an odd function is symmetric with respect to the origin.

EXAMPLE 1 Determine whether function f on the previous page is odd, even, or neither.

Solution For every real number x, $f(-x) = f(x)$. For example, $f(-5) = f(5) = -1$ and $f(-4) = f(4) = -2$. The graph is symmetric with respect to the y-axis.

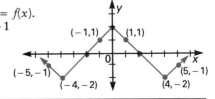

Therefore, f is an *even* function.

EXAMPLE 2 A portion of the graph of a periodic function f is shown at the right. The period is 6.

a. Complete the graph of f for the interval $-9 \le x \le 9$.

b. Find the amplitude of f.

c. Determine whether f is odd, even, or neither.

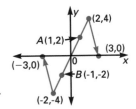

Solutions

a. Since the given part of the graph covers one cycle, repeat to the right and left until you reach -9 at the left and 9 at the right.

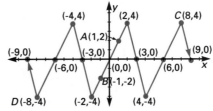

b. The maximum value of y is $M = 4$. The minimum value of y is $m = -4$. Therefore, the amplitude of f is
$$\frac{M - m}{2} = \frac{4 - (-4)}{2} = \frac{8}{2}, \text{ or } 4.$$

c. For every real number x, $f(-x) = -f(x)$. For example, $f(-1) = -2 = -f(1)$, $f(-8) = -4 = -f(8)$, and so on. The graph is symmetric with respect to the origin.

Thus, f is an *odd* function.

Additional Example 1

Determine whether the function graphed below is even, odd, or neither. Odd

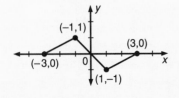

Additional Example 2

A portion of the graph of a periodic function f is shown. The period is 4.

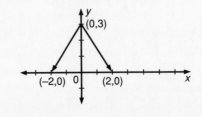

a. Complete the graph of f for the interval $-2 \le x \le 6$. See below.

b. Find the amplitude of f. 1.5

c. Is f odd, even, or neither? Even

a.

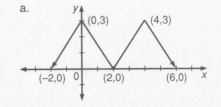

Classroom Exercises

The figure at the right shows the graph of a periodic function f.

1. What is the period of function f? 2
2. What is the amplitude of f? $\frac{3}{2}$
3. Is f odd, even, or neither? Even

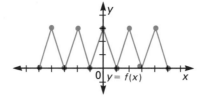

The figure at the right shows the graph of a periodic function g.

4. What is the period of function g? 4
5. What is the amplitude of g? 1
6. Is g odd, even, or neither? Odd

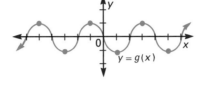

The figure at the right shows the graph of a periodic function h.

7. Tell how to graph h in the interval $4 \le x \le 12$.
8. What is the amplitude of h? $\frac{3}{2}$
9. Is h odd, even, or neither? Neither

7. Repeat the pattern for the section where $-4 \le x \le 4$.

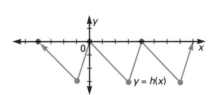

Written Exercises

In Exercises 1–6, a portion of the graph of a function is shown. Use the given period p to graph the function for the indicated interval. Determine whether the function is odd, even, or neither. Find the amplitude of each function.

1.

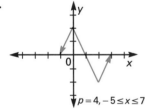

$p = 4, -5 \le x \le 7$

2.

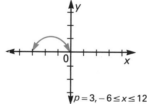

$p = 3, -6 \le x \le 12$

Critical Thinking Questions

Analysis: Since an odd function is symmetrical around the origin and an even function is symmetrical about the y-axis, ask students what restrictions must exist to have a function that is symmetric about the x-axis. Usually a curve that is symmetrical about the x-axis is not the graph of a function at all. However, if the only permitted value of y were 0, then such a function could exist.

Checkpoint

A portion of the graph of a function with period 4 is shown below.

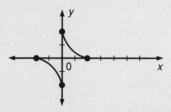

1. Complete the graph for the interval $2 \le x \le 6$.

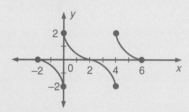

2. Determine whether f is even, odd, or neither. Odd
3. Find the amplitude of f. 2

Closure

Have students give the formal definiton of a periodic function, then give an informal definition in their own words. If the maximum and minimum values of a periodic function are known, ask students how they would find its amplitude. $\frac{M - m}{2} = $ amplitude

Guided Practice

Classroom Exercises 1–9

Independent Practice

A Ex. 1–4, **B** Ex. 5–13

Basic: WE 1–5
Average: WE 1–9
Above Average: WE 5–13

Additional Answers

Written Exercises

1. Even; 2

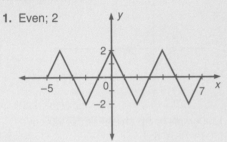

2. Even; $\frac{1}{2}$

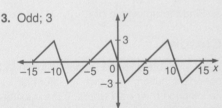

3. Odd; 3

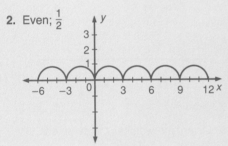

See page 709 for the answers to Written Ex. 4–7, 9, 11, 13.

3.

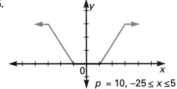

$p = 10, -15 \le x \le 15$

4.

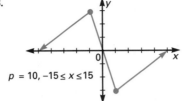

$p = 4, -6 \le x \le 10$

5.

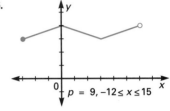

$p = 10, -25 \le x \le 5$

6.

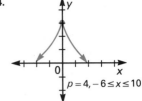

$p = 9, -12 \le x \le 15$

7. Draw the graph of $y = |2x|$, $-4 \le x \le 4$.

8. The graph of Exercise 7 represents one cycle of a periodic function f. The domain of f is the set of real numbers. What is the period of the function? 8

9. Graph the periodic function of Exercise 8 for $-12 \le x \le 12$. Determine the amplitude and whether the function is odd, even, or neither.

10. The graph of a periodic function g over one cycle is defined by
$y = \begin{cases} x + 1, & -1 \le x \le 2 \\ -2x + 7, & 2 < x \le 5 \end{cases}$. The domain of g is the set of real numbers. Determine the period of the function. 6

11. Graph the periodic function of Exercise 10 for $-1 \le x \le 17$.

12. The graph of a periodic function h over one cycle is defined by $y = |2x - 4| + 5$, $0 \le x \le 4$. The domain of h is the set of real numbers. Determine the period of the function. 4

13. Graph the periodic function of Exercise 12 for $-4 \le x \le 8$.

Mixed Review

Write each trigonometric ratio in simplest radical form. 17.5, 17.7, 18.5

1. $\sin 270$ -1
2. $\cos 150$ $\dfrac{-\sqrt{3}}{2}$
3. $\sin 780$ $\dfrac{\sqrt{3}}{2}$
4. $\sin \pi$ 0
5. $\cos \dfrac{3\pi}{4}$ $\dfrac{-\sqrt{2}}{2}$
6. $\sin 2\pi$ 0

Enrichment

On the chalkboard, draw a cube with side of length s and two of the diagonals forming angle θ, as shown below.

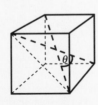

Have the students, without the use of tables or calculator, determine the cosine of the angle formed by the two diagonals of the cube. Suggest that the students first find the following.

1. The length of a diagonal of a face of the cube $s\sqrt{2}$
2. The length of a diagonal of the cube $s\sqrt{3}$ $\cos \theta = \frac{1}{3}$

18.7 Graphing the Sine and Cosine Functions

Objectives

To graph the sine and cosine functions for given domains
To determine $\sin(-\theta)$ and $\cos(-\theta)$ for given values of $\sin \theta$ and $\cos \theta$

As shown in a trigonometry table, there is one and only one sine (or cosine) ratio corresponding to any angle measure. Therefore, the sine and cosine relations are *functions*. The graph of $T = \sin \theta$ can be drawn by setting up a table of sample values and then drawing a smooth curve through the points corresponding to these table values.

EXAMPLE 1   Graph $T = \sin \theta$ for $0 \le \theta \le 2\pi$.

Solution   Set up a table of special radian measures for θ and find corresponding values for $T = \sin \theta$. The completed table is shown below with both exact and decimal values.

θ	0	$\frac{\pi}{6}$	$\frac{\pi}{3}$	$\frac{\pi}{2}$	$\frac{2\pi}{3}$	$\frac{5\pi}{6}$	π	$\frac{7\pi}{6}$	$\frac{3\pi}{2}$	$\frac{11\pi}{6}$	2π
$T = \sin \theta$	0	$\frac{1}{2}$	$\frac{\sqrt{3}}{2}$	1	$\frac{\sqrt{3}}{2}$	$\frac{1}{2}$	0	$\frac{-1}{2}$	-1	$\frac{-1}{2}$	0
	0	0.5	0.87	1	0.87	0.5	0	-0.5	-1	-0.5	0

If a calculator is used to obtain the table values, the calculator should be in the radian (RAD) mode, as illustrated below for $\theta = \frac{7\pi}{6}$.

$$7 \; \boxed{\times} \; \boxed{\pi} \; \boxed{\div} \; 6 \; \boxed{=} \; \boxed{\sin} \;\Rightarrow\; -0.5$$

Plot the ordered pairs in the table. Then, draw a smooth curve through the points.

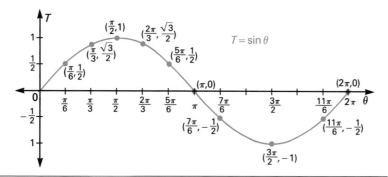

Additional Example 1

Graph $T = \sin \theta$ using the following five sample values for θ:

$-\pi$, $-\frac{\pi}{2}$, 0, $\frac{\pi}{2}$, and π.

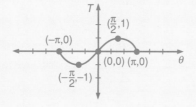

Teaching Resources

Quick Quizzes 139
**Reteaching and Practice
Worksheets** 139

GETTING STARTED

Prerequisite Quiz

Find each trigonometric ratio.

1. $\sin 0$ 0

2. $\cos \frac{\pi}{6}$ $\frac{\sqrt{3}}{2}$

3. $\sin \frac{\pi}{2}$ 1

4. $\sin \frac{2\pi}{3}$ $\frac{\sqrt{3}}{2}$

5. $\cos \frac{3\pi}{2}$ 0

6. $\sin \frac{11\pi}{6}$ $-\frac{1}{2}$

Motivator

Ask the class to construct a table of values for $y = \sin A$ using the following as domain: {0, 30, 90, 150, 180, 210, 270, 330, 360, 390, 450, 510, 540, 570, 630, 690, 720}. Have them graph these points and draw a smooth curve through them. Ask them if the function is periodic.   Yes.   Have them determine its amplitude. $\frac{1 - (-1)}{2} = \frac{2}{2} = 1$

Highlighting the Standards

Standard 9e: The graphing of trigonometric functions is seen by the Standards as a central element in the study of trigonometry.

Lesson Note

Students may find it easier in the beginning to use degree measures and multiples of 90 (rather than radian measures and multiples of $\frac{\pi}{2}$) when they graph the sine and cosine functions.

Point out that the graph of the cosine function is the same as that of the sine function, but shifted $\frac{\pi}{2}$ units to the left. Later on, this will be described in terms of a phase shift.

Math Connections

Formulas for the sine and cosine: The sine and cosine, like the tangent (see *Math Connections* for Lesson 18.5) may each be represented by an infinite series. The two series are as follows.

$$\cos x = 1 - \frac{x^2}{2}! + \frac{x^4}{4}! - \frac{x^6}{6}! + \cdots$$

$$\sin x = x - \frac{x^3}{3}! + \frac{x^5}{5}! - \frac{x^7}{7}! + \cdots$$

Here, x must be in radians.

Critical Thinking Questions

Analysis: The sine function is odd, the cosine function even. Ask students whether a function may be both odd and even. This question is related to the *Critical Thinking Questions* of Lesson 18.6 and has a similar answer. In order for both kinds of symmetry to be satisfied, it is necessary that the value of y for such a function be restricted to just the number 0. Thus, the function $y = 0$, with the x-axis as its graph, satisfies the given conditions.

Recall that if for every θ, $f(\theta) = f(\theta + p)$, where p is constant, then f is a periodic function. The reference triangles below illustrate that the sine function is periodic.

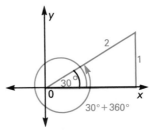 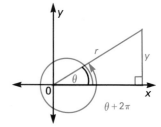

$\sin 30 = \sin(30 + 360) = \frac{1}{2}$, or $\sin \theta = \sin(\theta + 2\pi) = \frac{y}{r}$.

$\sin \frac{\pi}{6} = \sin(\frac{\pi}{6} + 2\pi)$ $\qquad f(\theta) = f(\theta + 2\pi)$

Therefore, $T = \sin \theta$ is periodic. The period is 2π.

The periodicity of the sine function can be visualized more clearly by extending the graph of Example 1.

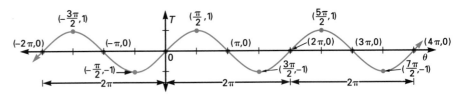

The basic shape is repeated every 2π units: $p = 2\pi$.
The maximum value M of $\sin \theta$ is 1, which occurs at $\theta = -\frac{3\pi}{2}$, $\theta = \frac{\pi}{2}$, $\theta = \frac{5\pi}{2}$, and so on.
The minimum value m of $\sin \theta$ is -1, which occurs at $\theta = -\frac{\pi}{2}$, $\theta = \frac{3\pi}{2}$, $\theta = \frac{7\pi}{2}$, and so on.
The amplitude of $\sin \theta$ is $\frac{M - m}{2} = \frac{1 - (-1)}{2} = \frac{2}{2} = 1$.

Notice in the graph of $T = \sin \theta$ that

as θ increases from 0 to $\frac{\pi}{2}$ (Quadrant I), T *increases* from 0 to 1,

as θ increases from $\frac{\pi}{2}$ to π (Quadrant II), T *decreases* from 1 to 0,

as θ increases from π to $\frac{3\pi}{2}$ (Quadrant III), T *decreases* from 0 to -1,

as θ increases from $\frac{3\pi}{2}$ to 2π (Quadrant IV), T *increases* from -1 to 0.

This pattern of increasing and decreasing is repeated every 2π units.

The sine function $T = \sin \theta$ is an odd function (and so is symmetric with respect to the origin), since $\sin\left(-\frac{\pi}{2}\right) = -1 = -\sin\frac{\pi}{2}$, $\sin(-\pi) = 0 = -\sin \pi$, $\sin\left(-\frac{5\pi}{2}\right) = -1 = -\sin\frac{5\pi}{2}$, and so on. In general, for any angle of measure θ,

$$\sin(-\theta) = -\sin\theta.$$

EXAMPLE 2 Find $\sin(-\theta)$ if $\sin\theta = 0.7071$.

Solution Use the property $\sin(-\theta) = -\sin\theta$.

$$\sin(-\theta) = -0.7071$$

The cosine function is also periodic, as illustrated in the reference triangles below.

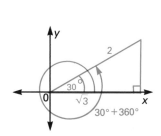 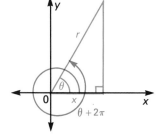

$$\cos 30 = \cos(30 + 360) = \frac{\sqrt{3}}{2}, \quad \text{or} \quad \cos\theta = \cos(\theta + 2\pi) = \frac{x}{r}.$$

Therefore, $\cos\theta$ is a periodic function with period 2π.

The periodicity is illustrated more clearly in the graph of $T = \cos\theta$. The table and graph are shown below for $-2\pi \le \theta \le 2\pi$.

θ	-2π	$-\frac{3\pi}{2}$	$-\pi$	$-\frac{\pi}{2}$	$-\frac{\pi}{3}$	0	$\frac{\pi}{3}$	$\frac{\pi}{2}$	π	$\frac{3\pi}{2}$	2π
$T = \cos\theta$	1	0	-1	0	$\frac{1}{2}$	1	$\frac{1}{2}$	0	-1	0	1

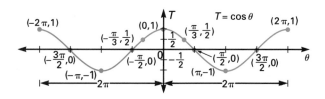

Common Error Analysis

Error: Students tend to memorize the graph of the sine function, but they have trouble remembering that the graph of the cosine function does *not* pass through the origin.

Suggest that the students use multiples of $\frac{\pi}{2}$ or 90 for θ in order to visualize each graph quickly.

Checkpoint

Use the following values of θ to make the indicated tables and graphs.

$-\frac{\pi}{2}$, 0, $\frac{\pi}{2}$, π, $\frac{3\pi}{2}$, 2π, $\frac{5\pi}{2}$, 3π

1. Make a table for $T = \sin\theta$.

θ	$-\frac{\pi}{2}$	0	$\frac{\pi}{2}$	π	$\frac{3\pi}{2}$	2π	$\frac{5\pi}{2}$	3π
T	-1	0	1	0	-1	0	1	0

2. Graph $T = \sin\theta$.

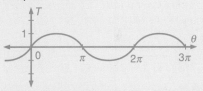

3. As θ increases from $-\frac{\pi}{2}$ to 3π, for what values of θ is $\sin\theta$ increasing?

$-\frac{\pi}{2} < \theta < \frac{\pi}{2}, \frac{3\pi}{2} < \theta < \frac{5\pi}{2}$

Closure

Have students give the period of the sine function. 2π The cosine function. 2π Ask them to find the amplitude of each of the functions above. 1 As the domain increases, beginning with 0, have students describe the difference in the appearance between the sine and cosine functions. The basic shape is the same; however, the sine function is an odd function while the cosine function is an even function.

Additional Example 2

Find $\sin(-\theta)$ if $\sin\theta = -0.4471$.
0.4471

Guided Practice

Classroom Exercises 1–10

Independent Practice

A Ex. 1–16, **B** Ex. 17–18, **C** Ex. 19–21

Basic: WE 1–17 odd, Problem Solving
Average: WE 5–19 odd, Problem Solving
Above Average: WE 9–21 odd, Problem Solving

Additional Answers

Written Exercises

1. $0, -\frac{1}{2}, -1, -\frac{1}{2}, 0, \frac{1}{2}, 1, \frac{1}{2}, 0$

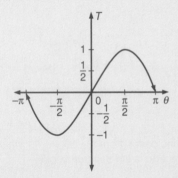

See pages 690–691 for the answers to Written Ex. 2, 4, 9, 10, 13–16, 18, 19, 21, and Mixed Review 1–8.

For the graph on the previous page, the basic shape is repeated every 2π units: $p = 2\pi$.

The maximum value M of $T = \cos \theta$ is 1.

The minimum value m of $T = \cos \theta$ is -1.

The amplitude of $T = \cos \theta$ is $\frac{M - m}{2} = \frac{1 - (-1)}{2} = \frac{2}{2} = 1$.

The cosine function $T = \cos \theta$ is an even function, since
$\cos\left(-\frac{\pi}{3}\right) = \cos \frac{\pi}{3} = \frac{1}{2}$, $\cos(-\pi) = \cos \pi = -1$,
$\cos\left(-\frac{3\pi}{2}\right) = \cos \frac{3\pi}{2} = 0$, and so on.
In general, for any angle of measure θ,

$$\cos(-\theta) = \cos \theta.$$

Notice also that the cosine function, like all even functions, is symmetric with respect to the vertical axis.

■■ Summary

$T = \sin \theta$
period: 2π
amplitude: 1
$\sin(-\theta) = -\sin \theta$
(an odd function)

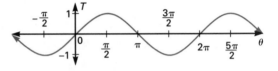

$T = \cos \theta$
period: 2π
amplitude: 1
$\cos(-\theta) = \cos \theta$
(an even function)

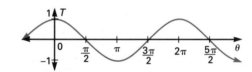

Classroom Exercises

Find each of the following.

1. the amplitude of the sine function 1
2. the period of the cosine function 2π
3. the minimum value of $T = \cos \theta$ -1
4. the period of $T = \sin \theta$ 2π
5. the maximum value of $T = \sin \theta$ 1
6. the amplitude of the cosine function 1

Tell whether $T = \cos \theta$ increases or decreases as θ increases in the given interval. (HINT: Use the graph of $T = \cos \theta$ above.)

7. from 0 to $\frac{\pi}{2}$
   Decreases
8. from $\frac{\pi}{2}$ to π
   Decreases
9. from π to $\frac{3\pi}{2}$
   Increases
10. from $\frac{3\pi}{2}$ to 2π
    Increases

Enrichment

Have the students draw graphs of $T = \sin x$ and $T = \cos x$ on a single pair of axes for $0 \le x \le 2\pi$.

Discuss with the students some of the properties of the functions and the interrelationships that can be observed from the graphs.

Here are a few examples.

1. Both functions are positive in quadrant I; both functions are negative in quadrant III.
2. The sine and cosine values are equal at $\frac{\pi}{4}$ and $\frac{5\pi}{4}$.
3. Since $\tan x = \frac{\sin x}{\cos x}$, and since $\cos x = 0$ at $\frac{\pi}{2}$ and $\frac{3\pi}{2}$, the tangent function is undefined at these values of x.

Have the students list as many of these observations as they can.

Some other observations are:

1. As x increases, both functions decrease in quadrant II; both functions increase in quadrant IV.
2. $\sin x < \cos x$ for $0 \le x < \frac{\pi}{4}$ and $\frac{5\pi}{4} < x \le 2\pi$, and
3. the cotangent function is undefined at 0, π, and 2π.

Written Exercises

In Exercises 1 and 2, complete the table of values for each function.
Use the results to graph the function.

1. $T = \sin \theta$

2. $T = \cos \theta$

θ	$-\pi$	$-\dfrac{5\pi}{6}$	$-\dfrac{\pi}{2}$	$-\dfrac{\pi}{6}$	0	$\dfrac{\pi}{6}$	$\dfrac{\pi}{2}$	$\dfrac{5\pi}{6}$	π
T	?	?	?	?	?	?	?	?	?

3. As θ increases, for what values of θ is T increasing in Exercise 1? $-\dfrac{\pi}{2} \le \theta \le \dfrac{\pi}{2}$

4. As θ increases, for what values of θ is T decreasing in Exercise 1?

5. As θ increases, for what values of θ is T increasing in Exercise 2? $-\pi \le \theta \le 0$

6. As θ increases, for what values of θ is T decreasing in Exercise 2? $0 \le \theta \le \pi$

7. Find $\sin(-\theta)$ if $\sin \theta = 0.5446$.
-0.5446

8. Find $\cos(-\theta)$ if $\cos \theta = 0.7431$.
0.7431

In Exercises 9–18, use the following values of θ to make the indicated tables and graphs. Graph the indicated functions in the same coordinate plane.

$$\left\{ -2\pi,\ -\frac{11\pi}{6},\ -\frac{3\pi}{2},\ -\frac{7\pi}{6},\ -\pi,\ -\frac{5\pi}{6},\ -\frac{\pi}{2},\ -\frac{\pi}{6},\ 0 \right\}$$

9. Make a table of values for $T = \cos \theta$.

10. Graph $T = \cos \theta$.

11. As θ increases, for what values of θ is $T = \cos \theta$ increasing? $-\pi \le \theta \le 0$

12. As θ increases, for what values of θ is $T = \cos \theta$ decreasing? $-2\pi \le \theta \le -\pi$

13. Make a table of values for $T = \sin \theta$.

14. Graph $T = \sin \theta$.

15. As θ increases, for what values of θ is $T = \sin \theta$ increasing?

16. As θ increases, for what values of θ is $T = \sin \theta$ decreasing?

17. For what values of θ does it appear that $\sin \theta = \cos \theta$? $\dfrac{\pi}{4} + k\pi$, where k is an integer.

18. Write briefly why the sine function is an odd periodic function.

In Exercises 19–21, $T = |\sin \theta|$ for $-2\pi \le \theta \le 4\pi$.

19. Graph $T = |\sin \theta|$.

20. Give the period of $T = |\sin \theta|$. π

21. Determine whether the function is odd or even. Give a reason.

Mixed Review

Graph each function in the same coordinate plane. *11.3*

1. $y = |x|$ **2.** $y = -4|x|$ **3.** $y = 2|x - 4|$ **4.** $y = 3|x + 6| - 4$

Graph each function in the same coordinate plane. *11.4*

5. $y = x^2$ **6.** $y = 2x^2$ **7.** $y = \dfrac{1}{2}x^2$ **8.** $y = x^2 + 4$

17.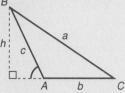

The area of any triangle is $\dfrac{1}{2}bh$. Now from the figure, $\dfrac{h}{c} = \sin(180 - A)$, or $h = c\sin(180 - A)$. But $\sin A = \sin(180 - A)$, so if the area is K, then $K = \dfrac{1}{2}b[c\sin(180 - A)] = \dfrac{1}{2}bc\sin A$.

Mixed Review

1.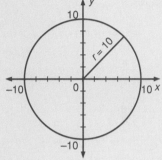

2.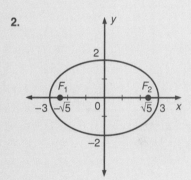

3.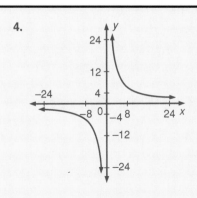

4.

Problem Solving

1. 0.0000312 au ≈ 2,900 mi or 4668 km;
   This is less than the earth's radius, so the center of mass is below the earth's surface.

2. d ≈ 571,000,000 mi or 918,700,000 km;
   The trailing zeros in these answers suggest that they have been rounded off. So, the converted values should be rounded off to the same level of precision. A distance of 0.0000312 au (≈ 2900 mi or 4668 km) added to these values would not significantly affect them, but would be lost in rounding off.

3. d = 9.1521657 au (estimated);
   difference from actual value: 0.0000301 au

Additional Answers, page 688

2. $-1, -\frac{\sqrt{3}}{2}, 0, \frac{\sqrt{3}}{2}, 1, \frac{\sqrt{3}}{2}, 0, -\frac{\sqrt{3}}{2}, -1$

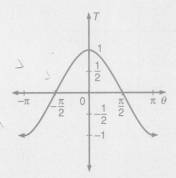

4. $-\pi \le \theta \le -\frac{\pi}{2}, \frac{\pi}{2} \le \theta \le \pi$

9. $T = \cos \theta$

θ	-2π	$-\frac{11\pi}{6}$	$-\frac{3\pi}{2}$	$-\frac{7\pi}{6}$	$-\pi$
T	1	$\frac{\sqrt{3}}{2}$	0	$-\frac{\sqrt{3}}{2}$	-1

θ	$-\frac{5\pi}{6}$	$-\frac{\pi}{2}$	$-\frac{\pi}{6}$	0
T	$-\frac{\sqrt{3}}{2}$	0	$\frac{\sqrt{3}}{2}$	1

Problem Solving Strategies: *Find an Estimate*

In many real-world situations, it may not be necessary to find an exact answer to a problem. Often, an estimate may be appropriate. In such cases, however, it is important to know how much the estimate may vary from the true value.

Suppose, for example, that you want to find the distance from the Earth to Jupiter on August 19, 1990, using the table below.

SUN-CENTERED COORDINATES OF MAJOR PLANETS FOR AUG 17, 1990 IN ASTRONOMICAL UNITS

	x	y	z
Earth/Moon	0.8189001	−0.5462572	−0.2368435
Mars	1.3894497	0.1451882	0.0290001
Jupiter	−2.238098	4.328104	1.909756
Saturn	3.934523	−8.443261	−3.656253

The table gives 3-dimensional, x-y-z coordinates for the planets using a coordinate system with the Sun at the origin. The x-axis is defined as lying in the direction of the Earth's position at the first moment of Spring, the vernal equinox.

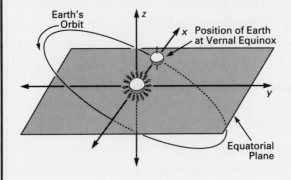

The x-y plane of the system is the **equatorial plane**, which corresponds to the plane of the equator of the Earth at the time of the equinox.

The unit of measure used in the table is the **astronomical unit** (**au**), which equals the mean distance of the Earth from the Sun, about 92,900,000 miles or 149,600,000 kilometers.

690 Chapter 18 Solving Triangles & Graphing Functions

10.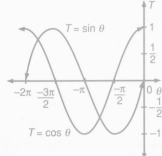

13. $T = \sin \theta$

θ	-2π	$-\frac{11\pi}{6}$	$-\frac{3\pi}{2}$	$-\frac{7\pi}{6}$	$-\pi$
T	0	$\frac{1}{2}$	1	$\frac{1}{2}$	0

θ	$-\frac{5\pi}{6}$	$-\frac{\pi}{2}$	$-\frac{\pi}{6}$	0
T	$-\frac{1}{2}$	-1	$-\frac{1}{2}$	0

14.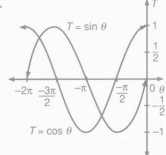

The values given for the Earth/Moon are actually for a point called the **center of mass** of the two bodies, which is located between their centers. This point may be used as an *approximation* of the position of the Earth, since it is not far from the Earth's center.

EXAMPLE   Estimate the distance between the centers of the Earth and Jupiter.

Solution   Use the distance formula for 3 dimensions.

$$d = \sqrt{(x_2 - x_1)^2 + (y_2 - y_1)^2 + (z_2 - z_1)^2}$$

Begin by subtracting the coordinates of Earth-Moon and Jupiter.

0.8189001	−0.5462572	−0.2368435
−2.238098	4.328104	1.909756
3.0569981	−4.8743612	−2.1465995

Substitute the results into the formula.

$$d = \sqrt{(3.0569981)^2 + (-4.8743612)^2 + (-2.1465995)^2}$$
$$= \sqrt{37.7125239} = \mathbf{6.1410523 \ au}$$

The center of mass of the Earth/Moon system is 0.0000312 au from the center of the Earth, so this is the *maximum amount that the above estimate can be in error.* Thus, the true value of the distance, to 7 decimal places, may be as *low* as 6.1410523 − 0.0000312 = 6.1410211 au, or as *high* as 6.1410523 − 0.0000312 = 6.1410835 au.

The actual value of the distance turns out to be 6.1410804 au. This value is within the bounds of the estimate. It differs from the estimate by 0.0000281 au.

Exercises

1. Use the values given at the bottom of the previous page to express the center of mass distance 0.0000312 au in miles and kilometers. How does this compare with the radius of the Earth? (The Earth's radius = 3,959 mi or 6,371 km.)

2. Express d in the above example in miles and kilometers. Do you think that a difference of 0.0000312 au would be significant in these figures? (HINT: Do the values for the Earth–Sun distances seem to be estimates themselves?)

3. Use the values in the table to estimate the distance between the centers of the Earth and Saturn. Compare this with the actual distance of 9.1521356 au.

15. $-2\pi \le \theta \le -\dfrac{3\pi}{2}$ and $-\dfrac{\pi}{2} \le \theta \le 0$

16. $-\dfrac{3\pi}{2} \le \theta \le -\dfrac{\pi}{2}$

18. Answers will vary. One possible answer: $T(-\theta) = T(\theta)$, so it is odd, and the function repeats every 2π units, so it is periodic.

19.

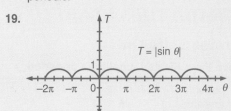

21. Even; $|\sin(-\theta)| = |-\sin\theta| = |\sin\theta|$

Mixed Review

1–4.

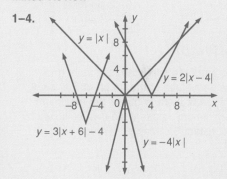

5–8.

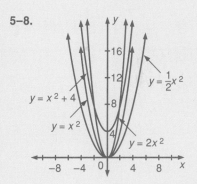

■■■GETTING STARTED

Prerequisite Quiz

Give the indicated amplitude or period.

1. period of $T = \sin \theta$ 2π
2. amplitude of $T = \sin \theta$ 1
3. amplitude of $T = \cos \theta$ 1
4. period of $T = \cos \theta$ 2π

Evaluate for $\theta = \dfrac{\pi}{2}$.

5. $\sin \theta$ 1   6. $2 \sin \theta$ 2
7. $5 \sin \theta$ 5   8. $-3 \sin \theta$ -3

Motivator

Have students recall the maximum and
minimum values of the graph of $T = \sin \theta$.
1; −1   Ask them to guess the maximum
and minimum values if $\sin \theta$ is multiplied by
the constant 3.   Max = 3; min = −3

18.8   Changing the Amplitude and Period

Objectives

To graph functions of the form $T = c + a \cos b\theta$ and
$T = c + a \sin b\theta$

To determine the amplitude, the period, and the maximum and
minimum values of the above functions

Periodic functions occur in various
fields, including medicine. For example, the electrocardiogram (EKG) at
the right shows a heart's electrical
impulses. The horizontal axis represents time in seconds, and the vertical axis represents the strength of the
electrical impulses.

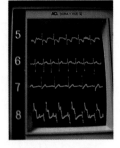

Recall that the amplitude of $T = \sin \theta$ and of $T = \cos \theta$ is 1. If $\sin \theta$
or $\cos \theta$ is multiplied by a constant a, then the amplitude will change.
This can be seen for the sine function by graphing $T = \sin \theta$,
$T = 2 \sin \theta$, and $T = -\frac{1}{2} \sin \theta$ on the same set of axes.

θ	$-\dfrac{\pi}{2}$	0	$\dfrac{\pi}{2}$	π	$\dfrac{3\pi}{2}$	2π	$\dfrac{5\pi}{2}$	3π
$T = \sin \theta$	-1	0	1	0	-1	0	1	0
$T = 2 \sin \theta$	-2	0	2	0	-2	0	2	0
$T = -\dfrac{1}{2} \sin \theta$	$\dfrac{1}{2}$	0	$-\dfrac{1}{2}$	0	$\dfrac{1}{2}$	0	$-\dfrac{1}{2}$	0

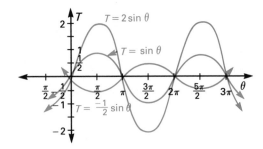

The graphs suggest that the period of $T = a \sin \theta$ is 2π regardless of
the value of a.

Highlighting the Standards

Standard 6e: Progress in understanding
how parameter changes affect functions is a
sign of mathematical maturity for students.

The amplitudes of $T = 2 \sin \theta$ and $T = -\frac{1}{2} \sin \theta$ are calculated below.

$T = 2 \sin \theta$

maximum: $M = 2$

minimum: $m = -2$

amplitude: $\dfrac{M - m}{2}$

$= \dfrac{2 - (-2)}{2} = 2$

$T = -\frac{1}{2} \sin \theta$

maximum: $M = \frac{1}{2}$

minimum: $m = -\frac{1}{2}$

amplitude: $\dfrac{M - m}{2}$

$= \dfrac{\frac{1}{2} - \left(-\frac{1}{2}\right)}{2} = \frac{1}{2}$

Amplitude and Period for $T = a \sin \theta$ and $T = a \cos \theta$
If $T = a \sin \theta$ or $T = a \cos \theta$, then for each real number $a \neq 0$,
(1) the *amplitude* is $|a|$, and
(2) the *period* is 2π.

The graph of $T = \sin \theta$ or $T = \cos \theta$ is
(1) *widened* vertically if $|a| > 1$, and
(2) *narrowed* vertically if $|a| < 1$.

The graph of $T = a \sin \theta$ or $T = a \cos \theta$ can be translated (moved) upward or downward by adding or subtracting a constant. This is illustrated for the cosine function in Example 1 below.

EXAMPLE 1   Graph $T = 4 + 2 \cos \theta$ for $0 \le \theta \le 2\pi$. Find the amplitude and period.

Plan   Make a table of values using multiples of $\frac{\pi}{2}$ for five sample values of θ.

Solution

θ	0	$\frac{\pi}{2}$	π	$\frac{3\pi}{2}$	2π
$\cos \theta$	1	0	-1	0	1
$4 + 2 \cos \theta$	6	4	2	4	6

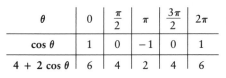

maximum value: $M = 6$
minimum value: $m = 2$
amplitude: $\dfrac{M - m}{2} = \dfrac{6 - 2}{2} = 2$

Thus, for $T = 4 + 2 \cos \theta$, the amplitude is 2 and the period is 2π.

Example 2 on the next page shows the effect of multiplying the angle measure θ by a constant. Recall that the period of $T = \sin \theta$ is 2π. Therefore, one cycle of the sine function is completed every 2π units.

Lesson Note

Have students discover the effects of changes in amplitude and period. Ask them to graph the following three functions on the same set of axes: $T = \sin \theta$, $T = 4 \sin \theta$, and $T = \frac{1}{2} \sin \theta$. For ease and speed, use as a domain multiples of $\frac{\pi}{2}$ from 0 through 2π. Then ask the students to describe and explain the differences in the graphs.

Use a similar procedure for changes in the period of a sine curve.

Math Connections

Sound: *Sound* is created by vibrations, for example, the vibrating column of air in a flute. If the sound is a pure tone, then its effect can be visualized as a sine wave on the screen of an oscilloscope. The equation of the sine wave will depend on the pitch of the tone. For example, the equation of "Middle C" on a piano might have as its equation $y = \sin 528\pi t$, where t is the time. The frequency of the vibrations is $528 \div 2$, or 264 vibrations (cycles) per second. (The frequency of a sine function is the reciprocal of its period.)

Critical Thinking Questions

Analysis: Mention to students that the frequency of a sine or cosine function is the number of cycles that occur over an interval that is one unit long. Then ask them to discover how the frequency f is related to the period p. They should be able to see that the two measures are reciprocals of each other (see *Math Connections*).

Additional Example 1

Graph $T = 3 - 4 \sin \theta$ for $0 \le \theta \le 2\pi$. Determine the amplitude and period.
Amplitude: 4; period: 2π

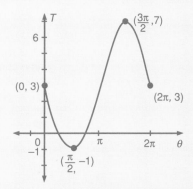

Common Error Analysis

Error: Students see that for $T = a \sin b\theta$, as $|a|$ grows larger, there is a vertical widening of the graph of $\sin \theta$. They then expect that the same will hold for $|b|$, except in a horizontal direction.

Emphasize that the relationship is *reversed*. As $|b|$ increases, there is a horizontal *narrowing* of the graph of $\sin \theta$.

Checkpoint

Graph one cycle of each function without using a table of values. Determine the period and amplitude.

1. $T = 4 \sin \theta$ *pd:* 2π; *amp:* 4

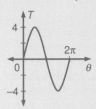

2. $T = 2 \cos \theta$ *pd:* 2π; *amp:* 2

Closure

Ask students to describe the effect of changing the value of each of the following on the graph of $T = c + a \sin b\theta$.

1. **a** $|a| > 1$ widens graph vertically; $|a| < 1$ narrows graph vertically.
2. **b** $|b| > 1$ narrows graph horizontally; $|b| < 1$ widens graph horizontally.
3. **c** $c > 0$ translates graph upward; $c < 0$ translates graph downward.

EXAMPLE 2   Graph $T = \sin 2\theta$ and $T = \sin \theta$ in the same coordinate plane for $0 \le \theta \le 2\pi$. Determine the amplitude and the period of $T = \sin 2\theta$.

Plan   For $T = \sin 2\theta$, use multiples of $\frac{\pi}{4}$ to get nine sample values.

Solution

θ	0	$\frac{\pi}{4}$	$\frac{\pi}{2}$	$\frac{3\pi}{4}$	π	$\frac{5\pi}{4}$	$\frac{3\pi}{2}$	$\frac{7\pi}{4}$	2π
2θ	0	$\frac{\pi}{2}$	π	$\frac{3\pi}{2}$	2π	$\frac{5\pi}{2}$	3π	$\frac{7\pi}{2}$	4π
$\sin 2\theta$	0	1	0	-1	0	1	0	-1	0

The graph of $T = \sin 2\theta$ completes 2 cycles from 0 to 2π. It completes one cycle every $\frac{2\pi}{2}$, or π units.

Therefore, the period is π and the amplitude is 1.

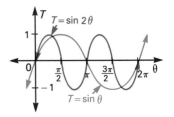

Notice that the graph of $T = \sin 2\theta$ above is a *horizontal narrowing* of the graph of $T = \sin \theta$. The cosine function behaves similarly.

Period for sin $b\theta$ and cos $b\theta$

If $T = \sin b\theta$, then the period is $\frac{2\pi}{|b|}$ for each real number $b \ne 0$.

If $T = \cos b\theta$, then the period is $\frac{2\pi}{|b|}$ for each real number $b \ne 0$.

The graph of $T = 4 \cos \frac{1}{3}\theta$ can be drawn without constructing a table of values. This is done by using the amplitude and the period of the graph, as illustrated in Example 3.

EXAMPLE 3   Graph one cycle of $T = 4 \cos \frac{1}{3}\theta$ in the same coordinate plane as the graph of $T = \cos \theta$. Do not use a table of values.

Plan   For $T = 4 \cos \frac{1}{3}\theta$, first determine the amplitude and the period. Then graph the function.

Solution $T = 4 \cos \frac{1}{3}\theta$: amplitude = 4, period = $\frac{2\pi}{\frac{1}{3}} = 2\pi(3) = 6\pi$

One cycle of the graph is completed every 6π units, as shown on the next page.

Additional Example 2

Graph $T = \cos \frac{1}{2}\theta$ and $T = \cos \theta$ in the same coordinate plane for $0 \le \theta \le 4\pi$. Determine the amplitude and period of $T = \cos \frac{1}{2}\theta$.

Amplitude: 1; period: 4π

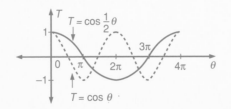

Additional Example 3

Graph one cycle of $T = 3 \sin 2\theta$ in the same coordinate plane as the graph of $T = \sin \theta$. Do not use a table of values.

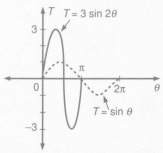

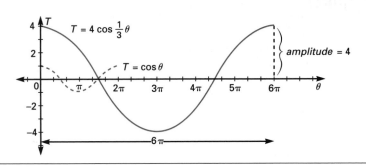

Notice that the graph of $T = 4 \cos \frac{1}{3}\theta$ is a *horizontal widening* and a *vertical widening* of the graph of $T = \cos \theta$.

EXAMPLE 4 Graph one cycle of the function $T = -1 + 2 \sin 3\theta$ without using a table of values. Determine the amplitude, the period, and the maximum and minimum values of $T = -1 + 2 \sin 3\theta$.

Solution First, graph $T = 2 \sin 3\theta$.

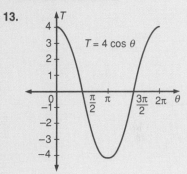

amplitude: 2   period: $\frac{2\pi}{3}$

maximum: 2   minimum: -2

Second, translate the graph of $T = 2 \sin 3\theta$ one unit downward.

maximum: $-1 + 2 = 1$

minimum: $-1 + (-2) = -3$

amplitude: $\dfrac{1 - (-3)}{2} = 2$

Therefore, for $T = -1 + 2 \sin 3\theta$, the amplitude is 2, the period is $\frac{2\pi}{3}$, the maximum value is 1, and the minimum value is -3.

Graph of $T = a \sin b\theta$ or $T = a \cos b\theta$

Amplitude	Vertical Effect of a on Graph of $\sin \theta$ or $\cos \theta$	Period	Horizontal Effect of b on Graph of $\sin \theta$ or $\cos \theta$
$|a|$	widens if $|a| > 1$ narrows if $|a| < 1$	$\dfrac{2\pi}{|b|}$	widens if $|b| < 1$ narrows if $|b| > 1$

The graph of $T = c + a \sin b\theta$ or $T = c + a \cos b\theta$ is a translation of the graph upward c units if $c > 0$, or downward $|c|$ units if $c < 0$.

18.8 Changing the Amplitude and Period   **695**

FOLLOW UP

Guided Practice

Classroom Exercises 1–16

Independent Practice

A Ex. 1–17, **B** Ex. 18–21, **C** Ex. 22–24

Basic: WE 1–15 odd
Average: WE 5–19 odd
Above Average: WE 9–23 odd

Additional Answers

Classroom Exercises

13.

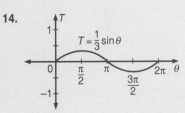

14.

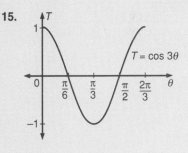

15.

Additional Example 4

Graph one cycle of the function $T = 3 - \frac{1}{2} \sin 2\theta$ without using a table of values.

Determine the amplitude, period, and maximum and minimum values of $T = 3 - \frac{1}{2} \sin 2\theta$.   Amplitude: $\frac{1}{2}$; period: π; maximum: 3.5; minimum: 2.5

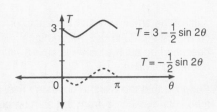

16.

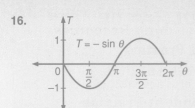

$T = -\sin\theta$

Written Exercises

1.

θ	$\frac{\pi}{8}$	$\frac{\pi}{4}$	$\frac{3\pi}{8}$	$\frac{\pi}{2}$	$\frac{5\pi}{8}$	$\frac{3\pi}{4}$	$\frac{7\pi}{8}$
T	1	0	−1	0	1	0	−1

θ	π	$\frac{9\pi}{8}$	$\frac{5\pi}{4}$	$\frac{11\pi}{8}$	$\frac{3\pi}{2}$	$\frac{13\pi}{8}$
T	0	1	0	−1	0	1

θ	$\frac{7\pi}{4}$	$\frac{15\pi}{8}$	2π
T	0	−1	0

$T = \sin 4\theta$

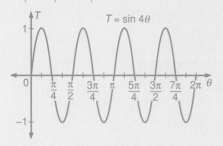

2.

θ	0	$\frac{\pi}{2}$	π	$\frac{3\pi}{2}$	2π
T	$\frac{1}{2}$	0	$-\frac{1}{2}$	0	$\frac{1}{2}$

$T = \frac{1}{2}\cos\theta$

See pages 826–827 for answers to Written Ex. 3–23.

Classroom Exercises

Give the amplitude and the period of each function.

1. $T = 4\cos\theta$ 4, 2π
2. $T = \frac{1}{3}\sin\theta$ $\frac{1}{3}$, 2π
3. $T = \cos 3\theta$ 1, $\frac{2\pi}{3}$
4. $T = -\sin\theta$ 1, 2π
5. $T = \sin 4\theta$ 1, $\frac{\pi}{2}$
6. $T = \frac{1}{2}\cos\theta$ $\frac{1}{2}$, 2π
7. $T = -\sin 2\theta$ 1, π
8. $T = 2\cos\frac{1}{3}\theta$ 2, 6π
9. $T = 1 + \sin\theta$ 1, 2π
10. $T = 3 - \cos\theta$ 1, 2π
11. $T = -4 + 4\sin\theta$ 4, 2π
12. $T = 3 + 5\sin\theta$ 5, 2π

13–16. Graph one cycle of each function of Classroom Exercises 1–4 without using a table of values.

Written Exercises

Graph each function for $0 \le \theta \le 2\pi$. Use a table of values.

1–8. Use the functions of Classroom Exercises 5–12 above.

Graph one cycle of each function without using a table of values. Determine the period and the amplitude.

9. $T = \sin 3\theta$
10. $T = \frac{1}{2}\sin 2\theta$
11. $T = 5\cos 4\theta$
12. $T = 3\cos 6\theta$
13. $T = -5\sin 2\theta$
14. $T = 2\cos\frac{1}{2}\theta$
15. $T = \frac{1}{4}\sin\frac{1}{3}\theta$
16. $T = 5\cos\frac{1}{4}\theta$
17. $T = -6\cos 9\theta$

Graph one cycle of each function without using a table of values. Determine the period, the amplitude, and the maximum and minimum values.

18. $T = 6 + 3\cos 2\theta$
19. $T = -6 + 5\sin 4\theta$
20. $T = -2\frac{1}{2} + \frac{1}{2}\sin 3\theta$

21. Graph $T = \cos\theta$ and $T = -\sin\theta$ in the same coordinate plane for $-2\pi \le \theta \le 2\pi$. For what values of θ does $\cos\theta = -\sin\theta$?
22. Graph $T = \theta + \sin\theta$ for $-2\pi \le \theta \le 2\pi$.
23. Graph $T = |3\sin 4\theta|$ for $-2\pi \le \theta \le 2\pi$.
24. The graph of a sine function has a period of $\frac{\pi}{2}$, a maximum value of 6, and a minimum value of 2. Write an equation for the function. $T = 4 + 2\sin 4\theta$

Mixed Review

Find each trigonometric ratio in simplest radical form.   18.5

1. $\sec\pi$ −1
2. $\tan\frac{3\pi}{4}$ −1
3. $\csc\frac{7\pi}{6}$ −2
4. $\cot\frac{5\pi}{3}$ $-\frac{\sqrt{3}}{3}$

Enrichment

Draw this picture of a weight suspended by a spring. Explain that if the weight is pulled down to point P and released, it will spring upward to point Q, then back to P. If there were no energy loss, this process would continue forever.

Have the students imagine a horizontally moving screen behind the weight, with the shadow of the weight on the screen. The path of the shadow would be a sine or cosine curve. Point out that motion described by such a curve is called *simple harmonic motion.*

Suppose that the weight moves so that at time t, its distance d from the starting point is given by the equation $d = -0.4\cos\pi t$. Have the students graph one cycle of this curve on axes labeled t and d.

18.9 Graphing the Tangent, Cotangent, Secant, and Cosecant Functions

Objective   To graph the tangent, cotangent, secant, and cosecant functions

Recall that $\tan \frac{\pi}{2}$, or $\tan 90°$, is undefined since division by zero is undefined. Similarly, $\tan \left(-\frac{\pi}{2}\right) = \frac{-1}{0}$ is undefined. More generally, the tangent ratio $\tan \theta$ is undefined for $\frac{\pi}{2}, \frac{3\pi}{2}, \frac{5\pi}{2}, \frac{7\pi}{2}, \cdots$, that is, for $\theta = \frac{\pi}{2} + k \cdot \pi$, where k is an integer.

However, as θ increases from 0 to $\frac{\pi}{2}$, $\tan \theta$ increases without bound. This is illustrated on a calculator by first selecting the radian mode (RAD), and then noticing that $\tan 1.50 = 14.10142$, $\tan 1.53 = 24.49841$, and $\tan 1.57 = 1{,}255.7653$.

To graph $T = \tan \theta$ for $-\frac{\pi}{2} \le \theta \le \frac{\pi}{2}$, make a table of values.

θ	$-\frac{\pi}{2}$	$-\frac{\pi}{3}$	$-\frac{\pi}{4}$	$-\frac{\pi}{6}$	0	$\frac{\pi}{6}$	$\frac{\pi}{4}$	$\frac{\pi}{3}$	$\frac{\pi}{2}$
$\tan \theta$	undef	-1.7	-1	-0.6	0	0.6	1	1.7	undef

Notice that as θ approaches $\frac{\pi}{2}$ from the left, $\tan \theta$ increases without bound. As θ approaches $-\frac{\pi}{2}$ from the right, $\tan \theta$ decreases without bound. The vertical lines $\theta = -\frac{\pi}{2}$ and $\theta = \frac{\pi}{2}$ are *asymptotes* of the graph of $T = \tan \theta$. There is no maximum or minimum value. Since there is no maximum or minimum, the amplitude is not defined.

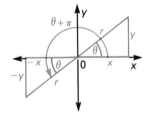

As seen in the figure at the right, the tangent function, like the sine and cosine functions, is periodic. To see this, notice that $\tan \theta = \frac{y}{x}$ and also $\tan (\theta + \pi) = \frac{-y}{-x} = \frac{y}{x}$.

Thus, $\tan \theta = \tan (\theta + \pi)$.

Therefore, the tangent function $T = \tan \theta$ is periodic with period π.

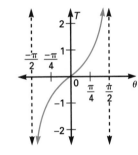

Teaching Resources

Problem Solving Worksheet 18
Quick Quizzes 141
**Reteaching and Practice
  Worksheets** 141
Transparencies 40, 41A, 41B, 42A, 42B, 43A, 43B, 44

▬▬ GETTING STARTED

Prerequisite Quiz

Find the value of each trigonometric ratio.

1. $\sec \frac{3\pi}{4}$ $-\sqrt{2}$ **2.** $\tan \left(-\frac{\pi}{3}\right)$ $-\sqrt{3}$

3. $\csc \pi$ Undef   **4.** $\cot \frac{5\pi}{6}$ $-\sqrt{3}$

5. $\tan \frac{\pi}{2}$ Undef   **6.** $\cot \frac{\pi}{4}$ 1

Motivator

Try to get the students to discover the asymptotic nature of the graph of the tangent function by relating the new to the familiar.

For the graph of $y = \frac{1}{x}$, ask students what happens to y-values as x approaches 0 from the right.   Values of y increase without bound.

Similarly, have students use a calculator to determine what happens to T as θ approaches 90 from the left for $T = \tan \theta$. Values for $\tan \theta$ increase without bound.

Highlighting the Standards

Standard 9e: This lesson completes the work of graphing the trigonometric functions. Students should now see clearly that trigonometry means much more than triangles.

Lesson Note

Review the reciprocal relationships between the sine and cosecant, cosine and secant, and tangent and cotangent functions.

It may be helpful to review the idea of an asymptote by sketching the graph of $f(x) = \frac{1}{x}$. Suggest that students draw the asymptotes first before they graph the tangent, cotangent, secant, and cosecant functions.

Math Connections

Historical Note: Trigonometry may be said to have really come of age in the period of the Renaissance. Finally, all six trigonometric functions came into full use with the appearance of a two-volume work *Opus palatinum de triangulis* by Georg Joachim Rheticus (1514–1576), a student of Copernicus. In this work, Rheticus calculated elaborate tables of the six functions.

Critical Thinking Questions

Analysis: The graphs of this lesson are characterized by the presence of asymptotes. Ask students whether this is the first such appearance. They should recall that the graph of the exponential functions of Lesson 13.2 had the x-axis as a horizontal asymptote, and the logarithmic functions of Lesson 13.4 had the y-axis as an asymptote. Both the x- and y-axes were asymptotes for the rectangular hyperbolas of Lesson 12.5.

This periodicity can be visualized more clearly by extending the graph of the tangent function.

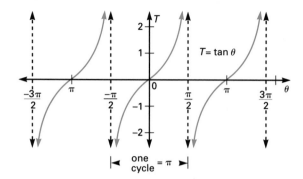

Notice that the function is an odd function because, for example, $\tan\left(-\frac{\pi}{4}\right) = -\tan\frac{\pi}{4} = -1$, $\tan\left(-\frac{\pi}{3}\right) = -\tan\frac{\pi}{3}$, and so on.

In general, for any angle measure θ for which $T = \tan(\theta)$ is defined,

$$\tan(-\theta) = -\tan\theta.$$

Recall that the period for $T = a\sin b\theta$ and of $T = a\cos b\theta$ is $\frac{2\pi}{|b|}$. However, for $T = a\tan b\theta$, the period is $\frac{\pi}{|b|}$.

EXAMPLE 1 Graph $T = \tan 2\theta$ for $-\frac{3\pi}{4} < \theta < \frac{3\pi}{4}$.

Solution Find the period: $p = \frac{\pi}{|b|} = \frac{\pi}{2}$. Then find and draw the asymptotes.

Sketch one cycle from $-\frac{1}{2}p$ to $\frac{1}{2}p$, that is, from $-\frac{1}{2}\left(\frac{\pi}{2}\right)$ to $\frac{1}{2}\left(\frac{\pi}{2}\right)$, or from $-\frac{\pi}{4}$ to $\frac{\pi}{4}$. Repeat the cycle from $-\frac{3\pi}{4}$ to $-\frac{\pi}{4}$ and from $\frac{\pi}{4}$ to $\frac{3\pi}{4}$.

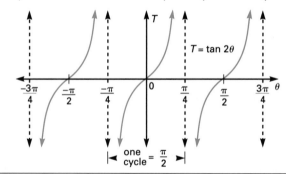

Additional Example 1

Graph $T = \tan\frac{1}{2}\theta$ for $-\pi < \theta < \pi$.

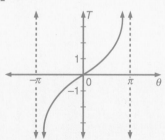

The cotangent ratio is the reciprocal of the tangent ratio. It is undefined where $\tan \theta = 0$, and 0 where $\tan \theta$ is undefined. Its period is the same as the tangent function, $\frac{\pi}{|b|}$, and its amplitude is not defined.

EXAMPLE 2 Graph $T = \sin \theta$ and $T = \csc \theta$ for $0 \le \theta \le 2\pi$ in the same coordinate plane. Determine the amplitude and the period of $T = \csc \theta$.

Plan Use the fact that the trigonometric ratios $\sin \theta$ and $\csc \theta$ are reciprocals of each other for $\sin \theta \ne 0$ to make a table of values.

Solution

θ	0	$\frac{\pi}{6}$	$\frac{\pi}{2}$	$\frac{5\pi}{6}$	π	$\frac{7\pi}{6}$	$\frac{3\pi}{2}$	$\frac{11\pi}{6}$	2π
$\sin \theta$	0	$\frac{1}{2}$	1	$\frac{1}{2}$	0	$\frac{-1}{2}$	-1	$\frac{-1}{2}$	0
$\csc \theta$	undef	2	1	2	undef	-2	-1	-2	undef

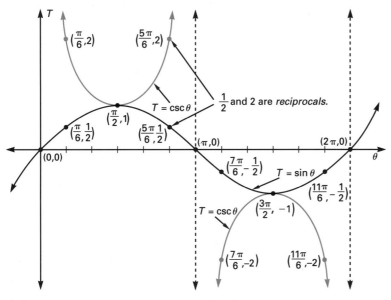

There is no maximum or minimum value for $T = \csc \theta$. Therefore, the cosecant function has no amplitude defined. The period of $T = \csc \theta$ is 2π, the same as that of $T = \sin \theta$.

Like sine and cosecant, the cosine and secant ratios are reciprocals of each other. The graph of $T = \sec \theta$ can be graphed on the same set of axes as the graph of $T = \cos \theta$, as done above for $\sin \theta$ and $\csc \theta$.

Common Error Analysis

Error: Students are so used to finding the amplitude of a periodic function that some of them may indicate an amplitude of "infinity" for the tangent, cotangent, secant, and cosecant functions.

Emphasize that there is no amplitude for a function that has no maximum or minimum value.

Checkpoint

Give the period and amplitude of each function.

1. $T = \tan 4\theta$ Period: $\frac{\pi}{4}$; no amplitude
2. $T = 6 \sec 2\theta$ Period: π; no amplitude
3. $T = 3 \csc \pi\theta$ Period: 2; no amplitude

Graph one cycle of the following function.

4. $T = \tan 4\theta$

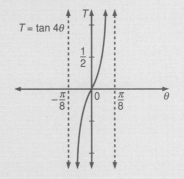

Closure

Ask students how they would graph $T = a \tan b\theta$. See Example 1. Have students give the period of the function above. $\frac{\pi}{b}$ the amplitude. Undefined

Additional Example 2

Graph $T = \sin 2\theta$ and $T = \csc 2\theta$ for $0 < \theta < 2\pi$ in the same coordinate plane. Determine the amplitude and period of $T = \csc 2\theta$.

No amplitude; period: π

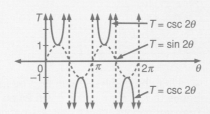

Guided Practice

Classroom Exercises 1–8

Independent Practice

Ⓐ Ex. 1–10, Ⓑ Ex. 11–16, Ⓒ Ex. 17, 18

Basic: WE 1–8
Average: WE 1–5, 11–15 odd
Above Average: WE 1–5, 13–17 odd

Additional Answers

Written Exercises

6. $T = \sec \theta$

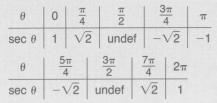

θ	0	$\frac{\pi}{4}$	$\frac{\pi}{2}$	$\frac{3\pi}{4}$	π
$\sec \theta$	1	$\sqrt{2}$	undef	$-\sqrt{2}$	-1

θ	$\frac{5\pi}{4}$	$\frac{3\pi}{2}$	$\frac{7\pi}{4}$	2π
$\sec \theta$	$-\sqrt{2}$	undef	$\sqrt{2}$	1

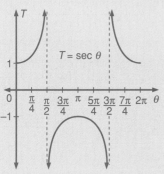

See pages 827–828 for the answers to Written Ex. 7–16.

Classroom Exercises

Give the period of each function.

1. $T = \sec 3\theta$ $\frac{2\pi}{3}$ 2. $T = 4 \tan 5\theta$ $\frac{\pi}{5}$ 3. $T = 2 \csc \frac{\pi}{2}\theta$ 4   4. $T = 6 \cot 2\theta$ $\frac{\pi}{2}$

5. $T = \csc \frac{1}{3}\theta$ 6π 6. $T = 3 \cot \pi\theta$ 1   7. $T = \sec(-\theta)$ 2π 8. $T = \tan 3\pi\theta$ $\frac{1}{3}$

Written Exercises

Complete the chart.

	$\sec \theta$	$\csc \theta$	$\tan \theta$	$\cot \theta$
1. Amplitude	None	None	None	None
2. Period	2π	2π	π	π
3. Maximum value	None	None	None	None
4. Minimum value	None	None	None	None
5. Values of θ, $0 \le \theta \le 2\pi$, for which the function is undefined	$\frac{\pi}{2}, \frac{3\pi}{2}$	$0, \pi, 2\pi$	$\frac{\pi}{2}, \frac{3\pi}{2}$	$0, \pi, 2\pi$

6. Graph $T = \sec \theta$ for $0 \le \theta \le 2\pi$ using a table of values. Use multiples of $\frac{\pi}{4}$ for θ.

Graph one cycle of each of the following functions.

7. $T = \tan 3\theta$ 8. $T = \cot \theta$ 9. $T = \cot \frac{1}{2}\theta$ 10. $T = \tan \frac{1}{3}\theta$

11. $T = \sec 4\theta$ 12. $T = \csc 2\theta$ 13. $T = \sec \frac{1}{3}\theta$ 14. $T = \csc 3\theta$

15. Graph $T = 2 \tan 3\theta$ and $T = \tan \theta$ in the same coordinate plane for $-\frac{\pi}{2} < \theta < \frac{\pi}{2}$. Compare the graphs.

16. Graph $T = \frac{1}{2} \sec 2\theta$ and $T = \sec \theta$ in the same coordinate plane for $0 \le \theta \le 2\pi$. Compare the graphs.

17. Determine whether $T = \sec \theta$ is an odd function, an even function, or neither. Even

18. Determine whether $T = \csc \theta$ is an odd function, an even function, or neither. Odd

Mixed Review

In Exercises 1–4, points $P(-2,4)$ and $Q(6,8)$ are given.

1. Write an equation of $\overleftrightarrow{PQ}$ in point-slope form. *3.4* $y - 4 = \frac{1}{2}(x + 2)$

2. Find the slope of a line perpendicular to $\overleftrightarrow{PQ}$. *3.7* -2

3. Find the coordinates of the midpoint of $\overline{PQ}$. *11.1* (2,6)

4. Find the distance between P and Q. *11.1* $4\sqrt{5}$

Enrichment

Explain that periodic functions such as $T = \sin \theta$ and $T = \sin 2\theta$ can be combined to form a new periodic function, $T = \sin \theta + \sin 2\theta$.

Have the students use a table of values to graph $T = \sin \theta + \sin 2\theta$ for $0 \le \theta \le 2\pi$. On the table, have the students show headings as follows, and values of T at intervals of $\frac{\pi}{6}$.

θ	$\sin \theta$	$\sin 2\theta$	T
0	0	0	0
$\frac{\pi}{6}$	0.5	0.87	$1.37 \approx 1.4$
⋮	⋮	⋮	⋮

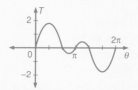

Note that the graph of T can also be formed by drawing the graphs of $T = \sin \theta$ and $T = \sin 2\theta$ on the same set of axes. Then, for various values of θ, the ordinates of the two curves can be added, taking signs of the ordinates into account.

18.10 Inverse Trigonometric Functions

Objectives To graph inverse trigonometric functions
To determine the inverse of a trigonometric function

So far, the trigonometric functions have been defined in terms of θ and T and graphed using θ and T axes. Now, the x- and y-axes will be used.

The relation $y = \sin x$ is a function. Therefore, for each angle measure x, there is only *one* sine value. The *inverse* of the sine function is not a function, since there is not a unique angle measure for a given sine value. For example, angles of 30, 150, 390. . . . (in radians, $\frac{\pi}{6}$, $\frac{5\pi}{6}$, $\frac{13\pi}{6}$ . . . .) all have a sine value of $\frac{1}{2}$.

However, if we restrict the domain of a function so that its range is a set of unique y values, then its inverse will be a function. This is shown in Example 1 below, where the function $y = \sin x$, $-\frac{\pi}{2} \le x \le \frac{\pi}{2}$, is graphed together with its inverse. Recall the method for graphing the inverse of a function (See Lesson 13.1).

EXAMPLE 1 Graph $y = \sin x$, $-\frac{\pi}{2} \le x \le \frac{\pi}{2}$. Then graph its inverse.

Solution
(1) On the same set of axes, graph $y = \sin x$ and $y = x$ for $-\frac{\pi}{2} \le \frac{\pi}{2}$.

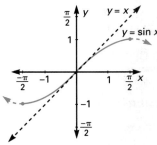

(2) Graph the mirror image of the solid portion of $y = \sin x$ with respect to the line $y = x$.

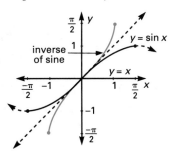

No y-value occurs more than once in the solid portion of the graph.

The domain of the inverse function is $\{x \mid -1 \le x \le 1\}$. There is just one y-value for every x-value.

Notice that the *range* of the resulting inverse function is numerically the same as the domain of the original function. The values of y in the range of an inverse function are called its **principal values**.

Additional Example 1

Graph $y = \cos x$ for $0 \le x \le \pi$. Then graph the inverse of the function.

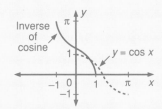

Math Connections

A Formula for the Inverse Tangent: The following formula represents the inverse tangent for its principal values.

$$\tan^{-1} x = x - \frac{x^3}{3} + \frac{x^5}{5} - \frac{x^7}{7} + \cdots$$

Here, $|x| < 0$.

Similar infinite-series formulas for other inverse trigonometric functions also exist.

Critical Thinking Questions

Analysis: In the lesson, students are cautioned to interpret $\text{Sin}^{-1} x$ as the inverse sine of x, not as the reciprocal of $\sin x$. Ask students how they think that the reciprocal of $\sin x$ should be represented using an exponent. They will probably conjecture, correctly, that the appropriate notation for the reciprocal of $\sin x$ is $(\sin x)^{-1}$. Earlier in the text expressions such as $\sin^2 x$ were first written as $(\sin x)^2$, so the reversion to this notation should not seem unnatural.

Checkpoint

Find the value of y for each of the following. Write the answer in terms of radians.

1. $y = \text{Sin}^{-1} 0$ 0
2. $y = \text{Sin}^{-1} \left(-\frac{\sqrt{3}}{2}\right)$ $-\frac{\pi}{3}$
3. $y = \text{Csc}^{-1} (-\sqrt{2})$ $-\frac{\pi}{4}$

Find the value of each angle measure to the nearest degree.

4. $\text{Cos}^{-1} (-0.4562)$ 117
5. $\text{Tan}^{-1} (-3.4516)$ -74

To indicate that the domain of the sine function is restricted to the interval from $-\frac{\pi}{2}$ to $\frac{\pi}{2}$, inclusive, write the function as $y = \text{Sin } x$ (with capital S) rather than $y = \sin x$.

The inverse of $y = \text{Sin } x$ is written $y = \text{Sin}^{-1} x$. Read "y is the angle measure whose sine is x." Note that $\text{Sin}^{-1} x$ does *not* mean $\frac{1}{\text{Sin } x}$.

The other trigonometric functions also have inverses that are functions if their domains are restricted. Two of these are graphed.

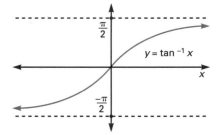

Range of Cos^{-1}
Principal Values: $\{y \mid 0 \le y \le \pi\}$ Range of Tan^{-1}
Principal Values: $\{y \mid -\frac{\pi}{2} < y < \frac{\pi}{2}\}$

EXAMPLE 2   Find y if $y = \text{Cos}^{-1} \left(-\frac{\sqrt{3}}{2}\right)$.

Plan   Sketch a reference triangle. Since y represents an angle measure, x and y cannot be used to label the coordinate axes of the reference-triangle diagram. Use u and v instead.

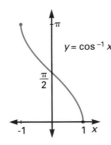

Solution $y = \text{Cos}^{-1} \left(-\frac{\sqrt{3}}{2}\right)$ means $\text{Cos } y = -\frac{\sqrt{3}}{2}$.

Principal values for inverse cosine are in Quadrants I and II. Draw the angle in Quadrant II, since the cosine is negative. The reference angle has measure 30.

Therefore, $y = 150$, or $\frac{5\pi}{6}$ in radian measure.

Often, an inverse function can be found by using a calculator.

EXAMPLE 3   Find A, to the nearest degree, if $A = \text{Tan}^{-1} (-0.9008)$.

Solution   Calculator steps: 0.9008 [+/−] [INV] [tan] $\Rightarrow -42.012527$
Therefore, A is -42 to the nearest degree.

Additional Example 2

Find y if $y = \text{Sin}^{-1} \left(-\frac{\sqrt{2}}{2}\right)$. $-\frac{\pi}{4}$

Additional Example 3

Find A, to the nearest degree, if $A = \text{Cos}^{-1} (-0.5454)$.   123

It can be shown that if $x > 0$, then $\mathrm{Cot}^{-1}\,x = \mathrm{Tan}^{-1}\left(\frac{1}{x}\right)$. (You are asked to show this in Exercise 20.) This relationship follows from the fact that the tangent and the cotangent are reciprocals of one another. The following definitions are based on the reciprocal relationship between the secant and cosine and between the cosecant and sine. For both definitions, $|x| > 1$.

$$y = \mathrm{Sec}^{-1}\,x = \mathrm{Cos}^{-1}\left(\frac{1}{x}\right) \qquad \leftarrow 0 \le y \le \pi,\ y \ne \frac{\pi}{2}$$

$$y = \mathrm{Csc}^{-1}\,x = \mathrm{Sin}^{-1}\left(\frac{1}{x}\right) \qquad \leftarrow -\frac{\pi}{2} \le y \le \frac{\pi}{2},\ y \ne 0$$

EXAMPLE 4 Evaluate $\cos\left[\mathrm{Csc}^{-1}(-3)\right]$.

Plan Let $y = \mathrm{Csc}^{-1}(-3) = \mathrm{Sin}^{-1}\left(\frac{1}{-3}\right)$. Then $\mathrm{Sin}\,y = -\frac{1}{3}$.

Solution Locate y in the proper quadrant. Principal values of y are in Quadrants I and IV. Since $\mathrm{Sin}\,y$ is negative (as is $\mathrm{Csc}\,y$), angle y terminates in Quadrant IV. Find the missing leg of the reference triangle.

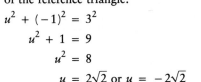

$$u^2 + (-1)^2 = 3^2$$
$$u^2 + 1 = 9$$
$$u^2 = 8$$
$$u = 2\sqrt{2}\ \text{or}\ u = -2\sqrt{2}$$

Since y terminates in Quadrant IV, $u = 2\sqrt{2}$.

Therefore, $\cos\left[\mathrm{Csc}^{-1}(-3)\right] = \cos y = \frac{2\sqrt{2}}{3}$, or approximately 0.9428.

Calculator steps: 1 ÷ 3 = +/− INV sin cos ⟹ 0.942809

 Focus on Reading

Indicate whether each statement is always true, sometimes true, or never true.

1. The inverse of a trigonometric function is a function. Sometimes
2. The inverse of $y = \cos x$ is $x = \cos y$. Always
3. If $B = \mathrm{Tan}^{-1}\,4$, then $\tan B = 4$. Always
4. The principal values of $y = \mathrm{Sin}^{-1}\,x$ are $0 \le y \le \pi$. Never
5. $\mathrm{Sin}^{-1}\,x$ is more conveniently written as $\dfrac{1}{\mathrm{Sin}\,x}$. Never

703

15.

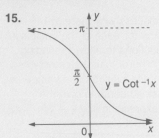

19. True. Since $-1 \leq x \leq 1$, so is $-x$. Thus, the domains are the same. Let $y = \text{Sin}^{-1}(-x)$. Then $\sin y = -x$. But $\sin y$ is an odd function, so $\sin(-y) = -(-x)$, or $\sin(-y) = x$. This means that $-y = \text{Sin}^{-1} x$, or that $y = -\text{Sin}^{-1} x$. So, $\text{Sin}^{-1}(-x) = -\text{Sin}^{-1}x$.

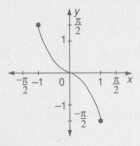

20. Let $y = \text{Cot}^{-1}x$. Then $\text{Cot } y = x$, or $\text{Tan } y = \frac{1}{x}$. So, $\text{Tan}^{-1}\left(\frac{1}{x}\right) = y = \text{Cot}^{-1}x$.

See page 705 for the answers to Mixed Review Ex. 1–4.

Classroom Exercises

Give the meaning of each statement. Is the statement true?

1. $30 = \text{Sin}^{-1}\frac{1}{2}$ **2.** $\frac{\pi}{4} = \text{Tan}^{-1} 1$ **3.** $-90 = \text{Sin}^{-1} 1$ **4.** $180 = \text{Cos}^{-1}(-1)$

Rewrite each sentence in terms of an inverse function.

5. $\text{Sin } A = b$ **6.** $\text{Cos } P = \frac{3}{5}$ **7.** $\text{Tan } M = 1$ **8.** $\text{Tan}\left(-\frac{3\pi}{4}\right) = -1$

Find the value of y for each of the following. Write the answer in radians.

9. $y = \text{Sin}^{-1} 0$ 0   **10.** $y = \text{Cos}^{-1} 1$ 0   **11.** $y = \text{Tan}^{-1} 0$ 0   **12.** $y = \text{Sin}^{-1}\left(-\frac{1}{2}\right) -\frac{\pi}{6}$

Written Exercises

Find the value of y for each of the following. Write the answer in radians.   **3.** $\frac{2\pi}{3}$

1. $y = \text{Cos}^{-1} 0$ $\frac{\pi}{2}$ **2.** $y = \text{Sin}^{-1}\frac{1}{2}$ $\frac{\pi}{6}$ **3.** $y = \text{Cos}^{-1}\left(-\frac{1}{2}\right)$ **4.** $y = \text{Tan}^{-1}\sqrt{3}$ $\frac{\pi}{3}$

5. $y = \text{Sin}^{-1} 1$ $\frac{\pi}{2}$ **6.** $y = \text{Tan}^{-1}(-1)$ $-\frac{\pi}{4}$ **7.** $y = \text{Sec}^{-1}\frac{2}{\sqrt{3}}$ $\frac{\pi}{6}$ **8.** $y = \text{Csc}^{-1}(-\sqrt{2})$ $-\frac{\pi}{4}$

Find the value of each angle measure to the nearest degree.

9. $\text{Sin}^{-1}(-0.4617)$ -27 **10.** $\text{Tan}^{-1} 4.3635$ 77   **11.** $\text{Cos}^{-1}(-0.7290)$ 137

Evaluate each expression.

12. $\cos\left[\text{Sin}^{-1}\left(-\frac{\sqrt{2}}{2}\right)\right]$ $\frac{\sqrt{2}}{2}$ **13.** $\sin\left(\text{Cos}^{-1}\frac{3}{5}\right)$ $\frac{4}{5}$ **14.** $\tan\left[\text{Sin}^{-1}\left(-\frac{5}{13}\right)\right]$ $-\frac{5}{12}$

15. Graph $y = \text{Cot}^{-1} x$ for the principal values $0 < y < \pi$.

Simplify.

16. $\sin(\text{Cos}^{-1} x)$ $\sqrt{1-x^2}$ **17.** $\csc\left(\text{Tan}^{-1}\frac{1}{x}\right)$ $\sqrt{x^2+1}$ **18.** $\cot\left(\text{Csc}^{-1}\frac{\sqrt{x^2+1}}{x}\right)$ $\frac{1}{x}$

19. Graph the function $y = \text{Sin}^{-1}(-x)$. Determine whether $\text{Sin}^{-1}(-x) = -\text{Sin}^{-1} x$ is true for $-1 \leq x \leq 1$. Explain.

20. Show that if $x > 0$, then $\text{Cot}^{-1} x = \text{Tan}^{-1}\left(\frac{1}{x}\right)$.

Mixed Review

Graph each equation of a conic section.   *11.5, 12.1, 12.2, 12.3*

1. $y = x^2 - 4x$ **2.** $x^2 + y^2 = 25$ **3.** $4x^2 + 9y^2 = 36$ **4.** $x^2 - y^2 = 25$

Enrichment

Write the area formula, $K = \frac{1}{2}bc \sin A$, on the chalkboard. Then write the Law of Sines and solve for b.

$$\frac{b}{\sin B} = \frac{c}{\sin C} \qquad b = \frac{c \sin B}{\sin C}$$

Have the students use this formula to substitute for b in the area formula. This gives a new formula for the area of a triangle:

$$K = c^2 \frac{\sin A \cdot \sin B}{2 \sin C}.$$

The area of any triangle is the product of the square of any side and the sines of the adjacent angles, divided by twice the sine of the opposite angle.

Now have the students use the new formula to find the area of triangle ABC for the given data.

1. m $\angle A = 30$, m $\angle B = 30$, $c = 6$. Give the answer in simplest radical form. $3\sqrt{3}$ sq units

2. m $\angle A = 52$, m $\angle B = 46$, $c = 18$. Give the correct answer to two significant digits. 93 sq units

Application: *Sine Curves*

As a part of their records, **meteorologists** compute the daily average temperature for a given city or location by averaging the high and low temperatures for the day. The graph below shows the average temperature on the first day of each month for a given city.

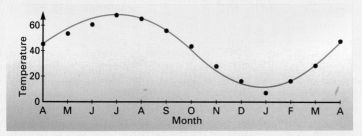

Notice that a sine curve drawn through the points models the graph. A model for a given site in the northern hemisphere can be written as

$$T(\theta) = c + a \sin \frac{2\pi}{365}(\theta - d).$$

The formula gives the average temperature for day θ, where θ is the number of days past March 21. The quantity c is the mean yearly temperature for the site, and a is the difference between the highest or lowest daily average and c. So, c translates the graph up or down and a is the amplitude. The period of the graph is $2\pi \div (2\pi/365)$ or 365 days. The quantity d is the number of days past March 21 that the daily average first exceeds the yearly mean. It translates the graph d units to the right.

EXAMPLE In Seattle, Washington, the mean yearly temperature is 50°F, the highest average temperature is 62°F, and the daily average first reaches 50°F on April 27. What is the modeling equation for Seattle?

Solution $c = 50 \qquad a = 62 - 50 = 12 \qquad d = \text{April 27} - \text{March 21} = 37$
So, $T(\theta) = 50 + 12 \sin \dfrac{2\pi}{365}(\theta - 37).$

In Exercises 1–3, write the modeling equation for each city given the mean yearly temperature, the highest daily average, and the date the daily temperature first reaches the yearly mean.
1. Detroit: 47, 71, Apr. 23   2. Miami: 75, 84, Apr. 25   3. Boston: 50, 72, Apr. 25
4. Give Detroit's average temperature for June 21. (Use the radian mode.)

Application

1. $47 + 24 \sin \frac{2\pi}{365}(\theta - 33)$

2. $75 + 9 \sin \frac{2\pi}{365}(\theta - 35)$

3. $50 + 22 \sin \frac{2\pi}{365}(\theta - 35)$

4. 67.4 degrees

Additional Answers, page 704

Mixed Review

1.
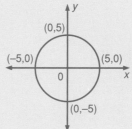

2.
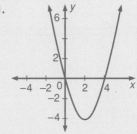

3. Foci: $(\pm\sqrt{5}, 0)$

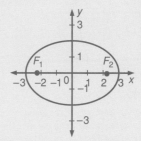

4. Foci: $(\pm 5\sqrt{2}, 0)$

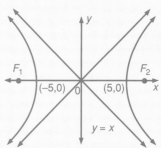

4. m ∠A = 33, m ∠B = 47, c = 22

5. m ∠A = 49, m ∠B = 56, m ∠C = 75

6. m ∠B = 46, m ∠C = 74,
b = 15 *or* m ∠B = 14,
m ∠C = 106, b = 5

7. Not possible

14.

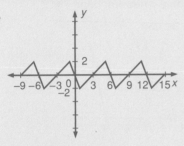

17. $p = \pi$; $a = 3$

θ	0	$\frac{\pi}{4}$	$\frac{\pi}{2}$	$\frac{3\pi}{4}$	π
T	3	0	−3	0	3

θ	$\frac{5\pi}{4}$	$\frac{3\pi}{2}$	$\frac{7\pi}{4}$	2π
T	0	−3	0	3

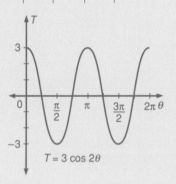

$T = 3 \cos 2\theta$

Chapter 18 Review

Key Terms

amplitude (p. 681)	Law of Cosines (p. 661)	periodic function (p. 681)
bearing (p. 669)	Law of Sines (p. 668)	principal value (p. 701)
cycle (p. 681)	oblique triangle (p. 661)	radian (p. 677)
Heron's Formula (p. 666)	period (p. 681)	semiperimeter (p. 666)

Key Ideas and Review Exercises

18.2 To find the area of a triangle, use $K = \frac{1}{2}bc \sin A$ for SAS, or use
$K = \sqrt{s(s - a)(s - b)(s - c)}$ for SSS, where $s = \frac{1}{2}(a + b + c)$.

1. Find the area of triangle *ABC* to two significant digits for $a = 12$, $c = 14$, and m ∠B = 125. 69

2. Find the area of triangle *ABC* in simplest radical form for $a = 4$, $b = 6$, and $c = 6$. $8\sqrt{2}$

3. Find, to two significant digits, the area of a regular hexagon inscribed in a circle of radius 12. 370

18.1, 18.3, 18.4 To solve an oblique (nonright) triangle, use the following laws.
- Law of Cosines: $a^2 = b^2 + c^2 - 2bc \cos A$, for SSS or SAS
- Law of Sines: $\frac{\sin A}{a} = \frac{\sin B}{b} = \frac{\sin C}{c}$, for ASA or SSA
To determine the number of solutions for SSA, see pages 672 and 673.

Solve triangle *ABC*, if possible. Find lengths to two significant digits and angle measures to the nearest degree.

4. $a = 12$, $b = 16$, m ∠C = 100

5. $a = 11$, $b = 12$, $c = 14$

6. m ∠A = 60, $c = 20$, $a = 18$

7. m ∠A = 80, $c = 56$, $a = 12$

8. Juan determines that the angle of elevation to the top of a building measures 42. At 120 ft closer to the building, the angle of elevation measures 48. Find the height of the building to two significant digits. 570 ft

18.5 To convert degrees to radians, use $1° = \frac{\pi}{180}$ radians.
To convert radians to degrees, use π radians = 180°, or 1 radian = 57.3°.

9. Express 240° as a radian measure in terms of π. $\frac{4\pi}{3}$

10. Express $-\frac{3\pi}{2}$ in degree measure. −270

11. Express 1.7 radians in degree measure. 97.4

12. Simplify $\tan \frac{5\pi}{6}$. $-\frac{\sqrt{3}}{3}$

13. Simplify $\frac{1}{3} \tan^2 \frac{4\pi}{3}$. 1

---

18. $p = 2\pi$; $a = 2$

θ	0	$\frac{\pi}{2}$	π	$\frac{3\pi}{2}$	2π
T	−3	−1	−3	−5	−3

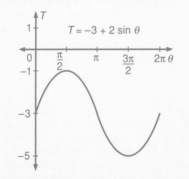

$T = -3 + 2 \sin \theta$

19.

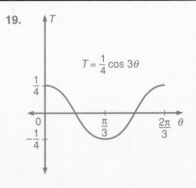

$T = \frac{1}{4} \cos 3\theta$

18.6 To determine the *period* of a periodic function, find the smallest value of p for which $f(x) = f(x + p)$.

To determine the *amplitude* of a periodic function, find $\dfrac{M - m}{2}$, where M is the maximum value and m is the minimum value of the function.

Part of the graph of a periodic function $y = f(x)$ with period $p = 6$ is shown at the right.

14. Graph $y = f(x)$ in the interval $-9 \le x \le 15$.

15. Find the amplitude of $y = f(x)$. 2

16. Determine whether f is even, odd, or neither. Odd

18.7, To graph trigonometric functions first determine the amplitude, if any, and
18.8, the period using the summary table below.
18.9

	Amplitude	Period	Odd or Even	Shape of Graph
$T = \sin \theta$	1	2π	Odd: $\sin(-\theta) = -\sin \theta$	See page 685.
$T = \cos \theta$	1	2π	Even: $\cos(-\theta) = \cos \theta$	See page 687.
$T = \tan \theta$	None	π	Odd: $\tan(-\theta) = -\tan \theta$	See page 697.
$T = \cot \theta$	None	π	Odd: $\cot(-\theta) = -\cot \theta$	See Ex. 8, page 700.
$T = \sec \theta$	None	2π	Even: $\sec(-\theta) = \sec \theta$	See Ex. 6, page 700.
$T = \csc \theta$	None	2π	Odd: $\csc(-\theta) = -\csc \theta$	See page 699.

For changes in amplitude and period, see pages 695 and 697–699.

For an example of a vertical translation of a graph, see page 695.

Graph each function for $0 \le \theta \le 2\pi$ using a table of values. Give the amplitude and the period of each function.

17. $T = 3 \cos 2\theta$ **18.** $T = -3 + 2 \sin \theta$

Graph one cycle of each function without using a table of values.

19. $T = \frac{1}{4} \cos 3\theta$ **20.** $T = 5 \sin \frac{1}{4}\theta$ **21.** $T = \tan 4\theta$ **22.** $T = 3 \csc 6\theta$

23. If $\sin \theta = 0.7431$, find $\sin(-\theta)$.
    -0.7431

24. If $\cos \theta = -0.8192$, find $\cos(-\theta)$.
    -0.8192

18.10 An equation such as $y = \operatorname{Sin}^{-1} x$ is read "y is the angle whose sine is x" (y is the principal value of the angle whose sine is x).

25. Find the value of y in terms of radians if $y = \operatorname{Cos}^{-1}\left(-\dfrac{\sqrt{3}}{2}\right)$. $\frac{5\pi}{6}$

26. Find $\operatorname{Tan}^{-1}(-0.3640)$ to the nearest degree. -20

27. Evaluate $\sin\left(\operatorname{Cos}^{-1}\dfrac{4}{5}\right)$. $\frac{3}{5}$

20.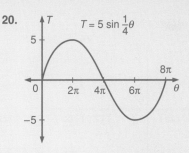

21.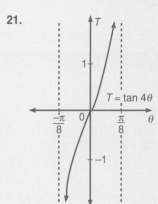

22.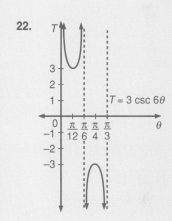

Chapter Test

13.

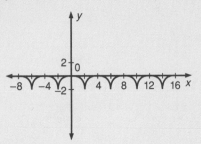

16. $p = \frac{2\pi}{3}$; $a = 5$

17. $p = 2\pi$; $a = \frac{1}{2}$

θ	0	$\frac{\pi}{2}$	π	$\frac{3\pi}{2}$	2π
T	-6	$-5\frac{1}{2}$	-6	$-6\frac{1}{2}$	-6

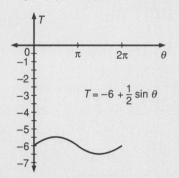

$T = -6 + \frac{1}{2}\sin\theta$

18.

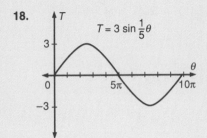

$T = 3\sin\frac{1}{5}\theta$

Find the area of triangle *ABC* to two significant digits.

1. $a = 12$, $b = 8$, m $\angle C = 85$ 48   **2.** $a = 10$, $b = 8$, $c = 8$ 31

3. Find, to two significant digits, the area of a regular octagon inscribed in a circle of radius 16.   720 sq units

In each triangle *ABC*, find lengths to two significant digits and angle measures to the nearest degree. Show all possible solutions.

4. $a = 20$, $b = 16$, m $\angle C = 105$, $c =$ __?__ 29   **5.** $a = 12$, $b = 20$, $c = 12$, m $\angle B =$ __?__ 113

6. $a = 14$, m $\angle A = 40$, m $\angle B = 80$, $c =$ __?__ 19   **7.** m $\angle A = 50$, $b = 11$, $a = 10$, m $\angle B =$ __?__ 57 or 123

8. A navigator sailing due north sights a lighthouse at a bearing of 28. After sailing another 55 km, the bearing of the lighthouse is then 115. Find his distance from the lighthouse, to two significant digits, at the time of the second measurement.   26 km

9. Express $-240°$ as a radian measure in terms of π. $-\frac{4\pi}{3}$

10. Express $-\frac{2\pi}{3}$ in degree measure. -120

11. Simplify cot $\frac{4\pi}{3}$. $\frac{\sqrt{3}}{3}$ **12.** Simplify $1 - \cos\left(-\frac{\pi}{3}\right)$. $\frac{1}{2}$

Part of the graph of a periodic function $y = f(x)$ with period $p = 4$ is shown at the right. (Exercises 13–15)

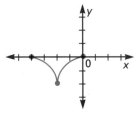

13. Graph $y = f(x)$ in the interval $-8 \le x \le 16$.

14. Find the amplitude of $y = f(x)$.   1

15. Determine whether f is even, odd, or neither.   Even

16. Give the amplitude and the period of $T = 5\cos 3\theta$.

17. Graph $T = -6 + \frac{1}{2}\sin\theta$ for $0 \le \theta \le 2\pi$, using a table of values. Give the amplitude and the period of the function.

Graph one cycle of each function without using a table of values.

18. $T = 3\sin\frac{1}{5}\theta$ **19.** $T = \tan 3\theta$

20. If $\sin\theta = -0.6691$, find $\sin(-\theta)$.   0.6691

21. Find the value of y in terms of radians if $y = \text{Sin}^{-1}\left(-\frac{\sqrt{3}}{2}\right)$. $-\frac{\pi}{3}$

22. The sides of a triangle are in arithmetic sequence, the shortest side being 5.0 in. long, and the perimeter of the triangle being 24.0 in. Find the length of the altitude to the 5.0-in. side to two significant digits.   7.3 in.

19.

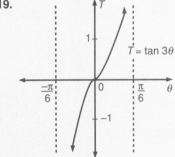

$T = \tan 3\theta$

College Prep Test

Choose the *one* best answer to each question or problem.

1. From which of the following statements, taken separately or together, can it be determined that m is greater than n? B

I. $2m + n > 12$ II. $m + n = 7$

(A) I alone, but not II
(B) I and II taken together, but neither taken alone
(C) II alone, but not I
(D) Neither I nor II nor both
(E) Both I alone and II alone

2. Three points on a line are X, Y, and Z, in that order. If $XZ - YZ = 6$, what is the ratio $\frac{YZ}{XZ}$? E

(A) 1:5   (B) 1:4
(C) 1:3   (D) 1:2
(E) It cannot be determined from the given information.

3. $2 \cos \frac{\pi}{4} = \underline{\quad?\quad}$ D

(A) 0   (B) $\frac{1}{2}$ (C) 1
(D) $\sqrt{2}$ (E) 2

4. $\text{Sin}^{-1} \frac{\sqrt{3}}{2} + \text{Cos}^{-1} \frac{\sqrt{3}}{2} = \underline{\quad?\quad}$ D

(A) 0   (B) $\frac{\pi}{6}$ (C) $\frac{\pi}{3}$
(D) $\frac{\pi}{2}$ (E) π

5. If $\frac{a + b}{a} = 3$ and $\frac{c + b}{c} = 5$, what is the value of $\frac{c}{a}$? E

(A) 2   (B) $\frac{5}{3}$ (C) $\frac{3}{5}$
(D) 15   (E) $\frac{1}{2}$

6. In triangle ABC below, if $BC < BA$, which of the following is true? C

(A) $x > y$ (B) $y < z$ (C) $y > x$
(D) $y = x$ (E) $z < x$

7. If $\frac{13a}{4}$ is an integer, then a could be any of the following except $\underline{\quad?\quad}$ C

(A) 8   (B) -32 (C) 6
(D) 0   (E) -112

8. If the area of a circle is 49π, what is the circumference of the circle? D

(A) 7   (B) 7π (C) 14
(D) 14π (E) 49

9. If v, w, x, y, and z are whole numbers, then $v[w(x + y) + z]$ will be an even number when which of the following is even? A

(A) v (B) w (C) x
(D) y (E) z

10. $\sin\left[\text{Tan}^{-1}\left(-\frac{3}{4}\right)\right] = \underline{\quad?\quad}$ C

(A) -1 (B) -0.75 (C) -0.6
(D) 0.6   (E) 0.75

11. If $P = \frac{h(a + b)}{2}$, what is the average of a and b when $P = 30$ and $h = 5$? A

(A) 6   (B) 150   (C) 15
(D) $\frac{35}{2}$ (E) 2

College Prep Test   **709**

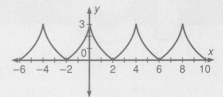

5. Even; $\frac{3}{2}$

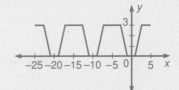

6. Neither; $\frac{1}{2}$

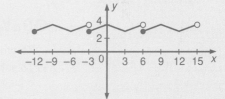

7.

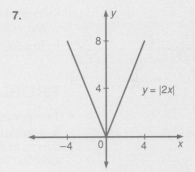

$y = |2x|$

9. Even; 4

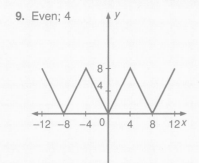

11.

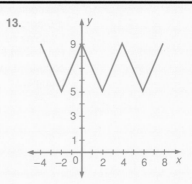

13.

709

19 TRIGONOMETRIC IDENTITIES AND EQUATIONS

OVERVIEW

This chapter of the text covers two topics, trigonometric identities and trigonometric equations. These topics are important both for their own mathematical value and for the opportunity they provide students to practice the algebraic skills they learned earlier in the course.

The last three lessons continue the study of vectors, which first appeared in Chapter 9. This includes the treatment of complex numbers as vectors, which is extended to the geometric interpretation of multiplying and dividing complex numbers.

OBJECTIVES

- To prove trigonometric identities
- To solve linear and quadratic trigonometric equations involving a single angle
- To apply multiple-angle formulas for the sine, cosine, and tangent
- To write a complex number in polar form and vice versa
- To apply DeMoivre's Theorem

PROBLEM SOLVING

Throughout the text, numerous problem-solving opportunities have been provided, but not all of these have been formally identified as problem-solving situations. An example of this is the *Brainteaser* on page 737, in which a student is required to use a variety of strategies such as *Logical Reasoning*, *Guess and Check*, or *Drawing a Diagram* to find two noncongruent triangles such that five of the six elements (angles and sides) of one triangle are congruent to five elements of the other triangle.

READING AND WRITING MATH

The *Focus on Reading* on page 714 asks students to determine the truth or falsity of mathematical statements. If the statement is false, students are asked to explain in their own words why it is false. Exercise 31 in the *Chapter Review* asks students to write an explanation of how to express (r, θ) in rectangular form.

TECHNOLOGY

Calculator: When students solve a trigonometric equation using a calculator, they will have to determine how it handles negative angle measures and angles with measures greater than 90 degrees. Some calculators will provide inverse functions for these measures, but others will display an error message.

SPECIAL FEATURES

Mixed Review pp. 715, 719, 723, 726, 731, 737, 741, 744, 748
Focus on Reading p. 714
Brainteaser pp. 715, 731, 737
Midchapter Review p. 735
Application: Solar Radiation p. 749
Key Terms p. 750
Key Ideas and Review Exercises pp. 750–751
Chapter 19 Test p. 752
College Prep Test p. 753
Cumulative Review (Chapters 17–19) pp. 754–755

PLANNING GUIDE

Lesson	Basic	Average	Above Average	Resources
19.1 pp. 714–715	FR all, CE all WE 1–17 odd Brainteaser	FR all, CE all WE 7–23 odd Brainteaser	FR all, CE all WE 11–25 odd Brainteaser	Reteaching 143 Practice 143
19.2 pp. 718–719	CE all WE 1–19 odd	CE all WE 9–27 odd	CE all WE 15–33 odd	Reteaching 144 Practice 144
19.3 p. 723	CE all WE 1–23 odd	CE all WE 7–29 odd	CE all WE 11–33 odd	Reteaching 145 Practice 145
19.4 p. 726	CE all WE 1–15 odd	CE all WE 5–19 odd	CE all WE 11–25 odd	Reteaching 146 Practice 146
19.5 pp. 730–731	CE all, WE 1–17 odd Brainteaser	CE all, WE 7–23 odd Brainteaser	CE all, WE 13–29 odd Brainteaser	Reteaching 147 Practice 147
19.6 pp. 734–735	CE all, WE 1–17 odd Midchapter Review	CE all, WE 5–21 odd Midchapter Review	CE all, WE 11–27 odd Midchapter Review	Reteaching 148 Practice 148
19.7 p. 737	CE all WE 1–13 odd Brainteaser	CE all WE 5–17 odd Brainteaser	CE all WE 5, 11–21 odd Brainteaser	Reteaching 149 Practice 149
19.8 pp. 740–741	CE all WE 1–13 odd	CE all WE 3–15 odd	CE all WE 7–19 odd	Reteaching 150 Practice 150
19.9 pp. 744	CE all WE 1–29 odd	CE all WE 3–31 odd	CE all WE 9–37 odd	Reteaching 151 Practice 151
19.10 pp. 748–749	CE all WE 1–15 odd Application	CE all WE 3, 7–19 odd Application	CE all, WE 3, 7, 9, 11, 19–25 odd Application	Reteaching 152 Practice 152
Chapter 19 Review pp. 750–751	all	all	all	
Chapter 19 Test p. 752	all	all	all	
College Prep Test p. 753	all	all	all	
Cumulative Review pp. 754–755	1–51 odd	5–55 odd	7–57 odd	

CE = Classroom Exercises   WE = Written Exercises   FR = Focus on Reading
NOTE: For each level, all students should be assigned all Mixed Review exercises.

▰INVESTIGATION

In this investigation, students are introduced to graphic representations of trigonometric equations.

Have the students consider the trigonometric equation $\sin x = \cos x$. One way to approach the solution of this equation is to graph the sine function and cosine function on the same coordinate plane and find the points where they are equal. On a graphing calculator, have the students use the built-in calculator function to graph the sine function.

$$\boxed{\text{Graph}}\ \boxed{\text{sin}}\ \boxed{\text{EXE}}$$

Next have the students graph the cosine function over the sine function using the keystrokes which will not erase the graph of the sine.

$$\boxed{\text{Graph}}\ \boxed{\text{cos}}\ \boxed{\text{Alpha}}\ \boxed{+}\ \boxed{\text{EXE}}$$

Ask the students to locate the points where the two functions are equal. The functions are equal for $x = -315, -135, 45,$ and 225.

Another way to approach the problem is to find the zeros of the function

$$y = \sin x - \cos x.$$

The zeros of this equation are the values of x for which $\sin x = \cos x$. Have them graph this equation over the previous two and report their observations.

$$\boxed{\text{Graph}}\ \boxed{\text{sin}}\ \boxed{\text{Alpha}}\ \boxed{+}\ \boxed{-}\ \boxed{\text{cos}}\ \boxed{\text{Alpha}}$$
$$\boxed{+}\ \boxed{\text{EXE}}$$

The graph of the function has zeros (crosses the x-axis) at the points $x = -315,$ $-135, 45,$ and 225.

A third way to approach the problem is to divide both sides of the equation by $\cos x$, where $\cos x \neq 0$, as shown below.

$$\frac{\sin x}{\cos x} = \frac{\cos x}{\cos x}$$
$$\tan x = 1$$

Have the students graph $y = \tan x$ and find the values for which $\tan x = 1$.

$$\boxed{\text{Graph}}\ \boxed{\text{tan}}\ \boxed{\text{Alpha}}\ \boxed{+}$$

$-315, -135, 45, 225$

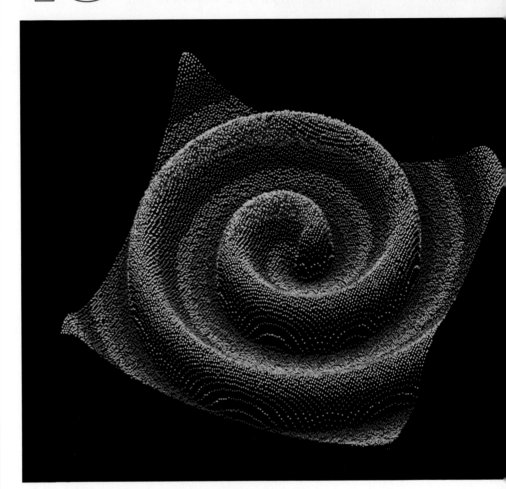

Certain systems that interact with their environment have the ability to generate *order out of chaos.* The image shown here is a computer-generated study of a chemical reaction that creates a spiral within a thoroughly mixed solution.

More About the Image

In some chemical reactions, one molecule can't be made unless it is in the presence of another molecule of the same type. Such reactions are *self-catalyzing.* An iterative process can occur in which the products of some steps feed back into the process either as further production or as inhibition. The resulting systems exhibit many fascinating behaviors, such as intermittency and self-organization. In 1958, B. P. Belousov discovered a chemical reaction that oscillates with clocklike precision, from yellow to colorless and back to yellow again, twice a minute. A. M. Zhabotinsky's study of Belousov's work led to the discovery of the reaction that produces the spiral waves in the computer study in the illustration.

19.1 Basic Trigonometric Identities

Objectives

To verify the basic trigonometric identities for given values of the variables

To prove basic trigonometric identities

An equation that is true for all permissible values of its variables is called an **identity**. For example, the equation $\dfrac{4}{3x - 9} = \dfrac{4}{3(x - 3)}$ is an identity since it is true for all $x \neq 3$.

Recall that $\sin \theta$ and $\csc \theta$ are reciprocals of each other for $\sin \theta \neq 0$, as shown below.

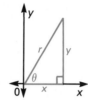

$$\sin \theta \csc \theta = \frac{y}{r} \cdot \frac{r}{y} = 1,$$

where $y \neq 0, r > 0$

Note that $y \neq 0$ if and only if $\sin \theta \neq 0$. Therefore, the equation $\sin \theta \csc \theta = 1$ is an identity since it is true for all values of θ for which $\sin \theta \neq 0$.

Reciprocal identities can be written for all of the trigonometric ratios.

Reciprocal Identities

$\sin \theta \csc \theta = 1$, for $\sin \theta \neq 0$ $\left(\sin \theta = \dfrac{1}{\csc \theta} \text{ or } \csc \theta = \dfrac{1}{\sin \theta}\right)$

$\cos \theta \sec \theta = 1$, for $\cos \theta \neq 0$ $\left(\cos \theta = \dfrac{1}{\sec \theta} \text{ or } \sec \theta = \dfrac{1}{\cos \theta}\right)$

$\tan \theta \cot \theta = 1$, for $\tan \theta \neq 0$, $\left(\tan \theta = \dfrac{1}{\cot \theta} \text{ or } \cot \theta = \dfrac{1}{\tan \theta}\right)$

$\cot \theta \neq 0$

Recall that the acute angles of a right triangle are complementary. For example, in right triangle ABC at the right, $\angle A$ and $\angle B$ are complementary since $60 + 30 = 90$. Notice that in the triangle,

$$\sin 60 = \frac{\sqrt{3}}{2} \text{ and } \cos 30 = \frac{\sqrt{3}}{2}.$$

Therefore, $\sin 60 = \cos 30$.

This suggests that the sine of an angle equals the cosine of its complement.

Teaching Resources

Quick Quizzes 143
Reteaching and Practice
   Worksheets 143

▰▰▰ GETTING STARTED

Prerequisite Quiz

Write each expression in terms of two of the variables *x*, *y*, and *r*.

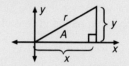

1. $\sin A$ $\dfrac{y}{r}$ 2. $\cos A$ $\dfrac{x}{r}$

3. $\tan A$ $\dfrac{y}{x}$ 4. $\sec A$ $\dfrac{r}{x}$

5. $\csc A$ $\dfrac{r}{y}$ 6. $\cot A$ $\dfrac{x}{y}$

Motivator

Lead students to discover the nature of an identity by asking them to evaluate $\sin^2 \theta + \cos^2 \theta$ for the following values of θ: 90, 30, 60, 45, 135, 240. Then ask them to generalize: $\sin^2 \theta + \cos^2 \theta = $?   1   Ask them if this equation appears to be true for all values of the variable.   Yes.

Highlighting the Standards

Standard 4c: With the derivations of the basic identities, students begin to see the increasingly large network of connections that flow through trigonometry.

Lesson Note

The basic identities are derived by expressing trigonometric ratios in terms of x, y, and r. In the next lesson, students will use the basic identities to prove other identities without reference to x, y, and r.

Try to lead students to discover some of the basic identities. Have them use a table to compare ratios such as sin 70 and cos 20. In this way they will notice that the sine of an angle is the same as the cosine of its complement.

Math Connections

Algebraic Identities: The trigonometric identities that are introduced in this lesson should remind students of the algebraic identities that have appeared earlier in the text. The term *identity* was introduced explicitly in the exercises of Lesson 1.4. The "identity" concept also occurs in several regular lessons, although not necessarily with the word "identity" mentioned. For example, in Chapter 5, the exponent rules such as

$$b^m \cdot b^n = b^{m+n}$$

are algebraic identities, as are the special-product rules such as

$$(a + b)(a - b) = a^2 - b^2.$$

The sine and cosine are said to be **cofunctions** of each other. The other cofunctions are tangent and cotangent, and secant and cosecant. The *cofunction identities* are stated below.

Cofunction Identities

$$\sin \theta = \cos (90 - \theta) \qquad\qquad \cos \theta = \sin (90 - \theta)$$
$$\tan \theta = \cot (90 - \theta) \qquad\qquad \cot \theta = \tan (90 - \theta)$$
$$\sec \theta = \csc (90 - \theta) \qquad\qquad \csc \theta = \sec (90 - \theta)$$

One of the cofunction identities was just illustrated for acute angles with positive measures. However, all of the identities are true for angles of all positive and nonpositive measures. This will be shown later in this chapter.

EXAMPLE 1 Express each function in terms of its cofunction.
   a. cos 74   **b.** sec 25

Solutions Use the cofunction and the complement.
   a. cos 74 = sin (90 − 74)   **b.** sec 25 = csc (90 − 25)
              = sin 16   = csc 65

The definitions of the basic trigonometric ratios in Lesson 17.6 can be used to prove the following *quotient identities*.

Quotient Identities

$$\tan \theta = \frac{\sin \theta}{\cos \theta}, \cos \theta \neq 0 \qquad\qquad \cot \theta = \frac{\cos \theta}{\sin \theta}, \sin \theta \neq 0$$

EXAMPLE 2 Prove the quotient identity $\tan \theta = \dfrac{\sin \theta}{\cos \theta}$.

Proof From Lesson 17.6, $\sin \theta = \dfrac{y}{r}$, $\cos \theta = \dfrac{x}{r}$, and $\tan \theta = \dfrac{y}{x}$ (see diagram).

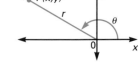

Therefore, $\dfrac{\sin \theta}{\cos \theta} = \dfrac{\frac{y}{r}}{\frac{x}{r}} = \dfrac{y}{x} = \tan \theta.$

Thus, $\tan \theta = \dfrac{\sin \theta}{\cos \theta}$ is an identity.

Additional Example 1

Express each function in terms of its cofunction.

a. tan 85   cot 5
b. csc 34   sec 56

Additional Example 2

Prove the quotient identity
$$\cot \theta = \frac{\cos \theta}{\sin \theta}.$$
$\cot \theta = \dfrac{x}{y}$, $\cos \theta = \dfrac{x}{r}$, $\sin \theta = \dfrac{y}{r}$,

so $\dfrac{\cos \theta}{\sin \theta} = \dfrac{\frac{x}{r}}{\frac{y}{r}} = \dfrac{x}{y} = \cot \theta.$

EXAMPLE 3 Verify that $\tan \theta = \dfrac{\sin \theta}{\cos \theta}$ for $\theta = \dfrac{5\pi}{6}$.

Solution $\dfrac{5\pi}{6}$ radians = 150 degrees (see diagram)

Therefore,

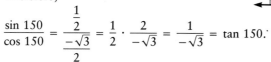

$$\frac{\sin 150}{\cos 150} = \frac{\frac{1}{2}}{\frac{-\sqrt{3}}{2}} = \frac{1}{2} \cdot \frac{2}{-\sqrt{3}} = \frac{1}{-\sqrt{3}} = \tan 150.$$

Thus, $\tan \dfrac{5\pi}{6} = \dfrac{\sin \dfrac{5\pi}{6}}{\cos \dfrac{5\pi}{6}}$.

The definitions of the trigonometric ratios can also be used to develop three special trigonometric identities called the *Pythagorean Identities*.

EXAMPLE 4 Prove that $\sin^2 \theta + \cos^2 \theta = 1$ is an identity.

Proof $\sin^2 \theta + \cos^2 \theta = \left(\dfrac{y}{r}\right)^2 + \left(\dfrac{x}{r}\right)^2 = \dfrac{y^2 + x^2}{r^2} = \dfrac{r^2}{r^2} = 1$

Therefore, $\sin^2 \theta + \cos^2 \theta = 1$ is an identity.

In Example 4, note that since $r > 0$, no denominator can be zero. Thus, all real values of the variable θ are possible values.

There are two other Pythagorean Identities. You are asked to prove one of these identities in Exercise 24. All three identities are summarized below.

Pythagorean Identities

For all values of θ for which $\tan \theta$, $\cot \theta$, $\sec \theta$, and $\csc \theta$ are defined,

$$\sin^2 \theta + \cos^2 \theta = 1$$
$$\tan^2 \theta + 1 = \sec^2 \theta$$
$$\cot^2 \theta + 1 = \csc^2 \theta$$

The Pythagorean Identities are often used in one of their alternate forms shown below.

$\sin^2 \theta = 1 - \cos^2 \theta$	$\tan^2 \theta = \sec^2 \theta - 1$	$\cot^2 \theta = \csc^2 \theta - 1$
$\cos^2 \theta = 1 - \sin^2 \theta$	$\sec^2 \theta - \tan^2 \theta = 1$	$\csc^2 \theta - \cot^2 \theta = 1$

19.1 Basic Trigonometric Identities **713**

Checkpoint

Express each function in terms of its cofunction.

1. sec 23   csc 67
2. sin $\frac{\pi}{3}$ cos $\left(\frac{\pi}{2} - \frac{\pi}{3}\right)$ = cos $\frac{\pi}{6}$

Verify each identity for the given value of θ.

3. sec θ · cos θ = 1; θ = $\frac{\pi}{4}$

   sec $\frac{\pi}{4}$ · cos $\frac{\pi}{4}$ = $\sqrt{2} \cdot \frac{\sqrt{2}}{2} = \frac{2}{2} = 1$

4. cos² θ = 1 − sin² θ; θ = 210

   cos² 210 = $\left(-\frac{\sqrt{3}}{2}\right)^2 = \frac{3}{4}$;

   1 − sin² 210 = $1 - \left(-\frac{1}{2}\right)^2 = 1 - \frac{1}{4} =$
   $\frac{3}{4}$; so cos² 210 = 1 − sin² 210

Closure

Have students define the following terms.

1. Identity   See page 711.
2. Reciprocal Identities   See page 711.
3. Quotient Identities   See page 712.
4. Pythagorean Identities   See page 713.
5. Cofunctions   See page 712.
6. Cofunction Identities   See page 712.

▄▄▄FOLLOW UP

Guided Practice

Classroom Exercises 1–15

Independent Practice

A Ex. 1–14, **B** Ex. 15–24, **C** Ex. 25, 26

Basic: FR, WE 1–17 odd, Brainteaser
Average: FR, WE 7–23 odd, Brainteaser
Above Average: FR, WE 11–25 odd, Brainteaser

Focus on Reading

Indicate whether each statement is true or false. If false, give a reason.

1. Trigonometric identities are equations that are true for all values of the variable θ.   False; they are true for all permissible values of θ.
2. sin θ = $\frac{1}{\csc \theta}$ is an example of a quotient identity.   False; reciprocal identity
3. The cofunction of cos θ is sec (90 − θ).   False; cofunction is sin (90 − θ).
4. Another form of the Pythagorean Identity sin² θ + cos² θ = 1 is cos² θ − 1 = − sin² θ.   True
5. tan θ = cot (90 − θ) is true only for acute angles.   False; it is true for all permissible values of θ.

Classroom Exercises
1. tan θ 2. cos θ 3. csc² θ − 1 4. 1
5. cos θ 6. 1 7. $\frac{\cos \theta}{\sin \theta}$ or $\frac{1}{\tan \theta}$ 8. tan θ 9. (90 − θ)

Complete each identity.

1. $\frac{\sin \theta}{\cos \theta}$ = _?_
2. sec θ · _?_ = 1
3. cot² θ = _?_
4. sin² θ + cos² θ = _?_
5. sin (90 − θ) = _?_
6. tan θ · cot θ = _?_
7. cot θ = _?_
8. $\frac{1}{\cot \theta}$ = _?_
9. sec _?_ = csc θ

Express each function in terms of its cofunction.

10. cos 60   sin 30
11. sin 45   cos 45
12. cot 5   tan 85
13. cos $\frac{\pi}{6}$ sin $\frac{\pi}{3}$

Verify each identity for the given value of θ.

14. sin² θ + cos² θ = 1; θ = 90
    sin² 90 + cos² 90 = 1² + 0² = 1

15. $\frac{\sin \theta}{\cos \theta}$ = tan θ; θ = 45
    $\frac{\sin 45}{\cos 45} = \frac{\frac{\sqrt{2}}{2}}{\frac{1}{\sqrt{2}}} = 1 = \tan 45$

Written Exercises

Express each function in terms of its cofunction.

1. sin 50   cos 40
2. tan 81   cot 9
3. csc 65.4   sec 24.6
4. cos $\frac{\pi}{8}$ sin $\frac{3\pi}{8}$

Verify each identity for the given value of θ.

5. sin θ csc θ = 1; θ = 135
6. tan θ = $\frac{1}{\cot \theta}$; θ = 135
7. sin (90 − θ) = cos θ; θ = 60
8. tan θ = cot (90 − θ); θ = 135
9. $\frac{1}{\sec \theta}$ = cos θ; θ = 30
10. $\frac{\sin \theta}{\cos \theta}$ = tan θ; θ = 150

Enrichment

To prove that an equation is *not* an identity, it is sufficient to use a *counterexample*.
For example, sin x = $\frac{1}{\cos x}$ is not an identity, since sin 30 = $\frac{1}{2}$, cos 30 = $\frac{\sqrt{3}}{2}$, and
$\frac{1}{2} \neq \frac{1}{\frac{\sqrt{3}}{2}}$.

Have students find counterexamples to show that the equations below are *not* identities.   Answers will vary.

1. csc a = sin a cos a
   (HINT: Let a = 0.)
2. sin $\frac{a}{b}$ = $\frac{\sin a}{\sin b}$
3. sin (a + b) = sin a + sin b
4. tan (a + b) = tan a + tan b
5. cos (a − b) = cos a − cos b

NOTE: Exercises 3–5 can be recalled when students study the sum and difference identities in Lesson 19.5.

11. $\dfrac{\cos \theta}{\sin \theta} = \cot \theta$; $\theta = 240$

12. $\dfrac{1}{\tan \theta} = \cot \theta$; $\theta = 30$

13. $\cos^2 \theta = 1 - \sin^2 \theta$; $\theta = 270$

14. $\sec^2 \theta - \tan^2 \theta = 1$; $\theta = 60$

15. $1 + \cot^2 \theta = \csc^2 \theta$; $\theta = \dfrac{2\pi}{3}$

16. $\sin^2 \theta = 1 - \cos^2 \theta$; $\theta = \dfrac{5\pi}{4}$

17. $\dfrac{\cos \theta}{\sin \theta} = \cot \theta$; $\theta = \dfrac{11\pi}{6}$

18. $\cot \theta = \dfrac{1}{\tan \theta}$; $\theta = \tan \theta$; $\theta = \dfrac{3\pi}{4}$

19. $1 + \tan^2 \theta = \sec^2 \theta$; $\theta = -\dfrac{7\pi}{3}$

20. $\cos^2 \theta = 1 - \sin^2 \theta$; $\theta = -\dfrac{3\pi}{2}$

Use the definitions of the trigonometric ratios to prove that each equation is an identity.

21. $\cos \theta \sec \theta = 1$

22. $\tan \theta \cot \theta = 1$

23. $\cot \theta = \dfrac{\cos \theta}{\sin \theta}$

24. $\tan^2 \theta + 1 = \sec^2 \theta$

Use identities from this lesson to prove the following.

25. $(1 - \sin^2 \theta)(1 + \tan^2 \theta) = 1$

26. $\dfrac{1}{\sec \theta - \tan \theta} - \dfrac{1}{\sec \theta + \tan \theta} = 2 \tan \theta$

Mixed Review

Simplify. *7.1, 7.4, 8.6*

1. $\dfrac{x^3 - 27}{4x - 12} \cdot \dfrac{x^2 + 3x + 9}{4}$

2. $\dfrac{-x - 9}{x^2 - 9} - \dfrac{2}{3 - x} \quad \dfrac{\frac{1}{x + 3}}{}$

3. $(-32)^{-\frac{4}{5}} \quad \dfrac{1}{16}$

Give the amplitude and period of each function. *18.8*

4. $T = 5 \cos \frac{1}{2}\theta$

5. $T = -6 \cos 4\theta$

6. $T = 3 \sin 3\theta$

Brainteaser

In the figure, the measure of $\angle A$ is 30. $\angle D$ is the angle formed by the bisectors of $\angle B$ and $\angle C$. Find the measure of $\angle D$. 105

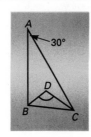

Additional Answers
Written Exercises

5. $\sin 135 \cdot \csc 135 = \dfrac{1}{\sqrt{2}} \cdot \sqrt{2} = 1$

6. $\dfrac{1}{\cot 135} = \dfrac{1}{-1} = -1 = \tan 135$

7. $\sin (90 - 60) = \sin 30 = \dfrac{1}{2} = \cos 60$

8. $\tan 135 = -1 = \cot (-45) = \cot (90 - 135)$

9. $\dfrac{1}{\sec 30} = \dfrac{\frac{1}{2}}{\frac{2}{\sqrt{3}}} = \dfrac{\sqrt{3}}{2} = \cos 30$

10. $\dfrac{\sin 150}{\cos 150} = \dfrac{1}{2} \div \left(\dfrac{-\sqrt{3}}{2}\right) = \dfrac{1}{2} \cdot \dfrac{-2}{\sqrt{3}} = \dfrac{-1}{\sqrt{3}}$
$= \tan 150$

11. $\dfrac{\cos 240}{\sin 240} = -\dfrac{1}{2} \div \left(\dfrac{-\sqrt{3}}{2}\right) = -\dfrac{1}{2} \cdot \dfrac{-2}{\sqrt{3}}$
$= \dfrac{1}{\sqrt{3}} = \cot 240$

12. $\dfrac{1}{\tan 30} = \sqrt{3} = \cot 30$

13. $\cos^2 270 = 0^2 = 1 - (-1)^2 = 1 - \sin^2 270$

14. $\sec^2 60 - \tan^2 60 = 2^2 - (\sqrt{3})^2 = 4 - 3 = 1$

15. $1 + \cot^2 \dfrac{2\pi}{3} = 1 + \left(\dfrac{-1}{\sqrt{3}}\right)^2 = 1 + \dfrac{1}{3} = \dfrac{4}{3}$
$= \left(\dfrac{2}{\sqrt{3}}\right)^2 = \csc^2 \dfrac{2\pi}{3}$

16. $\sin^2 \dfrac{5\pi}{4} = \left(\dfrac{-1}{\sqrt{2}}\right)^2 = 1 - \dfrac{1}{2} = \dfrac{1}{2}$
$= 1 - \left(\dfrac{-1}{\sqrt{2}}\right)^2 = 1 - \cos^2 \dfrac{5\pi}{4}$

17. $\dfrac{\cos \frac{11\pi}{6}}{\sin \frac{11\pi}{6}} = \dfrac{\sqrt{3}}{2} \div \left(\dfrac{-1}{2}\right) = \dfrac{\sqrt{3}}{2} \cdot \dfrac{-2}{1}$
$= -\sqrt{3} = \cot \dfrac{11\pi}{6}$

18. $\cot \dfrac{3\pi}{4} = -1 = \dfrac{1}{-1} = \dfrac{1}{\tan \frac{3\pi}{4}}$

19. $1 + \tan^2 \left(\dfrac{-7\pi}{3}\right) = 1 + (-\sqrt{3})^2$
$= 4 = 2^2 = \sec^2 \left(\dfrac{-7\pi}{3}\right)$

20. $\cos^2 \left(\dfrac{-3\pi}{2}\right) = 0^2 = 1 - 1^2$
$= 1 - \sin^2 \left(\dfrac{-3\pi}{2}\right)$

21. $\cos \theta \sec \theta = \dfrac{x}{r} \cdot \dfrac{r}{x} = 1$

22. $\tan \theta \cot \theta = \dfrac{y}{x} \cdot \dfrac{x}{y} = 1$

23. $\cot \theta = \dfrac{x}{y} = \dfrac{\frac{x}{r}}{\frac{y}{r}} = \dfrac{\cos \theta}{\sin \theta}$

24. $\tan^2 \theta + 1 = \left(\dfrac{y}{x}\right)^2 + 1 = \dfrac{y^2}{x^2} + 1$
$= \dfrac{x^2 + y^2}{x^2} = \dfrac{r^2}{x^2} = \sec^2 \theta$

25. $(1 - \sin^2 \theta)(1 + \tan^2 \theta)$
$= \cos^2 \theta \sec^2 \theta = \cos^2 \theta \left(\dfrac{1}{\cos^2 \theta}\right) = 1$

26. $\dfrac{1}{\sec \theta - \tan \theta} - \dfrac{1}{\sec \theta + \tan \theta}$
$= \dfrac{(\sec \theta + \tan \theta) - (\sec \theta - \tan \theta)}{(\sec \theta - \tan \theta)(\sec \theta + \tan \theta)}$
$= \dfrac{2 \tan \theta}{\sec^2 \theta - \tan^2 \theta}$
$= 2 \tan \theta$

Mixed Review

4. $p = 4\pi$; $a = 5$

5. $p = \dfrac{\pi}{2}$; $a = 6$

6. $p = \dfrac{2\pi}{3}$; $a = 3$

▬▬GETTING STARTED

Prerequisite Quiz

1. List the Reciprocal Identities.
$\sin \theta = \dfrac{1}{\csc \theta}$; $\cos \theta = \dfrac{1}{\sec \theta}$;
$\tan \theta = \dfrac{1}{\cot \theta}$

2. List the Quotient Identities.
$\dfrac{\sin \theta}{\cos \theta} = \tan \theta$; $\dfrac{\cos \theta}{\sin \theta} = \cot \theta$

3. List the Pythagorean Identities.
$\sin^2 \theta + \cos^2 \theta = 1$; $\tan^2 \theta + 1 = \sec^2 \theta$;
$1 + \cot^2 \theta = \csc^2 \theta$

Motivator

Write the following expression on the chalkboard.

$$\dfrac{\sin^2 \theta}{\cos \theta} + \cos \theta$$

Ask students to explain how they would find a common denominator. Multiply $\cos \theta$ by $\dfrac{\cos \theta}{\cos \theta}$. Now have students add the terms and simplify.

$$\dfrac{\sin^2 \theta}{\cos \theta} + \dfrac{\cos^2 \theta}{\cos \theta} = \dfrac{\sin^2 \theta + \cos^2 \theta}{\cos \theta} = \dfrac{1}{\cos \theta}$$

19.2 Proving Trigonometric Identities

Objectives To simplify trigonometric expressions
To prove trigonometric identities using basic identities

The identities of Lesson 19.1 can be used to simplify trigonometric expressions. One technique is to express all functions in terms of sines and cosines. This is illustrated in Example 1 below.

EXAMPLE 1 Simplify $\sin \theta \cot \theta$.

Plan First rewrite $\cot \theta$ in terms of $\sin \theta$ and $\cos \theta$.
Use the Quotient Identity, $\cot \theta = \dfrac{\cos \theta}{\sin \theta}$.

Solution $\sin \theta \cot \theta = \sin \theta \cdot \dfrac{\cos \theta}{\sin \theta} = \cos \theta$

Therefore, $\sin \theta \cot \theta = \cos \theta$.

You will find it helpful to make your own list of the *Reciprocal*, *Quotient*, and *Pythagorean Identities* of Lesson 19.1. Since these identities are used frequently, you should memorize them.

The Pythagorean and Reciprocal Identities are used in Example 2.

EXAMPLE 2 Simplify $\dfrac{\sin^2 \theta}{\cos \theta} + \cos \theta$.

Plan First combine the fractions by making the denominators alike.

Solution $\dfrac{\sin^2 \theta}{\cos \theta} + \dfrac{\cos \theta}{1} = \dfrac{\sin^2 \theta}{\cos \theta} + \dfrac{\cos \theta \cos \theta}{1 \cdot \cos \theta}$

$= \dfrac{\sin^2 \theta}{\cos \theta} + \dfrac{\cos^2 \theta}{\cos \theta}$

$= \dfrac{\sin^2 \theta + \cos^2 \theta}{\cos \theta}$

$= \dfrac{1}{\cos \theta} = \sec \theta$

Highlighting the Standards

Standard 9f: Verifying identities is not a practical use of trigonometry, but it sometimes helps develop mathematical intuition.

Additional Example 1

Simplify $\cos \theta \tan \theta$. $\sin \theta$

Additional Example 2

Simplify $\left(\dfrac{1}{\cos \theta} - \cos \theta\right) \csc^2 \theta$. $\sec \theta$

Of the two trigonometric equations below, one is an identity.

(1) $\sin\theta \cot\theta = \cos\theta$ (2) $\sin\theta = \cos\theta$

Equation (1) from Example 1 *is* an identity. Equation (2) is false for some values of θ (such as $\theta = 30$) and thus is *not* an identity.

EXAMPLE 3   Prove that $\dfrac{\sin\theta}{1 + \cos\theta} = \dfrac{1 - \cos\theta}{\sin\theta}$ is an identity.

Plan   The right side of the equation contains only $1 - \cos\theta$ in the numerator. Get $1 - \cos\theta$ in the numerator on the left by multiplying the numerator and the denominator of the left side of the equation by $1 - \cos\theta$. Then use basic algebraic and trigonometric identities to transform the expression on the left into the one on the right side.

Solution

$$\dfrac{\sin\theta}{1 + \cos\theta} \quad\Bigg|\quad \dfrac{1 - \cos\theta}{\sin\theta}$$

$$\dfrac{(1 - \cos\theta)\sin\theta}{(1 - \cos\theta)(1 + \cos\theta)}$$

Use $(a - b)(a + b) = a^2 - b^2$. $\dfrac{(1 - \cos\theta)\sin\theta}{1 - \cos^2\theta}$

Use the Pythagorean Identity $\dfrac{(1 - \cos\theta)\sin\theta}{\sin^2\theta}$
$1 - \cos^2\theta = \sin^2\theta$.

Simplify. $\dfrac{1 - \cos\theta}{\sin\theta} = \dfrac{1 - \cos\theta}{\sin\theta}$

The original equation will be an identity *unless* the transformation steps change the restrictions on the variable θ. Since this is not the case (only integral multiples of 180 are forbidden in the original and the transformed equations), the original equation is an identity.

In Example 4, each side of an equation is transformed into a third expression that is equivalent to each of the two sides.

EXAMPLE 4   Prove that $\cos^2\theta \tan^2\theta + 1 = \sec^2\theta + \sin^2\theta - \sin^2\theta \sec^2\theta$ is an identity.

Plan   Use the identity $\tan\theta = \dfrac{\sin\theta}{\cos\theta}$ to simplify part of the left side. The result, $\sin^2\theta + 1$, agrees with the right side in one term, $\sin^2\theta$. Therefore, ignore this term and concentrate on the rest of the right side, $\sec^2\theta - \sin^2\theta \sec^2\theta$.

Lesson Note

A review of algebraic skills may be necessary. When simplifying trigonometric expressions, compare them with algebraic analogs. For example, $\sin\theta \cdot \dfrac{\cos\theta}{\sin\theta}$ is similar to $a \cdot \dfrac{3}{a}$; and $(1 - \cos\theta)(1 + \cos\theta)$ is similar to $(1 - x)(1 + x)$. Thus, the given equation is an identity.

Math Connections

Line Functions: Some trigonometry texts of a generation ago represented the six trigonometric functions as so-called "line functions"; that is, as line segments associated with a circle of unit radius. The line-function diagrams geometrically illustrate the relationships among the six functions.

Some students may find it easier to memorize the basic identities using these diagrams. One of them appears in the first paragraph of Lesson 19.4. Here is another.

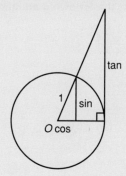

Additional Example 3

Prove that $\dfrac{1 - \sin\theta}{\cos\theta} = \dfrac{\cos\theta}{1 + \sin\theta}$ is an identity.

$\dfrac{1 - \sin\theta}{\cos\theta}$

$= \dfrac{(1 - \sin\theta)(1 + \sin\theta)}{\cos\theta\,(1 + \sin\theta)}$

$= \dfrac{1 - \sin^2\theta}{\cos\theta\,(1 + \sin\theta)} = \dfrac{\cos^2\theta}{\cos\theta\,(1 + \sin\theta)}$

$= \dfrac{\cos\theta}{1 + \sin\theta}$

Additional Example 4

Prove that $\cos^2\theta - \cos^2\theta \csc^2\theta + \csc^2\theta = 1 + \sin^2\theta \cot^2\theta$ is an identity.

$\cos^2\theta - \cos^2\theta \csc^2\theta + \csc^2\theta$
$= \cos^2\theta + \csc^2\theta(1 - \cos^2\theta)$
$= \cos^2\theta + \csc^2\theta \sin^2\theta$
$= \cos^2\theta + 1$
Also, $1 + \sin^2\theta \cot^2\theta = 1 + \cos^2\theta$.

Critical Thinking Questions

Analysis: Ask students to use the diagram of the *Math Connections* on page 717 to demonstrate the two identities $\tan \theta = \dfrac{\sin \theta}{\cos \theta}$ and $\sin^2 \theta + \cos^2 \theta = 1$.

Common Error Analysis

Error: Students may forget to use the Distributive Property with trigonometric expressions. Point out the similarity between expressions such as $x(\frac{1}{x} - x)$ and $\cos \theta \left(\dfrac{1}{\cos \theta} - \cos \theta\right)$. Remind students that $\cos \theta \left(\dfrac{1}{\cos \theta} - \cos \theta\right) = \cos \theta \dfrac{1}{\cos \theta} - \cos \theta \cos \theta$. Thus, the given equation is an identity.

Checkpoint

Simplify.

1. $\sin^2 \theta \csc^3 \theta$ $\csc \theta$

2. $\dfrac{\sin \theta}{\tan \theta}$ $\cos \theta$

3. $\dfrac{4 \cos \theta}{1 - \sin^2 \theta}$ $4 \sec \theta$

Prove that the following equation is an identity.

4. $\dfrac{\cos \theta + \cot \theta}{1 + \csc \theta} = \cos \theta$

$$\dfrac{\cos \theta + \cot \theta}{1 + \csc \theta} = \dfrac{\cos \theta + \dfrac{\cos \theta}{\sin \theta}}{1 + \dfrac{1}{\sin \theta}}$$

$$= \dfrac{\sin \theta \left(\cos \theta + \dfrac{\cos \theta}{\sin \theta}\right)}{\sin \theta \left(1 + \dfrac{1}{\sin \theta}\right)}$$

$$= \dfrac{\sin \theta \cos \theta + \cos \theta}{\sin \theta + 1}$$

$$= \dfrac{\cos \theta (\sin \theta + 1)}{\sin \theta + 1} = \cos \theta$$

Solution

$\cos^2 \theta \tan^2 \theta + 1$	$\sec^2 \theta + \sin^2 \theta - \sin^2 \theta \sec^2 \theta$
$\cos^2 \theta \dfrac{\sin^2 \theta}{\cos^2 \theta} + 1$	$\sin^2 \theta + \sec^2 \theta - \sin^2 \theta \sec^2 \theta$
$\sin^2 \theta + 1$	$\sin^2 \theta + \sec^2 \theta (1 - \sin^2 \theta)$
	$\sin^2 \theta + \sec^2 \theta \cos^2 \theta$

$$\sin^2 \theta + 1 = \sin^2 \theta + 1$$

Summary

To prove an identity,

(1) transform one side (the more complicated one, if possible) into the other side, or

(2) transform each side into the same expression.

The following guidelines should be observed when performing the above transformations.

(1) Apply the Reciprocal, Quotient, or Pythagorean Identities.

(2) Use basic algebraic identities such as $x^2 - y^2 = (x + y)(x - y)$.

(3) Combine the sum or difference of fractions into a single fraction.

(4) If all else fails, change all expressions into sines and cosines and simplify.

Classroom Exercises

Factor.

1. $\sin^2 \theta - \sin \theta$
$\sin \theta (\sin \theta - 1)$

2. $1 - \tan^2 \theta$
$(1 - \tan \theta)(1 + \tan \theta)$

3. $\sin \theta + \sin \theta \cos \theta$
$\sin \theta (1 + \cos \theta)$

Simplify.

4. $\sin \theta \dfrac{1}{\sin^2 \theta}$ $\csc \theta$

5. $(1 - \cos \theta)(1 + \cos \theta)$ $\sin^2 \theta$

6. $\dfrac{1}{1 - \sin \theta} + \dfrac{\sin \theta}{1 - \sin \theta}$

7. $\cos \theta \tan \theta$ $\sin \theta$

8. $\sin \theta \sec \theta$ $\tan \theta$

9. $\dfrac{1 - \sin^2 \theta}{\cos^2 \theta}$ 1

Written Exercises

Simplify.

1. $\tan \theta \cot \theta$ 1

2. $\cos \theta \left(\dfrac{1}{1 - \sin^2 \theta}\right)$ $\sec \theta$

3. $\cos \theta \csc \theta$ $\cot \theta$

4. $\csc^2 \theta - \csc^2 \theta \cos^2 \theta$
1

5. $\tan^2 \theta - \tan^2 \theta \sin^2 \theta$
$\sin^2 \theta$

6. $\dfrac{\cos \theta}{\cot \theta}$ $\sin \theta$

7. $\dfrac{\sin \theta - \sin \theta \cos \theta}{1 - \cos \theta}$ $\;\sin \theta\;$ **8.** $\dfrac{\cos \theta \sin^2 \theta + \cos^3 \theta}{\cos \theta}$ $\;1\;$ **9.** $\dfrac{2 \sin \theta}{1 - \cos^2 \theta}$ $\;2 \csc \theta$

Prove that each equation is an identity. (Exercises 10–30)

10. $\dfrac{\sin \theta}{\cos \theta} + \dfrac{\cos \theta}{\sin \theta} = \dfrac{1}{\sin \theta \cos \theta}$

11. $\sin \theta + \dfrac{\cos^2 \theta}{\sin \theta} = \csc \theta$

12. $(\sin \theta + \cos \theta)^2 = 1 + 2 \sin \theta \cos \theta$

13. $\cos^2 \theta \, (\cot^2 \theta + 1) = \cot^2 \theta$

14. $\cos^2 \theta - \sin^2 \theta = 1 - 2 \sin^2 \theta$

15. $\dfrac{\cos \theta}{\sin \theta} + \dfrac{\sin \theta}{1 + \cos \theta} = \csc \theta$

16. $\dfrac{\cot \theta + \tan \theta}{\csc^2 \theta} = \tan \theta$

17. $\dfrac{\sin \theta + \tan \theta}{\cot \theta + \csc \theta} = \sin \theta \tan \theta$

18. $\dfrac{\cot \theta}{1 + \cot^2 \theta} = \sin \theta \cos \theta$

19. $\dfrac{1 + \cot \theta}{\cot \theta \sin \theta + \dfrac{\cos^2 \theta}{\sin \theta}} = \sec \theta$

20. $\dfrac{1 + \sec \theta}{\sin \theta + \tan \theta} = \csc \theta$

21. $\dfrac{\csc \theta - \sin \theta}{\cot^2 \theta} = \sin \theta$

22. $\dfrac{1 + \sin \theta}{\cos \theta} = \dfrac{\cos \theta}{1 - \sin \theta}$

23. $\dfrac{1}{1 - \sin \theta} + \dfrac{1}{1 + \sin \theta} = 2 \sec^2 \theta$

24. $\dfrac{\csc \theta + \cot \theta}{\tan \theta + \sin \theta} = \cot \theta \csc \theta$

25. $\sin^2 \theta \sec^2 \theta + \sin^2 \theta \csc^2 \theta = \sec^2 \theta$

26. $\sin^4 \theta - \cos^4 \theta = 2 \sin^2 \theta - 1$

27. $\dfrac{\csc \theta}{\cot \theta + \tan \theta} = \cot \theta \sin \theta$

28. $\cos \theta \, (2 \sec \theta + \tan \theta)(\sec \theta - 2 \tan \theta) = 2 \cos \theta - 3 \tan \theta$

29. $(\sin \theta + \cos \theta)(\tan \theta + \cot \theta) = \csc \theta + \sec \theta$

30. $\dfrac{1}{\csc \theta - \cot \theta} - \dfrac{1}{\csc \theta + \cot \theta} = 2 \cot \theta$

31. Show that the determinant equation $\begin{vmatrix} \cos \theta & -\sin \theta \\ \sin \theta & \cos \theta \end{vmatrix} = 1$ is an identity.

32. For what value of t is $t \cos \theta \sin \theta - 1 + (\sin \theta + \cos \theta)^2 = 0$ an identity? $\;-2$

33. Simplify $\dfrac{1 - \sin \theta + \cos \theta - \sin \theta \cos \theta}{1 - \sin^3 \theta} \div \dfrac{1 + \cos \theta}{1 + \cos \theta + \cos^2 \theta} \cdot \dfrac{1 + \cos \theta + \cos^2 \theta}{1 + \sin \theta + \sin^2 \theta}$

Mixed Review

Solve. *1.4, 6.1*

1. $3x - 5(2 - x) = 4 - x$ $\;\dfrac{14}{9}$

2. $7x - 4 = -2x^2$ $\;-4, \dfrac{1}{2}$

Simplify. *17.6, 17.7, 18.5*

3. $\cos \dfrac{3\pi}{2}$ $\;0$

4. $\sin 120$ $\;\dfrac{\sqrt{3}}{2}$

5. $\tan \dfrac{3\pi}{4}$ $\;-1$

6. $\cot \dfrac{\pi}{4}$ $\;1$

19.2 Proving Trigonometric Identities **719**

Ask students to name the sets of identities of the previous lesson that can be used to simplify trigonometric expressions. Reciprocal, Quotient, and Pythagorean Identities   Have them identify two methods for simplifying trigonometric expressions taught in this lesson.   One method is to express all functions in terms of sines and cosines; the other is to use basic algebraic identities.

◤◢◤ FOLLOW UP

Guided Practice

Classroom Exercises 1–9

Independent Practice

A Ex. 1–15, **B** Ex. 16–27, **C** Ex. 28–33

Basic: WE 1–19 odd
Average: WE 9–27 odd
Above Average: WE 15–33 odd

Additional Answers

Classroom Exercises

6. $\dfrac{1 + \sin \theta}{1 - \sin \theta}$

Written Exercises

10. $\dfrac{\sin \theta}{\cos \theta} + \dfrac{\cos \theta}{\sin \theta} = \dfrac{\sin^2 \theta + \cos^2 \theta}{\sin \theta \cos \theta}$

$= \dfrac{1}{\sin \theta \cos \theta}$

11. $\sin \theta + \dfrac{\cos^2 \theta}{\sin \theta} = \dfrac{\sin^2 \theta}{\sin \theta} + \dfrac{\cos^2 \theta}{\sin \theta}$

$= \dfrac{1}{\sin \theta} = \csc \theta$

See pages 828–829 for the answers to Written Ex. 12–31.

Enrichment

Remind students that identities are valid only for permissible values of the variables (see Lesson 19.1). For example, the identity $\sin \theta = \dfrac{\sin \theta}{\cos \theta} \cdot \cos \theta$, or $\sin \theta = \tan \theta \cos \theta$, is not valid for angles coterminal with 90 or 270. For these angles, $\cos \theta = 0$ and $\tan \theta$ is undefined. Refer the students to Written Exercises 10–20. Have them determine which angles in the interval $0 \le \theta < 360$ are excluded.

10. 0, 90, 180, 270
11. 0, 180
12. None are excluded.
13. 0, 180
14. None are excluded.
15. 0, 180
16. 0, 90, 180, 270
17. 0, 90, 180, 270
18. 0, 180
19. 0, 90, 180, 270
20. 0, 90, 180, 270

Prerequisite Quiz

Find each trigonometric ratio.

1. $\sin 135$ $\dfrac{\sqrt{2}}{2}$ **2.** $\cos 270$ 0

3. $\tan 150$ $-\dfrac{\sqrt{3}}{3}$ **4.** $\sec 240$ -2

5. $\tan \dfrac{3\pi}{4}$ -1 **6.** $\csc \dfrac{11\pi}{6}$ -2

Motivator

Recall for students the methods for solving linear and quadratic equations. Then have students solve the following equations for x.

1. $5x + 6\sqrt{2} = 3x + 5\sqrt{2}$ $x = -\dfrac{\sqrt{2}}{2}$

2. $2x^2 = 3x - 1$
   $(2x - 1)(x - 1) = 0;\ x = \dfrac{1}{2};\ x = 1$

Lesson Note

Point out that Classroom Exercise 7 is a trigonometric equation in simplest form. This equation can be used to review the method of finding all values of θ, $0 \le \theta < 360$, given a trigonometric ratio of θ.

For each example in the lesson, compare the trigonometric equation with an algebraic equation of similar structure. For Example 1, a comparable algebraic equation could be $3x + 4 = x + 8$.

Highlighting the Standards

Standard 4a: The solution of trigonometric equations depends on the ability to see and use equivalent representations of a concept.

19.3 Introduction to Trigonometric Equations

Objective   To solve trigonometric equations involving one trigonometric ratio

An equation such as $\tan \theta = \dfrac{\sin \theta}{\cos \theta}$ is an identity since it is true for all values of θ for which $\tan \theta$ is defined. However, most equations encountered in algebra are *conditional*, that is, true for only some values of the variable. The equations $\cos \theta = -\dfrac{\sqrt{3}}{2}$, $\tan \theta = 1$, and $5 \sin \theta + 6\sqrt{2} = 3 \sin \theta + 5\sqrt{2}$ are conditional equations. Solving such equations often involves recognizing special angle relationships.

Special Angle Relationships (first-quadrant position)

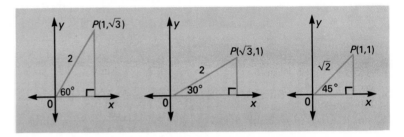

EXAMPLE 1   Solve $5 \sin \theta + 6\sqrt{2} = 3 \sin \theta + 5\sqrt{2}$, $0 \le \theta < 360$.

Solution   (1) Solve for $\sin \theta$.

$$5 \sin \theta + 6\sqrt{2} = 3 \sin \theta + 5\sqrt{2}$$
$$5 \sin \theta - 3 \sin \theta = 5\sqrt{2} - 6\sqrt{2}$$
$$2 \sin \theta = -\sqrt{2}$$
$$\sin \theta = -\dfrac{\sqrt{2}}{2}$$

(2) Sin θ is negative only in Quadrants III and IV.

(3) The reference angle measures 45.

$\left(\sin 45 = \dfrac{1}{\sqrt{2}} = \dfrac{\sqrt{2}}{2}\right)$

 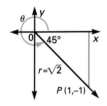

$\theta = 180 + 45$ or $\theta = 360 - 45$
$\quad = 225 \qquad\qquad\ = 315$

Thus, the solutions of the equation are 225 and 315.

Additional Example 1

Solve.

$4 \sin \theta + 3 = 6 \sin \theta + 4$, $0 \le \theta < 360$
210, 330

Solutions of a trigonometric equation in a specified interval, usually $0 \le \theta < 360$, are called *primary solutions*. Thus, 225 and 315 are primary solutions of the equation in Example 1. If there had been no specified interval, other solutions would differ from 225 and 315 by integral multiples of 360 as follows:

$\theta = 225 + k \cdot 360$ or $\theta = 315 + k \cdot 360$, where k is an integer.

For the next examples, the following figures review quadrantal angles.

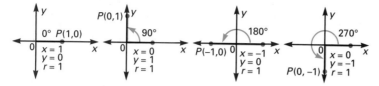

EXAMPLE 2   Solve $2 \cos^2 \theta = 3 \cos \theta - 1$, $0 \le \theta < 2\pi$.

Plan   The equation is of the form $2x^2 = 3x - 1$ and is thus in quadratic form (see Lesson 9.1). Solve as a quadratic equation.

Solution

$$2 \cos^2 \theta = 3 \cos \theta - 1$$
$$2 \cos^2 \theta - 3 \cos \theta + 1 = 0$$
$$(2 \cos \theta - 1)(\cos \theta - 1) = 0$$
$$2 \cos \theta - 1 = 0 \quad or \quad \cos \theta - 1 = 0$$
$$\cos \theta = \tfrac{1}{2} \quad or \quad \cos \theta = 1$$

$\cos \theta = \tfrac{1}{2}$: $\cos \theta$ is positive in Quadrants I and IV.
   The reference angle is 60.
$\cos \theta = 1$: quadrantal angle, $\theta = 0$

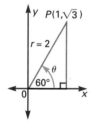   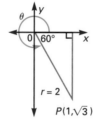   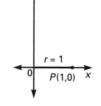

   $\theta = 60$ $\theta = 360 - 60 = 300$ $\theta = 0$

Write the solutions in radians.

$$60 \cdot \frac{\pi}{180} = \frac{\pi}{3} \qquad 300 \cdot \frac{\pi}{180} = \frac{5\pi}{3} \qquad 0 \cdot \frac{\pi}{180} = 0$$

Thus, the primary solutions are 0, $\frac{\pi}{3}$, and $\frac{5\pi}{3}$.

Math Connections

The Pendulum: There are many instances of conditional trigonometric equations in physics. A simple example is illustrated by a small weight tied to one end of a string that is attached to the ceiling at its other end. If the weight is pushed somewhat, so that it swings back and forth in a circular arc, then the effect produced is that of a pendulum. It can be shown that if, at a particular moment, the string forms an angle θ with an imaginary vertical line, then the force F that restores the weight to its equilibrium (resting) position is given by the equation $F = -mg \sin \theta$, where m is the mass of the small weight and g is the acceleration of a falling body under the influence of gravity.

Critical Thinking Questions

Analysis: Ask students to use the equation given in *Math Connections* to indicate the value of F for the case $\theta = 0$. Ask them to discuss the physical significance of this result. Students should see that for $\theta = 0$, the pendulum is vertical; that is, it is in its resting position. As would be supposed, $F = 0$ in this position.

Common Error Analysis

Error: Students may have difficulty with solving equations when a solution is a quadrantal angle, as in $\sin \theta = 1$, since the diagram does not involve a right triangle. Students should learn to recognize these special solutions.

For solutions that are nonquadrantal angles, the students tend to find only one of the two angles in the interval $0 \le \theta < 360$. Emphasize that for nonquadrantal angles, there will always be two solutions in this interval.

Additional Example 2

Solve $\sin \theta - 1 = -2 \sin^2 \theta$, $0 \le \theta < 2\pi$.
$\frac{\pi}{6}, \frac{5\pi}{6}, \frac{3\pi}{2}$

Checkpoint

Solve for θ, 0 ≤ θ < 360.

1. $\sin \theta = -\frac{\sqrt{3}}{2}$ 240, 300
2. $\cot \theta = -1$ 135, 315

Solve for θ, 0 ≤ θ < 2π.

3. $4 \cos \theta = -2\sqrt{3}$ $\frac{5\pi}{6}, \frac{7\pi}{6}$
4. $7 \tan \theta - 5 = 2 \tan \theta$ $\frac{\pi}{4}, \frac{5\pi}{4}$

Find solutions of each equation to the nearest degree, 0 ≤ θ < 360.

5. $2 \sin \theta - 1 = -8 \sin^2 \theta$
   14, 166, 210, 330
6. $2 \cos^2 \theta - \cos \theta = 0$ 60, 90, 270, 300

Closure

Ask students to give a definition of the term *primary solutions* (of a trigonometric equation). See page 721. Have students list the steps in solving a linear trigonometric equation. See Example 1. Then have them give the steps in solving a quadratic trigonometric equation. See Example 2.

▰▰▰FOLLOW UP

Guided Practice

Classroom Exercises 1–9

Independent Practice

A Ex. 1–22, **B** Ex. 23–30, **C** Ex. 31–34

Basic: WE 1–23 odd
Average: WE 7–29 odd
Above Average: WE 11–33 odd

If a trigonometric equation involves angles that are neither special nor quadrantal, then it is necessary to use a table or a calculator.

EXAMPLE 3 Solve $5 \tan^2 \theta - 4 \tan \theta = 0$, $0 \le \theta < 360$. Give answers to the nearest degree.

Solution

$5 \tan^2 \theta - 4 \tan \theta = 0$
$\tan \theta (5 \tan \theta - 4) = 0$ ← tan θ is the GCF.
$\tan \theta = 0$ *or* $5 \tan \theta - 4 = 0$
$\tan \theta = \frac{4}{5}$, or 0.8000

$\tan \theta = 0$:
quadrantal angle
Use $\tan \theta = \frac{0}{1} = 0$ and
$\tan \theta = \frac{0}{-1} = 0.$

$\tan \theta = 0.8000$:
tan θ is positive in Quadrants I and III. From a table or a calculator, the reference angle is 39 to the nearest degree. (By calculator, 0.8 [INV] [tan] $\Rightarrow$ 38.659808.)

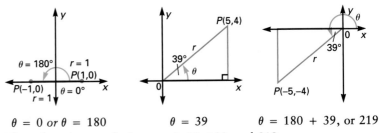

$\theta = 0 \text{ or } \theta = 180$ $\theta = 39$ $\theta = 180 + 39$, or 219

Thus, the primary solutions are 0, 39, 180, and 219.

Recall that sin θ can never be greater than 1. Thus, an equation such as $\sin \theta = 3$ has no solution.

EXAMPLE 4 Solve $-3 - 2 \sin \theta = -\sin^2 \theta$, $0 \le \theta < 2\pi$.

Solution

First rewrite in standard form. Then factor.

$-3 - 2 \sin \theta = -\sin^2 \theta$
$\sin^2 \theta - 2 \sin \theta - 3 = 0$
$(\sin \theta - 3)(\sin \theta + 1) = 0$

$\sin \theta = 3$ *or* $\sin \theta = -1$
sin θ cannot be $\theta = 270$
greater than 1. No $\theta = 270 \cdot \frac{\pi}{180}$, or
solution. $\frac{3\pi}{2}$ radians

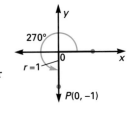

Thus, the primary solution is $\frac{3\pi}{2}$.

Additional Example 3

Solve $3 \sin^2 \theta = 2 \sin \theta$, $0 \le \theta < 360$.
0, 42, 138, 180

Additional Example 4

Solve $3 \cos \theta = 4 - \cos^2 \theta$, $0 \le \theta < 2\pi$.
0

Classroom Exercises

Give the steps for solving each equation. If factoring is necessary, identify the factoring as either common-monomial factoring or factoring into two binomials.

1. $3 \sin \theta + 6 = \sin \theta + 7$ **2.** $\cos^2 \theta - \cos \theta = 0$ **3.** $3 \sin^2 \theta = 2 \sin \theta + 1$
4. $\sin^2 \theta - 1 = 0$ **5.** $4 \cos \theta - 1 = 2$ **6.** $2 \tan^2 \theta - 3 \tan \theta = 0$

Solve for θ, $0 \le \theta < 360$.

7. $\cot \theta = 1$ 45, 225 **8.** $2 \cos \theta = \sqrt{3}$ 30, 330 **9.** $6 \sin \theta + 7 = 10$ 30, 150

Written Exercises

Solve for θ, $0 \le \theta < 360$.

1. $\cos \theta = -\dfrac{1}{2}$ 120, 240 **2.** $\tan \theta = 1$ 45, 225 **3.** $\sin \theta = \dfrac{\sqrt{3}}{2}$ 60, 120 **4.** $\tan \theta = \sqrt{3}$ 60, 240

5. $\csc \theta = 2$ 30, 150 **6.** $\cos \theta = \dfrac{1}{\sqrt{2}}$ 45, 315 **7.** $\tan \theta = -1$ 135, 315 **8.** $\cos \theta = -\dfrac{\sqrt{3}}{2}$ 150, 210

9. $\cos \theta = -1$ 180 **10.** $\sin \theta = 0$ 0, 180 **11.** $\tan \theta = 0$ 0, 180 **12.** $\sin \theta = 0.5299$ 32, 148

Solve for θ, $0 \le \theta < 2\pi$. **13.** $\frac{4\pi}{3}, \frac{5\pi}{3}$ **14.** $\frac{\pi}{3}, \frac{5\pi}{3}$ **15.** $\frac{3\pi}{4}, \frac{5\pi}{4}$ **16.** $\frac{3\pi}{2}$ **17.** $\frac{5\pi}{6}, \frac{7\pi}{6}$ **18.** $\frac{\pi}{4}, \frac{5\pi}{4}$

13. $2 \sin \theta = -\sqrt{3}$ **14.** $4 \cos \theta - 6 = -4$ **15.** $\sqrt{2} \cos \theta + 4 = 3$
16. $2 \sin \theta = 4 \sin \theta + 2$ **17.** $4\sqrt{3} \sec \theta = -8$ **18.** $5 \cot \theta - 3 = 2 \cot \theta$

Find solutions of each equation to the nearest degree, $0 \le \theta < 360$.

19. $4 \sin \theta + 1 = 3$ 30, 150 **20.** $3 \tan \theta - 4 = 4 \tan \theta + 2$ 99, 279
21. $7 \cos \theta + 3 = 2 \cos \theta - 1$ 143, 217 **22.** $\sqrt{2}(2 \sin \theta + \sqrt{2}) = 3\sqrt{2} \sin \theta + 1$
23. $2 \cos^2 \theta + \sqrt{3} \cos \theta = 0$ **24.** $2 \sin^2 \theta = -\sin \theta + 1$ 30, 150, 270
25. $3 \cos \theta + 2 = -\cos^2 \theta$ 180 **26.** $\tan^2 \theta = \tan \theta$ 0, 45, 180, 225
27. $\csc^2 \theta + \sqrt{2} \csc \theta = 0$ 225, 315 **28.** $\sec^2 \theta - 3 \sec \theta + 2 = 0$ 0, 60, 300
29. $4 \cot^2 \theta - 1 = 0$ 63, 117, 243, 297 **30.** $\tan^2 \theta + 5 \tan \theta + 6 = 0$
31. $2 \sin \theta \cos \theta - 2 \cos \theta + \sin \theta = 1$ **32.** $\cos^2 \theta + 2 \cos \theta - 1 = 0$ 66, 294
33. $2 \cos^3 \theta - \cos^2 \theta - 2 \cos \theta = -1$ **34.** $\csc^4 \theta - 4 \csc^2 \theta + 3 = 0$
    0, 60, 180, 300   35, 90, 145, 215, 270, 325

Mixed Review

Sketch one cycle of the graph of each function without using a table of values.   *18.8, 18.9*

1. $f(\theta) = 3 + 2 \sin \theta$ **2.** $f(\theta) = \cos \frac{1}{2}\theta$
3. $f(\theta) = 3 \cos 2\theta$ **4.** $f(\theta) = \tan 3\theta$

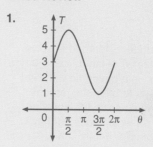
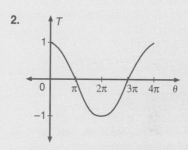

Enrichment

By using the Quotient and Pythagorean Identities, it is possible to express any of the six trigonometric functions in terms of each of the other functions. Show the students how each function can be expressed in terms of $\sin \theta$, as follows.

$\sin \theta = \sin \theta$;

$\csc \theta = \dfrac{1}{\sin \theta}$;

$\cos \theta = \pm\sqrt{1 - \sin^2 \theta}$

$\sec \theta = \dfrac{1}{\pm\sqrt{1 - \sin^2 \theta}}$

$\tan \theta = \dfrac{\sin \theta}{\pm\sqrt{1 - \sin^2 \theta}}$

$\cot \theta = \dfrac{\pm\sqrt{1 - \sin^2 \theta}}{\sin \theta}$

Now have the students express each function in terms of $\tan \theta$.

$\tan \theta = \tan \theta$

$\cot \theta = \dfrac{1}{\tan \theta}$;

$\cos \theta = \dfrac{1}{\sec \theta} = \dfrac{1}{\pm\sqrt{1 + \tan^2 \theta}}$

$\sec \theta = \pm\sqrt{1 + \tan^2 \theta}$

$\sin \theta = \pm\sqrt{1 - \cos^2 \theta} = \pm\sqrt{1 - \dfrac{1}{\sec^2 \theta}}$

$= \pm\sqrt{\dfrac{1 + \tan^2 \theta - 1}{1 + \tan^2 \theta}} = \dfrac{\pm\tan \theta}{\sqrt{1 + \tan^2 \theta}}$;

$\csc \theta = \dfrac{\pm\sqrt{1 + \tan^2 \theta}}{\tan \theta}$

Teaching Resources

Application Worksheet 19
Problem Solving Worksheet 19
Quick Quizzes 146
Reteaching and Practice
    Worksheets 146

19.4 Trigonometric Equations: Using Basic Identities

Prerequisite Quiz

1. Write $\sec^2 \theta$ in terms of $\tan \theta$. $\tan^2 \theta + 1$
2. Write $\cos^2 \theta$ in terms of $\sin \theta$. $1 - \sin^2 \theta$
3. Simplify $\sin \theta \cot \theta$. $\cos \theta$
4. Solve, to the nearest degree, $\tan \theta = -2$, $0 \le \theta < 360$.   117, 297

Motivator

In Lesson 19.3, students learned to solve an equation like $\tan^2 \theta + \tan \theta - 2 = 0$, since the equation is a quadratic with *like* terms. Ask students to offer suggestions about how they would solve the equation $\tan \theta + \sec^2 \theta - 3$, which does *not* have like terms.   Use $\sec^2 \theta = \tan^2 \theta + 1$ to rewrite the expression. Then solve as a quadratic equation.

■■TEACHING
    SUGGESTIONS

Lesson Note

When an equation contains different functions of an angle, students need to learn how to change the equation so that only one function is present. Encourage students to memorize the basic identities, in particular the Pythagorean Identities, which relate $\sin^2 \theta$ to $\cos^2 \theta$, $\tan^2 \theta$ to $\sec^2 \theta$, and $\cot^2 \theta$ to $\csc^2 \theta$.

Objective   To solve trigonometric equations using trigonometric identities

In medieval times, trigonometric ratios were represented in mathematics texts by line segments, as the figure at the right illustrates. The texts, which were in Latin, referred to the line segments as *touching* (tangent) or *cutting* (secant) a circle of radius 1. The figure suggests an identity that you already know and that appears in Example 1.

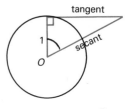

EXAMPLE 1   Solve, to the nearest degree, $\tan \theta + \sec^2 \theta = 3$, $0 \le \theta < 360$.

Plan   Use the Pythagorean Identity $\sec^2 \theta = \tan^2 \theta + 1$ to obtain expressions that involve only $\tan \theta$.

Solution

$$\tan \theta + \sec^2 \theta = 3$$
$$\tan \theta + (\tan^2 \theta + 1) = 3 \leftarrow \sec^2 \theta = \tan^2 \theta + 1 \text{ (See figure above.)}$$
$$\tan^2 \theta + \tan \theta - 2 = 0$$
$$(\tan \theta + 2)(\tan \theta - 1) = 0$$
$$\tan \theta + 2 = 0 \quad or \quad \tan \theta - 1 = 0$$
$$\tan \theta = -2 \quad or \quad \tan \theta = 1$$

$\tan \theta = -2$:
   $\tan \theta$ is negative in Quadrants II and IV. By calculator, the measure of the reference angle is 2 [INV] [tan] $\Rightarrow$ 63.434949, or 63 to the nearest degree.
   Then, either $\theta = 180 - 63$, or 117 (Quadrant II), or
   $\theta = 360 - 63$, or 297 (Quadrant IV).

$\tan \theta = 1$:
   $\tan \theta$ is positive in Quadrants I and III.
   Since $\tan 45 = 1$ the reference angle measures 45.
   Then, either $\theta = 45$ (Quadrant I)
   or $\theta = 180 + 45 = 225$ (Quadrant III).

Thus, the four primary solutions are 45, 117, 225, and 297.

Highlighting the Standards

Standard 9f: Solving the equations in this lesson demonstrates to students the interconnectedness of a great deal of the algebra and trigonometry they have learned.

Additional Example 1

Solve, to the nearest degree.

$2 - 2 \sin \theta = 3 \cos^2 \theta$, $0 \le \theta < 360$
90, 199, 341

EXAMPLE 2	Solve and check $\cos \theta + 2 \sin \theta = 2$, $0 \leq \theta < 360$. Express results to the nearest degree.
Plan	The equation can be written with only sines or cosines using the Pythagorean Identity. However, since this requires that both sides of the equation be squared, the equation must be checked for extraneous roots. To avoid a complicated solution, subtract $2 \sin \theta$ from each side before squaring.

Solution

$$\cos \theta + 2 \sin \theta = 2$$
$$\cos \theta = 2 - 2 \sin \theta$$
$$(\cos \theta)^2 = (2 - 2 \sin \theta)^2 \leftarrow \text{Square each side.}$$
$$\cos^2 \theta = 4 - 8 \sin \theta + 4 \sin^2 \theta$$
$$1 - \sin^2 \theta = 4 - 8 \sin \theta + 4 \sin^2 \theta \leftarrow \text{Use } \cos^2 \theta = 1 - \sin^2 \theta.$$
$$0 = 5 \sin^2 \theta - 8 \sin \theta + 3$$
$$0 = (5 \sin \theta - 3)(\sin \theta - 1)$$
$$5 \sin \theta - 3 = 0 \quad or \quad \sin \theta - 1 = 0$$
$$\sin \theta = \frac{3}{5} \quad or \quad \sin \theta = 1$$

$\sin \theta = 0.6$:
   $\sin \theta$ is positive in Quadrants I and II. By calculator, the measure of the reference angle is 37.
   Then, either $\theta = 37$ (Quadrant I)
   or $\theta = 180 - 37 = 143$ (Quadrant II).

$\sin \theta = 1$:
   Then $\theta = 90$ (quadrantal angle).

Checks

In Quadrant I, $\cos 37 = 0.8$. In Quadrant II, $\cos 143 = -0.8$.
$\cos 90 = 0$.

$\cos \theta + 2 \sin \theta = 2$	$\cos \theta + 2 \sin \theta = 2$	$\cos \theta + 2 \sin \theta = 2$
$\cos 37 + 2 \sin 37 \overset{?}{=} 2$	$\cos 143 + 2 \sin 143 \overset{?}{=} 2$	$\cos 90 + 2 \sin 90 \overset{?}{=} 2$
$0.8 + 2(0.6) \overset{?}{=} 2$	$-0.8 + 2(0.6) \overset{?}{=} 2$	$0 + 2 \cdot 1 \overset{?}{=} 2$
$2.0 = 2$	$0.4 = 2$	$2 = 2$
True	False	True
37 is a solution.	143 is not a solution.	90 is a solution.

Thus, the only primary solutions are 37 and 90.

Line Functions: Here is a modified and expanded version of the "line function" diagram on page 724 of the pupil's text. (See also *Math Connections* in Lesson 19.2.)

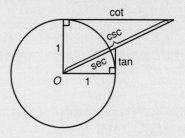

Critical Thinking Questions

Analysis: Ask students to use the diagram of *Math Connections* to demonstrate the identities $\cot^2 \theta + 1 = \csc^2 \theta$ and $\tan \theta \cdot \cot \theta = 1$. As a hint, tell them that they will have to use the fact that if two lines are parallel, then the measures of the alternate interior angles are equal.

Checkpoint

Solve each equation to the nearest degree, $0 \leq \theta < 360$. If, in your solution, you square each side of an equation, then check all the roots in the original equation.

1. $\sin \theta + 5 = 6 \cos^2 \theta$ 19, 161, 210, 330
2. $2 \sin \theta - \csc \theta = 1$ 90, 210, 330
3. $\tan \theta + 3 \cot \theta = 4$ 45, 72, 225, 252
4. $\sin \theta + \cos \theta = 1$ 0, 90

Additional Example 2

Solve and check $2 \cos \theta + \sin \theta = 1$.
Express results to the nearest degree.
90, 323

725

726

Closure

Have students give the first step in solving a trigonometric equation that contains sines and squares of cosines. Use the Pythagorean Identity $\cos^2 \theta = 1 - \sin^2 \theta$ to rewrite the equation. Suppose a trigonometric equation contains tangents and secants but no squares of these. Ask students to give the first step for getting an equation with all like terms. Square both sides of the equation.

◼◼◼FOLLOW UP

Guided Practice

Classroom Exercises 1–8

Independent Practice

A Ex. 1–12, **B** Ex. 13–20, **C** Ex. 21–25

Basic: WE 1–15 odd
Average: WE 5–19 odd
Above Average: WE 11–25 odd

Additional Answers

Classroom Exercises

1. Use $\cos^2 \theta = 1 - \sin^2 \theta$.
2. Use $\csc^2 \theta = 1 + \cot^2 \theta$.
3. Use $\tan^2 \theta = \sec^2 \theta - 1$.
4. Square both sides. Then use either alternate form of $\sin^2 \theta + \cos^2 \theta = 1$.

Written Exercises

10. 71, 120, 240, 289
12. 63, 135, 243, 315
21. 222, 318
22. 50, 130, 206, 334
23. 90, 180, 270

Classroom Exercises

Tell how to change each equation so that only one trigonometric ratio is present.

1. $2\cos^2 \theta = \sin \theta + 1$
2. $1 + \cot^2 \theta = 3\csc \theta$
3. $5\tan^2 \theta = 6\sec \theta - 5$
4. $\sin \theta = \cos \theta$

Solve each equation, $0 \le \theta < 360$.

5. $2\cos^2 \theta = 2 + \sin^2 \theta$ 0, 180
6. $\cos \theta \tan \theta = -1$ 270
7. $\sec \theta \cos^2 \theta - 1 = 0$ 0
8. $2\tan^2 \theta = \sec^2 \theta$ 45, 135, 225, 315

Written Exercises

Solve each equation to the nearest degree, $0 \le \theta < 360$. If, in your solution, you square each side of an equation, check all the roots in the original equation.

1. $2\cos^2 \theta = \sin \theta + 1$ 30, 150, 270
2. $2\cos^2 \theta - 2\cos \theta = -\sin^2 \theta$ 0
3. $1 + \cot^2 \theta - 3\csc \theta = 0$ 19, 161
4. $2\tan^2 \theta + 5\sec \theta + 4 = 0$ 120, 240
5. $3\sec^2 \theta - 4\tan \theta = 2$ 18, 45, 198, 225
6. $-\cos \theta - \sin^2 \theta = 1$ 180
7. $\cos^2 \theta = \sin^2 \theta + 1$ 0, 180
8. $\csc^2 \theta - 2\cot \theta = 0$ 45, 225
9. $\sec^2 \theta = 2\tan \theta + 4$ 72, 135, 252, 315
10. $6\sin^2 \theta - \cos \theta - 5 = 0$
11. $3\tan^2 \theta - 4\sec \theta = 4$ 65, 180, 295
12. $2\sec^2 \theta - \tan \theta = \tan^2 \theta + 4$
13. $\sin \theta = \cos \theta$ 45, 225
14. $3\sin \theta - \cos \theta = 1$ 37, 180
15. $\sec \theta - 1 = \tan \theta$ 0
16. $\sec \theta + \tan \theta = \sqrt{3}$ 30
17. $2\sin \theta + \cos \theta = 1$ 0, 127
18. $1 + \cos \theta = \sin \theta$ 90, 180
19. $3\cos \theta + 2\sin \theta = 2$ 90, 337
20. $\sin \theta + \cos \theta = \sqrt{2}$ 45
21. $3\cos \theta + 4\tan \theta + \sec \theta = 0$
22. $2\sin \theta - \cot \theta \cos \theta = 1$
23. $\sqrt{1 - \cos^2 \theta} - \cos \theta = 1$
24. $\sqrt{3 - 3\sin^2 \theta} + 1 = \sin \theta$ 90

25. Solve the system $\begin{array}{l} r\sin \theta = 4 \\ r\cos \theta = 3 \end{array}$ for r and θ, where $r > 0$ and $0 \le \theta < 360$. (HINT: Square each side and add.) $r = 5, \theta = 53$

Mixed Review

Simplify. 5.3, 5.5, 8.3, 8.4, 8.6, 9.2 1. $x^3 - 7x^2 + 8x + 4$

1. $(x - 2)(x^2 - 5x - 2)$
2. $\sqrt{8x^3} \cdot \sqrt{2x}$ $4x^2$
3. $(1 + \sqrt{3})(1 - \sqrt{3})$ -2
4. $\dfrac{1 - \sqrt{2}}{1 + \sqrt{2}}$ $2\sqrt{2} - 3$
5. $27^{-\frac{2}{3}}$ $\frac{1}{9}$
6. $\dfrac{20}{3 + i}$ $6 - 2i$

Enrichment

On the chalkboard, sketch the graphs of the sine and cosine curves on the same axes for $0 \le x < 360$.

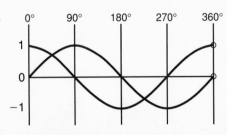

Then challenge the students to find the primary solutions of the following equations by inspection of the graph alone.

1. $\sin x + \cos x = 1$ 0, 90
2. $\sin x + \cos x = -1$ 180, 270
3. $\sin x + \cos x = 0$ 135, 315
4. $\sin x + \cos x = 2$ No solution
5. $\sin x - \cos x = 1$ 90, 180
6. $\cos x - \sin x = 1$ 0, 270

19.5 Sum and Difference Identities

Objectives

To simplify trigonometric expressions using sum and difference identities

To prove trigonometric identities using sum and difference identities

Even though 15 is not a special angle measure, cos 15 can be found in terms of the sines and cosines of two special angle measures whose difference is 15. This is done by applying the following trigonometric identity for the cosine of the difference of two angles θ (theta) and ϕ (phi): $\cos (\theta - \phi) = \cos \theta \cos \phi + \sin \theta \sin \phi$.

$$\cos 15 = \cos (60 - 45)$$
$$= \cos 60 \cos 45 + \sin 60 \sin 45$$
$$= \frac{1}{2} \cdot \frac{\sqrt{2}}{2} + \frac{\sqrt{3}}{2} \cdot \frac{\sqrt{2}}{2}$$
$$= \frac{\sqrt{2}}{4} + \frac{\sqrt{6}}{4}, \text{ or } \frac{\sqrt{2} + \sqrt{6}}{4}$$

To prove the above identity, refer to the figure at the right. M, N, P, and Q are points of a circle that has its center at the origin and that has a radius with a length of one unit. A circle with a radius of one unit is called a *unit circle*. The angles of rotation for M, N, and P are θ, ϕ, and $\theta - \phi$, respectively.

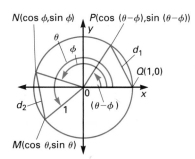

If (x,y) are the coordinates of point M, then the following are true.

$$\cos \theta = \frac{x}{r} = \frac{x}{1} = x \text{ and } \sin \theta = \frac{y}{r} = \frac{y}{1} = y$$

So, $(\cos \theta, \sin \theta)$ are the coordinates of point M of the unit circle. Similarly, it can be seen that $(\cos \phi, \sin \phi)$ and $(\cos (\theta - \phi), \sin (\theta - \phi))$ are the coordinates of points N and P, respectively.

Also in the figure, d_1 and d_2 are the lengths of chords $\overline{PQ}$ and $\overline{MN}$, respectively. Since both $\overparen{MN}$ (read "arc MN") and $\overparen{PQ}$ have $\theta - \phi$ as the measure of their central angle, these arcs are congruent. Therefore, the chords $\overline{MN}$ and $\overline{PQ}$ are also congruent. Thus, $d_1 = d_2$.

19.5 Sum and Difference Identities **727**

Teaching Resources

Project Worksheet 19
Quick Quizzes 147
Reteaching and Practice Worksheets 147

▬▬ GETTING STARTED

Prerequisite Quiz

Write each function of $(-\theta)$ in terms of a function of θ.

1. $\sin (-\theta)$ **2.** $\cos (-\theta)$ **3.** $\tan (-\theta)$
$-\sin \theta$ $\cos \theta$ $-\tan \theta$

Find each trigonometric ratio.

4. $\sin 30$ $\frac{1}{2}$ **5.** $\cos 240$ $-\frac{1}{2}$

6. $\tan 225$ 1 **7.** $\cos (-45)$ $\frac{\sqrt{2}}{2}$

Motivator

Ask the class if they think $\sin (A + B) = \sin A + \sin B$. Have them substitute the values $A = 90$ and $B = 30$, and emphasize that one *counterexample* shows that the generalization does not work.

▬▬ TEACHING SUGGESTIONS

Lesson Note

You might want to prove the tangent identities in class instead of assigning them as C-level exercises. Offer this hint: Rewrite $\tan (A + B)$ in terms of identities you already know, $\tan (A + B) = \frac{\sin (A + B)}{\cos (A + B)}$. Then ask the students, "What can you divide both numerator and denominator by to get the desired result?" Divide numerator and denominator by $\cos A \cos B$.

Highlighting the Standards

Standard 14e: Proof of the sum and difference identities illustrates to students that trigonometry is also an axiomatic system—with its own principles and rules.

Math Connections

The Formulas of Werner: The formula of Exercise 28 together with three similar formulas were once known as the "Formulas of Werner" because of their use by Johannes Werner (1468–1522) to simplify astronomical calculations. For example, an astronomer with the task of multiplying the two numbers 134,596,847,704,886 and 96,479,023,114,508 might let $2 \sin \theta$ equal the first number and $\cos \phi$ equal the second (while also adjusting the decimal point to get numerals that are less than 1). The astronomer could then use a table of sines and cosines to find the values of θ and ϕ, find the sines of their sum and difference, and add. After readjusting the decimal point (to compensate for the original adjustment), the original product would be obtained without the tedium of the actual multiplication. Before the invention of logarithms, this method of calculation, called *prosthaphaeresis* (see *Math Connections* for Lesson 13.7) was widely used by astronomers, including the famous Danish astronomer Tycho Brahe (1546–1601).

Critical Thinking Questions

Analysis: Ask students to prove one or more of the other three Werner Formulas below.

$\sin(\theta + \phi) - \sin(\theta - \phi) = 2 \cos \theta \sin \phi$
$\cos(\theta + \phi) + \cos(\theta - \phi) = 2 \cos \theta \cos \phi$
$\cos(\theta - \phi) - \cos(\theta + \phi) = 2 \sin \theta \sin \phi$

Now, using the Distance Formula, $d = \sqrt{(x_2 - x_1)^2 + (y_2 - y_1)^2}$, to express d_1 and d_2 gives the equations below.

$$d_1 = \sqrt{(\cos(\theta - \phi) - 1)^2 + (\sin(\theta - \phi) - 0)^2}$$
$$d_2 = \sqrt{(\cos \theta - \cos \phi)^2 + (\sin \theta - \sin \phi)^2}$$

Squaring and simplifying these equations gives the following.

$$d_1^2 = 2 - 2\cos(\theta - \phi)$$
$$d_2^2 = 2 - 2\cos \theta \cos \phi - 2 \sin \theta \sin \phi$$

Finally, set the values of d_1^2 and d_2^2 above equal to each other.

$$2 - 2\cos(\theta - \phi) = 2 - 2\cos \theta \cos \phi - 2 \sin \theta \sin \phi$$
$$-2\cos(\theta - \phi) = -2\cos \theta \cos \phi - 2 \sin \theta \sin \phi$$
$$\cos(\theta - \phi) = \cos \theta \cos \phi + \sin \theta \sin \phi$$

Recall from Lesson 18.7 that $\cos(-\theta) = \cos \theta$ and $\sin(-\theta) = -\sin \theta$. These two equations can be used to derive an identity for the cosine of the *sum* of the measures of two angles. This is shown below with a direct application of the new identity.

$$\cos(\theta + \phi) = \cos[\theta - (-\phi)]$$
$$= \cos \theta \cos(-\phi) + \sin \theta \sin(-\phi)$$
$$= \cos \theta \cos \phi + \sin \theta (-\sin \phi)$$
$$= \cos \theta \cos \phi - \sin \theta \sin \phi$$

Identities for the sine and tangent of the sum or difference of two angle measures can also be derived. The sum and difference identities are listed below. You are asked to prove the last four in Exercises 24–27.

Sum and Difference Identities

$\cos(\theta - \phi) = \cos \theta \cos \phi + \sin \theta \sin \phi$
$\cos(\theta + \phi) = \cos \theta \cos \phi - \sin \theta \sin \phi$
$\sin(\theta - \phi) = \sin \theta \cos \phi - \cos \theta \sin \phi$
$\sin(\theta + \phi) = \sin \theta \cos \phi + \cos \theta \sin \phi$

$$\tan(\theta - \phi) = \frac{\tan \theta - \tan \phi}{1 + \tan \theta \tan \phi} \qquad \tan(\theta + \phi) = \frac{\tan \theta + \tan \phi}{1 - \tan \theta \tan \phi}$$

EXAMPLE 1 Simplify $\sin 40 \cos 10 - \cos 40 \sin 10$ using an identity.

Solution The expression suggests the identity for $\sin(\theta - \phi)$.
$$\sin \theta \cos \phi - \cos \theta \sin \phi = \sin(\theta - \phi)$$
$$\sin 40 \cos 10 - \cos 40 \sin 10 = \sin(40 - 10) = \sin 30 = \frac{1}{2}$$

Additional Example 1

Use an identity to simplify $\cos 50 \cos 40 - \sin 50 \sin 40$.
$\cos(50 + 40) = \cos 90 = 0$

EXAMPLE 2 Find tan 105 without using a calculator or a table.

Plan Use the special angles 60 and 45 in the identity for $\tan(\theta + \phi)$.

Solution
$$\tan 105 = \tan(60 + 45)$$
$$= \frac{\tan 60 + \tan 45}{1 - \tan 60 \tan 45} \quad \leftarrow \text{THINK: Is this number positive or negative?}$$
$$\qquad\qquad\qquad\qquad \text{How do you know?}$$
$$= \frac{\sqrt{3} + 1}{1 - \sqrt{3} \cdot 1}$$
$$= \frac{(1 + \sqrt{3})(1 + \sqrt{3})}{(1 - \sqrt{3})(1 + \sqrt{3})} = \frac{1 + 2\sqrt{3} + 3}{1 - 3} = -2 - \sqrt{3}$$
Therefore, $\tan 105 = -2 - \sqrt{3}$.

EXAMPLE 3 Simplify $\cos\left(\frac{3\pi}{2} - \phi\right)$.

Solution $\frac{3\pi}{2}$ radians $= \frac{3}{2} \cdot 180 = 270$ degrees

Apply the identity $\cos(\theta - \phi) = \cos\theta\cos\phi + \sin\theta\sin\phi$.
$$\cos(270 - \phi) = \cos 270 \cos\phi + \sin 270 \sin\phi$$
$$= 0 \cdot \cos\phi + (-1) \cdot \sin\phi$$
$$= 0 - \sin\phi, \text{ or } -\sin\phi$$
Therefore, $\cos\left(\frac{3\pi}{2} - \phi\right) = -\sin\phi$.

EXAMPLE 4 Prove that $\frac{\sin(\theta + \phi)}{\cos\theta\cos\phi} = \tan\theta + \tan\phi$ is an identity.

Plan First use the identity for $\sin(\theta + \phi)$ to obtain sines and cosines of θ and ϕ. Then simplify.

Solution

$\dfrac{\sin(\theta + \phi)}{\cos\theta\cos\phi}$	$\tan\theta + \tan\phi$
$\dfrac{\sin\theta\cos\phi + \cos\theta\sin\phi}{\cos\theta\cos\phi}$	$\dfrac{\sin\theta}{\cos\theta} + \dfrac{\sin\phi}{\cos\phi}$
$\dfrac{\sin\theta\cos\phi}{\cos\theta\cos\phi} + \dfrac{\cos\theta\sin\phi}{\cos\theta\cos\phi}$	
$\dfrac{\sin\theta}{\cos\theta} + \dfrac{\sin\phi}{\cos\phi} = \dfrac{\sin\theta}{\cos\theta} + \dfrac{\sin\phi}{\cos\phi}$	

Therefore, the identity is proved.

Common Error Analysis

Error: Even after the identities have been proved, some students still make the error of interpreting sin $(A + B)$ as sin A + sin B. Use special angles to show why this is an error. For example, sin $(180 + 30) \neq$ sin 180 + sin 30, because sin $(180 + 30) =$ sin $210 = -\frac{1}{2}$, while sin 180 + sin 30 = $0 + \frac{1}{2} = \frac{1}{2}$, and $-\frac{1}{2} \neq \frac{1}{2}$.

Checkpoint

Simplify, using an identity. Give irrational answers in simplest radical form. (Ex. 1–4)

1. sin 35 cos 25 + cos 35 sin 25 $\frac{\sqrt{3}}{2}$
2. $\dfrac{\tan 144 - \tan 9}{1 + \tan 144 \tan 9}$ -1
3. sin 195 $\dfrac{\sqrt{2} - \sqrt{6}}{4}$
4. $\sin\left(\dfrac{\pi}{2} + \theta\right)$ $\cos\theta$
5. Prove the identity.
$$\tan(45 - \theta) = \dfrac{1 - \tan\theta}{1 + \tan\theta}$$
$$\tan(45 - \theta)$$
$$= \dfrac{\tan 45 - \tan\theta}{1 + \tan 45 \cos\theta}$$
$$= \dfrac{1 - \tan\theta}{1 + \tan\theta}$$

Closure

Have students give the sum and difference identities for each of the following.

1. cos $(A + B)$ 2. sin $(A + B)$
3. tan $(A + B)$ 4. cos $(A - B)$
5. sin $(A - B)$ 6. tan $(A - B)$
See page 728.

Additional Example 2

Find tan 165 without using a calculator or table. Express the result in simplest radical form.

$$\tan(120 + 45)$$
$$= \dfrac{-\sqrt{3} + 1}{1 - (-\sqrt{3} \cdot 1)} = \dfrac{1 - \sqrt{3}}{1 + \sqrt{3}}$$
$$= \dfrac{1 - \sqrt{3}}{1 + \sqrt{3}} \cdot \dfrac{1 - \sqrt{3}}{1 - \sqrt{3}} = \dfrac{4 - 2\sqrt{3}}{-2}$$
$$= -2 + \sqrt{3}$$

Additional Example 3

Simplify.

sin $(\pi + \theta)$. $-\sin\theta$

Additional Example 4

Prove that
sin $(\theta + \phi)$ − sin $(\theta - \phi)$ = 2 cos θ sin ϕ is an identity.
sin $(\theta + \phi)$ − sin $(\theta - \phi)$
= sin θ cos ϕ + cos θ sin ϕ −
(sin θ cos ϕ − cos θ sin ϕ)
= sin θ cos ϕ + cos θ sin ϕ −
sin θ cos ϕ + cos θ sin ϕ
= 2 cos θ sin ϕ

FOLLOW UP

Guided Practice

Classroom Exercises 1–8

Independent Practice

A Ex. 1–10, **B** Ex. 11–23, **C** Ex. 24–30

Basic: WE 1–17 odd, Brainteaser
Average: WE 7–23 odd, Brainteaser
Above Average: WE 13–29 odd, Brainteaser

Additional Answers

Classroom Exercises

1. $\cos 60 \cos 15 + \sin 60 \sin 15$
   $= \cos (60 - 15) = \cos 45 = \dfrac{\sqrt{2}}{2}$

2. $\sin 75 \cos 45 - \cos 75 \sin 45$
   $= \sin (75 - 45) = \sin 30 = \dfrac{1}{2}$

5. $\dfrac{\sqrt{6} - \sqrt{2}}{4}$

Written Exercises

8. $-\dfrac{\sqrt{2} - \sqrt{6}}{4}$

15. $\dfrac{\sin(\theta + \phi)}{\sin \theta \sin \phi}$
    $= \dfrac{\sin \theta \cos \phi + \cos \theta \sin \phi}{\sin \theta \sin \phi}$
    $= \dfrac{\cos \phi}{\sin \phi} + \dfrac{\cos \theta}{\sin \theta}$
    $= \cot \theta + \cot \phi$

16. $\cos(\theta - \phi) - \cos(\theta + \phi)$
    $= (\cos \theta \cos \phi + \sin \theta \sin \phi) -$
    $(\cos \theta \cos \phi - \sin \theta \sin \phi)$
    $= 2 \sin \theta \sin \phi$

17. $\sin(30 + \theta) + \sin(30 - \theta)$
    $= (\sin 30 \cos \theta + \cos 30 \sin \theta) +$
    $(\sin 30 \cos \theta - \cos 30 \sin \theta)$
    $= 2 \sin 30 \cos \theta$
    $= 2(\tfrac{1}{2}) \cos \theta$
    $= \cos \theta$

EXAMPLE 5   Given that $\sin \theta = \dfrac{12}{13}$, $\cos \phi = \dfrac{4}{5}$, θ is in Quadrant I and ϕ is in Quadrant IV, find $\cos(\theta + \phi)$.

Plan   First draw each angle in the proper quadrant.
Find the missing sides of each reference triangle.

$x^2 + 12^2 = 13^2$
$x^2 + 144 = 169$
$x^2 = 25$
$x = 5$
   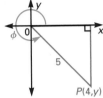
$4^2 + y^2 = 5^2$
$16 + y^2 = 25$
$y^2 = 9$
$y = -3$

$\cos(\theta + \phi) = \cos \theta \cos \phi - \sin \theta \sin \phi$

$\qquad\qquad = \dfrac{5}{13} \cdot \dfrac{4}{5} - \dfrac{12}{13} \cdot \left(-\dfrac{3}{5}\right)$ ← THINK: Can this number be greater than 1? How do you know?

$\qquad\qquad = \dfrac{20}{65} + \dfrac{36}{65}$, or $\dfrac{56}{65}$

Classroom Exercises

Use an identity to verify that the given equation is true.

1. $\cos 60 \cos 15 + \sin 60 \sin 15 = \dfrac{\sqrt{2}}{2}$

2. $\sin 75 \cos 45 - \cos 75 \sin 45 = \dfrac{1}{2}$

Simplify using an identity. Give your answers in simplest radical form.

3. $\sin 20 \cos 40 + \cos 20 \sin 40$ $\dfrac{\sqrt{3}}{2}$ 4. $\cos 70 \cos 50 - \sin 70 \sin 50$ $-\dfrac{1}{2}$

5. $\sin(45 - 30)$ 6. $\cos 75$ $\dfrac{\sqrt{6} - \sqrt{2}}{4}$ 7. $\sin 105$ $\dfrac{\sqrt{6} + \sqrt{2}}{4}$ 8. $\tan 15$ $2 - \sqrt{3}$

Written Exercises

Simplify using an identity. Give your answer in simplest radical form.

1. $\sin 80 \cos 55 + \cos 80 \sin 55$ $\dfrac{\sqrt{2}}{2}$ 2. $\cos 72 \cos 42 + \sin 72 \sin 42$ $\dfrac{\sqrt{3}}{2}$

3. $\dfrac{\tan 17 + \tan 43}{1 - \tan 17 \tan 43}$ $\sqrt{3}$ 4. $\dfrac{\tan 155 - \tan 5}{1 + \tan 155 \tan 5}$ $-\dfrac{\sqrt{3}}{3}$

5. $\cos 40 \cos 50 - \sin 40 \sin 50$ 0 6. $\cos 80 \cos 35 + \sin 80 \sin 35$ $\dfrac{\sqrt{2}}{2}$

7. $\tan 75$ $2 + \sqrt{3}$ 8. $\cos 195$ $\dfrac{-\sqrt{2} - \sqrt{6}}{4}$ 9. $\sin 165$ $\dfrac{\sqrt{6} - \sqrt{2}}{4}$ 10. $\tan 195$ $2 - \sqrt{3}$

Simplify.

11. $\cos(\pi + \phi)$ 12. $\tan(\pi - \phi)$ 13. $\cos\left(\dfrac{\pi}{2} + \phi\right)$ 14. $\sin\left(\dfrac{3\pi}{2} + \phi\right)$

   $-\cos \phi$ $-\tan \phi$ $-\sin \phi$ $-\cos \phi$

730   Chapter 19 Trigonometric Identities and Equations

Additional Example 5

Given that $\cos A = \dfrac{3}{5}$, $\tan B = \dfrac{5}{12}$, A is in Quadrant IV, and B is in Quadrant III, find $\tan (A - B)$.

$$\dfrac{-\dfrac{4}{3} - \dfrac{5}{12}}{1 + \left(-\dfrac{4}{3}\right)\dfrac{5}{12}} = -\dfrac{63}{16}$$

730

Prove each identity.

15. $\dfrac{\sin(\theta + \phi)}{\sin \theta \sin \phi} = \cot \theta + \cot \phi$ **16.** $\cos(\theta - \phi) - \cos(\theta + \phi) = 2 \sin \theta \sin \phi$

17. $\sin(30 + \theta) + \sin(30 - \theta) = \cos \theta$ **18.** $\cos\left(\dfrac{\pi}{3} + \theta\right) + \cos\left(\dfrac{\pi}{3} - \theta\right) = \dfrac{1}{\sec \theta}$

Find $\cos(\theta - \phi)$ and $\tan(\theta + \phi)$ for the conditions in Exercises 19–21.

19. $\cos \theta = \dfrac{3}{5}$, $\tan \phi = \dfrac{5}{12}$, where θ is in Quadrant I and ϕ is in Quadrant III $-\dfrac{56}{65}, \dfrac{63}{16}$

20. $\sin \theta = -\dfrac{8}{17}$, $\cos \phi = -\dfrac{4}{5}$, where θ is in Quadrant IV and ϕ is in Quadrant II $-\dfrac{84}{85}, -\dfrac{77}{36}$

21. $\cos \theta = \dfrac{1}{3}$, $\sin \phi = \dfrac{1}{2}$, where θ and ϕ are measures of acute angles

22. Show that $\cos(90 - \theta) = \sin \theta$ is an identity. (HINT: Use the difference identity for $\cos(\theta - \phi)$.)

23. Show that $\sin(90 - \theta) = \cos \theta$. (HINT: Use the identity $\cos(90 - \phi) = \sin \phi$ and let $\phi = 90 - \theta$.)

Prove the following identities. (Exercises 24–29)

24. $\sin(\theta - \phi) = \sin \theta \cos \phi - \cos \theta \sin \phi$ (HINT: Use the identities of Exercises 22 and 23.)

25. $\sin(\theta + \phi) = \sin \theta \cos \phi + \cos \theta \sin \phi$

26. $\tan(\theta - \phi) = \dfrac{\tan \theta - \tan \phi}{1 + \tan \theta \tan \phi}$ **27.** $\tan(\theta + \phi) = \dfrac{\tan \theta + \tan \phi}{1 - \tan \theta \tan \phi}$

28. $\sin(\theta + \phi) + \sin(\theta - \phi) = 2 \sin \theta \cos \phi$

29. $\sin A + \sin B = 2 \sin\left(\dfrac{A + B}{2}\right) \cos\left(\dfrac{A - B}{2}\right)$

30. Evaluate $\cos\left[\text{Tan}^{-1}\left(-\dfrac{3}{4}\right) - \text{Sin}^{-1}\dfrac{5}{13}\right]$. $\dfrac{33}{65}$

Mixed Review

Solve each equation or inequality. *5.1, 6.4, 7.6, 13.3*

1. $2^{3x+2} = 32$ 1   **2.** $x^2 - 8x - 20 \le 0$ $-2 \le x \le 10$ **3.** $\dfrac{4}{x} + \dfrac{5}{2x} = 13$ $\dfrac{1}{2}$ **4.** $\log_b 64 = 3$ 4

Brainteaser

Equilateral triangle CDE and square $ABCD$ share a common side $\overline{CD}$. $\overline{AE}$ intersects $\overline{DC}$ at point F.

Find the degree measure of angle DFE. 105

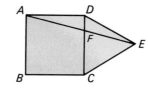

---

18. $\cos\left(\dfrac{\pi}{3} + \theta\right) + \cos\left(\dfrac{\pi}{3} - \theta\right)$
$= \left(\cos \dfrac{\pi}{3} \cos \theta - \sin \dfrac{\pi}{3} \sin \theta\right) + \left(\cos \dfrac{\pi}{3} \cos \theta + \sin \dfrac{\pi}{3} \sin \theta\right)$
$= 2 \cos \dfrac{\pi}{3} \cos \theta = \cos \theta = \dfrac{1}{\sec \theta}$

21. $\dfrac{\sqrt{3} + 2\sqrt{2}}{6}$, $-\dfrac{8\sqrt{2} + 9\sqrt{3}}{5}$

22. $\cos(90 - \theta)$
$= \cos 90 \cos \theta + \sin 90 \sin \theta$
$= 0 + 1 \cdot \sin \theta = \sin \theta$

23. $\cos(90 - \phi) = \sin \phi$
Let $\phi = 90 - \theta$.
Then $\cos(90 - (90 - \theta)) = \sin(90 - \theta)$.
So, $\cos \theta = \sin(90 - \theta)$.

24. $\sin(\theta - \phi)$
$= \cos[90 - (\theta - \phi)]$
$= \cos[(90 - \theta) + \phi]$
$= \cos(90 - \theta) \cos \phi - \sin(90 - \theta) \sin \phi$
$= \sin \theta \cos \phi - \cos \theta \sin \phi$

25. $\sin(\theta + \phi)$
$= \sin(\theta - (-\phi))$
$= \sin \theta \cos(-\phi) - \cos \theta \sin(-\phi)$
$= \sin \theta \cos \phi + \cos \theta \sin \phi$

26. $\tan(\theta - \phi)$
$= \dfrac{\sin(\theta - \phi)}{\cos(\theta - \phi)}$
$= \dfrac{\sin \theta \cos \phi - \cos \theta \sin \phi}{\cos \theta \cos \phi + \sin \theta \sin \phi} \cdot \dfrac{\frac{1}{\cos \theta \cos \phi}}{\frac{1}{\cos \theta \cos \phi}}$
$= \dfrac{\tan \theta - \tan \phi}{1 + \tan \theta \tan \phi}$

27. $\tan(\theta + \phi) = \tan(\theta - (-\phi))$
$= \dfrac{\tan \theta - \tan(-\phi)}{1 + \tan \theta \tan(-\phi)}$
$= \dfrac{\tan \theta + \tan \phi}{1 - \tan \theta \tan \phi}$

28. $\sin(\theta + \phi) + \sin(\theta - \phi)$
$= (\sin \theta \cos \phi + \cos \theta \sin \phi) + (\sin \theta \cos \phi - \cos \theta \sin \phi)$
$= 2 \sin \theta \cos \phi$

See page 748 for the answer to Written Ex. 29.

Enrichment

Have the students use the sum and difference identities to verify and review the following statements.

1. The sine function is an odd function. (HINT: Show that $\sin(-x) = -\sin x$, or that $\sin(0 - x) = -\sin x$.) $\sin(-x) = \sin(0 - x) = \sin 0 \cos x - \cos 0 \sin x = 0 \cdot \cos x - 1 \cdot \sin x = -\sin x$

2. The cosine function is an even function. $\cos(-x) = \cos(0 - x) = \cos 0 \cos x + \sin 0 \sin x = 1 \cdot \cos x + 0 \cdot \sin x = \cos x$

3. The sine of an acute angle equals the sine of its supplement. $\sin(180 - x) = \sin 180 \cos x - \cos 180 \sin x = 0 \cdot \cos x - (-1) \cdot \sin x = \sin x$

4. The period of the tangent function is 180. $\tan(x + 180) = \dfrac{\tan x + \tan 180}{1 - \tan x \tan 180} =$

$\dfrac{\tan x + 0}{1 - \tan x \cdot 0} = \dfrac{\tan x}{1} = \tan x$

5. The period of the sine function is 360. $\sin(x + 360)$
$= \sin x \cos 360 + \cos x \sin 360$
$= \sin x \cdot 1 + \cos x \cdot 0$
$= \sin x$

GETTING STARTED

Prerequisite Quiz

Simplify, using an identity.

1. sin 20 cos 70 + cos 20 sin 70   1
2. $1 - \sin^2 \theta$ $\cos^2 \theta$
3. $\sec^2 \theta - 1$ $\tan^2 \theta$
4. Find cos 75 in simplest radical form.
   $\frac{\sqrt{6} - \sqrt{2}}{4}$

Motivator

Ask the class if sin 2θ is the same as 2 sin θ. In other words, ask them if they can factor out 2 as a common monomial factor. To follow up their discussion have them evaluate sin 2θ and 2 sin θ for the following values of θ: 45, 30, 90. Ask them what this proves.   sin 2θ ≠ 2 sin θ

TEACHING SUGGESTIONS

Lesson Note

Emphasize the need to determine the correct sign when using the formulas for $\sin \frac{\theta}{2}$ and $\cos \frac{\theta}{2}$, while this is not necessary for $\tan \frac{\theta}{2}$. Note than there are only two trigonometric identities for which the sign is not provided by the identity itself.

Objectives

To simplify trigonometric expressions using double- and half-angle identities

To prove trigonometric expressions using double- and half-angle identities

The identity for sin 2θ can be derived from that for sin(θ + φ) by replacing φ by θ as shown below.

$$\sin(\theta + \phi) = \sin \theta \cos \phi + \cos \theta \sin \phi$$
$$\sin 2\theta = \sin(\theta + \theta) = \sin \theta \cos \theta + \cos \theta \sin \theta$$
$$= \sin \theta \cos \theta + \sin \theta \cos \theta$$
$$\sin 2\theta = 2 \sin \theta \cos \theta$$

Similarly, identities for cos 2θ and tan 2θ can be developed. You will be asked to derive these identities in Exercises 24 and 25.

The double-angle identities are listed below.

Double-Angle Identities

$\sin 2\theta =$	$\cos 2\theta = \cos^2 \theta - \sin^2 \theta$	$\tan 2\theta = \dfrac{2 \tan \theta}{1 - \tan^2\theta}$
$2 \sin \theta \cos \theta$	$\cos 2\theta = 1 - 2 \sin^2 \theta$	
	$\cos 2\theta = 2 \cos^2 \theta - 1$	

EXAMPLE 1   Simplify $2 \sin 22\frac{1}{2} \cos 22\frac{1}{2}$.

Solution
$$2 \sin \theta \cos \theta = \sin 2\theta$$
$$2 \sin 22\tfrac{1}{2} \cos 22\tfrac{1}{2} = \sin 2\left(22\tfrac{1}{2}\right) = \sin 45 = \frac{\sqrt{2}}{2}$$

EXAMPLE 2   Find tan 2θ if $\cos \theta = \frac{3}{5}$ and sin θ < 0. Simplify the result.

Solution   The cosine is positive when the sine is negative only in Quadrant IV.

$$3^2 + y^2 = 5^2 \quad \leftarrow \text{Solve for } y.$$
$$y = -4 \quad \leftarrow \text{Quadrant IV}$$
$$\tan \theta = -\frac{4}{3}$$

$$\tan 2\theta = \frac{2 \tan \theta}{1 - \tan^2 \theta} = \frac{2 \cdot \left(-\frac{4}{3}\right)}{1 - \left(-\frac{4}{3}\right)^2} = \frac{-\frac{8}{3}}{1 - \frac{16}{9}} = \frac{24}{7}$$

Therefore, $\tan 2\theta = \frac{24}{7}$.

Highlighting the Standards

Standard 4c: In these final lessons of the text, the connections between various mathematical topics become increasingly evident in examples and exercises.

Additional Example 1

Simplify 2 sin 15 cos 15.

$\sin 30 = \frac{1}{2}$

Additional Example 2

Find cos 2θ if $\sin \theta = -\frac{3}{5}$ and cos θ > 0.

$\frac{7}{25}$

EXAMPLE 3   Prove that $\cot \theta = \dfrac{1 + \cos 2\theta}{\sin 2\theta}$ is an identity.

Solution

$$\cot \theta \quad \bigg| \quad \dfrac{1 + \cos 2\theta}{\sin 2\theta}$$

$$\dfrac{\cos \theta}{\sin \theta} \quad \bigg| \quad \dfrac{1 + (2 \cos^2 \theta - 1)}{2 \sin \theta \cos \theta} \quad \leftarrow \text{Of the three identities for } \cos 2\theta, \text{ choose}$$
the one that eliminates the 1 in the numerator.

$$\bigg| \quad \dfrac{2 \cos^2 \theta}{2 \sin \theta \cos \theta}$$

$$\dfrac{\cos \theta}{\sin \theta} = \dfrac{\cos \theta}{\sin \theta} \quad \text{Therefore, the identity is proved.}$$

An identity for $\cos \frac{\theta}{2}$ can be derived from the double-angle identity $\cos 2\phi = 2 \cos^2 \phi - 1$. Solve for $\cos \phi$ and then replace ϕ by $\frac{\theta}{2}$.

$$2 \cos^2 \phi - 1 = \cos 2\phi$$

$$\cos^2 \phi = \dfrac{1 + \cos \phi}{2}$$

$$\cos \phi = \pm \sqrt{\dfrac{1 + \cos 2\phi}{2}}$$

Let $\phi = \dfrac{\theta}{2}$. $\cos \dfrac{\theta}{2} = \pm \sqrt{\dfrac{1 + \cos \left(2 \cdot \frac{\theta}{2}\right)}{2}}$, or $\pm \sqrt{\dfrac{1 + \cos \theta}{2}}$

An identity can be derived for $\sin \frac{\theta}{2}$ by solving $\cos 2\phi = 1 - 2 \sin^2 \phi$ for $\sin \phi$ and then replacing ϕ by $\frac{\theta}{2}$.

There are three identities for $\tan \frac{\theta}{2}$, one of which can be derived by using $\tan \dfrac{\theta}{2} = \dfrac{\sin \frac{\theta}{2}}{\cos \frac{\theta}{2}}$.

The half-angle identities are listed below.

Half-Angle Identities

$$\sin \dfrac{\theta}{2} = \pm \sqrt{\dfrac{1 - \cos \theta}{2}} \qquad\qquad \tan \dfrac{\theta}{2} = \pm \sqrt{\dfrac{1 - \cos \theta}{1 + \cos \theta}}$$

$$\cos \dfrac{\theta}{2} = \pm \sqrt{\dfrac{1 + \cos \theta}{2}} \qquad\qquad \tan \dfrac{\theta}{2} = \dfrac{\sin \theta}{1 + \cos \theta}$$

$$\tan \dfrac{\theta}{2} = \dfrac{1 - \cos \theta}{\sin \theta}$$

Math Connections

Multiple-Angle formulas: Double-angle formulas were known in the time of Ptolemy, but it was not until hundreds of years later that the French mathematician Francois Viete (1540–1603) derived the following formula for the cosine of the multiple angle $n\theta$.

$$\cos n\theta = \cos^n \theta - \dfrac{n(n-1)}{1 \cdot 2} \cos^{n-2} \theta \sin^2 \theta$$
$$+ \dfrac{n(n-1)(n-2)(n-3)}{1 \cdot 2 \cdot 3 \cdot 4} \cos^{n-4} \theta \sin^4 \theta - \cdots$$

He also derived a similar formula for the sine of the multiple angle.

Critical Thinking Questions

Analysis: Have students check the validity of Viete's formula by letting $n = 2$ in the formula and comparing the result with the first double-angle identity for $\cos 2\theta$ on page 732. They should observe that in the formula, every term after the first two terms is zero.

Common Error Analysis

Error: A typical error is to assign the same sign to $\cos \frac{\theta}{2}$ (or $\sin \frac{\theta}{2}$) as the sign of $\cos \theta$ (or $\sin \theta$).

Emphasize the importance of finding the quadrant in which $\frac{\theta}{2}$ terminates, based on the quadrant in which θ terminates. For example, if $0 \leq \theta < 360$ and θ terminates in quadrant IV, then θ is between 270 and 360. Therefore, $\frac{\theta}{2}$ must be between 135 and 180, so $\frac{\theta}{2}$ must terminate in quadrant II.

Additional Example 3

Prove that $\cos^4 \theta - \sin^4 \theta = \cos 2\theta$ is an identity.

$$\cos^4 \theta - \sin^4 \theta$$
$$= (\cos^2 \theta - \sin^2 \theta)(\cos^2 \theta + \sin^2 \theta)$$
$$= \cos 2\theta \cdot 1$$
$$= \cos 2\theta$$

Checkpoint

Use an identity to simplify each expression.

1. $\cos^2 67\frac{1}{2} - \sin^2 67\frac{1}{2}$ $-\frac{\sqrt{2}}{2}$

2. $\sin 22\frac{1}{2}$ $\frac{\sqrt{2 - \sqrt{2}}}{2}$

3. $\frac{2 \tan 75}{1 - \tan^2 75}$ $-\frac{\sqrt{3}}{3}$

In Exercises 4–6, $\cos \theta = \frac{12}{13}$, $\sin \theta < 0$, and $0 \le \theta < 360$. Find each of the following.

4. $\cos 2\theta$ $\frac{119}{169}$

5. $\cos \frac{\theta}{2}$ $-\frac{5\sqrt{26}}{26}$

6. $\sin \frac{\theta}{2}$ $\frac{\sqrt{26}}{26}$

Closure

Have students give each of the double-angle identities. See page 732. Ask them how many versions there are for $\cos 2\theta$.   3
Now have them give the half-angle identities. See page 733. Ask students how they determine the sign for $\sin \frac{1}{2}\theta$ and $\cos \frac{1}{2}\theta$.

Find the quadrant in which $\frac{\theta}{2}$ terminates.

EXAMPLE 4   Find $\cos \frac{\theta}{2}$ if $\tan \theta = \frac{4}{3}$, $\cos \theta < 0$, and $0 \le \theta < 360$. Express the result in simplest radical form.

Solution   The only quadrant in which the tangent is positive and the cosine negative is Quadrant III.

$r = \sqrt{(-3)^2 + (-4)^2} = 5$. Thus, $\cos \theta = -\frac{3}{5}$.
Since θ is in Quadrant III, $180 < \theta < 270$.

$$90 < \frac{\theta}{2} < 135$$

Therefore, $\frac{\theta}{2}$ is in Quadrant II and $\cos \frac{\theta}{2}$ is negative.

$$\cos \frac{\theta}{2} = -\sqrt{\frac{1 + \cos \theta}{2}} = -\sqrt{\frac{1 + \left(-\frac{3}{5}\right)}{2}} = -\frac{1}{\sqrt{5}} = -\frac{\sqrt{5}}{5}$$

P(-3,-4)   *r = 5*

EXAMPLE 5   Find $\tan 22\frac{1}{2}$. Express the result in simplest radical form.

Solution $22\frac{1}{2}$ is half of the special angle measure 45.
Use $\tan \frac{\theta}{2} = \frac{\sin \theta}{1 + \cos \theta}$, where $\theta = 45$.

$$\tan 22\frac{1}{2} = \frac{\sin 45}{1 + \cos 45}$$

$$= \frac{\frac{\sqrt{2}}{2}}{1 + \frac{\sqrt{2}}{2}} = \frac{2\left(\frac{\sqrt{2}}{2}\right)}{2\left(1 + \frac{\sqrt{2}}{2}\right)} = \frac{\sqrt{2}(2 - \sqrt{2})}{(2 + \sqrt{2})(2 - \sqrt{2})} = \sqrt{2} - 1$$

Classroom Exercises

Use an identity to verify that each equation is true.

1. $2 \sin 15 \cos 15 = \frac{1}{2}$ 2. $\cos^2 15 - \sin^2 15 = \frac{\sqrt{3}}{2}$

Given that $180 \le \theta < 360$, find each of the following.

3. $\sin \frac{\theta}{2}$ if $\cos \theta = 0.28$ 0.6   4. $\cos \frac{\theta}{2}$ if $\cos \theta = -0.28$ -0.6

Use an identity to simplify each expression.

5. $\frac{2 \tan 67\frac{1}{2}}{1 - \tan^2 67\frac{1}{2}}$ -1   6. $1 - 2 \sin^2 \frac{\pi}{8}$ $\frac{\sqrt{2}}{2}$ 7. $\sin 22\frac{1}{2}$ $\frac{\sqrt{2 - \sqrt{2}}}{2}$ 8. $\cos 67\frac{1}{2}$ $\frac{\sqrt{2 - \sqrt{2}}}{2}$

Additional Example 4

Find $\cos \frac{\theta}{2}$ if $\cos \theta = \frac{5}{13}$, $\sin \theta < 0$, and $0 \le \theta < 360$. Express the result in simplest radical form. $-\frac{3\sqrt{13}}{13}$

Additional Example 5

Find $\tan 67\frac{1}{2}$ in simplest radical form.
$\sqrt{2} + 1$

Written Exercises

Use an identity to simplify each expression.

1. $2 \sin 67\frac{1}{2} \cos 67\frac{1}{2}$ $\frac{\sqrt{2}}{2}$ **2.** $\cos^2 22\frac{1}{2} - \sin^2 22\frac{1}{2}$ $\frac{\sqrt{2}}{2}$ **3.** $\frac{2 \tan 15}{1 - \tan^2 15}$ $\frac{\sqrt{3}}{3}$

4. $\cos 22\frac{1}{2}$ **5.** $\tan 67\frac{1}{2}$ **6.** $\sin 157.5$ **7.** $\sin \frac{5\pi}{8}$ **8.** $\tan \frac{5\pi}{8}$ **9.** $\cos \frac{\pi}{8}$

In Exercises 10–15, $\cos \phi = \frac{4}{5}$, $\tan \phi < 0$, and $0 \le \phi < 360$. Find each of the following.

10. $\cos 2\phi$ $\frac{7}{25}$ **11.** $\sin \frac{\phi}{2}$ $\frac{\sqrt{10}}{10}$ **12.** $\tan 2\phi$ $-\frac{24}{7}$ **13.** $\tan \frac{\phi}{2}$ $-\frac{1}{3}$ **14.** $\cos \frac{\phi}{2}$ $\frac{3\sqrt{10}}{10}$ **15.** $\sin 2\phi$ $-\frac{24}{25}$

Prove each identity.

16. $\frac{\cos 2\theta}{\cos \theta - \sin \theta} = \cos \theta + \sin \theta$

17. $\frac{\tan \theta - \sin \theta}{2 \tan \theta} = \sin^2 \frac{\theta}{2}$

18. $(\cos \theta + \sin \theta)^2 = 1 + \sin 2\theta$

19. $\cos^4 \theta - \sin^4 \theta = \cos 2\theta$

20. $\frac{1 - \tan^2 \theta}{1 + \tan^2 \theta} = \cos 2\theta$

21. $\sin \theta = \frac{\cos \theta \sin 2\theta}{1 + \cos 2\theta}$

22. $\sin 3\theta = 4 \sin \theta \cos^2 \theta - \sin \theta$

23. $\cos 3\theta = 4 \cos^3 \theta - 3 \cos \theta$

Derive the identities that were given in the lesson for the following.

24. $\cos 2\theta$ (3 forms) **25.** $\tan 2\theta$ **26.** $\sin \frac{\theta}{2}$ **27.** $\tan \frac{\theta}{2}$ (3 forms)

28. The angle of elevation of a flagpole was measured at distances of 45 ft and 14.4 ft from the flagpole. The second measure of the angle of elevation was twice the first. Find the height of the flagpole to two significant digits. 27 ft

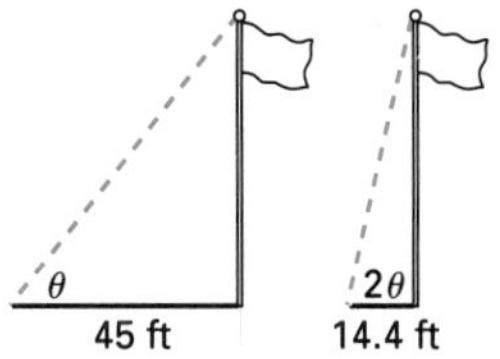

Midchapter Review

Verify each identity for the given value of θ. ***19.1***

1. $\cos(90 - \theta) = \sin \theta$; $\theta = 30$ **2.** $\cot \theta = \tan(90 - \theta)$; $\theta = 120$

3. Prove that $\tan \theta \sin \theta = \sec \theta - \cos \theta$ is an identity. ***19.2***

Solve for θ, $0 \le \theta < 2\pi$. ***19.3, 19.4***

4. $6 \csc \theta - 5 = 7$ $\frac{\pi}{6}, \frac{5\pi}{6}$ **5.** $1 - \cos \theta = \sin \theta$ $0, \frac{\pi}{2}$

Simplify. Give your answers in radical form. ***19.5, 19.6***

6. $\cos 177\frac{1}{2} \cos 57\frac{1}{2} + \sin 177\frac{1}{2} \sin 57\frac{1}{2}$ $-\frac{1}{2}$ **7.** $\tan \frac{\pi}{8}$ $\sqrt{2} - 1$

FOLLOW UP

Guided Practice

Classroom Exercises 1–8

Independent Practice

A Ex. 1–15, **B** Ex. 16–21, **C** Ex. 22–28

Basic: WE 1–17 odd, Midchapter Review

Average: WE 5–21 odd, Midchapter Review

Above Average: WE 11–27 odd, Midchapter Review

Additional Answers

Classroom Exercises

1. $2 \sin 15 \cos 15 = \sin 2(15)$
$= \sin 30 = \frac{1}{2}$

2. $\cos^2 15 - \sin^2 15 = \cos 2(15)$
$= \cos 30 = \frac{\sqrt{3}}{2}$

Written Exercises

4. $\frac{\sqrt{2 + \sqrt{2}}}{2}$ **5.** $\sqrt{2} + 1$

6. $\frac{\sqrt{2 - \sqrt{2}}}{2}$ **7.** $\frac{\sqrt{2 + \sqrt{2}}}{2}$

8. $-1 - \sqrt{2}$ **9.** $\frac{\sqrt{2 + \sqrt{2}}}{2}$

16. $\frac{\cos 2\theta}{\cos \theta - \sin \theta}$
$= \frac{\cos^2 \theta - \sin^2 \theta}{\cos \theta - \sin \theta}$
$= \frac{(\cos \theta + \sin \theta)(\cos \theta - \sin \theta)}{\cos \theta - \sin \theta}$
$= \cos \theta + \sin \theta$

17. $\frac{\tan \theta - \sin \theta}{2 \tan \theta} = \frac{\tan \theta}{\tan \theta}\left(\frac{1 - \cos \theta}{2}\right)$
$= \frac{1 - \cos \theta}{2}$
$= \sin^2 \frac{\theta}{2}$

See pages 752–753 for answers to Written Ex. 18–27 and Midchapter Review Ex. 1–3.

Enrichment

Draw these triangles on the chalkboard.

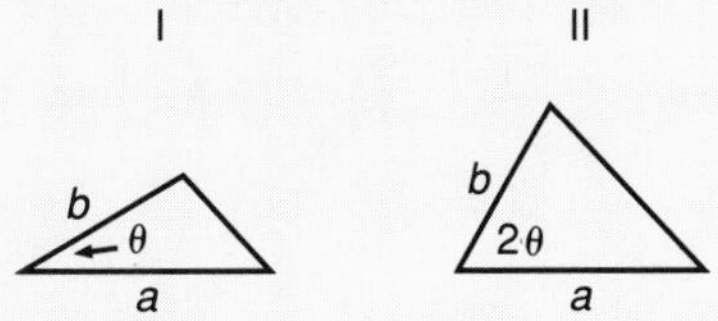

Point out that two sides of one triangle have the same lengths, *a* and *b*, as two sides of the second triangle, but the included angle in triangle II is twice as large as the included angle in triangle I.

Challenge students to find the ratio of the area of triangle I to the area of triangle II, using the formula $A = \frac{1}{2}ab \sin C$, where *C* is the measure of the included angle.

Note that the ratio will be a function of θ. $\frac{1}{2 \cos \theta}$

▉▉▉GETTING STARTED

Prerequisite Quiz

Solve for θ, 0 ≤ θ < 360.

1. $2 \cos \theta + 1 = 0$ 120, 240
2. $2 \cos^2 \theta + \cos \theta - 1 = 0$ 60, 180, 300
3. $3 \tan^2 \theta = 1$ 30, 150, 210, 330

Motivator

In the equation $\sin 2\theta + \sin \theta = 0$, ask students to change $\sin 2\theta$ so that the resulting equation will contain only trigonometric ratios that do not involve multiple angles.
$2 \sin \theta \cos \theta + \sin \theta = 0$

▉▉▉TEACHING SUGGESTIONS

Lesson Note

Review the solution of quadratic equations by factoring. Point out the similarity between equations such as $2x^2 + x - 1 = 0$ and $2 \cos^2 \theta + \cos \theta - 1 = 0$.

Math Connections

Snell's Law: In optics, the relationship between the angles of incidence (θ) and refraction (θ′) formed by a ray of light and the normal to the surface of the transparent substance that the ray is striking is given by the following trigonometric equation in two variables: $\sin \theta = k \sin \theta'$, where k is a nonzero constant that depends on the substance. This is called *Snell's Law*.

19.7 Trigonometric Equations: Multiple Angles

Objective To solve trigonometric equations involving multiple angles

To solve equations involving multiples of angle measures, it is sometimes helpful first to use identities to change each multiple-angle expression to a function of a single-angle measure as shown below.

EXAMPLE 1 Solve $\sin 2\theta + \sin \theta = 0, 0 \le \theta < 2\pi$.

Plan Use the double-angle identity $\sin 2\theta = 2 \sin \theta \cos \theta$.

Solution
$$\sin 2\theta + \sin \theta = 0$$
$$2 \sin \theta \cos \theta + \sin \theta = 0 \leftarrow \text{Factor out the common term.}$$
$$\sin \theta (2 \cos \theta + 1) = 0 \leftarrow \text{If } mn = 0, \text{ then } m = 0 \text{ or } n = 0.$$
$$\sin \theta = 0 \quad or \quad 2 \cos \theta + 1 = 0$$
$$\theta = 0, \pi \qquad \cos \theta = -\frac{1}{2} \text{ Quadrants II, III}$$
$$\theta = \frac{2\pi}{3}, \frac{4\pi}{3}$$

The primary solutions in radian measure are $0, \frac{2\pi}{3}, \pi,$ and $\frac{4\pi}{3}$.

EXAMPLE 2 Solve $\sin^2 \frac{\theta}{2} = \cos^2 \theta, 0 \le \theta < 360$.

Solution Substitute $\pm\sqrt{\dfrac{1 - \cos \theta}{2}}$ for $\sin \frac{\theta}{2}$.
$$\left(\pm\sqrt{\frac{1 - \cos \theta}{2}}\right)^2 = \cos^2 \theta \leftarrow (\sqrt{a})^2 = a$$
$$\frac{1 - \cos \theta}{2} = \cos^2 \theta$$
$$1 - \cos \theta = 2 \cos^2 \theta$$
$$2 \cos^2 \theta + \cos \theta - 1 = 0 \leftarrow \text{Factor the quadratic.}$$
$$(2 \cos \theta - 1)(\cos \theta + 1) = 0$$
$$2 \cos \theta = 1 \text{ or } \cos \theta = -1$$
$$\text{Quadrants I, IV} \rightarrow \quad \cos \theta = \frac{1}{2} \qquad \theta = 180$$
$$\theta = 60, 300$$

Therefore, the primary solutions are 60, 180, and 300.

736 Chapter 19 Trigonometric Identities and Equations

Additional Example 1

Solve.

$\cos 2\theta = \sin \theta, 0 \le \theta < 2\pi$.

$\dfrac{\pi}{6}, \dfrac{5\pi}{6}, \dfrac{3\pi}{2}$

Additional Example 2

Solve $\cos^2 \frac{\theta}{2} = \cos^2 \theta, 0 \le \theta < 360$.

0, 120, 240

Classroom Exercises

Tell what substitution is necessary to solve each equation.

1. $\sin 2\theta - \cos \theta = 0$ **2.** $\cos 2\theta + \sin \theta - 1 = 0$ **3.** $\sin^2 2\theta = \sin 2\theta$

4. $\cos^2 \frac{\theta}{2} = 1 - \frac{1}{2}\cos \theta$ **5.** $\cos 2\theta + 4\sin^2 \frac{\theta}{2} = 1$ **6.** $3\tan^2 \frac{\theta}{2} = 1$

7–12. Solve each of the equations of Classroom Exercises 1–6, where $0 \le \theta < 2\pi$.

Written Exercises

Solve each equation, where $0 \le \theta < 2\pi$ ($0 \le 2\theta < 4\pi$).

1. $\sin 2\theta - \sin \theta = 0$ **2.** $\cos 2\theta = \cos \theta$ **3.** $\sin 2\theta = 0$

4. $\sin 2\theta = 2\sin \theta$ **5.** $\cos 2\theta = -\sin \theta$ **6.** $\cos \theta = \sin \frac{\theta}{2}$

Solve each equation, where $0 \le \theta < 360$.

7. $\cos 2\theta = 3\sin \theta - 1$ **8.** $\sin 2\theta = \cos \theta$ **9.** $\cos 2\theta + 5\cos \theta = 2$

10. $\tan 2\theta = 0$ **11.** $\cos 2\theta = 0$ **12.** $\cos \theta + \sin 2\theta = 0$

13. $\tan^2 \frac{\theta}{2} = 0$ **14.** $2\cos^2 \frac{\theta}{2} = 3\cos \theta$ **15.** $\sin 2\theta = \tan \theta$

16. $\cos 2\theta \sin \theta + \sin \theta = 0$ **17.** $2\cos 2\theta + 2\sin^2 \theta = \cos \theta$

18. $\cos \theta = \cos \frac{\theta}{2}$ **19.** $\cos(2\theta - \pi) = \sin \theta$

20. $2\tan \frac{\theta}{2} - \csc \theta = 0$ **21.** $\sin 2\theta \cos 2\theta - 1 + \cos 2\theta = \sin 2\theta$

Mixed Review

Graph each equation.   *11.3, 12.3, 12.5, 12.7*

1. $y = |x| - 1$ **2.** $4x^2 - 25y^2 = 100$ **3.** $y = -\frac{8}{x}$ **4.** $y^2 + 8x - 6y = -1$

▰▰▱ Brainteaser

Every triangle has three sides and three angles. Euclid proved three cases, such as side-included angle-side (SAS), in which two triangles are congruent if only three of six elements are congruent.
Yet, two triangles can have *five* of the six elements congruent and still not be congruent. Which five elements can be congruent and still produce two noncongruent triangles? Try to find two noncongruent triangles that have five congruent elements. 2 sides and 3 angles equal; 1 possible pair: $\triangle ABC$, m $\angle A = 127$, m $\angle B = 32$, m $\angle C = 21$, $a = 27$, $b = 18$, $c = 12$; $\triangle DEF$, m $\angle D = 127$, m $\angle E = 32$, m $\angle F = 21$, $d = 18$, $e = 12$, $f = 8$

Critical Thinking Questions

Analysis: Ask students to show that if $\theta = 2\theta'$ and the incident ray is not perpendicular to the substance's surface, then $k < 2$. See discussion in *Math Connections*.

Checkpoint

Solve each equation, where $0 \le \theta < 360$.

1. $\sin 2\theta + 2\sin \theta = 0$ 0, 180

2. $\sin^2 \frac{\theta}{2} = \sin^2 \theta$ 0, 120, 240

3. $3\cos^2 \frac{\theta}{2} = \sin^2 \theta$ 120, 180, 240

Closure

Ask students to discuss the main strategy in solving equations involving trigonometric functions of multiple angles. Use identities to change multiple angle expressions to single angle measures. Then solve the equation.

▰▰▰▱ FOLLOW UP

Guided Practice

Classroom Exercises 1–12

Independent Practice

A Ex. 1–12, **B** Ex. 13–17, **C** Ex. 18–21

Basic: WE 1–13 odd, Brainteaser

Average: WE 5–17 odd, Brainteaser

Above Average: WE 5, 11–21 odd, Brainteaser

Additional Answers

See page 749 for the answers to Classroom Ex. 1–12, Written Ex. 1–21, and Mixed Review Ex. 1–4.

Enrichment

Write this equation on the chalkboard and ask the students to find all the primary solutions.

$\sin 3x = \frac{1}{2}$

Students should discover that for $0 \le x < 360$, they must solve for $3x$ in the interval $0 \le 3x < 1{,}080$.

10, 50, 130, 170, 250, 290

When students have understood this procedure, have them solve $\cos 5x = \frac{1}{2}$.

12, 60, 84, 132, 156, 204, 228, 276, 300, 348

■■■GETTING STARTED

Prerequisite Quiz

Write the complex number that corresponds to vector $\vec{v}$ with the origin as its initial point and the given point P as its terminal point.

1. $P(3,4)$ $3 + 4i$
2. $P(5,-9)$ $5 - 9i$
3. $P(-6,-3)$ $-6 - 3i$

4–6. Find the length of each vector described in Exercises 1–3.

4. 5   5. $\sqrt{106}$ 6. $3\sqrt{5}$

Motivator

Have students consider the following problem.

You are on a skiing trip in the mountains. On one downhill section you suddenly encounter a strong crosswind coming from your right. Will your downhill path change? If so, in what direction and by how much will your path be deflected?   You will be pushed to your left by an amount depending on the force of the wind.   Will your resulting path be straight or curved?   Curved

Highlighting the Standards

Standard 14f: Vectors in themselves form an axiomatic system and help students understand how a few rules of operation can support a large structure.

19.8 Vectors and Trigonometry

Objectives   To find the magnitude of a vector
                 To resolve a vector into two perpendicular vectors

In the figure at the right, a winch is attempting to pull a marble slab up a slippery ramp. The winch can exert a force of only 1,200 lb. Can the slab be moved up the ramp? In this lesson you will learn how to use vectors and trigonometry to answer this question.

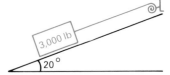

A vector $\vec{v}$ with initial point at the origin and terminal point at (a,b) is in *standard position*. It can be represented as (a,b), which is called the *rectangular form* of $\vec{v}$ (see Lesson 9.8).

In the figure below, $\vec{v_1} + \vec{v_2} = \vec{v}$. The sum, or *resultant*, of $\vec{v_1}$ and $\vec{v_2}$ is $\vec{v}$, and $\vec{v}$ is said to be *resolved* into the horizontal and vertical vector components, $\vec{v_1}$ and $\vec{v_2}$.

$$\vec{v_1} + \vec{v_2} = (-4.5,0) + (0,3)$$
$$= (-4.5 + 0,0 + 3)$$
$$= (-4.5,3)$$
$$= \vec{v}$$

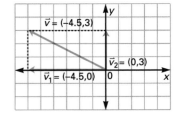

The numbers -4.5 and 3 are the *x-component* and *y-component*, respectively, of v.

■■■
Definition

> If $\vec{v} = (x,y)$, in standard position, is resolved into a horizontal vector $(x,0)$ and a vertical vector $(0,y)$, then the real numbers x and y are called the **x-component** and the **y-component** of $\vec{v}$.

The **magnitude**, or length, of a vector $\vec{v}$ in standard position is represented by $\| \vec{v} \|$. The distance formula can be used to determine the magnitude of $\vec{v}$ in terms of its x- and y-components.

If $\vec{v} = (x,y)$, then $\| \vec{v} \| = \sqrt{x^2 + y^2}$.

In the figure above, $\vec{v} = (-4.5,3)$ and

$$\| \vec{v} \| = \sqrt{(-4.5)^2 + 3^2} = \sqrt{29.25} \approx 5.4.$$

The magnitude of a vector $\vec{v}$ is also called its **norm**.

738   Chapter 19 Trigonometric Identities and Equations

EXAMPLE 1 Let $\vec{v} = (5, -3)$.

    a. Draw $\vec{v}$ in standard position. Then draw $\vec{v_1}$ and $\vec{v_2}$, the horizontal and vertical vector components of $\vec{v}$.

    b. Find the x- and y-components of $\vec{v}$. Then find $\| \vec{v} \|$.

Solutions   **a.**

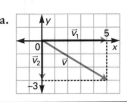

    b. x-component: 5
        y-component: -3

$$\| \vec{v} \| = \sqrt{5^2 + (-3)^2}$$
$$= \sqrt{34}$$
$$\approx 5.8$$

The absolute value of the x- and y-components of a vector $\vec{v} = (x,y)$ is related to the vector's magnitude $\| \vec{v} \|$ and to its reference angle α (alpha) with the x-axis.

$$\cos \alpha = \frac{|x|}{\| \vec{v} \|} \text{ and } \sin \alpha = \frac{|y|}{\| \vec{v} \|}$$

Thus, $|x| = \| \vec{v} \| \cdot \cos \alpha$ and
      $|y| = \| \vec{v} \| \cdot \sin \alpha$.

In physics, it is often useful to resolve a gravitational force vector into two perpendicular vectors. For example, a force $\overrightarrow{OG}$ of 800 pounds ($\| \overrightarrow{OG} \| = 800$) can be resolved into $\overrightarrow{OA}$ and $\overrightarrow{OB}$ so that $\overrightarrow{OA}$ makes an angle of measure 25 with the horizontal and is perpendicular to $\overrightarrow{OB}$.

Draw an x-axis through O at angles of 25 to the horizontal and 65 to $\overrightarrow{OG}$.

Draw a y-axis through O and perpendicular to the x-axis.

Resolve $\overrightarrow{OG}$ into $\overrightarrow{OA}$ and $\overrightarrow{OB}$.

The magnitudes of $\overrightarrow{OA}$ and $\overrightarrow{OB}$ are the solutions of the equations

$$\cos 65 = \frac{\| \overrightarrow{OA} \|}{800} \quad \text{and} \quad \sin 65 = \frac{\| \overrightarrow{OB} \|}{800}.$$

Lesson Note

When students are asked to find $|x|$ and $|y|$, given a vector $\vec{v}$ and the reference angle α, they do not need to memorize the formulas for $\cos \alpha$ and $\sin \alpha$. They can simply draw the reference angle in the first quadrant, as shown below, where $r = \| \vec{v} \|$. The x and y values will be positive, and therefore equal to $|x|$ and $|y|$ for the given vector.

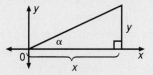

Math Connections

Banking of Curves: Turning a car requires *centripetal* force—force directed toward the center of the circular path. On a flat road, the friction of the highway against the tires provides this force. Higher speeds require more force for turning, and friction may not be enough, resulting in a skid.

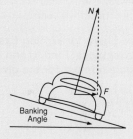

This is why roads are banked. The figure illustrates the *normal* force, the force exerted by the roadway against the car. On a banked curve, the horizontal component of the normal force supplies all or part of the centripetal force needed to turn the car.

Additional Example 1

Let $\vec{v} = (-2, 4)$.

a. Draw $\vec{v}$ in standard position. Then draw $\vec{v_1}$ and $\vec{v_2}$, the horizontal and vertical vector components of $\vec{v}$.

b. Find the x- and y-components of $\vec{v}$. Then find $\| \vec{v} \|$ to one decimal place.

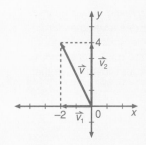

x-component: -2
y-component: 4
$\| \vec{v} \| = \sqrt{20} \approx 4.5$

Critical Thinking Questions

Analysis: In Example 2, explain to students that the inclined ramp exerts a normal force along the positive *y*-axis to resist an equal and opposite force exerted by the *y*-component of the slab's weight. (This is a consequence of Newton's Third Law.) Ask them to find the magnitude of the normal force.

They should see that it is 3,000 sin 70.

Common Error Analysis

Error: Students tend to misinterpret the symbol $\|\vec{v}\|$. For example, if $\overrightarrow{OA}$ is $(-5, -4)$, then some students think that $\|\overrightarrow{OA}\|$ means $(5, 4)$. Emphasize that $\|\overrightarrow{OA}\|$ is the length or magnitude of $\overrightarrow{OA}$, which is $\sqrt{(-5)^2 + (-4)^2}$, or $\sqrt{41}$.

Checkpoint

Find the *x*-component and the *y*-component. Then find $\|\vec{v}\|$ correct to one decimal place.

1. $\vec{v} = (-7.5, 3.0)$
   x: −7.5; *y*: 3.0; $\|\vec{v}\| \approx 8.1$
2. $\vec{v} = (-6.0, -2.0)$
   x: −6.0; *y*: −2.0; $\|\vec{v}\| \approx 6.3$

Refer to the figure for Written Exercises 9–12 in the textbook. Find |x| and |y| to the nearest whole number for the given data.

3. $\alpha = 40$, $\|\vec{v}\| = 300$
   |x| = 230, |y| = 193
4. $\alpha = 65$, $\|\vec{v}\| = 165$
   |x| = 70, |y| = 150

EXAMPLE 2 In the figure at the right, a marble slab is being pulled up a slippery ramp by a winch. Find the number of pounds of force required to prevent the slab from sliding down the ramp.

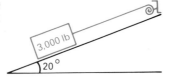

Plan Draw a force diagram with the *x*-axis in line with the force exerted by the winch. The slab's weight is pushing downward with a magnitude of 3,000 lb, due to gravity. Resolve this downward vector into two force vectors along the *x*- and *y*-axes as shown below.

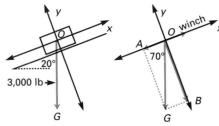

The winch must exert a force with a magnitude equal to $\|\overrightarrow{OA}\|$, since the force $\overrightarrow{OA}$ tends to pull the slab down the ramp.

Solution To find $\|\overrightarrow{OA}\|$, use right triangle *AOG*.

$$\cos 70 = \frac{\|\overrightarrow{OA}\|}{\|\overrightarrow{OG}\|} = \frac{\|\overrightarrow{OA}\|}{3,000}$$

$$\|\overrightarrow{OA}\| = 3,000 \cos 70 \approx 3,000(0.3420) = 1,026$$

Thus, the winch must exert a force of 1,026 lb.

Classroom Exercises

Use the figure at the right for Exercises 1–8. Express the vector in rectangular form.

1. $\vec{v}$ (8,6)   **2.** $\vec{w}$ (8, −6)

Find the *x*- and *y*-components of the vector.

3. $\vec{v}$ 8, 6   **4.** $\vec{w}$ 8, −6
5. Find $\|\vec{v}\|$. 10   **6.** Find $\|\vec{w}\|$. 10
7. What is the magnitude of the horizontal vector component of $\vec{v}$? 8
8. What is the magnitude of the vertical vector component of $\vec{w}$? 6

Additional Example 2

A weight of 600 lb is being pulled up a slippery ramp that forms an angle measuring 40 with the horizontal. Find, to the nearest pound, the force necessary to prevent the weight from sliding down the ramp. 386 lb

Written Exercises

Draw $\vec{v}$ in standard position. Then draw $\vec{v_1}$ and $\vec{v_2}$, the horizontal and vertical components of $\vec{v}$.

1. $\vec{v} = (4.5, 6.0)$ **2.** $\vec{v} = (-8.0, 6.0)$ **3.** $\vec{v} = (-24, -10)$ **4.** $\vec{v} = (15, 0)$

5–8. For the $\vec{v}$ of Exercises 1–4, find the x-component and y-component. Then find $\| \vec{v} \|$.

In Exercises 9–12, use the formulas $\cos \alpha = \dfrac{|x|}{\| \vec{v} \|}$ and $\sin \alpha = \dfrac{|y|}{\| \vec{v} \|}$ and the figure at the right to find $|x|$ and $|y|$ to the nearest whole number for the given data.

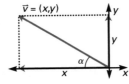

9. $\alpha = 30$, $\| \vec{v} \| = 200$ 173, 100   **10.** $\alpha = 50$, $\| \vec{v} \| = 425$ 273, 326

11. $\alpha = 45$, $\| \vec{v} \| = 300$ 212, 212   **12.** $\alpha = 62$, $\| \vec{v} \| = 510$ 239, 450

A mass is pulled up a slippery ramp that forms an angle α with the horizontal. For the given data, find the force required to prevent the mass from sliding down the ramp.

13. mass of 750 lb, $\alpha = 30$ 375 lb

14. mass of 2,500 lb, $\alpha = 25$ 1,057 lb

15. mass of 840 lb, $\alpha = 36$ 494 lb

16. mass of 1,200 lb, $\alpha = 28$ 563 lb

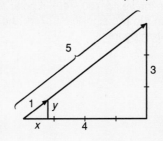

17. Prove that $\|(cx, cy)\| = |c| \cdot \|(x, y)\|$ for all real numbers c, x, and y.

In Exercises 18 and 19, use the following definition: If $\vec{v} = (a, b)$ and $\vec{w} = (c, d)$, then $\vec{v} = \vec{w}$ if and only if $a = c$ and $b = d$.

18. If $\vec{P} = (3, 5)$ and $\vec{Q} = (6m, 4 - n)$, find m and n so that $\vec{P} = \vec{Q}$. $m = \frac{1}{2}, n = -1$

19. Find h and k so that $(7 - h, 3k) = (k + 6, -2h)$. $h = 3, k = -2$

20. Draw the graph of all ordered pairs (x, y) such that $\|(x, y)\| = 5$.

Mixed Review

Simplify.   *5.2, 9.1, 9.2*   1. $\dfrac{b^6}{8a^9}$ 2. $2 - 4i$

1. $(2a^3 b^{-2})^{-3}$ **2.** $(6 - i) - (4 + 3i)$ **3.** $(3 + 2i)^2$ $5 + 12i$ **4.** $4 \div (3i)$ $-\dfrac{4i}{3}$

5. Find two numbers whose sum is 16 and the sum of whose squares is a minimum.   *11.6* 8, 8

6. Is the function $f(x) = 2x^3 - x^2 + x + 4$ odd, even, or neither?   *11.7* Neither

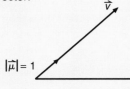
741

▬GETTING STARTED

Prerequisite Quiz

Use the figure below to find |x| and |y| to the nearest whole number for the given data.

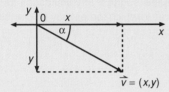

$$\vec{v} = (x,y)$$

1. $\alpha = 70$, $\|\vec{v}\| = 250$
   |x| = 86, |y| = 235
2. $\alpha = 53$, $\|\vec{v}\| = 450$
   |x| = 271, |y| = 359

Motivator

Have students consider the following problem. An object is moving in a plane such that the path it takes is a spiral. Resolve the path taken by the object into its separate components. There are two components: translation, and rotation about an axis. What common household appliance is suggested by the prior example. Phonograph or compact disc player Tell students that the polar coordinate system was devised to describe systems such as these.

19.9 Polar Form

Objective   To change a vector in standard position from rectangular form to polar form and vice versa

The point P in the first figure below has the **rectangular coordinates** $(-2\sqrt{3}, 2)$. In the second figure, the same point P is 4 units from the origin on $\overrightarrow{OP}$, which forms an angle of 150 with the positive x-axis. In the second case, P has the **polar coordinates** $(4, 150°)$. The positive x-axis is sometimes called the **polar axis**.

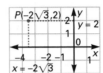 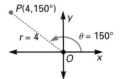

Since point P is the terminal point of $\overrightarrow{OP}$ in standard position,
(1) the rectangular form of $\overrightarrow{OP}$ is (x,y), and
(2) the **polar form** of $\overrightarrow{OP}$ is (r,θ), where
$$r = \sqrt{x^2 + y^2} = \|\overrightarrow{OP}\|,$$
$$\cos \theta = \frac{x}{r} \text{ and } \sin \theta = \frac{y}{r},$$
$$\text{or } x = r \cos \theta \text{ and } y = r \sin \theta.$$

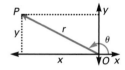

Notice that $r > 0$ and that θ is an angle of rotation.

In order to distinguish between the rectangular form and the polar form of a vector, degree symbols will be used.

EXAMPLE 1   Express $\overrightarrow{OP} = (-45, -35)$ in polar form (r, θ).

Plan   Sketch and label $\overrightarrow{OP}$ as shown. Find r and θ.

Solution
$$r = \sqrt{x^2 + y^2} = \sqrt{(-45)^2 + (-35)^2} \approx 57$$
$$\cos \theta = \frac{x}{r} = \frac{-45}{57} = -0.7895$$

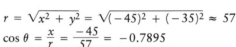

Find θ by calculator as follows:
$$45 \;\boxed{+/-}\; \boxed{\div}\; 57 \;\boxed{=}\; \boxed{INV}\; \boxed{cos} \Rightarrow 142.13635$$
So, $\theta = 142$ *or* $\theta = 360 - 142 = 218$.

Choose $\theta = 218$ since θ is in Quadrant III.
Thus, $\overrightarrow{OP} = (57, 218°)$ in polar form.

Highlighting the Standards

Standards 9g, 14d: At the secondary level, it is important that college intending students at least be introduced to polar coordinates and the trigonometric representations of complex numbers.

Additional Example 1

Express the vector $(60, -40)$ in polar form (r, θ) with r to two significant digits and θ to the nearest degree.   (72,326)

EXAMPLE 2   Express the polar form $(240, 140°)$ in rectangular form (x, y).

Solution

$$x = r \cos \theta \qquad\qquad y = r \sin \theta$$
$$\quad= 240 \cos 140 \qquad\quad = 240 \sin 140$$
$$\quad= 240(-0.7660) \qquad = 240(0.6428)$$
$$\quad= -183.84 \qquad\qquad = 154.272$$
$$\quad\approx -180 \qquad\qquad\quad \approx 150$$

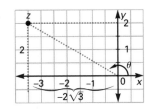

Thus, $(-180, 150)$ is the rectangular form of $(240, 140°)$.

A complex number $x + yi$ corresponds to a vector (x, y) (see Lesson 9.8). Therefore, the graph of $x + yi$ also has the polar coordinates (r, θ). Hence, $x + yi = r \cos \theta + i \cdot r \sin \theta = r(\cos \theta + i \sin \theta)$.

Definitions

> The notation $r(\cos \theta + i \sin \theta)$ is called the **polar form** of the complex number $z = x + yi$, where r is the *absolute value*, or **modulus**, of z, and θ is an **argument** of z.

EXAMPLE 3   Change $-2\sqrt{3} + 2i$ to polar form, $r(\cos \theta + i \sin \theta)$.

Solution

If $z = x + yi$, then $x = -2\sqrt{3}$ and $y = 2$.
Find $r = \sqrt{x^2 + y^2}$:

$$r = \sqrt{(-2\sqrt{3})^2 + 2^2} = \sqrt{16} = 4$$

Find θ.

$$\sin \theta = \frac{y}{r} = \frac{2}{4} = \frac{1}{2}$$

So, $\theta = 150$ and $r(\cos \theta + i \sin \theta) = 4(\cos 150 + i \sin 150)$.

Thus, $-2\sqrt{3} + 2i$ is $4(\cos 150 + i \sin 150)$ in polar form.

The polar form $r(\cos \theta + i \sin \theta)$ is sometimes abbreviated as r cis θ. Thus, $4(\cos 150 + i \sin 150)$ may be written as 4 cis 150.

The argument 150 is not unique in 4 cis 150, since multiples of 360 can be added to 150. For example, 4 cis 150 is the same number as 4 cis 510 and 4 cis (-210).

EXAMPLE 4   Change $8(\cos 240 + i \sin 240)$ to rectangular form.

Solution

$$8(\cos 240 + i \sin 240) = 8\left(\frac{-1}{2} + i \cdot \frac{-\sqrt{3}}{2}\right) = -4 - 4i\sqrt{3}$$

19.9 Polar Form   **743**

■■TEACHING SUGGESTIONS

Lesson Note

In the preceding lesson, the formulas for $\sin \alpha$ and $\cos \alpha$ were given in terms of $|x|$ and $|y|$, since α is a reference angle, and the sine and cosine of an acute angle are positive. In this lesson, the formulas for $\sin \theta$ and $\cos \theta$ do not involve absolute values, since θ is an angle of rotation and the signs of x and y are needed to determine its quadrant.

Math Connections

Vector Notations: This is a good opportunity to review the various kinds of vectors that have appeared in the text, as well as the notations that have been used to represent a vector.

1. a vector visualized as an arrow (Lesson 9.7)
2. a vector represented as an ordered pair in one of the following ways:
   a. in rectangular form (Lesson 9.8)
   b. as a complex number in the form $a + bi$ (Lesson 9.8)
   c. as a complex number in polar form (Lesson 19.9)

Also, a 2×1 column matrix (Lesson 16.5) may be thought of as a vector.

Critical Thinking Questions

Analysis: The definition of *polar form* of a complex number states that θ is *an* argument (not *the* argument) of z. Ask students why the indefinite article is used. They should see that, as later pointed out in the text, there are several arguments, not just one. In fact, if an integral multiple of 360 degrees is added to an argument, then the result is also an argument.

Additional Example 2

Express the polar form $(315, 205°)$ in rectangular form, (x, y), with x and y to two significant digits. $(-290, -130)$

Additional Example 3

Change $-1 - i\sqrt{3}$ to polar form, $r(\cos \theta + i \sin \theta)$. $2(\cos 240 + i \sin 240)$

Additional Example 4

Change $5\sqrt{2}(\cos 315 + i \sin 315)$ to rectangular form. $5 - 5i$

Checkpoint

Express each vector in polar form, (r,θ), with r to two significant digits and θ to the nearest degree.

1. (3.0,4.0) (5.0,53°)
2. (−70,80) (110,131°)

Express each vector in rectangular form, (x,y), with x and y to two significant digits.

3. (13,85°) (1.1,13)
4. (25,320°) (19,−16)

Express the complex number in polar form.

5. −6i 6(cos 270 + i sin 270)

Closure

To describe a circle using the polar coordinate system, ask students how they represent the polar counterpart of the *x*-value and the polar counterpart of the *y*-value. The "x-value" would be a fixed radius, and the "y-value" would be an angle of rotation.

◤FOLLOW UP

Guided Practice

Classroom Exercises 1–7

Independent Practice

A Ex. 1–16, **B** Ex. 17–34, **C** Ex. 35–38

Basic: WE 1–29 odd

Average: WE 3–31 odd

Above Average: WE 9–37 odd

Additional Answers

Written Exercises

10. (−23,130)   11. (−6.6, −4.6)
12. (10,−12)

32. $-\dfrac{3\sqrt{2}}{2} + \dfrac{3i\sqrt{2}}{2}$

Classroom Exercises

Match each item at the left with one of items a–i at the right.

1. (5,−2) e   2. magnitude of (5,−2) a   **a.** $\sqrt{29}$ **e.** rectangular form
3. (5,30°) d   4. x-component of (5,30°) i   **b.** 720   **f.** argument
5. y-component of (5,30°) g   **c.** 770   **g.** 5 sin 30
6. the 315 in $3\sqrt{2}$ cis 315 f   **d.** polar form   **h.** (5,−30°)
7. 50 + k · 360 if k = 2 c   **i.** 5 cos 30

Written Exercises

Express each vector in polar form (r,θ) with r to two significant digits and θ to the nearest degree.
   (7.2,236°)   (17,340°)   (23,218°)

1. (2.0,5.0) (5.4,68°)   2. (−4.0,−6.0)   3. (16,−7.0)   4. (−18,−15)
5. (−11,12)   6. (120,200)   7. (14,−130)   8. (0,−140)
 (16,133°)   (230,59°)   (130,276°)   (140,270°)

Express in rectangular form, (x,y), with x and y to two significant digits.

9. (12,80°) (2.1,12)   10. (130,100°)   11. (8.0,215°)   12. (16,310°)
13. (15,270°) (0,−15) 14. (260,204°)   15. (30,290°)   16. (44,140°)
      (−240,−106)   (10,−28)   (−34,28)

Express each complex number in polar form. **18.** 10 cis 135 **19.** 8 cis 210 **20.** $2\sqrt{2}$ cis 330

17. $3 + 3i\sqrt{3}$ 6 cis 60   18. $-5\sqrt{2} + 5i\sqrt{2}$ 19. $-4\sqrt{3} - 4i$ 20. $\sqrt{6} - i\sqrt{2}$
21. 4 + 0i 4 cis 0   22. −3i 3 cis 270   23. −4 4 cis 180   24. 1 + i $\sqrt{2}$ cis 45

Express each complex number in rectangular form.   **25.** $-\dfrac{3\sqrt{2}}{2} - \dfrac{3i\sqrt{2}}{2}$

25. 3(cos 225 + i sin 225)   26. 12(cos 30 + i sin 30) $6\sqrt{3} + 6i$
27. 10(cos 120 + i sin 120) $-5 + 5i\sqrt{3}$ 28. $4\sqrt{2}$(cos 300 + i sin 300) $2\sqrt{2} - 2i\sqrt{6}$
29. 7[cos(−90) + i sin(−90)] −7i   30. $2\sqrt{5}$[cos(−180) + i sin(−180)] $-2\sqrt{5}$
31. 2 cis 45 $\sqrt{2} + i\sqrt{2}$ 32. 3 cis 135   33. cis 70   34. 4 cis(−15)
                            0.34 + 0.94i   3.9 − 1.0i

Given $\overrightarrow{OP}$ and $\overrightarrow{OQ}$ in standard position, find m ∠POQ.

35. $\overrightarrow{OP}$ = (10,35°), $\overrightarrow{OQ}$ = (8,145°) 110   36. $\overrightarrow{OP}$ = (12,70°), $\overrightarrow{OQ}$ = (15,320°) 110
37. $\overrightarrow{OP}$ = (4,4), $\overrightarrow{OQ}$ = (−3,2) 101   38. $\overrightarrow{OP}$ = (−$\sqrt{3}$,−1), $\overrightarrow{OQ}$ = (−4,2) 57

Mixed Review

Simplify.   *18.5, 19.1, 19.6*

1. $\cos \dfrac{\pi}{2}$ 0   2. $\csc \dfrac{4\pi}{3}$ $-\dfrac{2\sqrt{3}}{3}$ 3. $\sin^2 \theta + \cos^2 \theta$ 1   4. $\dfrac{\sin 2\theta}{\cos \theta}$ 2 sin θ

Enrichment

In advanced courses, students are required to graph equations expressed in terms of polar coordinates, such as r = 2 + 4 cos x and r = 2 sin 4x. These graphs take on interesting shapes. Have the students do library research to discover figures such as the *cardioid*, the *limaçon*, and the *four-leafed rose*.

19.10 Multiplication and Division of Complex Numbers: DeMoivre's Theorem

Teaching Resources

Quick Quizzes 152
Reteaching and Practice
Worksheets 152

Objectives   To find products, quotients, and powers of complex numbers in polar form

To find the n different nth roots of a complex number

The polar form of a complex number has advantages over the rectangular form when finding products, quotients, and powers.

Given $z_1 = r_1(\cos \theta_1 + i \sin \theta_1)$ and $z_2 = r_2(\cos \theta_2 + i \sin \theta_2)$,

$$z_1 \cdot z_2 = [r_1(\cos \theta_1 + i \sin \theta_1)] \cdot [r_2(\cos \theta_2 + i \sin \theta_2)]$$
$$= r_1 r_2(\cos \theta_1 \cos \theta_2 + i^2 \sin \theta_1 \sin \theta_2 + i \sin \theta_1 \cos \theta_2 + i \cos \theta_1 \sin \theta_2)$$
$$= r_1 r_2[(\cos \theta_1 \cos \theta_2 - \sin \theta_1 \sin \theta_2) + i(\sin \theta_1 \cos \theta_2 + \cos \theta_1 \sin \theta_2)]$$
$$= r_1 r_2[\cos(\theta_1 + \theta_2) + i \sin(\theta_1 + \theta_2)], \text{ or } r_1 r_2 \text{ cis } (\theta_1 + \theta_2)$$

Product Rule
$$r_1(\cos \theta_1 + i \sin \theta_1) \cdot r_2(\cos \theta_2 + i \sin \theta_2)$$
$$= r_1 r_2[\cos(\theta_1 + \theta_2) + i \sin(\theta_1 + \theta_2)]$$

EXAMPLE 1   Given $z_1 = 4.5(\cos 25 + i \sin 25)$ and $z_2 = 8.2(\cos 115 + i \sin 115)$, find $z_1 \cdot z_2$ in polar form and in rectangular form.

Solution
$$z_1 \cdot z_2 = 4.5 \text{ cis } 25 \cdot 8.2 \text{ cis } 115$$
$$= 4.5 \cdot 8.2 \text{ cis}(25 + 115)$$
$$= 36.9 \text{ cis } 140$$
$$= 36.9(\cos 140 + i \sin 140) \leftarrow \text{polar form}$$
$$= 36.9(-0.7660 + 0.6428i)$$
$$= -28.3 + 23.7i \leftarrow \text{rectangular form}$$

The *quotient rule* is similar to the product rule.

Quotient Rule
$$\frac{r_1(\cos \theta_1 + i \sin \theta_1)}{r_2(\cos \theta_2 + i \sin \theta_2)} = \frac{r_1}{r_2}[\cos(\theta_1 - \theta_2) + i \sin(\theta_1 - \theta_2)]$$

Historical Note: Born a French Huguenot, Abraham DeMoivre (1667–1754) moved permanently to England after the revocation of the Edict of Nantes in 1685. There he made his living as a private teacher of mathematics. He made the acquaintance of Newton and Halley and was elected to the Royal Society in 1697. Much of his professonal life was devoted to the development of laws of probability and to the analytic side of trigonometry. It was in the latter field that his work led to the theorem that bears his name.

Critical Thinking Questions

Analysis: Ask students to review the definition of *principal nth root* on page 284. Then ask them how (or whether) the three cube roots of 5 can be expressed using the radical symbol only. They should see that, by the definition, only one of the roots, $\sqrt[3]{5}$, can be so represented since the other two cube roots are not real numbers. (The two complex roots are $\sqrt[3]{5}$ cis 120 and $\sqrt[3]{5}$ cis 240.)

Checkpoint

Express each product, quotient, or power in polar form and rectangular form.

1. $2(\cos 50 + i \sin 50) \cdot 9(\cos 120 + i \sin 120)$
   polar form: $18(\cos 170 + i \sin 170)$
   rectangular form: $-17.7 + 3.1i$

2. $\dfrac{\cos 90 + i \sin 90}{2(\cos 40 + i \sin 40)}$
   polar form: $0.5(\cos 50 + i \sin 50)$
   rectangular form: $0.3 + 0.4i$

3. $[3(\cos 45 + i \sin 45)]^2$
   polar form: $9(\cos 90 + i \sin 90)$
   rectangular form: $0 + 9i$, or $9i$

The quotient rule can be developed as follows.

$$\frac{r_1 \text{ cis } \theta_1}{r_2 \text{ cis } \theta_2} = \frac{r_1 \text{ cis } \theta_1 \cdot \text{cis}(-\theta_2)}{r_2 \text{ cis } \theta_2 \cdot \text{cis}(-\theta_2)} = \frac{r_1}{r_2} \cdot \frac{\text{cis}(\theta_1 - \theta_2)}{\text{cis } 0} = \frac{r_1}{r_2} \cdot \text{cis}(\theta_1 - \theta_2)$$

EXAMPLE 2 Express $\dfrac{18(\cos 40 + i \sin 40)}{6(\cos 100 + i \sin 100)}$ in polar form.

Plan Use $\dfrac{r_1 \text{ cis } \theta_1}{r_2 \text{ cis } \theta_2} = \dfrac{r_1}{r_2} \cdot \text{cis}(\theta_1 - \theta_2)$.

Solution $\dfrac{18 \text{ cis } 40}{6 \text{ cis } 100} = \dfrac{18}{6} \cdot \text{cis}(40 - 100) = 3 \text{ cis}(-60)$, or $3 \text{ cis } 300$

Thus, the quotient is $3(\cos 300 + i \sin 300)$.

Powers of complex numbers in polar form can be found as special cases of multiplication. For example, if $z = r(\cos \theta + i \sin \theta)$, then

$$z^2 = z \cdot z = r^2(\cos 2\theta + i \sin 2\theta),$$
$$z^3 = z \cdot z^2 = r^3(\cos 3\theta + i \sin 3\theta).$$

This pattern can be generalized to give *DeMoivre's Theorem*.

Theorem 19.1 **DeMoivre's Theorem:** For each complex number $z = r(\cos \theta + i \sin \theta)$ and for each positive integer n, $z^n = [r(\cos \theta + i \sin \theta)]^n = r^n(\cos n\theta + i \sin n\theta)$.

EXAMPLE 3 Express $(-\sqrt{2} + i\sqrt{2})^7$ in polar form and in rectangular form.

Plan Change $z = x + yi = -\sqrt{2} + i\sqrt{2}$ to polar form.

Solution $x = -\sqrt{2}$, $y = \sqrt{2}$, $r = \sqrt{x^2 + y^2} = \sqrt{4} = 2$
$\cos \theta = \dfrac{x}{r} = \dfrac{-\sqrt{2}}{2}$, $\sin \theta = \dfrac{y}{r} = \dfrac{\sqrt{2}}{2}$, $\theta = 135$

$z = r \text{ cis } \theta = 2 \text{ cis } 135$

$z^7 = (2 \text{ cis } 135)^7$
$= 2^7 \text{ cis } (7 \cdot 135) = 128 \text{ cis } 945 = 128 \text{ cis}(945 - 2 \cdot 360)$
$= 128 \text{ cis } 225 \leftarrow$ polar form
$128(\cos 225 + i \sin 225) = 128 \left(\dfrac{-\sqrt{2}}{2} + i \cdot \dfrac{-\sqrt{2}}{2} \right)$
$= -64\sqrt{2} - 64i\sqrt{2} \leftarrow$ rectangular form

Additional Example 2

Express in polar form.

$\dfrac{32(\cos 200 + i \sin 200)}{4(\cos 305 + i \sin 305)}$

$8(\cos 255 + i \sin 255)$

Additional Example 3

Express $(\sqrt{3} - i)^5$ in polar form and in rectangular form.

polar form: $32(\cos 210 + i \sin 210)$
rectangular form: $-16\sqrt{3} - 16$

A polar form of $-4\sqrt{3} + 4i$ is 8 cis 150. If multiples of 360 are added to the argument (150), equivalent polar forms can be written as follows.

$$8 \text{ cis } 150, \ 8 \text{ cis } 510, \ 8 \text{ cis } 870, \ 8 \text{ cis } 1{,}230, \ \ldots$$

If the first three of these polar forms and DeMoivre's Theorem are used, the three distinct cube roots of $-4\sqrt{3} + 4i$ can be found.

Let $z = r \text{ cis } \theta$ be a cube root of $-4\sqrt{3} + 4i$. Then $z^3 = -4\sqrt{3} + 4i$ and $(r \text{ cis } \theta)^3 = r^3 \text{ cis } 3\theta = 8 \text{ cis}(150 + k \cdot 360)$, where k is an integer.

Thus, $r^3 = 8$ and $3\theta = 150, 510, 870, \ldots$.
So, $r = 2$ and $\theta = 50, 170, 290, \ldots$.

The three cube roots in the form $r \text{ cis } \theta$ are thus as follows.

$$z_1 = 2 \text{ cis } 50 \qquad z_2 = 2 \text{ cis } 170 \qquad z_3 = 2 \text{ cis } 290$$

The graphs of the three cube roots are on a circle of radius 2 and separated by arcs of 120°. If $3\theta = 1{,}230$, then $\theta = 410$; there is *not* a fourth distinct cube root since 410 and 50 are coterminal angles.

It can be shown that each nonzero complex number has n distinct nth roots, where n is an integer greater than 1.

EXAMPLE 4 Find the 5 fifth roots of $-2\sqrt{2} - 2i\sqrt{2}$. Express in polar form.

Plan First, change $-2\sqrt{2} - 2i\sqrt{2}$ to polar form. Then, find five distinct numbers z such that z^5 equals the number in polar form.

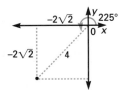

Solution $-2\sqrt{2} - 2i\sqrt{2} = 4 \text{ cis}(225 + k \cdot 360)$ in polar form.

Let $z = r \text{ cis } \theta$ be a fifth root of $4 \text{ cis}(225 + k \cdot 360)$.
$(r \text{ cis } \theta)^5 = r^5 \text{ cis } 5\theta = 4 \text{ cis}(225 + k \cdot 360)$

$$r^5 = 4 \ and \ 5\theta = 225 + k \cdot 360$$
$$r = \sqrt[5]{4} \qquad \theta = 45 + k \cdot 72$$

Let $k = 0, 1, 2, 3, 4$. Then $\theta = 45, 117, 189, 261, 333$.

The 5 fifth roots in polar form are $z_1 = \sqrt[5]{4} \text{ cis } 45$, $z_2 = \sqrt[5]{4} \text{ cis } 117$, $z_3 = \sqrt[5]{4} \text{ cis } 189$, $z_4 = \sqrt[5]{4} \text{ cis } 261$, and $z_5 = \sqrt[5]{4} \text{ cis } 333$.

Closure

Ask student to discuss the reasons why there are some advantages to finding products, powers, and quotients of complex numbers in polar form, as opposed to rectangular form. Polar forms involve sines and cosines. A cosine or sine function for any number cis a to the nth power = cis na, not a^n.

◼◼◼FOLLOW UP

Guided Practice

Classroom Exercises 1–8

Independent Practice

Ⓐ Ex. 1–9, Ⓑ Ex. 10–19, Ⓒ Ex. 20–25

Basic: WE 1–15 odd, Application
Average: WE 3, 7–19 odd, Application
Above Average: WE 3, 7, 9, 11, 19–25 odd, Application

Additional Answers

Written Exercises

8. 8 cis 150; $-4\sqrt{3} + 4i$
9. 4 cis 300; $2 - 2i\sqrt{3}$
10. 32 cis 300; $16 - 16i\sqrt{3}$
11. 64 cis 180; -64
12. 64 cis 90; $64i$
13. 16 cis 0; 16
14. 2 cis 15, 2 cis 135, 2 cis 255
15. 2 cis 20, 2 cis 140, 2 cis 260
16. 2 cis 20, 2 cis 110, 2 cis 200, 2 cis 290
17. 3 cis 30, 3 cis 150, 3 cis 270
18. 2 cis 30, 2 cis 120, 2 cis 210, 2 cis 300
19. $\sqrt[5]{6}$ cis 9, $\sqrt[5]{6}$ cis 81, $\sqrt[5]{6}$ cis 153, $\sqrt[5]{6}$ cis 225, $\sqrt[5]{6}$ cis 297

Additional Example 4

Express the 3 cube roots of 8 in polar form.
2(cos 0 + i sin 0),
2(cos 120 + i sin 120),
2(cos 240 + i sin 240), or
2 cis 0, 2 cis 120, 2 cis 240

3.

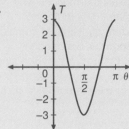

4.

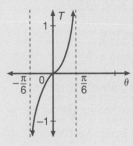

29. Let $A = \theta + \phi$, $B = \theta - \phi$

$\sin A + \sin B$
$= \sin (\theta + \phi) + \sin (\theta - \phi)$
$= 2 \sin \theta \cos \phi$ (Ex. 28)
Also, $A + B = 2\theta$ and $A - B = 2\phi$.
So, $2 \sin \left(\dfrac{A + B}{2}\right) \cos \left(\dfrac{A - B}{2}\right) =$
$2 \sin \theta \cos \phi$.

Classroom Exercises

If $z_1 = \text{cis } 30$, $z_2 = \text{cis } 60$, $z_3 = 2 \text{ cis } 165$, and $z_4 = 3 \text{ cis } 15$, find each of the following in polar form.

1. $z_1 \cdot z_2$ cis 90   **2.** $z_2 \div z_1$ cis 30   **3.** $z_3 \cdot z_4$ 6 cis 180   **4.** z_4^2 9 cis 30

5–8. For each expression of Exercises 1–4 above, find the rectangular form.   **5.** i **6.** $\dfrac{\sqrt{3} + i}{2}$ **7.** -6 **8.** $\dfrac{9\sqrt{3} + 9i}{2}$

Written Exercises

Express each product in polar form and in rectangular form.

1. $3(\cos 40 + i \sin 40) \cdot 5(\cos 120 + i \sin 120)$ 15 cis 160; $-14.1 + 5.13i$

2. $\frac{1}{4}(\cos 100 + i \sin 100) \cdot 12(\cos 150 + i \sin 150)$ 3 cis 250; $-1.03 - 2.82i$

3. $5[\cos(-20) + i \sin(-20)] \cdot 4(\cos 70 + i \sin 70)$ 20 cis 50; $12.9 + 15.3i$

Express each quotient in polar form with $0 \le \theta < 360$.

4. $\dfrac{8(\cos 150 + i \sin 150)}{2(\cos 50 + i \sin 50)}$ 4 cis 100   **5.** $\dfrac{6(\cos 75 + i \sin 75)}{3(\cos 225 + i \sin 225)}$ 2 cis 210

6. $\dfrac{9(\cos 100 + i \sin 100)}{\cos(-40) + i \sin(-40)}$ 9 cis 140   **7.** $\dfrac{\cos(-20) + i \sin(-20)}{4[\cos(-220) + i \sin(-220)]}$ $\frac{1}{4}$ cis 200

Express each power in polar form ($0 \le \theta < 360$) and rectangular form.

8. $[2(\cos 50 + i \sin 50)]^3$ **9.** $[\sqrt{2}(\cos 75 + i \sin 75)]^4$

10. $(1 + i\sqrt{3})^5$ **11.** $(2 - 2i)^4$ **12.** $(-2\sqrt{3} + 2i)^3$ **13.** $(-1 + i)^8$

Express the indicated roots of each complex number in polar form.

14. $8(\cos 45 + i \sin 45)$; 3 cube roots   **15.** $4 + 4i\sqrt{3}$; 3 cube roots

16. $16(\cos 80 + i \sin 80)$; 4 fourth roots   **17.** $27i$; 3 cube roots

18. $-8 + 8i\sqrt{3}$; 4 fourth roots   **19.** $3\sqrt{2} + 3i\sqrt{2}$; 5 fifth roots

The conjugate of $z = x + yi$ is $\overline{z} = x - yi$. Express the conjugate of each complex number in polar form.

20. $\cos 30 + i \sin 30$ cis 330   **21.** $\cos 210 + i \sin 210$ cis 150

22. $4(\cos 45 + i \sin 45)$ 4 cis 315   **23.** $2\sqrt{3}(\cos 300 + i \sin 300)$ $2\sqrt{3}$ cis 60

24. $r(\cos \theta + i \sin \theta)$ r cis$(360 - \theta)$ **25.** $r[\cos(-\theta) + i \sin(-\theta)]$ r cis θ

Mixed Review

Solve.   *2.3, 6.4, 9.5*

1. $|2x - 5| = 9$ **2.** $|2x - 5| < 9$ **3.** $x^2 - 4 < 0$ **4.** $x^2 + 2x = -4$
  $-2, 7$ $\{x \mid -2 < x < 7\}$ $\{x \mid -2 < x < 2\}$ $x = -1 \pm i\sqrt{3}$

748 Chapter 19 Trigonometric Identities and Equations

Enrichment

In the domain of real numbers, there is only one cube root of -1, namely, -1. In the domain of complex numbers, there must be three cube roots of -1. Challenge the students to find them by writing -1 as a complex number in polar form and then using DeMoivre's Theorem. Have them give the answers in both polar and rectangular form.

polar form:
cis 60, cis 180, cis 300

rectangular form:
$\frac{1}{2} + i \cdot \dfrac{\sqrt{3}}{2}$, $-1 + 0i$, $\frac{1}{2} - i \cdot \dfrac{\sqrt{3}}{2}$

Application: *Solar Radiation*

The amount of sunlight received by any given spot on
the earth's surface varies with the time of year. Let H
represent the amount of solar energy intercepted by a
one-meter-square patch when the sun is directly overhead.
Then $H \sin \theta$ represents the energy intercepted by the
same patch when the sun forms an angle of θ degrees with
the horizon, as shown at the right. To understand this,
notice that $H \sin \theta$ is the vertical component of the
sun's light with respect to the earth. That is, it represents the amount of energy beaming down on the earth.

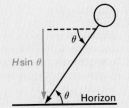

To find θ, consider the tilt of the earth. The plane containing the
earth's equator forms an angle of about $23\frac{1}{2}°$ with the plane containing
the earth's orbit around the sun. So, if $\propto$ is the angle of latitude of
a given point in the northern hemisphere, then the angle of inclination of the sun at high noon on the longest day of the year (around
June 21) equals $90° - (\propto - 23\frac{1}{2}°)$, where $23\frac{1}{2}° \leq \propto \leq 90°$. This is shown
below. Similarly, the angle of inclination at high noon on the
shortest day of the year (around December 22) equals $90° - (\propto + 23\frac{1}{2}°)$,
where $0 \leq \propto \leq 66\frac{1}{2}°$. For example, in a city at $35°$ north latitude on
December 22, $\theta = 90° - (35° + 23\frac{1}{2}°) = 31\frac{1}{2}°$. So $H \sin \theta \approx 0.52H$.

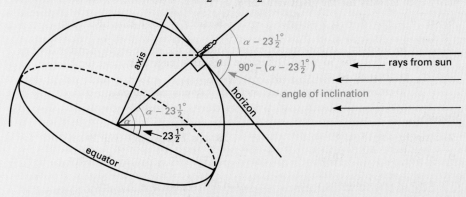

For Exercises 1–4, give the solar energy in terms of H received by a
one-meter-square patch at each northern hemisphere location at the
given latitude for June 21 and for December 22.

1. Houston: $30°$ **2.** Denver: $40°$ **3.** Juneau: $58°$ **4.** The Arctic Circle: $66\frac{1}{2}°$

5. Havana, Cuba sits near $23\frac{1}{2}°$ north latitude. Express the energy
received on December 22 as a percent of that received on June 21.

1. $0.99H$; $0.59H$ 2. $0.96H$; $0.45H$ 3. $0.82H$; $0.15H$ 4. $0.73H$; 0   5. 68%

Application **749**

Mixed Review

1.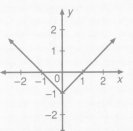

2.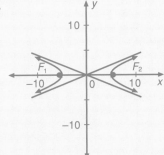

3.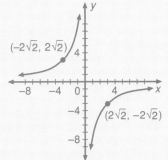

4.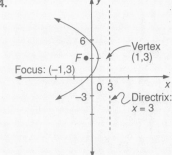

749

Chapter Review

5. $\sec^2\left(\dfrac{4\pi}{3}\right) = (-2)^2 = 4$

$\qquad = (\sqrt{3})^2 + 1$

$\qquad = \tan^2\left(\dfrac{4\pi}{3}\right) + 1$

6. $\sin 240 = -\dfrac{\sqrt{3}}{2}$

$\qquad = \left(-\dfrac{1}{2}\right)(\sqrt{3}) = \cos 240 \tan 240$

17. $\dfrac{\sec^2\theta}{\cot\theta + \tan\theta} = \dfrac{\sec^2\theta}{\dfrac{\cos\theta}{\sin\theta} + \dfrac{\sin\theta}{\cos\theta}}$

$\qquad = \dfrac{\sec^2\theta}{\dfrac{\cos^2\theta + \sin^2\theta}{\sin\theta\cos\theta}}$

$\qquad = \sec^2\theta \cdot \sin\theta \cdot \cos\theta$

$\qquad = \dfrac{\cos\theta}{\cos^2\theta} \cdot \sin\theta = \tan\theta$

18. $\dfrac{\cos 2\theta}{\cos\theta - \sin\theta}$

$\qquad = \dfrac{\cos^2\theta - \sin^2\theta}{\cos\theta - \sin\theta}$

$\qquad = \dfrac{(\cos\theta + \sin\theta)(\cos\theta - \sin\theta)}{\cos\theta - \sin\theta}$

$\qquad = \cos\theta + \sin\theta$

19. $\dfrac{4\pi}{3}, \dfrac{5\pi}{3}$ **20.** π

21. $0, \dfrac{\pi}{2}, \pi$ **22.** $\dfrac{\pi}{4}, \dfrac{3\pi}{4}, \dfrac{5\pi}{4}, \dfrac{7\pi}{4}$

25.

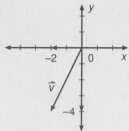

31. Answers will vary. One possible answer:
Use the formula $x = r\cos\theta$ to get the
x-coordinate and $y = r\sin\theta$ to get the
y-coordinate.

750

Chapter 19 Review

Key Terms

argument (p. 743)
cofunction (p. 712)
components of a vector (p. 738)
DeMoivre's Theorem (p. 746)
identity (p. 711)
modulus (p. 743)

norm of a vector (p. 738)
polar axis (p. 742)
polar coordinates (p. 742)
polar form (p. 742)
rectangular coordinates (p. 742)
rectangular form (p. 738)

Key Ideas and Review Exercises

19.1   To express a function in terms of its cofunction, use the cofunction identities on page 712.

To verify an identity, show that it is true for a given value of the variable. Basic trigonometric identities are the Reciprocal Identities, Quotient Identities, and Pythagorean Identities (pages 711–713).

Express each function in terms of its cofunction.

1. $\sin 28$ cos 62   **2.** $\cot 75$ tan 15   **3.** $\cos\dfrac{\pi}{3}$ $\sin\dfrac{\pi}{6}$ **4.** $\csc\dfrac{\pi}{4}$ $\sec\dfrac{\pi}{4}$

Verify each identity for the given value of θ.

5. $\sec^2\theta = \tan^2\theta + 1; \theta = \dfrac{4\pi}{3}$ **6.** $\sin\theta = \cos\theta\tan\theta; \theta = 240$

19.2,   To simplify a trigonometric expression, use the single-angle identities of Les-
19.5,   son 19.1, the Sum and Difference Identities (page 728), and the Double- and
19.6   Half-Angle Identities (page 732–733).

To prove an identity, use the methods summarized on page 718.

Simplify.

7. $2\sin 15\cos 15$ $\frac{1}{2}$ **8.** $\cos 20\cos 70 - \sin 20\sin 70$ 0

9. $\sin 22\frac{1}{2}$ $\dfrac{\sqrt{2 - \sqrt{2}}}{2}$ **10.** $\sin 165$ $\dfrac{\sqrt{6} - \sqrt{2}}{4}$ **11.** $\tan 105$ $-2 - \sqrt{3}$ **12.** $\sin(\pi - \theta)$ $\sin\theta$

Find each of the following if $\sin\theta = -\dfrac{4}{5}$, $\tan\theta < 0$, $\tan\phi = \dfrac{5}{12}$, and $\sin\phi > 0$.

13. $\tan(\theta - \phi)$ $-\frac{63}{16}$ **14.** $\cos\dfrac{\theta}{2}$ $-\dfrac{2\sqrt{5}}{5}$ **15.** $\sin 2\theta$ $-\frac{24}{25}$ **16.** $\cos 2\theta$ $-\frac{7}{25}$

Prove that each equation is an identity.

17. $\dfrac{\sec^2\theta}{\cot\theta + \tan\theta} = \tan\theta$ **18.** $\dfrac{\cos 2\theta}{\cos\theta - \sin\theta} = \cos\theta + \sin\theta$

750   Chapter 19 Review

750

19.3, To solve a trigonometric equation, use the basic identities to rewrite the
19.4, equation in terms of one function, if possible. If both sides must be
19.7 squared, check for extraneous roots.

Solve for θ, $0 \leq \theta < 2\pi$, if θ is a special or quadrantal angle. Otherwise find solutions to the nearest degree, $0 \leq \theta < 360$.

19. $\sin \theta = -\dfrac{\sqrt{3}}{2}$ **20.** $\cos \theta = -1$ **21.** $\sin^2 \theta = \sin \theta$ **22.** $\sin 2\theta = \cot \theta$

23. $2 + 2 \sin \theta = 3 \cos^2 \theta$ 19, 161, 270   **24.** $\sec \theta = \sqrt{3} + \tan \theta$ $\dfrac{11\pi}{6}$

19.8 To find the magnitude of $\vec{v}$, where $\vec{v} = (x,y)$, use $\|\vec{v}\| = \sqrt{x^2 + y^2}$.
To resolve a vector into two perpendicular vectors, see page 738.

25. If $\vec{v} = (-2.0, -4.0)$, draw $\vec{v}$ in standard position. Find the x- and y-components of $\vec{v}$. Then find $\|\vec{v}\|$ correct to one decimal place.

26. For $\theta = 40$ and $\|\vec{v}\| = 300$, find $|x|$ and $|y|$ to the nearest whole number.   230, 193

27. A mass of 820 lb is being pulled up a ramp at an angle that measures 20 to the horizontal. Find, to the nearest pound, the number of pounds required to stop the mass from sliding down the ramp.   280 lb

19.9 To convert the rectangular form $\vec{v} = (x,y)$ to the polar form (r,θ) and conversely, use $r = \sqrt{x^2 + y^2}$, and $\cos \theta = \dfrac{x}{r}$ and $\sin \theta = \dfrac{y}{r}$.
The polar form (r,θ) can be written as $r(\cos \theta + i \sin \theta)$, or r cis θ.

28. Express the vector $(-15, 3.0)$ in polar form (r,θ), with r to two significant digits and θ to the nearest degree.   (15,169°)

29. Express the vector $(14, 60°)$ in rectangular form (x,y). $(7, 7\sqrt{3})$

30. Express the complex number $2\sqrt{3}(\cos 240 + i \sin 240)$ in rectangular form. $-\sqrt{3} - 3i$

31. Write briefly how to express the polar form (r,θ) in the rectangular form (x,y).

19.10 For complex numbers in polar form, refer to the pages indicated below.
(1) To find products or quotients, use the rules on page 745.
(2) To find powers or nth roots, use DeMoivre's Theorem (pages 746–747).

Write each product, quotient, or power in rectangular and polar form, $0 \leq \theta < 360$.   **33.** $0.68 + 1.9i$; 2 cis 70

32. $4(\cos 25 + i \sin 25) \cdot 7(\cos 150 + i \sin 150)$ $-27.9 + 2.4i$; 28 cis 175

33. $\dfrac{10(\cos 120 + i \sin 120)}{5(\cos 50 + i \sin 50)}$ **34.** $[3(\cos 40 + i \sin 40)]^3$ $\dfrac{-27 + 27i\sqrt{3}}{2}$; 27 cis 120

35. Express the fourth roots of $3\sqrt{2} + 3i\sqrt{2}$ in polar form. $\sqrt[4]{6}$ cis 11.25, $\sqrt[4]{6}$ cis 101.25, $\sqrt[4]{6}$ cis 191.25, $\sqrt[4]{6}$ cis 281.25

Chapter 19 Review   **751**

 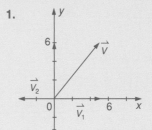
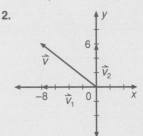
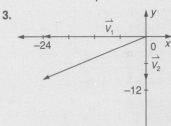
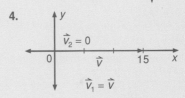
17. $\|(cx, cy)\|$
$= \sqrt{(cx)^2 + (cy)^2}$
$= \sqrt{c^2(x^2 + y^2)}$
$= |c|\sqrt{x^2 + y^2}$
$= |c| \cdot \|(x, y)\|$

20.

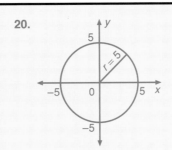

2. $\dfrac{\cos 150}{\sin 150} = \dfrac{\frac{-\sqrt{3}}{2}}{\frac{1}{2}} = \dfrac{-\sqrt{3}}{2} \cdot \dfrac{2}{1}$

$\qquad\qquad = -\sqrt{3} = \cot 150$

13. $\dfrac{1}{\tan^2 \theta} + \cot \theta \tan \theta = \cot^2 \theta + 1$

$\qquad\qquad = \csc^2 \theta$

14. $\dfrac{1}{1 - \sin \theta} + \dfrac{1}{1 + \sin \theta}$

$\quad = \dfrac{1 + \sin \theta + 1 - \sin \theta}{1 - \sin^2 \theta} = \dfrac{2}{\cos^2 \theta}$

21.

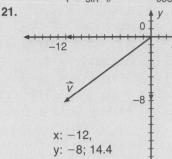

x: −12,
y: −8; 14.4

Additional Answers, page 735

18. $(\cos \theta + \sin \theta)^2$
$\quad = \cos^2 \theta + 2 \sin \theta \cos \theta + \sin^2 \theta$
$\quad = 1 + \sin 2\theta$

19. $\cos^4 \theta - \sin^4 \theta$
$\quad = (\cos^2 \theta + \sin^2 \theta)(\cos^2 \theta - \sin^2 \theta)$
$\quad = \cos^2 \theta - \sin^2 \theta = \cos 2\theta$

20. $\dfrac{1 - \tan^2 \theta}{1 + \tan^2 \theta} = \dfrac{1 - \tan^2 \theta}{\frac{1}{\cos^2 \theta}}$

$\qquad\quad = \cos^2 \theta \,(1 - \tan^2 \theta)$
$\qquad\quad = \cos^2 \theta - \sin^2 \theta$
$\qquad\quad = \cos 2\theta$

21. $\dfrac{\cos \theta \sin 2\theta}{1 + \cos 2\theta}$

$\quad = \dfrac{\cos \theta \,(2 \sin \theta \cos \theta)}{1 + (2 \cos^2 \theta - 1)}$

$\quad = \dfrac{(2 \cos^2 \theta) \sin \theta}{2 \cos^2 \theta} = \sin \theta$

---

1. Express sec 40 in terms of its cofunction. csc 50

2. Verify the identity $\cot \theta = \dfrac{\cos \theta}{\sin \theta}$ for $\theta = 150$.

Simplify.

3. $\dfrac{\sec^2 \theta - 1}{\tan \theta} \tan \theta$ **4.** $\dfrac{\cos \theta}{\sin \theta \cot \theta}$ 1   **5.** $\tan 2\theta \cdot \dfrac{\frac{2}{1 - \tan \theta}}{\frac{1 + \tan \theta}{\tan \theta}}$

6. $\cos^2 15 - \sin^2 15$ $\frac{\sqrt{3}}{2}$ **7.** $\tan 22\frac{1}{2}$ $\sqrt{2} - 1$ **8.** $\cos\!\left(\frac{3\pi}{2} - \theta\right)$ $-\sin \theta$

Find each of the following if $\cos \theta = \frac{5}{13}$, $\tan \theta < 0$, $\sin \phi = \frac{3}{5}$, and $\cos \phi < 0$.

9. $\cos(\theta - \phi)$ $-\frac{56}{65}$ **10.** $\cos \frac{\theta}{2}$ $\frac{3\sqrt{13}}{13}$ **11.** $\tan 2\phi$ $-\frac{24}{7}$ **12.** $\sin 2\theta$ $-\frac{120}{169}$

Prove that each equation is an identity.

13. $\dfrac{1}{\tan^2 \theta} + \cot \theta \tan \theta = \csc^2 \theta$ **14.** $\dfrac{1}{1 - \sin \theta} + \dfrac{1}{1 + \sin \theta} = \dfrac{2}{\cos^2 \theta}$

Solve for θ, $0 \le \theta < 2\pi$, if θ is a special or quadrantal angle. Otherwise, find solutions to the nearest degree, $0 \le \theta < 360$.

15. $\sin \theta + 1 = 4 \sin \theta$ 19, 161   **16.** $\tan \theta - 2 = 4 \tan \theta$ 146, 326

17. $-1 + 5 \cos \theta = 2 \sin^2 \theta$ $\frac{\pi}{3}, \frac{5\pi}{3}$ **18.** $\sin \theta - 1 = \cos \theta$ $\frac{\pi}{2}, \pi$

19. $2 + \cos 2\theta = 1 + \sin \theta$ 51, 129   **20.** $2 \sin^2 \theta - \sqrt{3} \sin \theta = 0$ $0, \frac{\pi}{3}, \frac{2\pi}{3}, \pi$

21. If $\vec{v} = (-12, -8)$, draw $\vec{v}$ in standard position. Find the x- and y-components of $\vec{v}$. Then find $\|\vec{v}\|$, correct to one decimal place.

22. For $\theta = 32$, $\|\vec{v}\| = 250$, find $|x|$ and $|y|$ to the nearest whole number. 212, 132

23. A mass of 650 lb is being pulled up a ramp at an angle that measures 30 to the horizontal. Find, to the nearest pound, the number of pounds required to stop the mass from sliding down the ramp.
325 lb

Write each product, quotient, or power in rectangular and polar form, $0 \le \theta < 360$.

24. $3(\cos 50 + i \sin 50) \cdot 9(\cos 10 + i \sin 10)$ $\frac{27 + 27i\sqrt{3}}{2}$, 27 cis 60

25. $\dfrac{15(\cos 110 + i \sin 110)}{5(\cos 40 + i \sin 40)}$ $1 + 2.8i$, 3 cis 70 **26.** $[2(\cos 30 + i \sin 30)]^3$ $8i$, 8 cis 90

27. Express in polar form the 5 fifth roots of $-4\sqrt{2} - 4i\sqrt{2}$. $\sqrt[5]{8}$ cis 45, $\sqrt[5]{8}$ cis 117,

28. Simplify $\tan(A + 2B)$ if $\cot A = \tan B = 2$. $-\frac{1}{2}$ $\sqrt[5]{8}$ cis 189, $\sqrt[5]{8}$ cis 261, $\sqrt[5]{8}$ cis 333

---

22. $\sin 3\theta$
$= \sin (\theta + 2\theta)$
$= \sin \theta \cos 2\theta + \cos \theta \sin 2\theta$
$= \sin \theta \,(2 \cos^2 \theta - 1) + 2 \cos^2 \theta \sin \theta$
$= 2 \cos^2 \theta \sin \theta + 2 \cos^2 \theta \sin \theta - \sin \theta$
$= 4 \sin \theta \cos^2 \theta - \sin \theta$

23. $\cos 3\theta$
$= \cos (\theta + 2\theta)$
$= \cos \theta \cos 2\theta - \sin \theta \sin 2\theta$
$= \cos \theta \,(2 \cos^2 \theta - 1) -$
$\quad \sin \theta \,(2 \sin \theta \cos \theta)$
$= 2 \cos^3 \theta - \cos \theta - 2 \sin^2 \theta \cos \theta$
$= 2 \cos^3 \theta - \cos \theta -$
$\quad 2(1 - \cos^2 \theta) \cos \theta$
$= 2 \cos^3 \theta - \cos \theta - 2 \cos \theta + 2 \cos^3 \theta$
$= 4 \cos^3 \theta - 3 \cos \theta$

24. $\cos 2\theta = \cos (\theta + \theta)$
$= \cos \theta \cos \theta - \sin \theta \sin \theta$
$= \cos^2 \theta - \sin^2 \theta$

$\cos 2\theta = \cos^2 \theta - \sin^2 \theta$
$= \cos^2 \theta - (1 - \cos^2 \theta)$
$= 2 \cos^2 \theta - 1$

$\cos 2\theta = \cos^2 \theta - \sin^2 \theta$
$= (1 - \sin^2 \theta) - \sin^2 \theta$
$= 1 - 2 \sin^2 \theta$

Choose the *one* best answer to each question or problem.

1. A train traveling 90 mi/h for 1 h covers the same distance as a train traveling 60 mi/h for how many hours? C
(A) $\frac{2}{3}$ (B) 1 (C) $1\frac{1}{2}$
(D) 3 (E) $\frac{1}{3}$

2. In a basket of 80 pears, exactly 4 were rotten. What percent of the pears were good? B
(A) 5 (B) 95 (C) 4
(D) 96 (E) 20

3.
```
P    Q    R    S    T
•────•────•────•────•
```
On the line segment above, if $PQ > QR > RS > ST$, which of the following must be true? C
(A) $PR > QT$ (B) $PQ > QT$
(C) $PS > QT$ (D) $QR > RT$
(E) $PR > QT$

4. If $a = 7b$, then the average of a and b, in terms of b, is ___?___ D
(A) $2b$ (B) $3b$ (C) $3\frac{1}{2}b$
(D) $4b$ (E) $8b$

5. If $\frac{p}{q} = 5$ and $p = 15$, then $2p - q = $ ___?___ B
(A) 10 (B) 27 (C) 33
(D) 25 (E) -5

6. The lengths of the sides of a rectangle are 6 cm and 8 cm. Which of the following equations can be used to find θ, the angle that a diagonal makes with the longer side? C
(A) $\sin \theta = \frac{3}{4}$ (B) $\cos \theta = \frac{3}{4}$
(C) $\tan \theta = \frac{3}{4}$ (D) $\tan \theta = \frac{4}{3}$
(E) $\tan \theta = \frac{3}{5}$

Questions 7–8 refer to the operation represented by * and defined by the equation $x * y = xy + x - y$ for all numbers x and y. For example, $6 * 8 = 6 \cdot 8 + 6 - 8 = 46$.

7. $-5 * \frac{1}{2} = $ ___?___ B
(A) -7 (B) -8
(C) 7 (D) 8 (E) 6

8. If $5 * x = 13$, then $x = $ ___?___ E
(A) 5 (B) 8 (C) $2\frac{3}{5}$
(D) 65 (E) 2

9. $\sin \frac{\pi}{3} \cos \frac{\pi}{6} - \cos \frac{\pi}{3} \sin \frac{\pi}{6} = $ ___?___ B
(A) $-\frac{1}{2}$ (B) $\frac{1}{2}$ (C) $\frac{3}{4}$
(D) 1 (E) $\frac{5}{4}$

10. $\dfrac{2}{\tan \theta + \cot \theta} = $ ___?___ C
(A) $\sin \theta$ (B) $\cos \theta$ (C) $\sin 2\theta$
(D) $\cos 2\theta$ (E) $2 \sin \theta$

11.
In right triangle SPQ above, if $\overline{QR}$ bisects $\angle PQS$, then $y = $ ___?___ C
(A) 15 (B) 30 (C) 60
(D) 5 (E) 65

12. $\dfrac{1}{10^{30}} - \dfrac{1}{10^{29}} = $ ___?___ D
(A) $\dfrac{9}{10^{30}}$ (B) $-\dfrac{9}{10^{29}}$
(C) $\dfrac{1}{10}$ (D) $-\dfrac{9}{10^{30}}$ (E) $\dfrac{9}{10^{29}}$

Additional Answers, page 735

25. $\tan 2\theta = \tan(\theta + \theta)$
$= \dfrac{\tan \theta + \tan \theta}{1 - \tan \theta \tan \theta}$
$= \dfrac{2 \tan \theta}{1 - \tan^2 \theta}$

26. $\sin \frac{\theta}{2}$: Let $\frac{\theta}{2} = \phi$.
Then $\cos 2\phi = 1 - 2 \sin^2 \phi$.
So, $2 \sin^2 \phi = 1 - \cos 2\phi$,
or $\sin^2 \phi = \dfrac{1 - \cos 2\phi}{2}$.
Thus, $\sin \phi = \pm \sqrt{\dfrac{1 - \cos 2\phi}{2}}$
and $\sin \frac{\theta}{2} = \pm \sqrt{\dfrac{1 - \cos \theta}{2}}$

27. $\tan \dfrac{\theta}{2} = \dfrac{\sin \frac{\theta}{2}}{\cos \frac{\theta}{2}}$

$= \dfrac{\pm \sqrt{\dfrac{1 - \cos \theta}{2}}}{\pm \sqrt{\dfrac{1 + \cos \theta}{2}}} \cdot \dfrac{\sqrt{2}}{\sqrt{2}}$

$= \pm \sqrt{\dfrac{1 - \cos \theta}{1 + \cos \theta}}$

$\tan \dfrac{\theta}{2}$
$= \pm \sqrt{\dfrac{1 - \cos \theta}{1 + \cos \theta}} \left(\dfrac{\sqrt{1 + \cos \theta}}{\sqrt{1 + \cos \theta}} \right)$
$= \pm \dfrac{\sqrt{1 - \cos^2 \theta}}{(\sqrt{1 + \cos \theta})^2} = \pm \dfrac{\sqrt{\sin^2 \theta}}{1 + \cos \theta}$
$= \dfrac{\sin \theta}{1 + \cos \theta}$

$\tan \dfrac{\theta}{2} = \pm \sqrt{\dfrac{1 - \cos \theta}{1 + \cos \theta}} \left(\dfrac{\sqrt{1 - \cos \theta}}{\sqrt{1 - \cos \theta}} \right)$
$= \pm \dfrac{(\sqrt{1 - \cos \theta})^2}{\sqrt{1 - \cos^2 \theta}} = \pm \dfrac{1 - \cos \theta}{\sqrt{\sin^2 \theta}}$
$= \dfrac{1 - \cos \theta}{\sin \theta}$

For the last two forms of the identity, the $\pm$ sign is not needed because $\tan \frac{\theta}{2}$ and $\sin \theta$ have the same sign.

Midchapter Review

1. $\cos(90 - 30) = \cos 60 = \frac{1}{2} = \sin 30$

2. $\tan(90 - 120) = \tan(-30)$
$= -\dfrac{\sqrt{3}}{3} = \cot 120$

3. $\tan \theta \sin \theta = \dfrac{\sin \theta}{\cos \theta} \sin \theta$
$= \dfrac{\sin^2 \theta}{\cos \theta} = \dfrac{1 - \cos^2 \theta}{\cos \theta}$
$= \sec \theta - \cos \theta$

For additional standardized test practice, see the SAT/ACT Practice Tests for Cumulative Tests, Chapters 17–19.

8. $\cos \theta = \frac{-12}{13}$; $\tan \theta = \frac{5}{12}$; $\csc \theta = \frac{-13}{5}$;
$\sec \theta = \frac{-13}{12}$; $\cot \theta = \frac{12}{5}$

12. m $\angle A = 47$, m $\angle B = 29$, m $\angle C = 104$

14. $c = 17$, m $\angle B = 53$, m $\angle C = 107$ or $c = 9.6$, m $\angle B = 127$, m $\angle C = 33$

18. $a = 4$, $p = \frac{2\pi}{3}$

19.

θ	0	$\frac{\pi}{8}$	$\frac{\pi}{4}$	$\frac{3\pi}{8}$	$\frac{\pi}{2}$
$f\theta$	-4	$-3\frac{2}{3}$	-4	$-4\frac{1}{3}$	-4

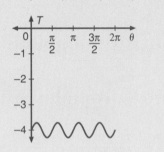

20.

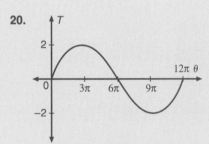

Cumulative Review (Chapters 17–19)

In right triangle ABC with right angle at C, find angle measures to the nearest degree and lengths to two significant digits.

1. $c = 6$, $b = 2$, m $\angle A = $ ___?___ 71 **17.3**

2. $a = 20$, m $\angle A = 72$, $b = $ ___?___ 6.5

Find each trigonometric ratio in simplest radical form. (Exercises 3–6)

3. csc 150 2 **17.7**

4. $\tan(-240)$ $-\sqrt{3}$

5. $\sin\left(\frac{3\pi}{4}\right)$ $\frac{\sqrt{2}}{2}$ **18.5**

6. $\sin\left(\frac{3\pi}{2}\right)$ -1

7. If $\sin \theta = -0.4377$, find all values of θ, $0 \le \theta < 360$, to the nearest degree. 206, 334 **17.8**

8. Find the values of each of the other five trigonometric ratios of θ if $\sin \theta = -\frac{5}{13}$ and $\cos \theta < 0$.

9. Find the area of triangle ABC to two significant digits if $a = 10$, $b = 6$, and m $\angle C = 35$. 17 square units **18.2**

10. Find, to two significant digits, the area of a regular octagon inscribed in a circle of radius 10. 280 square units

Solve each triangle ABC. Find lengths to two significant digits and angle measures to the nearest degree. (Exercises 11–14)

11. $a = 12$, $b = 10$, $c = 21$, m $\angle C = 150$ m $\angle A = 16$, m $\angle B = 14$ **18.1**

12. $a = 6$, $b = 4$, $c = 8$

13. $a = 24$, m $\angle A = 70$, m $\angle B = 50$ $b = 20$, $c = 22$, m $\angle C = 60$ **18.3**

14. $b = 14$, $a = 6$, m $\angle A = 20$ **18.4**

15. Evaluate $1 - \cos\left(-\frac{2\pi}{3}\right)$. $\frac{3}{2}$ **18.5**

16. Evaluate $2 \sin^2 \frac{4\pi}{3} - \cos \frac{4\pi}{3}$. 2

17. Find $\sin(-\theta)$ if $\sin \theta = -0.6754$. 0.6754 **18.7**

18. Give the amplitude and the period of $f(\theta) = 4 \sin 3\theta$. **18.8**

19. Sketch the graph of $f(\theta) = -4 + \frac{1}{3} \sin 4\theta$ for $0 \le \theta < 2\pi$, using a table of values.

Sketch one cycle of each graph without using a table of values.

20. $f(\theta) = 2 \sin \frac{1}{6}\theta$ **18.8**

21. $f(\theta) = \tan 6\theta$ **18.9**

22. Find $\text{Sin}^{-1}\left(-\frac{1}{2}\right)$. -30 **18.10**

23. Find $\cot\left[\text{Cos}^{-1}\left(-\frac{3}{5}\right)\right]$. $-\frac{3}{4}$

Verify each identity for the given value of θ.

24. $\cot \theta = \frac{\cos \theta}{\sin \theta}$, $\theta = 240$ **19.1**

25. $\tan^2 \theta + 1 = \sec^2 \theta$; $\theta = \frac{3\pi}{4}$

26. Express sin 20 in terms of its cofunction. cos 70

Simplify.

27. $\frac{\cos \theta}{1 - \sin^2 \theta}$ sec θ **19.2**

28. $\sin^2 \theta + \cot^2 \theta + \frac{1}{\sec^2 \theta}$ csc² θ

29. $\cos 70 \cos 10 + \sin 70 \sin 10$ **19.5**

30. $\sin\left(\frac{3\pi}{2} - \theta\right)$ $-\cos \theta$

31. $\sin 22.5$ $\frac{\sqrt{2 - \sqrt{2}}}{2}$ **19.6**

32. $\frac{\csc \theta \sin 2\theta}{\cos \theta}$ 2

21.

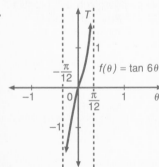

24. $\frac{\cos 240}{\sin 240} = -\frac{1}{2} \cdot \frac{-2}{\sqrt{3}} = \frac{\sqrt{3}}{3} = \cot 240$

In Exercises 33–35, $\cos \theta = \frac{4}{5}$,
$\sin \theta < 0$; $\tan \phi = -\frac{12}{5}$, $\cos \phi < 0$.
Find each of the following.

33. $\sin (\theta + \phi)$ $\frac{63}{65}$ *19.5*

34. $\cos \frac{\theta}{2}$ $-\frac{3\sqrt{10}}{10}$ *19.6*

35. $\sin 2\phi$ $-\frac{120}{169}$

Prove that each equation is an identity.

36. $\cot \theta \sin \theta = \cos \theta$ *19.2*

37. $\sin^2 \theta (1 + \tan^2 \theta) = \tan^2 \theta$

38. $\dfrac{\cot^2 \theta - 1}{1 + \cot^2 \theta} = 1 - 2 \sin^2 \theta$

39. $\dfrac{2 \tan \theta - \sin 2\theta}{2 \sin^2 \theta} = \dfrac{1}{\cot \theta}$ *19.6*

Solve for θ, $0 \le \theta < 2\pi$, if θ is a special or quadrantal angle. Otherwise, find solutions to the nearest degree, $0 \le \theta < 360$.

40. $\sin \theta = 5 \sin \theta + 3$ 229, 311   *19.3*

41. $4 \cos \theta + 3 = 5$ $\frac{\pi}{3}, \frac{5\pi}{3}$

42. $1 + \cos \theta = 2 \cos^2 \theta$ $0, \frac{2\pi}{3}, \frac{4\pi}{3}$

43. $\sin^2 \theta - \sqrt{2} \sin \theta = 0$ $0, \pi$

44. $3 - 2 \sin^2 \theta = 3 \cos \theta$ *19.4*

45. $1 + \sin \theta = \cos \theta$ $0, \frac{3\pi}{2}$

46. $\sin 2\theta + \cos \theta = 0$ *19.6*

47. Draw $\vec{v} = (6, -2)$ in stan-   *19.8*
dard position. Find the x- and
y-components of $\vec{v}$. Then
find $\| \vec{v} \|$.

48. For $\theta = 42$, $\| \vec{v} \| = 320$,
find $|x|$ and $|y|$ to the near-
est whole number. 238, 214

49. Express the vector $(40, 220°)$ *19.9*
in rectangular form, with x
and y to two significant
digits. $(-31, -26)$

50. Change $\frac{1}{2} - \frac{i\sqrt{3}}{2}$ to polar   *19.9*
form, $r(\cos \theta + i \sin \theta)$. 1 cis 300

Express each product, quotient, or
power in rectangular and polar form,
$0 \le \theta < 360$.

51. $4(\cos 20 + i \sin 20) \cdot$ *19.10*
$3(\cos 70 + i \sin 70)$ 12i, 12 cis 90

52. $\dfrac{12(\cos 75 + i \sin 75)}{3(\cos 15 + i \sin 15)}$

53. $[3(\cos 40 + i \sin 40)]^3$

54. Express in polar form the
fourth roots of $-8 + 8i\sqrt{3}$.

55. From the top of a lighthouse   *17.3*
180 ft above sea level, the
angle of depression of a ship
at sea is 32°20′. Find, to
three significant digits, the
distance of the ship from the
base of the lighthouse. 114 ft

56. A 20-ft ladder is placed
against a building so that the
lower end is 5 ft from the
base of the building. What
angle, to the nearest degree,
does the ladder make with
the ground? 76

57. The diagonals of a rectangle   *18.1*
are each 12 cm long, and
they intersect at an angle of
100. Find, to two significant
digits, the lengths of the $l = 9.2$ cm,
sides of the rectangle. $w = 7.7$ cm

58. A navigator sailing north   *18.3*
sights a lighthouse at a bear-
ing of 36. After sailing another 67 km, he finds that
the bearing of the lighthouse
is 122. How far, to two significant digits, is the ship
now from the lighthouse? 39 km

25. $\tan^2 \frac{3\pi}{4} + 1 = (-1)^2 + 1 = 2$
$= \sec^2 \frac{3\pi}{4}$

29. $\frac{1}{2}$

36. $\cot \theta \sin \theta$
$= \dfrac{\cos \theta}{\sin \theta} \cdot \sin \theta = \cos \theta$

37. $\sin^2 \theta (1 + \tan^2 \theta)$
$= \sin^2 \theta \cdot \sec^2 \theta$
$= \tan^2 \theta$

38. $\dfrac{\cot^2 \theta - 1}{1 + \cot^2 \theta} = \dfrac{\cot^2 \theta - 1}{\csc^2 \theta}$
$= \dfrac{\cot^2 \theta}{\csc^2 \theta} - \sin^2 \theta$
$= \cos^2 \theta - \sin^2 \theta$
$= (1 - \sin^2 \theta) - \sin^2 \theta$
$= 1 - 2 \sin^2 \theta$

39. $\dfrac{2 \tan \theta - \sin 2\theta}{2 \sin^2 \theta}$
$= \dfrac{2 \dfrac{\sin \theta}{\cos \theta} - 2 \sin \theta \cos \theta}{2 \sin^2 \theta}$
$= \dfrac{1}{\sin \theta \cos \theta} - \dfrac{\cos \theta}{\sin \theta} = \dfrac{1 - \cos^2 \theta}{\sin \theta \cos \theta}$
$= \dfrac{\sin^2 \theta}{\sin \theta \cos \theta}$ $\tan \theta = \dfrac{1}{\cot \theta}$

44. $0, \frac{\pi}{3}, \frac{5\pi}{3}$

46. $\frac{\pi}{2}, \frac{7\pi}{6}, \frac{3\pi}{2}, \frac{11\pi}{6}$

47.

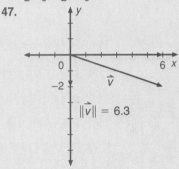

$\| \vec{v} \| = 6.3$

52. $2 + 2i\sqrt{3}$, 4 cis 60

53. $\dfrac{-27 + 27i\sqrt{3}}{2}$, 27 cis 120

54. 2 cis 30, 2 cis 120, 2 cis 210, 2 cis 300

Computer Investigations:
Algebra with Trigonometry

The following is a listing of the computer investigations that are included in *Holt Algebra with Trigonometry*. The related textbook page for each investigation is also indicated. For each investigation, you will need the *Investigating Algebra with the Computer* software.

Title	Appendix Page	Related Text Page
RATIONAL FUNCTIONS	757	233
LOCATING REAL ZEROS	760	361
THE ROLE OF a IN $y = ax^2$	762	391
THE ROLE OF c IN $y = ax^2 + c$	764	392
THE ROLE OF b IN $y = ax^2 + bx + c$	766	392
THE CIRCLE	768	409
THE ELLIPSE	770	413
THE HYPERBOLA	772	418
THE ELLIPSE (center not at the origin)	774	423
THE HYPERBOLA (center not at the origin)	776	423
THE PARABOLA	779	436
SYSTEMS OF FIRST– AND SECOND–DEGREE EQUATIONS	781	443
SYSTEMS OF TWO SECOND–DEGREE EQUATIONS	783	447
EXPONENTIAL FUNCTIONS AND EQUATIONS	785	465
THE LOGARITHMIC FUNCTION	787	469

Investigation: Rational Functions

Program: RATIONAL

Objectives: To use the computer to graph rational functions

To learn that the *x* intercepts of such a graph occur at *x* values which make the numerator equal to 0

To learn that vertical asymptotes occur at *x* values which make the denominator of the function equal to 0

To learn that a rational function often has a horizontal asymptote

1. a. With BASIC loaded into the computer and the disk in drive 1 (or drive A), run the program RATIONAL by typing the appropriate command.

Apple: RUN RATIONAL   IBM: RUN "RATIONAL"

Be sure to press the RETURN key after typing the command.

b. Read the opening messages of the program.

2. Use the program to graph the function $y = \dfrac{1}{x + 2}$.

a. Enter 0 for the degree of the numerator.

b. The numerator has only the DEGREE 0 TERM. Enter 1 for this term.

c. Enter 1 as the degree of the denominator.

d. For the denominator, enter 1 as the coefficient of the DEGREE 1 TERM, and 2 for the DEGREE 0 TERM.

e. Enter −4 for the MINIMUM X VALUE and 2 for the MAXIMUM X VALUE.

f. The computer draws the graph. Does the function have any zeros (*x* intercepts) over this interval?

g. For one *x* value the denominator of the function equals zero and therefore the function is undefined. The computer marks this point with a vertical dashed line (asymptote). What is this *x* value?

h. The graph also has a horizontal asymptote, which is a line the graph approaches but never touches. Which line is the horizontal asymptote?

i. What is the *y* intercept? (HINT: To check your estimate from the graph, substitute 0 for *x* in the equation defining the function.)

j. Answer N (no) to ANOTHER INTERVAL FOR SAME FUNCTION?

k. Answer Y (yes) to ANOTHER FUNCTION?

3. Have the computer graph $y = \dfrac{2x^2 + 3x - 2}{x}$.

   a. Enter 2 as the degree of the numerator. For the coefficients enter 2 as the DEGREE 2 TERM, 3 as the DEGREE 1 TERM, and -2 as the DEGREE 0 TERM.

   b. Enter 1 as the degree of the denominator; enter 1 and then 0 as the coefficients of the denominator.

   c. Enter -3 as the MINIMUM X VALUE and 2 as the MAXIMUM X VALUE.

   d. The computer draws the graph. What x value has a vertical asymptote?

   e. There is one negative zero and one positive zero. List each (to the nearest tenth).

   f. Write the factors of the numerator of the function. (?) (?)

   g. What is the relationship between the factors of the numerator and the zeros of the function?

   h. Answer N to ANOTHER INTERVAL? and Y to ANOTHER FUNCTION?

4. Graph the function, $y = \dfrac{x^2 - x - 2}{x^2 + 2x - 3}$.

   a. The numerator is degree 2 with coefficients 1, -1, and -2.

   b. The denominator is degree 2 with coefficients 1, 2, and -3.

   c. Enter -4 and 3 for the MINIMUM and MAXIMUM x values.

   d. List the zeros (x intercepts) of the function.

   e. Write the factors of the numerator of the function.

   f. What is the relationship between the factors of the numerator and the zeros of the function?

   g. List the two x values where the function is undefined.

   h. Write the factors of the denominator.

   i. What is the relationship between the factors of the denominator and the x values where the function is undefined?

   j. The function has a horizontal asymptote (not drawn by the computer). This may not be clear from the interval graphed so far. So answer Y to ANOTHER INTERVAL FOR SAME FUNCTION?

   k. Enter 1 and 8 for the MINIMUM and MAXIMUM x values.

   l. The graph now appears to get closer and closer to a horizontal line. What is the equation of this line? $y = \underline{\ ?\ }$

   m. Enter N to ANOTHER INTERVAL? and Y to ANOTHER FUNCTION?

5. Consider this function, $y = \dfrac{5x^2 - 11x}{2x^2 - 5x - 7}$.

   a. Before graphing this function, use algebra to make some predictions about the graph. First, factor the numerator.

b. Use the factors to predict the zeros of the function.

c. Factor the denominator.

d. Use the factors to predict the x values where vertical asymptotes will occur.

e. Now have the computer graph the function. The numerator is a polynomial of degree 2 with coefficients 5, -11, and 0. The denominator is a polynomial of degree 2 with coefficients 2, -5, -7.

f. Have the computer graph the function from $x = -4$ to $x = 6$.

g. Which of your predictions were correct?

h. It is not clear whether the function has a horizontal asymptote. Answer Y to ANOTHER INTERVAL FOR SAME FUNCTION?

i. Have the computer graph the function from $x = -8$ to $x = -1$.

j. The leftmost section of the graph may appear to be a straight line. However, it is actually a curved line that approaches a horizontal asymptote. What is the equation of this asymptote? (HINT: What fraction is formed by the first coefficient in the numerator and the first coefficient in the denominator of the function?)

k. Have the computer graph the same function from $x = 4$ to $x = 10$.

l. As x increases in the positive direction, does the graph appear to approach the same horizontal asymptote you listed in step **j**?

m. Enter N to ANOTHER INTERVAL FOR SAME FUNCTION?

n. Enter Y to ANOTHER FUNCTION?

6. a. Before having the computer graph each function below, predict the zeros and the vertical asymptotes. If necessary, factor the numerator and the denominator.

b. Graph the function over the indicated interval.

c. Check whether your predictions of the zeros and asymptotes were correct.

d. List the equation of any horizontal asymptote of the function. (To verify your estimate of a horizontal asymptote, regraph each function to the left and/or the right of the first interval listed below.)

Function	Interval	Function	Interval
i. $y = \dfrac{3}{x - 1}$	-4 to 5	**ii.** $y = \dfrac{4x}{x^2 + x - 12}$	-6 to 6
iii. $y = \dfrac{2x}{x^2 - 4}$	-3 to 3	**iv.** $y = \dfrac{4x - 3}{2x + 1}$	-5 to 5
v. $y = \dfrac{10x^2 + x - 42}{10x}$	-6 to 6	**vi.** $y = \dfrac{2x^2 - 4x - 6}{2x^2 + 3x - 5}$	-5 to 5

Investigation: Locating Real Zeros

Program: SOLVEPOL

Objectives: To use the computer to graph polynomial functions

To read the zeros of a function from its graph, and to estimate non-integral zeros to the nearest tenth.

1. **a.** With BASIC loaded into the computer and the disk in drive 1 (or drive A), run the program SOLVEPOL by typing the appropriate command.

   Apple: RUN SOLVEPOL   IBM: RUN "SOLVEPOL"

   Be sure to press the RETURN key after typing the command.

   b. Read the opening messages of the program.

2. Use the program to graph the function $f(x) = x^3 + 3x^2 - 2x - 6$.

   a. Enter 3 for THE DEGREE OF THE POLYNOMIAL.

   b. Enter the COEFFICIENTS one at a time: 1, 3, -2, -6. Press RETURN after typing each number.

   c. Enter -4 for the MINIMUM X VALUE and 3 for MAXIMUM X VALUE.

3. Use the graph on the screen to answer these questions.

   a. List one integer which is a <u>zero</u> of the function.

   b. Notice the function values for the integers -2 and -1. Is $f(-2)$ positive or negative?

   c. Is $f(-1)$ positive or negative?

   d. The function has a zero between -2 and -1. From the units marked on the x axis, estimate this zero to the nearest tenth.

4. **a.** There are two consecutive positive integers where the function changes from negative to positive. List these two integers.

   b. Estimate the positive zero of the function to the nearest tenth.

   c. Enter N (no) to ANOTHER INTERVAL FOR SAME FUNCTION?

   d. Enter Y (yes) to ANOTHER FUNCTION?

5. Have the computer graph the function $f(x) = 3x^4 + x^3 + 7x^2 + 3x - 6$.

   a. Enter 4 for the DEGREE OF THE POLYNOMIAL.

   b. Enter the COEFFICIENTS: 3, 1, 7, 3, -6.

   c. Enter -2 for MINIMUM X VALUE.

   d. Enter 2 for MAXIMUM X VALUE.

6. Use the graph to answer these questions.
   a. What negative integer is a zero of the function?
   b. List two consecutive integers which have a zero between them. _?_ , _?_

7. Focus on the interval between the two integers you listed in Exercise **6b**.
   a. Enter Y to ANOTHER INTERVAL FOR SAME FUNCTION?
   b. Graph the function between the two integers you listed in Exercise **7b**.
   c. Estimate the zero of the function to the nearest tenth.
   d. Enter N to ANOTHER INTERVAL FOR SAME FUNCTION?
   e. Enter Y to ANOTHER FUNCTION?

8. Have the computer graph $f(x) = 2x^3 - x^2 - 10x$.
   a. Enter 3 for the DEGREE, and the COEFFICIENTS: 2, -1, -10, 0.
   b. Enter -3 for MINIMUM X VALUE and 3 for MAXIMUM X VALUE.
   c. List two integers which are zeros of the function.
   d. List two consecutive integers which are not zeros but which have a zero between them. _?_ and _?_
   e. Enter N to ANOTHER INTERVAL?

9. Enter Y to ANOTHER FUNCTION?

10. a. Have the computer graph each function over the listed interval.
    b. From the graph list the zeros of each function. Estimate non-integer zeros to the nearest tenth.

Function	Interval
i. $f(x) = 3x^3 - x^2 - 6x + 2$	-2 to 2
ii. $f(x) = 8x^3 - 27x^2 + 25x - 6$	-1 to 3
iii. $f(x) = 4x^4 - 5x^2 + 2x + 1$	-2 to 2
NOTE: Enter 0 for the degree 3 coefficient.	
iv. $f(x) = 3x^5 - 8x^2$	-1 to 5
NOTE: Enter 0's for the missing coefficients.	
v. $f(x) = 2x^3 - x^2 + 2x - 1$	-2 to 2

Investigation: The Role of a in $y = ax^2$

Program: PARAB1

Objectives: To use the computer to graph parabolas of the form $y = ax^2$

To predict whether the parabola will open upward or downward

To predict whether the parabola will be more or less "steep" than the graph of $y = x^2$

1. a. With BASIC loaded into the computer and the disk in drive 1 (or drive A), run the program PARAB1 by typing the appropriate command.

   Apple: RUN PARAB1   IBM: RUN "PARAB1"

   Be sure to press the RETURN key after typing the command.

b. Read the opening messages of the program.

2. a. Use the program to draw the graph of $y = x^2$. Type 1 for A and press RETURN.

b. Is the graph a straight line?

c. The graph of $y = x^2$ is a <u>parabola</u>. Does the parabola open upward or downward?

3. Have the computer draw the graph of $y = -x^2$.

a. Answer Y (yes) to ANOTHER GRAPH? and N (no) to CLEAR THE SCREEN?

b. Type -1 for A and press RETURN.

c. Is the graph a parabola?

d. Does the graph open upward or downward?

4. Have the computer draw the graph of $y = 2x^2$.

a. Answer Y to ANOTHER GRAPH? and N to CLEAR THE SCREEN?

b. Enter 2 for A.

c. Does the parabola open upward or downward?

d. Is this parabola "steeper" (more narrow) than the graph of $y = x^2$?

5. Have the computer draw the graph of $y = -.5x^2$.

a. Answer Y to ANOTHER GRAPH? and N to CLEAR THE SCREEN?

b. Enter $-.5$ for A.

c. Does the parabola open upward or downward?

d. Is this parabola "steeper" (more narrow) than the graph of $y = -x^2$?

6. Consider the equation $y = .25x^2$. Before graphing this function, answer these questions.

a. Will the graph be a parabola?

b. Will it open upward or downward?

c. Will the graph be more steep or less steep than the graph of $y = x^2$?

7. Have the computer draw the graph of $y = .25x^2$. Leave the previous graphs on the screen. Were your predictions in Exercise 6 correct?

8. Consider the equation $y = 4x^2$. Before graphing this function, answer these questions.

   a. Will the graph be a parabola?

   b. Will it open upward or downward?

   c. Will the graph be more steep or less steep than the graph of $y = x^2$?

9. Have the computer draw the graph of $y = 4x^2$. Leave the previous graphs on the screen. Were your predictions in Exercise 8 correct?

10. Complete each sentence below.

    a. The graph of $y = ax^2$ will open upward if $a > \underline{\ ?\ }$.

    b. The graph of $y = ax^2$ will open downward if $\underline{\ ?\ }$.

    c. If $|a| > 1$, the graph of $y = ax^2$ will be $\underline{\ ?\ }$ (more steep/less steep) than the graph of $y = x^2$.

    d. If $|a| < 1$, the graph of $y = ax^2$ will be $\underline{\ ?\ }$ (more steep/less steep) than the graph of $y = x^2$.

11. Enter Y to ANOTHER GRAPH? Then enter Y for CLEAR THE SCREEN?

12. **a.** As a reference, graph $y = x^2$. (Enter 1 for A.)

    b. Before graphing each function below, predict whether the graph will open upward or downward and whether it will be more steep or less steep than $y = x^2$.

    c. Have the computer draw the graph. Do not clear the screen.

    d. Determine whether your predictions were correct.

    i. $y = 3x^2$ **ii.** $y = -2x^2$ **iii.** $y = -.8x^2$ **iv.** $y = .2x^2$

Investigation: The Role of c in $y = ax^2 + c$

Program: PARAB2

Objectives: To use the computer to graph parabolas of the form $y = ax^2 + c$
To discover that the vertex of such a parabola is $(0, c)$.

1. **a.** With BASIC loaded into the computer and the disk in drive 1 (or drive A),
run the program PARAB2 by typing the appropriate command.

Apple: RUN PARAB2   IBM: RUN "PARAB2"

Be sure to press the RETURN key after typing the command.

b. Read the opening messages of the program.

2. Use the program to graph $y = x^2$.

a. Type 1 for A and press RETURN.

b. Enter 0 for C (and press RETURN).

c. What is the y intercept?

d. What are the coordinates of the vertex? (? , ?)

3. Have the computer graph $y = x^2 + 3$.

a. Answer Y (yes) to ANOTHER GRAPH?

b. Keep the first graph on the screen. Answer N (no) to CLEAR THE SCREEN?

c. Enter 1 for A and 3 for C.

d. Does the graph of $y = x^2 + 3$ open upward or downward?

e. Where does the graph intersect the y axis?

f. What are the coordinates of the vertex? (? , ?)

g. What was the effect on the graph of adding the 3 behind the x^2?

4. Have the computer graph $y = x^2 - 2$.

a. Answer Y to ANOTHER GRAPH? and N to CLEAR THE SCREEN?

b. Enter 1 for A and -2 for C.

c. Does the graph open upward or downward?

d. What are the coordinates of the vertex of this graph?

e. What was the effect on the graph of the "-2" in the equation?

5. **a.** Before having the computer graph $y = x^2 - 5$, predict the coordinates of the vertex.

b. Have the computer draw the graph. Do not clear the screen. Was your prediction correct?

6. Answer Y to ANOTHER GRAPH? and Y to CLEAR THE SCREEN? Then graph $y = -2x^2$.

   a. Enter -2 for A and 0 for C.

   b. Does the graph open upward or downward?

   c. What is the vertex of this graph?

7. Have the computer graph $y = -2x^2 + 5$. Do not clear the screen.

   a. Enter -2 for A and 5 for C.

   b. Does this graph open upward or downward?

   c. What are the coordinates of the vertex of this graph?

   d. What was the effect on the graph of the "$+5$" in the equation?

8. a. Before having the computer graph $y = -2x^2 - 3$, predict the vertex.

   b. Now graph the equation. Do not clear the screen. Was your prediction correct?

9. *Complete:* The coordinates of the vertex of the graph of $y = ax^2 + c$ ($a \neq 0$) are ($\underline{\ ?\ }$, $\underline{\ ?\ }$).

10. Enter Y to ANOTHER GRAPH? Then enter Y for CLEAR THE SCREEN?

11. a. Predict the coordinates of the vertex of each of the two functions listed in each exercise.

    b. Have the computer graph the two functions on the same axes. (Enter 0 for C.)

    c. Check your predictions of the coordinates of the vertices.

    i. $y = 3x^2$
       $y = 3x^2 + 2$

    ii. $y = -4x^2$
        $y = -4x^2 - 3$

    iii. $y = 2x^2 - 8$
         $y = 2x^2 + 1$

    iv. $y = -.5x^2 + 5$
        $y = -.5x^2 - 4$

Investigation: The Role of b in $y = ax^2 + bx + c$

Program: PARAB3

Objectives: To use the computer to graph parabolas of the form $y = ax^2 + bx + c$

To discover that the coefficient b in the equation can move the parabola right or left and up or down

1. **a.** With BASIC loaded into the computer and the disk in drive 1 (or drive A), run the program PARAB3 by typing the appropriate command.

   Apple: RUN PARAB3   IBM: RUN "PARAB3"

   Be sure to press the RETURN key after typing the command.

   b. Read the opening messages of the program.

2. Use the program to graph the function $y = x^2$.

   a. Enter 1 for A, 0 for B, and 0 for C.

   b. The computer draws the <u>parabola</u> and the <u>axis of symmetry</u>. It also prints the equation of the axis of symmetry and the coordinates of the <u>vertex</u> of the parabola.

   c. What is the equation of the axis of symmetry? $x = $ _?_

   d. What are the coordinates of the vertex? (_?_ , _?_)

3. Have the computer graph $y = x^2 - 3x$.

   a. Answer Y (yes) to ANOTHER GRAPH? and N (no) to CLEAR THE SCREEN?

   b. Enter 1 for A, -3 for B, and 0 for C.

   c. Is the second parabola further to the right than the first one?

   d. What is the axis of symmetry of the second parabola (to the nearest tenth)? $x = $ _?_

   e. List the coordinates of the vertex of the second parabola.

4. Repeat Exercise 3 for $y = x^2 - 5x$. (A = 1, B = -5, C = 0)

   a. Is this parabola further to the right than the previous two?

   b. What is the axis of symmetry?

   c. What are the coordinates of the vertex?

5. Repeat Exercise 3 for $y = x^2 - 6x$. (A = 1, B = -6, C = 0)

   a. Is this parabola further to the right than all the others?

   b. What is the axis of symmetry?

   c. What are the coordinates of the vertex?

6. Clear the screen and graph the following parabolas on the same axes. In each case, list the axis of symmetry and the coordinates of each vertex.

   a. $y = -2x^2$ ($b = 0$, $c = 0$)   **b.** $y = -2x^2 + 4x$ ($c = 0$)

   c. $y = -2x^2 + 7x$ **d.** $y = -2x^2 + 9x$

7. Compare the four graphs of Exercise 6. What happened to the parabolas as the coefficient of the x term became larger and larger?

8. Clear the screen and graph the following parabolas on the same axes. List each axis of symmetry and the coordinates of each vertex.

   a. $y = 3x^2 + 6x$ $(c = 0)$
   b. $y = 3x^2 + 6x - 4$
   c. $y = 3x^2 + 6x - 7$
   d. $y = 3x^2 + 6x + 4$
   e. $y = 3x^2 + 6x + 7$

9. What happened to the parabolas of Exercise 8 as the value of c changed?

10. *Complete:* For the function $y = ax^2 + bx + c$ $(a \neq 0)$, changes in the value of b move the parabola to the __?__ or __?__ as well as __?__ or __?__ .

11. Graph each set of functions on the same screen. List each axis of symmetry and the coordinates of each vertex.

<table>
<tr><td colspan="2">**Set A**</td><td colspan="2">**Set B**</td></tr>
<tr><td>**a.** $y = 4x^2$; $(b = 0, c = 0)$</td><td></td><td>**a.** $y = -.5x^2$; $(b = 0, c = 0)$</td><td></td></tr>
<tr><td>**b.** $y = 4x^2 - 4x$ $(c = 0)$</td><td></td><td>**b.** $y = -.5x^2 - x$ $(c = 0)$</td><td></td></tr>
<tr><td>**c.** $y = 4x^2 - 8x$</td><td></td><td>**c.** $y = -.5x^2 - 2x$</td><td></td></tr>
<tr><td>**d.** $y = 4x^2 - 12x$</td><td></td><td>**d.** $y = -.5x^2 - 4x$</td><td></td></tr>
<tr><td colspan="2">**Set C**</td><td colspan="2">**Set D**</td></tr>
<tr><td>**a.** $y = 5x^2 + 10x$</td><td></td><td>**a.** $y = -x^2 + 4x$</td><td></td></tr>
<tr><td>**b.** $y = 5x^2 + 10x - 4$</td><td></td><td>**b.** $y = -x^2 - 4x$</td><td></td></tr>
<tr><td>**c.** $y = 5x^2 + 10x + 3$</td><td></td><td>**c.** $y = -x^2 + 4x + 5$</td><td></td></tr>
<tr><td>**d.** $y = 5x^2 + 10x + 7$</td><td></td><td>**d.** $y = -x^2 - 4x - 5$</td><td></td></tr>
</table>

Investigation: The Circle

Program: CIRCLE

Objectives: To use the computer to graph circles of the form $(x - h)^2 + (y - k)^2 = r^2$
To discover that the center of the circle is (h, k) and the radius is r

1. **a.** With BASIC loaded into the computer and the disk in drive 1 (or drive A), run the program CIRCLE by typing the appropriate command.

       Apple: RUN CIRCLE   IBM: RUN "CIRCLE"

   Be sure to press the RETURN key after typing the command.

   b. Read the opening messages of the program.

2. Use the program to graph the equation $x^2 + y^2 = 36$. Each equation must be in the form $(X - H)\wedge 2 + (Y - K)\wedge 2 = R\wedge 2$.

   a. The computer asks WHAT IS H? Type 0 and press RETURN.

   b. To WHAT IS K?, enter 0.

   c. To WHAT IS R^2?, enter 36.

   d. The graph should be a <u>circle</u>. It may not be perfectly round on your monitor. The <u>center</u> of the circle is marked by an X. What are the coordinates of the center? (_?_ , _?_)

   e. At what values does the circle intersect the x axis? _?_ and _?_

   f. At what values does the circle intersect the y axis? _?_ and _?_

   g. What is the length of the radius of the circle?

   h. Enter Y to the question ANOTHER GRAPH?

3. Have the computer graph this equation: $(x - 4)^2 + y^2 = 9$

   a. Enter 4 (not -4) for H and 0 for K.

   b. Enter 9 for R^2.

   c. The graph is a circle, although it may be distorted on your screen. What are the coordinates of the center? (_?_ , _?_)

   d. What is the length of the radius?

   e. What effect did the "-4" in the equation have on the graph?

   f. Answer Y to ANOTHER GRAPH?

4. Have the computer graph this equation: $x^2 + (y + 3)^2 = 16$

   a. Enter 0 for H.

   b. Think of $(y + 3)^2$ as $[y - (-3)]^2$. So enter -3 (not 3) for K.

   c. Enter 16 for R^2.

d. What are the coordinates of the center?

e. What is the length of the radius?

f. What effect did the " $+3$ " in the equation have on the graph?

g. Answer Y to ANOTHER GRAPH?

5. Have the computer graph this equation: $(x - 2)^2 + (y + 4)^2 = 25$

a. Enter 2 for H, -4 for K, and 25 for R^2.

b. What is the center?

c. What is the length of the radius?

6. Consider this equation: $(x + 5)^2 + (y - 1)^2 = 49$
Before graphing the equation, predict the center and the length of the radius.

7. Have the computer graph the equation in Exercise 6.

a. Enter -5 for H, 1 for K, and 49 for R^2.

b. Was your prediction of the center correct?

c. Was your prediction of the length of the radius correct?

8. Repeat Exercises **6** and **7** for this equation: $(x - 1)^2 + (y - 3)^2 = 16$
Be sure to enter the correct values for H, K, and R^2.

9. *Complete:* For an equation of the form $(x - h)^2 + (y - k)^2 = r^2 (r > 0)$,

a. the graph is a __?__ ;

b. the center is at the point (__?__ , __?__);

c. the length of the radius is __?__ .

10. Enter Y to ANOTHER GRAPH?

11. a. Before graphing each equation, predict the center and the length of the radius.

b. Have the computer graph each equation.

c. Check whether your predictions were correct.

d. The values of H, K, and R^2 are listed for the first four equations.

    i. $x^2 + y^2 = 64$ (H = 0, K = 0, R^2 = 64)

    ii. $(x - 3)^2 + y^2 = 81$ (H = 3, K = 0, R^2 = 81)

    iii. $x^2 + (y - 4)^2 = 25$ (H = 0, K = 4, R^2 = 25)

    iv. $(x + 3)^2 + (y - 2)^2 = 36$ (H = -3, K = 2, R^2 = 36)

    v. $(x - 5)^2 + (y + 1)^2 = 49$

    vi. $(x - 6)^2 + (y - 7)^2 = 4$

    vii. $(x + 8)^2 + (y + 5)^2 = 9$

Investigation: The Ellipse

Program: ELLIPSE1

Objectives: To use the computer to graph ellipses of the form $\dfrac{x^2}{a^2} + \dfrac{y^2}{b^2} = 1$, connecting each point of the ellipse to the two foci

To discover that such an ellipse is centered at the origin, that the foci are on the x axis if $a > b$ and on the y axis if $b > a$, that the x intercepts are a and $-a$, and that the y intercepts are b and $-b$

1. **a.** With BASIC loaded into the computer and the disk in drive 1 (or drive A), run the program ELLIPSE1 by typing the appropriate command.

    Apple: RUN ELLIPSE 1   IBM: RUN "ELLIPSE1"

    Be sure to press the RETURN key after typing the command.

    b. Read the opening messages of the program.

2. Use the program to graph the equation $\dfrac{x^2}{16} + \dfrac{y^2}{4} = 1$.

    a. To the question WHAT IS A^2? type 16 and press RETURN.

    b. Enter 4 for B^2?

    c. The computer calculates the coordinates of the <u>foci</u> of the <u>ellipse</u>. It marks the foci with X's. What are the coordinates of the foci?

    d. Which axis, x or y, are the foci on?

    e. Read the messages at the bottom of the screen. Then press any key to begin graphing.

    f. As the computer draws the graph, it connects each point of the ellipse to the two foci. At the bottom of the screen it shows the distance between the point and the foci and the sum of the two distances. What do you notice about the sum of the distances?

    g. What are the x intercepts of the graph?   **h.** What are the y intercepts?

    i. Answer Y (yes) to ANOTHER GRAPH?

3. Have the computer graph $\dfrac{x^2}{25} + \dfrac{y^2}{9} = 1$.

    a. Enter 25 for A^2 and 9 for B^2.   **b.** What are the coordinates of the foci?

    c. Which axis are the foci on?

    d. Press any key to begin graphing. What is true about the sum of the distances from each point plotted to the foci?

    e. What are the x intercepts of the ellipse?

    f. What are the y intercepts?

    g. Enter Y to ANOTHER GRAPH?

4. Have the computer graph $\dfrac{x^2}{16} + \dfrac{y^2}{36} = 1$.

   a. Enter 16 for A^2 and 36 for B^2.

   b. What are the coordinates of the foci?

   c. Which axis are the foci on?

   d. Press any key to begin graphing. What is the constant sum of the distances from each point plotted to the foci?

   e. What are the x intercepts of the ellipse?

   f. What are the y intercepts?

   g. Enter Y to ANOTHER GRAPH?

5. Consider the equation $\dfrac{x^2}{9} + \dfrac{y^2}{49} = 1$.

   Before graphing it, predict the following.

   a. Which axis will the foci be on?   **b.** x intercepts?   **c.** y intercepts?

6. Graph the equation.

   a. Did you predict the correct axis for the foci?

   b. What are the coordinates of the foci?

   c. Was your prediction of the x intercepts correct?

   d. Was your prediction of the y intercepts correct?

   e. Answer Y to ANOTHER GRAPH?

7. Repeat Exercises **5** and **6** for the equation $\dfrac{x^2}{100} + \dfrac{y^2}{64} = 1$.

   a. foci on which axis?   **b.** coordinates of foci?

   c. x intercepts?   **d.** y intercepts?

8. What determines which axis contains the foci of the ellipse?

9. *Complete:* For an equation of the form $\dfrac{x^2}{a^2} + \dfrac{y^2}{b^2} = 1$,

   a. the graph is an __?__ ;

   b. the x intercepts are __?__ and __?__ ;

   c. the y intercepts are __?__ and __?__ .

10. Enter Y to ANOTHER GRAPH?

11. a. Predict the x and y intercepts of the graph of each equation below. Also state which axis the foci are on.

   b. Have the computer graph the equation.

   c. Check whether your predictions were correct.

   d. List the coordinates of the foci.

     i. $\dfrac{x^2}{4} + \dfrac{y^2}{25} = 1$ **ii.** $\dfrac{x^2}{81} + \dfrac{y^2}{9} = 1$ **iii.** $\dfrac{x^2}{36} + \dfrac{y^2}{36} = 1$

     iv. In Exercise **iii**, $a^2 = b^2$. What special kind of ellipse results in this case?

Investigation: The Hyperbola

Program: HYPERB1

Objectives: To use the computer to graph hyperbolas centered at the origin

To discover that the foci of the hyperbola are on the x axis and the branches of the hyperbola open horizontally if the x term is first (the positive term) in the equation, and that the foci are on the y axis and the hyperbola opens vertically if the y term is positive

To discover that the x intercepts are a and −a if the x term is positive and that the y intercepts are a and −a if the y term is positive.

1. a. With BASIC loaded into the computer and the disk in drive 1 (or drive A), run the program HYPERB1 by typing the appropriate command.

Apple: RUN HYPERB1   IBM: RUN "HYPERB1"

Be sure to press the RETURN key after typing the command.

b. Read the opening messages of the program.

2. Use the program to graph the equation $\frac{x^2}{9} - \frac{y^2}{4} = 1$.

a. Enter 1 for the form of the equation, then 9 for A^2 and 4 for B^2.

b. The computer marks the <u>foci</u> of the <u>hyperbola</u>. Which axis are they on?

c. What are the coordinates of the foci?

d. Read the messages on the screen. Then press any key to begin graphing.

e. The computer connects each point to the two foci and prints these distances and the difference between them. What is true about the difference of the distances for all the points?

f. Do the branches of the hyperbola open vertically or horizontally?

g. What are the x intercepts?   **h.** Does the hyperbola have any y intercepts?

i. Press a key to see the <u>asymptotes</u>. What are their equations?

j. The asymptotes intersect at the <u>center</u> of the hyperbola. What are the coordinates?

k. Press a key and the computer draws a rectangle with the asymptotes as its diagonals. This rectangle contains the <u>vertices</u> of the hyperbola.

l. Answer Y (yes) to ANOTHER GRAPH?

3. Have the computer graph $\frac{y^2}{16} - \frac{x^2}{4} = 1$.

a. Choose 2 for the form of the equation; enter 16 for A^2 and 4 for B^2.

b. Which axis are the foci on?   **c.** What are the coordinates of the foci?

d. Press any key to begin graphing. What is the difference of the distances to the foci?

e. Do the branches of this hyperbola open vertically or horizontally?

f. Are there any x intercepts?   **g.** What are the y intercepts?

h. List the equations of the asymptotes.   **i.** Enter Y to ANOTHER GRAPH?

4. Have the computer graph $\frac{y^2}{9} - \frac{x^2}{64} = 1$. (Form 2; $A\hat{}2 = 9$; $B\hat{}2 = 64$.)

    a. Which axis contains the foci?   **b.** What are the coordinates of the foci?

    c. Press any key to begin graphing. What is the difference of the distances from any point on the graph to the foci?

    d. Do the branches of this hyperbola open vertically or horizontally?

    e. What are the y intercepts?   **f.** What are the asymptotes' equations?

    g. Enter Y to ANOTHER GRAPH?

5. Have the computer graph $\frac{x^2}{25} - \frac{y^2}{16} = 1$. (Form 1; $A\hat{}2 = 25$; $B\hat{}2 = 16$.)

    a. Which axis contains the foci?   **b.** What are the coordinates of the foci?

    c. Press any key to see the graph. What is the constant difference between the distances to the foci?

    d. Do the branches of this hyperbola open vertically or horizontally?

    e. What are the x intercepts?   **f.** What are the equations of the asymptotes?

    g. Enter Y to ANOTHER GRAPH?

6. Before graphing $\frac{x^2}{81} - \frac{y^2}{25} = 1$, make predictions about the graph.

    a. Which axis will contain the foci?

    b. Will the branches open vertically or horizontally?

    c. x intercepts (if any)?   **d.** y intercepts (if any)?

7. Have the computer graph the equation in Exercise 6. (Form 1; $A\hat{}2 = 81$; $B\hat{}2 = 25$.)

    a. Which of your predictions were correct?

    b. What are the coordinates of the foci?

    c. What are the equations of the asymptotes?

8. Make predictions about the graph of the equation $\frac{y^2}{9} - \frac{x^2}{49} = 1$.

    a. Which axis will contain the foci?

    b. Will the branches open vertically or horizontally?

    c. x intercepts (if any)?   **d.** y intercepts (if any)?

9. Have the computer graph the equation in Exercise 8. (Form 2, $A\hat{}2 = 9$, $B\hat{}2 = 49$)

    a. Were your predictions correct?   **b.** What are the coordinates of the foci?

    c. What are the equations of the asymptotes?

10. *Complete:* For an equation of the form $\frac{x^2}{a^2} - \frac{y^2}{b^2} = 1$,

    a. the graph is a __?__ ;   **b.** the foci are on the __?__ axis;

    c. the branches open __?__ (vertically or horizontally?);

    d. the x intercepts are __?__ and __?__ ;   **e.** there are no __?__ intercepts.

11. *Complete:* For an equation of the form $\dfrac{y^2}{a^2} - \dfrac{x^2}{b^2} = 1$,

    a. the graph is a __?__ ;
            b. the foci are on the __?__ axis;

    c. the branches open __?__ (vertically or horizontally?);

    d. the __?__ intercepts are __?__ and __?__ ;   **e.** there are no __?__ intercepts.

12. Enter Y to ANOTHER GRAPH?

13. **a.** For each equation, predict whether the foci will be on the *x* or *y* axis and whether the branches will open vertically or horizontally. Also predict the *x* and *y* intercepts (if any).

    b. Graph the equation.
            c. Check whether your predictions were correct.

    d. List the coordinates of the foci and the equations of the asymptotes.

    i. $\dfrac{x^2}{36} - \dfrac{y^2}{9} = 1$
        ii. $\dfrac{y^2}{36} - \dfrac{x^2}{36} = 1$
        iii. $\dfrac{y^2}{16} - \dfrac{x^2}{64} = 1$

    iv. $\dfrac{x^2}{121} - \dfrac{y^2}{4} = 1$
        v. $\dfrac{x^2}{144} - \dfrac{y^2}{25} = 1$
        vi. $\dfrac{y^2}{25} - \dfrac{x^2}{49} = 1$

Investigation: The Ellipse (center not at the origin)

Program: ELLIPSE2

Objectives: To use the computer to graph ellipses not centered at the origin

               To discover that the center of the ellipse is (h, k), that the major axis has length $2a$ (where $a > b$), and the minor axis has length $2b$ (where $a > b$).

1. **a.** With BASIC loaded into the computer and the disk in drive 1 (or drive A), run the program ELLIPSE2 by typing the appropriate command.

           Apple: RUN ELLIPSE2   IBM: RUN "ELLIPSE2"

    Be sure to press the RETURN key after typing the command.

    b. Read the opening messages of the program.

2. Use the program to graph the equation $\dfrac{(x - 2)^2}{25} + \dfrac{y^2}{16} = 1$.

    a. Enter 2 (not -2) for H and 0 for K.

    b. Enter 25 for the FIRST DENOMINATOR and 16 for the SECOND DENOMINATOR.

    c. The computer marks the <u>center</u> of the ellipse with an X and then graphs the ellipse. What are the coordinates of the center?

    d. Two <u>vertices</u> of the ellipse are $(-3, 0)$ and $(7, 0)$. What is the distance from each of these points to the center?

    e. What is the relationship between the answer to Exercise **2d** and the first denominator of the equation?

    f. The <u>major axis</u> of the ellipse connects the two vertices listed in Exercise **2d**. What is the length of the major axis?

g. Two other vertices of the ellipse are (2, 4) and (2, −4). What is the distance from each of these points to the center?

h. What is the relationship between the answer to Exercise **2g** and the second denominator of the equation?

i. The <u>minor axis</u> connects the two vertices listed in Exercise **2g**. The major and minor axes of the ellipse intersect at its center. What is the length of the minor axis?

j. Answer Y (yes) to ANOTHER GRAPH?

3. Graph $\dfrac{x^2}{36} + \dfrac{(y-4)^2}{9} = 1$.

a. Enter 0 for H and 4 (not −4) for K.

b. Enter 36 for the FIRST DENOMINATOR and 9 for the SECOND.

c. What are the coordinates of the center of this ellipse?

d. What are the coordinates of the end points of the major and minor axes?

e. What is the length of the major axis?

f. What is the length of the minor axis?

g. Answer Y to ANOTHER GRAPH?

4. Have the computer graph $\dfrac{(x+4)^2}{49} + \dfrac{(y-2)^2}{16} = 1$.

a. Think of $(x+4)^2$ as $[x-(-4)]^2$. So enter −4 for H and 2 for K.

b. Enter 49 for the FIRST DENOMINATOR and 16 for the SECOND.

c. What are the coordinates of the center of the ellipse?

d. What is the length of the major axis?

e. What is the length of the minor axis?

f. Answer Y to ANOTHER GRAPH?

5. Have the computer graph $\dfrac{(x-5)^2}{9} + \dfrac{(y+3)^2}{36} = 1$.

a. Enter 5 for H, −3 for K, and 9 and 36 for the denominators.

b. What are the coordinates of the center?

c. The major (longer) axis is vertical for this ellipse. What is the length of the major axis?

d. What is the length of the minor (shorter) axis?

e. Answer Y to ANOTHER GRAPH?

6. Before graphing $\dfrac{(x+3)^2}{64} + \dfrac{(y+1)^2}{49} = 1$, make predictions about the graph.

a. What are the coordinates of the center?

b. Is the major axis vertical or horizontal?

c. What is the length of the major axis?

d. What is the length of the minor axis?

7. Have the computer graph the equation in Exercise 6. (H = −3, K = −1, A^2 = 64, B^2 = 49) Were your predictions correct?

8. Repeat Exercises **6** and **7** for $\dfrac{(x+7)^2}{4} + \dfrac{(y-3)^2}{25} = 1$. (H = −7, K = 3, A^2 = 4, B^2 = 25)

9. Enter Y to ANOTHER GRAPH?

10. a. Before graphing each ellipse, repeat Exercise **6**.

    b. Graph the ellipse.

    c. Check your predictions.

      i. $\dfrac{x^2}{81} + \dfrac{y^2}{49} = 1$ (H = 0, K = 0, A^2 = 81, B^2 = 49)

      ii. $\dfrac{(x+1)^2}{100} + \dfrac{y^2}{64} = 1$ (H = −1, K = 0, A^2 = 100, B^2 = 64)

      iii. $\dfrac{x^2}{121} + \dfrac{(y-5)^2}{9} = 1$ (H = 0, K = 5, A^2 = 121, B^2 = 9)

      iv. $\dfrac{(x-6)^2}{16} + \dfrac{(y+2)^2}{25} = 1$ (H = 6, K = −2, A^2 = 16, B^2 = 25)

      v. $\dfrac{(x-8)^2}{36} + \dfrac{(y-1)^2}{49} = 1$

Investigation: The Hyperbola (center not at the origin)

Program: HYPERB2

Objectives: To use the computer to graph hyperbolas not centered at the origin

To discover that the center of the hyperbola is (h, k), that the major axis has length $2a$ and the minor axis has length $2b$

1. a. With BASIC loaded into the computer and the disk in drive 1 (or drive A), run the program HYPERB2 by typing the appropriate command.

            Apple: RUN HYPERB2   IBM: RUN "HYPERB2"

     Be sure to press the RETURN key after typing the command.

    b. Read the opening messages of the program.

2. Use the program to graph $\dfrac{(x-3)^2}{16} - \dfrac{y^2}{9} = 1$.

    a. Choose form 1. (Type a 1 and press RETURN.)

    b. Enter 3 (not −3) for H and 0 for K.

    c. Enter 16 for the FIRST DENOMINATOR and 9 for the SECOND DENOMINATOR.

    d. The computer graphs the <u>hyperbola</u>. One <u>vertex</u> is at the point (−1, 0). What are the coordinates of the other vertex?

e. Press any key and the computer draws the asymptotes. They intersect at the center of the hyperbola. What are the coordinates of this point?

f. What is the distance from each vertex to the center of the hyperbola?

g. What is the relationship between your answer in Exercise **2f** and the first denominator of the equation?

h. What are the equations of the asymptotes?

i. Press any key and the computer draws a rectangle between the branches of the hyperbola. Notice that the asymptotes contain the diagonals of this rectangle. Also, the length of the rectangle is the distance between the vertices. What is the width (the shorter dimension) of this rectangle?

j. What is the relationship of the answer to Exercise **2i** and the second denominator of the equation?

k. Answer Y (yes) to ANOTHER GRAPH?

3. Have the computer graph $\dfrac{x^2}{25} - \dfrac{(y-2)^2}{16} = 1$.

a. This is form 1; H = 0; K = 2; A^2 = 25; B^2 = 16.

b. What are the coordinates of the vertices of the hyperbola?

c. Press any key to see the asymptotes. What are the coordinates of the center where they intersect?

d. What is the distance from each vertex to the center?

e. What are the equations of the asymptotes?

f. Press a key to see the dashed rectangle. What is the width (the vertical dimension) of the rectangle?

g. Enter Y for ANOTHER GRAPH?

4. Graph $\dfrac{(y+2)^2}{4} - \dfrac{(x-1)^2}{36} = 1$.

a. Since the y term is first in this equation, choose form 2.

b. Enter 1 for H. Since $(y + 2)^2 = [y - (-2)]^2$, enter -2 for K.

c. Enter 4 for the FIRST DENOMINATOR and 36 for the SECOND.

d. One vertex of the hyperbola is (1, 0). What are the coordinates of the other vertex?

e. Press a key to see the asymptotes and press again to see the rectangle.

f. What are the coordinates of the center?

g. What is the distance from each vertex to the center and what is the relationship of this number to the first denominator of the equation?

h. What is the horizontal dimension of the rectangle and what is the relationship of this number to the second denominator of the equation?

i. What are the equations of the asymptotes?

j. Enter Y for ANOTHER GRAPH?

5. Before graphing $\dfrac{(x+4)^2}{49} - \dfrac{(y+3)^2}{25} = 1$, make predictions about the graph.

    a. Will the branches of the hyperbola open vertically or horizontally?

    b. What are the coordinates of the center?

    c. What are the coordinates of the vertices?

    d. What is the distance from each vertex to the center?

    e. What is the other (vertical) dimension of the rectangle between the branches?

6. Have the computer graph the hyperbola in Exercise 5.

    a. This is form 1; $H = -4$; $K = -3$; $A\hat{}2 = 49$; $B\hat{}2 = 25$.

    b. Were your predictions correct?

    c. What are the equations of the asymptotes?

7. Repeat Exercises **5** and **6** for $\dfrac{(y-4)^2}{9} - \dfrac{(x+5)^2}{64} = 1$.

    (Form 2, $H = -5$, $K = 4$, $A\hat{}2 = 9$, $B\hat{}2 = 64$)

8. Enter Y to ANOTHER GRAPH?

9. Before graphing each hyperbola predict the following.

    a. the coordinates of the center;

    b. whether the hyperbola opens vertically or horizontally;

    c. the coordinates of the two vertices;

    d. the dimensions of the rectangle between the branches.

    e. Graph the hyperbola and check your predictions.

    f. List the equations of the asymptotes.

      i. $\dfrac{x^2}{36} - \dfrac{y^2}{25} = 1$ (Form 1, $H = 0$, $K = 0$, $A\hat{}2 = 36$, $B\hat{}2 = 25$)

      ii. $\dfrac{(x+1)^2}{81} - \dfrac{y^2}{49} = 1$ (Form 1, $H = -1$, $K = 0$, $A\hat{}2 = 81$, $B\hat{}2 = 49$)

    iii. $\dfrac{(y-3)^2}{9} - \dfrac{x^2}{100} = 1$ (Form 2, $H = 0$, $K = 3$, $A\hat{}2 = 9$, $B\hat{}2 = 100$)

    iv. $\dfrac{(y-1)^2}{16} - \dfrac{(x+2)^2}{36} = 1$ (Form 2, $H = -2$, $K = 1$, $A\hat{}2 = 16$, $B\hat{}2 = 36$)

     v. $\dfrac{(x-8)^2}{4} - \dfrac{(y+1)^2}{25} = 1$

    vi. $\dfrac{(y+5)^2}{4} - \dfrac{(x-6)^2}{64} = 1$

Investigation: The Parabola

Program: PARAB4

Objectives: To use a computer to graph parabolas of the form $y = ax^2 + bx + c$ by showing each point of the parabola as equidistant from the focus and the directrix

To discover that the focus is above the directrix when $a > 0$ and below the directrix when $a < 0$

1. **a.** With BASIC loaded into the computer and the disk in drive 1 (or drive A), run the program PARAB4 by typing the appropriate command.

  Apple: RUN PARAB4   **IBM: RUN "PARAB4"**

  Be sure to press the RETURN key after typing the command.

  b. Read the opening messages of the program.

2. Use the program to graph the equation $y = .25x^2 + x - 4$.

  a. Type .25 for A and press the RETURN key.   **b.** Enter 1 for B.   **c.** Enter -4 for C.

  d. The computer draws a horizontal line called the <u>directrix</u>. What is the equation of the directrix? $y = \underline{\quad ? \quad}$

  e. The <u>focus</u> is marked with an X. What are the coordinates of the focus?

  f. Is the focus above or below the directrix?

  g. Press any key to begin graphing.

  h. As it plots the function, the computer connects each point to the focus and to the directrix. The segment connecting each point to the focus is the same length as the segment connecting the point to the directrix.

  i. When the parabola is complete, the computer prints the equation of the <u>axis of symmetry</u>. What is it? $x = \underline{\quad ? \quad}$

  j. What are the coordinates of the <u>vertex</u>?

  k. The vertex and the focus are both on which line?

  l. Is the vertex above the focus or below the focus?

  m. Where is the vertex in relation to the focus and directrix?

  n. Answer Y (yes) to ANOTHER GRAPH?

3. Have the computer graph $y = -.5x^2 + 3x + 2$.

  a. Enter $-.5$ for A, 3 for B, and 2 for C.

  b. What is the equation of the directrix?

  c. What are the coordinates of the focus?

  d. Is the focus above or below the directrix?

  e. Press any key to begin graphing.

  f. What is the equation of the axis of symmetry?

  g. What are the coordinates of the vertex?

h. Is the vertex above or below the focus?

i. Answer Y to ANOTHER GRAPH?

4. Have the computer graph $y = .4x^2 - 2x - 4$.

    a. Enter .4 for A, -2 for B, and -4 for C.

    b. What is the equation of the directrix?

    c. What are the coordinates of the focus?

    d. Is the focus above or below the directrix?

    e. Press any key to start graphing.

    f. What is the equation of the axis of symmetry?

    g. What are the coordinates of the vertex?

    h. Is the vertex above or below the focus?

    i. Enter Y to ANOTHER GRAPH?

5. Make predictions about the graph of $y = -.6x^2 + 4x - 2$.

    a. Will the focus be above or below the directrix?

    b. Will the vertex be above or below the focus?

6. Have the computer graph the function in Exercise 5. $(A = -.6, B = 4, C = -2)$

    a. Which of your predictions were correct?

    b. List this information about the parabola.
        axis of symmetry: $x = \underline{\ ?\ }$ vertex: $(\underline{\ ?\ }, \underline{\ ?\ })$
        directrix: $y = \underline{\ ?\ }$ focus: $(\underline{\ ?\ }, \underline{\ ?\ })$

    c. Enter Y to ANOTHER GRAPH?

7. Repeat Exercises **5** and **6** for the function $y = .2x^2 + 1.5x - 3$.

8. Complete: For the parabola defined by $y = ax^2 + bx + c\ (a \neq 0)$,

    a. when $a > 0$, the focus is $\underline{\ ?\ }$ (above/below) the directrix and the vertex is $\underline{\ ?\ }$ (above/ below) the focus;

    b. when $a < 0$, the focus is $\underline{\ ?\ }$ (above/below) the directrix and the vertex is $\underline{\ ?\ }$ (above/ below) the focus.

9. Enter Y to ANOTHER GRAPH?

10. For each function below,

    a. predict whether the focus will be above or below the directrix;

    b. predict whether the vertex will be above or below the focus;

    c. have the computer graph the function;

    d. check whether your predictions were correct;

    e. list the focus, directrix, axis of symmetry, and vertex.

        i. $y = x^2 - 3$ [Enter 0 for B.]   **ii.** $y = .75x^2 + 4x$ [Enter 0 for C.]

        iii. $y = -.25x^2 - 2x - 1$ **iv.** $y = -.3x^2 + 1.2x + .8$

Investigation: Systems of First– and Second–Degree Equations

Program: QUADLIN

Objectives: To use the computer to graph systems consisting of one linear and one quadratic equation

To discover that such systems may have none, one, or two solutions

1. **a.** With BASIC loaded into the computer and the disk in drive 1 (or drive A), run the program QUADLIN by typing the appropriate command.

Apple: RUN QUADLIN   IBM: RUN "QUADLIN"

Be sure to press the RETURN key after typing the command.

   b. Read the opening messages of the program.

2. Use the program to graph this system, $\begin{cases} x^2 + y^2 = 64 \\ x + y = 8 \end{cases}$

   a. For the first equation choose form 1 and press the RETURN key.

   b. Enter 1 for A (and press RETURN); enter 1 for B; enter 64 for C.

   c. Now enter the coefficients of the linear equation: A = 1, B = 1, C = 8.

   d. What figure is the graph of the quadratic equation?

   e. The straight line for the second equation intersects the first graph in how many points?

   f. List any solutions of the system: (_?_ , _?_), (_?_ , _?_)

   g. Enter Y (yes) to ANOTHER GRAPH?

3. Have the computer graph and solve this system, $\begin{cases} 4x^2 + 25y^2 = 100 \\ x + 2y = 4 \end{cases}$

   a. Enter 1 for the form of the quadratic equation.

   b. Enter the coefficients of the first equation: A = 4, B = 25, C = 100.

   c. Enter the coefficients of the linear equation: A = 1, B = 2, C = 4.

   d. What kind of figure is the graph of the first equation?

   e. How many points do the graphs of the two equations have in common?

   f. List any solutions of the system.

   g. Enter Y to ANOTHER GRAPH?

4. Have the computer graph this system, $\begin{cases} x^2 - 9y^2 = 81 \\ 2x + 3y = -6 \end{cases}$

   a. The first equation is form 1, with A = 1, B = −9, and C = 81.

   b. For the second equation, A = 2, B = 3, C = −6.

   c. What kind of figure is the graph of the quadratic equation?

d. The line intersects the quadratic graph in how many points?

e. List any solutions of the system.

f. Enter Y to ANOTHER GRAPH?

5. Have the computer graph this system, $\begin{cases} 2x^2 + y = 5 \\ y = -3x + 1 \end{cases}$.

  a. The first equation is form 2; $A = 2$; $B = 1$; $C = 5$.

  b. The second equation must be converted to the form required by the computer. So add $3x$ to both sides to obtain $3x + y = 1$.

  c. Enter 3 for A, 1 for B, and 1 for C.

  d. What kind of figure is the graph of the first equation?

  e. How many points of intersection do the two graphs have?

  f. List any solutions of the system.

  g. Enter Y to ANOTHER GRAPH?

6. Have the computer graph this system, $\begin{cases} 4x - 3 = y^2 \\ 9 = 6y - 3x \end{cases}$.

  a. The first equation is of form 3. To match the form $AX + BY^2 = C$, add 3 to both sides and subtract y from both sides. This gives $4x - y^2 = 3$.

  b. For the first equation, enter 4 for A, -1 for B, and 3 for C.

  c. Change the second equation to the form $Ax + By = C$ and write the result.

  d. Enter A, B, and C for the second equation.

  e. What kind of figure is the graph of the first equation?

  f. How many points of intersection do the two graphs have?

  g. List any solutions of the system.

7. Enter Y to ANOTHER GRAPH?

8. a. Use the computer to graph each system. If necessary, convert each equation to the required form before entering the coefficients.

  b. List the type of graph for each quadratic equation.

  c. List any solutions (to the nearest tenth) of each system.

  i. $\begin{cases} x^2 - y^2 = 4 \\ 3x + y = 6 \end{cases}$

  ii. $\begin{cases} 25x^2 + 16y^2 = 400 \\ y = 5 \end{cases}$

  iii. $\begin{cases} x^2 + 3y = 16 \\ 2y = 7 - 3x \end{cases}$

  iv. $\begin{cases} y = -2x^2 + 1 \\ y = x + 5 \end{cases}$

  v. $\begin{cases} y^2 = 5x \\ 10x + 3y = 5 \end{cases}$

  vi. $\begin{cases} 81y^2 - 16x^2 = 1296 \\ 13 = 11x - 7y \end{cases}$

Investigation: Systems of Two Second–Degree Equations

Program: QUADQUAD

Objectives: To use the computer to graph systems consisting of two quadratic equations

To discover that such systems may have none, one, two, three, or four solutions.

1. a. With BASIC loaded into the computer and the disk in drive 1 (or drive A), run the program QUADQUAD by typing the appropriate command.

    Apple: RUN QUADQUAD   IBM: RUN "QUADQUAD"

    Be sure to press the RETURN key after typing the command.

b. Read the opening messages of the program.

2. Use the program to graph and solve this system, $\begin{cases} 9x^2 + 4y^2 = 36 \\ x^2 - y^2 = 9 \end{cases}$.

a. The first equation is form 1. So type a 1 and press the RETURN key.

b. Enter 9 for A, 4 for B, and 36 for C.

c. The second equation is also of form 1. $A = 1$, $B = -1$, and $C = 9$.

d. What type of figure is the graph of the first equation?

e. What type of figure is the graph of the second equation?

f. The two graphs intersect in how many points?

g. Is there any solution to the system?

h. Enter Y (yes) to ANOTHER?

3. Use the computer to solve this system, $\begin{cases} 2x^2 + y = 4 \\ x + y^2 = 5 \end{cases}$.

a. The first equation is form 2. $A = 2$, $B = 1$, and $C = 4$.

b. The second equation is form 3. $A = 1$, $B = 1$, and $C = 5$.

c. What type of figure is the graph of the first equation?

d. What type of figure is the second graph?

e. How many points do the two graphs have in common?

f. List all solutions of the system.

g. Enter Y to ANOTHER?

4. Use the computer to solve this system, $\begin{cases} 25y^2 - x^2 = 100 \\ y = 4x^2 \end{cases}$.

a. The first equation is form 1. However, it must be transformed into the required form, which is $-x^2 + 25y^2 = 100$. So enter -1 for A, 25 for B, and 100 for C.

b. The second equation is form 2. Rearrange it by subtracting $4x^2$ from both sides to give $-4x^2 + y = 0$. So A $= -4$, B $= 1$, and C $= 0$.

c. What type of figure is the graph of each equation?

d. List all solutions to the system.

e. Enter Y to ANOTHER?

5. Use the computer to solve this system, $\begin{cases} y^2 = 8 - 4x \\ x^2 + y^2 = 81 \end{cases}$.

a. The first equation is form 3. Add $4x$ to both sides to obtain $4x + y^2 = 8$. So enter 4 for A, 1 for B, and 8 for C.

b. The second equation is in form 1, with A $= 1$, B $= 1$, and C $= 81$.

c. What type of figure is the graph of each equation?

d. List all solutions of the system.

6. Enter Y to ANOTHER?

7. a. Use the computer to solve each system. If necessary, convert each equation to the required form before entering the coefficients.

b. List the type of figure that forms the graph of each equation.

c. List any solutions of each system.

i. $\begin{cases} x^2 + 10y = 40 \\ 25x^2 - 36y^2 = 900 \end{cases}$

ii. $\begin{cases} x - 2y^2 = 10 \\ x = 3y^2 + 5 \end{cases}$

iii. $\begin{cases} x^2 + y^2 = 64 \\ y^2 = x - 9 \end{cases}$

iv. $\begin{cases} 3y^2 + 2x = 6 \\ y = 5x^2 \end{cases}$

v. $\begin{cases} 4x^2 = 3 - 2y \\ 4 = 3x^2 - 4y \end{cases}$

vi. $\begin{cases} 9x^2 = 25y^2 + 225 \\ 9x^2 + 144y^2 = 1296 \end{cases}$

Investigation: Exponential Functions and Equations

Program: EXPGRAPH

Objectives: To use the computer to graph exponential functions

To learn that the graph of an exponential function lies in quadrants I and II

To discover that the graph of $y = a^x$ rises from left to right if $a > 1$ and falls from left to right if $a < 1$

1. a. With BASIC loaded into the computer and the disk in drive 1 (or drive A), run the program EXPGRAPH by typing the appropriate command.

      Apple: RUN EXPGRAPH   IBM: RUN "EXPGRAPH"

Be sure to press the RETURN key after typing the command.

b. Read the opening messages of the program.

2. Use the program to graph $y = 2^x$.

a. Type 2 for A and press RETURN.

b. Is the graph a straight line?

c. The graph lies in which two quadrants? _?_ and _?_

d. Does the graph have an x intercept?

e. What is the y intercept?

f. As x increases (left to right), does the graph rise or does it fall?

g. Answer Y (yes) to the question ANOTHER GRAPH?

h. Answer N (no) to the question CLEAR THE SCREEN?

3. Have the computer graph $y = 3^x$.

a. Type 3 for A and press RETURN.

b. Is the graph a straight line?

c. The graph lies in which two quadrants?

d. Does the graph have an x intercept?

e. What is the y intercept?

f. For positive values of x, does the second graph rise more steeply than the first graph?

g. What point(s) do the two graphs have in common?

h. Answer Y to the question ANOTHER GRAPH? and N to CLEAR THE SCREEN?

4. Graph $y = .5^x$. That is, enter .5 for A.

a. Does the graph lie in the same quadrants as the first two graphs?

b. What point does the graph have in common with the first two graphs?

c. As *x* increases, does the graph rise or fall?

d. Answer Y to ANOTHER GRAPH? and Y for CLEAR THE SCREEN?

5. Consider the function $y = 4^x$. Before graphing the function, make predictions about it.

a. Will the function rise or fall from left to right?

b. Will the function intersect the *x* axis?

c. Where will the function intersect the *y* axis?

6. a. Graph the function (enter 4 for A). Were your predictions correct?

b. Answer Y to ANOTHER GRAPH? and N to CLEAR THE SCREEN?

7. Consider $y = .25^x$. Before seeing the graph, make predictions about it.

a. Will the function rise or fall from left to right?

b. What point will it have in common with $y = 4^x$?

8. a. Have the computer graph the function. Were your predictions correct?

b. Enter Y to ANOTHER GRAPH? and Y to CLEAR THE SCREEN?

9. Consider these three functions.

A. $y = 5^x$ **B.** $y = .2^x$ **C.** $y = 7^x$

Before having the computer graph them, answer these questions. Then graph all three functions on the same screen and check your predictions.

i. Which functions decrease (fall) from left to right?

ii. Which functions increase (rise) from left to right?

iii. Which function has the steepest slope; that is, which one will rise or fall the fastest?

iv. Which point do all three functions have in common?

Investigation: The Logarithmic Function

Program: LOGGRAPH

Objectives: To use the computer to graph logarithmic functions

To learn that the graph of a logarithmic function is the inverse of the corresponding exponential function

To discover that the graph of a logarithmic function lies in quadrants I and IV

1. **a.** With BASIC loaded into the computer and the disk in drive 1 (or drive A), run the program LOGGRAPH by typing the appropriate command.

   Apple: RUN LOGGRAPH   IBM: RUN "LOGGRAPH"

   Be sure to press the RETURN key after typing the command.

   b. Read the opening messages of the program.

2. Use the program to graph $y = \log_{10} x$. ◄——— This is read "log base 10 of x."

   a. Enter 10 for B and press RETURN.

   b. The computer draws the line $y = x$. Then it graphs $y = \log_{10} x$ to the right of this line and $y = 10^x$ to the left. The two curved graphs are mirror images across this line. Graphs like these are called <u>inverse functions</u>.

   c. In what quadrants does the graph of $y = \log_{10} x$ lie? _?_ and _?_

   d. What is the x intercept of $y = \log_{10} x$?   **e.** Does $y = \log_{10} x$ have a y intercept?

3. **a.** To the question ANOTHER GRAPH?, type Y (yes) and press RETURN.

   b. Have the computer graph $y = \log_2 x$. That is, enter 2 for B.

   c. The computer graphs $y = \log_2 x$ to the right of the line $y = x$ and $y = 2^x$ to the left of the line.

   d. The graph of $y = \log_2 x$ is in which quadrants?

   e. What is the x intercept of $y = \log_2 x$?

   f. Does $y = \log_2 x$ have a y intercept?   **g.** Answer Y to ANOTHER GRAPH?

4. Consider the function $y = \log_4 x$. Before having the computer graph this function and its inverse, write the equation of the inverse function.

   a. In what quadrants will the graph of $y = \log_4 x$ lie?

   b. What will be the x intercept of $y = \log_4 x$?

   c. Graph the function. Were your predictions correct?

   d. Answer Y to ANOTHER GRAPH?

5. Have the computer graph $y = \log_{.5} x$. That is, enter .5 for B.

   a. What is the inverse of $y = \log_{.5} x$?

   b. How does the graph of $y = \log_{.5} x$ differ from the earlier log graphs?

   c. What is the x intercept?   **d.** Answer N to ANOTHER GRAPH?

Formulas From Geometry

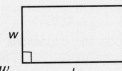

Rectangle

area: $A = \ell w$

perimeter: $P = 2\ell + 2w$

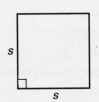

Rectangular Prism

volume: $V = \ell w h$

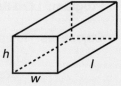

Square

area: $A = s^2$

perimeter: $P = 4s$

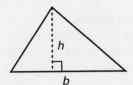

Cube

volume: $V = s^3$

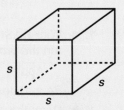

Triangle

area: $A = \frac{1}{2}bh$

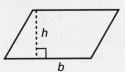

Right Circular Cylinder

volume: $V = \pi r^2 h$

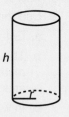

Parallelogram

area: $A = bh$

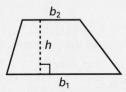

Rectangular Pyramid

volume: $V = \frac{1}{3}\ell w h$

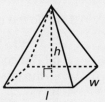

Trapezoid

area: $A = \frac{1}{2}h(b_1 + b_2)$

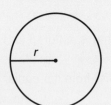

Right Circular Cone

volume: $V = \frac{1}{3}\pi r^2 h$

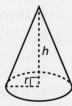

Circle

area: $A = \pi r^2$

circumference: $C = 2\pi r$

Sphere

volume: $V = \frac{4}{3}\pi r^3$

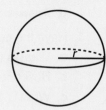

Table of Squares, Cubes, Square and Cube Roots

No.	Squares	Cubes	Square Roots	Cube Roots	No.	Squares	Cubes	Square Roots	Cube Roots
1	1	1	1.000	1.000	51	2,601	132,651	7.141	3.708
2	4	8	1.414	1.260	52	2,704	140,608	7.211	3.733
3	9	27	1.732	1.442	53	2,809	148,877	7.280	3.756
4	16	64	2.000	1.587	54	2,916	157,464	7.348	3.780
5	25	125	2.236	1.710	55	3,025	166,375	7.416	3.803
6	36	216	2.449	1.817	56	3,136	175,616	7.483	3.826
7	49	343	2.646	1.913	57	3,249	185,193	7.550	3.849
8	64	512	2.828	2.000	58	3,364	195,112	7.616	3.871
9	81	729	3.000	2.080	59	3,481	205,379	7.681	3.893
10	100	1,000	3.162	2.154	60	3,600	216,000	7.746	3.915
11	121	1,331	3.317	2.224	61	3,721	226,981	7.810	3.936
12	144	1,728	3.464	2.289	62	3,844	238,328	7.874	3.958
13	169	2,197	3.606	2.351	63	3,969	250,047	7.937	3.979
14	196	2,744	3.742	2.410	64	4,096	262,144	8.000	4.000
15	225	3,375	3.873	2.466	65	4,225	274,625	8.062	4.021
16	256	4,096	4.000	2.520	66	4,356	287,496	8.124	4.041
17	289	4,913	4.123	2.571	67	4,489	300,763	8.185	4.062
18	324	5,832	4.243	2.621	68	4,624	314,432	8.246	4.082
19	361	6,859	4.359	2.668	69	4,761	328,509	8.307	4.102
20	400	8,000	4.472	2.714	70	4,900	343,000	8.367	4.121
21	441	9,261	4.583	2.759	71	5,041	357,911	8.426	4.141
22	484	10,648	4.690	2.802	72	5,184	373,248	8.485	4.160
23	529	12,167	4.796	2.844	73	5,329	389,017	8.544	4.179
24	576	13,824	4.899	2.884	74	5,476	405,224	8.602	4.198
25	625	15,625	5.000	2.924	75	5,625	421,875	8.660	4.217
26	676	17,576	5.099	2.962	76	5,776	438,976	8.718	4.236
27	729	19,683	5.196	3.000	77	5,929	456,533	8.775	4.254
28	784	21,952	5.292	3.037	78	6,084	474,552	8.832	4.273
29	841	24,389	5.385	3.072	79	6,241	493,039	8.888	4.291
30	900	27,000	5.477	3.107	80	6,400	512,000	8.944	4.309
31	961	29,791	5.568	3.141	81	6,561	531,441	9.000	4.327
32	1,024	32,768	5.657	3.175	82	6,724	551,368	9.055	4.344
33	1,089	35,937	5.745	3.208	83	6,889	571,787	9.110	4.362
34	1,156	39,304	5.831	3.240	84	7,056	592,704	9.165	4.380
35	1,225	42,875	5.916	3.271	85	7,225	614,125	9.220	4.397
36	1,296	46,656	6.000	3.302	86	7,396	636,056	9.274	4.414
37	1,369	50,653	6.083	3.332	87	7,569	658,503	9.327	4.431
38	1,444	54,872	6.164	3.362	88	7,744	681,472	9.381	4.448
39	1,521	59,319	6.245	3.391	89	7,921	704,969	9.434	4.465
40	1,600	64,000	6.325	3.420	90	8,100	729,000	9.487	4.481
41	1,681	68,921	6.403	3.448	91	8,281	753,571	9.539	4.498
42	1,764	74,088	6.481	3.476	92	8,464	778,688	9.592	4.514
43	1,849	79,507	6.557	3.503	93	8,649	804,357	9.644	4.531
44	1,936	85,184	6.633	3.530	94	8,836	830,584	9.695	4.547
45	2,025	91,125	6.708	3.557	95	9,025	857,375	9.747	4.563
46	2,116	97,336	6.782	3.583	96	9,216	884,736	9.798	4.579
47	2,209	103,823	6.856	3.609	97	9,409	912,673	9.849	4.595
48	2,304	110,592	6.928	3.634	98	9,604	941,192	9.899	4.610
49	2,401	117,649	7.000	3.659	99	9,801	970,299	9.950	4.626
50	2,500	125,000	7.071	3.684	100	10,000	1,000,000	10.000	4.642

Table of Common Logarithms

N	0	1	2	3	4	5	6	7	8	9
1.0	0000	0043	0086	0128	0170	0212	0253	0294	0334	0374
1.1	0414	0453	0492	0531	0569	0607	0645	0682	0719	0755
1.2	0792	0828	0864	0899	0934	0969	1004	1038	1072	1106
1.3	1139	1173	1206	1239	1271	1303	1335	1367	1399	1430
1.4	1461	1492	1523	1553	1584	1614	1644	1673	1703	1732
1.5	1761	1790	1818	1847	1875	1903	1931	1959	1987	2014
1.6	2041	2068	2095	2122	2148	2175	2201	2227	2253	2279
1.7	2304	2330	2355	2380	2405	2430	2455	2480	2504	2529
1.8	2553	2577	2601	2625	2648	2672	2695	2718	2742	2765
1.9	2788	2810	2833	2856	2878	2900	2923	2945	2967	2989
2.0	3010	3032	3054	3075	3096	3118	3139	3160	3181	3201
2.1	3222	3243	3263	3284	3304	3324	3345	3365	3385	3404
2.2	3424	3444	3464	3483	3502	3522	3541	3560	3579	3598
2.3	3617	3636	3655	3674	3692	3711	3729	3747	3766	3784
2.4	3802	3820	3838	3856	3874	3892	3909	3927	3945	3962
2.5	3979	3997	4014	4031	4048	4065	4082	4099	4116	4133
2.6	4150	4166	4183	4200	4216	4232	4249	4265	4281	4298
2.7	4314	4330	4346	4362	4378	4393	4409	4425	4440	4456
2.8	4472	4487	4502	4518	4533	4548	4564	4579	4594	4609
2.9	4624	4639	4654	4669	4683	4698	4713	4728	4742	4757
3.0	4771	4786	4800	4814	4829	4843	4857	4871	4886	4900
3.1	4914	4928	4942	4955	4969	4983	4997	5011	5024	5038
3.2	5051	5065	5079	5092	5105	5119	5132	5145	5159	5172
3.3	5185	5198	5211	5224	5237	5250	5263	5276	5289	5302
3.4	5315	5328	5340	5353	5366	5378	5391	5403	5416	5428
3.5	5441	5453	5465	5478	5490	5502	5514	5527	5539	5551
3.6	5563	5575	5587	5599	5611	5623	5635	5647	5658	5670
3.7	5682	5694	5705	5717	5729	5740	5752	5763	5775	5786
3.8	5798	5809	5821	5832	5843	5855	5866	5877	5888	5899
3.9	5911	5922	5933	5944	5955	5966	5977	5988	5999	6010
4.0	6021	6031	6042	6053	6064	6075	6085	6096	6107	6117
4.1	6128	6138	6149	6160	6170	6180	6191	6201	6212	6222
4.2	6232	6243	6253	6263	6274	6284	6294	6304	6314	6325
4.3	6335	6345	6355	6365	6375	6385	6395	6405	6415	6425
4.4	6435	6444	6454	6464	6474	6484	6493	6503	6513	6522
4.5	6532	6542	6551	6561	6571	6580	6590	6599	6609	6618
4.6	6628	6637	6646	6656	6665	6675	6684	6693	6702	6712
4.7	6721	6730	6739	6749	6758	6767	6776	6785	6794	6803
4.8	6812	6821	6830	6839	6848	6857	6866	6875	6884	6893
4.9	6902	6911	6920	6928	6937	6946	6955	6964	6972	6981
5.0	6990	6998	7007	7016	7024	7033	7042	7050	7059	7067
5.1	7076	7084	7093	7101	7110	7118	7126	7135	7143	7152
5.2	7160	7168	7177	7185	7193	7202	7210	7218	7226	7235
5.3	7243	7251	7259	7267	7275	7284	7292	7300	7308	7316
5.4	7324	7332	7340	7348	7356	7364	7372	7380	7388	7396

Table of Common Logarithms

N	0	1	2	3	4	5	6	7	8	9
5.5	7404	7412	7419	7427	7435	7443	7451	7459	7466	7474
5.6	7482	7490	7497	7505	7513	7520	7528	7536	7543	7551
5.7	7559	7566	7574	7582	7589	7597	7604	7612	7619	7627
5.8	7634	7642	7649	7657	7664	7672	7679	7686	7694	7701
5.9	7709	7716	7723	7731	7738	7745	7752	7760	7767	7774
6.0	7782	7789	7796	7803	7810	7818	7825	7832	7839	7846
6.1	7853	7860	7868	7875	7882	7889	7896	7903	7910	7917
6.2	7924	7931	7938	7945	7952	7959	7966	7973	7980	7987
6.3	7993	8000	8007	8014	8021	8028	8035	8041	8048	8055
6.4	8062	8069	8075	8082	8089	8096	8102	8109	8116	8122
6.5	8129	8136	8142	8149	8156	8162	8169	8176	8182	8189
6.6	8195	8202	8209	8215	8222	8228	8235	8241	8248	8254
6.7	8261	8267	8274	8280	8287	8293	8299	8306	8312	8319
6.8	8325	8331	8338	8344	8351	8357	8363	8370	8376	8382
6.9	8388	8395	8401	8407	8414	8420	8426	8432	8439	8445
7.0	8451	8457	8463	8470	8476	8482	8488	8494	8500	8506
7.1	8513	8519	8525	8531	8537	8543	8549	8555	8561	8567
7.2	8573	8579	8585	8591	8597	8603	8609	8615	8621	8627
7.3	8633	8639	8645	8651	8657	8663	8669	8675	8681	8686
7.4	8692	8698	8704	8710	8716	8722	8727	8733	8739	8745
7.5	8751	8756	8762	8768	8774	8779	8785	8791	8797	8802
7.6	8808	8814	8820	8825	8831	8837	8842	8848	8854	8859
7.7	8865	8871	8876	8882	8887	8893	8899	8904	8910	8915
7.8	8921	8927	8932	8938	8943	8949	8954	8960	8965	8971
7.9	8976	8982	8987	8993	8998	9004	9009	9015	9020	9025
8.0	9031	9036	9042	9047	9053	9058	9063	9069	9074	9079
8.1	9085	9090	9096	9101	9106	9112	9117	9122	9128	9133
8.2	9138	9143	9149	9154	9159	9165	9170	9175	9180	9186
8.3	9191	9196	9201	9206	9212	9217	9222	9227	9232	9238
8.4	9243	9248	9253	9258	9263	9269	9274	9279	9284	9289
8.5	9294	9299	9304	9309	9315	9320	9325	9330	9335	9340
8.6	9345	9350	9355	9360	9365	9370	9375	9380	9385	9390
8.7	9395	9400	9405	9410	9415	9420	9425	9430	9435	9440
8.8	9445	9450	9455	9460	9465	9469	9474	9479	9484	9489
8.9	9494	9499	9504	9509	9513	9518	9523	9528	9533	9538
9.0	9542	9547	9552	9557	9562	9566	9571	9576	9581	9586
9.1	9590	9595	9600	9605	9609	9614	9619	9624	9628	9633
9.2	9638	9643	9647	9652	9657	9661	9666	9671	9675	9680
9.3	9685	9689	9694	9699	9703	9708	9713	9717	9722	9727
9.4	9731	9736	9741	9745	9750	9754	9759	9763	9768	9773
9.5	9777	9782	9786	9791	9795	9800	9805	9809	9814	9818
9.6	9823	9827	9832	9836	9841	9845	9850	9854	9859	9863
9.7	9868	9872	9877	9881	9886	9890	9894	9899	9903	9908
9.8	9912	9917	9921	9926	9930	9934	9939	9943	9948	9952
9.9	9956	9961	9965	9969	9974	9978	9983	9987	9991	9996

Table of Values of the Trigonometric Functions

Deg.(θ)	Rad.(θ)	Sin θ	Cos θ	Tan θ	Cot θ	Sec θ	Csc θ		
0° 00'	.0000	.0000	1.0000	.0000		1.000		1.5708	90° 00'
10'	.0029	.0029	1.0000	.0029	343.77	1.000	343.8	1.5679	50'
20'	.0058	.0058	1.0000	.0058	171.89	1.000	171.9	1.5650	40'
30'	.0087	.0087	1.0000	.0087	114.59	1.000	114.6	1.5621	30'
40'	.0116	.0116	.9999	.0116	85.940	1.000	85.95	1.5592	20'
50'	.0145	.0145	.9999	.0145	68.750	1.000	68.76	1.5563	10'
1° 00'	.0175	.0175	.9998	.0175	57.290	1.000	57.30	1.5533	89° 00'
10'	.0204	.0204	.9998	.0204	49.104	1.000	49.11	1.5504	50'
20'	.0233	.0233	.9997	.0233	42.964	1.000	42.98	1.5475	40'
30'	.0262	.0262	.9997	.0262	38.188	1.000	38.20	1.5446	30'
40'	.0291	.0291	.9996	.0291	34.368	1.000	34.38	1.5417	20'
50'	.0320	.0320	.9995	.0320	31.242	1.001	31.26	1.5388	10'
2° 00'	.0349	.0349	.9994	.0349	28.636	1.001	28.65	1.5359	88° 00'
10'	.0378	.0378	.9993	.0378	26.432	1.001	26.45	1.5330	50'
20'	.0407	.0407	.9992	.0407	24.542	1.001	24.56	1.5301	40'
30'	.0436	.0436	.9990	.0437	22.904	1.001	22.93	1.5272	30'
40'	.0465	.0465	.9989	.0466	21.470	1.001	21.49	1.5243	20'
50'	.0495	.0494	.9988	.0495	20.206	1.001	20.23	1.5213	10'
3° 00'	.0524	.0523	.9986	.0524	19.081	1.001	19.11	1.5184	87° 00'
10'	.0553	.0552	.9985	.0553	18.075	1.002	18.10	1.5155	50'
20'	.0582	.0581	.9983	.0582	17.169	1.002	17.20	1.5126	40'
30'	.0611	.0610	.9981	.0612	16.350	1.002	16.38	1.5097	30'
40'	.0640	.0640	.9980	.0641	15.605	1.002	15.64	1.5068	20'
50'	.0669	.0669	.9978	.0670	14.924	1.002	14.96	1.5039	10'
4° 00'	.0698	.0698	.9976	.0699	14.301	1.002	14.34	1.5010	86° 00'
10'	.0727	.0727	.9974	.0729	13.727	1.003	13.76	1.4981	50'
20'	.0756	.0756	.9971	.0758	13.197	1.003	13.23	1.4952	40'
30'	.0785	.0785	.9969	.0787	12.706	1.003	12.75	1.4923	30'
40'	.0814	.0814	.9967	.0816	12.251	1.003	12.29	1.4893	20'
50'	.0844	.0843	.9964	.0846	11.826	1.004	11.87	1.4864	10'
5° 00'	.0873	.0872	.9962	.0875	11.430	1.004	11.47	1.4835	85° 00'
10'	.0902	.0901	.9959	.0904	11.059	1.004	11.10	1.4806	50'
20'	.0931	.0929	.9957	.0934	10.712	1.004	10.76	1.4777	40'
30'	.0960	.0958	.9954	.0963	10.385	1.005	10.43	1.4748	30'
40'	.0989	.0987	.9951	.0992	10.078	1.005	10.13	1.4719	20'
50'	.1018	.1016	.9948	.1022	9.7882	1.005	9.839	1.4690	10'
6° 00'	.1047	.1045	.9945	.1051	9.5144	1.006	9.567	1.4661	84° 00'
10'	.1076	.1074	.9942	.1080	9.2553	1.006	9.309	1.4632	50'
20'	.1105	.1103	.9939	.1110	9.0098	1.006	9.065	1.4603	40'
30'	.1134	.1132	.9936	.1139	8.7769	1.006	8.834	1.4573	30'
40'	.1164	.1161	.9932	.1169	8.5555	1.007	8.614	1.4544	20'
50'	.1193	.1190	.9929	.1198	8.3450	1.007	8.405	1.4515	10'
7° 00'	.1222	.1219	.9925	.1228	8.1443	1.008	8.206	1.4486	83° 00'
10'	.1251	.1248	.9922	.1257	7.9530	1.008	8.016	1.4457	50'
20'	.1280	.1276	.9918	.1287	7.7704	1.008	7.834	1.4428	40'
30'	.1309	.1305	.9914	.1317	7.5958	1.009	7.661	1.4399	30'
40'	.1338	.1334	.9911	.1346	7.4287	1.009	7.496	1.4370	20'
50'	.1367	.1363	.9907	.1376	7.2687	1.009	7.337	1.4341	10'
8° 00'	.1396	.1392	.9903	.1405	7.1154	1.010	7.185	1.4312	82° 00'
10'	.1425	.1421	.9899	.1435	6.9682	1.010	7.040	1.4283	50'
20'	.1454	.1449	.9894	.1465	6.8269	1.011	6.900	1.4254	40'
30'	.1484	.1478	.9890	.1495	6.6912	1.011	6.765	1.4224	30'
40'	.1513	.1507	.9886	.1524	6.5606	1.012	6.636	1.4195	20'
50'	.1542	.1536	.9881	.1554	6.4348	1.012	6.512	1.4166	10'
9° 00'	.1571	.1564	.9877	.1584	6.3138	1.012	6.392	1.4137	81° 00'
		Cos θ	Sin θ	Cot θ	Tan θ	Csc θ	Sec θ	Rad.(θ)	Deg.(θ)

Tables

Table of Values of the Trigonometric Functions

Deg.(θ)	Rad.(θ)	Sin θ	Cos θ	Tan θ	Cot θ	Sec θ	Csc θ		
9° 00′	.1571	.1564	.9877	.1584	6.3138	1.012	6.392	1.4137	81° 00′
10′	.1600	.1593	.9872	.1614	6.1970	1.013	6.277	1.4108	50′
20′	.1629	.1622	.9868	.1644	6.0844	1.013	6.166	1.4079	40′
30′	.1658	.1650	.9863	.1673	5.9758	1.014	6.059	1.4050	30′
40′	.1687	.1679	.9858	.1703	5.8708	1.014	5.955	1.4021	20′
50′	.1716	.1708	.9853	.1733	5.7694	1.015	5.855	1.3992	10′
10° 00′	.1745	.1736	.9848	.1763	5.6713	1.015	5.759	1.3963	80° 00′
10′	.1774	.1765	.9843	.1793	5.5764	1.016	5.665	1.3934	50′
20′	.1804	.1794	.9838	.1823	5.4845	1.016	5.575	1.3904	40′
30′	.1833	.1822	.9833	.1853	5.3955	1.017	5.487	1.3875	30′
40′	.1862	.1851	.9827	.1883	5.3093	1.018	5.403	1.3846	20′
50′	.1891	.1880	.9822	.1914	5.2257	1.018	5.320	1.3817	10′
11° 00′	.1920	.1908	.9816	.1944	5.1446	1.019	5.241	1.3788	79° 00′
10′	.1949	.1937	.9811	.1974	5.0658	1.019	5.164	1.3759	50′
20′	.1978	.1965	.9805	.2004	4.9894	1.020	5.089	1.3730	40′
30′	.2007	.1994	.9799	.2035	4.9152	1.020	5.016	1.3701	30′
40′	.2036	.2022	.9793	.2065	4.8430	1.021	4.945	1.3672	20′
50′	.2065	.2051	.9787	.2095	4.7729	1.022	4.876	1.3643	10′
12° 00′	.2094	.2079	.9781	.2126	4.7046	1.022	4.810	1.3614	78° 00′
10′	.2123	.2108	.9775	.2156	4.6382	1.023	4.745	1.3584	50′
20′	.2153	.2136	.9769	.2186	4.5736	1.024	4.682	1.3555	40′
30′	.2182	.2164	.9763	.2217	4.5107	1.024	4.620	1.3526	30′
40′	.2211	.2193	.9757	.2247	4.4494	1.025	4.560	1.3497	20′
50′	.2240	.2221	.9750	.2278	4.3897	1.026	4.502	1.3468	10′
13° 00′	.2269	.2250	.9744	.2309	4.3315	1.026	4.445	1.3439	77° 00′
10′	.2298	.2278	.9737	.2339	4.2747	1.027	4.390	1.3410	50′
20′	.2327	.2306	.9730	.2370	4.2193	1.028	4.336	1.3381	40′
30′	.2356	.2334	.9724	.2401	4.1653	1.028	4.284	1.3352	30′
40′	.2385	.2363	.9717	.2432	4.1126	1.029	4.232	1.3323	20′
50′	.2414	.2391	.9710	.2462	4.0611	1.030	4.182	1.3294	10′
14° 00′	.2443	.2419	.9703	.2493	4.0108	1.031	4.134	1.3265	76° 00′
10′	.2473	.2447	.9696	.2524	3.9617	1.031	4.086	1.3235	50′
20′	.2502	.2476	.9689	.2555	3.9136	1.032	4.039	1.3206	40′
30′	.2531	.2504	.9681	.2586	3.8667	1.033	3.994	1.3177	30′
40′	.2560	.2532	.9674	.2617	3.8208	1.034	3.950	1.3148	20′
50′	.2589	.2560	.9667	.2648	3.7760	1.034	3.906	1.3119	10′
15° 00′	.2618	.2588	.9659	.2679	3.7321	1.035	3.864	1.3090	75° 00′
10′	.2647	.2616	.9652	.2711	3.6891	1.036	3.822	1.3061	50′
20′	.2676	.2644	.9644	.2742	3.6470	1.037	3.782	1.3032	40′
30′	.2705	.2672	.9636	.2773	3.6059	1.038	3.742	1.3003	30′
40′	.2734	.2700	.9628	.2805	3.5656	1.039	3.703	1.2974	20′
50′	.2763	.2728	.9621	.2836	3.5261	1.039	3.665	1.2945	10′
16° 00′	.2793	.2756	.9613	.2867	3.4874	1.040	3.628	1.2915	74° 00′
10′	.2822	.2784	.9605	.2899	3.4495	1.041	3.592	1.2886	50′
20′	.2851	.2812	.9596	.2931	3.4124	1.042	3.556	1.2857	40′
30′	.2880	.2840	.9588	.2962	3.3759	1.043	3.521	1.2828	30′
40′	.2909	.2868	.9580	.2994	3.3402	1.044	3.487	1.2799	20′
50′	.2938	.2896	.9572	.3026	3.3052	1.045	3.453	1.2770	10′
17° 00′	.2967	.2924	.9563	.3057	3.2709	1.046	3.420	1.2741	73° 00′
10′	.2996	.2952	.9555	.3089	3.2371	1.047	3.388	1.2712	50′
20′	.3025	.2979	.9546	.3121	3.2041	1.048	3.356	1.2683	40′
30′	.3054	.3007	.9537	.3153	3.1716	1.049	3.326	1.2654	30′
40′	.3083	.3035	.9528	.3185	3.1397	1.049	3.295	1.2625	20′
50′	.3113	.3062	.9520	.3217	3.1084	1.050	3.265	1.2595	10′
18° 00′	.3142	.3090	.9511	.3249	3.0777	1.051	3.236	1.2566	72° 00′
		Cos θ	Sin θ	Cot θ	Tan θ	Csc θ	Sec θ	Rad.(θ)	Deg.(θ)

Tables

Table of Values of the Trigonometric Functions

Deg.(θ)	Rad.(θ)	Sin θ	Cos θ	Tan θ	Cot θ	Sec θ	Csc θ		
18° 00′	.3142	.3090	.9511	.3249	3.0777	1.051	3.236	1.2566	72° 00′
10′	.3171	.3118	.9502	.3281	3.0475	1.052	3.207	1.2537	50′
20′	.3200	.3145	.9492	.3314	3.0178	1.053	3.179	1.2508	40′
30′	.3229	.3173	.9483	.3346	2.9887	1.054	3.152	1.2479	30′
40′	.3258	.3201	.9474	.3378	2.9600	1.056	3.124	1.2450	20′
50′	.3287	.3228	.9465	.3411	2.9319	1.057	3.098	1.2421	10′
19° 00′	.3316	.3256	.9455	.3443	2.9042	1.058	3.072	1.2392	71° 00′
10′	.3345	.3283	.9446	.3476	2.8770	1.059	3.046	1.2363	50′
20′	.3374	.3311	.9436	.3508	2.8502	1.060	3.021	1.2334	40′
30′	.3403	.3338	.9426	.3541	2.8239	1.061	2.996	1.2305	30′
40′	.3432	.3365	.9417	.3574	2.7980	1.062	2.971	1.2275	20′
50′	.3462	.3393	.9407	.3607	2.7725	1.063	2.947	1.2246	10′
20° 00′	.3491	.3420	.9397	.3640	2.7475	1.064	2.924	1.2217	70° 00′
10′	.3520	.3448	.9387	.3673	2.7228	1.065	2.901	1.2188	50′
20′	.3549	.3475	.9377	.3706	2.6985	1.066	2.878	1.2159	40′
30′	.3578	.3502	.9367	.3739	2.6746	1.068	2.855	1.2130	30′
40′	.3607	.3529	.9356	.3772	2.6511	1.069	2.833	1.2101	20′
50′	.3636	.3557	.9346	.3805	2.6279	1.070	2.812	1.2072	10′
21° 00′	.3665	.3584	.9336	.3839	2.6051	1.071	2.790	1.2043	69° 00′
10′	.3694	.3611	.9325	.3872	2.5826	1.072	2.769	1.2014	50′
20′	.3723	.3638	.9315	.3906	2.5605	1.074	2.749	1.1985	40′
30′	.3752	.3665	.9304	.3939	2.5386	1.075	2.729	1.1956	30′
40′	.3782	.3692	.9293	.3973	2.5172	1.076	2.709	1.1926	20′
50′	.3811	.3719	.9283	.4006	2.4960	1.077	2.689	1.1897	10′
22° 00′	.3840	.3746	.9272	.4040	2.4751	1.079	2.669	1.1868	68° 00′
10′	.3869	.3773	.9261	.4074	2.4545	1.080	2.650	1.1839	50′
20′	.3898	.3800	.9250	.4108	2.4342	1.081	2.632	1.1810	40′
30′	.3927	.3827	.9239	.4142	2.4142	1.082	2.613	1.1781	30′
40′	.3956	.3854	.9228	.4176	2.3945	1.084	2.595	1.1752	20′
50′	.3985	.3881	.9216	.4210	2.3750	1.085	2.577	1.1723	10′
23° 00′	.4014	.3907	.9205	.4245	2.3559	1.086	2.559	1.1694	67° 00′
10′	.4043	.3934	.9194	.4279	2.3369	1.088	2.542	1.1665	50′
20′	.4072	.3961	.9182	.4314	2.3183	1.089	2.525	1.1636	40′
30′	.4102	.3987	.9171	.4348	2.2998	1.090	2.508	1.1606	30′
40′	.4131	.4014	.9159	.4383	2.2817	1.092	2.491	1.1577	20′
50′	.4160	.4041	.9147	.4417	2.2637	1.093	2.475	1.1548	10′
24° 00′	.4189	.4067	.9135	.4452	2.2460	1.095	2.459	1.1519	66° 00′
10′	.4218	.4094	.9124	.4487	2.2286	1.096	2.443	1.1490	50′
20′	.4247	.4120	.9112	.4522	2.2113	1.097	2.427	1.1461	40′
30′	.4276	.4147	.9100	.4557	2.1943	1.099	2.411	1.1432	30′
40′	.4305	.4173	.9088	.4592	2.1775	1.100	2.396	1.1403	20′
50′	.4334	.4200	.9075	.4628	2.1609	1.102	2.381	1.1374	10′
25° 00′	.4363	.4226	.9063	.4663	2.1445	1.103	2.366	1.1345	65° 00′
10′	.4392	.4253	.9051	.4699	2.1283	1.105	2.352	1.1316	50′
20′	.4422	.4279	.9038	.4734	2.1123	1.106	2.337	1.1286	40′
30′	.4451	.4305	.9026	.4770	2.0965	1.108	2.323	1.1257	30′
40′	.4480	.4331	.9013	.4806	2.0809	1.109	2.309	1.1228	20′
50′	.4509	.4358	.9001	.4841	2.0655	1.111	2.295	1.1199	10′
26° 00′	.4538	.4384	.8988	.4877	2.0503	1.113	2.281	1.1170	64° 00′
10′	.4567	.4410	.8975	.4913	2.0353	1.114	2.268	1.1141	50′
20′	.4596	.4436	.8962	.4950	2.0204	1.116	2.254	1.1112	40′
30′	.4625	.4462	.8949	.4986	2.0057	1.117	2.241	1.1083	30′
40′	.4654	.4488	.8936	.5022	1.9912	1.119	2.228	1.1054	20′
50′	.4683	.4514	.8923	.5059	1.9768	1.121	2.215	1.1025	10′
27° 00′	.4712	.4540	.8910	.5095	1.9626	1.122	2.203	1.0996	63° 00′
		Cos θ	Sin θ	Cot θ	Tan θ	Csc θ	Sec θ	Rad.(θ)	Deg.(θ)

Tables

794

Table of Values of the Trigonometric Functions

Deg.(θ)	Rad.(θ)	Sin θ	Cos θ	Tan θ	Cot θ	Sec θ	Csc θ		
27° 00′	.4712	.4540	.8910	.5095	1.9626	1.122	2.203	1.0996	63° 00′
10′	.4741	.4566	.8897	.5132	1.9486	1.124	2.190	1.0966	50′
20′	.4771	.4592	.8884	.5169	1.9347	1.126	2.178	1.0937	40′
30′	.4800	.4617	.8870	.5206	1.9210	1.127	2.166	1.0908	30′
40′	.4829	.4643	.8857	.5243	1.9074	1.129	2.154	1.0879	20′
50′	.4858	.4669	.8843	.5280	1.8940	1.131	2.142	1.0850	10′
28° 00′	.4887	.4695	.8829	.5317	1.8807	1.133	2.130	1.0821	62° 00′
10′	.4916	.4720	.8816	.5354	1.8676	1.134	2.118	1.0792	50′
20′	.4945	.4746	.8802	.5392	1.8546	1.136	2.107	1.0763	40′
30′	.4974	.4772	.8788	.5430	1.8418	1.138	2.096	1.0734	30′
40′	.5003	.4797	.8774	.5467	1.8291	1.140	2.085	1.0705	20′
50′	.5032	.4823	.8760	.5505	1.8165	1.142	2.074	1.0676	10′
29° 00′	.5061	.4848	.8746	.5543	1.8040	1.143	2.063	1.0647	61° 00′
10′	.5091	.4874	.8732	.5581	1.7917	1.145	2.052	1.0617	50′
20′	.5120	.4899	.8718	.5619	1.7796	1.147	2.041	1.0588	40′
30′	.5149	.4924	.8704	.5658	1.7675	1.149	2.031	1.0559	30′
40′	.5178	.4950	.8689	.5696	1.7556	1.151	2.020	1.0530	20′
50′	.5207	.4975	.8675	.5735	1.7437	1.153	2.010	1.0501	10′
30° 00′	.5236	.5000	.8660	.5774	1.7321	1.155	2.000	1.0472	60° 00′
10′	.5265	.5025	.8646	.5812	1.7205	1.157	1.990	1.0443	50′
20′	.5294	.5050	.8631	.5851	1.7090	1.159	1.980	1.0414	40′
30′	.5323	.5075	.8616	.5890	1.6977	1.161	1.970	1.0385	30′
40′	.5352	.5100	.8601	.5930	1.6864	1.163	1.961	1.0356	20′
50′	.5381	.5125	.8587	.5969	1.6753	1.165	1.951	1.0327	10′
31° 00′	.5411	.5150	.8572	.6009	1.6643	1.167	1.942	1.0297	59° 00′
10′	.5440	.5175	.8557	.6048	1.6534	1.169	1.932	1.0268	50′
20′	.5469	.5200	.8542	.6088	1.6426	1.171	1.923	1.0239	40′
30′	.5498	.5225	.8526	.6128	1.6319	1.173	1.914	1.0210	30′
40′	.5527	.5250	.8511	.6168	1.6212	1.175	1.905	1.0181	20′
50′	.5556	.5275	.8496	.6208	1.6107	1.177	1.896	1.0152	10′
32° 00′	.5585	.5299	.8480	.6249	1.6003	1.179	1.887	1.0123	58° 00′
10′	.5614	.5324	.8465	.6289	1.5900	1.181	1.878	1.0094	50′
20′	.5643	.5348	.8450	.6330	1.5798	1.184	1.870	1.0065	40′
30′	.5672	.5373	.8434	.6371	1.5697	1.186	1.861	1.0036	30′
40′	.5701	.5398	.8418	.6412	1.5597	1.188	1.853	1.0007	20′
50′	.5730	.5422	.8403	.6453	1.5497	1.190	1.844	.9977	10′
33° 00′	.5760	.5446	.8387	.6494	1.5399	1.192	1.836	.9948	57° 00′
10′	.5789	.5471	.8371	.6536	1.5301	1.195	1.828	.9919	50′
20′	.5818	.5495	.8355	.6577	1.5204	1.197	1.820	.9890	40′
30′	.5847	.5519	.8339	.6619	1.5108	1.199	1.812	.9861	30′
40′	.5876	.5544	.8323	.6661	1.5013	1.202	1.804	.9832	20′
50′	.5905	.5568	.8307	.6703	1.4919	1.204	1.796	.9803	10′
34° 00′	.5934	.5592	.8290	.6745	1.4826	1.206	1.788	.9774	56° 00′
10′	.5963	.5616	.8274	.6787	1.4733	1.209	1.781	.9745	50′
20′	.5992	.5640	.8258	.6830	1.4641	1.211	1.773	.9716	40′
30′	.6021	.5664	.8241	.6873	1.4550	1.213	1.766	.9687	30′
40′	.6050	.5688	.8225	.6916	1.4460	1.216	1.758	.9657	20′
50′	.6080	.5712	.8208	.6959	1.4370	1.218	1.751	.9628	10′
35° 00′	.6109	.5736	.8192	.7002	1.4281	1.221	1.743	.9599	55° 00′
10′	.6138	.5760	.8175	.7046	1.4193	1.223	1.736	.9570	50′
20′	.6167	.5783	.8158	.7089	1.4106	1.226	1.729	.9541	40′
30′	.6196	.5807	.8141	.7133	1.4019	1.228	1.722	.9512	30′
40′	.6225	.5831	.8124	.7177	1.3934	1.231	1.715	.9483	20′
50′	.6254	.5854	.8107	.7221	1.3848	1.233	1.708	.9454	10′
36° 00′	.6283	.5878	.8090	.7265	1.3764	1.236	1.701	.9425	54° 00′
		Cos θ	Sin θ	Cot θ	Tan θ	Csc θ	Sec θ	Rad.(θ)	Deg.(θ)

Table of Values of the Trigonometric Functions

Deg.(θ)	Rad.(θ)	Sin θ	Cos θ	Tan θ	Cot θ	Sec θ	Csc θ		
36° 00′	.6283	.5878	.8090	.7265	1.3764	1.236	1.701	.9425	**54° 00′**
10′	.6312	.5901	.8073	.7310	1.3680	1.239	1.695	.9396	**50′**
20′	.6341	.5925	.8056	.7355	1.3597	1.241	1.688	.9367	**40′**
30′	.6370	.5948	.8039	.7400	1.3514	1.244	1.681	.9338	**30′**
40′	.6400	.5972	.8021	.7445	1.3432	1.247	1.675	.9308	**20′**
50′	.6429	.5995	.8004	.7490	1.3351	1.249	1.668	.9279	**10′**
37° 00′	.6458	.6018	.7986	.7536	1.3270	1.252	1.662	.9250	**53° 00′**
10′	.6487	.6041	.7969	.7581	1.3190	1.255	1.655	.9221	**50′**
20′	.6516	.6065	.7951	.7627	1.3111	1.258	1.649	.9192	**40′**
30′	.6545	.6088	.7934	.7673	1.3032	1.260	1.643	.9163	**30′**
40′	.6574	.6111	.7916	.7720	1.2954	1.263	1.636	.9134	**20′**
50′	.6603	.6134	.7898	.7766	1.2876	1.266	1.630	.9105	**10′**
38° 00′	.6632	.6157	.7880	.7813	1.2799	1.269	1.624	.9076	**52° 00′**
10′	.6661	.6180	.7862	.7860	1.2723	1.272	1.618	.9047	**50′**
20′	.6690	.6202	.7844	.7907	1.2647	1.275	1.612	.9018	**40′**
30′	.6720	.6225	.7826	.7954	1.2572	1.278	1.606	.8988	**30′**
40′	.6749	.6248	.7808	.8002	1.2497	1.281	1.601	.8959	**20′**
50′	.6778	.6271	.7790	.8050	1.2423	1.284	1.595	.8930	**10′**
39° 00′	.6807	.6293	.7771	.8098	1.2349	1.287	1.589	.8901	**51° 00′**
10′	.6836	.6316	.7753	.8146	1.2276	1.290	1.583	.8872	**50′**
20′	.6865	.6338	.7735	.8195	1.2203	1.293	1.578	.8843	**40′**
30′	.6894	.6361	.7716	.8243	1.2131	1.296	1.572	.8814	**30′**
40′	.6923	.6383	.7698	.8292	1.2059	1.299	1.567	.8785	**20′**
50′	.6952	.6406	.7679	.8342	1.1988	1.302	1.561	.8756	**10′**
40° 00′	.6981	.6428	.7660	.8391	1.1918	1.305	1.556	.8727	**50° 00′**
10′	.7010	.6450	.7642	.8441	1.1847	1.309	1.550	.8698	**50′**
20′	.7039	.6472	.7623	.8491	1.1778	1.312	1.545	.8668	**40′**
30′	.7069	.6494	.7604	.8541	1.1708	1.315	1.540	.8639	**30′**
40′	.7098	.6517	.7585	.8591	1.1640	1.318	1.535	.8610	**20′**
50′	.7127	.6539	.7566	.8642	1.1571	1.322	1.529	.8581	**10′**
41° 00′	.7156	.6561	.7547	.8693	1.1504	1.325	1.524	.8552	**49° 00′**
10′	.7185	.6583	.7528	.8744	1.1436	1.328	1.519	.8523	**50′**
20′	.7214	.6604	.7509	.8796	1.1369	1.332	1.514	.8494	**40′**
30′	.7243	.6626	.7490	.8847	1.1303	1.335	1.509	.8465	**30′**
40′	.7272	.6648	.7470	.8899	1.1237	1.339	1.504	.8436	**20′**
50′	.7301	.6670	.7451	.8952	1.1171	1.342	1.499	.8407	**10′**
42° 00′	.7330	.6691	.7431	.9004	1.1106	1.346	1.494	.8378	**48° 00′**
10′	.7359	.6713	.7412	.9057	1.1041	1.349	1.490	.8348	**50′**
20′	.7389	.6734	.7392	.9110	1.0977	1.353	1.485	.8319	**40′**
30′	.7418	.6756	.7373	.9163	1.0913	1.356	1.480	.8290	**30′**
40′	.7447	.6777	.7353	.9217	1.0850	1.360	1.476	.8261	**20′**
50′	.7476	.6799	.7333	.9271	1.0786	1.364	1.471	.8232	**10′**
43° 00′	.7505	.6820	.7314	.9325	1.0724	1.367	1.466	.8203	**47° 00′**
10′	.7534	.6841	.7294	.9380	1.0661	1.371	1.462	.8174	**50′**
20′	.7563	.6862	.7274	.9435	1.0599	1.375	1.457	.8145	**40′**
30′	.7592	.6884	.7254	.9490	1.0538	1.379	1.453	.8116	**30′**
40′	.7621	.6905	.7234	.9545	1.0477	1.382	1.448	.8087	**20′**
50′	.7650	.6926	.7214	.9601	1.0416	1.386	1.444	.8058	**10′**
44° 00′	.7679	.6947	.7193	.9657	1.0355	1.390	1.440	.8029	**46° 00′**
10′	.7709	.6967	.7173	.9713	1.0295	1.394	1.435	.7999	**50′**
20′	.7738	.6988	.7153	.9770	1.0235	1.398	1.431	.7970	**40′**
30′	.7767	.7009	.7133	.9827	1.0176	1.402	1.427	.7941	**30′**
40′	.7796	.7030	.7112	.9884	1.0117	1.406	1.423	.7912	**20′**
50′	.7825	.7050	.7092	.9942	1.0058	1.410	1.418	.7883	**10′**
45° 00′	.7854	.7071	.7071	1.0000	1.0000	1.414	1.414	.7854	**45° 00′**
		Cos θ	Sin θ	Cot θ	Tan θ	Csc θ	Sec θ	Rad.(θ)	Deg.(θ)

Glossary

abscissa: The x-coordinate of an ordered pair of real numbers. (p. 73)

absolute value: The absolute value of any real number x, written $|x|$, is x if $x > 0$, and $-x$ if $x < 0$. On a number line, $|x|$ is the distance between the graph of x and the origin. (pp. 1, 51)

absolute value of a complex number: The absolute value of the complex number $a + bi$, written $|a + bi|$, equals $\sqrt{a^2 + b^2}$. (p. 305)

amplitude: The amplitude of a periodic function is one-half the distance between its maximum and minimum values. (p. 681)

angle of elevation (depression): The angle between a horizontal line and the line of sight. (p. 629)

angle of rotation: An angle of rotation in standard position is the angle formed by rotating a ray counterclockwise from the positive x-axis (the *initial* side) to a new position (the *terminal* side). (p. 633)

antilogarithm: If $\log_b q = r$, then q is the base-b antilogarithm (antilog) of r. (p. 481)

arithmetic means: The terms between the first and last terms of an arithmetic sequence. *The* arithmetic mean of x and y is their average, $\dfrac{x + y}{2}$. (p. 549)

arithmetic sequence: A sequence in which the difference between successive terms (the *common difference*) is constant. (p. 545)

arithmetic series: The indicated sum of the terms of an arithmetic sequence. (p. 553)

asymptote: A line that the graph of a function approaches more and more closely without ever touching. (p. 419)

augmented matrix: A matrix representing a system of equations, in which the constants of the system form the right-hand column. (p. 603)

axiom: A statement that is assumed to be true. Also called a *postulate*. (p. 31)

base: In the power b^3, b is the base, or the repeated power. (p. 161)

bearing: An angle of rotation measured clockwise from due north. (p. 669)

binomial: A polynomial that has two terms. (p. 170)

binomial expansion: The sum of the terms of the nth power of a binomial. (p. 576)

characteristic: The integer part of a base-10 logarithm. (p. 482)

circle: The set of all points in a plane that are a fixed distance, the *radius*, from a given point, the *center*. (p. 409)

coefficient: The constant factor in a term having one or more variables. (p. 9)

combination: A subset of r objects from a set of n objects is a combination of n elements taken r at a time. (p. 511)

combined variation: A relation containing both direct and inverse variation. (p. 432)

common difference: The constant difference between successive terms in an arithmetic sequence. (p. 545)

common logarithm: A base-10 logarithm. (p. 481)

common ratio: The constant ratio between successive terms in a geometric sequence. (p. 558)

complex conjugates: Two complex numbers of the form $a + bi$ and $a - bi$. (p. 308)

complex number: A number of the form $a + bi$, where a and b are real numbers and $i = \sqrt{-1}$. (p. 303)

complex rational expression: A rational expression in which the numerator, the denominator, or both contain at least one rational expression. (p. 249)

composite of two functions: Given the functions f and g such that the range of f is in the domain of g, the composite of the two functions is the function whose value at x is $g(f(x))$. (p. 190)

compound sentence: The combination of two simple sentences by *and* or by *or*. (p. 46)

conic section: The intersection of a plane and a right circular cone. (p. 440)

conjugates: Two real numbers of the form $a + b$ and $a - b$. (p. 281)

conjunction: A compound sentence containing the connector *and*. (p. 46)

consistent system: A system of equations that has at least one solution. (p. 119)

constant of variation: The constant k in functions of the forms $y = kx$ (direct variation) or $xy = k$ (inverse variation). (pp. 105, 427)

corollary: A statement that can be inferred from a proven theorem. (p. 477)

cosecant: The cosecant of angle θ (csc θ) is the reciprocal of sin θ. (p. 643)

cosine: For an angle θ in standard position with a point $P(x,y)$ on the terminal side at a distance $r > 0$ from the origin, the cosine of angle θ (cos θ) $= \frac{x}{r}$. (p. 643)

cotangent: The cotangent of angle θ (cot θ) is the reciprocal of tan θ. (p. 643)

coterminal angles: Angles that share initial and terminal sides. (p. 635)

degree of a monomial: The sum of the exponents of its variables. (p. 170)

degree of a polynomial: The degree of the term of highest degree. (p. 170)

dependent events: Events for which the outcome of one affects the outcome of the other. (p. 525)

dependent system: A system of equations that has infinitely many solutions. (p. 120)

determinant: A real number associated with a square matrix. (p. 594)

direct variation: A linear function defined by an equation of the form $y = kx$ ($k \neq 0$). (p. 105)

discriminant: The discriminant of the quadratic equation $ax^2 + bx + c$ is $b^2 - 4ac$. (p. 343)

disjunction: A compound sentence containing the connector *or*. (p. 46)

domain: The set of first coordinates of the ordered pairs of a relation. (p. 74)

ellipse: The set of all points in a plane such that for each point, the sum of the distances from two fixed points (the *foci*) is constant. (p. 413)

empty set: The set containing no elements. (p. 48)

equivalent equations: Equations that have the same solution set. (p. 13)

equivalent expressions: Expressions that are equal for all values of the variables for which they have meaning. (p. 10)

equivalent matrices: Augmented matrices which can be obtained from each other by any of a series of operations called *elementary matrix transformations*. (p. 604)

equivalent systems: Systems of equations with the same solution set. (p. 124)

equivalent vectors: Vectors with the same magnitude and direction. (p. 323)

even function: A function for which $f(-x) = f(x)$ for all x in the domain of f. (p. 401)

event: Any subset of the sample space. (p. 517)

exponent: In the power b^n, n is the exponent. If n is a positive integer, it tells how many times the base b is a factor in the power b^n. (p. 161)

exponential function: An exponential function with *base b* is defined by $y = b^x$, where $b > 0$ and $b \neq 1$. (p. 464)

extraneous root: A solution of a derived equation that is not a solution of the original equation. (p. 253)

factorial notation: For a positive integer n, $n!$ (read "n factorial") is defined as $n(n - 1)(n - 2) \ldots 3 \cdot 2 \cdot 1$. (p. 500)

field: A set of numbers, along with two operations, that satisfies a specified list of eleven properties. (p. 8)

frequency distribution: A display giving a set of data and the frequency of occurence of each score. (p. 529)

function: A relation in which no two ordered pairs have the same first coordinate. (p. 75)

geometric means: The terms between the first and last terms of a geometric sequence. *The* geometric mean of x and y is xy or $-xy$ ($x \neq 0$, $y \neq 0$). (p. 562)

geometric sequence: A sequence in which the ratio of successive terms (the *common ratio*) is constant. (p. 558)

geometric series: The indicated sum of the terms of a geometric sequence. (p. 566)

greatest common factor: The expression of greatest degree and greatest constant factor that is a factor of two expressions. (p. 175)

half-plane: All points of a plane that lie on one side of a given line. (p. 110)

harmonic means: The terms between the first and last terms of a harmonic sequence. (p. 550)

harmonic sequence: The reciprocal of an arithmetic sequence that has no zero terms. (p. 550)

histogram: A bar graph that displays a frequency distribution. (p. 529)

hyperbola: The set of all points in a plane such that for each point, the difference of the distances to two fixed points (the *foci*) is constant. (p. 418)

identity: An equation true for all possible values of its variables. (pp. 16, 711)

imaginary number: Any complex number, $a + bi$, for which $a = 0$ and $b \neq 0$. (p. 303)

inclusive events: Two events which can occur simultaneously. (p. 521)

inconsistent system: A system of equations that has no solution. (p. 120)

independent events: Events for which the outcome of one event doesn't affect the outcome of the other. (p. 523)

independent system: A system of equations that has at most one solution. (p. 119)

intersection: The set of all elements common to two sets. (p. 148)

inverse relations: Relations in which the order of the elements of each ordered pair is reversed. (p. 459)

inverse variation: A function defined by an equation of the form $xy = k$ or $y = \frac{k}{x}$ ($k \neq 0$). (p. 427)

irrational number: A nonterminating, nonrepeating decimal. (p. 1)

joint variation: Direct variation of a quantity as the product of two or more other quantities. (p. 431)

least common denominator: The common multiple of the denominators of two fractions of least degree and least positive constant factor. (p. 14)

linear function: A function whose graph is a line. Its equation is called a *linear equation in two variables* and has the standard form $ax + by = c$, where a and b are not both equal to zero. (p. 78)

logarithm: A logarithm is an exponent. For positive numbers b and y ($b \neq 1$) such that $b^y = x$, the *base-b logarithm* of x ($\log_b x$) is equal to y. (p. 467)

logarithmic function: The logarithmic function with *base-b* is the inverse of the exponential function, and is defined by $y = \log_b x$ ($x > 0$, $b > 0$, $b \neq 1$). (p. 469)

mantissa: The decimal portion of a base-10 logarithm. (p. 482)

matrix: A rectangular array of numbers (*elements*) enclosed by brackets. (p. 585)

mean: The number found by adding a group of scores and then dividing the sum by the number of scores. (p. 529)

median: For an odd number of scores in an ordered distribution, the middle score; for an even number of scores, the mean of the two middle scores. (p. 529)

minor: The minor of an element of a determinant is the determinant found by deleting the column and row in which the element lies. (p. 599)

mode: The score in a distribution that occurs most frequently. (p. 529)

monomial: A number, a variable, or the product of a number and one or more variables. (p. 170)

multiplicity: The number of times that a factor is repeated in the factorization of a polynomial. (p. 224)

mutually exclusive events: Events that cannot occur simultaneously. (p. 521)

normal distribution: A distribution of data with a bell-shaped curve. (p. 533)

norm of a vector: The length or *magnitude* of a vector. (p. 738)

nth root: A solution of the equation $x^n = b$, for n a positive integer. (p. 284)

odd function: A function for which $f(-x) = -f(x)$ for all x in the domain of f. (p. 402)

odds: The odds that an event will occur are defined as the ratio of the number of successful outcomes to the number of failures. (p. 519)

opposite vectors: Vectors of equal length but opposite direction. (p. 323)

ordinate: The y-coordinate of an ordered pair of real numbers. (p. 73)

parabola: The set of all points in a plane that are equidistant from a given point (the *focus*) and a given line (the *directrix*). (pp. 391, 435)

periodic function: A function such that for all x in its domain and for some positive constant p, $f(x) = f(x + p)$. The smallest such value of p is the *period*. (p. 681)

permutation: An ordered arrangement of elements of a set. (p. 503)

polar axis: The positive x-axis. (p. 742)

polar coordinates: The polar coordinates of the point P are given by the ordered pair (r,θ), where r is the distance from P to the origin and θ is the angle of rotation from the polar axis to P. (p. 742)

polar form: The polar form of the complex number $z = x + yi$ is $r(\cos \theta + i \sin \theta)$, where r is the absolute value (*modulus*) of z, and θ is an *argument* of z. (p. 743)

polynomial: A monomial or the sum of two or more monomials. (p. 170)

principal values: The range of an inverse trigonometric function when the domain is restricted to a given set. (p. 701)

probability: The number of successful outcomes divided by the total number of possible outcomes. (p. 517)

proportion: An equation of the form $\frac{a}{b} = \frac{c}{d}$. (p. 14)

quadrant: One of the four regions, labeled counterclockwise I-IV, into which a coordinate plane is divided by its axes. (p. 73)

quadrantal angle: An angle with terminal side on a coordinate axis. (p. 633)

quadratic equation: An equation of the form $ax^2 + bx + c$, where $a \neq 0$. (p. 199)

quadratic formula: The quadratic formula, $x = -b \pm \dfrac{\sqrt{b^2 - 4ac}}{2a}$, gives the solutions of the quadratic equation $ax^2 + bx + c = 0$. (p. 317)

radian: The measure of the central angle of a circle intercepting an arc of length equal to the circle's radius. (p. 677)

radical: The symbols $\sqrt{x}$ and $\sqrt[n]{x}$ are called radicals; n is the *index*, and x is the *radicand*. (pp. 267, 284))

radical equation: An equation that contains a radical with a variable as the radicand. (p. 347)

range: The set of all second coordinates of the ordered pairs of a relation. Also, in statistics, the difference between the highest and lowest values of a set of data. (pp. 74, 530)

rational expression: A quotient of two polynomials. (p. 233)

rational number: A number that can be written as the quotient of two integers, with the divisor not zero. (p. 1)

real number: Any rational or irrational number. (p. 1)

reciprocal: A number such that the product of any nonzero number and its reciprocal is equal to 1. (p. 8)

rectangular coordinates: The rectangular coordinates of the point P are given by the ordered pair (x,y), and represent the distances to P along the x- and y-axes. (p. 742)

rectangular hyperbola: The graph of the equation $xy = k$, where k is a nonzero real number. (p. 426)

reference triangle: The triangle formed by drawing a perpendicular from the terminal side of an angle in standard position to the x-axis. The acute angle formed with vertex at the origin is the *reference angle*. (pp. 639, 640)

relation: A set of ordered pairs. (p. 74)

resultant: The sum of two vectors. (p. 324)

root: Any value of a variable that satisfies an equation. (p. 13)

sample space: The set of all possible outcomes of an experiment. (p. 517)

scalar multiple: The product of a real number (*scalar*) and a vector. (p. 325)

scientific notation: The representation of a number in the form $a \times 10^c$, where $1 \leq a < 10$ and c is an integer. (p. 166)

secant: The secant of angle θ (sec θ) is the reciprocal of cos θ. (p. 643)

sequence: An ordered list of numbers. (p. 545)

sigma notation: A shorthand method of writing a series, using the summation symbol Σ (sigma). (p. 555)

simple sentence: A statement of equality or inequality of two algebraic expressions or numbers. (p. 46)

sine: For an angle θ in standard position with a point $P(x,y)$ on the terminal side at a distance $r > 0$ from the origin, the sine of angle θ (sin θ) $= \frac{y}{r}$. (p. 643)

slope: The slope of a nonvertical line containing the points $A(x_1,y_1)$ and $B(x_2,y_2)$ is given by the formula $\frac{y_2 - y_1}{x_2 - x_1}$. (p. 84)

solution set: The set of all solutions of an open sentence which belong to its domain. (p. 42)

square root: A square root of the number b is a solution of the equation $x^2 = b$. The *principal square root* of b is the positive square root. (p. 267)

standard deviation: The principal square root of the variance. (p. 530)

synthetic division: A method using only the coefficients to divide a polynomial in x by $x - a$. (p. 217)

system of equations: A set of two or more equations. (p. 119)

tangent: For an angle θ in standard position with a point $P(x,y)$ on the terminal side at a distance $r > 0$ from the origin, the tangent of θ (tan θ) $= \frac{y}{x}$, $x \neq 0$. (p. 643)

theorem: A statement that can be proved from a set of axioms. (p. 31)

translation: Moving a graph in a coordinate plane without changing its shape or rotating it. (p. 388)

trinomial: A polynomial that has three terms. (p. 170)

union: The set of all elements in either of two sets or in both of them. (p. 148)

value of a function: The range element of a function for a given value of the domain. (p. 79)

variable: A symbol that can stand for any element of a given set. (p. 3)

variance: The sum of the squares of the deviations from the mean of a set of scores, divided by the number of scores. (p. 530)

vector: A directed line segment, having magnitude and direction. (p. 323)

y-intercept: The y-coordinate of the point at which a line crosses the y-axis. (p. 91)

zero of a function: Any value of x in the domain of the function f which satisfies the equation $f(x) = 0$. (p. 234)

Additional Answers

5.

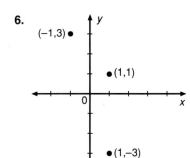

6.

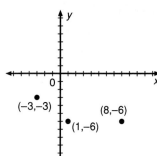

Written Exercises

1. D: { 1, 2}, R: { 1, 2}; function

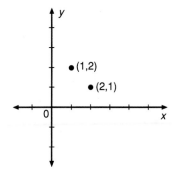

2. D: { −1, 0}, R: { 0, 4}; function

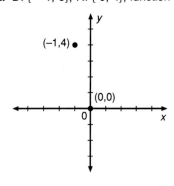

3. D: {−1}, R: {2, 6}; not a function

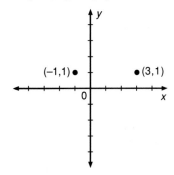

4. D: { −1, 3}, R: { 1}; function

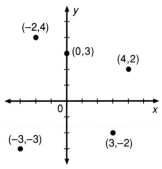

5. D: { −3, −2, 0, 3, 4},
R: { −3, −2, 2, 3, 4}; function

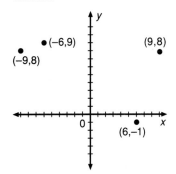

6. D: { −9, −6, 6, 9}, R: { −1, 8, 9};
function

7. D: { −5, −1, 7, 8}, R: { −1}; function

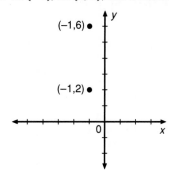

8. D: { 6}, R: −5, −1, 0, 2}; not a function

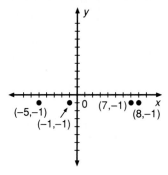

9. {(−2,3), (−1,−3), (1,−2), (4,1)}
D: {−2,−1,1,4};
R: {−3,−2,1,3}; function
10. {(−4,−2), (0,0), (2,1)}; D: {−4,0,2};
R: {−2,0,1}; function
11. {(0,0), (1,−1), (1,1) (2,−2), (2,2),
(3,−3), (3,3)}; D {0,1,2,3};
R: {−3,−2,−1,0,1,2,3}; not a function
12. Function   **13.** Not a function
14. Function
15. D: {x | −3 < x ≤ 4}
R: {y | −1 ≤ y < 3}; function
16. D: {x | −2 ≤ x < 3}
R: {y | −3 ≤ y ≤ 2}; function
17. D: {x | −3 ≤ x ≤ 3}
R: {y | −2 ≤ y ≤ 2}; not a function
18. D: {x | −3 ≤ x < 3}
R: {y | −3 ≤ y < 3}; not a function
19. D: {x | −3 ≤ x ≤ −2} together with
{x | −1 < x ≤ 3}; R: {−1,0,2}; function
20. D: {x | −4 ≤ x < −2} together with
{x | −1 ≤ x < 2}; R: {y | −3 < y ≤ 3};
function
21. D: {...,−7,−4,−1, 2,5,8,...}
R: {...,−5,−3,−1, 1,3,5,...}; function
22. D: {square of all integers ≥ 1};
R: {one plus squares of all
integers ≥ 2}; function

803

Mixed Review

1. $\{x \mid x > -3\}$

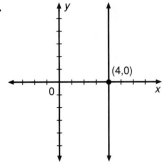

2. $\{x \mid x < 11\}$

3. $\{-1\frac{1}{2}, 2\frac{1}{2}\}$

4. $\{x \mid -4 < x < 7\}$

PAGE 81

6. $(-3,17)$, $(-2,2)$, $(-1,-7)$, $(0,-10)$, $(1,-7)$, $(2,2)$, $(3,17)$

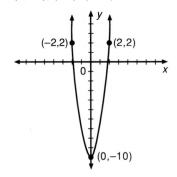

10. $\{-19, -13, 17\}$
11. $\{-8, -\frac{8}{3}, -2\}$
17. $-4a^2 + 12a - 1$
19. $W(t^3) = W(ttt)$
$\qquad = W(t) + W(t) + W(t)$
$\qquad = 3W(t)$

Mixed Review

1. Comm Prop Mult
2. Add Inverse Prop
3. Distr Prop
4. $-6xy^2 + 4xy$
5. -11

804

PAGE 94

18.

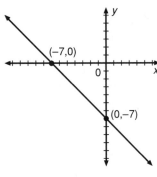

19.

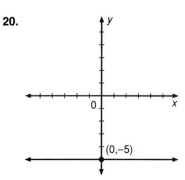

20.

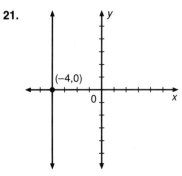

21.

22.

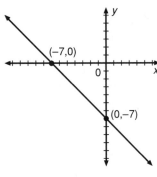

23.

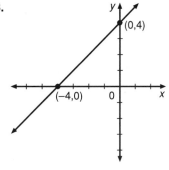

24.

Written Exercises

9.

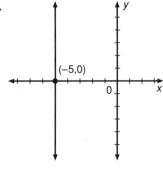

10.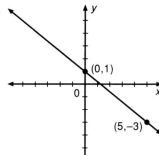

11.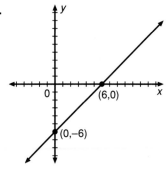

12.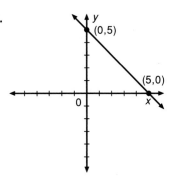

13.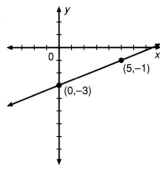

14.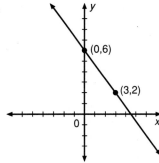

15.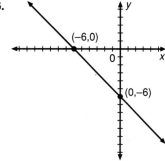

16.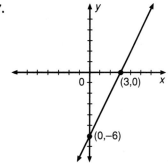

17.

18.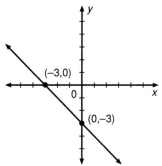

19.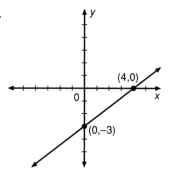

20.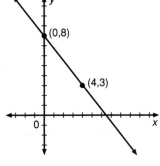

21.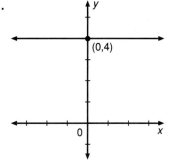

22.

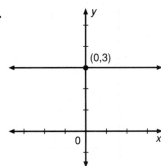

(0,3)

23.

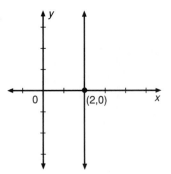

(2,0)

24.

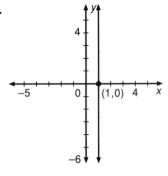

(1,0)

25.

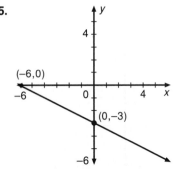

(−6,0)

(0,−3)

26.

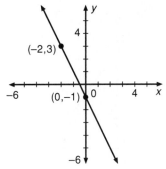

(−2,3)

(0,−1)

31.

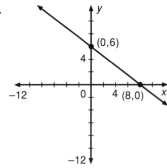

(0,6)

(8,0)

32.

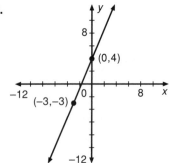

(0,4)

(−3,−3)

Midchapter Review

1.

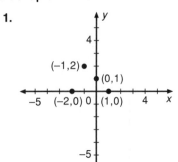

(−1,2)

(0,1)

(−2,0)

(1,0)

2.

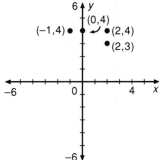

(0,4)

(−1,4)

(2,4)

(2,3)

3.

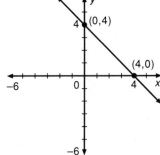

(0,2)

(2,0)

4.

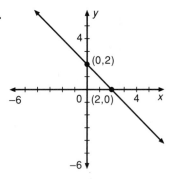

(0,4)

(4,0)

5.

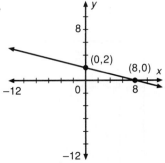

(0,2)

(8,0)

6.

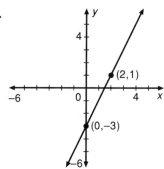

5.

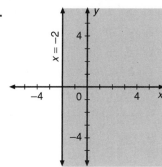

9.

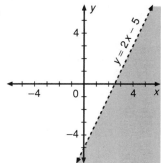

2.

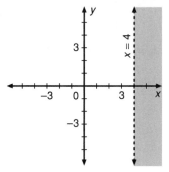

6.

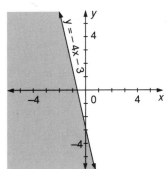

10.

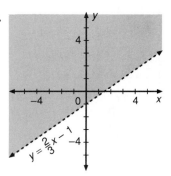

3.

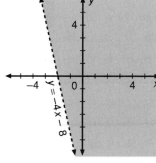

7.

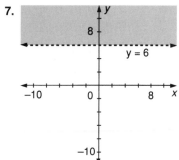

11.

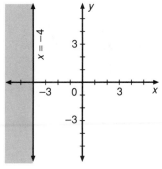

4.

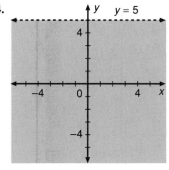

8.

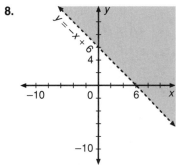

12.

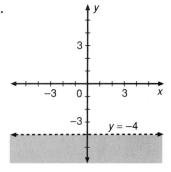

13.

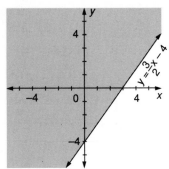

14.

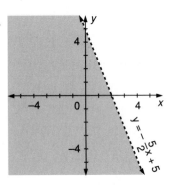

15.

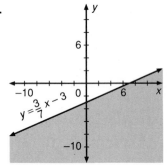

16.

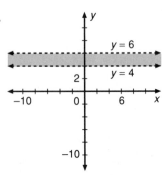

17.

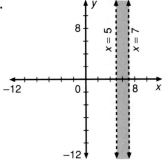

18.

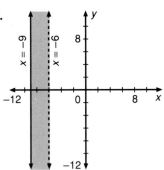

PAGE 122

4. Consistent, independent; (1,2)

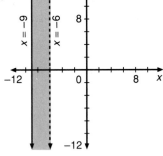

5. Inconsistent, independent; No solution

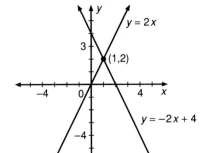

6. Consistent, dependent;
$\{(x,y) \mid y = -x + 6\}$

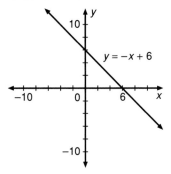

7. $\{(x,y) \mid y = -3x + 2\}$

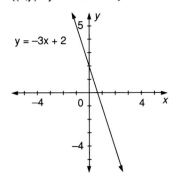

8. (1,7)

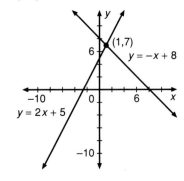

9. (1,2)

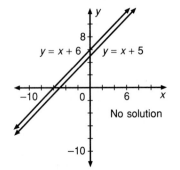

10. No solution

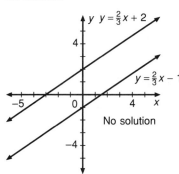

$y = \frac{2}{3}x + 2$

$y = \frac{2}{3}x - 1$

No solution

11. (1,0)

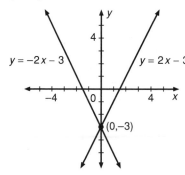

$y = -5x + 5$

(1,0)

$y = -3x + 3$

12. (0,−3)

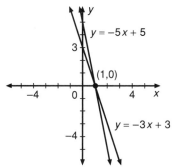

$y = -2x - 3$ $y = 2x - 3$

(0,−3)

13. (2,0)

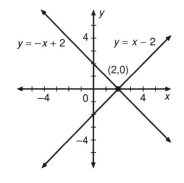

$y = -x + 2$ $y = x - 2$

(2,0)

14. Answers may vary. One possible answer: Graph both equations on the same coordinate plane. If there is a single point of intersection, the coordinates of that point are the system's solution. If the graphs are parallel, there is no solution. If the graphs are coincident, all the points on the line are solutions.

Mixed Review

4.

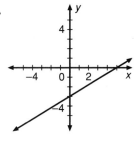

Application

2. Answers may vary. One possible answer: Supply could fall below the equilibrium point because of a shortage of raw materials, a strike, one-time production costs passed on as a price increase, and so on. The equilibrium point could shift because of long-term price increases, because of new foreign competition, and so on.

Statistics

4. $y = 0.10x + 0.28$; 7.28

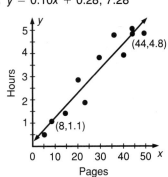

(44,4.8)

(8,1.1)

Pages / Hours

10. (3,2), (6,2), (6,8)

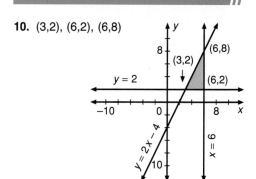

$y = 2$ (3,2)   (6,8)   (6,2)

$y = 2x - 4$ $x = 6$

11. (1,1), (1,3), (2,1)

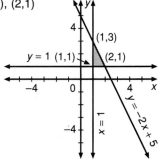

(1,3)

$y = 1$ (1,1)   (2,1)

$x = 1$ $y = -2x + 5$

12. (0,0), $(0,\frac{7}{2})$, (3,2), (4,0)

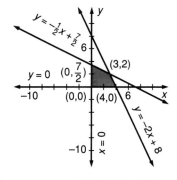

$y = -\frac{1}{2}x + \frac{7}{2}$

$y = 0$ $(0,\frac{7}{2})$ (3,2)

(0,0)   (4,0)

$x = 0$ $y = -2x + 8$

13. $y \le -4x + 7$ *and* $y \le x + 2$
and $y \ge -\frac{2}{3}x + \frac{1}{3}$

14. $y \ge 2$ *and* $-1 \le x \le 2$ *and*
$y \le -\frac{1}{3}x + \frac{8}{3}$

15. $y \ge 3x - 4$ *and* $x \le 2$ *and* $y \le 3$ *and*
$y \ge -\frac{4}{3}x + \frac{1}{3}$

11. $(x + y + 4)(x + y - 4)$
12. $(c - d + 5)(c - d - 5)$
13. $(x + y - 2)(x - y + 2)$
14. $(m + n + 3)(m - n - 3)$
15. $(x + 3 + y)(x + 3 - y)$
16. $(m - n + 3)(m - n - 3)$
17. $(a^2 - 8)^2$ **18.** $(y - 2)(y^2 + 2y + 4)$

24. $(7ab + 10)(7ab - 10)$

25. $(6x^3 + y^2)(6x^3 - y^2)$

26. $(2c - 5d)^2$

27. $(5a^2 + 6b^2)^2$

29. Not possible

30. $(3x + 2y)(3x + 8y)$

31. $(m - 4n)(m^2 + 4mn + 16n^2)$

32. $(5c + d)(25c^2 - 5cd + d^2)$

34. $(3x^2 - 10y)(9x^4 + 30x^2y + 100y^2)$

35. $(2x + 3y + 5)(2x + 3y - 5)$

36. $(3x - y + 4)(3x - y - 4)$

38. $(2x^{3m+2} + 3y^n)^2$

39. The inequality $x^2 + y^2 \geq 2xy$ is true for
all real numbers x and y, because it
can be rewritten as $x^2 - 2xy + y^2 \geq 0$,
or $(x - y)^2 \geq 0$, and the square of any
number $(x - y)$ is always nonnegative.

40. $[(a + b) + c]^2$
$= [(a + b) + c][(a + b) + c]$
$= (a + b)^2 + c(a + b) + c(a + b) + c^2$
$= a^2 + 2ab + b^2 + 2c(a + b) + c^2$
$= a^2 + 2ab + b^2 + 2ac + 2bc + c^2$
$= a^2 + b^2 + c^2 + 2ab + 2ac + 2bc$

810

14.

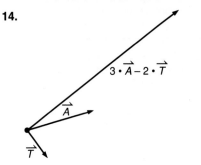

$3 \cdot \vec{A} - 2 \cdot \vec{T}$

$\vec{A}$

$\vec{T}$

15.

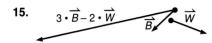

$3 \cdot \vec{B} - 2 \cdot \vec{W}$ $\vec{B}$ $\vec{W}$

16.

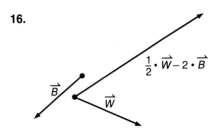

$\frac{1}{2} \cdot \vec{W} - 2 \cdot \vec{B}$

$\vec{B}$

$\vec{W}$

17.

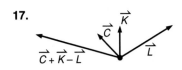

$\vec{K}$

$\vec{C}$

$\vec{L}$

$\vec{C} + \vec{K} - \vec{L}$

18.

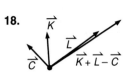

$\vec{K}$

$\vec{L}$

$\vec{C}$ $\vec{K} + \vec{L} - \vec{C}$

19.

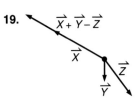

$\vec{X} + \vec{Y} - \vec{Z}$

$\vec{X}$

$\vec{Z}$

$\vec{Y}$

20.

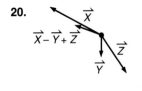

$\vec{X}$

$\vec{X} - \vec{Y} + \vec{Z}$

$\vec{Z}$

$\vec{Y}$

21.

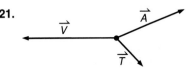

$\vec{A}$

$\vec{V}$

$\vec{T}$

22.

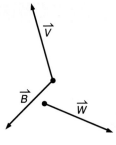

$\vec{V}$

$\vec{B}$ $\vec{W}$

23.

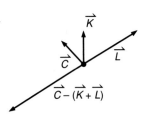

$\vec{K}$

$\vec{C}$ $\vec{L}$

$\vec{C} - (\vec{K} + \vec{L})$

24.

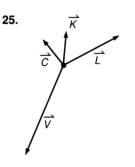

$2 \cdot (\vec{C} + \vec{K} + \vec{L})$

$\vec{K}$

$\vec{C}$ $\vec{L}$

25.

$\vec{K}$

$\vec{C}$ $\vec{L}$

$\vec{V}$

26.

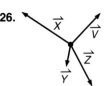

$\vec{X}$ $\vec{V}$

$\vec{Z}$

$\vec{Y}$

Mixed Review

1. Closure for Add
2. Comm Prop Add
3. Assoc Prop Add
4. Add Inverse Prop
5. Distr Prop
6. Mult Identity Prop

Classroom Exercises

9.

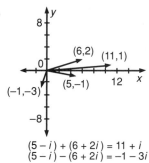

$$(5 - i) + (6 + 2i) = 11 + i$$
$$(5 - i) - (6 + 2i) = -1 - 3i$$

Written Exercises

1.

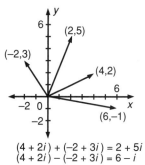

$$(4 + 2i) + (-2 + 3i) = 2 + 5i$$
$$(4 + 2i) - (-2 + 3i) = 6 - i$$

2.

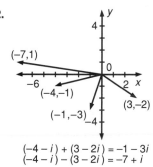

$$(-4 - i) + (3 - 2i) = -1 - 3i$$
$$(-4 - i) - (3 - 2i) = -7 + i$$

3.

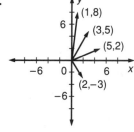

$$(3 + 5i) + (2 - 3i) = 5 + 2i$$
$$(3 + 5i) - (2 - 3i) = 1 + 8i$$

4.

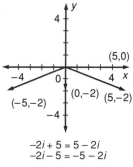

$$-2i + 5 = 5 - 2i$$
$$-2i - 5 = -5 - 2i$$

5.

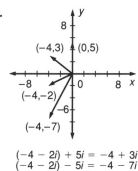

$$(-4 - 2i) + 5i = -4 + 3i$$
$$(-4 - 2i) - 5i = -4 - 7i$$

6.

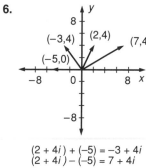

$$(2 + 4i) + (-5) = -3 + 4i$$
$$(2 + 4i) - (-5) = 7 + 4i$$

7.

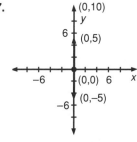

$$5i + (-5i) = 0 + 0i$$
$$5i - (-5i) = 0 + 10i$$

8.

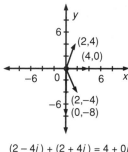

$$(2 - 4i) + (2 + 4i) = 4 + 0i$$
$$(2 - 4i) - (2 + 4i) = 0 - 8i$$

9.

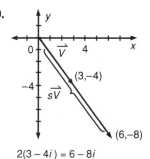

$$2(3 - 4i) = 6 - 8i$$

10.

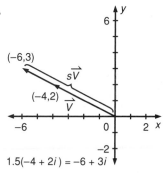

$$1.5(-4 + 2i) = -6 + 3i$$

11.

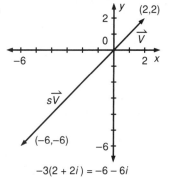

$$-3(2 + 2i) = -6 - 6i$$

811

12.

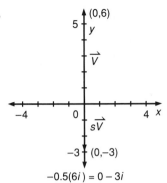

$-0.5(6i) = 0 - 3i$

13.

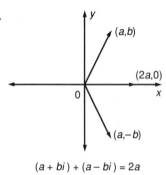

$(a + bi) + (a - bi) = 2a$

14.

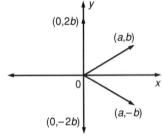

$(a + bi) - (a - bi) = 2bi$
$(a - bi) - (a + bi) = -2bi$

15.

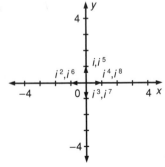

Mixed Review

1. Closure for Mult
2. Assoc Prop Mult
3. Comm Prop Mult
4. Mult Inverse Prop

PAGE 389

3.

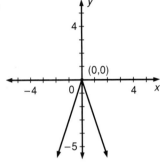

4.

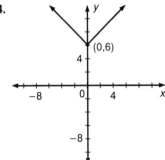

5.

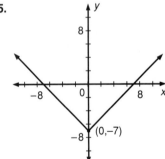

6.

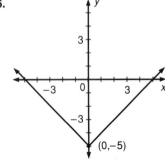

7.

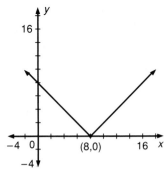

8.

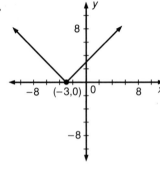

9.

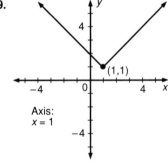

Axis:
$x = 1$

10.

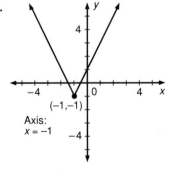

Axis:
$x = -1$

11.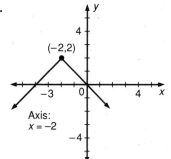

(-2,2)

Axis:
$x = -2$

12.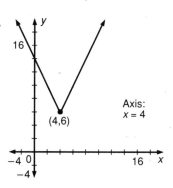

(4,6)

Axis:
$x = 4$

13.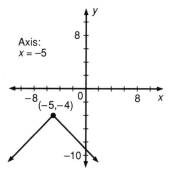

Axis:
$x = -5$

(-5,-4)

14.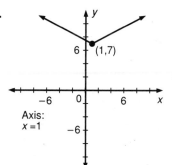

(1,7)

Axis:
$x = 1$

15.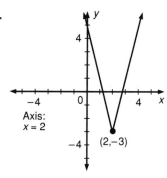

Axis:
$x = 2$

(2,-3)

16.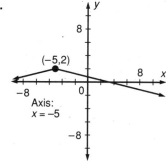

(-5,2)

Axis:
$x = -5$

17.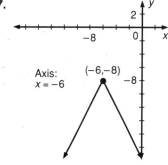

Axis:
$x = -6$

(-6,-8)

18. $(2, -10)$; $x = 2$
19. $(4, -5)$; $x = 4$
20. $(-3, -10)$; $x = -3$
21. $(4, -6)$; $x = 4$
22. $(-3, -1)$; $x = -3$
23. $(\frac{3}{2}, 6)$; $x = \frac{3}{2}$

24.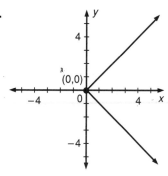

(0,0)

25.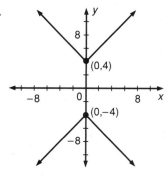

(0,4)

(0,-4)

26.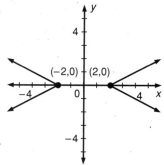

(-2,0) (2,0)

PAGE 394

Written Exercises

1.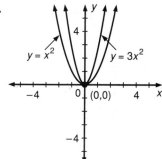

$y = x^2$ $y = 3x^2$

(0,0)

2.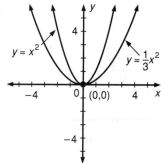

$y = x^2$ $y = \frac{1}{3}x^2$

(0,0)

3.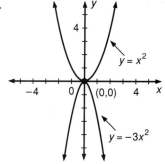

$y = x^2$

(0,0)

$y = -3x^2$

813

4.

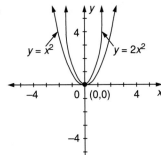

$y = x^2$ $y = 2x^2$

(0,0)

5.

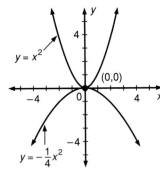

$y = x^2$

(0,0)

$y = -\dfrac{1}{4}x^2$

6.

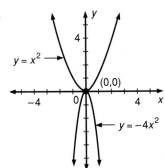

$y = x^2$

(0,0)

$y = -4x^2$

7.

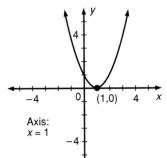

(1,0)

Axis:
$x = 1$

8.

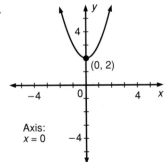

(0, 2)

Axis:
$x = 0$

9.

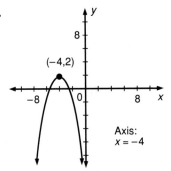

(2,1)

Axis:
$x = 2$

10.

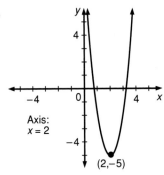

(−4,2)

Axis:
$x = -4$

11.

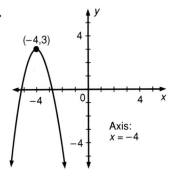

Axis:
$x = 2$

(2,−5)

12.

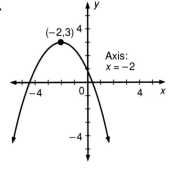

(−4,3)

Axis:
$x = -4$

13.

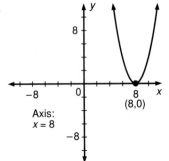

(8,0)

Axis:
$x = 8$

14.

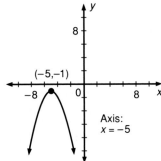

(−5,−1)

Axis:
$x = -5$

15.

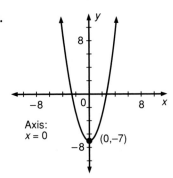

Axis:
$x = 0$

(0,−7)

16.

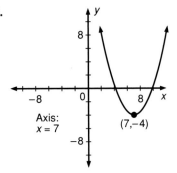

Axis:
$x = 7$

(7,−4)

17.

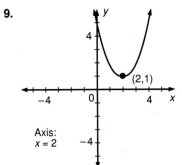

(−2,3)

Axis:
$x = -2$

18.

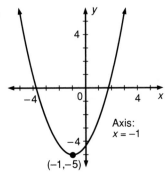

Axis:
$x = -1$

(−1,−5)

19. $y - 1 = -\frac{1}{4}(x + 4)^2$

20. $y + 3 = \frac{5}{9}(x + 1)^2$

21. $y + 1 = (x - 1)^2$

22. Answers will vary. One possible answer: Plot the vertex, (3,1), and draw the axis of symmetry, $x = 3$. Then make a table of values for $x = 3$, $x = 3 \pm 1$, and $x = 3 \pm 2$, or $x = 1,2,3,4,5$. Use these points to graph the parabola.

Midchapter Review

3.

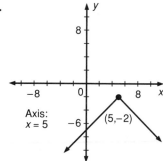

Axis:
$x = -3$

4.

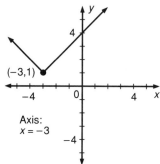

Axis:
$x = 5$

5.

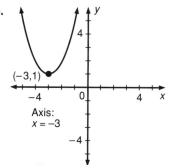

Axis:
$x = -3$

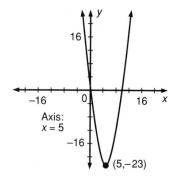

Classroom Exercises

10. $y + 23 = (x - 5)^2$; (5,−23); $x = 5$;

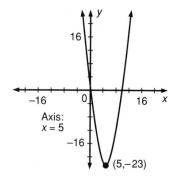

Axis:
$x = 5$

11. $y + 15 = (x + 4)^2$; (−4,−15); $x = -4$;

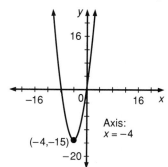

Axis:
$x = -4$

12. $y + 4 = (x - 2)^2$; (2,−4); $x = 2$;

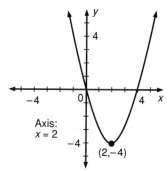

Axis:
$x = 2$

Written Exercises

1. $y + 5 = (x + 2)^2$; (−2,−5); $x = -2$;

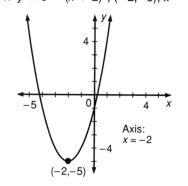

Axis:
$x = -2$

2. $y + 7 = (x - 3)^2$; (3,−7); $x = 3$;

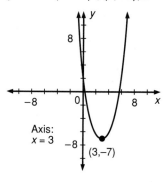

Axis:
$x = 3$

3. $y + 10 = (x - 5)^2$; (5,−10); $x = 5$;

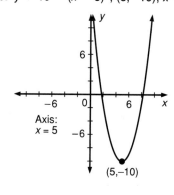

Axis:
$x = 5$

4. $y - 3 = (x - 3)^2$; (3,3); $x = 3$;

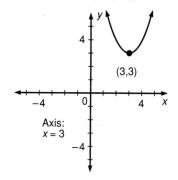

Axis:
$x = 3$

5. $y - 4 = (x + 4)^2$; (−4,4); $x = -4$;

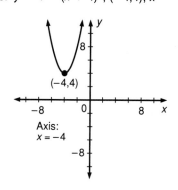

Axis:
$x = -4$

6. $y + 9 = (x + 2)^2$; $(-2,-9)$; $x = -2$;

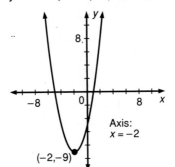

Axis:
$x = -2$

$(-2,-9)$

10. $y + 4 = (x - 2)^2$; $(2,-4)$; $x = 2$;

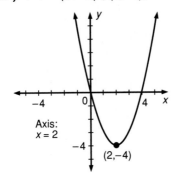

Axis:
$x = 2$

$(2,-4)$

14. $y + 8 = 3(x - 1)^2$; $(1,-8)$; $x = 1$;

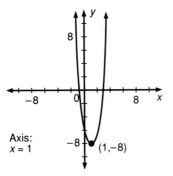

Axis:
$x = 1$

$(1,-8)$

7. $y - 2 = (x + 5)^2$; $(-5,2)$; $x = -5$;

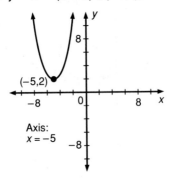

$(-5,2)$

Axis:
$x = -5$

11. $y + 16 = (x - 4)^2$; $(4,-16)$; $x = 4$;

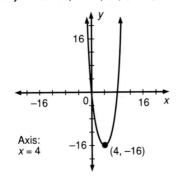

Axis:
$x = 4$

$(4,-16)$

15. $y + 3 = 4(x - 1)^2$; $(1,-3)$; $x = 1$;

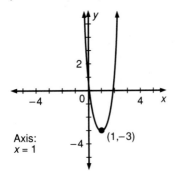

Axis:
$x = 1$

$(1,-3)$

8. $y + 9 = (x - 3)^2$; $(3,-9)$; $x = 3$;

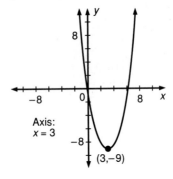

Axis:
$x = 3$

$(3,-9)$

12. $y + 9 = (x + 3)^2$; $(-3,-9)$; $x = -3$;

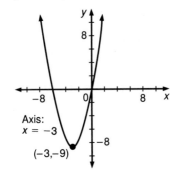

Axis:
$x = -3$

$(-3,-9)$

16. $y - 12 = -5(x + 1)^2$; $(-1,12)$; $x = -1$;

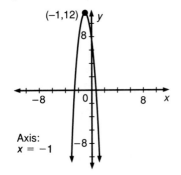

$(-1,12)$

Axis:
$x = -1$

9. $y + 1 = (x - 1)^2$; $(1,-1)$; $x = 1$;

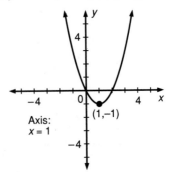

$(1,-1)$

Axis:
$x = 1$

13. $y + 2 = 2(x - 2)^2$; $(2,-2)$; $x = 2$;

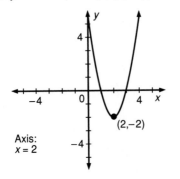

$(2,-2)$

Axis:
$x = 2$

17. $y - 8 = -2(x + 1)^2$; $(-1,8)$; $x = -1$;

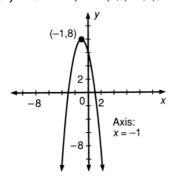

$(-1,8)$

Axis:
$x = -1$

18. $y - 35 = -2(x + 4)^2$; $(-4,35)$; $x = -4$;

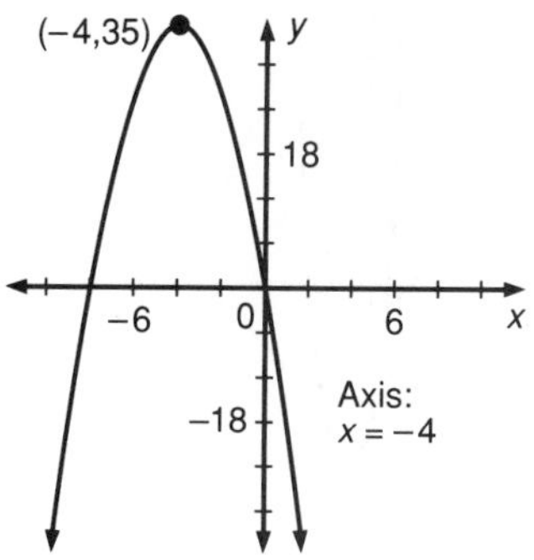

19. $y - 5 = -(x + 3)^2$; $(-3,5)$; $x = -3$;

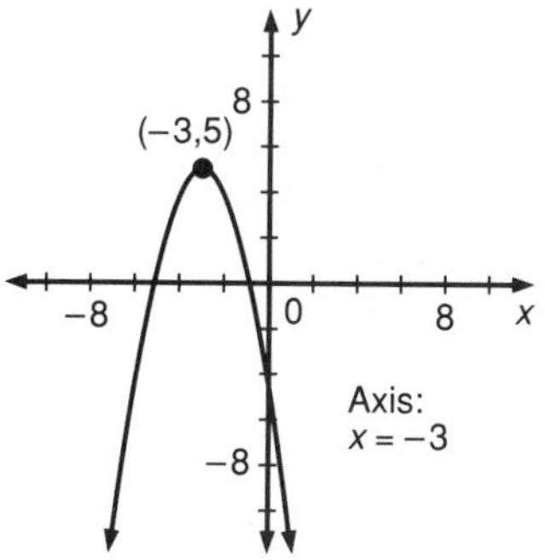

20. $y - 27 = -3(x - 3)^2$; $(3,27)$; $x = 3$;

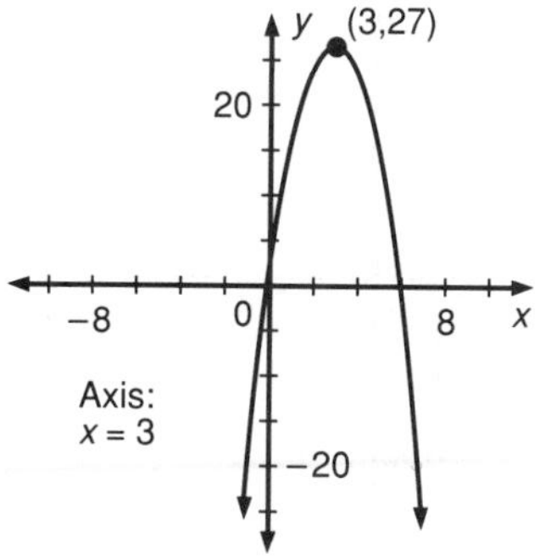

21. $y - 1 = -(x + 1)^2$; $(-1,1)$; $x = -1$;

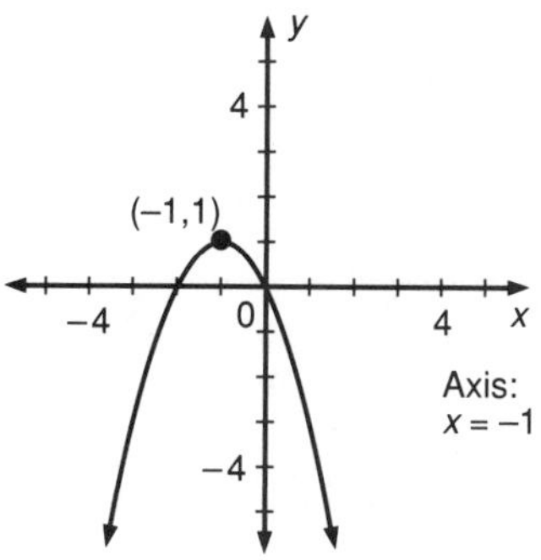

22.

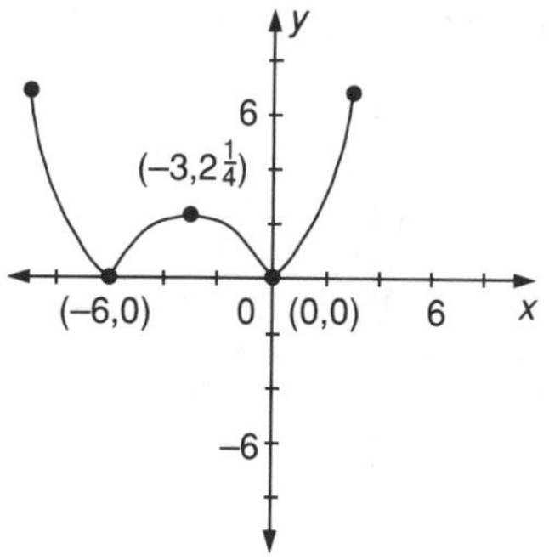

23.

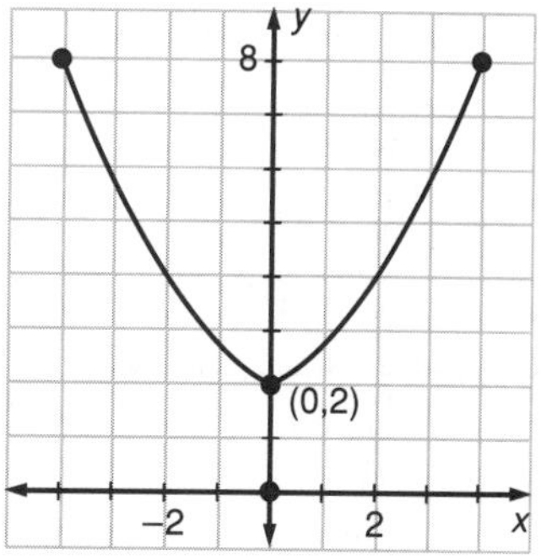

PAGE 412

Written Exercises

1. $x^2 + y^2 = 169$
2. $x^2 + y^2 = 9.61$
3. $(x - 2)^2 + (y - 3)^2 = 45$
4. $(x - 4)^2 + (y - 2)^2 = 52$
5. $(x + 2)^2 + (y - 8)^2 = 81$
6. $(x - 1)^2 + (y - 1)^2 = 3$
7. $(x - 1)^2 + (y + 1)^2 = \frac{17}{4}$
8. $\left(x - \frac{1}{4}\right)^2 + \left(y - \frac{1}{4}\right)^2 = \frac{17}{8}$

9.

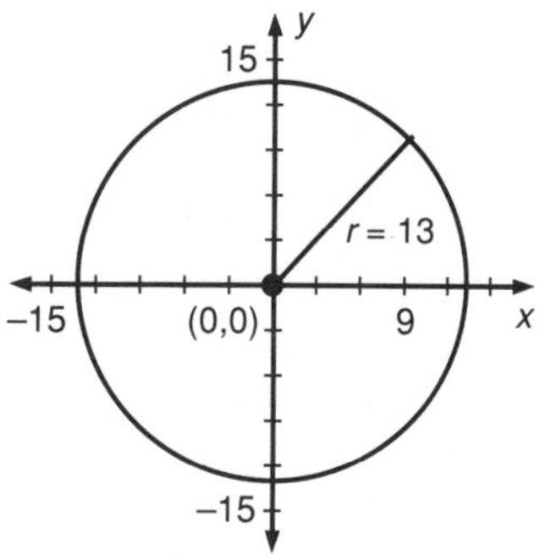

10.

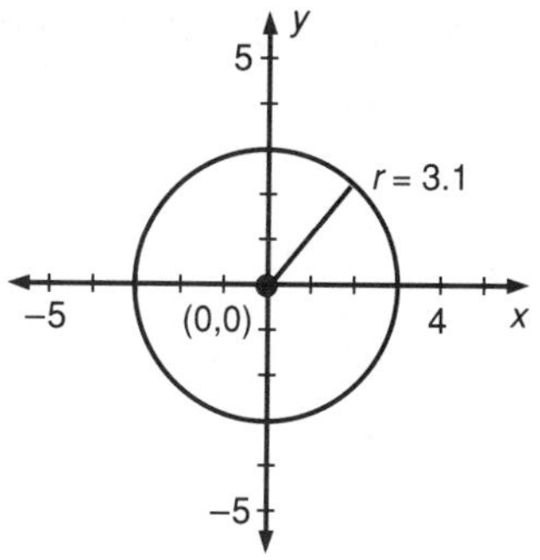

11.

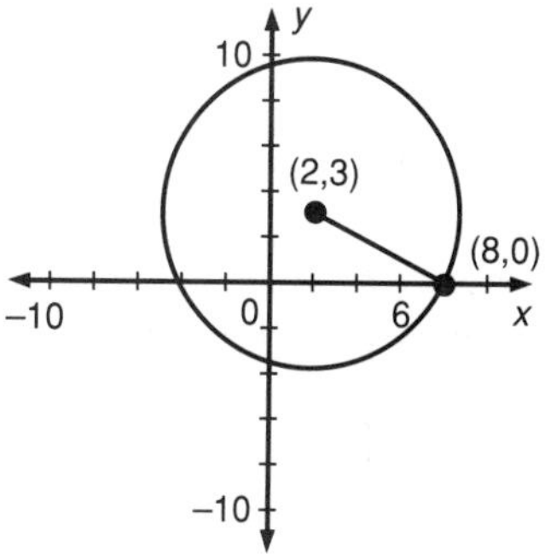

12.

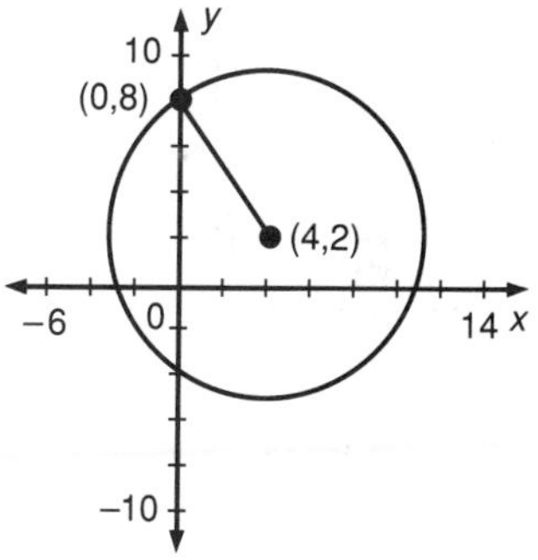

13.

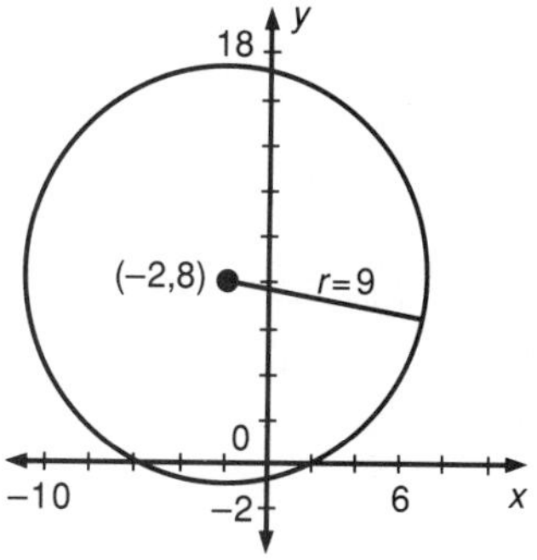

14.

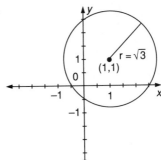

15.

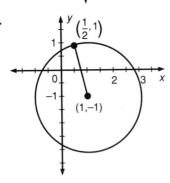

16.

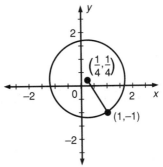

17.

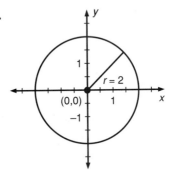

18.

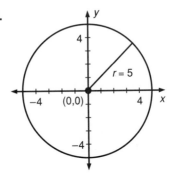

19.

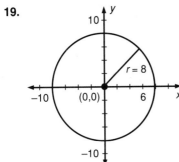

20.

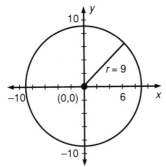

21.

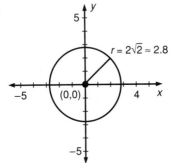

22.

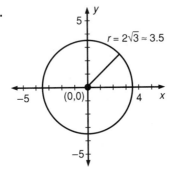

23.

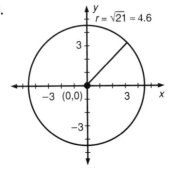

24.

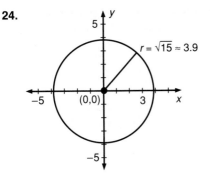

25.

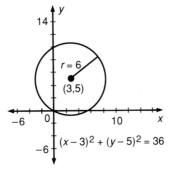

$(x-3)^2 + (y-5)^2 = 36$

26.

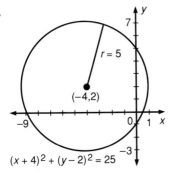

$(x+4)^2 + (y-2)^2 = 25$

27. $(x+3)^2 + (y-7)^2 = 100$

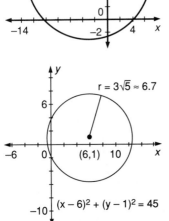

28.

$(x-6)^2 + (y-1)^2 = 45$

818

30. By definition, $CP = r$, where $P(x,y)$ is a point on the circle, C is its center, and r is its radius. C is given as $(0,0)$.

$CP = \sqrt{(x-0)^2 + (y-0)^2} = r$, so $x^2 + y^2 = r^2$.

31. By definition, $CP = r$, where $P(x,y)$ is a point on the circle, C is its center, and r is its radius. C is given as (h,k).

$CP = \sqrt{(x-h)^2 + (y-k)^2} = r$.

So, $(x-h)^2 + (y-k)^2 = r^2$.

36. $(-6)^2 + (8)^2 = 36 + 64 = 100$, so A is a point on the circle.

$(0)^2 + (10)^2 = 0 + 100 = 100$, so B is a point on the circle.

$(6)^2 + (-8)^2 = 36 + 64 = 100$, so C is a point on the circle.

$m\,\overline{AB} = \dfrac{8-10}{-6-0} = \dfrac{1}{3}$, $m\,\overline{BC} = \dfrac{10-(-8)}{0-6}$ $= -3$, and $m\,\overline{AB}(m\,\overline{BC}) = -1$, so $\overline{AB}$ is perpendicular to $\overline{BC}$, $\angle ABC$ is a right angle, and triangle ABC is a right triangle.

38. $m\,\overline{AB} = \dfrac{1}{3}$, so the slope of the perpendicular bisector is -3. Midpoint of $\overline{AB}$:

$\left(\dfrac{-6+0}{2}, \dfrac{8+10}{2}\right) = (-3,9)$

Equation of the line through $(-3,9)$ with slope -3: $y = -3x$

The center of the circle is $(0,0)$, which satisfies $y = -3x$, so $y = -3x$ passes through the center of the circle.

Mixed Review

1.

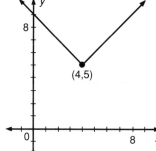

2.

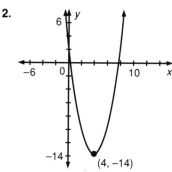

6.

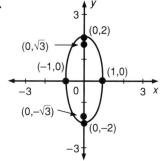

7.

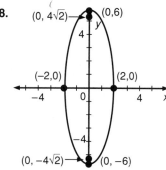

8.

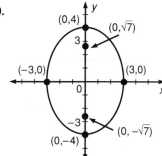

9.

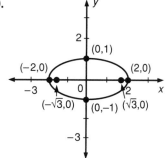

10.

11.

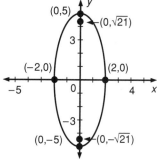

12.

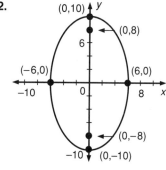

13. $\dfrac{x^2}{9} + y^2 = 1$

14. $\dfrac{x^2}{36} + \dfrac{y^2}{25} = 1$

15. $\dfrac{y^2}{25} + \dfrac{x^2}{9} = 1$

28. Let $d_1 + d_2 = 2a$. Then $PF_1 + PF_2 = 2a$, and the distance formula gives the following equation.

$$\sqrt{[x-(-f)]^2 + (y-0)^2} + \sqrt{(x-f)^2 + (y-0)^2} = 2a$$

$$\sqrt{(x+f)^2 + y^2} + \sqrt{(x-f)^2 + y^2} = 2a$$

$$\sqrt{(x+f)^2 + y^2} = 2a - \sqrt{(x-f)^2 + y^2}$$

$$x^2 + 2fx + f^2 + y^2$$
$$= 4a^2 - 4a\sqrt{x^2 - 2fx + f^2 + y^2} +$$
$$x^2 - 2fx + f^2 + y^2$$

$$4fx - 4a^2 = -4a\sqrt{x^2 - 2fx + f^2 + y^2}$$

$$fx - a^2 = -a\sqrt{x^2 - 2fx + f^2 + y^2}$$

$$f^2x^2 - 2a^2fx + a^4$$
$$= a^2x^2 - 2a^2fx + a^2f^2 + a^2y^2$$

$$a^4 - a^2f^2 - a^2x^2 + f^2x^2 = a^2y^2$$

$$a^2(a^2 - f^2) - x^2(a^2 - f^2) = a^2y^2$$

$$(a^2 - f^2)(a^2 - x^2) = a^2y^2$$

$$b^2(a^2 - x^2) = a^2y^2$$

$$b^2a^2 - b^2x^2 = a^2y^2$$

$$b^2x^2 + a^2y^2 = b^2a^2$$

$$\dfrac{x^2}{a^2} + \dfrac{y^2}{b^2} = 1$$

29. Same as for Exercise 28, but interchange x and y.

819

Mixed Review

1.

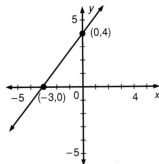

2.

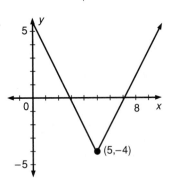

3.

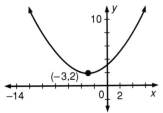

4.

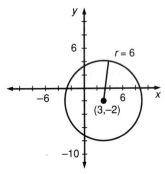

PAGE 422

11.

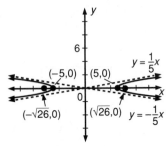

12.

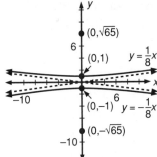

Written Exercises

1.

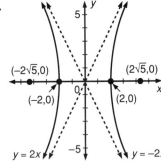

2.

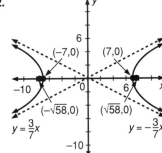

3.

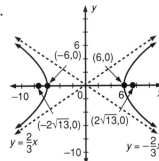

4.

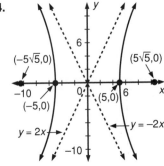

5.

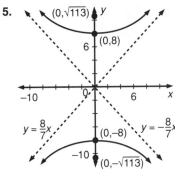

6.

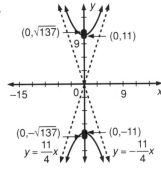

7.

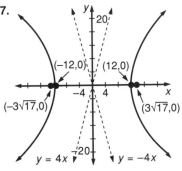

8.

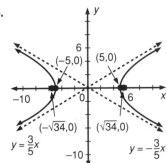

9.

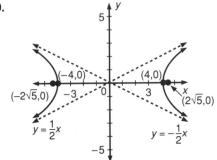

820

10.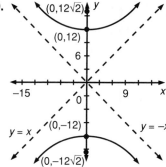

$(0,12\sqrt{2})$
$(0,12)$
$(0,-12)$
$(0,-12\sqrt{2})$
$y = x$ $y = -x$
-15 9 x

11.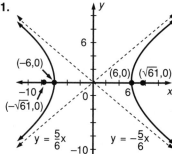

$(-6,0)$ $(6,0)$ $(\sqrt{61},0)$
$(-\sqrt{61},0)$
-10
$y = \frac{5}{6}x$ $y = -\frac{5}{6}x$

12.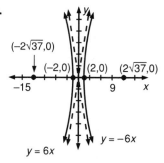

$(-2\sqrt{37},0)$
$(-2,0)$ $(2,0)$ $(2\sqrt{37},0)$
-15 9 x
$y = 6x$ $y = -6x$

19. $\dfrac{x^2}{100} - \dfrac{y^2}{16} = 1$

20. $\dfrac{y^2}{196} - \dfrac{x^2}{16} = 1$

22. Let $d_1 - d_2 = 2a$. Then $PF_1 - PF_2 = 2a$, and the distance formula gives the following equation.

$$\sqrt{[x - (-f)]^2 + (y - 0)^2} - \sqrt{(x - f)^2 + (y - 0)^2} = 2a$$

$$\sqrt{(x + f)^2 + y^2} - \sqrt{(x - f)^2 + y^2} = 2a$$

$$\sqrt{(x + f)^2 + y^2} = 2a + \sqrt{(x - f)^2 + y^2}$$

$$x^2 + 2fx + f^2 + y^2$$
$$= 4a^2 + 4a\sqrt{x^2 - 2fx + f^2 + y^2} + x^2 - 2fx + f^2 + y^2$$

$$4fx - 4a^2 = 4a\sqrt{x^2 - 2fx + f^2 + y^2}$$

$$fx - a^2 = a\sqrt{x^2 - 2fx + f^2 + y^2}$$

$$f^2x^2 - 2a^2fx + a^4$$
$$= a^2x^2 - 2a^2fx + a^2f^2 + a^2y^2$$

$$f^2x^2 - a^2x^2 - a^2f^2 + a^4 = a^2y^2$$

$$x^2(f^2 - a^2) - a^2(f^2 - a^2) = a^2y^2$$

$$(f^2 - a^2)(x^2 - a^2) = a^2y^2$$

$$b^2(x^2 - a^2) = a^2y^2$$

$$b^2x^2 - a^2y^2 = b^2a^2$$

$$\frac{x^2}{a^2} - \frac{y^2}{b^2} = 1$$

23. Same as Exercise 28, but interchange x and y.

Mixed Review

1.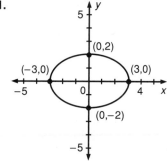

$(0,2)$
$(-3,0)$ $(3,0)$
$(0,-2)$
-5 4 x

2.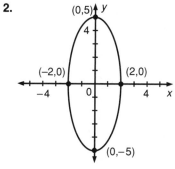

$(0,5)$
$(-2,0)$ $(2,0)$
-4 4 x
$(0,-5)$

Written Exercises

1.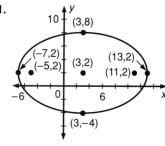

$(3,8)$
$(-7,2)$
$(-5,2)$ $(3,2)$ $(13,2)$
$(11,2)$
-6 6 x
$(3,-4)$

2.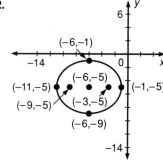

$(-6,-1)$
$(-6,-5)$
$(-11,-5)$ $(-3,-5)$ $(-1,-5)$
$(-9,-5)$
$(-6,-9)$
-14 x

3.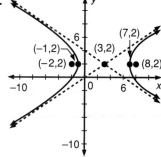

$(-1,2)$ $(3,2)$ $(7,2)$
$(-2,2)$ $(8,2)$
-10 6 x

4.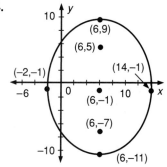

$(6,9)$
$(6,5)$
$(-2,-1)$ $(14,-1)$
$(6,-1)$
$(6,-7)$
$(6,-11)$
-6 10 x

821

5.

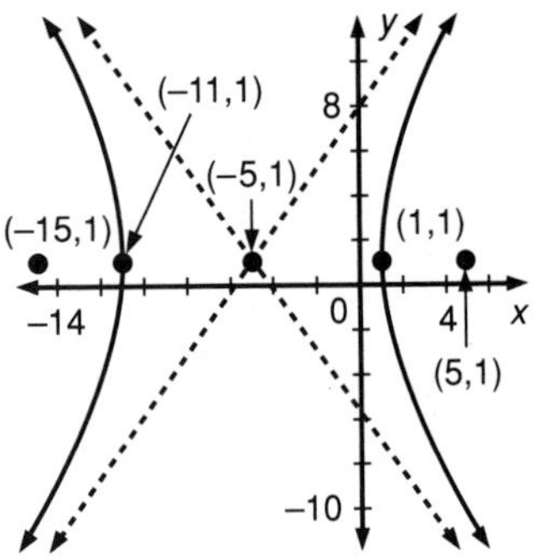

y
(−11,1)
8
(−5,1)
(−15,1)
(1,1)
−14
0
4
x
(5,1)
−10

6.

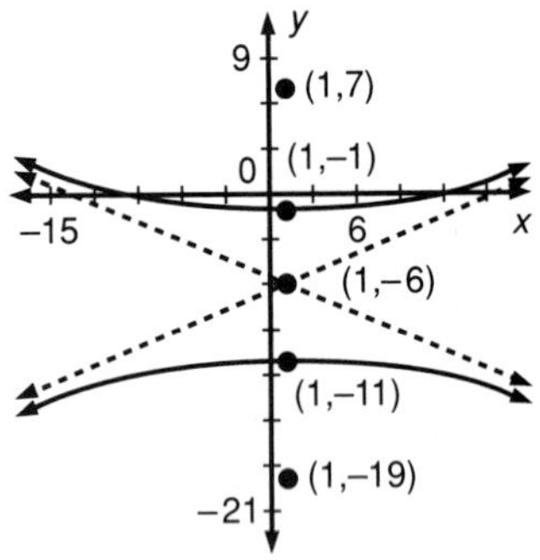

y
9
(1,7)
0
(1,−1)
−15
6
x
(1,−6)
(1,−11)
(1,−19)
−21

7.

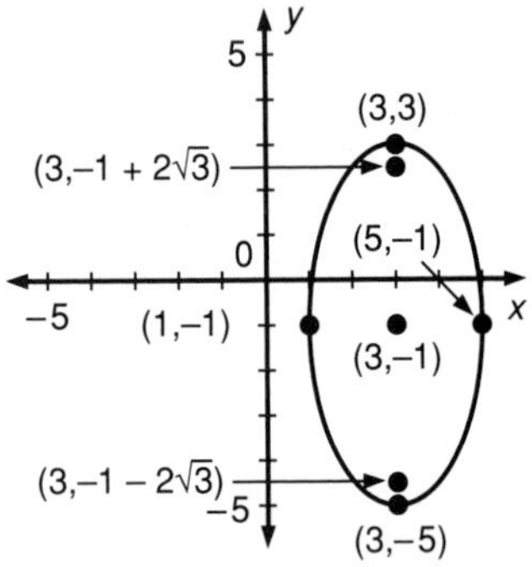

y
5
(3,3)
$(3,-1+2\sqrt{3})$
0
(5,−1)
−5
(1,−1)
x
(3,−1)
$(3,-1-2\sqrt{3})$
−5
(3,−5)

8.

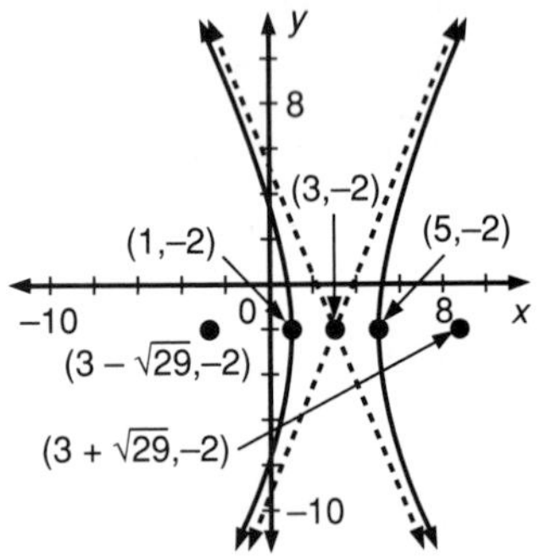

y
8
(3,−2)
(1,−2)
(5,−2)
−10
0
8
x
$(3-\sqrt{29},-2)$
$(3+\sqrt{29},-2)$
−10

9.

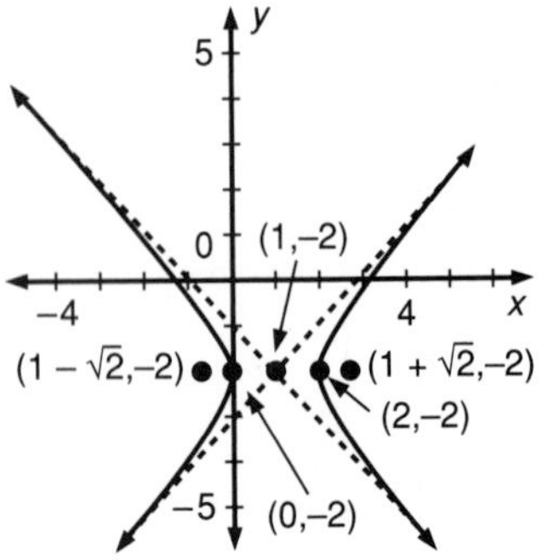

y
5
0
(1,−2)
−4
4
x
$(1-\sqrt{2},-2)$
$(1+\sqrt{2},-2)$
(2,−2)
−5
(0,−2)

10.

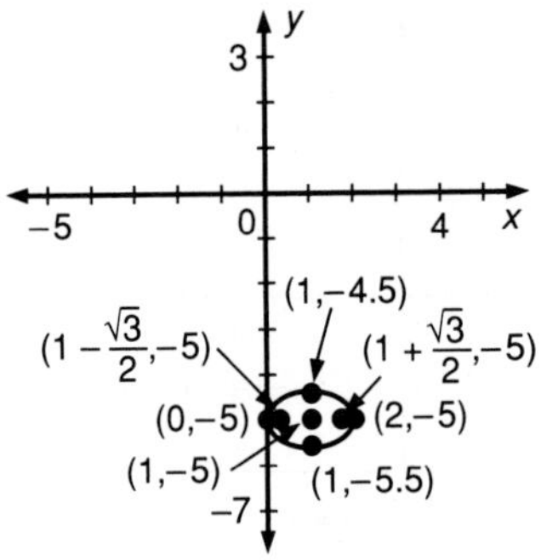

y
3
−5
0
4
x
(1,−4.5)
$(1-\frac{\sqrt{3}}{2},-5)$
$(1+\frac{\sqrt{3}}{2},-5)$
(0,−5)
(2,−5)
(1,−5)
−7
(1,−5.5)

11.

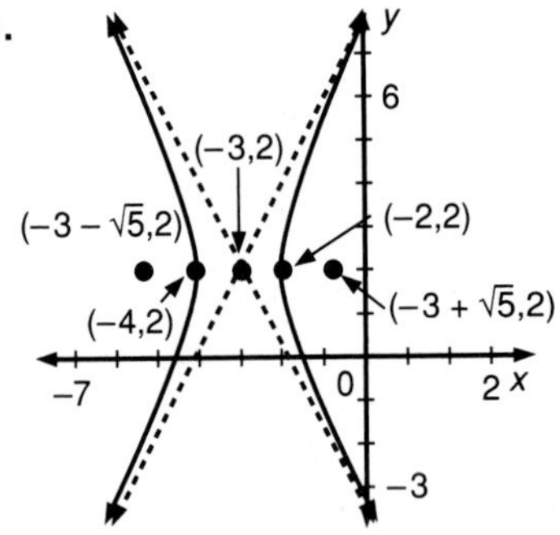

y
6
(−3,2)
$(-3-\sqrt{5},2)$
(−2,2)
(−4,2)
$(-3+\sqrt{5},2)$
−7
0
2
x
−3

12.

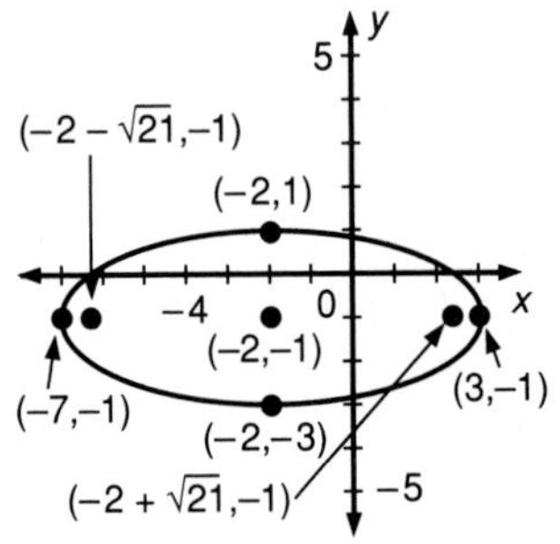

y
5
$(-2-\sqrt{21},-1)$
(−2,1)
−4
0
x
(−2,−1)
(−7,−1)
(3,−1)
(−2,−3)
$(-2+\sqrt{21},-1)$
−5

13.

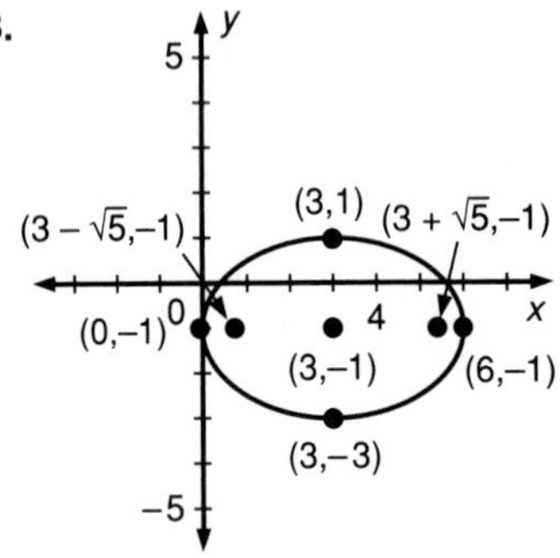

y
5
(3,1)
$(3-\sqrt{5},-1)$
$(3+\sqrt{5},-1)$
0
4
x
(0,−1)
(3,−1)
(6,−1)
(3,−3)
−5

14.

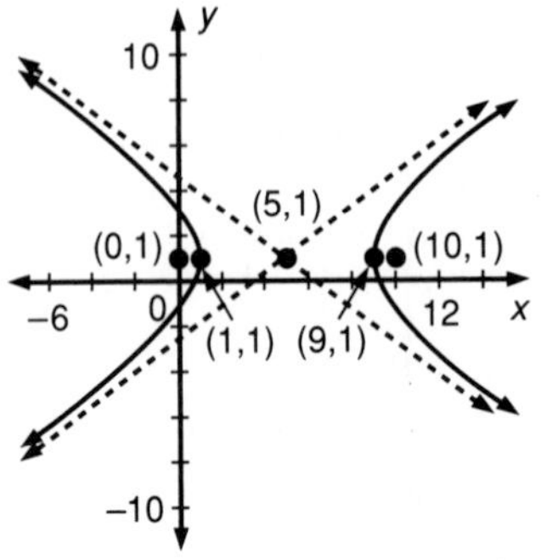

y
10
(5,1)
(0,1)
(10,1)
−6
0
12
x
(1,1)
(9,1)
−10

18.

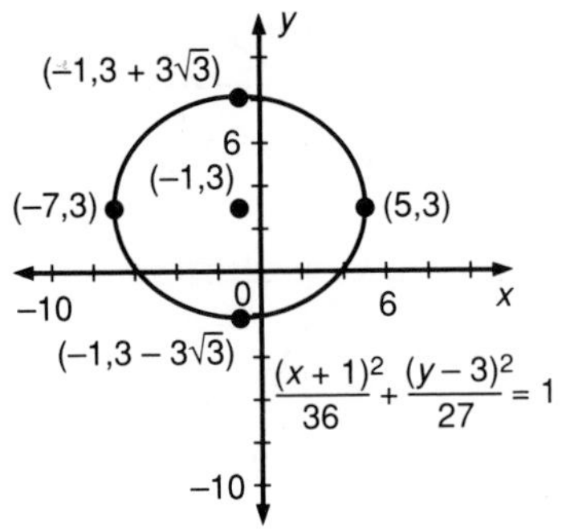

y
$(-1,3+3\sqrt{3})$
6
(−1,3)
(−7,3)
(5,3)
−10
0
6
x
$(-1,3-3\sqrt{3})$
$\frac{(x+1)^2}{36}+\frac{(y-3)^2}{27}=1$
−10

PAGE 439

9.

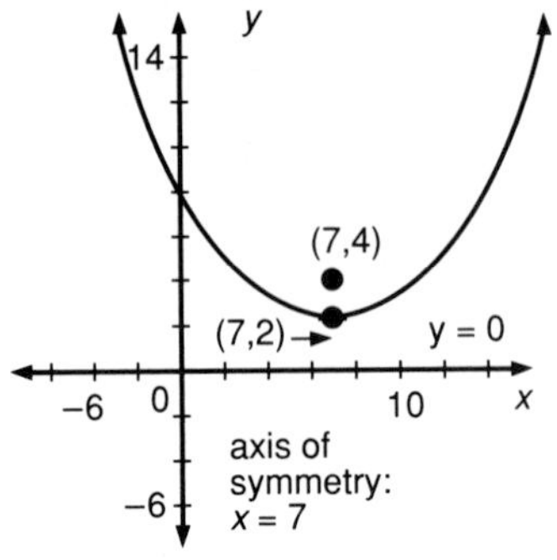

y
14
(7,4)
(7,2)
y = 0
−6
0
10
x
axis of symmetry:
x = 7
−6

10.

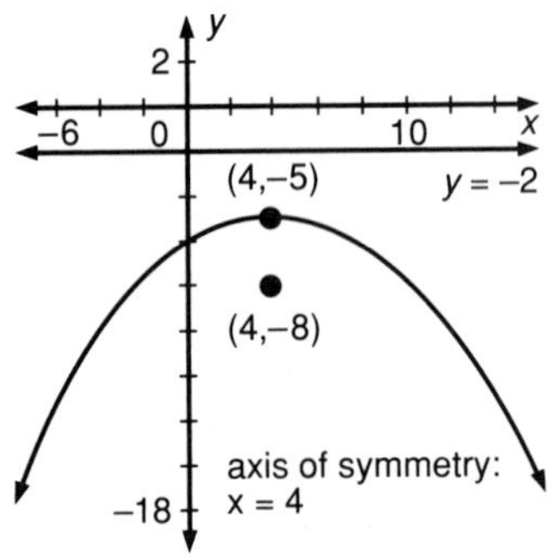

y
2
−6
0
10
x
(4,−5)
y = −2
(4,−8)
axis of symmetry:
x = 4
−18

11.

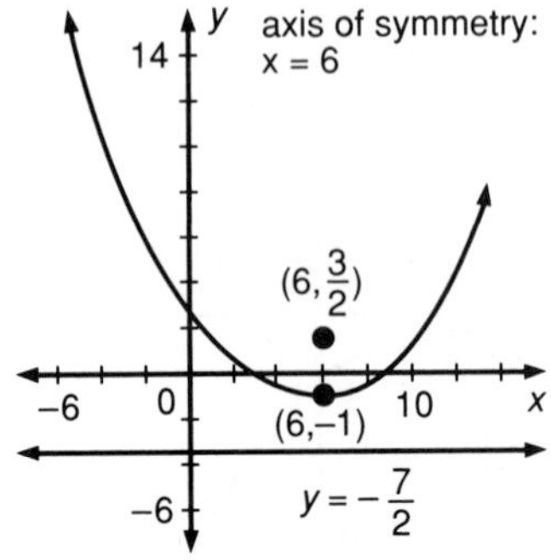

y
14
axis of symmetry:
x = 6
$(6,\frac{3}{2})$
−6
0
10
x
(6,−1)
$y=-\frac{7}{2}$
−6

12.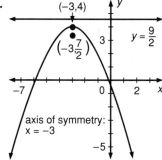

$(-3,4)$

$(-3\frac{7}{2})$

$y = \frac{9}{2}$

axis of symmetry:
$x = -3$

13.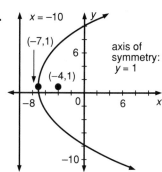

axis of
symmetry:
$y = 6$

$x = 7$

$(3,6)$ $\rightarrow$ $\leftarrow (5,6)$

14.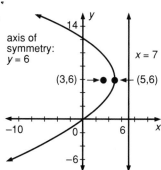

$x = -10$

$(-7,1)$

axis of
symmetry:
$y = 1$

$(-4,1)$

15.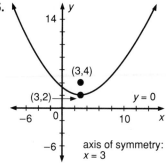

$(3,4)$

$(3,2)$

$y = 0$

axis of symmetry:
$x = 3$

16.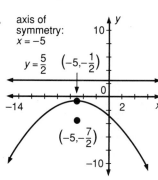

axis of
symmetry:
$x = -5$

$y = \frac{5}{2}$ $(-5,-\frac{1}{2})$

$(-5,-\frac{7}{2})$

17.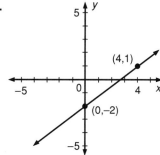

axis of
symmetry:
$x = 1$

$y = -\frac{11}{2}$

$(1,-\frac{9}{2})$ $(1,-5)$

18.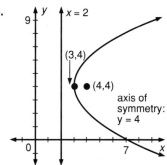

$x = 2$

$(3,4)$

$(4,4)$

axis of
symmetry:
$y = 4$

19.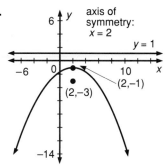

$x = -\frac{1}{2}$

axis of
symmetry:
$y = 5$

$(-\frac{3}{2},5)$ $(-1,5)$

20.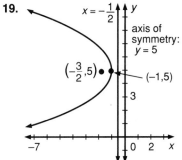

axis of
symmetry:
$x = 2$

$y = 1$

$(2,-1)$

$(2,-3)$

22. Let (x,y) be a point of the parabola. By definition, the distance to the focus, given as $(p,0)$, must be equal to the distance to the directrix, given as $x = -p$.

$$\sqrt{(x - p)^2 + (y - 0)^2}$$
$$= \sqrt{[x - (-p)]^2 + (y - y)^2}$$

$x^2 - 2px + p^2 + y^2 = x^2 + 2px + p^2$
$y^2 = 4px$

Mixed Review

1.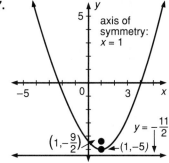

$(4,1)$

$(0,-2)$

2.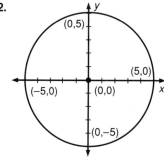

$(0,5)$

$(5,0)$

$(-5,0)$ $(0,0)$

$(0,-5)$

3.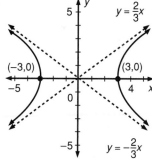

$y = \frac{2}{3}x$

$(-3,0)$ $(3,0)$

$y = -\frac{2}{3}x$

4.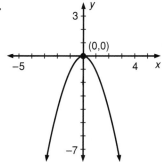

$(0,0)$

823

8.

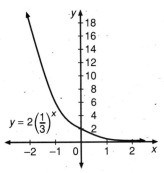

Domain: {real numbers}; Range: $\{y \mid y > 0\}$;
Decreasing; y-intercept $(0,2)$

9.

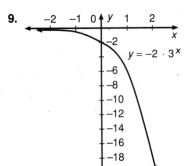

Domain: {real numbers}; Range: $\{y \mid y < 0\}$;
Decreasing; y-intercept $(0,-2)$

10.

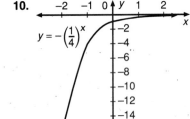

Domain: {real numbers}; Range: $\{y \mid y < 0\}$;
Increasing; y-intercept $(0,-1)$

11.

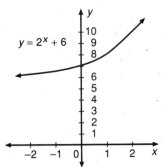

Domain: {real numbers}; Range: $\{y \mid y > 6\}$;
Increasing; y-intercept $(0,7)$

12.

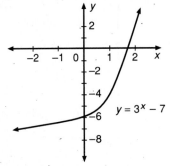

Domain: {real numbers}; Range: $\{y \mid y > -7\}$;
Increasing; y-intercept $(0,-6)$

17.

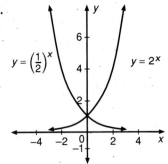

The graph of $y = \left(\frac{1}{2}\right)^x$ is the reflection
about the y-axis of the graph of $y = 2^x$.

18.

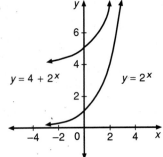

The graph of $y = 4 + 2^x$ is that of
$y = 2^x$ shifted 4 units up.

19.

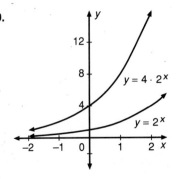

The graph of $y = 4 \cdot 2^x$ rises more
steeply than that of $y = 2^x$.

20.

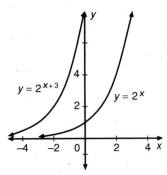

The graph of $y = 2^{x+3}$ is that of $y = 2^x$
shifted 3 units left.

21.

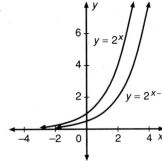

The graph of $y = 2^{x-1}$ is that of $y = 2^x$
shifted 1 unit right.

22.

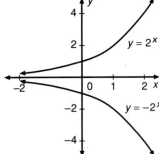

The graph of $y = -2^x$ is the reflection
about the x-axis of the graph of $y = 2^x$.

23.

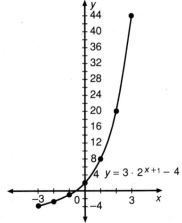

24.

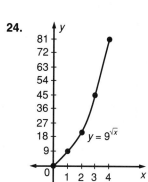

$y = 9^{\sqrt{x}}$

25. When $a = 1$, $y = a^x$ is the constant function $y = 1$.

26. No; any power of a positive number is positive.

27. The line $y = -c$

PAGE 593

3. $\begin{bmatrix} 16 & 13 & 2 \\ 5 & -1 & -8 \end{bmatrix}$

4. Does not exist: dimensions do not fit

5. $\begin{bmatrix} 14 & 8 & 18 \\ 29 & 15 & 39 \\ -11 & 83 & 9 \end{bmatrix}$

6. $\begin{bmatrix} 9 & -4 \\ -2 & 1 \end{bmatrix}$ **7.** $\begin{bmatrix} \frac{-2}{3} & \frac{-1}{3} \\ \frac{-2}{3} & \frac{-4}{3} \end{bmatrix}$

8. $\begin{bmatrix} \frac{3}{4} & \frac{5}{8} \\ \frac{1}{4} & \frac{3}{8} \end{bmatrix}$ **9.** $\begin{bmatrix} \frac{1}{23} & \frac{7}{46} \\ \frac{4}{23} & \frac{5}{46} \end{bmatrix}$

10. Not possible

11. $\begin{bmatrix} \frac{-8}{59} & \frac{1}{118} & \frac{-13}{59} \\ \frac{14}{59} & \frac{13}{118} & \frac{8}{59} \\ \frac{-13}{59} & \frac{9}{118} & \frac{1}{59} \end{bmatrix}$

12. $\begin{bmatrix} \frac{-2}{27} & \frac{-8}{27} & \frac{11}{54} \\ \frac{-19}{27} & \frac{-22}{27} & \frac{37}{54} \\ \frac{-8}{27} & \frac{-5}{27} & \frac{17}{54} \end{bmatrix}$

13. Not possible

14. $B \cdot I$

$= \begin{bmatrix} -1(1) + -3(0) & -1(0) + -3(1) \\ 2(1) + 1(0) & 2(0) + 1(1) \end{bmatrix}$

$= \begin{bmatrix} -1 & -3 \\ 2 & 1 \end{bmatrix}$

$I \cdot B$

$= \begin{bmatrix} 1(-1) + 0(2) & 1(-3) + 0(1) \\ 0(-1) + 1(2) & 0(-3) + 1(1) \end{bmatrix}$

$= \begin{bmatrix} -1 & -3 \\ 2 & 1 \end{bmatrix}$

Thus, $B \cdot I = I \cdot B$.

15. $A \cdot C = \begin{bmatrix} 2(-2) + 3(4) & 2(3) + 3(2) \\ -1(-2) + 4(4) & -1(3) + 4(2) \end{bmatrix}$

$= \begin{bmatrix} 8 & 12 \\ 18 & 5 \end{bmatrix}$

$C \cdot A = \begin{bmatrix} -2(2) + 3(-1) & -2(3) + 3(4) \\ 4(2) + 2(-1) & 4(3) + 2(4) \end{bmatrix}$

$= \begin{bmatrix} -7 & 6 \\ 6 & 20 \end{bmatrix}$

Thus, $A \cdot C \neq C \cdot A$.

16. $A \cdot (B \cdot C)$

$= \begin{bmatrix} 2 & 3 \\ -1 & 4 \end{bmatrix} \cdot \left(\begin{bmatrix} -1 & -3 \\ 2 & 1 \end{bmatrix} \cdot \begin{bmatrix} -2 & 3 \\ 4 & 2 \end{bmatrix} \right)$

$= \begin{bmatrix} 2 & 3 \\ -1 & 4 \end{bmatrix} \cdot \begin{bmatrix} -10 & -9 \\ 0 & 8 \end{bmatrix} = \begin{bmatrix} -20 & 6 \\ 10 & 41 \end{bmatrix}$

$(A \cdot B) \cdot C$

$= \left(\begin{bmatrix} 2 & 3 \\ -1 & 4 \end{bmatrix} \cdot \begin{bmatrix} -1 & -3 \\ 2 & 1 \end{bmatrix} \right) \cdot \begin{bmatrix} -2 & 3 \\ 4 & 2 \end{bmatrix}$

$= \begin{bmatrix} 4 & -3 \\ 9 & 7 \end{bmatrix} \cdot \begin{bmatrix} -2 & 3 \\ 4 & 2 \end{bmatrix} = \begin{bmatrix} -20 & 6 \\ 10 & 41 \end{bmatrix}$

Thus, $A \cdot (B \cdot C) = (A \cdot B) \cdot C$.

Brainteaser

The water level will drop. While the cobblestones are in the boat, they displace an amount of water equivalent to their weight. After the stones sink, they displace an amount of water equivalent to their volume. This volume is less, since the stones have a higher density than water. Otherwise they wouldn't sink.

PAGE 596

Classroom Exercises

5. $\begin{vmatrix} -3 & 5 \\ 2 & -3 \end{vmatrix}$ **6.** $\begin{vmatrix} -3 & 5 \\ 2 & -3 \end{vmatrix}$

7. $\begin{vmatrix} 6 & 5 \\ 9 & -3 \end{vmatrix}$ **8.** $\begin{vmatrix} -3 & 6 \\ 2 & 9 \end{vmatrix}$

Written Exercises

22. $\left(\dfrac{3b_2 + 4b_1}{a_1 b_2 - a_2 b_1}, \dfrac{-4a_1 - 3a_2}{a_1 b_2 - a_2 b_1} \right)$

24. $\left(\dfrac{p_1 n_2 + p_2 n_1}{m_1 n_2 + m_2 n_1}, \dfrac{m_1 p_2 - m_2 p_1}{m_1 n_2 + m_2 n_1} \right)$

25. $|-A| = \begin{vmatrix} -a_1 & -b_1 \\ -c_1 & -d_1 \end{vmatrix} = a_1 d_1 - c_1 b_1$

$|A| = \begin{vmatrix} a_1 & b_1 \\ c_1 & d_1 \end{vmatrix} = a_1 d_1 - c_1 b_1$

Thus, $|-A| = |A|$.

26. $|kA| = \begin{vmatrix} ka_1 & kb_1 \\ kc_1 & kd_1 \end{vmatrix}$

$= k^2 a_1 d_1 - k^2 c_1 b_1$

$= k^2(a_1 d_1 - c_1 b_1)$

$= k^2 |A|$

27. $|A \cdot B| = \begin{vmatrix} a_1 a_2 + b_1 c_2 & a_1 b_2 + b_1 d_2 \\ c_1 a_2 + d_1 c_2 & c_1 b_2 + d_1 d_2 \end{vmatrix}$

$= (a_1 a_2 + b_1 c_2)(c_1 b_2 + d_1 d_2) - (c_1 a_2 + d_1 c_2)(a_1 b_2 + b_1 d_2)$

$= a_1 a_2 b_2 c_1 + b_1 b_2 c_1 c_2 + a_1 a_2 d_1 d_2 + b_1 c_2 d_1 d_2 - [a_1 a_2 b_2 c_1 + a_2 b_1 c_1 d_2 + a_1 b_2 c_2 d_1 + b_1 c_2 d_1 d_2]$

$= b_1 b_2 c_1 c_2 + a_1 a_2 d_1 d_2 - a_2 b_1 c_1 d_2 - a_1 b_2 c_2 d_1$

$= (a_1 d_1 - c_1 b_1)(a_2 d_2 - c_2 b_2)$

$= |A| \cdot |B|$

Midchapter Review

1. $\begin{bmatrix} -2 & 4 \\ 1 & 10 \end{bmatrix}$

2. Does not exist; dimensions are unequal.

3. $\begin{bmatrix} 6 & -3 \\ 0 & -8 \end{bmatrix}$ **4.** $\begin{bmatrix} 8 & -3 \\ -1 & 0 \end{bmatrix}$

5. $\begin{bmatrix} 2 & 0 \\ -1 & 8 \end{bmatrix} = A$ **6.** $\begin{bmatrix} 8 & -8 \\ -4 & -4 \end{bmatrix}$

7. $\begin{bmatrix} -8 & 8 \\ 20 & 12 \end{bmatrix}$ **8.** $\begin{bmatrix} -12 & 32 \\ 2 & 16 \end{bmatrix}$

9. Does not exist; dimensions do not fit.

10. $\begin{bmatrix} -6 & 3 \\ 0 & 8 \end{bmatrix} = C$ **11.** $\begin{bmatrix} -7 & 7 \\ 1 & -1 \end{bmatrix}$

12. $\begin{bmatrix} -14 \\ 15 \end{bmatrix}$

PAGE 664

Classroom Exercises

1. $x^2 = 40^2 + 70^2 - 2(40)(70) \cos 50$

2. $(4.0)^2 = (2.8)^2 + (3.0)^2 - 2(2.8)(3.0) \cos A$

3. $x^2 = 50^2 + 60^2 - 2(50)(60) \cos 35$

4. $(40)^2 = (50)^2 + (70)^2 - 2(50)(70) \cos A$

Written Exercises

11.

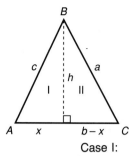

Case I:

For △ I, $c^2 = h^2 + x^2$.

For △ II, $a^2 = h^2 + (b - x)^2$
$$a^2 = h^2 + b^2 - 2bx + x^2$$

Subtract 1st equation from second equation.

$$
\begin{aligned}
a^2 &= h^2 + b^2 - 2bx + x^2 \\
c^2 &= h^2 + x^2 \\
\hline
a^2 - c^2 &= b^2 - 2bx
\end{aligned}
$$

But $\frac{x}{c} = \cos A$, or $x = c \cos A$,
so, $a^2 - c^2 = b^2 - 2b(c \cos A)$,
or $a^2 = b^2 + c^2 - 2bc \cos A$
by substitution.

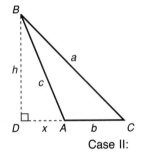

Case II:

For △ BDA, $c^2 = h^2 + x^2$.

For △ BDC, $a^2 = h^2 + (b + x)^2$
$$a^2 = h^2 + b^2 + 2bx + x^2$$

Subtract 1st equation from second equation.

$$
\begin{aligned}
a^2 &= h^2 + b^2 + 2bx + x^2 \\
c^2 &= h^2 + x^2 \\
\hline
a^2 - c^2 &= b^2 + 2bx
\end{aligned}
$$

But $\frac{x}{c} = \cos(180 - A)$,
so $x = c \cos(180 - A)$,
or $x = -c \cos A$.
Thus, by substitution,
$a^2 - c^2 = b^2 + 2b(-c \cos A)$,
or $a^2 = b^2 + c^2 - 2bc \cos A$.

826

12.

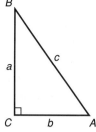

By the Law of Cosines,
$c^2 = a^2 + b^2 - 2ab \cos C$. But m ∠C = 90,
so cos C = 0. Thus, $c^2 = a^2 + b^2$.

PAGE 696

3.

θ	0	$\frac{\pi}{4}$	$\frac{\pi}{2}$	$\frac{3\pi}{4}$	π	$\frac{5\pi}{4}$
T	0	−1	0	1	0	−1

θ	$\frac{3\pi}{2}$	$\frac{7\pi}{4}$	2π
T	0	1	0

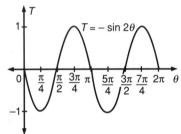

4.

θ	0	$\frac{\pi}{2}$	π	$\frac{3\pi}{2}$	2π
T	2	$\sqrt{3}$	1	0	−1

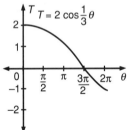

5.

θ	0	$\frac{\pi}{2}$	π	$\frac{3\pi}{2}$	2π
T	1	2	1	0	1

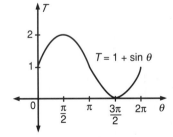

6.

θ	0	$\frac{\pi}{2}$	π	$\frac{3\pi}{2}$	2π
T	2	3	4	3	2

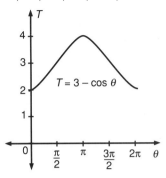

7.

θ	0	$\frac{\pi}{2}$	π	$\frac{3\pi}{2}$	2π
T	−4	0	−4	−8	−4

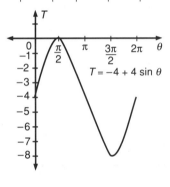

8.

θ	0	$\frac{\pi}{2}$	π	$\frac{3\pi}{2}$	2π
T	3	8	3	−2	3

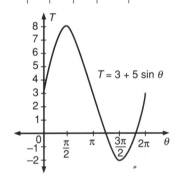

9. $p = \frac{2\pi}{3}$; $a = 1$

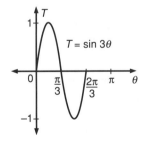

10. $p = \pi$; $a = \frac{1}{2}$

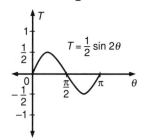

$T = \frac{1}{2}\sin 2\theta$

11. $p = \frac{\pi}{2}$; $a = 5$

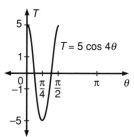

$T = 5\cos 4\theta$

12. $p = \frac{\pi}{3}$; $a = 3$

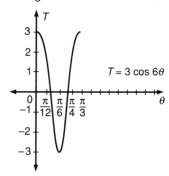

$T = 3\cos 6\theta$

13. $p = \pi$; $a = 5$

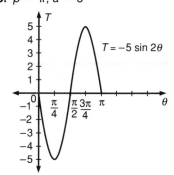

$T = -5\sin 2\theta$

14. $p = 4\pi$; $a = 2$

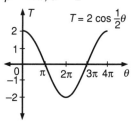

$T = 2\cos \frac{1}{2}\theta$

15. $p = 6\pi$; $a = \frac{1}{4}$

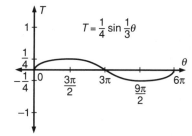

$T = \frac{1}{4}\sin \frac{1}{3}\theta$

16. $p = 8\pi$; $a = 5$

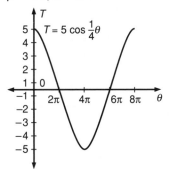

$T = 5\cos \frac{1}{4}\theta$

17. $p = \frac{2\pi}{9}$; $a = 6$

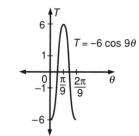

$T = -6\cos 9\theta$

18. $p = \pi$; $a = 3$; max $= 9$; min $= 3$

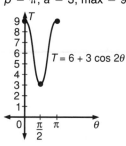

$T = 6 + 3\cos 2\theta$

19. $p = \frac{\pi}{2}$; $a = 5$; max $= -1$; min $= -11$

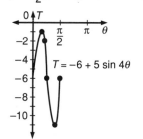

$T = -6 + 5\sin 4\theta$

20. $p = \frac{2\pi}{3}$; $a = \frac{1}{2}$; max $= -2$; min $= -3$

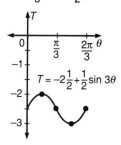

$T = -2\frac{1}{2} + \frac{1}{2}\sin 3\theta$

21. $-\frac{5\pi}{4}, -\frac{\pi}{4}, \frac{3\pi}{4}, \frac{7\pi}{4}$

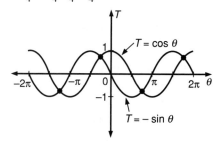

$T = \cos \theta$

$T = -\sin \theta$

22.

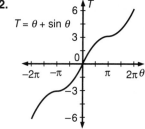

$T = \theta + \sin \theta$

23.

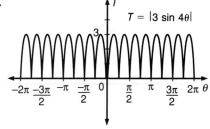

$T = |3\sin 4\theta|$

PAGE 700

7.

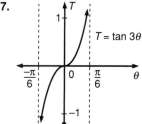

$T = \tan 3\theta$

8.

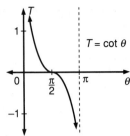

$T = \cot \theta$

9.

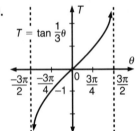

$T = \cot \frac{1}{2}\theta$

10.

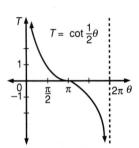

$T = \tan \frac{1}{3}\theta$

11.

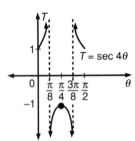

$T = \sec 4\theta$

12.

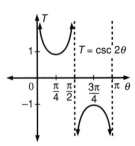

$T = \csc 2\theta$

13.

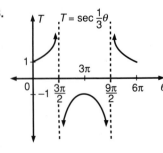

$T = \sec \frac{1}{3}\theta$

14.

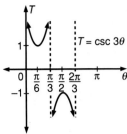

$T = \csc 3\theta$

15. One possible answer: $T = 2 \tan 3\theta$ is steeper, and has a period that is one-third as great as that of $T = \tan \theta$.

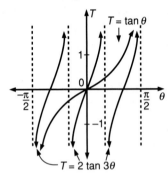

$T = \tan \theta$

$T = 2 \tan 3\theta$

16. One possible answer; $T = \frac{1}{2} \sec 2\theta$ is steeper, and has a period that is one-half as great as that of $T = \sec \theta$. It also extends to $y = \pm\frac{1}{2}$, whereas $T = \sec \theta$ extends only to $y = \pm 1$.

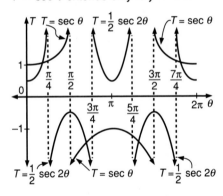

$T = \sec \theta$ $T = \frac{1}{2} \sec 2\theta$ $T = \sec \theta$

$T = \frac{1}{2} \sec 2\theta$ $T = \sec \theta$ $T = \frac{1}{2} \sec 2\theta$

PAGE 719

12. $(\sin \theta + \cos \theta)^2$
$= \sin^2 \theta + 2 \sin \theta \cos \theta + \cos^2 \theta$
$= 1 + 2 \sin \theta \cos \theta$

13. $\cos^2 \theta (\cot^2 \theta + 1)$
$= \cos^2 \theta (\csc^2 \theta)$
$= \dfrac{\cos^2 \theta}{\sin^2 \theta} = \cot^2 \theta$

14. $\cos^2 \theta - \sin^2 \theta = (1 - \sin^2 \theta) - \sin^2 \theta$
$= 1 - 2 \sin^2 \theta$

15. $\dfrac{\cos \theta}{\sin \theta} + \dfrac{\sin \theta}{1 + \cos \theta}$

$= \dfrac{\cos \theta (1 + \cos \theta)}{\sin \theta (1 + \cos \theta)} + \dfrac{\sin^2 \theta}{\sin \theta (1 + \cos \theta)}$

$= \dfrac{\cos \theta + \cos^2 \theta + \sin^2 \theta}{\sin \theta (1 + \cos \theta)}$

$= \dfrac{\cos \theta + 1}{\sin \theta (1 + \cos \theta)} = \dfrac{1}{\sin \theta} = \csc \theta$

16. $\dfrac{\cot \theta + \tan \theta}{\csc^2 \theta}$

$= \left(\dfrac{\cos \theta}{\sin \theta} + \dfrac{\sin \theta}{\cos \theta} \right) \div \dfrac{1}{\sin^2 \theta}$

$= \left(\dfrac{\cos \theta}{\sin \theta} + \dfrac{\sin \theta}{\cos \theta} \right) \cdot \sin^2 \theta$

$= \cos \theta \sin \theta + \dfrac{\sin^3 \theta}{\cos \theta}$

$= \dfrac{\cos^2 \theta \sin \theta + \sin^3 \theta}{\cos \theta}$

$= \dfrac{\sin \theta (\cos^2 \theta + \sin^2 \theta)}{\cos \theta}$

$= \dfrac{\sin \theta}{\cos \theta} = \tan \theta$

17. $\dfrac{\sin \theta + \tan \theta}{\cot \theta + \csc \theta}$

$= \dfrac{\sin \theta + \dfrac{\sin \theta}{\cos \theta}}{\dfrac{\cos \theta}{\sin \theta} + \dfrac{1}{\sin \theta}}$

$= \dfrac{\dfrac{\cos \theta \sin \theta}{\cos \theta} + \dfrac{\sin \theta}{\cos \theta}}{\dfrac{\cos \theta + 1}{\sin \theta}}$

$= \dfrac{\cos \theta \sin \theta + \sin \theta}{\cos \theta} \cdot \dfrac{\sin \theta}{\cos \theta + 1}$

$= \dfrac{\sin \theta (\cos \theta + 1)}{\cos \theta} \cdot \dfrac{\sin \theta}{\cos \theta + 1}$

$= \dfrac{\sin \theta}{\cos \theta} \cdot \sin \theta = \sin \theta \tan \theta$

18. $\dfrac{\cot \theta}{1 + \cot^2 \theta} = \dfrac{\cot \theta}{\csc^2 \theta}$

$= \cot \theta \cdot \dfrac{1}{\csc^2 \theta}$

$= \dfrac{\cos \theta}{\sin \theta} \cdot \sin^2 \theta = \sin \theta \cos \theta$

19. $\dfrac{1 + \cot \theta}{\cot \theta \sin \theta + \dfrac{\cos^2 \theta}{\sin \theta}}$

$= \dfrac{\dfrac{\sin \theta}{\sin \theta} + \dfrac{\cos \theta}{\sin \theta}}{\dfrac{\cos \theta \sin \theta}{\sin \theta} + \dfrac{\cos^2 \theta}{\sin \theta}}$

$= \dfrac{\sin \theta + \cos \theta}{\sin \theta} \cdot \dfrac{\sin \theta}{\cos \theta \sin \theta + \cos^2 \theta}$

$= \dfrac{\sin \theta + \cos \theta}{\cos \theta (\sin \theta + \cos \theta)}$

$= \dfrac{1}{\cos \theta} = \sec \theta$

20. $\dfrac{1 + \sec\theta}{\sin\theta + \tan\theta}$

$= \dfrac{\dfrac{\cos\theta}{\cos\theta} + \dfrac{1}{\cos\theta}}{\dfrac{\cos\theta \sin\theta}{\cos\theta} + \dfrac{\sin\theta}{\cos\theta}}$

$= \dfrac{\cos\theta + 1}{\cos\theta} \cdot \dfrac{\cos\theta}{\cos\theta \sin\theta + \sin\theta}$

$= \dfrac{\cos\theta + 1}{\sin\theta \,(\cos\theta + 1)}$

$= \dfrac{1}{\sin\theta} = \csc\theta$

21. $\dfrac{\csc\theta - \sin\theta}{\cot^2\theta}$

$= \dfrac{\dfrac{1}{\sin\theta} - \dfrac{\sin^2\theta}{\sin\theta}}{\dfrac{\cos^2\theta}{\sin^2\theta}}$

$= \dfrac{\dfrac{\cos^2\theta}{\sin\theta}}{\dfrac{\cos^2\theta}{\sin^2\theta}}$

$= \dfrac{\cos^2\theta}{\sin\theta} \cdot \dfrac{\sin^2\theta}{\cos^2\theta}$

$= \sin\theta$

22. $\dfrac{1 + \sin\theta}{\cos\theta} = \dfrac{(1 + \sin\theta)(1 - \sin\theta)}{(\cos\theta)(1 - \sin\theta)}$

$= \dfrac{1 - \sin^2\theta}{\cos\theta\,(1 - \sin\theta)}$

$= \dfrac{\cos^2\theta}{\cos\theta\,(1 - \sin\theta)}$

$= \dfrac{\cos\theta}{1 - \sin\theta}$

23. $\dfrac{1}{1 - \sin\theta} + \dfrac{1}{1 + \sin\theta}$

$= \dfrac{1 + \sin\theta + 1 - \sin\theta}{1 - \sin^2\theta}$

$= \dfrac{2}{\cos^2\theta} = 2\sec^2\theta$

24. $\dfrac{\csc\theta + \cot\theta}{\tan\theta + \sin\theta}$

$= \dfrac{\dfrac{1}{\sin\theta} + \dfrac{\cos\theta}{\sin\theta}}{\dfrac{\sin\theta}{\cos\theta} + \dfrac{\sin\theta\cos\theta}{\cos\theta}}$

$= \dfrac{1 + \cos\theta}{\sin\theta} \cdot \dfrac{\cos\theta}{\sin\theta\,(1 + \cos\theta)}$

$= \dfrac{1}{\sin\theta} \cdot \dfrac{\cos\theta}{\sin\theta}$

$= \cot\theta\,\csc\theta$

25. $\sin^2\theta\,\sec^2\theta + \sin^2\theta\,\csc^2\theta$

$= \dfrac{\sin^2\theta}{\cos^2\theta} + \dfrac{\sin^2\theta}{\sin^2\theta}$

$= \tan^2\theta + 1 = \sec^2\theta$

26. $\sin^4\theta - \cos^4\theta$

$= (\sin^2\theta - \cos^2\theta)(\sin^2\theta + \cos^2\theta)$

$= \sin^2\theta - \cos^2\theta$

$= \sin^2\theta - (1 - \sin^2\theta)$

$= 2\sin^2\theta - 1$

27. $\dfrac{\csc\theta}{\cot\theta + \tan\theta}$

$= \dfrac{\dfrac{1}{\sin\theta}}{\dfrac{\cos\theta}{\sin\theta} + \dfrac{\sin\theta}{\cos\theta}}$

$= \dfrac{\dfrac{1}{\sin\theta}}{\dfrac{\cos^2\theta + \sin^2\theta}{\sin\theta\cos\theta}}$

$= \dfrac{\dfrac{1}{\sin\theta}}{\dfrac{1}{\sin\theta\cos\theta}}$

$= \dfrac{1}{\sin\theta} \cdot \sin\theta\cos\theta$

$= \cos\theta$

$= \dfrac{\cos\theta}{\sin\theta}\,\sin\theta$

$= \cot\theta\,\sin\theta$

28. $\cos\theta\,(2\sec\theta + \tan\theta)(\sec\theta - 2\tan\theta)$

$= \cos\theta\,(2\sec^2\theta - 3\sec\theta\tan\theta - 2\tan^2\theta)$

$= \cos\theta\,[-3\sec\theta\tan\theta + 2(\sec^2\theta - \tan^2\theta)]$

$= \cos\theta\,(-3\sec\theta\tan\theta + 2)$

$= -3\cos\theta\!\left(\dfrac{1}{\cos\theta}\right)\tan\theta + 2\cos\theta$

$= 2\cos\theta - 3\tan\theta$

29. $(\sin\theta + \cos\theta)(\tan\theta + \cot\theta)$

$= \sin\theta\tan\theta + \cos\theta\tan\theta + \sin\theta\cot\theta + \cos\theta\cot\theta$

$= \dfrac{\sin^2\theta}{\cos\theta} + \sin\theta + \cos\theta + \dfrac{\cos^2\theta}{\sin\theta}$

$= \dfrac{1 - \cos^2\theta}{\cos\theta} + \sin\theta + \cos\theta + \dfrac{1 - \sin^2\theta}{\sin\theta}$

$= (\sec\theta - \cos\theta) + \sin\theta + \cos\theta + (\csc\theta - \sin\theta)$

$= \sec\theta + \csc\theta$

30. $\dfrac{1}{\csc\theta - \cot\theta} - \dfrac{1}{\csc\theta + \cot\theta}$

$= \dfrac{(\csc\theta + \cot\theta) - (\csc\theta - \cot\theta)}{(\csc\theta - \cot\theta)(\csc\theta + \cot\theta)}$

$= \dfrac{2\cot\theta}{\csc^2\theta - \cot^2\theta}$

$= 2\cot\theta$

31. $\begin{vmatrix} \cos\theta & -\sin\theta \\ \sin\theta & \cos\theta \end{vmatrix}$

$= \cos^2\theta + \sin^2\theta$

$= 1$

NCTM Evaluation Standards
Grades 9–12

STANDARD 1:
Mathematics as Problem Solving

In grades 9–12, the mathematics curriculum should include the refinement and extension of the methods of mathematical problem solving so that all students can—

1a. use, with increasing confidence, problem solving approaches to investigate and understand mathematical content;

1b. apply integrated mathematical problem-solving strategies to solve problems from within and outside mathematics;

1c. recognize and formulate problems from situations within and outside mathematics;

1d. apply the process of mathematical modeling to real-world problem situations.

STANDARD 2:
Mathematics as Communication

In grades 9–12, the mathematics curriculum should include the continued development of language and symbolism to communicate mathematical ideas so that all students can—

2a. reflect upon and clarify their thinking about mathematical ideas and relationships;

2b. formulate mathematical definitions and express generalizations discovered through investigations;

2c. express mathematical ideas orally and in writing;

2d. read written presentations of mathematics with understanding;

2e. ask clarifying and extending questions related to mathematics they have read or heard about;

2f. appreciate the economy, power, and elegance of mathematical notation and its role in the development of mathematical ideas.

STANDARD 3:
Mathematics as Reasoning

In grades 9–12, the mathematics curriculum should include numerous and varied experiences that reinforce and extend logical reasoning skills so that all students can—

3a. make and test conjectures;

3b. formulate counterexamples;

3c. follow logical arguments;

3d. judge the validity of arguments;

3e. construct simple valid arguments;

and so that, in addition, college-intending students can—

3f. construct proofs for mathematical assertions, including indirect proofs and proofs by mathematical induction.

STANDARD 4:
Mathematical Connections

In grades 9–12, the mathematical curriculum should include investigation of the connections and interplay among various mathematical topics and their applications so that all students can—

4a. recognize equivalent representations of the same concept;

4b. relate procedures in one representation to procedures in an equivalent representation;

4c. use and value connections among mathematical topics;

4d. use and value the connections between mathematics and other disciplines.

STANDARD 5:
Algebra

In grades 9–12, the mathematics curriculum should include the continued study of algebraic concepts and methods so that all students can—

5a. represent situations that involve variable quantities with expressions, equations, inequalities, and matrices;

5b. use tables and graphs as tools to interpret expressions, equations, and inequalities;

5c. operate on expressions and matrices, and solve equations and inequalities;

5d. appreciate the power of mathematical abstraction and symbolism;

and so that, in addition, college-intending students can—

5e. use matrices to solve linear systems;

5f. demonstrate technical facility with algebraic transformations, including techniques based on the theory of equations.

STANDARD 6:
Functions

In grades 9–12, the mathematics curriculum should include the continued study of functions so that all students can—

6a. model real-world phenomena with a variety of functions;

6b. represent and analyze relationships using tables, verbal rules, equations, and graphs;

6c. translate among tabular, symbolic, and graphical representations of functions;

6d. recognize that a variety of problem situations can be modeled by the same type of function;

6e. analyze the effects of parameter changes on the graphs of functions;

and so that, in addition, college-intending students can—

6f. understand operations on, and the general properties and behavior of, classes of functions.

STANDARD 7:
Geometry from a Synthetic Perspective

In grades 9–12, the mathematics curriculum should include the continued study of the geometry of two and three dimensions so that all students can—

7a. interpret and draw three-dimensional objects;

7b. represent problem situations with geometric models and apply properties of figures;

7c. classify figures in terms of congruence and similarity and apply these relationships;

7d. deduce properties of, and relationships between, figures from given assumptions;

and so that, in addition, college-intending students can—

7e. develop an understanding of an axiomatic system through investigating and comparing various geometries.

STANDARD 8:
Geometry from an Algebraic Perspective

In grades 9–12, the mathematics curriculum should include the study of the geometry of two and three dimensions from an algebraic point of view so that all students can—

8a. translate between synthetic and coordinate representations;

8b. deduce properties of figures using transformations and using coordinates;

8c. identify congruent and similar figures using transformations;

8d. analyze properties of Euclidean transformations and relate translations to vectors;

and so that, in addition, college-intending students can—

8e. deduce properties of figures using vectors;

8f. apply transformations, coordinates, and vectors in problem solving.

STANDARD 9:
Trigonometry

In grades 9–12, the mathematics curriculum should include the study of trigonometry so that all students can—

9a. apply trigonometry to problem situations involving triangles;

9b. explore periodic real-world phenomena using sine and cosine functions;

and so that, in addition, college-intending students can—

9c. understand the connection between trigonometric and circular functions;

9d. use circular functions to model periodic real-world phenomena;

9e. apply general graphing techniques to trigonometric functions;

9f. solve trigonometric equations and verify trigonometric identities

9g. understand the connections between trigonometric functions and polar coordinates, complex numbers, and series.

STANDARD 10:
Statistics

In grades 9–12, the mathematics curriculum should include the continued study of data analysis and statistics so that all students can—

10a. construct and draw inferences from charts, tables, and graphs that summarize data from real-world situations;

10b. use curve fitting to predict from data;

10c. understand and apply measures of central tendency, variability, and correlation;

10d. understand sampling and recognize its role in statistical claims;

10e. design a statistical experiment to study a problem, conduct the experiment, and interpret and communicate the outcomes;

10f. analyze the effects of data transformations on measures of central tendency and variability;

and so that, in addition, college-intending students can—

10g. transform data to aid in data interpretation and prediction;

10h. test hypotheses using appropriate statistics.

STANDARD 11:
Probability

In grades 9–12, the mathematics curriculum should include the continued study of probability so that all students can—

11a. use experimental or theoretical probability, as appropriate, to represent and solve problems involving uncertainty;

11b. use simulations to estimate probabilities;

11c. understand the concept of a random variable;

11d. create and interpret discrete probability distributions;

11e. describe, in general terms, the normal curve and use its properties to answer questions about sets of data that are assumed to be normally distributed;

and so that, in addition, college-intending students can—

11f. apply the concept of a random variable to generate and interpret probability distributions including binomial, uniform, normal, and chi square.

STANDARD 12:
Discrete Mathematics

In grades 9–12, the mathematics curriculum should include topics from discrete mathematics so that all students can—

12a. represent problem situations using discrete structures such as finite graphs, matrices, sequences, and recurrence relations;

12b. represent and analyze finite graphs using matrices;

12c. develop and analyze algorithms;

12d. solve enumeration and finite probability problems;

and so that, in addition, college-intending students can—

12e. represent and solve problems using linear programming and difference equations;

12f. investigate problem situations that arise in connection with computer validation and the application of algorithms.

STANDARD 13:
Conceptual Underpinnings of Calculus

In grades 9–12, the mathematics curriculum should include the informal exploration of calculus concepts from both a graphical and a numerical perspective so that all students can—

13a. determine maximum and minimum points of a graph and interpret the results in problem situations;

13b. investigate limiting processes by examining infinite sequences and series and areas under curves;

and so that, in addition, college-intending students can—

13c. understand the conceptual foundations of limit, the area under a curve, the rate of change, and the slope of a tangent line, and their applications in other disciplines;

13d. analyze the graphs of polynomial, rational, radical, and transcendental functions.

STANDARD 14:
Mathematical Structure

In grades 9–12, the mathematics curriculum should include the study of mathematical structure so that all students can—

14a. compare and contrast the real number system and its various subsystems with regard to their structural characteristics;

14b. understand the logic of algebraic procedures;

14c. appreciate that seemingly different mathematical systems may be essentially the same;

and so that, in addition, college-intending students can—

14d. develop the complex number system and demonstrate facility with its operations;

14e. prove elementary theorems within various mathematical structures, such as groups and fields;

14f. develop an understanding of the nature and purpose of axiomatic systems.

Correlation of Holt Algebra with Trigonometry to the NCTM Evaluation Standards

NCTM Standard	Holt Algebra with Trigonometry Lesson
1a	4.5, 6.2, 7.7, 17.8, 18.1
1b	2.4, 6.3, 9.4, 9.6, 10.3
1c	12.9
1d	1.5, 7.3, 12.6, 16.6
2a	1.4, 3.3, 7.6, 8.4, 10.1, 14.1, 17.1
2c	15.1
2d	5.5, 9.5, 9.7, 12.2, 14.6
2f	2.6, 16.3
3a	6.6, 11.7, 17.7
3c	1.6, 4.6, 5.2, 6.4, 7.4, 8.2, 9.3, 12.4, 14.2, 15.4, 16.4
3d	5.6, 14.3, 15.8
3e	8.5
4a	2.1, 4.1, 10.6, 12.7, 13.6, 18.2, 19.3
4b	9.1
4c	1.7, 4.7, 8.7, 9.8, 11.1, 12.5, 12.10, 13.3, 15.6, 19.1, 19.6
4d	5.8, 13.8
5a	2.3, 2.5, 4.4, 7.2
5b	1.3, 2.2, 3.1, 3.4, 3.9
5c	4.2, 16.1
5d	1.1, 3.5, 5.1, 6.5, 8.1, 10.2, 10.5, 13.5, 14.5, 18.5
5e	16.5
6a	3.6
6b	5.9, 7.1, 13.1, 13.7, 17.2, 18.6
6c	3.2, 13.4
6d	3.8
6e	11.3, 13.2, 18.8
6f	18.10
8a	12.1
8b	11.2, 11.4, 17.5, 17.6
8c	17.4
9a	17.3, 18.3, 18.4
9e	18.7, 18.9
9f	19.2, 19.4, 19.7
10c	14.10
11a	14.7
11b	14.8
11c	14.9
11e	14.11
12a	15.2, 15.5
13a	11.6
13b	15.3, 15.7
14a	8.6, 10.4
14b	1.2, 1.8, 3.7, 4.3, 5.3, 5.7, 6.1, 7.5, 8.3, 11.5, 12.3
14c	5.4, 16.2
14d	9.2, 10.7, 19.9, 19.10
14e	19.5
14f	19.8

Index

Boldfaced numerals indicate the pages that contain definitions.